AN INTRODUCTION TO SEISMOLOGY, EARTHQUAKES, AND EARTH STRUCTURE

To future generations of earth scientists — may their enthusiasm and creativity keep seismology vibrant and exciting

I cannot give any scientist of any age better advice than this: the intensity of the conviction that a hypothesis is true has no bearing on whether it is true or not. The importance of the strength of our conviction is only to provide a proportionally strong incentive to find out if the hypothesis will stand up to critical examination.

Sir Peter Medawar, *Advice to a Young Scientist*, 1979

An Introduction to Seismology, Earthquakes, and Earth Structure

Seth Stein
Department of Geological Sciences, Northwestern University, Evanston, Illinois

Michael Wysession
Department of Earth and Planetary Sciences, Washington University, St Louis, Missouri

BLACKWELL PUBLISHING
350 Main Street, Malden, MA 02148-5020, USA
9600 Garsington Road, Oxford OX4 2DQ, UK
550 Swanston Street, Carlton, Victoria 3053, Australia

First published 2003 by Blackwell Publishing Ltd.

14 2014

Library of Congress Cataloging-in-Publication Data

Stein, Seth.
An introduction to seismology, earthquakes, and earth structure / Seth Stein, Michael Wysession.
p. cm.
Includes bibliographical references and index.
ISBN 978-0-86542-078-6 (pb. : alk. paper)
1. Seismology. 2. Geology, Structural. 3. Earthquakes.
I. Wysession, Michael. II. Title.

QE534.3 .S74 2002
551.22—dc21

2001052639

A catalogue record for this title is available from the British Library.

Set in 9½/11½ pt Sabon
by Graphicraft Ltd, Hong Kong
Printed and bound in Singapore
by Ho Printing Singapore Pte Ltd

The publisher's policy is to use permanent paper from mills that operate a sustainable forestry policy, and which has been manufactured from pulp processed using acid-free and elementary chlorine-free practices. Furthermore, the publisher ensures that the text paper and cover board used have met acceptable environmental accreditation standards.

For further information on
Blackwell Publishing, visit our website:
www.blackwellpublishing.com

Contents

Preface

Science is only worth doing if it is interesting and fun. Hence the goal of a textbook is to interest students in a subject, convince them it is worth the effort required to learn about it, and help them do so. We have tried here to do all three.

For seismology, these should be easy. It is hard to imagine topics more interesting than the structure and evolution of a planet, as manifested by phenomena as dramatic as earthquakes. Our goal is to address them via an introduction to seismology, which is one of the cornerstones of the modern earth sciences. Seismology has been defined as the study of earthquakes and associated phenomena, or the study of elastic waves propagating in the earth. By integrating techniques and data from physics, mathematics, and geology, seismology has produced a remarkably sharp picture of the earth's interior that is a primary datum for studying the formation and evolution of terrestrial planets. Seismologists have also learned much about the nature of earthquakes and the tectonic processes responsible for them. These studies are not of purely academic interest; seismology is the major tool for earthquake hazard assessment, hydrocarbon exploration, and the peacekeeping role of nuclear test monitoring.

We thus believe that seismology should be part of the education of every solid earth scientist, rather than a specialized course for those whose primary interest is seismology or other branches of geophysics. The subject has much to offer mineralogists or petrologists studying the composition of the earth's interior, students of tectonics interested in processes of the lithosphere, geologists interested in the nature and evolution of the crust, engineers concerned with seismic hazards, and planetologists interested in the evolution of the terrestrial planets. As the earth sciences become increasingly more integrated and interdisciplinary, the advantages of understanding seismology will continue to grow.

Many students have been deterred from the subject because it requires confronting, often for the first time, both the physics of a continuous medium and wave propagation. We view these concerns as manageable. In fact, we believe that seismology is a good way to introduce these topics, because it applies what might otherwise seem abstract ideas. Seismic waves illustrate effects like reflection, refraction, diffraction, and dispersion by using them to study the earth. Earthquakes demonstrate concepts like rigid tectonic plates, stress and strain, and viscous mantle flow. Thus seismology is a natural way to discuss fundamental processes.

Our goal is to introduce key concepts and their application in present research. This twofold goal places several limitations on the text. First, time and space restrictions require a trade-off between the range of topics and the level of presentation. The resulting choices are, of necessity, subjective. Second, we end discussions when material, however fascinating, seems more appropriate for advanced classes or courses in a related field.[1] Third, these limitations preclude an account of the historical development of the subject, or a systematic assignment of credit for ideas and results. Fourth, in introducing topics of current research, we try to give our sense of issues while recognizing that others' views may differ. The danger in presenting the "current state of knowledge" in a text is that the field changes so rapidly that accounts can soon be out of date. We thus try to focus not on "what we know," but on "how we seek to find out," and highlight current findings in the context of studying interesting questions.

Given these limitations, suggestions for further reading are provided. When possible, the readings are texts or reviews rather than specialized research papers. In many cases, the sources of the figures used to illustrate a concept provide additional information. We also give some references to sites on the World Wide Web, recognizing the trade-off between the wealth of information there and the fact that the Web is volatile and sites can change locations or vanish.

The material is designed for advanced undergraduates and first-year graduate students. Readers are assumed to be familiar with ordinary differential equations and introductory physics. Further background, including basic earth science courses, is helpful but not essential. Material beyond this level is derived as needed. Thus, we seek a balance between presenting the mathematics like magic pulled from a hat and deriving so much so that the thematic flow is disrupted. Hence we

[1] Because subfields in the earth sciences overlap, the divisions between them are not sharp, and a given topic draws on several. As John Muir, an early member of the Seismological Society of America better known for founding the Sierra Club, pointed out, "when we look at anything in isolation we realize it is hitched to the rest of the universe."

review some useful mathematics in an Appendix, to which we refer. Other mathematical concepts, notably topics in Fourier analysis, are used as needed and then presented in more depth when appropriate.

Our goal is to introduce some concepts about seismology and its application to such studies of earth structure and earthquakes. Doing this requires developing basic ideas about wave propagation in a continuous solid medium, so the material of greatest interest to geologically oriented readers is somewhat postponed. Readers are urged to enjoy rather than endure the introductory material on elasticity and wave propagation. They risk only discovering the appeal of these topics and finding themselves taking subsequent advanced courses.

Part of the delights of the earth sciences is that they are less structured than some other sciences. There is no single set of topics covered in specific courses, which instead reflect the instructor's and students' interests. Certainly this is the case here. The topics we have chosen contain about a year's worth of class material, which we ourselves divide into several courses. Many students, of course, take only one. We have experimented with different groupings, all of which seemed to work well. We usually do not cover the Appendix in lectures, but assign its problems to identify areas for study or review.

We have found that the homework problems are helpful for understanding the topics. Given the nature of the modern earth sciences, many problems are designed to be done on computers. In our teaching, we expect that most will be done by writing programs, and hence require programming, beginning with simple problems in the Appendix and building to more complex ones in the chapters. A secondary motive is to ensure that students learn the skills of scientific programming, which are often not stressed in computer classes. Some of the problems can be done using spreadsheets, and most can be done with specialized mathematical software.

Some matters of style are worth mentioning. We illustrate interconnections between topics by referring both forward and backward to other sections. Figures are labeled with hyphens (e.g. 5.6-2), and equations with periods (e.g. 5.3.2). Footnotes generally cover side observations which we note in class but are not essential. We use both SI units (those based on the meter, kilogram, and second) and cgs units (those based on the centimeter, gram, and second) because both are common in the literature, although SI units are slowly superseding cgs. We also use other units when customary: seismic velocities are given in km/s and plate motions are given in the more intuitive mm/yr (e.g., 48 mm/yr rather than 1.5×10^{-9} m/s), following Emerson's dictum that "a foolish consistency is the hobgoblin of little minds."

We have enjoyed writing this book. It is a pleasure to try to summarize this diverse and fascinating discipline. We hope readers have as much fun as we did, and that our discussions prompt them to raise interesting and provocative questions as well as learn the material. We also hope that some readers are motivated to continue study of and research on these topics. Much remains to be learned about the earth and earthquake processes, and the opportunities for contributions are great for those with the energy and imagination to go beyond our current knowledge and ideas. Three hundred years after Isaac Newton's work in mechanics and optics laid what would become seismology's foundations, it is worth recalling his words: "I seem to have been only like a boy playing on the seashore, and diverting myself in now and then finding a smoother pebble or a prettier shell than ordinary, whilst the great ocean of truth lay all still undiscovered before me."

Acknowledgments

This book has evolved over many years, with much assistance from students, colleagues, and many who started off as one and became the other. Any effort to thank them all will omit some who should be credited, but is better than none. Similarly, we received much good advice, although not all of it could be accommodated, given the practical issues of manuscript length and level of material.

The text is better for thoughtful questions and assistance from students whose courses used its preliminary forms. In particular, Gary Acton, Don Argus, Craig Bina, John Brodholt, Po-Fei Chen, John DeLaughter, Charles DeMets, George Helffrich, Eryn Klosko, Lisa Leffler, Paul Lundgren, Frederick Marton, Andrew Michael, Andrew Newman, Phillip Richardson, Thomas Shoberg, Paul Stoddard, John Werner, Dale Woods, and Mark Woods helped in developing this material. Many of the figures result from their assistance, as well as the artistic talents of Ranjini Mahinda and Megan Murphy. Cheril Cheverton and Will Kazmeier provided valuable help in manuscript preparation.

We also benefited from suggestions and assistance by colleagues, especially Craig Bina, Raymon Brown, Wang-Ping Chen, Ken Creager, Robert Crosson, Joseph Engeln, Edward Flinn, Yoshio Fukao, Robert Geller, William Holt, Stephen Kirby, Simon McClusky, Emile Okal, Gary Pavlis, Aristeo Pelayo, Steve Roecker, Giovanni Sella, Tetsuzo Seno, Anne Sheehan, Zhang-Kang Shen, Robert Smalley, Robert Smith, Carol Stein, John Vidale, and Douglas Wiens. Jean van Altena, Cameron Laux, John Staples, and Nancy Duffy of Blackwell provided invaluable help.

Indirect support was provided by the National Science Foundation through PECASE award #NSF–EAR–9629018.

Most crucially, Carol Stein and Joan Wysession encouraged the project and allowed us the time required.

1 Introduction

I cannot help feeling that seismology will stay in the place at the center of solid earth science for many, many years to come.
The joy of being a seismologist comes to you, when you find something new about the earth's interior from the observation of seismic waves obtained on the surface, and realize that you did it without penetrating the earth or touching or examining it directly.

Keiiti Aki, presidential address to the Seismological Society of America, 1980

1.1 Introduction

This book is an introduction to seismology, the study of elastic waves or sound waves in the solid earth. Conceptually, the subject is simple. Seismic waves are generated at a *source*, which can be natural, such as an earthquake, or artificial, such as an explosion. The resulting waves propagate through the *medium*, some portion of the earth, and are recorded at a *receiver* (Fig. 1.1-1). A *seismogram*, the record of the motion of the ground at a receiver called a *seismometer*, thus contains information about both the source and the medium. This information can take several forms. The waves provide information on the location and nature of the source that generated them. If the *origin time* when the waves left the source is known, their *arrival time* at the receiver gives the *travel time* required to pass through the medium, and hence information about the speed at which they traveled, and thus the physical properties of the medium. In addition, because the amplitude and shape of the wave pulses that left the source are affected by propagation through the medium, the signals observed on seismograms provide additional information about the medium.

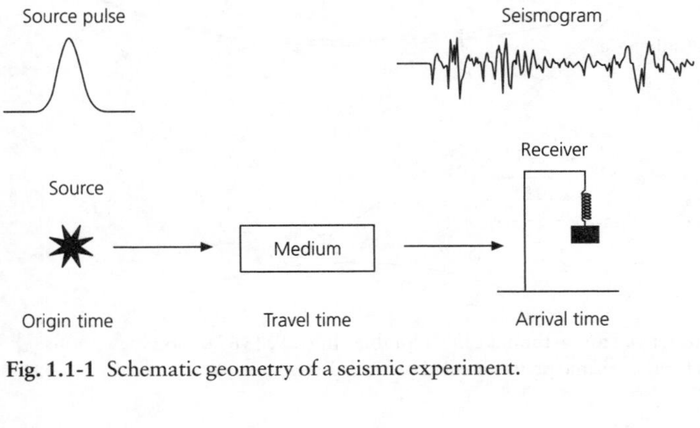

Source pulse

Seismogram

Receiver

Source

Medium

Origin time

Travel time

Arrival time

Fig. 1.1-1 Schematic geometry of a seismic experiment.

1.1.1 Overview

Before embarking on our studies, it is worth briefly outlining some of the ways in which seismology is used to study the earth, and some of the methods used. Seismology is the primary tool for the study of the earth's interior because little of the planet is accessible to direct observation. The surface can be mapped and explored, and drilling has penetrated to depths of up to 13 kilometers, though at great expense. Information about deeper depths, down to the center of the earth (approximately 6371 km), is obtained primarily from indirect methods. Seismology, the most powerful such method, is used to map the earth's interior and study the distribution of physical properties. The existence of the earth's shallow crust, deeper mantle, liquid outer core, and solid inner core are inferred from variations in seismic velocity with depth. Our ideas about their chemical compositions, including the presumed locations of changes in mineral structure due to the increase of pressure with depth, are also based on seismological data. Near the surface, seismology provides detailed crustal images that reveal information about the locations of economic resources like oil and minerals. Deeper in the earth, seismology provides the basic data for understanding earth's dynamic history and evolution, including the process of mantle convection.

Seismology is also the primary method for studies of earthquakes. Most of the information about the nature of faulting during an earthquake is determined from the resulting seismograms. These observations are useful for several purposes. Because earthquakes generally result from the motions of the plates making up the earth's lithosphere, which are the surface expression of convection within earth's mantle, knowledge of the direction and amount of motion is valuable for describing plate motions and the forces giving rise to them. Analysis of seismograms also makes it possible to investigate the physical processes that occur prior to, during, and after faulting. Such studies are helpful in assessing the societal hazards posed by earthquakes.

Our purpose here is to discuss some basic ideas about seismology and its applications. To do this, we first introduce several concepts about waves in a solid medium. We will see that a few simple but powerful ideas give a great deal of insight into how waves propagate and respond to variations in physical properties in the earth. Fortunately, most of these ideas are analogous to familiar concepts in the propagation of light and sound waves. As a result, studying the earth with seismic waves is conceptually similar to sensing the world around us using light and sound. For example, you are reading this by receiving light reflected off the paper. We see color because light has different wavelengths; the sky is blue because certain wavelengths are scattered preferentially. An even closer analogy is the use of sound waves by bats, dolphins, and submarines to "see" their surroundings. Seismology gives detailed images of earth structure, much as sound waves (ultrasound) and electromagnetic waves (X-rays) are used in medicine to study human bodies.

A familiar property of light is that it bends when traveling between materials in which its speed differs. Objects inserted into water appear crooked, because light waves travel more slowly in water than in air. Prisms and lenses use this effect, called *refraction*. This phenomenon occurs in the earth because seismic wave velocities generally increase with depth. Wave paths bend away from the vertical as they go deeper into the earth, eventually become horizontal ("bottom"), turn upward, and return to the surface (Fig. 1.1-2). The wave paths are thus used to infer the variation of seismic velocity, and hence the composition and physical properties of material, with depth in the earth.

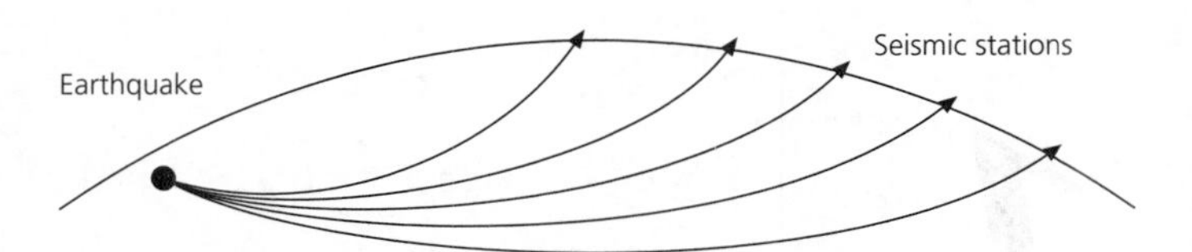

Fig. 1.1-2 Seismic ray paths in the earth, showing the effect of an increase in seismic velocity with increasing depth. The waves travel in curved paths between the earthquake and seismic stations.

Just as light waves *reflect* at a mirror, seismic waves reflect at interfaces across which physical properties change, such as the boundary between the earth's mantle and core. Because the amplitudes of the reflected and transmitted seismic waves depend on the velocities and densities of the material on either side of the boundary, analysis of seismic waves yields information on the nature of the interface. In addition to refraction and reflection, waves also undergo *diffraction*. Just as sound diffracts around the corner of a building, allowing us to hear what we cannot see, seismic waves bend around "obstacles" such as the earth's core.

The basic data for these studies are seismograms, records of the motion of the ground resulting from the arrival of refracted, reflected, and diffracted seismic waves. Seismograms incorporate precise timing, so that travel times can be determined. The seismometer's response is known, so the seismogram can be related to the actual ground motion. Because ground motion is a vector, three different components (north–south, east–west, and up–down) are typically recorded. Hence, although seismograms at first appear to be simply wiggly lines, they contain interesting and useful information.

To illustrate the use of seismology for the study of earth structure, consider a seismogram from a magnitude 6 earthquake in Colombia, recorded about 4900 kilometers away in Colorado (Fig. 1.1-3). Several seismic wave arrivals, called *phases*, are identified using a simple nomenclature that describes the path each followed from the source to the receiver.

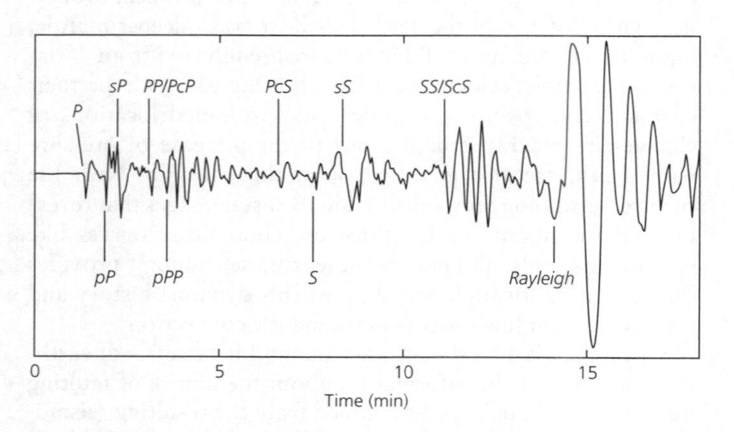

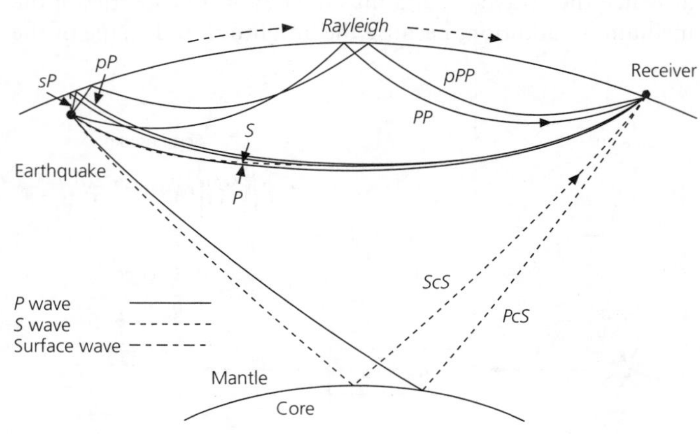

Fig. 1.1-3 *Left*: Long-period vertical component seismogram at Golden, Colorado, from an earthquake in Colombia (July 29, 1967), showing various seismic phases. The distance from earthquake to station is 44°. *Right*: Ray paths for the seismic phases labeled on the seismogram.

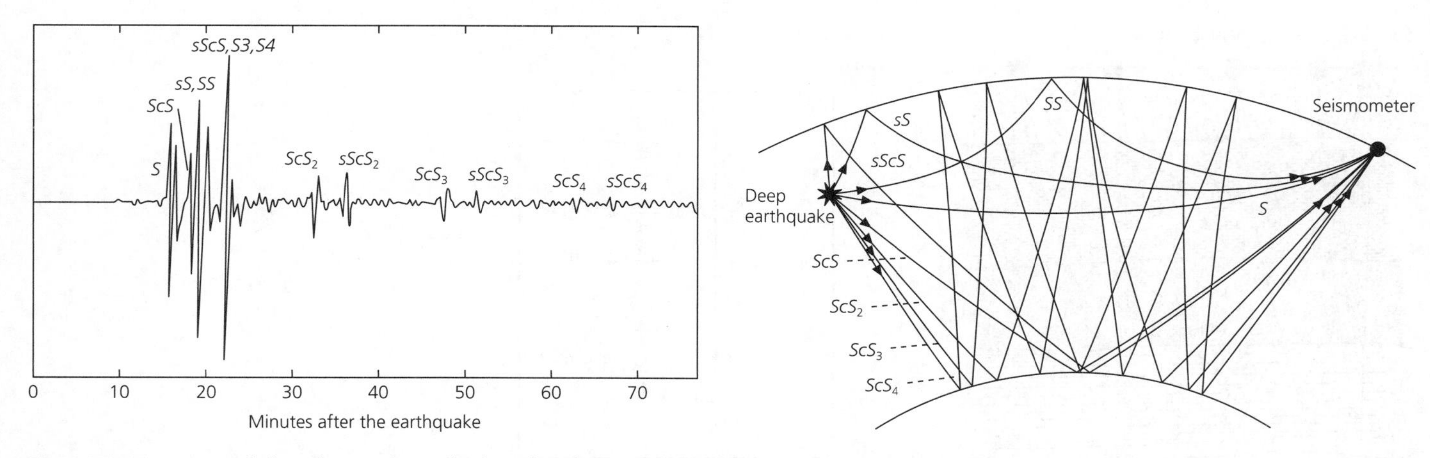

Fig. 1.1-4 Seismogram (*left*) and ray paths (*right*) for a deep focus earthquake in Tonga, recorded at Oahu (Hawaii), showing multiple core reflections.

We will see that seismic waves are divided into two types. In one type, *P* or compressional waves, material moves back and forth in the direction in which the wave propagates. In the other, *S* or shear waves, material moves at right angles to the propagation direction. *P* waves travel faster than *S* waves, so the first arriving pulse, labeled "*P*," is a *P* wave that followed a direct path from the earthquake to the seismometer.[1] Soon afterwards, a pulse labeled *pP* appears, which went upward from the earthquake, reflected off the earth's surface, and then traveled to the seismometer as a *P* wave. If the distribution of seismic velocity near the source is known, the depth of the earthquake below the earth's surface can be found from the time difference between the direct *P* and *pP* phases, because the primary differences between their ray paths are the *pP* segments that first go up to and then reflect off the surface. The phase marked *PP* is a compressional wave that went downward from the source, "bottomed," reflected at the surface, and repeated the process. Among the later arrivals on the seismogram are shear wave phases, including the direct shear wave arrival, *S*, and a shear phase *SS* that reflected off the surface, analogous to *PP*. All these phases, which traveled through the earth's interior, are known as *body waves*. The large amplitude wave train that arrives later, marked "Rayleigh," is an example of a different type of wave. Such *surface waves* propagate along paths close to the earth's surface.

Figure 1.1-4 shows a seismogram from an earthquake at a depth of 650 km in the Tonga subduction zone recorded in Hawaii. The seismometer is oriented such that all the arrivals are shear waves. In addition to *S* and *SS*, phases reflected at the core–mantle boundary appear. *ScS* went down from the source, reflected at the core–mantle boundary (hence "*c*"), and came back up to the seismometer. Its travel time gives the depth to the core if the velocity in the mantle is known. Alternatively, if the depth to the core is known, the travel time gives a vertical average of velocity with depth in the mantle. In addition, the large amplitude of these reflections constrains the contrast in physical properties between the solid rock-like lower mantle and the fluid iron outer core. Multiple reflections also occur: *ScSScS*, or ScS_2, reflects twice at the core–mantle boundary, ScS_3 reflects three times, and ScS_4 four times. Similar to the phase *SS*, the S_3 wave reflects twice off the surface, and S_4 reflects three times. By analogy to *pP*, *sScS* went upward from the source and was reflected first at the surface and then at the core–mantle boundary. Most of the multiple *SS* and *ScS* phases also have observable surface reflected phases (e.g., $sScS_2$, $sScS_3$, etc.).

These examples indicate some of the ways in which seismological observations are used to study earth structure. By collecting many such records, seismologists have compiled travel time and amplitude data for many seismic phases. Because the different phases have different paths, they provide multiple types of information about the distribution of seismic velocities, and therefore physical properties within the earth. Seismology can also be used to study the internal structure of other planets; seismometers were deployed on the lunar surface by each of the Apollo missions, and the Viking spacecraft that landed on Mars carried a seismometer.

An important use of seismology is the exploration of near-surface regions for scientific purposes or resource extraction. Figure 1.1-5 shows a schematic version of a common technique used. An artificial source at or near the surface generates seismic waves that travel downward, reflect off interfaces at depth, and are detected by seismometer arrays. The resulting data are processed using computers to enhance the arrivals corresponding to reflections and to estimate the velocity structure. Seismograms from different receivers are then displayed side by side, with the travel time increasing downward, to yield an image of the vertical structure. Reflections that match between seismograms give near-horizontal arrivals that often correspond to interfaces at depth. The vertical axis can be converted from time to depth using the estimated velocities, and reflectors

[1] The labels *P* and *S* come from the early days of seismology, when *P* stood for *primary* and *S* stood for *secondary*.

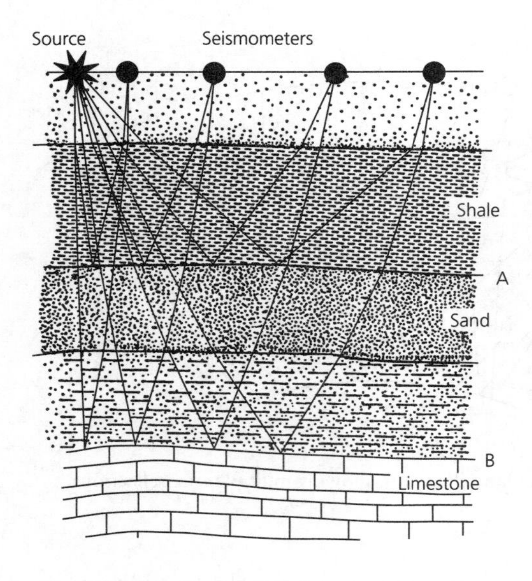

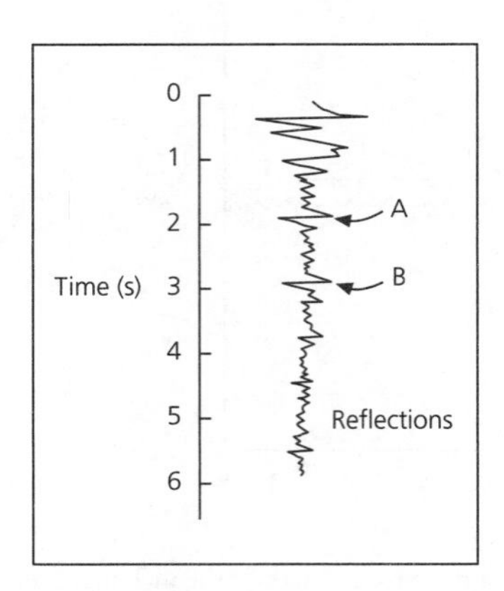

Fig. 1.1-5 Schematic example of the seismic reflection method, the basic tool of hydrocarbon exploration.

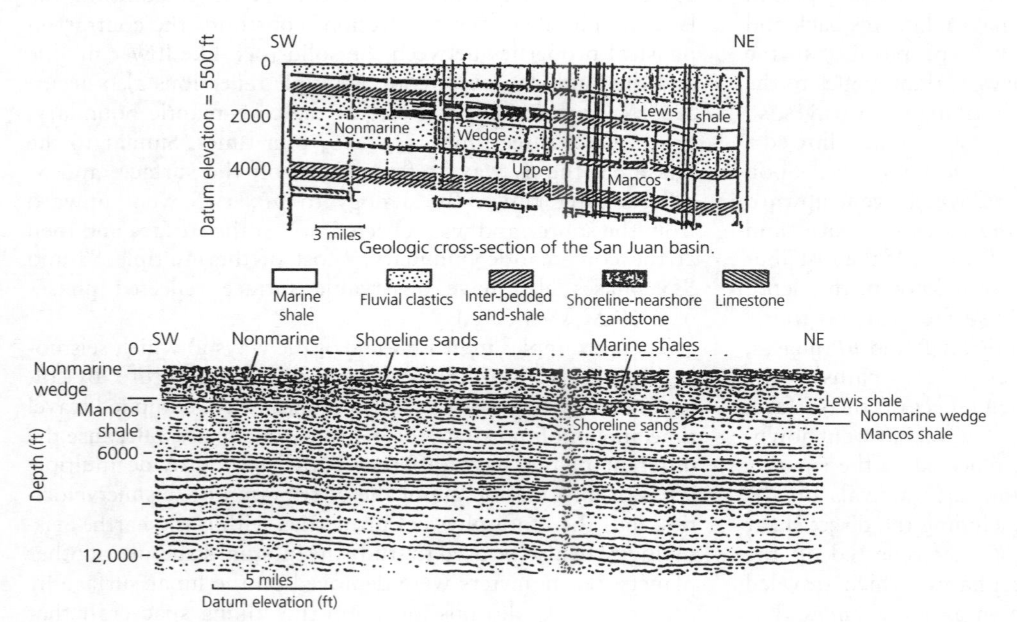

Fig. 1.1-6 Data from a reflection seismic survey across the San Juan Basin, New Mexico (*bottom*) and the resulting geological interpretation (*top*). [Sangree and Widmier, 1979. Reprinted by permission of the Society of Exploration Geophysicists.)

can be identified using geological information from the surface and drill holes (Fig. 1.1-6). Such seismic images of the subsurface provide a powerful tool for structural and stratigraphic studies. Although applications of seismology to exploration have traditionally been treated in universities as distinct from those dealing with earthquakes and the large-scale structure of the earth, this distinction is largely historical.[2] These applications draw on a common body of seismological principles, and the techniques used have considerable overlap.

Seismic sources — typically earthquakes — are also a major topic of seismological study. The location of an earthquake, known as the *focus* or *hypocenter*, is found from the arrival times of seismic waves recorded on seismometers at different sites. This location is often shown by the *epicenter*, the point on the earth's surface above the earthquake. The size of earthquakes is measured from the amplitude of the motion recorded on seismograms, and given in terms of *magnitude* or *moment*.[3] In addition, the geometry of the fault on which an earthquake

[2] This book follows this tradition and focuses on earthquakes and large-scale earth structure because of the existence of an excellent introductory literature dealing with exploration seismology and the inflexibility of university curricula.

[3] Magnitude is given as a dimensionless number measured in various ways, including the body wave magnitude m_b, surface wave magnitude M_s, and moment magnitude M_w, as discussed in Section 4.6. The seismic moment has the dimensions of energy, dyn-cm or N-m.

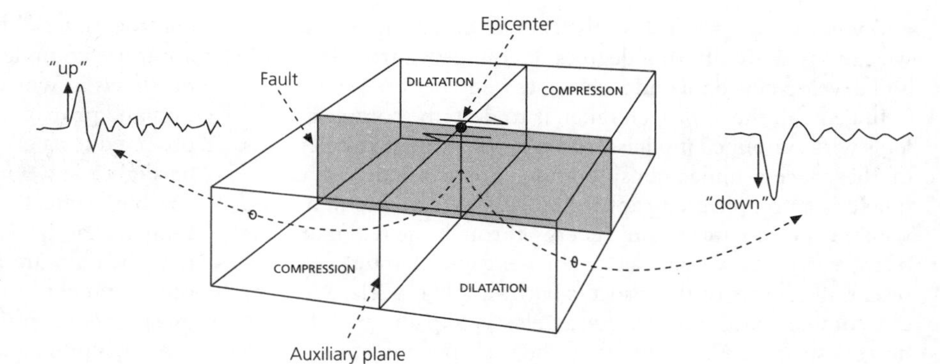

Fig. 1.1-7 First motions of seismic *P* waves observed at seismometers located in various directions about the earthquake allow the fault orientation to be determined.

occurred is inferred from the three-dimensional pattern of radiated seismic waves. Figure 1.1-7 illustrates the method used for an earthquake in which the material on one side of a vertically dipping fault moves horizontally with respect to that on the other side. This motion generates seismic waves that propagate away in all directions. In some directions the ground first moves away from the source (toward a seismic station), whereas in other directions the ground first moves toward the source (away from a receiver). The seismograms thus differ between stations. In the "toward" (called compressional) quadrants the first ground motion recorded is toward the receiver, whereas in the "away" (called dilatational) quadrants the first ground motion is away from the receiver. Because the seismic waves go down from the source, turn, and arrive at a distant seismographic station from below, the first motion is upward in a compressional quadrant and downward in a dilatational quadrant.[4] The compressional and dilatational quadrants can be identified using seismograms recorded at different azimuths around the source. The fault orientation and a surface perpendicular to it can then be found, because in these directions the first motion changes polarity. With the use of additional data we can often tell which of these surfaces was the actual fault. Given the fault orientation, the direction of motion can also be found; note that the compressional and dilatational quadrants would be interchanged if the fault had moved in the opposite direction. The pulse radiated from the earthquake also gives some information about the amount of slip that occurred, the size of the area that slipped, and the slip process.

Such observations of the location of earthquakes and the fault motion that occurred in them are among the most important data we have for understanding *plate tectonics*, the primary process shaping our planet. The earthquake analyzed in Fig. 1.1-7, for example, is like those that occur along the San Andreas fault in northern California, part of the boundary along which the Pacific plate moves northward with respect to the North American plate. The fault is visible at the earth's surface, so geological and geodetic observations also show the motion that occurs in earthquakes. In less accessible areas seismological observations provide most of the data used to identify the boundary along which motion occurs and to demonstrate its nature. This is the case for most plate boundaries, which occur in the oceans, beneath several kilometers of water. Similarly, in subduction zones, where lithospheric plates descend deep into the mantle and earthquakes can occur to depths of 660 km, direct observations are not possible, but analyses of seismograms reveal the motions and give insight into their tectonic causes.

1.1.2 Models in seismology

As summarized in the previous section, seismology provides a great deal of information about seismic sources, the structure of the earth, and the relation of earthquakes to the tectonic processes that produce them. Even so, we will see that there are major limitations on what the present seismological observations and other data tell us. For example, although we have good models of seismic velocity in the earth, we know much less about the composition of the earth and have only general ideas about the deep physical processes, such as convection, thought to be taking place. Similarly, although seismology provides a great deal of detail about the slip that occurs during an earthquake, we still have only general ideas about how earthquakes are related to tectonics, little understanding of the actual faulting process, no ability to predict earthquakes on time scales shorter than a hundred years, and only rudimentary methods to estimate earthquake hazards. This situation is typical of the earth sciences,[5] largely because of the complexity of the processes being studied and the limits of our observations. Our best response seems to be to show humility in face of the complexity of nature, recognize what we presently know

[4] These terms are not the same as compressional and shear waves; as often occurs in science, words have multiple meanings.

[5] In discussing analogous issues Sarewitz and Pielke (2000) note than even after billions of dollars spent on climate research, a senior scientist observes, "This may come as a shock to many people who assume that we do know adequately what's going on with the climate, but we don't," and the National Academy of Sciences states that deficiencies in our understanding "place serious limitations on the confidence" of climate modeling results.

and what we do not, use statistical techniques to assess what we can say with differing degrees of confidence from the data, and develop new data and techniques to do better.

In general, the approach taken is to describe complex problems with simplified models that seek to represent key elements of the process under consideration. For example, an earthquake is a complicated rupture process that occurs in a finite volume and radiates seismic energy through the real materials of the earth. As we will see in the next few chapters, we represent all aspects of this process with simple models. We treat the complex faulting process as elastic slip on an infinitely narrow surface. We further treat the rock around it as a simple elastic material, and thus describe the complex seismic wave disturbance that propagates through it, using a number of simplifications.

It is important to bear in mind that these models are only approximations to a more complicated reality. For example, although the radiated seismic energy is real (it can destroy buildings), the mathematical descriptions used to understand it are human constructs. *P* waves, *S* waves, seismic phases like *ScS*, seismic ray paths, surface waves, or the earth's normal modes are all approximations that make the radiated energy easier to conceptualize. Similarly, we model a fault as a planar slip surface and use seismological observations to characterize the slip geometry and history. However, although this process nicely replicates the seismic observations, it only approximates the actual physics of earthquake rupture.

We often use a hierarchy of different approximations, as appropriate. For example, we might first predict the approximate time when a packet of seismic energy arrives by treating it as a seismic ray, and then use a more sophisticated wave or normal mode calculation to predict its amplitude and hence learn more about the properties of the parts of the earth it traversed. Similarly, we first describe the earth as isotropic (having the same properties in all directions) and purely elastic (no seismic energy is lost to heat by friction) and then confront the deviations from these simplifications.

A similar approach is often followed when discussing the tectonic context of earthquakes. Although faults, earthquakes, volcanoes, and topography are real, we associate these with the boundaries of plates that are human approximations. We will see that the questions of when to regard a region as a plate and how to characterize its boundaries are not simple. The simplest analyses assume that plates are rigid and divided by narrow boundaries. Later, we treat the boundaries as broad zones, and eventually we confront the fact that plates are not perfectly rigid, but in fact deform internally, as shown by earthquakes that occur within them.

We often choose a type of model to represent the earth and then use seismological and other data to estimate the parameters of this model. Thus a characteristic activity of seismology, and of the earth sciences in general, is solving *inverse problems*. We start with the end result, the seismograms, and work backwards using mathematical techniques to characterize the earthquakes that generated the seismic waves and the material the waves passed through. Inverse problems are more complicated than the conceptually simpler *forward problems* in which we use the theory of seismic wave generation and propagation to predict the seismogram that would be observed for a given source and medium. Inverse problems are harder to solve for several reasons. Seismograms reflect the combined effect of the source and medium, neither of which is known exactly. There are often aspects of the inverse problem that the data are insufficient to resolve. Thus seismology and other branches of the earth sciences, to a greater extent than most other scientific disciplines, often infer a "big picture" from grossly limited and insufficient data. For example, our images of the earth from seismic waves suffer from the fact that the severely limited geographical distributions of both earthquakes and seismometers leave most of earth's interior unsampled. This situation is like a doctor examining a possible broken bone with only a few scattered bursts of x-rays from random directions.

Moreover, although the forward problem typically can be solved in a straightforward way, giving a unique solution, the inverse problem often has no unique solution. In fact, the data are generally somewhat inconsistent due to errors, so no model can exactly describe the data. Finally, the fact that solving the inverse problem yields a set of model parameters that describe the observations well does not necessarily mean that the resulting model actually reflects physical reality. This nonuniqueness reflects the logical tenet that because *a* implies *b*, *b* does not necessarily imply *a*. In fact, we often have no way of determining what the reality is. For example, we will never truly know the composition and temperature of the earth's core because we cannot go there. This limitation remains in spite of the fact that over time our models of the core have become increasingly consistent with seismological data, experimental results about materials at high pressure and temperature, and other data including inferences from meteorites about the composition of the solar system.[6]

A consequence of this approach is the need to consider issues of precision, accuracy, and uncertainty. Estimates of quantities like the magnitude or depth of an earthquake depend both on the precision, or repeatability, with which data like seismic wave arrival times and amplitudes are measured, and on the accuracy, or extent to which the resulting inferences correctly describe the earth. For example, earthquake magnitudes are simple measures of earthquake size, estimated in various ways from seismograms without accounting for effects like the geometry of the earthquake source or lateral variations in seismic velocities. Hence measurements at different sites yield various estimates, so it is of little value to argue whether an earthquake had magnitude 5.2 or 5.4. Similarly, focal depths are derived from seismic wave arrival times by assuming a velocity structure near the earthquake, which is often not well known. For

[6] Similar difficulties afflict most of the earth sciences. Field geologists will never know whether their inferences about the past history and environment of a region are correct; paleontologists will never know how realistic their models of ancient life are, etc.

example, the depth is sometimes estimated (Section 4.3.3) from half the product of the time difference between the direct *P* and *pP* phases (see Fig. 1.1-3) and the velocity. If the time difference is measured to 0.25 s, and the velocity is 8 km/s, the method of propagation of errors (Section 6.5.1) shows that the uncertainty in depth is about 1 km, so it makes little sense to report the depth to greater precision. In reality the uncertainty will be greater, because the velocity also has some uncertainty. It is important to bear in mind that assigning a single value to an earthquake depth may exceed the relevant accuracy because faulting extends over a finite area that may be large (on the order of 10 km for a magnitude 6 earthquake). Moreover, when we have alternative models with which to estimate a parameter (for example, the earthquake stress drop estimated from body waves depends on the assumed geometry of the fault), the uncertainty associated with an estimate using any particular model underestimates the uncertainty due to the fact that we do not know which model is best. It is thus useful to examine how the estimate depends on the precision of the observation, the model parameters, and the choice of models.

Seismologists generally assume that the best estimates of values and uncertainties come from studies by different investigators using multiple datasets and techniques. Ideally, studies using the same data increase precision by reducing random errors, and studies using different data and techniques increase accuracy by reducing the effect of systematic errors. For example, for the well-studied Loma Prieta earthquake, seismic moment estimates vary by about 25%, and M_s values vary by about 0.1 units.

However, statisticians have long noted the difficulties in assessing probabilities and uncertainties. Two famous examples are the *Titanic*, described as "unsinkable" (probability zero) and the space shuttle, which was lost on its twenty-fifth launch, surprisingly soon given the estimated probability of accident of 1/100,000. Other examples come from the history of measurements of physical constants, which shows that the reported uncertainties underestimate the actual errors. For example, the 27 successive measurements of the speed of light between 1875 and 1958 are shown by subsequent analysis to be consistently in error by much more than the assigned uncertainty. It appears that assessments of the formal or random uncertainty often significantly underestimate the systematic error, so the overall uncertainty is dominated by the unrecognized systematic error and thus larger than expected. As a result, measurements of a quantity often remain stable for some time, and then change by much more than the previously assumed uncertainty. One possible explanation, termed the "bandwagon effect," is the tendency to discount data that are inconsistent with previous ideas, but later prove more accurate than those included. Another effect appears to be the discarding of outliers: for example, although R. Millikan reported using all the observations in his Nobel prize-winning (1910) study of the charge of the electron, his notebooks show that he discarded 49 of 107 oil drops that appeared discordant, increasing the apparent precision of the result. Until a method is developed that excludes obviously erroneous data without discarding real disconforming evidence, making realistic uncertainty estimates will remain a challenge. Although such analyses are more difficult in the earth sciences — for example, an earthquake is a nonrepeatable experiment — they are useful to bear in mind.

This discussion brings out the fact that although we often speak of "finding" or "determining" quantities like earthquake source parameters or velocity structure, it might be better to speak of "estimating" or "inferring" these quantities. There is no harm in the common and more upbeat phrasing so long as we remember that these values reflect uncertainties due to random noise and errors of measurement (sometimes called *aleatory* uncertainty, after the Latin word for dice) and systematic (sometimes called *epistemic*) uncertainty due to our choice of model to describe the phenomenon under consideration.

Although these caveats sound worrisome, seismological models are far from useless. We can usually develop models that not only describe the data used to develop them, but to predict other data. For example, earthquake source models derived only from seismology often predict the observations made using field geology and geodesy (ground deformation), both for the specific earthquake studied and for others in the same region. Moreover, the seismological results often give useful insight that is consistent with other lines of evidence. For example, seismology, gravity, and geomagnetism all favor the earth having a dense liquid iron core chemically different from the rocky mantle. This idea is also consistent with the fact that meteorites — thought to be fragments of small planets — are divided into stony and iron classes. Hence seismologists use this modeling approach to understand the earth, while recognizing its limitations.

For several reasons, our models usually improve with time. First, the data improve in both quantity and quality. Second, new observational and analytical techniques are introduced. As a result, long-standing problems such as the velocity structure of the earth are repeatedly reassessed. Successive generations of models seek to explain additional types of data, and often contain more model parameters in the hope of better representing the earth. Using statistical tests, we find that in some cases the resulting improvements are significant, whereas in others the new model improves only slightly on earlier ones. An important point is that more complicated models can always fit data better, because they contain more free parameters, just as a set of points in the x–y plane can be better fit by a quadratic polynomial than by a straight line. Thus we can statistically test models to see whether a new model reduces the misfit to the data more than would be expected purely by chance due to the additional parameters. Another useful test is whether the new or old models do a better job of describing data that were not used in deriving either, a process called pure prediction. When new models pass these tests, we can accept them — and then look again to see which data are still not described well and try to do better.

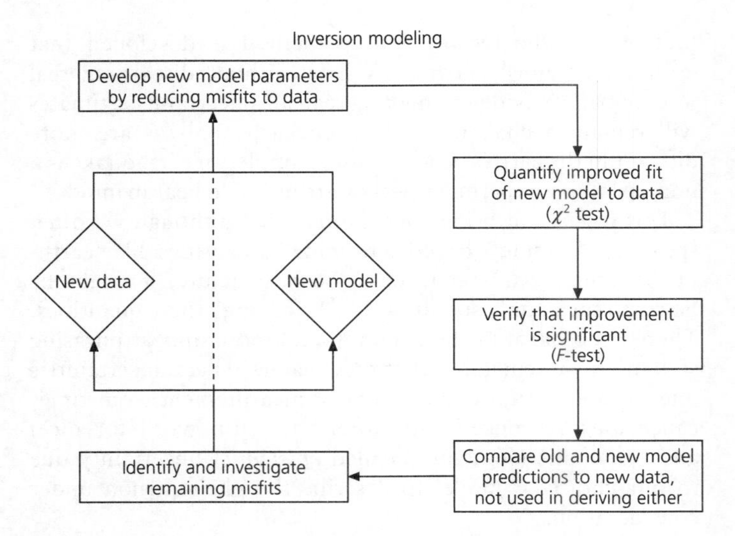

Fig. 1.1-8 Schematic illustration of how models of earth processes advance with time due to additional data and improved model parameterizations.

Over the years this process leads to a better understanding of how the earth works (Fig. 1.1-8). For example, Fig. 1.1-9 summarizes the development of global plate motion models, discussed in Chapter 5, that give the motion of the dozen or so major plates. The models are derived by inverting data consisting of the directions of plate motions along transform faults, the directions of plate motions during earthquakes, and the rates of plate motions shown by sea floor magnetic anomalies. Since 1972, when the first such model was made, the amount of available data has increased, and the data have become better, due to advances in seismology, sea floor imaging, and marine magnetic measurements. Similarly, the fit to the data has improved (or the misfit reduced) due both to the higher data quality and to improvements in the model, such as treating India and Australia as separate plates. Similar patterns of increased data and improved fit occur for many applications, including seismic velocity structure in the earth.

Many of the same issues surface when considering the models used to describe earth processes. For example, we will see that there are various models for what occurs at the core–mantle boundary or what causes earthquakes within downgoing plates at subduction zones. Such models assume that a particular set of physical processes occur, and show that for apparently plausible values of the (often unknown) relevant physical parameters, some behavior like that observed might be expected. Although these simple models attempt to reflect key aspects of the complex natural system, we often have no way of telling if and how well they succeed. Typically, various plausible models are suggested, all of which may in part be true and offer interesting insights into what may be occurring. The data often do not allow discrimination between them, so the model one prefers depends on one's geological instincts and prejudices, and models go in and out of vogue. A common scenario is for a model to become the consensus of the small group of researchers most interested in a problem, and then be challenged by fresh ideas or data from the outside. Hence, critically examining conventional wisdom often leads to discarding or modifying it, and so making progress in keeping with the

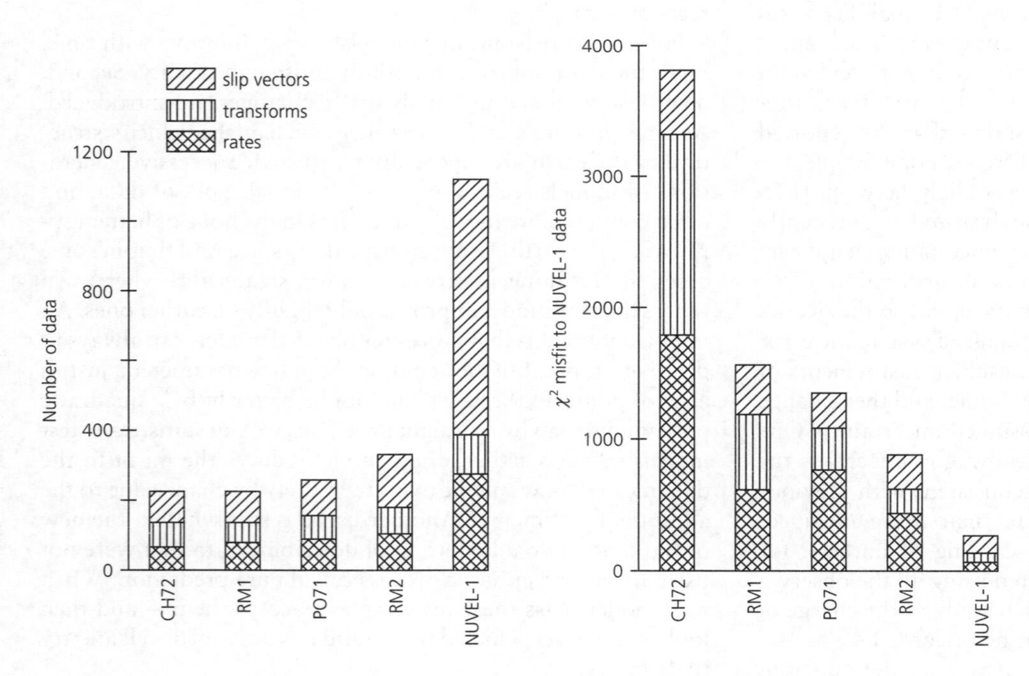

Fig. 1.1-9 Evolution of successive global plate motion models, as the amount of data increases and the misfit is reduced. *Left*: Number of data used to derive the models. Three types of data are inverted: earthquake slip vector azimuths, transform fault azimuths, and spreading rates. *Right*: The misfit to NUVEL-1 data for the various models. The vertical bars showing total misfit are separated into segments giving the misfit to each type of data. (DeMets *et al.*, 1990. *Geophys. J. Int.*, *101*, 425–78.)

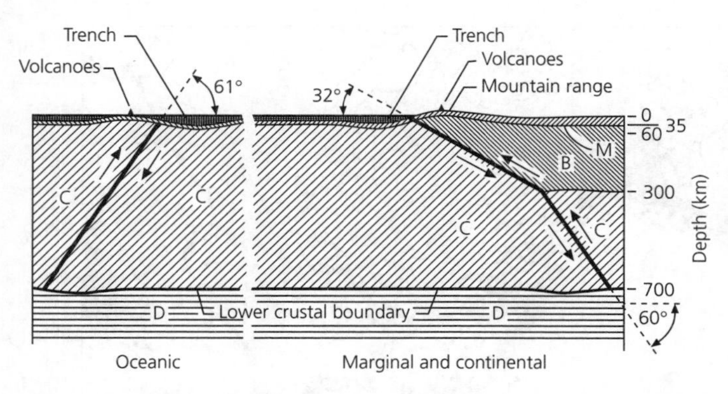

Fig. 1.1-10 Tectonic cartoon for oceanic and continental margin trenches, prior to the acceptance of plate tectonics. The association of dip-slip earthquakes with trenches, volcanism, and mountain ranges was recognized. Note the exaggeration of surface relief. (Benioff, 1955. From *Crust of the Earth*, ed. A. Poldervaart. Reproduced with permission of the publisher, the Geological Society of America, Boulder, CO. Copyright © 1955 Geological Society of America.)

ancient Jewish sages' observation that "the rivalry of scholars increases wisdom."[7] This process requires a constant cycle of learning and unlearning in which old models are discarded, even by those who helped create them, in favor of new models.

The classic geological example of advancing beyond conventional thinking is the plate tectonic revolution of the late 1960s. Although the idea of continental drift had been around for a long time and was strongly advocated by Alfred Wegener in 1915, it was not accepted by most of the geological community in the USA and Europe,[8] in part because seismological pioneer Harold Jeffreys argued that it was impossible. As a result, although it was recognized in the 1950s that earthquakes occurred on mid-ocean ridges that were young volcanic features and at deep sea trenches in association with volcanoes and mountain ranges (Fig. 1.1-10), their underlying nature was not understood. However, once paleomagnetic and marine geophysical data led to the recognition that oceanic lithosphere formed at mid-ocean ridges and subducted at trenches, the seismological observations made sense.

Thus, as in other sciences, progress in understanding seismological problems is typically incremental during "normal science" periods, in which we make small steady advances. Occasionally, however, exciting "paradigm shifts" occur when important new ideas change our views from our previous conventional thinking and permit great advances. This concept, developed by philosopher of science Thomas Kuhn (1962) for science-wide conceptual revolutions like the theory of plate tectonics, also describes progress in subfields. It is particularly apt in seismology, because many major faults move at most slightly for many years — and then break dramatically in large earthquakes.

1.2 Seismology and society

Seismology impacts society through applications including seismic exploration for resources, earthquake studies, and nuclear arms control. These topics involve both scientific and public policy issues beyond our focus on using seismic waves to study earth structure, earthquakes, and plate tectonics. However, given the natural interest of these societal applications, we briefly discuss some issues in earthquake hazard analysis and nuclear test monitoring, in part to motivate our discussions of the basic science.

These topics have the interesting feature that the state of seismological knowledge influences policy, so scientific uncertainties have broad implications. The choice of earthquake preparedness strategies depends in part on how well earthquake hazards can be assessed, and nations' willingness to negotiate test ban treaties depend in part on their confidence that compliance can be verified seismologically. Seismology thus faces the challenge, familiar in other applications like global warming or biotechnology, of explaining both knowledge and its limits. Failure to do so can have embarrassing consequences. For example, since the 1960s the Japanese government has spent more than $1 billion on an earthquake prediction program premised on the idea that large earthquakes will be preceded by observable precursory phenomena, despite the fact that (as discussed shortly) many seismologists increasingly doubt that such phenomena exist. This approach has so far failed to predict destructive earthquakes, like that which struck the Kobe area in 1995, and has focused most of its efforts on areas other than those where these earthquakes occurred. Critics have thus argued that the program is scientifically weak, diverts resources that could be more usefully employed for basic seismology and earthquake engineering, and gives the public the misleading impression that earthquakes can currently be predicted. Based on the program's record to date, the government would have been wiser to listen to these critics and to have been more candid with the public.[1]

[7] Alternative formulations of this idea include David Jackson's observation, (Fischman, 1992); "as soon as I hear 'everybody knows' I start asking 'does everybody know this, and how do they know it?'" the quotation used as the epigraph to this book by Nobel Laureate Peter Medewar; and the adage attributed to 1960s political activist Abbie Hoffman that "sacred cows make the best hamburger."

[8] Interestingly, many geologists in Southern Hemisphere countries like Australia and South Africa accepted continental drift early on and never abandoned it.

[1] Such issues were eloquently summarized by Richard Feynman's (1988) admonition after the loss of the space shuttle *Challenger*: "NASA owes it to the citizens from whom it asks support to be frank, honest, and informative, so these citizens can make the wisest decisions for the use of their limited resources. For a successful technology, reality must take precedence over public relations, because nature cannot be fooled."

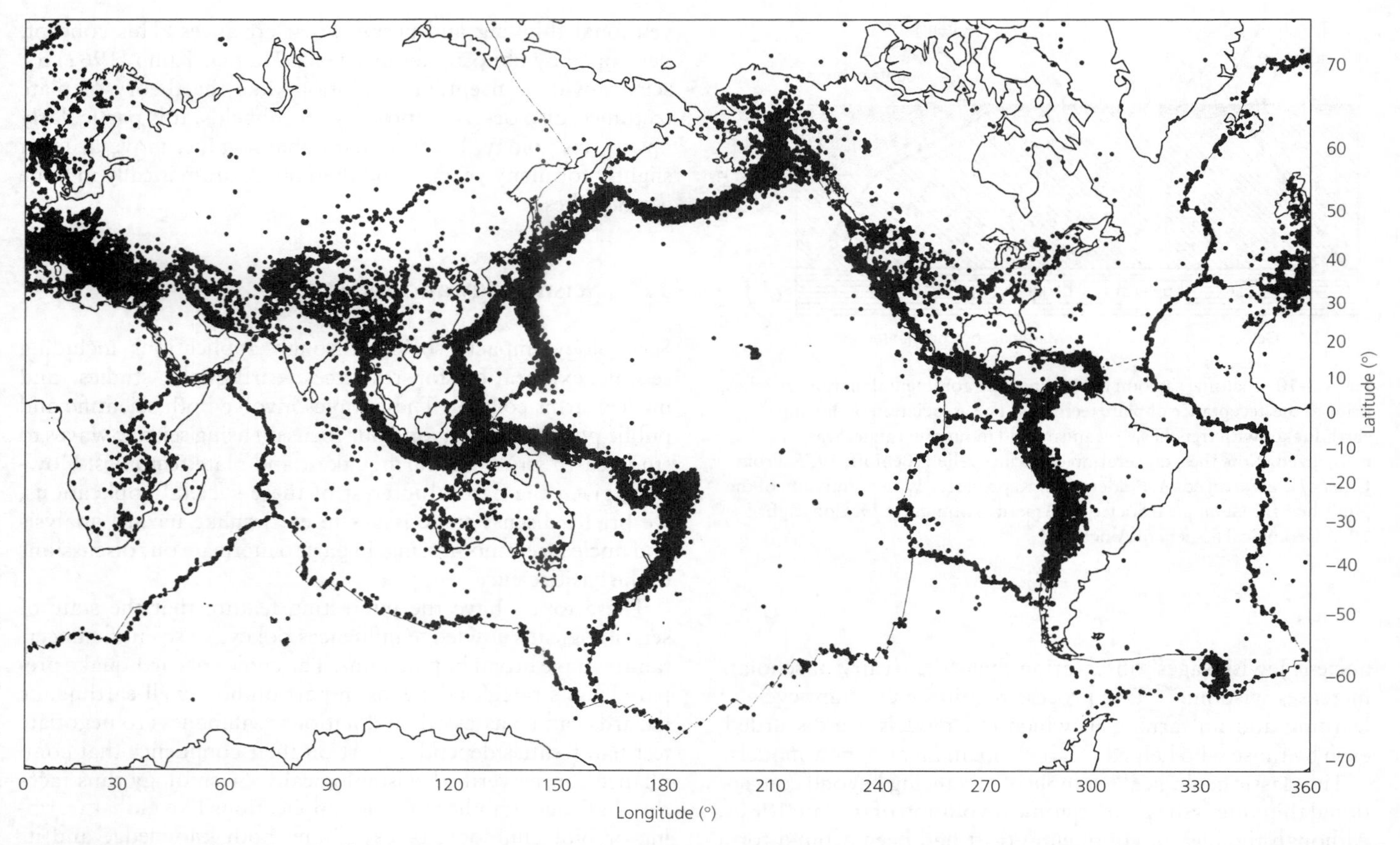

Fig. 1.2-1 Map showing epicenters of all earthquakes during 1963–95 with magnitudes of $m_b \geq 4$. Most earthquakes occur along the boundaries between tectonic plates. Where these boundaries are distinct, the earthquakes occur within narrow bounds. More diffuse plate boundaries, like the Himalayan plateau between India and China, show a much broader distribution of epicenters.

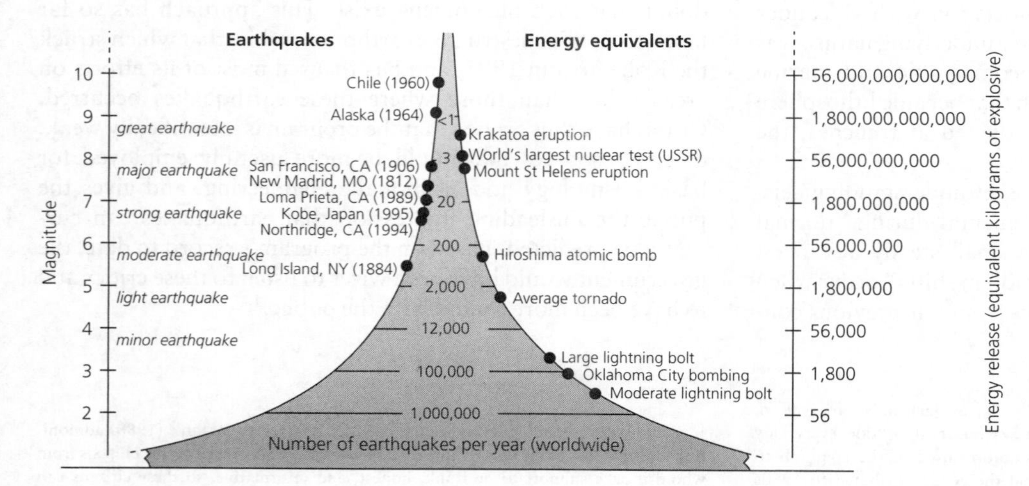

Fig. 1.2-2 Comparison of frequency, magnitude, and energy release of earthquakes and other phenomena. The magnitude used is moment magnitude, M_w. (After Incorporated Research Institutions for Seismology.)

1.2.1 *Seismic hazards and risks*

One of the primary motivations for studying earthquakes and seismology is the destruction caused by large earthquakes. In many parts of the world, seismic risks are significant, whether they are popularly recognized (as in Japan, where schools conduct earthquake drills) or not. Much of the challenge in assessing and addressing seismic hazards is that in any given area large earthquakes are relatively rare on human time scales, but can cause great destruction when they occur.

Earthquakes primarily occur at the boundaries where the 100 km-thick tectonic plates converge, diverge, or slide past each other. Although the plates move steadily, their boundaries are often "locked," and do not move most of the time. However, on time scales of a few hundred years, the boundary slips suddenly, and the accumulated motion is released in an earthquake. Figure 1.2-1 shows the locations of $m_b \geq 4$ earthquakes between 1963 and 1995. The earthquakes nicely define the plate boundaries, although some earthquakes also occur in *intraplate* regions, away from plate boundaries.

The energy released by large earthquakes is striking (Fig. 1.2-2). For example, the 1906 San Francisco earthquake involved about 4 m of slip on a 450 km-long fault, releasing about 3×10^{16} Joules[2] of elastic energy. This energy is equivalent to a 7 megaton nuclear explosion, much larger than the 0.012 megaton bomb dropped on Hiroshima. The largest recorded earthquake, the 1960 Chilean event in which about 21 m of slip occurred on a fault 800 km long and 200 km across, released about 10^{19} J of elastic energy, more than a 2000 Mt bomb. This earthquake released more energy than all the nuclear bombs ever exploded, the largest of which was 58 Mt. For comparison, the total global human annual energy consumption is about 3×10^{20} J.

Fortunately, the largest earthquakes are infrequent, because the energy released accumulates slowly over a long time. The San Francisco earthquake occurred on the San Andreas fault in northern California, part of the boundary along which the Pacific plate moves northward relative to the North American plate. Studies using the Global Positioning System satellites show that away from the plate boundary the two plates move by each other at a speed of about 45 mm/yr. Most parts of the San Andreas fault are "locked" most of the time, but slip several meters in a large earthquake every few hundred years. A simple calculation suggests that such earthquakes should occur on average about every 4000 mm/(45 mm/yr) or 90 years. The real interval is not uniform, for reasons that are unclear, and is longer, because some of the motion occurs on other faults.

Because plate boundaries extend for more than 150,000 km, and some earthquakes occur in plate interiors, earthquakes occur frequently somewhere on earth. As shown in Table 1.2-1, an earthquake of magnitude 7 occurs approximately monthly, and an earthquake of magnitude 6 or greater occurs on average every three days.[3] Earthquakes of a given magnitude occur about ten times less frequently than those one magnitude smaller. Because the magnitude is proportional to the logarithm of the energy released, most of the energy released seismically is in the largest earthquakes. A magnitude 8.5 event releases more energy than all the other earthquakes in a given year combined. Hence the hazard from earthquakes is due primarily to large (typically magnitude greater than 6.5) earthquakes.

In assessing the potential danger posed by earthquakes or other natural disasters, it is useful to distinguish between *hazards* and *risks*. The hazard is the intrinsic natural occurrence of earthquakes and the resulting ground motion and other effects. The risk is the danger the hazard poses to life and property. Hence, although the hazard is an unavoidable geological fact, the risk is affected by human actions. Areas of high hazard can have low risk because few people live there, and areas of modest hazard can have high risk due to large populations and poor construction. Earthquake risks can be reduced by human actions, whereas hazards cannot (hence the US government's National Earthquake Hazards Reduction Program is, strictly speaking, misnamed).

These ideas are illustrated by Table 1.2-2, which lists some significant earthquakes and their societal consequences. As shown, some very large earthquakes caused no fatalities because of their remote location or deep focal depth. In general, the most destructive earthquakes occur where large populations live near plate boundaries. The highest property losses occur in developed nations where more property is at risk, whereas fatalities are highest in developing nations. Although the statistics are often imprecise, the impact of major earthquakes can be enormous. Estimates are that the 1990 Northern Iran shock killed 40,000 people, and that the 1988 Spitak

Table 1.2-1 Numbers of earthquakes per year.

Earthquake magnitude (M_s)	Number per year	Energy released (10^{15} J/yr)
≥8.0	0–1	0–1,000
7–7.9	12	100
6–6.9	110	30
5–5.9	1,400	5
4–4.9	13,500	1
3–3.9	>100,000	0.2

Based upon data from the US Geological Survey National Earthquake Information Center. Energy estimates are based upon an empirical formula of Gutenberg and Richter (Gutenberg, 1959), and the magnitude scaling relations of Geller (1976), and are very approximate.

[2] The SI unit of energy is 1 Joule (J) = 1 Newton meter (N-m) = 10^7 ergs = 10^7 dyn-cm. Nuclear explosions are often described in megatons (Mt), equivalent to 1,000,000 tons of TNT or 4.2×10^{15} J.

[3] As part of his incorrect prediction of a magnitude 7 earthquake in the Midwest in 1990, I. Browning claimed that he had successfully predicted the 1989 Loma Prieta earthquake. In fact, he had said that near the date in question there would be an earthquake somewhere in the world with magnitude 6, a prediction virtually guaranteed to be true.

Table 1.2-2 Some notable and destructive earthquakes. (Values in this table are compiled from various sources, and different estimates have been reported, especially for older earthquakes.)

Location and date	Strength	Effects
Kourion, Cyprus July 21, 365	X MMI	Total destruction of this Greco-Roman city. Very large tsunami in the Mediterranean.
Basel, Switzerland October 18, 1356	XI MMI	Eighty castles destroyed over a wide area. 300 killed. Toppled cooking hearths caused fires that burned for many days.
Shansi, China January 23, 1556	8 M_s (est.)	Collapse of cave dwellings carved into bluffs of soft glacial loess. 830,000 reported killed (worst ever). Near the 1920 Kansu earthquake (see below).
Port Royal, Jamaica June 7, 1692	8 M_s (est.)	Widespread liquefaction caused one-third of Port Royal to spread and sink 4 m beneath the ocean surface. 2500 killed.
Lisbon, Portugal November 1, 1755	≥8 M_s (est.)	Large tsunamis seen all around the Atlantic. Felt over 1,600,000 km^2. Algiers destroyed. 70,000 killed. Largest documented earthquake in Europe (though several Italian quakes have killed >150,000 in past 500 years).
New Madrid, MO Dec. 1811 to Feb. 1812	7–7.4 M_s (est.)	Three large quakes (Dec. 16, 1811, Jan. 23, 1812, Feb. 7, 1812). Vertical movements up to 7 m. Widespread liquefaction. Changed course of Mississippi River. Felt over 5,000,000 km^2.
Charleston, SC August 31, 1886	7.2 M_s (est.)	No previous seismicity observed in this area between 1680 and 1886. Felt over 5,000,000 km^2. 14,000 chimneys damaged or destroyed. 90% of buildings damaged/destroyed. 60 killed.
Sanriku, Japan June 15, 1896	8.5 M_s (est.)	Tsunamis 35 m high washed away 10,000 houses and killed 26,000 along the Sanriku coast of Honshu. A similar Sanriku quake on March 2, 1933, killed 3000 with a 25 m high tsunami.
Assam, India June 12, 1897	8.7 M_s (est.)	One of the largest quakes ever felt. 1500 killed. Extremely violent ground shaking. Other Himalayan events on April 4, 1905 (20,000 killed), January 15, 1934 (10,000 killed), and August 15, 1950 (Ms = 8.6, 1526 killed).
San Francisco, CA April 18, 1906	7.8 M_s	About 4 m of slip on a 450 km-long fault. 28,000 buildings destroyed, largely by fires that burned for 3 days. 2500–3000 killed by fires (worst in USA).
Kansu, China December 16, 1920	8.5 M_s	180,000 killed, largely by downslope flow of liquefied soil over more than 1.5 km.
Tokyo, Japan September 1, 1923	8.2 M_s	Occurred in Sagami Bay, 80 km south of Tokyo. 134 separate fires merged to become a giant firestorm. 12 m tsunami hit shores of Sagami Bay. 143,000 killed.
Aleutian Islands, Alaska April 1, 1946	7.4 M_s	Large tsunami destroyed a power station and caused $25 million in damage in Hilo, Hawaii, where it rose to 7 m in height.
Lituya Bay, Alaska July 10, 1958	7.0 M_s	Massive landslides that slid into a local bay created a 60 m-high wave that washed up mountain sides as far as 540 m.
Hebgen Lake, MT August 17, 1959	7.5 M_s	Extensive landslides, including one that dammed a river and created a lake. Reactivated 160 Yellowstone geysers. Vertical displacement up to 6.5 m. 28 killed.
Chile May 21, 1960	9.5 M_w	Largest quake ever recorded. Fault area: 800 by 200 km. Slip: 21 m. Triggered eruption of Puyehue volcano. Massive landslides in Andes. Giant tsunami. 2000–3000 killed.
Alaska March 27, 1964	9.1 M_w	2nd largest quake ever recorded. Fault area: 500 by 300 km. Slip: 7 m. Large tsunamis, and widespread liquefaction. 200,000 km^2 of crustal surface deformed. 131 killed.
Peru May 31, 1970	7.8 M_s	Quake offshore caused large landslides. 30,000 killed, largely by 100,000,000 m^3 of rock and ice flowing down Andes mountain sides.
San Fernando Valley, CA February 9, 1971	6.6 M_s	Felt over more than 200,000 mi^2. 65 killed. 1000 injured. More than $500 million in direct losses.
Haicheng, China February 4, 1975	7.4 M_s	Successful prediction said to have led to an evacuation on the morning of the quake that possibly saved 100,000s of lives. 300–1200 killed.
Kalapana, Hawaii November 29, 1975	7.1 M_s	South flank of Kiluea volcano slid seaward. 14.6 m-high tsunami on Hawaiian shores. Largest Hawaiian earthquake since a 1868 quake that caused 22 m-high tsunamis and killed 148.
Tangshan, China July 27, 1976	7.6 M_s	Of a city of 1 million, >250,000 killed and 50,000 injured. Exact numbers speculative: fatalities may have exceeded the 1556 earthquake. In contrast to the 1975 Haicheng quake, this had no precursory behaviors.
Mexico City, Mexico September 19, 1985	7.9 M_s	Strong shaking lasted for 3 minutes due to sedimentary lake-fill oscillations. 10,000 killed. 30,000 injured. $3 billion in damage.
Spitak, Armenia December 7, 1988	6.8 M_s	Surface faulting showed 1.5 m of slip along a 10 km fault. 25,000 killed. 19,000 injured. 500,000 homeless. $6.2 billion in damages.
Loma Prieta, CA October 17, 1989	7.1 M_s	Slip along San Andreas segment south of San Francisco. 63 killed, most from the collapse of an elevated freeway in Oakland. About $6 billion in damages. Disrupted 5th game of World Series.
Caspian Sea, Iran June 20, 1990	7.7 M_s	100,000 structures damaged or destroyed. 40,000 killed. 60,000 injured. 500,000 left homeless. Over 700 villages destroyed, and another 300 damaged.
Luzon, Philippines July 16, 1990	7.8 M_s	Major rupture of Digdig fault, causing many landslides and major surface faulting. Extensive soil liquefaction. 1621 killed. 3000 injured.
Landers, CA June 28, 1992	7.3 M_w	Up to 6 m of horizontal displacement and 2 m of vertical displacement along a 70 km fault segment. 1 killed. 400 injured.

Table 1.2-2 (*cont'd*).

Location and date	Strength	Effects
Flores Island, Indonesia December 12, 1992	7.8 M_s	Tsunami heights reached 25 m. Extensive shoreline damage, where tsunami run-up was up to 300 m. 2200 killed. 30,000 buildings destroyed.
Northridge, CA January 17, 1994	6.7 M_w	Rupture on a blind thrust fault beneath Los Angeles. Many rock slides, ground cracks, and soil liquefaction. 58 killed. 7000 injured. 20,000 homeless. About $20 billion in damages.
Northern Bolivia June 9, 1994	8.2 M_s	Largest deep earthquake ever (depth was 637 km). Felt as far away as Canada.
Kobe, Japan January 16, 1995	6.8 M_s	5502 killed. 36,896 injured. 310,000 homeless. Massive destruction to world's 3rd largest seaport: 193,000 buildings, $100 billion in damages (highest to date).
NW of Balleny Islands March 25, 1998	8.2 M_w	Largest oceanic intraplate earthquake ever. Occurred west of Australia–Pacific–Antarctic plate triple junction in a region that was previously aseismic.
Izmit, Turkey August 17, 1999	7.4 M_s	5 m slip. 120 km rupture. 30,000 killed. $20 billion in economic loss. 12 major (M > 6.7) events this century have broken a total of 1000 km of the North Anatolian fault, including a 7.2 Mw aftershock on Nov. 12, 1999.
Chi-Chi, Taiwan September 21, 1999	7.6 M_w	150 km south of Taipei. 2333 killed. 10,000 injured. >100,000 homeless. Extensive seismic monitoring in Taiwan makes this one of the best seismically sampled earthquakes. One of largest observed surface thrust scarps.

(Armenia) earthquake killed 25,000. Even in Japan, where modern construction practices are used to reduce earthquake damage, the 1995 Kobe earthquake caused more than 5000 deaths and $100 billion of damage. On average during the past century earthquakes have caused about 11,500 deaths per year. As a result, earthquakes have had a significant effect upon the history and culture of many regions.

The earthquake risk in the United States is much less than in many other countries because large earthquakes are relatively rare in most of the country and because of earthquake-resistant construction.[4] The most seismically active area is southern Alaska, a subduction zone subject to large earthquakes. However, the population there is relatively small, so the 1964 earthquake (the second largest ever recorded instrumentally) caused far fewer deaths than a comparable earthquake would have in Japan. The primary earthquake impact in recent years has been in California. The 1994 Northridge earthquake killed 58 people and caused about $20 billion worth of damage in the Los Angeles area, and the 1989 Loma Prieta earthquake that shook the San Francisco area during a 1989 World Series baseball game killed 63 people and did about $6 billion worth of damage. Both these earthquakes were smaller (magnitude 6.8 and 7.1, respectively) than the largest known to occur on the San Andreas fault, such as the 1906 San Francisco earthquake, which had a magnitude of about 7.8.

Compared to other risks, earthquakes are not a major cause of death or damage in the USA. Most earthquakes do little harm, and even those felt in populated areas are commonly more of a nuisance than a catastrophe. Since 1811, US earthquakes have claimed an average of nine lives per year (Table 1.2-3), putting earthquakes at the level of in-line skating

[4] Many seismologists have faced situations like explaining to apprehensive telephone callers that the danger of earthquakes is small enough that the callers' upcoming family vacations to Disneyland are not suicidal ventures.

Table 1.2-3 Some causes of death in the United States, 1996.

Cause of death	Number of deaths
Heart attack	733,834
Cancer	544,278
Stroke	160,431
Lung disease	106,143
Pneumonia/influenza	82,579
Diabetes	61,559
Motor vehicle accidents	43,300
AIDS	32,655
Suicide	30,862
Liver disease/cirrhosis	25,135
Kidney disease	24,391
Alzheimer's	21,166
Homicide	20,738
Falling	14,100
Poison	10,400
Drowning	3,900
Fires	3,200
Suffocation	3,000
Bicycle accidents	695
Severe weather[1]	514
In-line skating[2]	25
Football[2]	18
Skateboards[2]	10
Earthquakes (1811–1983),[3] per year	9
Earthquakes (1984–98), per year	9

[1] From the National Weather Service (property loss due to severe weather is $10–15 billion/yr, comparable to the Northridge earthquake, and that from individual hurricanes can go up to $25 billion).
[2] From the Consumer Product Safety Commission.
[3] From Gere and Shah (1984).
All others from the National Safety Council and National Center for Health Statistics.

or football,[5] but far less than bicycles, for risk of loss of life. Similarly, the $20 billion worth of damage from the Northridge earthquake, though enormous, is about 10% of the annual loss due to automobile accidents. As a result, earthquakes pose an interesting challenge to society because they cause infrequent, but occasionally major, fatalities and damage. Society seems better able to accept risks that are more frequent but where individual events are less destructive.[6]

Similar issues surface when society must decide the costs, benefits, and appropriateness of various measures to reduce earthquake risks. Conceptually, the issues are essentially those faced in daily life. For example, a home security system costing $200 per year makes sense if one anticipates losing $1000 in property to a burglary about every five years ($200/year), but not if this loss is likely only once every 25 years ($40/year). However, the analysis is difficult, because the limited historical record of earthquakes makes it hard to assess their recurrence and potential damage.

Seismology is used in various ways to try to mitigate earthquake risks. Studies of past earthquakes are integrated with other geophysical data to forecast the location and size of future earthquakes. These estimates help engineers design earthquake-resistant structures, and help engineers and public authorities estimate and prepare for future damage by developing codes for earthquake-resistant construction. Seismology is also used by the insurance industry to develop rates for earthquake insurance, which can reduce the financial losses due to earthquakes and provide the resources for economic recovery after a damaging earthquake. Rates can be based on factors including the nature of a structure, its location relative to active faults, and soil conditions. Homeowners and businesses then decide whether to purchase insurance, depending on their perceived risk and the fact that damages must exceed a deductible amount (10–15% of the insured value) before the insurance company pays. A complexity for the insurer is that, unlike automobile accidents, whose occurrence is relatively uniform, earthquakes or other natural disasters are rare but can produce concentrated damage so large as to imperil the insurer's ability to pay claims. Approaches to this problem include limits on how much a company will insure in a given area, the use of reinsurance by which one insurance company insures another, catastrophe bonds that spread the financial risk into the global capital market, and government insurance programs.

1.2.2 *Engineering seismology and earthquake engineering*

Most earthquake-related deaths result from the collapse of buildings, because people standing in an open field during a large earthquake would just be knocked down. Thus it is often stated that in general "earthquakes don't kill people; buildings kill people." As a result, proper construction is the primary method used to reduce earthquake risks. This issue is addressed by engineering seismology and earthquake engineering, disciplines at the interface between seismology and civil engineering. Their joint goal is to understand the earthquake ground motions that can damage buildings and other critical structures, and to design structures to survive them or at least ensure the safety of the inhabitants.

These studies focus on the strong ground motion near earthquakes that is large enough to do damage, rather than the much smaller and often imperceptible ground motions used in many other seismological applications. Two common measures are used to characterize the ground motion at a site. One is the *acceleration*, or the second time derivative of the ground motion. Accelerations are primarily responsible for building destruction. A house would be unharmed on a high-speed train going along a straight track, where there is no acceleration. However, during an earthquake the house will be shaken and could be damaged if the accelerations were large enough. These issues are investigated using seismometers called *accelerometers* that can operate during violent shaking close to an earthquake but are less sensitive to the smaller ground motion from distant earthquakes. The seismic hazard to a given area is often described by numerical models that estimate how likely an area is to experience a certain acceleration in a given time. For example, the hazard map in Fig. 1.2-3 predicts the maximum acceleration expected at a 2% probability in the next 50 years, or at least once during the next 2500 (50/0.02) years. These values are given as a fraction of "g," the acceleration of gravity (9.8 m/s^2).

A second way to characterize strong ground motion uses *intensity*, a descriptive measure of the effects of shaking. Table 1.2-4 shows values for the commonly used Modified Mercalli intensity (MMI) scale, which uses roman numerals ranging from I (generally unfelt) to XII (total destruction). Intensity is not uniquely related to acceleration, which is a numerical parameter that seismologists compute for an earthquake and engineers use to describe building effects. The table shows an approximate correspondence between intensity and acceleration, but this can vary. However, intensity has the advantage that it is inferred from human accounts, and so can be determined where no seismometer was present and for earthquakes that occurred before the modern seismometer was invented (about 1890). Although intensity values can be imprecise (a fallen chimney can raise the value for a large area), they are often the best information available about historic earthquakes. For example, intensity data provide much of what is known about the New Madrid earthquakes of 1811 and 1812 (Fig. 1.2-4). These large earthquakes are interesting in that they occurred in the relatively stable continental interior of the North American plate (Section 5.6). Historical accounts show that houses fell down (intensity X) in the tiny Mississippi river town of New Madrid, and several chimneys toppled (intensity VII) near St Louis. Intensities can be used to infer earthquake magnitudes, albeit with significant uncertainties. These data have been used to infer the magnitude (about 7.2 ±

[5] These figures are for American football; in other countries soccer, termed football there, is safer for players but more dangerous for spectators.

[6] For example, although considerable attention is paid to aviation disasters and safety, far more lives could be saved at far less cost by enforcing automobile seat belt laws.

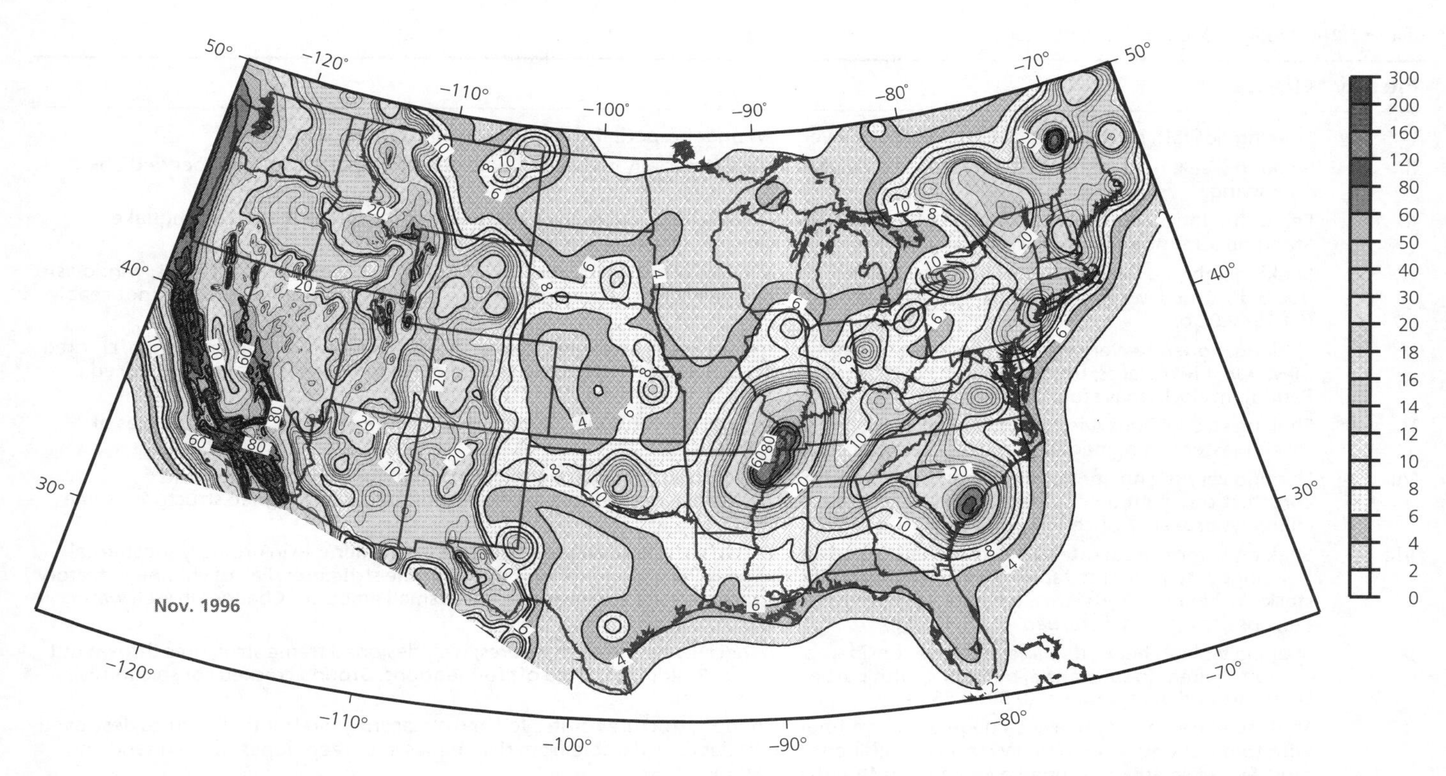

Fig. 1.2-3 A map of estimated earthquake hazards in the United States. The predicted hazards are plotted as the maximum acceleration of ground shaking expected at a 2% probability over a 50-year period. Although the only active plate boundaries are in the western USA, other areas are also shown as having significant hazards. (Courtesy of the US Geological Survey.)

0.3 in the study shown) and fault geometry of the historic earthquakes and to give insight into the effects of future ones.

The variation in ground motion with distance from an earthquake can be seen by plotting lines of constant intensity, known as *isoseismals*. Typically, as illustrated in Fig. 1.2-4, the intensity decays with distance from the earthquake. Similarly, strong motion data show that the variation in acceleration a with earthquake magnitude M and distance r from the earthquake can be described approximately by relations like

$$a(M, r) = b10^{cM}r^{-d}, \qquad (1)$$

where b, c, and d are constants that depend on factors including the geology of the area in question, the earthquake depth and fault geometry, and the frequency of ground motion. Hence the predicted ground acceleration increases with earthquake magnitude and falls off rapidly with distance at a rate depending on the rock type. For example, rocks in the USA east of the Rocky Mountains transmit seismic energy better than those in the western USA (Section 3.7.10), so earthquakes in the East are felt over a larger area than earthquakes of the same size in the West (Fig. 1.2-5). Because the shaking decays rapidly with distance, nearby earthquakes can do more damage than larger ones further away.

The damage resulting from a given ground motion depends on the types of buildings. As shown in Fig. 1.2-6, reinforced concrete fares better during an earthquake than a timber frame, which does better than brick or masonry. Hence, as also shown in Table 1.2-4, serious damage occurs for about 10% of brick buildings starting above about intensity VII (about 0.2 g), whereas reinforced concrete buildings have similar damage only around intensity VIII–IX (about 0.3–0.5 g). Buildings designed with seismic safety features do even better. The worst earthquake fatalities, such as the approximately 25,000 deaths in the 1988 Spitak (Armenia) earthquake, occur where many of the buildings are vulnerable (Fig. 1.2-7). Hence a knowledgeable observer[7] estimated that an earthquake of this size would cause approximately 30 deaths in California. This estimate proved accurate for the 1989 Loma Prieta earthquake, which was slightly larger and killed 63 people.

Designing buildings to withstand earthquakes is a technical, economic, and societal challenge. Research is being directed to better understand how buildings respond to ground motion and how they should be built to best survive it. Because such design raises construction costs and thus diverts resources from other uses, some of which might save more lives at less cost or otherwise do more societal good, the issue is to assess the seismic hazard and choose a level of earthquake-resistant

[7] Ambraseys (1989).

Table 1.2-4 Modified Mercalli intensity scale.

Intensity	Effects
I	Shaking not felt, no damage: not felt except by a very few under especially favorable circumstances.
II	Shaking weak, no damage: felt only by a few persons at rest, especially on upper floors of buildings. Delicately suspended objects may swing.
III	Felt quite noticeably indoors, especially on upper floors of buildings, but many people do not recognize it as an earthquake. Standing automobiles may rock slightly. Vibration like passing of truck. Duration estimated.
IV	Shaking light, no damage: during the day felt indoors by many, outdoors by very few. At night some awakened. Dishes, windows, doors disturbed; walls make creaking sound. Sensation like heavy truck striking building. Standing automobiles rocked noticeably. (0.015–0.02 g)
V	Shaking moderate, very light damage: felt by nearly everyone, many awakened. Some dishes, windows, and so on broken; cracked plaster in a few places; unstable objects overturned. Disturbances of trees and poles, and other tall objects sometimes noticed. Pendulum clocks may stop. (0.03–0.04 g)
VI	Shaking strong, light damage: felt by all, many frightened and run outdoors. Some heavy furniture moved; a few instances of fallen plaster and damaged chimneys. Damage slight. (0.06–0.07 g)
VII	Shaking very strong, moderate damage: everybody runs outdoors. Damage negligible in buildings of good design and construction; slight to moderate in well-built ordinary structures; considerable in poorly built or badly designed structures; some chimneys broken. Noticed by persons driving cars. (0.10–0.15 g)
VIII	Shaking severe, moderate to heavy damage: damage slight in specially designed structures; considerable in ordinary substantial buildings with partial collapse; great in poorly built structures. Panel walls thrown out of frame structures. Fall of chimneys, factory stacks, columns, monuments, walls. Heavy furniture overturned. Sand and mud ejected in small amounts. Changes in well water. Persons driving cars disturbed. (0.25–0.30 g)
IX	Shaking violent, heavy damage: damage considerable in specially designed structures; well-designed frame structures thrown out of plumb; great in substantial buildings, with partial collapse. Buildings shifted off foundations. Ground cracked conspicuously. Underground pipes broken. (0.50–0.55 g)
X	Shaking extreme, very heavy damage: some well-built wooden structures destroyed; most masonry and frame structures destroyed with foundations; ground badly cracked. Rails bent. Landslides considerable from river banks and steep slopes. Shifted sand and mud. Water splashed, slopped over banks. (More than 0.60 g)
XI	Few, if any, (masonry) structures remain standing. Bridges destroyed. Broad fissures in ground. Underground pipelines completely out of service. Earth slumps and land slips in soft ground. Rails bent greatly.
XII	Damage total. Waves seen on ground surfaces. Lines of sight and level destroyed. Objects thrown into the air.

Note: Parentheses show the average peak acceleration in terms of g (9.8 m/s), taken from Bolt (1999).

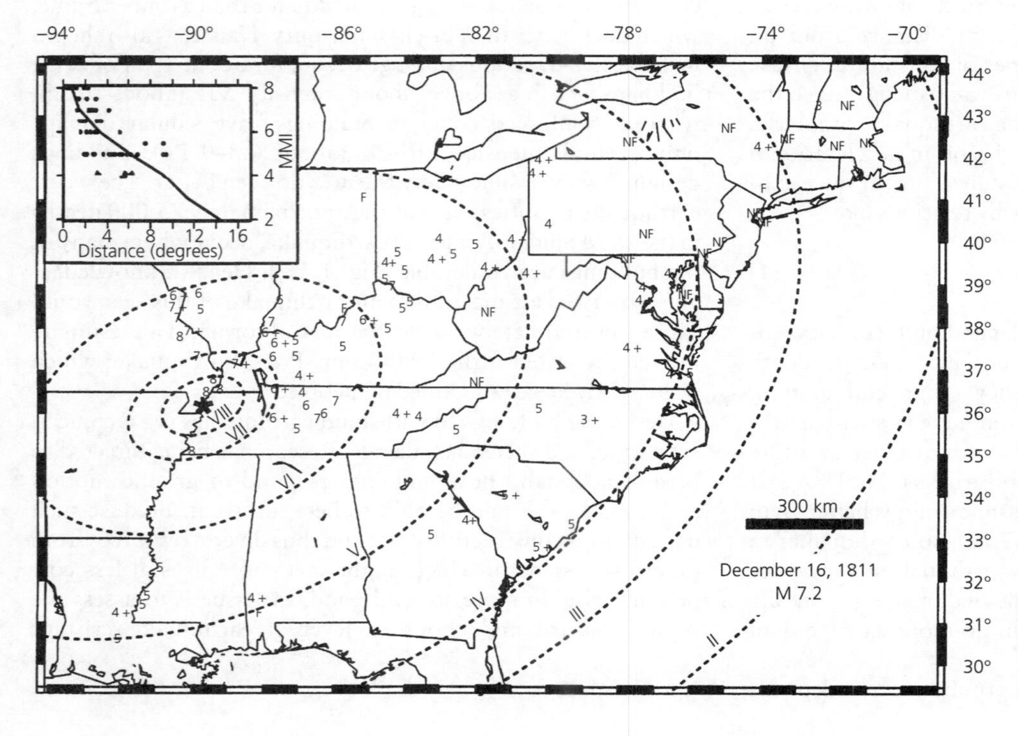

Fig. 1.2-4 Isoseismals for the first of the three largest earthquakes of the 1811–12 New Madrid earthquake sequence. Such plots, though based on sparse data, often provide the best assessment of historical earthquakes and of the effects of future ones. (After Hough *et al.*, 2000. *J. Geophys. Res.*, *105*, 23,839–64, Copyright by the American Geophysical Union.)

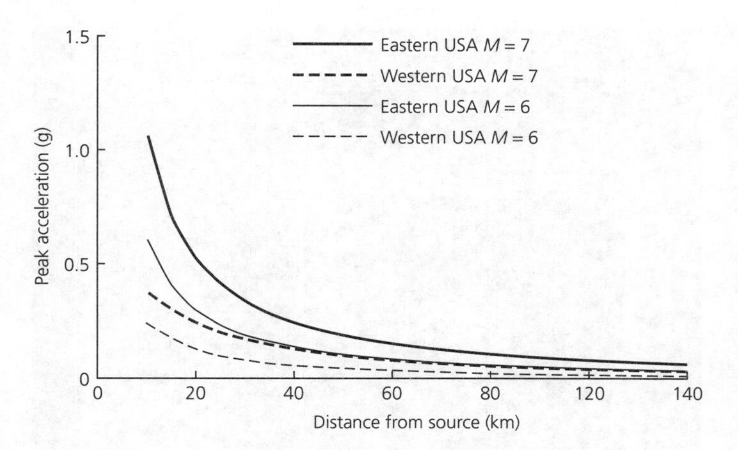

Fig. 1.2-5 Comparison of the predicted strong ground motion as a function of distance from magnitude 7 and 6 earthquakes in the eastern and western United States. Shaking from an earthquake in the east is comparable to that from one a magnitude unit larger in the west. The curves are computed from models by Atkinson and Boore (1995) and Sadigh *et al.* (1997).

Fig. 1.2-7 Five-story building in Spitak, Armenia, destroyed during the December 7, 1988, earthquake. The building was made from precast concrete frames that were inadequately connected. The failure of such buildings contributed greatly to the loss of 25,000 lives. (Courtesy of the US Geological Survey.)

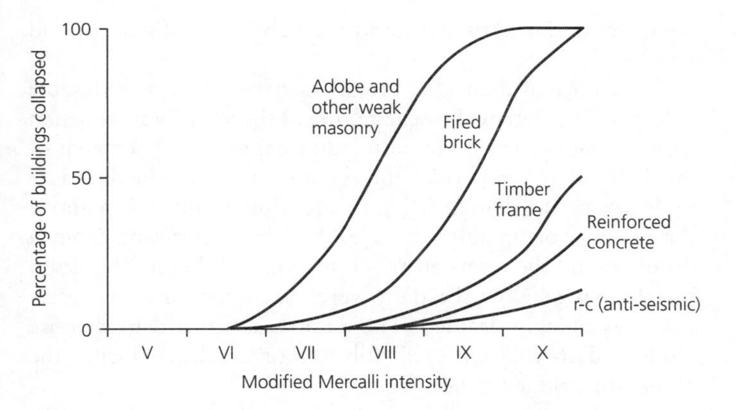

Fig. 1.2-6 Approximate percentage of buildings that collapse as a function of the intensity of earthquake-related shaking. The survival of buildings differs greatly for constructions of weak masonry, fired brick, timber, and reinforced concrete (with and without anti-seismic design). (After Coburn and Spence, *Earthquake Protection*, © 1992. Reproduced by permission of John Wiley & Sons Limited.)

construction that makes economic sense. Countries like the USA and Japan have the financial resources to study the effects of shaking on buildings, develop codes of appropriate building construction, and build structures to meet those codes. The task for building codes is to not be too weak, permitting unsafe construction and undue risks, or too strong, imposing unneeded costs and encouraging their evasion. Deciding where to draw this line is a complex policy issue for which there is no unique answer. Making the appropriate decisions is even more difficult in developing nations, many of which face serious hazards but have even larger alternative demands for resources that could be used for seismic safety. A classic example is the choice between building schools for towns without them or making existing schools earthquake-resistant.

A related issue is ensuring that buildings are built to the codes, given the tendency to evade expensive regulations designed to deal with events that are infrequent on a human time scale. For example, much damage occurred during large earthquakes in Turkey in 1999 because the building codes were not enforced. It has been reported that walls crumbled, revealing empty olive oil cans inserted during construction to save the costs of concrete.

Much of what has been learned about safe construction has been via trial and error. In California, the first major set of building codes was enacted following the 1933 Long Beach earthquake, which did $41 million worth of damage and killed 120 people. With successive destructive earthquakes, engineers have acquired a better sense of what works best, and building codes have been modified. For instance, buildings have become more resistant to the lateral shear that accompanies horizontal shaking with the use of shear walls consisting of concrete reinforced with steel. Similarly, measures have been developed to retrofit older buildings to increase their earthquake resistance.

An important factor for earthquake engineers is that structures resonate at different periods. Although the resonant period or periods depend on the specific building geometry and materials, they generally increase with an increase in the height or base width of a building. For example, typical houses or small buildings have periods of about 0.2 s, whereas a typical 10-story building has a period around 1 s. If the peak energy of ground motion is close to a building's resonant period, and the shaking continues long enough, the building may undergo large oscillations and be seriously damaged. This effect is like a swing — pushing at random intervals will likely stop the swing, whereas pushing repeatedly at its resonant period gives the person on it a good ride. Through this mechanism, an earthquake can destroy certain buildings and not others. Similarly, a building might collapse after a magnitude 7 earthquake, but remain standing after a magnitude 8 event with peak energy at a lower frequency. Sometimes damage occurs because adjacent buildings resonate out of phase, making their tops collide.

Another crucial factor for earthquake-resistant construction is the ground material of the site. Loose sediments and other weak rocks at the surface enhance ground motion compared to bedrock sites. As shown in Section 2.4.5, near-surface sediments can increase ground displacements by more than an order of magnitude. For instance, during the 1989 Loma Prieta earthquake, areas that sustained the worst damage corresponded to ones of high risk identified on the basis of subsurface geology. The failures of buildings in the Marina district, the Bay Bridge, and the Nimitz freeway all occurred on sedimentary layers.

An example of these effects occurred in 1985 in Mexico City, which is built on the sedimentary fill of an ancient lake that has dried up since the time of the Aztecs. A magnitude 7.9 earthquake at the subduction zone to the west caused the sedimentary basin to shake for more than 3 minutes (an unusually long time) at a dominant period of about 2 s. The worst damage was sustained by buildings with 6–15 stories, which had resonant periods of 1–3 s. Shorter or taller buildings were less damaged because they did not resonate with the ground shaking. This damage pattern has repeated for successive earthquakes.

1.2.3 *Highways, bridges, dams, and pipelines*

Buildings are not the only challenge for earthquake-resistant construction. Highways, bridges, parking structures, landfills, dams, pipelines, and power plants present additional problems. Many of these structures are crucial to society, so considerable effort is made to ensure that they will survive earthquakes.

Elevated highways often fail during earthquakes. Most of the lives lost during the 1989 Loma Prieta earthquake were due to the collapse of the Nimitz freeway in Oakland. In Los Angeles, the I-5 freeway was built to withstand a large earthquake, but parts were destroyed during the 1971 San Fernando earthquake. These were rebuilt, but parts collapsed again during the 1994 Northridge shock. A dramatic highway failure occurred during the 1995 Kobe earthquake, when a 20 km length of an expressway supported by large concrete piers fell over, crushing many cars and trucks.

Similar problems beset bridges, as illustrated in the 1989 Loma Prieta earthquake. The Bay Bridge connecting San Francisco and Oakland is a double-deck bridge built in 1936 with little flexibility and rests on sedimentary rocks. A large piece of the upper span collapsed during the earthquake (Fig. 1.2-8), and the bridge was closed for months for repairs. By contrast, the Golden Gate Bridge, a suspension bridge built into bedrock, was designed to withstand a large amount of shaking and fared well.

Fig. 1.2-8 Damage to the Bay Bridge, connecting San Francisco and Oakland, from the October 17, 1989, Loma Prieta earthquake. The bridge is of old construction (1936), and its supports rest in sedimentary fill that amplifies ground shaking. (Courtesy of the US Geological Survey.)

The failure of dams due to earthquakes poses considerable risk, as illustrated by the near-failure of the lower Van Norman dam during the 1971 San Fernando earthquake. A segment of the dam 600 m long broke and slid into the reservoir (Fig. 1.2-9), lowering the dam by 10 m and leaving it only 1.5 m above the water. Fortunately, the area had been suffering from a drought, and the reservoir was only half full. Eighty thousand people living below the dam were evacuated, and the reservoir was quickly drained. The dam was replaced by a more modern dam that suffered only minor cracking during the 1994 Northridge earthquake.

Dams have the special problem that they can cause earthquakes. This seems counter-intuitive, because the added weight of the water should increase the pressure on the rock below and inhibit faulting, because the two sides of the fault are pressed together harder, requiring a greater force to overcome the friction. However, it seems that the water impounded by dams sometimes flows into the rock, lowering the friction across faults and making rupture easier. The effect can be noticeable; seismicity associated with the man-made lake in Koyna, India, seems to follow a seasonal curve, being more active following the rainy season when reservoir levels are higher. One earthquake in 1967 was large enough to kill 200 people. The possibility of reservoir-induced earthquakes is thus considered when designing dams.

The greatest cause of earthquake-related death and destruction, other than the collapse of buildings, is fire. An important contributor to this problem is that water pipelines can rupture, making fire fighting harder. In the 1906 San Francisco earthquake, many buildings were damaged by the shaking, but fires that lasted three days are thought to have done ten times more

Fig. 1.2-9 Failure of the lower Van Norman dam that occurred during the February 9, 1971, San Fernando valley earthquake. Flooding did not occur because the region had been experiencing a drought, and the water level was low. (Courtesy of the US Geological Survey.)

damage (Fig. 1.2-10). Following the 1923 Tokyo earthquake, fires caused by overturned cooking stoves spread rapidly through the city and were unstoppable, due to ruptured water pipes. Many of the over 140,000 deaths resulted from fire, including a fire storm that engulfed 40,000 people who fled to an open area to escape collapsing buildings. In modern cities, natural gas pipelines can rupture, allowing flammable gas to escape and ignite. After the 1994 Northridge and 1995 Kobe earthquakes, both of which happened at night, the wide outbreaks of fires were the first way that rescue efforts could identify the areas that sustained the greatest damage. People in earthquake-prone areas are taught to turn off the gas supply to their homes if they smell gas after a large earthquake.

Fig. 1.2-11 Aerial view of Valdez, Alaska, showing the inundation of the coastline following the great 1964 earthquake. The resulting tsunami was as high as 32 m in places. (National Geophysical Data Center. Courtesy of the US Department of the Interior.)

1.2.4 *Tsunamis, landslides, and soil liquefaction*

Spectacular exceptions to the truism that "earthquakes don't kill people, buildings kill people" include tsunamis, landslides, avalanches, and soil liquefaction. Earthquake hazard planning thus includes identifying sites where these risks are present.

Tsunamis are large water waves that occur when portions of the sea floor are displaced by volcanic eruptions, submarine landslides, or underwater earthquakes (Fig. 1.2-11). Tsunamis are not noticeable as they cross the ocean, but can be amplified dramatically upon reaching the shore. The 1896 Sanriku (Japan) earthquake caused 35 m-high tsunamis that washed away 10,000 homes and killed 26,000 people. Hawaii is especially susceptible to tsunamis from earthquakes around the Pacific rim. Tsunamis from the 1960 Chilean earthquake killed 61 people in Hawaii, and the 1946 Alaska earthquake created a 7 m-high tsunami that washed over and short-circuited a power station, plunging Hilo into darkness. To address these risks, tsunami warning systems have been developed that assess

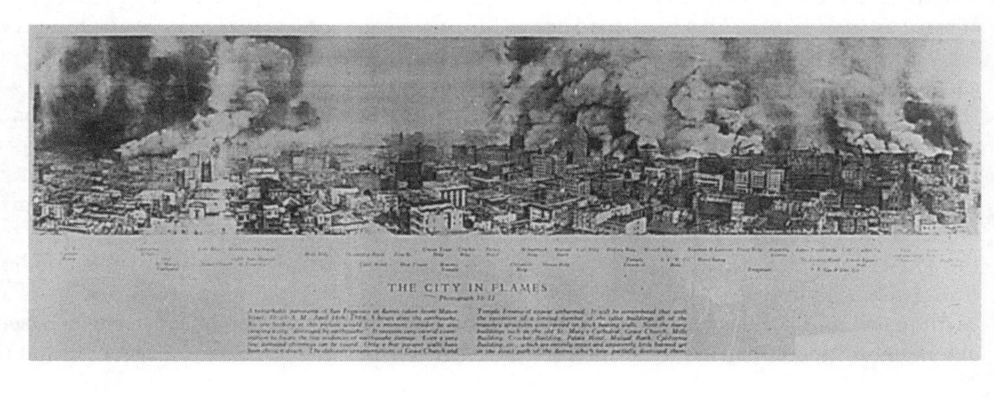

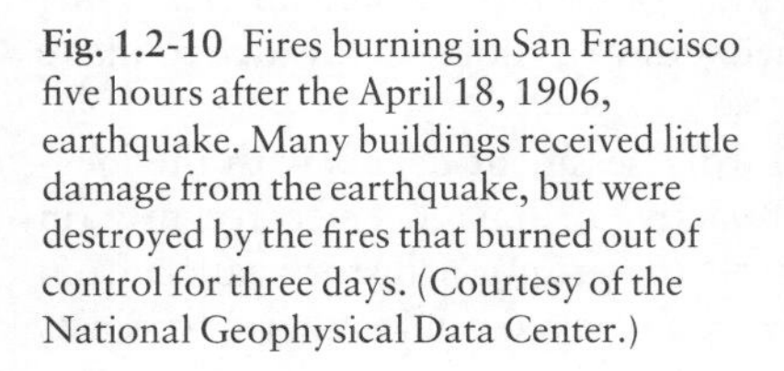

Fig. 1.2-10 Fires burning in San Francisco five hours after the April 18, 1906, earthquake. Many buildings received little damage from the earthquake, but were destroyed by the fires that burned out of control for three days. (Courtesy of the National Geophysical Data Center.)

Fig. 1.2-12 Landslide along California State Highway 17 in the Santa Cruz mountains, caused by shaking from the 1989 Loma Prieta earthquake. The landslide blocked the major commuter route between Santa Cruz and San Jose. (Courtesy of the US Geological Survey.)

Fig. 1.2-13 Damage to apartment buildings caused by soil liquefaction during the June 16, 1964, Niigata (Japan) earthquake. About a third of the city sank by as much 2 m as a result of sand compaction. (Courtesy of the National Geophysical Data Center.)

the likelihood that a large earthquake will generate a tsunami and issue warnings before the tsunami reaches distant areas.

Ground shaking in areas with steep topography can cause destructive landslides and avalanches (Fig. 1.2-12). For example, a 1970 earthquake in Peru caused rock and ice landslides that traveled downhill at speeds of 300 km/hr, burying villages and killing 30,000 people.

Another earthquake hazard involves *liquefaction*, a process by which loose water-saturated sands behave like liquids when vigorously shaken. Under normal conditions, the sand grains are in contact with each other, and water fills the pore spaces between them. Strong shaking moves the grains apart, so the soil behaves like a fluid slurry similar to "quicksand." Buildings can sink, otherwise undamaged, during the few seconds of peak ground shaking, and end up permanently stuck when the shaking stops and the soil resolidifies. A classic example is the tilting and sinking of buildings in Niigata, Japan, during a 1964 earthquake (Fig. 1.2-13).

Ground consisting of loose wet sediment is most susceptible to liquefaction. Sometimes the sand is ejected out of the surface as *sand blows*. This happened in the Marina district of the San Francisco waterfront during the 1989 Loma Prieta earthquake. Ironically, some of the material that erupted from the ground was building rubble from the 1906 San Francisco earthquake that had been bulldozed into the bay to make new waterfront property.

Liquefaction can be widespread and devastating, involving large downslope movements of soil called *lateral spreading*. In the 1920 Kansu, China, earthquake, downslope flows traveled over 1.5 km, killing 180,000 people. During the 1964 Alaska earthquake, parts of the Turnagain Heights section of Anchorage liquefied and collapsed. A dramatic example occurred on the island of Jamaica due to a magnitude 8 earthquake in 1692, where much of the town of Port Royal, built upon sand, sank about 4 m beneath the ocean. For years afterward, people on boats in the harbor could see houses below.

1.2.5 Earthquake forecasting

Reducing earthquake risks via resistant construction relies on identifying regions prone to earthquakes and estimating, even if crudely, how likely earthquakes are to occur and what shaking they might produce. Thus earthquake forecasting involves both scientific issues and the related question of how society can best use what seismology can provide.

Before addressing the predictions of earthquakes, it is useful to consider predictions for other geophysical processes. For example, severe storms are predicted in several ways. The first are long-term average forecasts: Chicagoans expect winter snowstorms, whereas Miamians expect fall hurricanes. Public authorities, power companies, homeowners, and businesses use the historical record of storms to prepare for them. Although surprises occur, long-term forecasting is generally adequate to ensure that needed resources (snow plows, salt) are available, whereas funds are not wasted on unneeded preparations (snow plows in Miami). Second, short-term weather forecasting often can identify conditions under which a storm is likely to form soon. Third, once formed, storms are tracked in *real time*, so people are often warned a day or more in advance to make preparations.

Similarly, volcanic hazard assessment begins with the location of volcanoes that are active or have been so recently (in geological terms). Based on the eruption history taken from historical accounts and the geologic record, long-term forecasts can be made. Short-term predictions are made using various phenomena that precede major eruptions: rising magma causes ground deformation, small earthquakes, and the release of volcanic gases. Finally, small eruptions usually precede a large one, making it possible to issue real-time warnings. Hence the record of volcanic predictions, though not perfect,[8] is reasonably good. The area around Mt St Helens was evacuated before the giant eruption of May 18, 1980, reducing the loss of life to only 60 people, including a geologist studying the volcano and citizens who refused to leave. The largest eruption of the second half of the twentieth century, Mt Pinatubo in the Philippines, destroyed over 100,000 houses and a nearby US Air Force base, yet only 281 people died because of evacuations during the preceding days.

Seismologists would like to do as well for earthquakes. We would like to be able to forecast where they are on average likely to occur in years to come, predict them a few years to hours before they occur, and issue real-time warnings after an earthquake has occurred in situations where such a warning would be useful. However, the record of seismology in these areas is mixed. To date there has been some success in long-term forecasting, little if any in short-term prediction, and some in real-time warning.

Earthquake forecasting, discussed in Section 4.7.3, estimates the probability that an earthquake of a certain magnitude will occur in a particular area during a specific time. For instance, a forecast might be a 25% probability of a magnitude 7 or greater earthquake occurring along the San Francisco segment of the San Andreas fault in the next 30 years. Forecasting uses the history of earthquakes on the fault and other geophysical information, such as the crustal motions measured using the Global Positioning System, to predict its likely future behavior. While forecasting is not relevant to short-term earthquake preparations, it is important in the enactment of building codes for earthquake-resistant construction, which are costly and require justification. Such forecasting is already successful in general ways; knowing that the San Andreas and nearby faults will be the sites of recurrent earthquakes has prompted building codes that are a major reason why the 1989 Loma Prieta and 1994 Northridge earthquakes caused few casualties.

Going beyond general forecasts is more difficult. For example, the probabilistic hazard map for the USA in Fig. 1.2-3 predicts a general pattern of higher hazards in areas of known past large earthquakes. Most of these, in California and Nevada, the Pacific Northwest, and Utah, are in the western USA, in the broad boundary zone between the Pacific and North American plates. In addition, high hazards are predicated in parts of the interior of the continent, near Charleston, South Carolina, and the New Madrid seismic zone in the Midwest. The map attempts to quantify this risk in terms of the maximum expected acceleration (recall that 0.2 g corresponds approximately to the onset of significant building damage) during a time interval. Such maps are made by assuming where and how often earthquakes will occur, how large they will be, and then using ground motion models like those in Fig. 1.2-5 to predict how much ground motion they will produce. Because these factors are not well understood, especially in intraplate regions where large earthquakes are rare, hazard estimates have considerable uncertainties.[9] For example, the high hazard predicted for parts of the Midwest, exceeding that in San Francisco or Los Angeles, results from specific assumptions, and alternative assumptions yield quite different estimates (Fig. 1.2-14).

Similarly, hazard estimates depend on the probability and hence recurrence time considered. Where the largest earthquakes are expected about every 200 years — for example, near a plate boundary as in California — a hazard map predicting the maximum acceleration expected at a 10% probability in the next 50 years, or at least once during the next 500 (50/0.1) years, will be similar to one for 2% probability in the next 50 years, or at least once during the next 2500 (50/0.02) years, because each portion of plate boundary is expected to rupture at least once in 500 years. However, the two maps would differ significantly where large earthquakes are less frequent — for example, in an intraplate region like the New Madrid zone (Sections 4.7.1, 5.6.3). This issue is important in choosing building codes because typical buildings have a useful life of about 50 years.

Because earthquakes are infrequent on a human time scale, it will be a long time before we know how well such estimates, which combine long-term earthquake forecasts and ground motion predictions, actually describe future earthquakes. Nonetheless, such estimates are used for purposes such as developing building codes and setting insurance rates. As a result, how to make meaningful predictions and hazard estimates, communicate their uncertainties to the public, and best use them for policy is a topic of discussion relevant not just to seismology but to the other earth sciences as well.

A key scientific challenge for hazard estimation is that the process determining when large earthquakes recur is unclear. The underlying basis for seismic forecasting is the principle of *elastic rebound* (Section 4.1). In this model, large-scale crustal motions, in most cases due to plate motions, slowly build up stress and strain across locked faults. When the stress reaches a critical threshold, seismic slip occurs along the fault, and the stress immediately drops. The process then begins again. The repeat time for these earthquakes depends on the rate at which crustal motions load the fault and the properties of the rocks that control when it slips.

[8] In 1982, uplift of the volcanic dome and other activity near the resort town of Mammoth Lakes, California, suggested that an eruption might be imminent. Geologists issued a volcano alert, resulting in significant tensions with local business leaders. When no eruption occurred, geologists were the target of much local anger, and the county supervisor who arranged for an escape route in the event of a volcanic eruption was recalled in a special election.

[9] Earthquake risk assessment has been described as "a game of chance of which we still don't know all the rules" (Lomnitz, 1989).

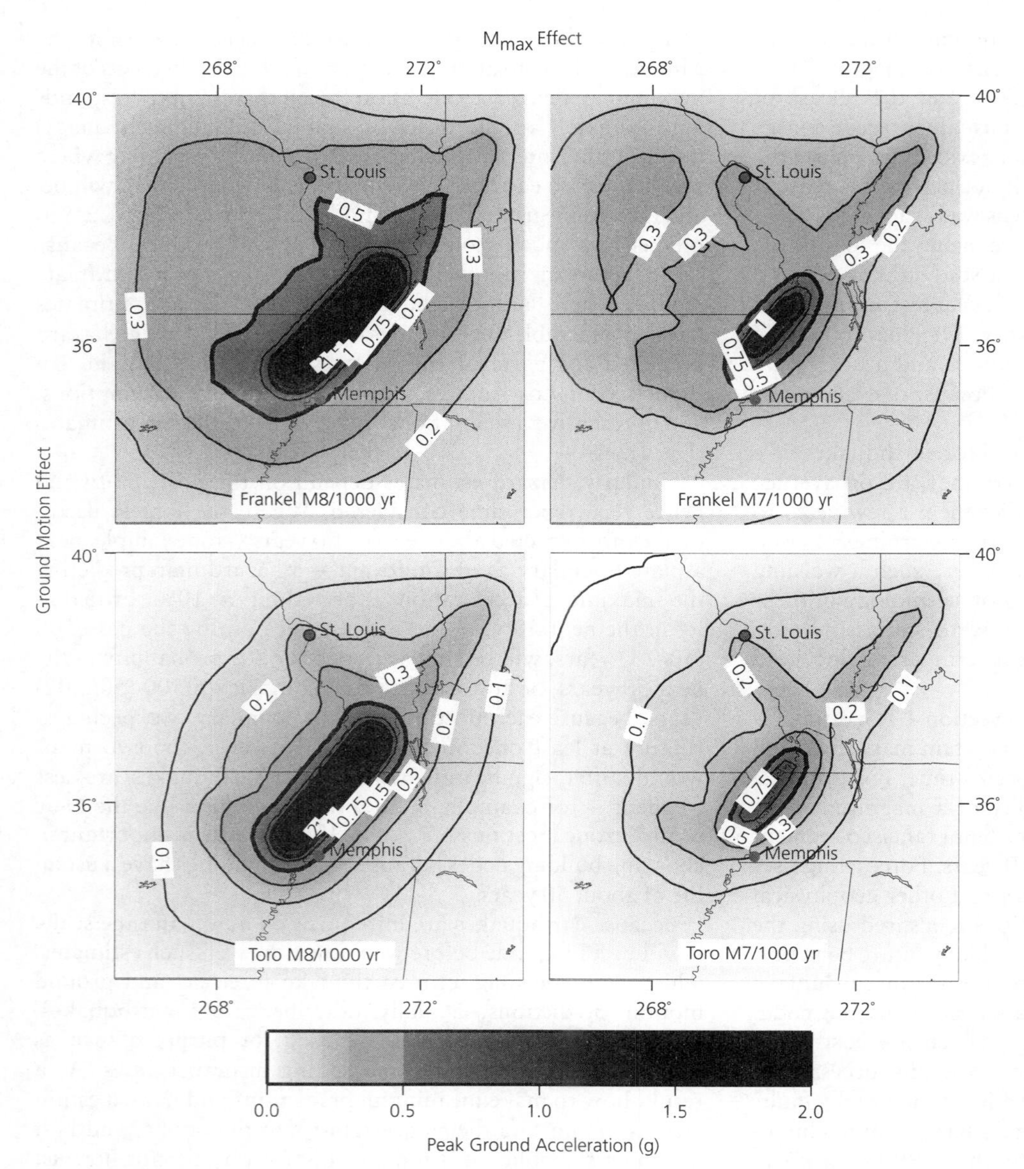

Fig. 1.2-14 Comparison of the predicted seismic hazard (peak ground acceleration expected at 2% probability in 50 years) from New Madrid seismic zone earthquakes for alternative parameter choices. Rows show the effect of varying the magnitude of the largest expected New Madrid fault earthquakes from 8 to 7, which primarily affects the predicted acceleration near the fault. Columns show the effect of two different ground motion models ("Frankel" and "Toro") which affect the predicted acceleration over a larger area. (Newman *et al.*, 2001.)

This idea implies that the history of large past earthquakes in an area should indicate the probable time of the next one. Naturally, the longer the history available, the better. Unfortunately, the duration of earthquake cycles is typically long compared to the approximately 100-year history of instrumental seismology. In some parts of the world, like China and Japan, historical records extend well into the past, whereas in the USA, the historic record is shorter. The earthquake history can be extended by *paleoseismology*, a branch of geology that studies the past history of faults. One of the best examples is the use of geological data to infer the history of large earthquakes on a major southern segment of the San Andreas fault. The last major earthquake recorded at a site at Pallett Creek, California, the 1857 Fort Tejon earthquake, is known from historical records to have caused shaking with an intensity of XI. The faulting is recorded by disruptions of sedimentary strata, including sand blows where material erupted during the earthquake. Sand blows and other structures from previous earthquakes were dated with radiometric carbon-14 methods, giving the dates of previous earthquakes. Despite the many uncertainties involved with these methods, including uncertainties in radiometric dating and the effects of climate variations and burrowing animals, the data show that faulting has recurred over the past thousands of years. However, assessing the size of past earthquakes and whether some earthquakes were missed is difficult.

The results can be surprising. For instance, large earthquakes near Pallett Creek appear to have occurred approximately in the years 1857, 1812, 1480, 1346, 1100, 1048, 997, 797, 734, and 671. Because the average time between events is 132 years,

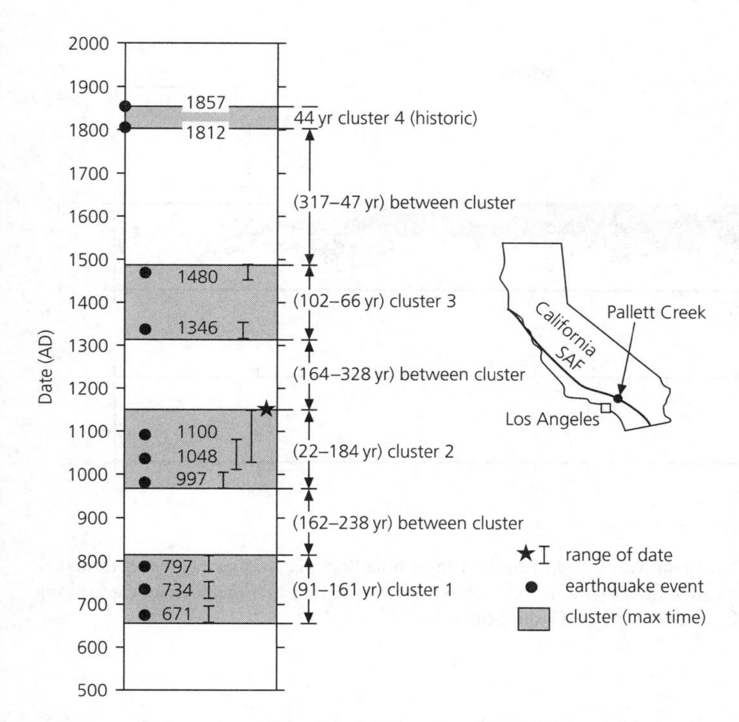

Fig. 1.2-15 Paleoseismic time series of earthquakes along the San Andreas fault near Pallett Creek, California, inferred from sedimentary deposits by Sieh *et al.* (1989). The sequence shows earthquake clusters separated by longer time intervals, illustrating the complexity of earthquake recurrence. (Keller and Pinter, *Active Tectonics: earthquakes, uplift, and the landscape*, © 1996. Reprinted by permission of Pearson Education.)

we might have expected the next large earthquake around the year 1989. However, the intervals between earthquakes vary from 45 years to 332 years, with a standard deviation of 105 years. Thus, given these data right after the 1857 earthquake, the simplest view would be that the earthquake would likely recur between 1885 and 2093. However, the time history suggests that something more complicated is going on (Fig. 1.2-15), as illustrated by the fact that the standard deviation of the recurrence time is similar to its mean. It looks as if the earthquakes are clustered: three earthquakes between 671 and 797, then a 200-year gap, then three between 997 and 1100, followed by a 246-year gap. Hence, using the earthquake history to forecast the next big earthquake is challenging, and the study's authors concluded in 1989 that one could estimate the probability of a similar earthquake before 2019 as only somewhere in the range 7–51%. For example, if the cluster that included the 1812 and 1857 earthquakes is over, then it may be a long time until the next big earthquake there.

The variability of recurrence times is striking because these data span for a long time history (10 earthquake cycles) on a plate boundary where the plate motion causing the earthquake is steady. The history of most faults is known only for the past few cycles, and the Pallett Creek data imply that these may not be representative of the long-term pattern. The recurrence may be even more complicated for earthquake zones within plates, many of which seem to act for only a few earthquake cycles, and others of which may be one-time events. Research, some of which is discussed in Section 5.7, is going on to investigate this complexity.

Even with the dates of previous major earthquakes, it is difficult to predict when the next will occur, as illustrated by the segment of the San Andreas fault near Parkfield, California. Compared to the southern segment just discussed, or the northern segment on which the 1906 earthquake occurred, the Parkfield segment is characterized by smaller earthquakes that occur more frequently and appear much more periodic. Earthquakes of magnitude 5–6 occurred in 1857, 1881, 1901, 1922, 1934, and 1966. The average recurrence interval is 22 years, and a linear fit to these dates made 1988 the likely date of the next event. In 1985, it was predicted at the 95% confidence level that the next Parkfield earthquake would occur before 1993, which was the USA's first official earthquake prediction. A comprehensive observing system was set up to monitor electrical resistivity, magnetic field strength, seismic wave velocity, microseismicity, ground tilting, water well levels and chemistry (especially radon content), and motion across the fault. The well-publicized experiment[10] hoped to observe precursory behavior, which seemed likely because surface cracks were observed 10 days before the 1966 earthquake and a pipeline ruptured 9 hours before the shock, and to obtain detailed records of the earthquake at short distances. As of 2002, the earthquake had not yet happened, making the current interval (35 years and growing) the longest yet observed between earthquakes there. The next Parkfield earthquake will eventually occur, but its non-arrival to date illustrates both the limitations of the statistical approaches used in the prediction (including the omission of the 1934 earthquake on the grounds that it was premature and should have occurred in 1944) and the fact that even in the best of circumstances nature is not necessarily cooperative or easily predicted. For that matter, it is unclear whether the Parkfield segment of the San Andreas fault shows such unusual quasi-periodicity because it differs from other parts of the San Andreas fault (in which case predicting earthquakes there might not be that helpful for other parts), or whether it results simply from the fact that, given enough time and different fault segments, essentially random seismicity can yield apparent periodicity somewhere. As is usual with such questions, only time will tell.

Such seismic forecasting involves the concept of *seismic gaps*, discussed further in Sections 4.7.3 and 5.4.3. The idea is that a long plate boundary like the San Andreas or an oceanic trench ruptures in segments. We would thus expect steady plate motion to cause earthquakes that fill in gaps and occur at relatively regular intervals. However, the Pallett Creek and

[10] The costs involved (more than $30 million) led *The Economist* magazine (Aug. 1, 1987) to argue that "Parkfield is geophysics' Waterloo. If the earthquake comes without warnings of any kind, earthquakes are unpredictable and science is defeated. There will be no excuses left, for never has an ambush been more carefully laid."

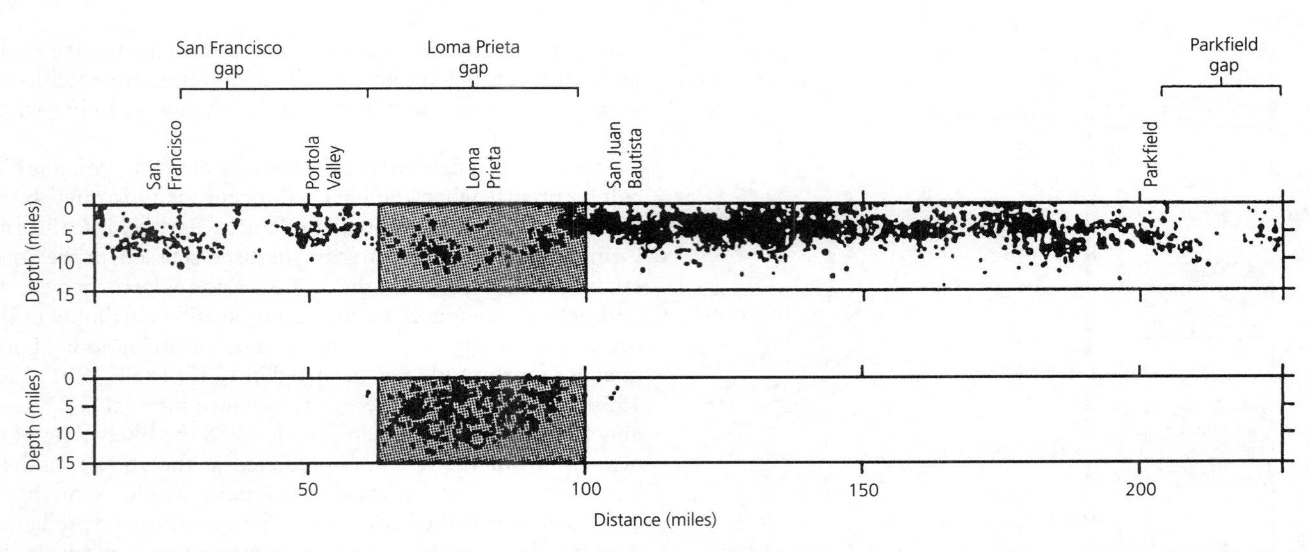

Fig. 1.2-16 Cross-section of the seismicity along the San Andreas fault before (top) and after (bottom) the 1989 Loma Prieta earthquake. This earthquake, whose rupture began at the large circle in the lower figure and is marked by the aftershocks (small circles), has been interpreted as filling a seismic gap along the San Andreas fault, although other interpretations have also been made. (Courtesy of the US Geological Survey.)

Parkfield examples show that the earth is more complicated. Some earthquakes may fit the gap idea; the 1989 Loma Prieta earthquake and its aftershocks have been interpreted as filling a gap along the San Andreas fault (Fig. 1.2-16), although the fact that the earthquake differed from the expected fault geometry has also been interpreted as making it different from the expected gap-filling earthquake. In other areas, however, the gap hypothesis has not yet proved successful in identifying future earthquake locations significantly better than random guessing. Faults deemed likely to rupture have not done so, and earthquakes sometimes occur on faults that were either unknown or considered seismically inactive. Understanding if, where, and when the gap hypothesis is useful is thus an active research area. Until it is resolved, it is unclear whether it is better to assume that all segments of a given fault are equally likely to rupture, making the probability of a major earthquake independent of time, or whether the segment that ruptured longest ago should have since accumulated the greatest elastic strain, and therefore be most likely to rupture next. This issue is important for hazard estimates.

In summary, several factors make earthquake forecasting difficult. In the meteorological case, storms occur frequently on human time scales, and we believe that we understand their basic physics. By contrast, the cycle of earthquakes on a given fault segment is long on a human time scale. Thus there are only a few places with a time history long enough to formulate useful hypotheses (recall that even the Pallett Creek 1000-year history shows major complexity). Moreover, because forecasts must be tested by their ability to predict future earthquakes, a long time will be needed to convincingly test models of earthquake recurrence and hazards. Even worse, the fundamental physics of earthquake faulting is not yet understood. Clearly, the process is complex. Earthquakes are at best only crudely periodic, and sometimes appear instead to cluster in time. Faults display a continuum of behavior from locking, to slow aseismic creep, to earthquakes. Thus the theoretical and experimental study of rock deformation and its application to earthquake faulting is an active field of research (Section 5.7).

1.2.6 *Earthquake prediction*

Earthquake prediction is defined as specifying within certain ranges the location, time, and size of an earthquake a few years to days before it occurs. Prediction is an even more difficult problem than long-term forecasting. A common analogy is that although a bending stick will eventually snap, it is hard to predict exactly when. To do so requires either a theoretical basis for knowing when the stick will break, given a history of the applied force, or observing some change in physical properties that immediately precedes the stick's failure.

Because little is known about the fundamental physics of faulting, many attempts to predict earthquakes have searched for *precursors*, observable behavior that precedes earthquakes. To date, as discussed next, this search has proved generally unsuccessful. As a result, it is unclear whether earthquake prediction is even possible. In one hypothesis, all earthquakes start off as tiny earthquakes, which happen frequently, but only a few cascade via a random failure process into large earthquakes.[11]

[11] This hypothesis draws on ideas from nonlinear dynamics or chaos theory, in which small perturbations can grow to have unpredictable large consequences. These ideas were posed in terms of the possibility that the flap of a butterfly's wings in Brazil might set off a tornado in Texas, or in general that minuscule disturbances do not affect the overall frequency of storms but can modify when they occur (Lorenz, 1993).

In this view, because there is nothing special about those tiny earthquakes that happen to grow into large ones, the interval between large earthquakes is highly variable, and no observable precursors should occur before them. If so, earthquake prediction is either impossible or nearly so.

Support for this view comes from the failure to observe a compelling pattern of precursory behavior before earthquakes. Various possible precursors have been suggested, and some may have been real in certain cases, but none have yet proved to be a general feature preceding all earthquakes, or to stand out convincingly from the normal range of the earth's variable behavior. Although it is tempting to note a precursory pattern after an earthquake based on a small set of data and to suggest that the earthquake might have been predicted, rigorous tests with large sets of data are needed to tell whether a possible precursory behavior is real and correlates with earthquakes more frequently than expected purely by chance. Most crucially, any such pattern needs to be tested by predicting future earthquakes.

One class of precursors involves *foreshocks*, earthquakes that occur before a main shock. Many earthquakes, in hindsight, have followed periods of anomalous seismicity. In some cases, there is a flurry of *microseismicity*: very small earthquakes like the cracking that precedes a bent stick's snapping. In other cases, there is no preceding seismicity. However, faults often show periods of either elevated or nonexistent microseismicity that are not followed by a large earthquake. Alternatively, the level of microseismicity before a large event can be unremarkable, occurring at a normal low level. The lack of a pattern highlights the problem with possible earthquake precursors: to date, no changes that might be associated with an upcoming earthquake are consistently distinguishable from the normal variations in seismicity that are not followed by a large earthquake.

Another class of possible precursors involves changes in the properties of rock within a fault zone preceding a large earthquake. It has been suggested that as a region experiences a buildup of elastic stress and strain, microcracks may form and fill with water, lowering the strength of the rock and eventually leading to an earthquake. This effect has been advocated based on data showing changes in the level of radon gas, presumably reflecting the development of microcracks that allow radon to escape. For example, the radon detected in groundwater rose steadily in the months before the 1995 Kobe earthquake, increased further two week before the earthquake, and then returned to a background level (Fig. 1.2-17).

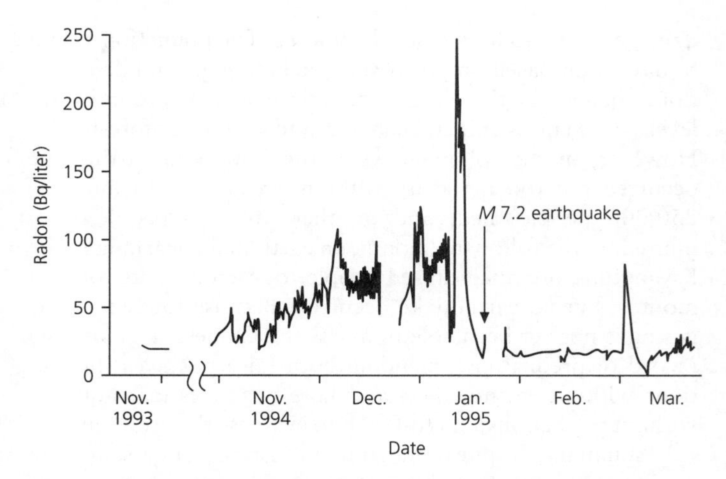

Fig. 1.2-17 Radon within groundwater before and after the January 16, 1995, Kobe earthquake in Japan. (Igarashi *et al.*, 1995. Reprinted with permission from *Science*, *269*, 60–1. Copyright 1995, American Association for the Advancement of Science.)

A variety of similar observations have been reported. In some cases, the ratio of *P*- and *S*-wave speeds in the region of an earthquake has been reported to have decreased by as much as 10% before an earthquake. Such observations would be consistent with laboratory experiments, and would reflect cracks opening in the rock (lowering wave speeds) due to increasing stress and later filling (increasing wave speeds). However, this phenomenon has not been substantiated as a general phenomenon. Similar difficulties beset reports of a decrease in the electrical resistivity of the ground before some earthquakes, consistent with large-scale microcracking. Changes in the amount and composition of groundwater have also been observed. For example, a geyser in Calistoga, California, changed its period between eruptions before the 1989 Loma Prieta and 1975 Oroville, California, earthquakes.

Efforts have also been made to identify ground deformation immediately preceding earthquakes. The most famous of these studies was the report in 1975 of 30–45 cm of uplift along the San Andreas fault near Palmdale, California. This highly publicized "Palmdale Bulge" was interpreted as evidence of an impending large earthquake and was a factor in the US government's decision to launch the National Earthquake Hazards Reduction Program aimed at studying and predicting earthquakes. However, the earthquake did not occur, and reanalysis of the data implied that the bulge had been an artifact of errors involved in referring the vertical motions to sea level via a traverse across the San Gabriel mountains. Subsequent studies, using newer and more accurate techniques including the Global Positioning System satellites, satellite radar interferometry, and borehole strainmeters have not yet convincingly detected precursory ground deformation.

An often-reported precursor that is even harder to quantify is anomalous animal behavior. What the animals are sensing (high-frequency noise, electromagnetic fields, gas emissions) is unclear. Moreover, because it is hard to distinguish "anomalous" behaviors from the usual range of animal behaviors, most such observations have been "postdictions," coming after rather than before an earthquake.

Despite these difficulties, Chinese scientists are attempting to predict earthquakes using precursors. Chinese sources report a successful prediction in which the city of Haicheng was evacuated in 1975, prior to a magnitude 7.4 earthquake that damaged more than 90% of the houses. The prediction is said to have been based on precursors, including ground deformation, changes in the electromagnetic field and groundwater levels, anomalous animal behavior, and significant foreshocks. However, in the following year, the Tangshan earthquake occurred not too far away without precursors. In minutes, 250,000 people died, and another 500,000 people were injured. In the following month, an earthquake warning in the Kwangtung province caused people to sleep in tents for two months, but no earthquake occurred. Because foreign scientists have not yet been able to assess the Chinese data and the record of predictions, including both false positives (predictions without earthquakes) and false negatives (earthquakes without predictions), it is difficult to evaluate the program.

In summary, despite tantalizing suggestions, at present there is still an absence of reliable precursors. The frustrations of this search have led to the wry observation that "it is difficult to predict earthquakes, especially before they happen." Most researchers thus feel that although earthquake prediction would be seismology's greatest triumph, it is either far away or will never happen. However, because success would be of enormous societal benefit, the search for methods of earthquake prediction will likely continue.

1.2.7 Real-time warnings

Some recent efforts are directed to the tractable goal of real-time warnings, where seismometers trigger an immediate warning if a set of criteria is met. For tsunamis, the warning may be several hours in advance, which is enough time for preparations. This is because tsunamis travel more slowly than seismic waves. A P wave travels from Alaska to Hawaii in about 7 minutes, whereas a tsunami traveling at about 800 km/hr across the ocean takes 5.5 hours. After the damage done to Hilo by the 1946 Alaska earthquake, the Seismic Sea Wave Warning System was organized for countries that rim the Pacific Ocean. Information from seismometers and tide gauges was phoned to the Tsunami Warning Center in Honolulu, Hawaii, which issued tsunami alerts if necessary.[12] Tsunami warning systems have since become more automated, using real-time digital seismic data to locate large earthquakes and derive information about their magnitudes, depths, and focal mechanisms. An assessment can be made of the likelihood of a tsunami, which usually results from vertical motion at the sea floor.

The situation is much more complicated with seismic waves. Although local seismic networks can automatically and immediately locate an earthquake and assess if it is hazardous, the warning time is short. For example, a warning after a major earthquake on the New Madrid fault system instantly relayed via Internet or radio to St Louis would arrive about 40 seconds before the first seismic waves. Seismologists, engineers, and public authorities are thus discussing what might be done with such short warning times. Although such times would not permit evacuations, certain steps might be useful. For example, real-time warnings are used in Japan to stop high-speed trains, and it may be practical to have gas line shut-off valves or other automatic responses connected to such a system. The questions are whether the improved safety justifies the cost and whether the risk of false alarms is serious.

A related approach is to provide authorities with near-real-time information, including data on the distribution of shaking, immediately after major earthquakes. Seismic networks are working to provide emergency management services with information that can help direct the needed response to the most affected areas during the chaotic few hours after a large earthquake, when the location and extent of damage are often still unclear.

1.2.8 Nuclear monitoring and treaty verification

Another important societal application of seismology is the monitoring of nuclear testing. Although atomic physics destabilized world politics through the invention of the atomic bomb, seismology has partially restabilized it. Throughout the cold war between the USA and the Soviet Union, seismology helped verify that treaties were being observed.

The role of seismology in nuclear monitoring began in 1957 when the USA detonated RAINIER, the first underground nuclear explosion. By the early 1960s it became clear that radioactive elements produced by atmospheric nuclear testing posed significant health threats. In 1963, 116 nations signed the Limited Test Ban Treaty, which banned nuclear testing in the atmosphere, in the oceans, and in space, and required testing to occur underground. At about this time, the US Air Force helped fund the deployment of the World Wide Standardized Seismographic Network (WWSSN). WWSSN stations provided important information for monitoring nuclear testing and a wealth of data that played a major role in modern geophysical seismology.

In 1976, countries began to abide by the Threshold Test Ban Treaty, which limited the size of underground nuclear tests to 150 kt (equivalent to 150 kilotons of TNT). Before then, the largest atmospheric test had been 58 Mt, and the largest underground test had been 4.4 Mt. Figure 1.2-18 shows the yields estimated seismologically for underground nuclear tests carried out by the Soviet Union. Although it was initially thought that some of the post-1976 explosions were greater than 150 kt, this turned out to reflect the different geologies of the western USA and central Asia. The conversion of seismic body wave magnitude m_b values into TNT yields was calibrated using the Nevada test site, but the western US crust is more seismically attenuating than the more stable Soviet sites in Kazakhstan and Novaya Zemlya (see Section 3.7.10). The yields of explosions in kilotons, Y, can be related to the observed seismic magnitudes by

[12] Serious or older television viewers may recall the episode of *Hawaii 5-0* in which criminals force the center to issue a spurious tsunami warning to prompt evacuation of downtown Honolulu and facilitate a robbery.

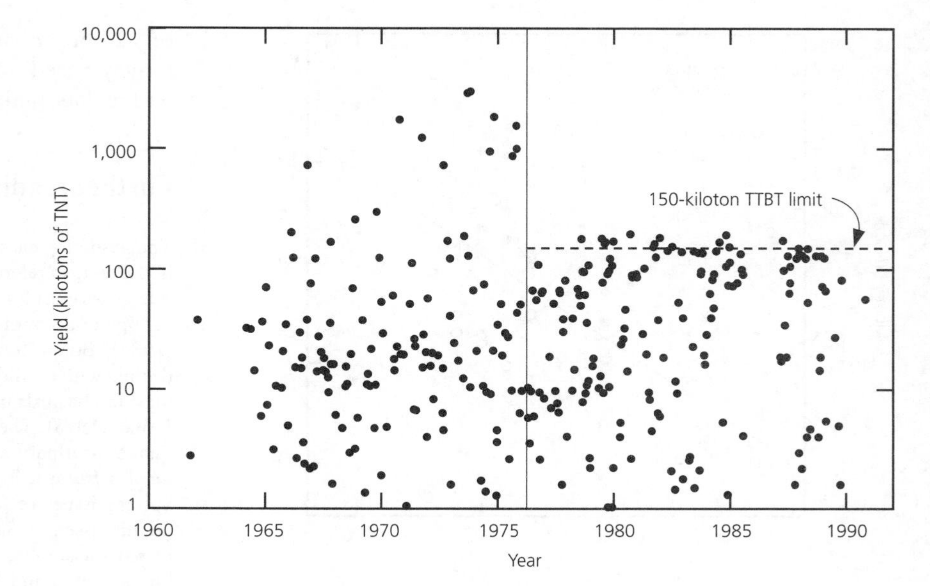

Fig. 1.2-18 Yields of underground nuclear tests carried out by the Soviet Union, determined through seismically observed m_b magnitudes. After the Threshold Test Ban Treaty (TTBT), seismology verified that the Soviet Union was in general compliance with the 150-kiloton limit. Data courtesy of P. Richards (personal communication).

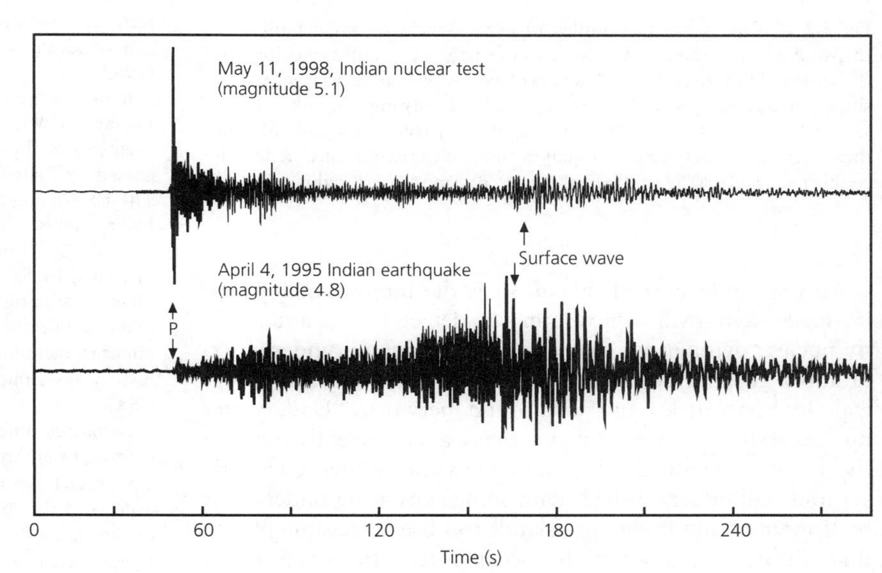

Fig. 1.2-19 Seismograms showing the differences between an earthquake and an explosion. For shallow earthquakes, in this case an m_b 4.8 shock in India, the *P* wave is much smaller than the surface waves. By contrast, the initial *P* wave is the largest arrival for explosions like this Indian nuclear test. Data recorded at Nilore, Pakistan. (Courtesy of the Incorporated Research Institutions for Seismology.)

$$m_b = C + 0.75 \log Y, \tag{2}$$

but the constant differs for Nevada ($C = 3.95$) and Kazakhstan ($C = 4.45$). With these corrections, it appears that the Soviet Union complied with the treaty.

Monitoring nuclear tests requires distinguishing them from earthquakes. Examples of the differences are shown in Fig. 1.2-19 for an earthquake and an explosion in India. Earthquakes occur by slip across a fault, generating large amounts of shear wave energy and hence large surface waves. By contrast, explosions involve motions away from the source, and so produce far less shear wave energy. Hence, for bombs the surface waves are dwarfed by the initial *P* wave. This difference is the basis for discrimination between earthquakes and explosions. A plot of M_s vs m_b (Fig. 1.2-20) separates earthquakes, which generate more surface wave energy (M_s), from the explosions, which generate more body (*P*) wave energy (m_b).

The challenge of seismic monitoring has increased in recent years. Since 1996 the USA has abided by the Comprehensive Test Ban Treaty (CTBT), which bans all nuclear testing, preventing the development of new nuclear weapons. Thus the focus of US monitoring efforts has expanded to include smaller countries around the world.[13] There is also the need to identify possible smaller nuclear tests, including those by terrorists. Hence seismic monitoring must identify explosions less than 1 kt, which have a magnitude of 4–4.5 (Eqn 2). This requires locating and identifying more than 200,000 earthquakes and additional mining explosions every year.

[13] A strategy described as "In God we trust, all others we verify."

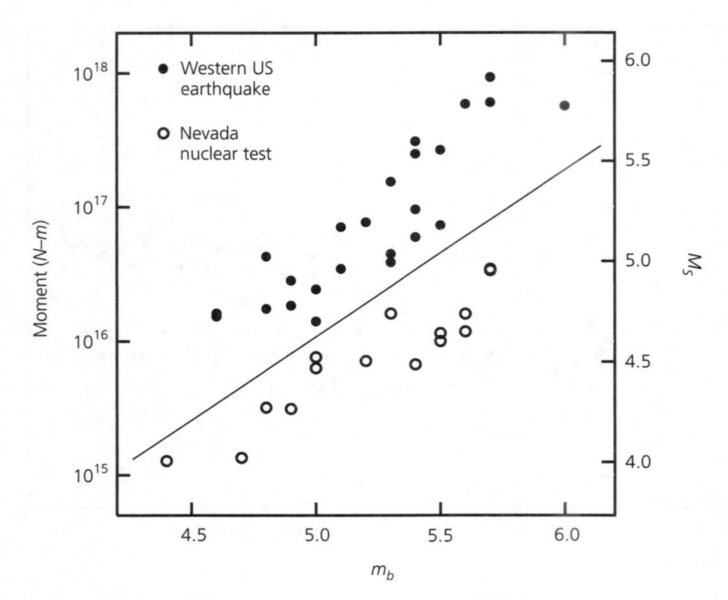

Fig. 1.2-20 Body wave magnitudes (m_b) versus surface wave magnitude (M_s) and seismic moment (M_0) for a set of earthquakes and explosions in the western USA. Because the *P* waves of explosions are very large, as shown in the previous figure, they have anomalously high m_b values for a given source energy (represented by M_0). A comparison of m_b and M_0 can thus discriminate between earthquakes and nuclear explosions. (After Al-eqabi *et al.*, 2001. © Seismological Society of America. All rights reserved.)

An important part of this effort is the International Monitoring System (IMS), whose aim is to detect, locate, and identify nuclear detonations that occur underground, underwater, or above ground. To do this, the IMS will combine seismological, hydroacoustic, and infrasound networks. Underwater nuclear tests create sound waves that travel efficiently through the ocean (Section 2.5.8), so a network of hydroacoustic stations will be established, with some sites using underwater hydrophones and others on islands to observe seismic phases that are generated when the oceanic acoustic waves reach land. Nuclear tests in the atmosphere will be detected by the infrasonic (frequencies less than 20 Hz, below the human hearing range) sound waves they generate. The IMS infrasound network will consist of small arrays of microphones that can determine the direction in which the infrasonic waves are traveling, so detection at multiple stations will identify the source of the waves.

Because most clandestine tests would likely occur underground, seismic stations will be a vital part of the IMS. The IMS seismic network will have 50 primary stations with three-component broadband seismometers. About half of these sites will be augmented with local arrays of short-period vertical-component sensors. Data will be telemetered in real time, so that there is no delay in monitoring. An auxiliary network of 120 broadband stations, distributed over 61 countries and largely based on existing networks, will aid in discrimination and replace malfunctioning primary stations.

Further reading

The seismological topics introduced in this chapter are discussed elsewhere in the text, so references are given in the appropriate sections. Many other references exist for the topics of societal interest discussed here.

Popular accounts of issues related to earthquakes include Gere and Shah (1984), Bolt (1999), and Brumbaugh (1999). Introductory treatments dealing with earthquakes and volcanoes from the point of view of the geology and hazards include Alexander (1993), Kovach (1995), and Sieh and LeVay (1998). The World Wide Web contains a wealth of general earthquake information; sites to start at include *http://www.scec.org*, *http://www.seismosoc.org*, *http://www.iris.edu*, and *http://earthquake.usgs.gov*. Specific issues related to volcano prediction studies at Mammoth Lakes are discussed by Sieh and LeVay (1998) and Hill (1998). For discussions of paleoseismology and geological effects of earthquakes, see Keller and Pinter (1996) and Yeats *et al.* (1997). The role of seismology in the plate tectonic revolution is discussed by Cox (1973) and Menard (1986); the general idea of scientific revolutions as "paradigm shifts" is given by Kuhn (1962).

Issues of assessing probabilities and uncertainties are discussed by Ekeland (1993); Henrion and Fischoff (1986) analyze the history of measurements of physical constants. Probabilistic seismic hazard analysis is discussed by Reiter (1990), Hanks and Cornell (1994), and Hanks (1997). The US Geological Survey National Seismic Hazard maps are described by Frankel *et al.* (1996), and a global hazard map is described by Shedlock *et al.* (2000). Uncertainties in earthquake probabilities for California are discussed by Savage (1991). Real-time seismology applications to earthquake risk mitigation are discussed by Kanamori *et al.* (1997). Sarewitz *et al.* (2000) discuss general issues of prediction and policy for the earth sciences, including earthquake prediction. Geschwind (2001) reviews the history of seismic risk mitigation and earthquake prediction policies in the USA.

A considerable volume of scientific literature addresses earthquake prediction, often arguing whether either a specific approach or any method can predict earthquakes. Turcotte (1991) gives a general review of many aspects of the topic, and Geller (1997) summarizes the history of earthquake prediction efforts, including that at Parkfield and the Palmdale Bulge. Geller *et al.* (1997) and Evans (1997) argue that earthquakes are unpredictable; Lomnitz (1994), Wyss *et al.* (1997), and Sykes *et al.* (1999) argue the other side. The Parkfield earthquake prediction experiment is summarized by Roeloffs and Langbein (1994); Davis *et al.* (1989) and Savage (1993) discuss the limitations of the statistical approach used. The controversy over the seismic gap hypothesis is discussed by Stein (1992); Kagan and Jackson (1991) and Jackson and Kagan (1993) argue against the hypothesis, and Nishenko and Sykes (1993) argue for it.

Earthquake engineering is discussed by Bray (1995), Chopra (1995), Krinitzsky *et al.* (1993), and Wiegel (1970). A good World Wide Web site to start at is *http://www.eeri.org*, which also provides an introduction to earthquake insurance. Issues in natural disaster insurance are discussed by Michaels *et al.* (1997).

Bolt (1976), Sykes and Davis (1987), Richards and Zavales (1990), and Lay (1992) discuss seismic verification of nuclear testing. More description of the Comprehensive Test Ban Treaty can be found at *http://pws.ctbto.org*.

2 Basic Seismological Theory

A very interesting example of sound waves in a solid, both longitudinal and transverse, are waves in the solid earth. Inside the earth, from time to time, there are earthquakes so sound waves travel around in the earth. Therefore if we place a seismograph at some location and watch the way the thing jiggles after there has been an earthquake somewhere else, we might get a jiggling, and a quieting down, and another jiggling . . . By using a large number of observations of many earthquakes at different places, we know what is inside the earth.

Richard Feynman, *The Feynman Lectures on Physics*, 1963

2.1 Introduction

We begin the study of seismic waves in the earth by addressing two basic questions. First, what in the physics of the solid earth allows waves to propagate through it? Second, how does the propagation of seismic waves depend on the nature of the material within the earth?

We will see that seismic waves propagate through the earth because the material within it, though solid, can undergo internal deformation. As a result, earthquakes and other disturbances generate seismic waves, which give information about both the source of the waves and the material they pass through.

To motivate these ideas, we first discuss a stretched string, a simple physical system that gives rise to waves analogous to seismic waves in the earth. As for the solid earth, deforming the string causes displacements that are functions of space and time satisfying the wave equation. The velocity of the propagating waves depends on the physical properties of the string in a way similar to that for waves in the earth, and the waves respond to changes in the physical properties of the string in ways analogous to what occurs for waves in the earth.

After discussing the string, we develop basic ideas about the mechanics of the solid earth. We introduce the stress tensor, which describes the forces acting within a deformable solid material, and the strain tensor, which describes the deformation. We then explore the relation between these tensors, and show that the displacements within the material can be described as functions of position and time satisfying the wave equation. Specifically, we will see how two types of seismic waves, *P* and *S*, propagate.

We then introduce concepts of wave propagation in the earth, with emphasis on how waves behave when they encounter changes in physical properties. These ideas give us the tools for Chapter 3, which discusses how seismic waves are used to study the interior of the earth, and Chapter 4, where we discuss how seismic waves are used to study earthquakes.

Although we focus on seismic waves, many of the concepts are similar to ones for other types of waves, so we will sometimes draw analogies to familar behavior of light, water, and sound waves.

2.2 Waves on a string

2.2.1 *Theory*

We consider an idealized mathematical string that extends in the x direction. Initially the string is straight in response to a tension force τ exerted along it, so u, the displacement from the equilibrium position in the y direction, is zero everywhere. After the string is plucked, portions of the string are displaced from their equilibrium positions and disturbances move along the string.

Our goal is to describe the displacement $u(x, t)$ as a function of both position along the string and of time. To do this, we apply Newton's second law of motion, $\mathbf{F} = m\mathbf{a}$, which states that the force vector equals the mass times the acceleration vector,[1] to a segment dx of the string. Once the string segment is displaced, the string is stretched and the tension directed along the

[1] Bold face is commonly used to denote vectors; see Section A.3.1.

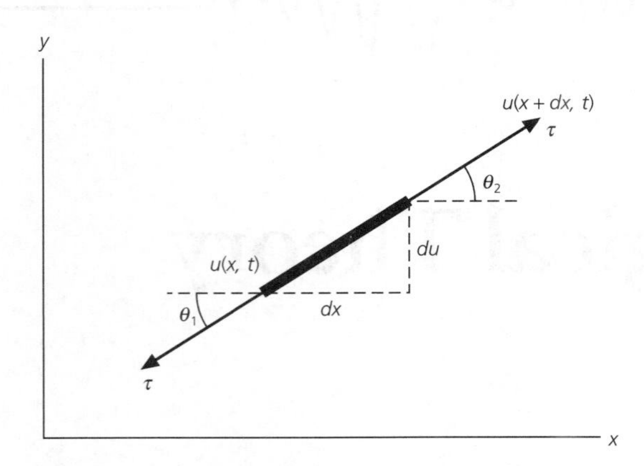

Fig. 2.2-1 Geometry of a segment of a string subject to a tension τ. A slight difference in the angles θ_1 and θ_2 provides a net force in the y direction of $F = \tau \sin \theta_2 - \tau \sin \theta_1$, which accelerates the string.

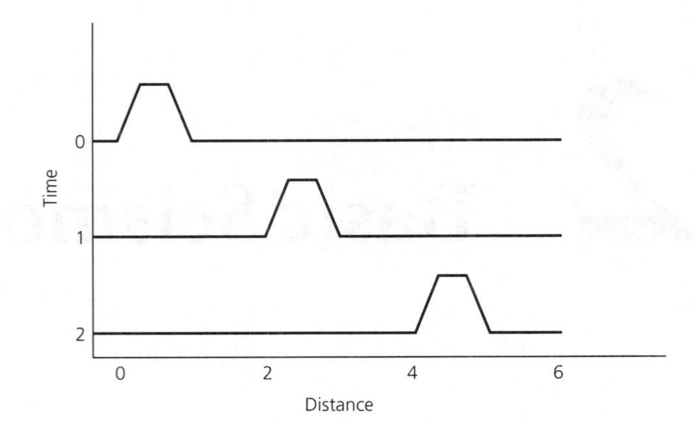

Fig. 2.2-2 "Snapshots" of a string showing a pulse $f(x - 2t)$ traveling to the right in the $+x$ direction. Because the velocity is 2, the pulse moves two distance units during each time unit. This pulse is one of many forms a traveling wave can take.

string gives rise to forces (Fig. 2.2-1) in the y direction of $\tau \sin \theta_2$ and $-\tau \sin \theta_1$ at the ends of the segment. The net force in the y direction equals the inertial term, which is the acceleration (second time derivative of the displacement) times the mass, where the mass is the product of the density ρ and dx. Hence, the vector equation $\mathbf{F} = m\mathbf{a}$ becomes the scalar equation

$$F(x, t) = \tau \sin \theta_2 - \tau \sin \theta_1 = \rho dx \frac{\partial^2 u(x, t)}{\partial t^2} \tag{1}$$

If the angles θ are small, $\sin \theta \approx \theta \approx \tan \theta$ can be approximated by the slope, so

$$\tau \left(\frac{\partial u(x + dx, t)}{\partial x} - \frac{\partial u(x, t)}{\partial x} \right) = \rho dx \frac{\partial^2 u(x, t)}{\partial t^2}, \tag{2}$$

which can be expanded by forming a Taylor series and discarding the higher-order terms:

$$\tau \left(\frac{\partial u(x, t)}{\partial x} + \frac{\partial^2 u(x, t)}{\partial x^2} dx - \frac{\partial u(x, t)}{\partial x} \right) = \tau \frac{\partial^2 u(x, t)}{\partial x^2} dx = \rho dx \frac{\partial^2 u(x, t)}{\partial t^2}, \tag{3}$$

yielding the *wave equation*:

$$\frac{\partial^2 u(x, t)}{\partial x^2} = \frac{1}{v^2} \frac{\partial^2 u(x, t)}{\partial t^2}, \tag{4}$$

where $v = (\tau/\rho)^{1/2}$.

This equation gives the relationship between the time and space derivatives of the displacement $u(x, t)$ along the string. We will see that the coupling between the two partial derivatives gives rise to waves propagating along the string with a velocity v. Because (4) describes the propagation of the scalar quantity $u(x, t)$ in one space dimension, it is called the *one-dimensional scalar wave equation*.

The wave equation is easily solved, because any function with the form $u(x, t) = f(x \pm vt)$ is a solution. To show this, note that the partial derivatives are

$$\frac{\partial^2 u(x, t)}{\partial x^2} = f''(x \pm vt) \quad \text{and} \quad \frac{\partial^2 u(x, t)}{\partial t^2} = v^2 f''(x \pm vt), \tag{5}$$

where f'' is the second derivative of f with respect to its argument. Thus, although we often think of solutions to the wave equation as sines and cosines, any function whose argument is $(x \pm vt)$ is a solution.

To see that a function $f(x - vt)$ describes a propagating wave, consider how it varies in space and time. As time increases by an increment dt, the argument stays constant provided that the distance increases by vdt. Because the function's value stays the same when its argument is constant, $f(x - vt)$ describes a wave of constant shape propagating with velocity v in the positive x direction (Fig. 2.2-2). Similarly, because $(x + vt)$ is constant if x decreases as time increases, $f(x + vt)$ describes a wave propagating with velocity v in the $-x$ direction. The sign relating the x and t terms thus shows which way the wave travels. We follow seismological convention and use the vector term "velocity" for v, although it is a scalar and thus better termed a "speed."

The velocity $v = (\tau/\rho)^{1/2}$ at which the waves propagate depends on two physical properties of the string: the tension with which it is stretched and its density. Equation 1 shows how these properties interact. Because the tension provides the force that tends to restore any displacement to the equilibrium position, greater tension gives higher acceleration and thus faster wave propagation. In contrast, because the density appears in the inertial term, higher density gives lower acceleration and slower wave propagation.

The fact that the velocity depends on the density illustrates one of the reasons why the string is a useful analogy for seismic waves in the earth. One goal of seismology is to study the composition of the earth. For this purpose, we measure the time that waves take to travel between sources and receivers, find the velocity at which the waves propagated, and thus learn about the properties of the earth.

2.2.2 *Harmonic wave solution*

Any function of the form $f(x \pm vt)$ describes a propagating wave as a function of time and distance. A particularly useful form is a *harmonic* or sinusoidal wave[2]

$$u(x, t) = Ae^{i(\omega t \pm kx)} = A\cos(\omega t \pm kx) + Ai\sin(\omega t \pm kx). \quad (6)$$

A harmonic wave is characterized by its amplitude A and two parameters, ω and k, which we will discuss shortly. Substituting into the wave equation (4) and canceling the exponential and constant show that the wave velocity is the ratio

$$v = \omega / k. \quad (7)$$

Although the exponential function $u(x, t)$ in Eqn 6 is complex, the physical displacement must be real. We thus describe the displacement as the real part of $u(x, t)$. The complex exponential form can be used for most purposes, because when a complex exponential appears in the solution of a physical problem, its conjugate also appears, so their sum yields a real displacement.

To understand the harmonic wave solution, consider the wave given by the real part of $u(x, t)$, which is $A\cos(\omega t - kx)$. Figure 2.2-3 shows how this function varies with both distance and time. The value of u is constant when the *phase* $(\omega t - kx)$ remains constant, as for a crest or a trough. Such lines of constant phase require that x increases when t increases. These lines indicate waves propagating in the $+x$ direction at a velocity shown by dx/dt, the slope of the line in the x–t plane.

Additional insight comes by examining $u(x, t)$ at a point in space, x_0. In terms of Fig. 2.2-3, this is a slice of the function on a plane parallel to the time axis, which intersects the distance axis at x_0. This gives a periodic function of time, $u(x_0, t) = A\cos(\omega t - kx_0)$ (Fig. 2.2-4, *top*). Because the function returns to the same value when ωt changes by 2π, the oscillation is characterized by the *period*, $T = 2\pi/\omega$, the time over which it repeats. The periodicity can also be described by the *frequency*, $f = 1/T = \omega/(2\pi)$, the number of oscillations within a unit time, or by the *angular frequency*, $\omega = 2\pi f$. The period has the dimensions of time, so the frequency and angular frequency have dimensions of time^{-1}. In Fig. 2.2-3, for example, $u(x, t) = A\cos(\pi t - 2\pi x)$, so the angular frequency is π (time units)$^{-1}$, the frequency is $1/2$ (time units)$^{-1}$, and the period is 2 time units.

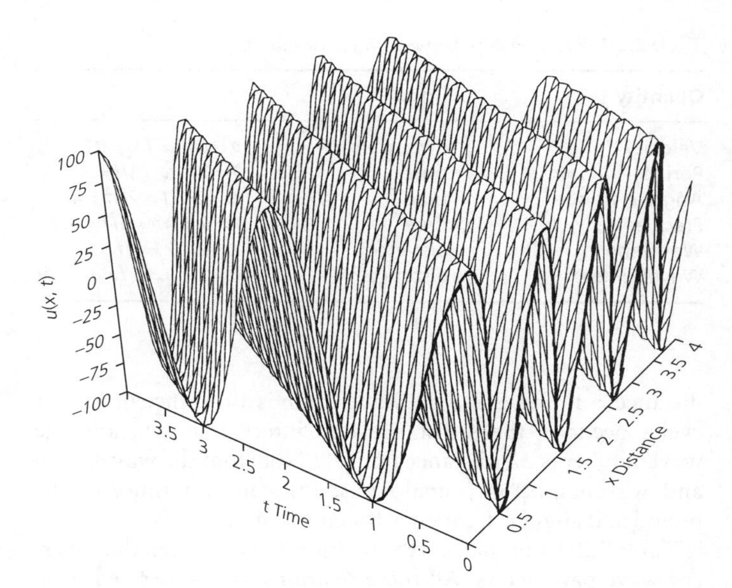

Fig. 2.2-3 Displacement as a function of position and time for the harmonic wave $u(x, t) = A\cos(\pi t - 2\pi x)$ propagating in the $+x$ direction. A line following a peak (or any part of the wave) in space and time represents the wave's velocity.

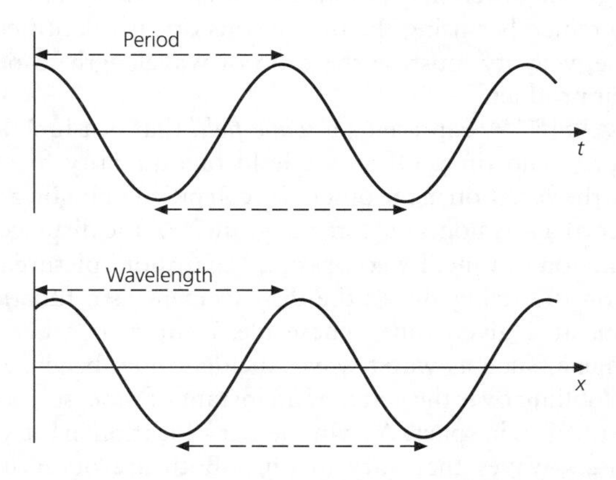

Fig. 2.2-4 A harmonic wave $u(x, t) = A\cos(\omega t - kx)$ shown at a fixed position as a function of time (*top*) and at a fixed time as a function of position (*bottom*).

Thus the interval shown, 4 time units, includes two full cycles of the oscillation. Equivalently, $1/2$ a cycle occurs in a unit time.

Alternatively, we can examine $u(x, t)$ at a fixed time, t_0, and plot $u(x, t_0) = A\cos(\omega t_0 - kx)$ as a function of position (Fig. 2.2-4, *bottom*). In terms of Fig. 2.2-3, this is a slice of the function on a plane parallel to the distance axis, which intersects the time axis at t_0. The displacement is periodic in space over a distance equal to the *wavelength*, $\lambda = 2\pi/k$, the distance between two corresponding points in a cycle. How the oscillation repeats in space can also be described by k, the *wavenumber* or *spatial frequency*, which is 2π times the number of cycles occurring in a unit distance. The wavelength has units of distance, so the wavenumber has dimensions of

[2] Properties of complex numbers are reviewed in Section A.2.

Table 2.2-1 Relationships between wave variables.

Quantity	Units	
Velocity	distance/time	$v = \omega/k = f\lambda = \lambda/T$
Period	time	$T = 2\pi/\omega = 1/f = \lambda/v$
Angular frequency	time^{-1}	$\omega = 2\pi/T = 2\pi f = kv$
Frequency	time^{-1}	$f = \omega/(2\pi) = 1/T = v/\lambda$
Wavelength	distance	$\lambda = 2\pi/k = v/f = vT$
Wavenumber	distance^{-1}	$k = 2\pi/\lambda = \omega/v = 2\pi f/v$

distance^{-1}. In Fig. 2.2-3 the wavelength is 1 distance unit, four cycles occur in the 4-distance unit interval shown, and the wavenumber is 2π (distance units)$^{-1}$. Note that the wavelength and wavenumber are analogous, for constant time, to the period and angular frequency for constant x.

Table 2.2-1 summarizes the relationships between the different wave parameters. All these relations can be derived from $v = \omega/k$ and the definitions of the other quantities. Note the analogy between period and angular frequency, which describe the wave in time at a fixed point in space, and wavelength and wavenumber, which describe the wave in space at a fixed time. Although the different relations may seem confusing, they are easy to remember using the dimensions of the quantities. For example, velocity must be the ratio of wavelength to period, not their product.

Thus $Ae^{i(\omega t \pm kx)}$ represents a *wave field* that is a function of both space and time. Often we hold one quantity fixed and observe the variation in the other. We can pick a point on a string and record a seismogram ("stringogram") of the displacement as a function of time. By contrast, a "snapshot" picture of the waves on the string shows the displacement as a function of position, at a given time. These ideas apply to other wave phenomena, such as water waves incident on a beach. A lifeguard, looking over the water at an instant of time, sees a wave field that varies in space. A swimmer, at a location in the water, encounters waves that vary in time. Both are observing, in different ways, a wave field that varies in both space and time. We will see that the same concept applies to seismic waves.

The harmonic wave solution describes a sinusoidal wave of a particular frequency. This might seem to make it a specific solution, not applicable to more complicated propagating waves. In particular, the sinusoid is defined for all times and distances, whereas in physical situations we deal with waves that exist only for a limited span in space and duration in time. Fortunately, as we will discuss later, an arbitrary wave shape can be decomposed into a set of harmonic waves using Fourier analysis. As a result, solutions describing the simple case of harmonic waves can be applied to more complicated cases.

2.2.3 *Reflection and transmission*

So far, we have discussed waves traveling along a string of uniform velocity. To use this as an analogy for the earth, within which physical properties vary with depth, we need to treat waves on a string with variable properties along its length. The simplest situation is a string composed of segments with uniform properties. If the segments are long enough, we treat the displacement in each segment as composed of propagating waves described by the solution for a uniform string with the appropriate properties, and then match solutions across the boundaries between segments.

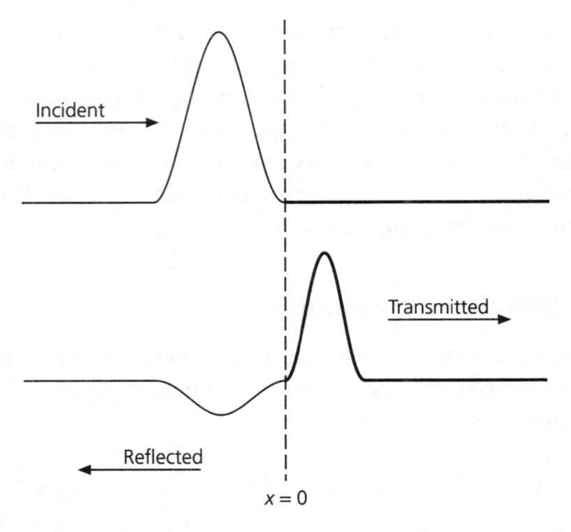

Fig. 2.2-5 A wave pulse incident from the left on a junction between two strings of different properties gives rise to transmitted and reflected wave pulses. The fact that the reflected wave is inverted shows that the impedance is greater in the right string. Similarly, the fact that the transmitted pulse has a smaller length shows that the velocity is lower in the right string.

To illustrate this approach, consider a junction between strings of different properties (Fig. 2.2-5). The junction at $x = 0$ separates string segment 1 on the left with density ρ_1 and velocity v_1 from string segment 2 on the right ($x > 0$) with density ρ_2 and velocity v_2. A wave arriving at the junction from the left yields two new waves. Some of the incident wave reflects from the junction, and thus travels to the left in string segment 1. The remainder of the incident wave is transmitted across the junction and travels to the right in string segment 2. We will show that the relative amounts of reflected and transmitted energy depend on the difference in properties across the interface.[3]

For the joined string segments, we write the total displacement in the left string segment as the sum of two harmonic waves

$$u_1(x, t) = Ae^{i(\omega t - k_1 x)} + Be^{i(\omega t + k_1 x)}. \tag{8}$$

The signs of the complex exponentials indicate that the incident wave, with amplitude A, travels in the $+x$ direction, whereas

[3] The wave's simultaneous reflection and transmission is analogous to shining a flashlight out of a window at night; you see the light reflected by the window, whereas someone outside sees the light transmitted through the window.

the reflected wave, with amplitude B, travels in the $-x$ direction. In the right-hand string segment there is only a transmitted wave going in the $+x$ direction

$$u_2(x, t) = Ce^{i(\omega t - k_2 x)}. \tag{9}$$

The waves in the two string segments have different wavenumbers because of the different velocities in the two segments.

The amplitudes of the reflected and transmitted waves are found using two *boundary conditions* that the physics of the string imposes on the solution at the junction $x = 0$. First, because the two segments at the junction stay joined, the displacement must always be continuous across the junction, so

$$u_1(0, t) = u_2(0, t),$$
$$Ae^{i\omega t} + Be^{i\omega t} = Ce^{i\omega t}. \tag{10}$$

For this to occur at all times, the angular frequency of the three waves must be the same, as we have assumed, and the amplitudes must satisfy

$$A + B = C. \tag{11}$$

Second, the y components of the tension forces acting on the two sides of the junction must always be equal, or the unequal forces would tear the string apart. Thus, by analogy to Eqn 2, we have another boundary condition

$$\tau \frac{\partial u_1(0, t)}{\partial x} = \tau \frac{\partial u_2(0, t)}{\partial x}. \tag{12}$$

Taking the derivatives and canceling terms gives

$$\tau k_1(A - B) = \tau k_2 C, \tag{13}$$

or, because the velocities on the two sides are $v_i = (\tau/\rho_i)^{1/2}$ and $k_i = \omega/v_i$,

$$\rho_1 v_1(A - B) = \rho_2 v_2 C. \tag{14}$$

We now have two equations (11 and 14) for the three constants A, B, and C, giving the amplitudes of the incident, reflected, and transmitted waves. We can eliminate C and find the ratio of the amplitudes of the reflected and incident waves, known as the *reflection coefficient*,

$$R_{12} = \frac{B}{A} = \frac{\rho_1 v_1 - \rho_2 v_2}{\rho_1 v_1 + \rho_2 v_2}. \tag{15}$$

Similarly, eliminating B yields the *transmission coefficient*, the ratio of transmitted and incident wave amplitudes,

$$T_{12} = \frac{C}{A} = \frac{2\rho_1 v_1}{\rho_1 v_1 + \rho_2 v_2}. \tag{16}$$

The "12" subscripts indicate that the reflection and transmission coefficients describe a wave incident from segment 1 upon segment 2; the corresponding coefficients for a wave incident from the right have subscripts "21." These can be derived by interchanging the subscripts, showing that

$$R_{12} = -R_{21}, \quad T_{12} + T_{21} = 2. \tag{17}$$

The reflection and transmission coefficients depend on the product of the density and velocity for each string, $\rho_i v_i$, a quantity called the *acoustic impedance*. Because the amount reflected depends on the difference in impedances between the two sides, the strongest reflections occur at boundaries where properties change significantly. One limiting case is if the materials on both sides of the junction are identical ($\rho_1 = \rho_2$ and $v_1 = v_2$), the reflection coefficient is zero and the transmission coefficient would be one. Hence, as expected, all the wave is transmitted, and none reflects. The other limiting case, total reflection and no transmission, occurs at the end of a string. The fixed end of a string, where no displacement occurs, can be treated as a junction with a string of infinite impedance. Hence the reflection coefficient is

$$R_{fixed} = \frac{\rho_1 v_1 - \infty}{\rho_1 v_1 + \infty} = -1, \tag{18}$$

so the entire incident wave pulse reflects with the opposite polarity. Similarly, a string whose end is free to move is described by the condition that the derivative $\partial u/\partial x$ is zero, because there is no force applied. This can be treated as a junction with a string of zero impedance, so the reflection coefficient is +1, and the entire incident pulse reflects with the same polarity. For values between the limiting cases, Eqn 15 shows that the polarity of the reflection depends upon whether the wave leaves or enters a string of greater impedance. If the impedance of segment 2 exceeds that of segment 1, waves going from segment 1 toward segment 2 reflect with reversed polarity, whereas waves going the other way reflect without changing polarity. Reflections at free and fixed ends are extreme cases of this property. Hence the amplitudes of reflections from boundaries can be used to infer changes in physical properties.

To illustrate these ideas, consider the reflection and transmission of waves on a string divided at $x = 10$ into two segments (Fig. 2.2-6). The left segment has $\rho_1 = 1$, $v_1 = 3$, and the right segment has $\rho_2 = 4$, $v_2 = 1.5$. At time 0 the string is plucked for a very short time by a source at the position marked by the triangle, so waves spread out in either direction.

At time 1, the first time shown, the wave traveling to the right has just encountered the junction (marked by a vertical dashed line). The reflection and transmission coefficients depend on the impedances $\rho_1 v_1 = 3$ and $\rho_2 v_2 = 6$. Thus for waves going from left to right $R_{12} = -0.33$ and $T_{12} = 0.67$. A small reflected pulse is generated, with a downward polarity opposite that of the incident pulse, because the reflection coefficient is negative. At

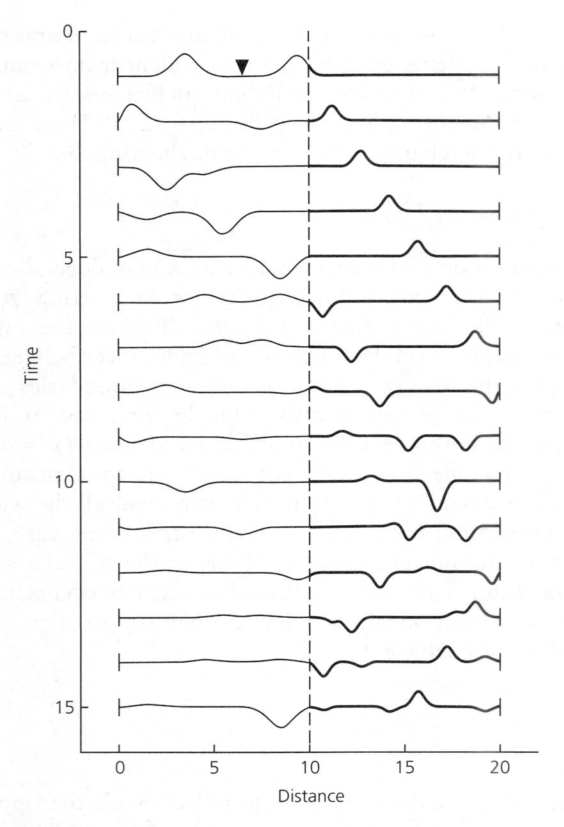

Fig. 2.2-6 Wave propagation on a string composed of two segments of different properties: the left (segment 1) with $\rho_1 = 1$, $v_1 = 3$, and the right (segment 2) with $\rho_2 = 4$, $v_2 = 1.5$. The triangle marks the position of the source (distance 6.5) that plucked the string at time 0. The traces are successive snapshots of the string one time unit apart. The vertical dashed line indicates the position of the junction. Both ends of the string are fixed, so reflections there have unchanged amplitude but reversed polarity.

time 2 we see this reflected wave traveling to the left and a larger transmitted wave traveling to the right. Note that, because of the different velocities, the reflected wave is further from the junction than the transmitted wave.

At time 2 the original pulse traveling to the left has reached the left end of the string. What happens to it depends on the boundary condition at the end. Here, we assumed that the ends were fixed, so at time 3 the pulse is inverted and reflected. Similarly at time 5 the first reflection off the junction has been inverted at the left end and now travels to the right.

When a pulse arrives at the junction, part is reflected and part is transmitted. For example, at time 6, the original pulse reflected from the left end has been converted at the junction into a transmitted wave with downward polarity and a reflected wave with positive polarity. As time goes by, many pulses develop, each with an amplitude that is the product of its history. Thus, if the initial pulses had unit amplitude, the first reflection has amplitude R_{12}. Once inverted by reflection off the fixed left end, this pulse has amplitude $R_{12}(-1)$. When it reaches the middle again (time 8), it gives rise to the small reflection with amplitude $R_{12}(-1)R_{12} = -0.11$ and a transmitted pulse with amplitude $R_{12}(-1)T_{12} = 0.22$.

By time 14, the original pulse that traveled to the right has been transmitted to segment 2, inverted by reflection off the right boundary (time 8), and is now incident on the junction from the right. The reflection and transmission coefficients for a wave incident from segment 2 are $R_{21} = 0.33$, $T_{21} = 1.33$. Thus the reflected and transmitted pulses have the same downward polarity as the incident wave and amplitudes $T_{12}(-1)R_{21} = -0.22$ and $T_{12}(-1)T_{21} = -0.89$.

It may seem curious that, because T_{21} is greater than 1, waves transmitted to the left have larger amplitude than the incident wave that generated them. This effect, although not appealing intuitively, is possible so long as the *energy* in the transmitted wave does not exceed that in the incident wave. We will show later that this is the case.

When a pulse is transmitted across the junction, its length as well as its amplitude changes. For example, the transmitted pulse at time 2 is shorter than the incident pulse. This results from the different velocities. To see this, recall that for a harmonic wave the angular frequencies of the transmitted and incident waves in the two strings are the same because the strings stay joined (Eqn 10). Thus

$$\omega = v_1 k_1 = v_2 k_2 = v_1 2\pi/\lambda_1 = v_2 2\pi/\lambda_2, \tag{19}$$

so the wavelength is shorter in the slower string. Another way to see this is from the time needed for an incident pulse to be transmitted (Fig. 2.2-7). If the pulse in segment 1 has length λ_1, it takes a time λ_1/v_1 to pass through the junction. The length of the transmitted pulse in segment 2 is the distance $v_2\lambda_1/v_1$ traveled by the leading edge of the transmitted pulse when the trailing edge of the incident pulse reaches the boundary.

A point worth noting is that the displacement at a point on the string is the sum of the displacements of all the waves passing by that point. For example, at time 10 (in Fig. 2.2-6) two waves, one traveling in either direction, add up to give a large pulse. At the next time step, the two waves have separated. Thus a wave has no lasting effect after crossing another; the waves "go through" each other. The concept that the waves can be added up without affecting each other is called *linear superposition*. This is generally assumed to be valid unless the amplitudes of the waves are so large that the material behaves *nonlinearly*, or differently from the simple elastic assumptions used to derive the propagating wave equation. Superposition allows us to form waves of arbitrary shape from harmonic waves of different frequencies using a Fourier series, as was done to form the pulses in this example. This posed no difficulty because in our derivation neither the velocity nor the reflection and transmission coefficients depended on frequency.

The fact that the amplitudes of waves on a string change as they are reflected and transmitted at interfaces where the properties of the string change illustrates a concept important for

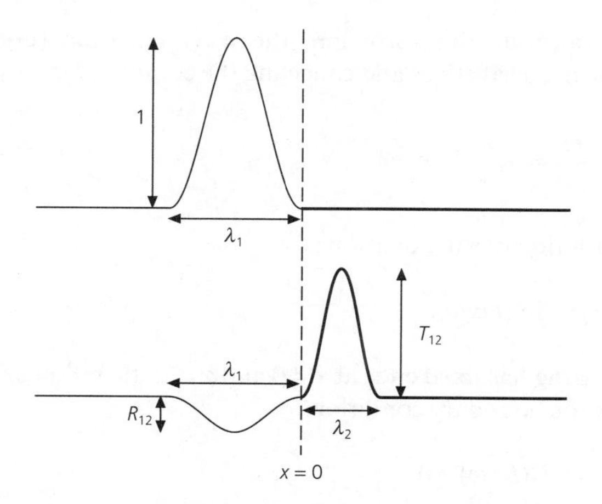

Fig. 2.2-7 An incident wave pulse of length λ_1 on a string with velocity v_1 generates a transmitted pulse of length λ_2 in a string with velocity v_2. The change in pulse length results from the distance the transmitted pulse travels while the incident pulse passes through the junction. If the amplitude of the incident pulse is 1, then the reflected and transmitted pulses have amplitudes R_{12} and T_{12}.

seismic waves in the earth. We will use this approach to show how we study changes in physical properties at depth in the earth from the amplitudes of reflected and transmitted waves.

2.2.4 *Energy in a harmonic wave*

We noted earlier that in some cases the transmission coefficient exceeds 1. To see how this occurs, we consider the energy transported by the traveling waves. It turns out that although amplitudes are easier to visualize, energy is often more useful for understanding wave behavior because energy is conserved, whereas amplitude is not. Hence, when a result for amplitudes is hard to understand, considering the energy can provide insight.

By analogy to the kinetic energy $mv^2/2$ of a point mass, the kinetic energy, KE, of a segment dx of the string is found from the velocity, the time derivative of the displacement, so

$$KE = \frac{\rho}{2}\left(\frac{\partial u}{\partial t}\right)^2 dx, \tag{20}$$

because the mass of the string is $m = \rho dx$.

The string also stores potential energy, because it is stretched, or deformed, from its equilibrium position. We will see shortly that a measure of the deformation is the *strain*, e, which for the string is the ratio of the change in the length to the original length. Hence for an element of the string (Fig. 2.2-1) with initial length dx, the strain due to the displacement du is

$$e = \frac{(dx^2 + du^2)^{1/2} - dx}{dx} = \left[1 + \left(\frac{du}{dx}\right)^2\right]^{1/2} - 1 = \frac{1}{2}\left(\frac{\partial u}{\partial x}\right)^2, \tag{21}$$

where the last step used the Taylor series $(1 + a^2)^{1/2} \approx 1 + a^2/2$ for small a. The potential energy stored in the string is the product of the tension and the strain integrated over the entire length L,

$$\frac{\tau}{2}\int_0^L \left(\frac{\partial u}{\partial x}\right)^2 dx, \tag{22}$$

so we can define the average potential energy, PE, in a segment dx as

$$PE = \frac{\tau}{2}\left(\frac{\partial u}{\partial x}\right)^2 dx. \tag{23}$$

We characterize the energy of a traveling wave by the kinetic and potential energy averaged over a wavelength. If $u(x, t) = A \cos(\omega t - kx)$, then the kinetic energy averaged over a wavelength is

$$KE = \frac{\rho}{2\lambda}\int_0^\lambda \left(\frac{\partial u}{\partial t}\right)^2 dx = \frac{\rho A^2\omega^2}{2\lambda}\int_0^\lambda \sin^2(\omega t - kx)dx. \tag{24}$$

The integral of the sinusoid squared over a period is

$$\int_0^\lambda \sin^2(\omega t - kx)dx = \lambda/2, \tag{25}$$

so the kinetic energy is

$$KE = A^2\omega^2\rho/4. \tag{26}$$

Similarly, the potential energy averaged over a wavelength is

$$PE = \frac{\tau}{2\lambda}\int_0^\lambda \left(\frac{\partial u}{\partial x}\right)^2 dx = \frac{\tau A^2 k^2}{2\lambda}\int_0^\lambda \sin^2(\omega t - kx)dx, \tag{27}$$

which, using Eqn 25, becomes

$$PE = \tau A^2 k^2/4 = A^2\omega^2\rho/4, \tag{28}$$

the same as the kinetic energy.

Hence the total energy transported, averaged over a wavelength, is the sum of the potential and kinetic energies:

$$E = PE + KE = A^2\omega^2\rho/2. \tag{29}$$

Another way to state this is in terms of the energy *flux*, the rate at which the wave transports energy past a point on the string. The average flux is just the averaged energy times the velocity

$$\dot{E} = A^2\omega^2\rho v/2. \tag{30}$$

For a string of a given density, the energy flux is proportional to the amplitude and angular frequency squared, so higher-frequency waves transport more energy.

Consideration of the energy explains how in Fig. 2.2-6 the transmitted wave can have higher amplitude than the incident wave. To see that an incident wave converting into reflected and transmitted waves conserves energy, assume that a wave in segment 1, described by $\cos(\omega t - k_1x)$, is incident on the junction. It gives rise to a reflected wave in segment 1, described by $R_{12}\cos(\omega t + k_1x)$, and a transmitted wave in segment 2, described by $T_{12}\cos(\omega t - k_2x)$. Using Eqns 15 and 16 for R_{12} and T_{12}, the net energy flux for the reflected and transmitted waves is the sum

$$\begin{aligned}\dot{E}_R + \dot{E}_T &= R_{12}^2\omega^2\rho_1v_1/2 + T_{12}^2\omega^2\rho_2v_2/2\\ &= (\omega^2/2)[R_{12}^2v_1\rho_1 + T_{12}^2v_2\rho_2]\\ &= \omega^2\rho_1v_1/2 = \dot{E}_I,\end{aligned} \tag{31}$$

which equals the energy flux in the incident wave. Thus, even if the amplitude of the transmitted wave exceeds that of the incident wave, the energy of the transmitted wave is less than that of the incident wave.[4]

2.2.5 *Normal modes of a string*

So far, we have discussed waves propagating along a string. Additional insight into propagating waves can be gained by considering standing waves, which are known as the *normal modes*, or *free oscillations*, of the string.

Recall that we began by applying Newton's second law to a string, and found that the displacement $u(x, t)$ as a function of position and time satisfied the scalar wave equation

$$\frac{\partial^2 u(x,t)}{\partial x^2} = \frac{1}{v^2}\frac{\partial^2 u(x,t)}{\partial t^2}. \tag{4}$$

We saw that this equation had solutions like

$$u(x, t) = A\cos(\omega t \pm kx), \tag{32}$$

which describes harmonic waves with angular frequency ω and wavenumber $k = 2\pi/\lambda$, propagating at velocity v such that $v = \omega/k$.

An alternative approach is to seek solutions of (4) with a $\cos(\omega t)$ time dependence, such that

$$u(x, t) = U(x, \omega)\cos(\omega t), \tag{33}$$

by substituting this form into the wave equation (Eqn 4).[5] Taking the derivatives and canceling the common factor yields

$$\frac{\partial^2 U(x,\omega)}{\partial x^2} = -\frac{\omega^2}{v^2}U(x,\omega). \tag{34}$$

One solution of this equation is

$$U(x, \omega) = \sin(\omega x/v). \tag{35}$$

If the string has fixed ends at $x = 0$ and $x = L$, then Eqn 35 must satisfy the boundary conditions

$$U(0, \omega) = U(L, \omega) = 0. \tag{36}$$

The solution already satisfies the boundary condition at $x = 0$, so all that is needed is to satisfy the boundary condition at $x = L$,

$$U(L, \omega) = \sin(\omega L/v) = 0, \tag{37}$$

which occurs for angular frequencies ω_n such that

$$\omega_n L/v = n\pi \quad \text{or} \quad \omega_n = n\pi v/L. \tag{38}$$

Thus the zero displacement boundary conditions at the string's ends require that it vibrate only at specific frequencies, called *eigenfrequencies*. The eigenfrequencies each correspond to a solution

$$U_n(x, \omega_n)\cos(\omega_n t), \tag{39}$$

where the spatial term

$$U_n(x, \omega_n) = \sin(\omega_n x/v) = \sin(n\pi x/L) \tag{40}$$

is known as the spatial *eigenfunction*.

To interpret these solutions physically, note that $\omega = vk = v2\pi/\lambda$, so the eigenfrequencies correspond to

$$\omega_n = n\pi v/L = 2\pi v/\lambda \quad \text{or} \quad L = n\lambda/2. \tag{41}$$

Thus each spatial eigenfunction has an integral number of half wavelengths along the string's length L, so the displacement at both ends is zero. The solutions are standing waves, known as the normal modes, or free oscillations, of the string, each of which has a characteristic spatial eigenfunction and vibrates at a characteristic eigenfrequency. Because the string is finite, it can vibrate only in these discrete modes that satisfy the boundary conditions. The eigenfrequencies are spaced $\pi v/L$ apart, so

[4] An analogous phenomenon occurs at beaches, where waves increase in amplitude as they approach the shore because the wave speed is proportional to the square root of water depth.

[5] This procedure amounts to taking the Fourier transform of the equation in frequency, and then using a Fourier series in space. Fourier analysis is discussed in chapter 6.

the longer the string is (i.e., the larger L gets), the closer the eigenfrequencies become.

A traveling wave can be expressed as the weighted sum of the string's normal modes, so it is the sum of the eigenfunctions, each weighted by the amplitude A_n and vibrating at its eigenfrequency ω_n,

$$u(x, t) = \sum_{n=0}^{\infty} A_n U_n(x, \omega_n) \cos(\omega_n t). \tag{42}$$

An important feature of this solution is that the modes are *orthogonal*, meaning that the integral over the string of the product of two different eigenfunctions is zero,

$$\int_0^L \sin\left(\frac{m\pi x}{L}\right) \sin\left(\frac{n\pi x}{L}\right) dx = \frac{L}{2}\delta_{mn}, \tag{43}$$

where δ_{mn}, the Kronecker delta symbol defined in Eqn A.3.37, is zero unless $m = n$. Each mode is independent and cannot be constructed by combining other modes. Thus we can think of the displacement of the string as a vector in a vector space (Section A.3.6) whose basis vectors are the eigenfunctions. Any particular set of waves is given by the amplitudes A_n, which are the weighting factors of the eigenfunctions or the components of the basis vectors.

The amplitude for each eigenfunction depends on the position of the source that generated the waves and on the behavior of the source as a function of time. The spatial part of A_n has the same form as U_n (Eqn 40), so

$$A_n = \sin(n\pi x_s/L) F(\omega_n), \tag{44}$$

where x_s is the position of the source, and $F(\omega_n)$ is a weighting factor describing how different frequencies contribute to the time history of the source. Thus the normal mode expression for the displacement (Eqn 42) can be written

$$u(x, t) = \sum_{n=0}^{\infty} \sin(n\pi x_s/L) F(\omega_n) \sin(n\pi x/L) \cos(\omega_n t). \tag{45}$$

Figure 2.2-8, computed in this way, illustrates how the first 40 modes of a string with fixed ends and a uniform velocity combine to give traveling waves. The source, at $x_s = 8$, is described by

$$F(\omega_n) = \exp[-(\omega_n \tau)^2/4] \tag{46}$$

with $\tau = 0.2$. The computer program used is similar to that discussed in Section A.8.1. The mode sum shows two waves, one propagating to the right and one propagating to the left, at the expected positions. Hence the mode sum correctly gives the propagating waves. In addition to the propagating waves, we see some small oscillations along the string because only the first 40 modes were summed.

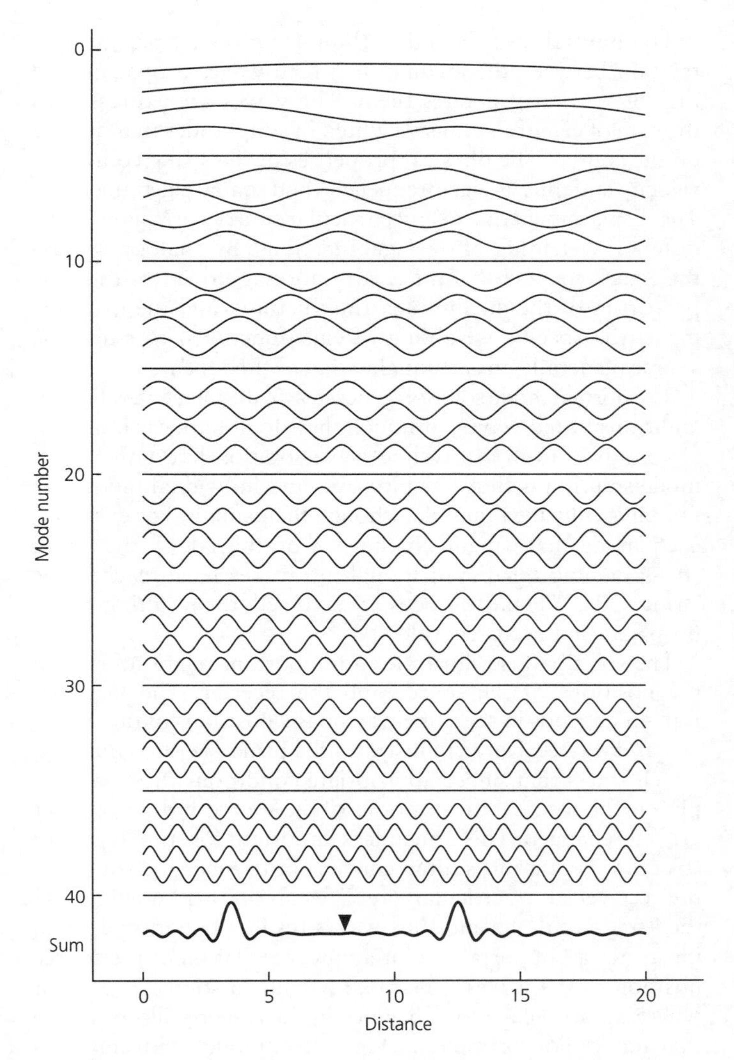

Fig. 2.2-8 Displacement of a string with fixed ends computed using the normal mode formulation. The string has length 20, velocity 3, and was plucked at time 0 by a source at position 8 (triangle). The bottom trace shows the displacement of the string at time 1.5, computed by summing the first 40 modes. The mode sum generates both the right- and the left-propagating waves at the appropriate positions. Spatial eigenfunctions for the individual modes, each of which corresponds to an integral number of half wavelengths, are also shown above the sum. The traces are normalized to unit amplitude.

We now have two ways to think of the displacement of the string as a function of time: either as propagating waves or as normal modes. Neither is more "real" — both are ways of representing how the displacement evolves. Thus comparing the two gives interesting insights. For example, consider studying the properties of the string. In the traveling wave formulation, we measure travel times and thus infer velocity. In the normal mode formulation, we measure eigenfrequencies and then infer velocities. Thus the eigenfrequencies are analogous to the travel times.

The normal mode solution (Eqn 45) gives insight into the relation between the medium in which waves propagate and the source that generates them. The waves are expressed as the sum of eigenfunctions weighted by amplitudes that depend on the source. The physical properties of the string control its velocity and thus its eigenfrequencies and spatial eigenfunctions. The displacement due to any particular source corresponds to a different weighting of the eigenfunctions. By analogy, we use the eigenfrequencies of the earth's normal modes to study the properties of the medium (earth structure), and the displacement (the specific weighting of eigenfunctions) to study the source (generally an earthquake) that excited them.

The normal mode solution generates all the incident, reflected, and transmitted waves, although they do not appear individually as they do in a traveling wave solution like Eqn 8. The mode solution is thus less intuitive, and individual modes are not physically meaningful, although their sum is. For example, each mode mathematically starts vibrating along the entire string at time zero, even though no waves have reached the string ends. When the modes are summed, the resulting waves propagate at the correct velocity.

The solution also illustrates an important relation between the positions of the source and the receiver. The fact that Eqn 45 depends in the same way on the positions of the source (x_s) and the receiver (x) illustrates the principle of *reciprocity*, which states that under appropriate conditions the same displacement occurs if the positions of the source and the receiver are interchanged. This principle is important for studying earth structure because it is often convenient to place the source or the receiver at a particular site. We can do this knowing that the same ray paths and thus waves result.[6] Equation 45 also illustrates an important point about the relation of the source position to the waves generated: namely, a source at a point where a particular mode has no displacement will not excite that mode. For example, in Fig. 2.2-8, modes with numbers that are multiples of five give zero displacement because the source term $\sin(n\pi x_s/20)$ is zero. Analogously, in the earth, surface waves whose displacements are largest near the surface are not excited well by deep earthquakes.

Finally, although we have discussed the normal modes of a uniform string, we could generalize these ideas to find the modes of a non-uniform string. One way to do this is to extend the method used to find the reflection and transmission coefficients (Section 2.2.3). We treat the string as a set of uniform pieces, use the harmonic wave solution in each piece, and impose displacement and traction boundary conditions at the junctions. We then numerically find eigenfrequencies that satisfy the fixed boundary condition at the string's end. The normal modes of the non-uniform string are then summed to give the traveling waves. The waves on the non-uniform string in Fig. 2.2-6 were calculated in this way.

[6] A familiar version for light waves, seen on the back of large trucks, warns other drivers that "If you can't see my mirrors, I can't see you."

2.3 Stress and strain

2.3.1 Introduction

By applying Newton's second law of motion, $\mathbf{F} = m\mathbf{a}$, to a string, we found that deforming the string gave rise to propagating waves. Similarly, deforming the solid earth produces seismic waves. We study these waves using concepts from *continuum mechanics*, which describes the behavior of a continuous deformable material made up of particles packed so closely together that density, force, and displacement can be thought of as continuous and differentiable functions. This approximation breaks down on an atomic distance scale, but is adequate for most seismological problems.

For these applications, we write Newton's second law in terms of the force per unit volume and the density, the mass per unit volume. If the density does not change with time, the force per unit volume $\mathbf{f}(\mathbf{x}, t)$ equals the inertial term, the product of the density ρ and the second derivative of the displacement vector $\mathbf{u}(\mathbf{x}, t)$ with time. Thus $\mathbf{F} = m\mathbf{a}$ becomes

$$\mathbf{f}(\mathbf{x}, t) = \rho \frac{\partial^2 \mathbf{u}(\mathbf{x}, t)}{\partial t^2}. \quad (1)$$

This vector equation can be written as a set of three equations, one for each component of the force and displacement vectors[1]

$$f_i(\mathbf{x}, t) = \rho \frac{\partial^2 u_i(\mathbf{x}, t)}{\partial t^2}. \quad (2)$$

In seismic wave propagation, both the displacement and the force vectors can vary in space and time. Although this dependence is generally not written explicitly, we will sometimes do so to remind ourselves that the solutions depend on space and time.

The goal of this section is to use Newton's second law to characterize a continuous medium and its response to applied forces. We first introduce the *stress tensor* that describes the forces acting on a deformable continuous medium. We then formulate the *equation of motion*, the version of Newton's law appropriate for a continuous medium, which relates the stress to the displacement. The variation in displacement within the material, described by the *strain tensor*, gives rise to internal deformation. This deformation is related to the stress via the *constitutive equation* that characterizes the properties of the material. Our brief discussion covers some basic results of continuum mechanics necessary for introductory seismology. The suggested reading listed at the end of the chapter provides further treatment of these and related topics.

[1] The three equations are written as one using index notation (Section A.3.5) in which the index i ranges from 1 to 3 over the coordinate axes. Index notation makes cumbersome vector equations shorter, clearer, and often easier to solve. These equations are often made even more compact using a dot superscript to indicate differentiation with respect to time, so the acceleration is $\ddot{u}_i$.

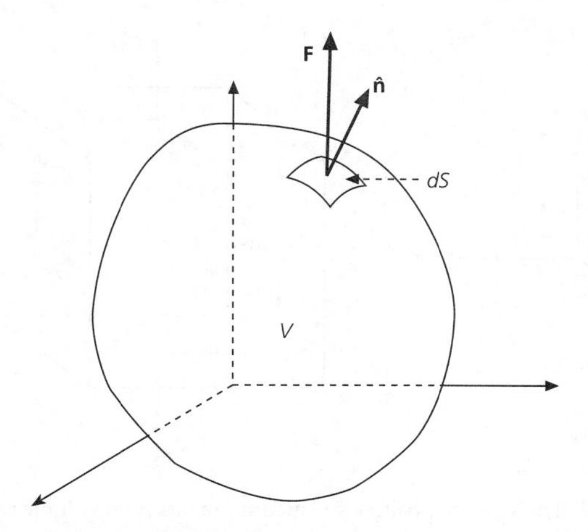

Fig. 2.3-1 Surface force on a volume element V within a material. The surface force $\mathbf{F}$ due to the material outside V acts on each element of surface dS, which has an outward-pointing unit normal vector $\hat{\mathbf{n}}$.

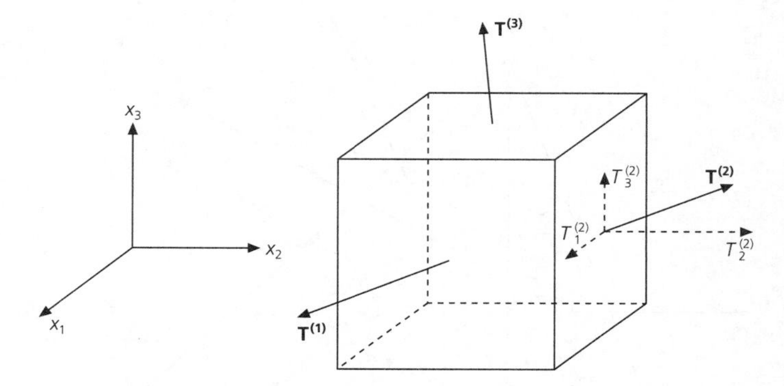

Fig. 2.3-2 Traction vectors acting on three faces of a volume element which are perpendicular to the coordinate axes. The superscript on $\mathbf{T}$ indicates the direction of the normal to the face on which $\mathbf{T}$ acts. The three components $T_i^{(2)}$ are shown.

2.3.2 Stress

Two types of forces can act on an object. The first is a *body force*, which acts everywhere within an object, resulting in a net force proportional to the volume of the object. A familiar example is the body force $\mathbf{g}$ due to gravity; the net force on an infinitesimal body with density ρ and volume dV is $\rho \mathbf{g} dV$. The units of a body force are force per unit volume.

A second type of force is a *surface force*, which acts on the surface of an object, yielding a net force proportional to the surface area of the object. For example, an object in a pool of fluid is subject to a pressure equal to the weight (a force) per unit area of the fluid above the object. At any point on the object's surface, the pressure is directed along the normal to the surface. Thus a surface force like pressure acts in different directions on different parts of an object, in contrast to gravity, which is a body force that always points down. Surface forces have units of force per unit area.

We now consider the forces acting on a small volume V, with surface S, within a larger continuous medium (Fig. 2.3-1). The material inside V is affected by body forces acting on everything inside V and surface forces, due to the material outside, acting on the surface S. If the surface force $\mathbf{F}$ acts on each element of surface dS, whose outward unit normal vector is $\hat{\mathbf{n}}$, we define the traction vector, $\mathbf{T}$, as the limit of the surface force per unit area at any point as the area becomes infinitesimal:

$$\mathbf{T}(\hat{\mathbf{n}}) = \lim_{dS \to 0} \frac{\mathbf{F}}{dS}. \tag{3}$$

The traction vector has the same orientation as the force, and is a function of the unit normal vector $\hat{\mathbf{n}}$ because it depends on the orientation of the surface.

The system of surface forces acting on a volume is described by three traction vectors. Each acts on a surface perpendicular to a coordinate axis (Fig. 2.3-2), and is thus parallel to the plane defined by the other two axes. We define $\mathbf{T}^{(j)}$ as the traction vector acting on the surface whose outward normal is in the positive $\hat{\mathbf{e}}_j$ direction. The components of the three traction vectors are $T_i^{(j)}$, where the upper index (j) indicates the surface and the lower (i) index indicates the component. For example, $T_3^{(1)}$ is the x_3 component of the traction on the surface whose normal is $\hat{\mathbf{e}}_1$.

This set of nine terms that describes the surface forces can be grouped into the *stress tensor*, σ_{ji}. The tensor's rows are the three traction vectors, such that

$$\sigma_{ji} = \begin{pmatrix} \sigma_{11} & \sigma_{12} & \sigma_{13} \\ \sigma_{21} & \sigma_{22} & \sigma_{23} \\ \sigma_{31} & \sigma_{32} & \sigma_{33} \end{pmatrix} = \begin{pmatrix} \mathbf{T}^{(1)} \\ \mathbf{T}^{(2)} \\ \mathbf{T}^{(3)} \end{pmatrix} = \begin{pmatrix} T_1^{(1)} & T_2^{(1)} & T_3^{(1)} \\ T_1^{(2)} & T_2^{(2)} & T_3^{(2)} \\ T_1^{(3)} & T_2^{(3)} & T_3^{(3)} \end{pmatrix}. \tag{4}$$

Thus the stress component σ_{ji} is the ith component of the traction vector acting on the surface whose outward normal points in the $\hat{\mathbf{e}}_j$ direction. The stress gives the force per unit area that the material on the outside (the side to which $\hat{\mathbf{n}}$ points) of the surface exerts on the material inside. In the special geometry of Fig. 2.3-2, where the surfaces are along coordinate axes, it is easy to see that $\sigma_{ji} = T_i^{(j)}$.

In some applications, it is more convenient to write the coordinate axes as x, y, and z, so the stress tensor is written

$$\sigma_{ji} = \begin{pmatrix} \sigma_{xx} & \sigma_{xy} & \sigma_{xz} \\ \sigma_{yx} & \sigma_{yy} & \sigma_{yz} \\ \sigma_{zx} & \sigma_{zy} & \sigma_{zz} \end{pmatrix}. \tag{5}$$

The stress tensor gives the traction vector $\mathbf{T}$ acting on any surface within the medium. To illustrate this, we examine the traction on an arbitrary element of surface dS, whose normal $\hat{\mathbf{n}}$ is not along a coordinate axis. Consider the material inside an infinitesimal tetrahedron of volume dV formed by this surface and three other faces, each perpendicular to a coordinate axis, with normal in the $-\hat{\mathbf{e}}_j$ direction (Fig. 2.3-3). The area of the

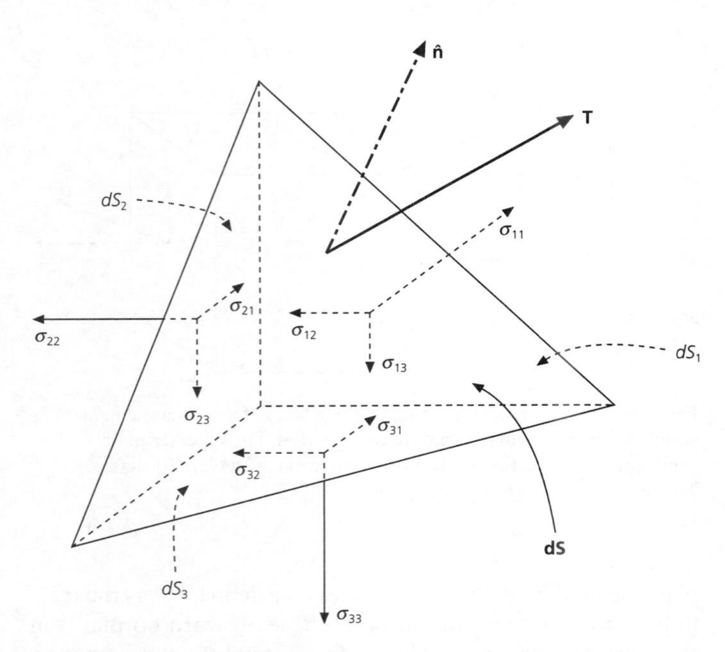

Fig. 2.3-3 Stress components on three faces of a tetrahedron, with normals parallel to coordinate axes. Summing the resulting forces yields the net force on the fourth (slanted) side.

face with its normal in the $-\hat{\mathbf{e}}_j$ direction is given by using the scalar product to find the cosine of the angle between $\hat{\mathbf{n}}$ and $\hat{\mathbf{e}}_j$,

$$(\hat{\mathbf{n}} \cdot \hat{\mathbf{e}}_j) dS = n_j dS. \tag{6}$$

Because traction is force per unit area, the net surface force in a given direction is found by multiplying each component of the traction by the area of the face it acts on and summing over the faces. Thus the total force in the $\hat{\mathbf{e}}_i$ direction is that due to this component of the traction, those resulting from the stress on the other three faces, and the component of the body force f in this direction. This total force equals the mass ρdV of the tetrahedron times the component of acceleration in the $\hat{\mathbf{e}}_i$ direction,

$$T_i dS - \sum_{j=1}^{3} \sigma_{ji} n_j dS + f_i dV = \rho \frac{\partial^2 u_i}{\partial t^2} dV. \tag{7}$$

Dividing by the area and letting dV/dS go to zero, we see that the stress tensor is related to the traction and normal vectors by

$$T_i = \sum_{j=1}^{3} \sigma_{ji} n_j = \sigma_{ji} n_j, \tag{8}$$

where the last form uses the index notation convention that a repeated index indicates summation (Section A.3.5). Because this equation gives the traction on an arbitrary surface, the stress tensor describes the surface forces acting on any volume within the material.

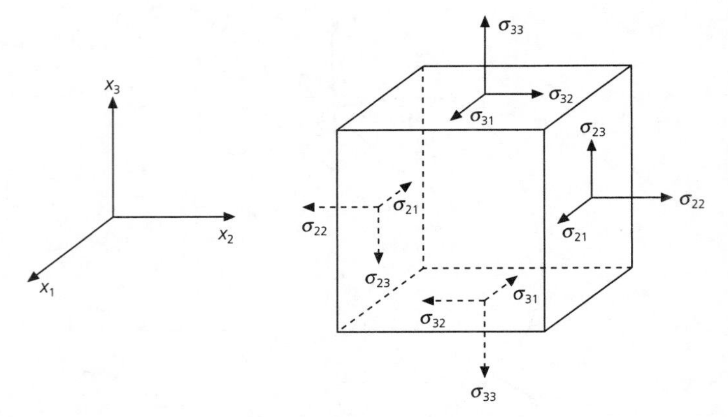

Fig. 2.3-4 The sense of positive stress components for a volume with faces perpendicular to the coordinate axes. σ_{ji} is the stress component acting in the $\hat{\mathbf{e}}_i$ direction on the face with outward normal in the $\hat{\mathbf{e}}_j$ direction.

The sign convention for stress components comes from the relation between the outward normal and the basis vectors. Figure 2.3-4 shows the positive stress components acting on a cube of material with faces perpendicular to the coordinate axes. For example, on the face with outward normal $\hat{\mathbf{e}}_3 = (0, 0, 1)$, σ_{33} is positive in the $\hat{\mathbf{e}}_3$ direction, and σ_{31} is positive in the $\hat{\mathbf{e}}_1$ direction. Because the tractions are $T_i = \sigma_{3i}$, positive σ_{33} and σ_{31} yield forces in the x_3 and x_1 directions. By contrast, on the opposite face with outward normal $-\hat{\mathbf{e}}_3 = (0, 0, -1)$, σ_{33} is positive in the $-x_3$ direction, and σ_{31} is positive in the $-x_1$ direction. Thus the tractions are $T_i = -\sigma_{3i}$, and positive σ_{33} and σ_{31} yield forces in the $-x_3$ and $-x_1$ directions.

The three diagonal components of the stress tensor, σ_{11}, σ_{22}, and σ_{33}, are known as *normal stresses*, and the six off-diagonal components are called *shear stresses*. The corresponding components of the traction vector are called normal and shear tractions. Figure 2.3-4 shows that positive normal stresses tend to expand the volume, whereas negative normal stresses make the volume smaller. Thus positive values of the normal tractions correspond to *tension*, whereas negative normal tractions correspond to *compression*. At most points within the earth, because material is under compression from the weight of rock above, the normal stress components are negative. Geophysicists thus often speak of the "maximum compressive stress," the most negative and largest in absolute value, and the "minimum compressive stress," the least negative and smallest in absolute value.

An important property of a stress tensor is that it is *symmetric*,

$$\sigma_{ij} = \sigma_{ji}. \tag{9}$$

To show this, consider the torque (Eqn A.3.32) τ_3 about the x_3 axis on a rectangle of material with sides dx_1, dx_2, along the coordinate axes (Fig. 2.3-5). If the torque is zero, the angular momentum of the block remains constant, so the block will not

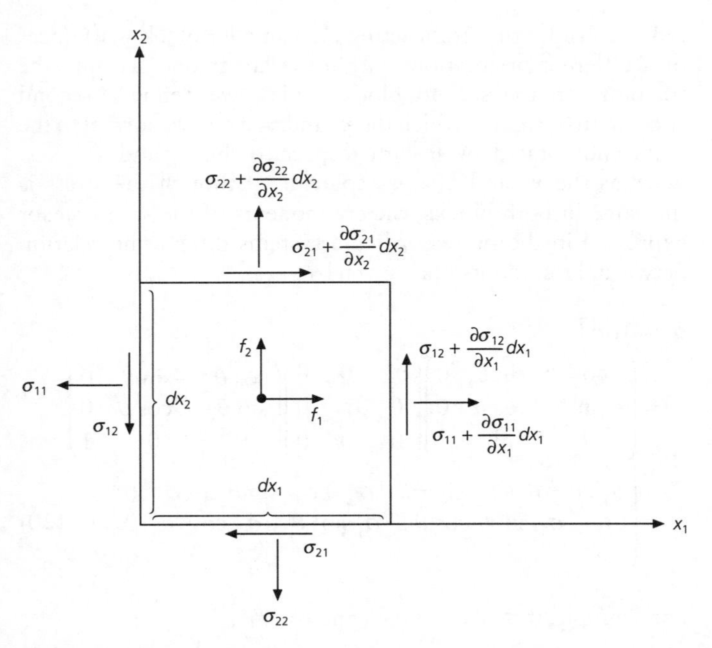

Fig. 2.3-5 Clockwise and counterclockwise torques about the x_3 axis on a rectangle due to the stress components and body forces. If the stress tensor were not symmetric, $\sigma_{12} = \sigma_{21}$, a net torque would arise.

start to rotate if it is not already doing so. The net body force, if any, is $f_i dx_1 dx_2$, where f_i is the force at the center of the block. Because a torque is the product of a force and a lever (or moment) arm, the shear stresses σ_{21} and σ_{12} acting on the faces along the x_1 and x_2 axes contribute no torque. The other stress components cause torques equal to the product of the lever arm and the traction, the stress component times the area of the face. Thus the total counterclockwise torque is the sum of that due to the shear tractions on the other two faces, with lever arms dx_1 and dx_2, the normal tractions on all four faces, with lever arms $dx_1/2$ and $dx_2/2$, and the two body force components acting at the center of the block, with lever arms $dx_1/2$ and $dx_2/2$:

$$\tau_3 = \left(\sigma_{12} + \frac{\partial \sigma_{12}}{\partial x_1} dx_1\right) dx_1 dx_2 - \left(\sigma_{21} + \frac{\partial \sigma_{21}}{\partial x_2} dx_2\right) dx_1 dx_2$$
$$- \left(\sigma_{11} + \frac{\partial \sigma_{11}}{\partial x_1} dx_1\right) dx_2 \frac{dx_2}{2} + \sigma_{11} dx_2 \frac{dx_2}{2}$$
$$+ \left(\sigma_{22} + \frac{\partial \sigma_{22}}{\partial x_2} dx_2\right) dx_1 \frac{dx_1}{2} - \sigma_{22} dx_1 \frac{dx_1}{2}$$
$$+ f_2 dx_1 dx_2 \frac{dx_1}{2} - f_1 dx_1 dx_2 \frac{dx_2}{2}. \quad (10)$$

Dividing by the area and letting dx_1 and dx_2 go to zero, we see that for there to be no torque, $\sigma_{12} = \sigma_{21}$. The same argument for the torque about the other two axes shows that $\sigma_{13} = \sigma_{31}$ and $\sigma_{23} = \sigma_{32}$. Thus, although the stress tensor has nine components, only the three normal ones and three of the six shear ones are independent.

Because the stress tensor is symmetric, we usually write (8) as

$$T_i = \sum_{j=1}^{3} \sigma_{ij} n_j = \sigma_{ij} n_j, \quad (11)$$

or, in terms of the vectors rather than their components,

$$\mathbf{T} = \sigma \hat{\mathbf{n}}. \quad (12)$$

Stress has units of force per area. In the cgs system of units based on the centimeter, gram, and second, force is given in dynes (dyn), with 1 dyn = 1 g-cm/s^2, so stress is given in dyn/cm^2, or *bars*, a unit equal to 10^6 dyn/cm^2. The bar has the convenient property that atmospheric pressure at sea level is 1.01 bars. In SI units based on the meter, kilogram, and second (mks), force is given in Newtons (N), with 1 N = 1 kg-m/s^2, so stress is given in *Pascals* (Pa), a unit equal to 1 N/m^2. The two sets of units can be related by noting that 1 Pa = 10^5 dyn/10^4 cm^2 = 10 dyn/cm^2 = 10^{-5} bars, so 1 MPa equals 10 bars.

2.3.3 Stress as a tensor

We have been using the term "tensor" without defining it. Already, we saw that it came from a relation between the traction and normal vectors, and is an entity with two subscripts that has properties similar to those of vectors. Vectors are entities that are independent of coordinate system, so that physical laws written using them do not depend on the coordinate system and can be analyzed using any convenient coordinate system. We now show that tensors are similar entities.

Specifically, a vector is an entity that remains the same in two coordinate systems (Section A.5.1), such that its components in two different Cartesian coordinate systems are related by the transformation matrix A. Hence, given two sets of axes (x_1, x_2, x_3) and (x'_1, x'_2, x'_3), the components of a vector $\mathbf{u}$ are related by

$$\mathbf{u}' = A\mathbf{u}. \quad (13)$$

The relation between the components of the stress tensor in two Cartesian coordinate systems can be found using the fact that it relates the traction and normal vectors in each coordinate system. The components of the traction and normal vectors in the two coordinate systems satisfy

$$\mathbf{T}' = A\mathbf{T}, \quad \hat{\mathbf{n}}' = A\hat{\mathbf{n}}. \quad (14)$$

The reverse transformation can be written using the inverse of A which, because A is orthogonal, equals its transpose:

$$\hat{\mathbf{n}} = A^{-1}\hat{\mathbf{n}}' = A^T\hat{\mathbf{n}}'. \quad (15)$$

In the primed coordinate system, the traction is related to the normal vector and the stress tensor by

$$\mathbf{T}' = \sigma'\hat{\mathbf{n}}', \tag{16}$$

so, by Eqns 14 and 15,

$$\mathbf{T}' = A\mathbf{T} = A\sigma\hat{\mathbf{n}} = A\sigma A^T\hat{\mathbf{n}}'. \tag{17}$$

Comparison of Eqn 16 and the last term in Eqn 17 shows that

$$\sigma' = A\sigma A^T. \tag{18}$$

This equation defines a tensor in Cartesian coordinates. Recall that what makes a vector more than a set of three numbers is its transformation properties: the numerical values of the components that describe it transform between coordinate systems in a way that preserves the vector as an entity independent of coordinate system. Similarly, a matrix of numbers is a tensor only if it transforms between coordinate systems according to Eqn 18. We derived this transformation by assuming that a tensor, in this case stress, is an operator relating two vectors, in this case the normal and traction, in a specific way regardless of coordinate system. The tensor's components transform between coordinate systems, so the tensor as an entity does not change. Because one application of the transformation matrix transforms a vector, two applications transform a tensor that relates two vectors. Unfortunately, tensors are harder to visualize than vectors. Although the stress tensor may seem puzzling, it is one of the easier tensors to interpret physically.

To illustrate these ideas, we consider an example of how a stress tensor's components change between coordinate systems. Assume that a block of material, with faces perpendicular to the x_1 and x_2 axes, is subject only to normal stresses σ_1 and σ_2 (Fig. 2.3-6), so the stress tensor is diagonal,

$$\sigma = \begin{pmatrix} \sigma_1 & 0 & 0 \\ 0 & \sigma_2 & 0 \\ 0 & 0 & 0 \end{pmatrix}. \tag{19}$$

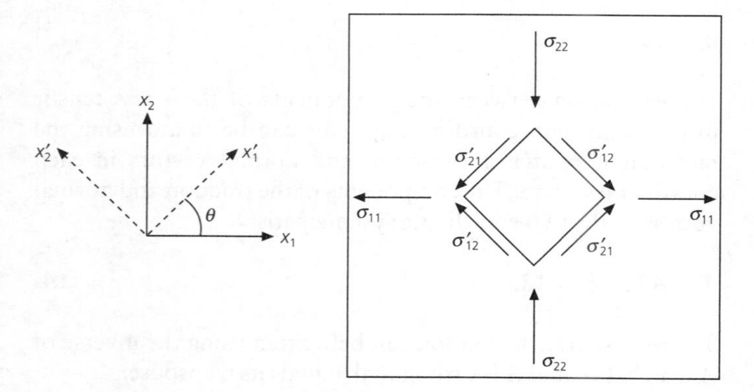

Fig. 2.3-6 An example of the stress tensor's different components in different coordinate systems. In the x_1, x_2 axis coordinate system, the stress tensor is diagonal. In contrast, shear stresses act on a volume with faces normal to the x_1' and x_2' coordinate axes, which are rotated by θ with respect to the x_1, x_2 axes.

Now, consider the stress acting on a smaller block, with faces of a different orientation, within the larger one. To find the tractions on the second block's sides, we define a second coordinate system in which the x_1' and x_2' axes are normal to the faces and rotated by θ with respect to the x_1 and x_2 axes, whereas the x_3 and x_3' axes coincide. Although the stress is the same in both blocks, the components of the stress tensor expressed in the two coordinate systems differ. The relation between the components is given by

$$\begin{aligned}\sigma' &= A\sigma A^T \\ &= \begin{pmatrix} \cos\theta & \sin\theta & 0 \\ -\sin\theta & \cos\theta & 0 \\ 0 & 0 & 1 \end{pmatrix}\begin{pmatrix} \sigma_1 & 0 & 0 \\ 0 & \sigma_2 & 0 \\ 0 & 0 & 0 \end{pmatrix}\begin{pmatrix} \cos\theta & -\sin\theta & 0 \\ \sin\theta & \cos\theta & 0 \\ 0 & 0 & 1 \end{pmatrix} \\ &= \begin{pmatrix} \sigma_1\cos^2\theta + \sigma_2\sin^2\theta & (\sigma_2 - \sigma_1)\sin\theta\cos\theta & 0 \\ (\sigma_2 - \sigma_1)\sin\theta\cos\theta & \sigma_1\sin^2\theta + \sigma_2\cos^2\theta & 0 \\ 0 & 0 & 0 \end{pmatrix}. \end{aligned} \tag{20}$$

For example, if $\sigma_1 = 1$, $\sigma_2 = -1$, and $\theta = 45°$,

$$\sigma' = \begin{pmatrix} 0 & -1 & 0 \\ -1 & 0 & 0 \\ 0 & 0 & 0 \end{pmatrix}. \tag{21}$$

Thus, although the large block is oriented such that the stress tensor causes only normal tractions, giving compression along the x_2 axis and tension along the x_1 axis, only shear tractions act on the smaller block because its sides are oriented differently. The negative shear stress values yield tractions in the $-x_2'$ direction on the face with normal $\hat{e}_1'$, and in the x_2' direction on the opposite face with normal $-\hat{e}_1'$, consistent with what we expect from the normal tractions on the larger block. Although the components of the stress tensor in the two coordinate systems differ, they represent the same entity, the physical state of stress.

2.3.4 *Principal stresses*

For a given state of stress, the traction vector acting on most surfaces within a material has components both normal to the surface and tangential to it. There are, however, some surfaces oriented such that the shear tractions on them vanish. These surfaces can be characterized by their normal vectors, called *principal stress axes*; the normal stresses on these surfaces are called *principal stresses*. The concept of principal stress axes is important for discussion of earthquake source mechanisms (Section 4.2).

To find the principal stresses, we use the concepts of eigenvalues and eigenvectors (Section A.5.2). The shear components of the traction will be zero if the traction and normal vectors are parallel, such that they differ only by a multiplicative constant, λ,

$$T_i = \sigma_{ij}n_j = \lambda n_i. \tag{22}$$

Thus the principal stress axes $\hat{\mathbf{n}}$ are the eigenvectors of the stress tensor, and the principal stresses λ associated with each one are the eigenvalues. The eigenvalues and eigenvectors can be found by solving the system of homogeneous linear equations

$$(\sigma_{ij} - \lambda\delta_{ij})n_j = 0$$

$$\begin{pmatrix} \sigma_{11} - \lambda & \sigma_{12} & \sigma_{13} \\ \sigma_{21} & \sigma_{22} - \lambda & \sigma_{23} \\ \sigma_{31} & \sigma_{32} & \sigma_{33} - \lambda \end{pmatrix} \begin{pmatrix} n_1 \\ n_2 \\ n_3 \end{pmatrix} = \begin{pmatrix} 0 \\ 0 \\ 0 \end{pmatrix}, \tag{23}$$

where the Kronecker delta symbol $\delta_{ij} = 0$ except when $i = j$, in which case it equals 1 (Eqn A.3.37). A nontrivial solution exists only for values of λ such that the matrix is singular (has no inverse), which occurs when its determinant is zero (Section A.4.3),

$$\det \begin{vmatrix} \sigma_{11} - \lambda & \sigma_{12} & \sigma_{13} \\ \sigma_{21} & \sigma_{22} - \lambda & \sigma_{23} \\ \sigma_{31} & \sigma_{32} & \sigma_{33} - \lambda \end{vmatrix} = 0. \tag{24}$$

Multiplying out the determinant gives the characteristic polynomial

$$\lambda^3 - I_1\lambda^2 + I_2\lambda - I_3 = 0, \tag{25}$$

whose coefficients, the invariants of the stress tensor, are independent of the coordinate system. In particular, I_1 is the *trace*, or sum of the diagonal elements, which has physical significance, as discussed in Section 2.3.6.

The roots λ of Eqn 25 are the eigenvalues or principal stresses, denoted σ_m, which are often ordered by decreasing value $\sigma_1 \geq \sigma_2 \geq \sigma_3$. In geology, where all stresses are compressive (negative), we usually order the principal stresses by magnitude, so $|\sigma_1| \geq |\sigma_2| \geq |\sigma_3|$. Each eigenvalue is then substituted into Eqn 23 to find the components of the associated eigenvector $\hat{\mathbf{n}}^{(m)}$. Because the stress tensor is symmetric, the three eigenvectors are automatically orthogonal if the roots are distinct (Section A.5.3), so there are three mutually perpendicular surfaces on which there is no tangential traction. Even if there are multiple roots, it is still always possible to find orthogonal $\hat{\mathbf{n}}^{(m)}$.

The principal stress axes are perpendicular and can be used as basis vectors for a useful coordinate system in which the stress tensor is diagonal. To transform vectors into this new coordinate system, we use a rotation matrix (Section A.5.1) whose rows are the components of the basis vectors of the new coordinate system written in the old coordinate system. In this case the rows are the eigenvectors, and the transformation matrix is

$$A = \begin{pmatrix} \hat{\mathbf{n}}^{(1)} \\ \hat{\mathbf{n}}^{(2)} \\ \hat{\mathbf{n}}^{(3)} \end{pmatrix} = \begin{pmatrix} n_1^{(1)} & n_2^{(1)} & n_3^{(1)} \\ n_1^{(2)} & n_2^{(2)} & n_3^{(2)} \\ n_1^{(3)} & n_2^{(3)} & n_3^{(3)} \end{pmatrix}. \tag{26}$$

Defining the diagonal matrix containing the eigenvalues as Λ,

$$\Lambda = \begin{pmatrix} \sigma_1 & 0 & 0 \\ 0 & \sigma_2 & 0 \\ 0 & 0 & \sigma_3 \end{pmatrix}, \tag{27}$$

we can describe all the eigenvalue–eigenvector pairs by writing Eqn 22 as a matrix equation,

$$\sigma A^T = A^T \Lambda \tag{28}$$

$$\sigma \begin{pmatrix} n_1^{(1)} & n_1^{(2)} & n_1^{(3)} \\ n_2^{(1)} & n_2^{(2)} & n_2^{(3)} \\ n_3^{(1)} & n_3^{(2)} & n_3^{(3)} \end{pmatrix} = \begin{pmatrix} n_1^{(1)} & n_1^{(2)} & n_1^{(3)} \\ n_2^{(1)} & n_2^{(2)} & n_2^{(3)} \\ n_3^{(1)} & n_3^{(2)} & n_3^{(3)} \end{pmatrix} \begin{pmatrix} \sigma_1 & 0 & 0 \\ 0 & \sigma_2 & 0 \\ 0 & 0 & \sigma_3 \end{pmatrix}.$$

Carrying out the tensor transformation (Eqn 18) shows that the stress tensor in the new coordinate system is now diagonal,

$$\sigma' = A\sigma A^T = \Lambda, \quad \sigma'_{ij} = \sigma_i\delta_{ij}, \tag{29}$$

where summation over i is not implied. To see why the stress tensor is diagonal, recall that each row of the stress tensor contains the components of the traction vector acting on a plane perpendicular to a coordinate axis. The new coordinate axes were chosen to be the principal stress axes, so on surfaces with these as normals the normal traction is the only nonzero component of the traction vector.

2.3.5 Maximum shear stress and faulting

An important seismological application of the principal stresses is that the simplest theory for rock fracture predicts that faulting will occur on the plane on which the shear stress is highest (Section 5.7.2). Although this is not exactly true, it gives insight into the relation between fault orientations and regional tectonics.

Given a state of stress, we can find the plane of maximum shear stress using the diagonalized stress tensor (Eqn 29), and thus a coordinate system whose basis vectors are the principal stress axes. By Eqn 11 the traction on a plane with normal vector $\hat{\mathbf{n}}$ is

$$T_i = \sigma'_{ij}n_j = \sigma_i\delta_{ij}n_j = \sigma_i n_i, \tag{30}$$

where summation over i is not implied. The squared magnitude of the traction normal to the surface is $(\mathbf{T} \cdot \hat{\mathbf{n}})^2 = (T_i n_i)^2$, so, using the triangular geometry (Fig. 2.3-7), the squared magnitude of τ, the tangential traction along the surface can be written as a function of the components of the normal vector

$$\begin{aligned} \tau^2(n_1, n_2, n_3) &= T_iT_i - (T_in_i)^2 \\ &= (\sigma_1 n_1)^2 + (\sigma_2 n_2)^2 + (\sigma_3 n_3)^2 \\ &\quad - (\sigma_1 n_1^2 + \sigma_2 n_2^2 + \sigma_3 n_3^2)^2. \end{aligned} \tag{31}$$

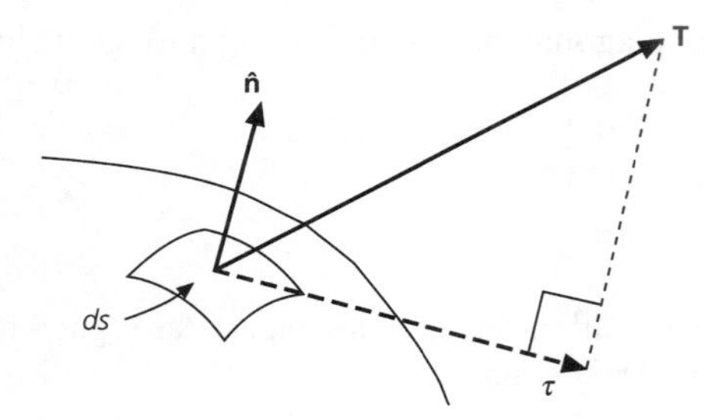

Fig. 2.3-7 Traction vector **T** acting on the surface dS, decomposed into two components. The *normal traction* is parallel to the normal, $\hat{\mathbf{n}}$, whereas τ is the *tangential traction* parallel to the surface.

This expression lets us find planes, characterized by their normal vectors $\hat{\mathbf{n}}$, on which τ^2 is a maximum. We eliminate n_3 using the fact that $n_3^2 = 1 - n_1^2 - n_2^2$, so

$$\tau^2(n_1, n_2) = n_1^2(\sigma_1^2 - \sigma_3^2) + n_2^2(\sigma_2^2 - \sigma_3^2) + \sigma_3^2 - [n_1^2(\sigma_1 - \sigma_3) + n_2^2(\sigma_2 - \sigma_3) + \sigma_3]^2. \quad (32)$$

At the maxima of τ^2, its derivatives with respect to n_1 and n_2 are zero:

$$\begin{aligned} 0 &= 2\tau\frac{\partial \tau}{\partial n_1} \\ &= 2n_1(\sigma_1 - \sigma_3)\{(\sigma_1 + \sigma_3) - 2[n_1^2(\sigma_1 - \sigma_3) + n_2^2(\sigma_2 - \sigma_3) + \sigma_3]\}, \\ 0 &= 2\tau\frac{\partial \tau}{\partial n_2} \\ &= 2n_2(\sigma_2 - \sigma_3)\{(\sigma_2 + \sigma_3) - 2[n_1^2(\sigma_1 - \sigma_3) + n_2^2(\sigma_2 - \sigma_3) + \sigma_3]\}. \end{aligned} \quad (33)$$

The first equation is satisfied if $n_1 = 0$, in which case $n_2^2 = 1/2$ satisfies the second equation because the term in braces is zero. For these values $n_3^2 = 1/2$, yielding a plane with unit normal $\hat{\mathbf{n}} = (0, 1/\sqrt{2}, 1/\sqrt{2})$. A second plane is found by setting $n_2 = 0$, so the first equation yields $\hat{\mathbf{n}} = (1/\sqrt{2}, 0, 1/\sqrt{2})$. Eliminating n_1 from Eqn 31 using the method used for n_3 yields two similar equations that can be solved for the third solution, $\hat{\mathbf{n}} = (1/\sqrt{2}, 1/\sqrt{2}, 0)$.

Each of these planes bisects the 90° angle between a pair of principal stress axes. Because two such planes can be defined for each pair of axes, there are other solutions. For example, because the condition for $n_1 = 0$ was that $n_2^2 = n_3^2 = 1/2$, $\hat{\mathbf{n}} = (0, -1/\sqrt{2}, 1/\sqrt{2})$ is also a solution.

To find the value of τ^2 as a function of $\hat{\mathbf{n}}$, we rewrite Eqn 31

$$\tau^2(n_1, n_2, n_3) = n_1^2 n_2^2[\sigma_1 - \sigma_2]^2 + n_2^2 n_3^2[\sigma_2 - \sigma_3]^2 + n_1^2 n_3^2[\sigma_1 - \sigma_3]^2. \quad (34)$$

This equation shows that of the three possible local maxima of the tangential traction, the largest value is

$$\tau = (\sigma_1 - \sigma_3)/2, \quad (35)$$

where σ_1 is the maximum principal stress and σ_3 is the minimum principal stress. This occurs on the planes with unit normal vectors

$$\hat{\mathbf{n}} = (1/\sqrt{2}, 0, 1/\sqrt{2}) \quad \text{and} \quad \hat{\mathbf{n}} = (-1/\sqrt{2}, 0, 1/\sqrt{2}). \quad (36)$$

Thus the planes of maximum shear stress are halfway between the maximum (1, 0, 0) and minimum (0, 0, 1) principal stress axes, and contain the intermediate principal stress axis. The derivatives (Eqn 33) are also zero at local minima, corresponding to the principal stress axes where $\tau^2 = 0$.

To apply this theory, consider an experiment in which a rock is compressed (Fig. 2.3-8) such that the principal stresses are negative, with $|\sigma_1| \geq |\sigma_2| \geq |\sigma_3|$. We expect fracture on the planes of maximum shear stress. By Eqn 36, there are two such planes, each 45° from the maximum and minimum principal stress axes and including the intermediate principal stress axis. Either plane is equally likely to fracture. Alternatively, if the experiment is conducted in a common laboratory situation known as uniaxial compression, where $|\sigma_1| \geq |\sigma_2| = |\sigma_3|$, failure should occur on any plane 45° from the maximum principal stress (σ_1) axis. Experiments (Section 5.7.2) support the idea that fracture is controlled by shear stress, but in a more complicated way such that the fracture plane is often

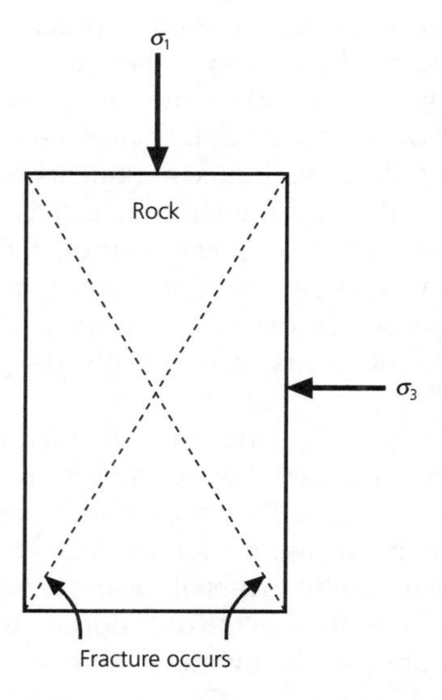

Fig. 2.3-8 Schematic illustration of an experiment in which a cylindrical rock sample is compressed along the direction of the maximum principal stress σ_1 until fracture occurs. The minimum principal stresses σ_2 and σ_3 are approximately equal. If fracture occurs on a plane of maximum shear stress, the rock breaks on a plane 45° from the direction of maximum principal stress.

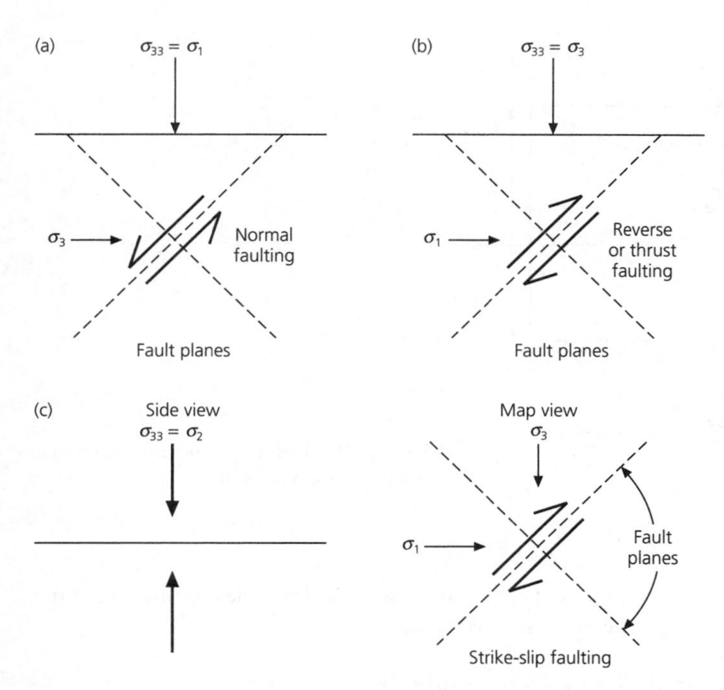

Fig. 2.3-9 Stress fields associated with three types of faulting, assuming that the earthquake occurred on a plane of maximum shear stress. Normal (a), reverse (b), and strike-slip (c) faulting involve different orientations of the principal stresses.

about 25°, rather than 45°, from the maximum principal stress direction.

For simplicity, however, assume that faults in the earth form on the planes of maximum shear stress. We will see (Section 2.3.10) that the earth's surface is a free surface, where tractions must be zero. Hence, at the surface one principal stress axis must be vertical, and the other two must be parallel to the surface. The three basic fault geometries — strike-slip, normal, and thrust — are related to the stress axes (Fig. 2.3-9). If the vertical principal stress is the most compressive, the fault dips at 45°, and normal faulting occurs. If, instead, the vertical principal stress is the least compressive, the fault geometry is the same, but reverse or thrust faulting occurs.[2] When the vertical principal stress is the intermediate principal stress, strike-slip motion occurs on a fault plane 45° from the maximum principal stress. Thus the geometry of faults, which can be mapped geologically or inferred from seismograms of earthquakes, can be used to study stress orientations. This model is subject to limitations, especially because earthquakes often occur on preexisting faults (Section 5.7.2). Nonetheless, the approach is useful, especially when integrated with other methods of estimating stress directions.

[2] Seismologists sometimes use the terms *reverse* and *thrust* fault interchangeably, whereas structural geologists reserve the term *thrust* for a shallow-dipping reverse fault.

2.3.6 *Deviatoric stresses*

Large compressive stresses occur at depth within the earth due to the weight of the overlying rock. It is convenient in many applications to remove the effect of the overall compressive stress and consider only the deviations from it. We thus define the *mean stress*

$$M = (\sigma_{11} + \sigma_{22} + \sigma_{33})/3 = \sigma_{ii}/3 \qquad (37)$$

as $\frac{1}{3}$ of the sum of the normal stresses, the trace of the stress tensor. The mean stress can be related to the principal stresses, because the trace of the stress tensor is independent of the coordinate system.

To see that the trace does not change, we write the transformation of the stress tensor between two coordinate systems (Eqn 18) in terms of the components, using the summation convention (Section A.3.5)

$$\sigma'_{ij} = A_{ik}\sigma_{kl}A^T_{lj} = A_{ik}\sigma_{kl}A_{jl}. \qquad (38)$$

The trace can be written

$$\sigma'_{ii} = \sigma'_{ij}\delta_{ij} = A_{ik}\sigma_{kl}A_{il} = \delta_{kl}\sigma_{kl} = \sigma_{kk}, \qquad (39)$$

because A is an orthogonal matrix, so that $A_{ik}A_{il} = \delta_{kl}$. Thus the trace is invariant under an orthogonal transformation, and so is known as the first invariant of a tensor. The other two invariants (Eqn 25) are also preserved by such transformations.

The mean stress can thus be written in terms of the trace of the diagonalized stress tensor (Eqn 29)

$$M = (\sigma_1 + \sigma_2 + \sigma_3)/3 \qquad (40)$$

as $\frac{1}{3}$ of the sum of the principal stresses. The *deviatoric* stress tensor is defined by removing the effect of the mean stress

$$D_{ij} = \sigma_{ij} - M\delta_{ij}$$

$$D = \begin{pmatrix} \sigma_{11} - M & \sigma_{12} & \sigma_{13} \\ \sigma_{21} & \sigma_{22} - M & \sigma_{23} \\ \sigma_{31} & \sigma_{32} & \sigma_{33} - M \end{pmatrix}. \qquad (41)$$

Thus, when the principal stresses are large and nearly equal, the deviatoric stress tensor removes their effect and indicates the remaining stress state. The deviatoric stress tensor can be diagonalized and has the same principal stress axes as the stress tensor.

This concept is important in discussing processes in the earth, because the deviatoric stresses result from tectonic forces and cause earthquake faulting and seismic wave propagation effects like anisotropy. At depths greater than a few kilometers, we often assume that a *lithostatic* state of stress exists, where the normal stresses are equal to minus the pressure of the overlying material and the deviatoric stresses are zero. Because

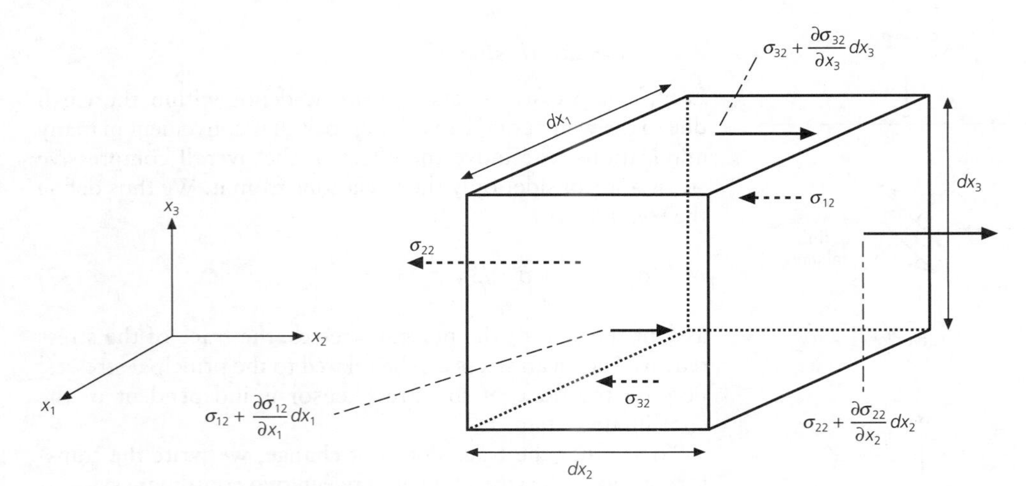

Fig. 2.3-10 Stress components contributing to force in the x_2 direction.

the weight of a column of material of height z and density ρ is ρgz, the pressure at a depth of 3 km beneath a column of rock with density 3 g/cm^3 is

$$P = (3\ \text{g/cm}^3)(980\ \text{cm/s}^2)(3 \times 10^5\ \text{cm})$$
$$\approx 9 \times 10^8\ \text{dyn/cm}^2 = 0.9\ \text{kbar}. \quad (42)$$

The approximation that the pressure at 3 km depth is about 1 kbar (100 MPa) is useful to remember.

The pressure causes compression and thus negative values of the principal stresses. If the state of stress at depth is lithostatic, the mean stress equals the negative of the pressure. Because deviatoric stresses exist, this relation is only approximate, but it is useful because the mean stress is usually thought to be much greater than the deviatoric stress.

2.3.7 *Equation of motion*

Now that we can describe the forces acting on the surface of a material element in terms of the stresses, we write Newton's second law (Eqn 1) in terms of body forces and stresses. This is the first step to deriving the equations describing seismic wave propagation.

Consider the forces acting on a block of material of density ρ and volume $dx_1 dx_2 dx_3$ with sides perpendicular to the coordinate axes (Fig. 2.3-10). The net body force, if any, is $f_i dx_1 dx_2 dx_3$, where f_i is the force per unit volume at the center of the block. The total force is the sum of the surface forces on each face plus the body force within the material.

For example, the net surface force in the x_2 direction is the sum of three terms, each of which describes the net force due to the difference in traction between opposing faces. The first term involves the difference between the traction in the $\hat{e}_2$ direction resulting from the stress on the face with normal $\hat{e}_2$ and that on the opposite face with normal $-\hat{e}_2$. Because stress is force per unit area, we multiply this difference by the area of the two faces, $dx_1 dx_3$, and use a Taylor series to obtain the net force due to these two faces,

$$[\sigma_{22}(\mathbf{x} + dx_2\hat{e}_2) - \sigma_{22}(\mathbf{x})]dx_1 dx_3$$
$$= \left[\sigma_{22}(\mathbf{x}) + \frac{\partial \sigma_{22}(\mathbf{x})}{\partial x_2} dx_2 - \sigma_{22}(\mathbf{x})\right] dx_1 dx_3$$
$$= \frac{\partial \sigma_{22}(\mathbf{x})}{\partial x_2} dx_1 dx_2 dx_3. \quad (43)$$

We then do the same for the force in the x_2 direction due to the pairs of faces with normals $\pm\hat{e}_1$ and $\pm\hat{e}_3$. Summing the three terms, adding the body force component, and equating this net force to the density times this component of the acceleration yields

$$\left[\frac{\partial \sigma_{12}}{\partial x_1} + \frac{\partial \sigma_{22}}{\partial x_2} + \frac{\partial \sigma_{32}}{\partial x_3}\right] dx_1 dx_2 dx_3 + f_2 dx_1 dx_2 dx_3$$
$$= \rho \frac{\partial^2 u_2}{\partial t^2} dx_1 dx_2 dx_3. \quad (44)$$

The first three terms give the net force from the tractions on opposite faces of the cube. As we saw, each stress component canceled with its value from the opposite face, so only the partial derivative of that component contributes to the net force. Hence the spatial variation of the stress field,[3] rather than the stress field itself, causes a net force. Dividing by the volume of the block yields

$$\frac{\partial \sigma_{12}}{\partial x_1} + \frac{\partial \sigma_{22}}{\partial x_2} + \frac{\partial \sigma_{32}}{\partial x_3} + f_2 = \sum_{j=1}^{3} \frac{\partial \sigma_{j2}}{\partial x_j} + f_2 = \rho \frac{\partial^2 u_2}{\partial t^2}. \quad (45)$$

[3] A field is a quantity that varies in space (Section A.6.1).

Similar equations apply for the x_1 and x_3 components of the force and acceleration. The set of three equations can be written simply using the summation convention

$$\frac{\partial \sigma_{ji}(\mathbf{x}, t)}{\partial x_j} + f_i(\mathbf{x}, t) = \rho \frac{\partial^2 u_i(\mathbf{x}, t)}{\partial t^2}. \quad (46)$$

Here the fact that the stresses, forces, and displacements can vary in both space and time is explicitly written. Alternatively, because the stress tensor is symmetric, we can write

$$\frac{\partial \sigma_{ij}(\mathbf{x}, t)}{\partial x_j} + f_i(\mathbf{x}, t) = \rho \frac{\partial^2 u_i(\mathbf{x}, t)}{\partial t^2}. \quad (47)$$

Note that the force in the i direction is obtained by summing over the faces j of the block. If the partial derivative with respect to x_j is denoted by a comma, Eqn 47 becomes

$$\sigma_{ij,j}(\mathbf{x}, t) + f_i(\mathbf{x}, t) = \rho \frac{\partial^2 u_i(\mathbf{x}, t)}{\partial t^2}. \quad (48)$$

This equation, called the *equation of motion*, is satisfied everywhere in a continuous medium. It expresses Newton's second law, $\mathbf{F} = m\mathbf{a}$, in terms of surface and body forces. The acceleration results from the body force and $\sigma_{ij,j}$, the divergence of the stress tensor. A stress field that does not vary with position has no divergence, and hence produces no force. It is interesting to note that the divergence of the stress tensor gives rise to a force, which is a vector, just as the divergence of a vector yields a scalar (Section A.6.3).

An important form of the equation of motion describes a body at equilibrium, whose acceleration is zero, so the divergence of the stress tensor exactly balances the body forces

$$\sigma_{ij,j}(\mathbf{x}, t) = -f_i(\mathbf{x}, t). \quad (49)$$

This *equation of equilibrium* must be satisfied for any static elasticity problem, such as finding the stresses due only to gravity.

Another important form, if no body forces are applied, is

$$\sigma_{ij,j}(\mathbf{x}, t) = \rho \frac{\partial^2 u_i(\mathbf{x}, t)}{\partial t^2}. \quad (50)$$

This is called the *homogeneous equation of motion*, where "homogeneous" refers to the lack of forces, as in the terminology of linear equations (Section A.4.4). This equation describes seismic wave propagation except at a source, such as an earthquake or an explosion, where a body force generates seismic waves.

2.3.8 Strain

If stresses are applied to a material that is not rigid, points within it move with respect to each other, and deformation results. The *strain tensor* describes the deformation resulting from the differential motion within the body.

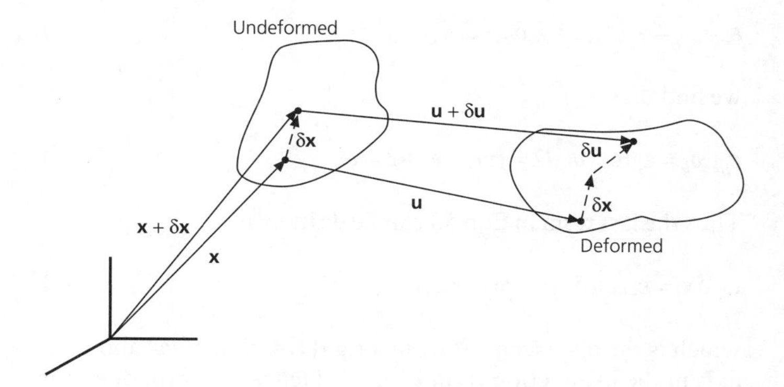

Fig. 2.3-11 Geometry showing how deformation arises from the relative displacement δu between two points originally separated by δx.

Consider an element of solid material within which displacements $\mathbf{u}(\mathbf{x})$ have occurred. If a point originally at $\mathbf{x}$ is displaced by $\mathbf{u}$ (Fig. 2.3-11), we describe the displacement of a nearby point originally at $\mathbf{x} + \delta\mathbf{x}$ by expanding the components of the displacement vector in a Taylor series,

$$u_i(\mathbf{x} + \delta\mathbf{x}) \approx u_i(\mathbf{x}) + \frac{\partial u_i(\mathbf{x})}{\partial x_j} \delta x_j = u_i(\mathbf{x}) + \delta u_i, \quad (51)$$

so that the relative displacement near $\mathbf{x}$, δu_i, is to the first order

$$\delta u_i = \frac{\partial u_i(\mathbf{x})}{\partial x_j} \delta x_j, \quad (52)$$

where the partial derivatives are evaluated at $\mathbf{x}$.

Although we are interested in deformation that distorts the body, there can also be a rigid body translation or a rigid body rotation, neither of which produces deformation. To distinguish these effects, we add and subtract $\partial u_j/\partial x_i$ to Eqn 52 and then separate it into two parts

$$\delta u_i = \frac{1}{2}\left(\frac{\partial u_i}{\partial x_j} + \frac{\partial u_j}{\partial x_i}\right)\delta x_j + \frac{1}{2}\left(\frac{\partial u_i}{\partial x_j} - \frac{\partial u_j}{\partial x_i}\right)\delta x_j = (e_{ij} + \omega_{ij})\delta x_j. \quad (53)$$

The ω_{ij} term corresponds to a rigid body rotation without deformation. To see this, note that because ω_{ij} is antisymmetric ($\omega_{ij} = -\omega_{ji}$), the diagonal terms are zero, and there are only three independent components. We can then form a vector ω with components

$$\omega_k = \varepsilon_{stk}\omega_{st}/2, \quad (54)$$

where ε_{stk} is the permutation symbol (Eqn A.3.39). Using the identity

$$\varepsilon_{ijk}\varepsilon_{stk} = \varepsilon_{kij}\varepsilon_{kst} = \delta_{is}\delta_{jt} - \delta_{it}\delta_{js}, \tag{55}$$

we find that

$$\varepsilon_{ijk}\omega_k = \varepsilon_{ijk}\varepsilon_{stk}\omega_{st}/2 = (\omega_{ij} - \omega_{ji})/2 = \omega_{ij}. \tag{56}$$

Thus the last term in Eqn 53 can be written as

$$\omega_{ij}\delta x_j = \varepsilon_{ijk}\omega_k \delta x_j = -\omega \times \delta \mathbf{x}, \tag{57}$$

which is the displacement from a rigid rotation of $|\omega|$ about an axis in the ω direction (Eqn A.3.31). Hence this term does not reflect deformation.

The other term in Eqn 53, e_{ij}, is the *strain tensor*, a symmetric tensor describing the internal deformation. Its tensor components

$$e_{ij} = \begin{pmatrix} \frac{\partial u_1}{\partial x_1} & \frac{1}{2}\left(\frac{\partial u_1}{\partial x_2} + \frac{\partial u_2}{\partial x_1}\right) & \frac{1}{2}\left(\frac{\partial u_1}{\partial x_3} + \frac{\partial u_3}{\partial x_1}\right) \\ \frac{1}{2}\left(\frac{\partial u_2}{\partial x_1} + \frac{\partial u_1}{\partial x_2}\right) & \frac{\partial u_2}{\partial x_2} & \frac{1}{2}\left(\frac{\partial u_2}{\partial x_3} + \frac{\partial u_3}{\partial x_2}\right) \\ \frac{1}{2}\left(\frac{\partial u_3}{\partial x_1} + \frac{\partial u_1}{\partial x_3}\right) & \frac{1}{2}\left(\frac{\partial u_3}{\partial x_2} + \frac{\partial u_2}{\partial x_3}\right) & \frac{\partial u_3}{\partial x_3} \end{pmatrix} \tag{58}$$

are spatial derivatives of the displacement field, $\mathbf{u}(\mathbf{x})$. If the displacement field does not vary, its derivatives are zero, so there is no deformation, only a rigid body translation.

The strain tensor can be written in terms of the x, y, z axes using the derivatives of the displacement vector components (u_x, u_y, u_z):

$$e_{ij} = \begin{pmatrix} \frac{\partial u_x}{\partial x} & \frac{1}{2}\left(\frac{\partial u_x}{\partial y} + \frac{\partial u_y}{\partial x}\right) & \frac{1}{2}\left(\frac{\partial u_x}{\partial z} + \frac{\partial u_z}{\partial x}\right) \\ \frac{1}{2}\left(\frac{\partial u_y}{\partial x} + \frac{\partial u_x}{\partial y}\right) & \frac{\partial u_y}{\partial y} & \frac{1}{2}\left(\frac{\partial u_y}{\partial z} + \frac{\partial u_z}{\partial y}\right) \\ \frac{1}{2}\left(\frac{\partial u_z}{\partial x} + \frac{\partial u_x}{\partial z}\right) & \frac{1}{2}\left(\frac{\partial u_z}{\partial y} + \frac{\partial u_y}{\partial z}\right) & \frac{\partial u_z}{\partial z} \end{pmatrix}. \tag{59}$$

The components of the strain tensor are dimensionless because they have units of length divided by length. The components are of two different types. The diagonal components show how the displacement in the direction of a coordinate axis varies along that axis. For example, if displacement occurs only in the x_1 direction ($u_2 = 0$, $u_3 = 0$) and u_1 changes only in that direction, then the only nonzero term in the tensor is e_{11}. Extension occurs along the x_1 axis if $\partial u_1/\partial x_1 > 0$ (Fig. 2.3-12a), whereas contraction occurs if it is negative (Fig. 2.3-12b). If e_{11} were constant within the material, it would equal the change in length per unit length along the x_1 axis. The other diagonal terms, e_{22} and e_{33}, represent similar strains along their coordinate axes.

The off-diagonal components describe changes along a coordinate axis of displacement in another direction. A simple case (Fig. 2.3-12c) is when only $u_1 \neq 0$, but u_1 changes only along the x_2 axis, so only e_{12} and e_{21} are nonzero. We can also have both $\partial u_1/\partial x_2$ and $\partial u_2/\partial x_1$ nonzero (Fig. 2.3-12d, e). Depending on the relative values of the derivatives, the strain components describe various deformations.

The strain tensor can be characterized by its eigenvectors, the principal strain axes, and associated eigenvalues, the principal strains. The strain tensor is diagonal when expressed in a coordinate system whose basis vectors are the principal strain axes. The trace or sum of diagonal terms of the strain tensor,

$$\theta = e_{ii} = \frac{\partial u_1}{\partial x_1} + \frac{\partial u_2}{\partial x_2} + \frac{\partial u_3}{\partial x_3} = \nabla \cdot \mathbf{u}, \tag{60}$$

known as the *dilatation*, equals the divergence of the displacement field $\mathbf{u}(\mathbf{x})$. The dilatation has physical significance because it gives the change in volume per unit volume associated with the deformation. To see this, note that in the principal strain axes coordinate system a block of material with initial volume $dx_1dx_2dx_3$ has a volume after deformation (Fig. 2.3-13) of

$$\left(1 + \frac{\partial u_1}{\partial x_1}\right) dx_1 \left(1 + \frac{\partial u_2}{\partial x_2}\right) dx_2 \left(1 + \frac{\partial u_3}{\partial x_3}\right) dx_3, \tag{61}$$

which, to first order,

$$\approx \left(1 + \frac{\partial u_1}{\partial x_1} + \frac{\partial u_2}{\partial x_2} + \frac{\partial u_3}{\partial x_3}\right) dx_1dx_2dx_3 = (1 + \theta)\, dx_1dx_2dx_3. \tag{62}$$

Thus, if we define the initial volume as $V = dx_1dx_2dx_3$,

$$V + \Delta V = (1 + \theta)V, \text{ so } \theta = \Delta V/V, \tag{63}$$

and the dilatation is the change in volume per unit volume.

It is worth noting that we have discussed the strain tensor in Cartesian coordinates. This tensor is more complicated when formulated in other coordinate systems, because it involves spatial derivatives of the basis vectors (Section A.7.4).

2.3.9 *Constitutive equations*

Various materials respond differently to an applied stress. For a given stress, a more rigid material responds with smaller strains than occur in a less rigid material. The relation between stress and strain is given by the material's *constitutive equation*.

The simplest types of materials are *linearly elastic*, such that there is a linear relation between the stress and strain tensors. We will see that when the earth behaves as linearly elastic, it gives rise to seismic waves. Linear elasticity is valid for the short time scale involved in the propagation of seismic waves, but not for longer time scales. On time scales of thousands of years or longer, the mantle rock flows as a viscous fluid (Section 5.7.3).

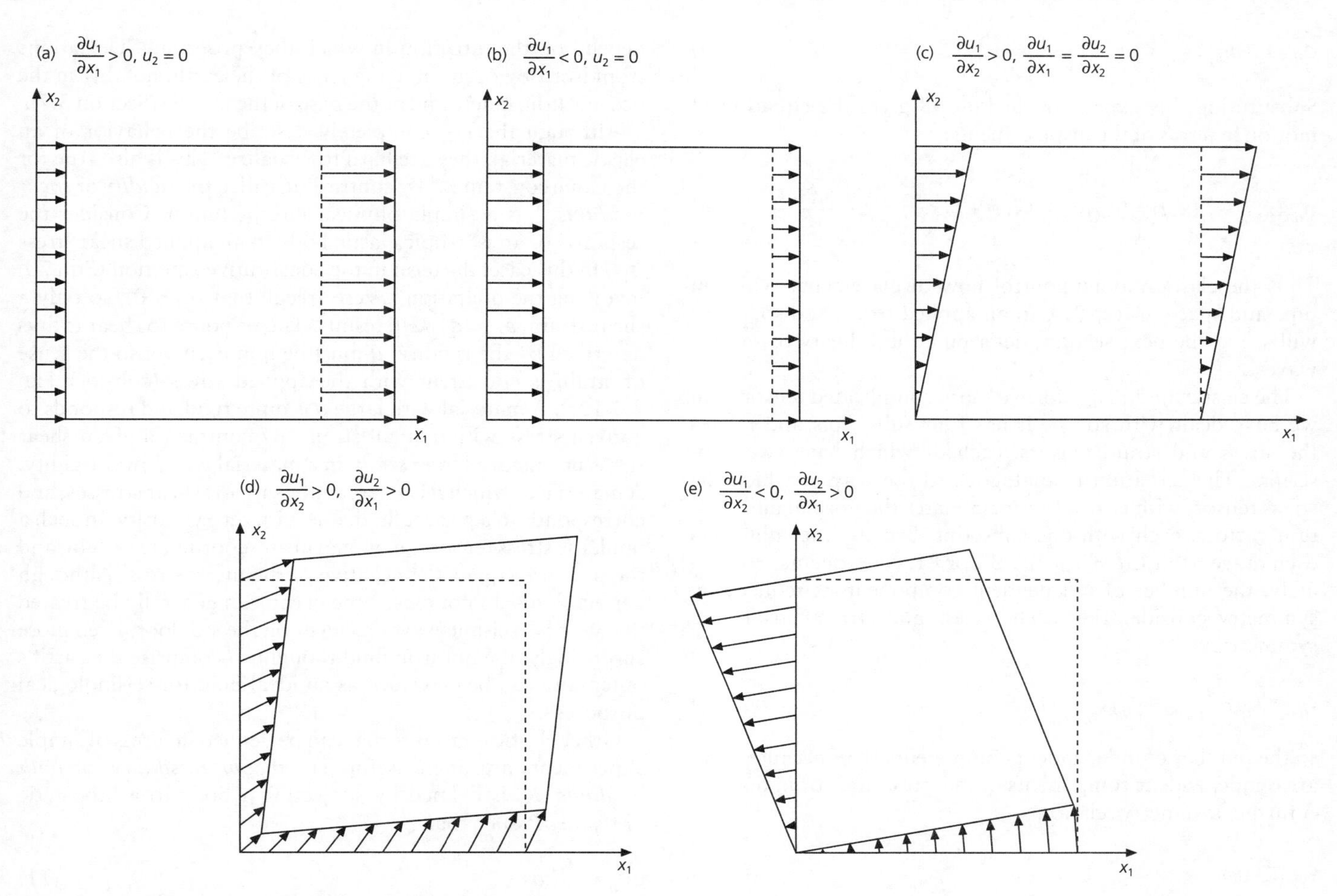

Fig. 2.3-12 Some possible strains for a two-dimensional element.

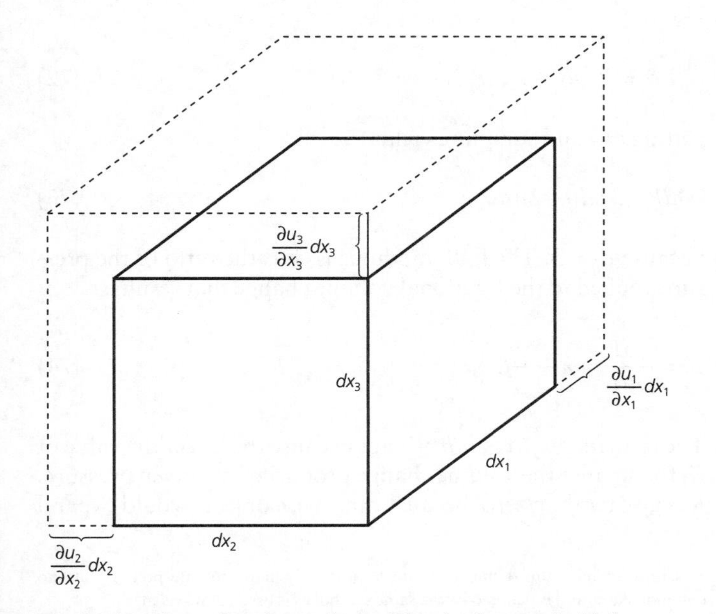

Fig. 2.3-13 Change in volume of a small block of material with faces normal to the coordinate axes, due to the principal strains. The fractional change in volume is the dilatation, the sum of the principal strains.

In assuming that material is elastic, we also assume that the displacements from an unstrained initial state are small. This assumption, known as *infinitesimal strain theory*, is generally valid for seismic waves. For example, a body wave may have a displacement on the order of 10 microns, and a wavelength on the order of 10 km. Expressing all quantities in meters, the resulting strain is about $(10^{-5}/10^4) = 10^{-9}$, certainly small enough for infinitesimal theory to be valid. However, for strains greater than about 10^{-4}, the linear relation between stress and strain fails. This occurs in regions of the earth's mantle under very high pressure, or when rocks break during an earthquake (Section 5.7.2).

The stress and strain for a linearly elastic material are related by a constitutive equation called *Hooke's law*,

$$\sigma_{ij} = c_{ijkl} e_{kl}, \tag{64}$$

written here using the summation convention. The constants c_{ijkl}, the *elastic moduli*, describe the properties of the material. To understand how the elastic moduli affect the equation of motion, we write the constitutive equation (64) using the fact that the strains are derivatives of the displacement,

$$\sigma_{ij} = c_{ijkl} u_{k,l}. \tag{65}$$

Substituting this expression in Eqn 48 gives the equation of motion in terms of the displacements:

$$\sigma_{ij,j}(\mathbf{x}, t) + f_i(\mathbf{x}, t) = (c_{ijkl} u_{k,l})_{,j}(\mathbf{x}, t) + f_i(\mathbf{x}, t) = \rho \frac{\partial^2 u_i(\mathbf{x}, t)}{\partial t^2}. \tag{66}$$

Thus the elastic moduli control how displacements evolve in time and space in response to an applied force, and so, as we will see in the next section, determine the velocity of seismic waves.

The elastic moduli c_{ijkl} form a more complicated tensor than we have dealt with so far. It has four subscripts and relates the stress and strain tensors, each of which have two subscripts. This situation is analogous to the way in which the stress tensor, with two subscripts, relates the normal and traction vectors, each with one subscript. Because the subscripts each range from 1 to 3, c_{ijkl} has 3^4, or 81, components. Fortunately, the number of independent components is reduced by symmetry considerations. The stress and strain tensors are symmetric

$$c_{ijkl} = c_{jikl}, \quad c_{ijkl} = c_{ijlk}, \tag{67}$$

so the number of independent components is 36 because there are 6 independent components of the stress and strain tensors. A further symmetry relation

$$c_{ijkl} = c_{klij}, \tag{68}$$

based on the idea of strain energy, which we will discuss later, reduces the number of independent components that characterize a general elastic medium to 21.

On a large scale, material within the earth has approximately the same physical properties regardless of orientation, a condition known as *isotropy*. For an isotropic material, the c_{ijkl} have further symmetries, so there are only two independent elastic moduli, which can be defined in various ways. One useful pair are the *Lamé constants* λ and μ, which are defined such that

$$c_{ijkl} = \lambda \delta_{ij} \delta_{kl} + \mu(\delta_{ik}\delta_{jl} + \delta_{il}\delta_{jk}). \tag{69}$$

In terms of the Lamé constants, the constitutive equation (Eqn 64) for an isotropic material is written

$$\sigma_{ij} = \lambda e_{kk} \delta_{ij} + 2\mu e_{ij} = \lambda \theta \delta_{ij} + 2\mu e_{ij}, \tag{70}$$

where θ is the dilatation. So, for example, $\sigma_{11} = \lambda\theta + 2\mu e_{11}$, and $\sigma_{12} = 2\mu e_{12}$. We will use this constitutive relation to study seismic waves in the next section. We will also see that the velocities of seismic waves depend on the elastic moduli, so in an isotropic material the velocities of seismic waves do not depend on the direction in which they propagate. Deviations from isotropy occur in many parts of the earth, notably in the oceanic lithosphere and at the base of the mantle (Section 3.7).

Although the c_{ijkl} completely describe the behavior of an elastic material, they are hard to visualize. This is also true for the Lamé constant λ.[4] By contrast, μ, called the *rigidity* or *shear modulus*, has a simple physical interpretation. Consider the response of an isotropic elastic body to an applied shear stress σ_{12}. In this case, the term in the constitutive equation (Eqn 70) involving the dilatation is zero (recall that $\delta_{12} = 0$), so only a shear strain, $e_{12} = \sigma_{12}/2\mu$, results. The response to shear is thus described by the rigidity. μ must be nonnegative, so the sense of strain is consistent with the applied stress (consider Fig. 2.3-12c). A material with large μ is quite rigid and responds to a given stress with a small strain. By contrast, a given shear stress produces a larger strain in a material with lower rigidity. A material in which μ is zero cannot support shear stresses, and corresponds to a perfect fluid, one with zero viscosity. In such a fluid, the stress tensor is diagonal in any coordinate system, and the pressure equals the negative of the mean stress. Although perfect fluids do not exist,[5] the ocean can generally be treated this way for seismic waves incident on the sea floor. Even more surprisingly, the hot iron fluid thought to comprise the earth's outer core can be described as an ideal fluid for seismological purposes.

Other elastic constants that can be defined in terms of simple experiments are often useful. The *incompressibility*, or *bulk modulus*, K, is defined by subjecting a body to a lithostatic pressure dP, such that

$$d\sigma_{ij} = -dP\delta_{ij}. \tag{71}$$

For an isotropic elastic body, the resulting strains, from Eqn 70, are

$$-dP\delta_{ij} = \lambda d\theta \delta_{ij} + 2\mu de_{ij}. \tag{72}$$

Setting $i = j$ and summing yields

$$-3dP = 3\lambda d\theta + 2\mu d\theta, \tag{73}$$

because $\delta_{ii} = 3$. The bulk modulus is thus the ratio of the pressure applied to the fractional volume change that results:

$$K = \frac{-dP}{d\theta} = \lambda + \frac{2}{3}\mu. \tag{74}$$

The term *incompressibility* is apt because the larger the value of K, the smaller the volume change produced by a given pressure. K is greater than zero, because otherwise objects would expand

[4] Unfortunately, this Lamé constant is not only hard to interpret; it has no common name and is denoted by the same symbol as is used for wavelength.

[5] Perfect fluids have been called "dry water" to illustrate that no real fluid behaves exactly this way.

when compressed.[6] In an ideal fluid, $K = \lambda$, so in this case λ has an easy physical interpretation.

Writing the constitutive equation (70) in terms of K and μ,

$$\sigma_{ij} = K\theta\delta_{ij} + 2\mu(e_{ij} - \theta\delta_{ij}/3) \tag{75}$$

shows that the response to an applied stress has two parts: a volume change characterized by K and a shear deformation, or change in shape, characterized by μ.

Two other elastic constants are defined by pulling the material along only one axis, leading to a state of stress called *uniaxial tension*. If the tension is applied along the x_1 axis, then by Equation 70,

$$\begin{aligned}\sigma_{11} &= (\lambda + 2\mu)e_{11} + \lambda e_{22} + \lambda e_{33}\\ \sigma_{22} &= 0 = \lambda e_{11} + (\lambda + 2\mu)e_{22} + \lambda e_{33}\\ \sigma_{33} &= 0 = \lambda e_{11} + \lambda e_{22} + (\lambda + 2\mu)e_{33}.\end{aligned} \tag{76}$$

Subtracting the last two equations shows that $e_{22} = e_{33}$, so

$$e_{22} = e_{33} = \frac{-\lambda}{2(\lambda + \mu)} e_{11} = -\nu e_{11}, \tag{77}$$

where ν, defined as *Poisson's ratio*, gives the ratio of the contraction along the other two axes to the extension along the axis where tension was applied. Substituting in the first line in Eqn 76 yields

$$\frac{\sigma_{11}}{e_{11}} = \frac{\mu(3\lambda + 2\mu)}{\lambda + \mu} = E, \tag{78}$$

where E is called *Young's modulus*, the ratio of the tensional stress to the resulting extensional strain.

The elastic constants E, ν, and K are often used in engineering because they are easily measured by simple experiments. However, for seismic wave propagation, λ, μ, and sometimes K are more natural constants.[7] Box 2.3-1 gives conversions between the various elastic constants.

Many seismological problems are simplified by assuming that $\lambda = \mu$. Such a material, called a *Poisson solid*, is often a good approximation for the earth. In this case, Poisson's ratio equals 0.25, Young's modulus $E = (5/2)\mu$, and the bulk modulus $K = (5/3)\mu$.

Because strain is dimensionless, the elastic constants λ, μ, E, and K all have dimensions of stress. For the earth's crust, μ is approximately 3×10^{11} dyn/cm^2. For comparison, the rigidity of steel is about 8×10^{11} dyn/cm^2. Young's modulus for the crust, assuming a Poisson solid, is 7.5×10^{11} dyn/cm^2, compared to 5×10^9 dyn/cm^2 for rubber.

[6] Such strange materials have been manufactured synthetically.

[7] In engineering the shear modulus μ is often termed G.

Box 2.3-1 Relations between moduli

$$\nu = \frac{\lambda}{2(\lambda + \mu)} = \frac{\lambda}{(3K - \lambda)} = \frac{E}{2\mu} - 1 = \frac{3K - 2\mu}{2(3K + \mu)} = \frac{3K - E}{6K}$$

$$E = \frac{\mu(3\lambda + 2\mu)}{\lambda + \mu} = \frac{\lambda(1 + \nu)(1 - 2\nu)}{\nu} = \frac{9K(K - \lambda)}{3K - \lambda} = 2\mu(1 + \nu)$$

$$= \frac{9K\mu}{3K + \mu} = 3K(1 - 2\nu)$$

$$K = \lambda + \frac{2}{3}\mu = \frac{\lambda(1 + \nu)}{3\nu} = \frac{2\mu(1 + \nu)}{3(1 - 2\nu)} = \frac{\mu E}{3(3\mu - E)} = \frac{E}{3(1 - 2\nu)}$$

$$\lambda = \frac{2\mu\nu}{1 - 2\nu} = \frac{\mu(E - 2\mu)}{3\mu - E} = K - \frac{2}{3}\mu = \frac{E\nu}{(1 + \nu)(1 - 2\nu)}$$

$$= \frac{3K\nu}{1 + \nu} = \frac{3K(3K - E)}{9K - E}$$

$$\mu = \frac{\lambda(1 - 2\nu)}{2\nu} = \frac{3}{2}(K - \lambda) = \frac{E}{2(1 + \nu)} = \frac{3K(1 - 2\nu)}{2(1 + \nu)} = \frac{3KE}{9K - E}.$$

2.3.10 *Boundary conditions*

For a string (Section 2.2.3), wave propagation across an interface depends on *boundary conditions* that relate the displacements and tractions across the interface. In the earth, we conduct similar analyses for three types of interface.

The boundary conditions at the earth's surface are derived for most seismological purposes by neglecting the atmosphere and treating the surface as a boundary between a solid and a vacuum. In this approximation, the earth's surface is a *free surface*, not subject to any force. At a free surface with normal $\hat{\mathbf{n}}$ the traction vector is zero, giving a constraint on those stress components that affect the components of the traction:

$$T_i = \sigma_{ij} n_j = 0. \tag{79}$$

Thus, in a coordinate system in which the surface is horizontal, the normal vector is $n_i = \delta_{i3}$, and $T_i = \sigma_{i3} n_3$, so

$$\sigma_{13} = \sigma_{23} = \sigma_{33} = 0. \tag{80}$$

The components of the stress tensor that do not affect the tractions, in this case σ_{11}, σ_{12}, and σ_{22}, are unconstrained. Similarly, no restriction is placed on the displacements. A free surface corresponds in the one-dimensional case to a string whose end is free to move.

There are also interfaces between two solids, a solid and a liquid, and between two liquids. Their boundary conditions are obtained by considering a volume, sometimes called a *Gaussian pill box*, along the interface between different materials (Fig. 2.3-14). The volume's long axis is along the interface, so the surface area, S, is large relative to the volume, V. We integrate the homogeneous equation of motion (Eqn 50) over the volume

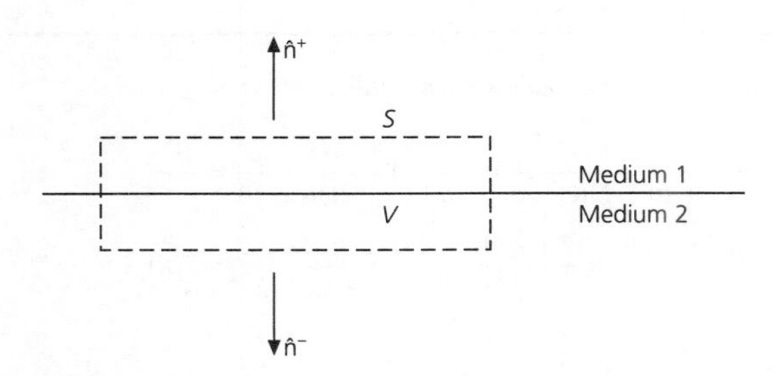

Fig. 2.3-14 "Gaussian pill box" used to formulate the boundary conditions across an interface. Application of the divergence theorem shows that the traction vector must be continuous across the interface, but that the entire stress tensor need not be.

$$\int \left(\sigma_{ij,j}(\mathbf{x}, t) - \rho \frac{\partial^2 u_i(\mathbf{x}, t)}{\partial t^2} \right) dV = 0, \tag{81}$$

and use the divergence theorem (Eqn A.6.10) to transform the first term to a surface integral, giving

$$\int \sigma_{ij}(\mathbf{x}, t) n_j dS - \int \rho \frac{\partial^2 u_i(\mathbf{x}, t)}{\partial t^2} dV = 0, \tag{82}$$

where n_j is the j component of the unit outward normal vector at each point on S. In the limit as the thickness approaches zero, the volume integral becomes negligible, so

$$\int \sigma_{ij}(\mathbf{x}, t) n_j dS = 0. \tag{83}$$

Because the thickness goes to zero, we neglect the ends, so that for the integral to be zero, the contributions from the top (+) and bottom (−) surfaces must satisfy

$$(\sigma_{ij} n_j)^+ + (\sigma_{ij} n_j)^- = 0. \tag{84}$$

Hence, because the unit normal on top is opposite that on the bottom ($n_j^+ = -n_j^-$), the three components of the traction vector, $T_i = \sigma_{ij} n_j$, must be continuous across the interface.

The continuity of traction leads to conditions on specific stress components, depending on the orientation of the interface. For example, if the interface is horizontal, then $n_j = \delta_{j3}$, so

$$T_i = \sigma_{ij} \delta_{j3} = \sigma_{i3} \tag{85}$$

must be continuous. If, instead, the boundary between two solids were vertical, then $n_j = \delta_{j1}$, so

$$T_i = \sigma_{ij} \delta_{j1} = \sigma_{i1} \tag{86}$$

would be continuous. Because the continuity conditions are for tractions rather than stresses, the stress components not involved in the traction condition need not be continuous.

Table 2.3-1 Boundary conditions.

Interface	Boundary conditions
solid–solid	$T_i^+ = T_i^-$
	$u_i^+ = u_i^-$
solid–liquid	$T_3^+ = T_3^-$
	$T_2 = T_1 = 0$
	$u_3^+ = u_3^-$
free surface	$T_i = 0$

At the interface between two solids, sometimes called a "welded" interface, all components of the displacement are continuous because no overlaps or tears occur. For the same reason, the tractions are continuous. This is the condition we used at the junction between two strings in Section 2.2.3.

At the interface between a solid and a perfect fluid the fluid can slip along the interface because its rigidity is zero, so it cannot support shear stress. Hence the components of traction tangential to the interface are zero in the fluid and, by the condition of continuity, in the solid as well. Thus the tangential displacement components need not be continuous, but the normal components of the traction and displacement are continuous.

Table 2.3-1 summarizes the boundary conditions for a horizontal interface between different media.

2.3.11 *Strain energy*

Because applying a force to an elastic material causes deformation, potential energy is stored within the material, as we saw for waves on a string (Section 2.2.4). To motivate this *elastic strain energy*, consider a spring with a restoring force $f = -kx$. Compressing the spring a distance dx requires work against the spring, equal to the integral of the force applied times the distance. If the spring is initially at equilibrium, the work is

$$W = \int_0^x kx dx = \frac{1}{2} kx^2, \tag{87}$$

which equals the potential energy stored in the spring.

By analogy, the strain energy stored in a volume is the integral of the product of stress and strain components summed

$$W = \frac{1}{2} \int \sigma_{ij} e_{ij} dV = \frac{1}{2} \int c_{ijkl} e_{ij} e_{kl} dV. \tag{88}$$

The strain energy is symmetric in ij and kl, providing the rationale for the statement (Eqn 68) that the tensor of elastic constants has the symmetry $c_{ijkl} = c_{klij}$.

2.4 Seismic waves

2.4.1 *The seismic wave equation*

The ideas of elasticity in the last section let us show that the equation of motion has solutions that describe the two types of propagating seismic (or elastic) waves, *compressional* and *shear* waves. We will see that these wave types propagate differently, with velocities that depend in different ways on the elastic properties of the material. Our approach to showing that the equations of elasticity have propagating wave solutions is conceptually similar to the way we showed (Section 2.2) that the physics of a string gives rise to traveling waves. In that analysis, we first demonstrated that waves occur on a uniform string, and then considered how waves propagate between strings of differing properties. That analysis considered propagating waves without regard to how they were generated.

Following that approach, we consider a homogeneous[1] region, one of uniform properties, within an elastic material. We assume that the region contains no source of seismic waves, which requires a body force. Once the waves propagate away from the source, the relation between the stresses and displacements is given by the homogeneous equation of motion, which includes no body force term, so $\mathbf{F} = m\mathbf{a}$ becomes

$$\sigma_{ij,j}(\mathbf{x}, t) = \rho \frac{\partial^2 u_i(\mathbf{x}, t)}{\partial t^2}. \qquad (1)$$

Before solving the equation, two points are worth noting. The equation of motion can be written and solved entirely in terms of displacements, because the stress is related to the strain, which is formed from derivatives of the displacement. The stress and strain are related by the constitutive relation, which characterizes the material. Thus, although the equation of motion does not depend on the elastic constants, the solution does. Second, the equation of motion relates spatial derivatives of the stress tensor to a time derivative of the displacement vector. The resulting solutions give the displacement vector and hence the strain and stress tensors as functions of both space and time. Often, for simplicity, these dependences are not explicitly written.

We solve Eqn 1 in Cartesian (x, y, z) coordinates, beginning with the x component,

$$\frac{\partial \sigma_{xx}(\mathbf{x}, t)}{\partial x} + \frac{\partial \sigma_{xy}(\mathbf{x}, t)}{\partial y} + \frac{\partial \sigma_{xz}(\mathbf{x}, t)}{\partial z} = \rho \frac{\partial^2 u_x(\mathbf{x}, t)}{\partial t^2}. \qquad (2)$$

To express this in terms of displacements, we use the constitutive law for an isotropic elastic medium (Eqn 2.3.70),

[1] Unfortunately, this word is used for two different concepts: a *homogeneous medium* has properties that do not vary with position, whereas a *homogeneous equation* has no forcing function or source term.

$$\sigma_{ij} = \lambda\theta\delta_{ij} + 2\mu e_{ij}, \qquad (3)$$

and write the strains in terms of displacements, which yields

$$\sigma_{xx} = \lambda\theta + 2\mu e_{xx} = \lambda\theta + 2\mu \frac{\partial u_x}{\partial x}$$

$$\sigma_{xy} = 2\mu e_{xy} = \mu\left(\frac{\partial u_x}{\partial y} + \frac{\partial u_y}{\partial x}\right)$$

$$\sigma_{xz} = 2\mu e_{xz} = \mu\left(\frac{\partial u_x}{\partial z} + \frac{\partial u_z}{\partial x}\right). \qquad (4)$$

We then take derivatives of the stress components

$$\frac{\partial \sigma_{xx}}{\partial x} = \lambda \frac{\partial \theta}{\partial x} + 2\mu \frac{\partial^2 u_x}{\partial x^2}$$

$$\frac{\partial \sigma_{xy}}{\partial y} = \mu\left(\frac{\partial^2 u_x}{\partial y^2} + \frac{\partial^2 u_y}{\partial y \partial x}\right)$$

$$\frac{\partial \sigma_{xz}}{\partial z} = \mu\left(\frac{\partial^2 u_x}{\partial z^2} + \frac{\partial^2 u_z}{\partial z \partial x}\right) \qquad (5)$$

using the fact that for a homogeneous material the elastic constants do not vary with position. Finally, substituting the derivatives into the equation of motion and using the definitions of the dilatation

$$\theta = \nabla \cdot \mathbf{u} = \frac{\partial u_x}{\partial x} + \frac{\partial u_y}{\partial y} + \frac{\partial u_z}{\partial z} \qquad (6)$$

and of the Laplacian (Section A.6.5)

$$\nabla^2(u_x) = \frac{\partial^2 u_x}{\partial x^2} + \frac{\partial^2 u_x}{\partial y^2} + \frac{\partial^2 u_x}{\partial z^2} \qquad (7)$$

yields

$$(\lambda + \mu)\frac{\partial \theta}{\partial x} + \mu\nabla^2(u_x) = \rho \frac{\partial^2 u_x}{\partial t^2} \qquad (8)$$

for the x component of the equation of motion (1).

Similar equations can be obtained for the y and z components of displacement. The three equations can be combined, using the vector Laplacian of the displacement field

$$\nabla^2\mathbf{u} = (\nabla^2 u_x, \nabla^2 u_y, \nabla^2 u_z), \qquad (9)$$

into a single vector equation:

$$(\lambda + \mu)\nabla(\nabla \cdot \mathbf{u}(\mathbf{x}, t)) + \mu\nabla^2\mathbf{u}(\mathbf{x}, t) = \rho \frac{\partial^2 \mathbf{u}(\mathbf{x}, t)}{\partial t^2}. \qquad (10)$$

This is the equation of motion for an isotropic elastic medium written entirely in terms of the displacements, with the dependence on position and time explicitly written to remind us that we seek a solution that varies in this way. Equation 10 can be rewritten using the vector identity (Eqn A.6.23)

$$\nabla^2\mathbf{u}=\nabla(\nabla\cdot\mathbf{u})-\nabla\times(\nabla\times\mathbf{u}) \tag{11}$$

to obtain

$$(\lambda+2\mu)\nabla(\nabla\cdot\mathbf{u}(\mathbf{x},t))-\mu\nabla\times(\nabla\times\mathbf{u}(\mathbf{x},t)) = \rho\frac{\partial^2\mathbf{u}(\mathbf{x},t)}{\partial t^2}. \tag{12}$$

Rather than solve Eqn 12 directly, we express the displacement field in terms of two other functions, ϕ and $\boldsymbol{\Upsilon}$, which are known as potentials;

$$\mathbf{u}(\mathbf{x},t)=\nabla\phi(\mathbf{x},t)+\nabla\times\boldsymbol{\Upsilon}(\mathbf{x},t). \tag{13}$$

In this representation, the displacement is the sum of the gradient of a *scalar potential*, $\phi(\mathbf{x},t)$, and the curl of a *vector potential*,[2] $\boldsymbol{\Upsilon}(\mathbf{x},t)$, both of which are functions of space and time. Although this decomposition appears to introduce complexity, it actually clarifies the problem, because the vector identities (Section A.6.4)

$$\nabla\times(\nabla\phi)=0 \quad \nabla\cdot(\nabla\times\boldsymbol{\Upsilon})=0 \tag{14}$$

separate the displacement field into two parts. The part associated with the scalar potential has no curl or rotation and gives rise to compressional waves. Conversely, the part associated with the vector potential has zero divergence, causes no volume change, and corresponds to shear waves. Because taking the curl discards any part of the vector potential that would give a nonzero divergence, we require that the vector potential satisfy $\nabla\cdot\boldsymbol{\Upsilon}(\mathbf{x},t)=0$.[3]

Substituting the potentials into Eqn 12 and rearranging terms using Eqn 14 yields

$$(\lambda+2\mu)\nabla(\nabla^2\phi)-\mu\nabla\times\nabla\times(\nabla\times\boldsymbol{\Upsilon})=\rho\frac{\partial^2}{\partial t^2}(\nabla\phi+\nabla\times\boldsymbol{\Upsilon}). \tag{15}$$

Using Eqn 11, the second term of Eqn 15 simplifies to

$$\begin{aligned}\nabla\times\nabla\times(\nabla\times\boldsymbol{\Upsilon})&=-\nabla^2(\nabla\times\boldsymbol{\Upsilon})+\nabla(\nabla\cdot(\nabla\times\boldsymbol{\Upsilon}))\\&=-\nabla^2(\nabla\times\boldsymbol{\Upsilon}),\end{aligned} \tag{16}$$

because the divergence of the curl is zero. After this substitution, the terms in Eqn 15 can be regrouped to give

$$\nabla\left[(\lambda+2\mu)\nabla^2\phi(\mathbf{x},t)-\rho\frac{\partial^2\phi(\mathbf{x},t)}{\partial t^2}\right]=-\nabla\times\left[\mu\nabla^2\boldsymbol{\Upsilon}(\mathbf{x},t)-\rho\frac{\partial^2\boldsymbol{\Upsilon}(\mathbf{x},t)}{\partial t^2}\right], \tag{17}$$

because the elastic constants do not vary with position, and the order of differentiation has no effect.

One solution of the equation can be found if both terms in brackets are zero. In this case, we have two wave equations, one for each potential. The scalar potential satisfies

$$\nabla^2\phi(\mathbf{x},t)=\frac{1}{\alpha^2}\frac{\partial^2\phi(\mathbf{x},t)}{\partial t^2}, \tag{18}$$

with the velocity

$$\alpha=[(\lambda+2\mu)/\rho]^{1/2}. \tag{19}$$

As we will see shortly, this solution corresponds to *P*, or compressional, waves. Similarly, the vector potential satisfies

$$\nabla^2\boldsymbol{\Upsilon}(\mathbf{x},t)=\frac{1}{\beta^2}\frac{\partial^2\boldsymbol{\Upsilon}(\mathbf{x},t)}{\partial t^2}, \tag{20}$$

with velocity

$$\beta=(\mu/\rho)^{1/2}, \tag{21}$$

and corresponds to *S*, or shear, waves.

Equations 18 and 20 are wave equations that are slightly different from those that we have previously encountered. Waves on a string (Section 2.2) satisfied the wave equation

$$\frac{\partial^2 u(x,t)}{\partial x^2}=\frac{1}{v^2}\frac{\partial^2 u(x,t)}{\partial t^2}, \tag{22}$$

describing the propagation of a scalar quantity in one space dimension. The scalar potential satisfies a similar scalar wave equation, with the difference that the space variable $\mathbf{x}$ is in three dimensions. The vector potential, a vector quantity, satisfies the analogous vector wave equation in three dimensions.

The wave equations in Eqns 18 and 20 are strictly valid only for a homogeneous medium because they were derived assuming that all derivatives of the elastic constants were zero. Although these equations were derived in Cartesian coordinates, they are valid in any coordinate system. We next discuss solutions of the wave equation, and then return to these two types of waves.

2.4.2 *Plane waves*

The scalar wave equation in three dimensions,

$$\nabla^2\phi(\mathbf{x},t)=\frac{1}{v^2}\frac{\partial^2\phi(\mathbf{x},t)}{\partial t^2}, \tag{23}$$

[2] Although $\boldsymbol{\Psi}$ is often used for the vector potential, we use $\boldsymbol{\Upsilon}$ (upsilon) to avoid confusion with the *SV* potential in the text section.

[3] This decomposition into scalar and vector potentials, known as *Helmholtz decomposition*, can be done for any vector field.

describes how the scalar field $\phi(\mathbf{x}, t)$ propagates in three dimensions. By analogy to the equation of motion (Eqn 2.3.50), Eqn 23 is a homogeneous wave equation, with no forcing function to act as a source of the waves. If there were, the inhomogeneous scalar wave equation in three dimensions with a source term $f(\mathbf{x}, t)$,

$$\nabla^2\phi(\mathbf{x}, t) - \frac{1}{v^2}\frac{\partial^2\phi(\mathbf{x}, t)}{\partial t^2} = f(\mathbf{x}, t), \tag{24}$$

would apply.

The harmonic wave solution to the scalar wave equation in one dimension (Eqn 2.2.6)

$$u(x, t) = Ae^{i(\omega t \pm kx)} \tag{25}$$

can be generalized to solve the three-dimensional scalar wave equation. This solution, known as a *harmonic plane wave*, is written[4]

$$\begin{aligned}\phi(\mathbf{x}, t) &= A \exp\,(i(\omega t \pm \mathbf{k}\cdot\mathbf{x})) \\ &= A \exp\,(i(\omega t \pm k_x x \pm k_y y \pm k_z z)),\end{aligned} \tag{26}$$

where $\mathbf{x}$ is now the position vector, and $\mathbf{k} = (k_x, k_y, k_z)$ is now the *wave vector*, sometimes also called the *wavenumber vector*. This solution describes a plane wave propagating in an arbitrary direction given by the wave vector, in contrast to the one-dimensional solution that describes propagation along a coordinate axis. To demonstrate this, we write $\mathbf{k} = |\,\mathbf{k}\,|\,\hat{\mathbf{k}}$, where $\hat{\mathbf{k}}$ is a unit vector in the direction of $\mathbf{k}$; so Eqn 26 becomes

$$\phi(\mathbf{x}, t) = A \exp\,(i[\omega t - |\,\mathbf{k}\,|(\hat{\mathbf{k}}\cdot\mathbf{x})]), \tag{27}$$

a plane wave propagating in the $\hat{\mathbf{k}}$ direction with velocity

$$v = \omega/|\,\mathbf{k}\,|. \tag{28}$$

Thus the wave vector describes two important features of a propagating wave. Its magnitude gives the wavenumber, the spatial frequency, and its direction gives the direction of propagation. The wave fronts, which at any time are surfaces of constant phase $(\omega t - \mathbf{k}\cdot\mathbf{x})$ and thus constant values of $\phi(\mathbf{x}, t)$, are planes perpendicular to the direction of propagation (Fig. 2.4-1). To see this, note that all points on a plane perpendicular to the wave vector have the same value of $\mathbf{k}\cdot\mathbf{x}$, because this scalar product is the projection of $\mathbf{k}$ on $\mathbf{x}$. The phase is periodic over a distance along the propagation direction equal to the wavelength, $2\pi/|\,\mathbf{k}\,|$. As for the waves on a string, we can use the complex exponential formulation so long as we ensure that the displacement is purely real, either by taking the real part of the complex exponential or by also using the complex conjugate.

[4] When the arguments of exponentials become lengthy, we sometimes use the notation $\exp\,(x) = e^x$ for clarity.

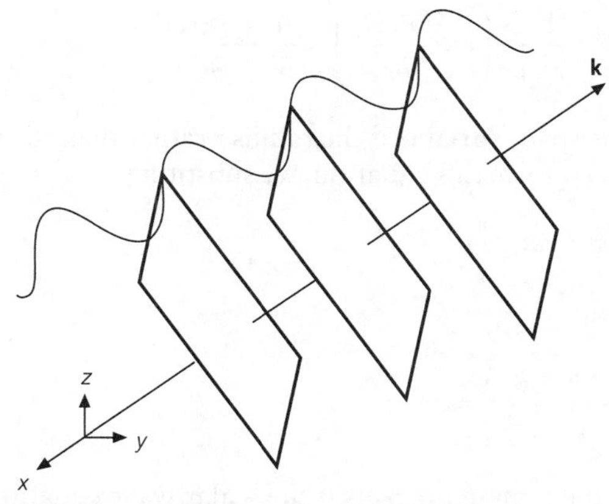

Fig. 2.4-1 Wave fronts for a harmonic plane wave traveling in the direction indicated by the wave vector **k**. The wavelength is $\lambda = 2\pi/|\,\mathbf{k}\,|$.

This solution to the three-dimensional scalar wave equation can be generalized to solve the vector wave equation in three dimensions,

$$\nabla^2\boldsymbol{\Upsilon}(\mathbf{x}, t) = \frac{1}{v^2}\frac{\partial^2\boldsymbol{\Upsilon}(\mathbf{x}, t)}{\partial t^2}, \tag{29}$$

which describes the propagation of a vector field. In Cartesian coordinates this breaks up into three scalar wave equations:

$$\begin{aligned}\nabla^2\Upsilon_x(\mathbf{x}, t) &= \frac{1}{v^2}\frac{\partial^2\Upsilon_x(\mathbf{x}, t)}{\partial t^2} \\ \nabla^2\Upsilon_y(\mathbf{x}, t) &= \frac{1}{v^2}\frac{\partial^2\Upsilon_y(\mathbf{x}, t)}{\partial t^2}, \\ \nabla^2\Upsilon_z(\mathbf{x}, t) &= \frac{1}{v^2}\frac{\partial^2\Upsilon_z(\mathbf{x}, t)}{\partial t^2}.\end{aligned} \tag{30}$$

The harmonic plane wave solution to the vector wave equation is then

$$\boldsymbol{\Upsilon}(\mathbf{x}, t) = \mathbf{A} \exp\,(i(\omega t - \mathbf{k}\cdot\mathbf{x})), \tag{31}$$

which is like Eqn 26 except that $\boldsymbol{\Upsilon}(\mathbf{x}, t)$ and the constant $\mathbf{A}$ are vectors.

2.4.3 Spherical waves

A second solution to the three-dimensional scalar wave equation yields waves with spherical, rather than planar, wave fronts. To obtain this solution, we express a scalar potential, $\phi(\mathbf{r}, t)$, and its Laplacian in spherical coordinates (Eqn A.7.17). We consider spherically symmetric solutions where ϕ is a function only of time and the radius r, so only the $\partial\phi/\partial r$ term in the Laplacian is nonzero. The spherically symmetric waves satisfy the homogeneous wave equation

$$\nabla^2\phi(r,t) = \frac{1}{r^2}\frac{\partial}{\partial r}\left(r^2\frac{\partial\phi(r,t)}{\partial r}\right) = \frac{1}{v^2}\frac{\partial^2\phi(r,t)}{\partial t^2}, \quad (32)$$

where the space variable is the radius r rather than the position vector $\mathbf{r}$. To solve this equation, we substitute

$$\phi(r,t) = \xi(r,t)/r \quad (33)$$

and obtain

$$\frac{1}{r}\left[\frac{\partial^2\xi}{\partial r^2} - \frac{1}{v^2}\frac{\partial^2\xi}{\partial t^2}\right] = 0. \quad (34)$$

Because the term in brackets is the scalar wave equation in one dimension, any function of the form $\xi = f(r \pm vt)$ satisfies Eqn 34 when $r \neq 0$. Thus any function of the form

$$\phi(r,t) = f(t \pm r/v)/r \quad (35)$$

is a spherically symmetric solution to the scalar wave equation.

This solution describes spherical wave fronts centered about the origin $\mathbf{r} = 0$, whose amplitude depends on the distance from the origin. When the minus sign is used, Eqn 35 represents waves diverging outward from a source at the origin, with the amplitude decaying as $1/r$. The plus sign yields an *incoming* spherical wave, growing in amplitude as $1/r$ and converging at the origin. It is common to impose a *radiation condition* that waves not enter the region of study from far away, and thus to discard the incoming wave solution.

However, Eqn 35 is not a solution to the homogeneous equation everywhere in space, because it is infinite at $\mathbf{r} = 0$. Physically this is because a wave spreading out from a point must have been generated by a seismic source there. Thus the outgoing wave, $\phi(r,t) = f(t - r/v)/r$, is actually a solution to the *inhomogeneous* wave equation

$$\nabla^2\phi(r,t) - \frac{1}{v^2}\frac{\partial^2\phi(r,t)}{\partial t^2} = -4\pi\delta(\mathbf{r})f(t). \quad (36)$$

This represents a point source at the origin with a time function $f(t)$. The delta function $\delta(\mathbf{r})$ (Section 6.2.5) is zero except at $\mathbf{r} = 0$, but its integral over a volume including the origin is 1. Thus, integrating over a volume including the origin shows that Eqn 35 is a solution to the inhomogeneous scalar wave equation (36) even at the origin. Hence, in seeking a solution to the homogeneous equation that yielded spherical waves, we have found a solution to the inhomogeneous equation which is used to study waves radiated by a seismic source.

The fact that the spherical wave solution (Eqn 35) represents an outgoing wave generated at the origin explains the distance-dependent amplitude factor $1/r$, which had no counterpart for the plane wave solution. As a spherical wave propagates away from its source, the area of the wave front, $4\pi r^2$, increases. Because, as we will see shortly, the energy per unit area of the wave front transported by a propagating wave

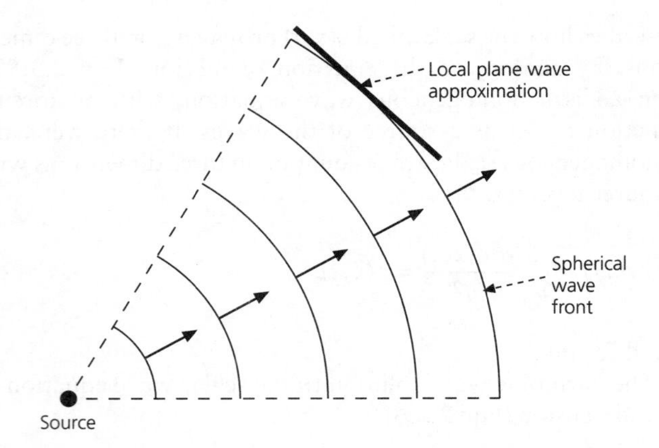

Fig. 2.4-2 As a spherical wave front moves far from the source, it can be locally approximated by a plane wave front due to the decreased curvature of the spherical wave.

is proportional to the amplitude squared, the energy per unit wave front decays as $1/r^2$. This decay, called *geometric spreading*, conserves energy. Similarly, the energy of spherical light waves decays with distance from a lamp as $1/r^2$.

A plane wave can be regarded as a limit of a spherical wave far from the source, because the spherical wave front becomes almost planar (Fig. 2.4-2). This approximation is often used in seismology when seismometers are far from an earthquake.

2.4.4 P and S waves

We found earlier in this section (Eqn 13) that the displacement can be separated into a scalar potential corresponding to P waves that satisfies the scalar wave equation

$$\nabla^2\phi(\mathbf{x},t) = \frac{1}{\alpha^2}\frac{\partial^2\phi(\mathbf{x},t)}{\partial t^2}, \quad (37)$$

and a vector potential corresponding to S waves that satisfies the vector wave equation

$$\boldsymbol{\nabla}^2\boldsymbol{\Upsilon}(\mathbf{x},t) = \frac{1}{\beta^2}\frac{\partial^2\boldsymbol{\Upsilon}(\mathbf{x},t)}{\partial t^2}. \quad (38)$$

To understand the displacements caused by the two types of waves, consider a plane wave propagating in the z direction. The scalar potential for a harmonic plane P wave satisfying Eqn 37 is

$$\phi(z,t) = A \exp(i(\omega t - kz)), \quad (39)$$

so the resulting displacement is the gradient

$$\mathbf{u}(z,t) = \boldsymbol{\nabla}\phi(z,t) = (0, 0, -ik)\, A \exp(i(\omega t - kz)), \quad (40)$$

which has a nonzero component only along the propagation direction z (Fig. 2.4-3). The corresponding dilatation is nonzero,

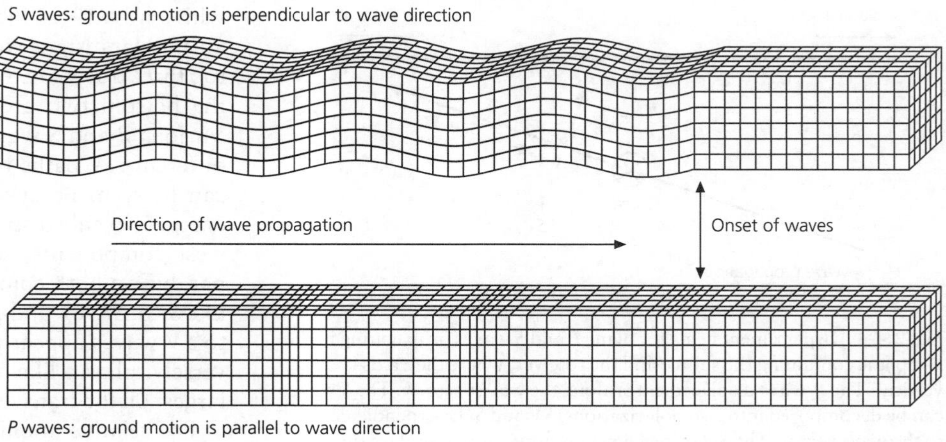

Fig. 2.4-3 Displacements produced by plane compressional and shear waves, shown by a "snapshot" in time. *P* waves produce displacement in the direction of wave propagation and a volume change. *S* waves produce displacement perpendicular to the direction of wave propagation and distort the material without any volume change.

$$\nabla \cdot \mathbf{u}(z, t) = -k^2 A \exp(i(\omega t - kz)), \tag{41}$$

so a volume change occurs. As the wave propagates, the displacements in the direction of propagation cause material to be alternately compressed and expanded. Thus the *P* wave generated by the scalar potential is called a compressional wave.

By contrast, for the *S* wave, or shear wave, described by the vector potential

$$\boldsymbol{\Upsilon}(z, t) = (A_x, A_y, A_z) \exp(i(\omega t - kz)), \tag{42}$$

the resulting displacement field is given by the curl

$$\mathbf{u}(z, t) = \nabla \times \boldsymbol{\Upsilon}(z, t) = (ikA_y, -ikA_x, 0) \exp(i(\omega t - kz)), \tag{43}$$

whose component along the propagation direction z is zero (Fig. 2.4-3). Thus the only displacement associated with a propagating shear wave is perpendicular to the direction of wave propagation. A shear wave causes no volume change, because the dilatation, $\nabla \cdot \mathbf{u}(z, t)$, is zero.

Comparison of the displacements for the *P* and *S* waves illustrates that a wave is characterized by two directions. One is the direction in which the wave propagates; the other is the direction in which the field that propagates changes. A compressional wave is an example of a *longitudinal* wave, because the propagating displacement field varies in the direction of propagation. A familiar example is a sound wave in air, which can be described as a compressional (elastic) wave in an ideal fluid. By contrast, a shear wave is an example of a *transverse* wave, because the propagating displacement field varies at right angles to the direction of propagation. The waves we considered on the string were transverse waves, because waves moved along the string, but their displacement was normal to the string. Electromagnetic waves are another familiar example of transverse waves.

The component of $\boldsymbol{\Upsilon}(z, t)$ in the direction of wave propagation (A_z) has no effect on the displacement field because taking the curl discards it. Thus, setting A_z to zero to satisfy the requirement that $\nabla \cdot \boldsymbol{\Upsilon}(z, t) = 0$ imposes no additional restriction on the displacement. Only A_x and A_y contribute to the displacement. Because each component of the displacement depends on only one of these terms, there can be two independent shear wave fields. For example, if A_x or A_y is zero, there will be only a y or an x component of displacement. Thus shear waves can have two independent polarizations, as is the case for other transverse waves, such as light.

In real applications, we often define the z axis as the vertical direction and orient the x–z plane along the great circle connecting a seismic source and a receiver. Plane waves traveling on the direct path between the source and the receiver thus propagate in the x–z plane. The shear wave polarization directions are defined as *SV*, for shear waves with displacement in the vertical (x–z) plane, and *SH*, for horizontally polarized shear waves with displacement in the y direction, parallel to the earth's surface. Both have displacements perpendicular to the propagation direction and the other polarization (Fig. 2.4-4, overleaf). Although we could choose any two orthogonal polarizations in the plane of the shear wave displacements, using *SV* and *SH* is particularly convenient. We will see that *P* and *SV* waves are coupled with each other when they interact with horizontal boundaries, whereas *SH* waves remain separate.

Seismometers record horizontal motions in the north–south and east–west directions, which rarely correspond exactly to the *SH* and *SV* polarizations. As a result, data from the horizontal components of seismometers are often *rotated*. The direction connecting the source and the receiver, corresponding to *SV* displacements, is called the *radial* direction, so a seismogram rotated to this direction is called the *radial* component. Similarly, the orthogonal direction corresponding to *SH* displacements is called the *transverse* direction, so a seismogram rotated to this direction is called the *transverse* component.

Because seismograms record components of the displacement vector, they can be rotated to give their components in a new coordinate system using Eqn A.5.9. If the back azimuth direction from the receiver to the source (Section A.7.2) is ζ',

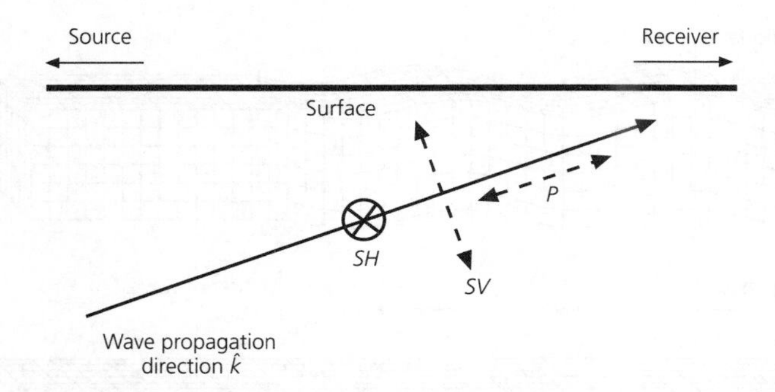

Fig. 2.4-4 Displacement fields for plane *P* and *S* waves propagating in the *x–z* plane containing the source and the receiver, where the *z* axis is vertical. The *P*-wave displacement is along the wave vector *k*. The *S* wave can be decomposed into two polarizations, *SV* and *SH*, perpendicular to the wave vector. The *SH* displacement is purely horizontal (in the *y* direction, out of the page), whereas the *SV* displacement is in the *x–z* plane.

we rotate the north–south (NS) and east–west (EW) components into radial (R) and transverse (T) components using

$$\begin{pmatrix} u_R \\ u_T \end{pmatrix} = \begin{pmatrix} \cos\theta & \sin\theta \\ -\sin\theta & \cos\theta \end{pmatrix} \begin{pmatrix} u_{EW} \\ u_{NS} \end{pmatrix} \quad (44)$$

with $\theta = 3\pi/2 - \zeta'$. Figure 2.4-5 shows seismograms recorded at an angular distance of 110° from a deep earthquake, where the top three traces are the components recorded at the station, and the bottom two are the radial and transverse components. Various *P* and *S* wave phases (Section 3.5), corresponding to different ray paths between the source and the seismometer, can be seen. Because the back azimuth is 323°, *SH* and *SV* energy is evenly distributed between the north–south and east–west components, so the S-wave phases are roughly comparable on both components. When rotated, however, phases like *SKS*, *SKKS*, and *PS* that involve conversions from *P* waves to *SV* waves appear primarily on the radial component. Conversely, phases like S_{diff} that involve primarily *SH* energy are largest on the transverse component.

The relative amplitudes on the radial and transverse components are shown by a *particle motion plot* of the amplitudes as a function of time (Fig. 2.4-6). As shown for two time segments from Fig. 2.4-5, the *SKS* and *SKKS* waves are primarily on the radial or *SV* component, whereas S_{diff} is primarily on the transverse or *SH* component.

The definitions of the *P*-wave velocity, termed α or v_P,

$$\alpha = [(\lambda + 2\mu)/\rho]^{1/2} = [(K + 4\mu/3)/\rho]^{1/2}, \quad (45)$$

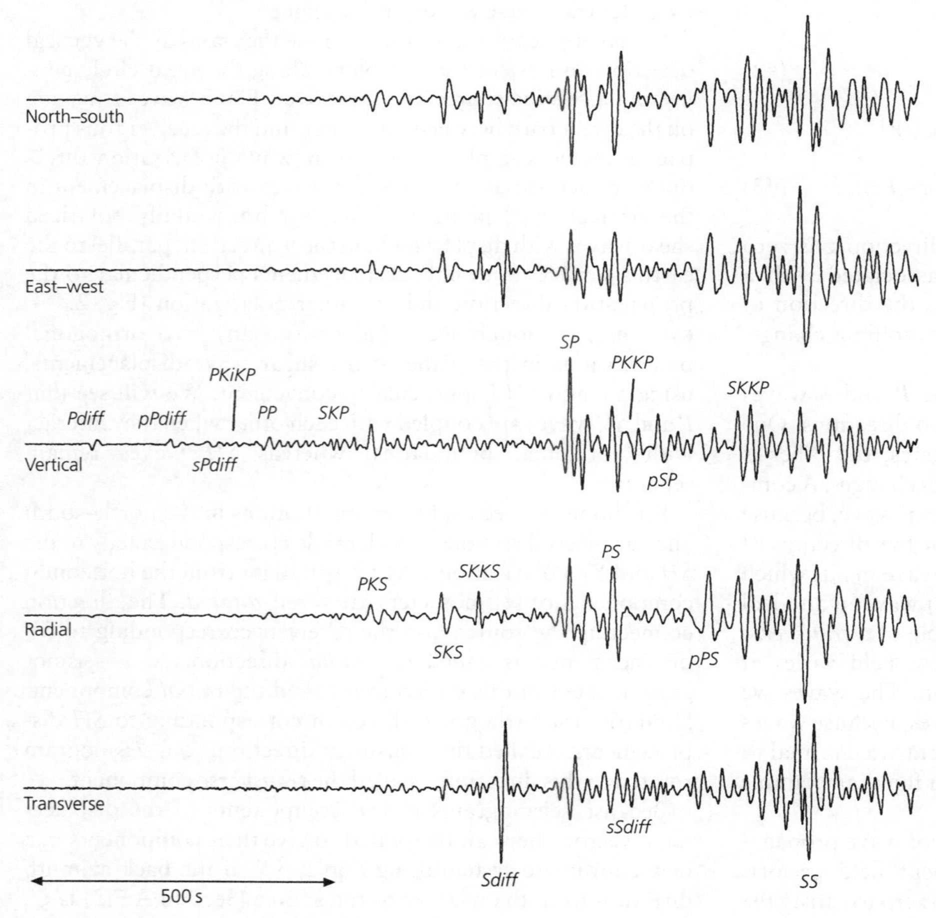

Fig. 2.4-5 Seismograms for a deep (597 km) earthquake on August 23, 1995, in the Mariana trench, recorded 110° away at Harvard, Massachusetts. *P*-wave phases are best seen on the vertical component, *SV*-wave phases are best seen on the radial component, and *SH*-wave phases are best seen on the transverse component.

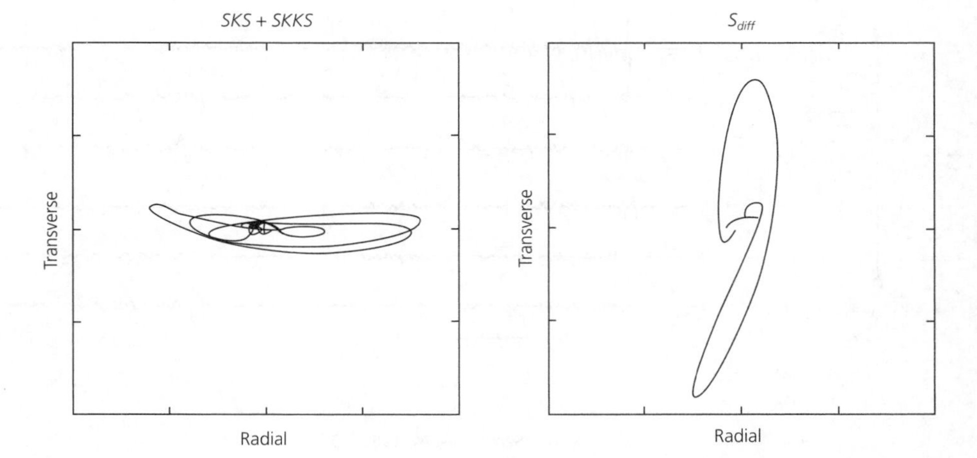

Fig. 2.4-6 Particle motion plots for two time segments of the radial and transverse components shown in Fig. 2.4-5. *SKS* and *SKKS*, which are primarily *SV* waves, are strongest on the radial component (*left*), whereas S_{diff} is primarily an *SH* wave, and so is strongest on the transverse component (*right*).

and *S*-wave velocity, termed β or v_S,

$$\beta = (\mu/\rho)^{1/2}, \tag{46}$$

show that the seismic velocities depend in different ways on the elastic constants of the material. Because the rigidity μ and the bulk modulus K (Eqn 2.3.74) are positive, *P* waves travel faster than *S* waves. Thus the first wave arriving from an earthquake is always a compressional wave. As a result, the nomenclature *P* originally denoted the first-arriving, "primary" wave, whereas *S* denoted the "secondary" wave.

Although both velocities depend on the rigidity, the shear velocity does not depend on the bulk modulus K, because these waves involve no volume changes. Because the shear velocity is proportional to the square root of the rigidity, shear waves cannot propagate through an ideal ($\mu = 0$) fluid. However, compressional waves propagate in an ideal fluid with a velocity proportional to $K^{1/2}$. Thus only compressional waves can travel through the earth's outer core or the ocean.[5]

To get a feel for these wave velocities, consider typical values for various parameters. The earth's crust is approximately a Poisson solid, with elastic constants $\lambda \approx \mu \approx 3 \times 10^{11}$ dyn/cm^2. Thus, for a density of 3 g/cm^3, the *P*-wave velocity is 5.5×10^5 cm/s, or 5.5 km/s. Similarly, the *S*-wave velocity is 3.2×10^5 cm/s, or 3.2 km/s. Hence a *P* wave propagating with a velocity of 5.5 km/s and a period of 2 s has a wavelength (Section 2.2) of (5.5 km/s × 2 s) or 11 km. The frequency is 0.5 s^{-1} (the unit s^{-1} is called a Hertz, or Hz), and the wavenumber is $2\pi/11 = 0.57$ km^{-1}. On the other hand, a wave with a period of 10 s and the same velocity has a wavelength of 55 km, and a frequency of 0.1 Hz. The longer-period wave has a longer wavelength and a lower frequency.

[5] The transverse waves we see at a beach are not seismic waves in the water, but instead propagate at the water surface and involve a rolling motion in two dimensions similar to Rayleigh waves (Section 2.7.2).

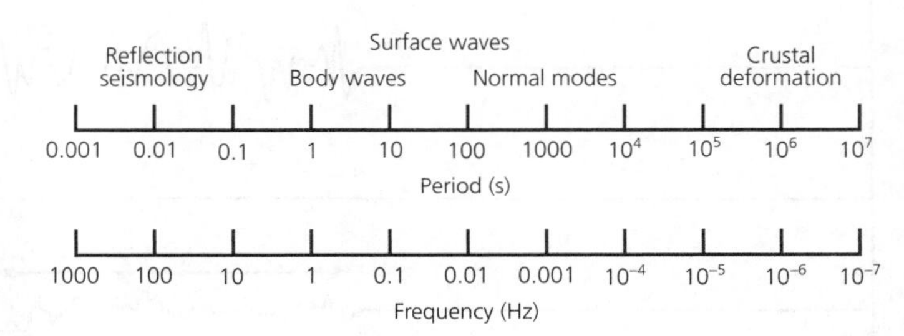

Fig. 2.4-7 Seismic spectrum showing the frequencies at which various analyses are conducted.

The "seismic spectrum," showing seismic waves of various frequencies and types, is shown in Fig. 2.4-7. Studies of earthquakes typically use the period range from approximately 0.1 s to more than 3000 s, or frequencies from 10 Hz to 3×10^{-4} Hz (0.1 mHz). Higher-frequency waves of 20–80 Hz generated by explosions or other artificial sources are used in reflection seismology to explore the earth's crust. Still higher frequencies, $3\text{–}12 \times 10^3$ Hz (3–12 kHz), propagating primarily in the ocean, are used by marine geophysicists to map the sea floor. At the other end of the spectrum, ground motions with periods longer than 10^4 s are due to slow crustal motions (Section 4.5) rather than propagating seismic waves.

Earthquake sources generate both *P* and *S* waves, with the *S* waves generally significantly larger. Figure 2.4-8 shows seismograms of the three components (vertical, or up–down, north–south, and east–west) of ground motion from seismic waves generated by an earthquake ~280 km beneath two seismic stations in Japan. The seismic waves are coming up vertically toward the surface. The first arrival, a *P* wave, has displacement along the direction of propagation, and therefore appears primarily on the vertical component. The large later arrival, a shear wave, has displacement perpendicular to the direction of propagation, and thus appears most on the horizontal components.

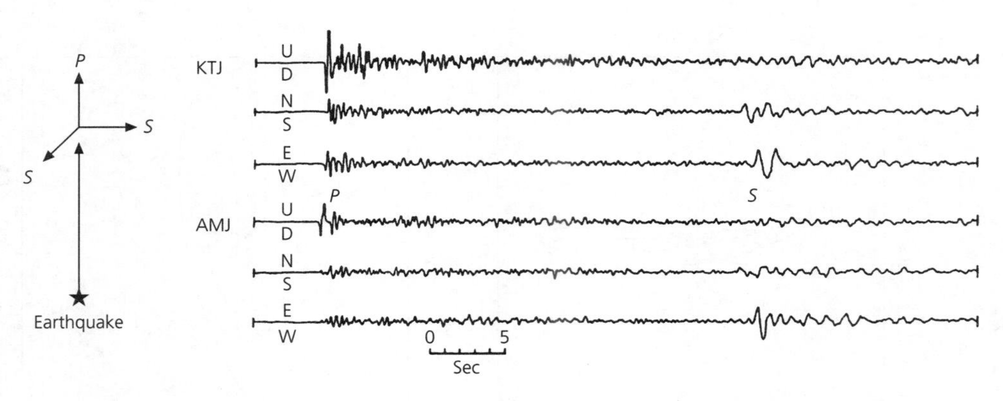

Fig. 2.4-8 Three-component seismograms at two stations from an earthquake beneath Japan. Because the stations are nearly above the earthquake, the *P* wave has its largest amplitude on the vertical (U–D, "Up"–"Down") components. (Ando *et al.*, 1983. *J. Geophys. Res.*, *88*, 5850–64, copyright by the American Geophysical Union.)

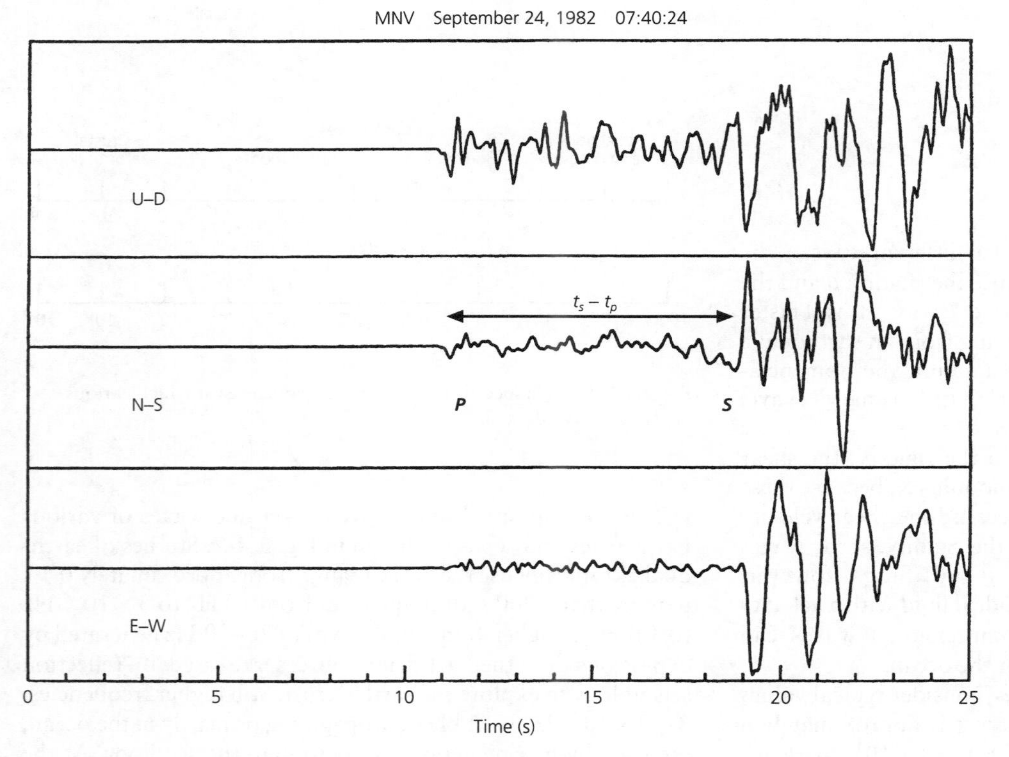

Fig. 2.4-9 Three-component seismogram of a magnitude 4.9 shallow-focus earthquake recorded 64 km away at Mina, Nevada. The difference in the arrival times of the *P* and *S* waves, $t_S - t_P$, can be used to estimate the distance between the earthquake and the seismometer.

These data also show an interesting effect. The *S* wave on the north–south components arrives earlier than on the east–west components. This observation has been interpreted as indicating that material beneath the seismic stations is ~5% anisotropic, such that in this region shear waves with displacements in the N–S direction propagate faster than those with displacements in the E–W direction. The anisotropy (Section 3.6) may reflect the presence of the mineral olivine, in which seismic waves propagate at different speeds depending on their direction with respect to the crystal structure. If enough olivine crystals are oriented in a consistent fashion, significant anisotropy can result. A second effect that could cause significant anisotropy is the presence of a region of aligned cracks.

Figure 2.4-9 shows a different type of seismogram: a record of a shallow earthquake in Nevada from a seismic station within 100 km of the source. The times when the *P* and *S* waves arrive can be measured from the seismograms. With a number of such observations at different locations, we will see (Chapter 7) that the location and origin time of the earthquake can be determined. Even with one seismic station, something about the location of the earthquake can be learned. Although the arrival times of the seismic waves cannot be converted to travel times without knowing when the earthquake occurred, we can learn something from the difference between the *P* and *S* arrival times. For typical values of the compressional and shear velocities in the crust, $\alpha = 5.5$ km/s and $\beta = 3.2$ km/s, the times required for *S* and *P* waves to travel a distance of x km are

$$t_s = x/3.2, \quad t_p = x/5.5. \tag{47}$$

The difference in travel times, which is also the difference in arrival times,

$$t_s - t_p = x(1/3.2 - 1/5.5) = x/7.6, \tag{48}$$

is thus a function of the distance between the source and the receiver. Because the S wave arrives about 8 s after the P wave, the earthquake is about 60 km away, in agreement with the distance found by an earthquake location program using arrival times from many seismic stations. This $S - P$ travel time technique gives an estimate of the distance from the seismometer to the earthquake, but does not yield the azimuth and hence the location.[1] Given $S - P$ times at several stations, the location can be found from the requirement that the earthquake must be a specific distance from each station. Schematically, this method can be thought of as locating the point on a map where arcs of circles with the appropriate radii intersect. The problem is actually more interesting, because the earthquake need not have occurred at the earth's surface.

2.4.5 *Energy in a plane wave*

Like waves on a string (Section 2.2.4), seismic waves transport energy both as kinetic energy and as strain, or potential, energy. To find this energy, consider harmonic plane S and P waves traveling in the z direction. An SH wave with displacement in the y direction is

$$u_y(z, t) = B \cos(\omega t - kz), \tag{49}$$

where this expression is written directly in terms of displacement, rather than potential. We will see shortly that this is a useful approach for SH waves.

The kinetic energy in a volume V is the integral of the sum of the kinetic energy associated with each component of the displacement

$$KE = \frac{1}{2}\int_V \rho\left(\frac{\partial u_i}{\partial t}\right)^2 dV, \tag{50}$$

because the mass is $m = \rho dV$. Hence for the plane wave (Eqn 49), the kinetic energy per unit wave front averaged over a wavelength λ is

$$KE = \frac{1}{2\lambda}\rho B^2\omega^2 \int_0^\lambda \sin^2(\omega t - kz)dz = \frac{1}{2\lambda}\rho B^2\omega^2 \frac{\lambda}{2} = B^2\omega^2\rho/4. \tag{51}$$

[1] An analogous method is used to estimate that a thunderstorm is a mile away for every 5 s between seeing lightning and hearing thunder, because light travels much faster than sound (about 330 m/s in air).

The strain energy (Eqn 2.3.88) is

$$W = \frac{1}{2}\int_V \sigma_{ij}e_{ij}dV. \tag{52}$$

Because the only nonzero strain components are

$$e_{32} = e_{23} = \frac{1}{2}\frac{\partial u_y}{\partial z} = Bk \sin(\omega t - kz)/2, \tag{53}$$

The only nonzero stress components are

$$\sigma_{32} = \sigma_{23} = \mu Bk \sin(\omega t - kz), \tag{54}$$

and the strain energy per unit area of wave front averaged over a wavelength in the propagation direction is

$$W = \frac{1}{2\lambda}\int_0^\lambda \mu B^2k^2 \sin^2(\omega t - kz)dz = \mu B^2k^2/4 = B^2\omega^2\rho/4, \tag{55}$$

where the last expression used the fact that $\mu = \beta^2\rho$ and $\beta k = \omega$. Thus the strain energy and kinetic energy averaged over a wavelength are equal, as we found for the string. Hence the total energy averaged over a wavelength is

$$E = KE + W = B^2\omega^2\rho/2, \tag{56}$$

and the average energy flux in the propagation direction is found by multiplying by the velocity

$$\dot{E} = B^2\omega^2\rho\beta/2. \tag{57}$$

The total energy and flux are proportional to the square of the amplitude and the frequency, so for waves of the same amplitude, the higher-frequency wave transports more energy.

Similarly, a plane P wave propagating in the z direction, described by the scalar potential

$$\phi(z, t) = A \exp(i(\omega t \pm kz)) \tag{58}$$

has a displacement which is the gradient of the potential,

$$\mathbf{u}(z, t) = \nabla\phi(z, t) = (0, 0, -ik) A \exp(i(\omega t - kz)), \tag{59}$$

with real part

$$u_z(z, t) = Ak \sin(\omega t - kz). \tag{60}$$

Using Eqn 50, the kinetic energy per unit wave front averaged over a wavelength is

$$KE = \frac{1}{2\lambda}\rho A^2k^2\omega^2 \int_0^\lambda \cos^2(\omega t - kz)dz = A^2\omega^2k^2\rho/4. \tag{61}$$

To find the strain energy (Eqn 52), we note that the only nonzero stress component is

$$\sigma_{zz} = (\lambda + 2\mu)e_{zz} = \rho\alpha^2 e_{zz}, \tag{62}$$

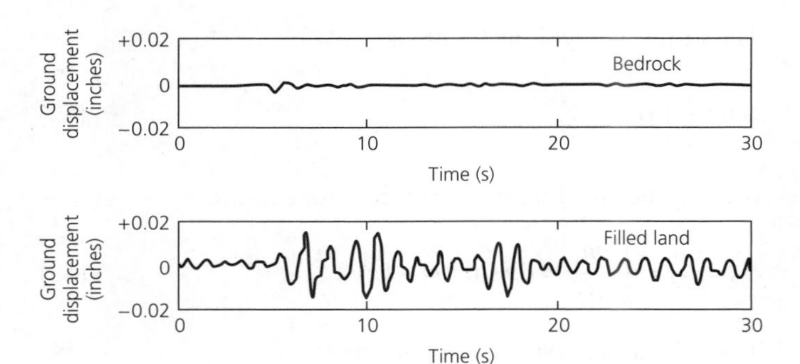

Fig. 2.4-10 Seismograms showing the ground displacement at two locations in the Marina district of San Francisco from a magnitude 5 aftershock of the 1989 Loma Prieta earthquake. The shaking on the filled land is about an order of magnitude larger than on bedrock. (Courtesy of the US Geological Survey.)

where the last form eliminates the Lamé constant λ and lets us reserve the symbol λ for wavelength. Thus the strain energy per unit wave front averaged over a wavelength is

$$W = \frac{1}{2\lambda}\int_0^\lambda \rho\alpha^2A^2k^4\cos^2(\omega t - kz)dz = A^2\omega^2k^2\rho/4, \qquad (63)$$

which equals the kinetic energy. Hence the total energy averaged over a wavelength is

$$E = KE + W = A^2\omega^2k^2\rho/2, \qquad (64)$$

and the average energy flux in the propagation direction is found by mutiplying by the P velocity

$$\dot{E} = A^2\omega^2k^2\rho\alpha/2. \qquad (65)$$

These expressions differ from those for the energy of the SH wave by a factor of k^2, because A is the amplitude of the potential, whereas in Eqns 56 and 57 B is the amplitude of the displacement. If we used the potential amplitude for a shear wave, the k^2 factor would be needed.

The energy flux gives insight into how waves behave when they change media. For example, as water waves travel into shallower water, their velocities decrease, so their amplitudes increase to conserve energy. Eventually the amplitudes exceed a critical level, and the wave breaks. Similarly, when seismic waves pass from bedrock into soft soil with lower velocity and density, their amplitudes increase. This effect is shown by Fig. 2.4-10, comparing seismograms of an aftershock of the Loma Prieta earthquake from the Marina district of San Francisco. The ground motion recorded by a seismometer located on a layer of soft landfill (*bottom*) is much larger than that on a nearby seismometer installed on bedrock (*top*). As a result, earthquake damage varies between structures built in soils and bedrock.

2.5 Snell's law

2.5.1 *The layered medium approximation*

In the last section, we saw that the equation of motion for a homogeneous elastic medium has solutions in which the displacement is described by potentials satisfying the wave equation. We now begin to use these solutions to describe seismic wave propagation in the earth. Applying results derived for an infinite homogeneous medium to a real planet with a complicated internal structure might seem like a large leap. Nonetheless, some significant problems can be explored using this approach.

For seismological purposes, we characterize the internal structure of the solid earth by the distribution of physical properties that affect seismic wave propagation and can be studied using seismic waves. We thus deal with the distribution of elastic properties and density, or, equivalently, of seismic velocities and density. A seismological model of elastic earth structure is the set of functions $\alpha(\mathbf{r})$, $\beta(\mathbf{r})$, $\rho(\mathbf{r})$ showing how the velocities and density depend on the position vector $\mathbf{r}$, and hence the radius, latitude, and longitude. Seismological results indicate that this distribution is complicated and difficult to characterize. For example, downgoing slabs of lithosphere extend to considerable depths at subduction zones. Fortunately, we can often make a series of useful approximations (Fig. 2.5-1). Because the solid earth's physical properties vary significantly more with depth than they do laterally, they can be approximated as spherically symmetric functions $\alpha(r)$, $\beta(r)$, $\rho(r)$ that depend only on the radius r. A medium whose properties vary only with depth is called *laterally homogeneous* or *stratified*, in contrast to a *laterally heterogeneous* medium where velocities vary laterally as well as with depth.

When the characteristic length of the region under consideration is small compared with the radius of the earth—as, for example, in local crustal studies—the earth's curvature can be neglected. The earth is thus further approximated as a laterally homogeneous halfspace, with velocities and density characterized by functions $\alpha(z)$, $\beta(z)$, $\rho(z)$ varying only with the depth z. A further useful simplification is to treat the earth as a halfspace consisting of finite thickness layers, each of uniform properties α_i, β_i, ρ_i.

An attractive feature of the layered model is that the solutions of the equation of motion discussed in the last section apply exactly only to a homogeneous medium. When a layered earth model is appropriate, it is possible to take the homogeneous medium solutions in each layer and "patch" them together at the interfaces to account for the propagation of seismic waves between layers. This can be done when plane waves adequately represent the wave fronts, an assumption that applies far enough away from the source that wave fronts can be considered planar. Treating a stratified medium as a set of uniform layers is analogous to the way we divided a string into uniform segments and matched solutions across their boundaries.

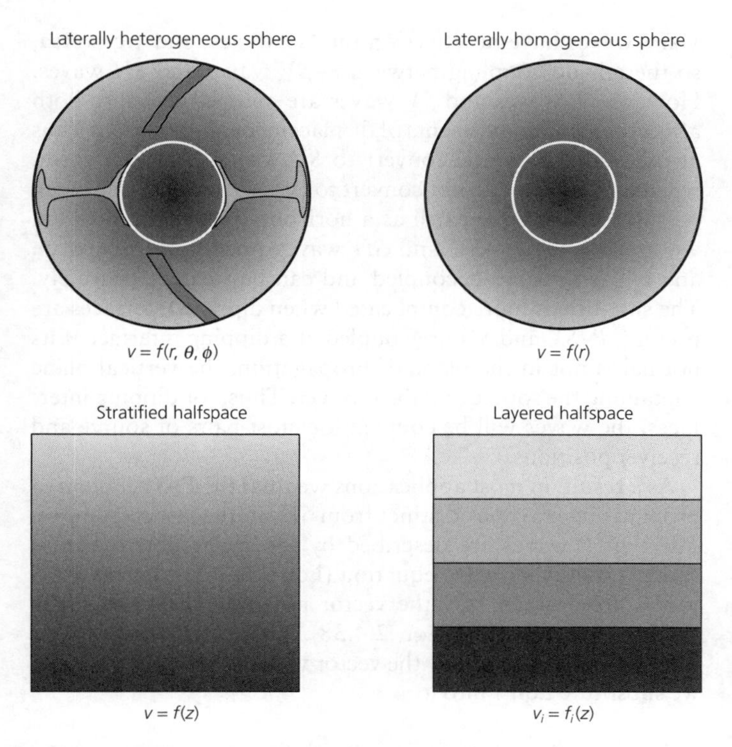

Fig. 2.5-1 Schematic illustration of some types of earth models used in seismology. The most accurate model, a laterally heterogeneous sphere, is often approximated as being spherically symmetric, with properties varying only with radius. A spherically symmetric model can be further approximated for many purposes as a stratified halfspace, in which properties vary only with depth, or as a layered halfspace composed of discrete uniform layers.

The real earth is not laterally homogeneous, much less composed of uniform layers, and seismic wave fronts do not extend as planes to infinity. The test of whether these approximations are useful is whether results derived by applying them to seismological data yield geologically meaningful inferences. We will see that this is surprisingly often the case. Laterally homogeneous models are thus useful both as representations of average earth structure and as starting models for more detailed investigations.

2.5.2 *Plane wave potentials for a layered medium*

Our first goal is to analyze what happens when a plane P or S wave is incident on the boundary between two halfspaces of homogeneous and isotropic elastic materials with different elastic constants and hence seismic velocities. We will derive Snell's law, the famous relation that describes the bending of wave fronts as a plane wave goes from one medium to the other. Once we can handle a single boundary, we generalize this solution to a stack of homogeneous layers. The layered approximation can be used, even when the elastic properties vary smoothly, by using a large number of thin layers.

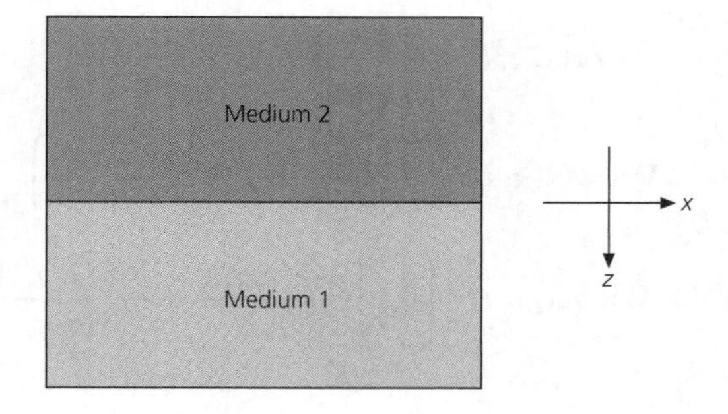

Fig. 2.5-2 Two halfspaces in contact, composed of materials with different elastic properties. The horizontal interface is in the x–y plane.

The geometry of the problem is shown in Fig. 2.5-2. We consider a plane wave with its direction of propagation, and thus wave vector, in the x–z plane. The displacements can be written using potentials that are functions only of x and z. Two halfspaces of different materials are in contact along a boundary that is the x–y plane, and the z axis, the normal to the interface, is positive downwards. This geometry has the attractive feature that the shear waves can be separated into the two polarizations discussed in the previous section: SV waves, whose displacement is only in the x–z plane, and SH waves, whose displacement has only a y component. Moreover, the displacement and hence potentials do not vary with y, and so can be written as functions of x, z, and t.

In Eqn 2.4.13 we saw that the displacement field can be decomposed into a scalar potential describing P waves and a vector potential for S waves. To separate the SV and SH waves, we split the vector potential $\mathbf{\Upsilon}$ into two terms,

$$\mathbf{\Upsilon}(x, z, t) = \mathbf{\Psi}(x, z, t) + \nabla \times \boldsymbol{\chi}(x, z, t). \quad (1)$$

The displacement vector can now be written using the scalar potential, $\phi(x, z, t)$, and the two vector potentials:

$$\begin{aligned}\mathbf{u}(x, z, t) &= \nabla\phi(x, z, t) + \nabla \times \mathbf{\Upsilon}(x, z, t) \\ &= \nabla\phi(x, z, t) + \nabla \times \mathbf{\Psi}(x, z, t) + \nabla \times \nabla \times \boldsymbol{\chi}(x, z, t). \quad (2)\end{aligned}$$

We choose the vector potentials to be

$$\mathbf{\Psi}(x, z, t) = (0, \psi(x, z, t), 0) \quad \text{and}$$
$$\boldsymbol{\chi}(x, z, t) = (0, \chi(x, z, t), 0). \quad (3)$$

Each potential has zero for its x and z components, and the y components are the scalar functions $\psi(x, z, t)$ for SV waves and $\chi(x, z, t)$ for SH waves. Thus the displacement vector is described by three scalar functions, one for each potential.

To find the resulting displacements, we carry out the vector operations in Eqn 2. Because the two vector potentials have only a y component, and neither ϕ, ψ, nor χ depend on y, the y derivatives are zero. Hence the P, SV, and SH terms give rise to displacement vectors with (x, y, z) components

$$(P) \qquad \nabla\phi(x, z, t) = \left(\frac{\partial\phi(x, z, t)}{\partial x}, 0, \frac{\partial\phi(x, z, t)}{\partial z}\right)$$

$$(SV) \qquad \nabla \times \boldsymbol{\psi}(x, z, t) = \left(\frac{-\partial\psi(x, z, t)}{\partial z}, 0, \frac{\partial\psi(x, z, t)}{\partial x}\right)$$

$$(SH)\ \nabla \times \nabla \times \boldsymbol{\chi}(x, z, t) = \left(0, -\left[\frac{\partial^2\chi(x, z, t)}{\partial x^2} + \frac{\partial^2\chi(x, z, t)}{\partial z^2}\right], 0\right). \tag{4}$$

Thus P and SV contribute to the x and z components of displacement, whereas SH contributes only to the y component. The divergences $\nabla \cdot \Psi$ and $\nabla \cdot \chi$ equal zero because only their y components are nonzero, and $\partial/\partial y$ of these components is zero. Hence, as expected, neither SH nor SV gives rise to a volume change.

The components of the displacement vector are found by grouping the components from Eqn 4:

$$u_x(x, z, t) = \frac{\partial\phi(x, z, t)}{\partial x} - \frac{\partial\psi(x, z, t)}{\partial z}$$

$$u_z(x, z, t) = \frac{\partial\phi(x, z, t)}{\partial z} + \frac{\partial\psi(x, z, t)}{\partial x}$$

$$u_y(x, z, t) = -\left(\frac{\partial^2\chi(x, z, t)}{\partial x^2} + \frac{\partial^2\chi(x, z, t)}{\partial z^2}\right) = -\nabla^2\chi(x, z, t). \tag{5}$$

These equations demonstrate that P–SV waves are independent of SH waves. The x and z components of displacement depend on both the P-wave potential ϕ and the SV-potential ψ. Thus for waves propagating in the x–z plane, the P and SV waves form a *coupled* system, which gives rise to two components of displacement. Neither the P nor the SV potentials contribute to the y component of displacement. Hence SH waves, which alone contribute to the y component of displacement, are *decoupled* from P and SV waves.

This coupling and decoupling persists when these waves interact with a horizontal interface parallel to the x–y plane. The boundary conditions at the interface constrain the displacements and tractions (Section 2.3.10). Because the normal to the interface has only a z component,

$$\hat{\mathbf{n}} = (0, 0, 1), \quad n_j = \delta_{j3}, \tag{6}$$

the tractions on the interface are given by

$$T_i = \sigma_{ij}n_j = \sigma_{i3} = (\sigma_{xz}, \sigma_{yz}, \sigma_{zz}). \tag{7}$$

The P–SV system gives rise to nonzero components of displacement u_x and u_z, and hence tractions σ_{xz} and σ_{zz}. For these waves, both $u_y = 0$ and $\sigma_{yz} = 0$. By contrast, the SH waves contribute only a y component of displacement, and their only nonzero traction component is σ_{yz}. Thus, at the interface, the P–SV waves have no effect on the SH waves, and vice versa, so there is no coupling between P–SV waves and SH waves. However, P waves and SV waves are coupled, because both affect the same components of displacement and traction. Thus at interfaces, P waves convert to SV waves, and vice versa, whereas SH waves do not convert to either P or SV waves.

When treating the earth as a horizontally layered medium, we assume that P–SV and SH waves propagating between any two points are decoupled and can be treated separately. The situation is more complicated when dipping interfaces are present. P–SV and SH are coupled at a dipping interface if its normal is not in the plane of propagation, the vertical plane containing the source and the receiver. Thus, for dipping interfaces, the waves will be coupled for most pairs of source and receiver positions.

As a result, in most applications we treat the P–SV system of propagating waves as distinct from SH. In the last section, we saw that P waves are described by the scalar potential that satisfies the scalar wave equation (Eqn 2.4.37), whereas the S waves are described by the vector potential $\boldsymbol{\Upsilon}$ satisfying the vector wave equation (Eqn 2.4.38). To see that the SV and SH potentials each satisfy the vector wave equation separately, we substitute Eqn 1 into it:

$$\nabla^2[\boldsymbol{\Psi}(x, z, t) + \nabla \times \boldsymbol{\chi}(x, z, t)] = \frac{1}{\beta^2}\frac{\partial^2}{\partial t^2}[\boldsymbol{\Psi}(x, z, t) + \nabla \times \boldsymbol{\chi}(x, z, t)], \tag{8}$$

and regroup the terms:

$$\nabla^2\boldsymbol{\Psi}(x, z, t) - \frac{1}{\beta^2}\frac{\partial^2\boldsymbol{\Psi}(x, z, t)}{\partial t^2}$$

$$= -\nabla^2[\nabla \times \boldsymbol{\chi}(x, z, t)] + \frac{1}{\beta^2}\frac{\partial^2}{\partial t^2}[\nabla \times \boldsymbol{\chi}(x, z, t)], \tag{9}$$

so the two potentials can be treated separately. Thus the P–SV system is described by

$$\nabla^2\phi(x, z, t) = \frac{1}{\alpha^2}\frac{\partial^2\phi(x, z, t)}{\partial t^2}, \quad \nabla^2\psi(x, z, t) = \frac{1}{\beta^2}\frac{\partial^2\psi(x, z, t)}{\partial t^2}. \tag{10}$$

Both of these are scalar wave equations, because ψ is the scalar function forming the y component of the SV vector potential (Eqn 3).

For SH waves we have two choices. Interchanging the curl and the other derivatives in the right side of Eqn 9 shows that the scalar function χ, the y component of the SH vector potential, satisfies a scalar wave equation. Alternatively, we can take the curl and recognize that by Eqns 4 and 5

$$u_y = \nabla \times \nabla \times \boldsymbol{\chi}(x, z, t), \tag{11}$$

so that

$$\nabla^2 u_y(x, z, t) = \frac{1}{\beta^2}\frac{\partial^2 u_y(x, z, t)}{\partial t^2}. \tag{12}$$

Thus the *SH*-wave displacement satisfies a scalar wave equation, and can be found without using the *SH* potential.

2.5.3 *Angle of incidence and apparent velocity*

We now consider *P–SV* waves propagating in the x–z plane that are described by harmonic plane wave solutions of the scalar wave equations (10),

$$(P) \quad \phi(x, z, t) = A \exp\,(i(\omega t - k_x x \pm k_{z_\alpha} z)) \tag{13}$$
$$(SV) \quad \psi(x, z, t) = B \exp\,(i(\omega t - k_x x \pm k_{z_\beta} z)).$$

The direction of wave propagation is described by the wave vector, which is the normal to the wave fronts. For propagation in the x–z plane, the direction is given by k_x and k_z because k_y is zero. Thus Eqn 13 represents waves propagating in the $+x$ direction (because of the negative sign in $-k_x x$), and in both the $+z$ and $-z$ directions.

Subscripts on **k** and k_z are needed because the magnitude of the wave vector differs for *P* and *SV* waves. We will see shortly that in this geometry k_x is the same for the *P* and the *SV* waves. The components of the wave vectors satisfy

$$|\mathbf{k}_\alpha|^2 = k_x^2 + k_{z_\alpha}^2 = \omega^2/\alpha^2 \quad |\mathbf{k}_\beta|^2 = k_x^2 + k_{z_\beta}^2 = \omega^2/\beta^2. \tag{14}$$

Because $k_y = 0$, k_x is the horizontal component of the wave vector.

The direction of propagation can also be expressed by the *angle of incidence* that the wave vector makes with the vertical (Fig. 2.5-3). Because the wave vectors, and therefore incidence angles, differ for *P* and *S* waves, we adopt the convention that i refers to *P*-wave incidence angles and j to *S*-wave incidence angles. Thus

$$\sin i = \frac{k_x}{(k_x^2 + k_{z_\alpha}^2)^{1/2}} = \frac{k_x}{|\mathbf{k}_\alpha|}, \quad \sin j = \frac{k_x}{(k_x^2 + k_{z_\beta}^2)^{1/2}} = \frac{k_x}{|\mathbf{k}_\beta|}. \tag{15}$$

We will see shortly that plane waves change direction when they cross an interface into a material with different seismic velocity (Fig. 2.5-4), so the orientation of the wave vector and the angle of incidence change. Hence the propagation of a plane wave is characterized by the changing orientations of the wave vector. We thus speak of a seismic *ray* that follows this *ray path*. Figures like Fig. 2.5-4 are often drawn showing only the ray paths and omitting the wave fronts that are normal to the ray.

It is useful to define the *apparent velocity*, c_x, the velocity at which a plane wave appears to travel along a horizontal surface. Figure 2.5-3 shows that in a time Δt a plane wave with

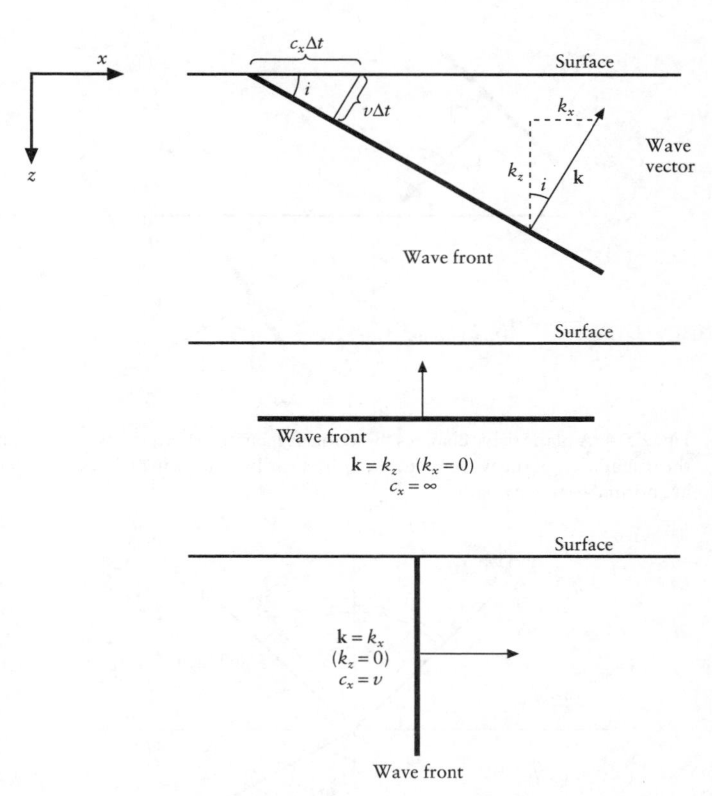

Fig. 2.5-3 The wave vector, **k**, is normal to the wave front and points in the direction of propagation. *Top*: For a plane wave traveling in the x–z plane, the propagation direction is given by the wave vector (k_x, k_z) or the incidence angle, i, between the wave vector and the vertical. In a time increment Δt the wave front moves a distance $v\Delta t$, where v is the medium velocity, and sweeps out a distance along the surface $c_x\Delta t$, where c_x is the apparent velocity along the surface. *Middle*: For a plane wave traveling vertically, the incidence angle $i = 0°$, **k** equals k_z, and c_x is infinite. *Bottom*: For a plane wave propagating horizontally, $i = 90°$, **k** equals k_x, and c_x equals the medium velocity.

incidence angle i in a medium with velocity v moves forward a distance $v\Delta t$ and moves across the horizontal surface a distance $c_x\Delta t$. Thus the horizontal apparent velocity is

$$c_x = v/\sin i. \tag{16}$$

The apparent velocity is always greater than or equal to the medium velocity, α for *P* waves and β for *S* waves. A horizontally propagating wave, with $i = 90°$, has an apparent velocity equal to the medium velocity. A vertically incident plane wave arrives everywhere on the surface at the same time, so it has an infinite apparent velocity.

The horizontal apparent velocity[1] can be written in terms of the horizontal component of the wave vector using Eqns 15 and 16:

[1] Because seismological observations are made at the earth's surface, the apparent velocity along the earth's surface is sometimes written as c rather than c_x, and k is sometimes used to denote k_x.

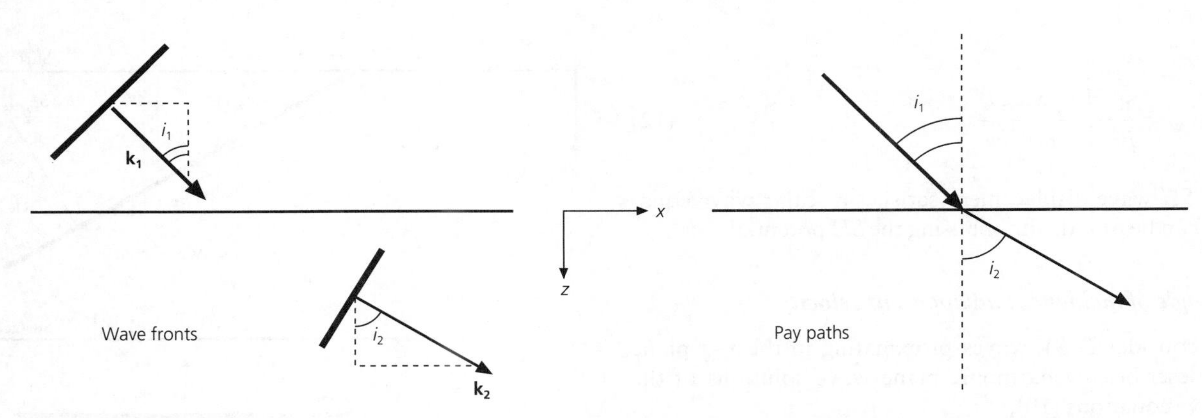

Fig. 2.5-4 A plane wave changes direction as it enters a material with different seismic velocity. The change in direction is represented by the change in the orientation of the wave vector **k**, or by a ray path showing successive orientations of the wave vector. The wave fronts, which are often not shown, are normal to the ray path.

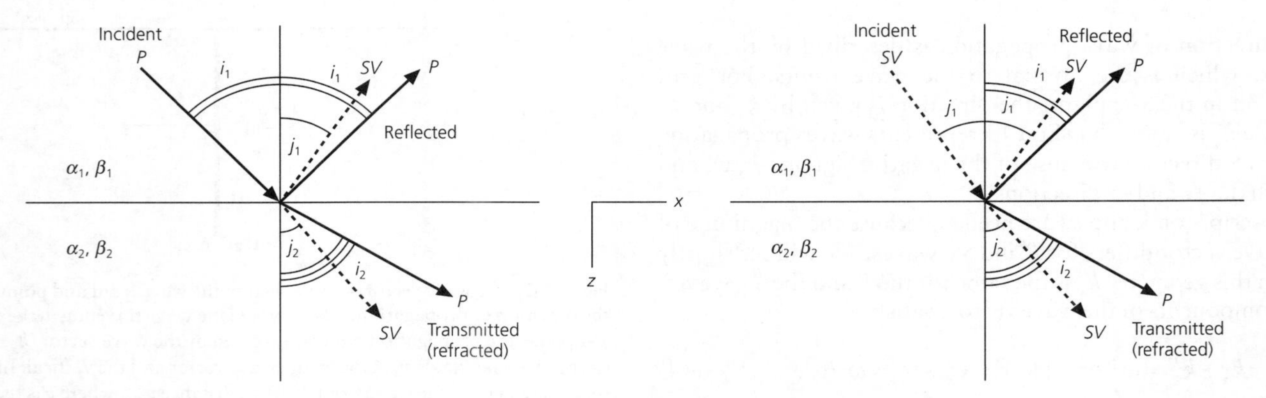

Fig. 2.5-5 Snell's law for plane waves propagating into a higher-velocity medium. *Left*: An incoming *P* wave generates transmitted and reflected *P* and *SV* waves. The reflected *P* wave has the same incidence angle, i_1, as the incoming *P* wave. Because in each medium the *P*-wave velocity exceeds the *S*-wave velocity, $j_1 < i_1$ and $j_2 < i_2$. *Right*: The same situation for an incoming *SV* wave. The incidence angles of the incoming and reflected *SV* waves, j_1, are equal. The relationships between the other incidence angles are the same as for an incident *P* wave.

$$c_x = \omega/k_x. \tag{17}$$

Thus we define the ratios of vertical to horizontal wavenumbers as

$$r_\alpha = k_{z_\alpha}/k_x = (c_x^2/\alpha^2 - 1)^{1/2} = \cot i,$$
$$r_\beta = k_{z_\beta}/k_x = (c_x^2/\beta^2 - 1)^{1/2} = \cot j, \tag{18}$$

so that the potentials (Eqn 13) can be written

$$(P) \quad \phi(x, z, t) = A \exp\,(i(\omega t - k_x x \pm k_x r_\alpha z))$$
$$(SV) \quad \psi(x, z, t) = B \exp\,(i(\omega t - k_x x \pm k_x r_\beta z)). \tag{19}$$

2.5.4 *Snell's law*

We now consider the relation between the angles of incidence for transmitted and reflected harmonic plane *P–SV* waves at an interface. In the geometry of Fig. 2.5-5, an interface at $z = 0$ separates medium 1 with *P* and *S* velocities α_1 and β_1 from medium 2 that has velocities α_2 and β_2. We first assume that $\alpha_1 < \alpha_2$ and $\beta_1 < \beta_2$.

A *P* wave incident from medium 1 generates reflected and transmitted *P* waves. In addition, part of the *P* wave is converted into a reflected *SV* wave and a transmitted *SV* wave. Each of these waves can be described by an appropriate potential. In medium 1 we have upgoing and downgoing *P* waves and an upgoing *SV* wave, so the potentials are

$$\phi(x, z, t) = \text{incident } P + \text{reflected } P$$
$$= A_1 \exp\,(i(\omega t - k_x x - k_x r_{\alpha_1} z))$$
$$+ A_2 \exp\,(i(\omega t - k_x x + k_x r_{\alpha_1} z))$$
$$\psi(x, z, t) = \text{reflected } SV = B_2 \exp\,(i(\omega t - k_x x + k_x r_{\beta_1} z)). \tag{20}$$

The form of each potential describes the wave. Terms like $k_x r_{\alpha_1}$, the z component of the wavenumbers, indicate which medium (1 or 2) and what wave type (*P* or *S*) this potential describes.

The direction of propagation for each wave is given by the components of the wave vector **k**. For example, the signs of the k_x and $k_x r_{\alpha_1}$ terms show that the incoming *P* wave with amplitude A_1 travels in the $+x$ and $+z$ directions as time increases. Similarly, the reflected *P* wave with amplitude A_2 and the reflected *SV* wave with amplitude B_2 travel in the $+x$ and $-z$ directions.

The downgoing *P* wave and *SV* waves in the second medium are given by the potentials

$$\phi(x, z, t) = \text{transmitted } P = A' \exp\,(i(\omega t - k_x x - k_x r_{\alpha_2} z))$$
$$\psi(x, z, t) = \text{transmitted } SV = B' \exp\,(i(\omega t - k_x x - k_x r_{\beta_2} z)). \quad (21)$$

A' and B' are the amplitudes of the transmitted *P* and *SV* waves, which travel in the $+x$ and $+z$ directions. We generally write the amplitudes of *P* waves as *A* and the amplitudes of *S* waves as *B*.

We can find the incidence angles of the transmitted and reflected waves from the incidence angle of the incoming wave. The boundary conditions for the solid–solid interface at $z = 0$ are that the components of the displacement and traction vectors are continuous (Section 2.3.10). Because all of the potentials contain the phase factor, $\exp\,(i(\omega t - k_x x))$ times a factor independent of *x* and *t*, all of the displacement and traction components have this phase factor. For the displacement and traction to be continuous at the interface for all *x* and all *t*, $(\omega t - k_x x)$ must be equal for each of the potentials. Thus the horizontal wavenumber k_x, and hence the apparent velocity along the interface $c_x = \omega / k_x$, must be the same for each wave. As a result, the waves travel along the interface at the same speed and stay in phase.

This condition and the definition of c_x (Eqn 16) give the familiar form of *Snell's law*:

$$c_x = \frac{\alpha_1}{\sin i_1} = \frac{\beta_1}{\sin j_1} = \frac{\alpha_2}{\sin i_2} = \frac{\beta_2}{\sin j_2}, \quad (22)$$

the ratio of the sine of the angle of incidence for each wave to the corresponding velocity is constant. Hence the incident and reflected *P* waves have the same incidence angle i_1. The transmitted *P* and *S* waves change direction by a factor depending on the velocities in the two media. A change in direction upon transmission into a medium with a different velocity is called *refraction*, so the waves in the second medium are called refracted or transmitted waves. Figure 2.5-5 illustrates the ray paths for the different waves.

The *S* wave reflected from the boundary satisfies

$$\sin j_1 = \sin i_1 (\beta_1 / \alpha_1). \quad (23)$$

Because in any medium *P* waves travel faster than *S* waves, Snell's law requires that $j_1 < i_1$. Hence the reflected *S* ray is closer to the vertical, or further from the interface, than the *P* ray in the same medium. Physically, this is because the *S* wave must be closer to the vertical than the *P* wave to have the same apparent velocity along the interface.

The angle of incidence for the refracted *P* wave is related to that for the incident *P* wave by

$$\sin i_2 = \sin i_1 (\alpha_2 / \alpha_1). \quad (24)$$

If the second medium has a higher velocity, then $i_2 > i_1$, so the transmitted ray is further from the vertical than the incident ray. It travels more horizontally, so the apparent velocities along the interface are equal. On the other hand, if $\alpha_1 > \alpha_2$, then the refracted *P* wave would be closer to normal incidence. (This effect, for light waves, makes a pencil appear to bend at the surface of a glass of water.)

The transmitted *S* wave satisfies

$$\sin j_2 = \sin i_1 (\beta_2 / \alpha_1). \quad (25)$$

Hence for $\beta_2 > \beta_1$, we get $j_2 > j_1$, so the transmitted *S* wave is more nearly horizontal than the reflected *S* wave. Similar relations apply for an incident *SV* wave (Fig. 2.5-5). The reflected *P* ray is bent further from the normal than the incident or reflected *SV* rays.

The fact that an incident *P* wave generates both *P* and *SV* waves, and vice versa, is a consequence of the displacement and traction boundary conditions at the interface, as we will see in Section 2.6. Some insight into why this should be can be obtained by considering Fig. 2.5-6, in which an incident *SV* wave disturbs the boundary, which then generates *P* waves in addition to the transmitted and reflected *SV* waves.

2.5.5 *Critical angle*

When a *P* wave impinges on a horizontal boundary, Eqn 24 shows that the incidence angle for the transmitted *P* wave in the second medium is

$$i_2 = \sin^{-1} [\sin i_1 (\alpha_2 / \alpha_1)], \quad (26)$$

where the notation $\sin^{-1}$ indicates the inverse sine function. If the second medium has a higher velocity, the transmitted *P* ray is further from the vertical than the incident ray. As the angle of incidence increases, the transmitted ray approaches the horizontal interface (Fig. 2.5-7, overleaf). Eventually, the incidence angle i_1 reaches a value i_c where $i_2 = 90°$ and the argument of the $\sin^{-1}$ term becomes 1, so

$$\sin i_c (\alpha_2 / \alpha_1) = 1 \quad \text{or} \quad \sin i_c = \alpha_1 / \alpha_2. \quad (27)$$

Thus for a wave incident at this *critical angle* of incidence, the transmitted wave grazes the interface.

Once the incidence angle exceeds the critical angle, which is a situation called *postcritical* incidence, no transmitted plane wave exists in the second medium. This phenomenon is sometimes called *total internal reflection*. In this case, as we will see

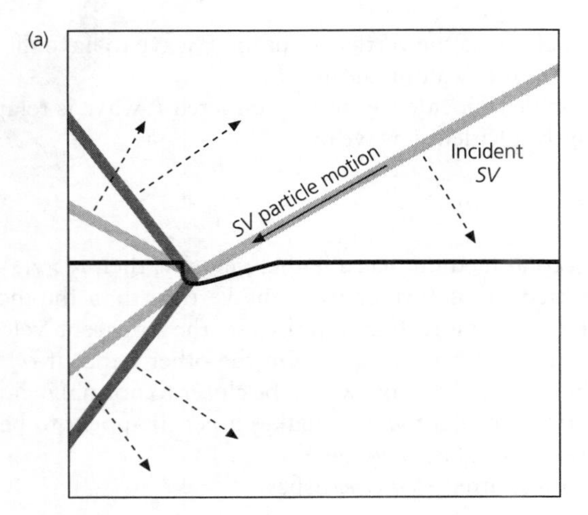

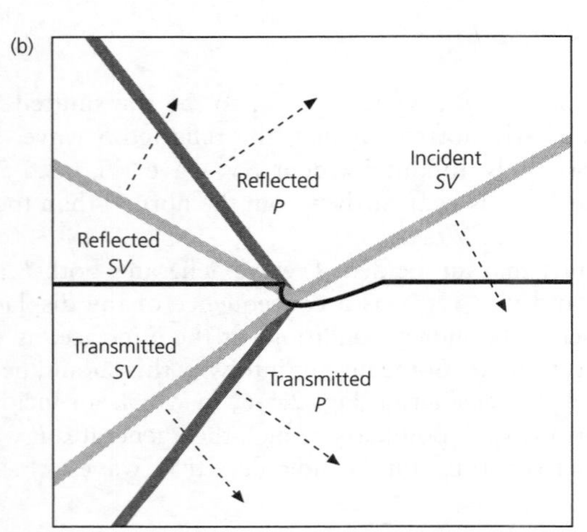

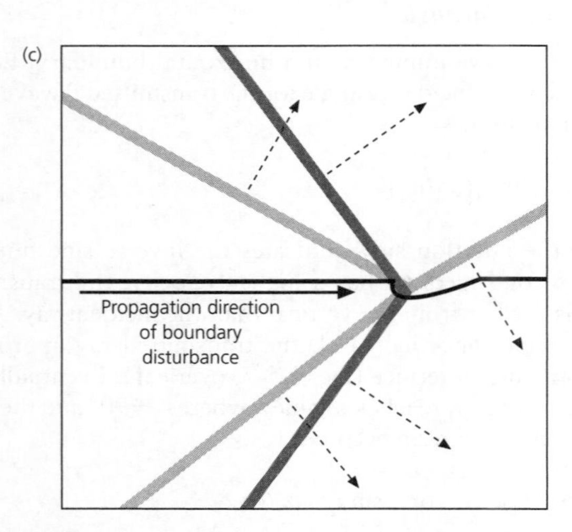

Fig. 2.5-6 Cartoon demonstrating how an *SV* wave (shown by the light grey wave front) incident at a boundary generates reflected and transmitted *P* (dark grey wave front) and *SV* waves, for the case shown in the bottom half of Fig. 2.5-5. a: The incident *SV* wave disturbs the boundary. b: The displaced boundary generates reflected and transmitted *P* and *SV* waves. c: As the incident *SV* wave advances, its intersection with the boundary moves, continuously generating reflected and transmitted waves.

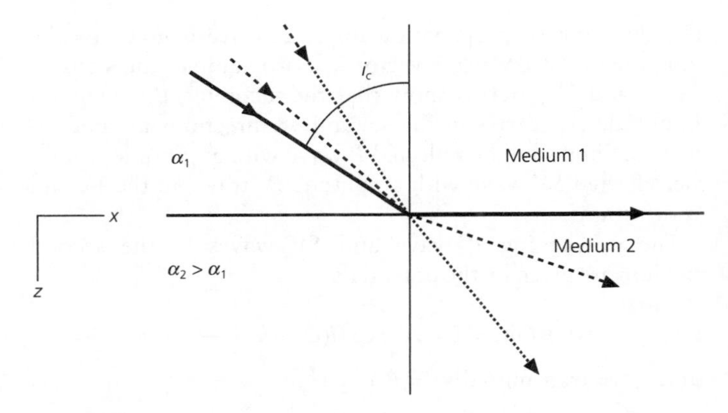

Fig. 2.5-7 Illustration of the critical angle i_c for *P* waves incident on a faster medium. The transmitted *S* and the reflected *P* and *S* waves are not shown. As the angle of incidence increases, the incoming waves become more nearly horizontal, and the refracted *P* waves approach the interface. For waves incident at an angle exceeding (more horizontal than) the critical angle, no traveling *P* wave is transmitted into medium 2.

in the next section, the *P*-wave potential for the second medium has a *z*-dependent real exponential term, $\exp(-k_z z)$, instead of a purely imaginary exponential term, $\exp(-ik_z z)$. Hence the displacement in the second medium is not a propagating plane wave, but occurs as an *evanescent wave* that travels along the interface and decays away from the interface.

Although for angles of incidence beyond the critical angle there is no transmitted *P* wave, there can still be a transmitted *S* wave. If the *S* velocity in medium 2 is greater than the *P* velocity in medium 1 there is a second critical angle

$$\sin i_{c_2} = \alpha_1/\beta_2 \tag{28}$$

beyond which no transmitted *P* or *S* waves occur.

2.5.6 *Snell's law for* SH *waves*

Snell's law also applies to *SH* waves. Because for *SH* waves the displacement satisfies the wave equation, *SH* waves in the first medium are described by

$$u_y(x, z, t) = B_1 \exp(i(\omega t - k_x x - k_x r_{\beta_1} z)) + B_2 \exp(i(\omega t - k_x x + k_x r_{\beta_1} z)), \tag{29}$$

where B_1 and B_2 are the amplitudes of the incoming and reflected *SH* waves (Fig. 2.5-8). In the second medium, the transmitted *SH* wave is

$$u_y(x, z, t) = B' \exp(i(\omega t - k_x x - k_x r_{\beta_2} z)). \tag{30}$$

As before, Snell's law

$$c_x = \beta_1/\sin j_1 = \beta_2/\sin j_2 \tag{31}$$

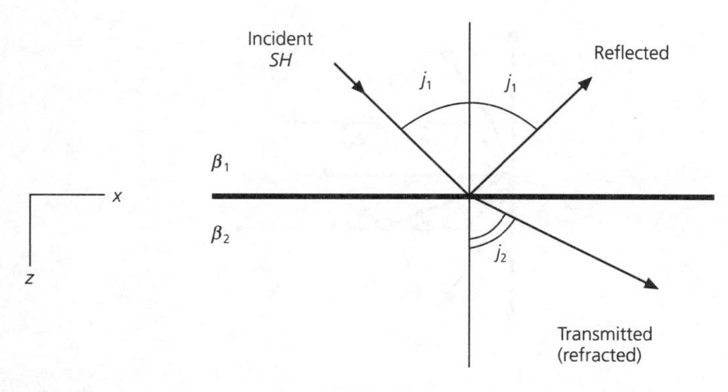

Fig. 2.5-8 An *SH* wave propagating in the x–z plane creates only transmitted and reflected *SH* waves when incident on a solid–solid interface in the x–y plane. The incident and reflected waves have the same incidence angle, j_1. For $\beta_2 > \beta_1, j_2 > j_1$.

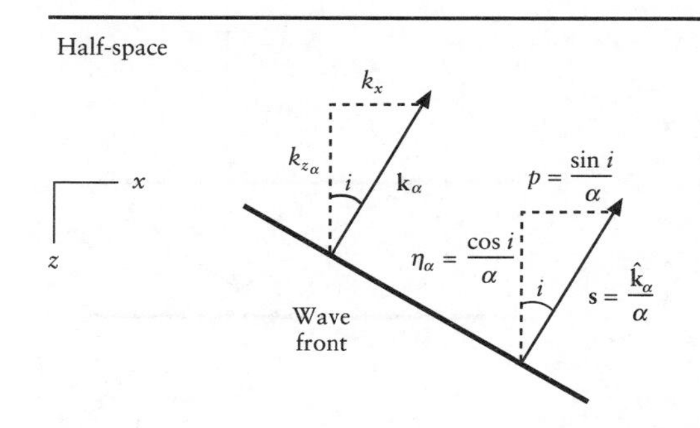

Fig. 2.5-9 Geometric interpretation of *P*-wave propagation in terms of the relation between the angle of incidence, i, the wave vector, $\mathbf{k}_\alpha$, the slowness vector, $\mathbf{s}$, the ray parameter or horizontal slowness, p, and the vertical slowness, η_α.

applies, because $(\omega t - k_x x)$ must be equal for all three waves for the traction and displacement to be continuous at the interface. The critical angle for *SH* waves is thus

$$\sin j_c = \beta_1/\beta_2. \tag{32}$$

2.5.7 *Ray parameter and slowness*

A useful way to characterize a wave's ray path is via its *ray parameter*, p, the reciprocal of the horizontal apparent velocity,

$$p = 1/c_x = \sin i/v = k_x/\omega, \tag{33}$$

where i is the incidence angle of either a *P* or an *S* wave, and v is the corresponding velocity. The harmonic plane wave solution can be written in terms of the ray parameter. To illustrate this, consider the potential for a *P* wave propagating in the x–z plane, and factor out the angular frequency:

$$\begin{aligned}\exp(i(\omega t - k_x x - k_x r_\alpha z)) &= \exp(i\omega(t - (k_x/\omega)x - (k_x/\omega)r_\alpha z)) \\ &= \exp(i\omega(t - px - \eta_\alpha z)) \\ &= \exp(i\omega(t - \mathbf{s}\cdot\mathbf{x})).\end{aligned} \tag{34}$$

Here we define the *slowness vector*,

$$\mathbf{s} = (p, \eta_\alpha), \tag{35}$$

whose components are the ray parameter p and $\eta_\alpha = (k_x/\omega)r_\alpha = pr_\alpha = r_\alpha/c_x = (1/\alpha^2 - p^2)^{1/2}$.

We can interpret η_α geometrically using the components of the wave vector, because by Eqn 18 $r_\alpha = k_{z_\alpha}/k_x$, so

$$\eta_\alpha = k_{z_\alpha}/\omega = k_{z_\alpha}/(|\mathbf{k}_\alpha|\alpha) = \cos i/\alpha. \tag{36}$$

η_α and the ray parameter p are closely related because both are functions of the angle of incidence divided by the velocity. Hence the magnitude of the slowness vector is

$$|\mathbf{s}| = (p^2 + \eta_\alpha^2)^{1/2} = (\sin^2 i/\alpha^2 + \cos^2 i/\alpha^2)^{1/2} = 1/\alpha. \tag{37}$$

Thus the reciprocal of the velocity, $1/\alpha$, is called the scalar *slowness*, an apt term because a low-velocity medium is very slow (has a high slowness), whereas a fast-velocity medium has low slowness. The slowness vector (Fig. 2.5-9) is directed along the ray (parallel to the wave vector) with a magnitude equal to the slowness, and can be written $\mathbf{s} = \hat{\mathbf{k}}_\alpha/\alpha$. Its components are the ray parameter p, also called *horizontal slowness*, and η_α, called the *vertical slowness*. Similarly, for *S* waves the slowness is

$$\begin{aligned}&\mathbf{s} = (p, \eta_\beta) = \hat{\mathbf{k}}_\beta/\beta, \\ &\eta_\beta = (1/\beta^2 - p^2)^{1/2} = \cos j/\beta = pr_\beta = r_\beta/c_x.\end{aligned} \tag{38}$$

Writing a harmonic plane wave in terms of slowness gives several insights. In the argument of the exponential in Eqn 34 $(i\omega(t - \mathbf{s}\cdot\mathbf{x}))$, the slowness term, $\mathbf{s}\cdot\mathbf{x}$, has the dimension of time, and shows the net travel time due to the vertical and horizontal propagation times, each of which is described by the corresponding component of the slowness. The slowness formulation also gives another view of Snell's law. We derived Snell's law by considering a harmonic plane wave incident on a horizontal interface and the resulting reflected and transmitted plane waves. The horizontal component of the wave vectors k_x, and hence the horizontal apparent velocity c_x, were continuous at the interface. By contrast, the terms related to the vertical component of the wave vectors like $k_z = k_x r_\alpha$ varied between layers and for *P* and *S* waves. The corresponding formulation in terms of slowness says that the ray parameter or horizontal slowness p is the same for the incident, reflected, and transmitted waves, whereas the vertical slowness depends on the medium and the wave type. Snell's law can thus be stated as: p is constant for a ray and any rays that it produces at interfaces.

An important application of the ray parameter is in describing the evolution of a ray that encounters a number of interfaces (Fig. 2.5-10). Each of the four rays generated at the first

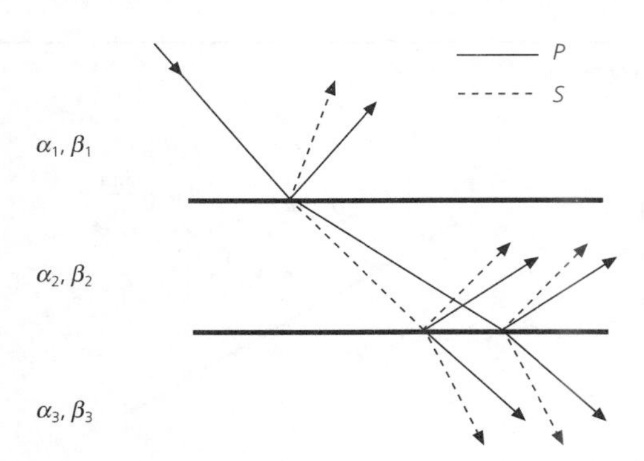

Fig. 2.5-10 A *P* wave incident on a stack of flat layers generates four waves, two reflected and two transmitted, at each interface. Each of these waves generates four more at each interface, and so on. All these waves have the same ray parameter, so their paths can be traced by applying Snell's law at each interface.

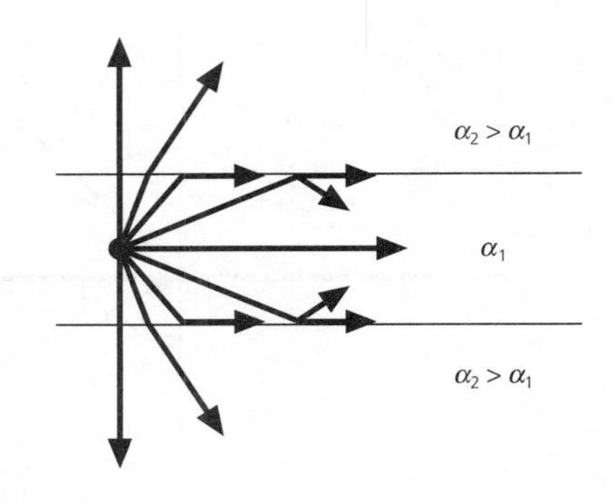

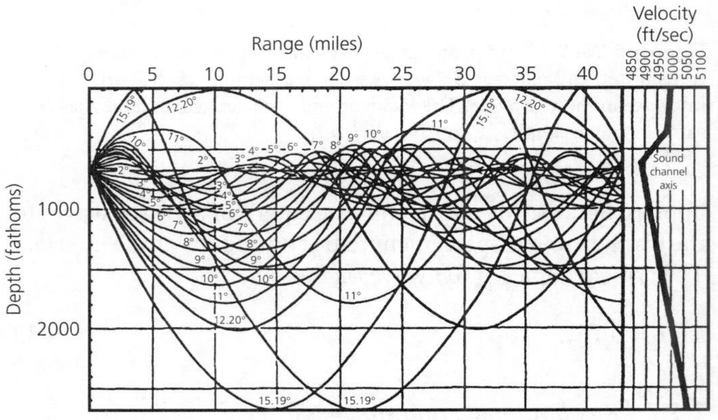

Fig. 2.5-11 *Top*: A low-velocity layer surrounded by high-velocity material acts as a waveguide. Rays incident on either interface at angles exceeding the critical angle undergo total internal reflection. *Bottom*: The SOFAR channel, a low-velocity zone (*right*) in the ocean, acts as a waveguide, as shown by ray paths from a source in the channel (*left*). Note the non-SI units for distance and velocity. (Ewing *et al.*, 1957)

interface in turn generates another four rays at the next interface, and so on. Because Snell's law applies at each interface, all these rays have the same ray parameter. As a result, p is constant along any ray path, no matter how many transmissions, reflections, or conversions the ray has undergone. This gives a way of tracing the ray path for a ray that began its travels with a certain ray parameter. In doing this on a computer, an advantage of the ray parameter is that it is zero for a vertically incident wave, whereas c_x is infinite.

2.5.8 *Waveguides*

Snell's law is one of seismology's most important tools, because seismic waves encounter variations in velocity due to changes in the physical properties of the materials, including the effects of composition, temperature, and pressure. In general, the velocity increases with depth, so seismic waves turn toward the horizontal as they go deeper. Eventually the ray "bottoms," turns upward, and reaches the surface (Fig. 1.1-3). Such ray paths can be modeled using Snell's law, either with many layers or with a version (Section 3.4) accommodating velocities that vary smoothly with depth and so give smooth ray paths. The ray path and the travel time along it thus provide information about the distribution of seismic velocities and physical properties with depth.

However, in some regions velocity decreases with depth, yielding a low-velocity medium between higher-velocity media (Fig. 2.5-11, *top*). If seismic waves are generated in the low-velocity medium, then total internal reflection will trap much of the seismic energy in the low-velocity channel, which acts as a *waveguide*.[2] One such waveguide occurs in the oceans, because the speed of sound in seawater is proportional to both temperature and pressure. The combination of temperature decreasing with depth and pressure increasing with depth produces a low-velocity region known as the SOFAR (SOund Fixing And Ranging) channel at a depth of ~1000 meters. Rays leaving a source in the channel at angles up to ±12° from the horizontal are internally reflected (Fig. 2.5-11, *bottom*). The ray paths are curved because of the smooth velocity structure. The SOFAR channel transmits sound very efficiently, allowing explosions, submarines, and whales to be detected at great distances. As a result, the speed of sound waves in the channel is being used to search for changes in ocean temperature that may be due to global warming. Similarly, earthquakes can be studied using seismic waves in the SOFAR channel that cause arrivals called T waves (Fig. 2.5-12, *top*), that can be detected by hydrophones in the water, or by seismometers when a T wave hits land. The ringing quality of T waves (Fig. 2.5-12, *bottom*) is due to the internal reflections within the SOFAR channel. Waveguides are also associated with fault zones due to their low velocities relative to the surrounding rocks.

2.5.9 *Fermat's principle and geometric ray theory*

As our discussions so far show, we can gain insight into the behavior of seismic waves by considering the ray paths associated

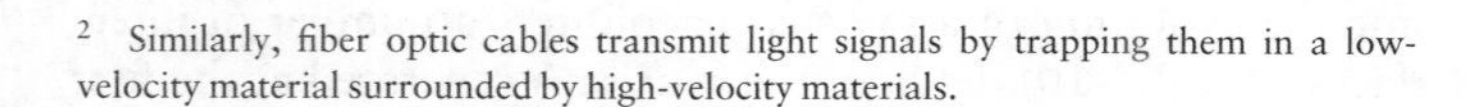
[2] Similarly, fiber optic cables transmit light signals by trapping them in a low-velocity material surrounded by high-velocity materials.

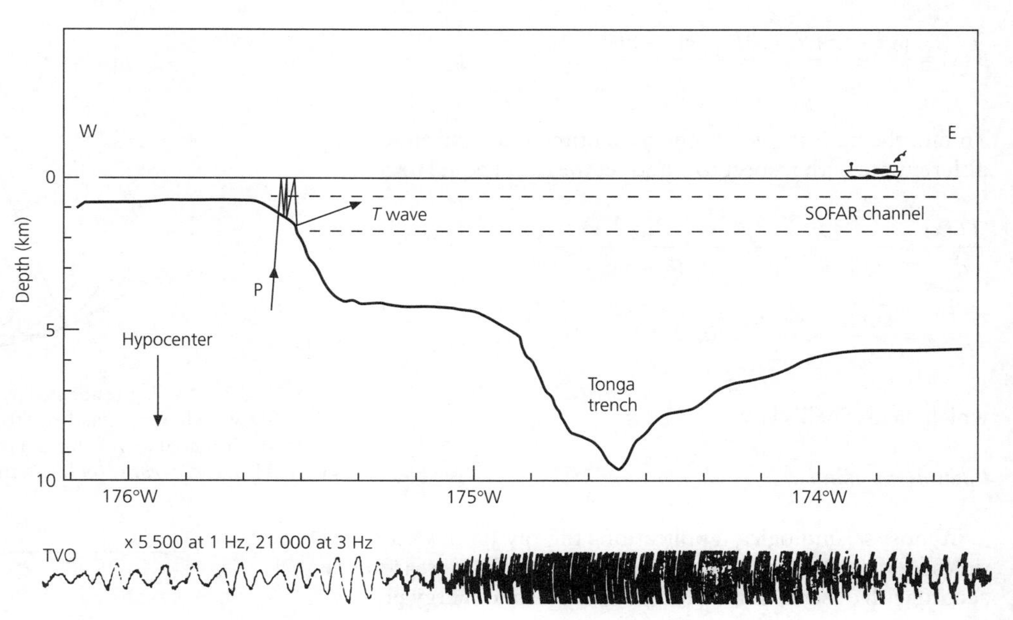

Fig. 2.5-12 *Top*: A *P* wave generated by an earthquake and reflected between the ocean floor and surface is trapped in the SOFAR channel and propagates as a *T* wave. *Bottom*: *T* waves recorded in Tahiti from an earthquake in Tonga. The amplitudes exceeded the gain on the seismometers, causing them to clip at the top and bottom. The high-frequency ringing of the *T* waves distinguishes them from body and surface waves. (Talandier and Okal, 1979.)

with them. This approach, studying wave propagation using ray paths, is called *geometric ray theory*. Although it does not fully describe important aspects of wave propagation, it is widely used because it often greatly simplifies the analysis and gives the correct answer or a good approximation.

The most obvious application of rays is for computing travel times. To find when a plane wave generated at one position will arrive at another, we use the travel time, which is the length of the ray path divided by the velocity. Thus, if waves follow complicated paths, their travel time is the sum of the travel times for each portion of the ray path. The travel time for a ray that has traveled through several media, sometimes as a *P* wave and sometimes as an *S* wave, is found using the appropriate path length and velocity for each segment.

The concept underlying this approach is *Fermat's principle*, a famous result from optics, the study of light. Fermat's principle states that the ray paths between two points are those for which the travel time is an extremum, a minimum or maximum, with respect to the nearby possible paths. The simplest case is two points in a homogeneous halfspace; the time needed to traverse the straight line connecting the points is less than for adjacent paths (Fig. 2.5-13). A second ray path for which the time is a minimum compared to adjacent paths is that of the reflected ray satisfying Snell's law. The direct ray path corresponds to an absolute minimum of the travel time, whereas the reflected ray corresponds to a local minimum.

Snell's law can be derived from Fermat's principle. Consider the possible ray paths (Fig. 2.5-14) between the point $(0, a)$ in medium 1, with velocity v_1, and the point $(b, -c)$ in medium 2, with velocity v_2. The ray paths can be parametrized by the point $(x, 0)$ where they cross the interface. The travel time as a function of x is

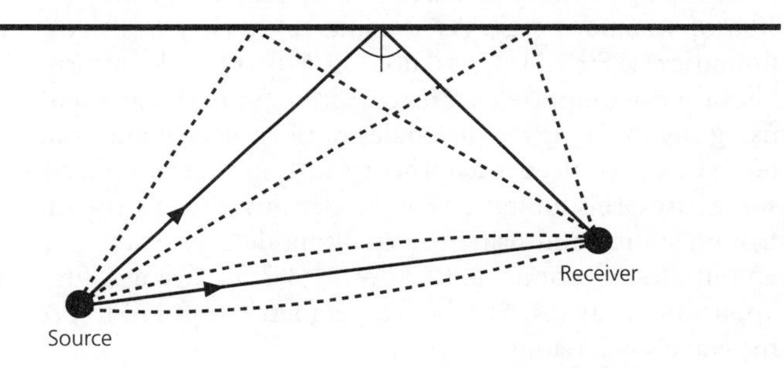

Fig. 2.5-13 Two ray paths (solid lines), one for the direct ray and one for the reflection obeying Snell's law, connecting two points in a homogeneous halfspace. The travel time for these paths is less than for nearby paths (dashed), in accord with Fermat's principle.

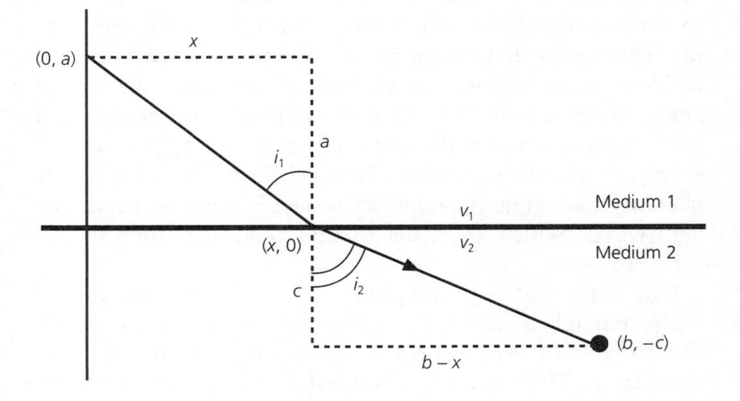

Fig. 2.5-14 Derivation of Snell's law for refraction using Fermat's principle. The ray path between points on opposite sides of the interface is that for which the travel time is a minimum.

$$T(x) = \frac{(a^2 + x^2)^{1/2}}{v_1} + \frac{((b - x)^2 + c^2)^{1/2}}{v_2}. \qquad (39)$$

To find the path for which the travel time is an extremum, we differentiate with respect to x and set the result equal to zero,

$$\frac{dT(x)}{dx} = \frac{x}{v_1(a^2 + x^2)^{1/2}} - \frac{(b - x)}{v_2((b - x)^2 + c^2)^{1/2}}$$

$$= \frac{\sin i_1}{v_1} - \frac{\sin i_2}{v_2} = 0, \qquad (40)$$

which yields Snell's law

$$v_1/\sin i_1 = v_2/\sin i_2. \qquad (41)$$

In most seismological applications the ray paths and travel times derived using Snell's law yield results in reasonable accord with observations, because most seismic energy propagates as though it followed ray paths. However, geometric ray theory is only an approximation to the solutions of the elastic equation of motion that describes the generation and propagation of seismic energy. As a result, ray theory has two major limitations. First, it does not directly provide information about wave amplitudes. Hence, although deriving Snell's law using ray theory gives the angles of the reflected and transmitted waves, we need wave theory to find their amplitudes. In some cases, this limitation can be circumvented by tracing rays from a source and using the resulting density of rays to infer amplitudes (Sections 2.8.4, 3.4.2, 3.7.3). Second, in other applications, as discussed next, geometric rays fail to describe the wave's behaviour.

2.5.10 Huygens' principle and diffraction

In some applications treating propagating waves as geometric rays fails to explain what we observe. For example, waves bend or *diffract* around the earth's core and so reach places to which Snell's law predicts no ray path. Similarly, although ray theory says that no energy is transmitted when a wave is incident on an interface at an angle greater than the critical angle, some energy is in fact transmitted. Addressing such issues requires explicitly considering the fact that seismic energy propagates as waves. To do this, we draw on results from both seismology and other wave phenomena, especially light waves, which are easier to study and have been investigated for many years.

One important approach, known as *Huygens' principle*, is illustrated in Fig. 2.5-15. Each point on a wave front is considered to be a *Huygens' source* that gives rise to another circular wave front. These wave fronts interfere constructively to give a circular wave front, and interfere destructively everywhere else. In three dimensions, the wave fronts are spherical.

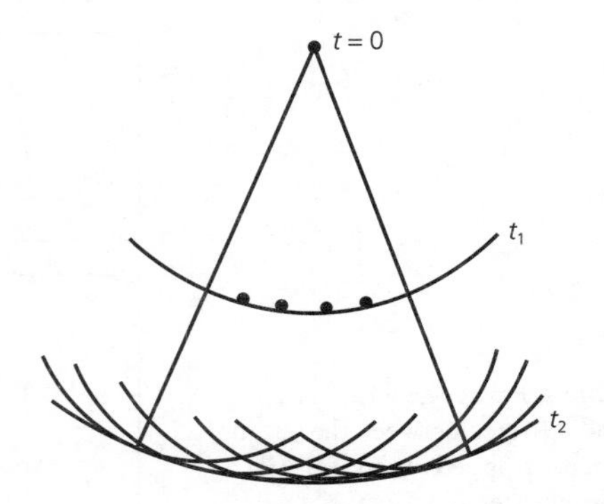

Fig. 2.5-15 Figure adapted from Huygens' original (1690) analysis showing how circular wave fronts can be generated by treating each point on the initial wave front as a point source of wave energy. (Reprinted from Huygens, *Treatise on Light*, trans. S. P. Thompson (Dover, New York).)

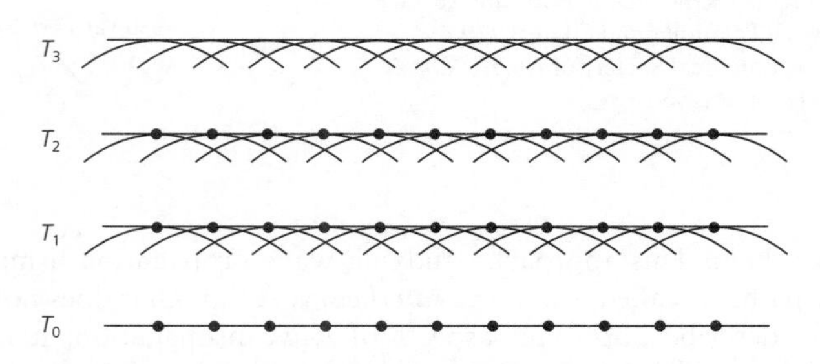

Fig. 2.5-16 Demonstration of Huygens' principle for the propagation of a straight wave front. Successive wave fronts are generated by drawing a circular wave from each point on the previous wave front and then drawing a line tangent to the circles. The circular wave fronts are assumed to interfere destructively everywhere else.

Although the point sources, known also as diffractors or scatterers, need not have a physical interpretation, in some cases they do. For example, heterogeneities in the crust and mantle scatter incident seismic waves. Hence, migration methods in exploration seismology (Section 3.3.7) improve images of the subsurface by undoing this scattering. Similarly, seismic energy that arrives before *PKP* waves that traverse the earth's core is thought to have been scattered by heterogeneities in the mantle.

Huygens' principle gives another way of thinking about phenomena we have discussed. It explains why a straight wave front generates subsequent straight wave fronts, as shown in Fig. 2.5-16. It is also another way of deriving Snell's law. Assume, as in Fig. 2.5-17, that a wave front A–A' in medium 1 is incident upon a boundary with medium 2. When the wave front reaches point A, energy begins to radiate outward, but if the velocity in the second medium is less, the radius of the circular wave front some time later is smaller in medium 2. Similarly, as the wave front reaches other points along the interface (for example, point B), circular wave fronts of different sizes spread out in the two media. By the time the initial wave front

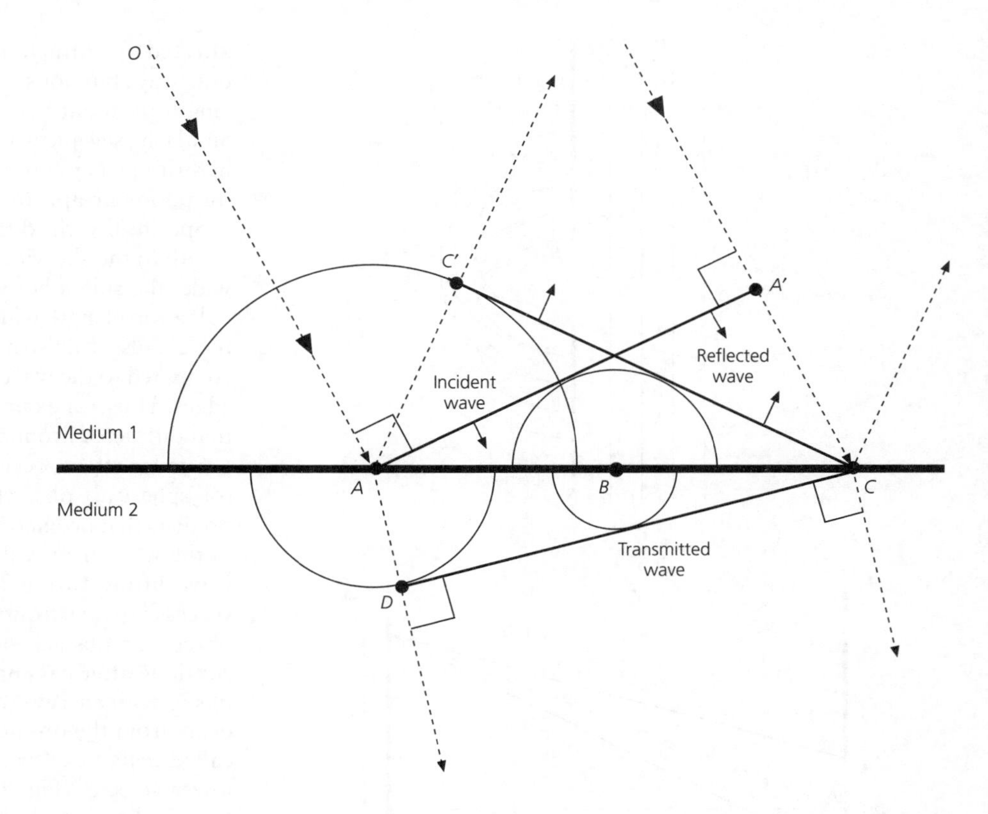

Fig. 2.5-17 Derivation of Snell's law using Huygens' principle. As an incident plane wave A–A' interacts with the boundary, the Huygens' sources combine to form a reflected wave front C–C' and a transmitted wave front C–D. Because the radii of the circular wave fronts are proportional to the velocity in each medium, the angles of the incident (O–A), reflected (A–C'), and transmitted (A–D) rays yield Snell's law.

reaches point C, one planar wave front, drawn as the tangent to the circular wave fronts in medium 1, is the reflected wave, and another gives the refracted wave. The directions of the waves, taken as the perpendiculars to the planar wave fronts, are those expected from Snell's law. Thus we have three ways of understanding how Snell's law comes about: Huygens' principle, Fermat's principle (Section 2.5.9), and the application of the interface boundary conditions to plane waves (Section 2.5.4). Each approach offers different insight into the phenomenon of reflection and refraction.

Huygens' principle also explains the phenomenon of *diffraction*, in which waves bend around obstacles. Although the phenomenon is complicated, the simple example of diffraction at a slit (Fig. 2.5-18, *top*) gives considerable insight. We assume that an incident planar wave front acts like a set of Huygens' sources, so the transmitted wave field is the superposition of waves from these sources. In front of the slit, the sources combine to give a planar transmitted wave front. In addition, energy propagates to the sides, and thus can be detected around the corners, although there is no geometric ray path to there. The analogous process occurs with shear waves that cannot pass through the liquid outer core, and so diffract around it (Section 3.5.2).

Although evaluating the amplitude of the diffracted waves requires going beyond Huygens' principle, a simple construction (Fig. 2.5-18, *middle*) shows some important aspects. If the slit has width d, then waves from either side of the slit will be out of phase by 90° and so interfere[3] destructively at distance D when the path difference is a half wavelength. Hence the amplitude will be zero at a distance x_0, or an angle θ, from the middle of the slit. By this condition

$$\lambda/2 = d \sin \theta \approx dx_0/D, \tag{42}$$

assuming $D >> d$. Thus the amplitude decays from its maximum at $\theta = 0$ to zero at $x_0 = \lambda D/2d$. A more sophisticated analysis[4] shows that the amplitude varies as

$$(\sin \zeta)/\zeta, \quad \text{where} \quad \zeta = 2\pi dx/\lambda D, \tag{43}$$

which is shown in Fig. 2.5-18 (*bottom*). This function has a central lobe of width $2x_0$ and a series of decreasing side lobes.

The slit illustrates general properties of diffraction, because diffraction around an obstacle is in many ways similar. An important point is that diffraction depends on the wavelength, so longer wavelengths have broader lobes and thus are more

[3] Interference and diffraction are terms for closely related wave phenomena between which there is no sharp distinction. Effects involving a few sources are typically called interference, whereas those involving many sources are often called diffraction.

[4] This analysis uses Fourier transforms, and so yields the $(\sin \zeta)/\zeta$ function that commonly appears in Fourier analysis, as we will see in Section 6.3.

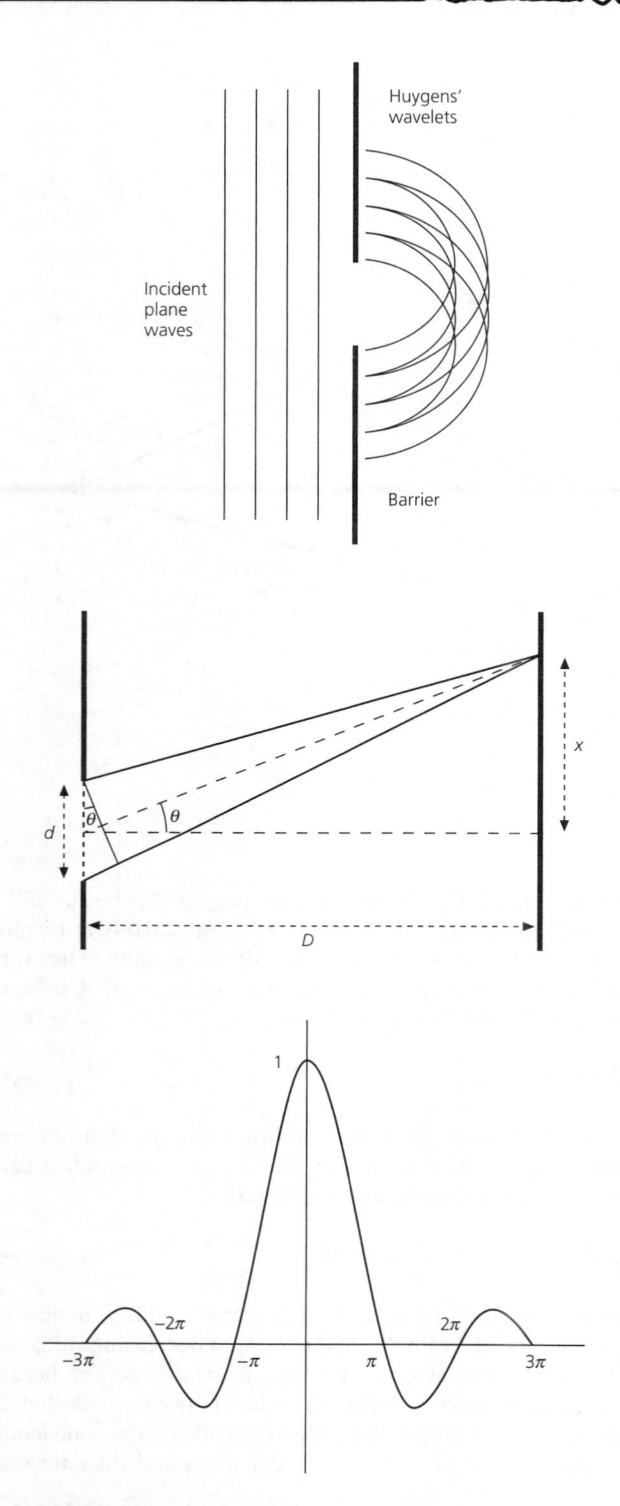

Fig. 2.5-18 *Top*: Use of Huygens' sources to describe waves diffracting at a slit. Energy diffracts around the corners to reach areas with no geometric ray paths leading to them. (Klein and Furtak, 1986. Copyright © 1986. Reprinted by permission of John Wiley & Sons, Inc.) *Middle*: Geometry for the analysis of diffraction by a slit of width d, observed at a distance D. *Bottom*: The $(\sin \zeta)/\zeta$ function describing the amplitude of the diffracted wave, showing the central lobe and side lobes.

affected by diffraction. For example, we can hear around open doorways but not see around them, because sound has a wavelength of about 0.1 m, compared to 10^{-7} m for visible light. Similarly, seismic waves that diffract around the core lose their high-frequency components. Hence the longer the wavelength, the poorer an approximation geometric ray theory becomes.

Specifically, the diffraction depends on the ratio of the wavelength to the slit width. If the slit is less than a half wavelength wide, the side lobes vanish. Hence, if an obstacle is less than half a wavelength wide, waves impinging on it are insensitive to the details of its structure. Conversely, if the slit is very wide compared to the wavelength, diffraction occurs only at the slit's edges. Thus, for example, seismic reflection images show waves that diffracted around the ends of interfaces (Section 3.3.7).

Similar effects occur when wave fronts encounter a circular (or spherical) obstacle (Fig. 2.5-19a). Geometric ray theory predicts that no energy will arrive behind the obstacle, so a hole in the wave front will develop and never close. In reality, the wave diffracts around the sphere, closing the gap behind it. The successive wave fronts illustrate why it is difficult to seismically observe an obstacle or a low-velocity zone. As the wave fronts continue after passing the sphere, the break in the wave front fills in with energy from either side until at large distances the delay from the obstacle is no longer observable. This process, called *waveform annealing*, also occurs if the obstacle has a lower velocity (Fig. 2.5-19b), so much of the energy arriving behind the obstacle diffracts around the obstacle rather than passing slowly through it. This effect can also be interpreted using Fermat's principle, because the resulting wave is that which traveled for the least time.

This example illustrates one possible reason why it has proved very difficult to seismologically observe plumes, upwellings from deep in the mantle that have been proposed to give rise to island chains like Hawaii. A seismic wave front encountering a narrow conduit of hot, slow rock diffracts around it, causing little travel time delay. By contrast, anomalously fast rock is easy to "see" seismologically. Hence seismology is very good at detecting subducting lithosphere at trenches (Section 5.4), because the cold material has a higher seismic velocity. This effect is illustrated by Fig. 2.5-19c, which shows a spherical anomaly faster than the surrounding material. By Fermat's principle, the anomaly is the fastest path between a source and a receiver. From the Huygens' principle view, the wave front moves further ahead through the fast material, and then spreads out laterally, advancing the rest of the wave front. The waves thus lose their planar appearance and appear to have emanated from a point source.

These analyses show that Huygens' principle describes the general features of diffraction. However, it does not provide direct information about amplitudes. For instance, although the wave fronts in Fig. 2.5-19 lose amplitude as they diffract around the sphere, this decay cannot be obtained from Huygens' principle. To go further requires an extension of Huygens' principle known as the Kirchhoff integral, which is beyond our scope.

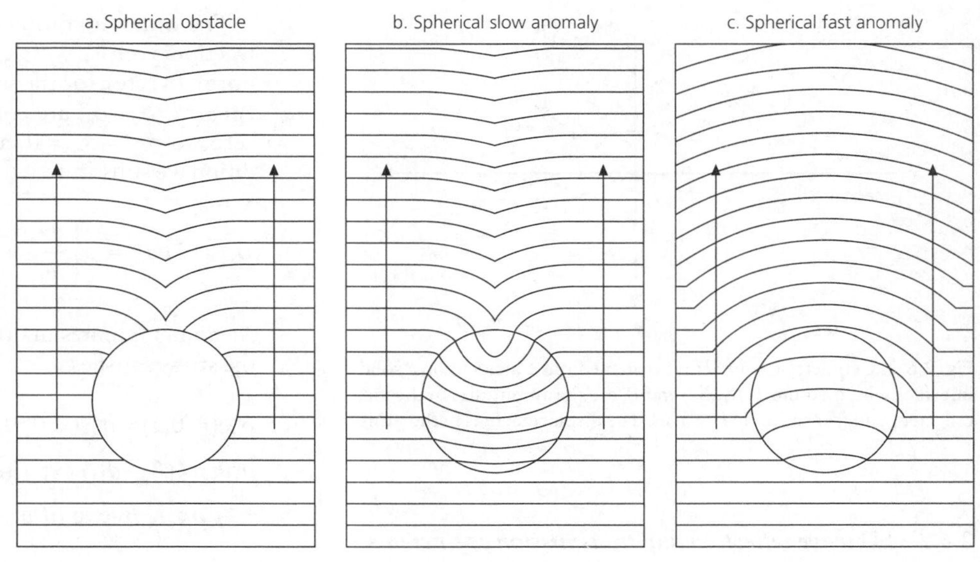

Fig. 2.5-19 Waves interacting with a spherical anomaly. a: A straight wave front diffracts around a circular or spherical obstacle, as described by Huygens' principle. Only the leading wave front is shown. This formulation shows the locations of the wave fronts, but not their amplitudes. b: Plane waves interacting with a low-velocity anomaly 30% slower than the surrounding material. The waves slow within the anomaly and diffract around it. After passing the obstacle, the wave front shows little perturbation, illustrating the difficulty of seismically observing low-velocity anomalies. c: Plane waves interacting with an anomaly 50% faster than the surrounding material. The overall speed of the wave field increases, demonstrating that seismically fast anomalies are easy to observe.

2.6 Plane wave reflection and transmission coefficients

2.6.1 *Introduction*

Seismic waves propagating in the earth encounter several types of interface (Fig. 2.6-1) at which physical properties change over short distances. For example, the earth's surface is a free surface, and the sea floor is a liquid–solid interface. Variations in velocity and density cause solid–solid interfaces such as the *Mohorovičić discontinuity*, or *Moho*, separating the crust and the mantle (Section 3.2). The upper and lower mantles are divided by regions of rapid velocity changes (Section 3.5), which can be described for many purposes as solid–solid interfaces. The core–mantle boundary is an interface between the solid mantle and fluid outer core, and the base of the outer core is an interface with the solid inner core. Nearly all our knowledge of these interfaces comes from observing their effects on seismic wave propagation.

In the last section we derived Snell's law, relating the bending of waves at an interface to the velocity contrast across it. We now discuss the amplitudes of the reflected and transmitted waves. We first consider two simple cases, *SH* waves at a boundary and *P–SV* waves at a free surface, and then outline how the same approach is applied for *P–SV* waves at an interface between solids. It turns out that although the angles of reflection and transmission, and hence the ray paths and travel times, depend only on the velocities, the amplitudes depend on the elastic constants in a more complicated way. As a result, the amplitudes of waves provide information beyond that conveyed by travel times, and so are valuable for studying the earth's interior.

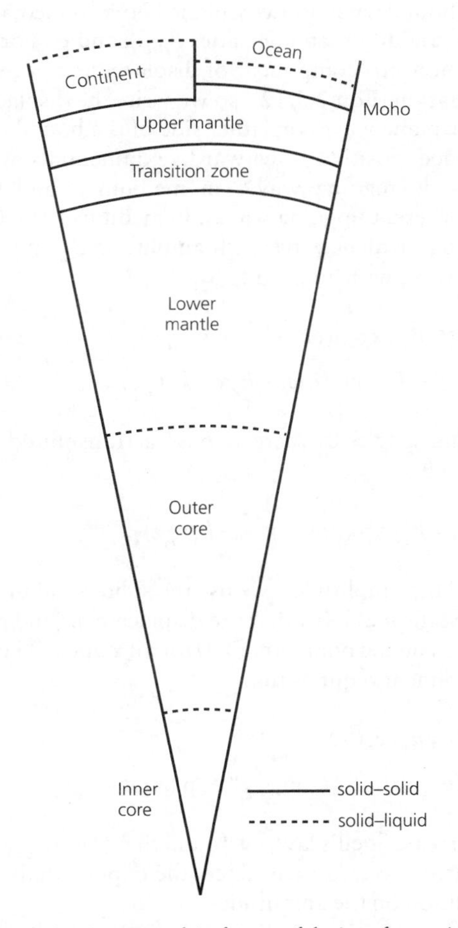

Fig. 2.6-1 Illustration (not to scale) of some of the interfaces within the earth that affect seismic waves.

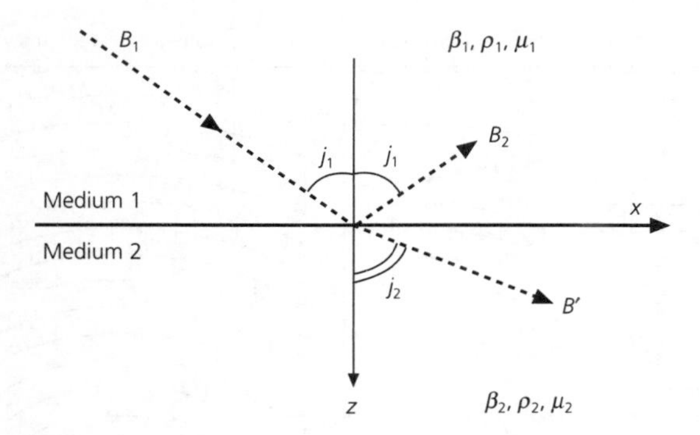

Fig. 2.6-2 Geometry for an *SH* wave in medium 1 incident on a solid–solid interface with medium 2. B_1, B_2, and B' are the amplitudes of the incident, reflected, and transmitted *SH* waves. The displacement is in the *y* direction.

2.6.2 SH *wave reflection and transmission coefficients*

We first consider the amplitudes of *SH* waves reflected and transmitted at a horizontal interface. Figure 2.6-2 illustrates the geometry of an *SH* wave propagating in the x–z plane incident on a boundary in the x–y plane between media with shear velocities, rigidities, and densities β_i, μ_i, and ρ_i. For *SH* waves, the only nonzero component of displacement, u_y, satisfies the wave equation (Eqn 2.5.12), so we write the displacements for harmonic plane waves on either side of the boundary. Because z is defined positive downward, exponentials with $-k_x r_{\beta_i} z$ represent downgoing waves in medium i, and those with $+k_x r_{\beta_i} z$ represent upgoing waves. In medium 1 ($z < 0$) there is a downgoing incident wave with amplitude B_1 and an upgoing reflected wave with amplitude B_2,

$$u_y^-(x, z, t) = B_1 \exp\left(i(\omega t - k_x x - k_x r_{\beta_1} z)\right) + B_2 \exp\left(i(\omega t - k_x x + k_x r_{\beta_1} z)\right). \quad (1)$$

In medium 2 ($z > 0$) there is only a transmitted wave with amplitude B',

$$u_y^+(x, z, t) = B' \exp\left(i(\omega t - k_x x - k_x r_{\beta_2} z)\right). \quad (2)$$

To find the amplitudes, we use the solid–solid interface conditions (Section 2.3.10) that the displacement and traction are continuous on the boundary $z = 0$ for all x and t. The continuity of displacement requires that

$$u_y^-(x, 0, t) = u_y^+(x, 0, t)$$
$$(B_1 + B_2) \exp\left(i(\omega t - k_x x)\right) = B' \exp\left(i(\omega t - k_x x)\right). \quad (3)$$

When deriving Snell's law, we found that $(\omega t - k_x x)$ is the same for all three waves, so we cancel the exponentials and obtain one condition on the amplitudes,

$$B_1 + B_2 = B'. \quad (4)$$

The other condition comes from the requirement that the traction vector, $T_i = \sigma_{ij} n_j$, be continuous. Because the unit normal vector for the interface is (0, 0, 1), the stress components σ_{xz}, σ_{yz}, σ_{zz} are continuous. For *SH* waves u_x and u_z are zero, so $\sigma_{xz} = \sigma_{zz} = 0$, and σ_{yz} is continuous. To use this condition we substitute

$$\sigma_{yz} = 2\mu e_{yz} = \mu\left(\frac{\partial u_y}{\partial z} + \frac{\partial u_z}{\partial y}\right) = \mu\left(\frac{\partial u_y}{\partial z}\right). \quad (5)$$

At points infinitesimally above and below the interface $z = 0$, the stress satisfies

$$\sigma_{yz}^-(x, 0, t) = \sigma_{yz}^+(x, 0, t),$$
$$\mu_1 i k_x r_{\beta_1}(B_2 - B_1) \exp\left(i(\omega t - k_x x)\right)$$
$$= -\mu_2 i k_x r_{\beta_2} B' \exp\left(i(\omega t - k_x x)\right). \quad (6)$$

Canceling the factors common to both sides gives the second condition

$$(B_1 - B_2) = B'(\mu_2 r_{\beta_2})/(\mu_1 r_{\beta_1}). \quad (7)$$

Solving Eqns 4 and 7 simultaneously yields the amplitudes of the reflected and transmitted waves. First, we eliminate B_2 and find the transmission coefficient,

$$T_{12} = \frac{B'}{B_1} = \frac{2\mu_1 r_{\beta_1}}{\mu_1 r_{\beta_1} + \mu_2 r_{\beta_2}}, \quad (8)$$

the ratio of the amplitude of the transmitted wave in medium 2 to that of the incident wave in medium 1. Similarly, eliminating B' from Eqns 4 and 7 gives the reflection coefficient

$$R_{12} = \frac{B_2}{B_1} = \frac{\mu_1 r_{\beta_1} - \mu_2 r_{\beta_2}}{\mu_1 r_{\beta_1} + \mu_2 r_{\beta_2}}, \quad (9)$$

the ratio of the amplitudes of the reflected and incident waves in medium 1.

The reflection and transmission coefficients depend on the angle of incidence because, by Eqn 2.5.38

$$r_{\beta_i} = c_x \cos j_i / \beta_i. \quad (10)$$

Hence, using Eqn 10 and recognizing that from the definition of the *S*-wave velocity, $\mu_i = \rho_i \beta_i^2$, the reflection and transmission coefficients can be written

$$T_{12} = \frac{2\rho_1 \beta_1 \cos j_1}{\rho_1 \beta_1 \cos j_1 + \rho_2 \beta_2 \cos j_2},$$
$$R_{12} = \frac{\rho_1 \beta_1 \cos j_1 - \rho_2 \beta_2 \cos j_2}{\rho_1 \beta_1 \cos j_1 + \rho_2 \beta_2 \cos j_2}. \quad (11)$$

Thus the reflection and transmission coefficients depend on the acoustic impedances $\rho_i\beta_i$, as did those for waves on a string (Section 2.2.3), but with an angle dependence that could not occur for a one-dimensional string. If the media are interchanged, the reflection coefficient reverses polarity, $R_{12} = -R_{21}$, and the transmission coefficients satisfy $T_{12} + T_{21} = 2$. Due to the displacement continuity condition (Eqn 3), $1 + R_{12} = T_{12}$. Large impedance contrasts favor reflection, whereas small contrasts favor transmission. In the limit of identical media there is no reflection ($R_{12} = 0$), and everything is transmitted ($T_{12} = 1$).

An interesting effect occurs for an *SH* wave incident on the earth's free surface. Because $\beta_2 = 0$, the reflection coefficient equals 1 regardless of the incidence angle, so the displacement is twice that of the upgoing wave. This also occurs at solid–liquid interfaces, such as the sea floor or the core–mantle boundary, which act as free surfaces for *SH* because no *SH* waves propagate in the liquid.

The transmission and reflection coefficients have a particularly simple form for vertical incidence ($j_1 = j_2 = 0$):

$$T_{12} = \frac{2\rho_1\beta_1}{\rho_1\beta_1 + \rho_2\beta_2}, \quad R_{12} = \frac{\rho_1\beta_1 - \rho_2\beta_2}{\rho_1\beta_1 + \rho_2\beta_2}. \tag{12}$$

These vertical incidence forms are easy to remember and are a useful approximation for nonvertical incidence.

The fact that the transmission and reflection coefficients depend on the contrast in both density and velocity, whereas the angles made by the waves depend only on velocity, makes the amplitudes valuable for studying elastic properties from seismological observations. Although each medium has three quantities of interest, β_i, μ_i, and ρ_i, only two are independent, because the velocities depend on the rigidities and densities. For example, if we regard the velocity and rigidity as independent, the angles of reflection and transmission give information about the velocity, and the amplitudes provide additional information about the rigidity.

2.6.3 *Energy flux for reflected and transmitted* SH *waves*

In some cases the transmission coefficient exceeds 1. For example, when an *SH* wave impinges on a higher-velocity medium at critical incidence, the transmitted wave becomes horizontal ($j_2 = 90°$) and Eqn 11 shows that the transmission coefficient is 2. As for the string (Section 2.2.4), this puzzling effect can be explained by examining how the incident wave energy divides between the reflected and transmitted waves.

We saw (Section 2.4.5) that the flux of energy per unit wave front in the propagation direction associated with a harmonic *SH* plane wave $u(x, t) = A \cos(\omega t - kx)$ is the product of the energy density and the velocity

$$\dot{E} = A^2\omega^2\rho\beta/2. \tag{13}$$

Because no energy accumulates at an interface, the flux of energy in the length of wave front incident on an element dx of

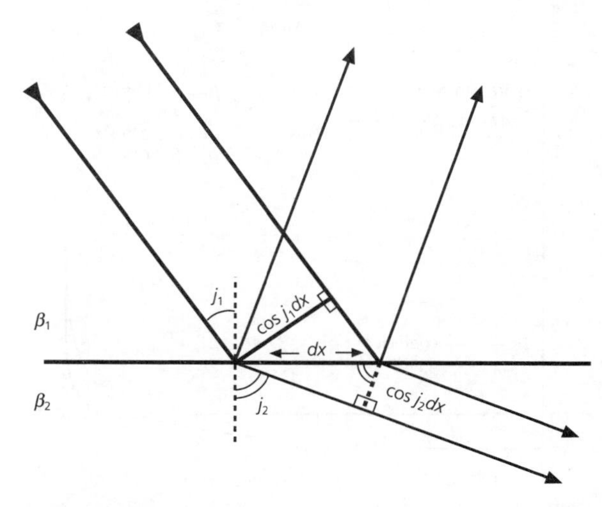

Fig. 2.6-3 The lengths of the incident, reflected, and transmitted wave fronts contributing to the energy flux though an element dx of an interface depend on the cosine of the angle of incidence for each wave.

the interface equals that of the reflected and transmitted waves removing energy from the interface. The length of the wave fronts contributing to the flux depends on the angles of incidence. Figure 2.6-3 shows that the relevant lengths are $\cos j_1 dx$ for the incident and reflected waves, and $\cos j_2 dx$ for the transmitted wave. Thus, for an incident wave of unit amplitude, the energy fluxes for the incident, reflected, and transmitted waves are

$$\dot{E}_I = \omega^2\rho_1\beta_1 \cos j_1 dx/2$$
$$\dot{E}_R = R_{12}^2\omega^2\rho_1\beta_1 \cos j_1 dx/2$$
$$\dot{E}_T = T_{12}^2\omega^2\rho_2\beta_2 \cos j_2 dx/2. \tag{14}$$

These satisfy the conservation of energy

$$\dot{E}_I = \dot{E}_R + \dot{E}_T, \tag{15}$$

as proved in one of this chapter's problems. The ratios of the transmitted and reflected energy fluxes to the incident energy flux are

$$\frac{\dot{E}_R}{\dot{E}_I} = R_{12}^2 \quad \text{and} \quad \frac{\dot{E}_T}{\dot{E}_I} = T_{12}^2\frac{\rho_2\beta_2 \cos j_2}{\rho_1\beta_1 \cos j_1}. \tag{16}$$

Because the energy ratios are proportional to the squares of the amplitudes, small amplitudes represent very small energies. For example, a reflected wave with $R_{12} = 0.1$ has an energy ratio of $\dot{E}_R/\dot{E}_I = 0.01$.

To see the angle dependence, consider an interface between media with $\beta_1 = 3.9$ km/s, $\rho_1 = 2.8$ g/cm^3, and $\beta_2 = 4.5$ km/s, $\rho_2 = 3.3$ g/cm^3, which approximates the continental Mohorovičić discontinuity. Figure 2.6-4 shows the reflection and transmission coefficients and the ratio of energy fluxes for angles of incidence between vertical and critical (58°). The energy flux ratios

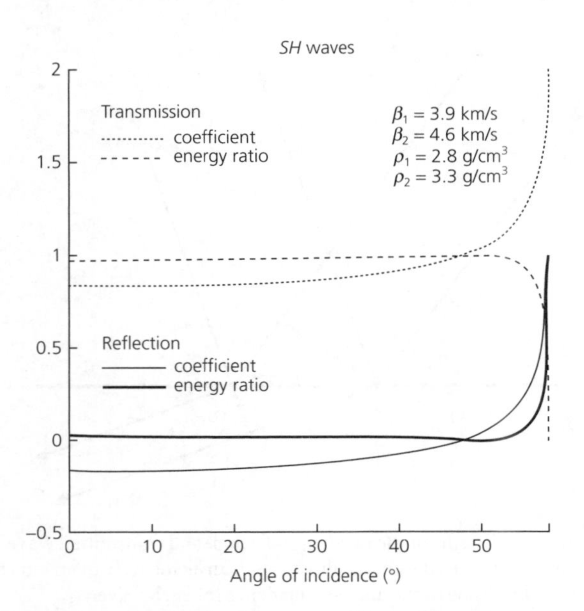

Fig. 2.6-4 For an *SH* wave incident on a solid–solid boundary, displacement reflection and transmission coefficients and the ratios of reflected and transmitted energy fluxes to that of the incident wave are given as functions of the angle of incidence of the incident wave. The critical angle for these values is 58°.

sum to one, so, as the reflected energy increases, the transmitted energy decreases.

At vertical incidence and for most of the range of incidence angles less than the critical angle, most of the energy is transmitted. In this range, the vertical incidence reflection and transmission coefficients and energy flux ratios are good approximations for nonvertical incidence. The behavior near the critical angle illustrates the value of considering the energies as well as the reflection and transmission coefficients. As the angle of incidence approaches the critical value, the transmission coefficient goes to 2, but the wave front factor $\cos j_2$ goes to zero, so the energy in the transmitted wave vanishes and all of the energy reflects.[1]

2.6.4 *Postcritical* SH *waves*

The transmitted and reflected waves behave differently for angles of incidence greater than the critical angle. Snell's law,

$$c_x = \beta_1/\sin j_1 = \beta_2/\sin j_2, \tag{17}$$

shows that for incidence angles less than the critical angle, the apparent velocity exceeds the velocity of the second medium, β_2. At critical incidence, $\sin j_2 = 1$, so the apparent velocity equals β_2. For incidence angles greater than the critical angle, $\sin j_1 > \sin j_c$, so the apparent velocity $c_x = \beta_1/\sin j_1$ is less than $\beta_1/\sin j_c = \beta_2$.

[1] The wave angles and amplitudes can be shown by a simple experiment using beams of light (Klosko *et al.*, 2000).

To see the effect on the transmitted wave of an apparent velocity less than that of medium 2, recall that the transmitted wave (Eqn 2) is described by

$$u_y^+(x, z, t) = B' \exp\,(i(\omega t - k_x x - k_x r_{\beta_2} z)). \tag{18}$$

If $c_x < \beta_2$, the quantity (Eqn 2.5.8)

$$r_{\beta_2} = (c_x^2/\beta_2^2 - 1)^{1/2} \tag{19}$$

becomes an imaginary number. As a result, $k_x r_{\beta_2}$, the z component of the wavenumber, also becomes imaginary, so Eqn 18 no longer describes a plane wave propagating in the $+z$ direction. The square root, which describes the imaginary number, has two possible signs. We pick the negative sign and define

$$r_{\beta_2} = -ir^*_{\beta_2}, \quad r^*_{\beta_2} = (1 - c_x^2/\beta_2^2)^{1/2} \tag{20}$$

so that the z term in the displacement,

$$\exp\,(-ik_x r_{\beta_2} z) = \exp\,(-k_x r^*_{\beta_2} z), \tag{21}$$

decays exponentially away from the interface in medium 2 as $z \to \infty$. Thus, instead of being a propagating wave, the transmitted wave becomes an *evanescent* or *inhomogeneous* wave "trapped" near the interface. Choosing the negative sign in Eqn 20 is a radiation boundary condition, because the opposite choice gives displacement increasing with depth as $z \to \infty$, as if energy originated there.

The behavior of the reflected wave for postcritical incidence results from the fact that the reflection coefficient (Eqn 9) becomes a complex number. Using Eqn 20 shows that

$$R_{12} = \frac{\mu_1 r_{\beta_1} + i\mu_2 r^*_{\beta_2}}{\mu_1 r_{\beta_1} - i\mu_2 r^*_{\beta_2}}. \tag{22}$$

This a complex number divided by its conjugate, so the magnitude of the reflection coefficient is 1, but there is a phase shift of 2ε:

$$R_{12} = e^{i2\varepsilon}, \quad \varepsilon = \tan^{-1} \frac{\mu_2 r^*_{\beta_2}}{\mu_1 r_{\beta_1}}. \tag{23}$$

The phase shift depends on the angle of incidence. At critical incidence, $c_x = \beta_2$, so $r^*_{\beta_2} = 0$ and $\varepsilon = 0°$. As the angle of incidence increases beyond critical, ε increases until grazing incidence, $j_1 = 90°$, where $c_x = \beta_1$, $r_{\beta_1} = 0$, and $\varepsilon = 90°$. A 90° phase shift turns a sine wave into a cosine wave, and vice versa, whereas a 180° phase shift is multiplication by −1. If the incident wave is made up of different frequencies, the phase shift affects each frequency, so the reflected wave can be computed using the Fourier transform. Figure 2.6-5 illustrates how the reflected wave would appear due to different phase shifts.

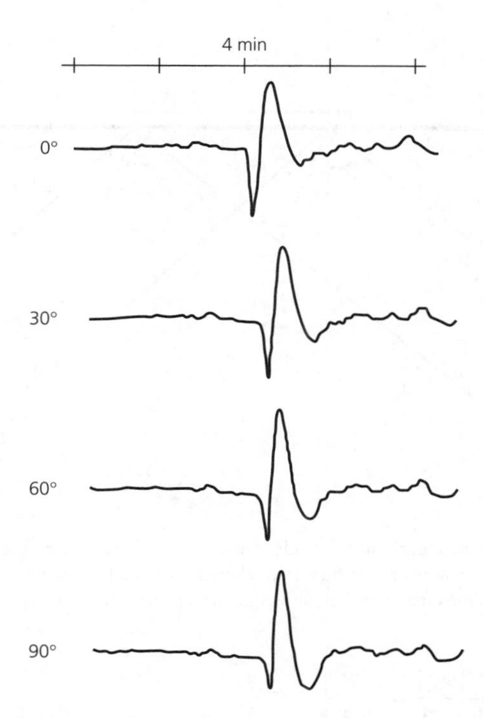

Fig. 2.6-5 The effect of phase shifts on a seismic waveform shown in the upper trace. (Choy and Richards, 1975.)

2.6.5 P–SV *waves at a free surface*

Determining the amplitudes of reflected and transmitted waves is more complicated for the *P–SV* system because waves convert from one type to the other. To illustrate this, we consider the simple case when a harmonic plane *P* wave incident on a free surface generates two reflected waves, one *P* and one *SV* (Fig. 2.6-6). To determine their amplitudes, we use potentials for both *P* and *SV*, in contrast to the *SH* case, where we used the displacements directly, and find solutions that satisfy the free surface boundary conditions.

There are two scalar potential terms, one for the upgoing incident *P* wave and one for the downgoing reflected *P* wave,

$$\phi_I(x, z, t) + \phi_R(x, z, t) = A_1 \exp(i(\omega t - k_x x + k_x r_\alpha z)) + A_2 \exp(i(\omega t - k_x x - k_x r_\alpha z)). \quad (24)$$

The downgoing reflected *SV* wave with amplitude B_2 is described by a vector potential with *y* component

$$\psi_R(x, z, t) = B_2 \exp(i(\omega t - k_x x - k_x r_\beta z)). \quad (25)$$

Using Eqn 2.5.5, the two nonzero components of the displacement are given by a combination of the *P* and *SV* potentials

$$u_x = \frac{\partial \phi}{\partial x} - \frac{\partial \psi}{\partial z}, \quad u_z = \frac{\partial \phi}{\partial z} + \frac{\partial \psi}{\partial x}. \quad (26)$$

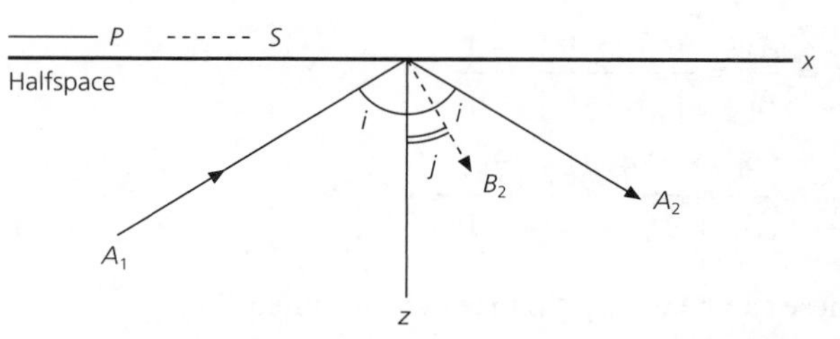

Fig. 2.6-6 Geometry for a *P* wave in a halfspace incident upon a free surface. A_1, A_2, and B_2 are the amplitudes of the incident *P*, reflected *P*, and reflected *SV* waves.

At the free surface, the traction vector, and hence the stress components σ_{xz}, σ_{yz}, σ_{zz}, must be zero for all *x* and *t*. σ_{yz} is automatically zero for *P–SV* waves in this geometry. Using Eqn 26, we express the other two stress components in terms of the potentials

$$\sigma_{xz} = 2\mu e_{xz} = \mu\left(\frac{\partial u_x}{\partial z} + \frac{\partial u_z}{\partial x}\right) = \mu\left(2\frac{\partial^2 \phi}{\partial x \partial z} + \frac{\partial^2 \psi}{\partial x^2} - \frac{\partial^2 \psi}{\partial z^2}\right)$$

$$\sigma_{zz} = \lambda\theta + 2\mu e_{zz} = \lambda\left(\frac{\partial^2 \phi}{\partial x^2} + \frac{\partial^2 \phi}{\partial z^2}\right) + 2\mu\left(\frac{\partial^2 \phi}{\partial z^2} + \frac{\partial^2 \psi}{\partial x \partial z}\right). \quad (27)$$

We then substitute the wave potentials from Eqns 24 and 25 into Eqn 27 and evaluate them at $z = 0$:

$$\sigma_{xz}(x, 0, t) = 0$$
$$= \mu[2r_\alpha(A_1 - A_2) + (r_\beta^2 - 1)B_2]k_x^2 \exp(i(\omega t - k_x x))$$
$$\sigma_{zz}(x, 0, t) = 0$$
$$= -[\lambda(1 + r_\alpha^2)(A_1 + A_2) + 2\mu(r_\alpha^2(A_1 + A_2) + r_\beta B_2)]k_x^2 \exp(i(\omega t - k_x x)). \quad (28)$$

Regrouping terms shows that the ratios of the amplitudes of the reflected *P* and *SV* waves to that of the incident *P* wave can be found by solving the two equations

$$2r_\alpha \frac{A_2}{A_1} + (1 - r_\beta^2)\frac{B_2}{A_1} = 2r_\alpha, \quad (29)$$

$$[(\lambda + 2\mu)(1 + r_\alpha^2) - 2\mu]\frac{A_2}{A_1} + 2\mu r_\beta \frac{B_2}{A_1} = 2\mu - (\lambda + 2\mu)(1 + r_\alpha^2). \quad (30)$$

Because $(1 + r_\alpha^2) = (c_x^2/\alpha^2) = c_x^2\rho/(\lambda + 2\mu)$, the last equation can be simplified to

$$(c_x^2\rho - 2\mu)\frac{A_2}{A_1} + 2\mu r_\beta \frac{B_2}{A_1} = 2\mu - c_x^2\rho. \quad (31)$$

Solving Eqns 29 and 31 using $(1 + r_\beta^2) = (c_x^2/\beta^2) = c_x^2\rho/\mu$ gives the amplitude ratios

$$R_P = \frac{A_2}{A_1} = \frac{4r_\alpha r_\beta - (r_\beta^2 - 1)^2}{4r_\alpha r_\beta + (r_\beta^2 - 1)^2},$$

$$R_{SV} = \frac{B_2}{A_1} = \frac{4r_\alpha(1 - r_\beta^2)}{4r_\alpha r_\beta + (r_\beta^2 - 1)^2}. \quad (32)$$

These can be written in many forms, including

$$R_P = \frac{A_2}{A_1} = \frac{4p^2\eta_\alpha\eta_\beta - (\eta_\beta^2 - p^2)^2}{4p^2\eta_\alpha\eta_\beta + (\eta_\beta^2 - p^2)^2},$$

$$R_{SV} = \frac{B_2}{A_1} = \frac{4p\eta_\alpha(p^2 - \eta_\beta^2)}{4p^2\eta_\alpha\eta_\beta + (\eta_\beta^2 - p^2)^2}. \quad (33)$$

The last form has the advantage that at vertical incidence the vertical slownesses are $\eta_\alpha = 1/\alpha$ and $\eta_\beta = 1/\beta$, whereas r_α and r_β are infinite (Eqn 2.5.36).

These amplitude ratios are the reflection coefficients for the *P* and *SV* potentials. In general, both reflected *P* and *SV* result. At vertical incidence the ray parameter p is zero, and Eqn 33 shows two interesting features. First, none of the incident *P* wave converts to reflected *SV* energy ($B_2 = 0$). Second, the reflected *P* wave is inverted because $A_2/A_1 = -1$. These effects also occur at grazing incidence, $i = 90°$, because η_α is zero.

The ratios of the displacement for the incident *P* and reflected *P* and *SV* waves can be found from the potentials using Eqn 26:

Incident *P*: $(u_x, u_z)_{PI} = (-ik_x, ik_x r_\alpha)\phi_I$

Reflected *P*: $(u_x, u_z)_{PR} = (-ik_x, -ik_x r_\alpha)\phi_R$

Reflected *SV*: $(u_x, u_z)_{SR} = (ik_x r_\beta, -ik_x)\psi_R$. (34)

Because the displacements are real numbers, they can be found by taking the real part of the complex expressions or by adding the complex conjugates.

Using these expressions, the amplitude of any component of the displacement can be found from the potential reflection and transmission coefficients. Thus the ratio of the displacements can differ by either a sign or a scale factor from the potential reflection and transmission coefficients. To see this, consider the ratios of the magnitudes of the displacements. Because the components of the wave vectors for *P* and *SV* waves satisfy

$$k_\alpha = [k_x^2 + (k_x r_\alpha)^2]^{1/2} = \omega/\alpha, \quad k_\beta = [k_x^2 + (k_x r_\beta)^2]^{1/2} = \omega/\beta, \quad (35)$$

the ratio of the magnitudes of the displacements for the reflected and incident *P* waves is

$$\frac{|u|_{PR}}{|u|_{PI}} = \frac{k_\alpha|\phi_R|}{k_\alpha|\phi_I|} = \frac{|A_2|}{|A_1|}, \quad (36)$$

and the ratio of the magnitudes of the reflected *SV* and incident *P* displacements is

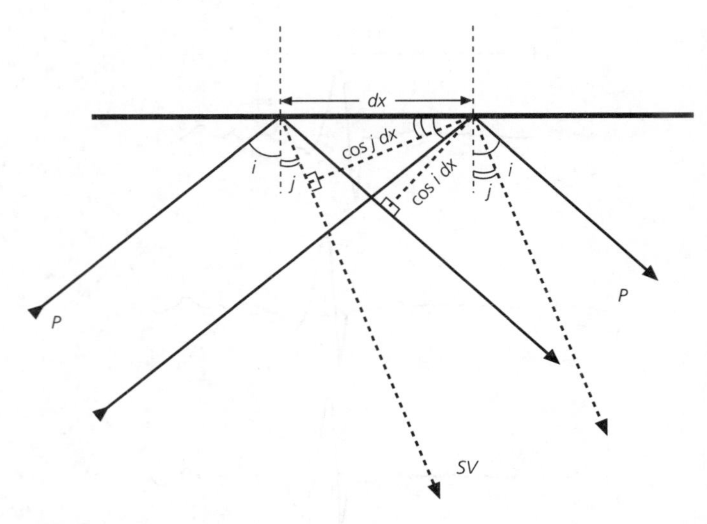

Fig. 2.6-7 The length of the incident and reflected wave fronts contributing to the energy flux at an element *dx* of a free surface depends on the cosine of the angle of incidence for each wave.

$$\frac{|u|_{SR}}{|u|_{PI}} = \frac{k_\beta|\psi_R|}{k_\alpha|\phi_I|} = \frac{\alpha|B_2|}{\beta|A_1|}. \quad (37)$$

We can gain further insight by considering how the incident wave's energy is partitioned between the two reflected waves. From Eqn 2.4.65, a harmonic plane *P* wave has an energy flux in the propagation direction

$$\dot{E} = A^2\omega^2k_\alpha^2\rho\alpha/2, \quad (38)$$

and a similar result applies for an *SV* wave. The lengths of wave fronts contributing to the flux at an element dx of the free surface (Fig. 2.6-7) are $\cos i\, dx$ for the *P* waves and $\cos j\, dx$ for the *S* wave. Thus the energy fluxes for the incident, reflected *P*, and reflected *SV* waves are

$$\dot{E}_{PI} = A_1^2\omega^2k_\alpha^2\rho\alpha\cos i\, dx/2$$

$$\dot{E}_{PR} = A_2^2\omega^2k_\alpha^2\rho\alpha\cos i\, dx/2$$

$$\dot{E}_{SR} = B_2^2\omega^2k_\beta^2\rho\beta\cos j\, dx/2, \quad (39)$$

so the ratios of the reflected energy fluxes to the incident energy flux are

$$\frac{\dot{E}_{PR}}{\dot{E}_{PI}} = \left(\frac{A_2}{A_1}\right)^2 \quad \frac{\dot{E}_{SR}}{\dot{E}_{PI}} = \left(\frac{B_2}{A_1}\right)^2\frac{\alpha\cos j}{\beta\cos i} = \left(\frac{B_2}{A_1}\right)^2\frac{\eta_\beta}{\eta_\alpha}. \quad (40)$$

Because energy does not accumulate at the free surface, these ratios always sum to 1.

Figure 2.6-8 shows an example of reflection coefficients and energy flux ratios as a function of the angle of incidence of the incoming *P* wave. Although there is no reflected *SV* wave at the limits, vertical and grazing incidence, there is a wide range

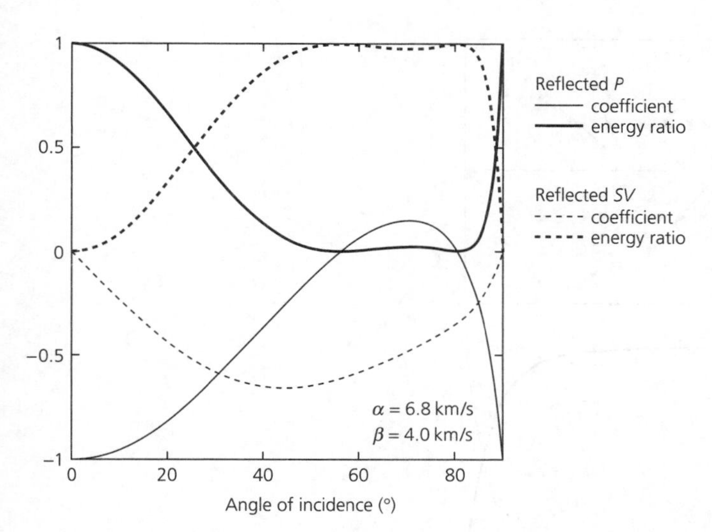

Fig. 2.6-8 For a *P* wave incident on a free surface, potential reflection and transmission coefficients and the ratios of reflected and transmitted energy fluxes to that of the incident wave are shown as functions of the angle of incidence of the incident *P* wave.

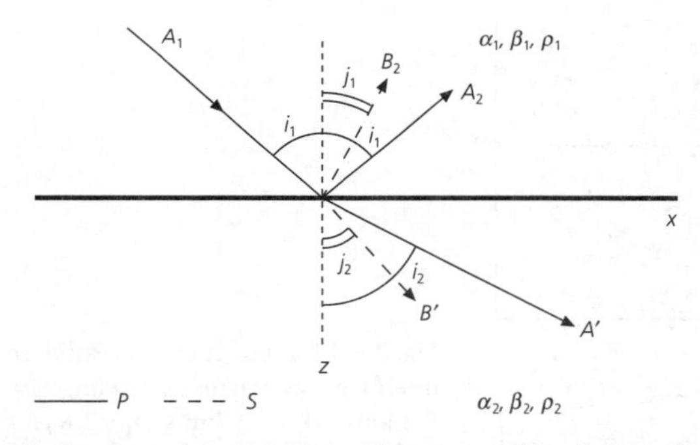

Fig. 2.6-9 Geometry for a *P* wave incident on a solid–solid interface. A_1, A_2, B_2, A', and B' are the amplitudes of the incident *P* wave, the reflected *P* and *SV* waves, and the transmitted *P* and *SV* waves.

of angles over which most of the energy reflects as *SV*. At two angles, the incident *P* wave converts entirely to *SV*.

2.6.6 *Solid–solid and solid–liquid interfaces*

The approach we used for *P–SV* waves at a free surface can be extended to a solid–solid interface. Consider the usual geometry (Fig. 2.6-9) in which *P–SV* waves propagating in the *x–z* plane interact with a horizontal interface at $z = 0$. An incident wave generates two reflected waves and two transmitted waves. The four ratios of the amplitudes of the reflected *P* and *SV* and transmitted *P* and *SV* waves to that of the incident wave are found from the boundary conditions. There are four equations because the *x* and *z* components of the displacement and traction are continuous at the interface. The resulting solutions are complicated and are not given here. Instead, we consider some general principles and examples.

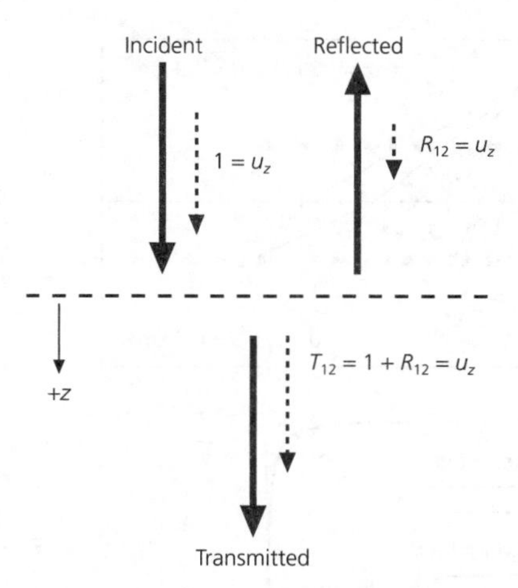

Fig. 2.6-10 Directions of propagation (solid line) and displacement amplitudes (dashes) for vertically incident, reflected, and transmitted *P* waves at a solid–solid interface.

The solutions are simple for vertical incidence. For a vertically incident *P* wave, no *SV* waves are generated. The displacement is only in the *z* direction, and the ratio of the displacement of the transmitted *P* wave to that of the incident wave is

$$\frac{(u_z)_T}{(u_z)_I} = T_{12} = \frac{2\rho_1\alpha_1}{\rho_1\alpha_1 + \rho_2\alpha_2}. \tag{41}$$

The corresponding ratio for the reflected *P* wave is

$$\frac{(u_z)_R}{(u_z)_I} = R_{12} = \frac{\rho_1\alpha_1 - \rho_2\alpha_2}{\rho_1\alpha_1 + \rho_2\alpha_2}. \tag{42}$$

These ratios, the vertical incidence transmission and reflection coefficients for displacements, satisfy $1 + R_{12} = T_{12}$, as required by continuity of displacements. As for the *SH* case (Section 2.6.2), the vertical incidence transmission and reflection coefficients depend only on the acoustic impedances. For the incident *SV* case, no *P* waves are generated, and the ratios of the displacement component u_x have the same form, but in terms of the shear velocity β.

Figure 2.6-10 illustrates an intriguing effect that occurs for a *P* wave vertically incident on an interface where $\rho_1\alpha_1 > \rho_2\alpha_2$, so R_{12} is positive. If the incident *P* wave is a pulse in the $+z$ direction of propagation with unit amplitude, then the reflected *P* wave is a pulse with amplitude R_{12} in the $+z$ direction. Hence the motion in the incident wave is in its direction of propagation ($+z$), whereas the motion in the reflected wave is opposite to its direction of propagation ($-z$). Often the motion in a *P* wave is called *compressional* if it is in the direction of propagation, and *dilatational* if it is opposite the direction of propagation. Thus an incident *P* wave with a compressional motion yields a reflection with dilatational motion. Sometimes the positive amplitude of motion for a *P* wave is defined to be in the

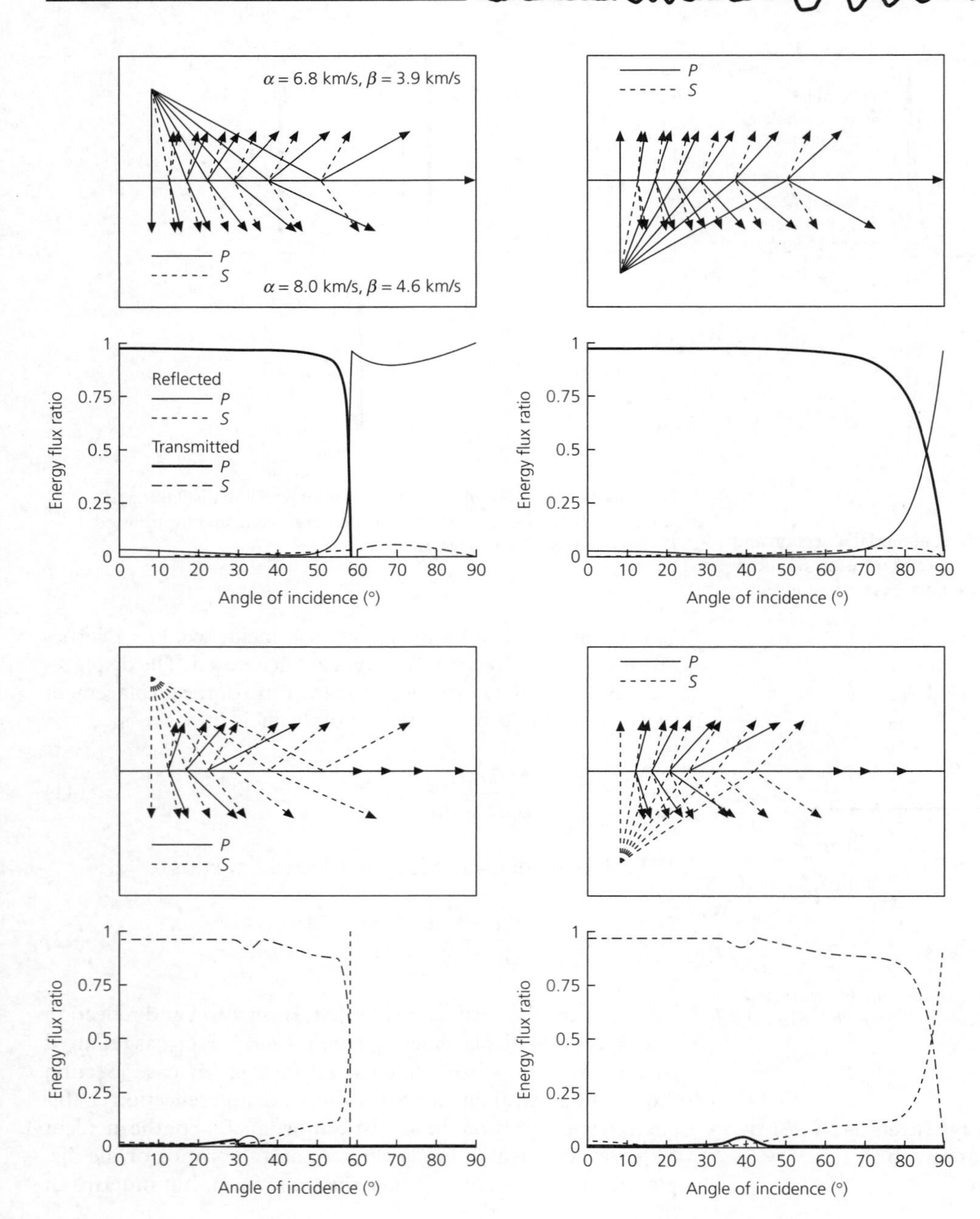

Fig. 2.6-11 Interactions at a solid–solid interface between media having $\alpha_1 = 6.8$ km/s, $\beta_1 = 3.9$ km/s, $\rho_1 = 2.8$ g/cm^3, and $\alpha_2 = 8.0$ km/s, $\beta_2 = 4.6$ km/s, $\rho_2 = 3.3$ g/cm^3. These values correspond approximately to the continental crust and mantle at the Mohorovičić discontinuity. Ray paths and ratios of reflected and transmitted energy fluxes to that of the incident wave are shown as a function of incidence angle for *P* and *SV* waves incident from above and below.

propagation direction, so the reflection coefficient is defined with the opposite sign from Eqn 42.

The amplitudes of the reflected and transmitted waves vary with the angle of incidence, as we illustrate by considering how the energy is partitioned between the four waves. Figure 2.6-11 shows an example for velocities and densities approximating the continental Mohorovičić discontinuity. Ray paths and energy flux ratios for *P* and *SV* waves incident from above and below are plotted. The four ratios are between 0 and 1, and sum to 1 because energy is conserved.

For a *P* wave vertically incident from above, the impedances $\rho_1\alpha_1 = 19.0$, $\rho_2\alpha_2 = 26.4$, yield reflection and transmission coefficients $R_{12} = -0.16$, $T_{12} = 0.84$, and energy flux ratios of

$$\frac{\dot{E}_R}{\dot{E}_I} = R_{12}^2 = 0.03, \quad \frac{\dot{E}_T}{\dot{E}_I} = T_{12}^2 \frac{\rho_2\alpha_2}{\rho_1\alpha_1} = 0.97. \tag{43}$$

These ratios are a good approximation for angles of incidence less than the critical angle $\sin^{-1}(\alpha_1/\alpha_2) = 58°$ because almost all the energy is transmitted as *P*. However, as the angle of incidence approaches the critical value, the transmitted P energy goes to zero, and most of the energy reflects as P. For most postcritical incidence angles, up to ~10% of the energy converts to SV, of which approximately half reflects and half is transmitted. In the limit of grazing incidence, however, all the energy is in the reflected P wave.

For a P wave incident from below, the situation is similar except that there is no critical angle behavior. For vertical incidence, the reflection and transmission coefficients are $R_{21} = 0.16$, $T_{21} = 1.16$, and the energy flux ratios are the same as before, because

$$\frac{\dot{E}_R}{\dot{E}_I} = R_{21}^2 = 0.03, \quad \frac{\dot{E}_T}{\dot{E}_I} = T_{21}^2 \frac{\rho_1 \alpha_1}{\rho_2 \alpha_2} = 0.97. \tag{44}$$

At high angles of incidence, >70°, the energy is increasingly in the reflected P wave.

The behavior of an S wave incident from above is analogous to that for a P wave incident from above. For this example, the S wave impedances are $\rho_1 \beta_1 = 10.9$, $\rho_2 \beta_2 = 15.2$, and the vertical incidence reflection and transmission coefficients are the same as for P waves. Hence at vertical incidence, almost all the energy is transmitted as S, a little reflects as S, and none converts to P. For near-vertical incidence, < ~20°, this pattern changes slowly. At shallower angles of incidence, however, the situation is more interesting, because there are three critical angles. Approaching the critical angle for the transmitted P wave, $\sin^{-1}(\beta_1/\alpha_2) = 29°$, the transmitted P energy increases somewhat. Beyond this angle there is no transmitted P wave, but the reflected P wave behaves in a similar way because it vanishes for $\sin^{-1}(\beta_1/\alpha_1) = 35°$. For larger angles of incidence, only the reflected and transmitted S waves exist, and the energy in the transmitted S wave falls off to zero at the critical angle $\sin^{-1}(\beta_1/\beta_2) = 58°$. Beyond this angle, the incident S wave undergoes total internal reflection.

The final case, an S wave incident from below, is analogous to that for a P wave incident from below. At vertical incidence almost all the energy is transmitted as S, a little is reflected as S, and none converts to P. There is a small reflected P wave near its critical angle, $\sin^{-1}(\beta_2/\alpha_2) = 35°$. More noticeably, the transmitted P wave is enhanced near the critical angle for the S-to-P conversion, $\sin^{-1}(\beta_2/\alpha_1) = 42°$. At higher angles of incidence, the transmitted S wave decreases as the reflected S wave increases.

This example bears out the complexity of interactions at a solid–solid interface. The detailed behavior depends on the four velocities and two densities. A useful approximation is that for media with similar impedances, most of the energy goes into the transmitted wave of the same type (P or S) as the incident wave. This makes sense, because if the materials were identical, all the energy would be transmitted. For a wave incident from a lower-velocity medium, this is approximately the case for angles of incidence less than the critical angle for those two waves. For a wave incident from the higher-velocity medium, most of the energy is transmitted in the same type of wave until near-grazing incidence. Because the incident wave is not seriously affected by small impedance changes, waves propagating through the earth change direction continuously according to Snell's law, but change amplitude significantly only at interfaces where the impedance contrasts are large. If this were not the case, we would not see distinct arrivals.

The approach used for the reflection and transmission coefficients at a solid–solid interface can be extended to a solid–liquid interface. Because there are no shear waves in the liquid, there

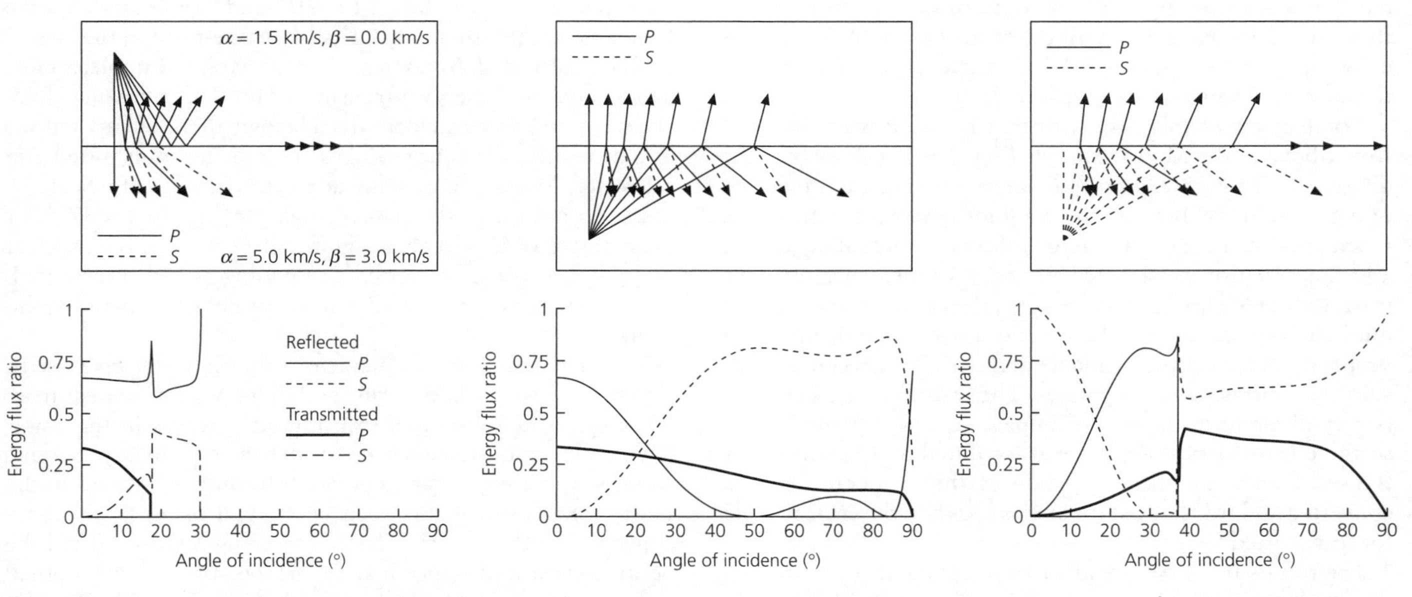

Fig. 2.6-12 Ray paths and energy flux ratios for an interface between the ocean, with $\alpha_1 = 1.5$ km/s, $\beta_1 = 0.0$ km/s, $\rho_1 = 1.0$ g/cm³, and an underlying crust with $\alpha_2 = 5.0$ km/s, $\beta_2 = 3.0$ km/s, $\rho_2 = 3.0$ g/cm³. Three cases, P waves incident from above and P and SV waves incident from below, are shown.

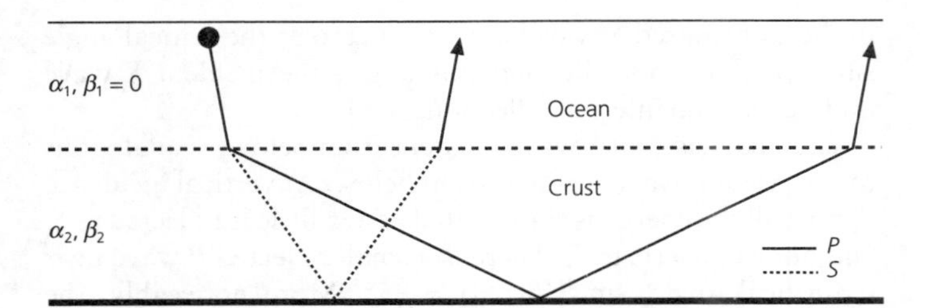

Fig. 2.6-13 Schematic illustration of a marine seismic experiment, in which a P wave generated in the water converts to P and S in the crust. The upgoing crustal S waves partially reconvert to P at the sea floor. Although no S waves travel through the water, the experiment can determine the S-wave properties of the crust. Not all reflected and transmitted waves are shown.

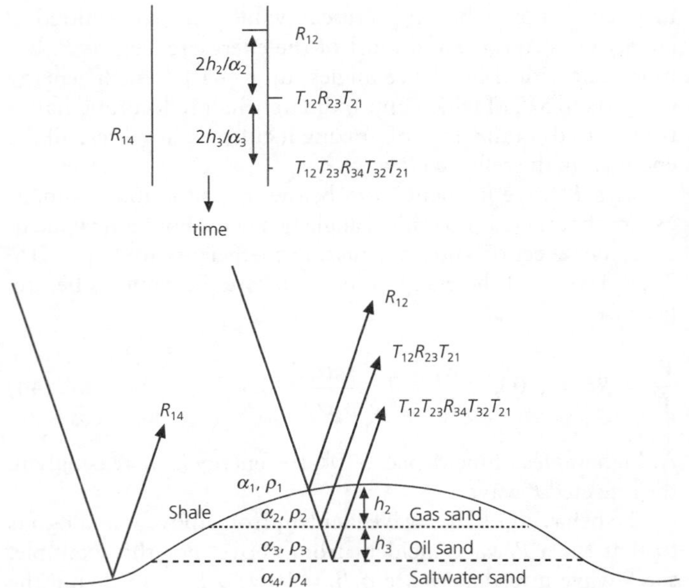

Fig. 2.6-14 Schematic illustration of a seismic reflection experiment, in which vertically incident waves reflect from a region with a variable velocity structure. The vertical ray paths are offset for clarity. The media have $\alpha_1 = 2.6$ km/s, $\rho_1 = 2.5$ g/cm^3, $\alpha_2 = 1.7$ km/s, $\rho_2 = 2.0$ g/cm^3, $\alpha_3 = 2.2$ km/s, $\rho_3 = 2.2$ g/cm^3, $\alpha_4 = 2.3$ km/s, $\rho_4 = 2.3$ g/cm^3. Impulse seismograms showing the arrivals resulting from an incident P-wave pulse of unit amplitude are plotted with time increasing downward. The resulting arrivals have amplitudes $R_{12} = 0.3$, $T_{12}R_{23}T_{21} = -0.2$, $T_{12}T_{23}R_{34}T_{32}T_{21} = -0.02$, and are separated by the time required to traverse the layers. The corresponding reflection from a point to one side of the region has amplitude $R_{14} = 0.1$. (After Dobrin, 1976.)

are three amplitude ratios. Similarly, because the fluid's shear velocity and rigidity are zero, there are three boundary conditions at the interface: continuity of vertical displacement and traction, and vanishing of the shear traction in the solid.

Figure 2.6-12 shows the three possible cases at the sea floor: P waves incident from above and P and SV waves incident from below. Because the impedance contrast at the sea floor is much greater than in the Mohorovičić discontinuity example, the relative amplitudes of the reflected and transmitted waves are quite different from those in Fig. 2.6-11. First, consider a P wave incident from above. At vertical incidence, $R_{12} = -0.82$, $T_{12} = 0.18$, so two-thirds of the incident energy reflects and only one-third is transmitted. As the angle of incidence increases, the fraction of reflected energy remains approximately the same, but the transmitted S wave grows at the expense of transmitted P. The first critical angle behavior occurs for transmitted P near $\sin^{-1}(\alpha_1/\alpha_2) = 17°$. Beyond this angle, a significant transmitted S wave exists until the critical angle for the P-to-S conversion, $\sin^{-1}(\alpha_1/\beta_2) = 30°$. For larger angles of incidence, the incident P wave is totally reflected.

Comparison of this case with that of the P wave incident from above in the Moho example (Fig. 2.6-11) shows several differences. In both examples P waves impinge on a medium of higher velocity. Because the sea floor impedance contrast is much greater, most of the energy reflects at vertical incidence, and this situation persists for all angles of incidence. By contrast, for the Moho example, most of the energy is transmitted until the critical angle. The critical angle for transmitted P occurs for much steeper incidence at the sea floor because the P-velocity contrast is much greater. The transmitted S behavior is very different in the two examples: $\alpha_1 > \beta_2$ for the Moho, so there is no critical angle for transmitted S. By contrast, at the sea floor a significant portion of the incident energy is converted and transmitted for angles less than the critical angle for transmitted S.

The results for waves incident from below also differ significantly between the two examples. A P wave incident on the sea floor from below is primarily reflected downward, largely as a P wave for angles less than ~20°, and largely as an S wave for angles greater than ~30°. Less than one-third of the energy is ever transmitted. By contrast, for the Moho example, almost all the incident P energy is transmitted until near-grazing incidence. For an S wave incident from below, all the energy reflects as S at vertical incidence, because there is no transmitted S in the water. At low angles of incidence, the fraction of reflected P increases until near the critical angle $\sin^{-1}(\beta_2/\alpha_2) = 37°$. For most angles of incidence, a significant portion of the incident upgoing S wave is converted to upgoing P and transmitted. This strong converted transmission does not occur in the Moho example.

The facts that P waves incident from the water give rise to significant S waves in the crust and that S waves incident from the crust yield substantial transmitted P waves in the water have important consequences for marine seismology. Seismic sources in the water can generate transmitted S waves in the crust, whose propagation can be studied using P waves reconverted at the sea floor from upcoming S waves. Thus the oceanic crust and upper mantle can be studied with both P waves and S waves, using sources that generate only P waves and receivers that detect only P waves (Fig. 2.6-13).

2.6.7 *Examples*

Using the amplitudes of reflected, converted, and transmitted waves to study interfaces is common in seismology, as we illustrate with two examples. In reflection seismology, *P* waves generated by near-surface sources and reflected from interfaces at depth are used to study the crust and uppermost mantle. We will see in the next chapter that the downgoing waves impinge on the reflectors at steep angles of incidence, and the data are often processed to simulate vertical incidence. Because the impedance contrasts are small, it is common to neglect *P*-to-*S* conversions and estimate the amplitudes of the reflected and transmitted *P* waves using vertical incidence reflection and transmission coefficients. The reflection and transmission coefficients inferred from seismic data are combined with the travel times to yield information about the subsurface geology.

Consider (Fig. 2.6-14) a hypothetical region where natural gas, oil, and saltwater are trapped in the pores of a sand unit. To describe the response of this region to a *P* wave impulse of unit amplitude, we consider only the first, or *primary*, reflection from each layer, because subsequent *multiple* reflections would be smaller. The resulting arrivals have amplitudes R_{12}, $T_{12}R_{23}T_{21}$, and $T_{12}T_{23}R_{34}T_{32}T_{21}$, and are separated by the time required to traverse the layers. By contrast, the corresponding reflection from a point to one side of the region would have amplitude R_{14}. The lateral variation in impedance contrasts causes a significant difference in the amplitude of the reflected waves.

For a second example, consider the downgoing slab of lithosphere at a subduction zone. As discussed in Chapter 5, the slab is colder than the surrounding mantle, and hence has higher seismic velocity. Seismic waves propagating in several geometries (Fig. 2.6-15) are used to study the upper surface of the slab. In one, *ScS*, an *S* wave reflected at the core–mantle boundary, is partially converted to a *P* wave, *ScSp*, at the slab surface. The ray paths can be found by using Snell's law at the dipping interface. Assume that the downgoing slab and overlying mantle have velocities α_1, β_1 and α_2, β_2 and the slab dips at angle θ. A vertically traveling *ScS* wave impinges on the interface at an angle $j_1 = \theta$, so the angles of incidence for transmitted *ScS* and *ScSp* are $j_2 = \sin^{-1}[(\beta_2/\beta_1)\sin j_1]$ and $i_2 = \sin^{-1}[(\alpha_2/\beta_1)\sin j_1]$. The amplitude of *ScSp* is enhanced because the *ScS* incidence angle is close to the critical angle for the conversion. *ScSp* travels faster than *ScS* and appears at seismometers primarily on the vertical component, whereas *ScS* arrives later and is primarily on the horizontal component. Additional information is obtained from *P* waves that reflect off the interface and appear at seismometers above the subduction zone later and with higher apparent velocity (steeper incidence) than the direct arrival. The travel times and

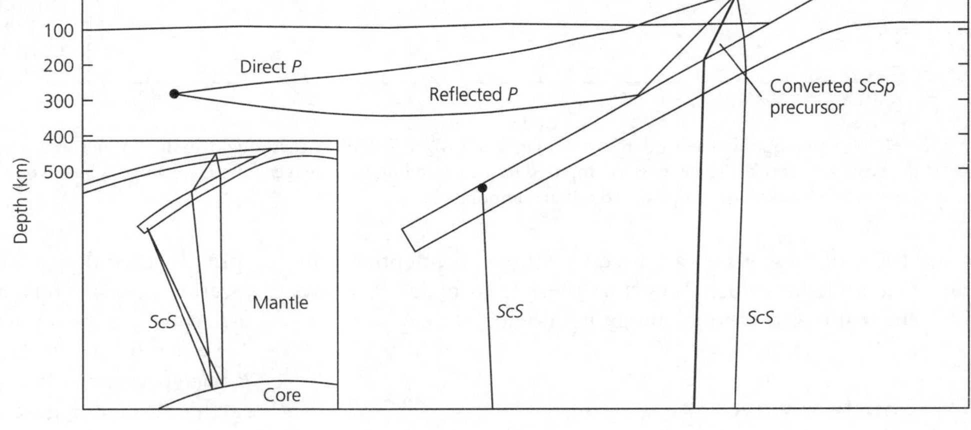

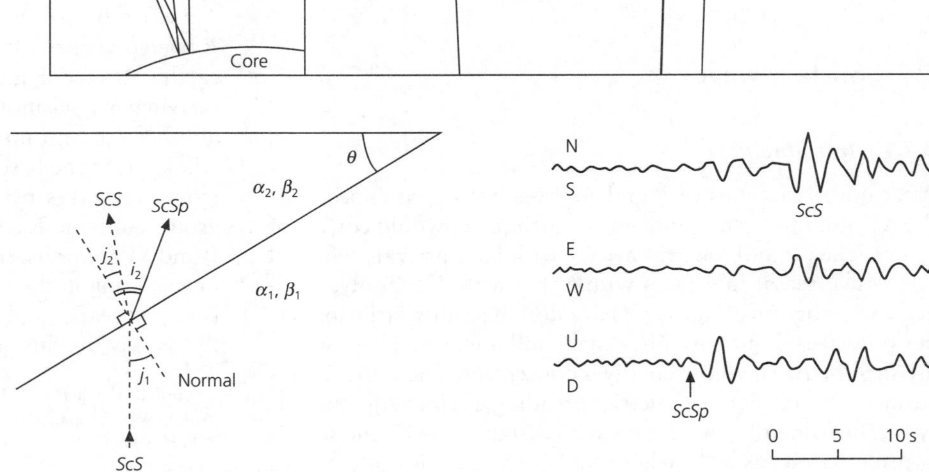

Fig. 2.6-15 Study of a subducting slab using seismic waves reflected and converted at its upper surface. *Top*: Ray paths for the conversion of upcoming *ScS* to *ScSp* and the reflection of *P* waves. (Helffrich *et al.*, 1989. *J. Geophys. Res.*, *94*, 753–63, copyright by the American Geophysical Union.) *Lower left*: Application of Snell's law at the dipping interface for the *ScS* to *ScSp* conversion. *Lower right*: Seismograms showing *ScS* and *ScSp* recorded in Hokkaido, Japan, for an earthquake in Honshu, Japan. *ScSp* arrives on the vertical component before *ScS* appears on the horizontal components. (Snoke *et al.*, 1979.)

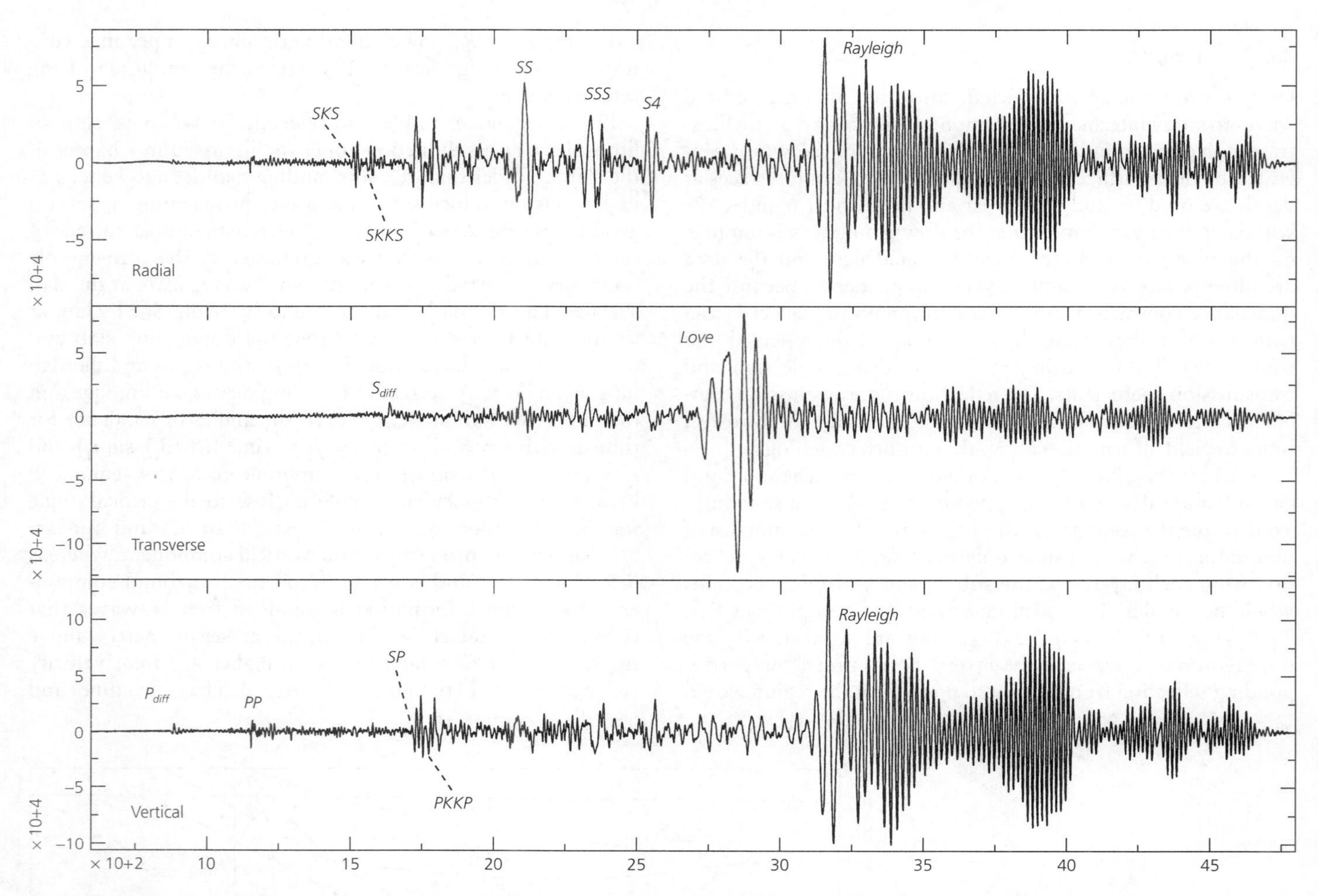

Fig. 2.7-1 Three-component seismogram of a magnitude M_w 7.7 shallow earthquake in the Vanuatu trench recorded 12,250 km away at station CCM. Note the large size of the surface waves compared to the preceding body waves. The Love wave is observed on the transverse component, and the Rayleigh wave is primarily seen on the vertical and radial components.

amplitudes of these waves are used to estimate the depth to the interface and the velocity contrast there, and hence to draw inferences about its thermal and mineralogical state.

2.7 Surface waves

2.7.1 *Introduction*

After our discussions of *P* and *S* waves, we might expect that the seismogram resulting from an earthquake would consist of pulses when *P* and *S* waves arrive, with later arrivals reflected and converted at interfaces within the earth. Generally, however, seismograms (Fig. 2.7-1) are dominated by large longer-period waves that arrive after the *P* and *S* waves. These waves are *surface waves* whose energy is concentrated near the earth's surface. As a result of geometric spreading, their energy spreads two-dimensionally and decays with distance r from the source approximately as r^{-1}, whereas the energy of body waves spreads three-dimensionally and decays approximately as r^{-2} (Section 2.4.3). Thus, at large distances from the source, surface waves are prominent on seismograms.

Two types of surface waves, known as *Love waves* and *Rayleigh waves* after their discoverers,[1] propagate near the earth's surface. Figure 2.7-1 shows a large surface wave train arriving on a seismometer's transverse component, followed by another wave group on the vertical and radial components. We will see that the first wave train contains Love waves resulting from *SH* waves trapped near the surface. The second wave group contains Rayleigh waves, which are a combination of *P* and *SV* motions. In our usual geometry (Fig. 2.7-2) of waves propagating in the x–z plane, the Rayleigh wave displacement is in this plane, and the Love wave displacement is parallel to the y axis. In this section, we examine the simplest cases of

[1] Lord Rayleigh (1842–1919), best known among seismologists for pioneering work in wave propagation, was awarded the Nobel prize for the discovery of argon. A. E. H. Love (1863–1940) made fundamental contributions to both seismology and geodynamics.

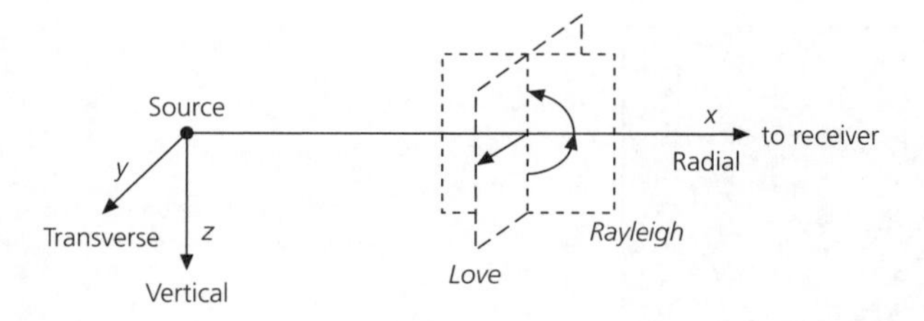

Fig. 2.7-2 Geometry for surface waves propagating in a vertical plane containing the source and receiver. Rayleigh (*P–SV*) waves appear on the vertical and radial components. Love (*SH*) waves appear on the transverse component.

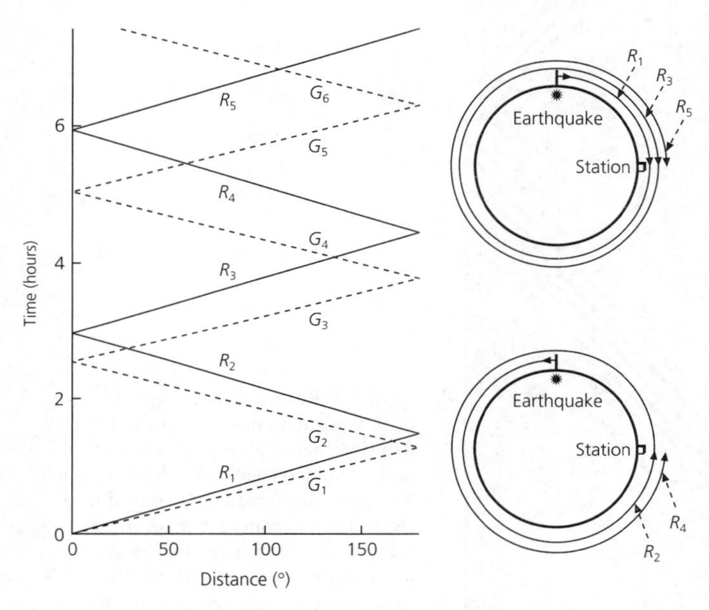

Fig. 2.7-3 Multiple surface waves circle the earth. *Right*: Odd-numbered arrivals (R_1, R_3, etc.) take the shortest path from the earthquake to the station, whereas even-numbered arrivals (R_2, R_4, etc.) travel in the opposite direction. *Left*: Travel times for multiple Rayleigh (R_n) and Love waves (G_n).

Rayleigh and Love waves, and use them to demonstrate some general ideas about surface waves.

An interesting difference between surface and body waves, due to their different rates of decay, is that surface waves can circle the globe many times after a large earthquake. Figure 2.7-3 shows such *multiple surface waves*, which are denoted as Rayleigh waves (R_n) and Love waves (G_n). The travel time plot (Fig. 2.7-3, *left*) illustrates the increasing time required for successive paths, indexed by n, from the earthquake to the station. An important feature of surface waves is *dispersion*, the fact that waves of different periods travel at different velocities. As a result, the surface wave arrivals are not sharp lines, but are spread out in time. These effects are shown in Fig. 2.7-4 (overleaf) by a record section composed of many vertical component seismograms at different distances from earthquakes, which yields an observed travel time plot. The data show the arrivals of R_1, R_2, R_3, and R_4, and a comparable 6-hour plot for the transverse component would show G_1 through G_5.

2.7.2 *Rayleigh waves in a homogeneous halfspace*

Rayleigh waves are a combination of P and SV waves that can exist at the top of a homogeneous halfspace. To describe them, we define the free surface as $z = 0$, measure z downward, and use potentials for waves propagating in the x–z plane. We consider only P and SV waves, because they can satisfy the free surface boundary conditions and do not interact with SH waves. The P and SV potentials are

$$\phi = A \exp\,(i(\omega t - k_x x - k_x r_\alpha z)),$$
$$\psi = B \exp\,(i(\omega t - k_x x - k_x r_\beta z)). \qquad (1)$$

For a combination of these potentials to describe energy trapped near the free surface, two conditions must apply. The solution must both ensure that the energy does not propagate away from the surface and satisfy the free surface boundary conditions.

For the energy to be trapped near the surface, the exponentials $\exp(-ik_x r_\alpha z)$ and $\exp(-ik_x r_\beta z)$ must have negative real exponents, so that the displacement will decay as $z \to \infty$. Because

$$r_\alpha = (c_x^2/\alpha^2 - 1)^{1/2}, \quad r_\beta = (c_x^2/\beta^2 - 1)^{1/2}, \qquad (2)$$

this radiation condition requires that $c_x < \beta < \alpha$, so that both square roots become imaginary, with a choice of sign such that

$$r_\alpha = -i(1 - c_x^2/\alpha^2)^{1/2}, \quad r_\beta = -i(1 - c_x^2/\beta^2)^{1/2}. \qquad (3)$$

Thus c_x, the apparent velocity along the surface, must be less than the shear velocity.

The other condition, the vanishing of traction at the free surface, arose for the P–SV reflection at a free surface (Section 2.6.5). The difference here is that the boundary conditions are satisfied with no incident wave. Using Eqn 2.6.28 without an incident wave shows that when the stress components are expressed in terms of the potentials, the amplitudes A and B must satisfy the continuity equations

$$\sigma_{xz}(x, 0, t) = 0 = 2r_\alpha A + (1 - r_\beta^2)B,$$
$$\sigma_{zz}(x, 0, t) = 0 = [\lambda(1 + r_\alpha^2) + 2\mu r_\alpha^2]A + 2\mu r_\beta B. \qquad (4)$$

Eliminating the Lamé constants from the second equation using $(1 + r_\alpha^2) = c_x^2/\alpha^2$ and the definitions of the velocities α and β gives a system of two homogeneous linear equations for A and B,

$$2(c_x^2/\alpha^2 - 1)^{1/2}A + (2 - c_x^2/\beta^2)B = 0,$$
$$(c_x^2/\beta^2 - 2)A + 2(c_x^2/\beta^2 - 1)^{1/2}B = 0. \qquad (5)$$

This system has nontrivial solutions if the determinant of the system is zero (Section A.4.4), such that

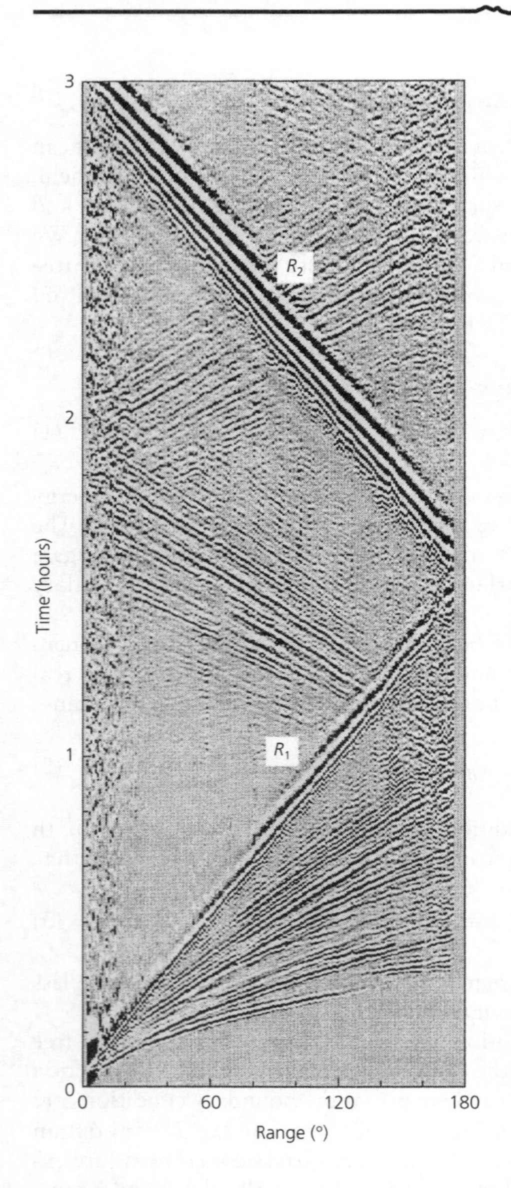

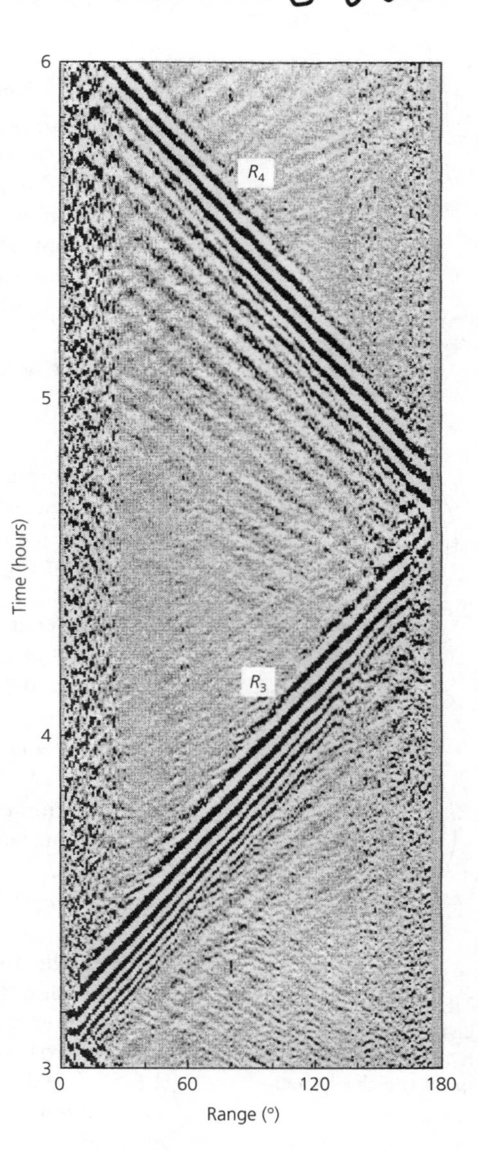

Fig. 2.7-4 Record section formed from vertical seismograms at stations of the IDA (International Deployment of Accelerometers) network. The R_1 through R_4 arrivals are spread out in time due to dispersion and contain lines of energy that cross the largest amplitudes at small angles. As discussed later, the lines show the phase velocity, and the overall amplitude pattern shows the group velocity. Body wave arrivals appear before and after R_1. (Shearer, 1994. *Eos*, *75*, 449, 451, 452. Copyright by the American Geophysical Union.)

$$(2 - c_x^2/\beta^2)^2 + 4(c_x^2/\beta^2 - 1)^{1/2}(c_x^2/\alpha^2 - 1)^{1/2} = 0. \quad (6)$$

For a halfspace with given velocities α and β, this equation gives the values of c_x that satisfy the free surface boundary condition. Of the four roots, one is zero, and only one is consistent with the requirement that $0 < c_x < \beta$. For a Poisson solid, in which $\alpha^2/\beta^2 = 3$, the determinant becomes

$$(c_x^2/\beta^2)[c_x^6/\beta^6 - 8c_x^4/\beta^4 + (56/3)c_x^2/\beta^2 - 32/3] = 0. \quad (7)$$

If we reject the trivial solution $c_x^2/\beta^2 = 0$, the equation is a cubic in c_x^2/β^2, with roots 4, $2 + 2/\sqrt{3}$ (≈ 3.155) and $2 - 2/\sqrt{3}$ (≈ 0.845). Only the last root satisfies $c_x < \beta$, the condition for waves to be trapped at the surface. Thus the apparent velocity of the Rayleigh wave in a halfspace that is a homogeneous Poisson solid is $c_x = (2 - 2/\sqrt{3})\beta = 0.92\,\beta$, slightly less than the shear velocity.

The coefficients of the potentials (Eqn 1), which can be found from Eqn 5, are

$$B = A(2 - c_x^2/\beta^2)/(2r_\beta) \quad (8)$$

and can be substituted into the potentials and used to find the displacements (Eqn 2.6.26). Taking the real parts of the exponentials and using the numerical values of c_x/β and c_x/α for a Poisson solid gives

$$u_x = Ak_x \sin(\omega t - k_x x)[\exp(-0.85\,k_x z) - 0.58\exp(-0.39\,k_x z)],$$
$$u_z = Ak_x \cos(\omega t - k_x x)[-0.85\exp(-0.85\,k_x z) + 1.47\exp(-0.39\,k_x z)]. \quad (9)$$

The displacement can be characterized by its variation in depth and distance along the surface. Both components are

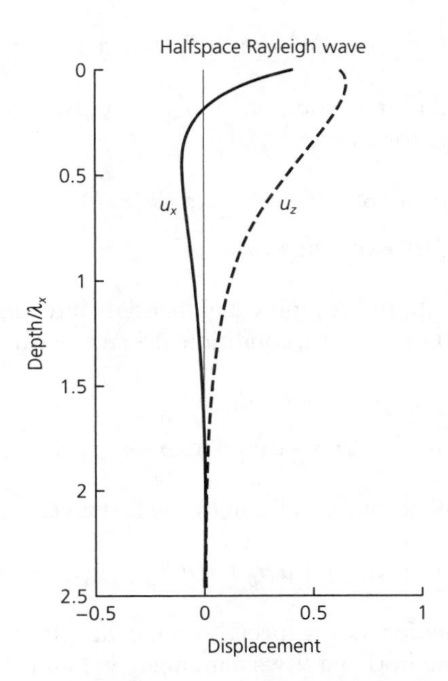

Fig. 2.7-5 Variation with depth of the x and z components of displacement for a Rayleigh wave in a halfspace composed of a Poisson solid. Both components decay with depth, plotted here normalized by the horizontal wavelength.

sinusoidal functions of $(\omega t - k_x x)$, and thus harmonic waves propagating in the $+x$ direction. Because the harmonic wave solution applies only in the x direction, the meaningful wavelength is the horizontal wavelength along the surface, $\lambda_x = 2\pi/k_x$. The displacement decays with depth as $\exp(-k_x z)$ (Fig. 2.7-5), so the depth to which a Rayleigh wave has significant displacement is proportional to its horizontal wavelength.

At the surface, $z = 0$, and the displacement components are

$$u_x = 0.42\, Ak_x \sin(\omega t - k_x x),$$
$$u_z = 0.62\, Ak_x \cos(\omega t - k_x x). \qquad (10)$$

To visualize these, consider the motion of a particle of material at $x = 0$ as a function of time. At $t = 0$, u_z is a maximum (z is positive downward), and $u_x = 0$. As time increases, the x and z displacements combine to give counterclockwise, or "retrograde", motion about an ellipse (Fig. 2.7-6, *left*). For a Poisson solid, the maximum vertical displacement at the surface is about 1.5 times the maximum horizontal displacement. The particle motion becomes "prograde" below a depth of about a fifth of the wavelength, because the decaying exponential term in u_x becomes negative.

The phase relation between the horizontal and vertical components of Rayleigh wave motion can be seen on seismograms, as shown in Fig. 2.7-6 (*right*). When the vertical displacement is at a negative maximum (e.g., about 785 s), the radial displacement is zero, corresponding to $t = 0$ in Fig. 2.7-6 (*left*). A quarter-period later (e.g., about 790 s) the vertical displacement is zero, and the radial displacement is at its positive maximum, corresponding to $t = T/4$.

Rayleigh waves also exist when the medium is more complicated than a homogeneous halfspace. In this case, rather than having a single apparent velocity for all frequencies, c_x is a function of frequency. We illustrate this idea next using Love waves.

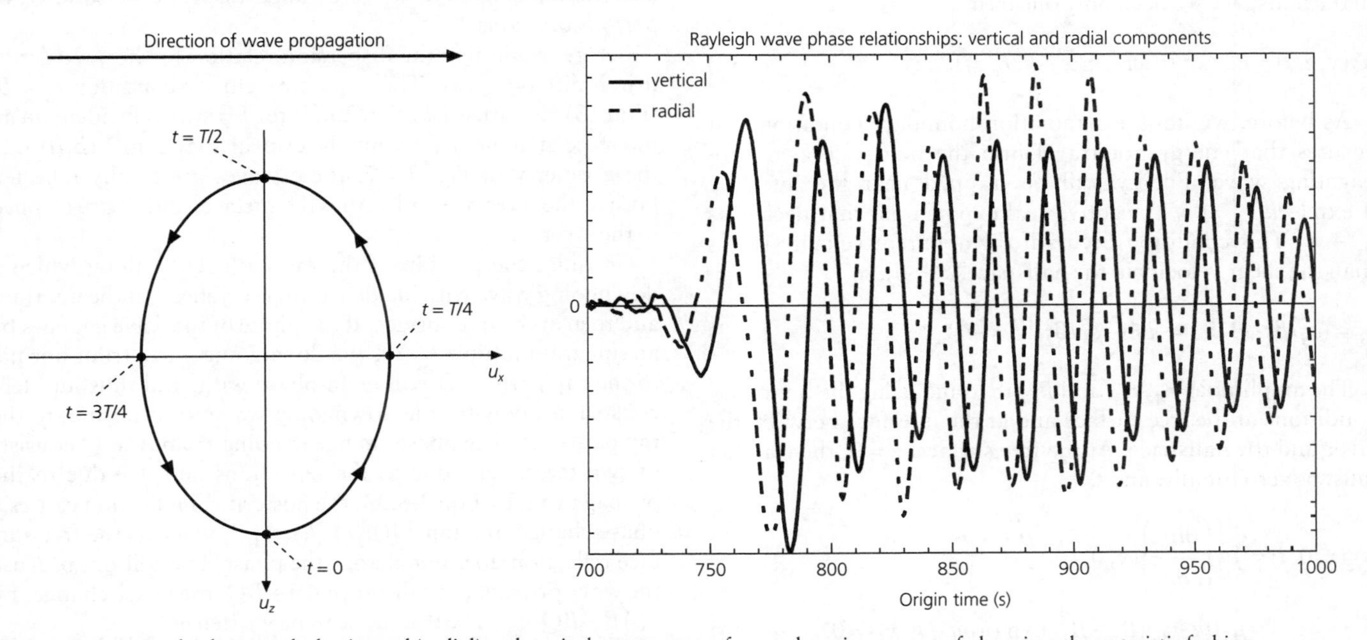

Fig. 2.7-6 For a Rayleigh wave, the horizontal (radial) and vertical components of ground motion are out of phase in a characteristic fashion. *Left*: Because the components are out of phase, the particle motion at a point on the free surface as a function of time is a retrograde ellipse. The particle moves opposite the direction of wave propagation at the top of the ellipse. *Right*: Comparison of the displacement components from seismograms of an earthquake in the Kuril Islands recorded in Micronesia, showing that one peaks when the other is zero.

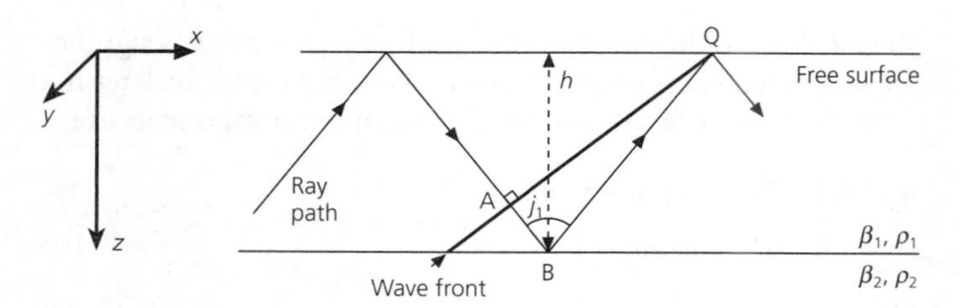

Fig. 2.7-7 Layer over a halfspace geometry for Love waves. Love waves exist if the layer's shear wave velocity is less than the halfspace velocity. The waves can be treated as constructive interference between *SH* waves incident on the interface beyond the critical angle.

2.7.3 *Love waves in a layer over a halfspace*

A second type of surface wave, a Love wave, results from the interactions of *SH* waves. The simplest geometry (Fig. 2.7-7) in which a Love wave occurs is a layer of thickness h of material with velocity β_1, underlain by a halfspace of material with a higher velocity β_2. Love waves require a velocity structure that varies with depth, and so cannot exist in a halfspace, in contrast to Rayleigh waves.

To describe the Love waves, we write the *SH*-wave displacement in the layer as the sum of an upgoing and a downgoing wave:

$$u_y^-(x,z,t) = B_1 \exp(i(\omega t - k_x x - k_x r_{\beta_1} z)) + B_2 \exp(i(\omega t - k_x x + k_x r_{\beta_1} z)). \quad (11)$$

In the halfspace we need only one term:

$$u_y^+(x,z,t) = B' \exp(i(\omega t - k_x x - k_x r_{\beta_2} z)). \quad (12)$$

As before, we impose a radiation boundary condition that ensures that energy not travel into the halfspace as a propagating wave. Energy will be trapped near the interface if $\exp(-ik_x r_{\beta_2} z)$ is a negative real exponential that decays as $z \to \infty$. This condition occurs if the apparent velocity is less than the shear velocity in the halfspace, $c_x < \beta_2$, so

$$r_{\beta_2} = (c_x^2/\beta_2^2 - 1)^{1/2} = -i(1 - c_x^2/\beta_2^2)^{1/2} = -ir_{\beta_2}^*. \quad (13)$$

The amplitudes B_1, B_2, and B' are found using the boundary conditions at the free surface and at the interface between the layer and the halfspace. At the free surface, $z = 0$, the traction must be zero for all x and t,

$$\sigma_{yz}(x,0,t) = \mu_1 \left(\frac{\partial u_y^-}{\partial z}\right)(x,0,t) = \mu_1(ik_x r_{\beta_1})(B_2 - B_1)\exp(i(\omega t - k_x x)) = 0, \quad (14)$$

so $B_1 = B_2$. At the interface $z = h$, the displacement must be continuous for all x and t, so

$$B_1[\exp(-ik_x r_{\beta_1} h) + \exp(ik_x r_{\beta_1} h)] = B' \exp(-ik_x r_{\beta_2} h). \quad (15)$$

Similarly, the stress component σ_{yz} must also be continuous at the interface for all x and t, so

$$\mu_1(-ik_x r_{\beta_1})B_1[\exp(-ik_x r_{\beta_1} h) - \exp(ik_x r_{\beta_1} h)] = \mu_2(-ik_x r_{\beta_2})B' \exp(-ik_x r_{\beta_2} h). \quad (16)$$

By combining the complex exponentials into sine and cosine functions (Eqn A.2.10), conditions 15 and 16 can be written

$$2B_1 \cos(k_x r_{\beta_1} h) = B' \exp(-ik_x r_{\beta_2} h),$$
$$2i\mu_1 r_{\beta_1} B_1 \sin(k_x r_{\beta_1} h) = -\mu_2 r_{\beta_2} B' \exp(-ik_x r_{\beta_1} h). \quad (17)$$

Dividing the second condition by the first gives

$$\tan(k_x r_{\beta_1} h) = (-\mu_2 r_{\beta_2})/(i\mu_1 r_{\beta_1}) = (\mu_2 r_{\beta_2}^*)/(\mu_1 r_{\beta_1}). \quad (18)$$

This equation has a special significance. It gives a relation between the horizontal wavenumber, k_x, and the horizontal apparent velocity, c_x, that must be satisfied for the Love wave to exist. Because $c_x = \omega/k_x$, this means that, for a given horizontal apparent velocity, Love waves must have specific horizontal wavenumbers and thus angular frequencies. Alternatively, for a particular period or angular frequency, Love waves can have only certain horizontal apparent velocities or wavenumbers. Hence different frequencies have different apparent velocities, a phenomenon that is called dispersion. Relations like Eqn 18, which give the apparent velocity, c_x, as a function of ω or k_x, are called *dispersion relations*, or *period equations*.

Before examining the dispersion relation further, we derive it in a different way. The apparent velocity condition $c_x < \beta_2$ (Eqn 13) also arose (Section 2.6.4) for *SH* waves incident on an interface at angles exceeding the critical angle, $\sin^{-1}(\beta_1/\beta_2)$. In the geometry of Fig. 2.7-7, these waves are totally reflected both at the interface and at the free surface, and so are trapped in the layer.

Consider the portion of the ray path ABQ along which a downgoing wave with incidence angle j_1 reflects at the interface and then at the free surface. If the phase of the wave changes by an integral multiple of 2π, the downgoing wave front normal to the ray path at Q will be in phase with, and thus interfere constructively with, the downgoing wave front normal to the ray path at A. The phase change in going from A to Q consists of two terms, one due to the reflections and one due to the propagation. By Eqn 2.6.23, the postcritical reflection causes a phase change of $2\tan^{-1}[(\mu_2 r_{\beta_2}^*)/(\mu_1 r_{\beta_1})]$, whereas the free surface reflection does not change the phase. In addition, because the wave propagated a distance $AB + BQ$, the phase changes by $-(AB + BQ)k_{\beta_1}$. The distance can be written as

$$AB + BQ = BQ\cos 2j_1 + h/\cos j_1 = (\cos 2j_1 + 1)(h/\cos j_1) = 2h\cos j_1, \quad (19)$$

using $\cos 2j_1 = 2\cos^2 j_1 - 1$. The condition for constructive interference is thus that the total phase change

$$-2k_{\beta_1} h \cos j_1 + 2\tan^{-1}[(\mu_2 r^*_{\beta_2})/(\mu_1 r_{\beta_1})] = 2n\pi, \tag{20}$$

or, because $\tan(n\pi) = 0$,

$$\tan(k_{\beta_1} h \cos j_1) = \tan(k_x r_{\beta_1} h) = (\mu_2 r^*_{\beta_2})/(\mu_1 r_{\beta_1}). \tag{21}$$

Thus the Love wave dispersion relation that we derived from the boundary conditions can also be viewed as an interference criterion for critically reflected *SH* waves, corresponding to propagating waves in the layer and an evanescent wave in the higher-velocity halfspace.

2.7.4 Love wave dispersion

The dispersion relation (Eqn 21) can be written as a function of any two of the three related parameters c_x, ω, and k_x. To find solutions, we write it in terms of frequency and apparent velocity as

$$\tan[(\omega h/c_x)(c_x^2/\beta_1^2 - 1)^{1/2}] = \frac{\mu_2(1 - c_x^2/\beta_2^2)^{1/2}}{\mu_1(c_x^2/\beta_1^2 - 1)^{1/2}}. \tag{22}$$

Because the tangent function is defined for real values, the square roots must be real, so the apparent velocity is bounded by $\beta_1 < c_x < \beta_2$. A graphical solution can be derived by defining a new variable,

$$\zeta = (h/c_x)(c_x^2/\beta_1^2 - 1)^{1/2}, \tag{23}$$

so that over the allowable range of the apparent velocity, $\zeta = 0$ at $c_x = \beta_1$, and $\zeta_{max} = h(1/\beta_1^2 - 1/\beta_2^2)^{1/2}$ at $c_x = \beta_2$. Hence, Eqn 23 becomes

$$\tan(\omega\xi) = \left(\frac{\mu_2(1 - c_x^2/\beta_2^2)^{1/2}}{\mu_1}\right)\left(\frac{h}{c_x\xi}\right). \tag{24}$$

As shown in Fig. 2.7-8, the left side of the equation, $\tan(\omega\zeta)$, has zeroes at $\zeta = n\pi/\omega$ and goes to infinity at $\zeta = \pi/2\omega$, $3\pi/2\omega$, etc. The right side of the equation, which has a hyperbolic appearance because of the $1/\zeta$ dependence, is infinite for $c_x = \beta_1$, where $\zeta = 0$, and decays monotonically to zero at $c_x = \beta_2$, where $\zeta = \zeta_{max}$. Solutions exist where the two curves intersect, giving the values of ζ and thus c_x for which a Love wave with a given ω occurs. The solutions are called *modes*, so that for a given frequency there are several modes, each with a different apparent velocity. The leftmost solution, with the lowest c_x, is called the *fundamental mode*; the others are *higher modes*, or *overtones*, numbered 1 through *n*.

Figure 2.7-8 illustrates Eqn 24 for three different periods using a model for the continental crust and mantle of a 40 km-thick layer with $\beta_1 = 3.9$ km/s and $\rho_1 = 2.8$ g/cm^3 underlain by a halfspace with $\beta_2 = 4.6$ km/s and $\rho_2 = 3.3$ g/cm^3. For waves with a period of 5 s, there are three solutions within the allowed apparent velocity range: $c_x = 3.92, 4.13$, and 4.55 km/s.

Consider now what happens for longer periods or lower frequencies. The zeroes of the tangent curve $\zeta = n\pi/\omega$ increase, so the spacing between the tangent curves, π/ω, also increases. As a result, there are fewer tangent curves within the allowable

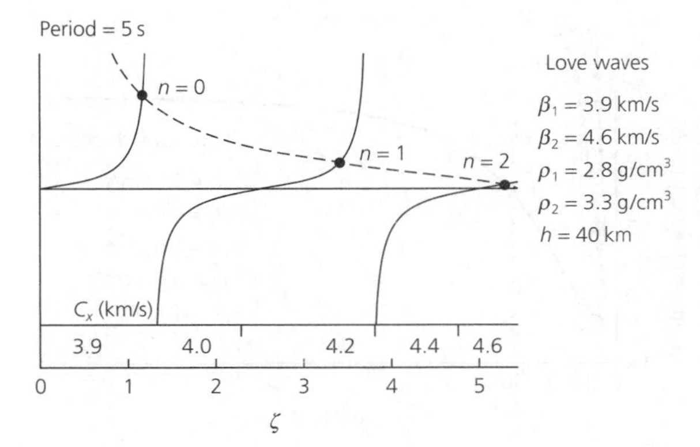

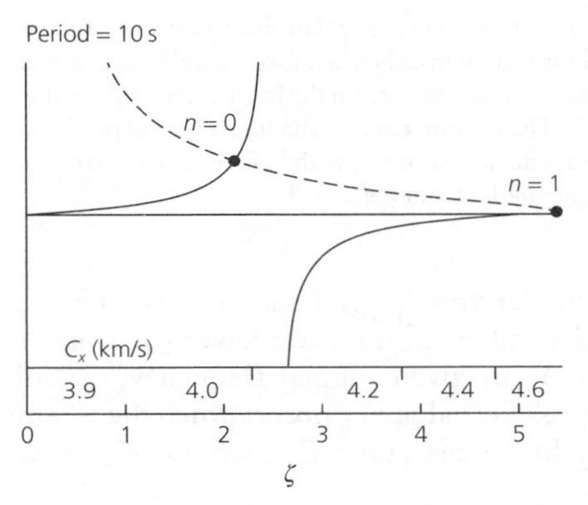

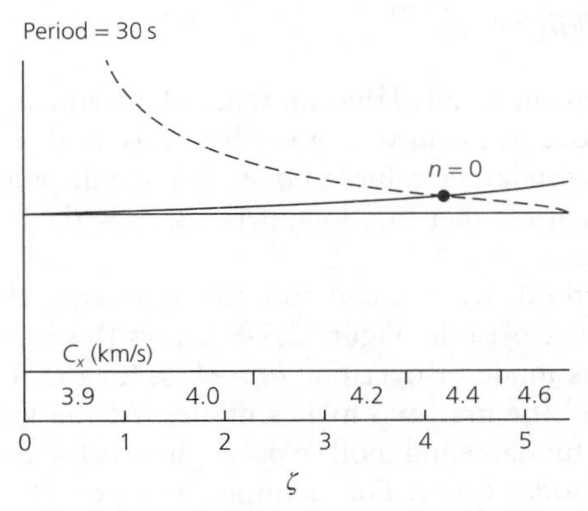

Fig. 2.7-8 Graphical solution of the dispersion relation for Love waves in a layer over a halfspace. The left side of Eqn 24 is represented by the solid curves, $\tan(\omega\zeta)$, with zeroes at $n\pi/\omega$. The decreasing dashed hyperbolas represent the right side of Eqn 24. The intersections of the curves (dots) are the roots of the equation and give the apparent velocities for a given period. The apparent velocities range between the shear velocities of the layer (β_1) and the halfspace (β_2). For longer periods there are fewer solutions and thus fewer modes.

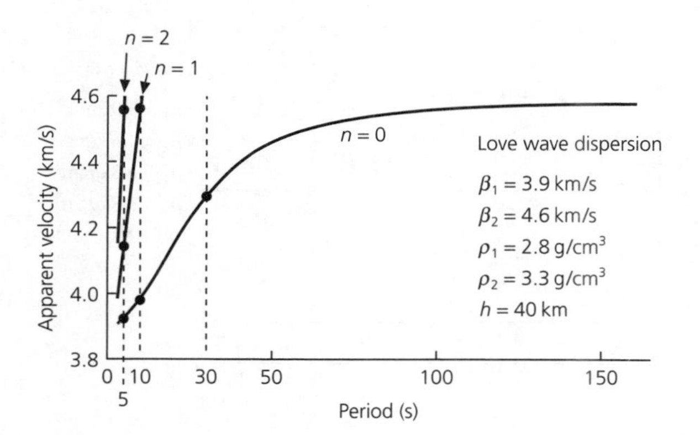

Fig. 2.7-9 Dispersion curves giving the relationship between apparent velocity and period for Love waves in a layer over a halfspace. For each mode, the apparent velocities range from the layer velocity β_1 to the halfspace velocity β_2. The bottom curve is the fundamental mode branch, and the overtone branches are above it, with higher velocities for any period. Dots show the modes from Fig. 2.7-8.

range of ζ, which is $n\pi/\omega < \zeta_{max}$. Thus, because the decaying curve does not depend on ω, there are fewer solutions, c_x, for longer periods. For any given angular frequency, the solution with the largest possible value of ζ occurs when the nth solution is ζ_{max}, so $c_x = \beta_2$. In this case, $\tan \omega\zeta_{max} = 0$, so $\omega\zeta_{max} = n\pi$, and

$$\omega = \omega_{cn} = n\pi/[h(1/\beta_1^2 - 1/\beta_2^2)^{1/2}]. \tag{25}$$

This angular frequency, called the cutoff angular frequency for the nth higher mode, is the lowest ω at which this mode exists. Tangent curves with larger values of n are beyond the allowed range of ζ. Thus, for sufficiently long periods, only the fundamental mode exists.

Using this method, we can compute the apparent velocity values for different periods. Figure 2.7-9 shows the resulting curves, known as mode or overtone *branches*, for the fundamental mode and the first two higher modes. At the longest periods only the fundamental mode exists, whereas for shorter periods higher modes occur. For example, at a period of 5 s there are three modes, for 10 s there are two modes, but at 30 s only the fundamental mode occurs. The longest-period modes for each branch have $c_x \to \beta_2$, so their apparent velocity depends on the shear velocity in the halfspace and is essentially unaffected by the shear velocity in the layer. Thus at long periods the branches in Fig. 2.7-9 approach the velocity in the halfspace, $\beta_2 = 4.6$ km/s. Similarly, the shortest-period modes for each branch have $c_x \to \beta_1 = 3.9$ km/s, so their apparent velocity approaches the layer velocity.

This variation in apparent velocity reflects differences in displacement among the modes. In the layer, because the amplitudes B_1 and B_2 of the upgoing and downgoing waves are equal, the displacement (Eqn 11) can be written

$$u_y^-(x, z, t) = 2B_1 \exp(i(\omega t - k_x x)) \cos(k_x r_{\beta_1} z). \tag{26}$$

In the halfspace, the displacement (Eqn 12) is

$$u_y^+(x, z, t) = B' \exp(i(\omega t - k_x x)) \exp(-k_x r_{\beta_2}^* z), \tag{27}$$

so, by the continuity of displacement at the interface $z = h$,

$$B' = 2B_1 \cos(k_x r_{\beta_1} h)/\exp(-k_x r_{\beta_2}^* h). \tag{28}$$

Thus, in both the layer and the halfspace, we have a wave propagating in the x direction, with horizontal wavenumber $k_x = 2\pi/\lambda_x = \omega/c_x$. In the layer, the displacement varies with depth as $\cos(k_x r_{\beta_1} z)$, and so oscillates. In the halfspace, the displacement decays exponentially with depth as $\exp(-k_x r_{\beta_2}^* z)$.

The variation in displacement in the x and z directions is illustrated in Fig. 2.7-10 for the three periods whose apparent velocities were found in Fig. 2.7-8. The horizontal variation is shown in the upper panels. Because the apparent velocity increases with period (Fig. 2.7-9), the horizontal wavelength increases with period for a given branch. Thus, for the fundamental mode ($n = 0$) cases shown, the longest period (30 s) has the highest apparent velocity and thus the longest horizontal wavelength. At a given period (Fig. 2.7-9), the higher the mode, the higher the apparent velocity, and thus the longer the horizontal wavelength. Hence for the three modes shown for period 5 s, $n = 2$ has the longest horizontal wavelength.

The variation with depth, known as the mode's vertical *eigenfunction*, is different for each mode. For a given branch, the depth of penetration in the halfspace increases with period, so, of the fundamental mode periods shown, the longest (30 s) "sees" deepest into the higher velocity halfspace, and thus has the highest apparent velocity. Conversely, the shortest period modes on a given branch penetrate to the shallowest depth, and thus have the lowest apparent velocity. At a given period, the higher modes oscillate more rapidly with depth in the layer, and so change sign more frequently. In the halfspace, however, the higher modes decay more slowly and penetrate deeper. The eigenfunction for a mode with order n has n zero crossings, or nodes, with depth.

The fact that the displacement behaves differently with depth for various modes and periods makes Love waves dispersive. In our derivation, the *intrinsic* shear velocities of the layer and halfspace do not depend on frequency. Nonetheless, the resulting *apparent* velocity along the free surface depends on frequency. This dispersion results from the fact that Love waves of different periods have different displacements with depth, and the intrinsic medium velocity varies with depth. As a result, surface wave dispersion is valuable for studying earth structure.

By contrast, the halfspace Rayleigh wave does not show this dispersion. This wave is a "true" surface wave because it can exist in a homogeneous halfspace due to the interaction of P and SV waves. By contrast, the Love wave in a layer over a halfspace exists because the properties of the medium vary with depth, and so cause interference between SH waves. Dispersive Love waves and Rayleigh waves also occur in media whose properties vary with depth in a more complicated way. The dispersion curves for Love and Rayleigh waves in such media can be calculated by several methods. One approach is to extend the method used in Section 2.7.3 by treating the medium as a set of homogeneous layers underlain by a halfspace. As for the

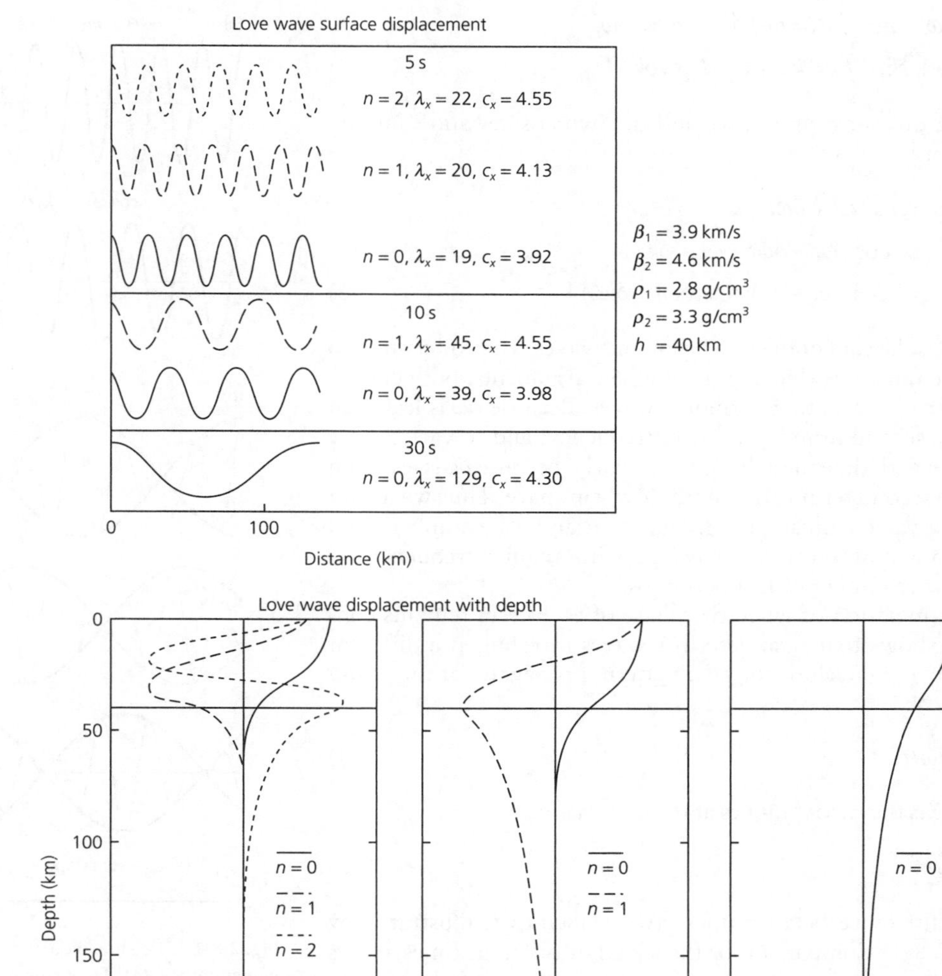

Fig. 2.7-10 Variation in displacement along the surface (*top*) and as a function of depth (*bottom*) for Love waves in a layer over a halfspace. The figure shows the modes for the three periods from Figs 2.7-8 and 9.

one layer case, we assume that the displacement in each layer is given by the exponential solutions, and find combinations of frequency and horizontal apparent velocity that satisfy the boundary conditions at the free surface, at each layer boundary, and in the halfspace. Another approach is to view surface waves as the normal modes of the spherical earth (Section 2.9).

2.8 Dispersion

2.8.1 *Phase and group velocity*

In the last section, we saw that the Love wave was dispersive, because its apparent velocity along the surface varied with frequency. To explore dispersion further, we first consider the simplest example, the net effect of two harmonic waves with slightly different frequencies and wavenumbers. We next consider dispersion in general terms, and discuss some features of surface wave and tsunami dispersion.

Consider the sum of two harmonic waves with slightly different angular frequencies and wavenumbers

$$u(x, t) = \cos(\omega_1 t - k_1 x) + \cos(\omega_2 t - k_2 x). \quad (1)$$

The angular frequencies and wavenumbers can be written in terms of the differences from their average values ω and k:

$$\omega_1 = \omega + \delta\omega, \quad \omega_2 = \omega - \delta\omega, \quad \omega >> \delta\omega,$$
$$k_1 = k + \delta k, \quad k_2 = k - \delta k, \quad k >> \delta k. \tag{2}$$

Using this substitution, we add the two cosines and simplify, yielding

$$u(x, t) = \cos(\omega t + \delta\omega t - kx - \delta kx) + \cos(\omega t - \delta\omega t - kx + \delta kx) = 2\cos(\omega t - kx)\cos(\delta\omega t - \delta kx). \tag{3}$$

Thus the sum of the two harmonic waves is a product of two cosine functions (Fig. 2.8-1). By their arguments, both correspond to propagating harmonic waves. Because $\delta\omega$ is less than ω, the second term has a lower frequency, and so varies more slowly with time than the first. Similarly, because δk is less than k, the second term varies more slowly in space. Thus we have a *carrier* wave with angular frequency ω and wavenumber k, on which a slower varying *envelope* with angular frequency $\delta\omega$ and wavenumber δk is superimposed.[1]

Examination of when the phase of each term remains constant shows that each describes waves traveling at a different speed. The envelope, or beat pattern, propagates at the *group velocity*

$$U = \delta\omega / \delta k, \tag{4}$$

whereas the carrier moves at the *phase velocity*,

$$c = \omega / k. \tag{5}$$

The difference between these two velocities is illustrated by Fig. 2.8-1. Comparison of the signal at different times shows that the envelope propagates at a different speed from the carrier. This difference explains why in the surface wave data of Fig. 2.7-4 individual lines had a slope (phase velocity) differing from the slope (group velocity) of the overall wave pattern.

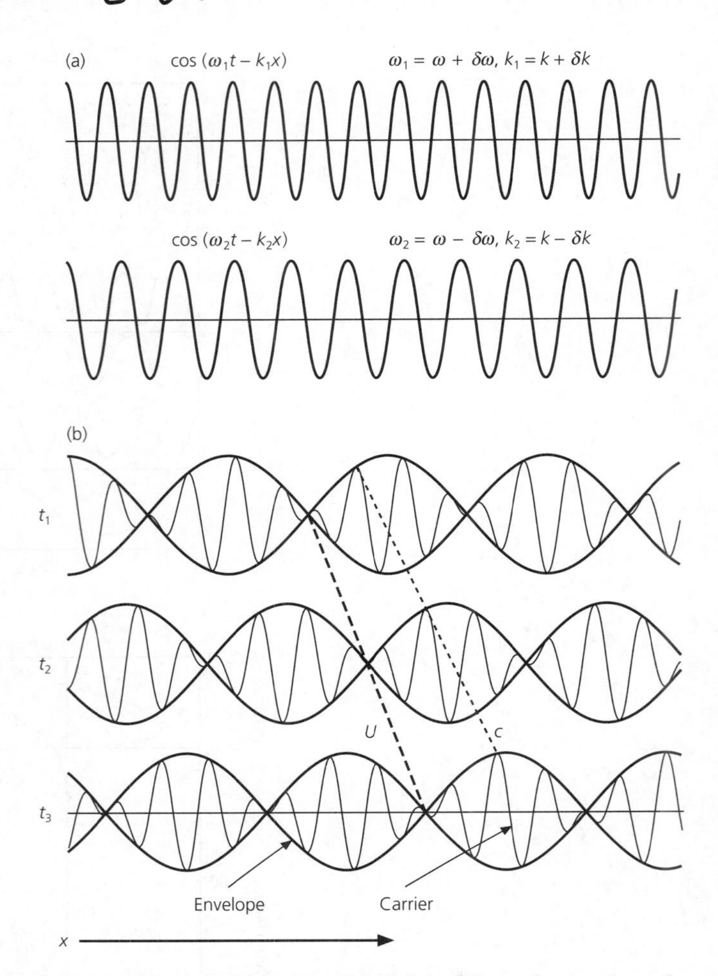

Fig. 2.8-1 Two sinusoidal waves with slightly different frequencies and wavenumbers (a). Their sum as a function of time (b) yields a beating pattern, or long-period envelope, which propagates at the group velocity, U. The carrier, the high-frequency oscillation whose amplitude is modulated by the envelope, propagates at the phase velocity, c.

2.8.2 Dispersive signals

Because dispersive waves of different frequencies propagate at different speeds, this process is best viewed by using Fourier analysis to decompose a wave into the frequencies that compose it. Hence, although we discuss Fourier analysis in Chapter 6, we introduce some key concepts here without proof. For a function of time $f(t)$, multiplication by the complex exponential $e^{-i\omega t}$ and integration over all time yields a function of angular frequency ω:

$$F(\omega) = \int_{-\infty}^{\infty} f(t)e^{-i\omega t}dt \tag{6}$$

known as the *Fourier transform* of $f(t)$. Because the integral involves a complex exponential, $F(\omega)$ is generally a complex function. Similarly, $f(t)$ and $F(\omega)$ are related by the *inverse Fourier transform*:

$$f(t) = \frac{1}{2\pi}\int_{-\infty}^{\infty} F(\omega)e^{i\omega t}d\omega. \tag{7}$$

Thus the time function $f(t)$ can be written as an integral over angular frequency of the complex exponentials $e^{i\omega t}$, weighted by the value of the transform at that angular frequency, $F(\omega)$. Because the Fourier transform is complex, it can be written

$$F(\omega) = A(\omega)e^{i\phi(\omega)} \tag{8}$$

in terms of its magnitude, $A(\omega) = |F(\omega)|$, and phase, $\phi(\omega)$. Thus the Fourier transform represents a time series by two real

[1] This derivation also describes the amplitude modulation (AM) transmission method used in radio, where the amplitude of the carrier is changed or *modulated* by the envelope, the signal of interest.

functions of angular frequency: the *amplitude spectrum*, $A(\omega)$, and the *phase spectrum*, $\phi(\omega)$.

The inverse Fourier transform lets us express a displacement field $u(x, t)$ as an integral over harmonic plane waves of all frequencies

$$u(x,t) = \frac{1}{2\pi}\int_{-\infty}^{\infty} A(\omega) \exp i[\omega t - k(\omega)x + \phi_i(\omega)]d\omega. \tag{9}$$

In this formulation, the wavenumber $k(\omega)$ and the amplitude $A(\omega)$ of each harmonic plane wave are functions of the angular frequency. At each angular frequency, the phase

$$\Phi(\omega) = \omega t - k(\omega)x + \phi_i(\omega) \tag{10}$$

has two parts. The term $\omega t - k(\omega)x$ gives the variation in the phase due to the propagation of the harmonic wave. Hence, as shown in Fig. 2.2-3, the propagation depends on both time (ωt) and space ($k(\omega)x$). Surfaces of constant phase travel with a *phase velocity*

$$c(\omega) = \omega / k(\omega) \tag{11}$$

that may vary as a function of angular frequency. The other phase term, $\phi_i(\omega)$, includes effects such as the initial phase of the wave when it was generated by a seismic source, which depends on the earthquake focal mechanism.

If the harmonic waves of different angular frequencies making up the displacement (Eqn 9) propagate with different phase velocities, the velocity at which a *wave group* propagates differs from the phase velocity at which individual harmonic waves travel. To find the group velocity of energy propagation in the angular frequency band between $\omega_0 - \Delta\omega$ and $\omega_0 + \Delta\omega$, we first approximate the wavenumber $k(\omega)$ by the first term of a Taylor series about ω_0,

$$k(\omega) \approx k(\omega_0) + \left.\frac{dk}{d\omega}\right|_{\omega_0} (\omega - \omega_0). \tag{12}$$

Substituting Eqn 12 in the inverse Fourier transform (Eqn 9) shows that the displacement due to harmonic waves with angular frequencies near ω_0 can be approximated by

$$u(x, t) \approx \frac{1}{2\pi}\int_{\omega_0-\Delta\omega}^{\omega_0+\Delta\omega} A(\omega) \exp\left[i\left(\omega t - k(\omega_0)x - \left.\frac{dk}{d\omega}\right|_{\omega_0}(\omega - \omega_0)x + \phi_i(\omega)\right)\right] d\omega. \tag{13}$$

Adding and subtracting $\omega_0 t$ and regrouping gives

$$u(x, t) \approx \frac{1}{2\pi}\int_{\omega_0-\Delta\omega}^{\omega_0+\Delta\omega} A(\omega) \exp\left[i\left((\omega - \omega_0)\left(t - \left.\frac{dk}{d\omega}\right|_{\omega_0} x\right) + (\omega_0 t - k(\omega_0)x) + \phi_i(\omega)\right)\right] d\omega. \tag{14}$$

The argument of the exponential has three terms, the first two of which describe traveling waves. The second term, $(\omega_0 t - k(\omega_0)x)$, describes a wave with average angular frequency ω_0 propagating at the phase velocity $c(\omega_0) = \omega_0 / k(\omega_0)$. By contrast, the first term describes a wave group with average angular frequency ω_0 propagating at a group velocity $U(\omega_0)$ given by the condition that

$$t - \left.\frac{dk}{d\omega}\right|_{\omega_0} x \tag{15}$$

remain constant, so

$$U(\omega_0) = \left(\left.\frac{dk}{d\omega}\right|_{\omega_0}\right)^{-1} = \left.\frac{d\omega}{dk}\right|_{\omega_0}. \tag{16}$$

If the signal has energy over a wide range of angular frequencies, similar expansions for each angular frequency band give the group velocity as a function of angular frequency

$$U(\omega) = \frac{d\omega}{dk}. \tag{17}$$

Although the group velocity can always be defined by Eqn 17, it does not always yield the velocity of energy propagation as a function of angular frequency. For example, if the wavenumber is a very rapidly varying function of angular frequency, then using only the first two terms in the Taylor series (Eqn 12) may not be adequate, and Eqn 17 may yield negative group velocities. In this case, the group velocity is no longer a useful concept. Fortunately, these approximations are generally valid for seismic surface waves.

At any angular frequency, the group velocity is related to the phase velocity by

$$U = \frac{d\omega}{dk} = \frac{d(ck)}{dk} = c + k\frac{dc}{dk}. \tag{18}$$

It is sometimes easier to think in terms of wavelength, restating Eqn 18 as

$$U = c - \lambda\frac{dc}{d\lambda}. \tag{19}$$

If a wave is not dispersive, different wavelengths travel at the same phase velocity, so $dc/d\lambda = 0$, and the phase and group velocities are equal.

For a dispersive wave, such as the Love wave in the previous section, the group velocity can be found from the dispersion relation. If the dispersion relation is

$$f(\omega, k) = 0, \tag{20}$$

then the change in f for a small change in ω and k is given by the Taylor series,

$$f(\omega + d\omega, k + dk) = f(\omega, k) + \left.\frac{\partial f}{\partial \omega}\right|_k d\omega + \left.\frac{\partial f}{\partial k}\right|_\omega dk. \tag{21}$$

Because ω and k define a mode, they satisfy the dispersion relation, $f(\omega, k) = 0$. If $\omega + d\omega$, $k + dk$, is also a solution, then $f(\omega + d\omega, k + dk)$ must also be zero, so the group velocity is given by

$$U = \frac{d\omega}{dk} = -\left(\frac{\partial f}{\partial k}\right)_\omega \bigg/ \left(\frac{\partial f}{\partial \omega}\right)_k. \tag{22}$$

2.8.3 *Surface wave dispersion studies*

It is useful to distinguish two types of dispersion. The familiar case is that of light, where the different frequencies travel through material such as a lens or a prism at different speeds. This phenomenon, known as *physical dispersion*, occurs in the earth but is a small effect (Section 3.7). In seismology, a more significant effect is that shown for Love waves in the previous section, where the apparent velocity along the surface varied with frequency although the intrinsic shear wave velocity in the layer and the halfspace did not. This type of dispersion, called *geometrical dispersion*, is noticeable and is frequently studied for surface waves. Because for surface waves the horizontal apparent velocity, c_x, and wavenumber, k_x, vary with frequency, these are sometimes written simply as c and k. Similarly, we usually speak of "phase velocity" or "group velocity" when we mean horizontal apparent phase or group velocity.

Figure 2.8-2 illustrates phase and group velocity curves for the fundamental mode Love wave in the layer over a halfspace geometry of the previous section. Although the phase velocity increases monotonically with period, as longer period waves "feel" the halfspace velocity, the group velocity curve has a minimum. This minimum occurs at a period (about 15 s) where the slope of the phase velocity curve becomes very steep. This is because, by Eqn 19, U decreases when the dispersion term $dc/d\lambda$ becomes large.

The fact that the surface wave velocities vary depending on the depth range sampled by each period makes surface wave dispersion valuable for studying earth structure. These studies are conducted both with Love waves, whose dispersion depends on the shear velocity, and Rayleigh waves, whose dispersion depends on both the compressional and the shear velocities.

Both phase and group velocity dispersion measurements are used. Group velocities are easier to measure because they are

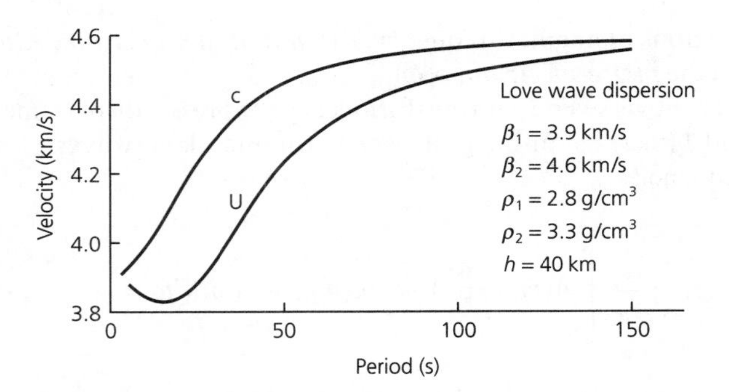

Fig. 2.8-2 Fundamental mode Love wave phase and group velocities for a model of the continental crust and mantle, a 40 km-thick layer with $\beta_1 = 3.9$ km/s, $\rho_1 = 2.8$ g/cm^3 underlain by a halfspace with $\beta_2 = 4.6$ km/s, $\rho_2 = 3.3$ g/cm^3. The group velocity has a minimum where the phase velocity curve becomes steep, as longer-period waves sample more of the velocity in the underlying halfspace.

the velocities at which a wave group visible on a seismogram travels. As shown by the Love waves in Fig. 2.8-3, the period can be measured from the time between successive peaks or troughs. Generally, the waves with longest periods travel fastest, and therefore appear first on seismograms. The group velocity is found by dividing the distance between the source and the receiver by the travel time of the wave group. Hence the wave group with a period of about 45 s arrived about 1145 s after the earthquake, and thus has a group velocity of about 3.7 km/s (4200 km in 1145 s). The later-arriving wave group with a period of about 35 s has a group velocity of about 3.6 km/s (4200 km in 1170 s). This method can be applied in a more sophisticated way by using the Fourier transform of a seismogram to isolate wave groups of different periods (Fig. 2.8-4). When the original record (*top*) is filtered at a succession of narrow frequency bands, energy is seen arriving at different group velocities.

To use such data, the results are typically plotted as a function of period and are compared to theoretical dispersion curves for different structures. For example, the group velocities for the seismogram in Fig. 2.8-3 are lower than predicted for the simple structure in Fig. 2.8-2. A better fit to the data is obtained for a model with lower layer and halfspace velocities.

This example illustrates a theme that we will encounter repeatedly: using seismological observations at the earth's surface (in this case dispersion curves), to study the velocity at depth. As noted in Section 1.1.2, this is an inverse problem, in contrast to the forward problem of predicting the observations expected for a given velocity structure. Although solving the forward problem is straightforward, it can be more difficult to find a model or models consistent with the observations. For the moment, we assume that such a model can be found, if only by trial and error, and defer more detailed discussion until Chapter 7.

Dispersion data are used to study more complicated velocity structures. Figure 2.8-5 shows the observed dispersion curves and inferred S-wave velocity structure for a study of the Walvis

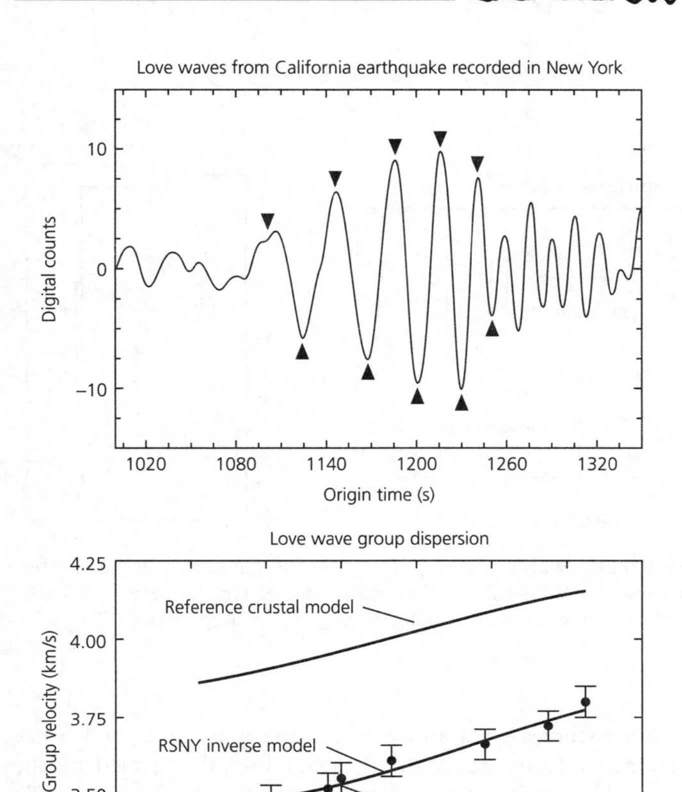

Fig. 2.8-3 *Top*: Love waves from an earthquake off the coast of California, recorded on the transverse component at station RSNY in New York, 4200 km away. Triangles indicate successive peaks and troughs of the waveform. *Bottom*: Observed (dots) and predicted (top line) group velocities for the reference structure in Fig. 2.8-2. The data are better fit by the predicted velocities (lower line) from a model with a 40 km-thick layer with shear velocity 3.6 km/s, overlying a halfspace with velocity 4.4 km/s.

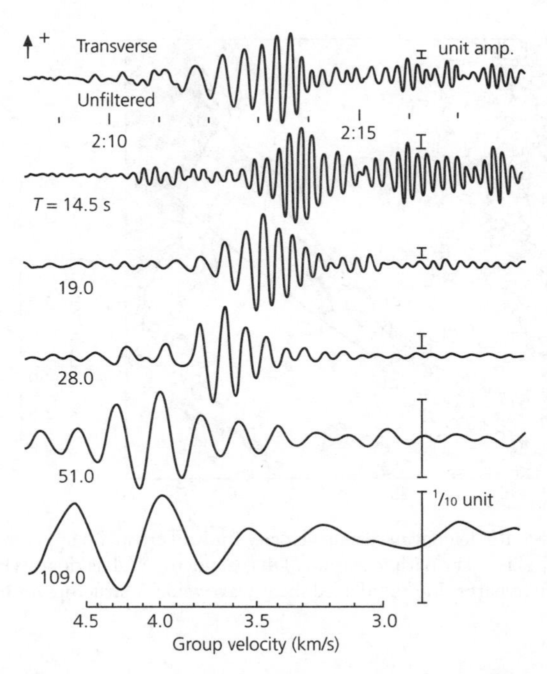

Fig. 2.8-4 Love wave group velocity dispersion shown by a seismogram from a Mongolian earthquake recorded in Japan (*top*). The data are filtered around five successive periods. Longer period energy arrives earlier, showing higher group velocity. (Kanamori and Abe, 1968.)

ridge, a linear elevated region in the South Atlantic. There are noticeable group velocity differences between two paths from an earthquake on the Mid-Atlantic ridge, one along the Walvis ridge and one off the ridge. For periods greater than about 20 s the off-ridge path is faster, indicating the presence of higher-velocity upper mantle material to a depth of about 45 km. This difference may reflect the processes that formed the Walvis ridge, which is thought to have been generated by a hot spot (Section 5.2.4), a fixed source of magma beneath the Mid-Atlantic ridge.

For periods less than about 50 s the group velocity increases with period, because the longer periods sample material whose velocity increases with depth. By contrast, for periods greater than about 50 s, the group velocity decreases with period. This decrease is interpreted as evidence for a low-velocity zone beneath the higher velocity "lid." The surface wave data thus provide evidence for the idea that the mechanically strong and cold (hence higher-velocity) plates of the earth's lithosphere are underlain by a low-velocity zone (Section 3.5.3) where temperatures approach the melting point of rock (Section 3.8.2).

Earth structure is also studied using phase velocities. These are more difficult to measure than group velocities, because they are defined for harmonic waves of a single frequency. Taking the Fourier transform of a seismogram yields the phase at each angular frequency, $\Phi(\omega)$. We assume that this phase, on a seismogram recorded at a distance x from an earthquake at time t after the earthquake, has three terms

$$\Phi(\omega) = [\omega t - k(\omega)x] + \phi_i(\omega) + 2n\pi$$
$$= [\omega t - \omega x/c(\omega)] + \phi_i(\omega) + 2n\pi. \tag{23}$$

The $\omega t - k(\omega)x$ term is the phase due to the propagation of the wave in time and space. The $\phi_i(\omega)$ term includes the initial phase at the earthquake and any phase shift introduced by the seismometer. The final term, $2n\pi$, reflects the periodicity of the complex exponential, because adding an integral multiple of 2π to the argument yields the same value.

The phase velocity can be found from observations in two ways. One method uses seismograms recorded at two stations, at distances x_1 and x_2 from an earthquake. If the waves arrive at times t_1 and t_2, taking the Fourier transform at each station gives the phase as a function of angular frequency:

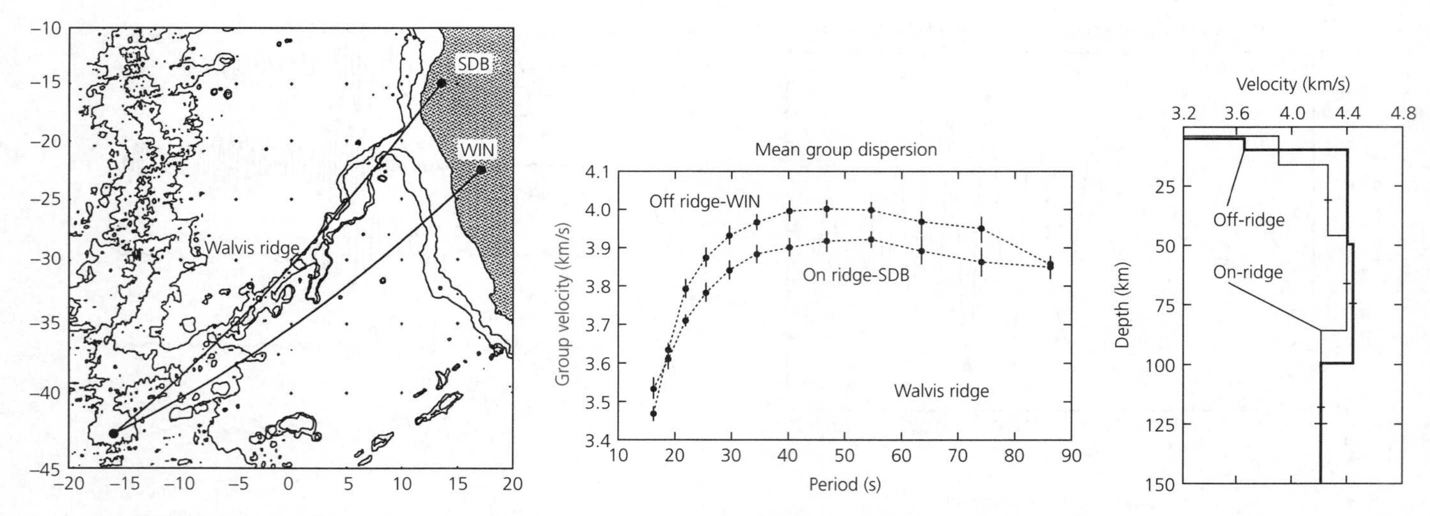

Fig. 2.8-5 Rayleigh wave group velocity study of crust and upper mantle structure along the Walvis ridge. *Left*: Ray paths from an earthquake on the Mid-Atlantic ridge. The path to station SDB is along the Walvis ridge, whereas the path to station WIN is similar, but off the ridge. *Center*: Dispersion curves for the two paths. *Right*: Inferred shear wave velocity structure for the two paths, showing lower velocities along the ridge. (Data from Chave, 1979.)

$$\Phi_1(\omega) = \omega t_1 - \omega x_1/c(\omega) + \phi_i(\omega) + 2n\pi,$$
$$\Phi_2(\omega) = \omega t_2 - \omega x_2/c(\omega) + \phi_i(\omega) + 2m\pi. \quad (24)$$

We then form the difference $\Phi_{21} = \Phi_2 - \Phi_1$, and solve for the phase velocity:

$$c(\omega) = \omega(x_2 - x_1)/[\omega(t_2 - t_1) + 2(m - n)\pi - \Phi_{21}(\omega)]. \quad (25)$$

The initial phase is common to both stations, so the $\phi_i(\omega)$ term drops out if the seismometers have the same response, and so contribute the same phase shift. If the seismometers have different responses, a correction term is added. The $2(m - n)\pi$ term is found empirically by ensuring that the phase velocity at long periods is reasonable.

Alternatively, a single-station measurement of phase velocity can be made by predicting the phase at the earthquake from its focal mechanism (Section 4.3). If $\phi_i(\omega)$ is assumed to be known, the phase velocity is

$$c(\omega) = \omega x/[\omega t + \phi_i(\omega) + 2n\pi - \Phi(\omega)]. \quad (26)$$

Figure 2.8-6 shows an example of using phase velocity data to study the evolution of the oceanic lithosphere. Various evidence shows that the oceanic lithosphere cools and thickens as it moves away from the spreading ridge where it formed (Section 5.3.2). As a result, surface wave velocities depend on the age of the lithosphere. Thus the Rayleigh wave phase velocity for the two paths shown is slowest for the path to TUC, approximately parallel to the East Pacific rise, which includes primarily young lithosphere. The other path to ARE, which includes older lithosphere, shows higher velocities. Similar effects are observed from group velocities.

Such studies yield an average dispersion curve, and hence average velocity along the great circle path traveled by the wave. However, the actual structure varies along the path. To study the evolution of the lithosphere, we would like to know the velocity of the lithosphere at each age. Unfortunately, the distribution of earthquakes and seismic stations is such that paths between earthquakes and seismic stations are rarely in lithosphere of a single age. Instead, we measure surface wave velocity on paths including different ages, as in Fig. 2.8-6.

Determination of the variable velocity structure along a path is a complicated inverse problem. The simplest approach, known as the "pure path" method, divides the study area into regions, in this case regions formed during age intervals, in which the velocity at each angular frequency is assumed to be constant. We then take a set of paths between individual earthquakes and seismic stations, such that the ith path has length L_i, and determine the phase or group velocity $v_i(\omega)$ for each path as a function of angular frequency. The total time required for the wave to travel the entire path is assumed to be the sum of the times required to traverse each of the regions along the path. Thus, if path i contains segments of lengths L_{ij} in each region j with velocity $v_j(\omega)$,

$$L_i/v_i(\omega) = \sum_{j=1}^{n} L_{ij}/v_j(\omega). \quad (27)$$

We find the velocity in each region $v_j(\omega)$ by writing this as a vector–matrix equation

$$\mathbf{d} = A\mathbf{m} \quad (28)$$

where the matrix $A_{ij} = L_{ji}$ and the data vector $d_i = L_i/v_i(\omega)$ are known, and the model vector $m_j = 1/v_j(\omega)$ is to be found.

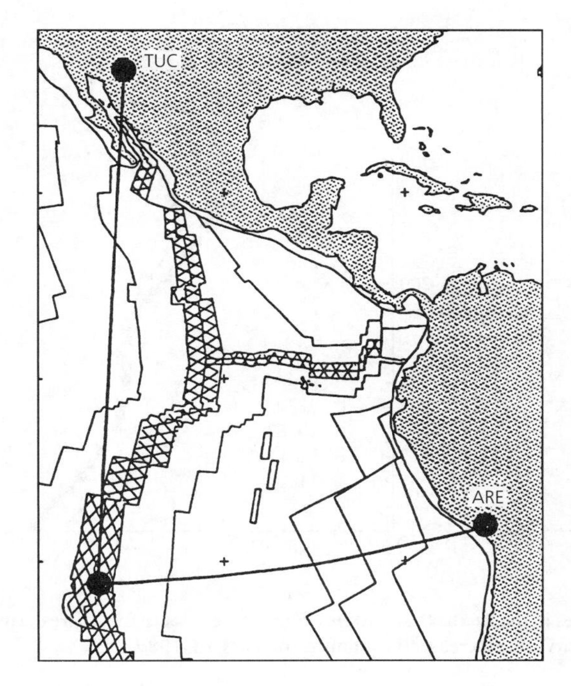

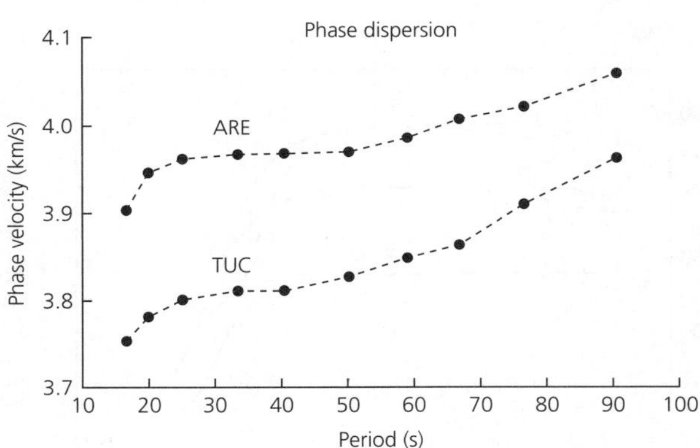

Fig. 2.8-6 Application of Rayleigh wave phase velocity data to study the evolution of the oceanic lithosphere. *Top*: Sample paths between earthquakes on the East Pacific Rise and seismic stations, which traverse lithosphere of various ages, as shown by the isochrons. The hatched regions are lithosphere younger than 3 million years. *Bottom*: Dispersion curves for the paths shown. The path to station TUC is through younger, hence lower-velocity, lithosphere than the path to ARE. (Data from Forsyth, 1975.)

Typically, because the study area is divided into a number of regions smaller than the number of paths, the number of observations exceeds the number of model parameters sought. Hence the data vector has more elements than the model vector, so the matrix A has more rows than columns and cannot be inverted. Such overdetermined systems of equations are common in seismology, especially in determining earth structure from observations. As we will see in Chapter 7, the best solution in a least squares sense to such systems of equations is found by premultiplying both sides, first by the transpose matrix and then by the inverse of A^TA,

$$\mathbf{m} = (A^TA)^{-1}A^T\mathbf{d}. \tag{29}$$

The results of such an analysis for Rayleigh wave phase velocity on many paths crossing the Pacific are shown in Fig. 2.8-7. As the lithosphere ages, the velocity and the depth to the low-velocity zone increase, presumably due to the cooling and thickening of the lithosphere.

Such studies, on both a global and a regional scale, have contributed greatly to our understanding of the earth's interior and processes. As we noted, finding velocity structure as a function of depth from dispersion data is an inverse problem, which exploits the fact that waves of different periods sample the structure at depth differently. The pure-path study illustrates a more complicated inverse problem, studying variations of velocity laterally as well as in depth. Our ability to study lateral structure comes from the fact that different source–receiver paths sample different regions. Hence these studies have the common feature of using observations on the boundaries of a region (either laterally or at depth) to learn about the structure within it, via observations resulting from sampling the region in different ways. Such approaches are examples of *tomography*, which we will discuss in Chapter 7.

2.8.4 *Tsunami dispersion*

Dispersion is also observed for tsunamis, the water waves generated by earthquakes that were discussed in Section 1.2.4. Tsunamis are like wind-driven water waves, in that they involve gravitational potential energy stored by vertical displacements of the water.[2] Although the underlying physics of the propagation differs, there are similarities in the way tsunamis and surface waves propagate.

As shown in Fig. 2.8-8 (*left*), tsunami dispersion is similar to that of Rayleigh and Love waves, in that the waves with longer periods travel faster and thus arrive earlier. The dispersion relations (Fig. 2.8-8, *right*) show two effects that depend on the period, and thus on the wavelength. At long periods, where the wavelengths are much greater than the ocean depth, d, the phase velocities are essentially nondispersive and are given by

$$c = \sqrt{gd}, \tag{30}$$

where g is the acceleration of gravity. Thus tsunami velocities depend on ocean depth, as shown. However, at shorter periods, where the wavelengths are much less than the ocean depth and so do not "feel" the ocean floor, the tsunami velocities depend on wavelength as

[2] Although tsunamis are often called "tidal waves," they have no connection to tides.

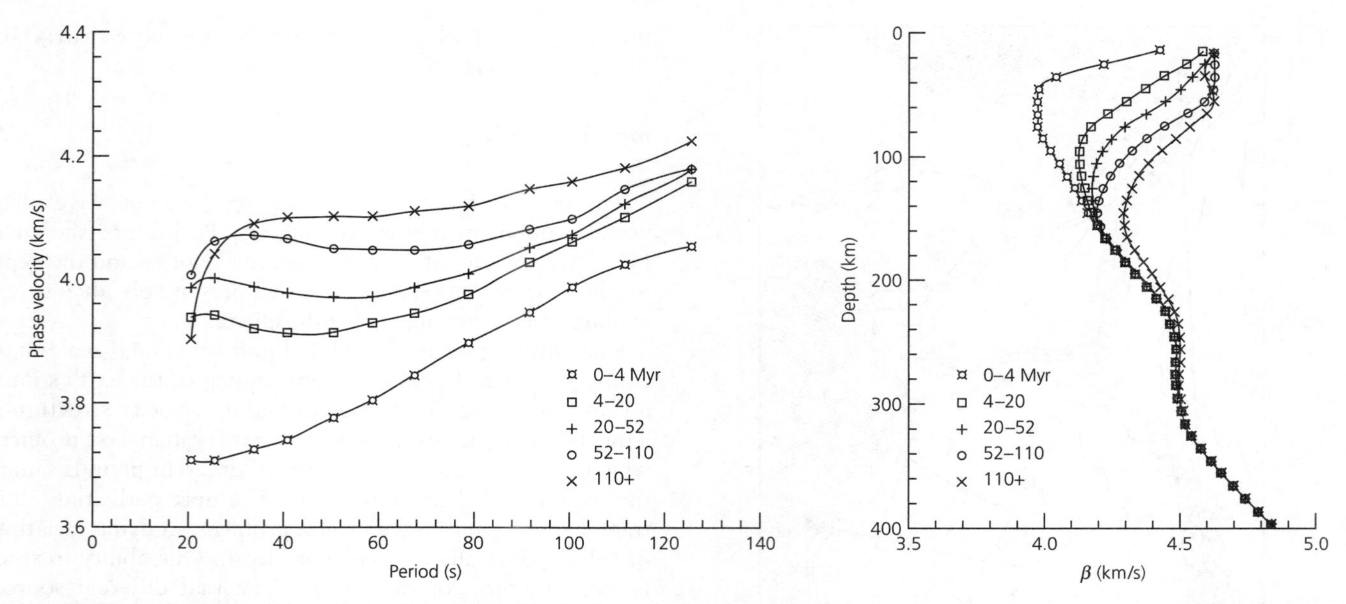

Fig. 2.8-7 *Left*: Rayleigh wave phase velocity dispersion results for five age provinces in the Pacific basin. *Right*: Shear wave velocity structure derived from the data. As the lithosphere ages, the phase velocity and depth to the low-velocity zone increase. (Nishimura and Forsyth, 1989.)

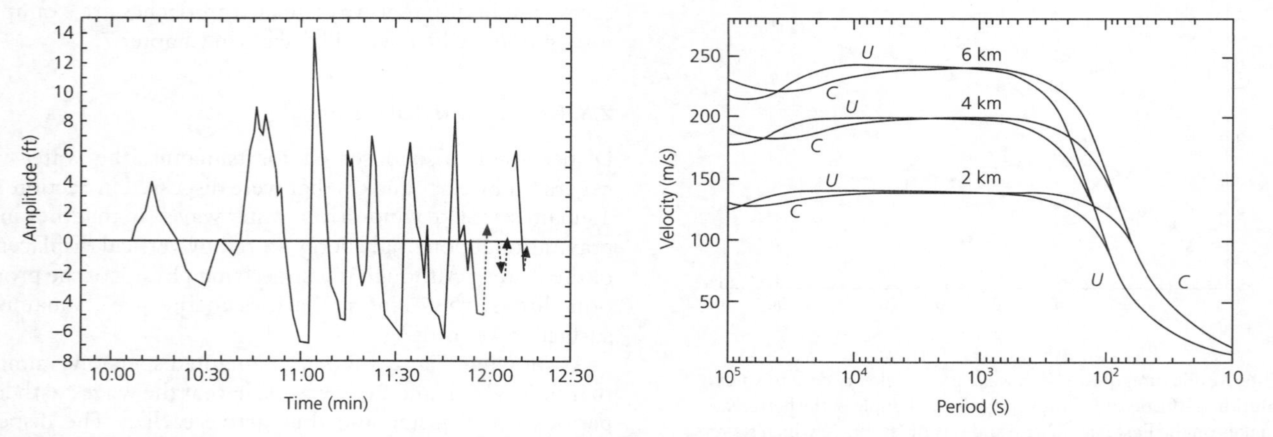

Fig. 2.8-8 *Left*: Tide gauge record of the tsunami at Hilo, Hawaii, from the great 1960 Chilean earthquake. Dispersion is seen, with the longer-period waves arriving first. (After Eaton *et al.*, 1961. © Seismological Society of America. All rights reserved.) *Right*: Theoretical tsunami dispersion curves for group (U) and phase (C) velocities for different ocean depths. At longer periods the velocity is roughly constant and controlled by the ocean depth, whereas at shorter periods, where the tsunami waves do not reach to the bottom, the velocities vary with period. (Ward, 1989.)

$$c = (\lambda g / 2\pi)^{1/2}, \tag{31}$$

so shorter-period waves travel more slowly.

Like surface waves, tsunamis travel across the earth's surface, so their amplitudes decay roughly according to $1/\sqrt{r}$ due to two-dimensional spreading. However, applying Snell's law to their horizontal propagation shows that the paths of surface waves and tsunamis deviate from the shortest great circle path if there are large lateral velocity variations. This effect, called *multipathing* because waves arrive at a receiver from several directions, can cause large changes in the waves' amplitudes due to the effects of focusing and defocusing (Section 3.7.3). As a result, the amplitude variations can be inferred from the concentration of ray paths that left the source uniformly spaced. Denser paths show rays focusing and increasing amplitudes, whereas sparser paths indicate defocusing and lower amplitudes. Figure 2.8-9 shows focusing and defocusing for the tsunami in Fig. 2.8-8 (*left*), due to variations in ocean depth. We will also use this method to study body wave amplitudes in Chapter 3.

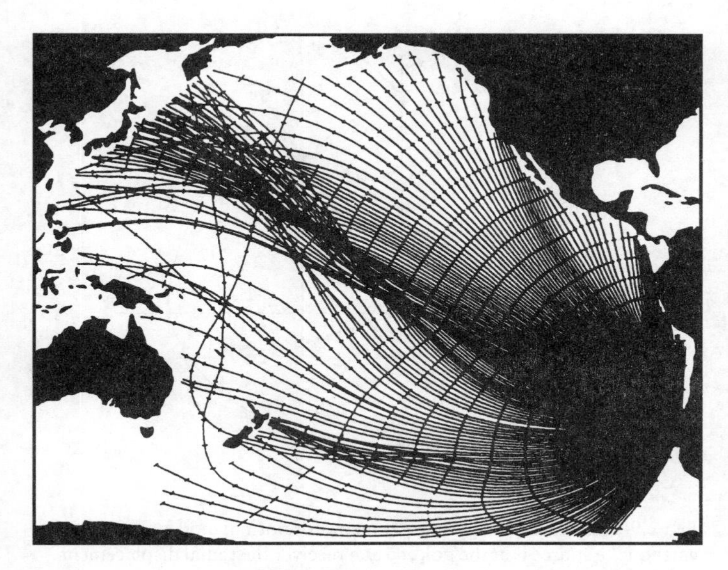

Fig. 2.8-9 Ray paths for the tsunami in Fig. 2.8-8 (*left*). Tick marks show the travel times in increments of hours. Variations in ocean depth, and therefore in tsunami velocities, cause multipathing that results in large variations in amplitudes. (Woods and Okal, 1987. *Geophys. Res. Lett.*, *14*, 765–8, copyright by the American Geophysical Union.)

2.9 Normal modes of the earth

2.9.1 *Motivation*

We started this chapter (Section 2.2) by considering the motion of a string that resulted from applying a force, and saw that the displacement could be viewed in two ways: either as waves propagating along the string or as the sum of standing waves, called normal modes. Both of these descriptions came from applying Newton's second law of motion, and are equivalent because all the features of wave propagation, such as the velocities and amplitudes of the reflected and transmitted waves, come out the same. This concept, called *mode–wave duality*, is useful in seismology because the two formulations provide different insights and jointly lead to deeper understanding. Neither formulation is more "real" — both are mathematical ways of representing the displacement, which is the physical quantity.

In a similar way, we end this chapter by extending the duality to the three-dimensional earth. We discuss how all body and surface waves can be described as the sums of the normal modes, also called free oscillations, of the spherical earth. These sums yield not only the reflections and transmissions from all boundaries, but also waves produced by effects like diffraction that are difficult to model because geometric optics fails (Section 2.5.10). However, when we discuss seismological investigations of earth structure in Chapter 3, it will turn out that most studies do not use a normal mode approach, for two reasons. First, normal mode calculations are more complicated than those for rays and plane waves. Second, by representing all seismic waves simultaneously, mode solutions do not select specific seismic phases. Hence a phase like *ScS* emerges from a computation summing many modes, whereas simpler ray or plane wave calculations often directly give the information (for example, travel times and amplitudes) that we seek. However, there are applications in which modal solutions are useful, making the topic worthy of study for reasons beyond its physical elegance, although the latter may well be what draws many seismologists (ourselves included) to it.

2.9.2 *Modes of a sphere*

The earth's modes show many features seen for the one-dimensional string, so we begin by recalling some basic results. We saw in Section 2.2.5 that once a one-dimensional string is excited, its motion can be described as

$$u(x, t) = \sum_{n=0}^{\infty} A_n\, U_n(x, \omega_n) \cos(\omega_n t), \tag{1}$$

which is the sum of standing waves or *eigenfunctions*, $U_n(x, \omega_n)$, each of which is weighted by the amplitude A_n and vibrates at its *eigenfrequency* ω_n. The eigenfunctions and eigenfrequencies depend on the physical properties of the string, whereas the amplitudes depend on the position and nature of the source that excited the motion. We saw that eigenfunctions that satisfy the wave equation in one dimension are sine and cosine functions. For a homogeneous (uniform) string of length L and velocity v, the boundary conditions of zero displacement at the fixed ends require that

$$U_n(x, \omega_n) = \sin(n\pi x/L) = \sin(\omega_n x/v), \tag{2}$$

so the eigenfrequencies are

$$\omega_n = n\pi v/L. \tag{3}$$

Because the frequency, velocity, and wavelength of a traveling wave are related by $\omega = 2\pi v/\lambda$ (Section 2.2.2), Eqn 3 requires that $L = n\lambda/2$, so each spatial eigenfunction has an integral number of half wavelengths along the string. A finite string can vibrate only in these discrete modes, which satisfy the boundary conditions. The eigenfrequencies are spaced $\pi v/L$ apart in frequency, so if the string were infinite, the eigenfrequencies would be continuous rather than discrete. Finally, we saw that the amplitudes depend on the value of the eigenfunction at the point where the source excited the motion.[1]

[1] Representing the displacement as a sum of sines and cosines, where the eigenfunctions have discrete eigenfrequencies, corresponds to a Fourier series, whereas a continuous distribution of eigenfrequencies corresponds to a Fourier transform. We use both concepts informally as needed throughout the text, and develop them more formally in Chapter 6.

Additional insight into the earth's modes comes from the two-dimensional problem of Love waves in a layer over a halfspace (Section 2.7.3). The medium was semi-infinite, extending vertically from the surface to all depths, and horizontally in both directions. We wrote a solution of the wave equation in both the layer and the halfspace as the product of separate terms describing the vertical and horizontal behaviors. We then used boundary conditions of zero traction at the free surface, continuity of traction and displacement at the interface, and energy decaying away from the interface downward, and found that these conditions require that Love waves have discrete eigenfrequencies that depend on the thickness of the layer and the shear velocity of the layer and the halfspace. Each of these eigenfrequencies thus corresponds to a vertical and horizontal eigenfunction. Interestingly, the eigenfrequencies form discrete overtone branches (Fig. 2.7-9), so that for a given apparent velocity there are several possible eigenfrequencies. Because the medium is two-dimensional, we need two parameters to list all the eigenfrequencies. One parameter, the overtone number, varies discretely (0, 1, 2, . . .) because the thickness of the layer gives a discrete dimension. The other parameter, the frequency, varies continuously along an overtone branch, because the horizontal dimension is infinite.

To extend one- and two-dimensional ideas to wave propagation in the three-dimensional spherical earth, we formulate the normal mode solution in spherical coordinates (Section A.7). Because waves propagate away from the seismic source, we put the pole of the coordinate system there (Fig. 2.9-1). We then write the displacement vector $\mathbf{u}(r, \theta, \phi) = (u_r, u_\theta, u_\phi)$ that satisfies the equation of motion (Eqn 2.4.10) as a function of radius r and surface position (θ, ϕ). A slight linguistic complication is that in spherical coordinates the radial direction is the vertical, whereas for plane waves the term "radial" (Fig. 2.7-2) denotes the horizontal direction in the vertical plane containing the source and the receiver. In this spherical geometry, u_θ is in the direction analogous to that of plane wave propagation, and u_ϕ is transverse to it.

By analogy to the string (Eqn 1), we write the displacement as a normal mode sum

$$\mathbf{u}(r, \theta, \phi) = \sum_n \sum_l \sum_m {}_nA_l^m \, {}_ny_l(r) \, \mathbf{x}_l^m(\theta, \phi) \, e^{i_n\omega_l^m t}. \qquad (4)$$

Because the medium is three-dimensional, each mode is described by its radial (depth) order n, and two surface orders l and m. All three indices have discrete integer values, because the earth is a finite body. The eigenfrequency depends on all three, and the spatial behavior is described by a radial (or vertical) eigenfunction ${}_ny_l(r)$, which is a scalar, and a surface eigen-function $\mathbf{x}_l^m(\theta, \phi)$, which is a vector. The sum depends on the weights for each eigenfunction, ${}_nA_l^m$, which are excitation amplitudes that depend on the seismic source. Thus a mode's displacement varies along the earth's surface depending on both the excitation of that mode and the location relative to the source, which combine to control the value of the surface

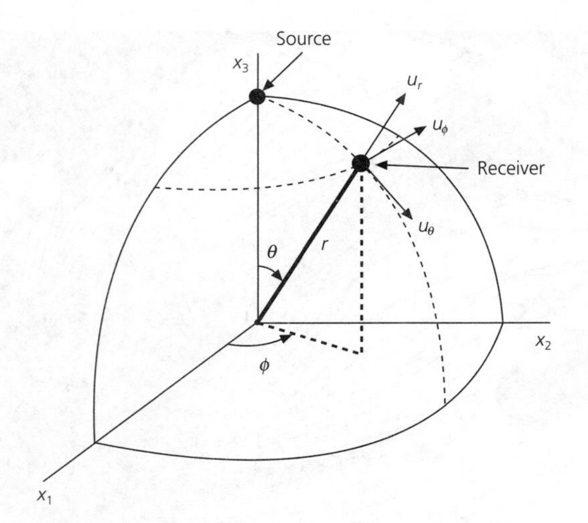

Fig. 2.9-1 Spherical coordinate geometry for normal modes. The earthquake source is at the pole, so at a receiver the radial displacement component u_r is vertical, u_θ is in the horizontal direction in the vertical plane containing the source and the receiver, and u_ϕ is in the transverse direction.

eigenfunction. As with modes on a string, we can think of the displacement as a vector in a vector space (Section A.3.6) whose basis vectors are the eigenfunctions, which are weighted and combined to describe the displacement.

Although Eqn 4 seems abstract, it turns out to be useful. If we take the Fourier transform of a long seismogram, which might extend for days or even weeks following a great earthquake, we find that the amplitude spectrum[2] (Eqn 2.8.8) is made up of normal modes that appear as peaks at certain distinct frequencies (Fig. 2.9-2). Hence thinking about a seismogram as a sum of modes gives additional insight into its nature.

Separating the radial and surface eigenfunctions in the normal mode sum (Eqn 4) has interesting consequences. The earth is close to being spherically symmetric (sometimes termed laterally homogeneous), because its structure varies much more with depth than it does laterally at a given depth. By analogy to Love waves, we expect the surface eigenfunction to be an analytic form related to the wave equation. Moreover, if the earth were laterally homogeneous (as assumed in our Love wave example), the surface eigenfunction would not affect the eigenfrequency. Thus, for a laterally homogeneous earth, we can write the eigenfrequencies as ${}_n\omega_l^m = {}_n\omega_l$. We will see later that this useful approximation also assumes that the earth is perfectly spherical and not rotating.

The eigenfrequency depends on the radial eigenfunction, which is found by solving the equation of motion in the spherical earth subject to boundary conditions at different depths. Although the boundary conditions (continuity of stress and tractions) do not sound unduly formidable, they turn out to be complicated because the tractions involve stresses and hence

[2] As discussed in Section 6.2, the amplitude spectrum is the magnitude of the Fourier transform, and its square shows how much energy is present at different frequencies.

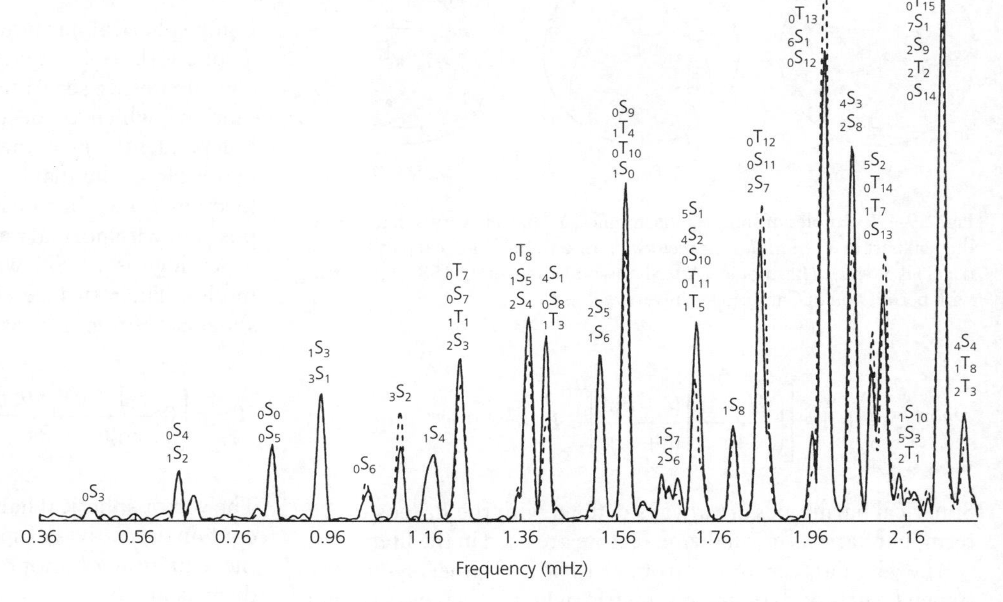

Fig. 2.9-2 Amplitude spectrum of the radial component of a 35-hour seismogram following the great June 9, 1994, deep focus Bolivia earthquake, recorded at Pasadena, California. Many peaks are labeled with several modes, indicating coupling between modes of similar frequencies. The solid line is the observed spectrum, and the dashed line is the spectrum predicted by a three-dimensional earth velocity model. (Dahlen and Tromp, 1998. Copyright © by Princeton University Press. Reprinted by permission of Princeton University Press.)

the gradients of displacements. As noted in Section A.7.4, gradients in spherical coordinates require taking the derivatives of the unit basis vectors that vary with position, unlike those in Cartesian coordinates that always point the same way. Thus we leave the problem of finding the radial eigenfunctions, and hence the eigenfrequencies, for advanced texts, just as we did for a string and for surface waves. As a result, we will also not address the issue of computing the excitation, which depends on the radial eigenfunctions at the source depth.

2.9.3 *Spherical harmonics*

The surface eigenfunctions are based on *spherical harmonics*, functions often used to expand a function on the surface of a sphere, much as sines and cosines are used in Cartesian coordinates. Because we use the seismic source as the pole, θ is the angular distance from the pole, or colatitude, and ϕ is the azimuth around the pole, or longitude (Fig. 2.9-1).

The angular variations are described by a set of functions called *Legendre polynomials*, which are indexed by the *degree*, or *angular order*, l,

$$P_l(x) = \frac{1}{2^l l!}\frac{d^l}{dx^l}(x^2 - 1)^l. \tag{5}$$

The first several polynomials are

$$P_0(x) = 1, \quad P_1(x) = x, \quad P_2(x) = (1/2)(3x^2 - 1),$$
$$P_3(x) = (1/2)(5x^3 - 3x), \tag{6}$$

and some examples are shown in Fig. 2.9-3. For a sphere, $x = \cos\theta$, so x ranges from $-1 \le x \le 1$. Legendre polynomials

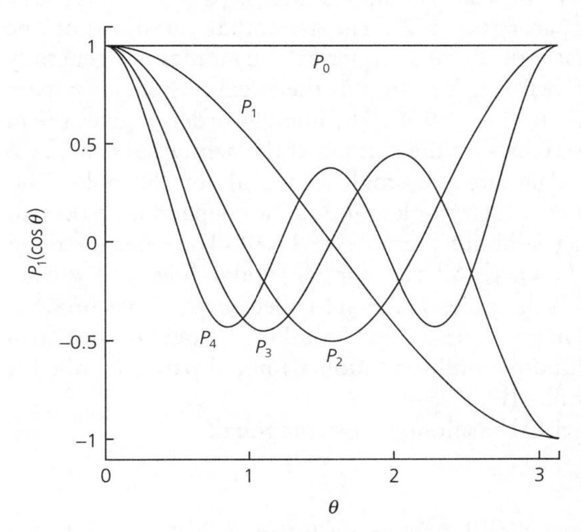

Fig. 2.9-3 Examples of Legendre polynomials for the interval 0–π used to describe the displacements associated with normal mode oscillations.

are orthogonal over this interval, and so are a suitable basis set for describing the angular variations.

The azimuthal variations are included by forming the *associated Legendre functions*,

$$P_l^m(x) = \left[\frac{(1 - x^2)^{m/2}}{2^l l!}\right]\left[\frac{d^{l+m}}{dx^{l+m}}(x^2 - 1)^l\right], \tag{7}$$

where the *azimuthal order*, m, varies over $-l \le m \le l$. The azimuthal functions $e^{im\phi}$ and associated Legendre functions are combined to give the *fully normalized spherical harmonics*,

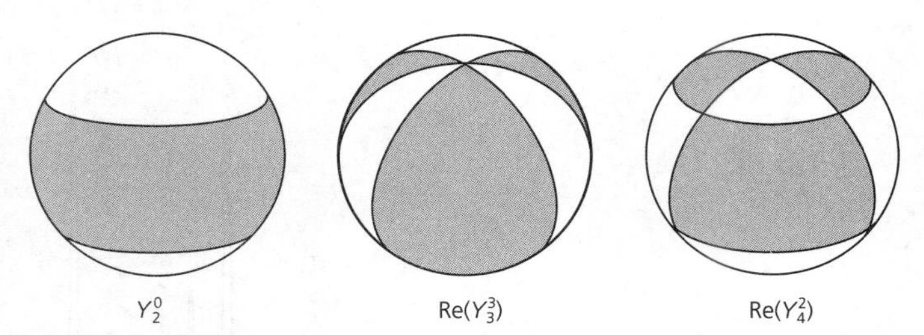

Fig. 2.9-4 Examples of spherical harmonics. Y_2^0 (*left*) is a *zonal* harmonic, the real part of Y_3^3 (*middle*) is a *sectoral* harmonic, and the real part of Y_4^2 (*right*) is a *tesseral* harmonic. (After Lapwood and Usami, 1981, reprinted with permission of Cambridge University Press.)

$$Y_l^m(\theta, \phi) = (-1)^m \left[\left(\frac{2l+1}{4\pi}\right)\frac{(l-m)!}{(l+m)!}\right]^{1/2} P_l^m(\cos\theta)e^{im\phi}. \qquad (8)$$

Spherical harmonics are always defined with the $P_l^m(\cos\theta)e^{im\phi}$ term, but various normalizing factors are used in the literature.

The angular variations from 0 to π are either symmetric (when $l+m$ is odd) or antisymmetric (when $l+m$ is even) about the equator ($\theta = \pi/2$). The azimuthal variations are periodic ($\phi + 2\pi = \phi$). Because spherical harmonics are generally complex functions, we can plot their real or imaginary parts over the sphere (Fig. 2.9-4). The angular order, l, gives the number of nodal lines on the surface. If the azimuthal order m is zero, the nodal lines are small circles about the pole. These are called *zonal* harmonics, and do not depend on ϕ (i.e., they are symmetric about the pole at $\theta = 0$). The other extreme is for $m = l$, where all the surface nodal lines are great circles through the pole. These are called *sectoral* harmonics. When $0 < |m| < l$, there are combined angular and azimuthal (colatitudinal and longitudinal) nodal patterns called *tesseral* harmonics (Fig. 2.9-4).

Spherical harmonics are orthogonal,

$$\int_0^{2\pi}\int_0^{\pi} \sin\theta\, Y_{l'}^{m'*}(\theta, \phi)\, Y_l^m(\theta, \phi) d\theta d\phi = \delta_{l'l}\delta_{m'm}, \qquad (9)$$

so that the integral of the product of one with the conjugate of another over the sphere is zero.[3] The spherical harmonics therefore form an orthogonal set of basis vectors that can be used to expand any function on the surface of a sphere, much as we used sines for the string (and would do so for any other Cartesian coordinate problem). Spherical harmonics are used to represent planetary quantities, including lateral variations in seismic velocity, surface topography, and gravitational and magnetic fields. The shape of the field represented depends on the amplitudes of the different spherical harmonic components.

[3] As defined in Eqn A.3.37, $\delta_{nm} = 0$ unless $n = m$.

2.9.4 *Torsional modes*

Using spherical harmonics, we can write the normal modes of a sphere (Eqn 4) explicitly. You may recall that in Cartesian coordinates we separated the displacements into *P–SV* and *SH* motions, which are decoupled in the sense that they propagate independently in a medium whose properties vary only in depth along the plane containing the source and the receiver (Section 2.5.2). In spherical geometry, we do a similar decomposition with normal modes.

Analogous to *SH* waves, we have *torsional*, or *toroidal*, modes. Their surface eigenfunctions are given by the *vector spherical harmonics* with (r, θ, ϕ) components

$$\mathbf{T}_l^m = \left(0, \frac{1}{\sin\theta}\frac{\partial Y_l^m(\theta, \phi)}{\partial\phi}, \frac{-\partial Y_l^m(\theta, \phi)}{\partial\theta}\right). \qquad (10)$$

The vector spherical harmonics are vectors whose components contain derivatives of spherical harmonics, which arise because the equation of motion involves spatial derivatives of the displacements.

The displacement vector $\mathbf{u} = (u_r, u_\theta, u_\phi)$ that corresponds to torsional modes is

$$\mathbf{u}^T(r, \theta, \phi) = \sum_n \sum_l \sum_{m=-l}^{l} {}_nA_l^m\, {}_nW_l(r)\, \mathbf{T}_l^m(\theta, \phi) e^{i_n\omega_l^m t}. \qquad (11)$$

The radial eigenfunction ${}_nW_l(r)$ varies with depth, even though the resulting displacement has no radial component because u_r is always zero. Thus torsional modes have only horizontal displacements and are analogous to *SH* waves. Similarly, their divergence is zero, so they cause no volume change.

Torsional modes are denoted ${}_nT_l^m$, where n is the radial order, l is the angular order, and m is the azimuthal order. For given radial and angular orders, the $2l + 1$ modes of different azimuthal orders $-l \le m \le l$ are called *singlets*, and the group of singlets is called a *multiplet*. If the earth were perfectly spherically symmetric, and not rotating, then all the singlets in a multiplet would have the same eigenfrequency. This condition is called *degeneracy*. For example, the period of ${}_nT_l^0$ would be the same for ${}_nT_l^{\pm1}$, ${}_nT_l^{\pm2}$, ${}_nT_l^{\pm3}$, etc. In the real earth, the singlet frequencies vary, which is an effect called *splitting*. However, the splitting is small enough that for most applications we ignore it, dropping the m superscript and referring to the entire ${}_nT_l^m$ multiplet as ${}_nT_l$, with eigenfrequency ${}_n\omega_l$.

For torsional modes, the horizontal displacements, u_θ and u_ϕ, are zero along nodal lines, because the angular displacements u_θ vanish where $\partial Y_l^m/\partial\phi = 0$ and the azimuthal displacements u_ϕ vanish where $\partial Y_l^m/\partial\theta = 0$. For example, consider the lowest-frequency (longest-period or gravest) torsional normal mode singlet, ${}_0T_2^0$ (Fig. 2.9-5). There are no radial motions, and the angular displacements are always zero, because $m = 0$. To see this, note, from Eqn 10, that u_θ is proportional to

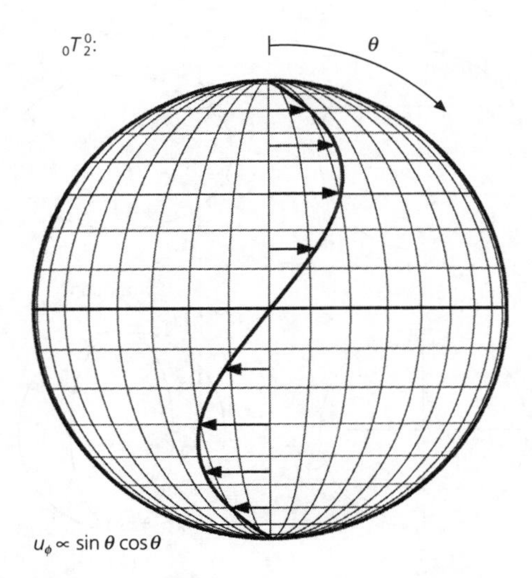

Fig. 2.9-5 Displacement associated with torsional mode $_0T_2^0$.

$$\frac{1}{\sin\theta} P_2^0(\cos\theta) \frac{\partial}{\partial\phi}(e^{im\phi}) = \frac{1}{\sin\theta} P_2^0(\cos\theta)(im)e^{im\phi} = 0. \qquad (12)$$

The only nonzero displacement component is the azimuthal one, u_ϕ, which is proportional to

$$e^{im\phi} \frac{\partial}{\partial\theta} P_2^0(\cos\theta) = 3\sin\theta\cos\theta. \qquad (13)$$

The azimuthal motions vanish at the poles ($\theta = 0°$ and $180°$) and at the equator ($\theta = 90°$). The motions are in opposite directions across the equator because $\sin\theta$ is an odd function. This node is the surface expression of a nodal plane that bisects the earth along the equator. The pattern of oscillations extends throughout the mantle.[4]

The radial order describes how the mode varies with radius, and the angular and azimuthal orders describe how it varies with latitude and longitude. For torsional modes, n gives the number of spherical nodal surfaces within the earth. If $n = 0$, there are no nodal surfaces, and the direction of motion at a given latitude and longitude is the same at all depths. For torsional modes, l equals one more than the number of nodal lines on the surface. The shape and distribution of these nodal lines varies according to the azimuthal order, m, which gives the number of vertical nodal planes that bisect the earth, passing through the pole. For $m = 0$, the nodal lines are small circles about the pole. If $m = l - 1$, the nodal lines are great circles through the pole.

The $_0T_2^1$ singlet has a longitudinal great circle node at the surface (Fig. 2.9-6). The motions are shear displacements

[4] Because the outer core is liquid, the core–mantle boundary is a free surface for torsional modes excited by earthquakes. These modes do not propagate into the outer core, and therefore never reach the inner core, which theoretically has its own set of torsional modes.

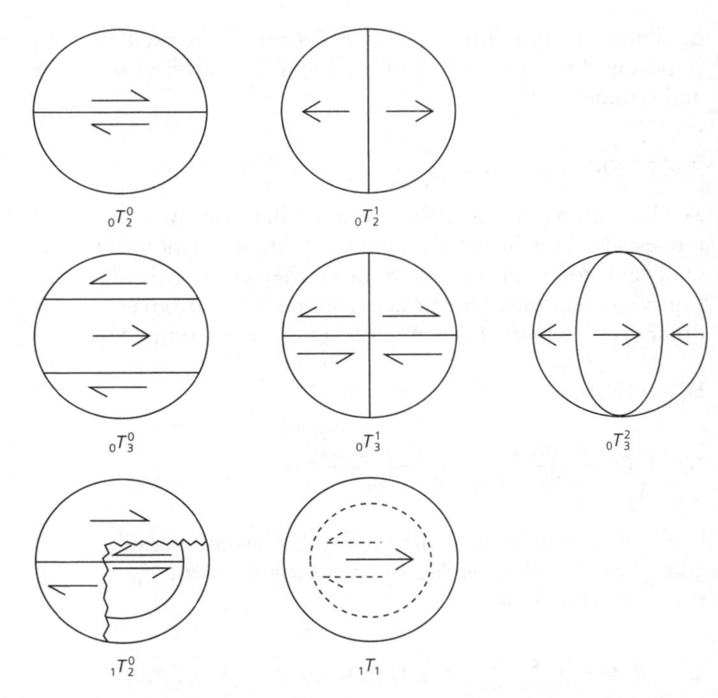

Fig. 2.9-6 Examples of the displacements for several torsional modes. The examples for $_1T_2^0$ and $_1T_1$ schematically show the variation with depth.

about the pole that oscillate toward and away from the nodal plane. The period of $_0T_2$ is 44 minutes: 22 minutes rotating in one direction, then 22 minutes rotating back again. For higher angular orders l, more nodal planes occur. $_0T_3^0$ has two latitudinal nodal lines at the surface, $_0T_3^1$ has one, and $_0T_3^2$ has none. As l increases, the number of divisions of the surface increases.

Torsional modes with $n = 0$ ($_0T_l^m$) are called *fundamental modes*, and have motions at depth in the same direction as at the surface. This is not true, however, for modes with $n > 0$, called *overtones*. As shown in the cutaway for $_1T_2^0$, there is a spherical nodal surface within the mantle across which displacements reverse. We will see shortly that an overtone of order n has n radially symmetric nodal surfaces at depths determined by the velocity structure of the mantle.

You may have wondered what happened to $_0T_1$ and $_0T_0$. Because the number of nodal planes equals $l - 1$, $_0T_1$ has no nodal planes. Physically, this corresponds to rigid body rotation. As we will discuss in Section 4.4.4, seismic waves generated by earthquakes are generally well described by treating the source as a double couple of body forces, which generates no net torque, and therefore no change in rotation. In rare cases, giant earthquakes may cause enough vertical displacement of rock to affect the rate of the earth's rotation. However, because torsional modes do not involve radial motions, even in these cases conservation of angular momentum demands that $_0T_1$ be zero. There are, however, overtones with $l = 1$ ($_1T_1$, $_2T_1$, etc.). These involve the entire top spherical shell of the earth oscillating in one direction, with deeper shells oscillating in opposing directions. The mode ${}_0T_0$ has no physical meaning and is undefined.

2.9.5 *Spheroidal modes*

P–SV motions are described in a similar way by *spheroidal* modes, also known as *poloidal* modes. These are more complicated than torsional modes, because they combine radial and transverse motions. The surface eigenfunctions are given by two other *vector spherical harmonics*, with (r, θ, ϕ) components

$$\mathbf{R}_l^m = (Y_l^m, 0, 0),$$

$$\mathbf{S}_l^m = \left(0, \frac{\partial Y_l^m(\theta, \phi)}{\partial \theta}, \frac{1}{\sin\theta}\frac{\partial Y_l^m(\theta, \phi)}{\partial \phi}\right). \tag{14}$$

Each corresponds to a different radial eigenfunction, ${}_nU_l(r)$ and ${}_nV_l(r)$, so the displacement vector $\mathbf{u} = (u_r, u_\theta, u_\phi)$ for spheroidal modes is

$$\mathbf{u}^S(r, \theta, \phi) = \sum_n \sum_l \sum_{m=-l}^{l} {}_nA_l^m[{}_nU_l(r)\mathbf{R}_l^m(\theta, \phi) + {}_nV_l(r)\mathbf{S}_l^m(\theta, \phi)]\, e^{i_n\omega_l^m t}. \tag{15}$$

Thus the radial eigenfunction ${}_nU_l(r)$ corresponds to radial motion, and ${}_nV_l(r)$ corresponds to horizontal motion.

To see that the mode formulation separates *P–SV* from *SH* and fully represents the displacement in three dimensions, note that the three vector spherical harmonics are orthogonal,

$$\mathbf{T}_l^m \cdot \mathbf{S}_l^m = \mathbf{T}_l^m \cdot \mathbf{R}_l^m = \mathbf{S}_l^m \cdot \mathbf{R}_l^m = 0. \tag{16}$$

Spheroidal modes ${}_nS_l^m$ follow a similar nomenclature as torsional modes. The fundamental modes, with no internal nodal surfaces, are described by $n = 0$. As n increases, the number of internal nodal surfaces increases, although, unlike for torsional modes, n is not the number of nodal surfaces. The angular order l equals the number of nodal lines at the surface (rather than $l - 1$ for torsional modes), and m represents the number of great circle nodal lines passing through the pole. The spheroidal *radial* modes, which have $l = 0$ and thus only radial motions, have no torsional analogue.

Some examples of spheroidal modes are shown in Fig. 2.9-7. The "breathing" mode ${}_0S_0$ involves radial motions of the entire earth that alternate between expansion and contraction. The gravest (lowest-frequency or longest-period) of earth's modes observed to date is ${}_0S_2$, which has a period of 3233 s, or 54 minutes.[5] The ${}_0S_2^0$ singlet alternates between an oblate (flat disk) and prolate (football) shape, and is accordingly referred to as the "football" mode. Displacements for the ${}_0S_2^1$ and ${}_0S_2^2$ singlets are also shown. There is no ${}_0S_1$ mode, which would correspond to a lateral translation of the planet. Increasing l results in more surface nodal lines, as shown for ${}_0S_3$, and increasing n results in more internal nodal surfaces.

[5] The ${}_1S_1$ Slichter mode due to lateral sloshing of the solid inner core through the liquid iron outer core, which has yet to be observed, should in theory have a period of about 5.5 hours.

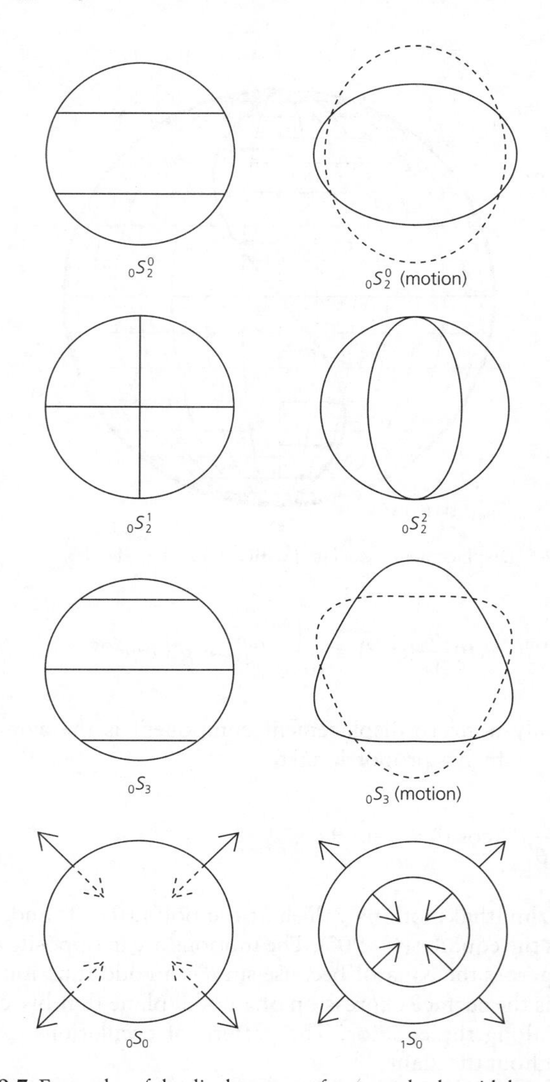

Fig. 2.9-7 Examples of the displacements for several spheroidal modes.

2.9.6 *Modes and propagating waves*

We can gain considerable insight into normal modes by considering their relation to traveling waves. To do this, we use a mathematical approximation (that we will not derive) for the associated Legendre functions. When the angular order l is much greater than the azimuthal order m,

$$P_l^m(\cos\theta) \approx (-1)^m l^m (2/l\pi \sin\theta)^{1/2} \cos[(l + 1/2)\theta + m\pi/2 - \pi/4)], \tag{17}$$

so the spherical harmonics behave approximately like

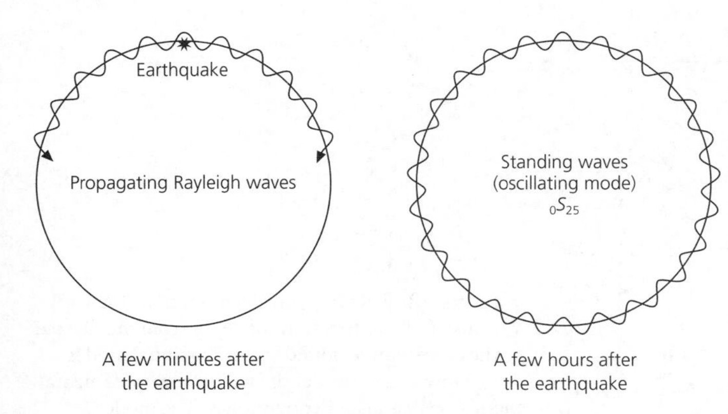

Fig. 2.9-8 Cartoon of the equivalence of surface waves and normal modes. Once surface waves from an earthquake make multiple passes around the earth, they can be viewed as standing waves, or normal modes, such that the mode with angular order l has $l + 1/2$ wavelengths around the earth. This example is for $_0S_{25}$.

$$Y_l^m(\theta, \phi) \approx A(2/l\pi \sin \theta)^{1/2} \cos [(l + 1/2)\theta] e^{im\phi}, \tag{18}$$

where A contains the remaining factors. Using this approximation and representing the cosine as complex exponentials shows that terms in the mode sums (Eqns 11 and 15), which involve the products $Y_l^m(\theta, \phi) e^{i_n\omega_l^m t}$, give rise to terms corresponding to propagating waves with horizontal wave vector (Section 2.4.2)

$$\mathbf{k}_x = (k_\theta, k_\phi), \quad k_\theta = (1/a)[(l + 1/2)^2 - m^2/\sin^2 \theta]^{1/2},$$
$$k_\phi = m/(a \sin \theta), \tag{19}$$

where the factor of the earth's radius a converts the angular terms to wavenumbers along the surface. Hence the mode with angular order l and frequency $_n\omega_l$ corresponds to a traveling wave with horizontal wavelength

$$\lambda_x = 2\pi/|\mathbf{k}_x| = 2\pi a/(l + 1/2) \tag{20}$$

that has l + 1/2 wavelengths around the earth (Fig. 2.9-8). These waves travel at a horizontal phase velocity

$$c_x = {}_n\omega_l/|\mathbf{k}_x| = {}_n\omega_l a/(l + 1/2). \tag{21}$$

This equivalence is easily visualized a while after an earthquake, where globe-circling surface waves can be viewed as standing waves, or modes. Waves corresponding to different singlets propagate in different directions, as shown by the various values of m.

This approximation also gives insight into the correspondence between spheroidal and torsional modes and P–SV and SH waves (or Rayleigh and Love waves). The spheroidal and torsional mode displacements depend on vector spherical harmonics, and thus on the derivatives of spherical harmonics. Taking derivatives of Eqn 18 shows the ratio of the partial derivatives,

$$\frac{\partial Y_l^m(\theta, \phi)}{\partial \theta} \Bigg/ \frac{\partial Y_l^m(\theta, \phi)}{\partial \phi} \gg 1, \tag{22}$$

because l was assumed to be much greater than m. For torsional modes, the $\mathbf{T}_l^m$ vector spherical harmonic (Eqn 10) generally has a ϕ component greater than its θ component, so its displacement is primarily perpendicular to the plane connecting the source and the receiver, like an SH or a Love wave (Fig. 2.9-1). By contrast, the spheroidal mode vector spherical harmonic $\mathbf{S}_l^m$ (Eqn 14) generally has a θ component greater than its ϕ component, and so causes displacement primarily in the plane connecting the source and the receiver, like a P–SV or a Rayleigh wave.

We can use these ideas to relate modes to specific body and surface wave phases. A good place to start is to recall that for Love waves in a layer over a halfspace, the boundary conditions at the free surface and the interface require that the Love wave have discrete eigenfrequencies that depend on the layer thickness and the shear velocity of the layer and the halfspace. We thus obtain a *dispersion relation* (Section 2.7.3) giving the phase velocity as a function of frequency for these modes. Because the dispersion relation depends on the earth structure assumed in computing it, we can compare the observed dispersion of surface waves to the predictions of different earth models, and invert the observations to derive earth models that better fit the data (e.g., Fig. 2.8-3).

Analogous computations for the spherical earth predict the normal mode eigenfunctions and eigenfrequencies, which depend on the earth model assumed. Figure 2.9-9 shows a plot of radial eigenfunctions for some modes. As for surface waves, modes with different eigenfrequencies sample different depths within the earth. For example, as noted in Fig. 2.9-6, Fig. 2.9-9A shows that a torsional overtone of order n has n nodal surfaces at depths determined by the velocity structure of the mantle. Thus the observed eigenfrequencies can be inverted to model the earth's radial velocity structure. This process yields earth models that match the observed eigenfrequencies quite well, as illustrated by the dashed line in Fig. 2.9-2. Moreover, the results can be checked by combining them with travel time observations. For instance, before *PKJKP* body waves[6] were observed, the shear velocity of the inner core was constrained using normal modes like $_{10}S_2$ that have large displacements in the inner core.

Figure 2.9-10 shows a plot of the eigenfrequency versus angular order for torsional modes. The modes plot along distinct lines, corresponding to overtone branches. The lowest line is the fundamental branch (radial order $n = 0$) with the lowest eigenfrequency (longest period) for any given angular order. The

[6] As discussed in Section 3.5, *PKJKP* is an elusive body wave phase that propagates in the inner core as a shear wave, and so provides information on the difficult-to-constrain shear velocity there.

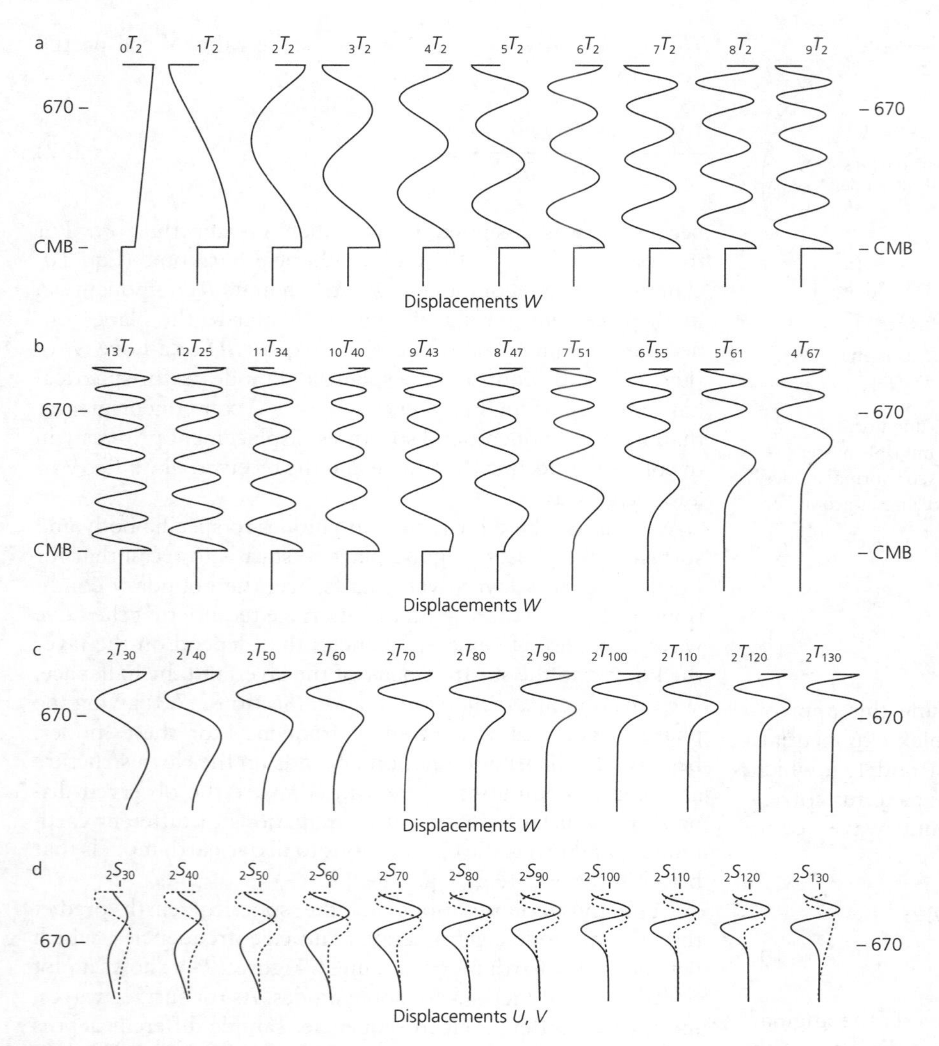

Fig. 2.9-9 Radial (vertical) eigenfunctions for various modes as functions of depth from the surface to the core–mantle boundary. a: Torsional modes with a low angular order of $l = 2$ for the fundamental mode ($n = 0$) and higher overtones. The modes sample fairly evenly across the whole mantle, with the radial order giving the number of times the displacements change sign. b: Torsional modes with about the same frequency (14 mHz). When $l < \sim 4n$, the modes correspond to ScS_{SH} waves, and the eigenfunctions span the whole mantle. When $l > \sim 4n$, the modes correspond to *SH* waves that bottom in the mid-mantle, and the eigenfunctions tail off before reaching the core–mantle boundary. c: Second-overtone branch of Love wave-equivalent torsional modes. Because the radial order is always $n = 2$, the curves always have two zero crossings, so the displacement directions are always divided into three regions. The eigenfunctions get shallower at higher angular orders. d: Second-overtone branch of Rayleigh wave-equivalent spheroidal modes. As with b, the eigenfunctions get shallower at higher angular orders. The solid lines show the eigenfunction for radial displacements, U, and the dashed lines show the eigenfunction for tangential displacements, V. (Dahlen and Tromp, 1998. Copyright © by Princeton University Press. Reprinted by permission of Princeton University Press.)

lines of successively higher eigenfrequencies (shorter periods) define overtone branches with increasing n. On any branch, the eigenfrequency increases for higher angular order l.

As we have seen, the angular order l relates modes to traveling waves of a specific wavelength (Eqn 20) or phase velocity (Eqn 21). Thus frequency–angular order plots for normal modes as in Fig. 2.9-10 correspond to dispersion (phase velocity–period) plots for surface waves (Fig. 2.7-8) and are sometimes called normal mode dispersion plots.

Various regions of the torsional mode dispersion plot in Fig. 2.9-10 correspond to different body and surface shear wave (*SH*) phases, which are discussed further in the next chapter. The horizontal phase velocity (Eqn 21) of the waves corresponding to a given mode can be related to the horizontal phase velocity of a surface wave or the apparent velocity of a body wave phase. The upper left of the figure, with high frequency and low angular order l, contains modes that contribute to body wave phases with high apparent velocities and thus near-vertical incidence (recall from Section 2.5.3) that $c_x = v/\sin i$), such as the core reflections (Figs 1.1-2 and 3.5-5) *ScS*, *sScS*, and ScS_2. The dashed line corresponds to modes with a phase velocity around 7.3 km/s, which is the apparent velocity of shear waves that diffract around the core. We will see that these SH_{diff} waves bottom and turn at the core–mantle boundary, and so represent the transition between direct *S* and *ScS*, which reflects at the core–mantle boundary. To the right of the dashed line are modes corresponding to *S* wave phases that bottom in the mantle, like *S*, *SS*, *sS*, *sSS*, and *SSS*. Modes further to the right (higher l) for a given frequency have lower phase velocity, and thus correspond to body wave phases (Section 3.4) that bottom at shallower depths in the mantle. The difference is shown by the radial eigenfunctions (Fig. 2.9-9b) for torsional modes of about the same frequency. Modes to the left of $_9T_{43}$ have significant displacement throughout the mantle, corresponding to phases that reach the core–mantle boundary, whereas those to the right increasingly correspond to phases that penetrate only to shallower depths.

We can also consider modes that are equivalent to surface waves, bearing in mind a slight notational complexity that the higher ($n > 0$) overtone branches are sometimes termed "higher

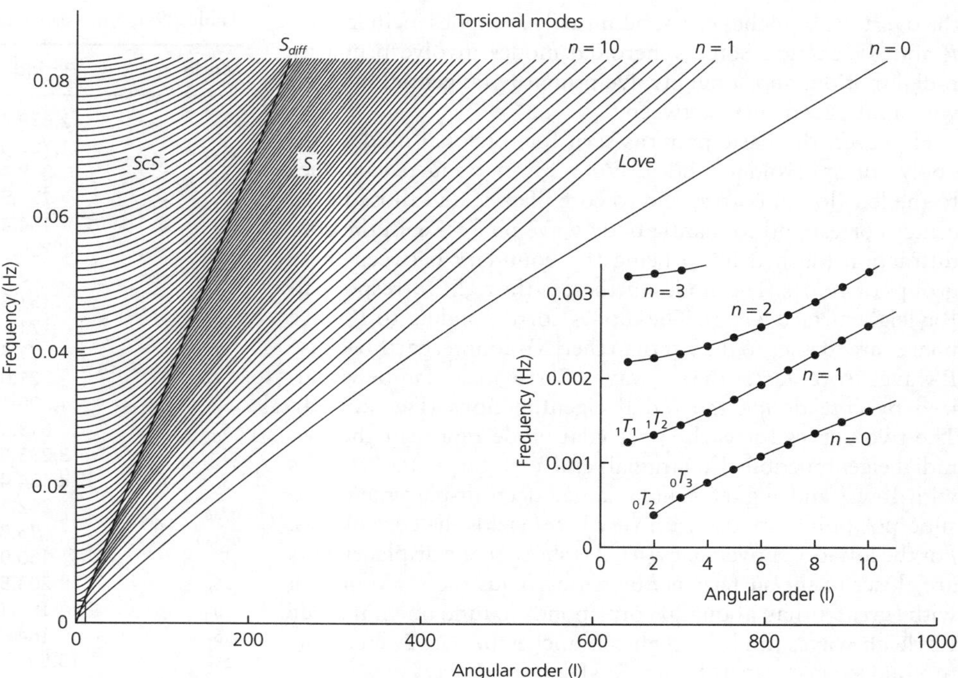

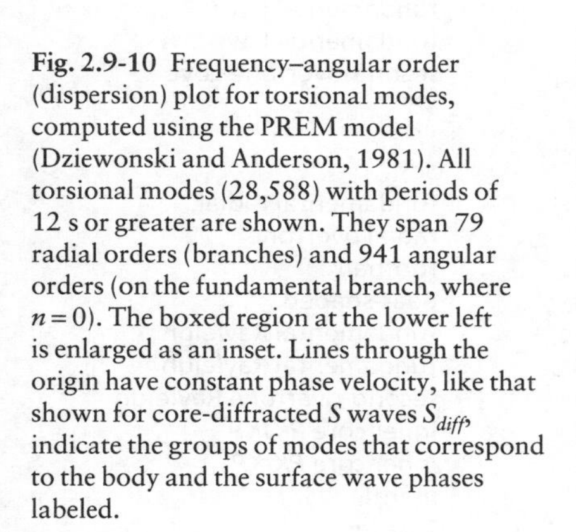

Fig. 2.9-10 Frequency–angular order (dispersion) plot for torsional modes, computed using the PREM model (Dziewonski and Anderson, 1981). All torsional modes (28,588) with periods of 12 s or greater are shown. They span 79 radial orders (branches) and 941 angular orders (on the fundamental branch, where $n = 0$). The boxed region at the lower left is enlarged as an inset. Lines through the origin have constant phase velocity, like that shown for core-diffracted S waves S_{diff}, indicate the groups of modes that correspond to the body and the surface wave phases labeled.

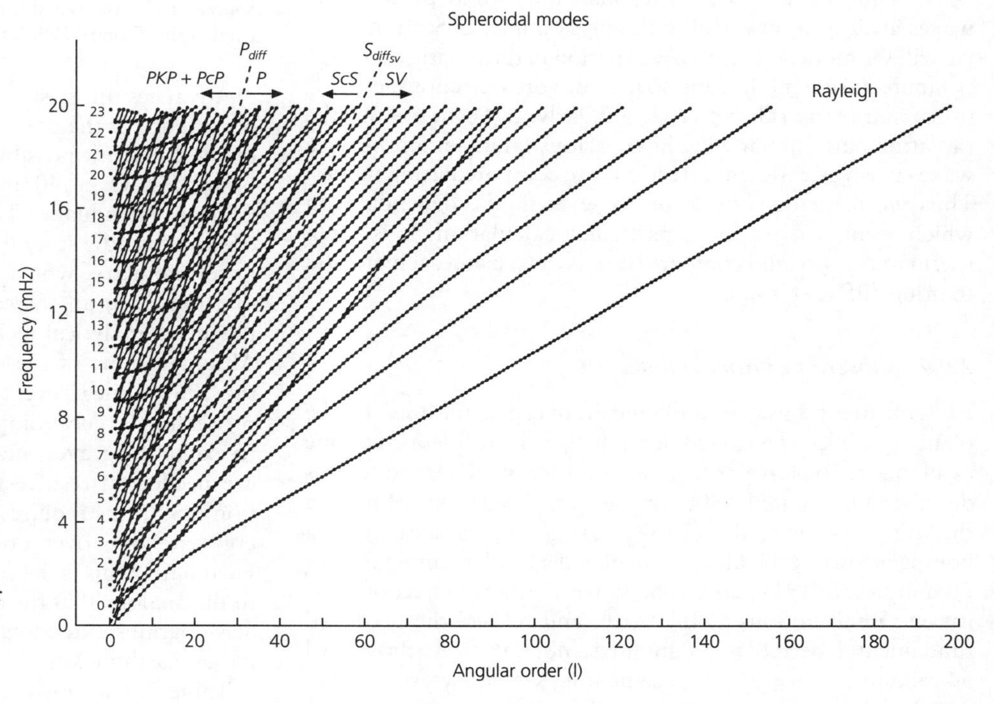

Fig. 2.9-11 Frequency–angular order (dispersion) plot for spheroidal modes, computed using the PREM model. All spheroidal modes with periods of 50 s or greater are shown. Note the complexity of the branches compared to the toroidal modes. The dashed lines show the phase velocities of modes corresponding to the core diffractions P_{diff} and $S_{diff_{SV}}$ (also called SV_{diff}). To the left of the P_{diff} line are modes corresponding to core reflected and transmitted phases like *PcP*, *PKiKP*, and the various branches of *PKP*. To the right of this line are modes corresponding to *P* waves that bottom in the mid-mantle. To the left of the $S_{diff_{SV}}$ line are modes corresponding to core reflected and transmitted phases like *ScS* and *SKS* (mixed in with the *PcP* and *PKP* modes to the left of the P_{diff} line). To the right of the $S_{diff_{SV}}$ line are modes corresponding to *SV* waves that bottom in the mid-mantle. The first few mode branches at the right correspond to Rayleigh waves. (After Dahlen and Tromp, 1998. Copyright © by Princeton University Press. Reprinted by permission of Princeton University Press.)

modes" when discussing surface waves (Section 2.7.4). The torsional modes furthest to the right in Fig. 2.9-10, which are the lowest-overtone branches, can be viewed as Love waves. The $n = 0$ branch with l greater than about 20 corresponds to fundamental mode Love waves, that for $n = 1$ corresponds to the first Love wave overtone, and so on. The radial eigenfunctions in Fig. 2.9-9c for the $_2T$ ($n = 2$) branch show that modes with successively higher l have displacements increasingly concentrated near the surface. This is consistent with our observation that higher-frequency (shorter-period) Love waves for a given overtone branch n have displacements closer to the surface (Fig. 2.7-10).

The situation for spheroidal modes is more complicated (Fig. 2.9-11). The fundamental branch remains distinct, but the overtone branches cross, because these modes include both *P* and *SV* energy. Some spheroidal modes involve primarily radial motion, and some involve primarily tangential motion, with a full spectrum in between.

However, the basic patterns seen for torsional modes also apply for spheroidal modes. For a given frequency, modes to the left (low l) correspond to core phases, and those in the center correspond to mantle body wave phases, with the core diffraction (dashed lines) being the boundary between these groups of modes. The modes furthest to the right correspond to Rayleigh surface waves. The modes corresponding to *P*-wave phases are further to the left than their *SV* counterparts because *P* waves travel faster than *S* waves. These ideas can be visualized by considering the radial eigenfunctions (Fig. 2.9-9d). The two curves for each spheroidal mode represent the two radial eigenfunctions $U(r)$ (radial) and $V(r)$ (tangential). Modes with low l and higher n have larger deep displacements, so superposition of modes with very low l yields the core phases. For the low-order overtones ($n = 2$ is shown), the displacements are closer to the surface as l increases. Thus the $n = 0$ branch with l greater than about 20 corresponds to fundamental mode Rayleigh waves, and the higher branches ($n = 1$, 2, etc.) correspond to successively higher Rayleigh wave overtones.

The equivalence between normal modes and propagating waves gives us a powerful tool. For example, in Section 3.5.5 we will see models of wave propagation in the mantle that were computed using modes, and so include core reflections, diffractions, and many other phases. Similarly, in Section 4.3.4 the radiation patterns showing how various faults radiate surface wave energy in different directions are computed using modes. Thus we use either mode or wave methods, depending on which seems easiest for a particular calculation. It is often useful to do both and compare the answers, using each method to bring different insight.

2.9.7 *Observing normal modes*

As with many basic seismological concepts, the idea of the planet's modes developed long before instruments became available to observe them. As the theory of elasticity was developed in the mid-1800s, there were discussions of finding the "pitch" of the earth.[7] In 1882, Lamb modeled the earth as a homogeneous steel ball, and calculated a fundamental mode of 78 minutes. In 1911, Love took into account the effect of gravity on radial motions of the earth, and revised the predicted fundamental period to 60 minutes, not far from the actual 54 minutes. However, because making seismometers that can detect such long-period motions is difficult (Section 6.6), it was only after the great 1952 Kamchatka earthquake that this mode was actually observed on a strainmeter recording.

Table 2.9-1 Some torsional and spheroidal modes.

Mode	Period	Description or associated phase
${}_0T_2$	2,639.4	fundamental torsional
${}_0T_3$	1,707.6	fundamental torsional
${}_1T_1$	808.4	radial overtone
${}_1T_2$	757.5	radial overtone
${}_9T_2$	104.4	radial overtone
${}_0T_{30}$	259.5	fundamental Love
${}_0T_{130}$	68.9	fundamental Love
${}_2T_{30}$	151.3	second-overtone Love
${}_4T_{67}$	71.3	*SH*
${}_{10}T_{40}$	71.4	SH_{diff}
${}_{13}T_7$	71.6	ScS_{SH}
${}_0S_0$	1,228.1	fundamental radial
${}_1S_0$	613.0	radial overtone
${}_0S_2$	3,233.5	football
${}_0S_3$	2,134.4	pear-shaped
${}_0S_{30}$	262.1	fundamental Rayleigh
${}_0S_{130}$	75.8	fundamental Rayleigh
${}_1S_{30}$	160.9	second-overtone Rayleigh
${}_{10}S_6$	203.5	inner core *PKJKP*
${}_{11}S_5$	197.1	inner core *PKIKP*
${}_{14}S_3$	184.9	mantle ScS_{SV}
${}_1S_1$	19,500	Slichter

Sources: Dziewonski and Anderson (1981); Wysession and Shore (1994); Dahlen and Tromp (1998).

Advances in seismic instrumentation, together with the occurrence of the great 1960 Chilean and 1964 Alaska earthquakes, made it possible to identify and study large numbers of modes. Over 40 modes were identified from the 1960 Chilean earthquake. The number of modes that have been observed is now several thousands, due to continued advances in seismometry, which permit recording at very long periods (Section 6.6), an increase in the number of stations, more powerful analytical techniques, and many large earthquakes. Although none of the earthquakes has come close in size to the 1960 Chilean event, the advances in instrumentation and processing largely compensate. Large earthquakes are needed to excite the gravest modes, and long lengths of seismograms are needed to resolve their properties. As discussed in Section 6.3.3, this requires a seismogram that has significant energy extending over a time much longer than a mode's period. Fortunately this is the case for the largest earthquakes, leading to the analogy that the earth rings like a bell after they occur.[8] Seismograms extending for many days are analyzed after the largest earthquakes.

Table 2.9-1 shows the periods of several modes, some of which have been discussed earlier. Note that for the fundamental (${}_0S$ and ${}_0T$) overtone branches, modes with angular orders greater than about 20 correspond to the fundamental mode Rayleigh and Love waves with those periods and are

[7] Earth's gravest observed mode, ${}_0S_2$ corresponds to a note of E, twenty octaves below middle E on a piano. Johannes Kepler, among others, wrote about the "music of the spheres," and thought that each planet's revolution around the sun corresponded to a musical note. The earth's 365.25 day revolution would correspond to a note of C#, 33 octaves below middle C#.

[8] Actually, because the earth vibrates at many frequencies, rather than just one, and is laterally heterogeneous, a better but less poetic analogy would be that the earth rattles like a dented garbage can.

often viewed as traveling waves. However, the longest-period modes, like ${}_0S_2$, ${}_0S_3$, ${}_0T_2$, ${}_0T_3$, etc. have such long periods that we think of them as modes. Higher-order modes are often thought of in terms of a body wave phase to which they contribute. Of course, the descriptions are equivalent.

2.9.8 *Normal mode synthetic seismograms*

As we will see in many places in this text, various techniques are used to create theoretical, often called synthetic, seismograms for the earth. One of these is normal mode summation, analogous to the way the propagating waves on the string were generated in Section 2.2. This summation is also the way that a music synthesizer creates a particular sound by summing the right combination of harmonic overtones (i.e., modes).[9]

For example, torsional mode displacements (Eqn 11) are synthesized by

$$\mathbf{u}^T(r_r, \theta_r, \phi_r) = \sum_n \sum_l \sum_{m=-l}^{l} {}_nA_l^m(r_s, r_r)\, {}_nW_l(r_r)\, \mathbf{T}_l^m(\theta_r, \phi_r)\, e^{i_n\omega_l^m t} e^{-\frac{{}_n\omega_l^m t}{2_nQ_l}}. \tag{23}$$

To do this, we need to know the modes' radial eigenfunctions, ${}_nW_l$ and eigenfrequencies ${}_n\omega_l^m$, which are determined by the earth's velocity and density structure. These modes are then weighted by excitation amplitudes ${}_nA_l^m$, determined by the depth, geometry, and time history of the seismic source and the depth of the receiver. We also need to know the attenuation, or quality, factor ${}_nQ_l$, discussed in Sections 3.7 and 7.4, which measures the rate at which the mode's seismic energy is lost by friction (without this effect, the earth would ring like a bell forever). This formulation assumes that all singlets in a multiplet have the same quality factor.

The modes of the earth are found by computing the radial eigenfunctions and the corresponding eigenfrequencies. Although this process is beyond our scope here, several techniques have been developed to do this. Some involve propagating the values of stresses and displacements from the center of the earth to the surface, layer by layer, while satisfying the boundary conditions at each layer. The frequency of the mode is iterated until the final surface values satisfy the free surface boundary conditions. This process is analogous to that used to determine the periods of the Love waves in the layer over the halfspace example (Section 2.7.3).

The amplitudes, or excitation coefficients, depend on the earthquake's fault geometry. One of the many advantages of evaluating the normal modes in a coordinate system whose pole is at the seismic source is that the radiated energy has strong symmetry. As noted in Section 1.1 and discussed further in Chapter 4, earthquakes radiate energy in a pattern with four-lobed symmetry about the fault plane. Thus any given fault geometry is reflected by various combinations of the $m = 0, \pm 1$, and ± 2 singlets. The excitation also depends on the depth of the source, much as that for the string depended on the source position.[10] An earthquake at 600 km depth strongly excites modes whose eigenfunctions are large at that depth, whereas other modes are barely excited. However, the relative excitations will be very different for an earthquake at 10 km depth. For example, as previously discussed, fundamental mode surface waves correspond to the fundamental ($n = 0$) branch of torsional and spheroidal modes, for angular orders greater than about 20. Because these modes' radial eigenfunctions are small at great depths, a 600 km-deep earthquake does not excite surface waves efficiently.

The modes are summed at a specific receiver location. Thus the displacements in Eqn 23 are expressed in terms of the radius of the source r_s, the radius of the receiver r_r, and the colatitude and azimuth of the receiver, θ_r and ϕ_r. A slightly disturbing feature of the mode sum (Eqn 23) is that both the time functions and the vector spherical harmonics are complex numbers. However, the sum gives the displacement as a real number. Similarly, although individual modes oscillate everywhere on earth at all times, even before a traveling wave from an earthquake could arrive, the mode sum yields waves that arrive after a finite time. Thus, although modes are mathematical objects that are hard to visualize, their sum gives rise to a meaningful physical displacement (Fig. 2.9-12).

Figure 2.9-13 shows a comparison of observed seismograms with synthetic seismograms created using normal mode summation. The fits are good enough that many studies use observed normal mode amplitudes to find the fault geometry and focal depth of earthquakes, especially when they are large and remote from seismometers. This process is an inverse problem, corresponding to the forward problem of generating a synthetic seismogram.

It is worth noting that while the synthetic receivers are usually placed at the surface (where seismometers are), they can also be computed for any depth within the earth. Figure 2.9-14 shows a record section that would be recorded at a distance of 70° from an earthquake if seismometers could be placed a depths ranging from the surface to the core–mantle boundary. We will use this idea shortly to visualize shear wave propagation (Section 3.5.5) by evaluating normal mode synthetic seismograms at 100,000 locations in the mantle.

2.9.9 *Mode attenuation, splitting, and coupling*

So far, we have discussed the modes of a spherically symmetric, nonrotating, purely elastic, and isotropic earth. This

[9] Although the fundamental notes for a clarinet, trumpet, and flute might be the same, playing each instrument excites a different suite of overtones, giving a different sound. For instance, because the open end of the clarinet allows only odd-numbered overtones, the absence of even-numbered overtones contributes to its warm, dark sound. Early synthesizers added only a few overtones, producing a false, tinny sound. Modern synthesizers sum overtones up to and beyond frequencies of 20,000 Hz, the limit of human hearing, so the synthesized sounds can be indistinguishable from those of the actual instruments.

[10] This effect is analogous to the way in which bowing at different locations on a violin string makes different sounds.

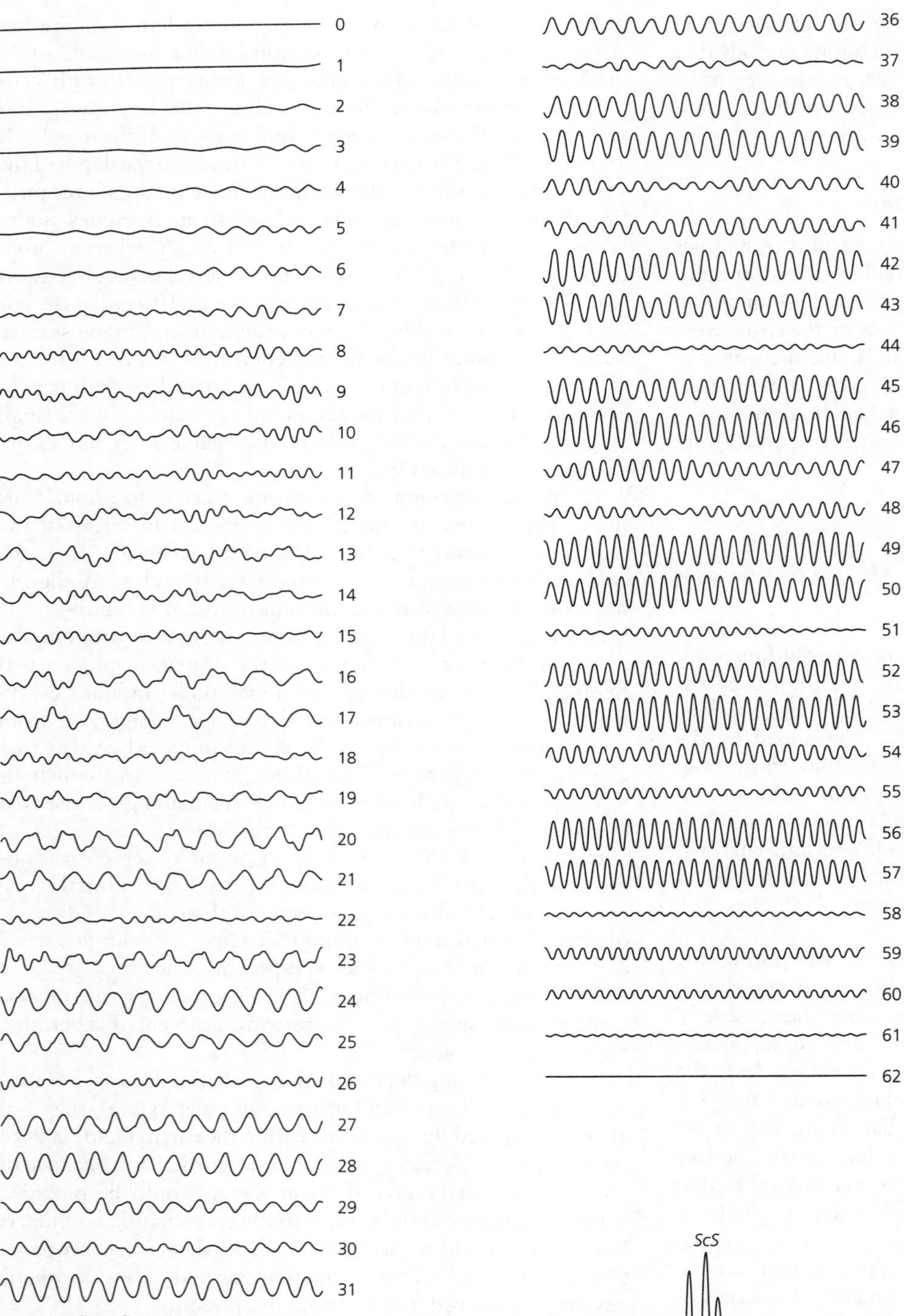

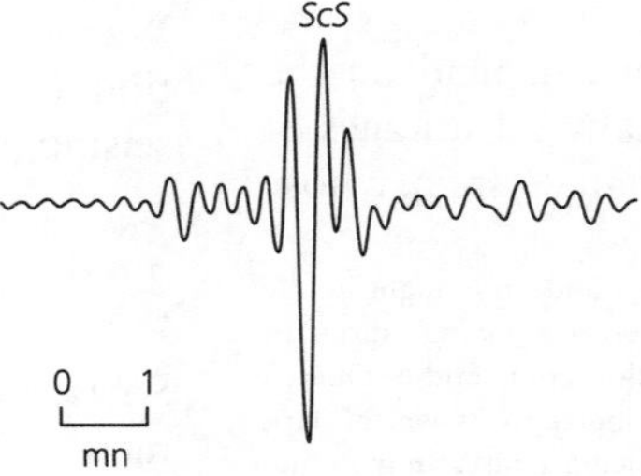

Fig. 2.9-12 Synthesis of a body wave seismogram using torsional normal modes. The numbered lines are mode sums for successive overtone branches, and their sum gives a seismogram including the core reflection *ScS*. (Figure by E. Okal. Reprinted courtesy of E. Okal.)

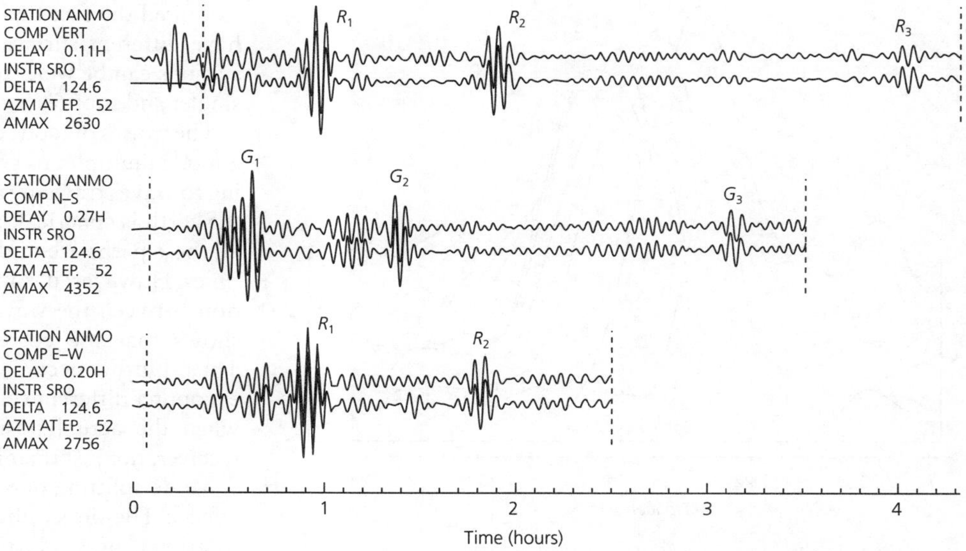

Fig. 2.9-13 Modeling data with normal mode synthetic seismograms. The three pairs are the vertical, north–south, and east–west traces recorded at station ANMO 124.6° from an earthquake in Indonesia. The top trace in each pair shows the data, and the bottom trace is the normal mode synthetic. (Woodhouse and Dziewonski, 1984. *J. Geophys. Res.*, *89*, 5953–86, copyright by the American Geophysical Union.)

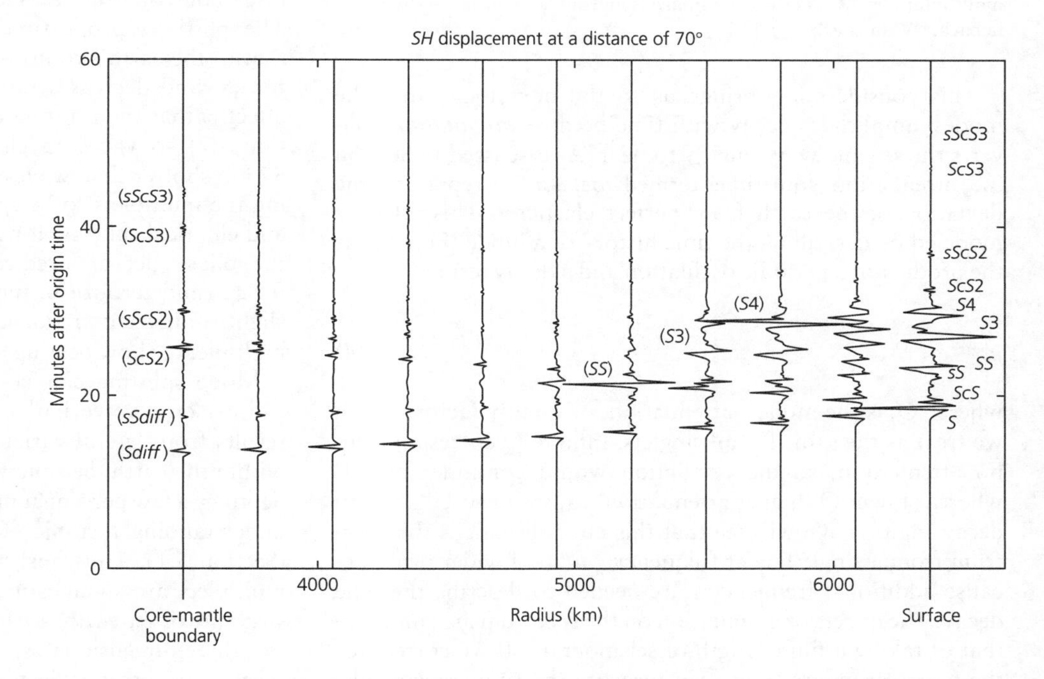

Fig. 2.9-14 Shear wave synthetic seismograms computed at a series of depths, all at a distance of 70° from a 600 km-deep hypothetical earthquake. (After Wysession and Shore, 1994. *Pure Appl. Geophys.*, *142*, 295–310, reproduced with the permission of Birkhauser.)

idealized body, sometimes called a SNREI ("sneery") earth, is a reasonable approximation, because the earth is approximately spherically symmetric and elastic, and its rotation period is long compared to those of the normal modes. In this case, we expect the normal mode spectrum of an earthquake to show sharp peaks for each mode. However, when we look at data like Fig. 2.9-2, we see that some peaks vary in width and that some mode peaks overlap with others. These features reflect the complexities of making measurements of the modes of the real earth.

The first effect worth noting is that seismograms are not infinitely long. Thus each mode's displacement is not a pure sinusoid of single frequency extending for infinite time, but instead stops when the seismogram ends. We will see in Section 6.3.3 that taking a finite portion of a sinusoid broadens its spectrum from a sharp spectral line (a delta function) to a wider peak. Physically, this is because other frequencies are needed to make the time function end rather than go on forever. The shorter the time we use, the worse the broadening is. This problem seems easy to solve, since we can take a seismogram for as long as we want, and thus make peaks narrow. However, we do not want to go on too long, because the longer we wait after an earthquake, the more the earthquake's signal will decay relative to the ground noise, which can include signals from other earthquakes.

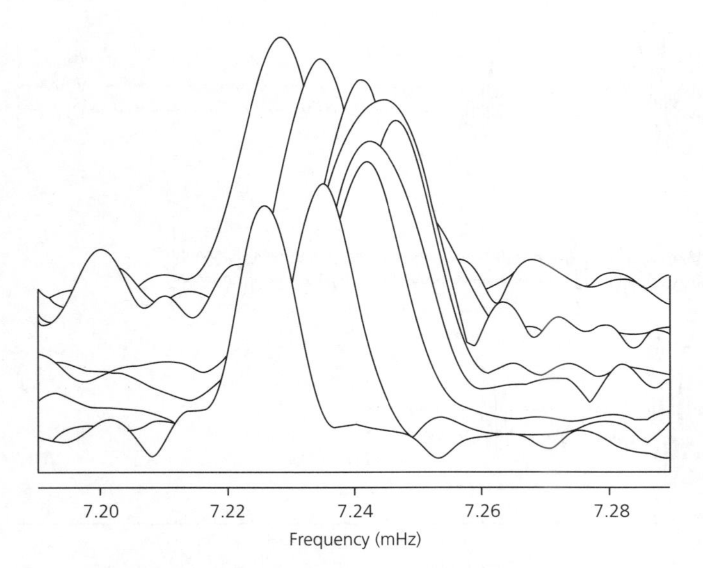

Fig. 2.9-15 Amplitude spectra of the nine singlets of the split spheroidal mode multiplet ${}_{18}S_4$. The $m = -4$ singlet is in front, and the $m = 4$ singlet is in back. (Widmer *et al.*, 1992.)

This consideration brings us to the next issue, that the modes' amplitudes decay with time because *attenuation* converts the seismic wave energy to heat. As discussed in Section 3.7, attenuation (sometimes termed *anelasticity*) represents the deviation of the earth from perfect elasticity. This effect is modeled by describing the time history of a mode (Eqn 23) as the product of a periodic oscillation and a decay term

$$e^{i_n\omega_l^m t} e^{-\frac{_n\omega_l t}{2_nQ_l}}, \tag{24}$$

where ${}_nQ_l$ is the mode's attenuation, or quality factor, which we treat as the same for all singlets. Infinite Q corresponds to no attenuation, so the oscillation would continue forever, whereas lower Q (higher attenuation) causes the oscillation to decay rapidly. We will see that this effect broadens the spectrum from a single line at frequency ${}_n\omega_l^m$ to a wider peak, because additional frequencies are needed to describe the time decay. The effects of attenuation on the spectrum are similar to that of taking a finite length of seismogram. If we correct for the finite seismogram, we can measure the Q of each mode. These data can then be used to determine how anelasticity within the earth varies with depth (Section 7.4).

Other factors can also affect spectral peaks. For a SNREI earth, a mode's frequency depends only on the radial order n and the angular order l, so the $2l + 1$ singlets of different azimuthal order $-l \leq m \leq l$ would have the same eigenfrequency. However, in the real earth, the singlet frequencies vary slightly, causing mode splitting. The split singlets broaden the peak produced by the entire multiplet. Peaks due to individual singlets can sometimes be resolved on high-quality long-period seismograms (Fig. 2.9-15). To identify singlets, the analysis shown used a stacking method (Section 6.5) that exploited the fact that individual singlets within the multiplet have different surface eigenfunctions, so spectra at different stations can be weighted and combined to enhance the desired singlet and suppress others.

The causes of mode splitting can be visualized by considering a mode multiplet to be a superposition of singlets corresponding to waves traveling along different paths around the earth. If the earth is spherical, nonrotating, and spherically symmetric, all these paths are of the same length and have the same travel times. However, if some paths take longer than others, the relation between the wave velocity and eigenfrequency (Eqn 21) shows that the corresponding eigenfrequencies will differ. Thus splitting occurs when waves traveling on different paths encounter different velocities. Put another way, splitting occurs when the actual positions on earth of the source and the receiver, not just their relative positions, matter.

Mode splitting due to the rotation of the earth reflects two effects. The direct effect is that the Coriolis force due to the rotation causes splitting, because waves traveling in the direction of the rotation travel faster than those going the other way. The splitting is proportional to the ratio of the mode's period to that of the earth's rotation (24 hours), so this effect is largest for ${}_0S_2$ and decreases for shorter-period modes. An indirect effect is that the rotating earth takes an elliptical shape (Section A.7), so waves traveling across the poles travel a distance 67 km shorter than waves traveling around the equator, causing the multiplets to be split. Figure 2.9-16 shows rotational and elliptical splitting for the ${}_0S_2$ multiplet. The amplitudes of the split singlets are predicted to be greatest for $m = \pm 1$, smaller for ± 2, and zero for 0. Interference between the singlets with slightly different frequencies causes the time series for the multiplet to show beating (Section 2.8.1).

Mode splitting can be caused by any other process that causes some wave paths to be faster than others. Splitting results from lateral variations in velocity, or inhomogeneity, within the earth. Seismic velocities vary laterally at any given depth by a few percent at most, but these variations are vital for understanding tectonic effects, including mantle convection (Section 5.1). Thus, just as the average frequencies of mode multiplets are significant for determining the radial velocity structure of the earth, so the frequencies of singlets help resolve the three-dimensional structure. Splitting also results from seismic anisotropy (Section 3.6), which occurs when waves traveling in different directions through a region travel at different velocities. For example, Fig. 3.6-13 shows splitting resulting from anisotropy in the inner core.

A related effect is called mode *coupling*. Recall that in the homogeneous string the modes were purely orthogonal and did not interact with each other. Similarly, in the ideal SNREI earth, energy is not transferred from the oscillations of one mode to another. However, real-earth effects like rotation, ellipticity, lateral inhomogeneity, and anisotropy affect not only the eigenfrequencies, but also the eigenfunctions. As a result, the eigenfunction of a given mode contains both the eigenfunction it would have for a SNREI earth and perturbations due to

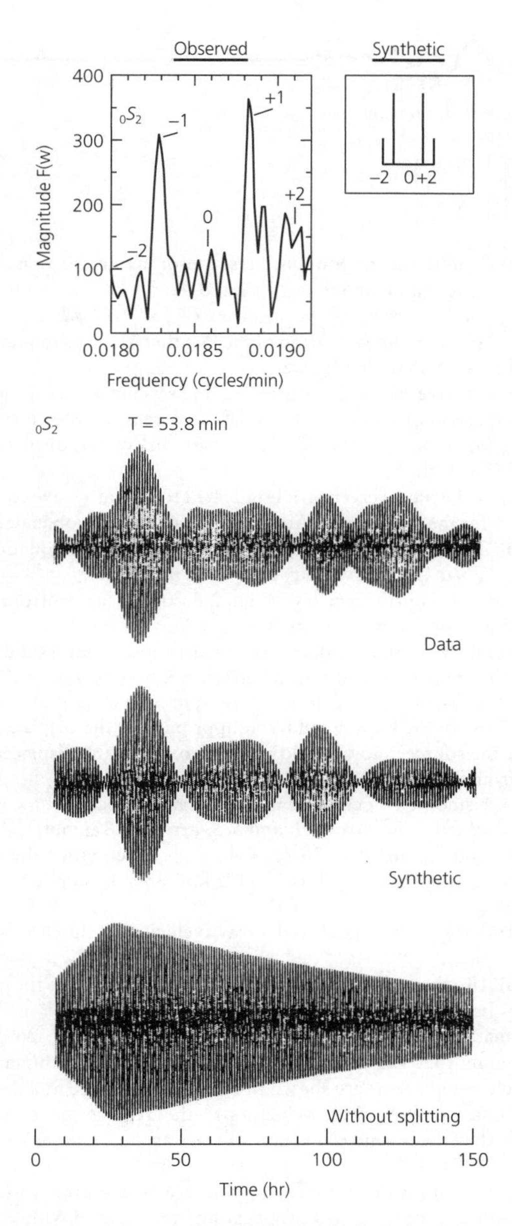

Fig. 2.9-16 Splitting observations for the football mode $_0S_2$ from the great 1960 Chilean earthquake, recorded at station Isabella (California). Splitting causes the singlets to stand out as distinct peaks in the spectrum and the time series to show beating due to interference between the singlets. A synthetic seismogram, computed by predicting the singlet amplitudes and combining them in the time domain with the effects of attenuation and finite seismogram length matches the data better than a similar synthetic seismogram without rotational splitting. (Geller and Stein, 1977; Stein and Geller, 1978. © Seismological Society of America. All rights reserved.)

contributions from the eigenfunctions of some other modes with very similar eigenfrequencies. Coupling can occur between modes on separate branches, between modes on the same branch, and even within a single mode multiplet between different azimuthal orders. Thus, although an earthquake should excite only the $m = 0, \pm 1$, and ± 2 singlets because it radiates energy in a pattern with fourfold symmetry about the fault plane, energy is transferred to the other singlets. Some coupling also occurs between torsional and spheroidal modes, much as plane *P–SV* and *SH* waves can be coupled at a dipping interface (Section 2.5.2). Hence torsional modes can contribute to the radial displacement, which would not be possible for a SNREI earth. As a result, some spectral peaks in Fig. 2.9-2 have several mode labels, corresponding to modes with similar frequencies that are coupled. These composite modes are called supermultiplets.

Although the theory of mode splitting and coupling is beyond our scope, it is worth noting that it is closely allied to conceptually similar problems in other branches of science. The splitting due to earth's rotation is similar to that for waves in a rotating bowl of water, or to the Zeeman effect in atomic physics, where spectral lines are split by a magnetic field. The normal mode problems are addressed by exploring how perturbations to the equation of motion due to rotation, ellipticity, lateral heterogeneity, etc. change the eigenfrequencies and eigenfunctions from those for an unperturbed (SNREI) earth.

In summary, the peaks in a normal mode spectrum reflect the combined effects of the earthquake, spherical and elastic earth structure, attenuation, rotation, ellipticity, lateral heterogeneity, and anisotropy. As a result of extensive studies, these effects are surprisingly well modeled, as shown by the good (though not perfect) agreement between the synthetic and observed spectra in Fig. 2.9-2. Thus, as is so often the case, data showing the deviations of the real earth from a simple model are used to explore these deviations and better describe the real earth.

Further reading

Further information about the topics of this chapter can be obtained from many sources, a few of which are listed here. Basic wave concepts are discussed in books on wave propagation (e.g., Bland, 1988; French, 1971; Main, 1978), classical mechanics (e.g., Feynman *et al.*, 1963; Marion, 1970), and applied mathematics (e.g., Butkov, 1968; Morse and Feshbach, 1953; Menke and Abbott, 1990; Snieder, 2001). Introductions to topics in continuum mechanics are given by Fung (1965, 1969) and Malvern (1969). Fermat's principle, Huygens' principle, and diffraction are discussed in optics texts like Baker and Copson (1950) and Klein and Furtak (1986).

Several introductory texts treat the seismological material in this chapter, including Ewing *et al.* (1957), Officer (1958), Richter (1958), Bullen and Bolt (1985), Lay and Wallace (1995), Shearer (1999), and Udias (1999). Advanced treatments beyond our discussions are given by Aki and Richards (1980), Hudson (1980), Ben-Menahem and Singh (1981), Lapwood and Usami (1981), Kennett (1983), Bath and Berkhout (1984), and Dahlen and Tromp (1998).

A number of sources discuss specific topics that we address. Geller and Stein (1978) discuss string examples like those used here, including of the source term and of the modes of a non-uniform string. Young and Braile (1976) review the solutions for reflection and transmission at a solid–solid interface, and give the computer program used to calculate the energies in Fig. 2.6-11 and 12. Madariaga (1972) derives the equivalence between modes and traveling waves.

Problems

1. What are the reflection and transmission coefficients for a junction between two identical strings? Give a physical interpretation of the result.
2. In Fig. 2.2-6, find the seismic velocities of the two different string segments by measuring the distance versus time slope of the wave pulses on the left and right sides of the figure. Are these velocities the same as the velocities given in the figure caption?
3. For the stress tensor

$$\sigma = \begin{pmatrix} 2 & 1 & 3 \\ 1 & -1 & -2 \\ 3 & -2 & 5 \end{pmatrix}$$

find the traction on
   (a) the x–y plane,
   (b) the y–z plane,
   (c) the plane with normal $(3, 2, -1)$.
4. To derive the reflection coefficient for the end of a string:
   (a) Express the total displacement due to incident and reflected harmonic waves of unknown amplitudes.
   (b) Find the relation between these amplitudes at a fixed string end, where the displacement is zero, and at a free end, where the traction is zero.
5. For the stress tensor

$$\sigma = \begin{pmatrix} 0 & 2 & 0 \\ 2 & 0 & 0 \\ 0 & 0 & 0 \end{pmatrix}$$

   (a) Find the principal stresses and their associated directions.
   (b) Find the surfaces on which the maximum tangential traction occurs, and the value of this traction.
6. Estimate the pressure expected at a depth of 1000 km in the earth.
7. Given the stress tensor, whose elements are in kbar:

$$\sigma = \begin{pmatrix} -150 & -2 & 1 \\ -2 & -155 & 3 \\ 1 & 3 & -145 \end{pmatrix}$$

   (a) What physical situation do the large negative values on the diagonal represent?
   (b) What is the mean stress?
   (c) What is the deviatoric stress tensor?
   (d) At what depth in the earth might this state of stress be found?
8. Give an example of a strain tensor for which there is
   (a) an increase in volume,
   (b) a decrease in volume,
   (c) shear strain but no volume change.
   Which of these strains could result from a P wave, and which could result from an S wave?
9. Estimate by what fraction the volume of a block of a Poisson solid with the rigidity of crustal rock will be compressed at a depth of 30 km relative to its volume at the earth's surface.
10. Determine whether the Lamé constant λ can be negative and, if so, under what conditions.
11. For the strain tensor

$$e = \begin{pmatrix} 3 & 0 & 0 \\ 0 & 1 & 1 \\ 0 & 1 & 2 \end{pmatrix}$$

   (a) Find the corresponding stress tensor, assuming an isotropic solid with Lamé constants λ and μ.
   (b) Find the stored elastic strain energy, $W = \sigma_{ij}e_{ij}/2$.
12. Give a physical interpretation of the fact that Young's modulus for rubber is less than that for steel.
13. An alternative to using potentials to find seismic wave solutions to the equation of motion in terms of displacements is to formulate wave equations for the dilatation and curl of the displacement field. To see this:
   (a) Take the divergence of Eqn 2.4.12 to obtain a wave equation for the dilatation θ. At what velocity does θ propagate?
   (b) Take the curl of Eqn 2.4.12 to obtain a wave equation for $\nabla \times \mathbf{u}$. At what velocity does $\nabla \times \mathbf{u}$ propagate?
14. Derive the constitutive law (Eqn 2.3.70) for an isotropic and linearly elastic material using the c_{ijkl} in Eqn 2.3.69.
15. Derive the ratio of P- and S-wave velocities in a Poisson solid.
16. Use the gradient operator in spherical coordinates (Eqn A.7.14) to find the displacement field from the spherical wave scalar potential $f(t - r/v)/r$. How would you approximate the displacements near the source? How would you approximate the displacements far from the source?
17. On a seismometer located at an earthquake hypocenter, the phases reflected from the core, PcP and ScS, arrive at 8 minutes, 31 seconds, and 15 minutes, 36 seconds, respectively after the earthquake. If the earth's radius is 6371 km, and the core's radius is 3480 km:
   (a) Find the average P- and S-wave velocities in the earth's mantle.
   (b) Use these average velocities to estimate how close the mantle is to a Poisson solid.
18. Estimate the P- and S-wave velocities in the upper mantle by assuming that it is a Poisson solid, and that the earthquake for which seismograms are shown in Fig. 2.4-8 occurred at a depth of 280 km. Compare these velocities to the average mantle values. Note that the seismograms do not start at the earthquake origin time.
19. To get a feel for the distance and time scales in seismic wave propagation, consider waves propagating in a material with velocity 8 km/s.
   (a) Find the wavelengths of waves with periods of 0.1 s, 1 s, and 100 s.
   (b) Find the periods and frequencies of waves with wavelengths of 1 m, 1 km, and 100 km.
20. For waves propagating in an arbitrary direction given by the wavenumber vector $\mathbf{k}$,
   (a) Show that the P-wave displacement due to the scalar potential

$$\phi(\mathbf{x}, t) = e^{i(\omega t - \mathbf{k}\cdot\mathbf{x})}$$

   is parallel to the propagation direction.

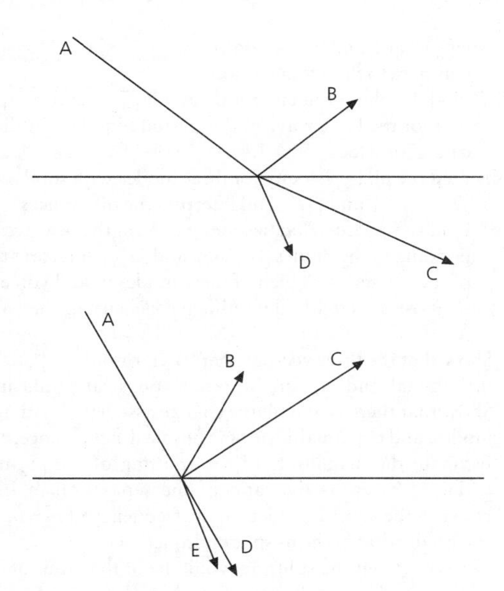

Fig. P2.1 See Problem 22.

(b) Show that the S-wave displacement due to the vector potential

$$\Upsilon(\mathbf{x}, t) = \mathbf{A}e^{i(\omega t - \mathbf{k}\cdot\mathbf{x})}, \quad \mathbf{A} = (A_x, A_y, A_z),$$

is perpendicular to the propagation direction.

21. For a medium composed of upper, middle, and lower layers with velocities of 6, 8, and 10 km/s, calculate the angle of incidence in the 8 and 10 km/s layers for a ray with an incidence angle of 10° in the 6 km/s layer. What is the smallest angle of incidence in the 6 km/s layer that causes total internal reflection at the 8 km/s–10 km/s interface?

22. For the two cases of an incident wave hitting a plane boundary between two media shown in Fig. P2.1,
    (a) Determine which waves are P waves and which are S waves.
    (b) Determine which media are liquid and which are solid.
    (c) For the two media in each case, determine which has the higher P-wave velocity.

23. Consider two rays that originate from a source at $x = 0$, $z = 0$, in a medium with velocity 1 km/s with angles of incidence 0° and 30° (Fig. P2.2). Assume that these rays cross an interface at $z = 2$ km into a medium with velocity 1.5 km/s and travel to the boundary at $z = 4$ km. For each of the ray paths:
    (a) Compute the angle of incidence in the upper layer, the ray path length in each layer, and the total travel time.
    (b) Compute the components and magnitude of the slowness vector $\mathbf{s} = (p, \eta)$ in each layer. Check that the magnitude is related to the velocity as expected.
    (c) Derive the total travel time from the scalar product of slowness and distance $(\mathbf{s} \cdot \mathbf{x})$ for the ray path. Remember to use the appropriate slowness components and horizontal and vertical distances in each layer. Check that these travel times agree with those from (a).

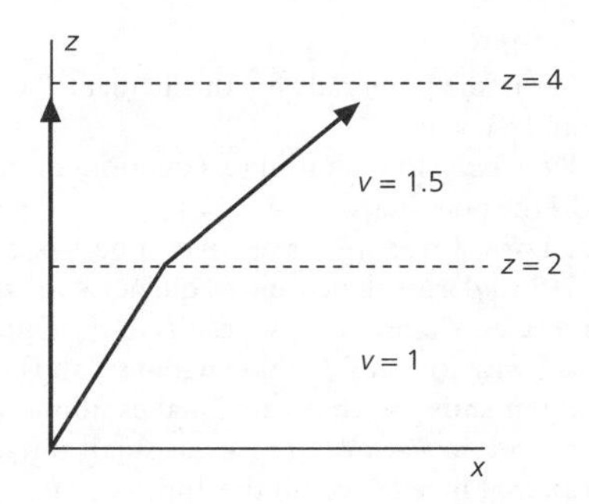

Fig. P2.2 See Problem 23.

24. Fermat's principle problems:
    (a) Use Fermat's principle to show that the angles of incidence for the incident and reflected waves at the surface of a homogeneous halfspace (Fig. 2.5-13) are equal.
    (b) Use the second derivative of the travel time to determine whether the ray path in (a) is a minimum- or a maximum-time path.
    (c) Use the second derivative of the travel time to show that the refracted ray path in Fig. 2.5-14 is a minimum-time path.

25. For an SV wave incident on a free surface:
    (a) Write the potentials for the incident SV wave and reflected P and SV waves.
    (b) Derive the continuity equations at the interface in terms of both the potentials and the amplitude coefficients.
    (c) Assume that the potential reflection coefficients, the ratios of the reflected SV and P potentials to that of the incident SV wave, are

$$\frac{B_2}{B_1} = \frac{4p^2\eta_\alpha\eta_\beta - (\eta_\beta^2 - p^2)^2}{4p^2\eta_\alpha\eta_\beta + (\eta_\beta^2 - p^2)^2}, \quad \frac{A_2}{B_1} = \frac{4p\eta_\beta(\eta_\beta^2 - p^2)}{4p^2\eta_\alpha\eta_\beta + (\eta_\beta^2 - p^2)^2}.$$

    Evaluate the potential reflection coefficients at vertical incidence, and explain the result physically.
    (d) Find the displacement magnitude ratios and energy flux ratios for the two reflected waves relative to the incident wave.
    (e) Show that the energy fluxes satisfy conservation of energy.

26. Show that conservation of energy is satisfied by:
    (a) The energy flux for the incident, reflected and transmitted SH waves at an interface (Eqn 2.6.14).
    (b) The energy flux for the incident P wave and reflected P and SV waves at a free surface (Eqn 2.6.39).

27. For the $ScSp$ conversion at the top of the downgoing slab (Fig. 2.6-15), assume that ScS is traveling vertically in the slab, which dips at 30°. Assume that the velocities in the slab are $\alpha = 9.3$ km/s and $\beta = 5.2$ km/s, and the overlying mantle between the slab and the surface has velocities $\alpha_2 = 8.0$ km/s and $\beta_2 = 4.6$ km/s.
    (a) Find the angle of incidence for ScS and $ScSp$ at the top of the slab and at the earth's surface.
    (b) Use this result and the seismograms shown to estimate the depth of the slab. Bear in mind that the $ScSp$ and ScS arrivals observed at a given station originated from different points on the slab.

28. For a P wave incident on a horizontal solid–solid interface

(Fig. 2.6-9):

(a) Write the potentials for the incident P wave and reflected P and SV waves.

(b) Derive the four continuity equations at the interface in terms of the potentials.

29. For the Love waves in a layer over a halfspace, use the model in Fig. 2.7-9 to derive the cutoff frequencies for the first and second higher modes. Compare these results to the figure.

30. A second way to study the downgoing slab is to use observations from Japan showing that earthquakes about 1300 km away can give rise to two P-wave arrivals, a small direct one and a larger one presumably reflected off the upper surface of the slab (Fig. 2.6-15). Using the geometry and velocities assumed in problem 27:

(a) Determine the angles of incidence at the surface if the apparent velocities of the direct and reflected arrivals are 8.5 and 16 km/s.

(b) Determine the angle of incidence at the slab top of the reflection. To see if the large amplitude of the reflection might occur because of near-critical incidence, compute this critical angle and compare the two.

(c) Suppose that a P-to-S wave conversion also occurred at the slab top. For the converted wave, find the angle of incidence at the slab and the angle of incidence and apparent velocity expected at the surface.

31. For Love waves in a layer over a halfspace, derive a vertical wavelength to show how the displacement oscillates with depth in the layer. Also, derive a vertical decay constant for the halfspace, a distance over which the displacement decays to e^{-1} of its value at the interface. Show how these quantities vary with apparent velocity for a given period. For different modes at a given period, interpret the result in terms of the rate at which the displacement oscillates in the layer and the depth of penetration in the halfspace.

32. For a dispersive wave, derive the following relations between group velocity, phase velocity, wavelength, frequency, and period:

$$\text{(a) } U = c - \lambda \frac{dc}{d\lambda}, \quad \text{(b) } U = c^2 \frac{dT}{d\lambda}, \quad \text{(c) } U = -\lambda^2 \frac{df}{d\lambda}.$$

33. Find the displacements for ${}_0T_3^0$ as functions of θ and ϕ in the manner done for ${}_0T_2^0$ in Eqns 2.9.12 and 2.9.13.

34. (a) Show that for $m = 0$,

$$Y_{l0}(\theta, \phi) = \left(\frac{2l+1}{4\pi}\right)^{1/2} P_l(\cos\theta).$$

(b) Use (a) to find the spherical harmonic Y_{00} associated with radial modes ${}_nS_0$.

(c) Evaluate the vector spherical harmonics associated with the radial modes and explain what the results imply for these modes' displacements.

35. Using the relation between modes and traveling waves and the data in Table 2.9-1:

(a) Because ${}_0T_2$ samples the mantle fairly uniformly (Fig. 2.9-9a), assume that the phase velocity appropriate for this mode is the average mantle shear wave velocity from problem 17 and find the period you would expect. How does this compare to the actual period?

(b) Find the phase velocity for the mode ${}_0T_{130}$ and compare it to that for the Love wave of this period found in the dispersion calculation (Section 2.7.4).

(c) Find the phase velocity for three modes with similar periods: ${}_4T_{67}$, ${}_{10}T_{40}$, and ${}_{13}T_7$, and interpret the differences.

(d) Find the phase velocities and wavelengths of waves corresponding to the modes ${}_0S_3$, ${}_0S_{30}$, and ${}_0S_{130}$. Interpret the trend of the velocities. Which of these modes would you expect to be most affected by lateral heterogeneity in the earth, and why?

36. (a) Show that the three vector spherical harmonics $\mathbf{T}_l^m$, $\mathbf{S}_l^m$, $\mathbf{R}_l^m$ are orthogonal, and explain this result's physical significance.

(b) Show that there is no volume change associated with torsional modes, and explain this result's physical significance.

37. (a) Estimate the magnitude of the splitting of the ${}_0S_2$ multiplet in Fig. 2.9-16a as the ratio of the separation in frequency between the $m = \pm 2$ singlets to the frequency of $m = 0$, which is essentially that of the unsplit multiplet.

(b) We expect that the splitting would be of the order of the ratio of the unsplit mode's period to that of the earth's rotation. Compute this ratio and compare the result to the results of (a).

Computer problems

C-1. Write a subroutine to generate the values of the function $\cos(\omega t - kx)$. Use it to plot the function as a

(a) function of time from $t = 0$ to 10, at $x = 1$, for $\omega = 1$, $k = 1$.

(b) function of time from $t = 0$ to 10, at $x = 0$, for $\omega = 4$, $k = 1$.

(c) function of position from $x = 0$ to 10, at $t = 0$, for $\omega = 1$, $k = 2$.

(d) function of position from $x = 0$ to 10, at $t = 0$, for $\omega = 1$, $k = 4$.

C-2. Write a subroutine that uses the P and S velocities on either side of a solid–solid interface and the angle of incidence for a wave of a specific type to find the angles of reflection and transmission for both P and S waves. The subroutine should calculate and list any possible critical angles for that incident wave, and indicate whether any of the reflected or transmitted waves are past the critical angle.

C-3. Write a program that takes the velocities and densities on either side of a solid–solid interface and finds the vertical incidence displacement reflection and transmission coefficients, and energy flux ratios, for P and S waves incident from either side. Use the program to estimate these quantities for the core–mantle boundary (although it is a solid–liquid boundary), if the lower mantle has $\alpha = 13.7$ km/s, $\beta = 7.2$ km/s, $\rho = 5.5$ g/cm^3, and the core has $\alpha_2 = 8.0$ km/s, $\beta_2 = 0.0$ km/s, $\rho_2 = 9.9$ g/cm^3.

C-4. Write a program, using the result of C-2, to generate figures like the ray paths in Fig. 2.6-11 for an interface with given velocities on either side. Use the program to show the ray paths for the possible incident wave types on either side of a planar interface with the properties of the core–mantle boundary (problem C-3).

3 Seismology and Earth Structure

Ordinary language undergoes modification to a high pressure form when applied to the interior of the earth; a few examples of equivalents follow:

Ordinary meaning:	***High pressure form:***
dubious	*certain*
perhaps	*undoubtedly*
vague suggestion	*positive proof*
trivial objection	*unanswerable argument*
uncertain mixture of all the elements	*pure iron*

Francis Birch, 1952

3.1 Introduction

A major application of seismology is the determination of the distribution of seismic velocities, and hence elastic properties, within the earth. This distribution, known as *earth structure*, gives the basic constraint on the mineralogical, chemical, and thermal state of the earth's interior. Seismological data are important for this purpose because their resolving power is generally superior to that of other geophysical methods. For example, although gravity and magnetic data indicate the presence of a dense fluid core at depth, they provide only relatively weak constraints on its density and size. By contrast, seismological data indicate the depth of the core–mantle boundary and the sharp change in properties that occurs there. Above the boundary, both *P* and *S* waves propagate in the solid mantle, whereas in the liquid outer core no *S* waves propagate and the *P*-wave velocity drops sharply. The observed velocities are the primary basis for our models of the physical properties and chemical composition of the material on either side of this boundary. Similarly, the distinction between the crust and the mantle and many inferences about their structure and composition come from seismological observations. More generally, by establishing the essentially layered structure of the earth, seismology provides the primary evidence for the process of differentiation whereby material within planets became compositionally segregated during their evolution. As a result, many crucial issues about the other terrestrial planets could be resolved if seismological data were available.

Constraints from seismology are crucial for other disciplines of the earth sciences, and vice versa. Seismology gives *earth models* describing the distribution of *P*- and *S*-wave velocities and density. Going from an earth model to a description of the chemical, mineralogical, thermal, and rheological state of the earth's interior requires additional information. There are thus two types of uncertainty in our knowledge of the earth's interior. In some cases, such as the structure of the inner core, the seismological results are still under discussion. In others — for example, the nature of the 660 km discontinuity in the mantle — the basic seismological results are generally accepted, but their mineralogic and petrologic interpretations remain under investigation. Given our scope here, we only summarize the implications of seismological data for models of the earth's interior.

The fundamental data for seismological studies of the earth's interior are the travel times of seismic waves. The measurements available are the arrival times of seismic waves at receivers. To convert these to travel times, the origin time and location of the source must be known. These parameters, which are known for artificial sources, must be estimated from the observations for earthquake sources. Hence travel time data include information about both the source and the properties of the medium, and separating the two is a challenge in many seismological studies.

The travel times are used to learn about the velocity structure between the source and the receiver. As we saw in the last chapter, waves follow paths that depend on the velocity structure. Hence the structure must be known to find the paths that the waves took. To illustrate this, consider the travel time between two points. If the velocity were constant, the ray path would be a straight line, and the velocity could be found by dividing the distance by the travel time. If, instead, an interface separates media with different velocities, the ray path would consist of two line segments, depending on the velocities, and the travel time would be the sum of the time spent along each segment. For a more complicated velocity distribution, the ray path would also be more complicated.

This problem can be posed mathematically by writing the travel time between the source (s) and receiver (r) as the integral of 1/velocity, or slowness, along the ray path

$$T(s, r) = \int_s^r \frac{1}{v(x)} dx. \quad (1)$$

In simple cases, where the ray path is a set of segments with constant velocity, the integral is just a sum over the time in each segment. Thus the travel time gives an integral constraint on the velocity distribution between the source and the receiver, but does not indicate which of the many paths satisfying the constraint the ray followed. As a result, an individual measurement is inadequate to show the distribution of velocities. Fortunately, as we shall see, a set of travel times between different sources and receivers provides much more information. In addition, useful information is derived from the amplitudes and waveforms of seismic waves.

This example illustrates an interesting feature of determining velocity structure from travel times. If the velocity structure is known, the forward problem of finding the travel times and amplitudes is straightforward. However, the inverse problem of using the travel times and amplitudes measured at the surface to find the velocity structure at depth is more difficult, and various methods are used. For example, in addition to using travel times directly, we have seen that velocity structure is studied using the dispersion of surface waves (Section 2.8) and the eigenfrequencies of normal modes (Section 2.9), quantities that correspond to travel times.

In this chapter, we follow the approach discussed in Section 1.1.2 of treating the earth with a series of progressively more complex and, hopefully, more accurate models. We begin with the homogeneous, isotropic, elastic, layered halfspace used in Chapter 2 to derive seismic wave propagation. This approximation of uniform flat layers is often used in crust and upper mantle studies, where the distance between source and receiver is less than a few hundred kilometers. We then consider larger source–receiver distances, for which spherical geometry is required, and then the anisotropic and anelastic behavior of the earth. Throughout these discussions, we will see that although velocity varies primarily with depth, there are important lateral variations, or heterogeneities. Finally, we consider the implications of the observed heterogeneous, anisotropic, and anelastic velocity structure for the composition of the earth. Later, in Chapter 7, we discuss further how seismic data can be used to study laterally variable velocity structure.

3.2 Refraction seismology

3.2.1 *Flat layer method*

The simplest approach to the inverse problem of determining velocity at depth from travel times treats the earth as flat layers of uniform-velocity material. We thus begin by deriving the travel time curves for such a model, which show when seismic waves arrive at a particular distance from a seismic source. The travel times, especially those of waves that are critically refracted at the interfaces, are used to find the velocities of the layers and underlying halfspace and the layer thicknesses. As a result, this technique is called *refraction seismology*.

Refraction seismology is used on vastly differing scales. Near-surface structure at depths less than 100 meters can be studied using a sledge hammer or a shotgun as a source and a single receiver. Similar methods are used to study the crust and the upper mantle, with earthquake or explosion sources and many receivers at distances of hundreds of kilometers.

The simplest situation, shown in Fig. 3.2-1, is a layer of thickness h_0, with velocity v_0, overlying a halfspace with a higher velocity, v_1. We write the velocities as "v" to indicate that the analysis applies for either P or S waves. There are three basic ray paths from a source on the surface at the origin to a surface receiver at x. The travel times for these paths can be found using Snell's law.

The first ray path corresponds to a *direct wave* that travels through the layer with travel time

$$T_D(x) = x/v_0. \quad (1)$$

This travel time curve (Fig. 3.2-2) is a linear function of distance, with slope $1/v_0$, that goes through the origin.

The second ray path is for a wave reflected from the interface. Because the angles of incidence and reflection are equal,

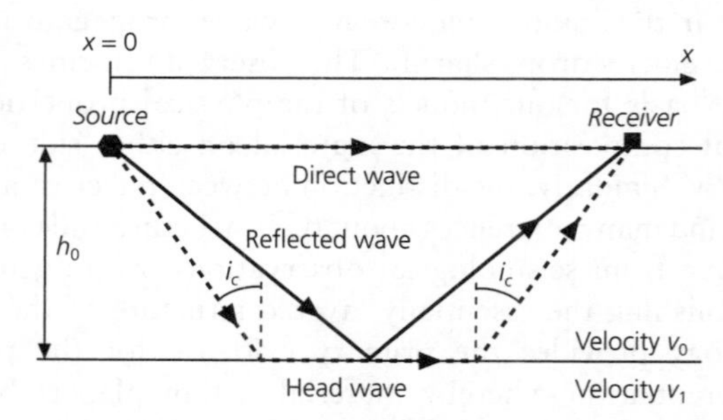

Fig. 3.2-1 Three basic ray paths for a layer over a halfspace model. The direct and reflected rays travel within the layer, whereas the head wave path also includes a segment just below the interface. For the head wave to exist, the layer velocity v_0 must be less than the halfspace velocity v_1.

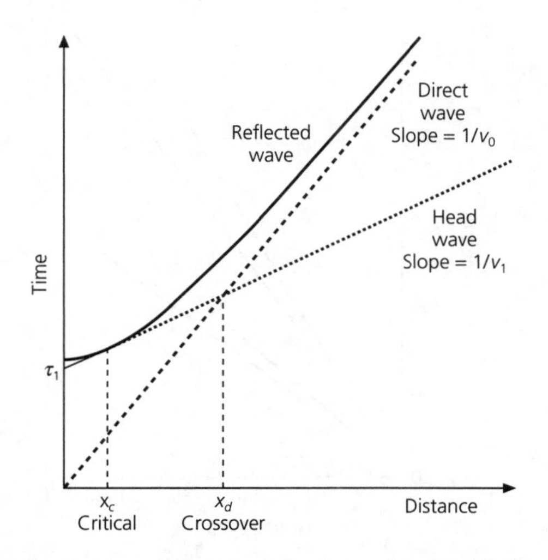

Fig. 3.2-2 Travel time versus source-to-receiver distance plot for the three ray paths in Fig. 3.2-1. The direct wave is the first arrival for receivers closer than the crossover distance x_d. Beyond x_d the head wave arrives first. The head wave exists only beyond the critical distance x_c.

the wave reflects halfway between the source and the receiver. The travel time curve can be found by noting that $x/2$ and h_0 form two sides of a right triangle, so

$$T_R(x) = 2(x^2/4 + h_0^2)^{1/2}/v_0. \tag{2}$$

This curve is a hyperbola, because it can be written

$$T_R^2(x) = x^2/v_0^2 + 4h_0^2/v_0^2. \tag{3}$$

For $x = 0$ the reflected wave goes straight up and down, with a travel time of $T_R(0) = 2h_0/v_0$. At distances much greater than the layer thickness ($x \gg h$), the travel time for the reflected wave asymptotically approaches that of the direct wave.

The third type of wave is the *head wave*, often referred to as a refracted wave. This wave results when a downgoing wave impinges on the interface at an angle at or beyond the critical angle. Its travel time can be computed by assuming that the wave travels down to the interface such that it impinges at the critical angle, then travels just below the interface with the velocity of the lower medium, and finally leaves the interface at the critical angle and travels upward to the surface. Thus the travel time is the horizontal distance traveled in the halfspace divided by v_1 plus that along the upgoing and downgoing legs divided by v_0:

$$T_H(x) = \frac{x - 2h_0 \tan i_c}{v_1} + \frac{2h_0}{v_0 \cos i_c}$$
$$= \frac{x}{v_1} + 2h_0\left(\frac{1}{v_0 \cos i_c} - \frac{\tan i_c}{v_1}\right). \tag{4}$$

The last step used the fact that the critical angle (Section 2.5.5) satisfies

$$\sin i_c = v_0/v_1. \tag{5}$$

To simplify Eqn 4, we use trigonometric identities showing that

$$\cos i_c = (1 - \sin^2 i_c)^{1/2} = (1 - v_0^2/v_1^2)^{1/2} \tag{6}$$

and

$$\tan i_c = \frac{\sin i_c}{\cos i_c} = \frac{v_0/v_1}{(1 - v_0^2/v_1^2)^{1/2}}, \tag{7}$$

so Eqn 4 can be written

$$T_H(x) = x/v_1 + 2h_0(1/v_0^2 - 1/v_1^2)^{1/2} = x/v_1 + \tau_1. \tag{8}$$

Thus the head wave's travel time curve is a line with a slope of $1/v_1$ and a time axis intercept of

$$\tau_1 = 2h_0(1/v_0^2 - 1/v_1^2)^{1/2}. \tag{9}$$

This intercept is found by projecting the travel time curve back to $x = 0$, although the head wave appears only beyond the *critical distance*, $x_c = 2h_0 \tan i_c$, where critical incidence first occurs.

Because $1/v_0 > 1/v_1$, the direct wave's travel time curve has a higher slope but starts at the origin, whereas the head wave has a lower slope but a nonzero intercept. At the critical distance the direct wave arrives before the head wave. At some point, however, the travel time curves cross, and beyond this point the head wave is the first arrival even though it traveled a longer path. The *crossover distance* where this occurs, x_d, is found by setting $T_D(x) = T_H(x)$, which yields

$$x_d = 2h_0\left(\frac{v_1 + v_0}{v_1 - v_0}\right)^{1/2}. \tag{10}$$

Hence the crossover distance depends on the velocities of the layer and the halfspace and the thickness of the layer.[1]

Thus we can solve the inverse problem of finding the velocity structure at depth from the variation of the travel times observed at the surface as a function of source–receiver distance. This simple structure is described by three parameters. The two velocities, v_0 and v_1, are found from the slope of the two travel time curves. We then identify the crossover distance and use Eqn 10 to find the third parameter, the layer thickness, h_0. Alternatively, the layer thickness can be found from the reflection time or the head wave intercept (Eqn 9) at zero distance. Each of these methods exploits the fact that there is more than one ray path between the source and the receiver.

[1] A simple analogy is driving to a distant point by a route combining streets and a highway. If the destination is far enough away, it is quicker to take a longer route including the faster highway than a direct route on slower streets. The point at which this occurs depends on the relative speeds and the additional distance required to use the highway.

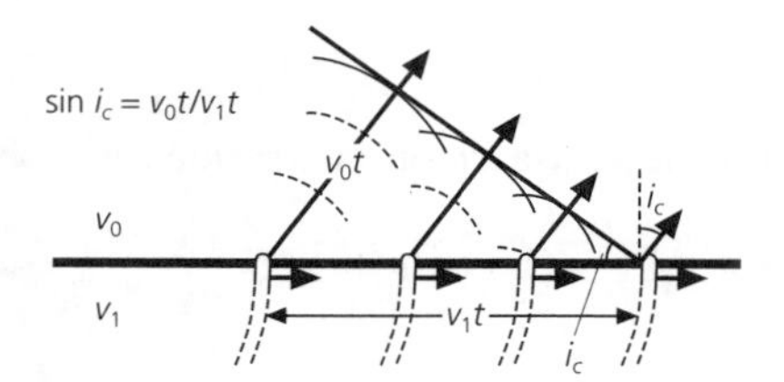

Fig. 3.2-3 Generation of an upgoing head wave by Huygens' sources due to a refracted pulse propagating along a boundary. The head wave travels in the upper layer at a slower velocity (v_0) than the refracted wave creating it, which travels in the layer below at velocity v_1. (After Griffiths and King, 1981.)

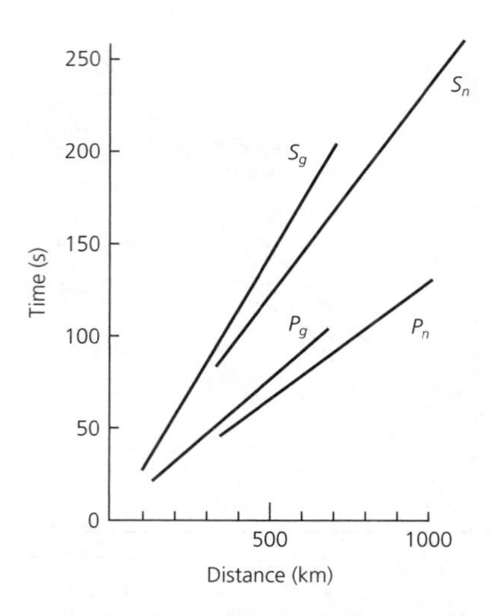

Fig. 3.2-4 Schematic of Mohorovičić's results showing the existence of a distinct crust and mantle. The travel time curves are labeled using modern nomenclature: the direct waves are P_g and S_g, and the head waves are P_n and S_n. (After Bonini and Bonini, 1979. *Eos*, 60, 699–701, copyright by the American Geophysical Union.)

Despite this solution's elegance, the basic assumption about the travel time of the head wave may seem unsatisfying, because it is unclear why energy should follow this path. However, the result conforms with observations — the experiment diagrammed in Fig. 3.2-1 yields an arrival whose travel time is given by Eqn 8. To understand why, we can view the head wave in several ways. As shown in this chapter's problems, it corresponds to a minimum time path between the source and the receiver, so, by Fermat's principle (Section 2.5.9), we expect such a wave. Another approach, using Huygens' principle (Section 2.5.10), is to consider the refracted wave traveling horizontally below the boundary at the velocity of the half-space, generating spherical waves that propagate upward in the lower-velocity layer (Fig. 3.2-3). The spherical waves interfere to produce upgoing plane waves that leave the interface at the critical angle.[2] However, our analysis of postcritical incidence (Section 2.6.4), which showed that an evanescent wave propagates along the interface, does not fully describe the head wave. A more sophisticated analysis than is appropriate here shows that the geometry in Fig. 3.2-1 gives the head wave's travel time, but not its amplitude, because geometrical optics are not applicable. Thus, although the energy propagation is more complicated than along the geometric ray path, the travel time predicted is correct.

Seismic refraction data led A. Mohorovičić[3] in 1909 to one of the most important discoveries about earth structure. Observing two P arrivals (Fig. 3.2-4), he identified the first as having traveled in a deep high-velocity (7.7 km/s) layer, and the second as a direct wave in a slower (5.6 km/s) shallow layer about 50 km thick. These layers, now identified around the world, are known as the *crust* and the *mantle*. The boundary between them is known as the Mohorovičić discontinuity, or Moho. We now denote the head wave as P_n and the direct wave as P_g ("g" for "granitic"). Corresponding arrivals are also observed for S waves. The Moho, which defines the boundary between the crust and the mantle, has been observed around the world. One of the first steps in studying the nature of the crust is characterizing the depth to Moho, or crustal thickness, and the variation in P_n velocity from site to site.

Travel time plots for refraction experiments can be made by displaying seismograms in *record sections*. Because seismograms are functions of time, aligning several as a function of distance yields a travel time plot showing the different arrivals. Figure 3.2-5 shows a record section of a profile of seismograms recorded in England from explosive sources. In addition to P_n and P_g, the reflection off the Moho, known as P_mP, is well recorded. As expected, the direct and head wave travel times are linear with distance, whereas the reflection has a hyperbolic curvature. The figure is plotted as a *reduced travel time plot*, in which the time shown is the true time minus the distance divided by a constant velocity. This reduces the size of the plot, and makes waves arriving at the reducing velocity appear as a line parallel to the distance axis.

The geometry discussed here can correspond to different physical experiments. A single source can be recorded simultaneously at receivers at different distances. Alternatively, multiple sources at different distances can be recorded by a single receiver at different times. A single receiver can be moved away from a fixed source, so the same source is recorded at different distances. Similarly, a source can be moved away from a fixed receiver. Results of various experiments can be combined, using the principle of *reciprocity*, which states that the travel time is unchanged if the source and the receiver are interchanged. As a result, we can use travel time measurements without considering whether the source was at one position and the receiver at another, or the reverse. Moreover, because earth structure presumably is not changing during the experiment, data collected at different times can be combined.

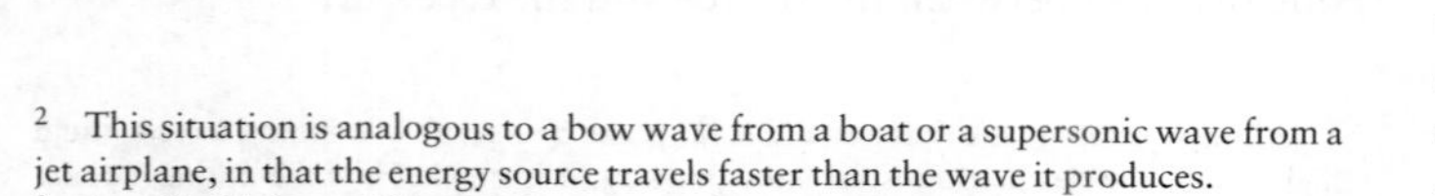

[2] This situation is analogous to a bow wave from a boat or a supersonic wave from a jet airplane, in that the energy source travels faster than the wave it produces.

[3] Andrija Mohorovičić (1857–1936), working in Zagreb, Croatia (then part of the Austro-Hungarian Empire), studied travel times from earthquakes in the region using recently invented pendulum seismographs.

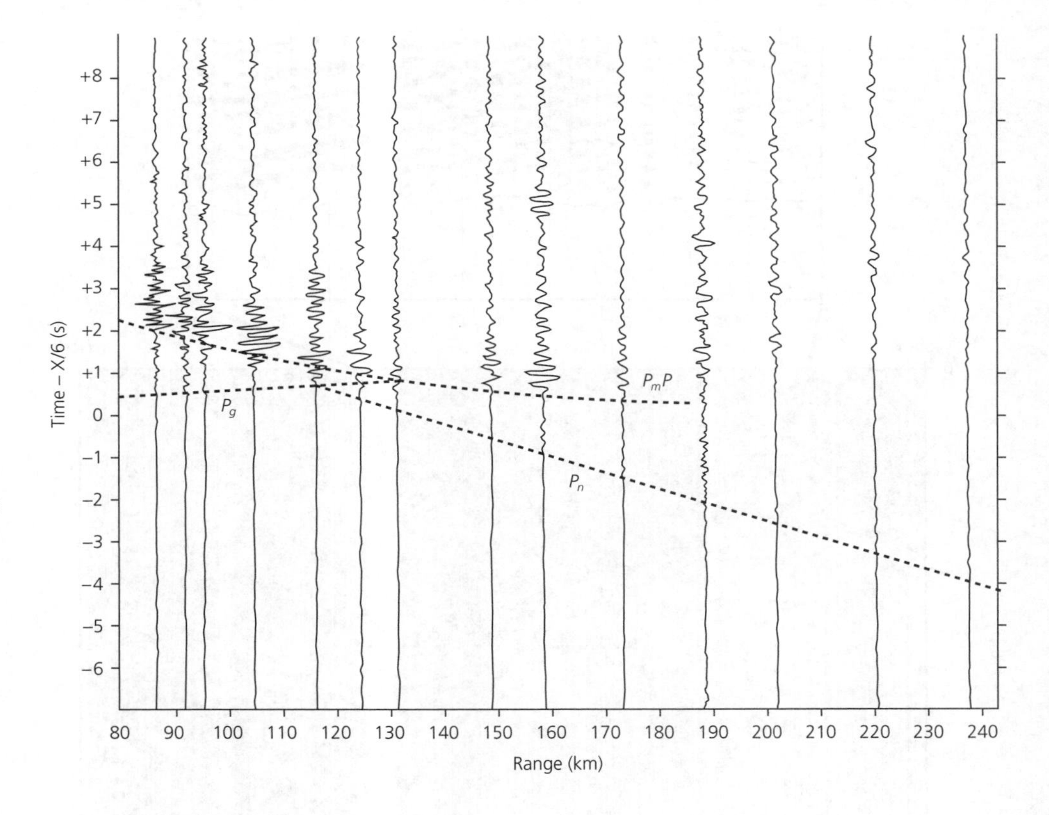

Fig. 3.2-5 Seismograms from a refraction profile, plotted with a reducing velocity of 6 km/s. The direct wave P_g, Moho head wave P_n, and Moho reflection P_mP are observed. P_g does not asymptotically approach P_mP as in Fig. 3.2-2 because the crust, instead of being homogeneous, has increasing velocity with depth. (Bott *et al.*, 1970. From *Mechanism of Igneous Intrusion*, ed. G. Newall and N. Rast, © 1970 by John Wiley & Sons Ltd. Reproduced by permission.)

Refraction data often show other arrivals in addition to P_g, P_n, and P_mP. Figure 3.2-6 shows a record section that also contains head waves P_i and P_n2 from boundaries within the crust and the mantle and P_iP, a reflection off a mid-crustal interface, which is analogous to the P_mP reflection off the Moho.

Such data require a model with multiple layers. Figure 3.2-7 shows a model in which a head wave arises at each interface where the velocity increases with depth. The travel time curve for a head wave at the top of the nth layer is a line with slope $1/v_n$, that can be extrapolated to its intercept on the t axis, τ_n, and written

$$T_{H_n}(x) = x/v_n + \tau_n, \tag{11}$$

where, by analogy to the layer over the halfspace case (Eqn 9),

$$\tau_n = 2\sum_{j=0}^{n-1} h_j(1/v_j^2 - 1/v_n^2)^{1/2}. \tag{12}$$

The thickness of successive layers can be found by starting with the top layer, whose thickness h_0 is given by Eqn 9 or 10, and continuing downward using the iterative formula

$$h_{n-1} = \frac{\tau_n - 2\sum_{j=0}^{n-2} h_j(1/v_j^2 - 1/v_n^2)^{1/2}}{2(1/v_{n-1}^2 - 1/v_n^2)^{1/2}}. \tag{13}$$

Thus for two layers over a halfspace, the thickness of the second layer is found by setting $n = 2$, so

$$h_1 = \frac{\tau_2 - 2h_0(1/v_0^2 - 1/v_2^2)^{1/2}}{2(1/v_1^2 - 1/v_2^2)^{1/2}}. \tag{14}$$

A few examples illustrate some other complexities of refraction experiments. If the velocity increases with depth, the travel time curve for the head wave at the top of each successive layer has a shallower slope. By contrast, a low-velocity layer (Fig. 3.2-8) does not cause a head wave, so the travel time curve does not have a first arrival with the corresponding velocity, and depths to interfaces calculated using Eqn 13 are incorrect. Another possible problem occurs if a layer is thin or has a small velocity contrast with the one below it. Although a head wave results, it may never appear as a first arrival (Fig. 3.2-9), causing a *blind zone* that can be missed in the interpretation.

3.2.2 *Dipping layer method*

The refraction method can also be applied if the interfaces between layers are not horizontal. Conducting a *reversed profile* yields the travel times for ray paths in both the down-dip and the up-dip directions. This can be done using receivers on either side of a source, sources on either side of a receiver, or both. In this geometry, the depths to the interface below the source and the receiver differ due to the dip angle, θ. Consider

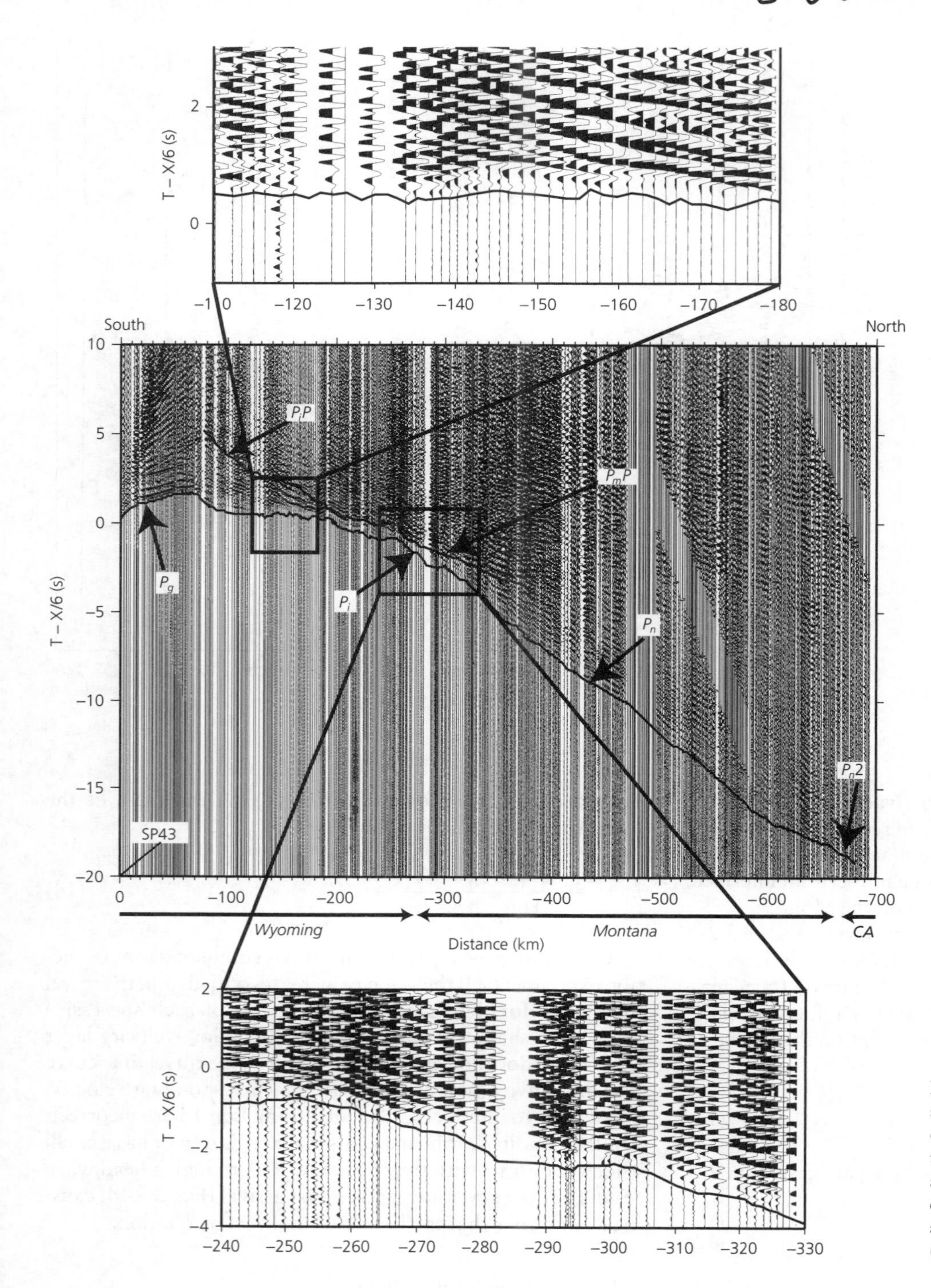

Fig. 3.2-6 Seismic refraction record section, plotted with a reducing velocity of 6 km/s. In addition to P_g, P_n, and P_mP, there are also arrivals P_i and P_n2 interpreted as head waves from boundaries within the crust and the mantle, and P_iP, interpreted as a reflection off a mid-crustal interface. (Snelson *et al.*, 1998.)

the down-dip ray path (Fig. 3.2-10) from a source, below which the perpendicular distance to the interface is h_d, to a receiver at a distance x, below which the perpendicular distance to the interface is $(h_d + x \sin \theta)$. The travel time for the head wave in the down-dip direction is the sum of the distance along the interface divided by v_1 plus that for the upgoing and downgoing legs divided by v_0

$$T_d(x) = \frac{x \cos \theta - (2h_d + x \sin \theta) \tan i_c}{v_1} + \frac{(2h_d + x \sin \theta)}{v_0 \cos i_c}. \quad (15)$$

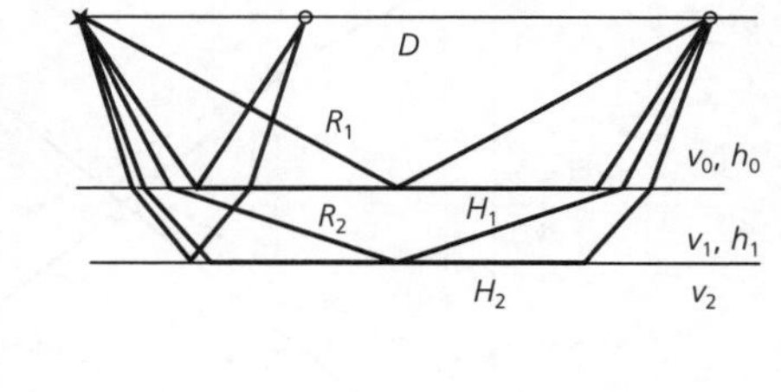

Fig. 3.2-7 Ray paths and travel times for a multilayered model in which velocity increases with depth. Each layer gives rise to a head wave H_i, whose intercept on the time axis is τ_i, and a reflection R_i. The direct wave arrival is also shown.

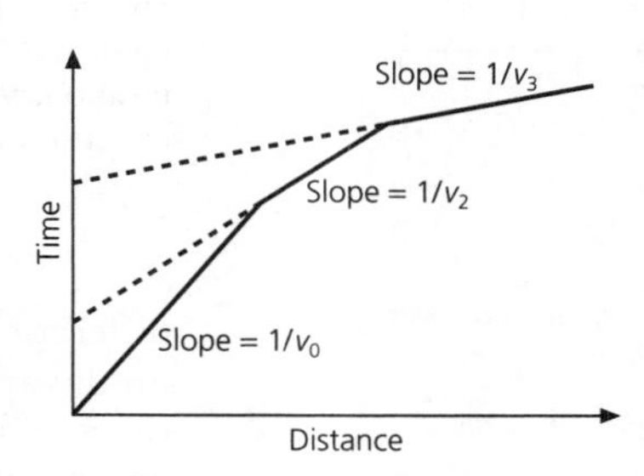

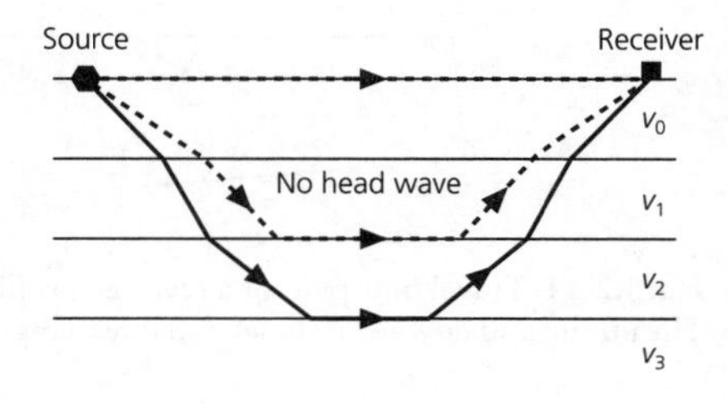

Fig. 3.2-8 Travel time curves, showing first arrivals only, for a model with three layers over a halfspace. Because the middle layer is a low-velocity layer with $v_1 < v_0$, no head wave arises at its top.

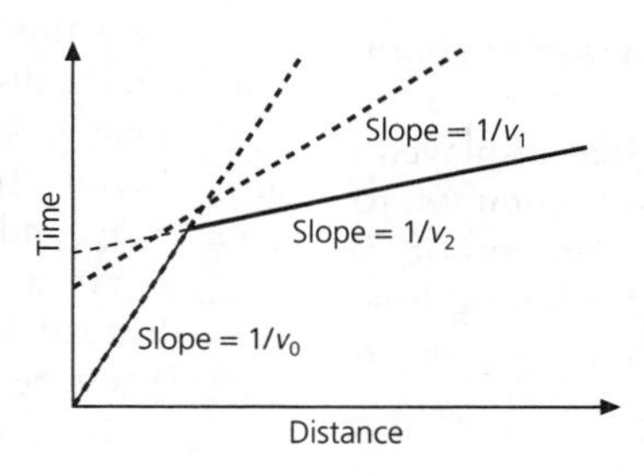

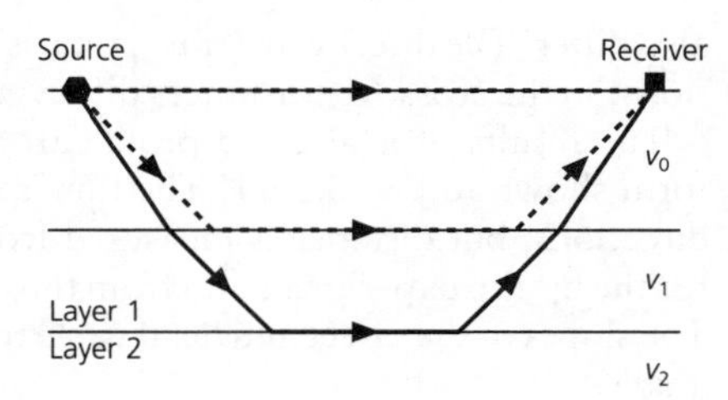

Fig. 3.2-9 Travel time curves, showing first arrivals only, for a blind zone geometry where the head wave from the top of layer 1 is never the first arrival because this layer is too thin.

Fig. 3.2-10 Head wave ray path in the down-dip direction for a dipping interface over a higher-velocity halfspace. The layer thickness is measured perpendicular to the interface.

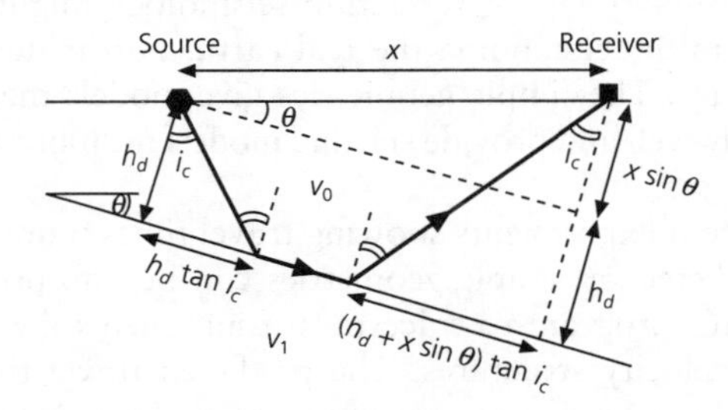

For the flat case, ($\theta = 0$), this is just Eqn 4. Simplifying using Eqns 5 and 7 yields

$$T_d(x) = \frac{x \cos\theta \sin i_c}{v_0} + \frac{(2h_d + x \sin\theta)(1 - \sin^2 i_c)}{v_0 \cos i_c}$$

$$= \frac{x \sin(i_c + \theta)}{v_0} + \frac{2h_d \cos i_c}{v_0} = \frac{x}{v_d} + \tau_d, \qquad (16)$$

which is a straight line with slope $1/v_d$ and intercept τ_d.

Similarly, the travel time for the head wave in the up-dip direction is

$$T_u(x) = \frac{x \sin(i_c - \theta)}{v_0} + \frac{2h_u \cos i_c}{v_0} = \frac{x}{v_u} + \tau_u, \qquad (17)$$

where h_u is the perpendicular distance to the interface below the receiver. Thus the apparent velocities, corresponding to the slopes of the head wave travel time curves, differ in the up-dip and down-dip directions by a factor depending on the dip angle,

$$v_u = v_0/\sin(i_c - \theta) \qquad v_d = v_0/\sin(i_c + \theta). \qquad (18)$$

The apparent velocity in the up-dip direction is greater than the halfspace velocity, and that in the down-dip direction is smaller. The time axis intercepts

$$\tau_u = 2h_u \cos i_c/v_0, \qquad \tau_d = 2h_d \cos i_c/v_0, \qquad (19)$$

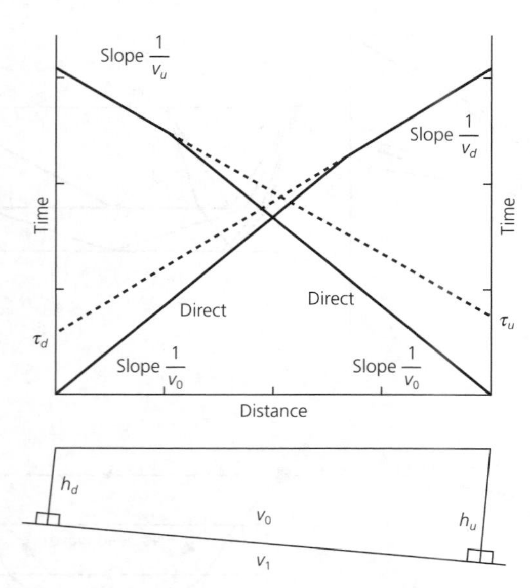

Fig. 3.2-11 Travel time plot for a reversed profile and its interpretation. The up-dip and down-dip slopes and intercepts differ.

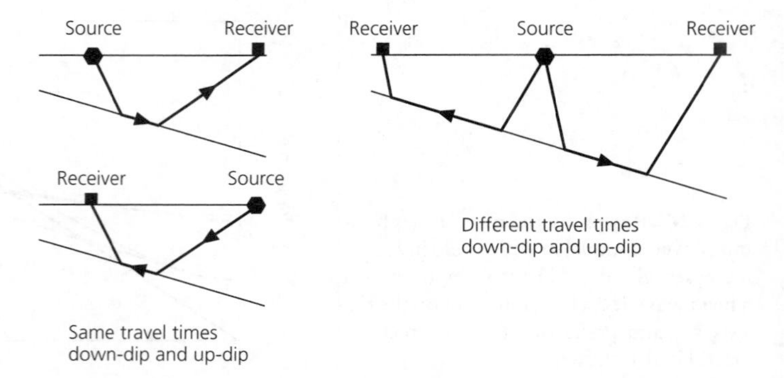

Fig. 3.2-12 *Left*: If the source and the receiver are interchanged on a reversed refraction profile, the travel time is unchanged. *Right*: Different up-dip and down-dip travel times occur because, for a given source position, waves going the same distance along the surface in opposite directions sample the dipping interface differently.

also differ. The direct wave travel time is the same in both directions, so the crossover distances differ.

The results of a reversed profile are often displayed in the form shown in Fig. 3.2-11. The time axis is common to both directions, but distance is measured from one end of the axis for the up-dip experiment and from the other for the down-dip. The slopes of the direct and head wave travel times yield the dip angle

$$\theta = \frac{1}{2}\left(\sin^{-1}\frac{v_0}{v_d} - \sin^{-1}\frac{v_0}{v_u}\right) \tag{20}$$

and the critical angle

$$i_c = \frac{1}{2}\left(\sin^{-1}\frac{v_0}{v_d} + \sin^{-1}\frac{v_0}{v_u}\right). \tag{21}$$

The halfspace velocity v_1 is found from the critical angle and v_0, and the intercept times then yield the layer thickness.

Two additional points about reversed profiles are worth noting. First, the different up-dip and down-dip head wave travel time curves do not imply that for a given pair of locations, it makes a difference whether the source is up-dip and the receiver down-dip, or the reverse (Fig. 3.2-12). By reciprocity, the two experiments give the same travel time. Thus, for a ray path connecting two points, it does not matter whether the wave travels up-dip or down-dip. By contrast, for two receivers at the same distance from a source, one up-dip and one down-dip, the travel times differ because the ray paths encounter the dipping interface at different depths. Similarly, the travel times differ for two sources at the same distance from a receiver, one up-dip and one down-dip. If the dip were zero, then the travel times would be the same for all these cases because all ray paths encounter the interface at the same depth. Another way to view this is that for a flat geometry the travel time depends only on the distance between the source and the receiver. For a dipping geometry, the position as well as the separation matters, because the depth to the interface varies.

Second, the dip found from a reversed profile is not a true dip if the profile is not perpendicular to the strike of the layer. Instead, the measured dip is an apparent dip along the profile. The true dip can be found from the apparent dips along two reversed profiles that cross at a reasonably large angle, using a standard technique in structural geology.

3.2.3 Advanced analysis methods

Because the analysis above has been for simple geometries and uniform-velocity layers, refraction seismology might seem of little use in understanding the real earth. Fortunately, this is not the case. The simple geometries give models that fit data reasonably well and provide starting models for more sophisticated analyses.

Data from experiments showing travel times more complex than predicted by simple geometries can be interpreted with a computer program to trace rays using Snell's law through possible velocity structures. The predicted travel time curve is found by taking rays that arrive at a given distance, and integrating the slowness along their paths (Eqn 3.1.1). Figure 3.2-13 shows a record section and the inferred velocity structure for a refraction survey in central California. Ray paths calculated through the structure shown yield a good fit to the complicated travel time data. For example, the late arrivals about 8 km from the source are interpreted as resulting from a low-velocity region associated with a set of faults. The model

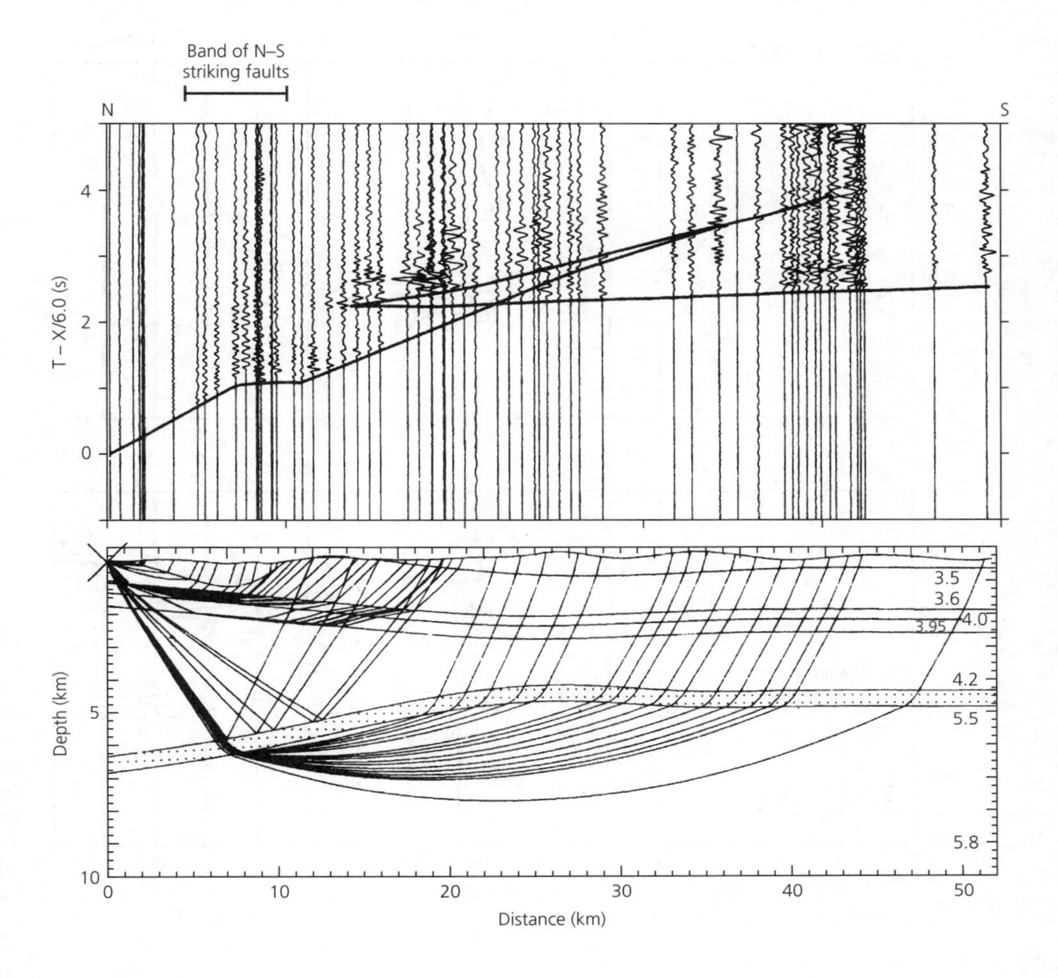

Fig. 3.2-13 Reduced travel time plot and ray tracing results for a seismic refraction survey. The solid line on the travel time plot shows the travel times predicted by the model. (Meltzer *et al.*, 1987. © Seismological Society of America. All rights reserved.)

also fits the travel times showing several velocity increases beyond this distance.

The restriction of uniform-velocity layers can also be surmounted. Geological instincts (a useful but occasionally unreliable tool) lead us to expect that rock types, and thus velocities, should often vary smoothly rather than in discrete jumps. Thus we expect velocity gradients with depth, rather than sharp interfaces. This possibility can be tested using advanced methods of analysis that predict both the travel times and the amplitudes of the expected arrivals. The amplitudes make it possible to distinguish gradients from uniform layers, even if the travel times predicted are the same. Although the methods are beyond our scope here, we discuss some results briefly.

To illustrate the relation between velocity structure and amplitudes, consider theoretical, or synthetic, seismogram record sections for the head wave, P_n, and Moho reflection, P_mP, predicted by two crustal models (Fig. 3.2-14). The seismograms were computed using a method known as reflectivity, which avoids the limitations of ray and plane wave analysis. The travel times are reduced at 8 km/s, and the direct wave is not shown. Both models have the same average velocity structure, a 30 km-thick layer of 6.5 km/s material over an 8 km/s halfspace, so the travel times are similar. However, the amplitudes of the arrivals differ noticeably because the models have different fine structure near the Moho.

For the sharp Moho model (Fig. 3.2-14, *top*) the reflected wave is small for distances less than the critical distance (*subcritical* reflection), largest near the critical distance, and large for distances greater than critical (*supercritical, postcritical*, or *wide angle* reflections). Because the boundary is sharp, this amplitude behavior is similar to that predicted for plane waves (Fig. 2.6-11). P_mP also shows the expected phase shift for reflection past critical incidence (Section 2.6.4). The head wave first appears near the critical distance, 83 km, and is small, as expected from the plane wave approximation that predicts no transmitted wave past the critical angle.

Figure 3.2-14 (*bottom*) shows the effect of velocity gradients above and below the Moho. Seismic energy trapped near the Moho yields larger P_n amplitudes than for the sharp Moho case. In addition, for subcritical distances, the reflection is smaller than without a gradient above the Moho, because it no longer reflects off a sharp interface. Hence the amplitudes

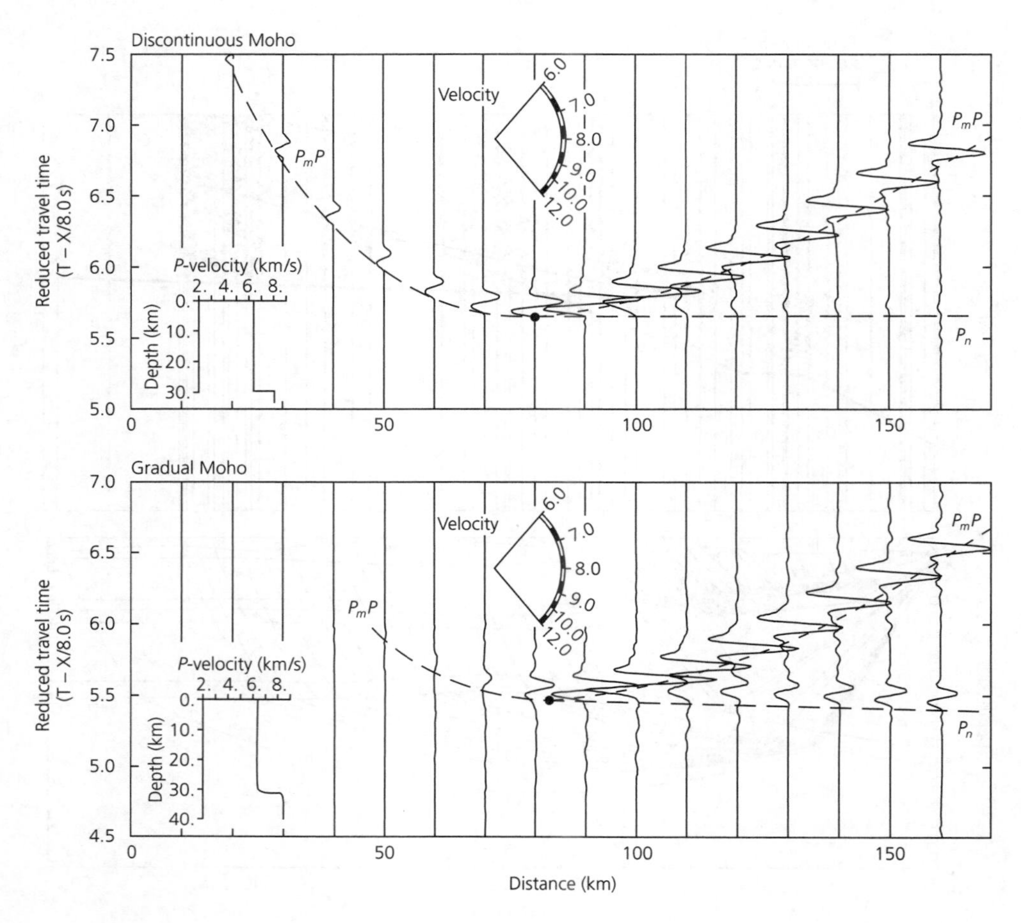

Fig. 3.2-14 Synthetic seismograms showing how the amplitudes of the head wave, P_n, and the reflected wave, P_mP, depend on the velocity structure at the Moho. Two cases with the same average-velocity structure are shown. At the *top* the Moho is a sharp transition, and at the *bottom* there are gradients above and below the Moho. The velocity scale shows the slopes of arrivals with different velocities. (After Braile and Smith, 1975.)

of P_n and P_mP indicate the presence or absence of gradients at the Moho.

Figure 3.2-15 illustrates these ideas for the oceanic crust and the mantle. Theoretical seismograms (Fig. 3.2-15, *center*) computed for a layered model that fits travel times predict strong reflections off the top of layer 3 (P_3P) and the Moho (P_mP). The observed data (Fig. 3.2-15, *bottom*) show strong P_mP reflections, suggesting a sharp Moho transition. However, strong P_3P reflections are not observed, implying that the transition between layers 2 and 3 is a gradient rather than a sharp jump. Thus, although the results of refraction studies are often reported as layered models that fit the travel times, amplitude studies are needed to show whether sharp interfaces exist.

An interesting point is that, because layers are distinguished from gradients by interpreting the amplitudes of seismic waves, this distinction depends on the wavelength of the wave used to study the structure. A reasonable approximation is that waves "see" only structures longer than their wavelengths. In other words, waves are affected by the medium properties averaged over their wavelengths. For example, the velocity structures in Fig. 3.2-16 appear identical to waves with a wavelength of 1 km, but look quite different for a wavelength of 1 m. Thus profile 3 appears as a sharp interface for waves with wavelength 1 km, a gradient for 100 m wavelength, and a stack of layers for 10 m wavelength. The velocity structure depends on the wavelengths under discussion, so a velocity "gradient" is a structure that cannot be distinguished, with the wavelengths used, from one in which velocity changes smoothly. Similarly, an "interface" is a region that cannot be distinguished from a sharp velocity change with the wavelengths used.

3.2.4 *Crustal structure*

Information about crust and upper mantle structure around the world has been acquired by refraction surveys conducted on different scales. The size of the sources and the source-to-receiver distances increase with the depth of the structures being studied. Earthquakes or large explosions, including nuclear weapons tests, have enough energy to reveal the Moho. For example, the profile in Fig. 3.2-5, which showed clear Moho arrivals, was almost 250 km long and used sources containing 136 kg of explosive. Shorter profiles are used to study structure within the crust, as in Fig. 3.2-13. The recording stations are either permanent seismic stations or, in most cases, portable seismometers. Refraction studies are also conducted

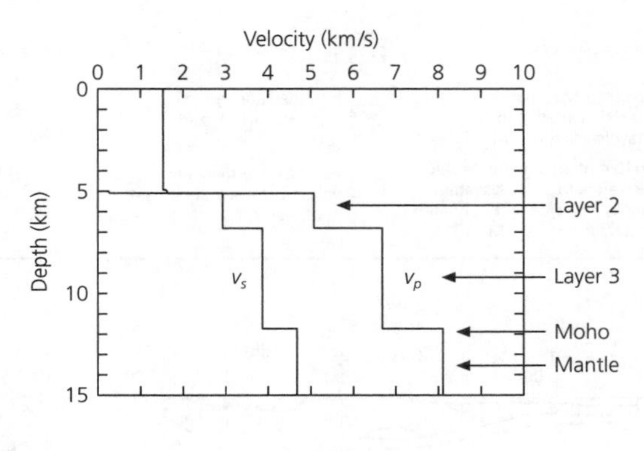

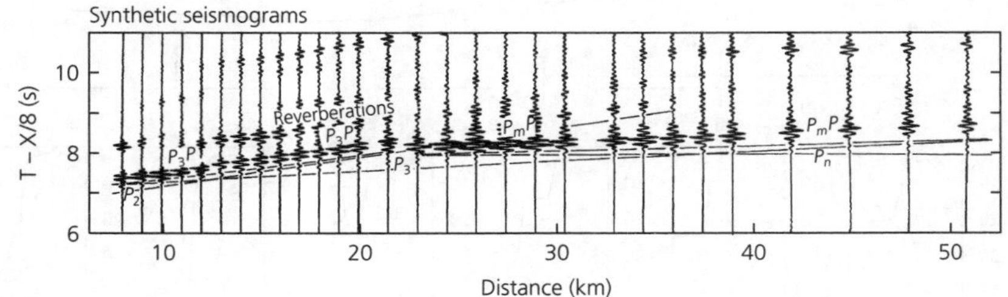

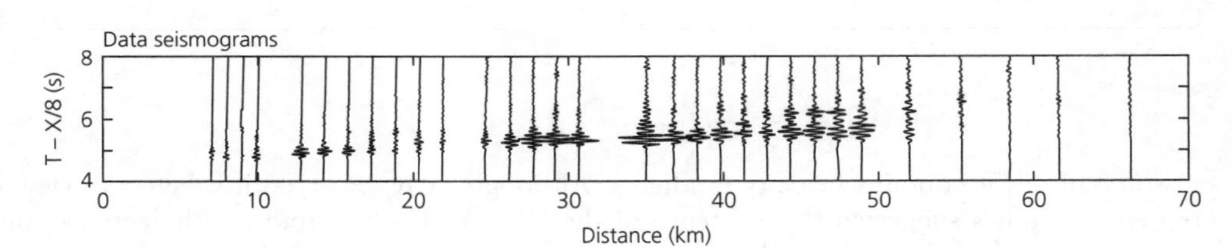

Fig. 3.2-15 *Top*: Oceanic crust model with sharp transitions between layer 1 (water), layer 2 (unconsolidated sediment), layer 3 (crustal rock), and the mantle. *Center*: Synthetic seismograms for this model. P_2, P_3, and P_n are head waves from layers 2, 3, and the mantle. P_3P and P_mP are reflections off the tops of layer 3 and the mantle. *Bottom*: Data showing an absence of the large P_3P arrivals predicted by the layered model. (After Spudich and Orcutt, 1980. *Rev. Geophys. Space Phys.*, *18*, 627–45, copyright by the American Geophysical Union.)

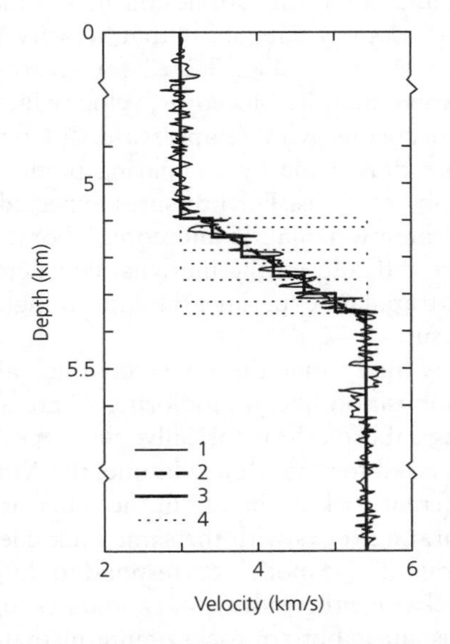

Fig. 3.2-16 Different velocity profiles that are indistinguishable when examined by using 1 km wavelength seismic waves, but distinguishable with much shorter wavelengths. (Spudich and Orcutt, 1980. *Rev. Geophys. Space Phys.*, *18*, 627–45, copyright by the American Geophysical Union.)

at sea. In some cases, disposable sonobuoys or retrievable ocean bottom seismometers are deployed, and a ship steams away firing "shots." In other cases, two ships are used. Marine refraction data (e.g., Fig. 3.2-15) are analyzed by treating the water as an upper layer of known velocity. The refraction results are combined with those from seismic reflection techniques, discussed in the next section, in which the velocity structure is derived from the travel times of subcritical reflections, rather than refractions. Refraction and reflection results are complementary and yield improved knowledge of structure.

The oceanic crust is about 7 km thick, and is relatively uniform from site to site, except at mid-ocean ridges. As a result, a single simple model like that in Fig. 3.2-15 is often applicable. By contrast, the continental crust is thicker and variable, as illustrated in Fig. 3.2-17 for a cross-section across the west coast of the United States. The thin crust beneath the Pacific Ocean thickens across the continent–ocean transition, such that beneath the coast ranges the Moho is about 25 km deep. Beneath the Sierra Nevada range, the depth to the Moho reaches 35–40 km. The refraction data also show complicated and variable-velocity structures within the crust. Thus the crust is not a uniform layer, or even a uniform set of layers, because

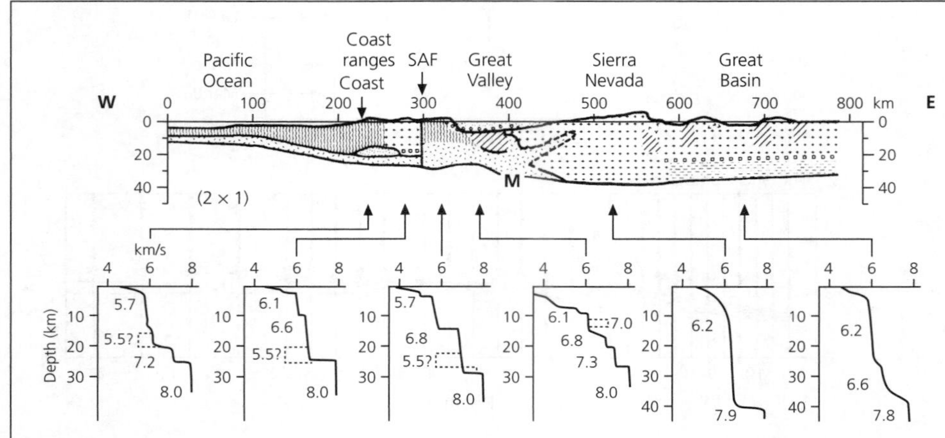

Fig. 3.2-17 Crustal velocity model and inferred geologic structure for a cross-section across the west coast of the USA. "SAF" denotes the San Andreas fault. Dashed lines indicate low-velocity zones. (After Mooney and Weaver, 1989. From *Geophysical Framework of the Continental United States*, ed. L. C. Pakiser and W. D. Mooney, with permission of the publisher, the Geological Society of America, Boulder, CO. © 1989 Geological Society of America.)

in some places it contains velocity gradients. Although early refraction studies suggested the existence of the Conrad discontinuity dividing the upper and lower crust, it now appears that high (greater than about 6.5 km/s)-velocity lower crust is present in some places but not in others. Furthermore, some areas show low-velocity zones within the crust.

Refraction studies show regional variations in crustal thickness and P_n velocities, as illustrated for North America in Fig. 3.2-18. East of ~104°W, the crust is typically thick (~42 km), and P_n velocities are high (~8.1 km/s). To the west, the crust is often thinner, with lower P_n velocities. The thin crust and low P_n velocities beneath the Basin and Range province may reflect hotter material near the surface, consistent with active extension. As seen here and globally (Fig. 3.2-19), mountain ranges often have thick crust. The thick crust is thought to be due to isostasy, whereby the excess mass of the mountains is at least partially compensated by a crustal root with density less than that of the mantle.

The continental Moho can be modeled as a simple interface for the wavelengths used in most refraction studies. However, seismic reflection studies, with shorter wavelengths, sometimes show a laminated structure of high- and low-velocity layers (Fig. 3.2-20). In other cases, however, the Moho is not observed in reflection data. Some of these complexities may reflect difficulties associated with seismic reflection studies in laterally varying media (Section 3.3). Nonetheless, the Moho appears to be a complicated transition zone 0–5 km wide, with properties varying between locations (Fig. 3.2-21). Rather than regarding the Moho as the base of a homogeneous crustal layer, it is better to view it as a zone where velocities increase rapidly with depth to values above about 7.7 km/s.

Velocity structures are often interpreted in terms of composition, as in Fig. 3.2-17. To do this, seismological results are combined with other geophysical data (e.g., gravity), geological fieldwork, and laboratory studies of the seismic velocities of rocks. The laboratory data show that velocity varies with composition, as shown in Fig. 3.2-22 for igneous rocks of the crust and upper mantle. Moreover, velocity increases with pressure and decreases with temperature. Inferences about composition are thus made by comparing predicted velocities to seismic observations. For pressures expected at greater depths, as for the lower mantle and core, laboratory experiments are more difficult, so thermodynamic calculations are also used to extrapolate experimental data to higher temperatures and pressures.

Such analyses imply that the upper continental crust has an average composition like granodiorite, whereas the upper oceanic crust is gabbroic.[4] Historically, two types of models have been suggested for the Moho. In one, the Moho divides chemically different rocks, whereas in the other, it is a phase boundary separating rocks with the same bulk chemistry but different minerals. These models correspond to different combinations of rocks on either side. Two candidates for the lower continental crust are gabbro or rocks of intermediate composition in the granulite facies. The most popular candidate for the upper mantle is peridotite, which would make the Moho a

[4] Some relevant rock and mineral nomenclature is summarized in Section 3.2.5.

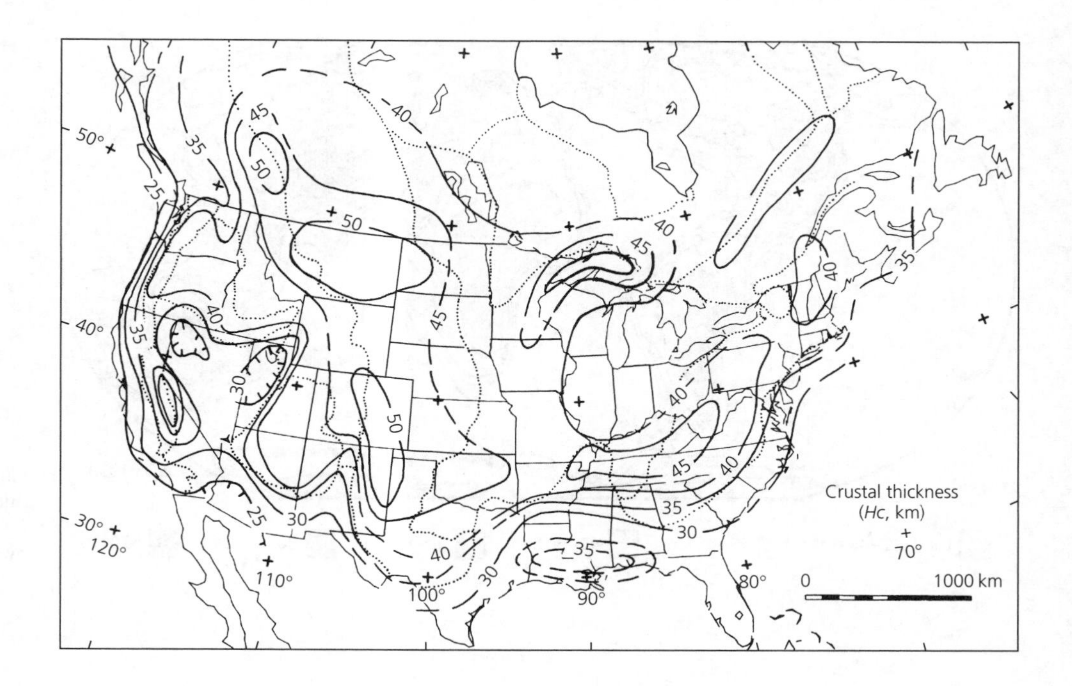

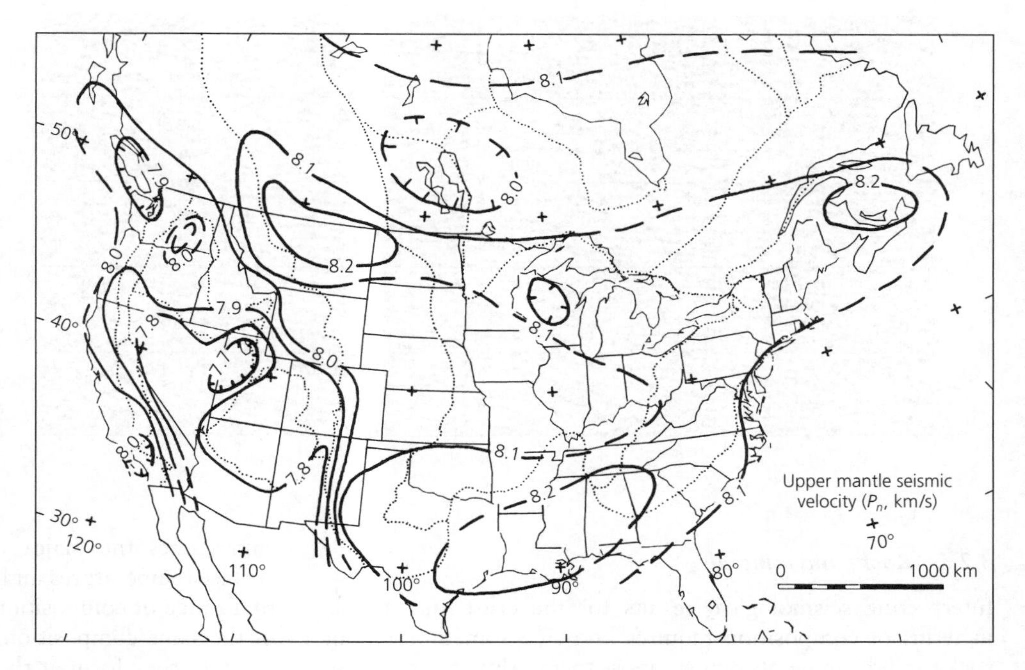

Fig. 3.2-18 Crustal thickness (depth to Moho) (*top*) and P_n velocity (*bottom*) maps for part of North America. Contour intervals are 5 km and 0.1 km/s. (Braile *et al*., 1989. From *Geophysical Framework of the Continental United States*, ed. L. C. Pakiser and W. D. Mooney, with permission of the Geological Society of America, Boulder, CO. © 1989 Geological Society of America.)

compositional boundary. Another candidate is eclogite, a rock with the same bulk chemistry as gabbro, but denser mineral phases. If the upper mantle were eclogite and the lower continental crust gabbroic, the continental Moho would be a phase boundary. However, although eclogite and peridotite have similar seismic velocities, peridotite seems a more likely composition for the upper mantle. One of the reasons is that olivine, a major component of peridotite, yields anisotropic seismic velocities due to its crystal structure. Such anisotropic P_n velocities are observed in the oceanic upper mantle and in some locations in the continental upper mantle (Section 3.6).

The status of the lower continental crust is more controversial. A granulite model is popular, but gabbro cannot be ruled out. Similarly, the origin of the laminated structure of the Moho is still unclear. Possible explanations include metamorphosed sediments, cumulate layering, tectonic banding, and lenses of partial melt. In any event, this structure seems to be laterally variable.

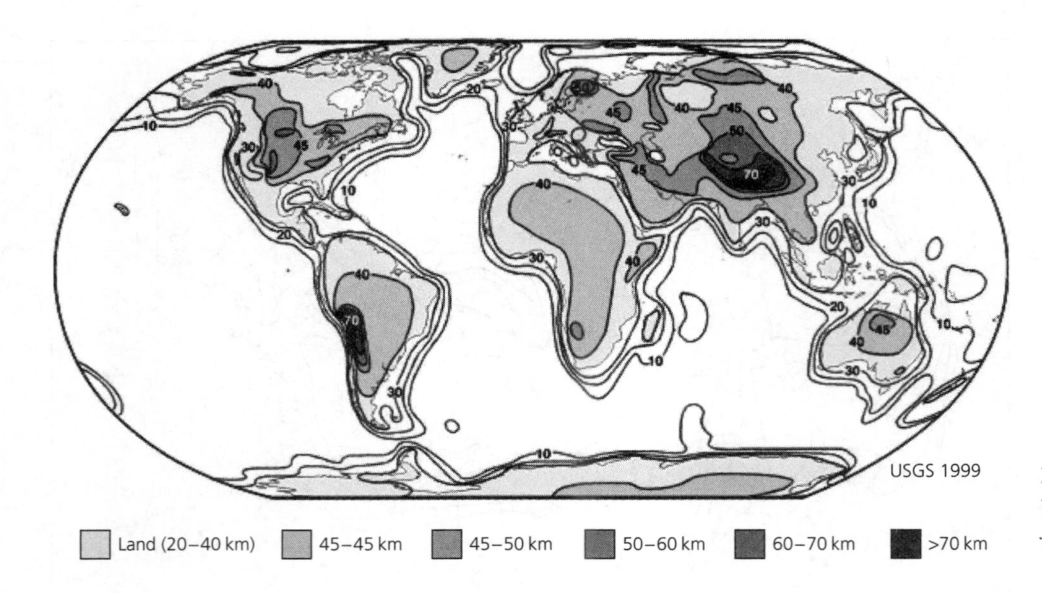

Fig. 3.2-19 Global map of crustal thickness. (Mooney *et al.*, 1998. *J. Geophys. Res.*, *103*, 727–47. Copyright by the American Geophysical Union.)

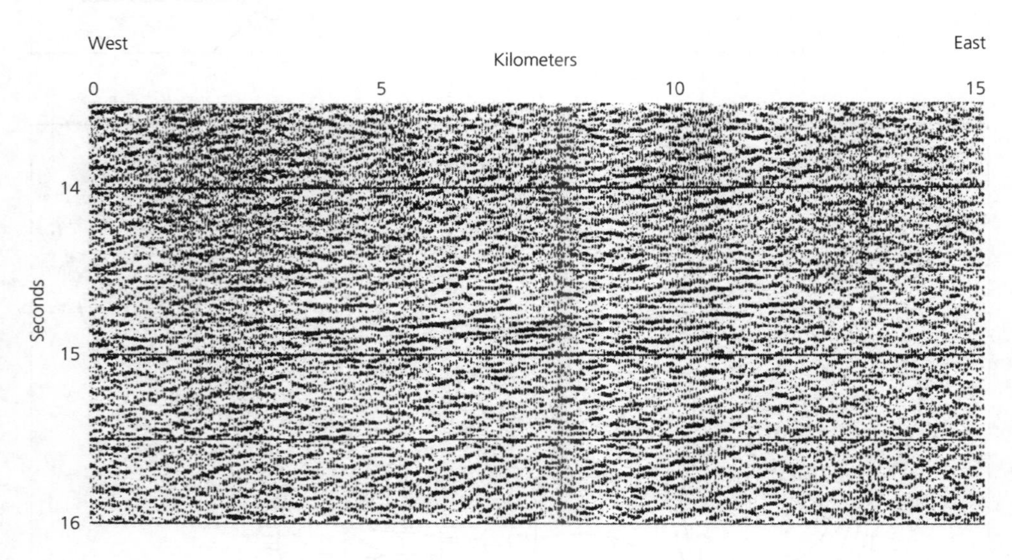

Fig. 3.2-20 Seismic reflection profile from the Wichita Mountains of southeastern Oklahoma. The "ringing" Moho reflections at 14.5–15 s in the middle of the section suggest that the Moho has a laminated velocity structure over several km. (Hale and Thompson, 1982. *J. Geophys. Res.*, *87*, 4625–35, copyright by the American Geophysical Union.)

3.2.5 Rocks and minerals

Interpreting seismological results for the crust and mantle in terms of composition requires knowing something about rocks and the minerals that compose them. Although these are complicated subjects, we summarize a few essential terms.

For our discussions of crust and upper mantle structure, the most important rocks are the igneous rocks formed by cooling a molten magma. These rocks are classified primarily by the weight percent of silica, SiO_2. A common nomenclature describes rocks as *acidic* or *silicic* for a weight percent of $SiO_2 >$ 66%, *intermediate* for 66–52%, *basic* or *mafic* for 52–45%, and *ultrabasic* or *ultramafic* for < 45%.

Physical properties of rocks, such as density and seismic velocity, depend on their mineral composition. Figure 3.2-23 summarizes the major minerals in various rocks at near-surface temperatures and pressures. Because rock names refer to a range of compositions, those shown are averages. Rocks of the same composition have different names depending on whether they form at the earth's surface (extrusive rocks) or below it (intrusive rocks). Hence an extrusive rock of gabbroic composition is a *basalt*.

Several important silicate (SiO_2-bearing) minerals are mentioned in the figure. *Quartz* is pure SiO_2. *Olivine* is a solid solution, $(Mg, Fe)_2SiO_4$, whose composition varies from pure Fe_2SiO_4 (*fayalite*) to pure Mg_2SiO_4 (*forsterite*). Due to its crystal structure, olivine has anisotropic seismic velocities. *Pyroxene* is a solid solution with end members $MgSiO_3$ (*enstatite*), $FeSiO_3$ (*ferrosilite*), $CaMg(SiO_3)_2$ (*diopside*), and $CaFe(SiO_3)_2$ (*hedenbergite*), though only certain ranges of

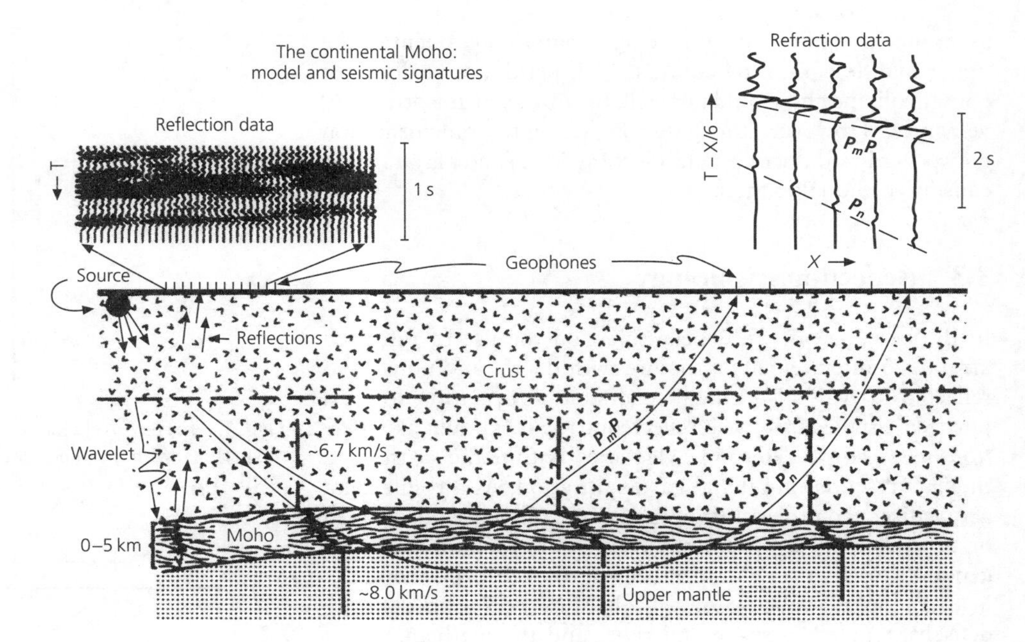

Fig. 3.2-21 Schematic model for the continental Moho as a laminated structure. Refraction studies using relatively longer wavelengths would show clear P_mP and P_n arrivals, whereas reflection studies using shorter wavelengths would show reverberations. (Braile and Chiang, 1986. *Reflection Seismology*, 257–72, copyright by the American Geophysical Union.)

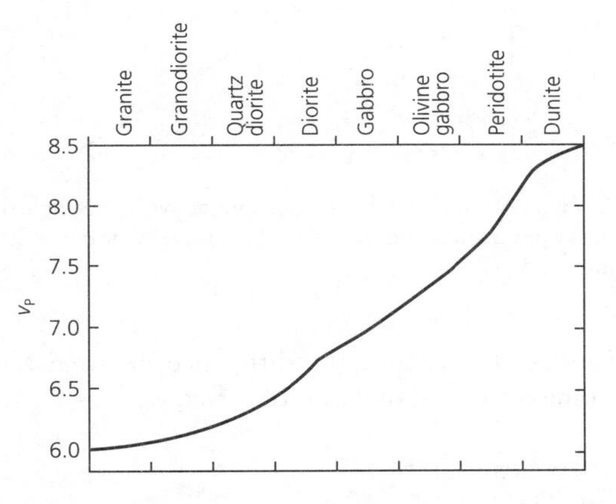

Fig. 3.2-22 Variation of *P*-wave velocity with lithology for crust and upper mantle rocks, at a pressure of 1.5 kbar (150 MPa). Velocity increases with decreasing silica content. (Fountain and Christensen, 1989. From *Geophysical Framework of the Continental United States*, ed. L. C. Pakiser and W. D. Mooney, with permission of the Geological Society of America, Boulder, CO. © 1989 Geological Society of America.)

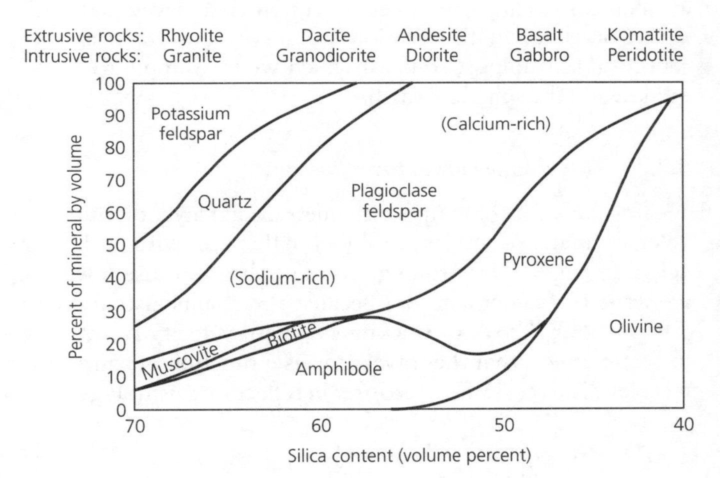

Fig. 3.2-23 Simplified igneous rock classification. Compositions are shown as the volume percent of major minerals for a rock of given silica content (horizontal axis). Thus a granodiorite of about 60% silica content contains about 20% amphibole, 5% biotite, 53% plagioclase feldspar, 17% quartz, and 5% potassium feldspar. Rock names are given for intrusive and extrusive forms.

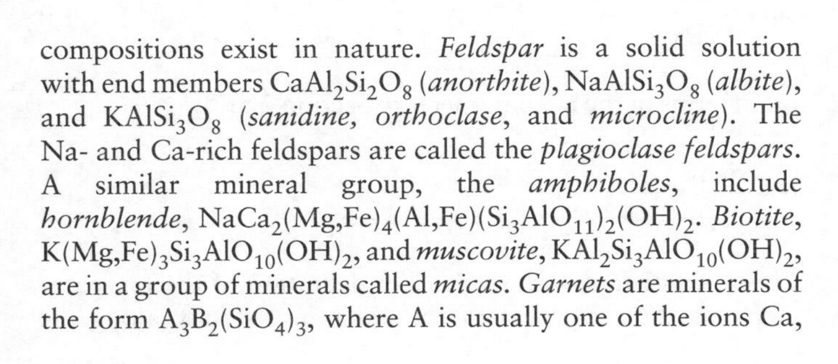
compositions exist in nature. *Feldspar* is a solid solution with end members $CaAl_2Si_2O_8$ (*anorthite*), $NaAlSi_3O_8$ (*albite*), and $KAlSi_3O_8$ (*sanidine, orthoclase*, and *microcline*). The Na- and Ca-rich feldspars are called the *plagioclase feldspars*. A similar mineral group, the *amphiboles*, include *hornblende*, $NaCa_2(Mg,Fe)_4(Al,Fe)(Si_3AlO_{11})_2(OH)_2$. *Biotite*, $K(Mg,Fe)_3Si_3AlO_{10}(OH)_2$, and *muscovite*, $KAl_2Si_3AlO_{10}(OH)_2$, are in a group of minerals called *micas*. *Garnets* are minerals of the form $A_3B_2(SiO_4)_3$, where A is usually one of the ions Ca, Mg, or Fe, and B is typically any of Al, Fe, or Cr. Garnets are comparatively dense, and thus significant for discussions of phase changes.

The figure describes rocks in terms of their mineralogy at surface conditions. With increasing pressure due to increasing depth in the earth, minerals transform to denser phases. Thus, for example, a gabbro containing plagioclase feldspar, pyroxene, and olivine transforms to a chemically identical eclogite rock containing quartz, pyroxene, and garnet. Hence an argument against eclogite being a major component of the upper mantle is that, by contrast with peridotite, it does not contain olivine and would not yield the observed anisotropic P_n velocities. However, the gabbro-to-eclogite transformation may occur in subducting slabs (Section 5.4.2) and play a role in causing earthquakes there.

3.3 Reflection seismology

In the last section, we concentrated on the use of refracted arrivals to infer velocity structure with depth, and noted that reflected arrivals also contain valuable information for this purpose. Studies using the reflected arrivals, known as *reflection seismology*, determine velocities within the crust, and thus are essential in oil and gas exploration. As a result, data acquisition and processing methods have often been developed first by reflection seismologists. For example, digital data were generally used in exploration before they became common in earthquake studies. Similarly, because reflection data are densely sampled in space and time, and the mathematics of wave propagation in a layered medium is simpler than for a spherical earth, techniques are often first developed with reflection data. In this section we survey basic concepts in reflection seismology, some of which we later apply to earthquakes and the spherical earth.

3.3.1 Travel time curves for reflections

We first consider the simplest geometry: a flat layer of uniform-velocity material underlain by a halfspace with a higher velocity (Fig. 3.2-1). Although most applications use P waves, we write the velocity as "v" because the results also apply to S waves. For a layer of thickness h_0 with velocity v_0, we saw in Section 3.2.2 that the travel time as a function of source-to-receiver distance, known as *offset* in reflection seismology, is

$$T(x)^2 = x^2/v_0^2 + 4h_0^2/v_0^2 = x^2/v_0^2 + t_0^2. \tag{1}$$

The travel time curve $T(x)$ is a hyperbola (Fig. 3.3-1) that intercepts the T axis at $t_0 = 2h_0/v_0$, the travel time at zero offset. This time is called the two-way vertical travel time, because the corresponding ray traveled vertically down to the reflector and back. Although this curve is the same as the "reflected wave" curve in Fig. 3.2-2, the convention in reflection seismology is to plot time increasing downward,[1] because later arrivals reflect deeper in the earth.

The layer velocity is found from the slope of the hyperbola. Because the slope decreases with increasing velocity, "flatter" travel time curves indicate higher velocities. To see this, note that a plot of $T(x)^2$ versus x^2 has slope $1/v_0^2$. Alternatively, the variation in travel time with offset is often stated in terms of *normal moveout (NMO)*, the difference between the travel time at some offset and that at zero offset,

$$T(x) - t_0 = (x^2/v_0^2 + t_0^2)^{1/2} - t_0. \tag{2}$$

Once the velocity is found, the layer thickness is given by the vertical travel time.

To see the relation between the travel time curve and ray paths, consider the ray paths to two points dx apart, which differ in travel time by dT (Fig. 3.3-2). Because the ray paths differ in length by vdT, the angle of incidence can be found using

$$\sin i = \frac{vdT}{dx} \tag{3}$$

or, in terms of the ray parameter p (Section 2.5.7),

$$p = \frac{\sin i}{v} = \frac{dT}{dx}. \tag{4}$$

This is consistent with our earlier definition of the ray parameter as the reciprocal of the apparent velocity along the

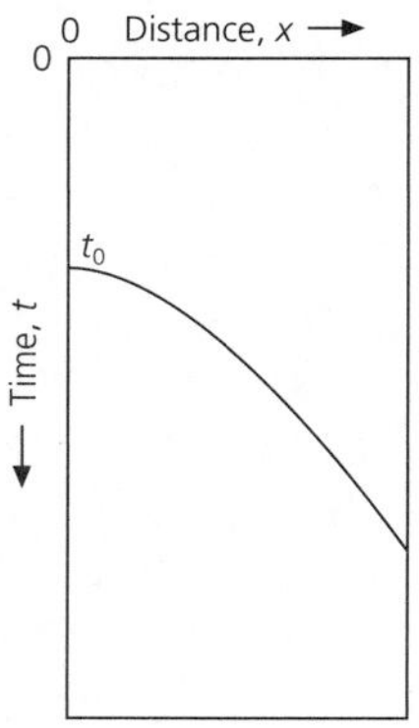

Fig. 3.3-1 The travel time curve for a reflection off a flat interface is a hyperbola, with the minimum at $x = 0$ corresponding to a vertical ray path. The slope is zero at $x = 0$ and increases with the offset distance.

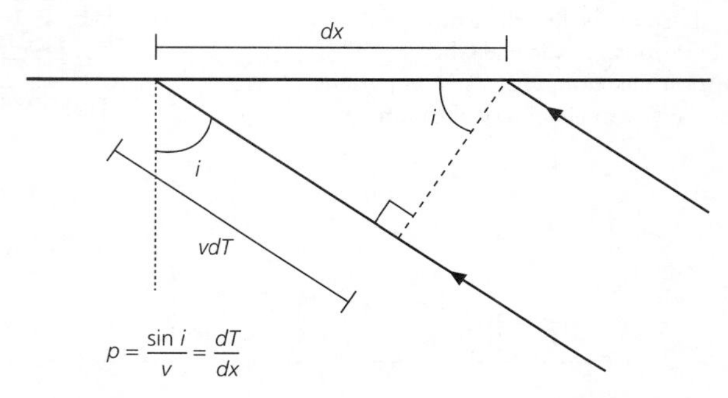

Fig. 3.3-2 Two rays showing the relationship between the angle of incidence, ray parameter, and the slope of the travel time curve for a flat medium.

[1] Earthquake seismologists generally follow the opposite convention.

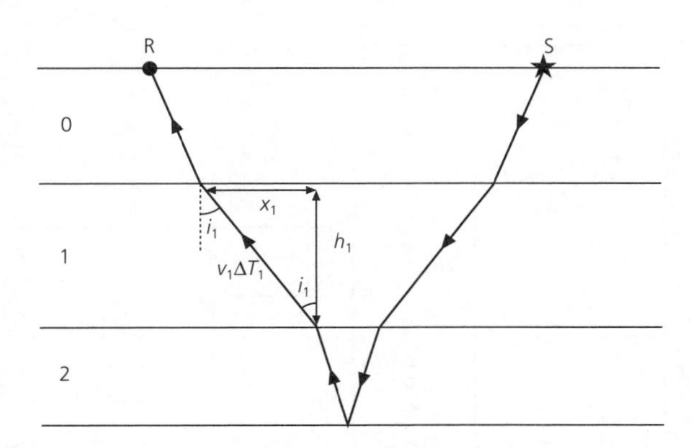

Fig. 3.3-3 Ray geometry for a reflection in a flat-layered medium. Layer thicknesses are h_j, horizontal distances traveled in the layers are x_j, and one-way travel times spent in the layers are ΔT_j.

surface of the wave front, which moves a distance dx in time dT, because

$$p = 1/c_x = 1/(dx/dT). \tag{5}$$

Thus the ray parameter and the angle of incidence of the ray emerging at a distance x can be found from dT/dx, the slope of the travel time curve evaluated at x. From Eqn 2, the slope is zero at $x = 0$ and then increases with offset; so the angle of incidence is nearly zero (vertical incidence) at short distances and becomes closer to 90° (horizontal) at larger distances (Fig. 3.3-1).

This lets us find the travel time curve for reflections in a geometry with multiple horizontal layers. Figure 3.3-3 shows that the reflection R_{n+1} from the top of the $(n+1)^{\text{th}}$ layer (or the bottom of the nth layer) has traveled through n layers, each of thickness h_j and velocity v_j. Such rays, which have been reflected only once, are known as primary reflections. Because, by Snell's law, the ray parameter p is constant along a ray, the incidence angles i_j in each layer can be found from the incidence angle i_0 in the top layer,

$$p = \frac{\sin i_j}{v_j} = \frac{\sin i_0}{v_0}. \tag{6}$$

A downgoing ray, which travels a horizontal distance x_j in the jth layer, spends a time ΔT_j in the layer. Thus, in going down and up again, the ray travels a total horizontal distance

$$x(p) = 2\sum_{j=0}^{n} x_j = 2\sum_{j=0}^{n} h_j \tan i_j \tag{7}$$

in a total time

$$T(p) = 2\sum_{j=0}^{n} \Delta T_j = 2\sum_{j=0}^{n} \frac{h_j}{v_j \cos i_j}. \tag{8}$$

We explicitly write $x(p)$ and $T(p)$, because the two sums are formulated in terms of the ray parameter. To see how they allow us to compute the corresponding travel time curve $T(x)$, consider a single layer, where x_0 $(= x/2)$ is the horizontal distance along each of the downgoing and upgoing legs. In this case, Eqn 8 becomes

$$T(x) = 2[(x/2)^2 + h_0^2]^{1/2}/v_0, \tag{9}$$

because

$$\cos i_0 = h_0(x_0^2 + h_0^2)^{-1/2}. \tag{10}$$

Hence Eqn 8 yields Eqn 9, which is equivalent to the relation we derived earlier showing that the travel time curve for the reflection is a hyperbola (Eqn 1).

For multiple layers, we approximate the travel time curve for the reflection R_{n+1} off the top of the $(n+1)^{\text{th}}$ layer as a hyperbola,

$$T(x)_{n+1}^2 = x^2/\bar{V}_n^2 + t_n^2, \tag{11}$$

and find the two parameters, $\bar{V}_n$ and t_n. t_n is the total two-way (up and down) vertical travel time at zero offset, which is twice the sum of the one-way vertical travel times Δt_j for each layer

$$t_n = 2\sum_{j=0}^{n} \Delta t_j = 2\sum_{j=0}^{n} (h_j/v_j). \tag{12}$$

The velocity term, $\bar{V}_n$, is a little trickier. From the geometry, the distance traveled by the downgoing ray in layer j is

$$x_j = v_j \Delta T_j \sin i_j = (v_j^2/v_0)\Delta T_j \sin(i_0), \tag{13}$$

where the last step used Snell's law (Eqn 6). Hence, by Eqn 7, the total distance, x, can be written

$$x = 2\sum_{j=0}^{n} x_j = 2\frac{\sin i_0}{v_0}\sum_{j=0}^{n} v_j^2 \Delta T_j. \tag{14}$$

Because the ray parameter is constant along a ray, the slope of the travel time curve is, by Eqn 4,

$$\frac{dT}{dx} = \frac{\sin i_0}{v_0} = x/(2\sum_{j=0}^{n} v_j^2 \Delta T_j). \tag{15}$$

For the hyperbolic approximation (Eqn 11), the slope of the travel time curve is

$$\frac{dT}{dx} = \frac{x}{\bar{V}_n^2 T}, \tag{16}$$

so we define

$$\bar{V}_n^2 = (2\sum_{j=0}^{n} v_j^2 \Delta T_j)/T. \tag{17}$$

Because this was derived for an arbitrary incidence angle, vertical incidence can be used for simplifications, so in each layer the travel time equals the one-way vertical travel time, $\Delta T_j = \Delta t_j$, and the total travel time is $T = 2\sum_{j=0}^{n}\Delta t_j$. Hence

$$\bar{V}_n^2 = \left(\sum_{j=0}^{n} v_j^2 \Delta t_j\right) \Big/ \left(\sum_{j=0}^{n} \Delta t_j\right). \tag{18}$$

$\bar{V}_n$, the appropriate average velocity for the travel time curve, is the time-weighted *root mean square*, or *rms*, velocity for the first n layers. This hyperbolic approximation and the exact solutions agree well except for large offsets.

These results let us find the layer velocities from the travel time curves. Given a reflection from the top of the nth layer, with vertical two-way travel time t_{n-1} and rms velocity $\bar{V}_{n-1}$, and a reflection from the top of the $(n+1)^{\text{th}}$ layer, with vertical two-way travel time t_n and rms velocity $\bar{V}_n$, the velocity in the nth layer is

$$v_n^2 = \frac{\bar{V}_n^2 t_n - \bar{V}_{n-1}^2 t_{n-1}}{t_n - t_{n-1}}. \tag{19}$$

This relationship is called the *Dix equation*.[2] The resulting velocity, called an *interval velocity*, is better determined for larger offsets, where the slope of the travel time curve is greater. Because the later reflections have higher velocities, and hence flatter travel time curves (Fig. 3.3-4), larger offsets are required to determine velocities at greater depths.

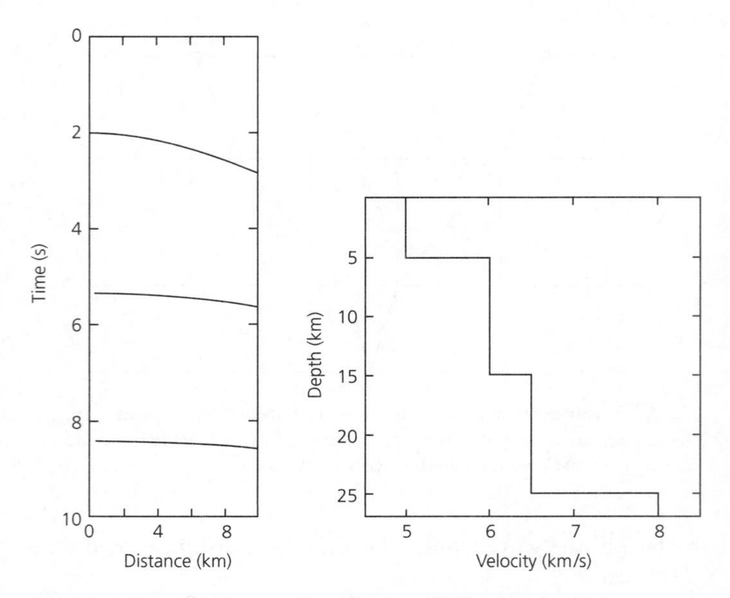

Fig. 3.3-4 Travel time curves for reflections (*left*) from a layered structure (*right*) corresponding to continental crust. Reflections from deeper interfaces are flatter, or have shallower slopes, due to the increase of velocity with depth.

Travel time calculations are more complicated for dipping layers. Figure 3.3-5 shows the geometry for a reflector of dip θ, whose depth along the perpendicular to the reflector below the origin is h. The travel times can by derived using an imaginary source on the line from the surface source normal to the reflector, at the same distance below the layer, so that travel times from the imaginary source to the receivers are the same as from the true source. Applying the law of cosines to triangle RIS shows that

$$\begin{aligned} T^2 &= [x^2 + 4h^2 - 4hx\cos(\theta + \pi/2)]/v_0^2 \\ &= [x^2 + 4h^2 + 4hx\sin\theta]/v_0^2. \end{aligned} \tag{20}$$

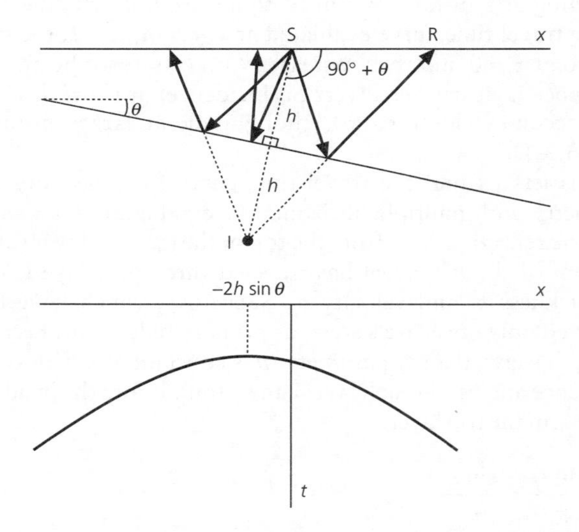

Fig. 3.3-5 The travel time curve for a reflection off a dipping interface can be derived using an imaginary source (I) at depth that gives the same travel times. The resulting hyperbola has a minimum at a nonzero offset. S and R denote the source and the receiver.

This travel time curve is a hyperbola with minimum at $-2h\sin\theta$, so it is not symmetric about $x = 0$. Reflections from a stack of dipping layers yield travel time curves of approximately this form.

It is sometimes useful to think of the earth as having a continuous distribution of velocity with depth, $v(z)$, rather than a stack of discrete layers, each with uniform velocity. The expressions for the ray path and travel time of a ray with ray parameter p for discrete layers can be generalized. The ray path (Fig. 3.3-6) is given by Snell's law, because the ray parameter,

$$p = \sin i / v(z), \tag{21}$$

is constant along a ray. If velocity increases with depth, $\sin i$ and thus i increase, so the ray bends away from the vertical on its way down. Once $i = 90°$, the ray turns, becomes horizontal, and then goes upward. At the deepest point, the turning, or bottoming, depth z_p, the velocity is the reciprocal of the ray parameter, $p = 1/v(z_p)$. If on some portion of the ray path the velocity decreases with depth, the ray bends toward the

[2] Named after its discoverer, pioneering exploration seismologist C. Hewitt Dix (1905–84).

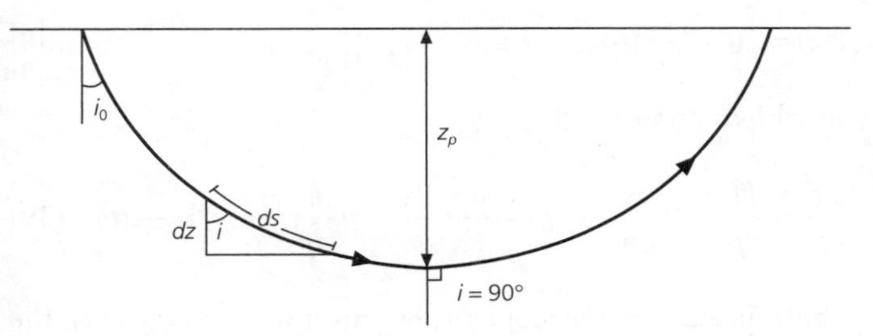

Fig. 3.3-6 Ray path in a medium with velocity increasing smoothly with depth. The ray parameter is constant along the ray path, so the angle of incidence changes as the velocity changes. The incidence angle is smallest at the surface, where velocity is lowest, and is 90° at the bottoming depth, z_p.

vertical. The ray does not turn upward until it gets below the low-velocity region.

We thus replace the sums over layer thickness h_j with integrals over depth, such that the expression for the distance traveled by the ray (Eqn 7) becomes

$$x(p) = 2\int_0^{z_p} \tan i \, dz = 2p \int_0^{z_p} \left(\frac{1}{v^2(z)} - p^2\right)^{-1/2} dz, \tag{22}$$

because

$$\sin i = pv(z) \quad \text{and} \quad \cos i = (1 - \sin^2 i)^{1/2} = (1 - p^2v^2(z))^{1/2}. \tag{23}$$

This is sometimes written in terms of the slowness, the reciprocal of velocity, as

$$u(z) = 1/v(z), \tag{24}$$

so that

$$x(p) = 2p\int_0^{z_p} \frac{dz}{(u^2(z) - p^2)^{1/2}}. \tag{25}$$

Similarly, the travel time sum (Eqn 8) becomes

$$T(p) = 2\int_0^{z_p} \frac{dz}{v(z)\cos i} = 2\int_0^{z_p} \frac{dz}{v(z)(1 - p^2v^2(z))^{1/2}}$$

$$= 2\int_0^{z_p} \frac{u^2(z)dz}{(u^2(z) - p^2)^{1/2}}. \tag{26}$$

This integral is valid everywhere except at the exact bottom of the curve, where $u(z)$ equals p. A useful way to view this is to note that the ray path (Fig. 3.3-6) can be written as an integral over ds, where $dz = \cos i \, ds$. The travel time is thus

$$T(p) = \int \frac{ds}{v(z)} = \int u(z)ds, \tag{27}$$

the integral of the slowness along the ray path. Slowness, though less intuitive to use than velocity,[3] can lead to simpler formulations.

3.3.2 *Intercept-slowness formulation for travel times*

So far, we have given travel time curves as $T(x)$, the travel time as a function of distance. We now develop an alternative formulation that offers interesting insights and is useful for data analysis. To do so, we note that ΔT_j, the one-way travel time in the jth layer with velocity v_j, is related to the thickness, h_j, and the horizontal distance traveled, x_j (Fig. 3.3-3), by

$$v_j\Delta T_j = (x_j^2 + h_j^2)^{1/2}. \tag{28}$$

The incidence angle i_j for this ray satisfies

$$\sin i_j = \frac{x_j}{(x_j^2 + h_j^2)^{1/2}} = \frac{x_j}{v_j\Delta T_j} \qquad \cos i_j = \frac{h_j}{(x_j^2 + h_j^2)^{1/2}} = \frac{h_j}{v_j\Delta T_j}. \tag{29}$$

We rewrite Eqn 28 as

$$v_j\Delta T_j = \frac{x_j^2 + h_j^2}{(x_j^2 + h_j^2)^{1/2}} = x_j \sin i_j + h_j \cos i_j, \tag{30}$$

or

$$\Delta T_j = \frac{x_j \sin i_j}{v_j} + \frac{h_j \cos i_j}{v_j} = p_j x_j + \eta_j h_j, \tag{31}$$

where

$$p_j = (\sin i_j)/v_j = \sin i_j u_j \quad \text{and} \quad \eta_j = (\cos i_j)/v_j = \cos i_j u_j. \tag{32}$$

Thus in layer j we have entities introduced in Section 2.5.7: p_j is the ray parameter, or horizontal slowness, and η_j is the vertical slowness. These are the components of the slowness vector that has magnitude equal to the slowness, and points in the direction of wave propagation. Hence u_j, the slowness in the layer, is

$$u_j^2 = 1/v_j^2 = p_j^2 + \eta_j^2. \tag{33}$$

By Eqn 31, the travel time a ray spends in a layer is the sum of the horizontal slowness times the horizontal distance traveled and the vertical slowness times the vertical thickness. The total travel time is the sum over all layers, with a factor of two to account for both downgoing and upgoing legs,

[3] It is somehow harder to think of a zone of high slowness than a low-velocity zone.

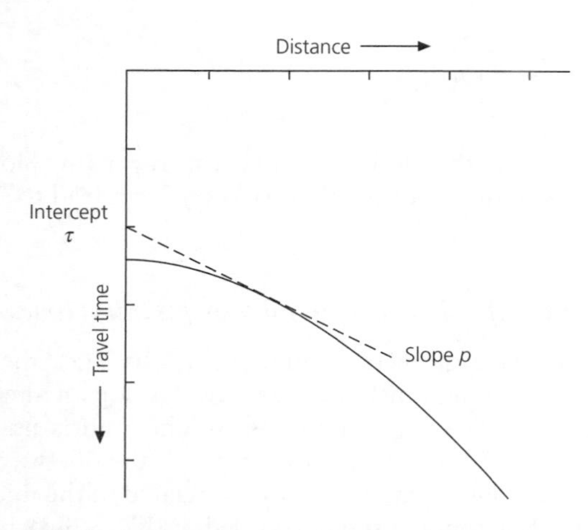

Fig. 3.3-7 Relation between the travel time curve $T(x)$ and the line tangential to a point on it, which has a slope, or slowness, p and a time axis intercept τ.

$$T(x) = 2\sum_{j=0}^{n} \Delta T_j = 2\sum_{j=0}^{n} p_j x_j + 2\sum_{j=0}^{n} \eta_j h_j. \quad (34)$$

By Snell's law, the horizontal ray parameter is constant along the ray path, so $p_j = p$, and

$$T(x) = px + 2\sum_{j=0}^{n} \eta_j h_j, \quad (35)$$

where $x = 2\sum_{j=0}^{n} x_j$ is the total horizontal distance traveled. This formulation is equivalent to the way we formulated the travel time as the scalar product of the distance and slowness vectors (Eqn 2.5.34).

Formulating the travel time curve in this way gives interesting insight. We define

$$T(x) = px + \tau(p), \quad (36)$$

where the function

$$\tau(p) = 2\sum_{j=0}^{n} \eta_j h_j = 2\sum_{j=0}^{n} (1/v_j^2 - p^2)^{1/2} h_j = 2\sum_{j=0}^{n} (u_j^2 - p^2)^{1/2} h_j. \quad (37)$$

Because p is the slope of the travel time curve (dT/dx) and hence of a line tangential to it at the point (T, x), τ is the intercept of the tangent line with the time axis (Fig. 3.3-7). In general τ and p differ for different points on the travel time curve, so the travel time curve can be described by the values of either (T, x) or (τ, p). Thus the function $\tau(p)$ is called the *intercept-slowness* representation of the travel time curve. Although less intuitive, the $\tau(p)$ formulation is equivalent to $T(x)$.

Given that the slope of the travel time curve $T(x)$ has special significance, it is natural to investigate the slope of the function $\tau(p)$. To do this, we write Eqn 36 with the ray parameter, rather than the distance, as the independent variable,

$$\tau(p) = T(p) - px(p), \quad (38)$$

and differentiate

$$\frac{d\tau}{dp} = \frac{dT}{dp} - p\frac{dx}{dp} - x(p) = \frac{dT}{dx}\frac{dx}{dp} - p\frac{dx}{dp} - x(p) = -x(p). \quad (39)$$

Thus, just as p is the slope of the travel time curve, $T(x)$, the distance, x, is minus the slope of the $\tau(p)$ curve.

To illustrate these ideas, we show that the $\tau(p)$ formulation gives the travel time curve for the reflected wave in a layer over a halfspace. Figure 3.3-3 shows that $x_0 = x/2$, so, using Eqn 32,

$$p = \frac{(x/2)}{v_0[(x/2)^2 + h_0^2]^{1/2}}, \quad \eta_0 = \frac{h_0}{v_0[(x/2)^2 + h_0^2]^{1/2}}. \quad (40)$$

Hence, by Eqns 36 and 37, the travel time curve is

$$T(x) = px + 2\eta_0 h_0 = \frac{(x^2/2) + 2h_0^2}{v_0[(x/2)^2 + h_0^2]^{1/2}}$$
$$= 2[(x/2)^2 + h_0^2]^{1/2}/v_0, \quad (41)$$

which is the familiar hyperbola (Eqn 9).

To see how this travel time curve appears when written as $\tau(p)$, we write Eqn 37 for a layer over a halfspace:

$$\tau(p) = 2(1/v_0^2 - p^2)^{1/2} h_0. \quad (42)$$

This can also be written as

$$(v_0^2\tau^2)/(4h_0^2) + v_0^2 p^2 = 1, \quad (43)$$

which is an ellipse whose axes are the τ and p axes (Fig. 3.3-8). It intersects the τ axis at ($\tau = t_0 = 2h_0/v_0, p = 0$), and the p axis at ($\tau = 0$, $p = 1/v_0$). Both these points have significance. The first, where the travel time curve has zero slope and the time axis intercept is the vertical two-way travel time, corresponds to the zero-offset point $x = 0$.

The second, where the travel time curve has slope $1/v_0$ and time axis intercept 0, is the $\tau(p)$ position of the linear travel time curve for the direct wave. Hence the line for the direct wave maps to a point in the $\tau(p)$ plane that is on the ellipse describing the reflected wave. To understand why this occurs, we use the fact that distance is minus the derivative of the $\tau(p)$ curve (Eqn 39) and differentiate Eqn 42, giving

$$x(p) = -d\tau/dp = 2ph_0(1/v_0^2 - p^2)^{-1/2}, \quad (44)$$

so at the point $p = 1/v_0$, $x = \infty$. This makes sense, because as $x \to \infty$, the reflected wave is asymptotic to the direct wave (Fig. 3.2-2).

The head wave is easily mapped into the $\tau(p)$ plane, because its travel time curve (Eqn 3.2.8) is

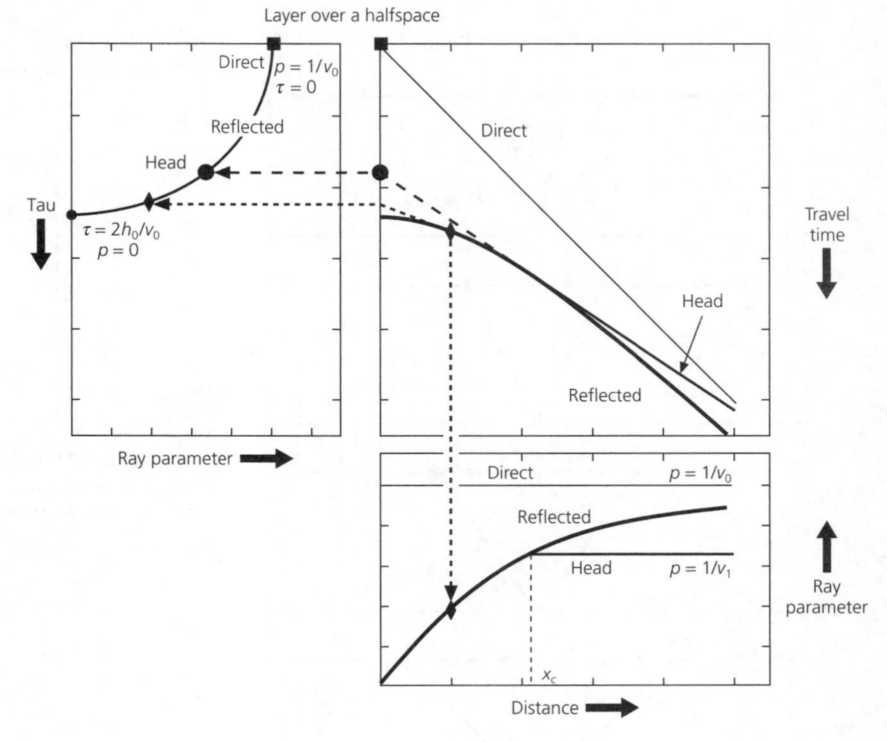

Fig. 3.3-8 Travel time curves $T(x)$ for a layer over a halfspace and their representation in the (τ, p) plane. Each point on the $T(x)$ curves has a slope (ray parameter) p and intercept τ. The linear travel time curves for the direct and head waves each map into a point (square and circle) in the (τ, p) plane. The hyperbolic travel time curve for the reflection maps into an ellipse in the (τ, p) plane. Note how an arbitrary point on the reflection's travel time curve, marked by the diamond, maps into the other two curves.

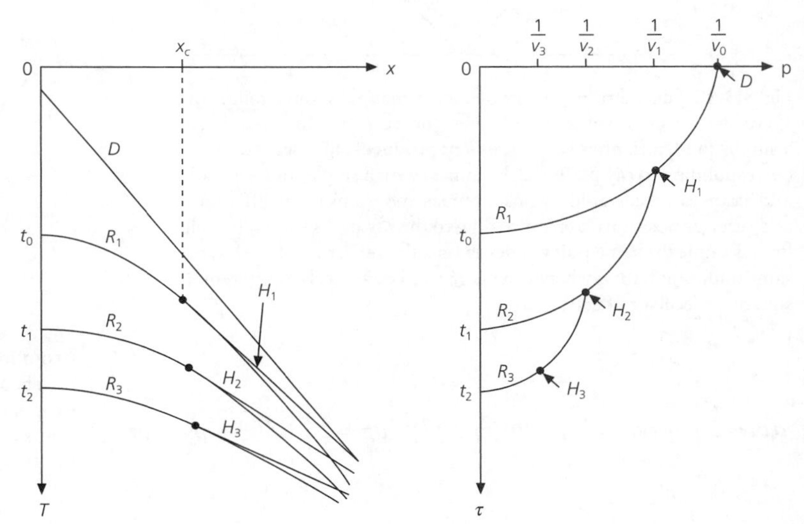

Fig. 3.3-9 Relation between the travel time curve $T(x)$ and the function $\tau(p)$ for multiple layers over a halfspace. D denotes the direct wave; R_i and H_i are reflections and head waves at the top of the ith layer; x_c is the critical distance for H_1. (After Diebold and Stoffa, 1981. Reproduced by permission of the Society of Exploration Geophysicists.)

$$T_H(x) = x/v_1 + 2h_0(1/v_0^2 - 1/v_1^2)^{1/2}$$
$$= x/v_1 + \tau_1, \qquad (45)$$

a line with slope equal to the reciprocal of the halfspace velocity, $p = 1/v_1$, and intercept τ_1. Thus the head wave maps into a point on the ellipse describing the reflected wave, corresponding to the critical distance x_c where the head and reflected waves are the same. To see this, note that for $p = 1/v_1$, Eqn 44 gives

$$x(p) = -d\tau/dp = 2h_0v_0(v_1^2 - v_0^2)^{-1/2} = x_c. \qquad (46)$$

This point divides the ellipse describing the reflected wave into a subcritical portion, between the τ axis and the head wave, and a postcritical portion, between the head wave and the p axis. We will see shortly that the fact that different arrivals have distinct locations in the $\tau(p)$ plane provides the basis for techniques that can separate these arrivals.

This analysis can be extended to more complex geometries. For multiple layers, the $\tau(p)$ curves corresponding to reflections off successive layers are all portions of different ellipses (Fig. 3.3-9). For a continuous velocity distribution, the summation for τ (Eqn 37) becomes an integral

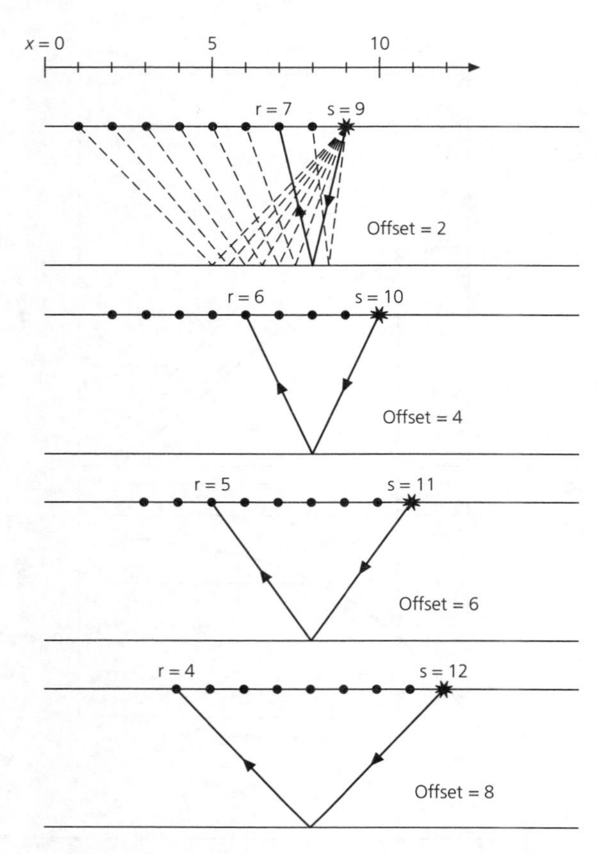

Fig. 3.3-10 Schematic geometry of a multichannel seismic reflection survey with a single source (star) and eight receivers (dots) moving along a survey line. Each physical experiment produces eight seismograms corresponding to ray paths (dashed lines) with a single source location and a range of receiver locations. Four seismograms from different source and receiver positions, corresponding to the ray paths shown by solid lines, sample the same point at depth on a flat reflector. These have the same midpoint halfway between source and receiver, but different source-to-receiver offsets.

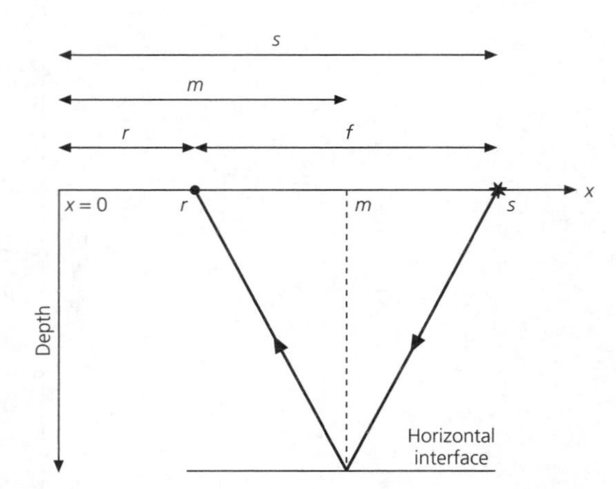

Fig. 3.3-11 Relation between source, receiver, midpoint, and offset coordinates measured along the survey line. Any two specify an individual seismogram.

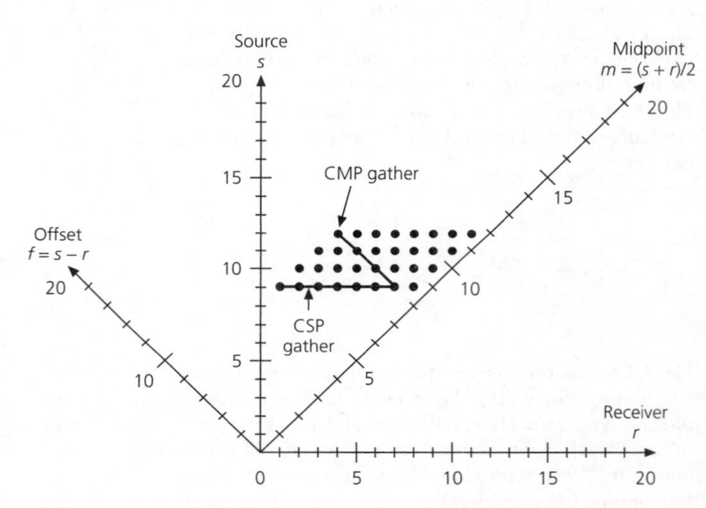

Fig. 3.3-12 An individual trace is characterized by its position in a two-dimensional diagram showing its source, receiver, midpoint, and offset coordinates. Dots show the traces indicated in Fig. 3.3-10. Physical experiments correspond to a common source point (CSP) gather; the four traces in Fig. 3.3-10 with the same midpoint form the common midpoint (CMP) gather shown.

$$\tau(p) = 2\int_0^{z_p} \eta(z)dz = 2\int_0^{z_p} (1/v^2(z) - p^2)^{1/2}dz = 2\int_0^{z_p} (u^2(z) - p^2)^{1/2}dz. \tag{47}$$

Formulating travel time curves as $\tau(p)$ is useful for some techniques that invert for velocity structure.

3.3.3 *Multichannel data geometry*

A feature of reflection seismology is *multichannel* geometry, the use of multiple source and receiver locations, so that points on reflecting interfaces are sampled repeatedly. Figure 3.3-10 illustrates how such coverage is accomplished by combining experiments performed with a seismic source and an array of eight receivers at fixed distances from the source. Each time the source is activated, eight seismograms, or *traces*, are recorded. The source and the receivers are then moved, and the experiment is repeated, giving eight more traces. Eventually each point on the reflector is sampled four times, producing "four-fold coverage."

We assume initially that the velocity structure is layered and varies only with depth. Even so, the four seismograms that sample the same point are not identical, because they correspond to different source and receiver positions, and thus different offset distances between the source and the receiver. Hence each trace is a record of displacement, or pressure, as a function of time, t, $u(s, r, t)$, characterized by the source and receiver positions.

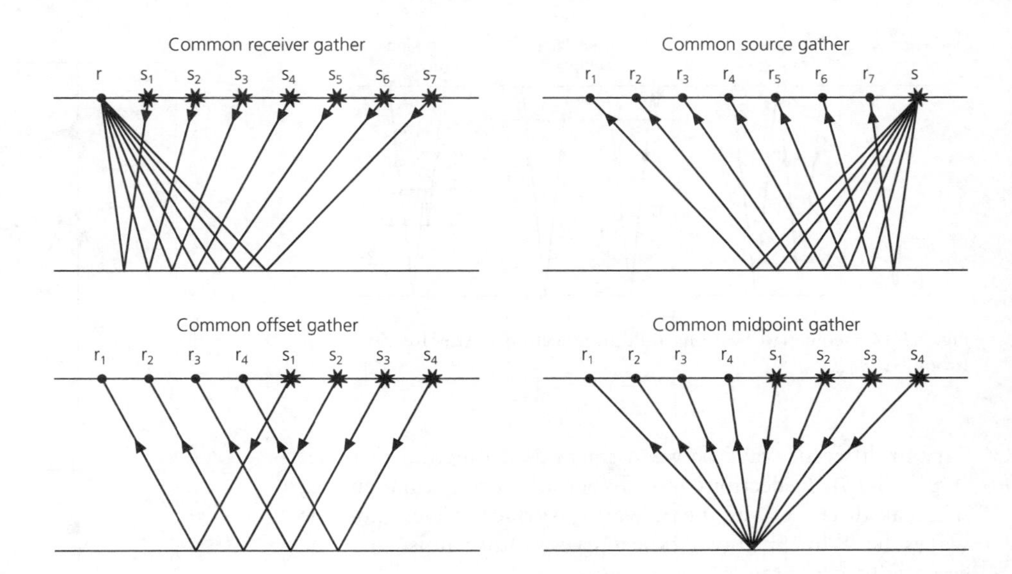

Fig. 3.3-13 Schematic of the four different gather types.

The data are analyzed by grouping the seismograms that sampled the same point on the reflector. In this flat-layered geometry, these seismograms have the same point, known as the *midpoint*, halfway between the source and the receiver. For each midpoint, there is a set of traces with different offsets. The midpoint m and offset f are defined in terms of the source location s and the receiver position r as

$$m = (s + r)/2, \quad f = (s - r). \tag{48}$$

Thus an individual seismogram is specified by either the source and receiver positions, or the midpoint and offset (Fig. 3.3-11). These are plotted using two perpendicular axes (Fig. 3.3-12), one for the source location and one for the receiver position. The midpoint and offset for each seismogram are indicated by distance along axes 45° from the s and r axes. Note that the scales on these axes differ from the other two.

To illustrate this relationship, consider the four experiments in Fig. 3.3-10, with eight receivers and a single source. Each experiment produced data at points, shown by dots, with constant source position and successive receiver positions. Successive experiments yielded data along a similar horizontal line, but displaced by the motion of the source and the receiver.

The data can be sorted and combined in various ways that need not correspond to an actual experiment (Fig. 3.3-13). Each experiment corresponds to a set of records with the same source position, a *common source point*, or *CSP*, gather. Traces with the same midpoint and different offsets can be grouped in a *common midpoint*, or *CMP*, gather. Similarly, common receiver point and common offset gathers can be formed.

Ordering traces by midpoint and offset makes no distinction between a source at position a and a receiver at position b, or the reverse. This assumption is justified by the principle of reciprocity, by which these two geometrices should produce identical seismograms. Thus a common receiver point gather can simulate a reversed profile (Section 3.2.2) because, by reciprocity, it gives the same data as a common source point gather shot in the opposite direction.

Later in this section, we will discuss a few aspects of the data collection process. The sources can be explosives, sound sources in water, or vibration sources on land. The source coordinate is thus sometimes referred to as a source point, shot point, or vibration point. The receivers are typically single-component vertical seismometers, known as geophones, for land applications, and pressure transducers, or hydrophones, for marine surveys. The receiver coordinate is thus often termed the geophone coordinate. Generally large numbers of receivers, which are themselves groups of receivers, are used. Increasingly, data are collected over two-dimensional areas, and so are processed to yield three-dimensional velocity structures.

3.3.4 *Common midpoint stacking*

Because the traces in a CMP gather have ideally sampled the same subsurface point with different offsets, they can be combined to enhance reflected arrivals. The process begins with a set of traces showing the data as a function of offset and time. The data contain "signals" of interest, primary reflections from interfaces that are used to determine velocity structure with depth. The data also contain "noise," arrivals of no interest, including direct waves, head waves,[4] surface waves (sometimes termed "ground roll"), and waves from the source that travel in the air. The data may also contain arrivals (Fig. 3.3-14) that have been reflected more than once, which are known as *multiples*, by contrast with the once-reflected primary reflections.

To enhance primary reflections and suppress everything else, we exploit the fact that the arrival times of various signals

[4] In the previous section we focused on direct and head waves, illustrating the adage that "one person's signal is another's noise."

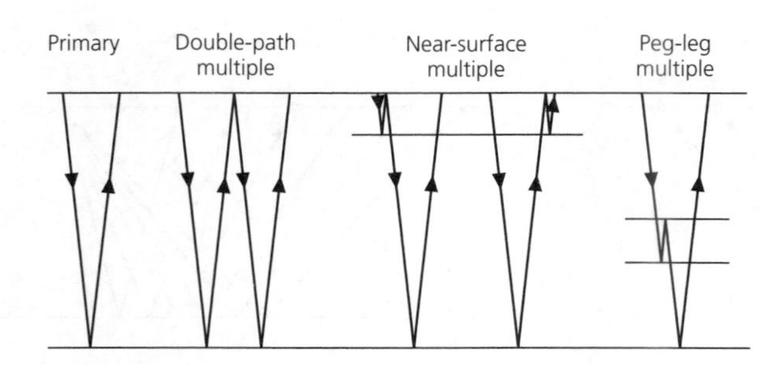

Fig. 3.3-14 Geometry of various multiple reflections. (After Kearey and Brooks, 1984.)

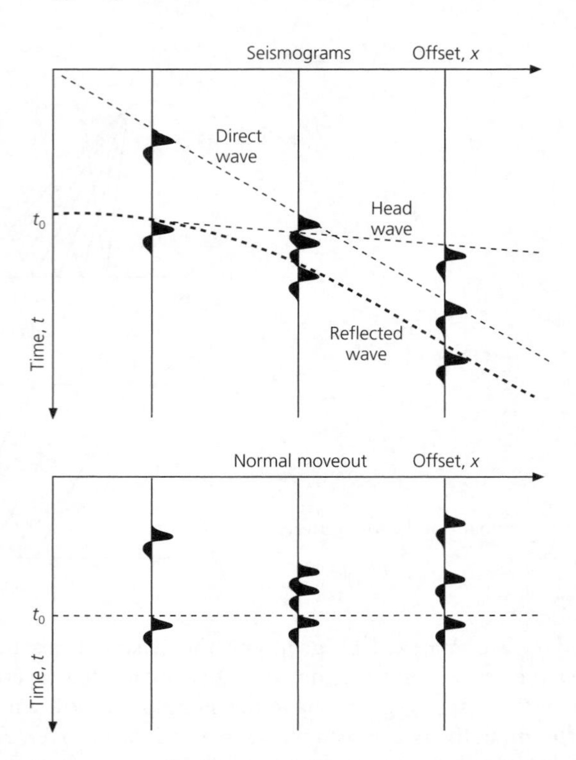

Fig. 3.3-15 Schematic example of the normal moveout correction, shown for the three arrivals for a single layer. NMO aligns all traces (lower panel) in a common midpoint gather by a time shift corresponding to the hyperbolic travel time curve of a reflection. The desired reflection is thus in phase between traces, whereas other arrivals are out of phase. CMP stacking, which adds the traces after this time shift, enhances the desired reflection and suppresses other arrivals.

vary in different ways between traces as a function of offset (Fig. 3.3-15). Reflections have hyperbolic travel time curves, whereas direct waves, head waves, surface waves, and air waves have linear travel time curves. Other noise may be essentially incoherent between traces.

Consider a reflection whose variation in travel time with offset is the normal moveout (NMO),

$$T(x) - t_0 = (x^2/\bar{V}^2 + t_0^2]^{1/2} - t_0, \tag{49}$$

where t_0 and $\bar{V}$ are the vertical two-way time and rms velocity. If each trace is shifted forward in time by the appropriate NMO, this reflection appears at the same time for all offsets (Fig. 3.3-15). By contrast, arrivals with different moveouts, such as the direct wave, do not align. Similarly, multiple reflections do not align, because they reflected off shallower interfaces than primary reflections with a similar arrival time, and thus have a lower rms velocity. This method is similar to forming reduced travel time plots (Section 3.2), where a linear time shift lines up direct or head waves whose linear travel time curve has apparent velocity equal to the reducing velocity. In this case, the hyperbolic time shift lines up reflections with hyperbolic travel time curves.

If the traces are added after this time shift, the resulting sum, in theory, is the single trace that would have been recorded at zero offset, with coincident source and receiver. The reflection that was aligned is in phase on all traces, and thus sums constructively and gives a strong arrival. By contrast, other arrivals will have been shifted such that they are sometimes out of phase, and thus sum destructively, yielding weaker arrivals. The process of time shifting and then summing the traces with different offsets for a given midpoint is called *common midpoint (CMP) stacking*.

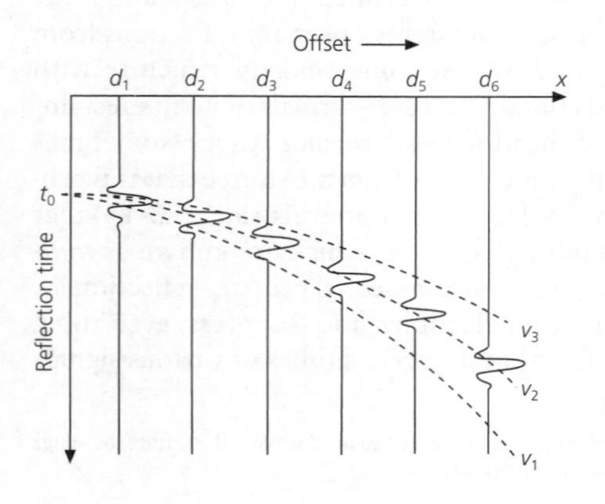

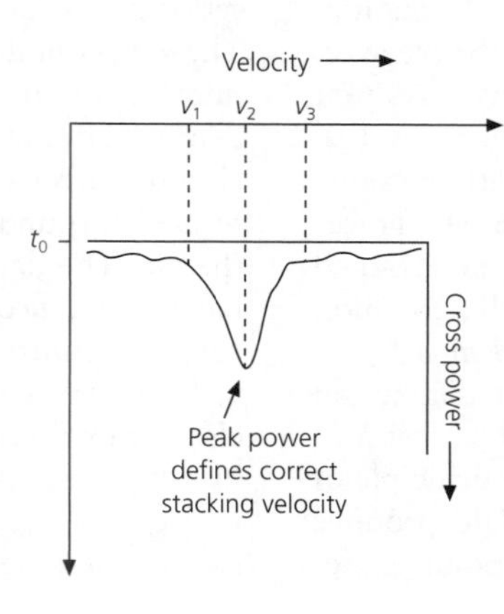

Fig. 3.3-16 Schematic of CMP stacking and velocity analysis. *Left*: Stacking is done for a range of stacking velocities, each corresponding to a different hyperbola in offset–time space. *Right*: The peak in the velocity spectrum, or power in the resulting stack, shows the best stacking velocity. (After Taner and Kohler, 1969. Reproduced by permission of the Society of Exploration Geophysicists.)

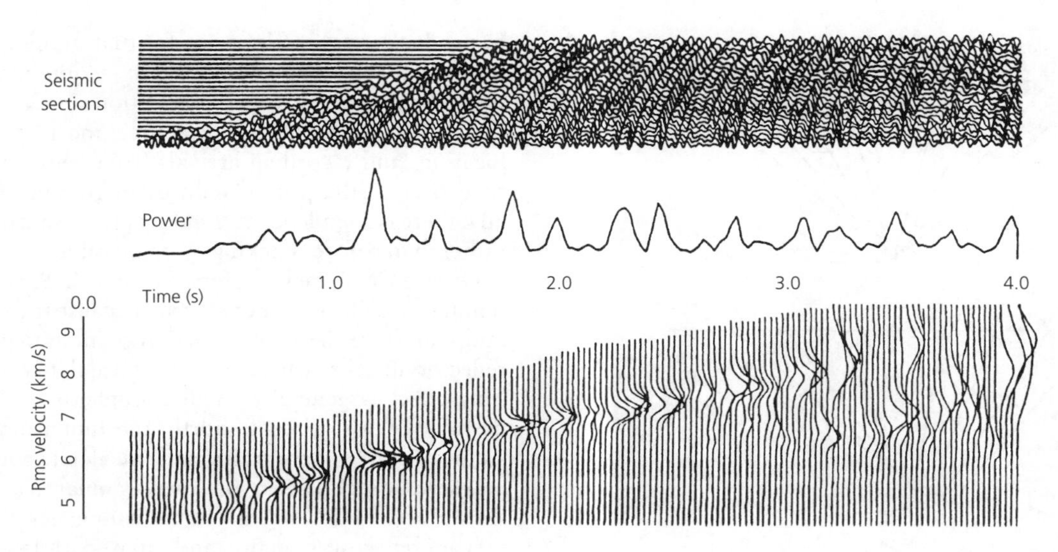

Fig. 3.3-17 Example of CMP stacking and velocity analysis. Velocity analysis at different times yields the best stacking velocity as a function of time (*bottom*). The stacking velocity increases with time because later arrivals reflected off deeper interfaces. (After Taner and Kohler, 1969. Reproduced by permission of the Society of Exploration Geophysicists.)

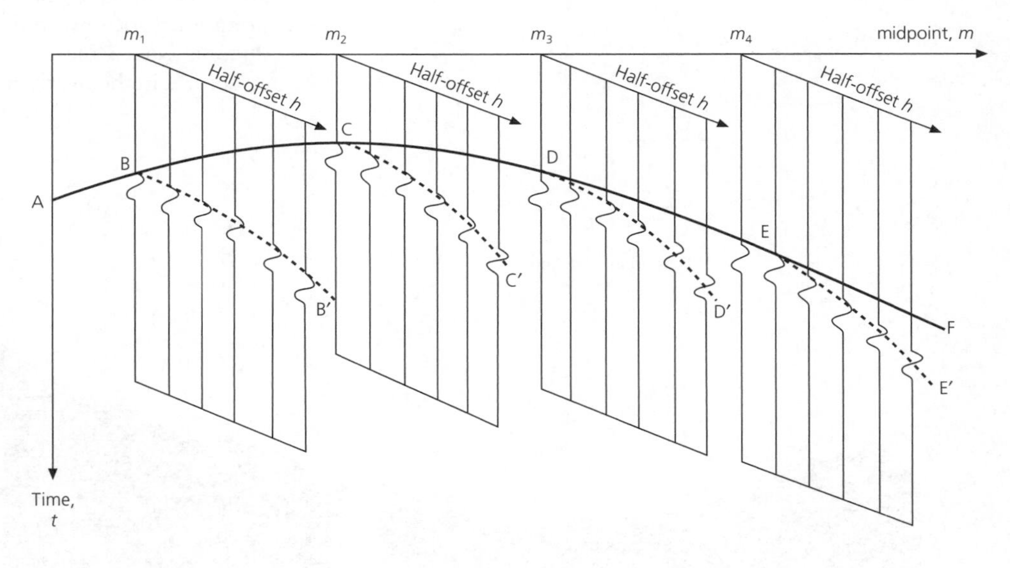

Fig. 3.3-18 Schematic geometry illustrating formation of a zero-offset section by common midpoint stacking. Each CMP gather is stacked over all offsets, as shown by the dashed lines like B–B′, to produce a single zero-offset trace for that midpoint. Taken together, these traces form a zero-offset section, a plane in midpoint–time space, containing arrivals like that shown by the solid curve A–F. (After Robinson, 1983. *Migration of Geophysical Data*, © 1983, p. 24. Reprinted by permission of Pearson Education.)

Real data contain more than one reflection, and the appropriate velocities are unknown. Thus the velocities are found by stacking with a range of velocities and determining which gives the best results. As illustrated in Fig. 3.3-16, traces are stacked along hyperbolas corresponding to different velocities. The stack output as a function of stacking velocity, known as a velocity spectrum, has peak amplitude at the velocity that best aligns arrivals on the different traces. This stacking velocity is close to the rms velocity if the data are reasonably good and the structure is approximately a set of flat layers.

Because later reflections have higher rms velocities, they yield higher stacking velocities. Thus velocity analysis is conducted as a function of time. In Fig. 3.3-17, the best stacking velocity, indicated by the maximum in the velocity spectrum, increases with time for deeper arrivals. This increase "tunes" the stacking to bring arrivals with various stacking velocities "into focus." Peaks in the power of the velocity spectrum show the arrival of strong coherent reflections. At later times there are several peaks, as multiples arrive. Using the stacking velocities, interval velocities for different depths are found from the Dix equation.

Figure 3.3-18 illustrates the CMP concept geometrically. The traces give displacement or pressure as a function of midpoint, offset, and time, $u(m, f, t)$. CMP gathers can be thought of as planes parallel to the offset and time axes, each with the appropriate midpoint. Each gather is stacked over all offsets

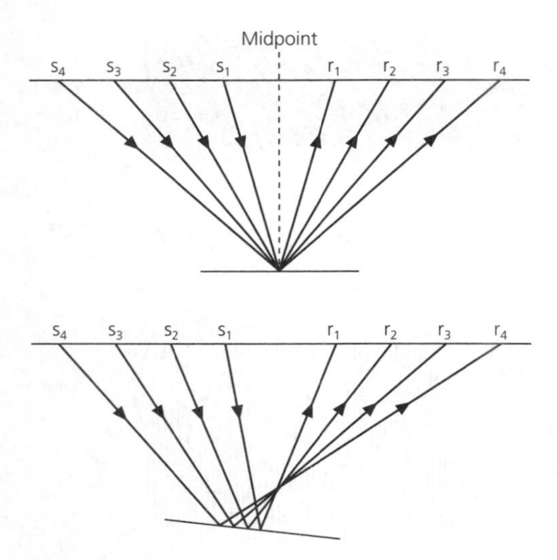

Fig. 3.3-19 *Top*: Traces with a common midpoint sample the same point on a reflector when a reflector and the structure above it are horizontal. *Bottom*: If the structure dips, traces with a common midpoint do not reflect at the same point. (After Kearey and Brooks, 1984.)

to produce a zero-offset trace for that midpoint. These traces together form a zero-offset *seismic section*, $u(m, 0, t)$, a function of midpoint and time. This section simulates moving along the survey line with a single source and receiver at the same location, and recording arrivals from below as a function of time. Because this process reduces the volume of data dramatically, there is a tendency to conduct processing operations after, rather than before, stacking when possible.

Often a CMP stack is referred to as a CDP, or common depth point, stack. CMP is a better term, because traces with the same midpoint have the same reflection point at depth only when a reflector and the structure above it are flat-lying (Fig. 3.3-19). This effect is generally small enough that CMP stacking is useful. We will discuss shortly the limitations on reflection studies due to deviations from the ideal flat geometry.

A seismic section is in some ways similar to a "picture" of the subsurface. Major arrivals in the data generally represent significant reflectors at depth, and can be correlated with geologic structure. As a result, analysis of seismic reflection data is a powerful geological tool. For example, Fig. 3.3-20 (*top*) shows a seismic section across the Peru trench. Data of one polarity are black, making coherent reflectors more visible. The interpretation (*bottom*) indicates the top of the crust of the subducting Nazca plate, including small grabens, and complex structures in the overlying accretionary prism.

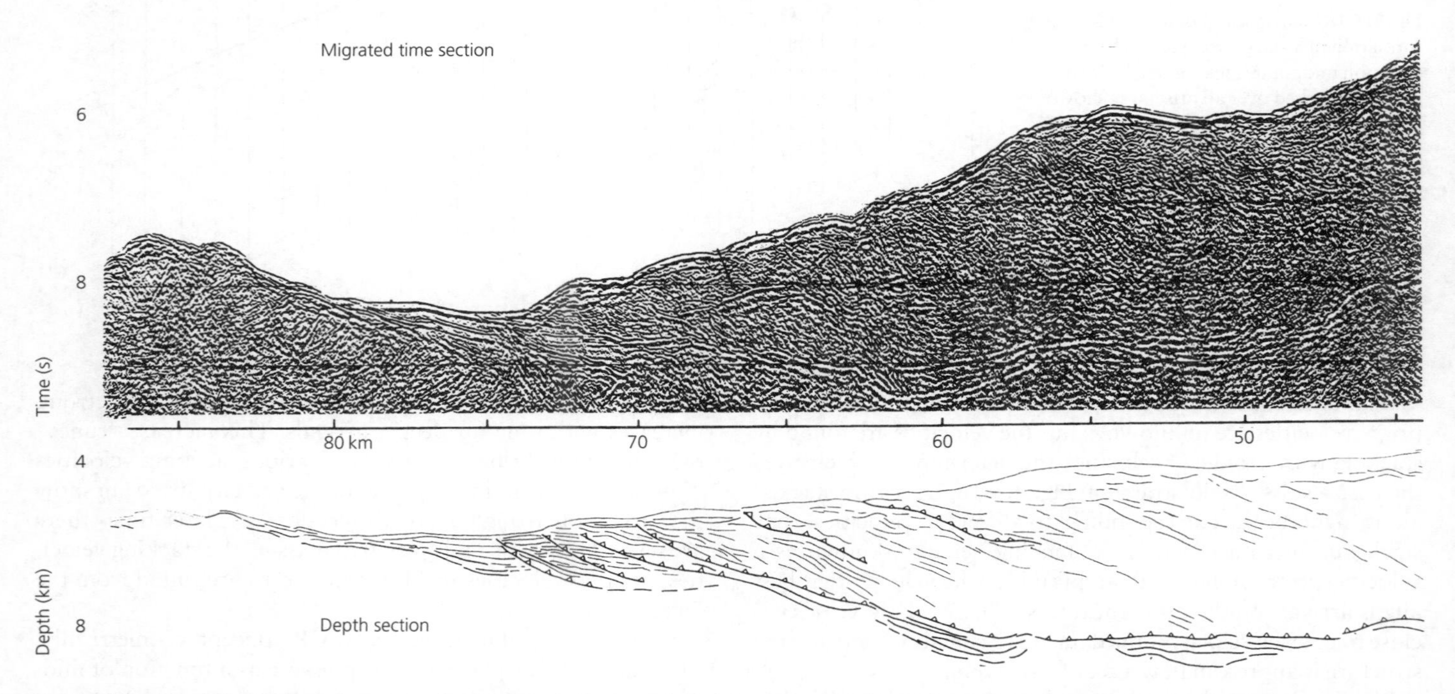

Fig. 3.3-20 Migrated seismic section across the Peru trench, showing the subducting Nazca plate dipping to the right. The data were collected with air gun sources shot at 35 m intervals and recorded by a 1600 m-long array with 24 hydrophone groups. The data were sampled every 4 milliseconds. (After Von Huene *et al.*, 1985. *J. Geophys. Res.*, *90*, 5429–42, copyright by the American Geophysical Union.)

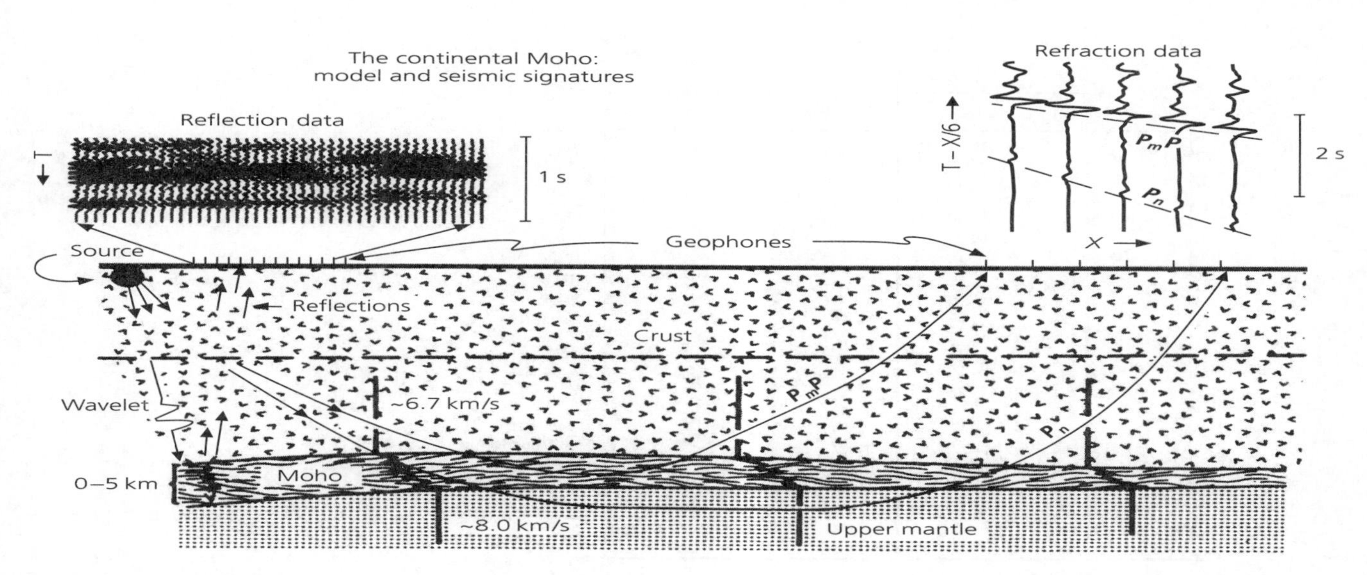

Fig. 3.3-21 Reflection data (*left*), showing muting (*right*) to eliminate the head waves that arrive first and the large surface waves that arrive later. (After Claerbout, 1985.)

3.3.5 Signal enhancement

The best hope of reducing artifacts in a seismic section due to noise and other difficulties is to exclude them before stacking. Thus, as in many signal processing applications, the idea is to identify characteristics of the "noise" we seek to reject, and use those characteristics to exclude it.

For example, variations in the thickness of a near-surface low-velocity layer due to weathering produce arrival time variations. Similar variations can result from sea floor topography, because the water is a low-velocity material of varying thickness, or from elevation changes along a land survey. These shifts can cause the travel time of reflections to deviate from the hyperbolic moveout with offset assumed in stacking, and hence degrade a stacked section and produce spurious relief on a deeper reflector. To minimize these problems, a *static time correction*, shifting traces back or forward in time, can be applied.

Direct waves, head waves, surface waves, air waves, and the like are often identifiable on CSP gathers from their arrival times and linear travel time curves. Data corresponding to the time–distance ranges in which the undesired arrivals appear can be set equal to zero, or *muted* before the gathers are stacked (Fig. 3.3-21).

Another approach to isolating reflections uses the fact that the apparent velocity along the surface,

$$c_x = 1/p = v/\sin i = \omega/k_x, \tag{50}$$

is higher for reflections, which have angles of incidence close to the vertical, than for surface or air waves. Hence the reflections have a longer apparent wavelength along the surface, $\lambda_x = 2\pi c_x/\omega$. Thus the effects of surface waves can be reduced by summing a *group* of receivers to produce a single trace. Arrivals with wavelengths shorter than the length of the group interfere destructively and are reduced in amplitude, enhancing the longer-wavelength reflections. Hence traces from a single source–receiver pair are often actually a sum of a number of geophones or hydrophones. In this way, the data collection process, rather than subsequent analysis, enhances the reflections.

Differences in the apparent velocity can also be used to enhance reflections after the data are collected. In this approach, arrivals with different apparent velocities on common source gathers are separated by *velocity filtering*, using a double Fourier transform. As we saw in Section 2.8.2, and discuss further is Chapter 6, the Fourier transform and inverse transform relate a function of time $f(t)$ and its transform $F(\omega)$, a function of angular frequency,

$$F(\omega) = \int_{-\infty}^{\infty} f(t)e^{-i\omega t}dt \qquad f(t) = \frac{1}{2\pi}\int_{-\infty}^{\infty} F(\omega)e^{i\omega t}d\omega. \tag{51}$$

Similarly, because the wavenumber is the spatial frequency (Section 2.2.2), it is related to the distance in the same way that angular frequency is related to time. Hence, a function of the horizontal distance $g(x)$ and its corresponding function of horizontal wavenumber $G(k_x)$ are related by the Fourier transform pair

$$G(k_x) = \int_{-\infty}^{\infty} g(x)e^{ik_x x}dx \qquad g(x) = \frac{1}{2\pi}\int_{-\infty}^{\infty} G(k_x)e^{-ik_x x}dk_x. \tag{52}$$

By convention, opposite signs are used in the exponentials for the time and space transforms.

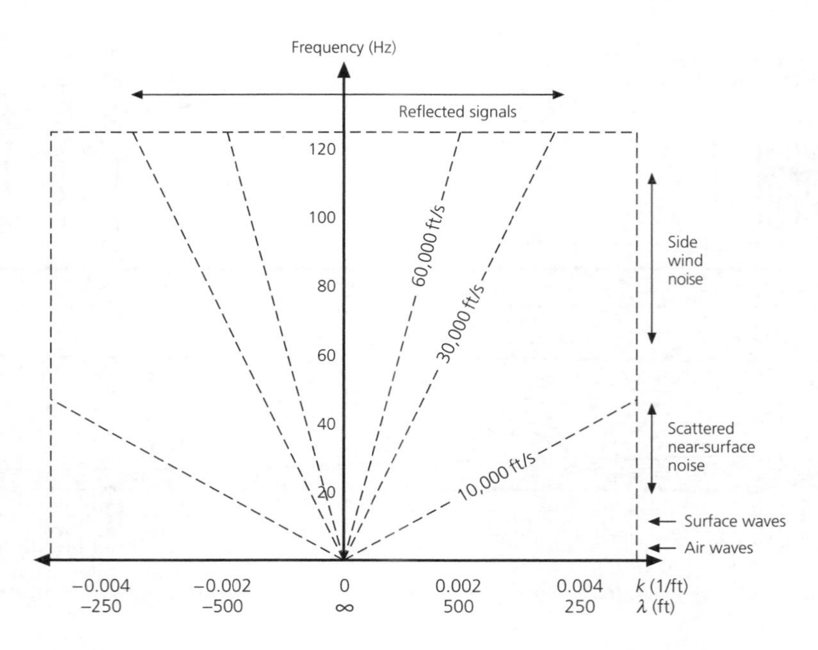

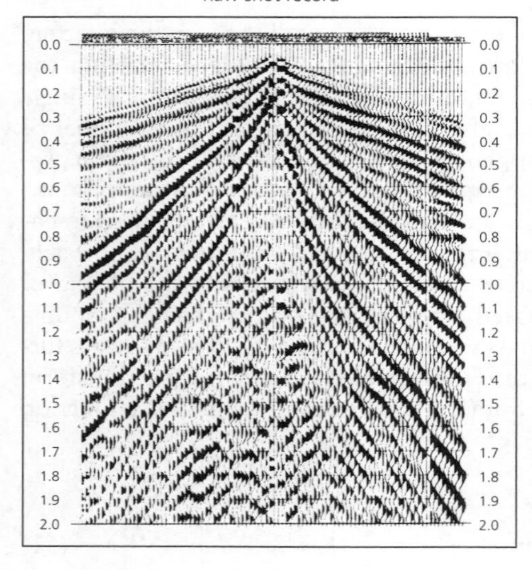

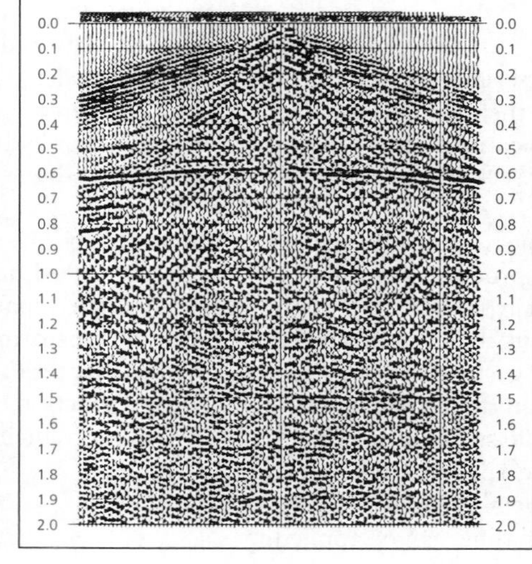

Fig. 3.3-22 Velocity filtering by Fourier transformation into the horizontal wavenumber and frequency domain. *Top*: Positions of reflected waves, noise, air waves, and surface waves in the (k_x, f) plane. Slopes correspond to lines of equal apparent velocity (in ft/s). (After Kanasewich, 1981.) *Bottom*: Common source gather before and after velocity filtering. Surface waves have been suppressed by removing low apparent velocity data, thus enhancing reflections. (Hosking Geophysical.)

A gather $u(x, t)$ is the displacement as a function of horizontal distance and time, so the double Fourier transform,

$$U(k_x, \omega) = \int_{-\infty}^{\infty}\int_{-\infty}^{\infty} u(x, t) \exp\,[i(-\omega t + k_x x)]dxdt, \quad (53)$$

converts it to the horizontal wavenumber and angular frequency domains. Plotting the transform as a function of k_x and ω (or, equivalently, k_x and frequency f) separates the data into portions of different apparent velocity, because a given velocity, $c_x = \omega/k_x$, plots as a straight line (Fig. 3.3-22). It is thus possible to suppress arrivals with a given range of apparent velocities by setting the data in some region of (k_x, ω) space to zero, and inverse transforming the data back to (x, t) space, using the inverse of the double Fourier transform

$$u(x, t) = \frac{1}{4\pi^2}\int_{-\infty}^{\infty}\int_{-\infty}^{\infty} U(k_x, \omega) \exp\,[i(\omega t - k_x x)]dk_x d\omega. \quad (54)$$

Rather than having an abrupt boundary, the data at the edges of the portion of the (k_x, ω) space of interest are tapered smoothly to zero for reasons discussed in Chapter 6.

Thus the double Fourier transform converts data containing arrivals that overlap in the (x, t) domain into the (k_x, ω) domain, where the arrivals have distinct properties that make it easy to separate them. This separation is exploited to filter

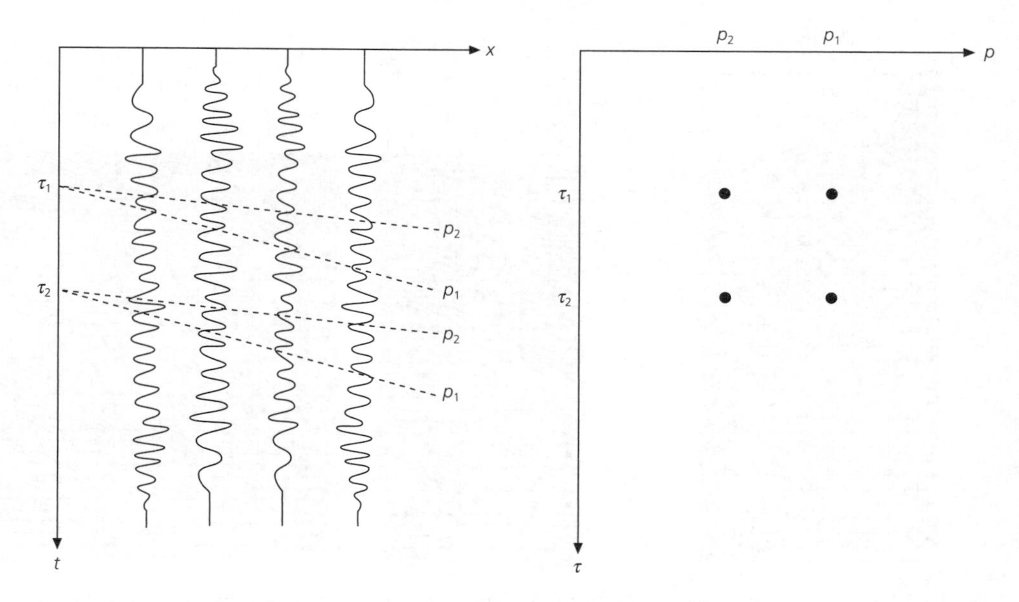

Fig. 3.3-23 Schematic illustration of slant stacking: data are summed along lines in the (x, t) plane (*left*) corresponding to values of intercept τ and slope p, and so yield points in the (τ, p) plane (*right*).

the data, which are then transformed back to (x, t). Velocity filters are also called *dip filters*, because they separate arrivals based on their slope (dip) in the (k_x, ω) domain. A variant of this method for data recorded in two spatial dimensions, $u(x, y, t)$, is to take the triple Fourier transform $U(k_x, k_y, \omega)$. Because the transform is in terms of both components of the horizontal wave vector, it can be filtered to suppress arrivals coming from certain directions.

Another approach to transforming data such that components are more easily separated uses the intercept-slowness formulation of travel time curves. As discussed in Section 3.3.2, the function $\tau(p)$ describes each point on a travel time curve $T(x)$ by the time axis intercept τ and the slope p of the line tangential to the curve at that point. Thus seismic data can be described as functions either of position and time, $u(x, t)$, or of slope and intercept, $\bar{u}(\tau, p)$. To transform from one representation to the other, the data $u(x, t)$ are summed along lines of constant slope in the (x, t) plane, which correspond to values of intercept τ and slope p (Fig. 3.3-23),

$$\bar{u}(\tau, p) = \int_{-\infty}^{\infty} u(x, \tau + px)dx. \tag{55}$$

This integral, which maps all the data along each slanted line in (x, t) to a point in (τ, p), is called a *slant stack*, or *Radon transform* of the data. It is also called a *plane wave decomposition*, because it decomposes the data according to p, the reciprocal of the apparent velocity of a plane wave. The inverse slant stack operation that transforms the slant stack back into the (x, t) space can be written[5]

[5] Claerbout (1985).

$$u(x, t) = 1/t^2 * \frac{1}{2\pi} \int_{-\infty}^{\infty} \bar{u}(t - px, p)dp, \tag{56}$$

where "$*$" is the convolution operation, discussed shortly. This expression is similar to a slant stack in the (τ, p) plane, because data are summed along a line of constant τ.

All the data are mapped from one domain into the other, so no data are lost by this transformation. Thus, after slant stacking, we can use the fact that the $\tau(p)$ representation of the travel time curve is in some ways simpler than the $T(x)$ representation. Because different arrivals fall in different parts of the (τ, p) plane (Fig. 3.3-8), undesired arrivals can be suppressed by zeroing portions of the data. For example, the gather in Fig. 3.3-24 shows a strong surface wave, the late-arriving linear arrival with an apparent velocity of about 1.35 km/s and intercept about 0. In the usual (x, t) space, it would be hard to filter out this arrival without suppressing the reflections. After slant stacking, this arrival shows up as a region of large amplitude with $\tau \approx 0$ and $p = 1/1350$ s/m $\approx$ 740 μs/m. Once the slant stack is filtered by eliminating all data with $p > 650$ μs/m and inverse transformed, the surface wave is significantly reduced. In practice, rather than having an abrupt boundary, the data at the edges of the portion of the (τ, p) space of interest are tapered smoothly to zero for reasons discussed in Chapter 6.

The slant stack and velocity filtering with the double Fourier transform are related, because both exploit properties of the data associated with the apparent velocity. As a result, slant stacking can be done by transforming data to the (k_x, ω) domain, evaluating the transform for constant values of the ray parameter, and then inverse transforming to the time domain.

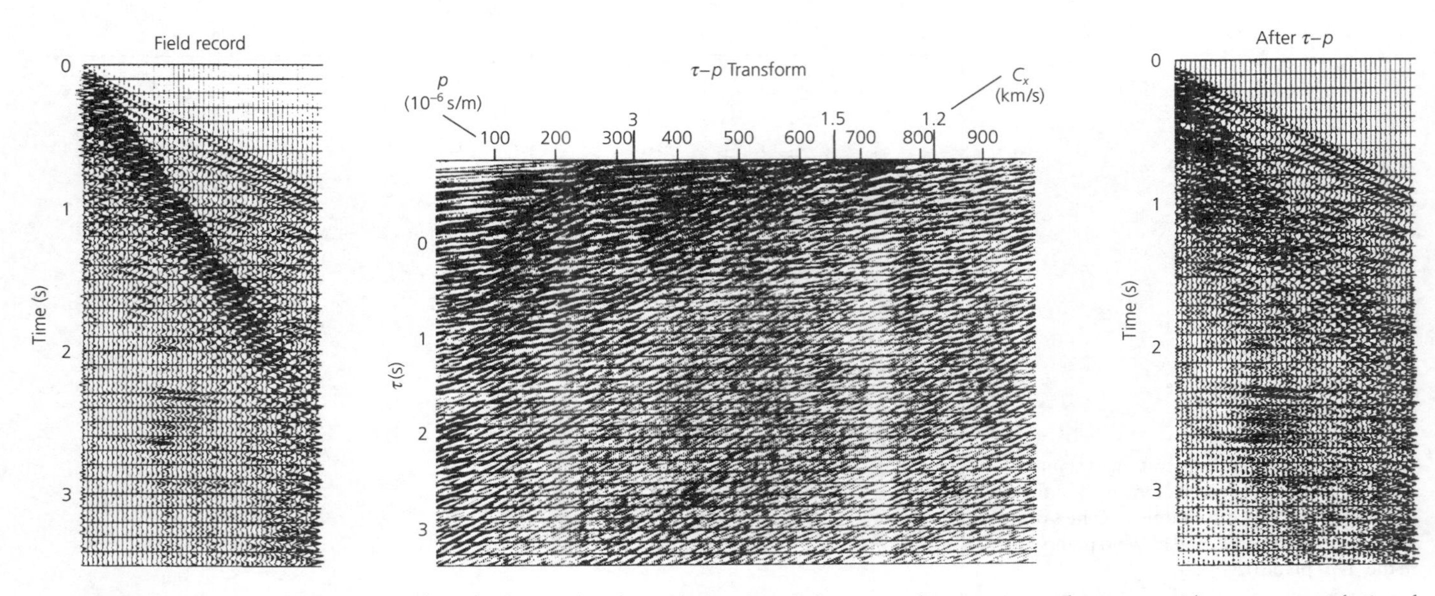

Fig. 3.3-24 *Left*: Common source point gather of Vibroseis data from Alaska, showing prominent late-arriving surface waves with an apparent velocity of about 1.35 km/s and intercept about 0. *Center*: Slant stack of the data. The p axis is labeled both with values of p (μs/m) and apparent velocity (km/s). The surface waves appear as a region of large amplitude with $\tau \approx 0$ and $p \approx 740$ μs/m. *Right*: The inverse slant stack, after suppression of data with $p > 650$ μs/m, shows the surface wave significantly reduced. (Tatham, 1989. With kind permission from Kluwer Academic Publishers.)

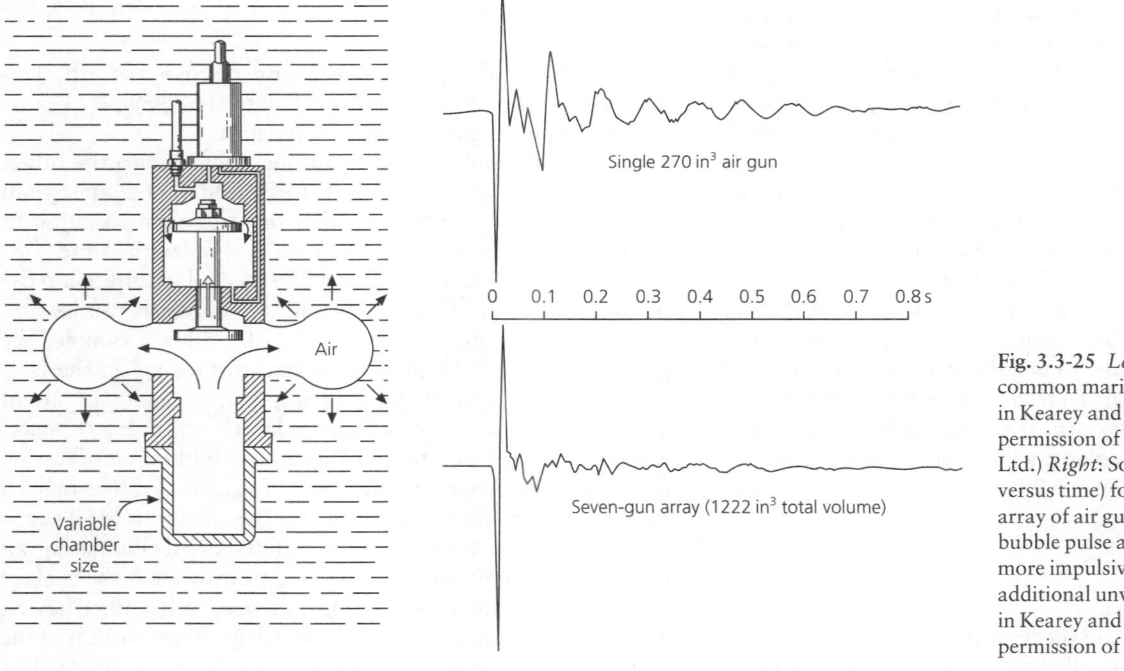

Fig. 3.3-25 *Left*: Schematic of an air gun, a common marine seismic source. (Fig. 3.18 in Kearey and Brooks, 1984, redrawn with permission of Bolt Associates and Sodera Ltd.) *Right*: Source wavelets (pressure versus time) for a single air gun and an array of air guns. The array reduces the bubble pulse and makes the wavelet more impulsive, though it still contains additional unwanted complexity. (Fig. 3.19 in Kearey and Brooks, 1984. Redrawn with permission of Bolt Associates.)

3.3.6 *Deconvolution*

Another useful technique, *deconvolution*, "sharpens" the reflections from interfaces. Ideally, each reflection would be a sharp pulse approximating a delta function, so the arrival time of the reflection and the depth of the reflector would be determined precisely. The sharpness of the reflected pulse determines vertical resolution: how close in travel time, and thus depth, two interfaces can be and still give distinct reflected arrivals.

Seismic sources do not generate delta function signals. Figure 3.3-25 shows the signal produced by an air gun, a common source used in marine surveys. The damped oscillation results

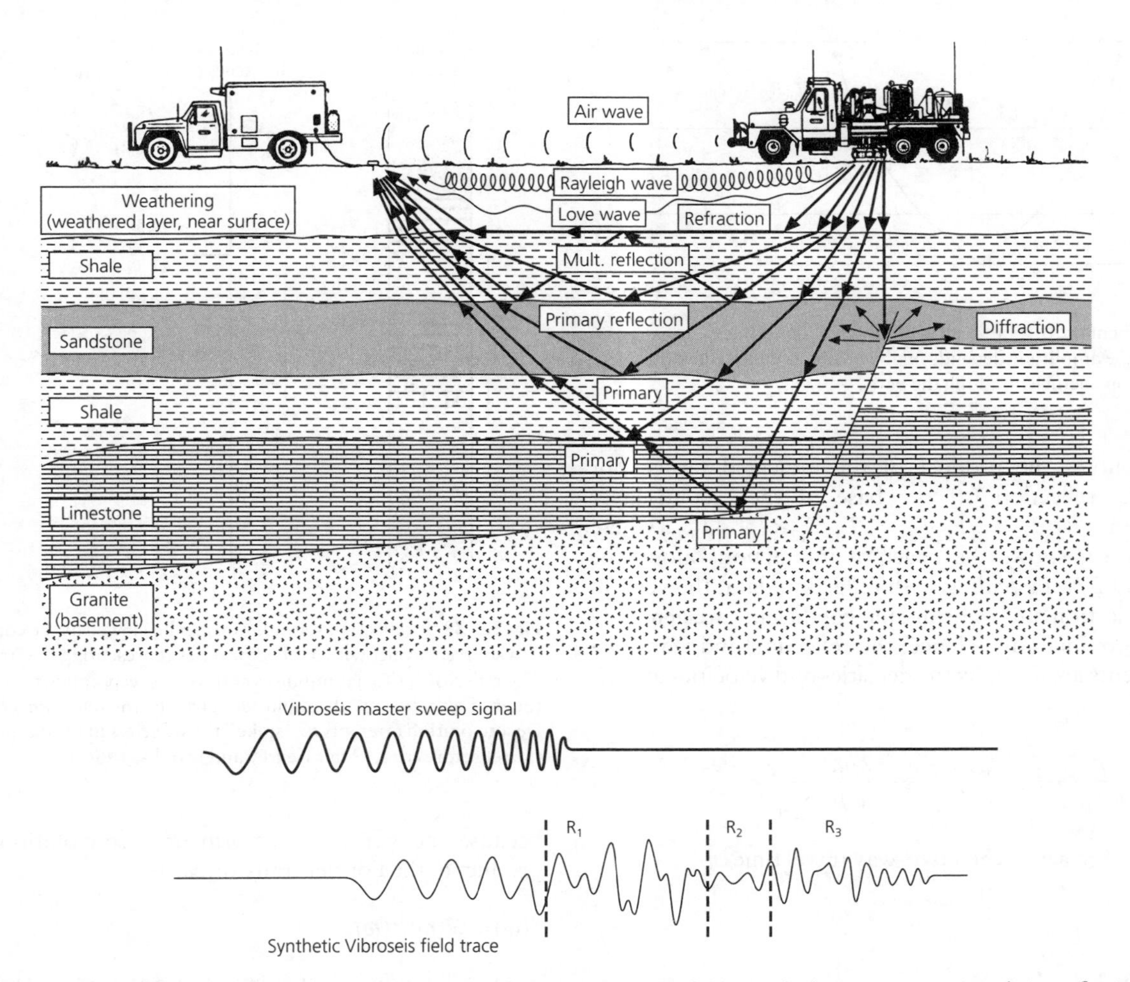

Fig. 3.3-26 Schematic geometry of a Vibroseis survey (*top*) and sweep signal (*center*). The field records (*bottom*) contain interfering reflections off various interfaces, and so require processing to identify individual reflections. (With permission of Conoco.)

from expansion and contraction of the air bubble that the gun injects into the water. The signal can be sharpened using multiple air guns offset in time, which interfere to give a sharper pulse. Figure 3.3-26 shows the "sweep" signal generated by a Vibroseis[6] unit, a truck-mounted seismic source used in land surveys. The signal extends for a period of time T (typically 7–35 s) over which the frequency varies through a range f_1–f_2, generally within 10–60 Hz. Such signals, also called "chirps," can be written

$$w(t) = \cos 2\pi \left(f_1 t + \frac{(f_2 - f_1)}{2T} t^2 \right). \tag{57}$$

Because the duration of the sweep is often longer than the difference in travel time between interfaces, the resulting seismogram is a complicated combination of sweep signals with different amplitudes and time delays reflected from different interfaces.

[6] Vibroseis is a trademark of the Continental Oil Company. The first such continuously operating variable frequency seismic source was invented by Selwyn Sacks in his Ph.D. thesis in 1961.

Thus reflection data, like any other seismograms, include the effects of both the source and the structure. Separating these effects is a basic theme in seismology, because we are usually interested in either the source (as for earthquakes) or the structure, as in this application. To separate source and structure, we describe a seismogram, $s(t)$, as resulting from the source pulse, known in reflection applications as a wavelet, $w(t)$, and a time series that describes the effects of the structure, in this case a reflector series, $r(t)$.

To find the reflector series, we recall from Section 2.6.7 that a wave with initial unit amplitude acquires an amplitude equal to the product of the reflection and transmission coefficients along its path. Thus, for a set of layers with velocity v_j and thickness h_j, the amplitude of the primary reflection from the bottom of the ith layer is the product of the reflection coefficient at the base of the layer times all the transmission coefficients for both the up and down parts of the path,

$$R_{i\,i+1} \prod_{j=0}^{i-1} T_{j\,j+1} T_{j+1\,j}, \tag{58}$$

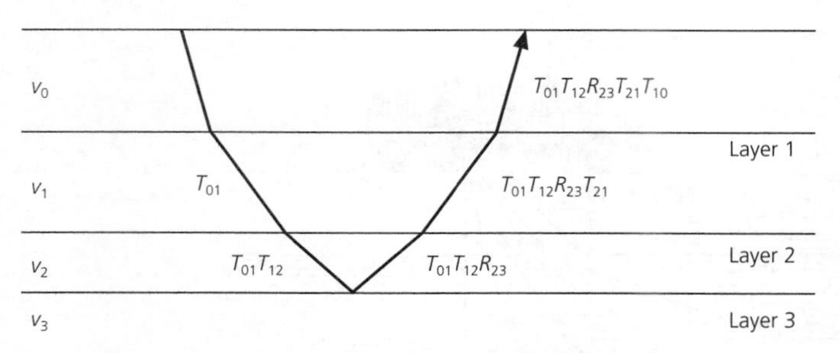

Fig. 3.3-27 Schematic of a ray path through several interfaces, showing how the amplitude depends on the product of the reflection and transmission coefficients along the path.

where Π denotes the product of the indicated terms. For example, the reflection off the base of the second layer has amplitude $R_{23}T_{01}T_{10}T_{12}T_{21} = T_{01}T_{12}R_{23}T_{21}T_{10}$, where the second form shows the order of interactions along the path (Fig. 3.3-27). In dealing with reflection data, the vertical incidence reflection and transmission coefficients are generally suitable approximations. Hence, the reflection and transmission coefficients are given by the densities and velocities at each interface

$$R_{i\,i+1} = \frac{\rho_i v_i - \rho_{i+1} v_{i+1}}{\rho_i v_i + \rho_{i+1} v_{i+1}}, \quad T_{i\,i+1} = \frac{2\rho_i v_i}{\rho_i v_i + \rho_{i+1} v_{i+1}}, \tag{59}$$

and the reflection arrives at a two-way travel time of

$$t_i = 2\sum_{j=0}^{i} \frac{h_j}{v_j}, \tag{60}$$

which is the sum of the vertical travel times in each of the layers. Thus the reflector series for primary reflections off a set of N layers is a sum of impulses, each corresponding to the reflection from the bottom of the ith layer,

$$r(t) = \sum_{i=0}^{N} \delta(t - t_i) R_{i\,i+1} \prod_{j=0}^{i-1} T_{j\,j+1} T_{j+1\,j}. \tag{61}$$

$\delta(t - t_i)$ is the delta function, a spike in time that is zero at all times except t_i, when it equals 1. The reflector series is thus a set of spikes with the appropriate amplitude and arrival time, each corresponding to a specific reflection.

We will see in Chapter 6 that the resulting seismogram is given by an operation known as the *convolution* of $w(t)$ and $r(t)$, which is written

$$s(t) = w(t) * r(t) \equiv \int_{-\infty}^{\infty} w(t - \tau) r(\tau) d\tau. \tag{62}$$

This equation defines convolution in the time domain. Convolution can also be described in the frequency domain,

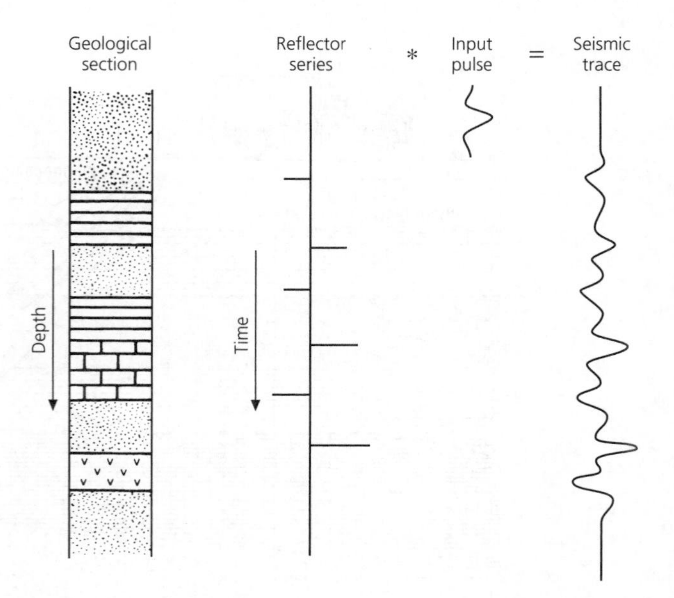

Fig. 3.3-28 A reflection seismogram can be viewed as the convolution of a source wavelet with a reflector series representing the structure. The reflector series has impulses at times corresponding to the arrival times of reflections with amplitudes given by the reflection coefficients. Deconvolution attempts to "spike" the wavelets in the data, revealing the reflector series. (After Kearey and Brooks, 1984.)

because the Fourier transform of a convolution equals the product of the Fourier transforms,

$$S(\omega) = W(\omega) R(\omega). \tag{63}$$

As shown schematically in Fig. 3.3-28, the convolution yields a trace in which the source wavelet appears at times corresponding to the spikes in the reflector series, with the appropriate amplitudes. If the time between the spikes corresponding to individual reflectors is shorter than the duration of the wavelet, interference can give a complicated signal.

These expressions show why it would be desirable to have a delta function source wavelet, because the Fourier transform of a delta function is simply 1. Thus, if $w(t) = \delta(t)$, the seismogram would equal the reflector series. Although a physical source wavelet is not a delta function, the seismograms can be manipulated mathematically to simulate such a wavelet. This can be done by creating an *inverse filter*[7] $w^{-1}(t)$, that, when convolved with the wavelet, yields a delta function

$$w^{-1}(t) * w(t) = \delta(t). \tag{64}$$

Applying this filter, which "spikes" the wavelet, leaves only the reflector series

$$w^{-1}(t) * s(t) = w^{-1}(t) * w(t) * r(t) = r(t). \tag{65}$$

[7] The notation $w^{-1}(t)$ does not mean $1/w(t)$.

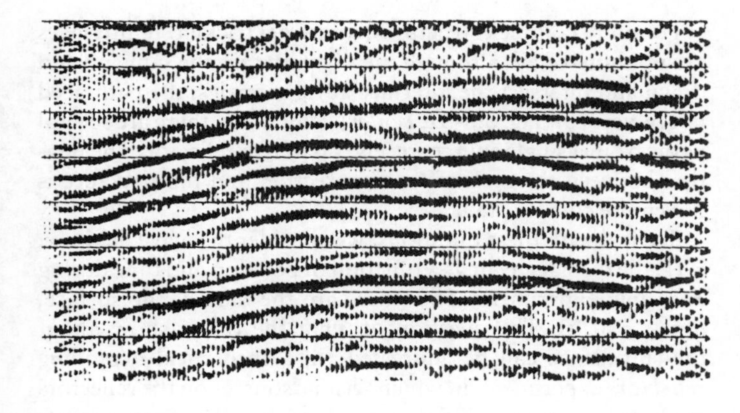

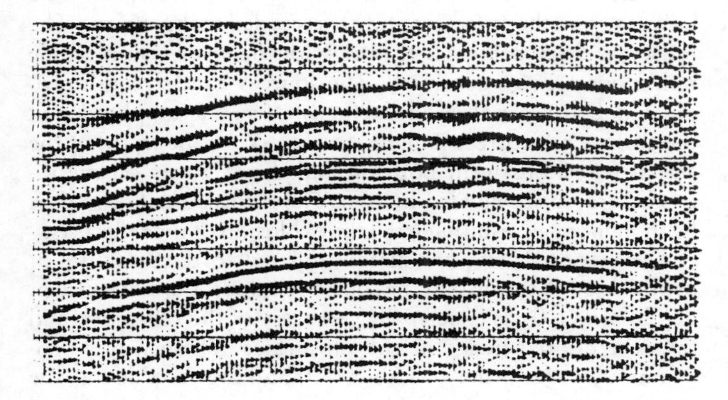

Fig. 3.3-29 *Top*: Seismic section before deconvolution. *Bottom*: Seismic section after deconvolution, showing sharper arrivals for the major reflections. (Yilmaz, 1987. Reproduced by permission of the Society of Exploration Geophysicists.)

Because this operation is the inverse of convolution, it is called *deconvolution*.

To create the inverse filter, note that the Fourier transform of the convolution (Eqn 64) yields

$$W^{-1}(\omega)W(\omega) = 1, \tag{66}$$

so the transform of the inverse filter is just $1/W(\omega)$. Hence deconvolution can be done by dividing the Fourier transforms

$$S(\omega)/W(\omega) = R(\omega). \tag{67}$$

This works well except at frequencies where the source wavelet's spectrum is small. Deconvolution makes the arrivals from reflectors stand out more distinctly (Fig. 3.3-29) and easier to interpret.

An alternative, but similar, approach is used with Vibroseis data for which the wavelet is very long. The goal is to identify times in the trace when the sweep signal arrives. Similarities between two time series $f(t)$ and $g(t)$ are shown by their *cross-correlation*, an operation (Section 6.3.4) defined by

$$c(L) = \lim_{T\to\infty} \frac{1}{T} \int_{-T}^{T} f(t + L)g(t)dt. \tag{68}$$

The cross-correlation is largest as a function of L, the lag time, when the series are most similar. For finite time series, the integration is over the times when f and g are nonzero. A special case is the *auto-correlation*, the cross-correlation of a function with itself

$$a(L) = \lim_{T\to\infty} \frac{1}{T} \int_{-T}^{T} f(t + L)f(t)dt, \tag{69}$$

which is always maximum at zero lag. The auto-correlation of a Vibroseis sweep, called a Klauder wavelet, is sharply peaked at zero lag (Fig. 3.3-30). Thus cross-correlating a sweep with the recorded trace is similar to using a spiking filter, because it produces sharp spikes when reflections arrive (Fig. 3.3-31). This similarity is not surprising, because cross-correlation and convolution are similar operations (compare Eqns 62 and 68).

Reflections can also be enhanced by filtering in the frequency domain to enhance certain frequency ranges and reject others. The frequency response of geophones varies, but the records may contain frequencies as low as a few Hz and in excess of 100 Hz. As a result, the signal-to-noise ratio can vary significantly as a function of frequency, so filtering often improves reflection quality. The appropriate frequencies may change with time in the record. For example, the later-arriving reflections have longer periods because high-frequency energy is lost to attenuation, the process by which seismic energy is converted to heat (Section 3.7).

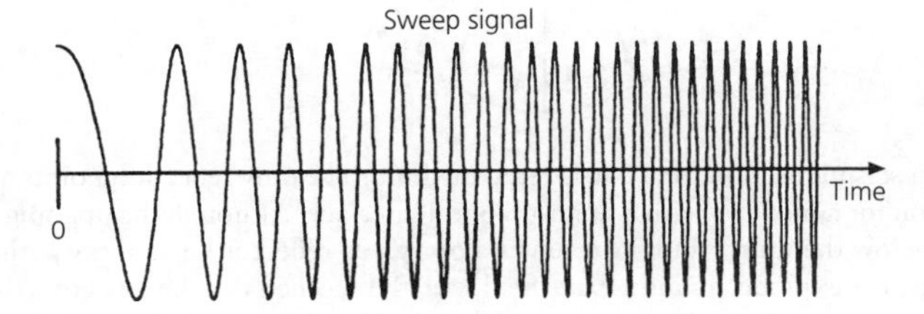

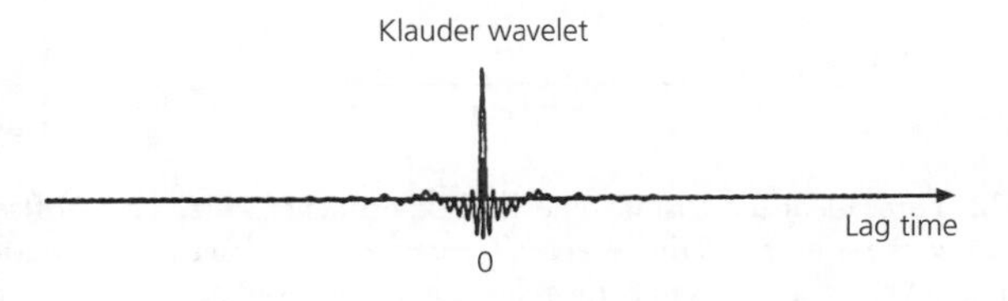

Fig. 3.3-30 The auto-correlation of a Vibroseis sweep signal is an impulsive Klauder wavelet.

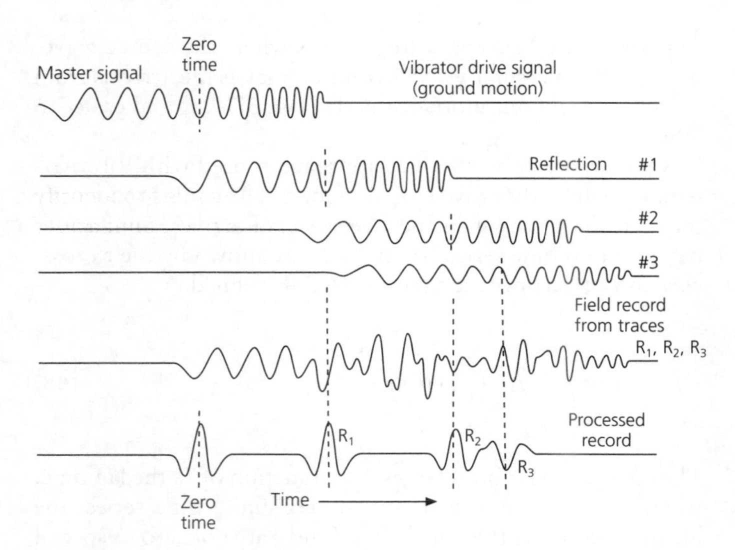

Fig. 3.3-31 A Vibroseis record is a sum of sweep signals reflected from various interfaces. Cross-correlation with a sweep signal produces Klauder wavelets at the reflection times. (Conoco.)

3.3.7 Migration

Given the "cleanest" possible seismic section, how good an image of the subsurface is it? Ideally, the section produced by CMP stacking is a zero-offset section, because the traces have been converted to what would be recorded for a coincident source and receiver. The ray path down to a reflector and back up must be the same, so Snell's law requires that this path be normally incident on the reflector. If the structure were composed of horizontal interfaces, the reflection paths would be vertical, and the time section could be converted to a depth section by using the velocities to scale the time axis (Fig. 3.3-32, *left*). In this case, a reflection's arrival time indicates depth to a reflector directly below the source and receiver.

For more complex structures, things get trickier. If interfaces are not horizontal, although the ray paths are the same up and down and intersect the interface at right angles, the path need not be vertical (Fig. 3.3-32, *center*). Moreover, there are several paths from a single source–receiver pair to a reflector. The relation between the zero-offset time section and the structure is thus more complicated.

To deal with these questions, we consider the wave field $u(x, z, t)$, the displacement as a function of position and time during a seismic experiment. The traces are the data at the surface, $u(x, z = 0, t)$. The question of what the traces show about the subsurface can be addressed via a theoretical *exploding reflectors* experiment, in which seismic sources on the reflectors explode at time zero (Fig. 3.3–32, *right*). Waves propagate upward from the reflectors and are recorded at the surface. The reflectors do not interact further with the waves, so multiple reflections are not generated. The sources have strength proportional to the reflection coefficients, so the amplitudes at the surface are correct. Finally, to correct for the fact that the actual reflections went both up and down, times on the recorded traces are divided by two. The recorded data can thus be thought of as resulting from the explosion of the reflectors.

The recorded data are directly related to the structure at depth. At $t = 0$, the instant the sources explode, the wave field at depth, $u(x, z, 0)$, is exactly the geometry of the reflectors, and thus the desired image of the subsurface. These waves propagate upward to the surface $z = 0$, and are recorded as the seismic section $u(x, 0, t)$. Hence the reflectors can be found from the section by removing the effects of propagation, using an operation called *migration*.

We first consider a constant-velocity medium in which a point source at (x_0, z_0) explodes at $t = 0$. The resulting displacement is a circular wave front (Section 2.4.3) that expands with time at a rate equal to the velocity (Fig. 3.3-33) and is described using a delta function

$$u(x, z, t) = \delta((x - x_0)^2 + (z - z_0)^2 - (vt)^2), \quad (70)$$

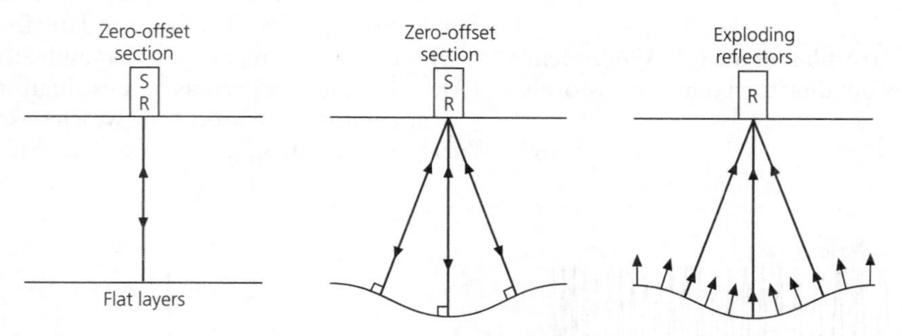

Fig. 3.3-32 Three idealized seismic reflection experiments. *Left*: A zero-offset seismic section for a flat-layered medium. The only reflection points are directly below the source and the receiver. *Center*: A zero-offset seismic section for a medium with a nonhorizontal interface. Although the upgoing and downgoing ray paths are the same, the reflection points need not be directly below the source and the receiver. For a given reflector, several ray paths can produce arrivals at a single receiver. *Right*: A conceptual model in which reflectors explode, giving a wave field with the geometry of the reflectors that propagates to the surface, producing the observed seismic section. Migration seeks to reverse this process and find the initial wave field from the seismic section. (After Claerbout, 1985.)

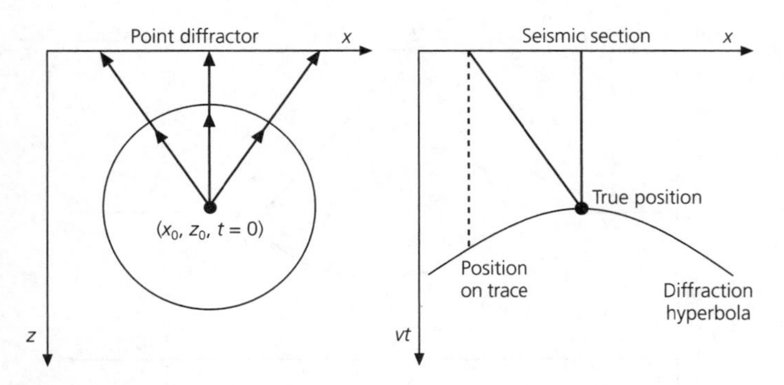

Fig. 3.3-33 Effect of a point source or diffractor. *Left*: The point diffractor acts as a source of spherical (circular in two dimensions) wave fronts. *Right*: The resulting seismic section with time scaled by velocity so that the vertical axis has the dimensions of distance. The diffraction appears as a hyperbola. The point at the apex corresponds to the true position of the diffractor. The other points are due to later arrivals from the diffractor at nonvertical incidence, so their positions on the section do not indicate a reflector directly below the receiver.

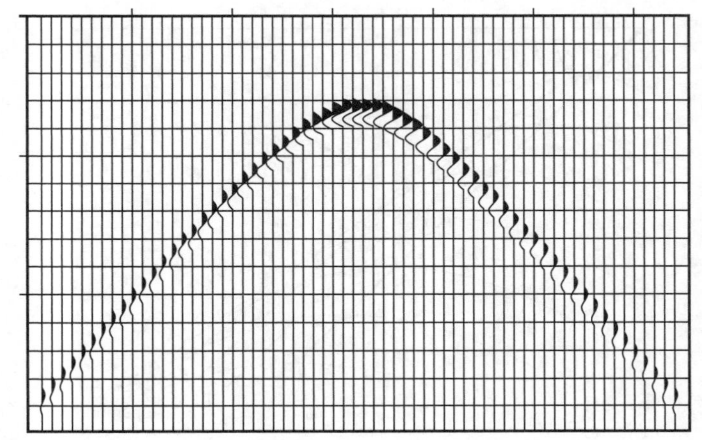

Fig. 3.3-34 Diffraction hyperbola with true amplitudes. (After Claerbout, 1985.)

whose downgoing half we ignore. The resulting seismic section is the wave field at the surface, $z = 0$,

$$u(x, 0, t) = \delta((x - x_0)^2 + (z_0)^2 - (vt)^2). \tag{71}$$

This is a hyperbola with apex at $(x = x_0,\ t = z_0/v)$, showing that the wave front arrives first directly above the source, and arrives later at points farther away. Thus the arrival seen on the seismic section is not equivalent to geologic structure with depth. A way to visualize the relation between the source position and the seismic section is to plot the time axis in units of vt, giving a time scale equal to the propagation distance. Thus an arrival time equals the distance along the true path from the source to the receiver. As illustrated, the depth of the source is shown correctly on the section only by the arrival time at a receiver directly above the source. For all other points on the surface, the arrival appears at a time corresponding to the travel time to that point, along a path that was not vertical. Hence, except above the source, the arrival on the section does not correspond to a source directly below the receiver, and the arrival time does not give the source depth directly.

The hyperbolic arrival on a seismic section due to a point source at depth is called a *diffraction hyperbola*. It lets us understand how complicated structures appear on seismic sections, because by Huygens' principle (Section 2.5.10) the reflection from an interface can be found by treating the interface as a set of point sources. The resulting reflection is found by summing the wave fronts from these Huygens' sources, which are also called point diffractors, or point scatterers. Because each source produces a diffraction hyperbola on the seismic section, the section resulting from a set of point diffractors is the sum of their diffraction hyperbolas. In considering this sum, we use the results of a more sophisticated analysis showing that the diffraction hyperbola's amplitude is largest at the apex and decays as the cosine of the angle off to the sides. Because Eqn 71 describes only the travel times, it does not include this term, whose effect is visible in Fig. 3.3-34.

To illustrate this approach, consider an interface dipping at an angle β. This should give the same reflections as a line of closely spaced point diffractors, so the seismic section will be a sum of the resulting hyperbolas. As shown in Fig. 3.3-35, the hyperbolas interfere constructively, causing an apparent interface. Interestingly, this apparent interface does not pass through the apex of each hyperbola, so it is displaced from the real interface and appears to have a shallower dip angle, α. Because the scaled travel time to the true interface equals the scaled arrival time on the trace, the real and apparent dips are related by $\sin \beta = \tan \alpha$.

Seismic sections from simple structures can appear quite different from the actual structure. For example, consideration of ray paths for a single reflector with a synclinal structure shows that several arrivals from different points on the reflector appear on a single zero-offset trace, each with a different travel time. As a result, an apparent anticlinal, or "bowtie," structure appears (Fig. 3.3-36). Another common effect is that the edges of sharp interfaces can give rise to long diffraction "tails" (Fig. 3.3-37). This effect is analogous to diffraction at the edges of a slit (Fig. 2.5-18).

The goal of migration is to undo the effects of diffraction and hence convert the data to realistic images of the subsurface. Migration can thus be thought of as an inverse scattering or inverse diffraction problem. Because this requires removing propagation effects, migration methods are derived using forward models of the propagation process. The idea that the section is the sum of diffractions suggests one approach known as *diffraction sum migration*, or *Kirchhoff migration*. Because point diffractors cause hyperbolas on the seismic section, the amplitude of each point on the migrated section is found by summing the unmigrated section with appropriate scaling along hyperbolic trajectories (Fig. 3.3-38). This operation should collapse all the signal in diffraction hyperbolas to points at

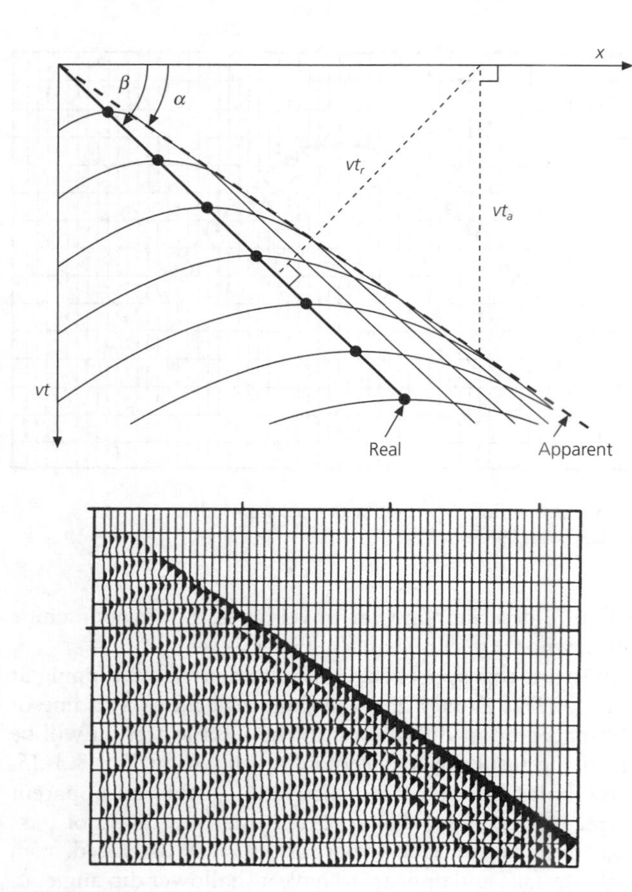

Fig. 3.3-35 A dipping layer can be modeled as a line of point diffractors. On a time section, interference between the diffraction hyperbolas produces an apparent reflector, as shown schematically (*top*) and with true amplitudes (*bottom*). Because the scaled travel time to the true interface (vt_r) equals the scaled arrival time on the trace (vt_a), the apparent dip α is shallower than the true dip β. (After Claerbout, 1985.)

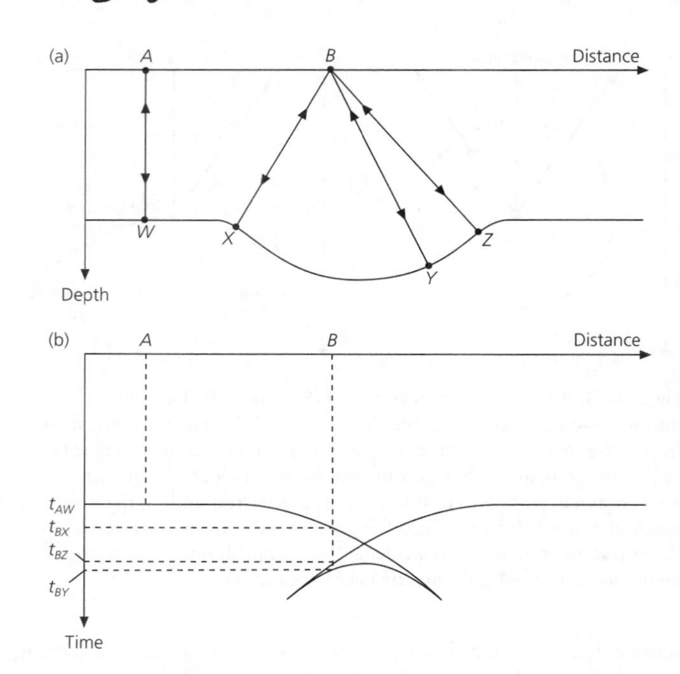

Fig. 3.3-36 *Top*: Illustration of several ray paths for reflections at zero offset from a reflector with a synclinal structure. *Bottom*: Because these arrivals, each from a different point on the reflector, have different travel times, the time section shows an apparent anticlinal, or "bowtie," structure. (Kearey and Brooks, 1984.)

their apexes, and thereby reconstruct the reflectors as a set of point diffractors. Thus diffraction artifacts like those in Figs 3.3-36 and 37 should be removed, and apparent interfaces with shallow dips should be converted to interfaces with the steeper true dips. Figure 3.3-39 shows the improvement to a seismic section from Kirchoff migration. The resulting migrated time section can be converted to a depth section using a velocity–depth function.

The appearance of a migrated section depends on the assumed velocity. Using a too-slow velocity reduces the length

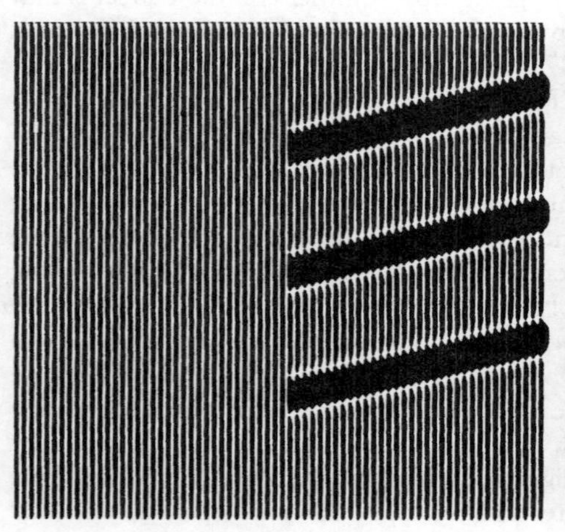

Fig. 3.3-37 The ends of truncated interfaces (*left*) act as diffractors, so a time section (*right*) shows spurious down-dip extensions of the interfaces. (Claerbout, 1976.) (http://sepwww.stanford.edu/sep/prof/)

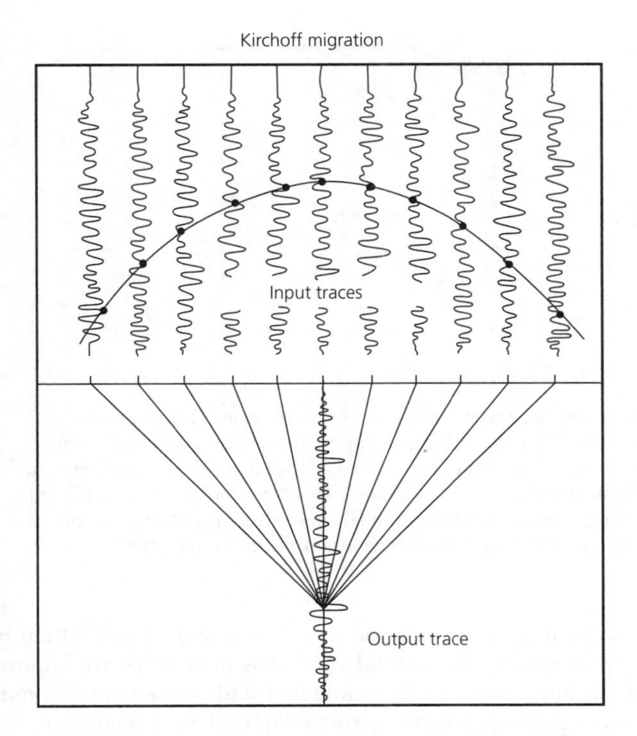

Fig. 3.3-38 Diffraction sum migration reverses the effects of diffraction by summing the time section along hyperbolic trajectories, thus collapsing hyperbolas to their apexes. (Schneider, 1971. Reproduced by permission of the Society of Exploration Geophysicists.)

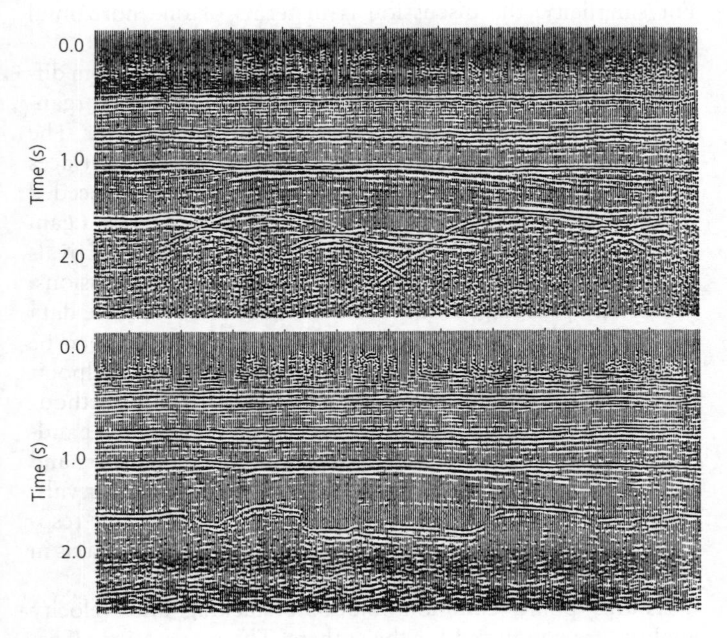

Fig. 3.3-39 Time section before (*top*) and after (*bottom*) migration. Elimination of the diffractions produces a better image of structures at depth, such as those at about 1.8 s, where bowties and diffraction "tails" have been suppressed. (Prakla-Seismos)

of hyperbolas' "tails," but does not fully collapse them, and so is termed undermigration. Similarly, a too-high velocity over-migrates the data, converting upward-pointing hyperbolas into downward-pointing ones. As a result, correct imaging of dipping structures depends on an accurate velocity model.

Other migration methods, called wave equation migration, use a double Fourier transform to map a wave field, $u(x, z, t)$, from the horizontal distance and time (x, t) domain to the horizontal wavenumber and angular frequency (k_x, ω) domain. The transform is

$$U(k_x, z, \omega) = \int_{-\infty}^{\infty}\int_{-\infty}^{\infty} u(x, z, t) \exp\,[i(-\omega t + k_x x)]dxdt, \tag{72}$$

with inverse transform

$$u(x, z, t) = \frac{1}{4\pi^2}\int_{-\infty}^{\infty}\int_{-\infty}^{\infty} U(k_x, z, \omega) \exp\,[i(\omega t - k_x x)]dk_x d\omega. \tag{73}$$

If we consider only P waves, the wave field $u(x, z, t)$ satisfies the wave equation in two dimensions:

$$\frac{\partial^2 u}{\partial x^2} + \frac{\partial^2 u}{\partial z^2} = \frac{1}{v^2}\frac{\partial^2 u}{\partial t^2}. \tag{74}$$

The corresponding condition on the transform $U(k_x, z, \omega)$ is found by substituting the inverse transform for u, taking the derivatives, and canceling, yielding

$$\frac{\partial^2 U}{\partial z^2} = \left(k_x^2 - \frac{\omega^2}{v^2}\right)U. \tag{75}$$

Because the components of the wavenumber vector are related by

$$|\mathbf{k}|^2 = k_x^2 + k_z^2 = \omega^2/v^2, \tag{76}$$

the transform satisfies

$$\frac{\partial^2 U}{\partial z^2} = -k_z^2 U. \tag{77}$$

If the velocity is constant with depth, k_z is independent of z, so integrating Eqn 77 yields

$$U(k_x, z, \omega) = U(k_x, 0, \omega) \exp\,[\pm ik_z z]. \tag{78}$$

This equation relates the wave field at the surface and at any depth. The operation of converting one to the other is called *downward* or *upward continuation* of the wave field. The sign of the exponential distinguishes upcoming from downgoing waves. Because z increases downward, upcoming waves occur when k_z and ω have the same sign. To ensure this, we define k_z as a function of ω,

$$k_z(\omega, k_x) = \text{sgn}(\omega)\sqrt{\omega^2/v^2 - k_x^2}, \tag{79}$$

where the function sgn (ω) is 1 when ω is positive, and is −1 when ω is negative. Using this definition, the inverse transform Eqn 73 becomes

$$u(x, z, t) = \frac{1}{4\pi^2}\int_{-\infty}^{\infty}\int_{-\infty}^{\infty} U(k_x, 0, \omega) \exp[i(\omega t - k_x x + k_z(\omega, k_x)z)]dk_x d\omega. \tag{80}$$

This integral relates the Fourier transform of the seismic section recorded on the surface, $z = 0$, $U(k_x, 0, \omega)$, and the upcoming wave field at depth at earlier times. By the exploding reflector model, the image of the subsurface is the wave field at $t = 0$, when the reflectors have just exploded. Thus the image can be found by setting $t = 0$:

$$u(x, z, 0) = \frac{1}{4\pi^2}\int_{-\infty}^{\infty}\int_{-\infty}^{\infty} U(k_x, 0, \omega) \exp[i(-k_x x + k_z(\omega, k_x)z)]dk_x d\omega. \tag{81}$$

Although this integral migrates the transform of the seismic section into the desired image, the integral over ω and k_x has to be done separately to find the image at every depth z. A way to get around this is to replace the ω integration with one over k_z, by expressing ω as a function of k_x and k_z,

$$\omega(k_x, k_z) = \text{sgn}(k_z)v\sqrt{k_x^2 + k_z^2}, \tag{82}$$

and changing variables using

$$\frac{d\omega}{dk_z} = \frac{k_z v}{\sqrt{k_x^2 + k_z^2}}. \tag{83}$$

This change converts Eqn 81 into an inverse Fourier transform from the wavenumber (k_x, k_z) domain to the space (x, z) domain

$$u(x, z, 0) = \frac{1}{4\pi^2}\int_{-\infty}^{\infty}\int_{-\infty}^{\infty} U(k_x, 0, \omega(k_x, k_z)) \exp[i(-k_x x + k_z z)]\frac{k_z v}{\sqrt{k_x^2 + k_z^2}}dk_x dk_z, \tag{84}$$

so inverting the double transform once gives the image for all x and z.

The application of migration methods to data involves various complexities. The time axis in the section can be scaled to account for the variation in velocity with depth. Often in complex geology horizontal variations in velocity are important, so migration can be conducted with numerical methods that can propagate waves through laterally varying media.

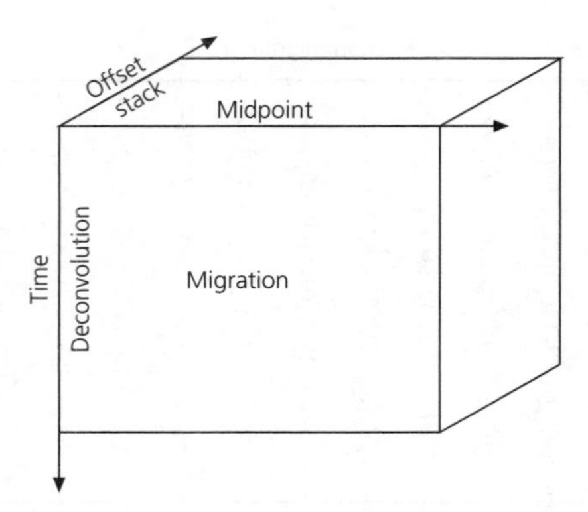

Fig. 3.3-40 Schematic illustration of the relation between processing operations for reflection data. Deconvolution applied along the time axis increases temporal resolution. CMP stacking along the offset axis collapses the data to the midpoint–time plane (compare to Fig. 3.3-18), yielding a seismic section and enhancing reflections. Migration applied in this plane improves lateral resolution. (After Yilmaz, 1987.)

3.3.8 Data processing sequence

The various processing operations for seismic reflection data can be combined in different ways. To illustrate this, we summarize a common sequence for some of the possible operations. For simplicity, the discussion is in terms of one horizontal dimension, but the approach applies to two dimensions.

Preprocessing consists of initial steps. Because data from different receivers are recorded simultaneously, they are reorganized (demultiplexed) to produce a trace for each receiver. The traces are then edited to eliminate effects such as noisy traces or recording errors. Static time shifts are applied when needed (Section 3.3.5). The amplitudes are then adjusted using a gain recovery function that corrects for the fact that the later arrivals have lower amplitudes because of reflections, transmissions, geometric spreading, and attenuation (Section 3.7). The data are combined into common source gathers, and can then be viewed as a volume defined by the time, offset, and midpoint axes (Fig. 3.3-40). They can then be filtered using methods (Section 3.3.5) including muting of undesired arrivals, band-pass filtering to enhance or suppress certain frequencies, and velocity or slant stack filtering to suppress certain arrivals. Deconvolution (Section 3.3.6), which improves the time resolution of the data, can be viewed as acting along the time axis in Fig. 3.3-40.

As this point, common midpoint stacking and velocity analysis are conducted for the gathers. These operations (Section 3.3.4) combine data for each midpoint to produce a seismic section that approximates what would be recorded at zero offset. Geometrically, this acts along the offset axis to collapse all the data to the midpoint–time plane. Migration (Section 3.3.7) in the midpoint–time plane seeks to eliminate artifacts due to diffractions and convert the seismic section to an image of the subsurface. The migrated section can then be converted, using assumptions about velocities, to a depth section. The depth section is then interpreted together with geological data and other types of geophysical data, in some cases from drill holes, to understand the subsurface geology.

This discussion of the processing sequence brings out the point that although it is natural to treat a seismic section as an accurate image of the subsurface, it is actually a display of a seismic wave field showing the energy arriving as a function of two-way travel time. Thus the quantity shown, vertical displacement or pressure, need not correspond to any geological reflector of interest. Large arrivals can result from interference between reflections from small impedance contrasts. Moreover, because a seismic section has been produced by mathematical operations, rather than the physical experiment it simulates, noise in the data and errors in the processing can produce spurious artifacts. For example, the conversion of time to depth is only as accurate as the velocities found by stacking or otherwise, perhaps from measurements in a drill hole.

As we have seen in discussing migration, seismic sections are most likely to deviate from the desired images when the medium has significant lateral variations. For example, a medium with random heterogeneities can yield spurious short layered segments, because the reflected energy depends on the vertical changes in impedance. Thus long-wavelength vertical variations in impedance are suppressed, whereas both short- and long-wavelength horizontal variations are preserved, and so can yield a structure with apparent horizontal layering. This effect can be viewed as a velocity filter (Section 3.3.5) that reduces horizontal resolution for structures with steep dips. Similar effects, which are prone to occur at large offsets, may contribute to the horizontally discontinuous layering observed in deep crustal reflection data (Section 3.2.4). Hence, as we will see in various contexts throughout our discussions (e.g., Section 7.3), studying three-dimensional velocity structure is an interesting and challenging enterprise.

3.4 Seismic waves in a spherical earth

In the previous sections, we developed the theory to use the travel times of seismic waves to study the velocity structure of a medium composed of flat layers. This analysis is useful when the ray paths between the source and the receiver are short enough that the earth's curvature can be neglected. Because this is the case for distances less than a few hundred kilometers, such analysis is used to study structure in the crust and the uppermost mantle. In this section we develop the corresponding theory for a spherical earth, which can be used for greater distances and thus greater depths. Application of these results, discussed in the next section, is our primary tool for studying the structure of the deep earth.

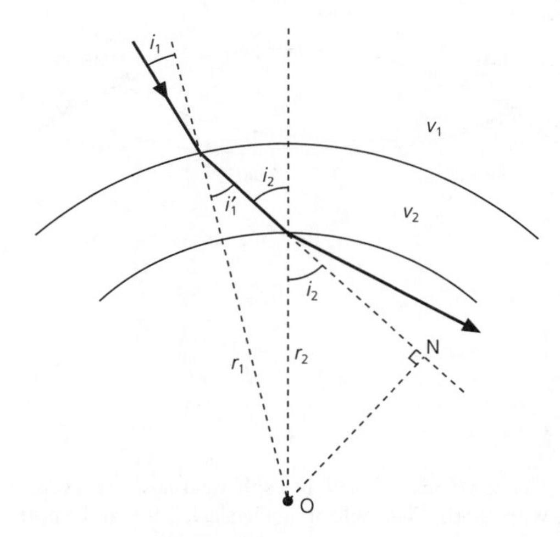

Fig. 3.4-1 Geometry of Snell's law for a spherical earth.

3.4.1 *Ray paths and travel times*

By analogy to the way we previously represented the earth using uniform flat layers, we now treat it as a series of concentric spherical shells of uniform-velocity material. The ray paths and travel times for the spherical geometry are described by expressions similar to those for flat layers (Section 3.3.1). Consider the portion of a seismic ray's path connecting points at radial distances r_1 and r_2 from the earth's center (Fig. 3.4-1). If v_1 and v_2 are the velocities above and below r_1, and i_1, i_1' and i_2 are the angles shown, then by Snell's law

$$\frac{r_1 \sin i_1}{v_1} = \frac{r_1 \sin i_1'}{v_2}. \tag{1}$$

However, $r_1 \sin i_1' = r_2 \sin i_2$ because both equal the length ON, so we rewrite Eqn 1 as

$$\frac{r_1 \sin i_1}{v_1} = \frac{r_2 \sin i_2}{v_2}. \tag{2}$$

Thus we define the ray parameter p for a spherical earth as

$$p = \frac{r \sin i}{v}, \tag{3}$$

where r is the radial distance from the center of the earth, v is the velocity at that point, and i is the incidence angle between the ray path and the radius vector. By reducing the thickness of the shells ever thinner, the velocity becomes a continuous function of radius, $v(r)$. Equation 3 is thus Snell's law for a spherical earth, which describes the ray path. As for the flat earth, the ray parameter is constant along the ray path, and thus identifies a particular ray.

It may seem strange that different forms of the ray parameter and Snell's law occur for a sphere. At any given depth, the flat

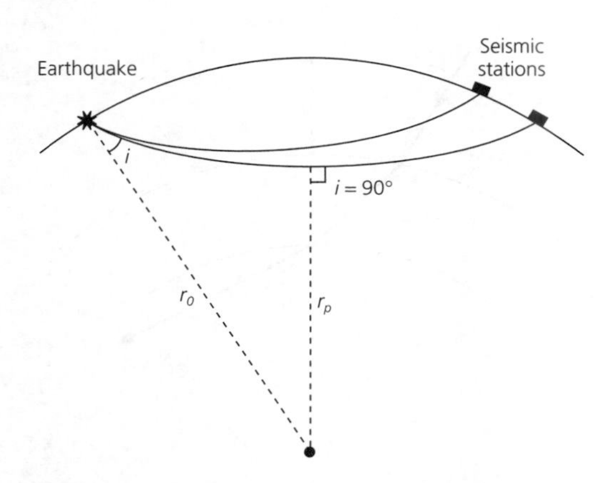

Fig. 3.4-2 Geometry of a ray path in a spherical earth with velocity increasing with depth. The angle of incidence, i, is 90° at the bottoming depth r_p.

layer formulation, that $p = \sin i/v$ is constant, is valid. The factor r corrects for the change along the path of the orientation of the normal to the interface, which is the radius. If r changes so slowly along the path that its variation can be ignored, we obtain the flat case. Thus the flat layer version is used for near-surface refraction and reflection studies.

The condition of constant ray parameter relates the ray path to the velocity structure. For a source at a radius r_0 (the earth's radius for a surface source) where the velocity is v_0,

$$p = r_0 \sin i/v_0. \tag{4}$$

Rays leaving the source at different angles thus have different ray parameters. As the ray travels downward, r decreases, and in general v increases, so sin i and thus i increase, because p is constant. The ray eventually "bottoms" and turns upward when $i = 90°$ (Fig. 3.4-2). At this bottoming depth, $r = r_p$, and

$$p = r_p/v_p. \tag{5}$$

From this point the ray returns to the surface. Different rays, with different p, thus bottom at different depths.

Consider two rays with ray parameters p and $p + dp$, that arrive at nearby points on the earth's surface (Fig. 3.4-3). The ray with ray parameter p takes a travel time T to travel a distance Δ, measured by the angle subtended at the earth's center, whereas the ray with $p + dp$ takes $T + dT$ to travel $\Delta + d\Delta$. In the limit, as the distance between the two points goes to zero,

$$\frac{v_0 dT}{r_0 d\Delta} = \sin i, \tag{6}$$

so

$$\frac{dT}{d\Delta} = \frac{r_0 \sin i}{v_0} = p. \tag{7}$$

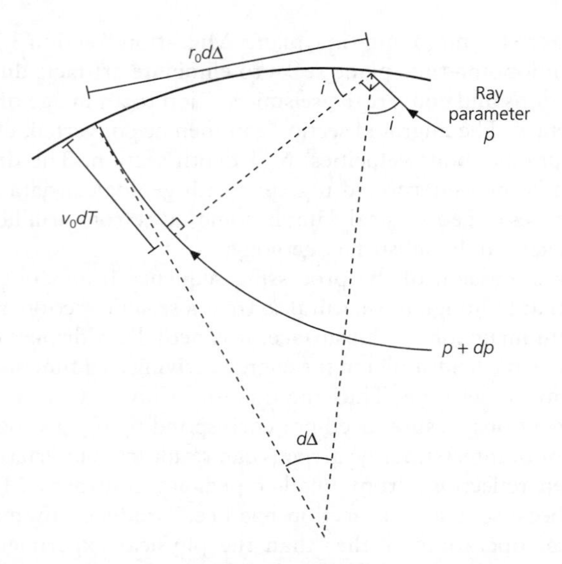

Fig. 3.4-3 Two rays with infinitesimally different ray parameters illustrating the relationship $p = dT/d\Delta$.

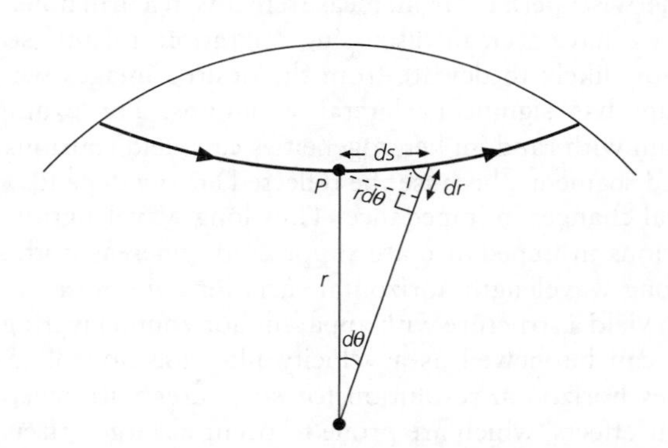

Fig. 3.4-4 Variables defining ds, a portion of the ray path subtending an angle $d\theta$.

Thus, as for the flat layer case (Section 3.3.1) the ray parameter is the reciprocal of the apparent velocity along the surface, c_x:

$$p = \frac{1}{c_x} = 1\bigg/\left(\frac{d\Delta}{dT}\right) = \frac{dT}{d\Delta}. \tag{8}$$

Hence the ray parameter can be measured from the difference in arrival times at nearby stations. Conversely, the slope of a travel time curve $T(\Delta)$ is the ray parameter of the ray emerging at a distance Δ.

Because the geometry is spherical, it is natural to describe the ray path in polar coordinates. Consider (Fig. 3.4-4) the point P on the ray path with polar coordinates (r, θ). A small portion of the ray path, ds, subtends an angle at the center of the earth $d\theta$, so

$$(ds)^2 = (dr)^2 + r^2(d\theta)^2 \quad \text{and} \quad \sin i = r\frac{d\theta}{ds}. \tag{9}$$

Substitution in Snell's law (Eqn 3) gives

$$p = \frac{r \sin i}{v} = \frac{r^2}{v}\frac{d\theta}{ds}. \tag{10}$$

We thus use Eqns 9 and 10 to form and equate two expressions for $(ds/d\theta)^2$,

$$\frac{r^4}{p^2v^2} = \left(\frac{dr}{d\theta}\right)^2 + r^2, \tag{11}$$

and manipulate them to obtain

$$d\theta = \frac{\pm p\,dr}{r(\zeta^2 - p^2)^{1/2}}, \tag{12}$$

where ζ is defined by $\zeta = r/v$. Integrating this expression from the source depth, which we assume to be the surface r_0, to the deepest point on the ray r_p, and doubling to account for the upward path, gives

$$\Delta(p) = \int d\theta = 2p\int_{r_p}^{r_0} \frac{dr}{r(\zeta^2 - p^2)^{1/2}}. \tag{13}$$

This integral gives the angular distance Δ traveled by the ray with ray parameter p in an earth with a velocity distribution $v(r)$.

A similar integral expression for the travel time of this ray comes from combining Eqns 9 and 10,

$$\frac{p^2v^2}{r^2} = r^2\left(\frac{d\theta}{ds}\right)^2 = 1 - \left(\frac{dr}{ds}\right)^2, \tag{14}$$

so that a portion of the ray path is

$$ds = \pm\frac{r}{v}\frac{dr}{(\zeta^2 - p^2)^{1/2}}. \tag{15}$$

Thus the travel time, defined by the integral of the slowness along the ray path, is given by

$$T(p) = \int \frac{ds}{v} = 2\int_{r_p}^{r_0} \frac{\zeta^2 dr}{r(\zeta^2 - p^2)^{1/2}}. \tag{16}$$

These integral expressions for the distance $\Delta(p)$ and travel time $T(p)$ of a ray in a spherical geometry are analogous to those for $x(p)$ (Eqn 3.3.25) and $T(p)$ (Eqn 3.3.26) in a layered material. As written, the integrals are from the surface to the bottoming depth, with the factor of 2 accounting for the return trip to the surface. If the source is not at the surface, the limits of integration are changed appropriately.

For the flat geometry, we found it useful to describe the travel time curve in terms of its slope, the ray parameter, p, and the time axis intercept of its tangent, τ (Section 3.3.2). To do the same for the spherical geometry, we write the travel time curve as

$$T(p) = p\Delta(p) + \tau(p). \tag{17}$$

We then evaluate the function

$$\tau(p) = T(p) - p\Delta(p) \tag{18}$$

using the integral expressions (Eqns 13 and 16), and find that

$$\tau(p) = 2\int_{r_p}^{r_0} \frac{(\zeta^2 - p^2)^{1/2}}{r} dr. \tag{19}$$

This formulation can be used to invert travel time curves for velocity structure.

3.4.2 *Velocity distributions*

Different distributions of velocity with depth produce characteristic travel time curves. Figure 3.4-5 (overleaf) shows the usual situation in which velocity increases slowly with depth. Given two rays, the one with a smaller angle of incidence at the source has a smaller p, thus bottoms deeper at a point with smaller r_p and larger v_p, and eventually emerges further from the source. Thus the ray parameter decreases, and travel time increases, monotonically with distance, Δ. The travel time curve, $T(\Delta)$, is concave downward because its slope, $p(\Delta)$, decreases with distance ($dp/d\Delta = d^2T/d\Delta^2 < 0$). The intercept-slowness curve, $\tau(p)$, is smooth. To show these relations in different ways, the plots in Fig. 3.4-5 are aligned so that the distance axis is common to the ray path, $T(\Delta)$, and $p(\Delta)$ plots, the depth axis is the same for the ray path and velocity–depth plots, and the time axis is the same for the $T(\Delta)$ and $\tau(p)$ plots.

A more complicated situation occurs when velocity increases rapidly with depth (Fig. 3.4-6). Rays that bottom either above or below the region of high velocity gradient behave as in Fig. 3.4-5, so the corresponding portions of the travel time and ray parameter curve show T increasing with Δ, and p decreasing with Δ. By contrast, rays that bottom in the region of high velocity gradient are bent upward more and emerge at smaller values of Δ than would otherwise be the case. As a result, three rays with different ray parameters emerge at the same distance Δ. Thus the $p(\Delta)$ and $T(\Delta)$ curves have three distinct branches. On the two normal forward branches $dp/d\Delta < 0$. However, on

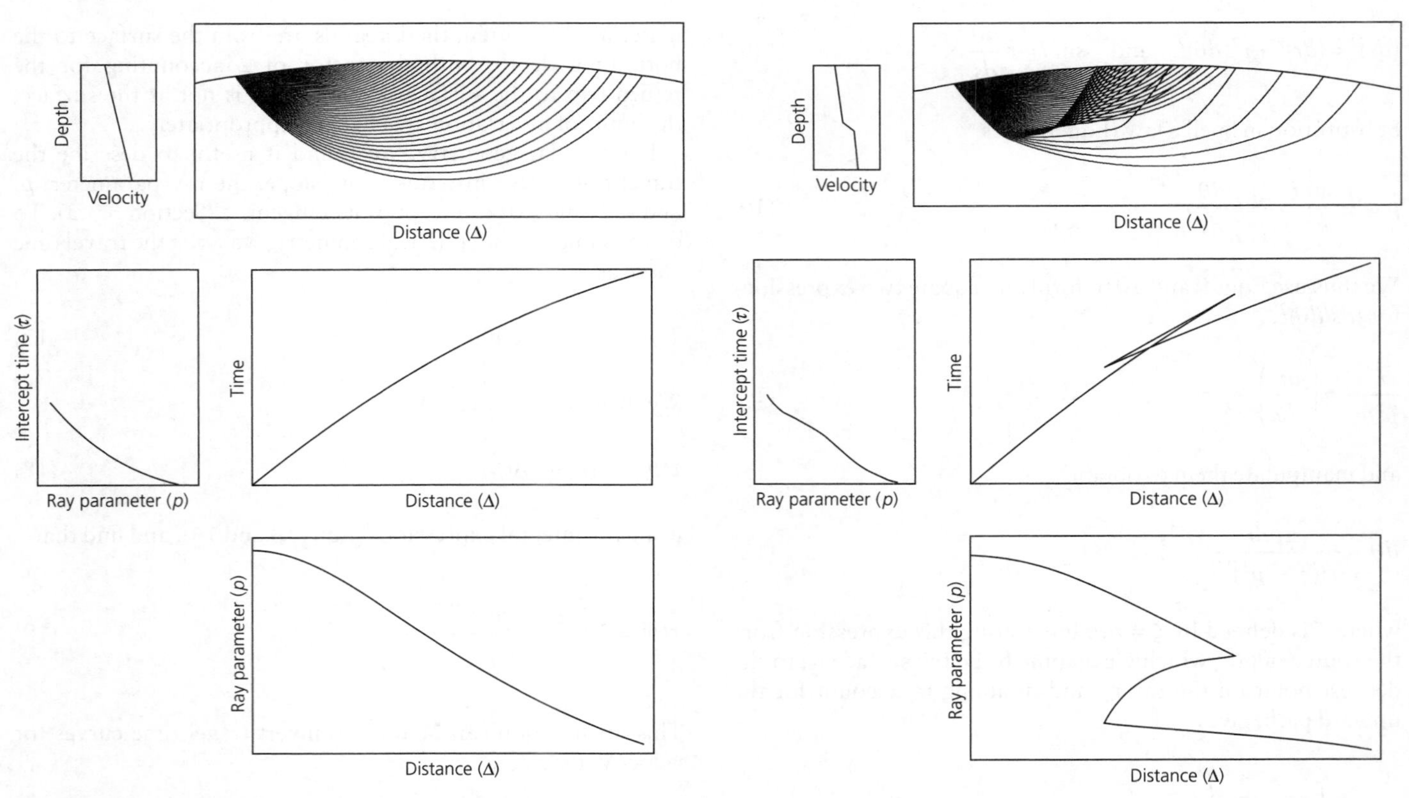

Fig. 3.4-5 Ray paths, $T(\Delta)$, $p(\Delta)$, and $\tau(p)$ relationships for velocity increasing slowly with depth.

Fig. 3.4-6 A triplication occurs if velocity increases rapidly, because at some distances three rays arrive. The triplication appears as the three branches in the $T(\Delta)$ and $p(\Delta)$ curves. The cusps on the travel time curve where the branches meet correspond to reversals in the $p(\Delta)$ plot.

the back branch, Δ decreases with decreasing p, so $dp/d\Delta > 0$. Thus rays with smaller incidence angles arrive closer to the source, giving a characteristic *triplication* in the travel time curve and a reversal in the $p(\Delta)$ curve. We will see in the next section that triplications are observed in the travel time curve for waves in the mantle, due to velocity increases that are thought to result from mineral phase transitions.

A triplication is similar to the travel time curves for the direct, reflected, and head waves for a layer over a halfspace (Fig. 3.2-2). The back branch of the triplication is analogous to the reflection, and the two forward branches are analogous to the direct and head waves. As the velocity increase becomes sharper and more like the sharp jump between a layer and halfspace, the back branch extends further in either direction, so the triplication looks increasingly like the travel times for a layer over a halfspace.

As we discussed in Section 2.8.4, geometric ray theory gives information about amplitudes as well as travel times. Because the rays plotted left the source at uniform increments of angle, the amplitude expected at some distance depends on *geometric spreading*, or the density of rays arriving. We expect high amplitudes where rays are concentrated, and low amplitudes where rays are sparse. Mathematically, the concentration of rays is proportional to $di/d\Delta$, the range of incidence angles for the rays that arrive in a given distance. To find this, we differentiate the definition of the ray parameter (Eqn 7),

$$\frac{d^2T}{d\Delta^2} = \frac{dp}{d\Delta} = \frac{d(r \sin i/v)}{d\Delta} = \frac{r}{v}\cos i\frac{di}{d\Delta}. \tag{20}$$

Thus the amplitude is proportional to the second derivative of the travel time curve, or the derivative of the $p(\Delta)$ curve. For a triplication, the back branch meets the two forward branches at two points on the travel time and $p(\Delta)$ curves. Here $dp/d\Delta = \infty$, so large amplitudes are expected. This situation is called a *caustic*.

A third important case is a low-velocity zone, where velocity decreases with depth and then increases (Fig. 3.4-7). Rays entering the low-velocity zone bend down, rather than up, so no rays bottom there. To see this, note that for a ray to bottom, it must turn upward (to a larger angle of incidence) as it goes deeper (to smaller values of r), so that $di/dr < 0$. Conversely, if $di/dr > 0$, the ray turns downward and cannot bottom. These conditions can be written in terms of the velocity–depth function by differentiating both sides of

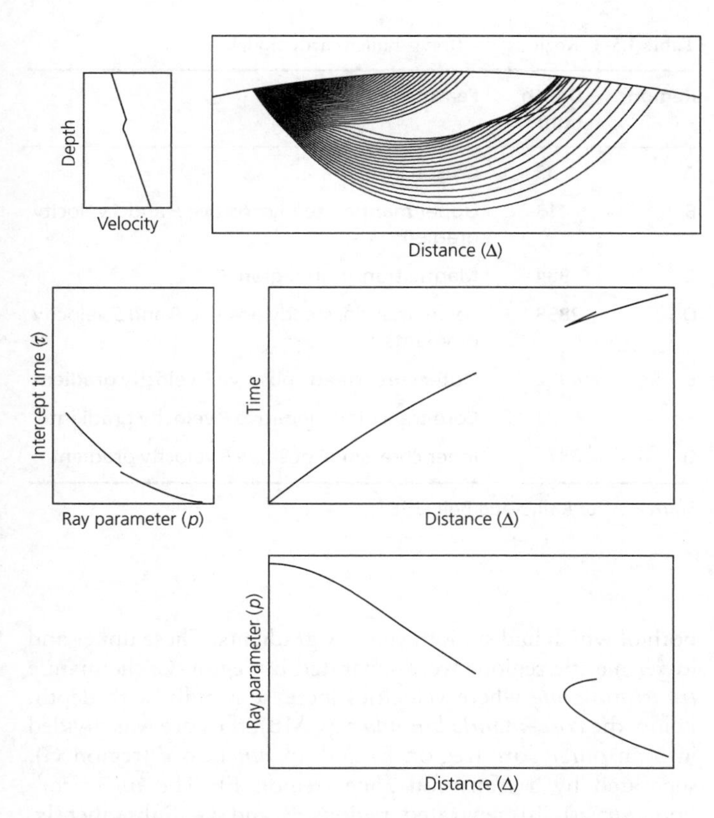

Fig. 3.4-7 A low-velocity zone gives rise to a shadow zone, a distance range where no direct geometric arrivals appear, and hence discontinuous $T(\Delta)$, $p(\Delta)$, and $\tau(p)$ curves.

$$\sin i = \frac{pv}{r}, \tag{21}$$

yielding

$$\cos i \frac{di}{dr} = p\left(\frac{1}{r}\frac{dv}{dr} - \frac{v}{r^2}\right) = \sin i\left(\frac{1}{v}\frac{dv}{dr} - \frac{1}{r}\right) \tag{22}$$

and thus

$$\frac{di}{dr} = \tan i\left(\frac{1}{v}\frac{dv}{dr} - \frac{1}{r}\right). \tag{23}$$

The condition that no rays bottom in a depth region where di/dr is positive implies that the velocity decreases fast enough that

$$\frac{dv}{dr} > \frac{v}{r}. \tag{24}$$

This situation causes a *shadow zone*, a region of the earth's surface where no rays arrive. Just below the low-velocity zone, rays reach a given Δ by two paths, giving two values of p and travel time for a given distance. The back branch, with $dp/d\Delta > 0$, corresponds to the rays that would have bottomed at the depth of the low-velocity zone, had the velocity there been high enough. The forward branch, which continues to greater distances, corresponds to the rays that bottom deeper. The concentration of rays just past the shadow zone corresponds to the point where the two branches meet. Here $dp/d\Delta = \infty$, so large amplitudes occur. We will see that this situation occurs as a result of the drop in velocity across the core–mantle boundary, which gives rise to a shadow zone.

3.4.3 *Travel time curve inversion*

To infer the distribution of velocity with depth, travel time curves are compiled from seismograms recorded at different source–receiver distances. The inverse problem of deriving velocity structure from the $T(\Delta)$ curves can be done in various ways. One is to use a computer program, based on Snell's law, to trace rays through different velocity structures and compute the corresponding travel time curves. Figures 3.4-5–7 were derived this way. This approach solves the inverse problem by solving the forward problem repeatedly until a satisfactory solution is found. An alternative is to solve the inverse problem directly by deriving $v(r)$ from $T(\Delta)$.

Various methods have been used to solve the inverse problem. A classic one is the *Herglotz–Wiechert integral.* This approach is based on Eqn 13, which gives the distance traveled by a ray with ray parameter p as a function of the velocity structure

$$\Delta(p) = 2p \int_{r_p}^{r_0} \frac{dr}{r(\zeta^2 - p^2)^{1/2}}, \tag{25}$$

where $\zeta = r/v$, and p is the ray parameter for the ray arriving at Δ. This can be converted to

$$\int_0^{\Delta_1} \cosh^{-1}\left(\frac{p(\Delta)}{\zeta_1}\right) d\Delta = \pi \ln\left(\frac{r_0}{r_1}\right), \tag{26}$$

where $\zeta_1 = r_1/v_1$ at radius r_1, the bottoming point of the ray that emerges at Δ_1.[1] This formula is used by starting with an observed travel time curve, $T(\Delta)$, and forming its derivative $dT/d\Delta = p(\Delta)$ numerically. The integral is done numerically from $\Delta = 0$ to $\Delta = \Delta_1$, using the fact that $\zeta_1 = dT/d\Delta$ at a distance Δ_1. The equation then gives the radius, r_1, at which the velocity is r_1/ζ_1.

This method sometimes fails when velocity decreases with depth, giving a low-velocity zone. In some such cases, it can still be applied using a method called "earth stripping." To do this, $v(r)$ is found down to the low-velocity zone using the Herglotz–Wiechert integral. Equations 13 and 16 are then used with r',

[1] Bullen and Bolt (1985).

the outer radius of the low-velocity zone, substituted for r_p, to find the distance and time the ray traveled on its way down to the low-velocity zone. Subtracting these from the known $T(\Delta)$ curve gives a $T'(\Delta)$ curve for a "mini-earth" with radius r'. Because in this "mini-earth" the velocity increases with depth, the Herglotz–Wiechert method is applied again.

3.5 Body wave travel time studies

We saw in the last section that travel time data can be used to determine seismic velocity as a function of depth. Beginning early in the 1900s, travel time tables were compiled by combining data from many earthquakes observed at various epicentral distances. These seismological observations provide the primary data for our view of the basic features of the earth's velocity structure. This picture, an essentially layered earth composed of a thin crust, a mantle, a liquid outer core, and a solid inner core, is key to our thinking about how the earth evolved and operates. This concept was largely developed by the 1940s, as illustrated in Fig. 3.5-1, showing the classic Jeffreys–Bullen[1] (JB) earth model. The JB model treated the earth as a series of shells, characterized by the behavior of the velocity with depth (Table 5.1-1). The mantle was divided into an *upper mantle* (region B) and a *lower mantle* (region D),

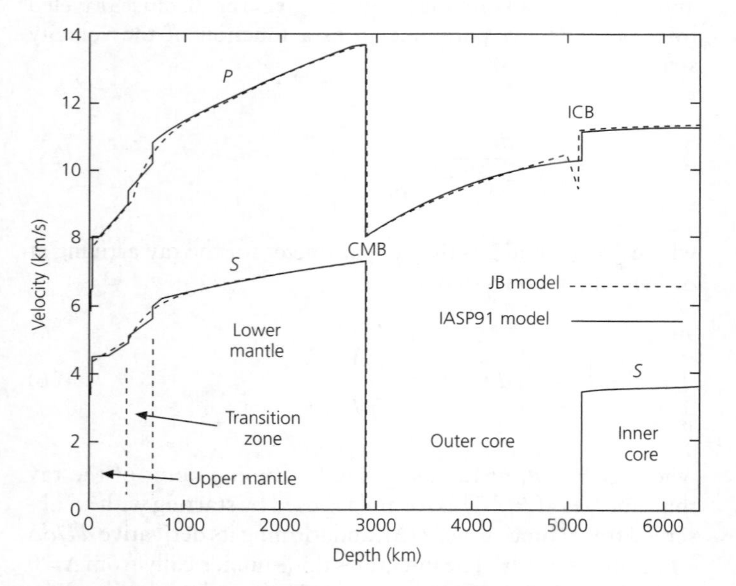

Fig. 3.5-1 Comparison of the classic Jeffreys–Bullen earth model (Jeffreys and Bullen, 1940) and a newer model, IASP91 (Kennett and Engdahl, 1991). Although IASP91 and its successor, AK135 (Kennett *et al.*, 1995), have improved resolution in the mantle transition zone and the core, the newer models are generally similar to that derived using hand-cranked calculators.

[1] The model was derived from extensive joint research into earth structure by Sir Harold Jeffreys (1891–1989), who established in 1926 that the core was liquid, and Keith Bullen (1906–76).

Table 3.5-1 Regions in Jeffreys–Bullen earth model.

Region	Depth (km)	Features of region
A	33	Crustal layers
B	413	Upper mantle: steady positive *P* and *S* velocity gradients
C	984	Mantle transition region
D	2898	Lower mantle: steady positive *P* and *S* velocity gradients
E	4982	Outer core: steady positive *P* velocity gradient
F	5121	Core transition: negative *P* velocity gradient
G	6371	Inner core: small positive *P* velocity gradient

Source: After Bullen and Bolt (1985).

both of which had smooth velocity gradients. These upper and lower mantle regions were separated by region C, the mantle *transition zone* where velocities increase rapidly with depth. Below the *core–mantle boundary* (CMB), the core was divided into an *outer core* (region E) and an *inner core* (region G), separated by a transition zone (region F). The *inner core boundary* (ICB) separated regions F and G. Subsequently, the lower mantle was divided into regions D′ (1000–2700 km depth), most of the lower mantle with a smooth velocity gradient, and D″ (2700–2900 km), the zone above the core–mantle boundary with a reduced velocity gradient.

Subsequent studies have derived models, such as the IASP91 model, also shown in Fig. 3.5-1, which confirm the basic structure of the JB model and provide better resolution of important regions. For example, the JB model did not resolve shear velocities in the inner core, whereas recent models have finite *S* velocity in the inner core, implying that it is solid. Similarly, recent models provide more details about the mantle transition zone and the core–mantle boundary, and do not include the velocity "notch" at the inner core–outer core boundary.

Jeffreys' and Bullen's derivation of a radially symmetric earth model from travel time observations converted the previous crude picture of the earth into one that has since changed only in detail. More recent radial velocity models do not differ much from each other, so they are likely to be converging on an accurate radial model for the earth. Such average, or *reference*, models and travel time curves, such as JB, IASP91, and PREM (for Preliminary Reference Earth Model, Section 3.8), are derived from data around the world and so average over local variations in structure. Regional differences can then be viewed as perturbations relative to a reference model.

However, lateral differences in structure can be significant and provide insight into tectonic processes. Thus a major current goal of seismology is to define the three-dimensional velocity structure that results from the fact that the earth is a

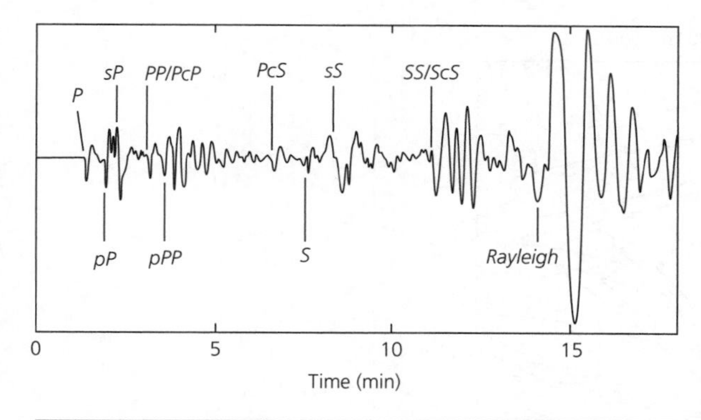

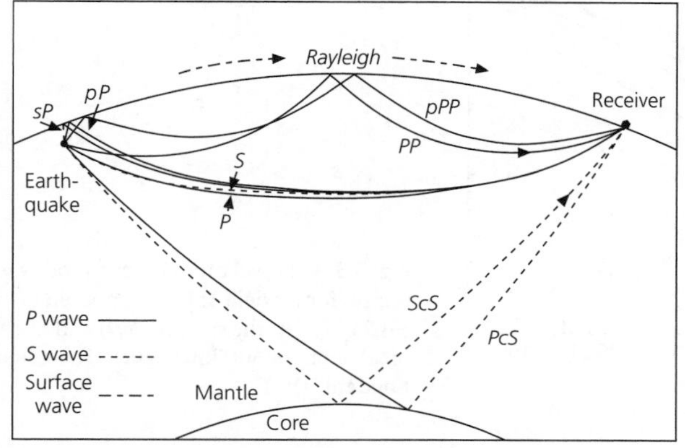

Fig. 3.5-2 *Top*: Long-period vertical component seismogram at Golden, Colorado, showing various seismic phases. *Bottom*: Ray paths for some of the seismic phases labeled on the seismogram. Paths taken as *P* waves are shown as solid lines; paths taken as *S* waves are shown as dashed lines. Although *P* and *S* are both direct phases, they do not travel the exact same path because their velocities differ. Similarly, the ray path for *PcS* is asymmetric, and *pP* and *sP* do not reflect off the surface at the same location.

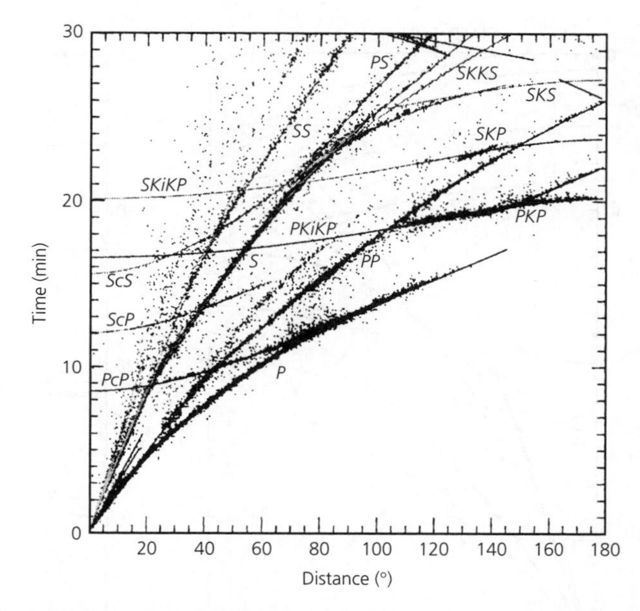

Fig. 3.5-3 Travel time data for various body wave phases and travel time curves for model IASP91. The travel times are corrected to those for an earthquake at the surface. The data are 57,655 travel times from 104 sources (earthquakes and explosions). (Kennett and Engdahl, 1991.)

geologically active planet. Convection in the earth causes three-dimensional temperature variations that result in observable velocity variations. In addition, mantle flow appears to generate seismic anisotropy at the top and the bottom of the mantle, and magnetic stresses due to outer core flow may cause inner core anisotropy. Resolving this three-dimensional structure requires sophisticated analysis techniques. For example, travel time studies are complemented by waveform modeling, and stacking techniques are applied to enhance seismic signals. The suggested reading provides some reviews of recent studies.

This section focuses on determining velocity structure, so we largely defer discussion of the chemical, mineralogical, thermal, and rheological factors that cause these variations for later sections.

3.5.1 *Body wave phases*

We have seen that seismic waves can travel between a source and a receiver along multiple paths. For example, increases in velocity can cause triplications, yielding three distinct arrivals at a receiver. Multiple reflections off various layers and diffractions can bring additional arrivals. Hence seismograms contain many arrivals, or *phases*, corresponding to different travel paths. This is illustrated by Fig. 3.5-2, discussed in Section 1.1, showing a few of the phases that are observed and some of the corresponding ray paths. All the phases shown, except for the Rayleigh surface wave, are *body waves* that travel through the earth's interior.

Such seismograms provide the observations that are combined to generate travel time tables. Figure 3.5-3 illustrates the process; the dots are travel times observed at various epicentral distances for a set of earthquakes and nuclear explosions. The data define lines giving the travel times of different phases. Such observations can be used to develop and test earth models giving *P* and *S* velocities as a function of depth. These models predict the observed travel times quite well, as shown by the fit of the theoretical travel times (lines in Fig. 3.5-3) to the observations. The travel times depend on the source depth, as shown in Fig. 3.5-4 for a surface source and a source at 600 km depth.

Although the details of an earth model depend on the specific data used to construct it, the key features of IASP91 are characteristic of recent models. The model represents a global average of the velocity structure that varies somewhat between locations. The crust is 35 km thick, an average between thin oceanic and thick continental crust (Fig. 3.2-17). Velocities increase smoothly through the upper mantle, to a depth of 410 km. The mantle transition zone, from about 400–700 km depth, contains depth intervals near 410 km and 660 km

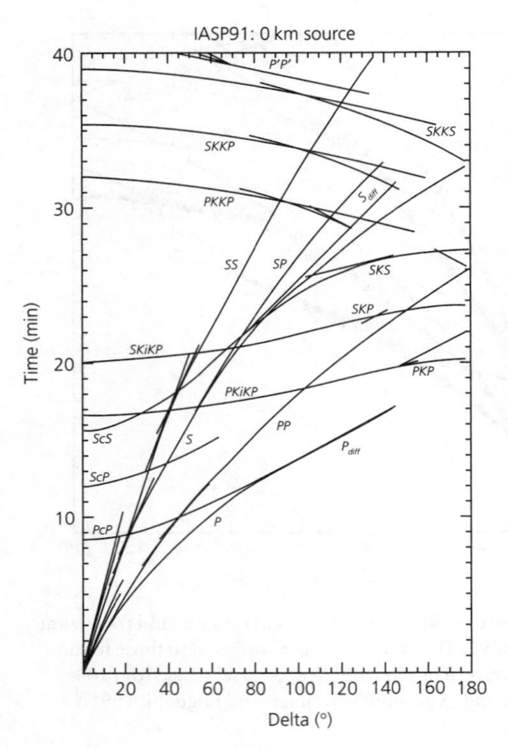

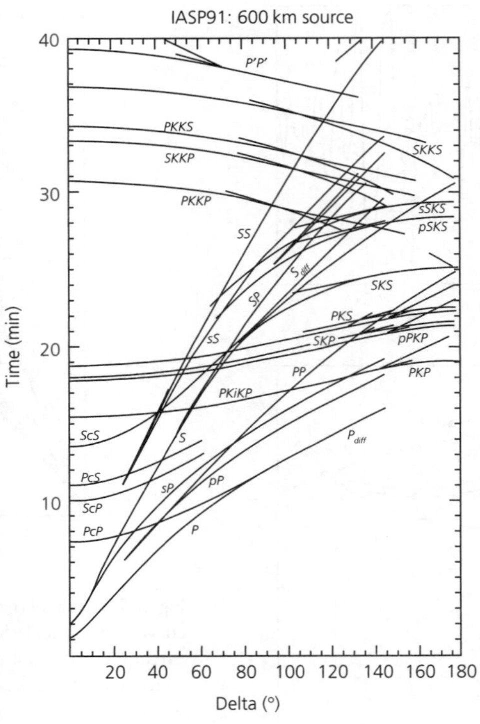

Fig. 3.5-4 Travel time curves for body wave phases for model IASP91 computed for an earthquake at the surface (*left*) and at a focal depth of 600 km (*right*). (Kennett and Engdahl, 1991.)

where the velocities increase rapidly.[2] Although these regions are often referred to as the 410 km and 660 km discontinuities, their exact depths vary from place to place. From about 700 to 2890 km depth the velocities increase smoothly throughout the lower mantle. At about 2890 km, the *P* velocity drops sharply, and the *S* velocity goes to zero, corresponding to the liquid outer core. The outer core extends to a depth of about 5150 km, beneath which the solid inner core has higher velocities, including a finite *S*-wave velocity. As we will see, these variations in velocity with depth are thought to reflect important changes in the physical, chemical, thermal, and mineralogical state of the materials present.

Seismic phases are named, based on their paths through the earth (Fig. 3.5-5, Table 3.5-2). The direct *P*-wave and *S*-wave arrivals are denoted "*P*" and "*S*." Another class of arrivals involve reflections at the earth's surface. The *P*-wave arrival corresponding to a single surface reflection is called *PP*, that for two reflections is *PPP*, and so on. Similarly, *SS* and *SSS* correspond to *S* waves reflected at the surface. Because *P* waves can convert to *S* waves, and vice versa, *PS* is a *P* wave converted to an *S* wave upon surface reflection, and *SP* is the reverse. Consideration of the ray paths shows that the travel time for *PP* at a given distance should be twice the travel time of *P* at half that distance — that is, to a point midway between the source and the receiver. Similarly, the travel time for *PPP* should be three times the travel time for one-third the distance.

Table 3.5-2 Body wave phase nomenclature.

Name	Description
P	Compressional wave
S	Shear wave
K	*P* wave through outer core
I	*P* wave through inner core
J	*S* wave through inner core
PP	*P* wave reflected at surface
PPP	*P* wave reflected at surface twice
SP	*S* wave reflected at surface as *P* wave
PS	*P* wave reflected at surface as *S* wave
pP	*P* wave upgoing from focus, reflected at surface
sP	*S* wave upgoing from focus, converted to *P* at surface
c	Wave reflected at core–mantle boundary (e.g., *ScS*)
i	Wave reflected at inner core–outer core boundary (e.g., *PKiKP*)
P′	Abbreviation for *PKP*
P_d or P_{diff}	*P* wave diffracted along core–mantle boundary

Source: After Bolt (1982).

The surface-reflected phases *PP* and *SS* (as well as *SSS*, *SSSS* or *S*4, etc.) have unusual characteristics. By Fermat's principle (Section 2.5.9), seismic phases have either minimum or maximum travel times with respect to adjacent paths. Most arrivals (*P*, *S*, *pP*, *ScS*, etc.) are minimum-time phases, but the surface

[2] Although early estimates put the locations of the discontinuities at depths of 400 and 670 km, recent revisions place them closer to 410 and 660 km. We will use the differing values interchangeably.

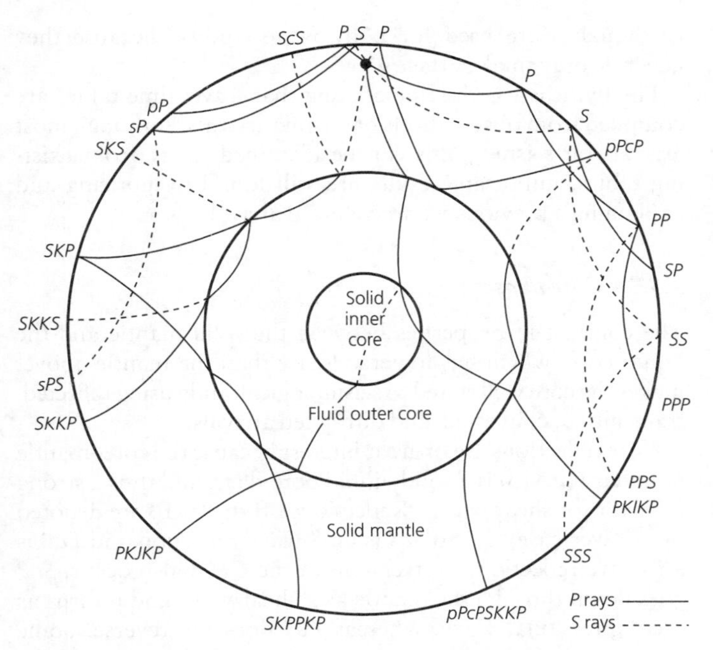

Fig. 3.5-5 Examples of body wave phases illustrating the nomenclature used. "*P*" and "*S*" designate direct ray paths, whereas "*p*" and "*s*" denote upgoing paths from the earthquake. Hence *SP* designates an *S* wave through the mantle reflected at the surface as *P*. "*c*" designates a reflection at the core–mantle boundary, so *PcP* is a *P* wave reflected at the core, and *PcS* is a *P* wave reflected as *S*. "*K*" and "*I*" denote *P* waves that traveled through the outer and inner cores, and "*i*" designates a reflection at the inner core's boundary. Hence *PKIKP* travels through the mantle, outer core, and inner core. *PKJKP*, which travels as *S* through the inner core, has only recently been conclusively observed. (After Bolt, 1982. From *Inside the Earth* by Bruce A. Bolt. © 1982 by W. H. Freeman and Company. Used with permission.)

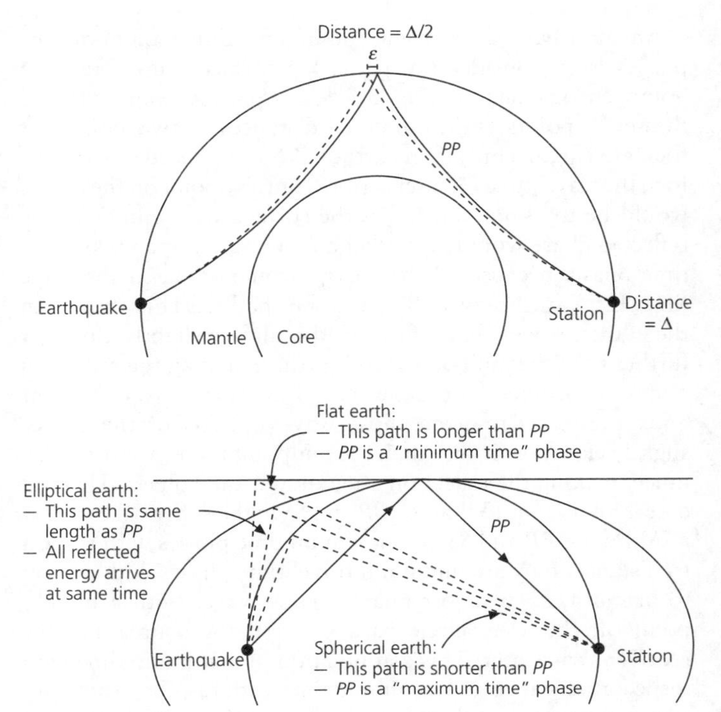

Fig. 3.5-6 *Top*: Ray path for a surface reflection. The reflection is a maximum-time phase, because the travel time for reflection at the midpoint Δ/2 is longer than on nearby alternative paths. *Bottom*: Ray paths for a surface reflection in a homogeneous medium, in which all reflections off the elliptical surface have the same travel time. The reflection off the midpoint is a minimum-time phase if the surface is flat, and a maximum-time phase if the surface is circular.

reflections are maximum-time phases with respect to distance. To see this, consider ray paths for a surface reflection that differ slightly from the true path, so the reflection bounces off the surface a small distance ε from the actual bounce point at $\Delta/2$, halfway between the source and the receiver (Fig. 3.5-6, *top*). Their travel time is thus the sum of the travel times for two legs

$$T(\Delta) = T(\Delta/2 + \varepsilon) + T(\Delta/2 - \varepsilon). \tag{1}$$

Using the first two terms of the Taylor series

$$T(\Delta/2 + \varepsilon) \approx T(\Delta/2) + \varepsilon\frac{dT}{d\Delta} + \frac{\varepsilon^2}{2}\frac{d^2T}{d\Delta^2},$$

$$T(\Delta/2 - \varepsilon) \approx T(\Delta/2) - \varepsilon\frac{dT}{d\Delta} + \frac{\varepsilon^2}{2}\frac{d^2T}{d\Delta^2}, \tag{2}$$

shows that

$$T(\Delta) \approx 2T(\Delta/2) + \varepsilon^2\frac{d^2T}{d\Delta^2}. \tag{3}$$

By Fermat's principle, the true ray path is that on which the derivative of travel time with respect to ε is zero,

$$\frac{dT}{d\varepsilon} = 2\varepsilon\frac{d^2T}{d\Delta^2} = 0, \tag{4}$$

so ε is zero, giving the expected bounce point. To see if this is a minimum or a maximum, we form the second derivative

$$\frac{d^2T}{d\varepsilon^2} = 2\frac{d^2T}{d\Delta^2}. \tag{5}$$

Figure 3.5-4 and Section 3.4.2 show that the direct *P* and *S* waves have travel time curves that are concave down, $d^2T/d\Delta^2 < 0$, so their surface reflections *PP* and *SS* are maximum-time phases. Thus *PP* or *SS* waves traveling along the same azimuth that reflect at the surface either closer or further than the point where *PP* reflects arrive earlier. By contrast, the core reflections like *ScS* have travel time curves that are concave upward, so in Eqn 5 $d^2T/d\Delta^2 > 0$, and its surface reflection *ScS2* is a minimum-time phase.

An intuitive way to view minimum- and maximum-time phases is to consider ray paths for surface reflections in a homogeneous medium (Fig. 3.5-6, *bottom*). An ellipse defines the set of points whose summed distances to two points, or foci, are equal. Thus, if an earthquake and a receiver were the foci, the travel time for a reflection from any point on the ellipse would be the same. Hence, if the surface were elliptical, the reflected phase would be neither a minimum- nor a maximum-time phase, because all the energy would arrive at the same time. If the surface were flat, and thus had less curvature than the ellipse, waves that reflect off the surface slightly closer or further than the midpoint travel further, making the reflection a minimum-time phase. However, if the surface were circular and more curved than the ellipse, waves reflected off the surface slightly closer or further than the midpoint travel a shorter distance, making the reflection a maximum-time phase. This last case is analogous to that for *PP* and *SS* in the spherical earth.

Although *PP* and *SS* are maximum time phases with respect to distance, they are minimum travel time phases with respect to azimuth, as are most phases. Thus waves with a bounce point off the great circle path between the source and the receiver arrive later. This combination of maximum time with respect to distance and minimum time with respect to azimuth makes the surface reflections sample an "X"-shaped region of the surface, known as the *Fresnel zone* (Section 3.7.3), near the bounce point. The fact that these are maximum-time phases also causes them to undergo a $\pi/2$ phase shift[3] (Fig. 2.6-5). Each successive bounce at the surface causes another $\pi/2$ phase shift, so *SSS* is phase-shifted by π and inverted with respect to direct *S*. *S4* undergoes a $3\pi/4$ phase shift, and *S5* has a 2π shift, giving it the same shape as the original *S*.

Figure 3.5-5 is drawn for an earthquake beneath the earth's surface. Because earthquakes occur to depths of 700 km, seismic ray paths go up from earthquakes as well as down. Lower-case "*p*" and "*s*" identify upgoing compressional and shear waves (Fig. 3.5-2). *pP* goes up as a *P* wave and reflects near the epicenter, whereas *sP* goes up as an *S* wave and converts to a *P* wave at the surface. These reflections are useful because the travel time difference between direct *P* and *pP*, for example, indicates the depth of the earthquake. After an upgoing wave reflects at the free surface, it can undergo later conversions, so *pPP*, *sPS*, etc. are possible arrivals.

Many other body wave phases have been identified and are included in travel time tables. In addition, some tables give arrival times for Love and Rayleigh surface waves. As shown in Fig. 2.7-4, these surface waves are dispersive, so different frequencies have different arrival times, making the time shown approximate. This time is still useful for various purposes, including allowing earth structure studies to avoid phases that may be obscured by surface waves. In many cases, deep earthquakes are used for body wave studies, because they generate only small surface waves.

Finally, it is worth remembering that travel time tables are compiled from observations of seismic arrivals. Although most arrivals on seismograms can be identified today from existing tables, important results are still found by noticing and explaining a previously unrecognized arrival.

3.5.2 *Core phases*

The contrast in properties between the solid mantle and the liquid core, which has lower velocity than the mantle above, makes the core well suited to seismological study using reflected, transmitted, converted, and diffracted arrivals.

Core reflections are of great interest because the core–mantle boundary (CMB) is a solid–liquid boundary, and thus a strong reflector for shear waves. Reflections off the CMB are denoted by a lower-case "*c*," so *ScS* is an *S*-wave reflection and *PcP* is a *P*-wave reflection. Conversions at the CMB also occur. *ScP* goes down through the mantle as a shear wave and returns as a compressional wave, whereas *PcS* does the reverse. Some phases undergo multiple reflections at both the core and the surface; *ScSScS* (or *ScS2*) bounces twice at the CMB and once at the surface. Such reflections, known as multiple *ScS*, are shown in Fig. 1.1-4.

ScS is a more distinct arrival than *PcP*, because the liquid core does not transmit shear waves. The *SH* part of the motion in the incident *ScS* cannot convert to *P* waves at the CMB, so is totally reflected. Hence *ScS* is often well recorded on the transverse component (Section 2.4.4) of a seismometer. By contrast, the *PcP* reflection is generally weak, because the impedance contrast (Section 2.6.6) is small, so most *P* energy incident on the CMB is transmitted. The small impedance contrast (about 5%) arises because the *P*-wave velocity decrease going from the mantle to the core (about 13.7 km/s to 8.1 km/s) is offset by the density increase (about 5.5 g/cm^3 to 9.9 g/cm^3).

Core reflections, especially *ScS*, are useful in studies of earth structure, because they give a vertical average velocity for the mantle. The travel time curves for these phases are concave upward (Fig. 3.5-4), like that for the reflection off the top of a layer in a flat geometry (Section 3.2.1). Similarly, they have finite travel time at zero distance because of the time needed to get down to the core and back.

The travel times and amplitudes of core phases are also used to study structure near and within the core, because their ray paths are sensitive to the structure. To illustrate this idea, consider *P*-wave ray paths (Fig. 3.5-7, *top left*) within the earth. Rays leaving the source at progressively smaller angles of incidence (closer to the vertical) bottom deeper in the mantle and so reach greater distances. As the bottoming depth approaches the core–mantle boundary, the travel times of *P* and *PcP* converge (Figs 3.5-3 and 4). Eventually, at about 98° (the precise distance depends on the depth of the earthquake and the exact velocity structure), *P* grazes the core–mantle boundary, and *P* and *PcP* are identical.

[3] This phase shift, also known as a Hilbert transform, can be viewed by thinking of the pulse as made up of sine and cosine functions and turning each into the other.

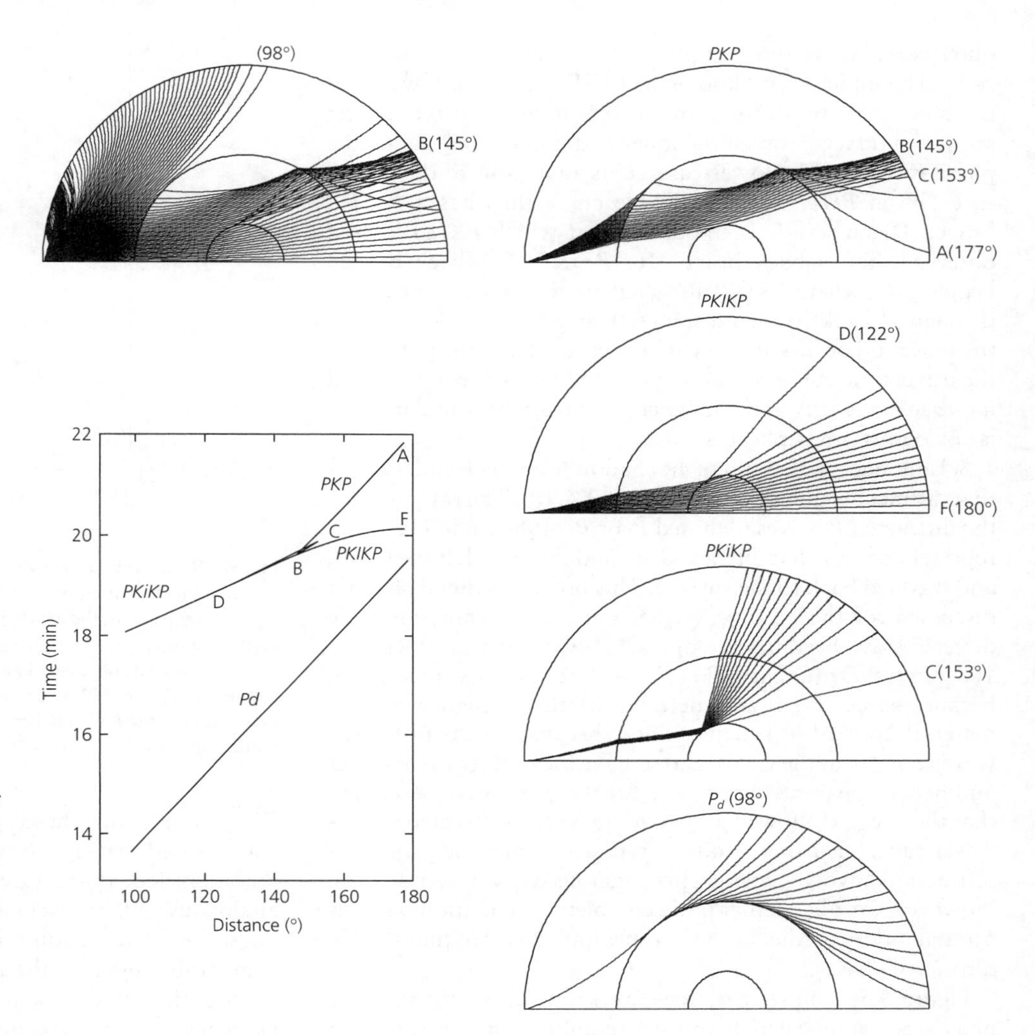

Fig. 3.5-7 Ray paths and travel times for major core phases, computed for earth model PREM. *Top left*: Paths for direct rays (i.e., excluding reflections and diffractions). *Right*: Ray paths for four other phases: *PKP* passes through the outer core, *PKIKP* also penetrates the inner core, *PKiKP* reflects from the boundary between the outer and inner cores, and P_d (also called P_{diff}) diffracts along the core–mantle boundary. *Lower left*: Travel time curves for these phases. Points on the earth's surface are labeled with their distances in degrees.

A ray with a slightly smaller angle of incidence, however, refracts downward at the CMB, because the core has a lower *P* velocity than the mantle. It thus enters the core, travels through it, refracts into the mantle, and reaches the surface. This phase is called *PKP*, where "*K*" denotes passage through the outer core.[4] For an angle of incidence slightly below grazing, *PKP* reaches the surface at point A (Fig. 3.5-7, *top right*), at a distance close to 180°. Rays with smaller angles of incidence penetrate deeper into the core, and thus arrive at distances successively less than 180°, down to a distance of about 145° (point B). At this point the pattern reverses, because rays with smaller angles of incidence arrive at successively greater distances. This goes on for rays reaching distances up to point C (~153°, depending on the earth model), corresponding to the ray that grazes the inner core–outer core boundary.

The ray paths show that the low velocity in the outer core gives rise to a geometrical shadow zone, where Snell's law predicts that no direct rays arrive.[5] We have seen (Fig. 3.4-7) that the corresponding travel time curve should have a break due to the shadow zone, and then two branches on the far side of the shadow zone. For the core, the shadow zone occurs for distances between ~98° to ~145° (point B, Fig. 3.5-7, *top left*). Beyond 145°, the travel time curve has two branches for *PKP*. The AB branch (sometimes labeled PKP_2) is the back branch, on which rays with smaller angles of incidence appear at smaller distances, whereas the BC branch is the forward branch on which rays with smaller angles of incidence appear at larger distances.

In reality, body waves are observed in the shadow zone. Much of the body wave energy arrives as surface-reflected (*PP*, *PPP*, *SS*, etc.) or multiply core-reflected (*ScS*2, etc.) arrivals. Other arrivals are due to *P* waves that encounter the inner core. Because the inner core has higher *P*-wave velocity than the

[4] "*K*" is from *Kern*, the German word for core.

[5] Although the core's existence had been inferred from the earth's gravity (Section 3.8), the discovery of this shadow zone in 1906 by Richard Oldham (1858–1936) provided the first direct evidence and set the paradigm for future core studies.

outer core, waves refract upward and emerge in the shadow zone. These phases are known as *PKIKP*, because *P* waves in the inner core are denoted by "*I*." In addition, waves reflect at the boundary between the inner and outer cores, giving the phase *PKiKP*.[6] (The lower-case "*i*" is analogous to the lower-case "*c*" in *PcP*.) The travel time curve thus has a *PKIKP* branch DF, where D is the distance at which *PKIKP* is first observed, and a back branch for *PKiKP*. The back branch begins at C, where *PKiKP* and *PKP* are the same, and extends through D back to zero distance (Figs 3.5-3 and 4), because the reflection occurs at vertical incidence. Hence the portion of the travel time curve containing CD and DF is due to the rapid increase of velocity at the inner core–outer core boundary, and is analogous to a triplication.

Seismic energy also enters the shadow zone via *P* and *S* waves that diffract around the core (Section 2.5.10). The ray paths for the diffracted *P* waves (denoted P_d or P_{diff}) shown in Fig. 3.5-7 represent energy that diffracted around the core, left the CMB, and traveled back to the surface. This process is much like that discussed for the head wave (Section 3.2.1). Thus, once the direct *P* wave becomes the diffracted wave at a distance near 100°, its travel time curve (Fig. 3.5-4) loses the curvature it had, because successive rays penetrated deeper to higher-velocity material. Instead, it becomes linear because all the diffracted waves bottom at the CMB, and so have the same ray parameter and hence apparent velocity. As for the head wave, assuming that the energy followed a ray path gives the diffracted wave's travel time but cannot fully describe its amplitude, because diffraction involves energy propagating as waves, not rays. However, we will see that more complete formulations such as normal modes predict both the times and the amplitudes of the diffracted phases.

Figure 3.5-7 shows that the travel time curve for the core phases is complicated because it combines the effects of a geometric shadow zone, which gives two *PKP* branches, a triplication-like feature containing the *PKIKP* and *PKiKP* branches, and a diffraction branch. In reality, even these models are simplifications of a more complex reality. Figure 3.5-8, showing the travel times of several million *PKP* arrivals, illustrates several significant deviations from the theoretical curves in Fig. 3.5-7. First, the arrivals do not fall along narrow lines. This is partly due to errors of observation, but also due to the heterogeneous structure of the crust, mantle, and core, which makes some arrivals early and others late. Second, the *PKP-BC* branch continues beyond its geometrically predicted limit of 153°. This is because the *PKP-BC* wave diffracts around the inner core, although its amplitude decreases rapidly in the process, so there are few observations beyond 160°.

Third, and most importantly, the *PKP* travel times show an additional branch not predicted by geometric ray theory. These arrivals, labeled *PKP precursors*, appear to be a continuation of the *PKP-AB* branch and arrive as much as 20 s before the

[6] Observations of this phase by Inge Lehmann (1888–1993) in 1936 provided the first evidence for the existence of the inner core.

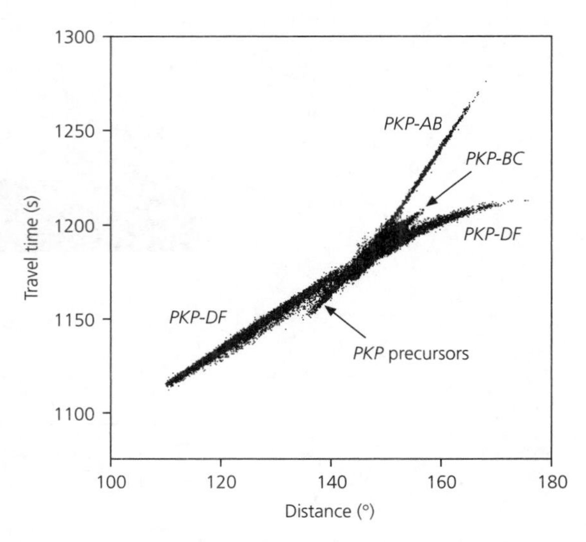

Fig. 3.5-8 Arrival times of *PKP* waves recorded by the International Seismological Centre during 1964–87. A point is plotted if there are at least 200 arrivals in the catalog for that time and distance. Although these arrival times are similar to the predicted travel time curves in Fig. 3.5-7, there are some differences. The *PKP-BC* branch is observed beyond its geometrical limit (153°) due to diffraction around the inner core, and precursors to the *PKP-DF* branch are observed that result from seismic scattering at the CMB and in the mantle. (Courtesy of K. Koper.)

PKP-DF branch. These arrivals puzzled seismologists until it was realized that they were waves reflected, or scattered, from inhomogeneous structures in the mantle. This scattering is analogous to that discussed in Section 3.3.7 in the context of migration in reflection seismology. Because the scatterers are comparable in size (about 10–15 km) to the wavelengths of short-period *P* waves in the lower mantle, they behave as Huygens' sources (Section 2.5.10). Thus a *PKP-AB* wave interacting with a scatterer at the CMB radiates waves in all directions (Fig. 3.5-9). Those arriving before *PKP-DF* are clearly observed, whereas those arriving afterwards are lost amid *PKP-DF*. The range of observable scattered *PKP* waves is shown as the shaded regions in Fig. 3.5-9, illustrating another way in which seismic energy reaches the shadow zone. Although most such scattering occurs near the CMB, modeling of the *PKP* precursors suggest that waves are also scattered by small reflectors throughout the mantle, as shown by the dark shaded region in Fig. 3.5-9.

Some core phases begin as *S* waves (Fig. 3.5-5). Although no *S* waves propagate in the liquid outer core, phases like *SKS* travel through the mantle as an *S* wave and through the core as a *P* wave. *SKKS* is similar to *SKS*, but also involves an underside reflection at the CMB. Because the *P* velocity of the uppermost core (about 8.1 km/s) is not much larger than the *S* velocity of the lowermost mantle (about 7.2 km/s), *SKS* and *SKKS* waves do not change direction significantly as they cross the CMB. Thus *SKS*, *SKKS*, *SKKKS*, etc. are the only waves that bottom near the top of the core and are used to constrain the outer core's velocity structure.

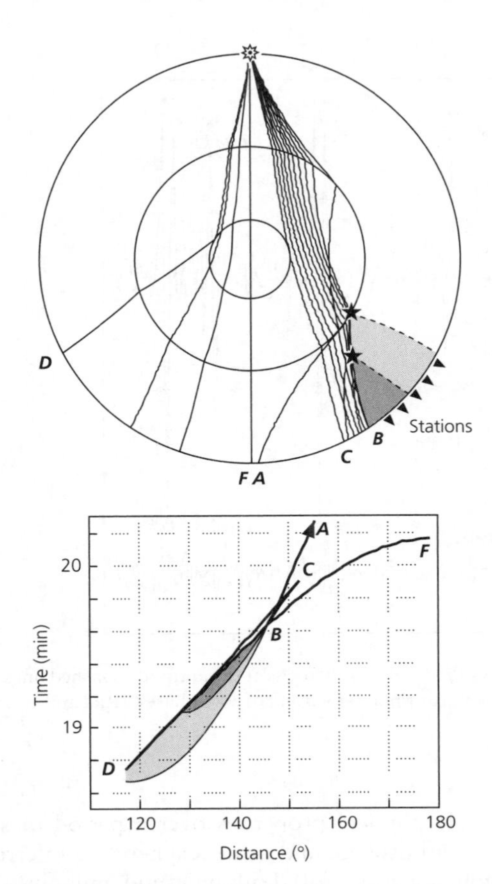

Fig. 3.5-9 A model for the *PKP* precursors shown in Fig. 3.5-8. *Top*: *PKP-AB* waves that interact with scatterers (stars) cause arrivals at distances less than the geometrically allowed *AB* range. As shown by the travel times (*bottom*), these arrivals precede the *PKP-DF* arrivals. Scatterers at the base of the mantle yield waves in the light and dark shaded regions, and others in the mid-mantle yield waves in the dark shaded region. (Hedlin *et al.*, 1997. Reproduced with permission from *Nature*.)

Other core phases, some of which are not included in the travel time plots in Fig. 3.5-4, have also been reported. These include *PKKP* (Fig. 3.5-10), a *P* wave that has undergone underside reflection at the CMB, *PKPPKP* (sometimes called P′P′), a *PKP* phase reflected at the surface, and *PKIIKP*, an underside reflection from the outer core–inner core boundary. An especially elusive phase has been *PKJKP*, which, by analogy to *PKIKP*, travels through the inner core as an *S* wave. The weak amplitude of this phase, combined with the fact that it arrives late in the seismogram amid other phases, has made it difficult to observe. *PKJKP* has been verified only recently, by stacking data from very large deep earthquakes that generate the large body waves needed to produce even small *PKJKP*, while not generating surface waves that mask the small core arrivals.[7]

[7] This observation and normal mode results are overcoming seismologists' prior reservations, exemplified by comments like "the inner core, which may exist, is said to have the following properties. . . ."

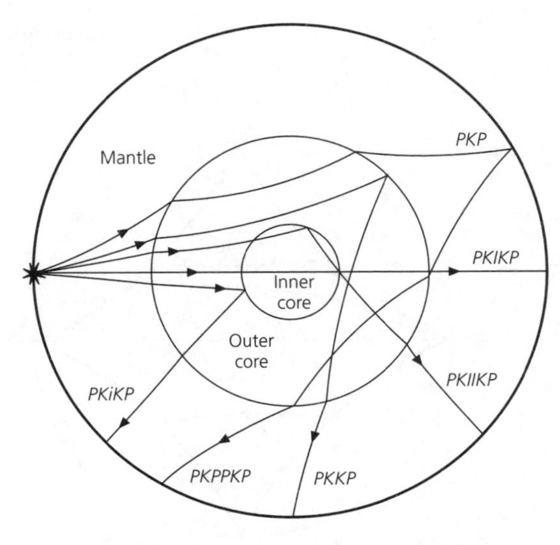

Fig. 3.5-10 Some additional core phases. *PKKP* and *PKIIKP* are underside reflections at the core–mantle boundary and outer core–inner core boundaries, and *PKPPKP* (*P′P′*) is an underside surface reflection.

Core phases can be challenging to study with travel time data because their travel time curves are complicated and some of the arrivals are small. Amplitude and waveform studies provide additional information. As we have seen (Section 3.2.3), amplitudes can be used to differentiate between structures that would give similar travel times. Some insight into the amplitudes can be obtained from the ray densities (Section 3.4.2). For example, the AB and BC branches of the *PKP* travel time curve meet at the far side of the shadow zone, at point B. Figure 3.5-7 shows that rays which left the source at uniform angle increments are concentrated there, so large amplitudes are expected at this caustic.

This discussion of amplitudes brings out another interesting point. Although the earth is approximately spherical, we have discussed only waves propagating in the plane containing the source, the receiver, and the center of the earth. One case in which sphericity is important is near the antipode, the point 180° from the source. Figure 3.5-11 shows seismograms recorded at PTO (Porto, Portugal) and MAL (Malaga, Spain) from an earthquake in New Zealand. Phases like *PP* and *PKP* are focused at the antipode, because paths in any direction from the source arrive at the same time. Note the larger arrivals at PTO, only 0.7° from the antipode.

3.5.3 *Upper mantle structure*

The velocity structure of the upper mantle shows two major effects. First, it has discontinuities and velocity gradients that are essentially radially symmetric, which are believed due to the effects of pressure on the minerals present. Second, it contains significant lateral heterogeneity that is primarily associated with temperature variations due to cold subducting oceanic lithosphere. We discuss the radial velocity structure here,

Antipodal focusing

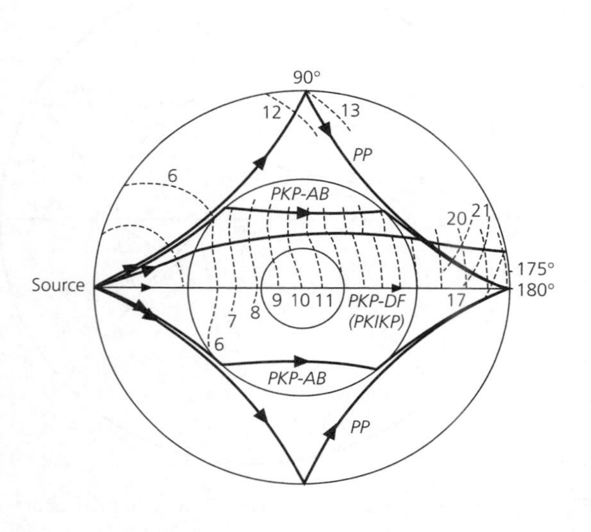

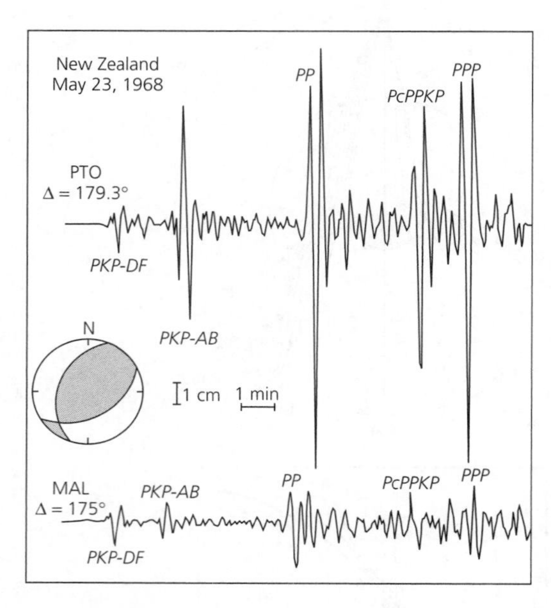

Fig. 3.5-11 Focusing of *P* waves at the antipode, 180° from an earthquake. *Left*: Seismic rays *PKP-AB* and *PP* focus at the antipode. Dashed lines represent wave fronts, whose propagation time in minutes is given. *Right*: Seismograms showing antipodal focusing of core phases. (Rial and Cormier, 1980. *J. Geophys. Res.*, *85*, 2661–8, copyright by the American Geophysical Union.)

explore its mineralogical causes in Section 3.8, and consider the effects of subducting lithosphere in Section 5.4.

We have already discussed the velocity structure of the uppermost mantle shown by surface wave dispersion (Section 2.8). Body wave analyses reveal a similar structure. The sub-crustal lithosphere shows generally fast *P*- and *S*-wave velocities of about 8.1 and 4.5 km/s. This high-velocity layer provides a way of defining the lithosphere, termed the *seismic lithosphere* or *lid*, from seismological observations. The thickness of the seismic lithosphere varies with location. At mid-ocean ridges, where oceanic plates are created, its thickness approaches zero. Beneath stable cratons, the fast lithospheric velocities extend to about 200 km. As a global average, the seismic lithosphere extends to about 80–100 km depth.

In most regions of the world, we find a seismic low-velocity zone (LVZ) beneath the seismic lithosphere. The LVZ approximately coincides with the expected mechanically weak asthenosphere underlying the stronger lithosphere. The lithosphere and asthenosphere are defined by their mechanical properties, such that plates of strong lithosphere slide over weaker asthenosphere. This contrast, as we will see, results from the fact that the lithosphere is the cold outer thermal boundary layer of the solid earth (Sections 3.8, 5.1). By contrast, the high-velocity seismic lithosphere and underlying LVZ are seismologically defined entities. The rough correspondence between the seismological and mechanical layers indicates that the two are closely related, and that seismic observations can be used to map mechanical structure. The two sets of layers are not identical for several reasons including the fact that seismic waves sample physical properties over a period of seconds, whereas the lithosphere and asthenosphere are inferred from data sampling periods of thousands and millions of years (Section 5.7).

The depth and magnitude of the LVZ vary regionally. In tectonically active regions like western North America, the LVZ is well developed and relatively shallow. In stable continental regions that have not experienced tectonism for a long time, the LVZ is deeper and less pronounced, and may not even be present. The thick, high-velocity layer under continents has led to the suggestion that it may reflect a chemically distinct *tectosphere*. For this hypothesis, continents behave differently from the oceanic lithosphere, where surface wave dispersion shows a pronounced LVZ for all ages (Fig. 2.8-7). This persistence may reflect the fact that oceanic lithosphere is never older than 180 Ma, tectonically young by continental standards, because older oceanic lithosphere is subducted away.

We will show in Section 5.7 that the contrast between the high-velocity seismic lithosphere and the asthenosphere LVZ is probably related to variations in material strength between the cold lithosphere and the warmer asthenosphere. There may also be some effects of partial melting. This situation differs from the velocity differences between the crust and the mantle, which result from their different compositions. Beneath the LVZ, which extends to an average depth of about 200 km, temperatures increase only slowly, but velocities increase significantly in response to the increasing pressure.

The transition zone between the upper and lower mantles is marked by the velocity discontinuities at depths of about 410

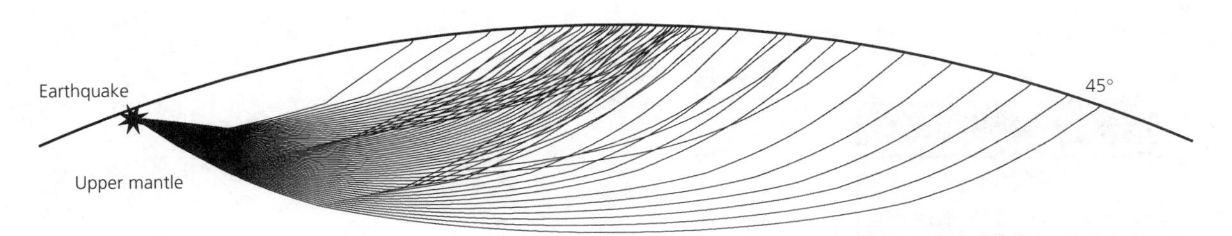

Fig. 3.5-12 Ray paths for *P* waves through the upper mantle, computed for earth model PREM, showing triplications due to mantle discontinuities.

and 660 km. We saw in the last section that a rapid velocity increase (Fig. 3.4-6) produces a triplication in the travel time curve. Upper mantle travel times show two triplications around 15° and 22° caused by the 410 and 660 km discontinuities. Ray paths for such a structure are shown in Fig. 3.5-12. Some reference models such as PREM also have a discontinuity at 220 km, and regional studies also often find discontinuities at other depths in the upper mantle.

One difficulty in studying the transition zone, or other regions of complex velocity structure, is that travel time curves are composites of data from many earthquakes at different distances. The process of combining the data can make the details of the triplication difficult to observe. Moreover, $dT/d\Delta$, the derivative of the travel time curve that is used in inverting for velocity, is uncertain due to the scattered data. These difficulties can be addressed in several ways. One is to derive information from the waveforms as well as the travel times. A second is to use arrays of seismometers spaced closely enough that it is possible to identify arrivals corresponding to the different branches of triplications and directly measure $dT/d\Delta$ by tracing them across the array. Such dense data also facilitate waveform studies.

Figure 3.5-13 (overleaf) illustrates these ideas with an array study of upper mantle *P*-wave structure under the Gulf of California spreading center. Data from ten earthquakes are combined into a record section for the epicentral distance range 9–40°. The travel time curves show two triplications, one near 15° due to the 410 km discontinuity, and another around 22° due to the 660 km discontinuity. Travel times and synthetic seismograms predicted by the velocity structure (GCA) derived from the data fit the data well, including the back branches (C-B and D-E) of the triplications. The effects of the discontinuities appear in the $p(\Delta)$ data as two groups of later arrivals for which p increases with Δ. These arrivals are the back branches of the triplications (Fig. 3.4-6). The remaining arrivals show p decreasing with Δ, and thus are the forward branches.

Figure 3.5-14 compares the GCA model to upper mantle models for other tectonic environments: ARC-TR (arc-trench) for the Japan subduction zone, T7 for the tectonically active western portion of North America, and K8 for the stable Eurasian shield. Above 200 km, all show a LVZ overlain by a higher-velocity lid, but the depth and extent of the LVZs differ. The shield model, for example, has the thickest lid. Below 200 km, GCA shows the lowest velocity. The depths of the 410 and 660 km discontinuities differ between the models. These differences are thought to reflect the fact that the mineral phase transformations causing the discontinuities occur at pressures (and hence depths) that depend on temperature. Thus lateral temperature changes, especially those associated with subduction zones, should change the depths at which these transitions occur (Section 5.4.2).

Waveform modeling provides additional information about the transition zone. For example, waveform modeling of intermediate-period *S* waves shows a discontinuity at about 520 km depth that is not observed with short-period *P* waves. The phase transition thought to cause this discontinuity may occur over a greater depth range than for the 410 and 660 km discontinuities, making it visible only to longer-period waves.

3.5.4 *Lower mantle structure*

Velocities increase rapidly with depth for roughly 100 km beneath the 660 km discontinuity, but then increase more slowly. The rapid increase implies that mineral transformations continue, whereas the slow increase implies that the mineralogy and composition of the material are not changing significantly, and that the velocity increases are primarily due to the material being compressed by higher pressure. However, weak seismic discontinuities have been reported at a variety of depths such as 900 and 1300 km. These may represent either global discontinuities like the 410 and 660 ones, or local velocity anomalies, perhaps due to fragments of old subducted slabs.

The situation changes dramatically in the D″ layer at the very base of the mantle, a fascinating and poorly understood region[8] that has a velocity structure whose complexity rivals that of the lithosphere. D″, the bottom few hundred kilometers of the mantle, was initially differentiated from the rest of the mantle (D′) because the velocity gradient with depth is lower. This lower gradient is expected, because D″ is a thermal boundary layer between the mantle and hotter core. The expected ~1000°C temperature difference across D″ would lower velocities and thus decrease the velocity gradient.

However, detailed velocity models show that at the top of this lower-gradient region the velocity increases sharply

[8] The uncertainties about D″ have been illustrated by describing its thickness as 250 ± 250 km (Jeanloz, 1990).

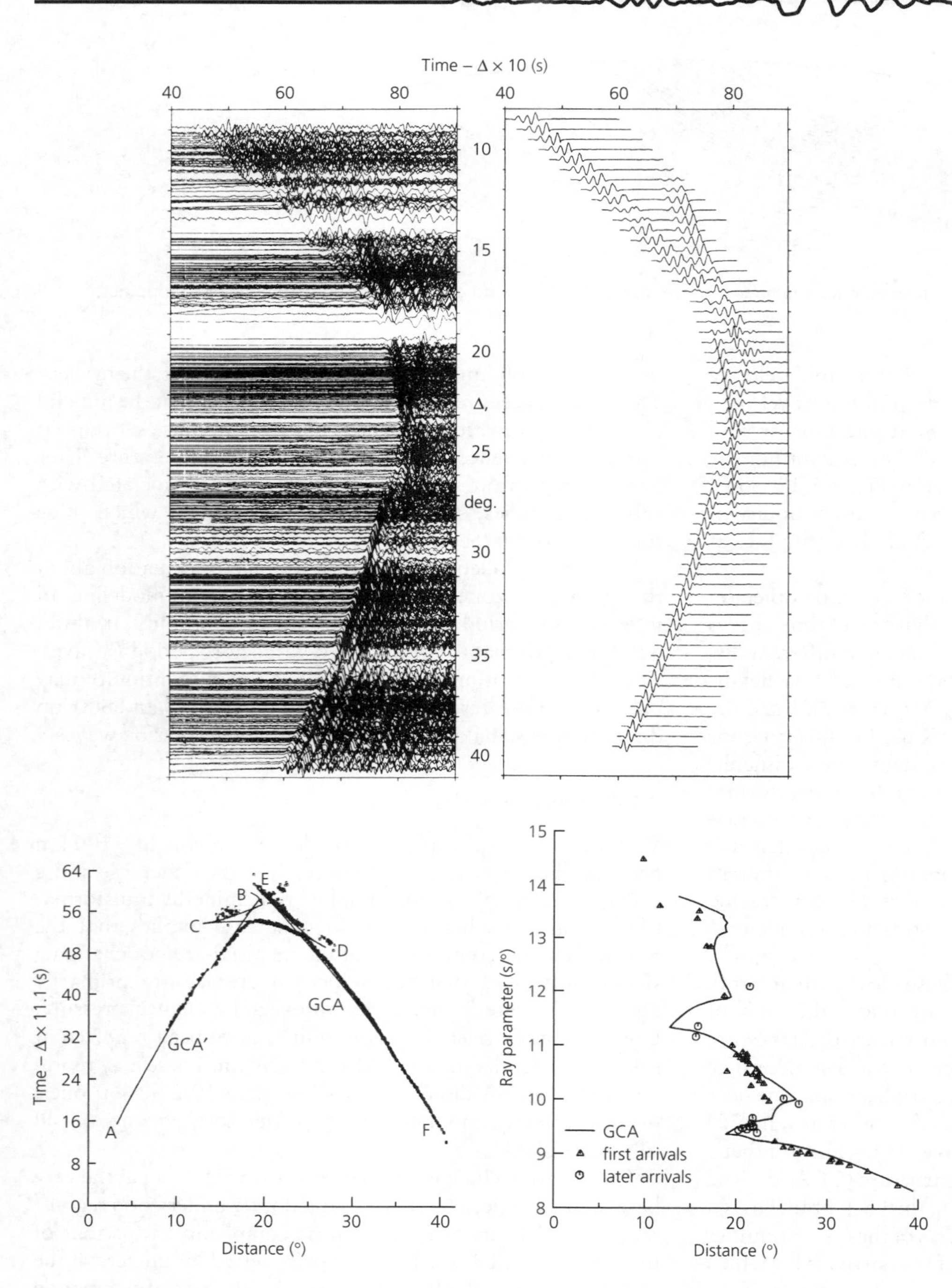

Fig. 3.5-13 Seismic array study of upper mantle structure. *Top*: Record sections, plotted with a reducing velocity of 10°/s, showing observed (*left*) and synthetic (*right*) seismograms. *Bottom left*: Reduced travel time plot, showing travel time data and model predictions. *Bottom right*: $p(\Delta)$ plot and model predictions. The two triplications are evident in the record sections, travel time plot, and $p(\Delta)$ plots. The slight break in the travel time curve at 13° is due to use of slightly different models (GCA′ versus GCA). (Walck, 1984.)

(Fig. 3.5-15). A feature of such models is that the high- and low-velocity regions trade off to give similar travel times as PREM, which does not contain the high-velocity region. Thus D″ is now often delineated by the location of the discontinuous velocity increase, which averages about 250 km above the CMB. This is ironic, in that D″ was first named for a region of lower than expected velocities.

Observations of the velocity increase, known as the D″ discontinuity, are usually made with the phases *PdP* and *SdS*, each of which combines waves that reflect off and refract just under the discontinuity (Fig. 3.5-16). *PdP* and *SdS* arrive between the direct (*P* and *S*) and core-reflected (*PcP* and *ScS*) phases, as shown. The discontinuity has been observed at many locations on the CMB, but other locations, even nearby, do not show a *PdP* or *SdS* arrival. Moreover, although the average depth of the discontinuity is 250 km above the CMB, the observed depths range from 100 to 450 km above the CMB.

One possible explanation for this variability is that the discontinuity has large topographic variations over small spatial wavelengths that focus and defocus waves. Another possibility

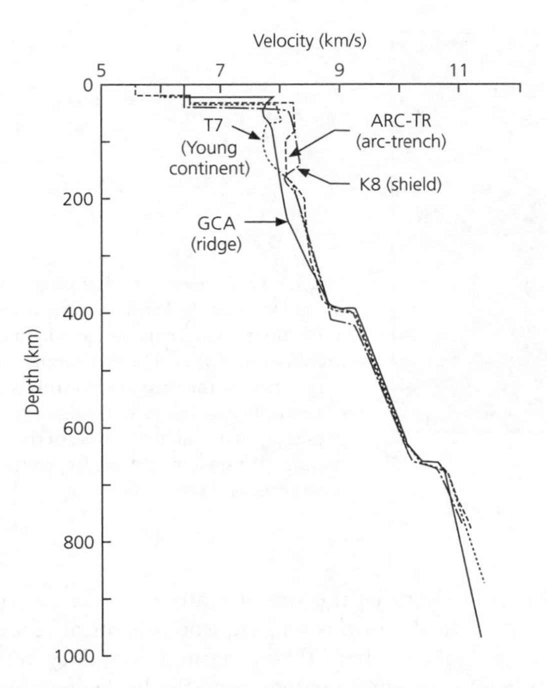

Fig. 3.5-14 Comparison of model GCA, derived from the data in Fig. 3.5-13, to *P*-wave velocity models for other tectonic regions. (Walck, 1984.)

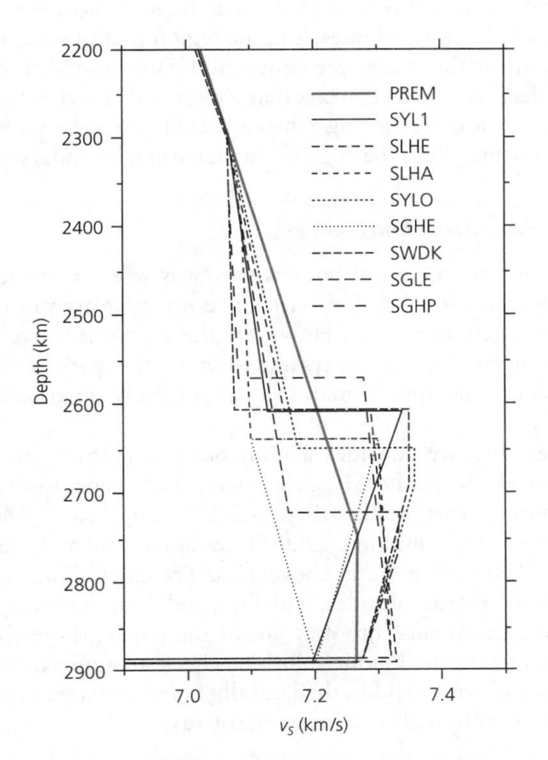

Fig. 3.5-15 Velocity structures from several studies showing an increase in velocity about 250 km above the core–mantle boundary, known as the D″ discontinuity. (Wysession *et al.*, 1998. *The Core–Mantle Boundary Region*, 273–97, copyright by the American Geophysical Union.)

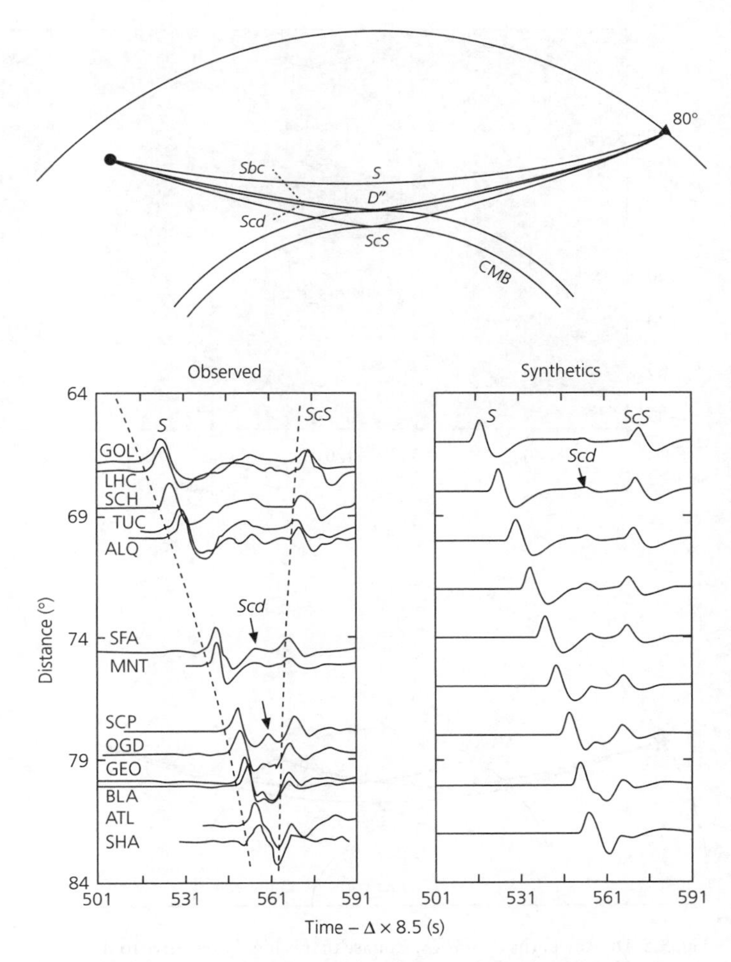

Fig. 3.5-16 *Top*: Schematic ray paths of the two arrivals making up the phase *SdS*. *Sbc* reflects off the D″ discontinuity, and *Scd* refracts just below it. (Wysession *et al.*, 1998. *The Core–Mantle Boundary Region*, 273–97, copyright by the American Geophysical Union). *Bottom*: Observed *Scd* arrivals (arrows) (*left*) compared to synthetic seismograms (*right*) computed using velocity model SYLO (Fig. 3.5-15). (After Young and Lay, 1990. *J. Geophys. Res.*, *95*, 17, 385–402, copyright by the American Geophysical Union.)

is that there is no actual discontinuity, but that complex three-dimensional velocity heterogeneities give the appearance of a discontinuity. This possibility is supported by observations of the increased scattering of seismic waves passing through D″. In either case, it is possible that the increase in velocity is associated with subducted lithosphere that sank to the bottom of the mantle. There is a correlation between regions of fast velocities in D″ and the projected locations of fossil slabs from ancient subduction zones (Fig. 3.5-17), which should retain a cold thermal anomaly for a long time after reaching the CMB (Section 5.4.1). Seismic modeling suggests that this mechanism could generate *PdP* and *SdS* phases.

D″ shows additional complexities. There is strong evidence for significant seismic anisotropy (Section 3.6.6). Large lateral

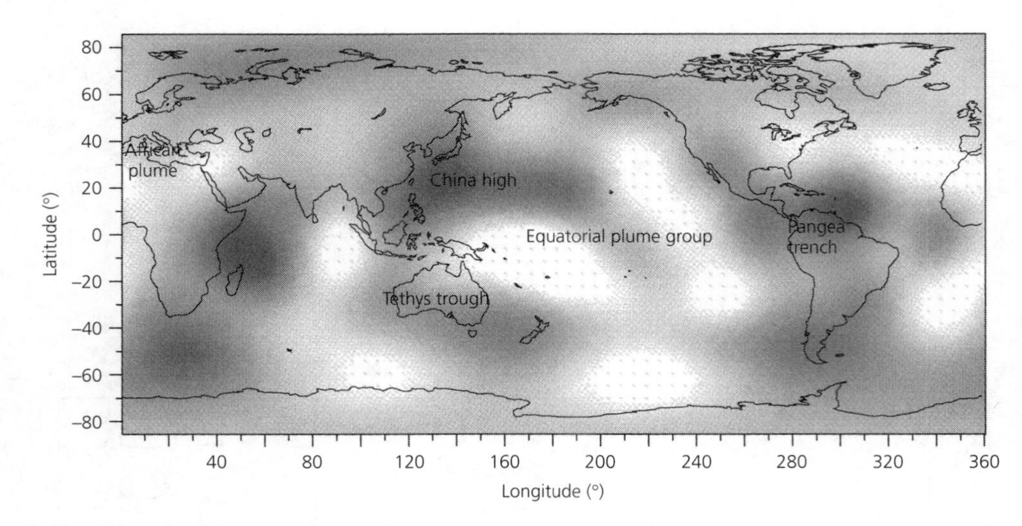

Fig. 3.5-17 *P*-velocity variations at the base of the mantle. Dark areas represent anomalously fast velocities, and light areas are slow anomalies. The fast anomalies correlate with the predicted locations of lithosphere subducted during the Mesozoic that sank to the base of the mantle. (Wysession, 1996b. Reproduced with permission from *Nature*.)

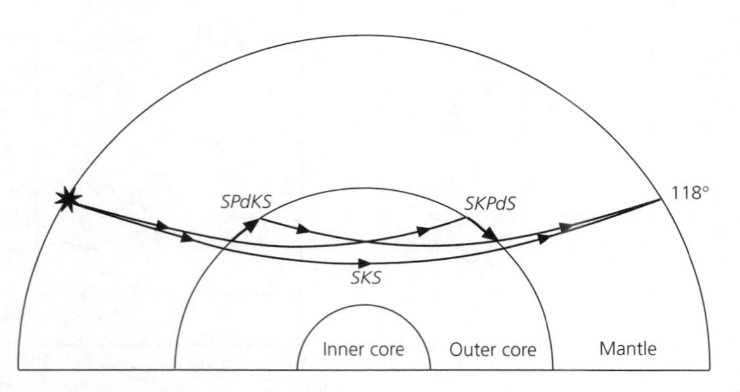

Fig. 3.5-18 Ray paths of *SPdKS*, a phase that is highly sensitive to the ultra-low-velocity zone at the base of the mantle. As with many studies of the deep mantle and core, it is analyzed using the difference between its travel time and another phase — in this case, *SKS*.

variations at both small and large spatial wavelengths occur for velocities within D″ and for topography on the CMB. There is also evidence for an ultra-low-velocity zone (ULVZ) at the very bottom 10–20 km of the mantle. The ULVZ is observed with an unusual body wave phase, *SPdKS*, which is similar to *SKS* but travels partly as a diffracted *P* wave at either or both of its entrance and exit points from the core. *SPdKS* appears as a shoulder of the *SKS* arrival, and is very sensitive to the *P*-wave velocity structure just above the CMB (Fig. 3.5-18). Modeling of *SPdKS* waveforms suggests that v_p may be 10% lower than in the rest of D″, and the reflection coefficients of *PcP* precursors that reflect off the top of the ULVZ suggest that v_s may decrease by 30%. The ULVZ may result from partial melt, because it is most prominent where D″ velocities are slowest, implying that the high temperatures causing the low velocities may also cause more partial melting.

In summary, much uncertainty remains about the detailed structure of D″ and its causes. This is hardly surprising, because the CMB is likely to be the site of many processes involving lateral and vertical motions and vigorous chemical reactions. An analogy might be that D″ is a thermal boundary between the mantle and the core, analogous to the lithosphere, which is the thermal boundary layer at the top of the mantle. The high-velocity layer at the base of D″ may be a chemical layer, analogous to the crust. These complexities have led the CMB to be called the graveyard of ancient ocean lithosphere, the birthplace of mantle plumes, and the region that most significantly controls the outer core convection patterns and thus the earth's magnetic field. The fact that we study this region largely via seismic "remote sensing" through 2890 km of heterogeneous mantle may limit the degree to which it can be understood.[9]

3.5.5 Visualizing body waves

To end our discussion of body waves, it is worth considering their physical nature. We have treated body wave arrivals like *S* and *ScS* as geometric rays. However, although it is convenient to describe these waves as rays and to show their paths through ray tracing, this approximation does not fully describe their behavior.

To see this, we consider a numerical simulation showing time snapshots of the *SH* shear wave field generated by a 600 km-deep earthquake (Fig. 3.5-19). The wave field is synthesized by summing 28,000 torsional normal modes (Section 2.8) with periods above 12 s. The calculations show accurate relative amplitudes, with light and dark shades representing displacements into and out of the paper, respectively. Although the normal mode solution is itself an approximation to the actual wave field in the laterally heterogeneous earth, it is much closer to reality than geometric rays.

[9] The geophysical significance of the CMB and the large uncertainties remaining about it are summarized by D. Stevenson's description of D″ as "the sum of all of our ignorance of the interior of the earth."

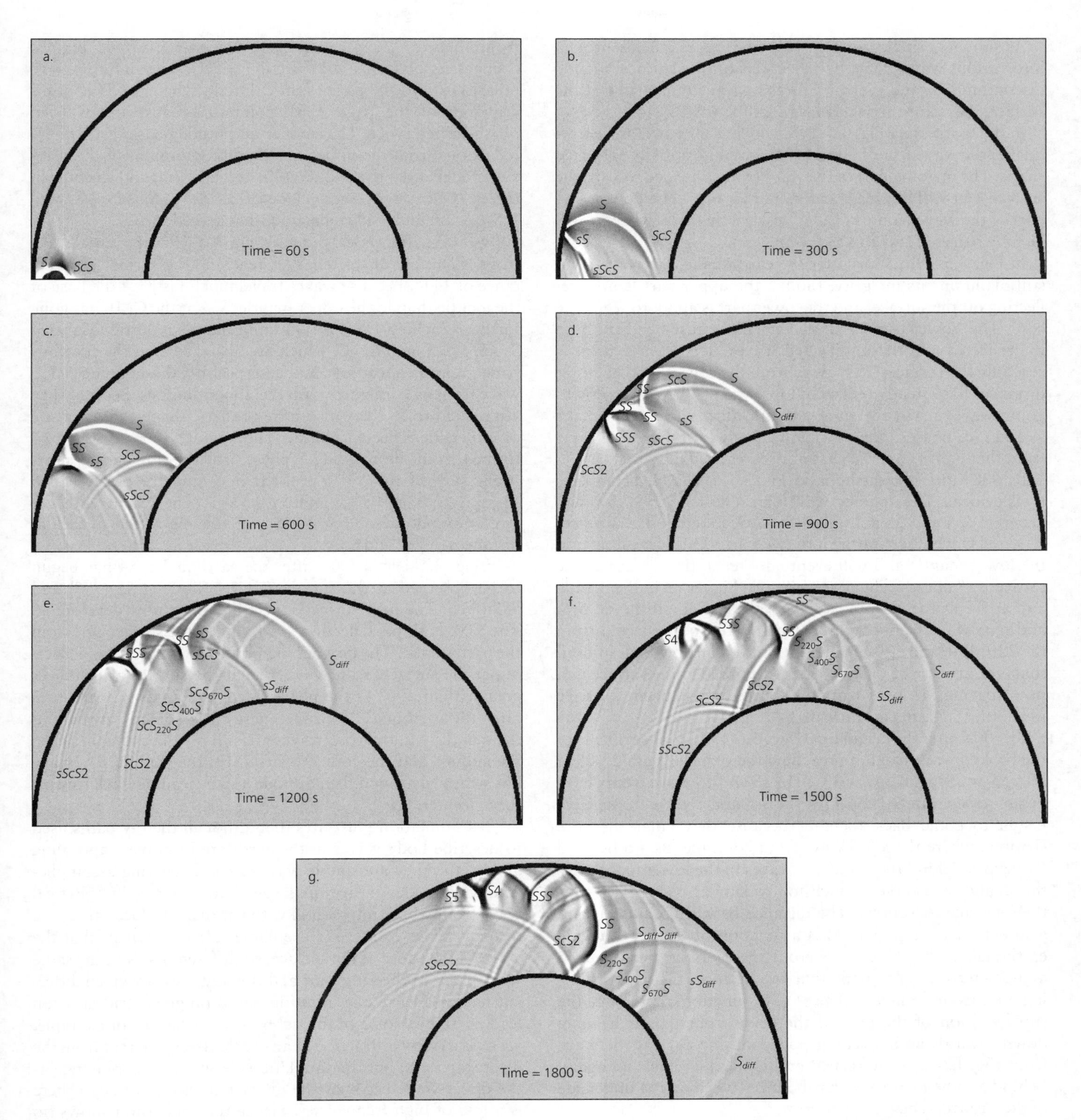

Fig. 3.5-19 Snapshots of a synthetic *SH* wave field showing the propagation of waves after a 600 km-deep earthquake. The initial wave front moves away from the source at the lower left side of the figures. The wave front develops complexity due to interactions with the surface, CMB, and internal discontinuities and velocity gradients. The wave field is computed using the spherically symmetric PREM velocity model. Amplitudes are raised to a power of 0.8 to enhance smaller signals. (After Wysession and Shore, 1994.)

As shown, a single spherical S wave front is quickly broken into various wave fronts by reflections off the surface, mantle discontinuities, and the core. As the wave fronts arrive at the surface, they cause arrivals that we call S, ScS, sS, etc.

In the first frame, Fig. 3.5-19a, which is 60 s after the earthquake, the wave front maintains much of its initially spherical shape. The upgoing part of the wave front is headed toward the surface, but will not reach it for another 67 s. The downgoing part of the wave front is headed toward the core, where it will be fully reflected and give rise to ScS.

In Fig. 3.5-19b, 300 s after the earthquake, the wave front still maintains its integrity, though the upper part is now reflecting off the surface, and the lower part is about to reach the core. The slower upper mantle velocities cause bends in both the reflected and the unreflected waves. The S wave front is reaching the surface 12.5° away from the source, and at closer distances has already reflected downward. When these downgoing waves reach the surface again, they will be called the sS and $sScS$ phases. The downgoing portion of the initial wave front that will become ScS has not yet reached the core.

By 600 s after the earthquake (Fig. 3.5-19c) added complexity is evident. The upgoing wave that reflected at the core will generate ScS and its multiples ($ScS2$, $ScS3$, etc.). The surface-reflected wave is separating into two parts. One is heading into the lower mantle and will eventually reach the surface as the $sScS$ and sS phases. The other will turn higher up in the mantle and arrive at the surface as the SS phase. Behind the sS, ScS, and $sScS$ wave fronts are upper mantle echoes reflected from the 220, 400, and 670 km discontinuities. However, despite all that is going on, the only phase yet recorded at the surface is S, now arriving 31° away from the source. sS will begin to arrive in another 63 s, at a distance of 24°.

By 900 s after the origin time (Fig. 3.5-19d), four segments of the broken wave front are reaching the surface: S at 52°, sS at 39°, SS at 38°, and ScS at 33°. The sS and SS wave fronts have begun to separate. In contrast, the ScS and S wave fronts have begun to come back together because they enter the core shadow, where the S/ScS wave front continues as a diffracted S_{diff} wave. Behind the S and ScS waves in the lower mantle are the sS and $sScS$ waves, which follow similar paths except for their surface reflections. The distance between S and sS (and also between ScS and $sScS$) is a function of the depth of the earthquake. The three wave front segments labeled SS form a characteristic "Y" shape that results from the waves turning in the mid-mantle. The "Y"'s junction represents the superposition of the part of the wave front that is heading down toward the bottoming point and the part of the wave front that has already turned and is heading back up again. Behind SS, the phase SSS that bounces twice on the underside of the surface, is beginning to form.

In Fig. 3.5-19e, 1200 s after the earthquake, most of the initial S wave front is actually S_{diff} because S grazes the core at about 100°. The surface-reflected sS wave now also diffracts around the core as sS_{diff}. SSS is now fully developed, and is reaching the surface behind SS. The polarity of SSS is different from that of SS, because each successive surface bounce changes its phase by $\pi/2$ (Section 3.5.1). The initial S-wave polarity is into the page (light-colored), whereas SSS is primarily out of the page (dark colored) because it has been phase-shifted twice. The smaller-amplitude phases evident are reflections from the upper mantle discontinuities in the velocity model at depths of 220, 400, and 670 km, and so come in threes. One set of these, labeled as $ScS_{220}S$, $ScS_{400}S$, and $ScS_{670}S$, are underside reflections that precede $ScS2$.

By 1500 s after the earthquake (Fig. 3.5-19f), the initial wave front is entirely diffracted S_{diff}, reaching the surface at a distance of 111°. Because waves travel much faster at the base of the mantle than in the upper mantle, S_{diff} at the CMB has gone further, reaching 152°. A set of mid-mantle reflections labeled $S_{220}S$, $S_{400}S$, and $S_{670}S$, which are also visible in the previous panel, appear ahead of SS. These peel off the upgoing S/S_{diff} wave front as it interacts with the discontinuities. Because they are related to SS, they also have the "Y"-shape characteristic of underside-reflected phases. The upgoing parts of the "Y" formed from the upgoing S phase, but the downgoing parts (right side of the "Y") peel off S_{diff} and are better called $S_{diff200}S_{diff}$, $S_{diff400}S_{diff}$, and $S_{diff670}S_{diff}$. The waves with the largest amplitudes, SS and SSS, are arriving at the surface at distances of 76° and 63°.

In Fig. 3.5-19g, 1800 s after the earthquake, $S4$ has begun to be observed at the surface (71°), following SS (97°) and SSS (83°). The next surface reflection, $S5$, is now developing. The $ScS2$ multiple reflection is arriving at the surface 36° from the earthquake. The downgoing part of SS is from S_{diff} reflecting at the surface, so it will arrive at the surface at distances greater than 200° as the phase $S_{diff}S_{diff}$. By now, 30 minutes after the earthquake, seismic energy has spread throughout the mantle. Multiple ScS waves are still reverberating between the surface and the core. At the CMB, the leading S_{diff} wave has wrapped around the antipode and is heading back toward the epicenter.

This simulation illustrates that although the ray paths used to describe body waves in the earth are intuitively appealing and useful, they are simple ways of characterizing a complicated wave field. An earthquake generates an initially spherical wave front whose interaction with various interfaces gives rise to many wave fronts. We use names for the arrivals that the wave fronts cause at the surface, so different parts of the same wave front, or the same part at different times, are given different names. Hence our intuition based on geometric rays can lead us to miss some of the richness that occurs. For example, we tend to view diffraction as an exotic effect different from the direct ray path, but the simulation shows no major change as the direct wave becomes the diffracted wave, although there is a loss of high frequencies. Hence the simulation shows no obvious core shadow zone, because seismic energy reaches the shadow zone by diffraction and multiple reflections. The essential point is that the wave fields are the physical entities, whereas rays are useful approximations whose limitations should be kept in mind.

3.6 Anisotropic earth structure

3.6.1 General considerations

So far in this chapter, we have considered a view of the earth developed from analyses of seismic waves assuming that they propagated through an earth made up of purely isotropic, linearly elastic material (Section 2.3.9). In such material, the stresses are linearly proportional to the strains via Hooke's law

$$\sigma_{ij} = c_{ijkl} e_{kl}, \tag{1}$$

and the 81-term tensor of elastic moduli, c_{ijkl}, reduces to two independent elastic constants, λ and μ. As a result, the material's elastic properties are the same in all directions. Although isotropy is a good first approximation in the earth, it is sometimes important to consider deviations from isotropy, or *anisotropy*. In such cases, Hooke's law applies, but the relation between the stresses and strains involves more than two elastic constants. Although there can be up to 21 independent elastic constants, any material in which more than two are needed is called *anisotropic*.

Having more than two elastic constants means that the material's properties differ depending on the direction. Because seismic wave velocities depend on the elastic constants, waves traveling through anisotropic material travel faster or slower depending on their direction, and complicated wave phenomena can occur. For example, a shear wave can be split into two pulses, each with a different polarity and traveling at a different speed (Figs 3.6-1, 2.4-8).

Anisotropy can result from a material's being non-uniform, a condition called *heterogeneity* or *inhomogeneity*. A common situation is when material has directionality in its structure. For instance, plywood is a superposition of thin layers of wood, so its strength (shear modulus) differs in different directions.

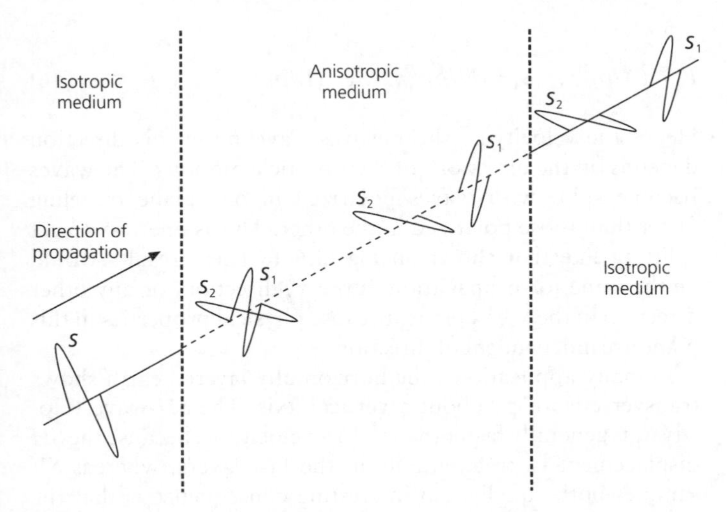

Fig. 3.6-1 Schematic of an initially polarized shear wave split along the fast and slow anisotropic directions, yielding pulses separated in time. The pulses remain split after leaving the anisotropic region.

Similarly, a stack of rock layers with different isotropic velocities can as a whole behave anisotropically, so seismic waves travel with different speeds parallel or perpendicular to the layers. This situation is called shape-preferred orientation (SPO) anisotropy. Anisotropy can also occur for homogeneous materials. For example, the crystal structure of the mineral olivine is homogeneous in that it is composed of the same repeating groups of atoms, but acts anisotropically because its acoustic properties vary in different directions relative to the crystal lattice. This situation is called lattice-preferred orientation (LPO) anisotropy.

The anisotropic variations of the seismic velocity of earth materials are small compared to the large changes in seismic velocity that occur radially from the surface to the core. Hence, in developing radial models of seismic velocity, anisotropy has traditionally been treated as a secondary effect. Nevertheless, recent efforts to better quantify three-dimensional velocity variations sometimes find that anisotropic perturbations are comparable to lateral velocity changes. It is often difficult, however, to distinguish between the effects of anisotropy and those of heterogeneity. For example, curvature on a refracting interface can simulate many of the effects associated with anisotropy.

An important reason to study anisotropy is that material flow at depth appears to preferentially orient olivine crystals within upper mantle rocks. Hence mapping the seismically "fast" direction lets us investigate the relation between plate motions and mantle flow at depth. Although anisotropy studies are ongoing, and both results and interpretations will change over time, they represent a major frontier in deep earth studies.

3.6.2 Transverse isotropy and azimuthal anisotropy

As discussed in Section 2.3.9, the symmetry of the stress and strain tensors and the idea of strain energy means that no more than 21 of the 81 elastic constants c_{ijkl} are independent. We can thus write the c_{ijkl} tensor as a matrix C_{mn}, where the indices m and n vary from 1 to 6 as the pairs of indices (i, j) or (k, l) take values of (1, 1), (2, 2), (3, 3), (2, 3), (1, 3) and (1, 2), respectively:

$$C_{mn} = \begin{pmatrix} c_{1111} & c_{1122} & c_{1133} & c_{1123} & c_{1113} & c_{1112} \\ c_{2211} & c_{2222} & c_{2233} & c_{2223} & c_{2213} & c_{2212} \\ c_{3311} & c_{3322} & c_{3333} & c_{3323} & c_{3313} & c_{3312} \\ c_{2311} & c_{2322} & c_{2333} & c_{2323} & c_{2313} & c_{2312} \\ c_{1311} & c_{1322} & c_{1333} & c_{1323} & c_{1313} & c_{1312} \\ c_{1211} & c_{1222} & c_{1233} & c_{1223} & c_{1213} & c_{1212} \end{pmatrix}$$

$$= \begin{pmatrix} C_{11} & C_{12} & C_{13} & C_{14} & C_{15} & C_{16} \\ C_{21} & C_{22} & C_{23} & C_{24} & C_{25} & C_{26} \\ C_{31} & C_{32} & C_{33} & C_{34} & C_{35} & C_{36} \\ C_{41} & C_{42} & C_{43} & C_{44} & C_{45} & C_{46} \\ C_{51} & C_{52} & C_{53} & C_{54} & C_{55} & C_{56} \\ C_{61} & C_{62} & C_{63} & C_{64} & C_{65} & C_{66} \end{pmatrix}. \tag{2}$$

For an isotropic material, the c_{ijkl} tensor can be written in terms of two independent elastic constants

$$c_{ijkl} = \lambda\delta_{ij}\delta_{kl} + \mu(\delta_{ik}\delta_{jl} + \delta_{il}\delta_{jk}), \tag{3}$$

so its matrix form is

$$C_{mn} = \begin{pmatrix} \lambda + 2\mu & \lambda & \lambda & 0 & 0 & 0 \\ \lambda & \lambda + 2\mu & \lambda & 0 & 0 & 0 \\ \lambda & \lambda & \lambda + 2\mu & 0 & 0 & 0 \\ 0 & 0 & 0 & \mu & 0 & 0 \\ 0 & 0 & 0 & 0 & \mu & 0 \\ 0 & 0 & 0 & 0 & 0 & \mu \end{pmatrix}. \tag{4}$$

However, the crystal structures of many earth materials require additional independent elastic coefficients. For example, ice, quartz, olivine, or plagioclase feldspar require 5, 6, 9, and 21 constants, respectively. In such cases, the matrix is more complicated.

One of the most important forms of anisotropy, known as *transverse isotropy* (also known as radial anisotropy, axisymmetry, and cylindrical symmetry), occurs for a stack of layered materials. Each layer is isotropic in its properties, but these properties differ between layers (as in plywood). Thus the elastic properties, and hence seismic velocities, of the stack as a whole are identical regardless of the amount of rotation about the axis of symmetry, which is perpendicular to the layers. However, these aggregate properties differ in the perpendicular directions.

A transversely isotropic material can be characterized by five independent elastic coefficients, A, C, F, L, N, that represent its aggregate properties. If the axis of symmetry is x_3, so properties in that direction differ from those in the x_1–x_2 plane, the elastic constant matrix (Eqn 4) becomes

$$C_{mn} = \begin{pmatrix} A & A - 2N & F & 0 & 0 & 0 \\ A - 2N & A & F & 0 & 0 & 0 \\ F & F & C & 0 & 0 & 0 \\ 0 & 0 & 0 & L & 0 & 0 \\ 0 & 0 & 0 & 0 & L & 0 \\ 0 & 0 & 0 & 0 & 0 & N \end{pmatrix}. \tag{5}$$

Comparisons with matrices 2 and 4 show that terms that were the same for an isotropic material (consider C_{11} and C_{33}, or C_{55} and C_{66}) now differ, because terms involving the x_3 direction differ from those in the x_1 or x_2 directions.

This matrix gives the velocities of waves propagating in different directions. First, consider waves propagating in the x_1 direction (Fig. 3.6-2, *top*). By analogy to the isotropic case, A corresponds to $\lambda + 2\mu$ for the x_1 direction, N corresponds to μ for the x_2 direction, and L corresponds to μ for the x_3 direction. Thus the P velocity and the two orthogonal S velocities are

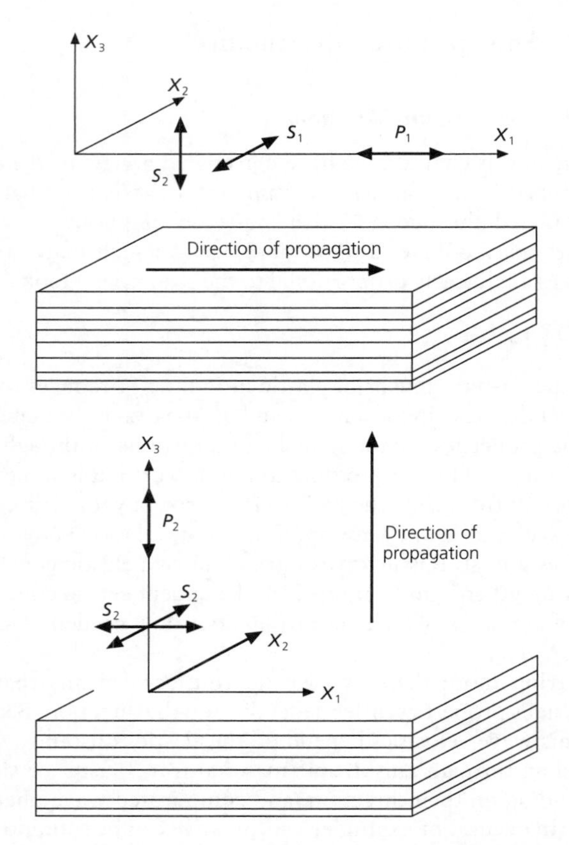

Fig. 3.6-2 Cartoon showing the effects of transverse isotropy due to layering. *Top*: Directions of oscillations for P and S waves propagating in the x_1 direction, in the plane of layering. The shear wave oscillating in the plane of the layering has velocity S_1, which is generally faster than that for the shear wave oscillating across the layers, S_2. *Bottom*: Directions of oscillations for P and S waves propagating in the x_3 direction, perpendicular to the layering. The compressional wave velocity, P_2, is generally less than P_1. Both shear waves have the same velocity, S_1.

$$P_1 = (A/\rho)^{1/2}, \quad S_1 = (N/\rho)^{1/2}, \quad S_2 = (L/\rho)^{1/2}. \tag{6}$$

Hence the velocity of shear waves traveling in this direction depends on the directions of their particle motions. The waves become split, with waves polarized in one plane traveling faster than those polarized in the other. This is one way to get splitting like that shown in Fig. 3.6-1. These results would be the same for propagation in the x_2 direction, or any other direction in the x_1–x_2 plane, because physical properties in this plane are independent of direction.

In many applications, the horizontally layered earth shows transverse isotropy about a vertical axis. The *SH*-wave velocity S_1 is generally faster than the *SV* velocity S_2, because the *SH* displacement is preferentially in the fast layers, whereas *SV* samples both equally. An interesting consequence is that the shear velocity inferred from the dispersion of Love waves, which are *SH* waves, would be higher than that from Rayleigh waves, which involve *SV*.

By contrast, for *P* and *S* waves propagating in the x_3 (axis of symmetry) direction (Fig. 3.6-2, *bottom*), both *S* velocities equal S_2 in Eqn 6. The *P* velocity reflects the fact that *C* corresponds to $\lambda + 2\mu$ for the x_3 direction, so

$$P_2 = (C/\rho)^{1/2}. \tag{7}$$

For layered materials, typically $P_1 > P_2$, so *P* waves propagate faster in the x_1 direction than in the x_3 direction. This is because the wave travels preferentially in the fast layers in the x_1 direction, whereas a *P* wave traveling in the x_3 direction must also traverse the slow layers.

Transverse isotropy is often characterized by three parameters:

$$\xi = N/L = (S_1/S_2)^2, \quad \phi = C/A = (P_2/P_1)^2, \quad \eta = F/(A - 2L). \tag{8}$$

If the material were isotropic, $\xi = \phi = \eta = 1$. For layered structures, generally $\xi > 1$ and $\phi < 1$.

A second common type of anisotropy is *azimuthal anisotropy*, in which velocities vary as a function of horizontal direction. One way to obtain this is to have transverse isotropy with the x_3 axis turned to horizontal, which is analogous to standing plywood vertically. In general, the *P*-wave velocity varies with azimuth as

$$P(\theta) = A_1 + A_2 \cos 2\theta + A_3 \sin 2\theta + A_4 \cos 4\theta + A_5 \sin 4\theta, \tag{9}$$

where the constants A_i depend on the 21 elastic constants.

3.6.3 *Anisotropy of minerals and rocks*

An important source of seismic anisotropy is minerals that are anisotropic due to their crystal structure. At microscopic levels the anisotropy can be enormous, with velocities along different mineralogical axes varying by more than 100%. Generally, however, the anisotropic mineral grains are randomly oriented, so seismic waves have wavelengths long enough to average out the anisotropic effects, leaving only weak anisotropy. However, in some cases the mineral grains are aligned, causing significant anisotropy.

Laboratory studies of the elastic moduli of minerals give insight into such LPO anisotropy. Some studies involve static methods like twisting or squeezing samples, but most use the vibrational properties of mineral samples as small as 1 mm. At very high pressures, a technique called Brillouin scattering, which measures how laser light passing through the mineral is distorted, yields elastic constants for samples smaller than 0.1 mm.

One of the most important anisotropic minerals is olivine (Fig. 3.6-3), which comprises much of the upper mantle (Section 3.8). For waves propagating in the fastest direction, the *P*-wave velocity is 9.89 km/sec and the *S* velocities are 4.89 km/s and 4.87 km/s. By contrast, the slowest *P* velocity in

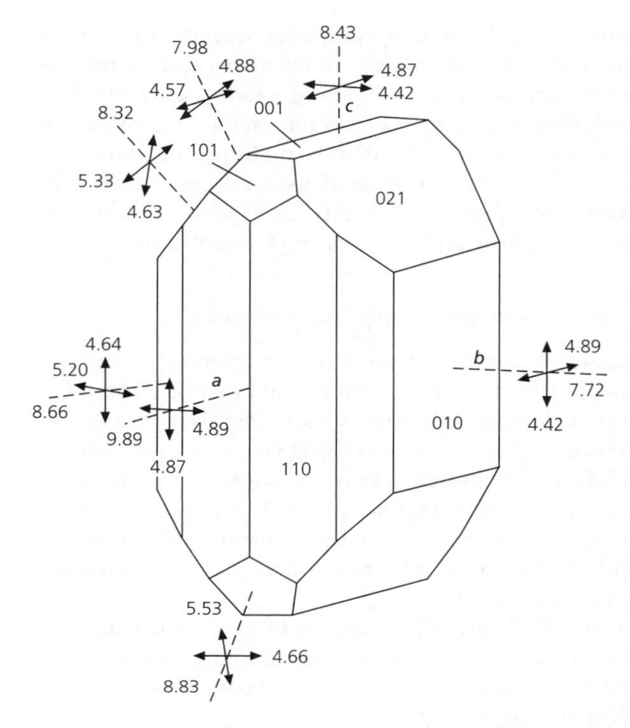

Fig. 3.6-3 *P* and *S* velocities (km/s) in different directions relative to the crystal structure of olivine. *P* velocities are in the directions of the dashed lines, and the *S* velocities are shown by the adjacent pairs of perpendicular lines. The *a* axis, corresponding to the [100] crystal face, is the fastest direction through the crystal. It is also the dominant slip direction, so olivine crystals align in the direction of plastic flow. (Babuska and Cara, 1991. With kind permission from Kluwer Academic Publishers.)

this example is 7.72 km/s. The magnitude of anisotropy is characterized by

$$k = (v_{max} - v_{min})/v_{mean}. \tag{10}$$

For *P*-waves in the olivine crystal, $\alpha_{max} = 9.89$ km/s, $\alpha_{min} = 7.72$ km/s, and $\alpha_{mean} = 8.81$ km/s, so $k = 25\%$. The maximum and minimum *S* velocities are 5.53 km/s and 4.42 km/s, so $k = 22\%$. Although for olivine the anisotropy of *P* and *S* waves is similar, they can differ greatly for other minerals.

Other important minerals range from nearly isotropic to extremely anisotropic. One of the most isotropic minerals is garnet, where *k* for both *P* and *S* waves is $\leq 1\%$. At the other extreme, sheet silicates like mica can have values of *k* up to 60% for *P* waves and 116% for *S* waves.

As a result, a major factor controlling a rock's anisotropy is the anisotropy of the minerals composing it and their relative proportions. Another important factor is the presence of deviatoric stresses, which can cause a preferred orientation of anisotropic mineral grains that might otherwise be randomly distributed. Crystals are generally oriented with their smallest widths in the direction of maximum compression. For example, in micas, which are important components of highly foliated schists, the flat crystals are oriented parallel to the plane of least compression. Thus slip occurs more easily parallel to the developing foliation, because the planar mica faces contain the weakest bonds. Shear in a preferred direction can also recrystallize different mineral assemblages, so the resulting anisotropy reflects a combination of the preferred orientation of anisotropic materials and the presence of laminar structures.

3.6.4 *Anisotropy of composite structures*

Anisotropy can also result from an asymmetric combination of materials. The upper continental crust often contains horizontally layered sedimentary rocks. Similarly, oceanic crust is comprised of sediments overlying layers of basalt and gabbro. Such layering can yield transverse isotropy, with the symmetry axis oriented vertically. On a regional scale, plate collisions often cause significant metamorphism, sometimes yielding transverse isotropy due to the preferred orientation of the foliation of gneisses and schists.

Fluid-filled cracks, for example in a volcanic region, can also cause anisotropy. For a material containing two-dimensional fluid-filled cracks whose normals are parallel to the x_1 axis, the anisotropy is given by

$$C_{ij} = \begin{pmatrix} \lambda+2\mu & \lambda & \lambda & 0 & 0 & 0 \\ \lambda & \lambda+2\mu & \lambda & 0 & 0 & 0 \\ \lambda & \lambda & \lambda+2\mu & 0 & 0 & 0 \\ 0 & 0 & 0 & \mu & 0 & 0 \\ 0 & 0 & 0 & 0 & \mu(1-\varepsilon) & 0 \\ 0 & 0 & 0 & 0 & 0 & \mu(1-\varepsilon) \end{pmatrix}, \quad (11)$$

where ε is the crack density given by $\varepsilon = Na^3/V$, N is the number of cracks in the volume V, and a is the half-width of a crack. If the cracks become infinitely small, $\varepsilon = 0$, yielding the isotropic case (Eqn 4). In general, the anisotropy depends on the geometry of the inclusions and their contrast in properties with the surrounding matrix. For computational ease, rods (prolate spheroids) and disks (oblate spheroids) are often assumed in seismic modeling.

3.6.5 *Anisotropy in the lithosphere and the asthenosphere*

Anisotropy in the lithosphere takes many forms, including that in glaciers whose flow aligns the ice crystals. Closer to our applications, several effects generate anisotropy in the oceanic crust. Horizontal sediment layers can create transverse isotropy of up to 15% with a vertical symmetry axis. In the upper crustal layer of vertical-sheeted basaltic dikes, azimuthal anisotropy is thought to exist with a horizontal axis perpendicular to the dikes and thus in the spreading direction.

Sub-crustal oceanic lithosphere shows strong azimuthal anisotropy. The flow processes associated with plate spreading

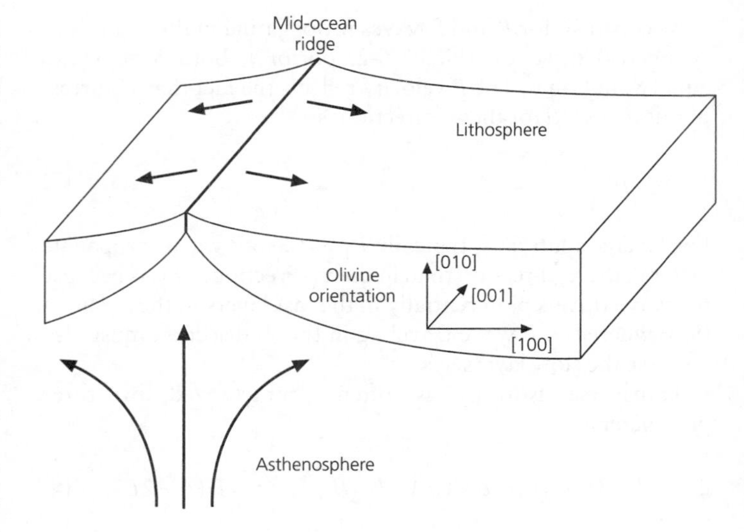

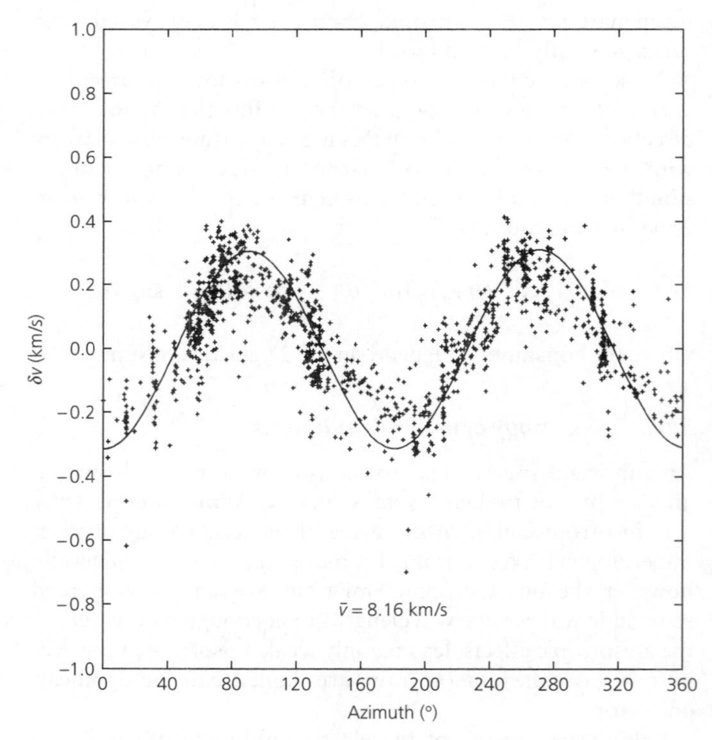

Fig. 3.6-4 *Top*: Illustration of how the spreading process yields a preferred orientation of olivine crystals in the oceanic lithosphere, with the fast axis of velocity ([100]) in the spreading direction. *Bottom*: Variations in P_n wave velocities near Hawaii. The azimuth is measured relative to the trend of the isochrons (90° from the spreading direction), so the maxima at 90° and 270° show that the fast direction of the azimuthal anisotropy is in the direction of spreading when the plate formed. (Morris *et al.*, 1969. *J. Geophys. Res.*, *74*, 4300–16, copyright by the American Geophysical Union.)

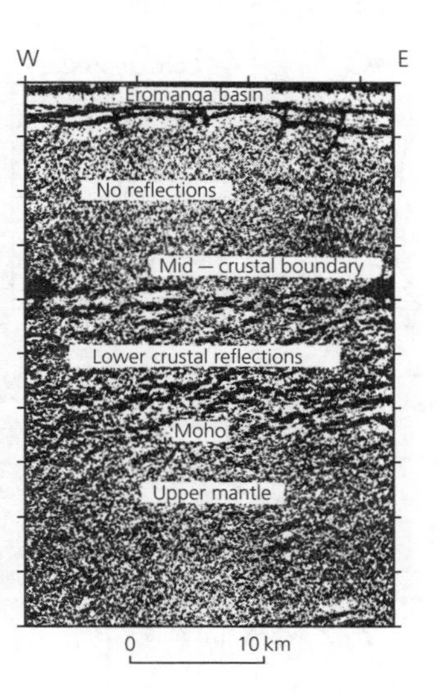

Fig. 3.6-5 *Left*: Seismic reflection profile of the crust and upper mantle in eastern Australia. The lower crust has multiple, discontinuous, and sub-horizontal reflectors possibly due to strain-induced fabrics, igneous layering, or free fluids. This structure yields vertical-axis transverse isotropy. (Finlayson *et al.*, 1989. *Properties and Processes of Earth's Lower Crust*, 1–16, by permission of Australian Geological Survey Organisation.) *Right*: Schematic cross-section of the crust in the northern Ruby Mountains of the North American Basin and Range. There is a strong tendency toward horizontally layered features, although the likely origins of such fabrics vary with depth. (Smithson, 1989. *Properties and Processes of Earth's Lower Crust*, 53–63, permission as above.)

(Fig. 3.6-4, *top*) appear to orient olivine crystals preferentially in the spreading direction, along their [100] slip axes.[1] Because P waves propagate fastest in this direction (Fig. 3.6-3), P_n head waves that sample the uppermost mantle just below the Moho (Section 3.2.1) show a strong azimuthal velocity dependence (Fig. 3.6-4, *bottom*). This variation is approximately described by the cos 2θ term in Eqn 9, where θ is measured from the spreading direction, so the velocity is highest in the spreading direction or 180° from it. This anisotropy is "frozen in" as the lithosphere ages, and so records the spreading direction.

Because continental crust is more complicated than oceanic crust, so is its anisotropy. A primary source of anisotropy in the upper crust is the presence of cracks, often fluid-filled. Such cracks often have a near-vertical orientation induced by regional stress fields parallel to the cracks. When these cracks occur in horizontal sediments that would by themselves have vertical-axis transverse isotropy, the combined result can be orthorhombic symmetry. The lower continental crust tends to have strong sub-horizontal layering, perhaps resulting from ductile deformation, which causes seismic anisotropy. Figure 3.6-5 shows such layering in a seismic reflection profile and a schematic diagram.

Anisotropy within and beneath continental lithosphere is often studied with a technique called *shear wave splitting*. When *SKS* waves convert from P waves in the outer core to S waves in the lower mantle, they are entirely polarized in the radial (*SV*) direction, because all the initial *SH* energy was reflected when the downgoing S wave encountered the core–mantle boundary. As these shear waves travel across the mantle and crust, however, they can be *split* when traveling through anisotropic media (Fig. 3.6-6). Assuming transverse isotropy with a horizontal axis of symmetry, the two polarized waves travel at different

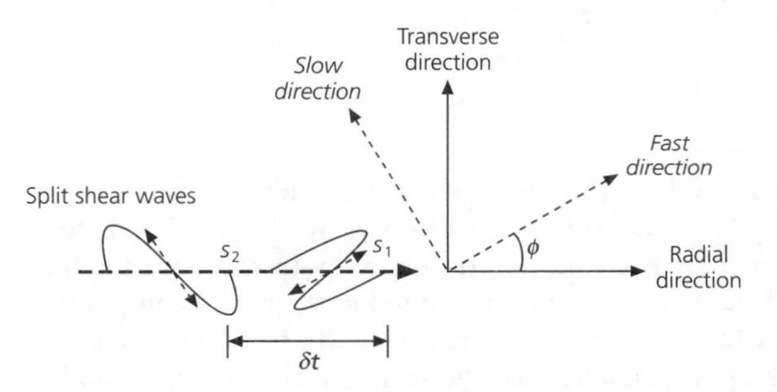

Fig 3.6-6 Splitting of an incoming shear wave into pulses oriented along the fast (s_1) and slow (s_2) directions of anisotropy. The polarization angle ϕ gives the rotation of the fast axis relative to the radial propagation direction, and δt is the time difference between the split pulses.

speeds and arrive at different times. Thus, if the *SKS* signal on the radial component in an isotropic earth is $s(t)$, its projection into the fast and slow polarizations is, respectively,

$$s_1(t) = s(t) \cos \phi, \quad -s_2(t) = s(t - \delta t) \sin \phi, \tag{12}$$

where ϕ is the polarization angle between the radial direction and the fast axis, and δt is the delay time between the fast and slow polarizations. We would normally not expect any *SKS* on the transverse component, but anisotropy yields a combination of both the fast and the slow polarizations on both the radial and the transverse components, given by

$$R(t) = s(t) \cos^2 \phi + s(t - \delta t) \sin^2 \phi,$$
$$T(t) = [(s(t) - s(t - \delta t))/2] \sin 2\phi. \tag{13}$$

For example, in Fig. 3.6-7a (*top*), *SKS* appears on the transverse component. The two components are rotated to yield the fast and slow polarizations, $s_1(t)$ and $s_2(t)$ (Fig. 3.6-7a, *middle*). The time shift δt is then applied, and the signals are rotated

[1] This representation of crystallographic axes is discussed in mineralogy texts like Klein and Hurlbut (1985).

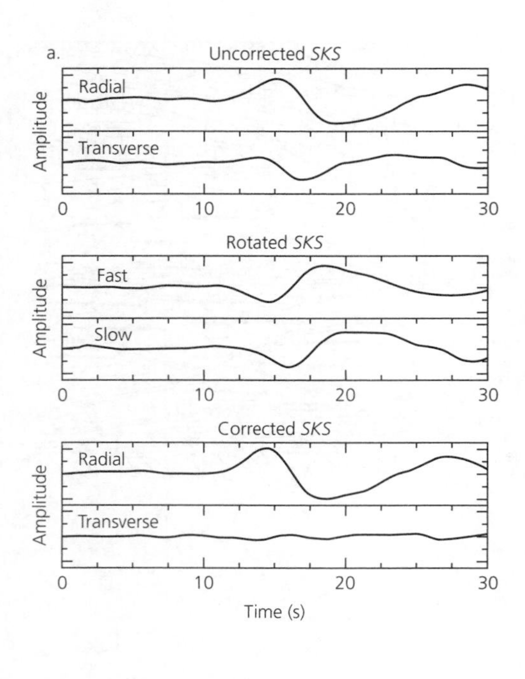

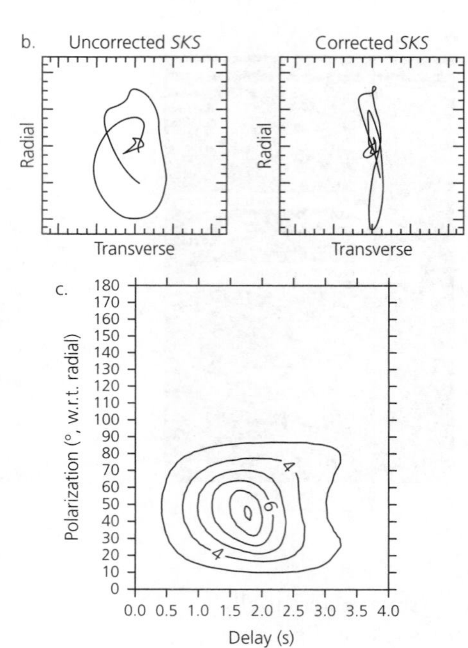

Fig. 3.6-7 Shear wave splitting of *SKS* waves for a Kuril Islands earthquake, stacked across an array of seismometers in New Zealand. a: *SKS* waveforms before and after processing. *Top*: radial and transverse components before processing. Note the large *SKS* signal on the transverse component, which should not be there for an isotropic earth. *Middle*: *SKS* waveforms after rotation into the fast and slow polarizations. *Bottom*: *SKS* waveform after the splitting has been removed so that all *SKS* is on the radial component. b: Particle motion plots (Section 2.4) of *SKS* on the radical and transverse components before and after removal of the transverse signal. c: Contour plot of the amplitude in the radial component as a function of the delay time and polarization angle. The minimum corresponds to the best-fitting value. (Gledhill and Gubbins, 1996. *Phys. Earth Planet. Inter.*, *95*, 227–36, with permission from Elsevier Science.)

again so that all of the signal appears on the radial component (Fig. 3.6-7a, *bottom*). As shown in Fig. 3.6-7b, before correction, particle motion occurs on both components, but after correction, the motion is limited to the radial component. The fact that this technique removes the transverse signal shows the appropriateness of the transversely isotropic model. The values of ϕ and δt are found by minimizing the transverse signal, as shown by the contour plot in Fig. 3.6-7c. Typical values for the magnitude of shear wave splitting, δt, are in the 0–2 s range.

Seismic anisotropy within continents is thought to reflect crystal alignment created during a tectonic episode and then "frozen in." The anisotropy is a result of the last episode of tectonism, which resets any previous anisotropy. Because continental rock can be as old as 4 Ga (the mean age is about 1.5 Ga), anisotropy in continental lithosphere can reveal information about very old tectonic events such as episodes of mountain building. For plate collisions the fast axis is usually sub-perpendicular to the principal stress axis, or parallel to the resulting orogenic belts. There may also be deeper anisotropy due to oriented olivine in the flowing asthenosphere. However, it is sometimes difficult to distinguish this effect from lithospheric anisotropy. For instance, in eastern North America the fast axis is oriented WSW–ENE, parallel to the direction of both absolute plate motion (Section 5.2.4) (and thus presumably asthenospheric flow) and major orogenic boundaries like the Appalachian Mountains (Fig. 3.6-8).

Surface wave observations indicate that anisotropy extends to a depth of about 300 km beneath oceans. The *S*-wave velocity inferred from Love waves, which are *SH* waves, is higher than inferred from Rayleigh waves, which involve *SV*. Figure 3.6-9 shows the squared *S*-wave velocity ratio ξ (Eqn 8) versus depth for several ages of oceanic lithosphere. The deviation of ξ from 1 reflects transverse isotropy with *SH* velocities faster than *SV* velocities. Because the oceanic lithosphere extends to a depth of about 100–125 km, anisotropy seems to extend into the asthenosphere.

In addition, Rayleigh wave velocities show azimuthal anisotropy similar to that found for P_n waves that sample the uppermost mantle at much shallower depths. Both types of anisotropy may reflect mantle flow (Fig. 3.6-4). The flow-induced preferred orientation of olivine would give azimuthal anisotropy in the spreading direction. Taking paths in different directions averages out the azimuthal effect, leaving a net transverse isotropy that is symmetric about the vertical. An interesting consequence of this model is that near the ridges, where mantle material is upwelling, transverse isotropy should be less significant, as the data show. At older ages, mantle flow will be more horizontal, increasing transverse isotropy.

3.6.6 *Anisotropy in the mantle and the core*

Although most of the mantle shows little or no anisotropy, this is not so for the D″ region at the base of the mantle, where complex interactions with the liquid outer core may occur (Section 3.5.4). Studying anisotropy in a narrow layer nearly 3000 km below the heterogeneous mantle and crust is challenging, but initial investigations suggest anisotropy on the order of several percent, comparable to the isotropic velocity variations. D″ anisotropy seems to fall in to two categories. Beneath regions of paleo-subduction, such as western Central America and the northern Pacific rim, *SH* waves in the form of *S*, *ScS*, or S_{diff} travel faster than their *SV* counterparts (Fig. 3.6-10). This behavior has been modeled as transverse isotropy. However,

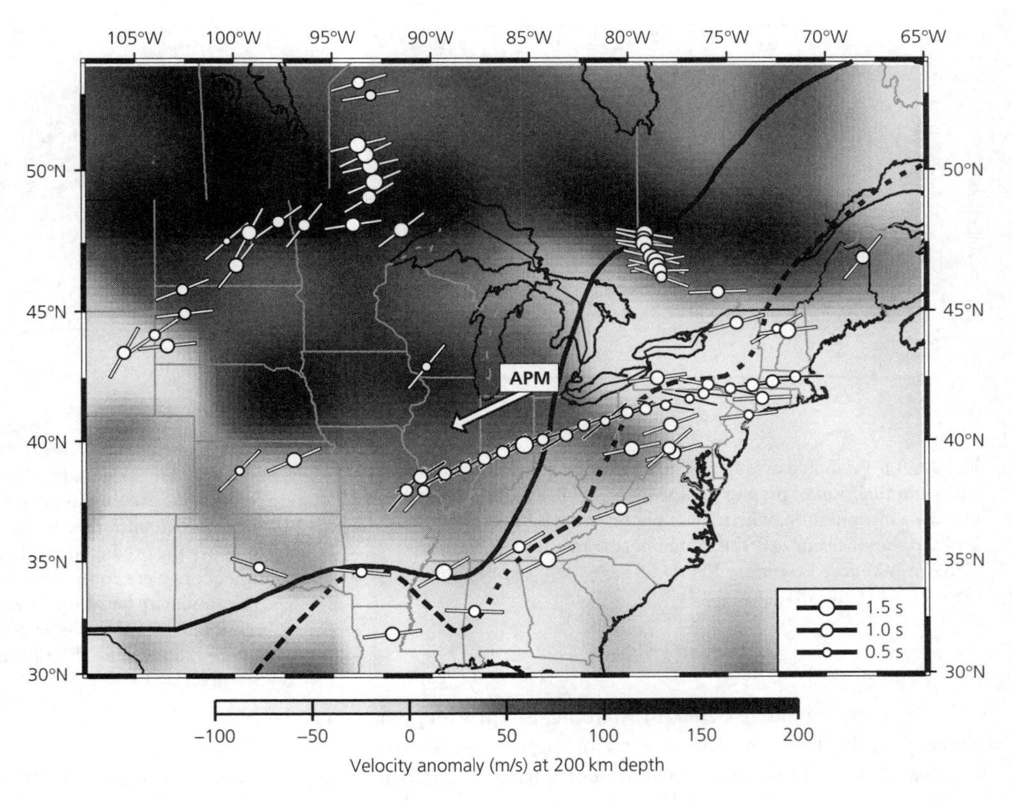

Fig. 3.6-8 Map of the eastern USA showing shear wave splitting results from *SKS* and *SKKS*. Lines point in the direction of the fast axis, assuming horizontally oriented transverse isotropy, and the sizes of the circles represent the magnitude of the splitting in seconds. The background is a map of the shear wave velocity anomalies at 200 km depth (van der Lee and Nolet, 1997. *J. Geophys. Res.*, *102*, 22, 815–38, copyright by the American Geophysical Union.) The splitting direction is approximately parallel to the Appalachian orogenic belts (dashed line) and aligned with the absolute plate motion (APM). Note the regional variations for different locations. (Fouch *et al.*, 2000. *J. Geophys. Res.*, *105*, 6255–76, copyright by the American Geophysical Union.)

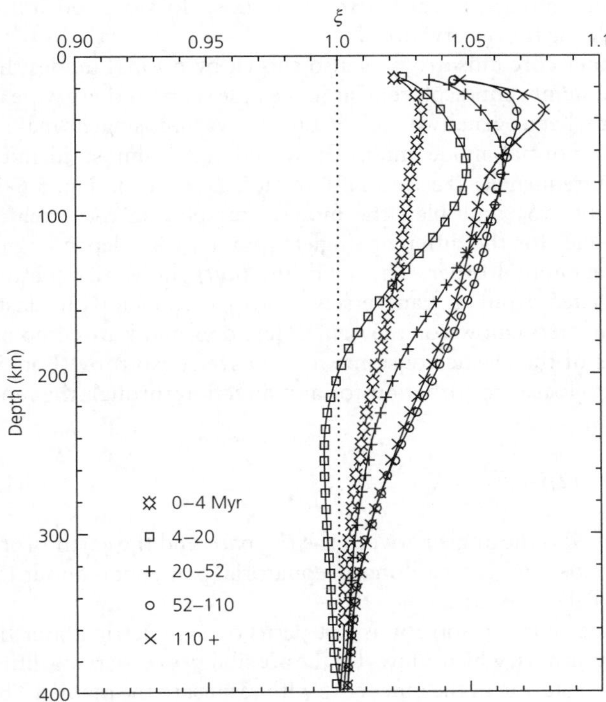

Fig. 3.6-9 Depth variations of ξ, the square of the V_{SH}/V_{SV} ratio, beneath the Pacific Ocean. ξ tends to exceed 1, meaning that *SH* is faster than *SV*, consistent with olivine in both the lithosphere and the asthenosphere being preferentially oriented by the spreading process. (Nishimura and Forsyth, 1989.)

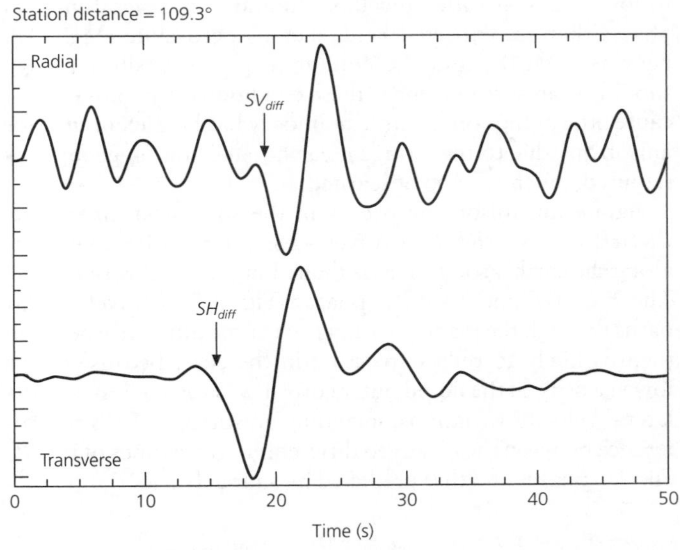

Fig. 3.6-10 Evidence for anisotropy at the base of the mantle, shown by diffracted arrivals for a South American earthquake recorded at Canadian station DAWY. Arrows show estimates of the onset times. Diffracted *SH* arrives before *SV*, suggesting transverse isotropy. (Kendall and Silver, 1996. Reproduced with permission from *Nature*.)

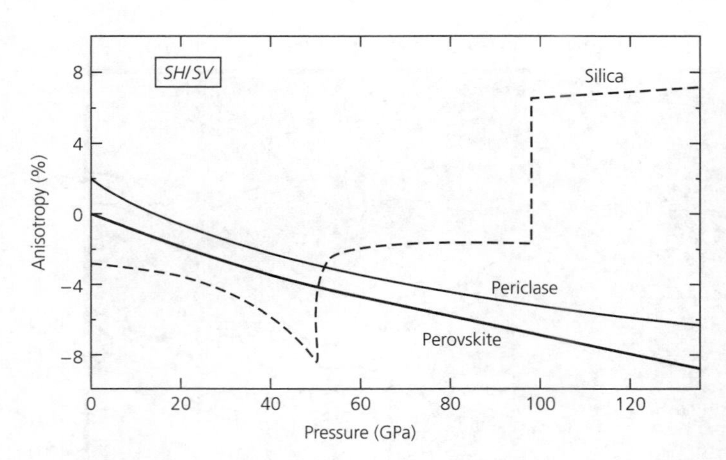

Fig. 3.6-11 Predicted anisotropic behavior for perovskite, periclase, and silica as a function of pressure in the mantle. The far right corresponds to the lowermost mantle, where these phases are major components. The kinks in the silica curve result from phase transitions. (Stixrude, 1998. *The Core–Mantle Boundary Region*, 83–96, copyright by the American Geophysical Union.)

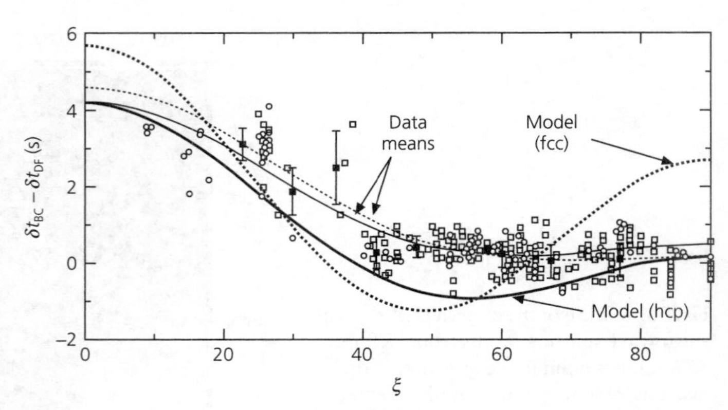

Fig. 3.6-12 *PKP-BC* – *PKP-DF* travel time residuals as a function of ξ, the angle between the *PKP-DF* ray and the earth's spin axis. Circles and thin solid line are for data from Song and Helmberger (1993); squares and thin dashed line are for data from Creager (1992). The thin solid and dashed lines are the smoothed fits to the residuals. The heavy solid and dashed lines are the predicted residuals for the transverse isotropy expected if the inner core were composed of iron in either the hcp and the fcc structures. The similarity between the hcp curve and the data support hcp as the crystal phase for the inner core. (Stixrude and Cohen, 1995. *Science*, *267*, 1972–5, copyright 1995 American Association for the Advancement of Science.)

D″ anisotropy beneath the mid-Pacific is variable, with *SH* waves usually but not always arriving before the accompanying *SV* waves. This effect may reflect vertical structures due to lower-most mantle upwelling. In addition, several mineral phases that are expected here, such as perovskite ($MgSiO_3$), periclase (MgO), and the columbite phase of silica (SiO_2), should be anisotropic under these conditions (Fig. 3.6-11). Because little of the core–mantle boundary has been examined for anisotropy due to the stringent earthquake-station geometries required, much is yet to be learned.

Significant anisotropy occurs in the solid iron inner core. *PKIKP* waves (*PKP-DF*) travel ~3 s faster in the inner core along the earth's rotation axis than along the equatorial plane. The *PKP-DF* and *PKP-BC* phases (Fig. 3.5-7) travel similar paths through the mantle, so any travel time difference between them is likely to reflect structure in the core. Because of the low viscosity of the liquid outer core, flow should eliminate any lateral velocity variations, including anisotropy. Thus the difference between the observed differential travel times of the *BC* and *DF* phases and that predicted by a model

$$\delta t_{BC} - \delta t_{DF} = (t_{BC} - t_{DF})_{\text{observed}} - (t_{BC} - t_{DF})_{\text{predicted}}, \tag{14}$$

is likely to be a function of inner core structure along the *DF* path.

Figure 3.6-12 shows *BC-DF* residuals versus ξ, the angle between the *PKP-DF* ray segment in the inner core and the earth's spin axis. Small values of ξ correspond to paths parallel to the spin axis, and the corresponding large residuals indicate that near-axial *PKP-DF* waves travel faster and arrive sooner. Also shown are theoretical predictions for the anisotropic behavior of solid iron in the hexagonal close-packed (hcp) and face-centered-cubic (fcc) structures. The hcp structure of iron, aligned along the earth's rotation axis, does a good job of modeling the observations.

Inner core anisotropy is also shown by normal modes that have significant displacement in the inner core. If there were no lateral heterogeneity or anisotropy, the various singlets making up a normal mode multiplet would have almost identical eigenfrequencies (Section 2.9). In fact, as shown in Fig. 3.6-13 for the ${}_{18}S_4$ multiplet, the modes are split, so the eigenfrequencies for the different singlets (points) vary depending on the azimuthal order. The solid line (*left*) shows the splitting predicted from a transversely isotropic model with elastic parameters (shown on the *right*). Here α, β, and γ are combinations of the elastic constants for transverse isotropy (Eqn 5). The velocity perturbation for any direction through the inner core is

$$\delta v/v = (2\beta - \gamma)\cos^2 \xi, \tag{15}$$

where ξ is the angle between the ray path and the earth's rotation axis. $\delta v/v$ is zero along an equatorial path, but is about 1% parallel to the axis.

Inner core anisotropy is not perfectly symmetric about the rotation axis, which allows for the possibility of observing differential rotation of the inner core with respect to the mantle. This phenomenon has been reported, seen as temporal variations of the *BC-DF* residuals (Eqn 14) for similar earthquake-station geometries. Quantification of such differential rotation and its implications for the generation of the magnetic field in the convecting outer core are active research areas.

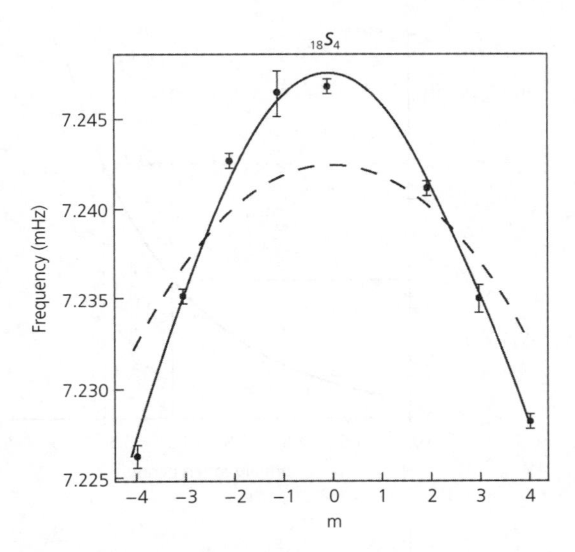

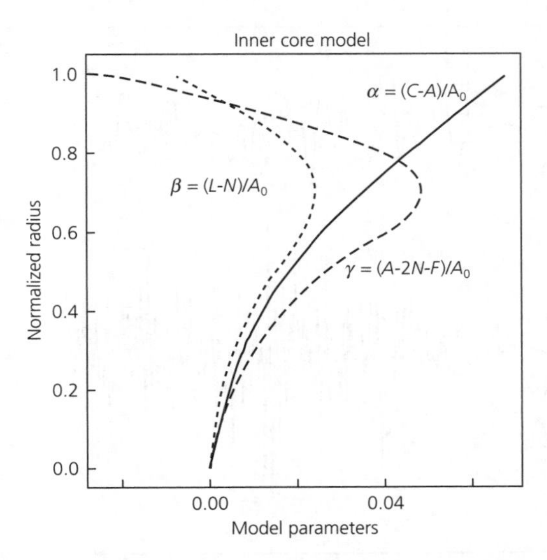

Fig. 3.6-13 Evidence for transverse isotropy in the inner core from the splitting of a spheroidal normal mode multiplet. Points represent the observed frequencies for the different azimuthal orders (*left*). The dashed curve is the prediction for the inner core including the effects of the earth's rotation and ellipticity but not anisotropy. The solid curve incorporates the anelistic model at the *right* by combinations of the elastic constants for transverse isotropy. In this formulation, α, β, and γ differ from their usual seismological definitions. (Tromp, 1993. Reproduced with permission from *Nature*.)

3.7 Attenuation and anelasticity

3.7.1 Wave attenuation

In the last section, we extended our view of the earth as an isotropic elastic medium to include the effects of anisotropy. We now consider *anelasticity*, or deviation from elasticity, which is one of the reasons why seismic waves *attenuate* or decrease in amplitude as they propagate. We have already discussed how the reflection and transmission of seismic waves at discrete interfaces reduce their amplitudes. Here, we consider four other processes that can reduce wave amplitudes: *geometric spreading*, *scattering*, *multipathing*, and anelasticity. The first three are elastic processes, in which the energy in the propagating wave field is conserved. By contrast, anelasticity, sometimes called *intrinsic attenuation*, involves conversion of seismic energy to heat.

As in many seismological applications, it is worth first considering familiar analogous behaviors for light. As you move away from a street lamp at night, the light appears dimmer for several reasons. The first is geometric spreading: light moves outward from the lamp in expanding spherical wave fronts (Section 2.4.3). By the conservation of energy, the energy in a unit area of the growing wave front decreases as r^{-2}, where r is the radius of the sphere or distance from the lamp.

Second, the light dims as it is scattered by air molecules, dust, and water in the air. As we have discussed, scattering results when objects acting as Huygens' sources scatter energy in all directions. This effect is dramatic on a foggy night because the scattered light causes a halo around the lamp.

Third, the light is focused or defocused by changes in the refractive properties of the air.[1] This effect is termed multipathing in seismology. Focusing and defocusing can be illustrated by looking at the street light through binoculars. Looking through binoculars the usual way, the waves are focused by the lenses, and the lamp appears closer and brighter. Reversing the binoculars makes the lamp appear further away and dimmer.

Fourth, some of the light energy is absorbed by the air and converted to heat. This process differs from the other three in that light energy is actually lost, not just moved onto a different path.

All four processes are important for seismic waves. The first three are described by elastic wave theory, and can increase or decrease an arrival's amplitude by shifting energy within the wave field. By contrast, anelasticity reduces wave amplitudes only because energy is lost from the elastic waves. So much of seismology is built upon the approximation that the earth responds elastically during seismic propagation that it is easy to forget that the earth is not perfectly elastic. However, without anelasticity, seismic waves from every earthquake that ever occurred would still be reverberating until the accumulating reverberations shattered the earth. Elasticity is a good approximation for the earth's response to seismic waves, but

[1] This process causes mirages, where light is refracted differently by hot air just above the ground. Similarly the distorted appearance of the setting sun results from seeing different parts of it through different levels of the atmosphere which refract light differently because of the vertical density gradient.

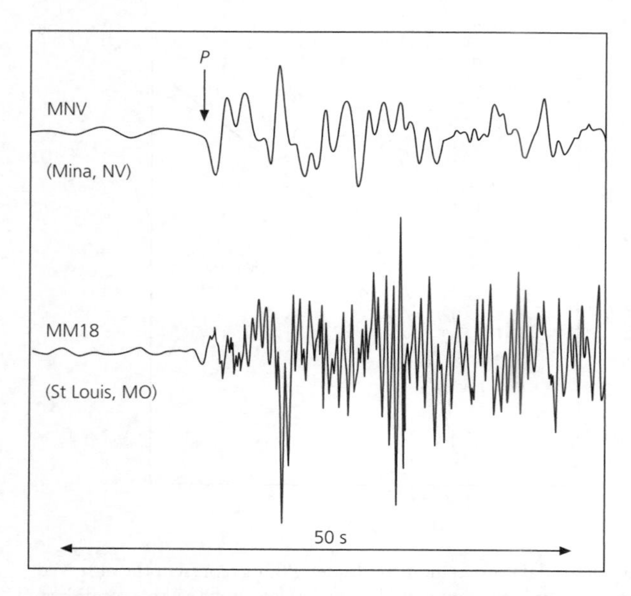

Fig. 3.7-1 Regional variations in attenuation seen in seismograms from an April 14, 1995, earthquake in Texas recorded in Nevada (MNV, $\Delta = 15°$) and Missouri (MM18, $\Delta = 14°$). The MNV record has less high frequency energy because the tectonically active western USA is more attenuating than the stable mid-continent.

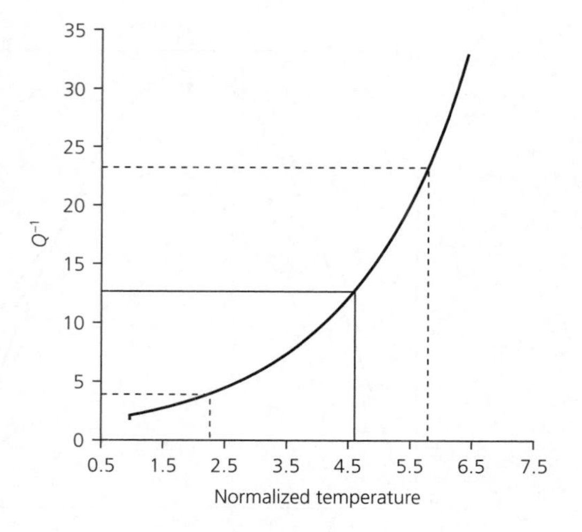

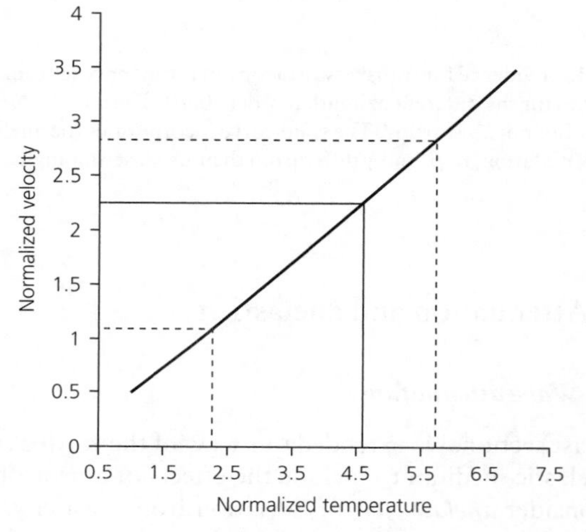

Fig. 3.7-2 Schematic representation of the variations of seismic attenuation (*top*) and normalized velocity (*bottom*) as a function of normalized temperature changes. Attenuation is more sensitive to increased temperature. (Romanowicz, 1995. *J. Geophys. Res., 100*, 12,375–94, copyright by the American Geophysical Union.)

there are many important implications and applications of anelasticity.

Anelasticity results because the kinetic energy of elastic wave motion is lost to heat by permanent deformation of the medium. The large-scale, or macroscopic, term for this process is *internal friction*. Among the smaller-scale, or microscopic, mechanisms that may cause this dissipation are stress-induced migration of defects in minerals, frictional sliding on crystal grain boundaries, vibration of dislocations, and the flow of hydrous fluids or magma through grain boundaries. Theoretical and experimental work is being carried out to examine possible mechanisms of seismic attenuation.

The study of anelasticity has lagged behind that of the elastic wave velocities because of the complexities involved in measuring attenuation and understanding its physical causes. Although measuring seismic wave amplitudes is straightforward, they depend on both the source, which is not perfectly known, and all the elastic and anelastic effects anywhere along the paths that the seismic energy traveled between the source and the receiver. Hence it can be hard to distinguish the effects of anelasticity from elastic processes.

This inherent uncertainty is somewhat compensated by the fact that variations in anelasticity are large, as illustrated by comparison of records of an earthquake in Texas at stations in Nevada and Missouri (Fig. 3.7-1). The Nevada seismogram has much less high-frequency energy, showing that the crust in the western USA is much more attenuating than that in the Midwest. By comparison, seismic velocity variations between these areas are generally less than ±10%. Even so, because of the difficulties in measuring attenuation, variations in attenuation at both regional and global scales are much less resolved than similar variations in velocity.

Attenuation is valuable for studying temperature variations within the earth. Many important geophysical processes (mantle convection, plate tectonics, magmatism, etc.) involve lateral variations in temperature. Elastic velocities are also sensitive to temperature, but are better for mapping cold (fast) anomalies like subducting slabs than hot (slow) material like that at midocean ridges (Section 2.5.10). As shown in Fig. 3.7-2, seismic velocities depend nearly linearly upon temperature, whereas attenuation depends exponentially on temperature. Thus combining velocity and attenuation studies can provide valuable information. Figure 3.7-3 shows the velocity and attenuation structure at a portion of the East Pacific rise axis, where a low-velocity, high-attenuation region is interpreted as a melt-filled magma chamber.

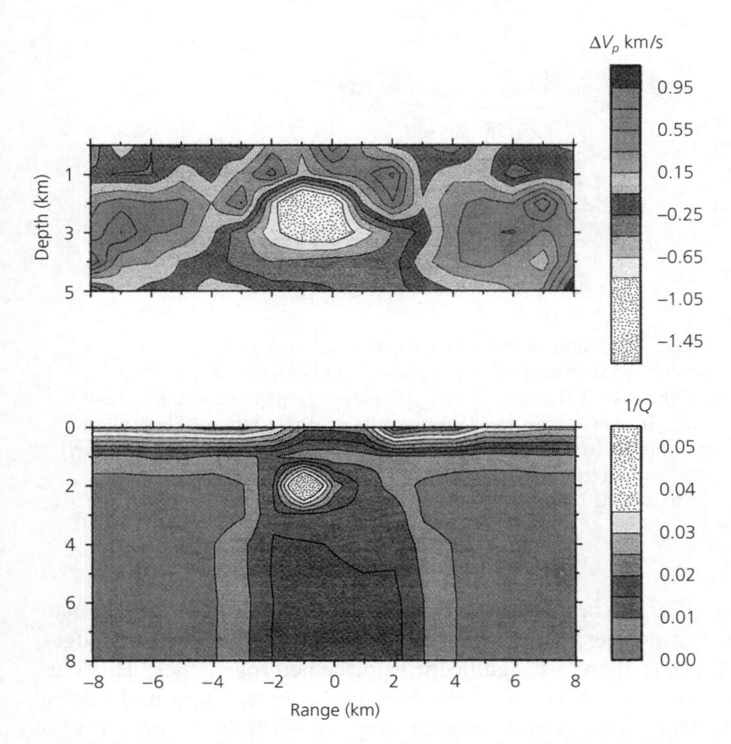

Fig. 3.7-3 Results of *P*-wave velocity (*top*) and attenuation (*bottom*) tomography across the axis of the East Pacific rise. (Solomon and Toomey, 1992, reproduced with the permission of Annual Reviews Inc.)

3.7.2 *Geometric spreading*

The most obvious effect causing seismic wave amplitudes to vary with distance is geometric spreading, in which the energy per unit wave front varies as a wave front expands or contracts. Geometric spreading differs for surface and body waves. For a homogeneous elastic spherical earth, a surface wave front would spread as it moved from the source to a distance 90° away, refocus as it approached the antipode on the other side of the earth from the source, and so on. The amplitudes would be largest at the source and antipode, where all the energy would be concentrated, and smallest halfway between, 90° from the source. On a homogeneous flat earth, the surface waves would spread out in a growing ring with circumference $2\pi r$, where r is the distance from the source. Conservation of energy[2] requires that the energy per unit wave front decrease as $1/r$, whereas the amplitudes, which are proportional to the square root of energy (Eqn 2.4.65), decrease as $1/\sqrt{r}$. However, because the earth is a sphere, the ring wraps around the globe (Fig. 3.7-4), making the energy per unit wavefront vary as

$$1/r = 1/(a \sin \Delta), \quad (1)$$

[2] As we saw in discussing wave reflection and transmission (Section 2.2.4), amplitudes are easier to visualize, but energy is conserved, and hence often more useful for understanding wave behavior.

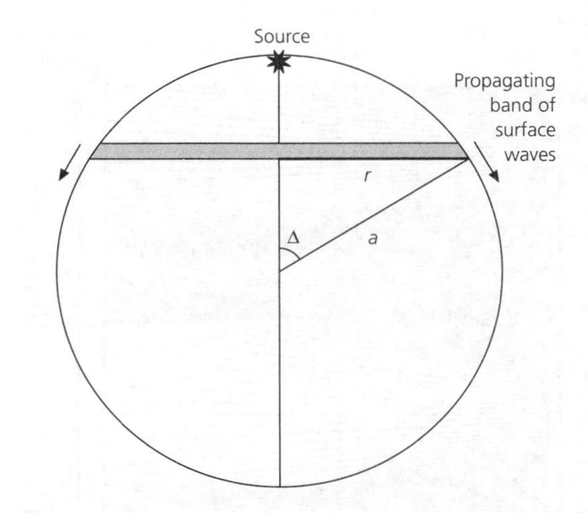

Fig. 3.7-4 Geometric spreading of surface waves for a laterally homogeneous earth yields a wave front that is a ring whose circumference varies as $a \sin \Delta$.

where Δ is the angular distance from the source. Thus the amplitudes decrease as $(a \sin \Delta)^{-1/2}$, with minimum at $\Delta = 90°$, and maxima at 0° and 180°. Actually, not all the energy would focus at the antipode and source even if the earth had no lateral variations in velocity, because some defocusing would result from the earth's ellipsoidal shape. Lateral heterogeneity, discussed next, further distorts the wavefront.

For body waves, consider a spherical wavefront moving away from a deep earthquake. Energy is conserved on the expanding spherical wavefront whose area is $4\pi r^2$, where r is the radius of the wavefront. Thus the energy per unit wave front decays as $1/r^2$, and the amplitude decreases as $1/r$. In reality, because body waves travel through an inhomogeneous earth, their amplitude depends on the focusing and defocusing of rays by the velocity structure. The effects of the variations in velocity with depth were shown in Section 3.4 by considering the density of rays with different incidence angles that arrive at a given distance. These amplitude variations are viewed as geometric spreading and described by the second derivative of the travel time curve (Eqn 3.4.20). Thus, although the phenomenon of geometric spreading is intuitive, quantification of its effects is complicated.

3.7.3 *Multipathing*

Seismic waves are also focused and defocused by lateral variations in velocity. Although physically this process is the same as the effects of vertical variations, it is often distinguished by the term *multipathing*. The distinction reflects our view of the earth as an essentially layered planet with secondary lateral variations.

As we discussed for tsunamis (Fig. 2.8-9), seismic waves refract towards low-velocity anomalies and away from high-velocity anomalies. Figure 3.7-5 illustrates this effect for a plane wave passing through a refracting layer of variable thickness.

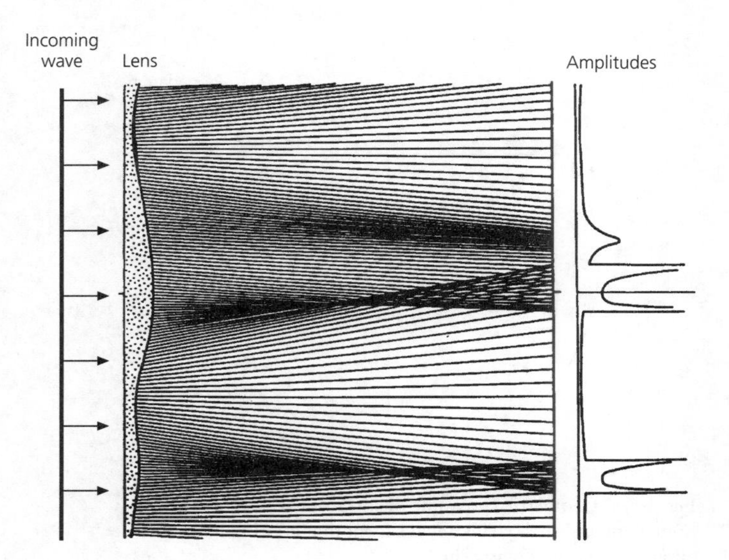

Fig. 3.7-5 An example of how velocity heterogeneities affect wave amplitudes. A plane wave impinging from the left is refracted by a layer of variable thickness. The amplitudes of the waves arriving at the right are shown. Regions of wide ray spacing have low amplitudes, and dense spacing yields large amplitudes. Concentrated lines, or caustics, cause very high amplitudes. (Hannay, 1986. Reproduced with permission from the Institute of Mathematics.)

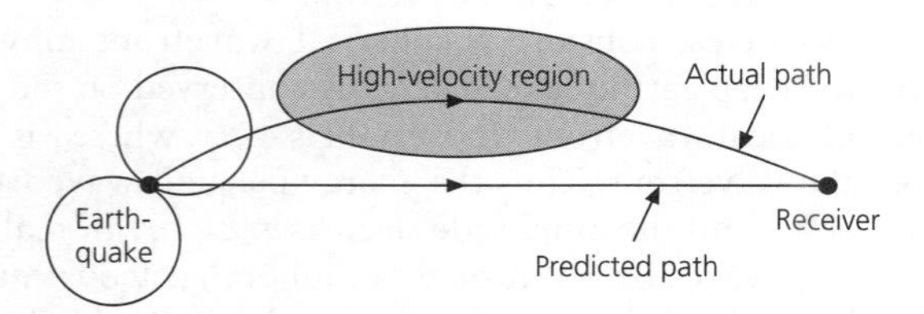

Fig. 3.7-6 Schematic example of how velocity heterogeneity can cause an erroneous estimate of either the focal mechanism or attenuation. The figure-eight structure at the earthquake shows the amplitude of a radiated surface wave as a function of azimuth, which depends on the focal mechanism (in this case dip-slip motion on a vertical fault). The predicted path would leave the source with a lower amplitude than the actual path, which is bent by the high-velocity region. Hence a focal mechanism study using these data without accounting for the perturbed ray path would be incorrect. Conversely, modeling the amplitudes without considering the high-velocity region would yield too-low estimates of attenuation.

The ray paths, which are normal to the local wave front, show how the initially planar wave is refracted. The ray spacing represents the energy density, so amplitudes are low where the rays are far apart, and high where they are close together. In some cases the energy focuses into caustics, areas of infinitely high energy density, which appear as solid black regions.

This example illustrates that velocity variations can affect the amplitudes of seismic waves some distance away. For example, small velocity heterogeneities near an earthquake can cause large amplitude variations at teleseismic distances. This effect can be important, because most earthquakes occur at plate boundaries, such as subduction zones or mid-ocean ridges, where there are significant velocity heterogeneities. This phenomenon can cause difficulties in the interpretation of seismic data. For example, assume (Fig. 3.7-6) that the actual wave path from an earthquake to a receiver differs from that predicted due to a region of anomalously fast velocities. If the amplitudes of these waves were used to study the earthquake's focal mechanism, the result would be biased because the waves left the source in a direction different from that expected if the velocity heterogeneity were not present. Conversely, if the focal mechanism were known, the observed amplitude would differ from that expected, so an estimate of the attenuation would be incorrect.

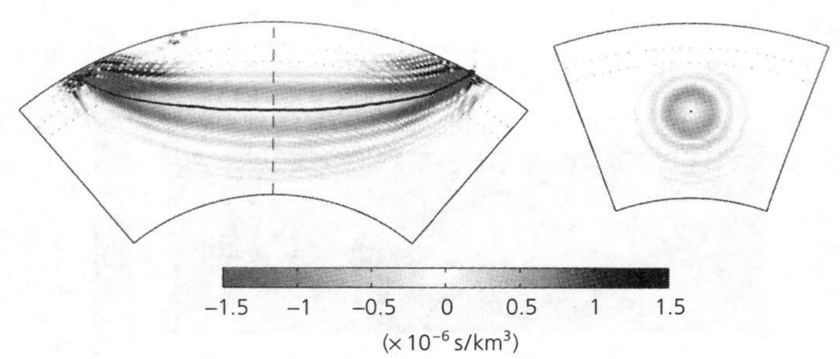

Fig. 3.7-7 Numerical simulation of the paths taken by seismic energy associated with the body wave phases *S* and *sS* for a 120 km-deep earthquake. The values shown, computed using normal modes, show the sensitivity of the travel time to velocity perturbations. These phases sample the structure in a banana-shaped region shown in side view (*left*) and end-on (*right*) surrounding the geometric ray path (solid line). (Zhao *et al.*, 2000.)

When multipathing occurs, the seismic waves arriving at a receiver can be viewed as having taken some ray paths in addition to the direct path, and so have sampled a larger region of the earth. A way to view this is that Fermat's principle giving the geometric ray path applies exactly only to waves of infinite frequency. For waves of finite frequency, we can view the seismic waveform as a coherent sum of energy that travels all possible paths that arrive within a half-period of the infinite-frequency wave, which took the shortest time. These paths form a volume called the first *Fresnel zone* around the infinite-frequency path. Successive half-periods correspond to higher-order Fresnel zones. For longer-period waves, the maximum time over which energy arrives coherently is longer, so the Fresnel zones are proportionately larger. For example, teleseismic body waves sample a banana-shaped region about the geometric ray path. Figure 3.7-7 shows Fresnel zones for a body wave phase in a laterally homogeneous earth, plotted in terms of how the travel time is affected by velocity perturbations. The curved ray path represents the effects of vertical variations in velocity on the infinite frequency ray, and the surrounding "banana" represents the effects of finite-frequency waves. Lateral heterogeneity would distort the "banana."

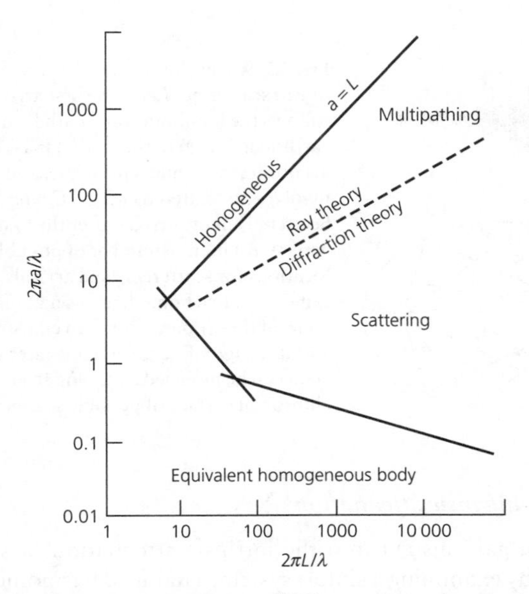

Fig. 3.7-8 Schematic representation of different approaches to seismic wave propagation in a medium with velocity heterogeneity. The approach depends on the ratio of the heterogeneity size a to the wavelength λ and the distance L the wave travels through the heterogeneous region. (After Aki and Richards, 1980. From *Quantitative Seismology*, © 1980 by W. H. Freeman and Company, used with permission.)

3.7.4 *Scattering*

A related effect to multipathing is the scattering of seismic waves. Both effects are complicated, and the distinction between them is gradational. As shown in Fig. 3.7-8, whether the effects of velocity heterogeneity are regarded as scattering depends on the ratio of the heterogeneity size to the wavelength and the distance the wave travels through the heterogeneous region. When the heterogeneity is large compared to the wavelength, we regard the wave as following a distinct ray path that is distorted by multipathing. When the velocity heterogeneities are closer in size to the wavelength, we think of scattered energy rather than distinct ray paths. However, when the heterogeneities are much smaller than the wavelength, they simply change the medium's overall properties. The further the wave travels in the heterogeneous region, the more useful the scattering description becomes. Hence for longer distances, the wavelength range viewed as scattering increases.[3]

Figure 3.7-8 also illustrates that diffraction can be viewed as behavior intermediate between scattering and multipathing. As we have seen, some of the behavior of diffracted waves can be derived either using a Huygens' source scattering representation (Section 2.5.10) or by using ray paths in a medium with variable velocity, as for the head wave (Section 3.2.1) or core diffraction (Section 3.5.2). These ray paths were not truly geometric, in that energy was required to follow paths that did not obey Snell's law. The distinction between ray theory and diffraction depends on wavelength, as discussed in Section 2.5.10, so waves diffracted around the core are depleted in the higher frequencies.[4]

Scattering can be viewed in different ways. In some situations we view the scattering as deterministic, and try to image distinct scatterers. For example, migration methods in reflection seismology (Section 3.3.7) seek to undo the effects of scattering and produce a clearer image of the subsurface. In other situations, we view the medium as containing many scatterers and consider their effects on the wave field statistically. This approach is taken to the scattering of *PKP* waves (Fig. 3.5-8), with a wavelength of about 10 km, by lower mantle heterogeneities of about that size.

Scattering is especially important in the continental crust, which has many small layers and reflectors resulting from billions of years of continental evolution. Although these structures do not significantly affect waves with wavelengths longer than tens of km, for shorter-wavelength waves they can act as point scatterers or Huygens' sources. Hence some of the scattered energy arrives at a receiver after the initial pulse that obeyed Fermat's principle and took the shortest path. This scattered energy causes an arrival to have a *coda*, a tail of incoherent energy that decays over a duration of seconds or minutes. The main arrival has a polarity related to the direction of propagation that can be observed on a three-component seismometer by forming particle motion plots (Fig. 2.7-6). By contrast, the scattered energy arrives from various directions and thus shows little or no preferred particle motion.

Figure 3.7-9 demonstrates the scattering for a seismic arrival. The unscattered wave travels the shortest distance and gives the initial arrival (*left*). Scattered energy lost from this arrival that instead arrives later could have been scattered from an infinite number of locations that would yield the observed travel time. In a constant-velocity medium, the locus of these possible scatterers forms an ellipsoid with the source and the receiver as foci (*center*). Larger ellipsoids define the possible scatterers for energy that arrives later (*right*). These ellipsoids are distorted by velocity heterogeneity and are analogous to the Fresnel volume used when we consider the waves as following distinct ray paths.

Scattering is especially noticeable on the moon. Figure 3.7-10 contrasts seismic records of an earthquake and the impact of a rocket on the moon. Most of the earthquake's energy arrives in the main *P*- and *S*-wave arrivals. By contrast, on the moon

[3] The fact that light scattering in the atmosphere depends on wavelength and the distance traveled has familiar consequences. Because the shortest wavelengths of visible light are the most scattered, blue light reaching us from all directions makes the sky appear blue. The loss of blue light makes the sun appear yellow, although it would appear white if observed from a spacecraft. At sunset, when the sunlight passes through a longer path in the atmosphere than at other hours, intermediate wavelengths are also scattered, leaving direct light from the sun enhanced in the longest visible wavelengths (red light) and making the sun appear red.

[4] This effect makes it hard to understand what someone is saying when they are standing around a corner, because the voice sounds muffled due to the loss of the higher frequencies.

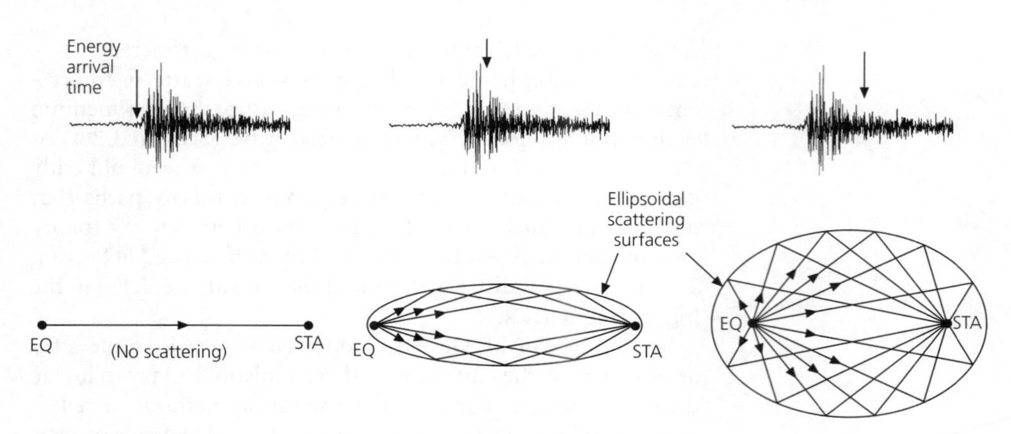

Fig. 3.7-9 Development of a *P*-wave coda due to scattering. *Left*: The first arrival follows the minimum-time path from the earthquake (EQ) to the station (STA) according to Fermat's principle, and involves no scattered energy. *Center*: Scattered energy arrives after the first arrival. An infinite number of possible locations for scatterers yield arrivals at this same time. In a homogeneous medium the locus of these points forms an ellipsoidal surface. *Right*: Energy arriving later in the coda can be modeled as arising from a larger ellipsoidal surface of possible scatterers.

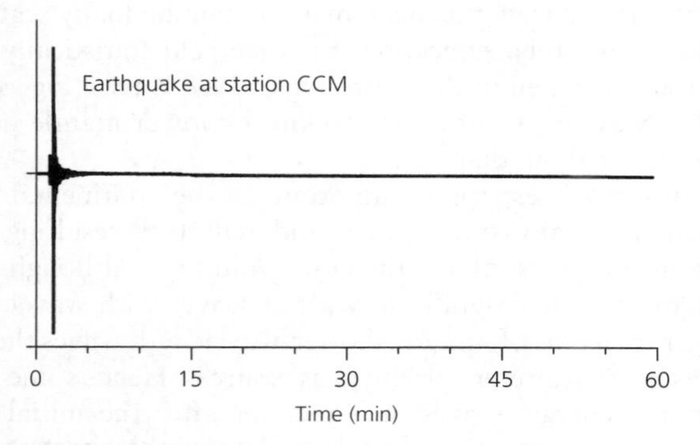

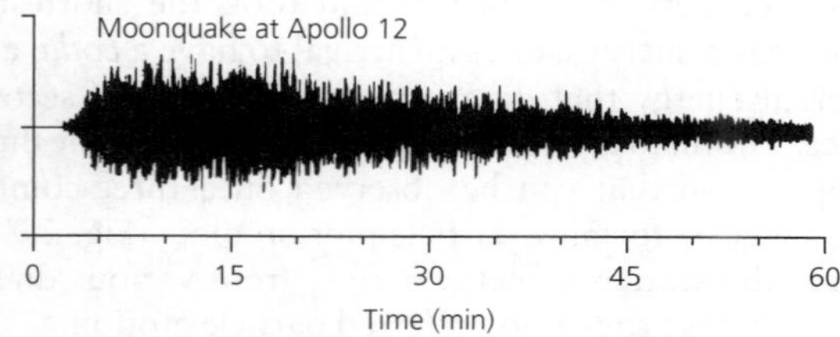

Fig. 3.7-10 Comparison of seismograms for the earth and the moon. *Top*: Seismogram recorded at Cathedral Cave, Missouri (CCM), from a small earthquake 183 km away. *Bottom*: Seismogram recorded by the Apollo 12 seismometer of the impact of the Apollo 14 Saturn booster rocket 147 km away. The terrestrial record shows high attenuation, whereas the lunar seismogram shows intense scattering due to the fractured regolith and very weak attenuation due to the lack of intergranular water. (Mitchell, 1995. *Rev. Geophys.*, *33*, 441–62, copyright by the American Geophysical Union.)

the energy is intensely scattered, and no main arrivals can be identified. This is probably because intrinsic attenuation is much larger in the earth's crust than on the moon. The movements of interstitial fluids in the earth's crust greatly reduce seismic wave amplitudes, whereas energy scattered by the moon's highly fractured near-surface regolith layer is poorly absorbed and reverberates. As a result, efforts to identify seismic phases and use them to study the moon's internal structure have been generally unsuccessful.

3.7.5 *Intrinsic attenuation*

We can gain insight into the intrinsic attenuation of seismic waves by examining a simple system, a damped harmonic oscillator composed of a spring and a dashpot. We use Newton's second law, $\mathbf{F} = m\mathbf{a}$, to describe the displacement $u(t)$ of a mass m. The restoring force of the spring is proportional to minus the spring constant k times the spring extension or displacement from the equilibrium positions, so

$$m\frac{d^2u(t)}{dt^2} + ku(t) = 0. \tag{2}$$

Once set in motion by an impulse, this frictionless system has a purely elastic response described by a perpetual harmonic oscillation

$$u(t) = Ae^{i\omega_0 t} + Be^{i\omega_0 t}, \tag{3}$$

where A and B are constants, and the mass moves back and forth with a natural frequency

$$\omega_0 = (k/m)^{1/2}. \tag{4}$$

One example of this general solution is

$$u(t) = A_0 \cos(\omega_0 t). \tag{5}$$

Once the motion is started, this undamped oscillation continues forever, because no energy is lost. However, this is no longer the case if the system contains a dashpot, or damping term. The damping force is proportional to the velocity of the mass and opposes its motion. Hence the equation of motion (Eqn 2) becomes

$$m\frac{d^2u(t)}{dt^2} + \gamma m\frac{du(t)}{dt} + k\,u(t) = 0, \tag{6}$$

where γ is the *damping factor*. To simplify this, we define the *quality factor*

$$Q = \omega_0/\gamma, \tag{7}$$

and rewrite Eqn 6 as

$$\frac{d^2u(t)}{dt^2} + \frac{\omega_0}{Q}\frac{du(t)}{dt} + \omega_0^2\, u(t) = 0. \tag{8}$$

This differential equation, which describes the damped harmonic oscillator, can be solved assuming that the displacement is the real part of a complex exponential

$$u(t) = A_0 e^{ipt}, \tag{9}$$

where p is a complex number. Substituting Eqn 9 into Eqn 8 yields

$$(-p^2 + ip\omega_0/Q + \omega_0^2)\, A_0 e^{i(pt)} = 0. \tag{10}$$

For this to be satisfied for all values of t,

$$-p^2 + ip\omega_0/Q + \omega_0^2 = 0. \tag{11}$$

Because p is complex, we break it into its real and imaginary parts,

$$p = a + ib, \quad p^2 = a^2 + 2iab - b^2, \tag{12}$$

so Eqn 11 gives

$$-a^2 - 2iab + b^2 + ia\omega_0/Q - b\omega_0/Q + \omega_0^2 = 0, \tag{13}$$

which can be split into equations for the real and imaginary parts and solved separately:

$$\text{Real:} \quad -a^2 + b^2 - b\omega_0/Q + \omega_0^2 = 0, \tag{14}$$

$$\text{Imaginary:} \quad -2ab + a\omega_0/Q = 0.$$

Solving the imaginary part for b gives

$$b = \omega_0/2Q, \tag{15}$$

and putting this into the equation for the real part gives

$$a^2 = \omega_0^2 - \omega_0^2/4Q^2 = \omega_0^2(1 - 1/4Q^2). \tag{16}$$

Thus we define

$$\omega = a = \omega_0(1 - 1/4Q^2)^{1/2}, \tag{17}$$

and rewrite Eqn 9 with separate real and imaginary parts,

$$u(t) = A_0 e^{i(\omega t + ibt)} = A_0 e^{-bt} e^{i\omega t}. \tag{18}$$

The real part is the solution for the damped harmonic displacement,

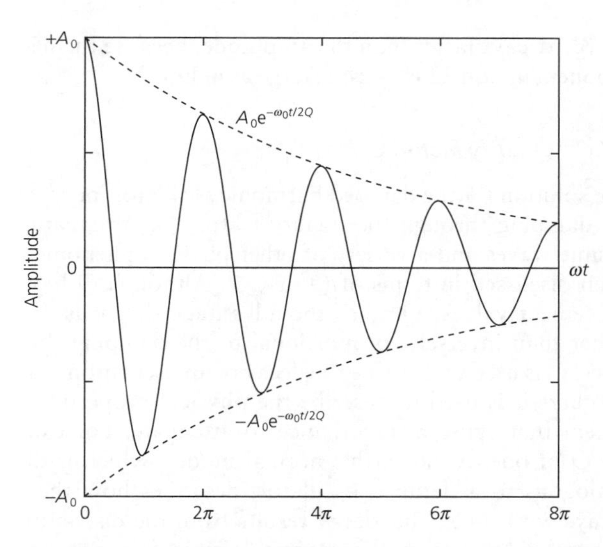

Fig. 3.7-11 For a damped harmonic oscillator, the envelope (dashed lines) amplitude is initially A_0, but decays with time at a rate determined by the quality factor, Q.

$$u(t) = A_0 e^{-\omega_0 t/2Q} \cos(\omega t). \tag{19}$$

This solution shows how the damped oscillator responds to an impulse at time zero (Fig. 3.7-11). It is no longer a simple harmonic oscillation because it differs in two ways from the undamped solution (Eqn 5). The exponential term expresses the decay of the signal's envelope, or overall amplitude,

$$A(t) = A_0 e^{-\omega_0 t/2Q}, \tag{20}$$

which is superimposed on the harmonic oscillation given by the cosine term. Moreover, the frequency of the harmonic oscillation (Eqn 17) is changed from the natural frequency of the undamped system, ω_0, by an amount depending on the quality factor. Q is inversely proportional to the damping factor, γ, so the smaller the damping, the greater Q is. For no damping, Q is infinite, and the damped solution reduces to the undamped one, because its amplitude does not decay with time (Eqn 20), and its frequency remains ω_0 (Eqn 17). As the damping increases, Q decreases, so the amplitude decays faster, and the frequency changes more from its undamped value. Equation 20 shows that the amplitude decays to e^{-1} (0.37) of its original value by the *relaxation time*

$$t_{1/e} = 2Q/\omega_0. \tag{21}$$

Because the energy in an oscillating system is proportional to the square of the amplitude, as we saw for a harmonic wave in Section 2.2.4, Eqn 20 gives the energy of the oscillator as

$$E(t) = \frac{1}{2}kA^2(t) = \frac{1}{2}kA_0^2 e^{-\omega_0 t/Q} = E_0 e^{-\omega_0 t/Q}. \tag{22}$$

Energy decays faster than the amplitude, because the negative exponent in Eqn 22 is twice as large as in Eqn 21.

3.7.6 *Quality factor, Q*

The solution for the damped harmonic oscillator incorporated the damping through the quality factor, Q. Attenuation for seismic waves and a variety of other physical phenomena are often discussed in terms of Q or Q^{-1}. Although Q has more convenient values, Q^{-1} has the advantage that it is directly, rather than inversely, proportional to the damping. In some cases, Q is used to describe the decay of an oscillation, whereas in others it is used to describe the physical properties of the system that cause a disturbance to attenuate. For example, the Q of one of the earth's normal modes, which is directly analogous to a damped oscillator, describes how the mode decays with time. This decay results from the distribution of material in the earth that causes seismic energy to be lost to heat. This distribution can be described in terms of a Q, or anelastic attenuation, structure analogous to the elastic velocity structure.

As a result, we speak of the Q of surface waves, body waves, and crustal phases like *Lg*. We also speak of the variation within the earth of Q_α and Q_β, which controls the attenuation of *P* and *S* waves. The anelastic structure of the earth, given by variations in Q_α and Q_β, is analogous to the elastic velocity structure because Q can be viewed mathematically as an imaginary part of the velocity. To see this, note that (9), which we used to derive the decaying oscillation, can be viewed as an oscillation with a complex frequency p

$$u(t) = A_0 e^{ipt} = A_0 e^{i(a+ib)t} \tag{23}$$

where the real and imaginary parts of the frequency are

$$a = \omega \qquad b = \omega^* = \omega_0/2Q \approx \omega/2Q \tag{24}$$

assuming that attenuation is small (Q large) enough that $\omega \approx \omega_0$. Hence we write

$$Q^{-1} = 2b/a = 2\omega^*/\omega. \tag{25}$$

Treating the attenuation as an imaginary part of the frequency and dividing by the wavenumber lets us treat the corresponding velocity for a propagating wave as complex,

$$c + ic^* = \omega/k + i\omega^*/k = \omega/k + i\omega Q^{-1}/2k \tag{26}$$

so

$$Q^{-1} = 2c^*/c. \tag{27}$$

Thus we can express the attenuation of *P*- and *S*-waves by using the quality factors Q_α and Q_β to give imaginary parts to the velocities. If there is no attenuation ($Q = \infty$) the frequency and the velocity have no imaginary parts. This formulation is useful because it means that methods used to invert surface wave velocities or normal mode eigenfrequencies to find velocity in the earth can also be used to invert observations of their attenuation to find the distribution of anelasticity.

We pose the complex parts of the velocities in terms of the properties of the material causing attenuation by treating the elastic moduli as having imaginary parts. For the shear velocity

$$\begin{aligned} \beta + i\beta^* &= \beta(1 + iQ_\beta^{-1}/2) \\ &= ((\mu + i\mu^*)/\rho)^{1/2} = \beta(1 + i\mu^*/\mu)^{1/2} \\ &\approx \beta(1 + i\mu^*/2\mu) \end{aligned} \tag{28}$$

where the last step used the first term of the Taylor series, because the attenuation and hence imaginary part is small. Comparing terms shows that

$$Q_\beta^{-1} = \mu^*/\mu. \tag{29}$$

A similar analysis shows that the quality factor for *P* waves is given by the imaginary parts of both the bulk and shear moduli

$$Q_\alpha^{-1} = (K^* + 4/3\mu^*)/(K + 4/3\mu). \tag{30}$$

Physically, it is useful to think of energy as being lost in either compressional or shear deformation, so we express their attenuation in terms of imaginary parts of the compressibility and rigidity

$$Q_K^{-1} = K^*/K \qquad Q_\mu^{-1} = \mu^*/\mu = Q_\beta^{-1}. \tag{31}$$

These quality factors are related to those for the velocities by

$$Q_\alpha^{-1} = LQ_\mu^{-1} + (1 - L)Q_K^{-1} \qquad L = (4/3)(\beta/\alpha)^2. \tag{32}$$

In general little energy is lost in compression, so Q_K^{-1} is very small, and thus most of the attenuation for *P* waves occurs in shear, making $Q_\alpha^{-1} \approx (4/9)Q_\beta^{-1}$.

Techniques for measuring Q in the earth follow from those used to measure Q for the decay of an oscillation. From Eqn 20, taking the natural logarithm of the envelope shows that

$$\ln A(t) = \ln A_0 - \omega_0 t/2Q, \tag{33}$$

so Q can be found from the slope of the logarithmic decay. Alternatively, if successive peaks one full period $T = 2\pi/\omega_0$ apart have amplitudes

$$\begin{aligned} A_1(t_1) &= A_0 \exp(-\omega_0 t_1/2Q) \quad \text{and} \\ A_2(t_1 + T) &= A_0 \exp(-\omega_0(t_1 + T)/2Q), \end{aligned} \tag{34}$$

their ratio is

$$A_1/A_2 = \exp[-\omega_0 t_1/2Q - \omega_0(t_1 + T)/2Q] = \exp(\pi/Q), \tag{35}$$

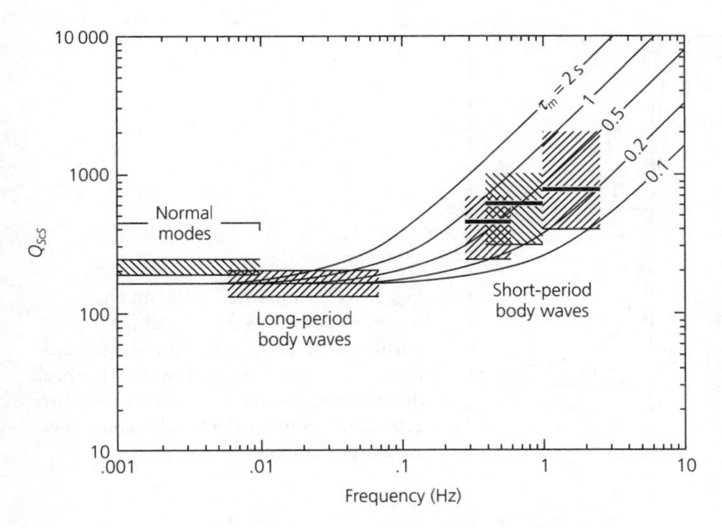

Fig. 3.7-12 Frequency dependence of attenuation for seismic waves in the mantle. Q is shown as though all measurements were for *ScS* waves, a good measure of the average mantle value because of their path from surface to core and back. (Sipkin and Jordan, 1979. © Seismological Society of America. All rights reserved.)

so

$$Q = \pi / \ln (A_1/A_2). \tag{36}$$

To illustrate this approach, note that in Fig. 3.7-11 the second peak, at $\omega t = 2\pi$, is about 2/3 of the first peak, at $\omega t = 0$. Thus $Q \approx \pi/\ln(3/2) \approx 8$. This is small compared to Q for mantle rocks, which is in the range of 200–500, but comparable to that for some sedimentary rocks. For example, *S* waves in shale have $Q \approx 10$.

Another way to view Q is as the number of cycles the oscillation takes to decay to a certain level. The number of cycles n, is

$$n = t/T = \omega t/2\pi \approx \omega_0 t/2\pi, \tag{37}$$

where the last approximation, based on Eqn 17, assumes that the attenuation is small enough ($Q >> 1$) so that $\omega \approx \omega_0$. The amplitude at time t_n, after n cycles, is

$$A(t_n) \approx A_0 e^{\frac{-n\pi}{Q}}, \tag{38}$$

so, if we define n as equal to Q,

$$A(t_n) \approx A_0 e^{-\pi} \approx 0.04 A_0. \tag{39}$$

Thus, after Q cycles, the amplitude drops to a level of $e^{-\pi}$ or 4% of the original amplitude. Hence, in Fig. 3.7-11, more than 95% of the amplitude is lost after $Q \approx 8$ cycles.

Q can describe the oscillation's decay in either time or space. For standing waves like normal mode oscillations, Q describes the decay of amplitudes with time. For traveling waves, we replace t with x/c, where x is the distance traveled and c is the velocity. Thus Eqn 20 becomes

$$A(x) = A_0 e^{\frac{-\omega_0 x}{2cQ}}, \tag{40}$$

which describes how the amplitude decays with the distance the wave propagates.

When these techniques are used to measure Q for seismic waves, we find that Q varies with frequency (Fig. 3.7-12). Q is essentially constant at low frequencies, about 0.001 to 0.1 Hz, but then increases with frequency. Thus Q values derived from normal mode analysis are lower than those obtained from higher-frequency waves. Although our first instinct might be that Q should be frequency-independent, such a situation imposes a stringent requirement. Because $Q = \omega/\gamma$, constant Q requires a physical mechanism in the earth with damping proportional to frequency. We will explore this issue shortly.

Before doing so, it also worth noting that our model of the damped oscillator assumes that the attenuation is *linear*, such that Q is independent of the amplitude of the wave. This is the same as assuming that the amplitudes are not too large. In most rocks this condition is satisfied if the strains involved with the wave propagation are less than about 10^{-6}. Although this is true at teleseismic distances, it is not the case near an earthquake or an explosion, where the elastic strain can exceed 10^{-4}. Large earthquakes can cause large strains, and hence a region of nonlinear attenuation.

3.7.7 *Spectral resonance peaks*

We are interested in understanding how anelasticity in the earth causes the attenuation of propagating waves. This behavior is an example of the general case of how a damped harmonic oscillator responds to a driving force that depends on frequency. To see this, we modify Eqn 8 by adding a harmonic driving force, and so have the inhomogeneous equation

$$\frac{d^2u}{dt^2} + \gamma \frac{du}{dt} + \omega_0^2 u = e^{i\omega t}. \tag{41}$$

The solution is found using a trial solution

$$u(t) = A(\omega)\, e^{i\phi(\omega)}\, e^{i\omega t}. \tag{42}$$

Substituting this in Eqn 41 yields the amplitude response, $A(\omega)$, and phase response, $\phi(\omega)$,

$$A(\omega) = [(\omega_0^2 - \omega^2)^2 + (\omega\gamma)^2]^{-1/2}, \quad \phi(\omega) = \tan^{-1}\left[\frac{-\gamma\omega}{\omega_0^2 - \omega^2}\right]. \tag{43}$$

As shown in Fig. 3.7-13, the amplitude and the phase responses depend on the damping factor γ and how far the forc-

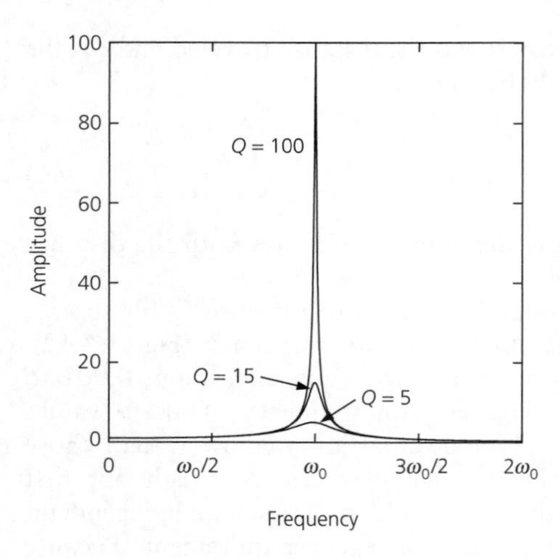

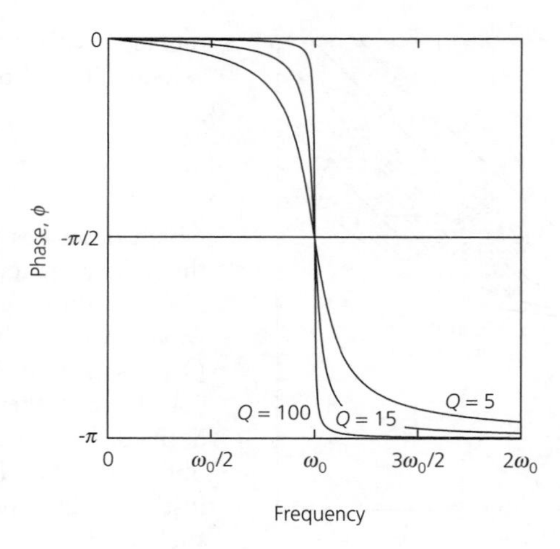

Fig. 3.7-13 Amplitude (*left*) and phase (*right*) response of a forced, damped harmonic oscillator with natural frequency ω_0. For greater damping (lower Q) the peak decreases and both it and the phase shift are broadened from the sharp values they have with little damping.

ing frequency ω is from the oscillator's natural or *resonant* frequency, ω_0. The resonance curve shows how the damped harmonic oscillator responds to the frequency-dependent driving force. The closer the driving frequency ω is to the oscillator's natural frequency ω_0, the more the oscillator responds.

The resonance curve can be viewed in terms of the frequency at which the peak occurs

$$\omega_p = (\omega_0^2 - \gamma^2/2)^{1/2} = \omega_0(1 - 1/2Q^2)^{1/2} \tag{44}$$

and the amplitude of the peak

$$A(\omega_p) = Q/(\omega_0^2(1 - 1/4Q^2)^{1/2}). \tag{45}$$

If the oscillator is undamped ($\gamma = 0$, $Q = \infty$) the peak occurs at its natural frequency and shows an infinite response. Adding damping lowers the amplitude of the peak and shifts it. However, the shift is very small unless the system is much more damped ($Q < 2$) than occurs for seismic wave attenuation. The damping also spreads out the peak in frequency, so the more the damping, the broader and lower the peak. To see why, recall that the more the damping, the faster the oscillation decays as a function of time (Fig. 3.7-11). As we will see in Chapter 6, the spectrum of an undamped sinusoid is a sharp line, or delta function, so additional frequencies, and thus a broader peak, correspond to the decaying sinusoid. The phase response also has significance, as we will see when we discuss seismometers (Section 6.6).

The resonance curve concept appears in a wide variety of applications, because many physical systems can be viewed as damped harmonic oscillators. Three commonly considered in seismology are the attenuation of the earth's normal modes, the behavior of a seismometer, and the response of a building to ground motion. An earthquake puts energy at various frequencies into the earth, exciting its normal modes (Section 2.9). These modes form a set of damped harmonic oscillators, so the amplitude spectrum of a long-period seismogram contains peaks that correspond to the net resonance curve for each mode multiplet. The width of a peak depends on the frequencies and amplitudes of the mode's singlets and the mode's damping. Seismometers can also be viewed as damped harmonic oscillators, whose natural frequency and damping control their response to ground motion. In addition, as mentioned in Section 1.2.2, buildings can be considered damped harmonic oscillators. This concept is important in designing earthquake-resistant structures, because buildings are most vulnerable to ground motion with frequencies close to their natural frequencies, so damping is added to reduce the resulting motion.

3.7.8 *Physical dispersion due to anelasticity*

An important consequence of seismic wave attenuation is *physical dispersion*, in which waves at different frequencies travel at different velocities. This differs from the *geometrical dispersion* discussed in Sections 2.7 and 2.8, in which surface waves of different frequencies have different apparent velocities at the surface because they sample different depths and hence encounter material of different velocities. Thus, although the intrinsic velocity of the rock at any depth is treated as frequency-independent, dispersion occurs because of the depth-variable velocity of the material. By contrast, with physical dispersion the intrinsic velocity of waves in the medium varies with frequency.[5]

To see how physical dispersion results from attenuation, consider how a seismic wave changes shape. Assume that a delta function wave, a pulse of infinite height and unit area (Fig. 3.7-14), propagates through a homogeneous elastic medium with intrinsic velocity c:

[5] The rainbow results from physical dispersion for light waves passing through water drops in the atmosphere or a prism. Different frequencies (colors) of light travel at different speeds through the water or prism, and thus refract at different angles, separating initially white light into different colors.

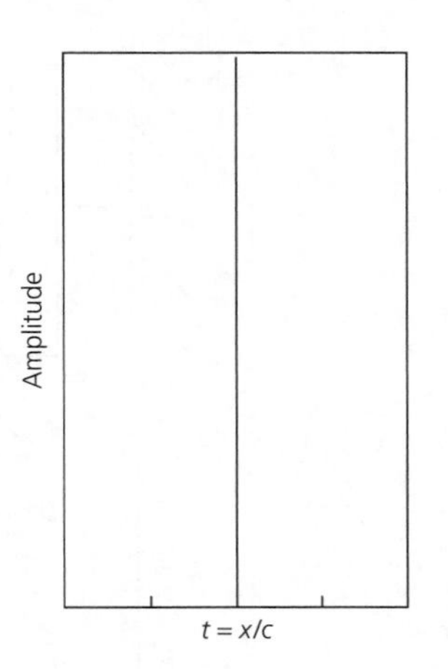

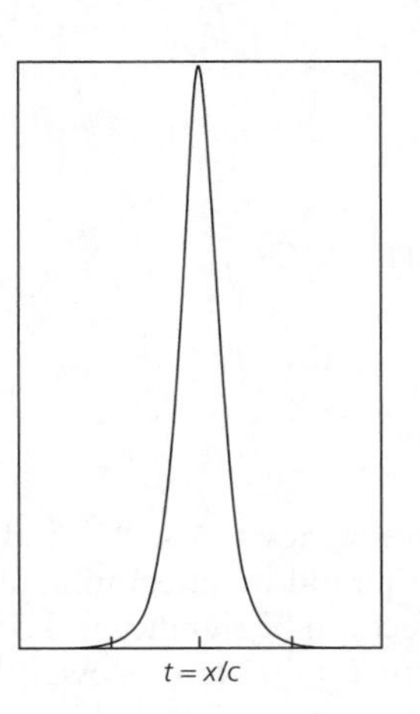

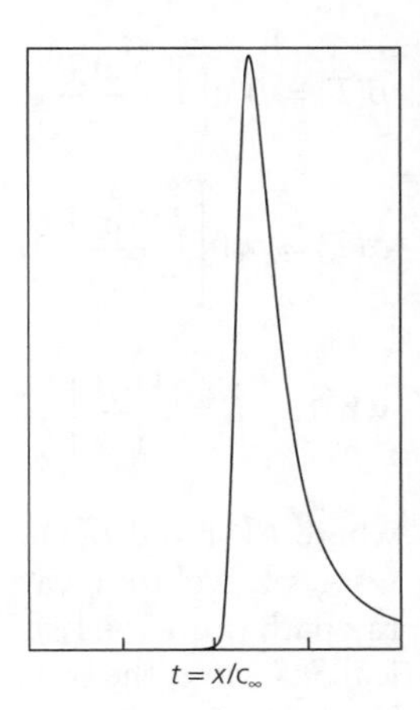

Fig. 3.7-14 *Left*: A propagating wave pulse composed of a delta function. With no dispersion, all frequencies arrive at the same time. *Center*: The delta function after broadening by attenuation, showing that energy arrives before the high-frequency arrival time. *Right*: The pulse including physical dispersion, which makes the lower-frequency waves travel more slowly, so that they do not arrive before the highest-frequency component.

$$u(x, t) = \delta(t - x/c). \tag{46}$$

The Fourier transform of the delta function,

$$F(\omega) = \int_{-\infty}^{\infty} u(x, t)e^{-i\omega t}dt = \int_{-\infty}^{\infty} \delta(t - x/c)e^{-i\omega t}dt = e^{(-i\omega x/c)}, \tag{47}$$

shows that the delta function is made up of waves of all frequencies, as we discuss further in Section 6.2.5. If there is no dispersion, all the frequencies travel at the same speed and arrive at the same time. The effect of attenuation as a function of distance is given by writing Eqn 40 as a function of frequency,

$$A(\omega) = e^{\frac{-\omega x}{2cQ}}, \tag{48}$$

which shows that if Q is constant, the rate at which the amplitude decays with distance increases strongly with frequency. To see how this attenuation affects the delta function wave, we multiply Eqn 47 by Eqn 48 and use the inverse Fourier transform to return to the time domain

$$u(x, t) = \frac{1}{2\pi}\int_{-\infty}^{\infty} A(\omega)F(\omega)e^{i\omega t}d\omega = \frac{1}{2\pi}\int_{-\infty}^{\infty} e^{\frac{-\omega x}{2cQ}} e^{\frac{-i\omega x}{c}} e^{i\omega t}d\omega. \tag{49}$$

Evaluating the integral yields

$$u(x, t) = [(x/2cQ)/((x/2cQ)^2 + (x/c - t)^2)]/\pi, \tag{50}$$

so the delta function is broadened by attenuation into a wavelet that is symmetric in time about its maximum at $t = x/c$ (Fig. 3.7-14, *center*).

A problem with this solution is that seismic energy arrives before the geometric arrival time of the delta function pulse, $t = x/c$, which is the arrival time of the infinite-frequency component. In fact, because the tails of the wavelet extend to infinity on both sides of $t = x/c$, some energy arrives before the earthquake occurred. This impossible situation, called *noncausality*, results from the fact that attenuation broadened the pulse by preferentially removing the high-frequency components.[6]

Thus the physical mechanisms that cause attenuation in the earth must prevent waves of all frequencies from traveling at the same speed. Instead, there must be dispersion, where the lower frequencies causing the tails travel more slowly and arrive later. We saw in Section 2.8 that in a dispersive medium we distinguish the phase velocity c, the speed of a wave of a single frequency, from the group velocity that describes the speed of a wave group. Thus the mathematical condition for causality is that $u(x, t) = 0$ for all $t < x/c_\infty$, where $c_\infty = c(\infty)$ is the phase velocity of the infinite-frequency waves that arrive first. One such dispersion relation for phase velocity as a function of frequency, called Azimi's attenuation law, is

$$c(\omega) = c_0\left[1 + \frac{1}{\pi Q}\ln\left(\frac{\omega}{\omega_0}\right)\right], \tag{51}$$

where c_0 is a reference velocity corresponding to a reference frequency ω_0.[7] This relation provides the needed causality, because the resulting pulse (Fig. 3.7-14, *right*) has high frequencies arriving at or soon after $t = x/c_\infty$, whereas the low frequencies arrive later over a duration depending on the value of Q. If there is no attenuation ($Q = \infty$), Eqn 51 yields no dispersion, and the delta function is not broadened.

From Eqn 51, the *P*- and *S*-wave velocities α and β vary as a function of period T, as

[6] We noted a similar effect in Section 2.9.8: namely, that individual normal modes of a single frequency appear to predict displacement before a wave could arrive, but their sum gives a wave at the correct time.

[7] Aki and Richards (1980).

$$\beta(T) = \beta(1)\left(1 - \frac{\ln T}{\pi} Q_\mu^{-1}\right),$$

$$\alpha(T) = \alpha(1)\left[1 - \frac{\ln T}{\pi}\left(LQ_\mu^{-1} + (1-L)Q_K^{-1}\right)\right],$$

$$\text{where} \quad L = \frac{4}{3}\left(\frac{\beta}{\alpha}\right)^2, \tag{52}$$

where $\alpha(1)$ and $\beta(1)$ are the velocities at 1 s. We find how a wave's travel time varies with period by integrating along its ray path (Eqn 3.4.16). The effect can be significant. For a vertical *ScS* wave, the travel time for $T = 40$ s is 5 s slower than for $T = 1$ s. Out of a total travel time of about 934 s, this is a difference of 0.5%. For vertical *PcP* waves at the same periods, the travel time difference is 1 s out of 511 s, or 0.2%.

This phenomenon causes a discrepancy between the seismic velocity structure found by inverting observations of long-period normal modes and short-period body waves. The velocities inferred from normal modes are consistently slower than those from body waves. The discrepancy reflects the fact that attenuation causes longer-period waves that are studied as normal modes to travel at lower velocities than the body waves. Failure to take this effect into account can cause errors in the predicted arrival times of body waves of several seconds.

The pulse in Fig. 3.7-14 (*right*) is also known as an *attenuation operator*, and can be used to model the effects of attenuation on seismic waveforms. As discussed in Section 3.3.6 and derived in Section 6.3, seismic signals can be modeled by convolving the source–time function with operators describing different effects. Thus a synthetic seismogram computed for an elastic earth can be convolved with the attenuation operator to create a more realistic pulse.

Body wave attenuation is often characterized using the parameter t^*. If a ray travels through a region of constant Q,

$$t^* = \frac{t}{Q} = \frac{\text{travel time}}{\text{quality factor}}. \tag{53}$$

Because Q varies within the earth, we derive t^* by integrating along the ray path,

$$t^* = \int \frac{dt}{Q} = \sum_{i=1}^{N} \frac{\Delta t_i}{Q_i}, \tag{54}$$

where Δt_i and Q_i are the travel time and Q values on the ith path segment. For P waves, t_α^* is often about 1 s, whereas S waves typically have t_β^* around 4 s. The values of t^* increase with increased distance, but are also affected by the number of passages through the asthenosphere (about 80–220 km depth). For example, *ScS* tends to have a higher t^* (greater attenuation) than S at the same distance because of the longer ray path, and S waves from deep earthquakes that only cross the asthenosphere once have lower t^* than S waves from shallow events.

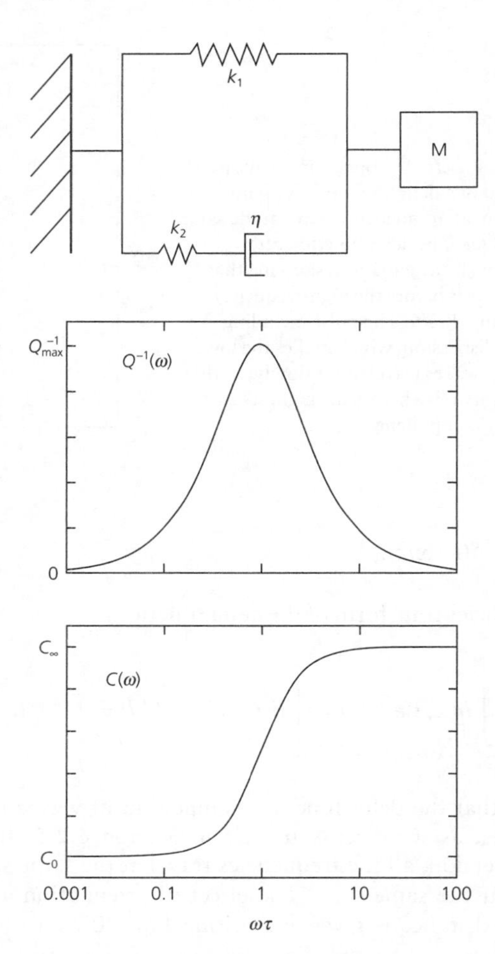

Fig. 3.7-15 *Top*: Schematic diagram of a standard linear solid, made up of a mass connected to two springs and a dashpot. This system responds elastically to waves with periods that are short compared to the relaxation time, τ, and viscously for periods longer than the relaxation time. *Center*: Absorption peak for this material. Q^{-1} approaches zero for large and small values of $\omega\tau$ and is greatest for $\omega\tau = 1$. *Bottom*: Phase velocity dispersion resulting from the attenuation. The velocity is c_0 at low frequencies and increases to c_∞ at high frequencies.

3.7.9 *Physical models for anelasticity*

A common model for the anelastic processes in the earth causing attenuation treats the material as a *viscoelastic* or *standard linear* solid, which combines elastic and viscous responses to an incident seismic wave. This model is represented by a spring with constant k_1 in parallel with a spring with constant k_2 and a dashpot with viscosity η (Fig. 3.7-15). If a step function strain $H(t)$ (0 for $t < 0$, 1 afterwards) is applied, the stress response includes an instantaneous elastic contribution from spring k_1 and a delayed response from the dashpot and spring k_2,

$$\sigma(t) = k_1 H(t) + k_2 e^{-t/\tau}, \tag{55}$$

where τ is the relaxation time constant $= \eta/k_2$.

The response to harmonic waves depends on the product of the angular frequency and the relaxation time. For wave periods that are very short compared to the relaxation time, the system responds mostly elastically, and there is little attenuation. For wave periods much longer than the relaxation time, the system responds mostly in a viscous manner, so there is no attenuative loss of energy. As shown in Fig. 3.7-15, the attenuation[8] varies as

$$Q^{-1}(\omega) = \frac{k_2}{k_1}\frac{\omega\tau}{1+(\omega\tau)^2}. \tag{56}$$

At very low or very high frequencies $Q^{-1}(\omega)$ approaches zero, so Q becomes infinite. The greatest attenuation, or *absorption peak*,[9] occurs at $\omega\tau = 1$, where

$$Q^{-1}_{max} = Q^{-1}(1/\tau) = k_2/2k_1. \tag{57}$$

The phase velocity also depends on $\omega\tau$:

$$c(\omega) = c_0\left[1 + \frac{k_2}{2k_1}\frac{(\omega\tau)^2}{1+(\omega\tau)^2}\right], \tag{58}$$

where $c_0 = (k_1/\rho)^{1/2}$. The phase velocity is lowest (c_0) at low frequencies, and reaches

$$c_\infty = c_0(1 + k_2/2k_1) = c_0(1 + Q^{-1}_{max}) \tag{59}$$

at high frequencies. This model thus has the key feature of the physical dispersion relation (Eqn 51) discussed earlier, that long-period waves travel more slowly than high-frequency waves.

Given this model, the fact that seismological observations find relatively constant Q over a large range of low frequencies from about 0.001 to 0.1 Hz (Fig. 3.7-12) is surprising. Moreover, theoretical and laboratory studies of the physical mechanisms thought to cause attenuation in the earth also suggest that Q should be strongly frequency-dependent. Hence the relatively constant value at low frequencies is thought to result from the superposition of many different mechanisms. A possible explanation comes from noting that a typical attenuation spectrum for a polycrystalline structure (Fig. 3.7-16, *top*) contains multiple attenuation peaks or absorption bands. The absorption bands depend on the material's composition and grain size and vary with temperature (recall Fig. 3.7-2) and pressure, such that higher pressure decreases attenuation, whereas higher temperature increases it. Waves of various frequencies traversing the earth may feel the net effect of absorption bands with different relaxation times, yielding a flat absorption spectrum (Fig. 3.7-16, *bottom*). The higher-frequency waves in Fig. 3.7-12 that show a frequency-dependent Q would be above the flat part of the absorption spectrum.

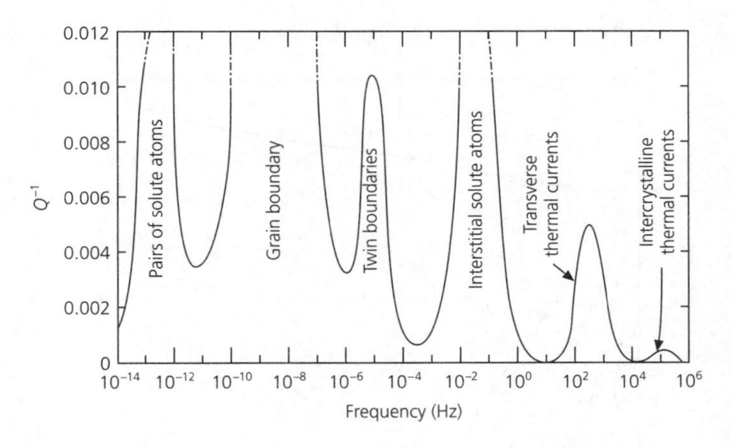

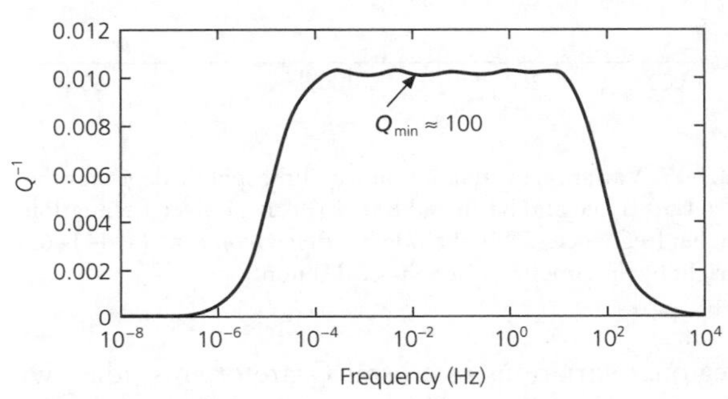

Fig. 3.7-16 *Top*: Relaxation spectrum for a polycrystalline material showing attenuation peaks at different frequencies due to different microscopic mechanisms. *Bottom*: Schematic model to explain the observation that Q is roughly constant over a wide range of frequencies. The superposition of absorption peaks for different compositions at different temperatures and pressures yields a flat absorption band. (Liu *et al.*, 1976.)

3.7.10 Q from crust to inner core

Attenuation is inferred in all regions of the earth except for the liquid iron outer core, and varies greatly both laterally and vertically. In the crust, the greatest attenuation (lowest Q or highest Q^{-1}) occurs near the surface (Fig. 3.7-17), presumably due to the presence of fluids. Attenuation is lowest at about 20–25 km depth, and then increases again, presumably due to increasing temperature. Attenuation decreases as a function of frequency, as in Fig. 3.7-12, and varies geographically. Q in the upper crust is roughly proportional to the time since the last major tectonic activity in a region, perhaps due to crack generation and fluid flow during tectonism and gradual crack annealing after tectonism ceased.

[8] Kanamori and Anderson (1977).

[9] This effect is like driving over a bump: at a high speed inertia keeps the car in line and the bump is not very noticeable. At low speed, we feel only a gradual swell in the road. However, at an intermediate speed the bump gives the maximum jolt.

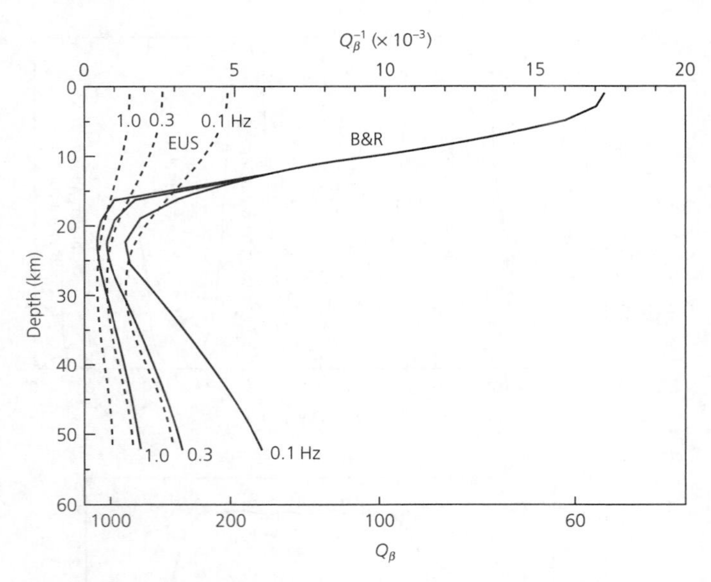

Fig. 3.7-17 Variation in attenuation with lithospheric depth for the eastern USA (EUS) and Basin and Range (B&R). Lower attenuation occurs for higher frequencies. (Mitchell, 1995. *Rev. Geophys.*, *33*, 441–62, copyright by the American Geophysical Union.)

Regional variations in crustal Q are often studied with Lg waves, a superposition of higher-mode surface waves that give prominent arrivals in continental regions. Q_{Lg} for the USA varies regionally (Fig. 3.7-18), with values as high as 750 in the stable East and as low as 250 in the tectonically active West. This regional difference in attenuation, also seen in Figs 3.7-1 and 3.7-17, has implications for seismic hazards (Section 1.2.2). Similarly, the fact that the USA tested nuclear weapons in the western USA, which is more attenuative than the areas used by the Soviet Union, is significant for verifying test ban treaties (Section 1.2.8).

Attenuation in the upper mantle varies with depth, with the lowest Q in the asthenosphere from about 80 to 220 km depth (Fig. 3.7-19). At these depths the temperature approaches, and perhaps exceeds, the melting temperatures of rock, so a small percentage of partial melt may exist. This pattern of attenuation is similar to that for seismic velocities, which are lowest in the asthenosphere. Hence both the elastic velocity and anelastic attenuation reflect the physical processes causing the mechanically weak asthenosphere. Beneath the asthenosphere, Q increases gradually with depth, presumably because temperature increases at a slower rate than pressure.

Q_μ increases with depth through the lower mantle, reaching values in excess of 500. There is some indication that attenuation is enhanced in the D″ region at the base of the mantle. Although no attenuation of P waves is detected for the outer core, there is significant attenuation of *PKIKP* waves traversing the inner core, yielding Q_K estimates in the range of 150–300.

Lateral variations in attenuation are studied using tomographic methods similar to those used for velocity (Sections 2.8.3, 7.3). Where temperatures vary over short distances, significant attenuation variations can occur, as shown in Fig. 3.7-3 for a mid-ocean ridge. Similarly, a cross-section through the back-arc spreading center above the Tonga subduction zone (Fig. 3.7-20) shows that Q_α exceeds 10,000 within the cold and rigid subducting slab, but is less than 75 beneath the hot back-arc basin. Such attenuation data, especially when combined with velocity data, are valuable for tectonic studies.

3.8 Composition of the mantle and the core

Seismology yields information about velocities within the earth. To derive inferences about the composition of the earth, the seismological data are combined with results from geology, geodesy, geomagnetism, cosmochemistry, and the physics and

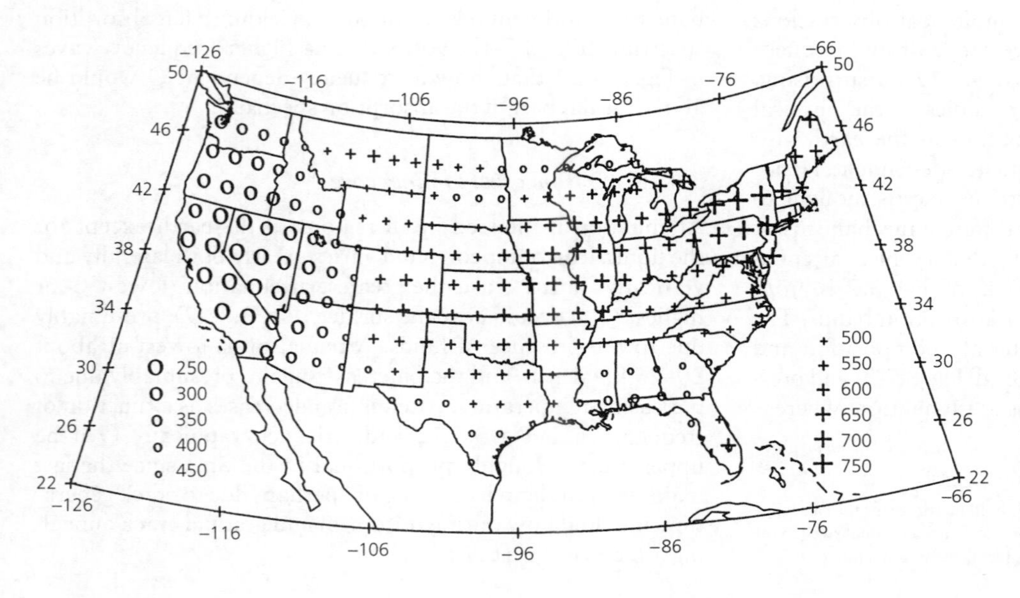

Fig. 3.7-18 Q_{Lg} for the USA mapped from the codas of 1 Hz Lg waves. Q_{Lg}, which reflects attenuation within the crust, shows higher attenuation in the tectonically active western USA and lower attenuation in the tectonically inactive east. (Mitchell *et al.*, 1997.)

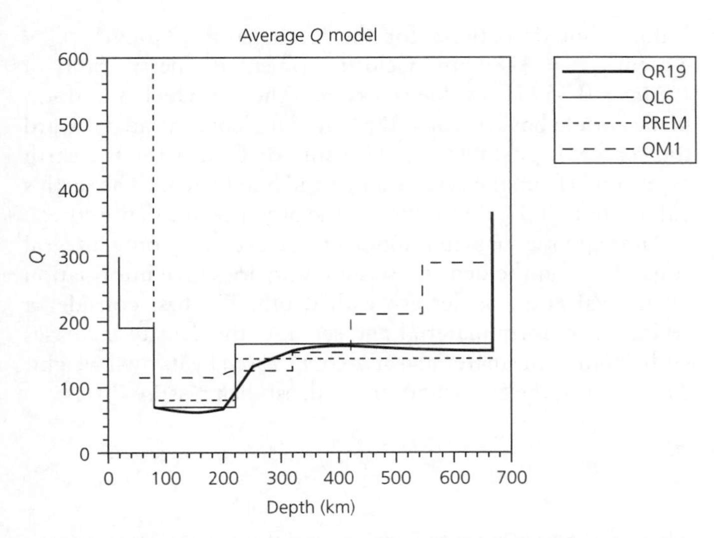

Fig. 3.7-19 Models of Q in the upper mantle showing that attenuation is highest at 80–220 km depth and then increases with depth. (Romanowicz, 1995. *J. Geophys. Res.*, *100*, 12,375–94, copyright by the American Geophysical Union.)

chemistry of materials at high temperature and pressure. A general view of the earth's composition has emerged, although aspects are still under investigation. This view is a cornerstone of our thinking about the evolution of the earth and other planets. We will summarize some basic ideas that are presently under discussion, and the suggested readings provide more information.

3.8.1 *Density within the earth*

A starting point for analysis of the earth's composition is a model of the variation in density with depth. The density is an important constraint on the nature of the material, and can be combined with velocities to derive elastic constants. Densities are less well known than velocities, and their estimation requires more inferences. As with velocities, we use a radially symmetric density model for most applications and consider lateral perturbations when needed.

The basic constraint on the earth's density is that its average is given by the earth's mass M, which can be found from the acceleration of gravity at the surface $r = a$ using the law of gravitation,

$$g = GM/a^2. \tag{1}$$

Because $g = 9.8$ m/s^2, $G = 6.67 \times 10^{-11}$ Nm^{2}kg^{-2}, and $a = 6371$ km, we find $M = 5.97 \times 10^{24}$ kg. The mass is the volume integral of the density, so if density varies only with depth

$$M = 4\pi \int_0^a \rho(r) r^2 dr, \tag{2}$$

the average density, ρ_o, is found by dividing the mass by the volume,

$$\rho_o = M/[(4/3)\pi a^3]. \tag{3}$$

The resulting average density of the earth is about 5.5 g/cm^3. The fact that this value is significantly higher than the density of the surface rocks (about 3 g/cm^3) is evidence for a core of much denser, and hence presumably different, material.

A second constraint on the density, which also indicates a dense core, comes from the moment of inertia about the rotation axis. This is defined by (Fig. 3.8-1) integrating over volumes dV, each at a distance $l = r \sin\theta$ from the spin axis,

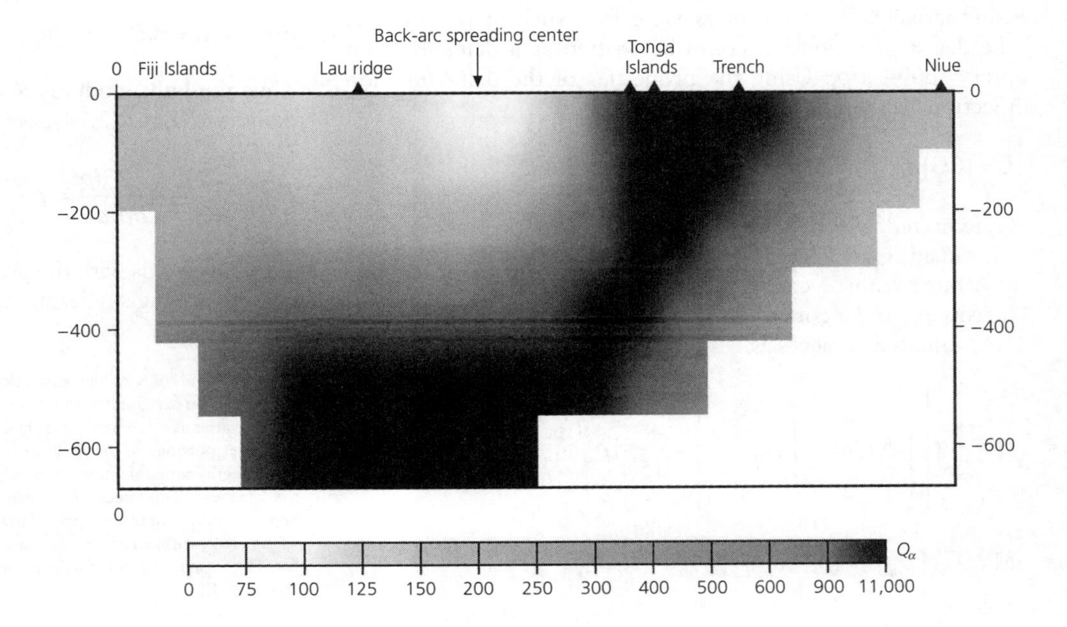

Fig. 3.7-20 Cross-section across the Tonga subduction zone, showing large lateral variations in Q_α between the cold subducting slab (black) and the hotter back-arc basin. (Roth *et al.*, 1999. *J. Geophys. Res.*, *104*, 4795–809, copyright by the American Geophysical Union.)

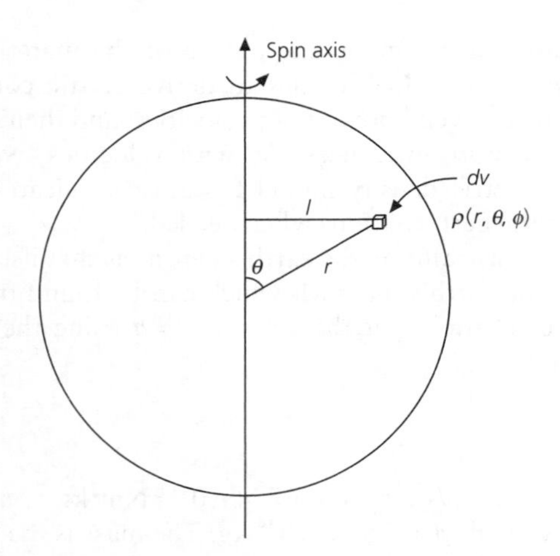

Fig. 3.8-1 A planet's moment of inertia is found by integrating about the spin axis. The moment arm, l, to a volume element, dV, is $r \sin \theta$.

$$C = \iiint l^2 \rho(r, \theta, \phi) dV = \int_0^{2\pi}\int_0^{\pi}\int_0^{a} \rho(r)(r^2 \sin^2 \theta) r^2 \sin \theta \, dr d\theta d\phi$$

$$= \frac{8}{3}\pi \int_0^a \rho(r) r^4 dr. \quad (4)$$

The ratio of the moment of inertia to the mass gives a scalar that depends on the density distribution. If the earth were homogeneous, the density everywhere would equal the average density, $\rho(r) = \rho_o$, and

$$C = (8/15)\pi a^5 \rho_o, \quad M = (4/3)\pi a^3 \rho_o, \quad C/Ma^2 = 0.4. \quad (5)$$

Alternatively, if all the mass were in a shell at the surface, the density distribution could be written as a delta function $\rho(r) = \delta(r - a)\rho_s$. Using the properties of the delta function (Section 6.2.5), Eqns 2 and 4 yield

$$C = (8/3)\pi \rho_s a^4, \quad M = 4\pi \rho_s a^2, \quad C/Ma^2 = 0.67. \quad (6)$$

As expected, a distribution with material concentrated toward the outside gives a larger ratio.

A more realistic case is a two-shell planet, with a mantle of density ρ_m and a core of density ρ_c and radius r_c. The integrals are evaluated in pieces as

$$C = \frac{8}{3}\pi \left[\int_0^{r_c} \rho_c r^4 dr + \int_{r_c}^{a} \rho_m r^4 dr \right] = \frac{8}{15}\pi[\rho_m a^5 + (\rho_c - \rho_m) r_c^5],$$

$$M = \frac{4\pi}{3}[\rho_m a^3 + (\rho_c - \rho_m) r_c^3]. \quad (7)$$

Values similar to those for the earth (ρ_c = 12 g/cm^3, ρ_m = 5 g/cm^3, r_c = 3480 km) yield a moment of inertia ratio of $C/Ma^2 = 0.35$. This value is less than the 0.4 which a uniform planet would have, because the material is concentrated toward the center. It is similar to the value of C/Ma^2 for the earth determined from the earth's shape and gravity field. The earth's value, about 0.33, thus indicates the presence of a dense core.

Although the mass and moment of inertia give only integral constraints on the density, seismic velocities give information on the variation of density with depth. We first consider a region of uniform material and see how the density increases with depth as the material is *self-compressed* by its own weight. At a radius r, the gradient of the hydrostatic pressure $P(r)$ is

$$\frac{dP}{dr} = -g\rho, \quad (8)$$

where $\rho(r)$ and $g(r)$ are the density and the acceleration of gravity at that depth. The derivative is negative, because pressure increases with depth. The local value of gravity, $g(r)$, depends on the total mass $m(r)$ within the sphere of radius r,[1]

$$g = Gm/r^2. \quad (9)$$

The pressure derivative can then be written as

$$\frac{dP}{dr} = \frac{-\rho G m}{r^2}. \quad (10)$$

The elastic constants of the material are introduced using the definitions of the density and the dilatation θ (Eqn 2.3.60),

$$\rho = m/V, \quad d\theta = dV/V, \quad (11)$$

so that differentiation yields

$$d\rho = -(m/V^2)dV = -\rho d\theta. \quad (12)$$

Thus the bulk modulus K can be expressed, starting with its definition (Eqn 2.3.74), as

$$K = -\frac{dP}{d\theta} = -\frac{dP}{d\rho}\frac{d\rho}{d\theta} = \rho\frac{dP}{d\rho}. \quad (13)$$

Combining this with the pressure derivative equation (Eqn 10) gives the change in density with depth

[1] $g(r)$ depends only on the mass below radius r, because a spherical shell of uniform density has no net gravitational effect inside the sphere. This situation arises because gravity varies as r^{-2}, whereas the shell's mass varies as r^2, so larger contributions from the closer portions of the shell are canceled by those from the rest. The fact that a sphere's gravitational attraction is the same as if all its mass were at the center arises in the same way. This effect is not a general property of the center of mass and does not apply for bodies of other shapes. However, it applies for the electric field, which also varies as r^{-2}, within a uniformly charged sphere. Deriving this result is said to have delayed Newton for years before presenting the theory of gravitation in 1686. (Feynman *et al.*, 1963.)

$$\frac{d\rho}{dr} = \frac{d\rho}{dP}\frac{dP}{dr} = \frac{-\rho^2 Gm}{Kr^2}. \tag{14}$$

To include the observations of seismic velocities, we define the *seismic parameter*, Φ, and *bulk sound speed*, $\Phi^{1/2}$, such that

$$\Phi = \alpha^2 - (4/3)\beta^2 = K/\rho. \tag{15}$$

Thus we can write the *Adams–Williamson equation* relating the velocity structure to the derivative of density with radius,

$$\frac{d\rho}{dr}(r) = \frac{-\rho(r)Gm(r)}{\Phi(r)r^2} = \frac{-\rho(r)g(r)}{\Phi(r)}, \tag{16}$$

where the dependences on radius are explicitly shown. This equation can be used to estimate the density structure by starting with the near-surface density, using the seismic velocities to find its derivative, and computing the density at a deeper point. The resulting density and value of $g(r)$ are then used in the next step.

However, density increases with depth as a result of mineral phase changes as well as of self-compression, so the Adams–Williamson equation is insufficient. This difficulty was identified in 1936 by K. Bullen, who used the Adams–Williamson approach to find the density throughout the mantle. He then computed the moment of inertia of the mantle and subtracted it from the moment of inertia of the earth, to find the moment of inertia of the core. Figure 3.8-2 shows the C/Ma^2 value calculated for the core as a function of the assumed density at the top of the mantle, which is the initial density for the Adams–Williamson calculation. For reasonable values of near-surface density, ≈ 3.3 g/cm^3, the core would have C/Ma^2 greater than 0.4, implying that density decreases with depth in the core. This seems unlikely, because the solid inner core should be denser than the liquid outer core. Only implausibly high near-surface densities could cure the problem.

This issue was resolved in the 1950s by F. Birch[2] in a classic series of papers showing that at least one of two assumptions underlying the method was inappropriate. One implicit assumption is that the temperature increases with depth along an *adiabatic gradient*, or "adiabat," such that if a piece of material moves vertically, the pressure-induced temperature change leaves the material at the same temperature as its new surroundings (Eqn 5.4.10). However, the temperature gradient in the mantle is thought to exceed the adiabatic gradient, because a superadiabatic gradient is required for the thermal convection expected in the mantle.[3] The superadiabatic gradient can be included by modifying the Adams–Williamson equation (16) to

$$\frac{d\rho}{dr} = -\frac{\rho g}{\phi} + g\alpha\tau, \tag{17}$$

where α is the coefficient of thermal expansion,[4] and τ is the portion of the temperature gradient exceeding the adiabatic gradient. This correction for higher temperature lowers the calculated mantle densities, and hence increases the calculated C/Ma^2 for the core, making the problem of the core density structure worse.

Hence the assumption of homogeneous material whose density changes only by self-compression must be incorrect. Birch showed that inhomogeneity can be identified using the function $1 - (1/g)d\phi/dr$. Figure 3.8-3 compares values of this function derived from seismic velocity data with values predicted for compression of homogeneous mantle material. Below 1000 km the mantle behaves as a homogeneous material, while at shallower depths it does not. This is because the mineral phase transitions expected at the 410 and 660 km discontinuities involve denser atomic packings, and therefore transitions to higher densities, than predicted by the Adams–Williamson equation.

As a result, density models of the earth include rapid changes in the transition zone. Figure 3.8-4 shows the velocity and density structure for earth model PREM (Table 3.8-1). Within the lower mantle, outer core, and inner core, density increases smoothly with depth according to the Adams–Williamson equation. At the boundaries between these regions, density

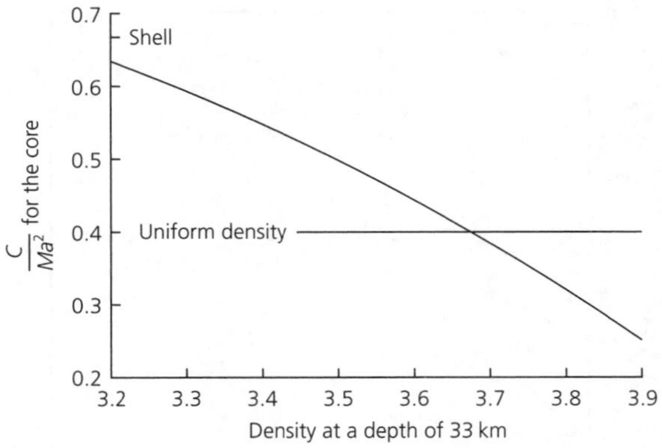

Fig. 3.8-2 Moment of inertia ratio of the earth's core as a function of density at the top of a uniform mantle. For any realistic upper mantle density the ratio would exceed 0.4, implying that the outer core is denser than the inner core. The alternative is that density increases beyond self-compression occur in the mantle. (After Birch, 1954. *Trans. Am. Geophys. Un.*, *35*, 79–85, copyright by the American Geophysical Union.)

[2] Francis Birch (1903–92) pioneered the use of rock and mineral physics in studies of the earth's composition.

[3] For an adiabatic gradient, rising material reaches the same temperature, and hence density, as its surroundings, and thus has no tendency to continue rising. However, for a superadiabatic gradient, the rising material remains hotter and less dense than its surroundings, and thus tends to continue rising.

[4] The coefficient of thermal expansion, which gives the change in density with temperature T, is $\alpha = (-1/\rho)\partial\rho/\partial T$.

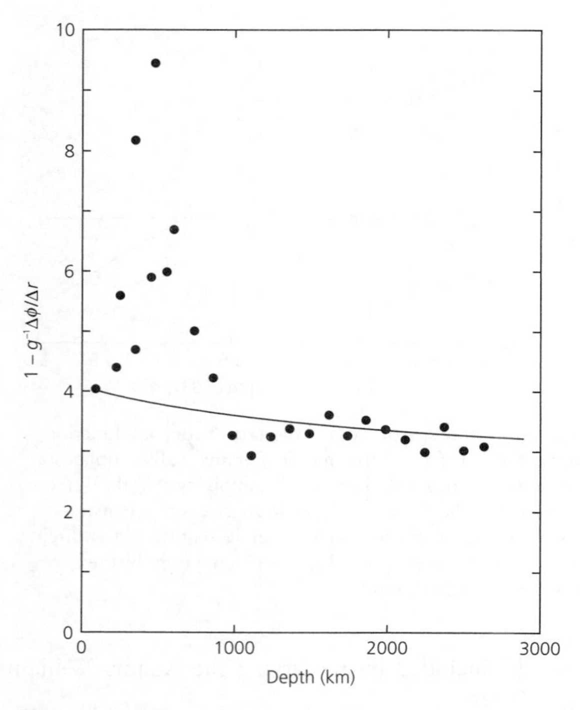

Fig. 3.8-3 Comparison of observed values (dots) of the function $1 - g^{-1}(\Delta\phi/\Delta r)$ for the mantle, with calculated values (line) of this function for compression of homogeneous material. In the upper mantle transition zone, self-compression alone cannot be occurring, motivating the expectation of mineralogical phase transformations. (After Birch, 1952. *J. Geophys. Res.*, *57*, 227–86, copyright by the American Geophysical Union.)

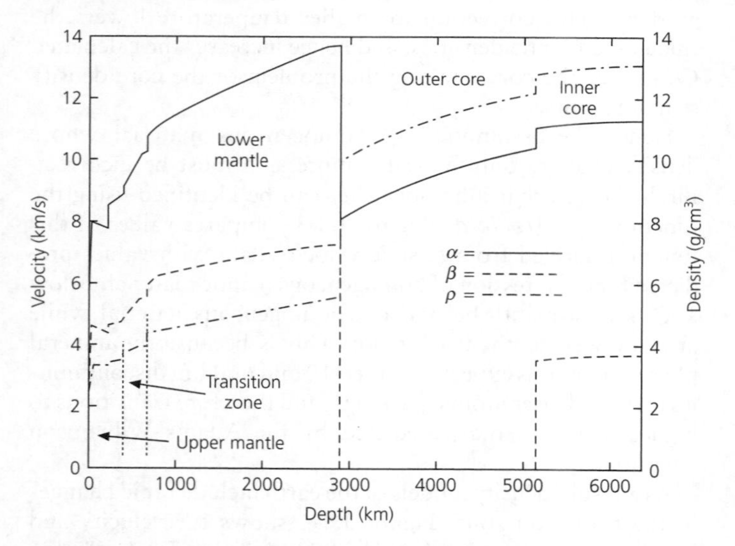

Fig. 3.8-4 Seismic velocities and density for the Preliminary Reference Earth Model (PREM). (Dziewonski and Anderson, 1981.)

changes sharply. The CMB is the most significant boundary with respect to density, with an increase from 5.57 g/cm^3 for mantle rock to 9.90 g/cm^3 for the liquid iron outer core. Density also changes sharply at the 410 and 660 km discon-

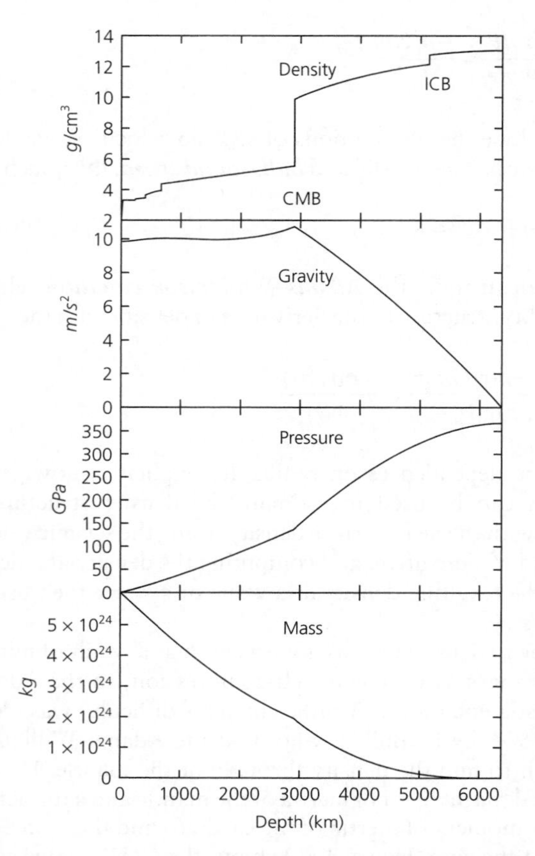

Fig 3.8-5 Density, gravity, pressure, and mass as functions of depth for the PREM model.

tinuities. Such models are developed to satisfy the travel time data, other seismological data including eigenfrequencies of the earth's normal modes, and the constraints on density.

A density profile lets us compute a pressure profile, and thus use the results of experiments showing which mineral phases exist at particular pressures. To do this we integrate both sides of Eqn 8,

$$P(r) = -\int_0^r g(r)\rho(r)dr, \tag{18}$$

using $\rho(r)$ and the resulting values of $g(r)$. As shown in Fig. 3.8-5, pressure starts at 1 bar at the surface, and rises to about 13.3 GPa (133 kbar) at the 410 km discontinuity, 23.8 GPa at the 660 km discontinuity, 136 GPa at the CMB, 329 GPa at the ICB, and 364 GPa at the center of the earth.

The curve for gravity is interesting. Gravity averages 9.8 m/s^2 at earth's surface,[5] and is zero at earth's center, where the mass of the earth pulls evenly in all directions. Gravity

[5] The value of gravity at the surface is a complicated function, varying laterally as a result of density anomalies within the earth, dynamic forces that lift up or pull down the surface, and a latitudinal effect due to the ellipsoidal shape of the earth (Section A.7.2).

Table 3.8-1 PREM Model.

	Depth (km)	ρ (g/cm³)	α (km/s)	β (km/s)
Ocean	0.0	1.020	1.450	0.000
	3.0	1.020	1.450	0.000
Crust	3.0	2.600	5.793	3.191
	15.0	2.600	5.793	3.191
	15.0	2.900	6.792	3.889
	25.0	2.900	6.792	3.889
Upper mantle	25.0	3.381	8.101	4.479
	40.0	3.379	8.091	4.473
	60.0	3.377	8.079	4.465
	80.0	3.375	8.067	4.457
	80.0	3.375	8.005	4.377
low-velocity zone	115.0	3.371	7.984	4.363
	150.0	3.367	7.963	4.350
	185.0	3.363	7.942	4.338
	220.0	3.359	7.920	4.325
	220.0	3.436	8.519	4.589
	265.0	3.463	8.606	4.620
	310.0	3.490	8.692	4.651
	370.0	3.516	8.778	4.683
	400.0	3.543	8.865	4.714
Transition zone	400.0	3.724	9.092	4.874
	450.0	3.787	9.347	5.019
	500.0	3.850	9.601	5.163
	550.0	3.913	9.856	5.307
	600.0	3.976	10.111	5.451
	635.0	3.984	10.165	5.478
	670.0	3.992	10.219	5.505
Lower mantle	670.0	4.381	10.727	5.913
	721.0	4.412	10.885	6.061
	771.0	4.443	11.040	6.207
	871.0	4.504	11.219	6.277
	971.0	4.563	11.390	6.344
	1071.0	4.621	11.552	6.407
	1171.0	4.678	11.707	6.469
	1271.0	4.735	11.856	6.527
	1371.0	4.790	11.998	6.583
	1471.0	4.844	12.135	6.637
	1571.0	4.898	12.266	6.689
	1671.0	4.951	12.394	6.739
	1771.0	5.003	12.518	6.788
	1871.0	5.055	12.638	6.836
	1971.0	5.106	12.757	6.882
	2071.0	5.157	12.873	6.928
	2171.0	5.207	12.988	6.973
	2271.0	5.257	13.103	7.017
	2371.0	5.307	13.218	7.061
	2471.0	5.357	13.333	7.106
	2571.0	5.407	13.450	7.150
	2671.0	5.457	13.568	7.195
	2741.0	5.491	13.652	7.227
	2771.0	5.506	13.659	7.226
	2871.0	5.556	13.684	7.226
	2891.0	5.566	13.689	7.225
Outer core	2891.0	9.903	8.065	0.000
	2971.0	10.029	8.199	0.000
	3071.0	10.181	8.360	0.000
	3171.0	10.327	8.513	0.000
	3271.0	10.467	8.658	0.000
	3371.0	10.602	8.795	0.000
	3471.0	10.730	8.926	0.000
	3571.0	10.853	9.050	0.000
	3671.0	10.971	9.167	0.000
	3771.0	11.083	9.278	0.000
	3871.0	11.191	9.384	0.000
	3971.0	11.293	9.484	0.000
	4071.0	11.390	9.579	0.000
	4171.0	11.483	9.668	0.000
	4271.0	11.571	9.754	0.000
	4371.0	11.655	9.835	0.000
	4471.0	11.734	9.912	0.000
	4571.0	11.809	9.985	0.000
	4671.0	11.880	10.055	0.000
	4771.0	11.947	10.123	0.000
	4871.0	12.010	10.187	0.000
	4971.0	12.069	10.249	0.000
	5071.0	12.125	10.309	0.000
	5149.5	12.166	10.355	0.000
Inner core	5149.5	12.764	10.987	3.434
	5171.0	12.775	10.995	3.440
	5271.0	12.825	11.030	3.465
	5371.0	12.871	11.063	3.487
	5471.0	12.912	11.092	3.508
	5571.0	12.949	11.119	3.526
	5671.0	12.982	11.142	3.542
	5771.0	13.010	11.162	3.556
	5871.0	13.034	11.179	3.568
	5971.0	13.054	11.193	3.578
	6071.0	13.069	11.204	3.585
	6171.0	13.080	11.212	3.590
	6271.0	13.086	11.217	3.594
	6366.0	13.088	11.218	3.595
	6371.0	13.088	11.218	3.595

Source: Dziewonski and Anderson (1981).

increases slightly across the mantle, reaching a maximum of 10.7 m/s^2 at the CMB because of the high density of the core relative to the mantle. Inside the core, gravity decreases nearly linearly toward the earth's center. The high density of the core is also shown by the mass distribution; the core has only 16% of the earth's volume, but has almost one-third of the mass.

3.8.2 *Temperature in the earth*

Seismology gives insight into the *geotherm*, the temperature as a function of radius, which both controls and reflects the composition, mineralogy, and evolution of the earth. The geotherm depends on the sources of heat and modes by which the heat is transferred upward in the earth. Thermal convection, heat transfer by the motions of material due to the density changes resulting from temperature, occurs in the mantle. The most obvious manifestations of this convection are the mid-ocean ridges, which are its hot upwelling limbs, and subducting plates, which are its cold downwelling limbs. A separate convection system in the fluid outer core is believed to cause the earth's magnetic field. In addition, heat is transferred by conduction through the lithosphere, the core–mantle boundary, and the inner core, which may also be convecting.

The geotherm is harder to estimate than the pressure profile and remains a subject of debate. A geotherm is inferred by modeling radioactive generation of heat in the crust and the mantle, conduction of heat across the lithosphere, CMB, and inner core, and adiabatic temperature gradients associated with convection in the mantle and the outer core. The predicted temperatures are required to match the expected temperatures of the phase transitions in the transition zone and the expected freezing point of iron at the ICB.[6] Given the uncertainties involved, estimates of the temperature at the center of the earth vary from 5000 K to almost 7000 K,[7] with recent work favoring the lower end of this range.

A sample geotherm for the mantle is shown in Fig. 3.8-6. The most striking feature is the contrast between the shallow temperature gradient in the mid-mantle and the steep gradients in the upper and lower thermal boundary layers, the lithosphere and D″. The difference reflects the assumptions that heat is conducted primarily through the boundary layers, giving the steep gradients, but is convected between them, yielding a shallower near-adiabatic gradient. The predicted temperature rises from about 0°C at the surface to about 1300°C at a depth of 100 km, giving an average thermal gradient of 13°C/km. From there to the base of the mantle the temperature rises only another 1600°C, corresponding to a low gradient of only about 0.6°C/km. Over the bottom few hundred kilometers of the mantle, however, the temperature rises another 1400°C to a CMB temperature of about 4000°C (~3700 K). Thus the temperature changes across the boundary layers at the surface and CMB are comparable. However, because the surface area of the CMB is only about 30% of the earth's surface, much more heat flows out of the earth than flows out of the core. Most of this extra heat is generated by the decay of radioactive isotopes in the mantle and the crust. An important caveat is that if there are additional thermal boundary layers in the mantle, or if the thermal conductivity of the mantle is higher than expected, the temperatures in the lower mantle will be elevated, and the temperature change across D″ will be less.

The geotherm gives insight into the variations with depth of seismic velocity and attenuation and the strength, or stress, the material can support (Section 5.7). Higher temperatures reduce seismic velocity and strength, but increase attenuation. Conversely, higher pressures increase the velocity and strength, but reduce attenuation. These properties thus depend on the balance between the temperature and the pressure. The cold lithosphere has high velocity and low attenuation, and behaves as rigid plates. However, the rapidly increasing temperature with depth brings the geotherm close to, if not above, the *solidus*, or melting temperature curve. This yields the low-velocity zone, where there is high attenuation and weak material that forms the asthenosphere underlying the moving plates. In the lower mantle, temperatures are only slightly greater than in the asthenosphere, so the higher pressures make the rock stronger. Hence the lower mantle is thought to have a viscosity that is about 100 times greater than that in the upper mantle. Temperatures increase rapidly in D″, causing velocities slower than expected from the lower mantle velocity gradient. The ultra-low-velocity zone at the base of the mantle may be due to partial melting, showing that the geotherm has intersected the solidus. As discussed later, the high temperatures in the core keep the outer core liquid, but the rapid increase in pressure due to the weight of the outer core makes the inner core freeze into a denser solid. The inner core is therefore close to the melting temperature of iron, so it has low shear velocities.

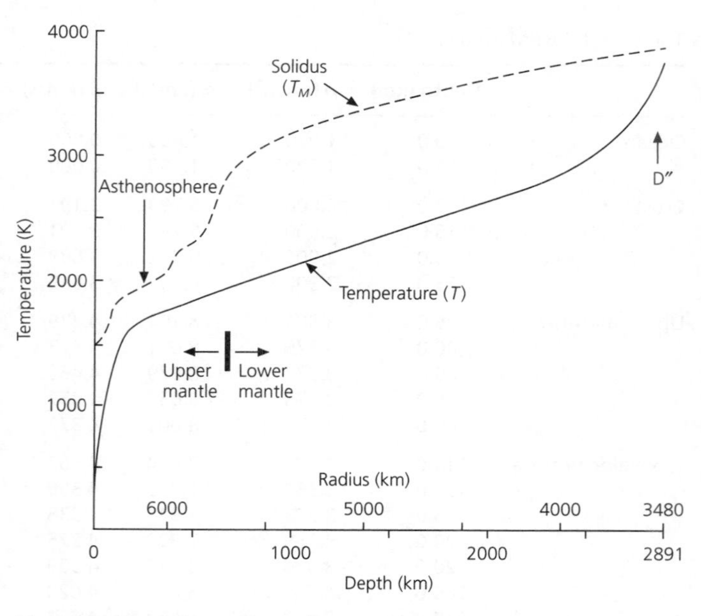

Fig. 3.8-6 A sample mantle geotherm with steep temperature gradients in the thermal boundary layers at the top and bottom of the mantle and a near-adiabatic gradient in the lower mantle. The melting curve, or solidus, is also shown. Temperatures are given in absolute temperature. (Stacey, 1992. From *Physics of the Earth*, 3rd edn, copyright © 1992 by John Wiley & Sons, Inc. (New York). Reprinted by permission.)

3.8.3 Composition of the mantle

Models of the composition of the mantle are derived by comparing the velocity and density (and therefore pressure) profiles

[6] Although our instincts based on water make it strange to think of temperatures near 5000° as "freezing," this occurs as the solid inner core forms from the liquid outer core.

[7] Temperatures in the deep earth are often given as absolute (Kelvin) temperatures, equal to the Celsius temperatures plus 273.15°.

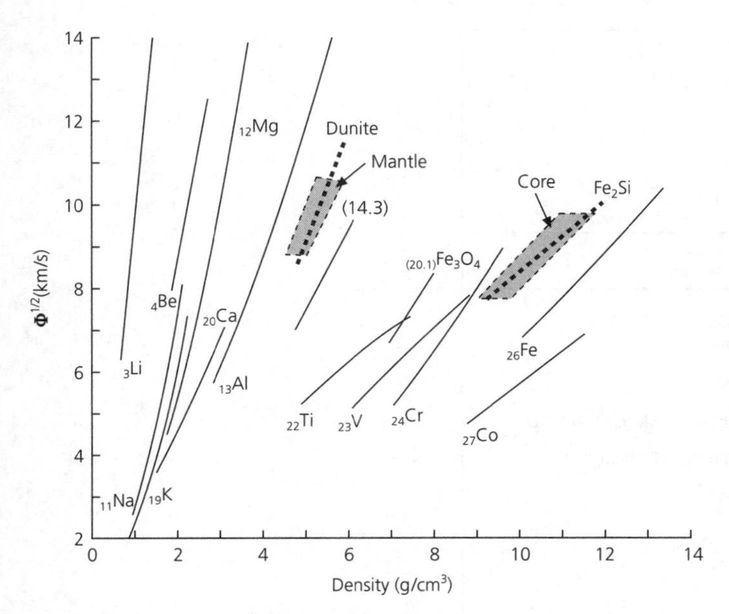

Fig. 3.8-7 Bulk sound speed as a function of density for various materials, obtained from experiments (lines) compared to the range for the mantle and the core from seismic observations and density models (shaded). Also shown are the results for a dunite rock and the composition Fe_2Si. The numbers shown are mean atomic numbers. (After Birch, 1968. *Phys. Earth Planet. Inter.*, *1*, 141–7, with permission from Elsevier Science.)

derived from seismic data to temperature profiles and results for earth materials at high pressure and temperature. A key result from experiments is that the bulk sound speed (Eqn 15) and the density for a material are approximately linearly related for a given mean atomic weight. The mean atomic weight is the mean molecular weight of a formula unit, such that forsterite (magnesian olivine) Mg_2SiO_4 has $\bar{m} = (2 \times 24 + 28 + 4 \times 16)/7 = 20$, and fayalite (iron olivine) Fe_2SiO_4 has $\bar{m} = (2 \times 56 + 28 + 4 \times 16)/7 = 29$. Figure 3.8-7 shows this result for various elements whose atomic numbers[8] are labeled. Also shown are ranges of density and bulk sound speed for the mantle and core derived from seismically based models. The mantle and the core occupy different parts of the plot.

This result suggests that the mantle and the core are chemically different, and provides a way of testing which chemical compositions are plausible. Dunite, a rock containing 92% olivine, which in turn is 90% forsterite, fits the mantle data. Curves for more iron-rich olivine would plot further to the right, such that olivine with more than 50% fayalite would be outside the range observed for the mantle.

The core data plot much further to the right, indicating that the core is composed of material of higher atomic number. The data are to the left of the curve for pure iron, suggesting that the core is composed of iron plus a lower atomic weight ("lighter")

[8] The atomic number is the number of protons, whereas the atomic weight is the number of protons and neutrons.

Table 3.8-2 Pyrolite model, mineralogy above transition zone.

Mineral	Composition	wt(%)
Olivine (Fo_{89})	$(Mg_{0.89}, Fe_{0.11})_2SiO_4$	57
Orthopyroxene	$(Mg, Fe)SiO_3$	17
Clinopyroxene	$(Ca, Mg, Fe)_2Si_2O_6 - NaAlSi_2O_6$	12
Pyrope-rich garnet	$(Mg, Fe, Ca)_3(Al, Cr)_2Si_3O_{12}$	14

Source: Ringwood (1979).

element. For example, the composition Fe_2Si (iron plus 20% weight Si) fits the core data.

Various chemical models for the mantle have been proposed. The concepts involved can be illustrated by considering a proposed composition called *pyrolite* that satisfies various petrological, cosmochemical, and geophysical constraints. Pyrolite is similar to natural peridotites (Fig. 3.2-23), which are acceptable source rocks for basaltic magmas that result from partial melting of mantle rock. The variation in seismic velocity and density with depth is assumed to result from transformations to denser phases as a result of increased pressure. Table 3.8-2 gives a composition whose density at surface temperature and pressure conditions would be 3.38 g/cm³ and has *P*- and *S*-wave velocities consistent with those observed for the upper mantle.

In the upper mantle, the model's major mineral component is olivine. Such a composition satisfies the density and bulk sound speed data (Fig. 3.8-7) and is consistent with the observed seismic anisotropy (Fig. 3.6-4). The transition zone corresponds to a series of solid state phase changes (Fig. 3.8-8). Olivine undergoes several transformations before converting to a *perovskite* structure in the lower mantle. Pyroxene first transforms to garnet, and somewhat deeper, the calcium-bearing component of the garnet transforms to a perovskite structure. Because of the predicted predominance of perovskite (~70%) in the voluminous lower mantle, perovskite is the most abundant material in the earth.

Figure 3.8-9 shows the predicted volume fraction of the major mineral phases as a function of depth. The α phase of olivine, which occurs in the crust and the upper mantle, transforms with increased pressure to its β phase *wadsleyite*, which has a modified spinel structure. This transformation is observed experimentally to occur at a pressure of about 12 GPa (120 kbar), corresponding to the 410 km discontinuity. The β phase transforms to a γ, or spinel, structure known as *ringwoodite* (Fig. 3.8-10) at a pressure of ~15 GPa, corresponding to the less dramatic seismic discontinuity at 520 km. At pressures above about 24 GPa, corresponding to the 660 km discontinuity, γ spinel breaks down to a perovskite structure and (Mg, Fe)O *magnesiowustite*.

The $(Mg,Fe)SiO_3$ pyroxene component also undergoes changes, beginning with a transformation to garnet below about 200 km. Below 600 km, some of the *Mg*-bearing garnet, majorite, transforms to a structure called *ilmenite*. Beneath about 660 km, the majorite/ilmenite transforms to perovskite. Some of the

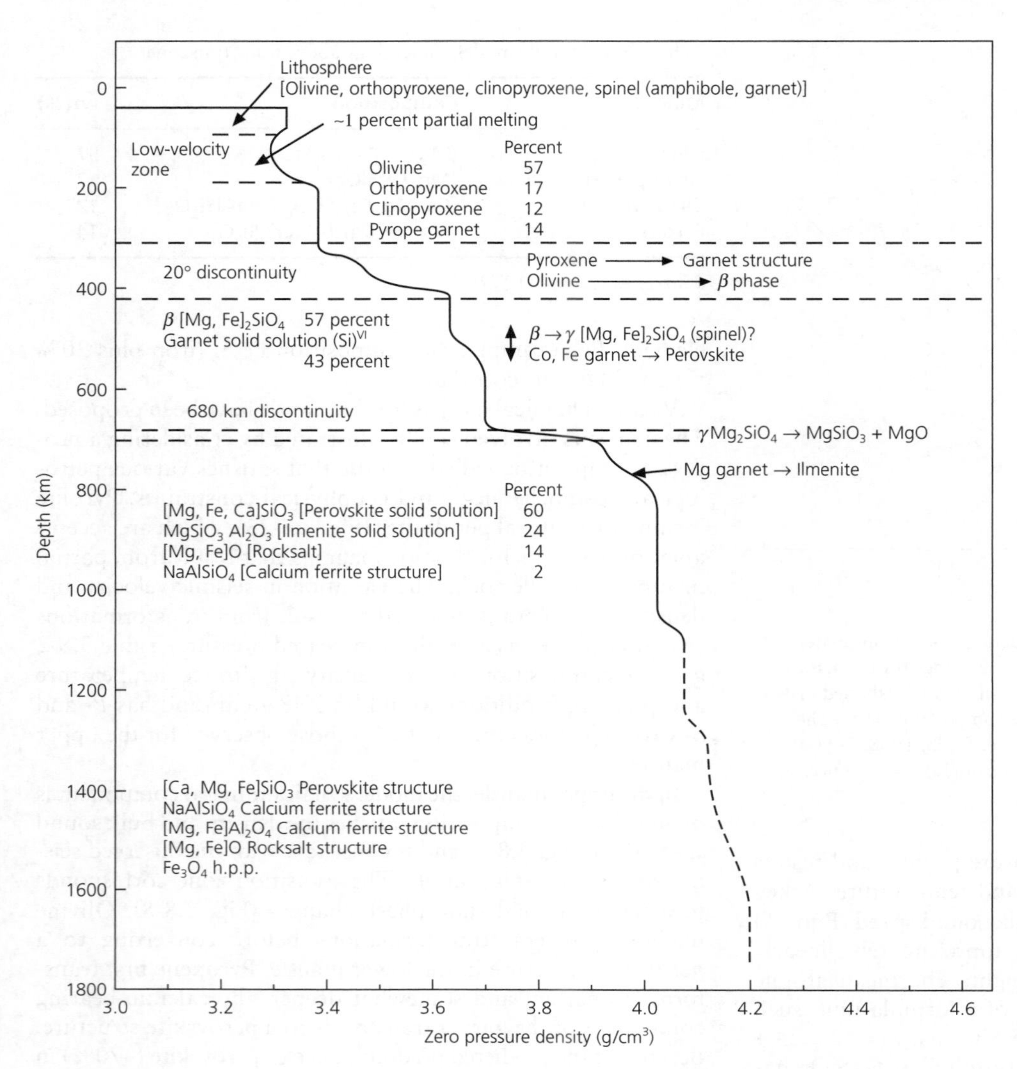

Fig. 3.8-8 Predicted mineral assemblages as a function of depth for a mantle of pyrolite composition. (Ringwood, 1979. Composition and origin of the earth, in *The Earth, Its Origin, Structure and Evolution*, ed. M. W. McElhinny, copyright 1979 by Academic Press, reproduced by permission of the publisher.)

majorite probably survives into the lower mantle as *stishovite*, a high-pressure phase of quartz, and an Al_2O_3-rich phase. Unlike the olivine transformations that cause distinct seismic discontinuites in the transition zone, the pyroxene and garnet transformations occur gradually and contribute to a high velocity gradient through the transition zone down to about 770 km (Section 3.5.4).

These phase changes are investigated using experiments that simulate the pressures, temperatures, and compositions in the earth. Because the experiments are difficult, extrapolations of lower pressure and temperature data via thermodynamic calculations are also used. An important factor for the velocity structure is that some phase transformations happen gradually over a range of depths (Fig. 3.8-11). A simple *univariant* phase change, in which material of a single composition changes completely from one phase to another as pressure increases, causes a sharp discontinuity in velocity. A more complicated *multivariant* phase change involving a system of variable compositions causes two or more phases to coexist over a broad region of pressure, and so produces a velocity gradient. Thus seismological studies that better define the velocity structure of the transition zone improve our understanding of its composition.

The mineralogical models agree with the depths of the seismic discontinuities and their other characteristics. The olivine α-to-β reaction should occur over a narrow depth range, as shown by the volume fractions in Fig. 3.8-9. This prediction is consistent with the sharpness of the seismic discontinuity, which is observed with high-frequency (short-wavelength) waves. The transformation is exothermic (releasing heat) and hence would occur at lower pressures in subducting slabs due to the colder temperatures (Section 5.4.2). This expectation agrees with seismic observations showing an elevation of the 410 km discontinuity in and around subducting lithosphere. By contrast, the β-to-γ transformation should occur over a broader depth range. This prediction agrees with seismic observations of the 520 km discontinuity, which is invisible to high-frequency waves and seen only with longer wavelengths.

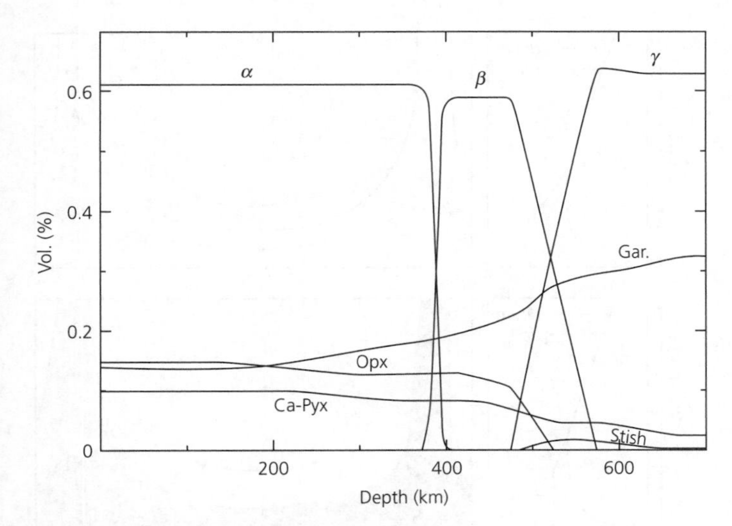

Fig. 3.8-9 A model for the relative proportions of major mineral phases as a function of depth in the upper mantle. The rapid changes between the olivine and spinel phases (α, β, γ) cause seismic discontinuities at depths of 410 km and 520 km, whereas the gradual transformation of pyroxene to garnet steepens the velocity gradient in the transition zone (410 km to 660 km). (Weidner, 1986. Reproduced with permission of Springer-Verlag.)

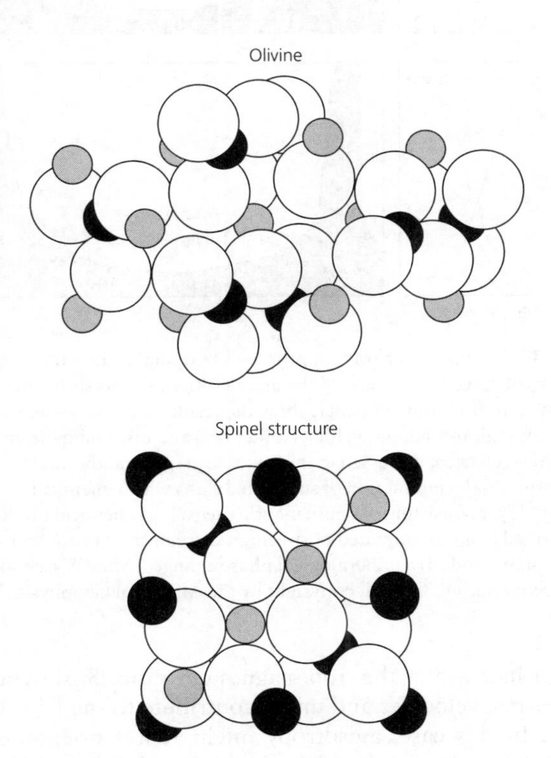

Fig. 3.8-10 Comparison of the crystal structures of $(Mg, Fe)_2SiO_4$, in its low-pressure α olivine phase (*top*) and its γ-spinel ringwoodite phase (*bottom*), which is about 10% denser. Spheres correspond to ions of oxygen (white), silicon (black), and magnesium/iron (grey). (After Press and Siever, 1982.)

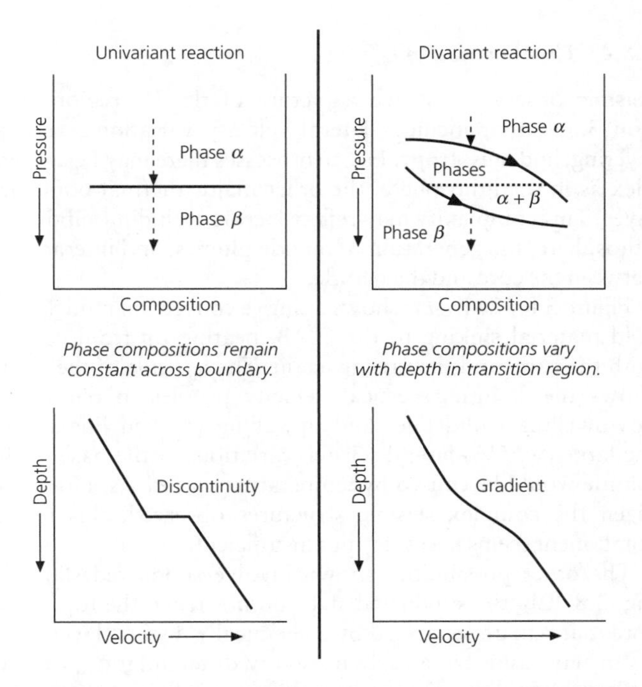

Fig. 3.8-11 Schematic phase diagrams showing the relation between the nature of a phase change and the corresponding velocity discontinuity. (Bina and Wood, 1987. *J. Geophys. Res.*, 92, 4853–66, copyright by the American Geophysical Union.)

However, the γ-spinel to perovskite and magnesiowustite transition should occur over a narrow depth range, consistent with the observed sharpness of the 660 km seismic discontinuity. The reaction is endothermic (absorbs heat) and so should occur at greater depths for colder temperatures. Studies have shown that the discontinuity is depressed to depths of 700 km or more in and around subducting lithosphere.

An unresolved question is whether the lower mantle is chemically distinct from the upper mantle, which has important implications for how the two have mixed during the earth's evolution. In models like those depicted in Fig. 3.8-8, the two are assumed to have the same bulk chemistry, and the increasing velocity and density in the lower mantle result from self-compression. The velocity data do not appear to require phase changes in the lower mantle. However, the lower mantle may be denser than expected for pyrolite, and hence perhaps enriched in iron and silica. The observation that some subducting lithosphere penetrates the 660 km discontinuity (Section 5.4) indicates that mixing occurs. However, even if all slabs reach the lower mantle, the earth may not be old enough for the lower and upper mantles to be well mixed.[9] Another possibility is that the early earth had distinct upper and lower mantle convection systems, and whole mantle convection began later.

[9] Consider a bowl of cake batter after only a few beats of a mixing spoon.

3.8.4 Composition of D″

Seismic observations give a picture of the D″ region (Section 3.5.4) that includes lateral velocity variations, vertical layering, and anisotropy. Hence processes there may be as complex as in the lithosphere, the other major thermal boundary layer. This complexity may reflect factors including subducted lithosphere, the generation of mantle plumes, and interactions between the core and the mantle.

Figure 3.8-12a (*right*) shows a simple convection model, with cold material sinking to the CMB, heating up from contact with the core, and then rising again. The *left* side of the figure shows the resulting vertical velocity profiles in regions of downwelling (solid line) and upwelling (dashed line). Thus the large (> ±5%) lateral seismic variations at the base of the mantle would be caused by temperature variations. However, given the complex seismic structures observed, this model component seems necessary but insufficient.

The other possibilities shown involve subducted slabs. In Fig. 3.8-12b, the subducted slabs do not reach the top of the core, but remain separated by a chemically distinct layer. This layer may result from early planetary differentiation, or may have grown by chemical reactions between the mantle and the core. High-pressure experiments imply that perovskite and magnesiowustite would react with iron. These mantle dregs might be thinned in regions of mantle downwelling, and thickened beneath upwellings. Layering in the dregs may explain observations of transverse isotropy in downwelling regions and azimuthal anisotropy in upwelling regions (Section 3.6.6). The velocity increase of the D″ discontinuity may be partly caused by ponded slab material, which will still be colder and have higher velocity than ambient rock. This discontinuity may be enhanced by dregs flowing up and over ponded slabs. The ultra-low-velocity zone (ULVZ) at the very bottom of the mantle may be due to the lower velocities of an iron-rich layer or to partial melting within it.

Another possibility is that the part of the subducted lithosphere that started as basaltic ocean crust and then transformed to eclogite transforms to a material that is seismically faster than the rest of the lower mantle (Fig. 3.8-12c). This phase could delaminate from the slabs and accumulate, forming a different chemical boundary layer. If it remained solid, it might partially explain the D″ discontinuity. Alternatively, if it melted, it might explain the ULVZ. Either way, its laminar nature might explain the observed seismic anisotropy. The lateral variations in velocity would correlate with anisotropy; *SH* waves would travel fast in downwelling regions because of transverse isotropy, but be slowed by the vertical laminations beneath upwellings.

D″ may also signify the bottom of the perovskite stability field (Fig. 3.8-12d). Large radial changes in temperature and/or composition at the base of the mantle could move perovskite or a secondary phase out of its range of stability, causing a phase transformation. One possibility is a transformation of perovskite to stishovite and magnesiowustite, which occurs

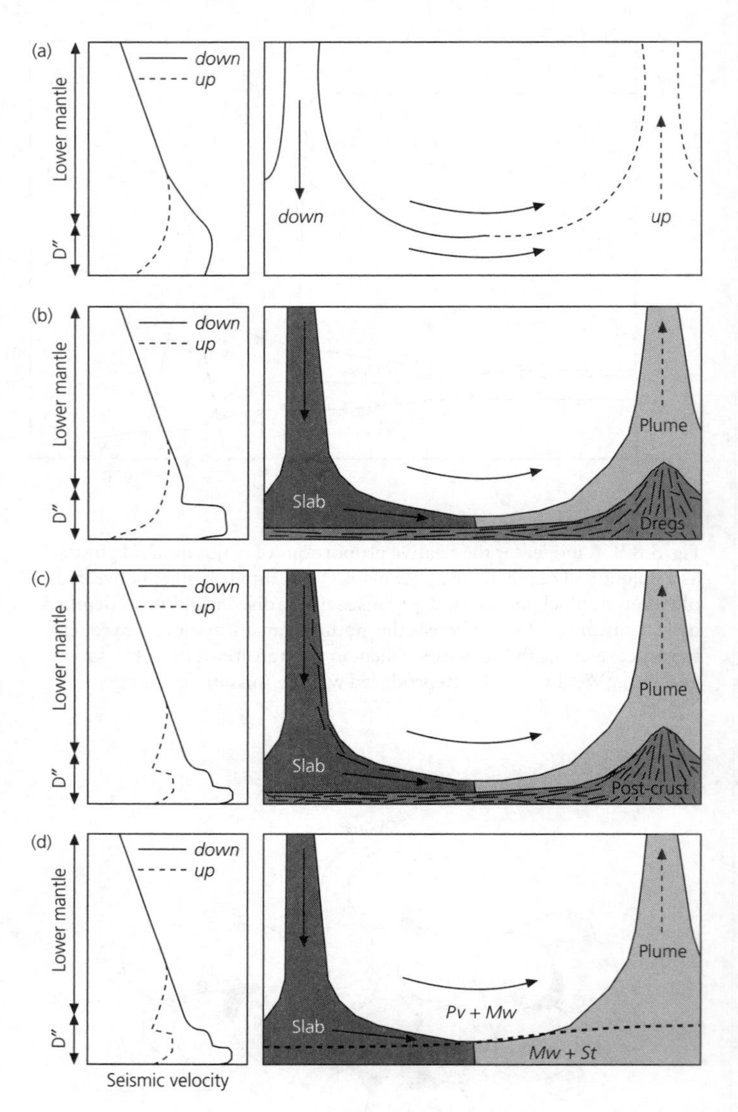

Fig. 3.8-12 Schematic diagram of processes that might cause the velocity structures observed at the base of the mantle. *Right* panels show the different scenarios, and *left* panels show the resulting velocity–depth profiles for regions of downwelling (solid lines) and upwelling (dashed lines). The scenarios, discussed in the text, are (a) general thermal convection, (b) the interaction of subducted slabs with a chemical boundary layer consisting of dense mantle dregs, (c) a chemical boundary layer formed from delaminated post-eclogitic ocean crust brought down with the slabs, and (d) a mineralogical phase change. (After Wysession, 1996a. *Subduction*, 369–84, copyright by the American Geophysical Union.)

with an increase in the iron/magnesium ratio. Stishovite has high seismic velocities and might contribute to the D″ discontinuity. In this case, anisotropy might reflect orientation of crystals due to lateral flow. The denser magnesiowustite might settle to the bottom, forming the ULVZ.

Given our limited knowledge, D″ may involve these and other effects. For example, if the vertical temperature difference across D″ is small (about 300°C), then convection should play

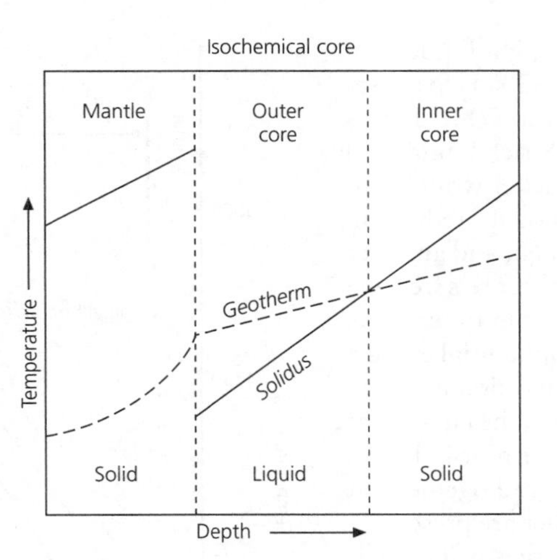

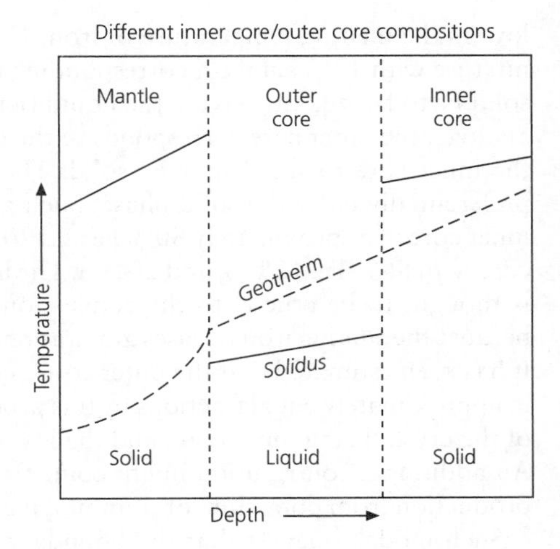

Fig. 3.8-13 Possible relationships between the geotherm (dashed line) and the solidus (solid line), for the inner and outer cores. *Left*: If the core is homogeneous, the solidus should be continuous across the inner and outer cores, so the gradient of the geotherm must be shallower than that of the solidus for the inner core to be solid and the outer core to be liquid. *Right*: If the inner and cores are chemically different, the solidus can differ between them, allowing a steeper gradient for the geotherm.

a lesser role relative to a chemical boundary layer. If the contrast is large, perhaps 1500°C, plume generation should be more significant, and it would be harder to maintain a distinct chemical layer.

3.8.5 *Composition of the core*

Interesting issues about the core also remain unresolved. The density and bulk sound speed data (Fig. 3.8-7) suggest that the core has a composition similar to that of iron, but with a less dense element of lower atomic number added. Other arguments for an iron core are from cosmochemistry. Meteorites are roughly divided into stony meteorites, resembling the mantle, and iron meteorites, composed of an iron–nickel alloy, which are thought to be similar to the core.[10] Convection of molten iron is also considered the only suitable mechanism for generating the earth's magnetic field. The light element lowering the core density is unknown: candidates include sulphur, silicon, oxygen, potassium, and hydrogen. Laboratory experiments suggest that 10–15% of a light element would yield an acceptable density.

It may seem surprising that the inner core is solid, because it should be at a higher temperature than the liquid outer core. Thus the effects of pressure favoring the denser solid phase must exceed those of temperature. From the ICB to the center of the earth, temperature is thought to increase by only 100–200°C, or about 3% of the inner core temperatures, which are about 5000°C. Pressure, however, is thought to increase about 11%, from about 329 GPa at the ICB to 364 GPa at earth's center (Fig. 3.8-5). The density inferred from the seismological data is consistent with that for solid iron expected from experiments and modeling.

This situation requires that the inner core geotherm be at temperatures below the melting temperature curve (solidus), whereas the outer core geotherm must be above the solidus.

[10] Iron meteorites — and thus presumably the solid inner core — are like steel, recalling legends in which swords forged from meteorites are very strong and have magical powers.

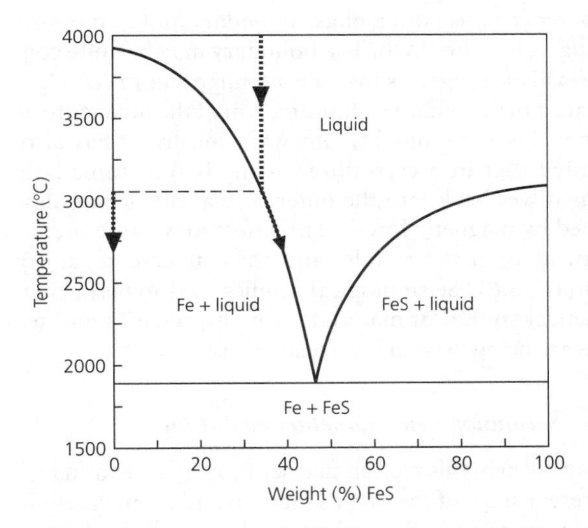

Fig. 3.8-14 Melting relations for the Fe–FeS system at the pressure of the core–mantle boundary (1.4 Mbar). When a cooling liquid with 33% FeS reaches the phase boundary, solid Fe freezes out, enriching the liquid in FeS. In this analogy, the inner core is freezing out from, and thus chemically different from, the outer core. (Data from Usselman, 1975.)

Two suggestions have been offered for this effect. If the inner and outer cores were chemically identical (Fig. 3.8-13, *left*), the solidus should rise smoothly with depth. The geotherm would be shallower than the solidus, so that they intersect at the ICB, but steeper than the adiabatic gradient required for convection in the outer core. However, some theoretical calculations suggest that the superadiabatic temperature gradient in the core required for convection would be steeper than the solidus. If so, the solid inner and liquid outer cores can be explained by assuming that the inner core is chemically different from the outer core, and thus has a different melting curve (Fig. 3.8-13, *right*). Thus, only in the inner core does the geotherm lie below the solidus and result in a solid phase.

Figure 3.8-14 illustrates this idea, assuming that the light element in the core is sulfur. In this phase diagram for the Fe–FeS system extrapolated to core conditions, sulphur significantly lowers the melting temperature of iron. Cooling a liquid iron mixture with 12% sulphur, corresponding to 33% FeS, causes solid Fe to freeze out, leaving the liquid richer in FeS.[11] In this analogy, the outer core corresponds to the FeS-rich liquid, and the inner core to the denser Fe solid. The nickel would also preferentially enter the solid phase. Such a model predicts an inner core of approximately 80% Fe and 20% Ni, and an outer core with 86% Fe, 12% S, and 2% Ni. The inner core's freezing is thought to be crucial to the convection in the outer core, because the sinking iron releases gravitational potential energy. It has been estimated that the outer core's convection is driven in approximately equal fractions by this process, the latent heat of the crystallizing inner core, and the loss of primordial heat. An additional contribution might come from radiogenic heat production from potassium or uranium, if either are present.

Such models suggest that the boundary between the inner and outer cores is both a phase boundary and a compositional boundary, like the CMB. The boundary may be quite complex. Some evidence suggests that the attenuation of *PKP-DF* waves is greatest in the outer few hundred km of the inner core, implying that this zone may be somewhat mushy. It has also been suggested that iron crystallizes at the ICB at some latitudes, and dissolves back into the outer core at other latitudes, constrained by magnetic forces. This effect may cause preferential alignment of iron crystals, and thus inner core anisotropy (Section 3.6.6). Seismological studies and experimental and theoretical studies of materials at high pressures and temperatures are being used to investigate these issues.

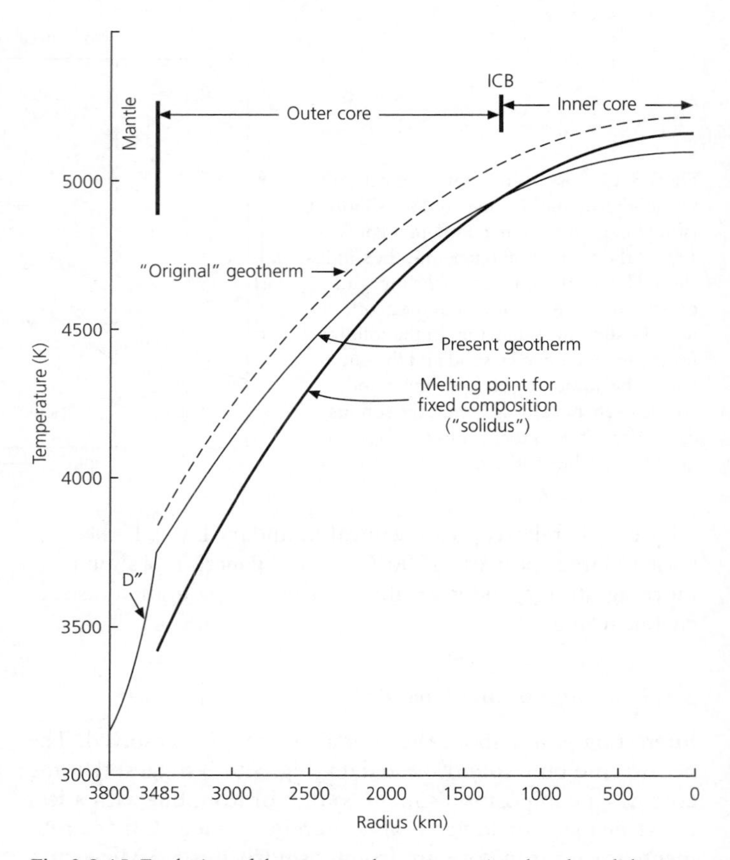

Fig. 3.8-15 Evolution of the core geotherm, assuming that the solidus is continuous between the inner and outer cores. Early in the earth's history the core geotherm (dashed line) was everywhere greater than the solidus, making the whole core molten. As the core cooled, the geotherm lowered, causing the growth of a frozen inner core. The ICB is the current intersection between the geotherm and the solidus. (After Stacey, 1992. From *Physics of the Earth*, 3rd edn, copyright © 1992 by John Wiley & Sons, Inc. (New York). Reprinted by permission.)

3.8.6 Seismology and planetary evolution

We have seen in this section that seismology gives a snapshot of the present stage of the earth's thermal and chemical evolution. Seismology shows the present thickness of the lithosphere, which may have increased with time, and provides much of our information about plate tectonic processes and mantle convection. Seismology similarly provides most of what we know about the core, including the present sizes of the inner and outer cores that reflect the progressive freezing of the solid inner core from the liquid outer core. Hence, as shown in Fig. 3.8-15, the core has been cooling with time, causing the inner core to grow.

What we know about the earth and our more limited knowledge of the moon and other planets suggest that although there are differences among the inner planets that reflect their initial compositions, there are also similarities in their evolution. As shown in Fig. 3.8-16, planets may follow a similar life cycle, with phases including their formation, early convection and core formation, plate tectonics, terminal volcanism, and quiescence. This evolution is driven by the available energy sources and reflects the planets' cooling with time. Thus, even though the planets formed at about the same time, they are at different stages in their life cycles.[12] The earth is in its middle age, characterized by active plate tectonics.

Hence the approaches used to study the earth's interior can be applied to other planets. A five-station seismological network deployed on the moon by the Apollo missions found a very low level of seismicity, of which most reflected meteoroid impacts or small moonquakes generated by tidal forces. Travel time studies yielded the velocity profile shown in Fig. 3.8-17, which has considerable uncertainty owing to the small number of seismometers and the difficulty of identifying arrivals due to scattering (Fig. 3.7-10). Various interpretations have been made of these results. Although it is tempting to correlate the low-velocity zone with an asthenosphere, thermal models predict that this region would be too cold. As a result, the zonation of the mantle is thought to represent compositional differences.

[11] This effect in which the composition of the liquid and the solid differ is called *fractional crystallization* and has many geological applications, including formation of rocks from a cooling magma. It can be illustrated with partially frozen apple juice, where the liquid tastes sweeter because it is enriched in sugar relative to the solid fraction.

[12] Consider a human and dog born on the same date.

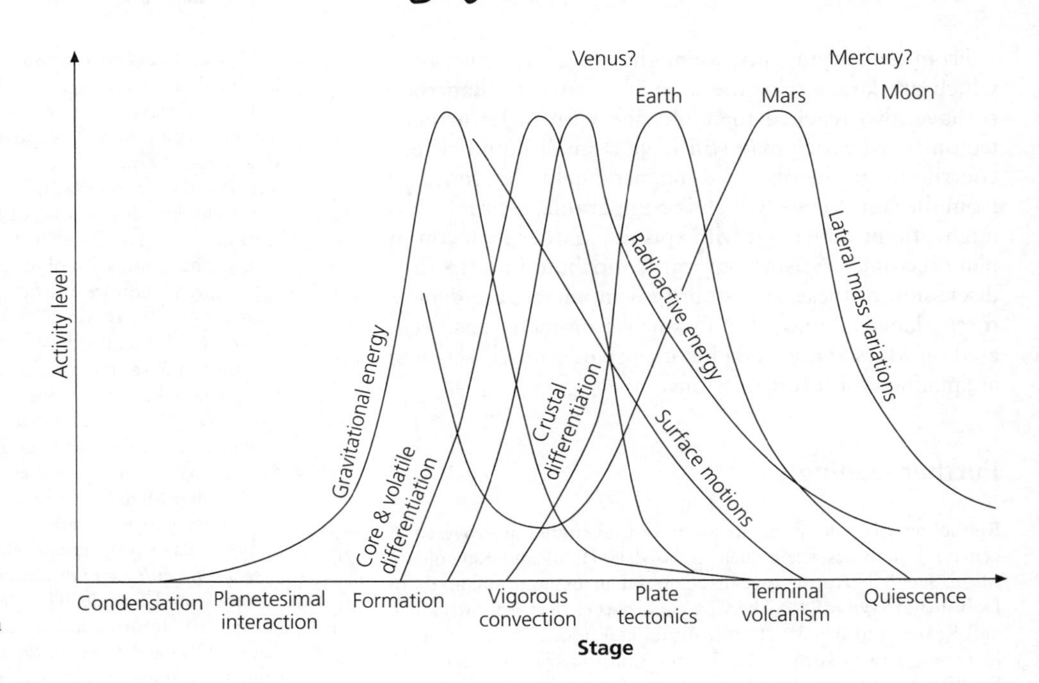

Fig. 3.8-16 A model for the evolution of the terrestrial planets, showing the energy sources at each stage, presented in text and verse by Kaula (1975). (© 1975 by Academic Press, reproduced by permission of the publisher.)

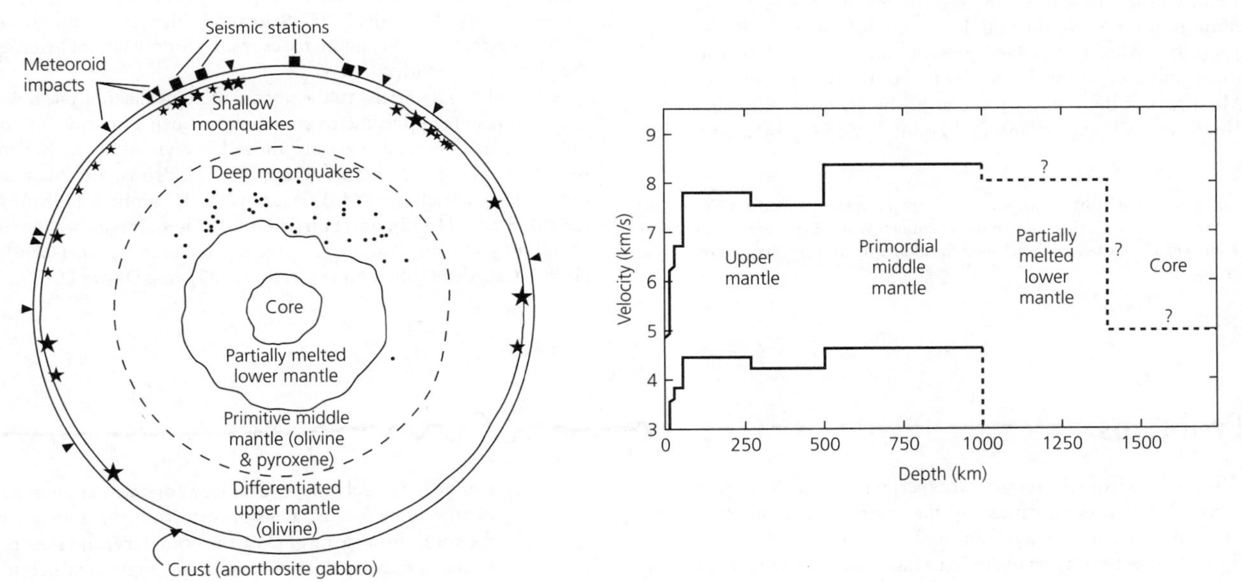

Fig. 3.8-17 Velocity model from lunar seismic data (*left*) and a possible compositional interpretation (*right*). Squares show seismometer locations, arrowheads show major meteoroid impacts, and large and small dots denote shallow and deep moonquakes. (After Nakamura, 1983 (*J. Geophys. Res.*, *88*, 677–86, copyright by the American Geophysical Union) and Hubbard, 1984.)

There is a suggestion of decreased velocity below 1000 km, which thermal models suggest could be consistent with an asthenosphere. Seismological efforts to detect a core are inconclusive, and the moment of inertia ratio of 0.39 allows for at most a small core.

Hence it appears that the moon now has a thick lithosphere and is tectonically inactive. It thus seems to have lost much of its heat, presumably because of its small size, which favors rapid heat loss. In general, we would expect the heat available from the gravitational energy of accretion and radioactivity to increase as the planet's volume, whereas the rate of heat loss through the surface should depend on its surface area. Hence the remaining heat should vary as

$$\text{remaining heat} = \frac{\text{available}}{\text{loss}} = \frac{(4/3)\pi r^3}{4\pi r^2} = \frac{r}{3}, \tag{19}$$

so larger planets would retain more heat and be more active.

From such arguments, we might expect Mercury and Mars, which are larger than the moon but smaller than the earth, to have also reached their old age with little further active tectonics. Mercury may still have a small liquid core, which contributes to the observed magnetic field, due to tidal forces from the sun. Venus, which is comparable in size to the earth, might still be active but with episodic, rather than continuous, plate tectonics. Seismology can contribute little to the active discussion of these topics until seismometers are deployed on these planets. Although only one seismometer has been operated on Mars and yielded inconclusive results,[13] seismometers are planned for future missions.

Further reading

Refraction seismology and its use in crustal studies are covered in many general geophysics texts, such as Fowler (1990) and Reynolds (1997). More detailed treatments can be found in exploration textbooks like Dobrin and Savit (1988), Sheriff and Geldart (1982), Telford *et al.* (1976), and Kearey and Brooks (1984). Additional information can be obtained in review papers such as Braile and Smith (1975), Kennett (1977), or Spudich and Orcutt (1980). A summary of crustal structure results and interpretations for the continental USA can be found in Pakiser and Mooney (1989). Meissner (1986) presents an integrated treatment of observations and models for the continental crust. Reviews on the nature of the Mohorovičić discontinuity are given by Jarchow and Thompson (1989), Braile and Chiang (1986), and Fountain and Christensen (1989). Gibson and Levander (1988) discuss posible artifacts in lower crustal reflection data.

The extensive literature on reflection seismology includes the introductory exploration texts listed above and advanced treatments, including Claerbout (1976, 1985), Robinson and Treitel (1980), Waters (1981), Sheriff and Geldart (1982), Robinson (1983), and Yilmaz (1987). The subject is closely allied to that of geophysical signal processing, discussed in texts including Kanasewich (1981) and Hatton *et al.* (1986).

Applications of seismology to earth structure are discussed in texts and the research literature. Introductory texts such as Bolt (1982), Bott (1982), Gubbins (1990), Doyle (1995), Lay and Wallace (1995), Lowrie (1997), Shearer (1999), and Udias (1999), have good overviews. Simon (1981) is a manual for seismogram interpretation, showing examples of records for earthquakes at various distances and depths. The classic texts by Gutenberg (1959) and Jeffreys (1976) are excellent starting points for further treatment of the data and methods. Bullen and Bolt (1985) has a detailed discussion of ray theory for the spherical earth. Aki and Richards (1980), Ben-Menahem and Singh (1981), and Kennett (1983) treat both ray theory and more advanced methods. The normal mode simulation of body wave propagation shown in Fig. 3.5-19 is available at *http://epsc.wustl.edu/seismology/michael/movie.html.*

Karato and Spetzler (1990) review the physical mechanisms causing anelasticity. Information about anisotropy can be found in Babuska and Cara (1991) and Silver (1996). For discussion of scattering and attenuation, see Kanamori and Anderson (1977), Brennan and Smylie (1981), Jackson (1993), Mitchell (1995), Sato and Fehler (1998), and Romanowicz (1998). Garnero (2000) summarizes results for the lateral heterogeneity of the lowermost mantle.

We alluded only briefly to the nonseismological geophysical data and to chemical results applicable to study of the earth's interior. In addition to journal articles, useful texts are those by Wyllie (1971), Bullen (1975), Ringwood (1975), Wood and Fraser (1977), Brown and Mussett (1993), Bott (1982), Melchior (1986), Jacobs (1987), Lambeck (1988), Anderson (1989), Stacey (1992), and Poirier (2000). Useful reviews can be found in McElhinny (1979), Ahrens (1995a, b, c), Boschi *et al.* (1996), Boehler (1996), Crossley (1997), Gurnis *et al.* (1998), and Davies (1999).

[13] Due to operational constraints, the seismometer was mounted on the lander portion of the spacecraft, rather than in direct contact with Mars. It is rumored that consideration was given to saving weight on the lander by moving the seismometer to the orbiter.

Problems

1. Use the data from the refraction experiment in Fig. 3.2-5 to find the crust and mantle velocities and the crustal thickness. Remember that this is a reduced travel time plot.
2. For a case of two layers overlying a halfspace, derive an expression for the thickness of the second (deeper) layer in terms of the second crossover distance.
3. Analyze the data from the marine refraction experiment (Lewis, 1978) shown in Fig. P3.1, assuming for simplicity that the structure consists of a water layer, a crustal layer, and a mantle halfspace.
   (a) Assuming that the first arrivals are described by two line segments, for head waves at the top of the crust and mantle, find the corresponding velocities.
   (b) Although the direct wave traveling in the water layer is not shown, the P velocity for water is 1.5 km/s. Use the time intercept for the crustal head wave to find the water depth.
   (c) Use the time intercept for the P_n wave to find the crustal thickness.
4. To show that the head wave is predicted by Fermat's principle, consider a layer of thickness h with velocity v_0, overlying a halfspace with a higher velocity, v_1.
   (a) Derive the travel time to distance x for a wave that is incident on the boundary at a distance y from the source, travels for some distance just below the boundary, and then returns to the surface at the same incidence angle at which it went down.
   (b) Find the y value giving an extremal travel time, and show that it corresponds to the critical angle of incidence.
   (c) Determine if this travel time is a minimum or a maximum.
5. Use the data for the reversed profile shown in Fig. P3.2 to find the crust and mantle velocities, Moho dip, and crustal thickness.
6. (a) Derive the travel time for the head wave on the up-dip path of a reversed profile with a dipping layer (Eqn 3.2.17).
   (b) Show that the equations for the travel time of the head wave for a dipping layer (Eqns 3.2.16 and 3.2.17) reduce to the flat layer result in the case of zero dip.
7. Derive the Dix equation for interval velocity (Eqn 3.3.19) from the formula for rms velocity.
8. Consider two pairs of seismograms. One pair have the same midpoint, but the offset for one record is the negative of the first. The other pair have the same source point, but the offset for one record

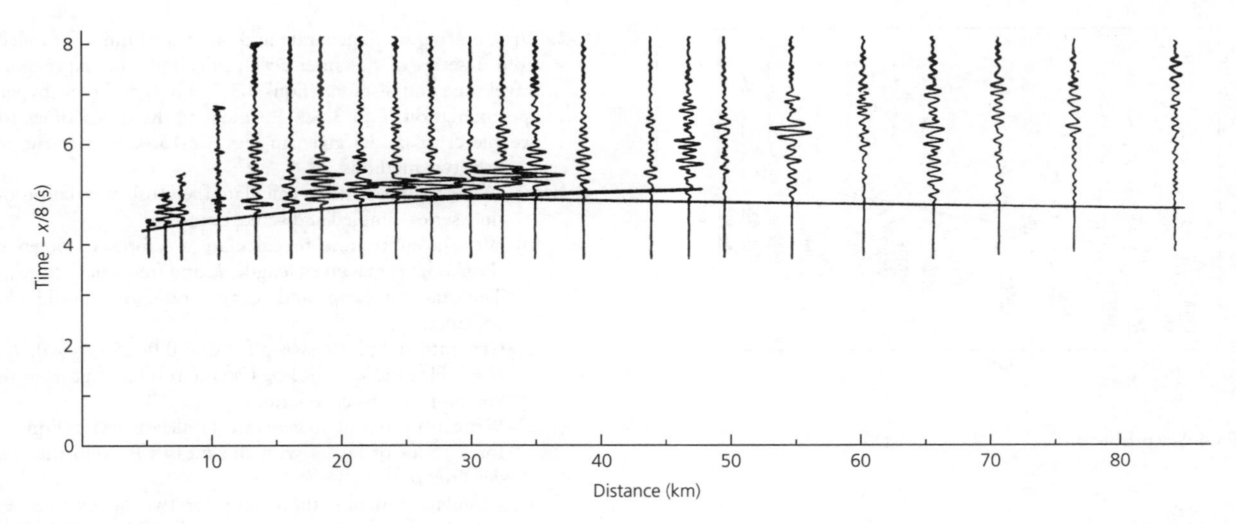

Fig. P3.1 See problem 3.

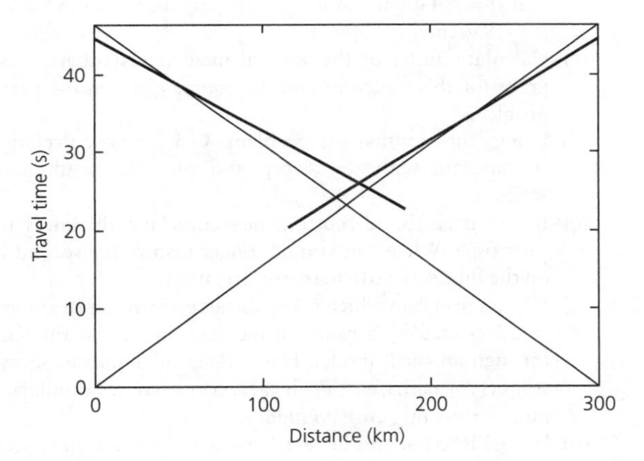

Fig. P3.2 See problem 5.

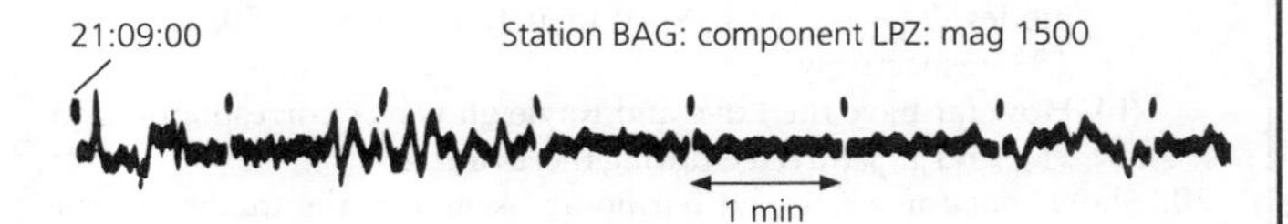

Fig. P3.3 See problem 16.

is the negative of the first. Sketch the ray paths for a single dipping layer, and explain which have the same travel times and why.

9. Define the cross-correlation (Eqn 3.3.68) for discrete time series. Such a series with N points can be written $f(t) = f(n\Delta t)$, where n goes from 0 to $N-1$ and Δt is the time increment between points.
10. Given a common offset gather, what can you tell about structure along a profile?
11. Assume that a 24-fold seismic survey records data sampled every 40 milliseconds, and that each trace is 10 s long. For a source spacing of 25 m, how many data points are recorded in a 100 km-long survey?
12. Given the definition of the travel time curve for a spherical earth $T(p) = p\Delta(p) + \tau(p)$, prove that $d\tau/dp = -\Delta(p)$.
13. (a) Use the travel times for *PcP* and *PKiKP* at vertical incidence (Fig. 3.5-4) to estimate the average *P*-wave velocity in the outer core.
    (b) Use the travel times for *PKiKP* and *PKIKP* at vertical incidence (Figs. 3.5-4 and 3.5-7 to estimate the average *P*-wave velocity in the inner core.
14. Compare the travel time curves (Fig. 3.5-4) for earthquakes at the surface and at a depth of 600 km. Identify and explain some of the differences.
15. Use the travel time curves (Fig. 3.5-4) for earthquakes at the surface and at a depth of 600 km to find p in s/degree for direct *P* waves at 40° and 60°. Find the angle of incidence at the earthquake for these rays by converting p to s/radian and using the velocities in Fig. 3.5-1. Explain how the angle of incidence of rays reaching a given distance depends on earthquake focal depth.
16. The seismogram in Fig. P3.3 for July 21, 1964, at Baguio (Philippines) contains arrivals from an earthquake that occurred in the Solomon Islands at 21 hours, 1 minute, 50 seconds. To analyze these data, which may be easier on an enlarged photocopy,
    (a) Measure the arrival time of the *P* wave and use the earthquake origin time to find its travel time.
    (b) Use the travel time curves to find how far from the station the earthquake occurred.
    (c) Trace the first 8 minutes of the seismogram after the *P* wave. Identify the *S* and *PP* phases on your tracing (use the travel time table for help). Can you identify other phases?
    (d) Identify the free surface reflections *pP* and *sP*. Measure their times after *P*, and use these times to estimate the depth.
17. The travel time curve for P_{diff}, the *P* wave diffracted along the core–mantle boundary, conveys information about the velocity at the base of the mantle. The travel time curve is linear, with ray parameter $p = dT/d\Delta = r_{cmb}/v_{cmb}$, where r_{cmb} is the radius of the core–mantle boundary and v_{cmb} is the velocity at the base of the mantle.
    (a) Measure the ray parameter in s/degree from the record section in Fig. P3.4, and compare it to the slope of the travel time curve in Fig. 3.5-4.
    (b) Convert p to s/radian, and find the velocity at the base of the mantle.
    (c) Imagine a location near the base of the mantle that is 180° away from an earthquake. The first *SH* wave to reach that spot will be SH_{diff}. What is the first *SV* wave (of nonzero amplitude) to reach that spot?

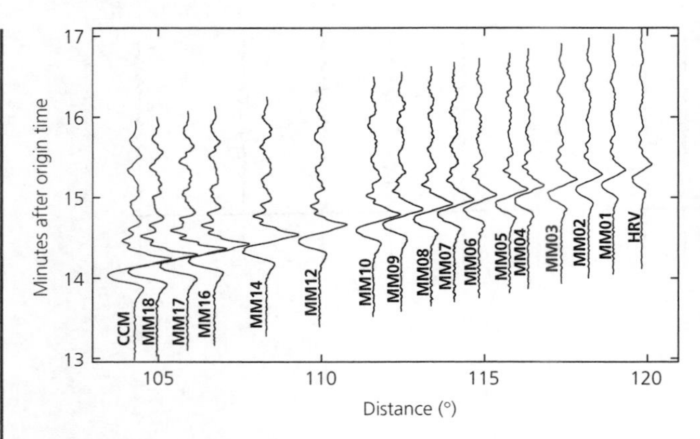

Fig. P3.4 See problem 17.

18. Derive $R(t)$ and $T(t)$ in Eqn 3.6.13.
19. (a) Use Table 2.9-1 to find the attenuation relaxation times for modes ${}_0T_2$, ${}_0T_{30}$, and ${}_0S_{30}$ if their Q values are 250, 130, and 183, respectively.
    (b) How far have the Love and Rayleigh waves corresponding to ${}_0T_{30}$ and ${}_0S_{30}$ traveled during these times?
20. Show that for a damped harmonic oscillator, the quality factor $Q = 2\pi E/(-\Delta E)$, where E is the energy in the oscillating system, and ΔE is the amount of energy lost during one cycle of the oscillation.
21. Find the percentage shear wave velocity differences due to physical dispersion between waves with periods of 1 and 10 s in the case of
    (a) a hot back-arc basin ($Q = 25$),
    (b) a cold lithospheric slab ($Q = 250$).
    Explain physically what causes the difference between the results for parts (a) and (b).
22. Show that $\alpha^2 - (4/3)\beta^2 = K/\rho$ and that $Q_\alpha^{-1} = LQ_\mu^{-1} + (1 - L)Q_K^{-1}$ where $L = (4/3)(\beta/\alpha)^2$
23. Use the acceleration of gravity at the core–mantle boundary ($g = 10.7$ m/s^2) to find the total mass and average density of earth's core.
24. Assuming that the earth is ellipsoidal, but otherwise homogeneous:
    (a) What source location results in the greatest amount of antipodal defocusing of surface waves?
    (b) What source location results in the least amount of antipodal defocusing of surface waves?
    (c) For (a), estimate the approximate time range for the earliest and latest arrival of a surface wave with a phase velocity of 4.0 km/s.

Computer problems

C-1. Write a program to trace direct, reflected, and head wave paths for a dipping layer over a halfspace. Have the program compute the travel time for each path from the length of the path in each material (i.e., rather than using the analytic expressions for travel time). Use the program to replicate the results of problem 5.

C-2. Write a program to generate and plot travel times for reflections from a series of flat interfaces using both the expressions for travel time and distance (Eqns 3.3.7, 3.3.8) and their hyperbolic approximation (Eqn 3.3.11). Calculate the travel times for the oceanic crust model given in Fig. 3.2-15. Compare the results from the two methods.

C-3. (a) Write a subroutine to calculate the cross-correlation of two time series sampled at discrete times.
(b) Write a subroutine to calculate a Vibroseis sweep signal (Eqn 3.3.57) of a given length, T, and frequency range (f_1, f_2). The start time, t_0, and sample rate, Δt, should also be parameters.
(c) Generate and plot a sweep for $\Delta t = 0.0025$ s, $t_0 = 0$, $T = 5$ s, $f_1 = 7$ Hz, and $f_2 = 14$ Hz. Use the results of part (a) to find and plot its auto-correlation.

C-4. (a) Write a subroutine to generate a reflector series (Eqn 3.3.61) for a series of layers with thicknesses h_i, velocities v_i, and densities ρ_i.
(b) Calculate and plot the results for two layers over a half-space, if the first layer is 3 km thick, with $v = 2.5$ km/s and $\rho = 2.1$ g/cm^3, the second is 4 km thick with $v = 3.2$ km/s and $\rho = 2.4$ g/cm^3, and the halfspace has $v = 4.5$ km/s and $\rho = 2.8$ g/cm^3.
(c) Calculate and plot the vertical incidence synthetic seismogram for this structure and the source given in the previous problem.
(d) Using the results of problem C-3, cross-correlate the seismogram with the sweep and plot the resulting time series.
(e) Repeat parts (b)–(d), cutting the second layer thickness in half each time. When can you no longer resolve the second layer on the time series after cross-correlation?

C-5. (a) Write a program which takes a source at any depth and traces rays, selected by a range of incidence angles at the source, through an earth model. Have the graphic output show the source, ray paths, earth's surface, core–mantle boundary, and inner core–outer core boundary.
(b) Using PREM or another earth model, trace rays for sources at the surface and at 300 km depth. Have the ray paths show the effects of upper mantle discontinuities and the core.
(c) Have the program produce a travel time plot. Can you resolve the upper mantle discontinuities in this plot?

C-6. (a) Write a program that computes the mass, M, moment of inertia about the polar axis, C, and C/Ma^2 ratio for a planet of radius a. To do this, treat the planet as a series of n shells whose densities you input.
(b) Determine models for the densities of

	a	M	C/Ma^2
earth	6371 km	5.977×10^{24} kg	0.331
moon	1738 km	7.352×10^{22} kg	0.395
Mars	3390 km	6.419×10^{23} kg	0.365

that satisfy the observed M and C/Ma^2. Can you satisfy the data for Mars and the earth without a dense core?

4 Earthquakes

Much of what is known about earthquakes follows from study of the motion of the ground.

Charles Richter, *Elementary Seismology*, 1958

4.1 Introduction

Seismology deals with the generation and propagation of seismic waves. Our initial focus has been on the propagation of seismic waves and how they can be used to study the interior of the earth. We now turn to the generation of seismic waves and how they are used to study earthquakes. This association is so strong that seismology is sometimes viewed as the science of earthquakes, rather than of elastic waves in the earth. Both definitions are used, but the latter has become more common because seismology is the primary tool used to investigate earth structure as well as earthquakes, whereas techniques other than elastic waves are also used to investigate earthquakes.

Earthquakes almost invariably occur on *faults*, surfaces in the earth on which one side moves with respect to the other. Typically, earthquakes occur on faults previously identified by geological mapping, which shows that motion across the fault has occurred in the past. Earthquakes that occur on land and close enough to the surface often leave visible ground breakage along the fault. For example, earthquakes occur along the San Andreas fault, which can be seen cutting across California for great distances (Fig. 4.1-1). One of these, the famous 1906 magnitude 7.8 San Francisco earthquake on the San Andreas fault was one of the first US earthquakes to be studied carefully. Contemporary accounts showed that several meters of relative motion occurred along several hundred kilometers of the San Andreas fault (Fig. 4.1-2).

Fig. 4.1-1 Aerial photograph of the San Andreas fault in the Carrizo Plain in California, seen from the south. Note the displacement of stream gullies as the Pacific plate (*near side*) has moved to the left (northwest) relative to North America. (Copyright John S. Shelton.)

The earthquake and the resultant fires did such damage (Fig. 1.2-10) that a study commission was formed. As part of the investigation, H. Reid proposed the *elastic rebound* theory of earthquakes on a fault. In this model, materials at distance on opposite sides of the fault move relative to each other, but friction on the fault "locks" it and prevents the sides from slipping (Fig. 4.1-3). Eventually the strain accumulated in the rock is more than the rocks on the fault can withstand, and the fault slips, resulting in an earthquake. The motion illustrated in this cartoon by an offset fence can sometimes be seen after earthquakes using other linear features, including rows of trees, railroad tracks, or roads (Fig. 4.1-4).

The elastic rebound idea was a major conceptual breakthrough, because the faulting seen at the surface had been previously regarded as an incidental side effect of an earthquake, rather than its cause. Subsequently, earthquake studies have been widely pursued for several reasons. One is to understand the large-scale geological processes causing earthquakes. It

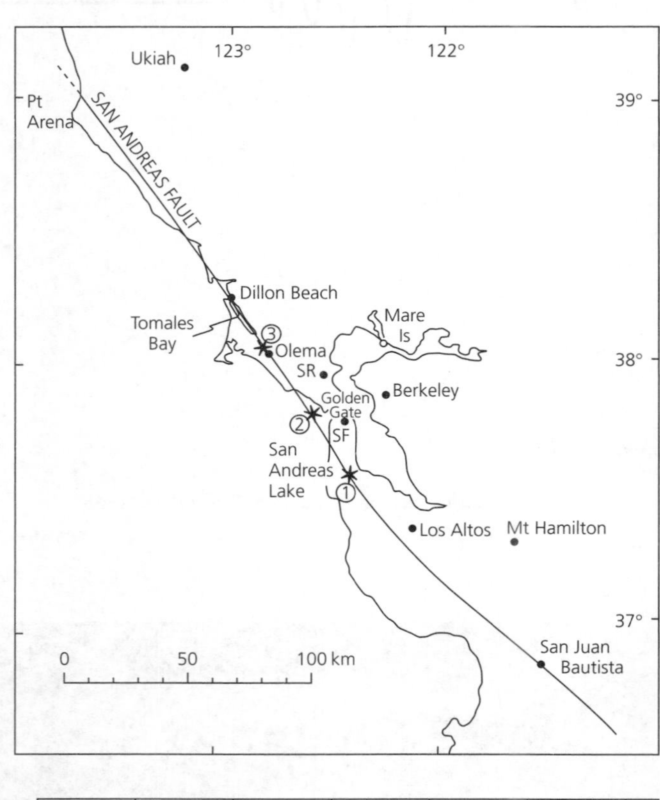

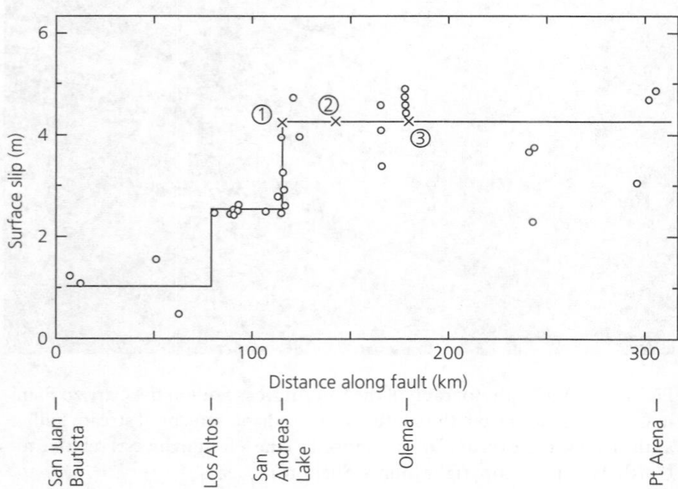

Fig. 4.1-2 Map of the portion of the San Andreas fault that slipped in the 1906 San Francisco earthquake (*top*) and the amount of surface slip reported at various points along it (*bottom*). This slip is the distance by which the earthquake displaced originally adjacent features on opposite sides of the fault. (Boore, 1977. © Seismological Society of America. All rights reserved.)

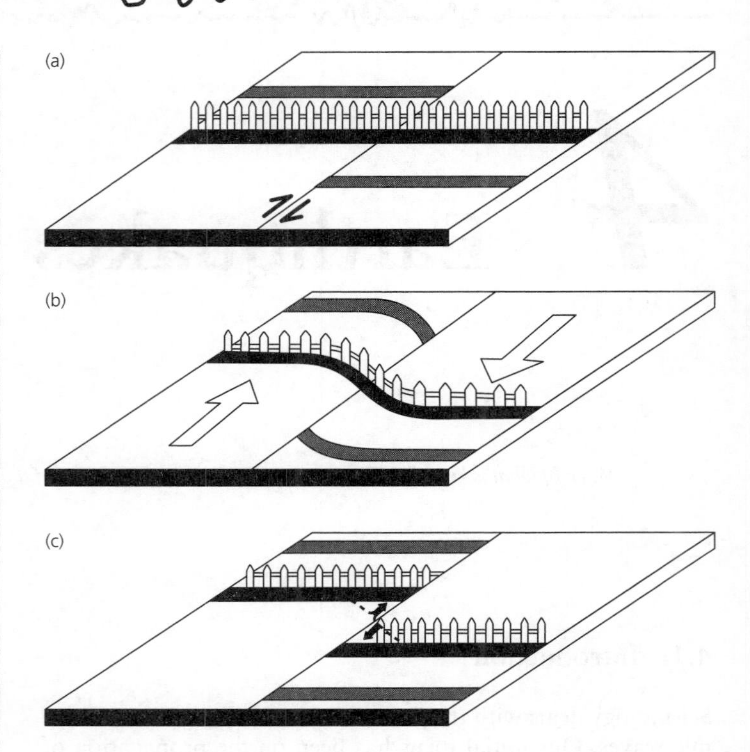

Fig. 4.1-3 The elastic rebound model of earthquakes assumes that between earthquakes, material on the two sides of a fault undergoes relative motion. Because the fault is locked, features across it that were linear at time (a), such as a fence, are slowly deformed with time (b). Finally the strain becomes so great that the fault breaks in an earthquake, offsetting the features (time c). (Courtesy of S. Wesnousky.)

Fig. 4.1-4 Displacement of crop rows resulting from slip along the Imperial fault, El Centro, California, on October 15, 1979. (Courtesy of the National Geophysical Data Center.)

turns out that earthquakes largely reflect the motions of lithospheric plates, and so provide valuable information about how and why plates move. For example, earthquakes on the San Andreas fault result from the steady motion between the North American and Pacific plates (Fig. 5.2-3). A second reason is to understand the fundamental physics of earthquake faulting. There are many unanswered questions about how and when faults break, even for earthquakes that occur near the earth's surface, where data are relatively easy to gather. These issues are important for society because, as discussed in Chapter 1, knowledge of where and when earthquakes are likely, and of the expected ground motion during them, can help mitigate the risk they pose.

The largest earthquakes typically occur at plate boundaries. Using elastic rebound theory, we think of them as reflecting the most dramatic part of a process called the *seismic cycle*, which takes place on segments of the plate boundary over hundreds to thousands of years. During the *interseismic* stage, which makes up most of the cycle, steady motion occurs away from the fault but the fault itself is "locked," although some aseismic creep can also occur on it. Immediately prior to rupture there is the *preseismic* stage that can be associated with small earthquakes (foreshocks) or other possible precursory effects. The earthquake itself marks the *coseismic* phase during which rapid motion on the fault generates seismic waves. During these few seconds, meters of slip on the fault "catch up" with the few mm/yr of motion that occurred over hundreds of years away from the fault. Finally, a *postseismic* phase occurs after the earthquake, and aftershocks and transient afterslip occur for a period of years before the fault settles into its steady interseismic behavior again.

Studying this cycle is difficult because it extends for hundreds of years, so we do not have observations of it in any one place. Instead, we have observations from different places, which we assume can be combined to give a complete view of the process. It is far from clear how good that view is and how well our models represent its complexity. As a result, earthquake physics remains an active research area that integrates a variety of techniques. Most faults are identified from the earthquakes on them, and seismology is the primary tool used to study the motion during the earthquakes and infer the long-term nature of motion on the faults. Moreover, because earthquakes are such dramatic events, historical records of earthquakes are often available and provide data on the earthquake cycle for a given fault or fault segment. Field studies, both on land and under water, also provide information about the location, geometry, and history of faults. Geodetic measurements are used to study ground deformation before, during, and after earthquakes, and thus the processes associated with fault locking and afterslip. For oceanic regions and deep earthquakes, where geodetic and geological observations are not available, almost all of what we know about the earthquakes themselves comes from seismology. The results for individual earthquakes are then combined and integrated with those from other techniques, as discussed in the next chapter, to better understand how earthquakes in a given region reflect the large-scale tectonic processes that cause them.

Of these approaches, our primary focus in this book is the information that seismology provides about earthquakes. The arrival time of seismic waves at seismometers at different sites is first used to find the location of an earthquake, known as the *focus*, or *hypocenter*, using techniques discussed in Chapter 7. Next, as discussed in this chapter, the amplitudes and shapes of the radiated seismic waves are used to study the size of the earthquake, the geometry of the fault on which it occurred, and the direction and amount of slip. We introduce these techniques and discuss their applications, while leaving their derivation and details for more advanced treatments listed at the chapter's end.

It is worth bearing in mind that learning about earthquake faulting from the seismic waves that are generated is an inverse problem, like learning about earth structure from seismic waves. As discussed in Section 1.1.2, this means that studying seismic waves alone is limited in what it can tell about the earthquake process. We will see that the seismic waves radiated from an earthquake reflect the geometry of the fault and the motion on it, and so can give an excellent picture of the *kinematics* of faulting. However, they contain much less information about the actual physics, or *dynamics*, of faulting. In the next chapter, we discuss how seismological results are being combined with experimental and theoretical studies of rock friction and fracture to explore the physics of earthquakes.

4.2 Focal mechanisms

4.2.1 *Fault geometry*

To describe the geometry of a fault, we assume that the fault is a planar surface across which relative motion occurred during an earthquake. Geological observations of faults that reach the surface show that this is often approximately the case (Fig. 4.2-1), although complexities are common. Similarly, we will see that this assumption is usually (but not always)

Fig. 4.2-1 Fault cutting across a moraine near Crowley Lake, California. The land in front has dropped relative to the background. (Copyright John S. Shelton.)

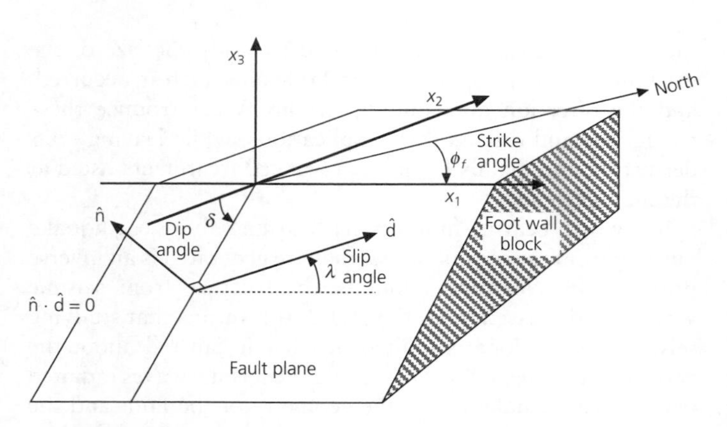

Fig. 4.2-2 Fault geometry used in earthquake studies. The fault plane, with normal vector $\hat{\mathbf{n}}$, separates the lower, or foot wall, block from the upper hanging wall block (not shown). The slip vector, $\hat{\mathbf{d}}$, describes the motion of the hanging wall block with respect to the foot wall block. The coordinate axes are chosen with x_3 vertical and x_1 oriented along the fault in the plane of the earth's surface, such that the fault dip angle, δ, measured from the $-x_2$ axis, is less than 90°. The slip angle λ is measured between the x_1 axis and $\hat{\mathbf{d}}$ in the fault plane. ϕ_f is the strike of the fault measured clockwise from north. (After Kanamori and Cipar, 1974. *Phys. Earth Planet. Inter.*, 9, 128–36, with permission from Elsevier Science.)

consistent with seismic data. Thus the fault geometry is described in terms of the orientation of the fault plane and the direction of slip along the plane.

The geometry of this model is shown in Fig. 4.2-2. The fault plane is characterized by $\hat{\mathbf{n}}$, its *normal vector*. The direction of motion is given by $\hat{\mathbf{d}}$, the *slip vector* in the fault plane. The slip vector indicates the direction in which the upper side of the fault, known as the *hanging wall block*, moved with respect to the lower side, the *foot wall block*. Because the slip vector is in the fault plane, it is perpendicular to the normal vector.

Several different coordinate systems are useful in studying faults. One is aligned such that the x_1 axis is in the fault *strike* direction, the intersection of the fault plane with the earth's surface. The x_3 axis points upward, and the x_2 axis is perpendicular to the other two. The dip angle δ gives the orientation of the fault plane with respect to the surface. Because the x_1 axis could be defined in two directions, 180° apart, it is chosen so that the dip measured from the $-x_2$ axis is less than 90°. The direction of motion is represented by the slip angle, λ, measured counterclockwise in the fault plane from the x_1 direction, which gives the motion of the hanging wall block with respect to the foot wall block. To orient this system relative to the geographic one, the fault strike ϕ_f is defined as the angle in the plane of the earth's surface measured clockwise from north to the x_1 axis.

Alternatively, the orientation of the fault and slip can be described by giving the normal and slip vectors in a geographic coordinate system with $\hat{\mathbf{x}}$ pointing north, $\hat{\mathbf{y}}$ pointing west, and $\hat{\mathbf{z}}$ pointing up. In this coordinate system, the unit normal vector to the fault plane is

$$\hat{\mathbf{n}} = \begin{pmatrix} -\sin\delta\sin\phi_f \\ -\sin\delta\cos\phi_f \\ \cos\delta \end{pmatrix}, \tag{1}$$

and the slip vector, a unit vector in the slip direction, is

$$\hat{\mathbf{d}} = \begin{pmatrix} \cos\lambda\cos\phi_f + \sin\lambda\cos\delta\sin\phi_f \\ -\cos\lambda\sin\phi_f + \sin\lambda\cos\delta\cos\phi_f \\ \sin\lambda\sin\delta \end{pmatrix}. \tag{2}$$

These two different coordinate systems, (ϕ_f, δ, λ) and ($\hat{\mathbf{n}}$, $\hat{\mathbf{d}}$), are useful for different purposes. Some calculations are more easily done with respect to the fault, whereas others are more easily done with respect to geographic directions.

Although the slip direction varies such that the slip angle ranges from 0° to 360°, several basic fault geometries, described by special values of the slip angle, are useful to bear in mind (Fig. 4.2-3). When the two sides of the fault slide horizontally by each other, pure *strike-slip* motion occurs. When $\lambda = 0°$, the hanging wall moves to the right, and the motion is called *left-lateral*. Similarly, for $\lambda = 180°$, *right-lateral* motion occurs. To tell which is which, look across the fault and see which way the other side moved. The other basic fault geometries describe *dip-slip* motion. When $\lambda = 270°$, the hanging wall slides downward, causing *normal faulting*. In the opposite case, $\lambda = 90°$, and the hanging wall goes upward, yielding *reverse*, or *thrust*, *faulting*.[1] Most earthquakes consist of some combination of these motions and have slip angles between these values. It is thus useful, when thinking about earthquake mechanisms, to remember the three basic faults. As discussed in Section 2.3.4, the basic fault types can be related to the orientations of the principal stress directions.

This discussion brings out the point that although texts typically show vertically dipping strike-slip faults,[2] they are by no means the norm. In fact, as discussed later, the largest earthquakes occur on shallow-dipping thrust faults at subduction zones. Although such faults are harder to study, because the fault trace is generally under water, the same basic principles apply.

Real faults, of course, have finite dimensions and complicated geometries. If we treat a fault as rectangular, the dimension along the strike is called the fault *length*, and the dimension in the dip direction is known as the fault *width*. Actual earthquake fault geometries can be much more complicated than a rectangle. The fault may curve and require a three-dimensional description. Rupture may occur over a long time and consist of several sub-events on different parts of the fault with different orientations. Such complicated seismic events, however, can be treated as a superposition of simple events. Thus, if we

[1] Seismologists often use the terms "reverse" and "thrust" fault interchangeably, whereas structural geologists reserve the term "thrust" for a shallow-dipping reverse fault.

[2] In part because many authors have spent time in California, and in part because they are easy to draw.

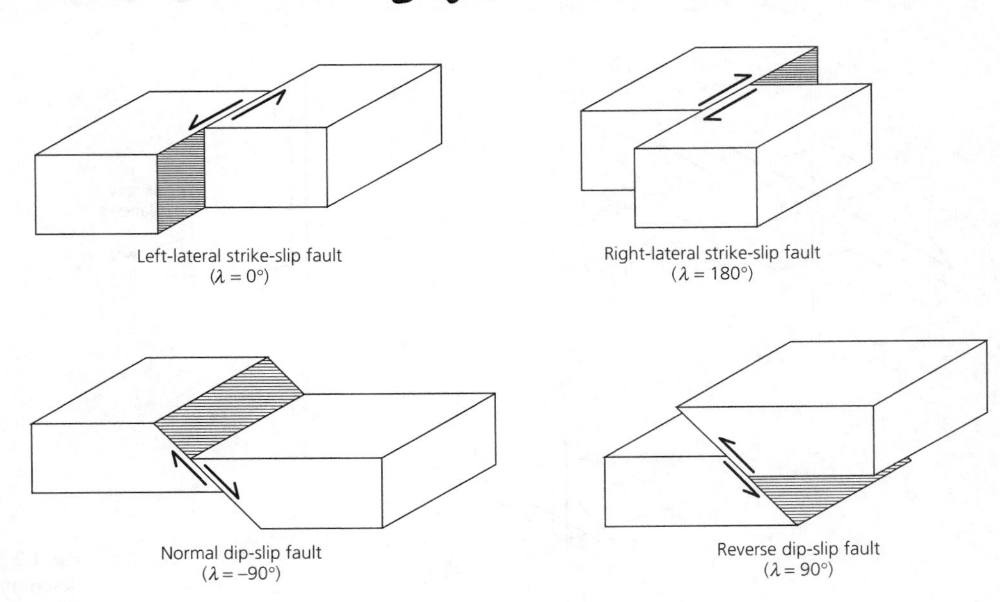

Fig. 4.2-3 Basic types of faulting. Strike-slip motion can be either right- or left-lateral. Dip-slip faulting can occur as either reverse (thrust) or normal faulting. (Eakins, 1987.)

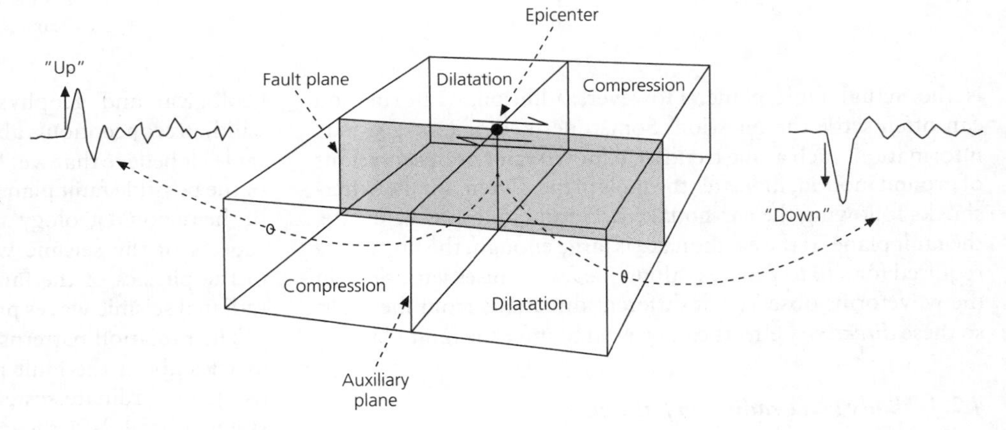

Fig. 4.2-4 First motions of *P* waves observed at seismometers located in various directions about the earthquake provide information about the fault orientation. The two nodal planes separate regions of compressional and dilatational first arrivals. One nodal plane is the fault plane, and the other is the auxiliary plane, but these data cannot distinguish which is the actual fault plane.

understand the seismic waves generated by a simple, two-dimensional, rectangular fault, we can model those resulting from a more complicated set of ruptures. This application of the principle of superposition is based on the assumption of linear elasticity and is analogous to the way we constructed seismic waves by summing normal modes (Sections 2.2.5 and 2.9).

4.2.2 *First motions*

Seismograms recorded at various distances and azimuths are used to study the geometry of faulting during an earthquake, known as the *focal mechanism*. This operation uses the fact that the pattern of radiated seismic waves depends on the fault geometry. The simplest method, which we discuss first, relies on the first motion, or polarity, of body waves. More sophisticated techniques, discussed in the next section, use the waveforms of body and surface waves.

The basic idea is that the polarity (direction) of the first *P*-wave arrival varies between seismic stations at different directions from an earthquake. Figure 4.2-4 illustrates this concept for a strike-slip earthquake on a vertical fault. The first motion is either *compression*, for stations located such that material near the fault moves "toward" the station, or *dilatation*, where the motion is "away from" the station. Thus when a *P* wave arrives at a seismometer from below, a vertical-component seismogram records an upward or downward first motion, corresponding to either compression or dilatation.

The first motions define four quadrants, two compressional and two dilatational. The division between quadrants occurs along the fault plane and a plane perpendicular to it. In these directions, because the first motion changes from dilatation to compression, seismograms show small or zero first motions. These perpendicular planes, called *nodal planes*, separate the compressional and dilatational quadrants. If these planes can be found, the fault geometry is known. A problem is that the first motions from slip on the actual fault plane and from slip on the plane perpendicular to it, the *auxiliary plane*, would be the same, so the first motions alone cannot resolve which plane

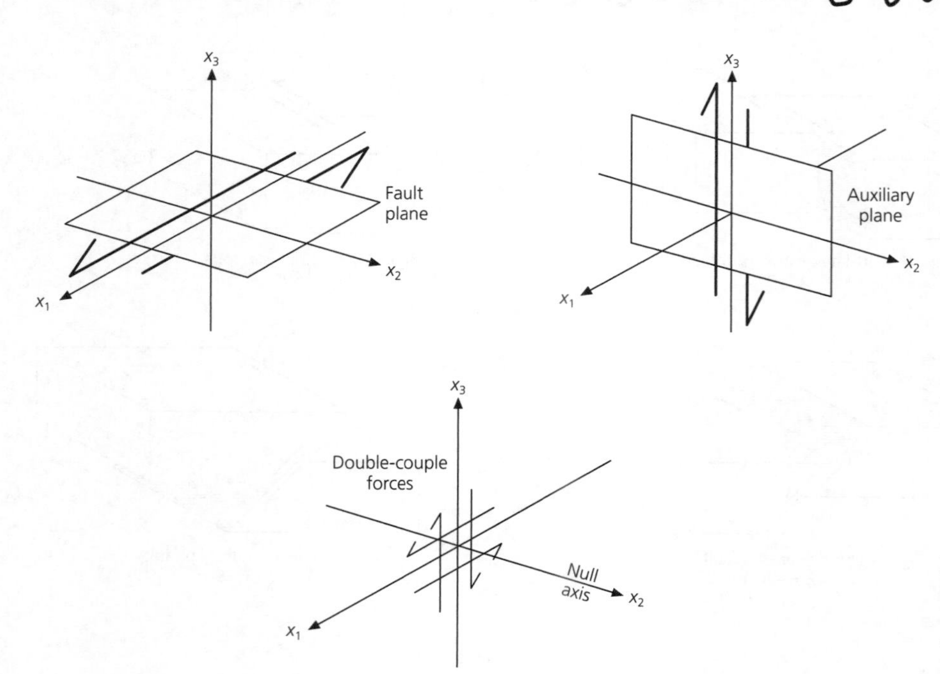

Fig. 4.2-5 A fault-oriented coordinate system for describing the radiation pattern of an earthquake. The body forces equivalent to the faulting are a pair of force couples acting about the null axis. (After Pearce, 1977.)

is the actual fault plane. However, additional information can often settle the question. Sometimes geologic or geodetic information, such as the trend of a known fault or observations of ground motion, indicates the fault plane. Often, smaller aftershocks following the earthquake occur on, and thus delineate, the fault plane. If the earthquake is large enough, the finite time required for slip to progress along the fault causes variations in the waveforms observed at different directions from the fault, so these *directivity* effects can be used to infer the fault plane.

4.2.3 *Body wave radiation patterns*

The radiation patterns of *P* and *S* waves, which we will not derive, can be obtained using the theory of seismic sources. The radiation patterns turn out to be those that would be generated by a set of forces with a corresponding geometry. Specifically, the radiation due to motion on the fault plane is what would occur for a pair of force *couples*, pairs of forces with opposite direction a small distance apart. If one couple was oriented in the slip direction with forces on opposite sides of the fault plane, the other couple would be oriented in the corresponding direction on opposite sides of the auxiliary plane. Thus the elastic radiation can be described as resulting from a *double couple*, and these forces are known as the *equivalent body forces* for the fault slip, discussed further in Section 4.4.

It is important to bear in mind that the equivalent forces are only a simple model representing the complex faulting process that actually took place. We can view the faulting as occurring within a "black box" about which the radiated seismic waves provide only limited information. The seismic waves tell us only that some processes within the box produced seismic waves described by the equivalent forces. Often we have other geological and geophysical data, together with (hopefully valid) preconceptions, about the source. In particular, we often (at least believe that we) have good reasons to favor slip on one of the possible fault planes and to interpret the faulting in terms of the regional geology and stress field. Similarly, we interpret aspects of the seismic wave field in terms of simple models of the physics of the faulting process, while recognizing that radiated seismic waves provide only a partial picture.

The radiation patterns of double couples have natural symmetries about the fault plane, and are thus normally written using a coordinate system oriented along the fault. In such a system (Fig. 4.2-5), the fault plane lies in the x_1–x_2 plane, so its normal is the x_3 axis. The slip vector is in the fault plane, parallel to the x_1 axis. The slip is such that material above the x_1–x_2 plane moves in the $+x_1$ direction with respect to the material on the other side. The radiation pattern would be the same if the slip in the x_3 direction occurred on the auxiliary plane, which lies in the x_2–x_3 plane and whose normal is the x_1 axis. Thus we can interchange the slip (x_1) and normal (x_3) directions, so the slip vector on one plane is the normal vector on the other, and vice versa. However, the direction orthogonal to both, known as the *null axis*, is distinct. In this geometry, the equivalent body force double couple acts about the x_2 axis, and the forces are oriented along the x_1 and x_3 directions.

To see how the radiation patterns vary with the direction of the receiver, consider the radiation field in spherical coordinates, where θ is measured from the x_3 axis and ϕ is measured in the x_1–x_2 plane (Figs 4.2-6 and 7). Seismic source theory shows that far from the source, the displacement due to compressional waves, which create the radial ($\hat{e}_r$) component of the displacement (u_r) because their motion is along the propagation direction, is

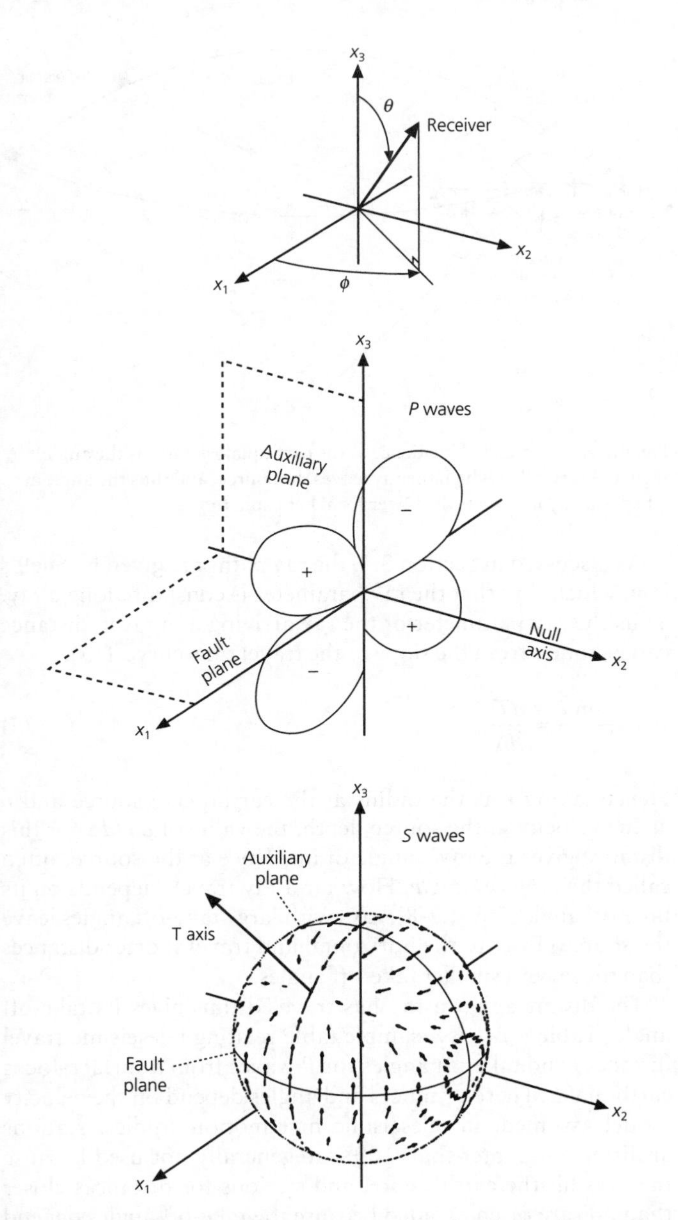

Fig. 4.2-6 The body wave radiation pattern for a double couple source has symmetry in the spherical coordinate system shown, corresponding to the axes in Fig. 4.2-5. θ is measured from the x_3 axis, the normal to the fault (x_1–x_2) plane, and ϕ is measured in the fault plane. The *P*-wave radiation pattern has four lobes that go to zero at the nodal planes, which are the fault and auxiliary (x_2–x_3) planes. The *S*-wave radiation pattern describes a vector displacement that does not have nodal planes but is perpendicular to the *P*-wave nodal planes. *S*-wave motion converges toward the T axis, diverges from the P axis, and is zero on the null axis. (After Pearce, 1977, 1980.)

$$u_r = \frac{1}{4\pi\rho\alpha^3 r}\dot{M}(t - r/\alpha) \sin 2\theta \cos \phi. \tag{3}$$

This expression has several parts. The first term is an amplitude term. In an infinite medium, for which this was derived, the amplitude would decay as $1/r$. The second term reflects the pulse radiated from the fault, $\dot{M}(t)$, which propagates away at the *P*-wave speed α and arrives at a distance r at time $t - r/\alpha$. $\dot{M}(t)$ is called the seismic moment rate function or *source time function*. It is the time derivative of the *seismic moment function*

$$M(t) = \mu D(t)S(t), \tag{4}$$

which describes the faulting process in terms of the rigidity of the material and history of the slip $D(t)$ and fault area $S(t)$. The latter terms are time-dependent, because they can vary during an earthquake. As discussed in Section 4.6, the best measure of earthquake size and energy release is the static (or scalar) *seismic moment*

$$M_0 = \mu \bar{D} S, \tag{5}$$

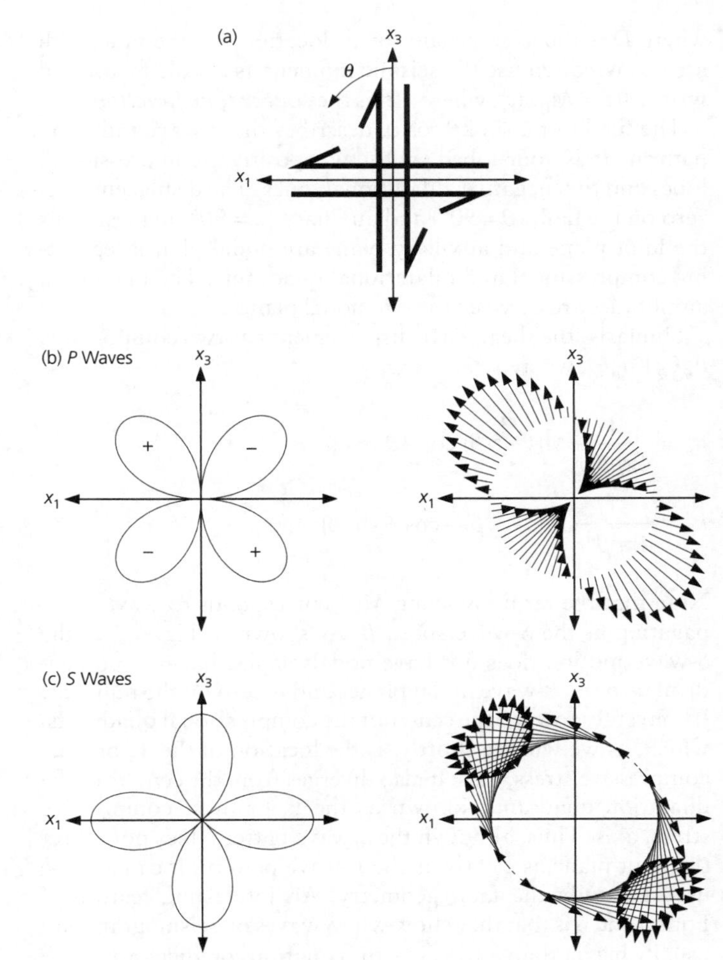

Fig. 4.2-7 Radiation amplitude patterns of *P* and *S* waves in the x_1–x_3 plane. a: Fault geometry, showing the symmetry of the double couple about the x_2 axis. b: Radiation pattern for *P* waves, showing the amplitude (*left*) and direction (*right*). c: Same as (b), but for *S* waves.

where $\overline{D}$ is the average slip (or dislocation) on the fault with area S. We often use the seismic moment as a scale factor and write $M(t) = M_0 x(t)$, where $x(t)$ is the *source time function*.

The final term, $\sin 2\theta \cos \phi$, describes the *P*-wave radiation pattern. It is four-lobed, with two positive, compressional, lobes and two negative, dilatational, ones. The displacement is zero on the fault ($\theta = 90°$) and auxiliary ($\phi = 90°$) planes. Thus the fault plane and auxiliary plane are nodal planes separating compressional and dilatational quadrants. The maximum amplitudes are between the two nodal planes.

Similarly, the shear wave displacement has two components, $u_\theta \hat{e}_\theta + u_\phi \hat{e}_\phi$, where

$$u_\theta = \frac{1}{4\pi\rho\beta^3 r} \dot{M}(t - r/\beta) \cos 2\theta \cos \phi,$$

$$u_\phi = \frac{1}{4\pi\rho\beta^3 r} \dot{M}(t - r/\beta)(-\cos \theta \sin \phi). \qquad (6)$$

Note that the term involving $\dot{M}(t)$ corresponds to waves propagating at the *S*-wave speed β. As shown in Fig. 4.2-6, the *S*-wave motion does not have nodal planes, but it is perpendicular to the *P*-wave nodal planes and is zero on the null axis. It converges toward the center of the compressional quadrants, which, as we will see shortly, is the location of the **T**, or least compressive stress, axis. It also diverges from the centers of the dilatation quadrants, known as the **P**, or most compressive stress, axis. Thus, although the *S*-wave pattern does not reflect the fault plane as clearly as the *P*-wave pattern, it can also be used to study the fault geometry. An interesting feature of Eqns 3 and 6 is that they show why *S* waves on seismograms are usually bigger than *P* waves — the equations predict an average ratio of α^3/β^3, or about 5.

Because the radiated seismic waves vary as a function of θ and ϕ, seismograms recorded at different directions from the earthquake can be used to find the fault geometry. The *P* wave is the first wave to arrive from an earthquake, so on a seismogram it is an isolated arrival whose polarity is often easy to identify. A set of *P*-wave first motions thus often makes it possible to locate the nodal planes that divide the regions of different polarity. The first *S* waves are harder to use, because they arrive later in the seismogram and can be buried in a complicated wave train. It is still possible, however, to use the *S*-wave information. One way to do this is to consider the relative amplitudes of the two *S*-wave components.

One additional concept is needed to determine fault plane solutions using the first motions from various seismic stations. The radiation patterns show the displacements that would occur on a sphere with infinitesimal radius about the source. The observations, of course, are at stations some finite distance from the source. We thus need to convert the observations at the stations to hypothetical ones surrounding the source. To do this, recall that seismic waves do not travel in straight lines from the earthquake to a station. Instead, because seismic velocities vary with depth, rays follow curved paths.

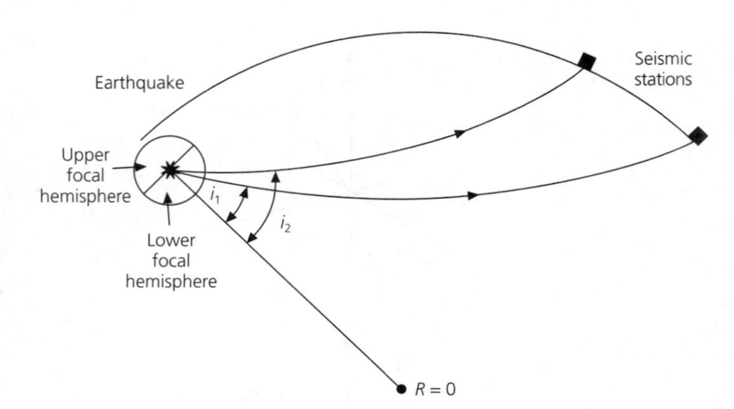

Fig. 4.2-8 The angle of incidence at the earthquake source is the angle from the vertical at which the ray leaves the source, and thus the angle at which the ray intersects the lower focal hemisphere.

As discussed in Section 3.4, the ray paths are given by Snell's law, which says that the ray parameter is constant along a ray. Thus the ray parameter of the ray arriving at a given distance can be found from the slope of the travel time curve $T(\Delta)$,

$$p = \frac{r \sin i}{v} = \frac{dT}{d\Delta}. \qquad (7)$$

Hence taking r as the radius at the earthquake source and v as the velocity at the source depth, the value of $dT/d\Delta$ for this distance gives the ray's angle of incidence at the source, often called the *take-off angle*. How far a ray travels depends on its take-off angle (Fig. 4.2-8); rays with large take-off angles leave the source closer to the horizontal and travel shorter distances than those with smaller take-off angles.

The distance that a ray has traveled thus gives its take-off angle. Table 4.2-1 is a sample table relating teleseismic travel distances and take-off angles for *P* waves from a surface-focus earthquake. These distances and angles depend on the velocity model assumed. In teleseismic first motion studies, stations at distances greater than 100° are generally not used because the rays hit the earth's core, and stations for distances closer than 30° are often avoided because the take-off angles depend strongly on the upper mantle velocity structure used. In local earthquake studies, care is taken to ensure that the velocity model is appropriate.

Using such tables, the distances to seismic stations can be converted to take-off angles. Thus the locations of compressions and dilatations can be converted to their positions on the surface of the *lower focal hemisphere*, a hemisphere with infinitesimal radius about the source. A similar approach can be used for data directly above a deep earthquake, where the upper focal hemisphere is a natural representation.

4.2.4 *Stereographic fault plane representation*

We have seen that fault geometry can be found from the distribution of data on a sphere around the focus. Because plotting

Table 4.2-1 *P*-wave take-off angles for a surface-focus earthquake.

Distance (°)	Take-off angle (°)	Distance (°)	Take-off angle (°)	Distance (°)	Take-off angle (°)
21	36	47	25	73	19
23	32	49	24	75	18
25	30	51	24	77	18
27	29	53	23	79	17
29	29	55	23	81	17
31	29	57	23	83	16
33	28	59	22	85	16
35	28	61	22	87	15
37	27	63	21	89	15
39	27	65	21	91	15
41	26	67	20	93	14
43	26	69	20	95	14
45	25	71	19	97	14

Source: After Pho and Behe (1972).

on a piece of paper is easier than plotting on a sphere, a *stereographic projection* that transforms a hemisphere to a plane is used to plot the data. The graphic construction that does this is a *stereonet* (Fig. 4.2-9).[3] On this net, the azimuth is shown by the numbers from 0° to 360° around the circumference. The dip angles are shown by the numbers from 90° to 0° along the net's equator. The angle 90°, straight down, hits the middle of the net, whereas 0°, the horizontal direction, is at the edge.

To see how to use this net, consider how planes through the center of the focal sphere appear (Fig. 4.2-10). A vertically dipping, N–S-striking, plane intersects the hemisphere such that it plots as a straight line through the center of the net. A N–S-striking plane with a different dip intersects the net edge at 0° and 180°, but intersects the equator at a position corresponding to the dip. For example, planes dipping 70°E and 60° W intersect the equator at the 70°E and 60°W marks. Thus, *meridians* on the net (the curves going from the top to the bottom) represent N–S-striking planes with different dips.

Planes striking in other azimuths are plotted in a similar way (Fig. 4.2-11) by rotating the stereonet.[4] Thus, a plane striking at an angle ϕ (measured clockwise from north) is plotted by rotating the stereonet so that the vertical (N–S) axis points in the ϕ direction. The plane with the desired dip is now a meridian, so it can be found using the scale along the equator. After plotting the plane by tracing the appropriate meridian, we rotate the net back to its original orientation. Hence planes striking in azimuths other than N–S appear as meridians relative to their strike direction, with the appropriate dip. All of these meridians are thus great circles, the curves formed when a plane through the center of the sphere intersects the surface of the sphere.

[3] Seismologists generally use an equal-area or Schmidt projection, rather than an equal-angle or Wulff projection. The techniques used are the same for the two.

[4] This can be done either by the traditional method, rotating a piece of tracing paper over a stereonet, or by using a computer program that plots points and planes on a stereonet.

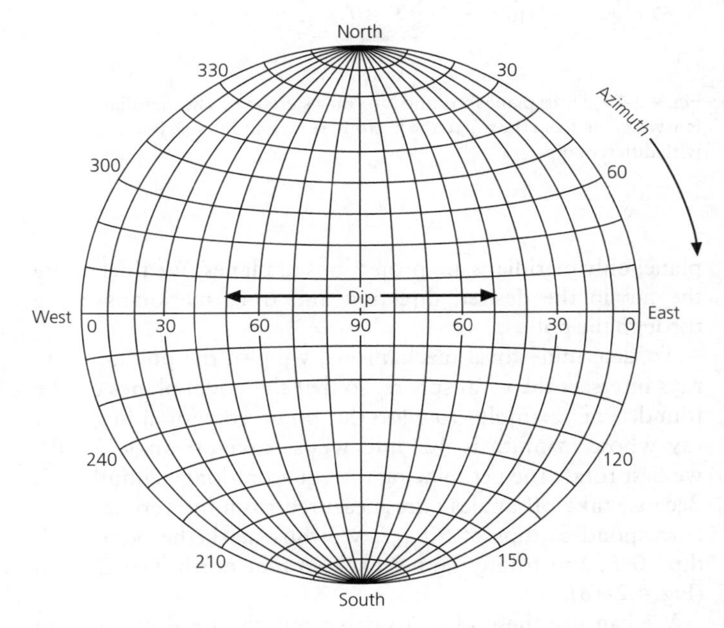

Fig. 4.2-9 A stereonet used to display a hemisphere on a flat surface. The azimuth is shown by the numbers around the circumference, and dip angles are shown by the numbers along the equator.

We can also plot planes perpendicular to a given plane. To do this, rotate the stereonet so that the plane lies on a meridian, and find the point on the equator 90° from the intersection of the plane with the equator (Fig. 4.2-12). This point is the *pole* for the plane, because it represents the point at which the normal to the plane intersects the sphere. Any plane perpendicular to the first plane contains the normal, and hence must pass through the pole. To draw such perpendicular planes, remember that an arbitrary curve on the stereonet does not represent a

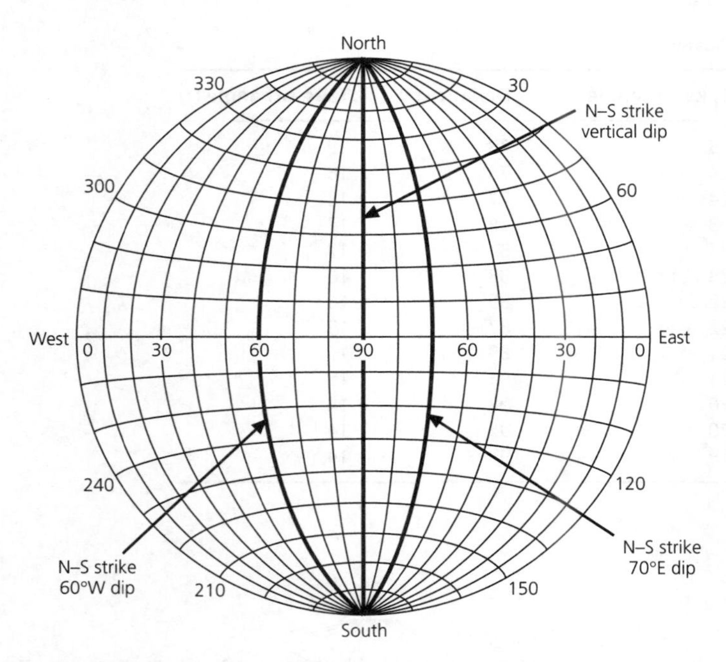

Fig. 4.2-10 Three planes striking N–S on a stereonet. The meridians (curves going from the top to the bottom) represent N–S-striking planes with different dips.

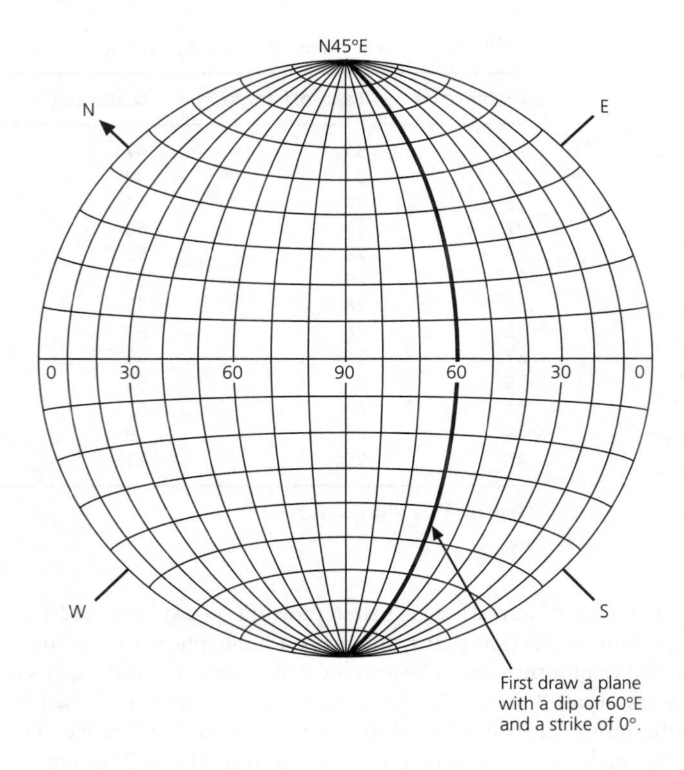

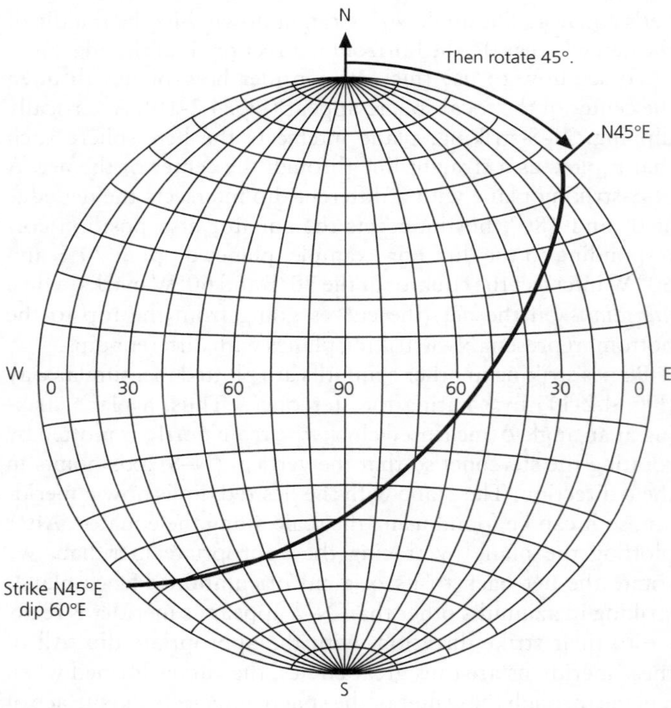

Fig. 4.2-11 To plot a plane striking N45°E and dipping 60°E, rotate the stereonet (or tracing paper above it) so that the strike is at the top and the dip can be measured along the equator. After plotting the appropriate meridian, rotate the net back to the geographic orientation with north at the top.

plane; only meridians are projections of planes. We thus rotate the net in the desired direction and trace meridians going through the pole.

To determine focal mechanisms, we plot the points where rays intersect the focal sphere, so that the nodal planes can be found. For example, to plot the point corresponding to a ray whose azimuth is 40° and whose take-off angle is 60°, we first rotate the net, placing the equator along azimuth 40°. Because take-off angles i are measured from the vertical, they correspond to dips of $90 - i$. We thus mark the point with dip 30°E, and rotate the net back so that north is at the top (Fig. 4.2-13).

We can use these ideas to determine the focal mechanism from a set of P-wave first motions. First, we find the polarities of the first arrivals at seismic stations. Each station corresponds to a point on the focal sphere with the same azimuth and an incidence angle corresponding to the ray that emerged there. We then plot the location of each station on the stereonet and mark whether the first motion is dilatation or compression. Next, by rotating the tracing paper or using a stereonet program, we find the nodal planes that best separate the compressions from the dilatations. In doing this, we ensure that the two planes are orthogonal, with each one passing through the pole to the other. Provided the distribution of stations on the focal sphere is adequate, we can find the nodal planes, which are the fault plane and the auxiliary plane.

Different types of faults appear differently on a stereonet (Fig. 4.2-14). The black and white quadrants, representing

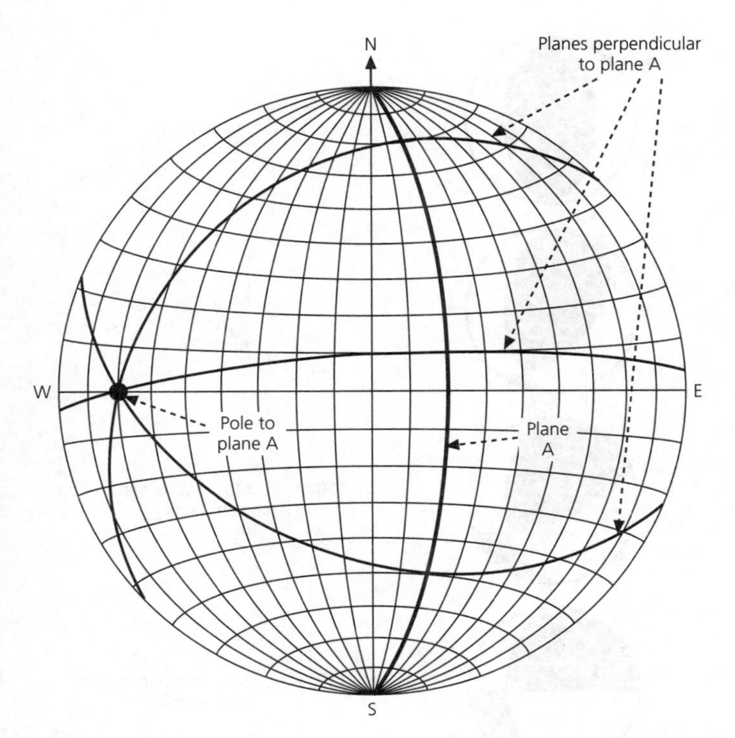

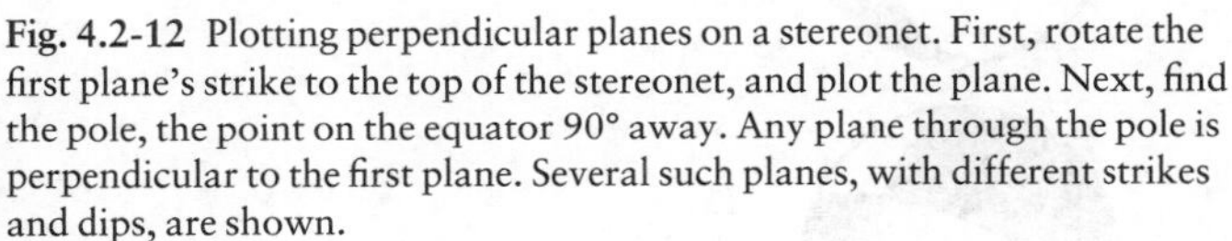
Fig. 4.2-12 Plotting perpendicular planes on a stereonet. First, rotate the first plane's strike to the top of the stereonet, and plot the plane. Next, find the pole, the point on the equator 90° away. Any plane through the pole is perpendicular to the first plane. Several such planes, with different strikes and dips, are shown.

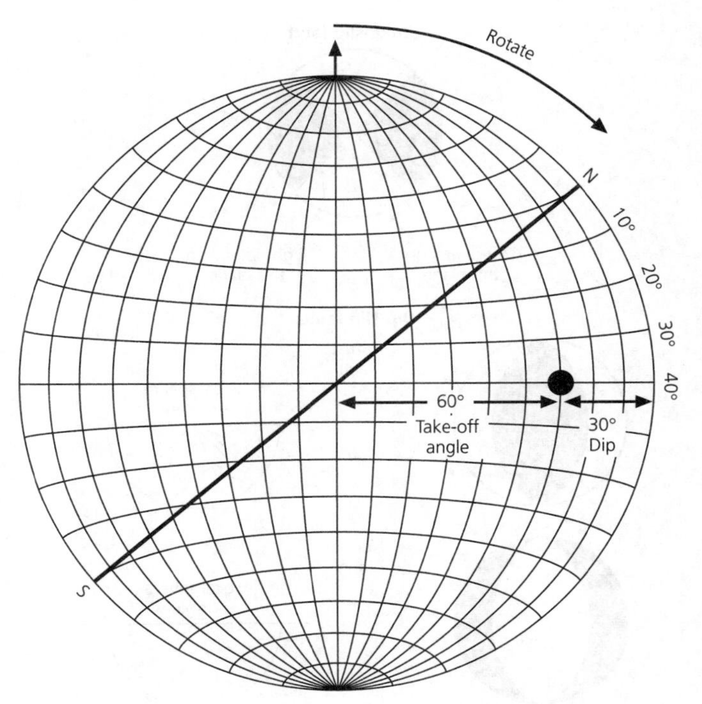

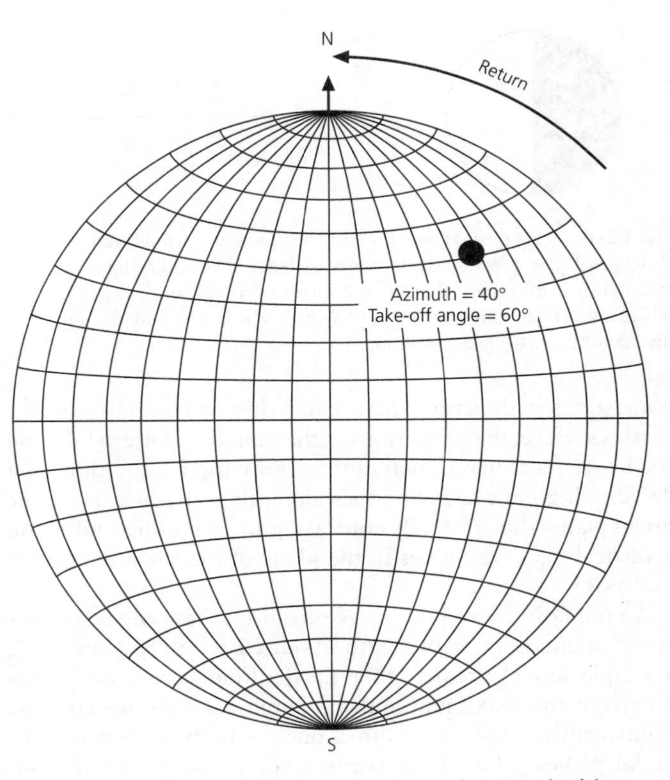

Fig. 4.2-13 To plot a point on a stereonet, rotate the azimuth of the point to the equator, measure the take-off angle from the vertical (or equivalently the dip from horizontal), plot the point and rotate back to the geographic orientation with north at the top.

compression and dilatation, show the fault geometry. A four-quadrant "checkerboard" indicates pure strike-slip motion on a vertical fault plane. The motion would be right-lateral if one plane is the fault plane, and left-lateral on the other. As we mentioned earlier, often the distribution of aftershocks or geologic information (or prejudices) is used to infer which was the actual fault plane, and thus the sense of slip. A pure dip-slip fault that dips at 45° (the fourth quadrant is on the upper focal hemisphere) gives a three-quadrant "beachball." The center region is compressional for a thrust fault, and dilatational for a normal fault. The difference reflects the different direction of fault motion, as the side-view cartoon shows. For a dip-slip rupture on a vertical fault, only two quadrants of the "beachball" are visible, because the others are on the upper focal hemisphere.

The pattern is a little more complicated for oblique-slip faults with a mixture of strike-slip and dip-slip motion. The mechanisms in Fig. 4.2-15 have the same N–S-striking, 45°E-dipping fault plane, but with slip directions varying from pure thrust, to pure strike-slip, to pure normal. Thus the auxiliary plane varies but always passes through the normal to the fault plane, and the slip vector can be found because it is the normal to the auxiliary plane, and thus is in the fault plane (Fig. 4.2-5).

It is important to bear in mind that although the focal mechanisms look different, they reflect the same four-lobed *P*-wave radiation pattern (Fig. 4.2-6). However, because the fault

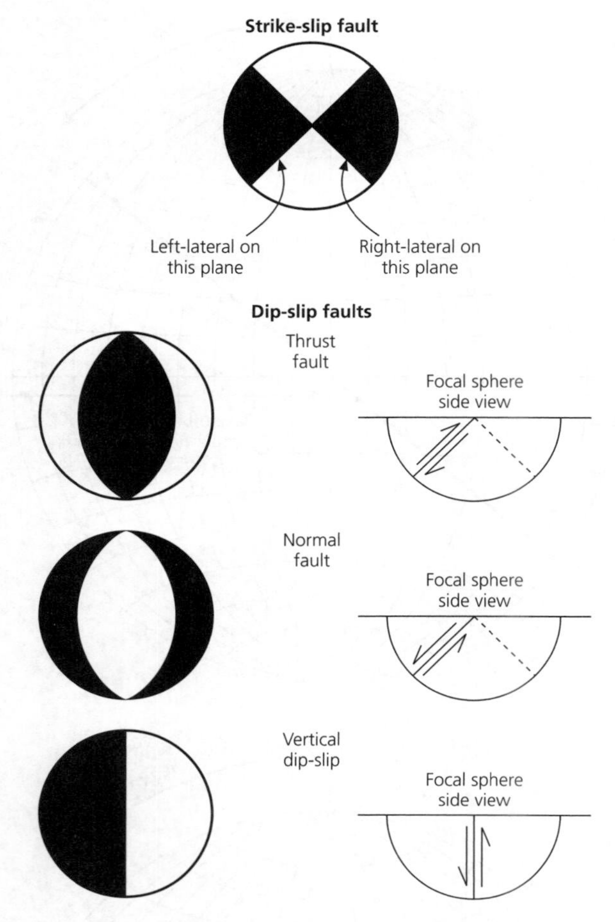

Fig. 4.2-14 Focal mechanisms for earthquakes with various fault geometries. Compressional quadrants are black. The strike-slip mechanism is for pure strike-slip motion on a vertical fault plane, which could be oriented either NE–SW or NW–SE. The pure dip-slip mechanisms are for faults striking N–S.

plane and slip direction are oriented differently relative to the earth's surface, the projections of the radiation pattern lobes on the lower focal hemisphere differ.[5] Pure dip-slip motion on a 45° dipping fault has two lobes along the vertical axis, so the nodal planes dip at 45°. By contrast, pure strike-slip motion on a vertical plane has lobes in the plane of the surface, and the null axis is vertical.

A common use of earthquake focal mechanisms is to infer stress orientations in the earth. As discussed in Section 2.3.4, a simple model predicts that the faulting occurs on planes 45° from the maximum and minimum compressive stresses. Equivalently, these stress directions are halfway between the nodal planes. Thus the maximum compressive (P) and minimum compressive stress (T) axes can be found by bisecting the dilatational and compressional quadrants, respectively (Fig. 4.2-16). Although T is called the "tension" axis, it is actually the minimum compressive stress, because compression occurs at depth in the earth. The intermediate stress axis, known as the B or null axis, is perpendicular to both the T and the P axes. This direction is also perpendicular to both the slip

[5] This concept can be seen by marking the *P*-wave quadrants on a ball and rotating it. For additional insight, the *S*-wave radiation pattern (Fig. 6) can also be marked on the ball.

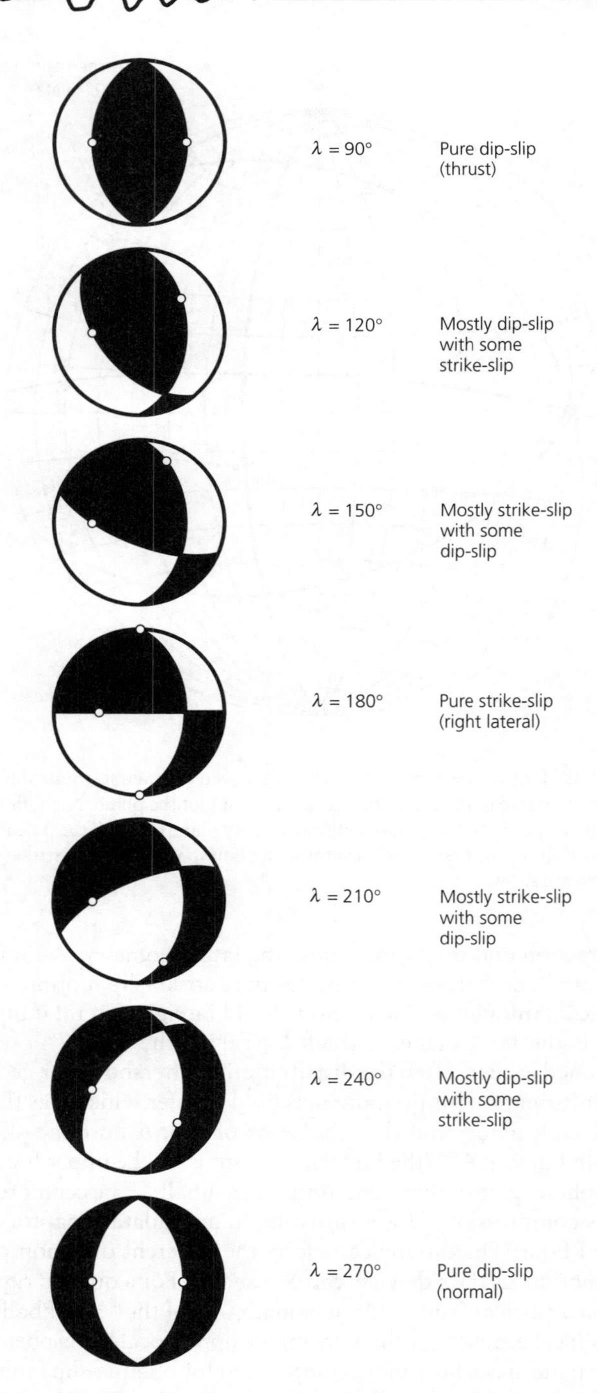

Fig. 4.2-15 Focal mechanisms for earthquakes with the same N–S-striking fault plane, but with slip angles varying from pure thrust, to pure strike-slip, to pure normal faulting.

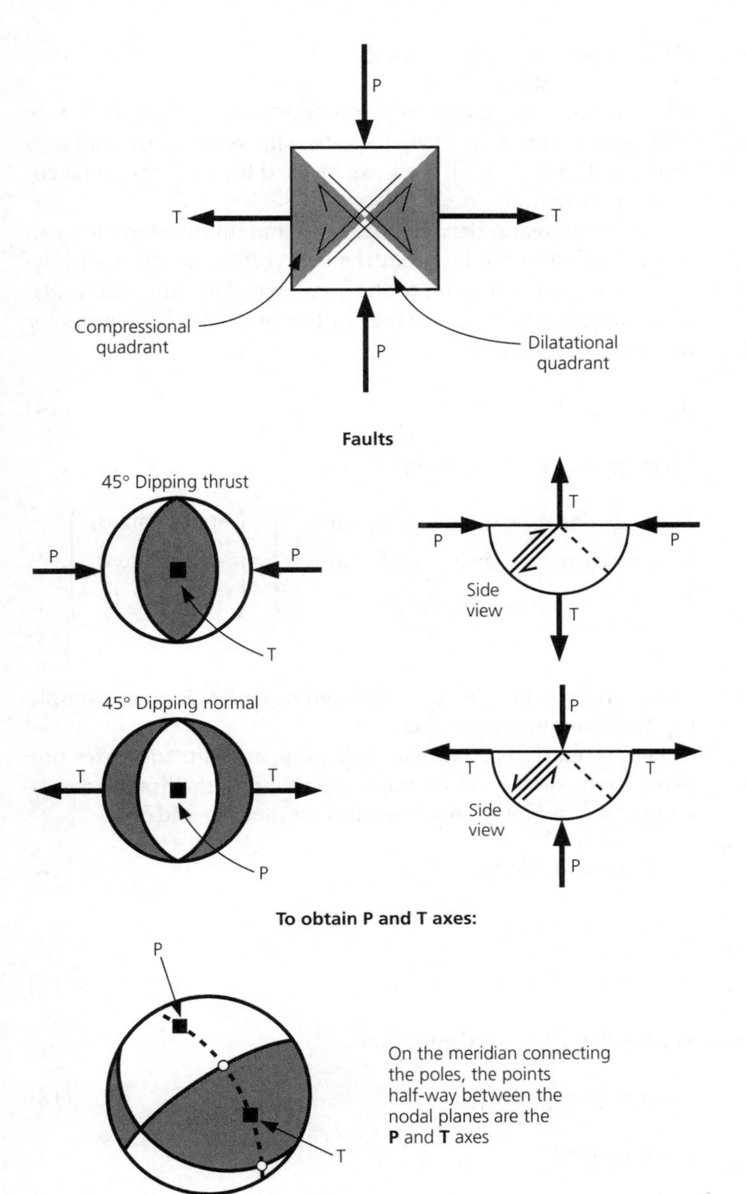

Fig. 4.2-16 Cartoon illustrating the relation between fault planes and the maximum compressive principal stress (P) and the minimum compressive stress (T) axes. The P and T axes can be found by bisecting the dilatational and compressional quadrants, respectively. On a stereonet, this is done by using the great circle (meridian) connecting the poles for the two nodal planes and finding the point halfway between them.

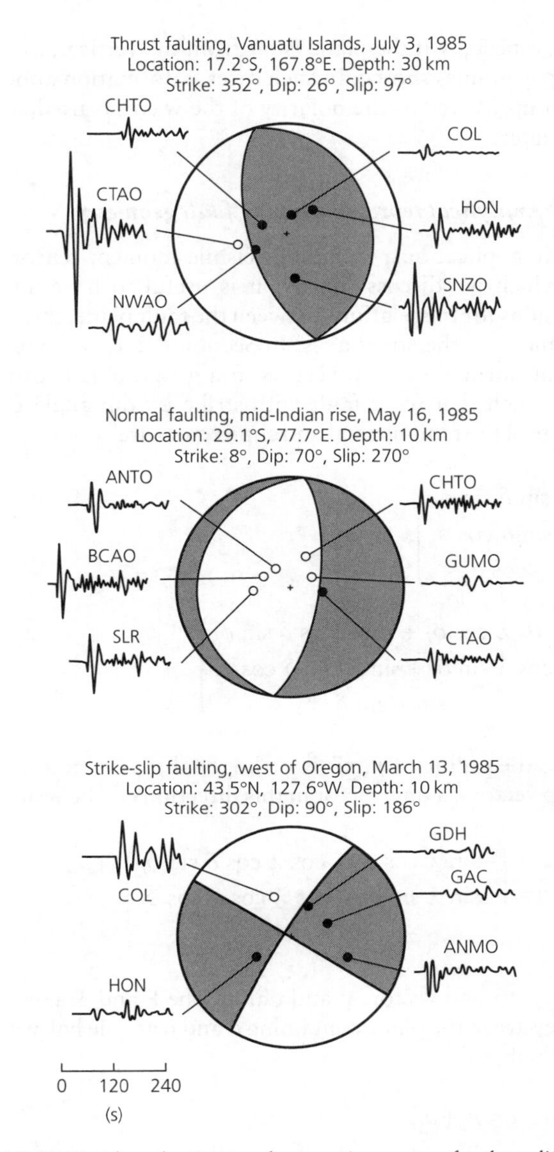

Fig. 4.2-17 Focal mechanisms and some seismograms for three different earthquakes. Compressional quadrants are shown shaded.

and the normal vectors, and is the intersection of the two nodal planes.

To bisect the angle between the two nodal planes on the stereonet, we find the poles for the two planes (each of which is in the other plane), draw the great circle (meridian) connecting them, and mark the point on it halfway between the poles (Fig. 4.2-16). We can thus infer stress directions from a focal mechanism. Different fault types correspond to different orientations of the stress axes, as noted in Fig. 2.3-9. If the P axis is vertical, the fault plane dips at 45°, and normal faulting occurs. If, instead, the T axis is vertical, the fault geometry is the same, but reverse faulting occurs. When the null axis is vertical, strike-slip motion occurs on a fault plane 45° from the maximum principal stresses, which are in the plane of the surface.

Figure 4.2-17 shows the focal mechanisms and a few of the seismograms for three earthquakes. Note that in some cases the first arrival is small and difficult to identify. This is especially likely when the station is near a nodal plane, where the amplitude is small. It is also worth noting that often many stations plot near the center of the focal sphere, because they are at large distances from the source, so rays to them have small angles of incidence. As a result, it is sometimes hard to constrain nodal planes, especially if the plane is far from the vertical, as in the dip-slip examples shown. In such cases, information about the waveforms as well as the polarity of the waves is used, as discussed later.

4.2.5 Analytical representation of fault geometry

In many applications, including seismic moment tensor analysis, which we discuss shortly, it is useful to have analytic expressions for the relations between the fault plane, the auxiliary plane, and the stress axes. In Section 4.2.1, we expressed the fault normal and slip vectors in a geographic coordinate system, such that for a fault with strike ϕ_f, dip angle δ, and slip angle λ the fault normal and slip vectors are

$$\hat{\mathbf{n}} = \begin{pmatrix} -\sin\delta \sin\phi_f \\ -\sin\delta\cos\phi_f \\ \cos\delta \end{pmatrix},$$

$$\hat{\mathbf{d}} = \begin{pmatrix} \cos\lambda\cos\phi_f + \sin\lambda\cos\delta\sin\phi_f \\ -\cos\lambda\sin\phi_f + \sin\lambda\cos\delta\cos\phi_f \\ \sin\lambda\sin\delta \end{pmatrix}. \quad (8)$$

Because the null (or B) axis is orthogonal to the fault normal and slip vectors, a unit vector in this direction can be written

$$\hat{\mathbf{b}} = \hat{\mathbf{n}} \times \hat{\mathbf{d}} = \begin{pmatrix} -\sin\lambda\cos\phi_f + \cos\lambda\cos\delta\sin\phi_f \\ \sin\lambda\sin\phi_f + \cos\lambda\cos\delta\cos\phi_f \\ \cos\lambda\sin\delta \end{pmatrix}. \quad (9)$$

Similarly, to find vectors $\mathbf{p}$ and $\mathbf{t}$ along the P and T axes, note that they are in the plane containing $\hat{\mathbf{d}}$ and $\hat{\mathbf{n}}$ and lie halfway between them, so

$$\begin{aligned} &\mathbf{t} = \hat{\mathbf{n}} + \hat{\mathbf{d}} \quad t_i = n_i + d_i, \\ &\mathbf{p} = \hat{\mathbf{n}} - \hat{\mathbf{d}} \quad p_i = n_i - d_i, \\ &\hat{\mathbf{b}} = \hat{\mathbf{n}} \times \hat{\mathbf{d}} \quad b_i = \varepsilon_{ijk} n_j d_k. \end{aligned} \quad (10)$$

It turns out that the null axis is perpendicular to both the P and the T axes. To see this, we use the cross-product (Eqn A.3.43) to form a vector perpendicular to both axes,

$$\begin{aligned} (1/2)(\mathbf{t} \times \mathbf{p}) &= (1/2)(\hat{\mathbf{n}} + \hat{\mathbf{d}}) \times (\hat{\mathbf{n}} - \hat{\mathbf{d}}) = (\varepsilon_{ijk}/2)(n_j + d_j)(n_k - d_k) \\ &= (\varepsilon_{ijk}/2)(n_j n_k - n_j d_k + d_j n_k - d_j d_k), \end{aligned} \quad (11)$$

and simplify, using

$$\begin{aligned} &\hat{\mathbf{n}} \times \hat{\mathbf{n}} = \varepsilon_{ijk} n_j n_k = 0, \quad \hat{\mathbf{d}} \times \hat{\mathbf{d}} = \varepsilon_{ijk} d_j d_k = 0, \\ &\varepsilon_{ijk} d_j n_k = -\varepsilon_{ijk} n_j d_k, \end{aligned} \quad (12)$$

to see that

$$(1/2)(\mathbf{t} \times \mathbf{p}) = -\varepsilon_{ijk} n_j d_k = -(\hat{\mathbf{n}} \times \hat{\mathbf{d}}), \quad (13)$$

which is just the negative of a unit vector along the null axis, **b**. Thus either the fault normal vector, slip vector, and null axis or the P, T, and B (null) axes can be used for an orthogonal coordinate system.

The relationship between the fault and auxiliary planes can be derived from the fact that the slip vector, which lies in the fault plane, is the normal to the auxiliary plane and vice versa. Thus if $\hat{\mathbf{n}}_1$, $\hat{\mathbf{d}}_1$ and $\hat{\mathbf{n}}_2$, $\hat{\mathbf{d}}_2$ are the fault normal and slip vectors for the two nodal planes,

$$\hat{\mathbf{d}}_1 = \hat{\mathbf{n}}_2 \quad \text{and} \quad \hat{\mathbf{d}}_2 = \hat{\mathbf{n}}_1. \quad (14)$$

Writing out $\hat{\mathbf{d}}_1 = \hat{\mathbf{n}}_2$ by components,

$$\begin{pmatrix} \cos\lambda_1\cos\phi_{f_1} + \sin\lambda_1\cos\delta_1\sin\phi_{f_1} \\ -\cos\lambda_1\sin\phi_{f_1} + \sin\lambda_1\cos\delta_1\cos\phi_{f_1} \\ \sin\lambda_1\sin\delta_1 \end{pmatrix} = \begin{pmatrix} -\sin\delta_2\sin\phi_{f_2} \\ -\sin\delta_2\cos\phi_{f_2} \\ \cos\delta_2 \end{pmatrix}. \quad (15)$$

The corresponding relation between $\hat{\mathbf{n}}_1$ and $\hat{\mathbf{d}}_2$ is found simply by interchanging subscripts.

These equations relate the strike, dip, and slip angles for one plane to the other. To use them, we multiply the first by $\cos\phi_{f_1}$ and the second by $\sin\phi_{f_1}$, and subtract them to find

$$\cos\lambda_1 = \sin\delta_2 \sin(\phi_{f_1} - \phi_{f_2}), \quad (16)$$

or, equivalently,

$$\cos\lambda_2 = \sin\delta_1 \sin(\phi_{f_2} - \phi_{f_1}). \quad (17)$$

We also have the third equation

$$\cos\delta_2 = \sin\lambda_1 \sin\delta_1, \quad (18)$$

or, equivalently,

$$\cos\delta_1 = \sin\lambda_2 \sin\delta_2. \quad (19)$$

An additional constraint comes from the fact that the two nodal planes are perpendicular:

$$\hat{\mathbf{n}}_1 \cdot \hat{\mathbf{n}}_2 = 0, \quad (20)$$

so

$$\begin{aligned} &\sin\delta_1\sin\phi_{f_1}\sin\delta_2\sin\phi_{f_2} + \sin\delta_1\cos\phi_{f_1}\sin\delta_2\cos\phi_{f_2} \\ &\quad + \cos\delta_1\cos\delta_2 = 0, \\ &\sin\delta_1\sin\delta_2\cos(\phi_{f_1} - \phi_{f_2}) + \cos\delta_1\cos\delta_2 = 0; \end{aligned} \quad (21)$$

or

$$\tan\delta_1 \tan\delta_2 \cos(\phi_{f_1} - \phi_{f_2}) = -1. \quad (22)$$

These equations allow us to find the the second nodal plane and the slip vector on it (ϕ_{f_2}, δ_2, λ_2) from the first nodal plane and the slip on it (ϕ_{f_1}, δ_1, λ_1). The hard part, getting the angles in the appropriate quadrants, can be done by first finding δ_2 from Eqn 18, and then finding sin λ_2 from Eqn 19 and cos λ_2 by combining Eqns 16 and 17. Given both sine and cosine, λ_2 can be placed in the correct quadrant. We then find ϕ_{f_2} from Eqns 22 and 16. Finally, if $90° < \delta_2 < 180°$, we change (ϕ_{f_2}, δ_2, λ_2) to ($180° + \phi_{f_2}$, $180° - \delta_2$, $360° - \lambda_2$).

If the nodal planes have been found from first motions using a stereonet, the situation differs because the strike and dip of both planes are known, but the slip angles are not. We then choose one nodal plane and find the slip angle on it. This can be done using Eqns 16 and 18 to find cos λ_1 and sin λ_1, and then placing λ_1 in the correct quadrant.

4.3 Waveform modeling

As noted in the previous section, *P*-wave first motions are often inadequate to constrain focal mechanisms. Additional information is obtained by comparing the observed body and surface waves to theoretical, or *synthetic*, waveforms computed for various source parameters, and finding a model that best fits the data, either by forward modeling or by inversion. Waveform analysis also gives information about the earthquake depths and rupture processes which cannot be extracted from the first motions. We discuss such analysis first for body waves and then for surface waves.

4.3.1 *Basic model*

To generate synthetic waveforms, we regard the ground motion recorded on a seismogram as a combination of factors: the earthquake source, the earth structure through which the waves propagated, and the seismometer. Each factor can be thought of as an operation whose effects depend on the frequency of the seismic waves. Hence it is often useful to think of the seismogram $u(t)$ in terms of its Fourier transform $U(\omega)$, which represents the contribution of the different frequencies:

$$u(t) = \frac{1}{2\pi}\int_{-\infty}^{\infty} U(\omega)e^{i\omega t}d\omega \qquad U(\omega) = \int_{-\infty}^{\infty} u(t)e^{-i\omega t}dt \tag{1}$$

As as in earlier discussions (Sections 2.8, 3.3, 3.7), we use the Fourier transform and related concepts while deferring more general treatment of Fourier analysis to Chapter 6. The essence of this approach is that we represent a seismogram or individual factors that make it up either as a time series or by its Fourier transform, depending on which is more convenient, and switch back and forth using the transform and inverse transform relations.

This approach to generating synthetic seismograms from earthquakes is conceptually the same as that discussed in Section 3.3.6 for reflection seismograms. There, we described the combined effect of various factors as the *convolution* of time series representing each factor. Recall that the convolution of two time series $w(t)$ and $r(t)$ is written

$$s(t) = w(t) * r(t) = \int_{-\infty}^{\infty} w(t-\tau)r(\tau)d\tau. \tag{2}$$

Thus a seismogram $u(t)$ can be written

$$u(t) = x(t) * e(t) * q(t) * i(t), \tag{3}$$

where $x(t)$ is the source time function, the "signal" the earthquake puts into the ground, $e(t)$ and $q(t)$ represent the effects of earth structure, and $i(t)$ describes the instrument response of the seismometer. We also noted (and will prove in Section 6.3.1) that convolution in the time domain is equivalent to multiplication in the frequency domain, so Eqn 3 can be written as the product of Fourier transforms of the four factors

$$U(\omega) = X(\omega)E(\omega)Q(\omega)I(\omega). \tag{4}$$

Each factor can be described in the time domain or the frequency domain. For example, the seismogram depends on how the seismometer responds to ground motion of different frequencies. Figure 4.3-1 (*top*) shows the instrument response, the amplification of a signal as a function of period, for a long-period seismometer. Ground motion with periods around the peak response ($T = 15$ s) is enhanced relative to that at longer or shorter periods. As discussed in Section 6.6, seismometer responses differ; some have peak response at short (e.g., 1 s) periods, whereas others have better response at longer periods. The seismometer response can also be described in the time domain by taking its inverse Fourier transform (Fig. 4.3-1, *bottom*). The resulting time series, $i(t)$, is the *impulse response*, describing how the seismometer responds to a sharp impulse. For the seismometer illustrated in Fig. 4.3-1, the impulse response has a sharp initial peak, followed by a smaller "backswing."

In this formulation, the effects of earth structure are divided into two factors. One, $e(t)$, gives the effect of reflections and conversions of seismic waves at different interfaces along the ray path and the effect of geometric spreading of the rays due to the velocity structure (Section 3.4.2). All these effects are elastic wave phenomena. There is also anelastic *attenuation* described by $q(t)$, whereby some of the seismic waves' mechanical energy is lost by conversion into heat. Attenuation, discussed in Section 3.7, is illustrated by the decay with time of a damped harmonic oscillation with frequency ω:

$$f(t) = Ae^{i\omega t}e^{-\omega t/2Q}. \tag{5}$$

The quality factor Q characterizes the attenuation: the amplitude decays by e^{-1} in a time $2Q/\omega$ (Fig. 3.7-11), so the higher

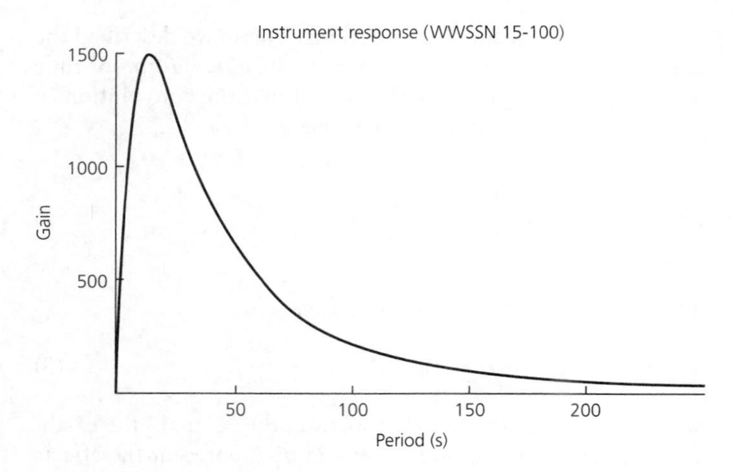

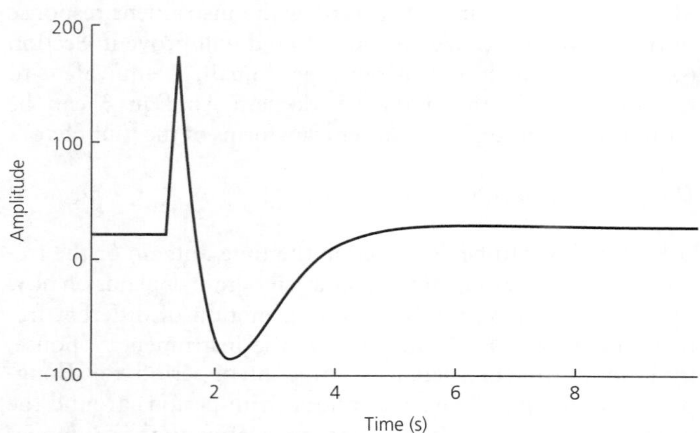

Fig. 4.3-1 The response of a long-period seismometer. *Top*: Gain, or magnification, of an arriving signal as a function of period. *Bottom*: Impulse response in the time domain. This seismometer is a long-period World Wide Standardized Seismographic Network (WWSSN) analog instrument, a type installed around the world in the 1960s that produced many crucial results prior to the advent of digital instrumentation.

that Q is, the slower the decay, and thus the lower the attenuation. The operators $q(t)$ or $Q(\omega)$ describe the effect of attenuation over the range of frequencies making up the seismogram being synthesized.

4.3.2 *Source time function*

The earthquake source signal, $x(t)$, is the *source time function* produced by the faulting. In the simplest case of a short fault that slips instantaneously, the seismic moment function (Eqn 4.2.4) is a step function whose derivative, a delta function (Section 6.2-5), is the source time function. Real faults, however, give rise to more complicated source time functions. Consider a simple case in which the rupture at each point on a rectangular fault radiates an impulse. However, the total radiated signal is not impulsive, because the finite fault does not all break at the same time. Instead, waves arrive first from the initial point of rupture, and later from points further along

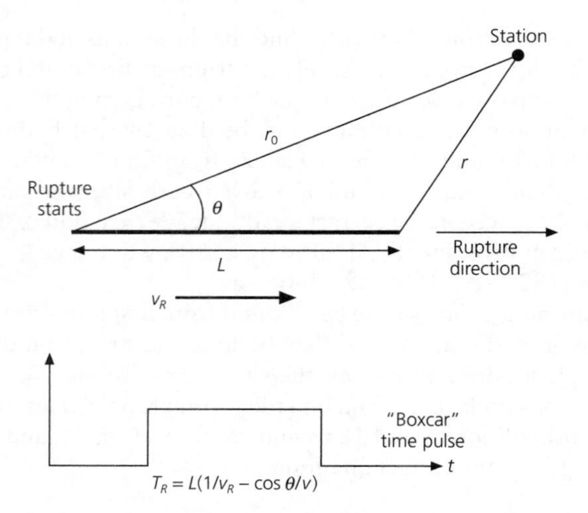

Fig. 4.3-2 For a fault of length L, the duration of the source time function varies as a function of azimuth, depending on the ratio of the rupture velocity v_R and the wave velocity v.

the fault. Assume (Fig. 4.3-2) that the rupture propagated at the *rupture velocity* v_R along a fault of length L. Consider a receiver at a distance r_o and azimuth θ from the initial point of rupture. The first seismic arrival is at time r_o/v where v is either α or β, for P or S waves, respectively. The far end of the fault ruptures a time L/v_R later, giving a seismic arrival at time $(L/v_R + r/v)$, where r is the distance from the far end to the receiver. The law of cosines shows that

$$r^2 = r_o^2 + L^2 - 2r_oL\cos\theta, \tag{6}$$

which, for points far from the fault ($r \gg L$), is approximately

$$r \approx r_o - L\cos\theta. \tag{7}$$

Thus the time pulse due to the finite fault length is a "boxcar" of duration

$$T_R = L(1/v_R - \cos\theta/v) = (L/v)(v/v_R - \cos\theta), \tag{8}$$

known as the *rupture time*. Because v_R is typically assumed to be about 0.7–0.8 times the shear velocity β, v/v_R is about 1.2 for shear waves and 2.2 for P waves. The maximum duration occurs 180° from the rupture direction, and the minimum is in the rupture direction.[1] These expressions can be modified for different fault shapes and rupture propagation directions, such as rupture propagating outward from the center of a circular fault.

[1] A familiar analogous effect occurs during thunderstorms. Thunder is generated by the sudden heating of air along a lightning channel in the atmosphere. Observers in positions perpendicular to the channel hear a brief, loud, thunder clap, whereas observers in the channel direction hear a prolonged rumble. Here the minimum duration occurs at azimuth 90°, and the maxima are at 0° and 180°, because the "rupture velocity" is much greater than the sound velocity, so v/v_R is approximately zero, and the time function duration varies as $\cos\theta$. (Few, 1980)

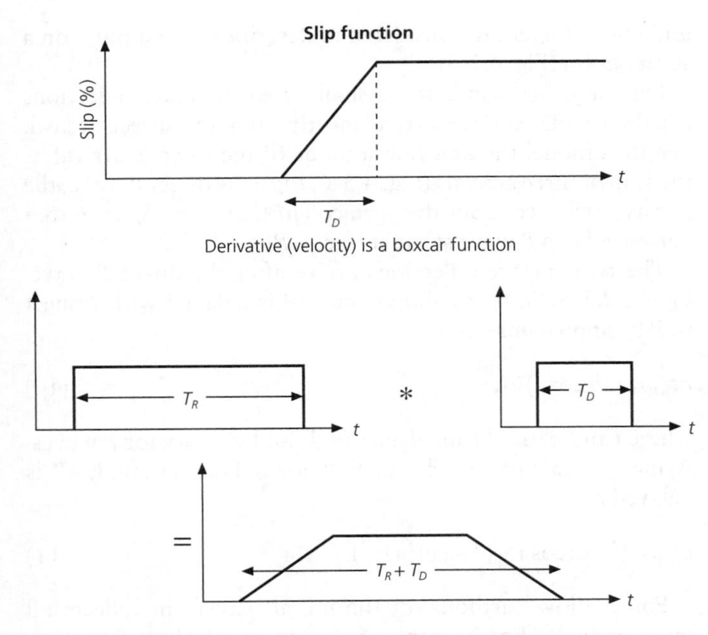

Fig. 4.3-3 The source time function depends on the derivative of the history of slip on the fault. A ramp time history (*top*) with duration T_D has a "boxcar" time derivative. When convolved with the "boxcar" time function due to rupture propagation (*center*), a trapezoidal source time function results (*bottom*).

A second effect lengthening the time function is that, even at a single location on the fault, slip does not occur instantaneously. The slip history is often modeled as a ramp function (Fig. 4.3-3) that begins at time zero and ends at the *rise time* T_D. The source time function depends on the derivative of the slip history, as noted in Section 4.2.3. For a ramp, the derivative is a "boxcar." Convolving the finiteness and rise time effects yields a trapezoid whose length is the sum of the rise and rupture times, which is often used to represent an earthquake source time function. Other shapes of comparable length, like triangles, are also used, because (as we will see) seismograms are often insensitive to the details of the source time function. However, we will also see that for large earthquakes, body wave modeling can resolve a more complicated time function corresponding to the variation in slip along the fault as a function of space and time.

The radiated pulse varies in time duration as a function of azimuth from the rupture direction, due to the finite rupture length (Eqn 8). Because the area of the pulse is the same at all azimuths, the magnitude of the source time function varies inversely with its duration (Fig. 4.3-4). In some cases these effects, called *directivity*, can be used to identify the fault plane (because no similar effect is associated with the auxiliary plane) and study the rupture propagation. Directivity is related to the Doppler effect for sound and light waves, which shifts the frequency of a moving oscillator to higher frequency when the oscillator moves toward an observer, and lower frequency when it moves away. However, directivity results from interference

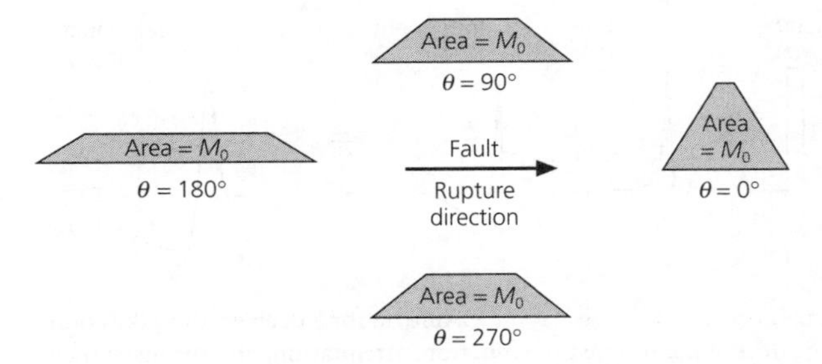

Fig. 4.3-4 Effects of rupture directivity on the source time function at different azimuths from the rupture. Because the same energy arrives, the area of each source time function, corresponding to the seismic moment, is the same. However, in the direction of rupture propagation more energy arrives in a shorter time, whereas in the opposite direction less energy arrives over a greater duration.

between different parts of a finite fault, whereas the Doppler effect in its simplest form occurs for a moving point source.[2]

An interesting question is when we need to consider the effects of a finite earthquake source. We have shown (Eqn 8) that the difference in the arrival time of waves traveling at velocity v from different parts of the fault with length L is the rupture time T_R, which is approximately L/v. If this difference is comparable to the period of the seismic wave, the arriving waveform will be significantly affected. Thus, when the ratio

$$\frac{T_R}{T} = \frac{L/v}{\lambda/v} = \frac{L}{\lambda} \tag{9}$$

is small, the fault length is short compared to the wavelength of the seismic waves, and we can neglect the finiteness of the source and treat it as a point. This criterion is similar to that noted in Section 3.2.3, that seismic waves cannot "see" earth structures much smaller than their wavelengths. For a finite fault, this occurs because the rupture velocity is comparable to the seismic velocity.

An interesting consequence of Eqn 9 is that a fault can seem finite for body waves, but not for surface waves. A 10 km-long fault, which we might expect for a magnitude 6 earthquake, is comparable to the wavelength of a 1 s body wave propagating at 8 km/s, but small compared to the 200 km wavelength of a 50 s surface wave propagating at 4 km/s. On the other hand, a 300 km-long fault for a magnitude 8 earthquake would be a finite source for both waves.

4.3.3 *Body wave modeling*

The elastic structure operator $e(t)$ representing the effects of reflections and transmissions along the ray path primarily reflects interactions near the earth's surface, where the largest change

[2] The Doppler effect is used to detect motion in applications ranging from police and weather radar to astronomical studies of "red-shifted" light that show the universe expanding. For discussion of the relation between directivity and the Doppler effect, see Douglas *et al.* (1988).

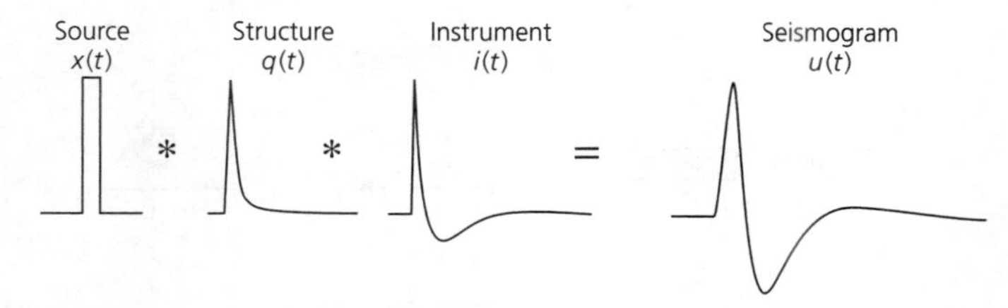

Fig. 4.3-5 The *P*-wave arrival waveform for a deep earthquake combines the effect of the source time function, attenuation, and the instrument. Near-source structure can be neglected because surface reflections arrive much later. (After Chung and Kanamori, 1980. *Phys. Earth Planet. Inter.*, *23*, 134–59, with permission from Elsevier Science.)

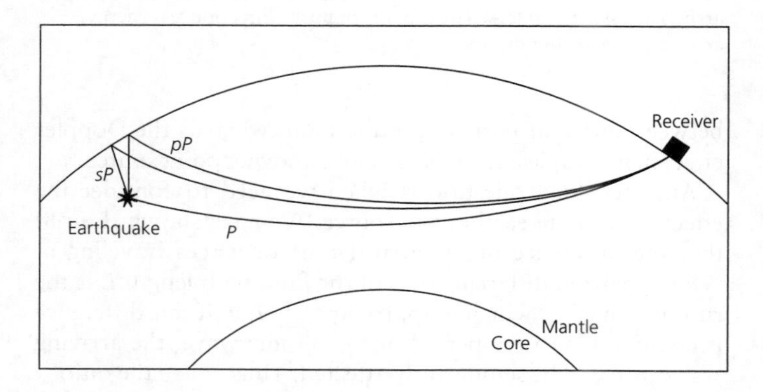

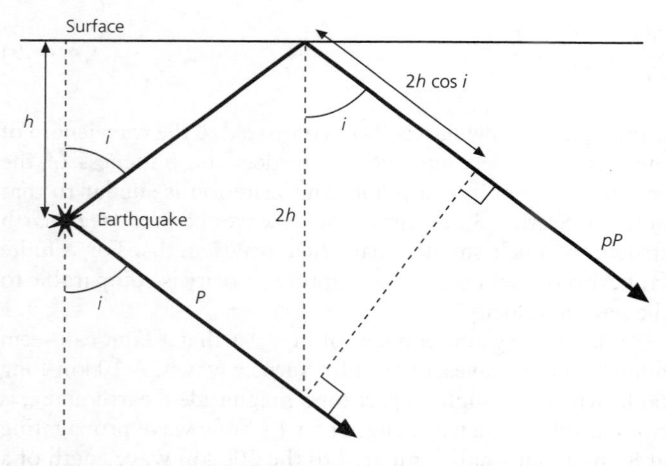

Fig. 4.3-6 *Top*: The *P*-wave arrival for a shallow earthquake at distance $30° < \Delta < 90°$ from the source is modeled as the sum of arrivals due to the direct *P* wave and the free surface reflections *pP* and *sP*. *Bottom*: Geometric construction used to derive the delay time of *pP* with respect to direct *P*.

in physical properties occurs. It is thus useful to consider two simple cases. For a deep earthquake, the surface reflections and other reflected, refracted, and diffracted arrivals arrive much later than the direct *P* wave, so we can describe the direct *P* wave without them. Moreover, at distances $30° < \Delta < 90°$ from the source, the effects of upper mantle triplications and core structure (Section 3.5) can be ignored. Thus, the structure operator can be neglected, and only the source, attenuation, and seismometer are considered to describe the first pulse on a seismogram (Fig. 4.3-5).

On the other hand, for a shallow earthquake, reflections off the earth's surface arrive shortly after the direct arrival. We thus model the first few seconds of the *P*-wave arrival as the sum of three arrivals (Fig. 4.3-6, *top*); the direct *P* wave, the *P* wave reflected from the surface (*pP*), and the *S* wave that converted to a *P* wave at the surface (*sP*).

The two surface reflections arrive after the direct *P* wave. Figure 4.3-6 (*bottom*) shows that *pP* is delayed with respect to *P* by approximately

$$\delta t_{pP} = (2h \cos i)/\alpha, \tag{10}$$

where i and α are the incidence angle and velocity for *P* waves. A messier calculation shows that for a Poisson solid, *sP* is delayed by

$$\delta t_{sP} = (h/\alpha)(\cos i + (3 - \sin^2 i)^{1/2}). \tag{11}$$

For shallow earthquakes the initial waveform reflects all three arrivals. For example, for a source 10 km deep in a medium with $\alpha = 6.8$ km/s, the time delays δt_{pP} and δt_{sP} are 2.7 s and 3.8 s at a distance $\Delta = 50°$, where the incidence angle is 24°. These arrivals are hard to resolve from the *P* arrival, because the seismometer's impulse response (Fig. 4.3-1) is long enough that it has not completely responded to the direct arrival before the others arrive.

The four factors in Eqn 3 can be combined to synthesize body waves. Although the derivation has some subtleties, the result reflects the basic ideas just discussed. The displacement as a function of time, distance, and azimuth, for an initial *P*-wave arrival at distances 30–90° from the source, is

$$u(t, \Delta, \phi) = i(t) * q(t) * \frac{M_0}{4\pi\rho_h\alpha_h^3} \frac{g(\Delta)}{a} C(i_0) \times \Big[R^P(\phi, i_h)x(t - \tau^P) + R^P(\phi, \pi - i_h) \prod^{PP}(i_h) x(t - \tau^{pP}) + R^{SV}(\phi, \pi - j_h) \frac{\alpha_h \cos i_h}{\beta_h \cos j_h} \prod^{SP}(j_h) x(t - \tau^{sP}) \Big]. \tag{12}$$

This formulation includes the seismometer and attenuation factors and a complicated-looking third term incorporating the source and structure factors. This term has distinct pieces, each with a physical interpretation. The amplitude scale factor $M_0/(4\pi\rho_h\alpha_h^3)$ contains the earthquake's seismic moment M_0 and the density and *P*-wave velocity at the source depth h. The $g(\Delta)/a$ factor, where a is the earth's radius, describes the amplitude variations due to geometric spreading of rays. The $C(i_0)$ factor corrects the amplitude for the effects of the free surface, where the rays arrive at the receiver at an angle of incidence, i_0.

The term in brackets has three parts, corresponding to *P*, *pP*, and *sP*. Each includes the source time function $x(t)$ lagged by

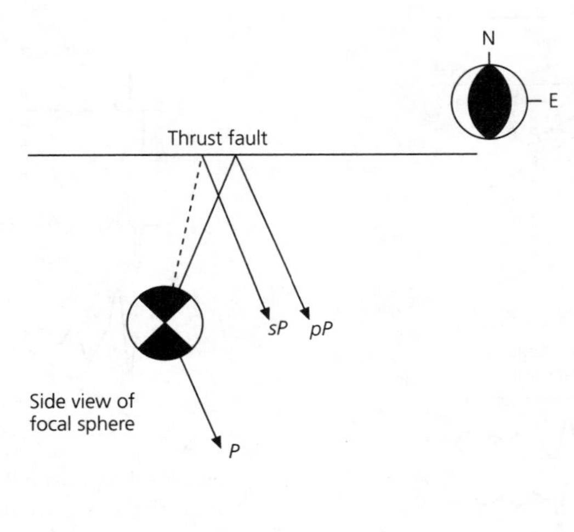

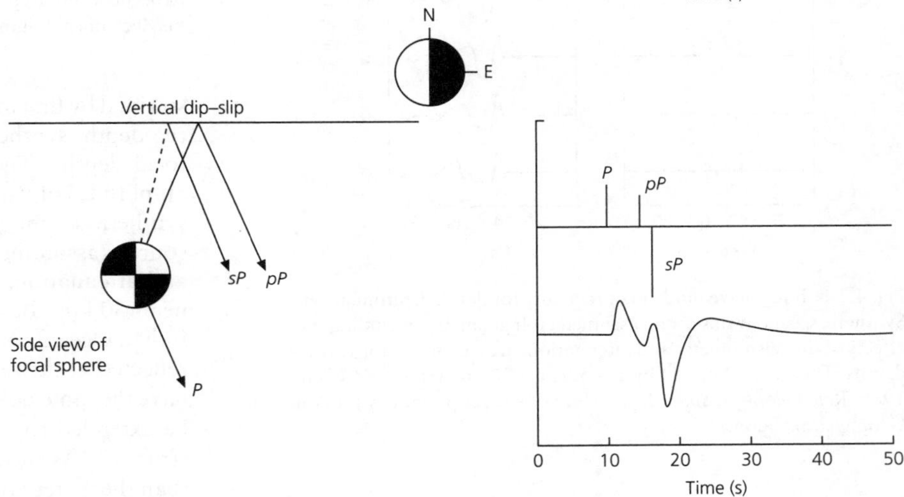

Fig. 4.3-7 Cartoon illustrating the relative polarities and amplitudes of the direct *P* wave and the near-source free surface reflections *pP* and *sP* for different focal mechanisms. The arrivals are shown as impulses, and then including the effects of attenuation and the seismometer. (Okal, 1992. © Seismological Society of America. All rights reserved.)

the travel time for that ray, τ^P, τ^{pP}, and τ^{sP}. Each arrival's amplitude depends on the body wave radiation pattern at the source for that wave type

$$R^P(\phi, i) = s_R(3\cos^2 i - 1) - q_R \sin 2i - p_R \sin^2 i,$$

$$R^{SV}(\phi, j) = \frac{3}{2} s_R \sin 2j + q_R \cos 2j + \frac{1}{2} p_R \sin 2j,$$

$$R^{SH}(\phi, j) = -q_L \cos j - p_L \sin j, \qquad (13)$$

which depend on the take-off angle (i for *P* waves and j for *S* waves) and a set of fault geometry factors which include the fault strike, dip and slip angles (Fig. 4.2-2) ϕ_f, δ, λ, and the azimuth ϕ (clockwise from north) to the station. For $P - SV$ waves these factors are

$$s_R = \sin\lambda \sin\delta \cos\delta,$$

$$q_R = \sin\lambda \cos 2\delta \sin(\phi_f - \phi) + \cos\lambda \cos\delta \cos(\phi_f - \phi),$$

$$p_R = \cos\lambda \sin\delta \sin 2(\phi_f - \phi) - \sin\lambda \sin\delta \cos\delta \cos 2(\phi_f - \phi),$$

and those for *SH* waves are

$$p_L = \sin\lambda \sin\delta \cos\delta \sin 2(\phi_f - \phi) + \cos\lambda \sin\delta \cos 2(\phi_f - \phi),$$

$$q_L = -\cos\lambda \cos\delta \sin(\phi_f - \phi) + \sin\lambda \cos 2\delta \cos(\phi_f - \phi). \qquad (14)$$

The reflected phases' amplitudes also include the plane wave potential reflection coefficients at the free surface, $\Pi^{PP}(i_h)$ and $\Pi^{SP}(j_h)$, which depend on the angles of incidence. Finally, the *sP* term is scaled by a factor $(\alpha_h \cos i_h)/(\beta_h \cos j_h)$ which incorporates several effects, including the fact that near the source the wave incident on the surface is better treated as a spherical wave than a plane wave.

We could similarly model the *SH* wave, which arrives much later, by summing direct *S* and *sS* using an expression analogous to Eqn 12, with the *S*-wave velocity, take-off angles, delay times, and the *SH*-wave radiation pattern R_{SH}.

This formulation shows how synthetic body wave seismograms depend on the assumed focal depth, which determines the time separation between arrivals, the mechanism, which determines the relative amplitudes of the arrivals, and the

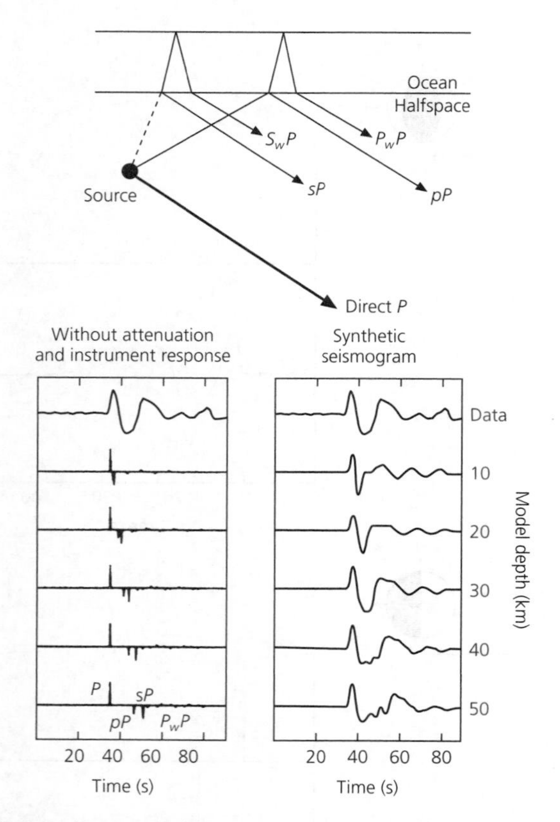

Fig. 4.3-8 Body wave modeling procedure for depth determination. Synthetic seismograms for an assumed fault geometry, including the effects of the seismometer and attenuation, are calculated for various depths. The data are best fit by a depth near 30 km. (Stein and Wiens, 1986. *Rev. Geophys. Space Phys.*, *24*, 806–32, copyright by the American Geophysical Union.)

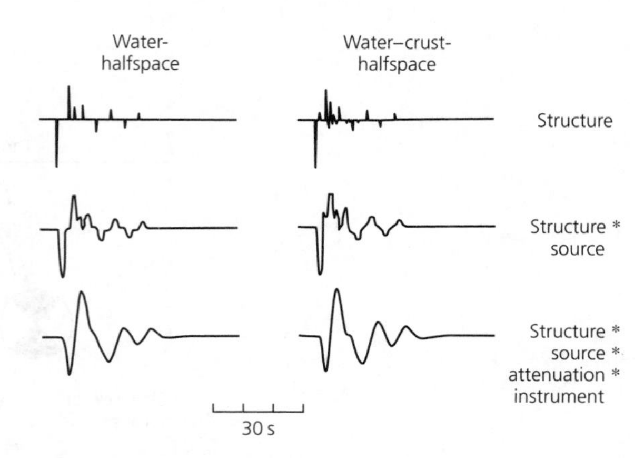

Fig. 4.3-9 Synthetic *P*-wave seismograms for an earthquake occurring beneath the ocean, modeled both without and with a distinct crustal layer. The crustal layer has a smaller effect than the water layer. (Stein and Kroeger, 1980. Reproduced with the permission of the American Society of Mechanical Engineers.)

source time function, which determines the pulse shape. Figure 4.3-7 illustrates this concept for *P* waves from two dip-slip faults, one dipping vertically and the other at 45°. The arrivals are shown first as impulses and then after convolution with the seismometer and attenuation operators. In one case *pP* leaves the focal sphere (shown in side view) with the same polarity as *P*, whereas in the other it leaves with opposite polarity. Its polarity then reverses at the free surface. Thus *pP* on a seismogram need not have the opposite polarity from *P*. Similar effects occur to *sP*. As a result, the relative polarities and amplitudes of the arrivals vary with the mechanism, making the seismogram a useful diagnostic.

Source parameters can be studied by generating synthetic seismograms for various values and finding the best fit to the data, either by forward modeling ("trial and error") or by inversion. Often first motion, body wave, and surface wave analyses (discussed next) are combined. Although first motion data are often consistent with various focal mechanisms, the different methods used together generally yield a consistent and better constrained result.

Figure 4.3-8 shows an example for an earthquake near the Sumatra trench, whose mechanism was reasonably well constrained by first motions. To check the mechanism and estimate the depth, synthetic seismograms were computed for various focal depths. The left panel shows the expected timing and amplitudes of various arriving phases, and the right shows the synthetic seismogram resulting from including the effect of the source (assuming a trapezoidal time function), seismometer, and attenuation. The data are fit well by a source at a depth near 30 km. Because the earthquake occurred beneath the Indian Ocean, some rays *reflected* at the sea surface, and others reflected at the sea floor. The sea floor reflection, p_wP, should have the same polarity (up) as *pP*, as observed. This method can be extended to include the effects of crust and upper mantle structure. As shown in Fig. 4.3-9, a crustal layer has less effect than the water layer, because the water layer has a greater contrast in velocity and density.

Such depth determinations from body wave modeling are often better than those provided by earthquake location programs using arrival times. For example, the International Seismological Center assigned the earthquake represented in Fig. 4.3-8 a depth of 0 ± 17 km. Even if the depth is restricted to be within the earth, the modeling shows that this solution is too shallow.

How well the details of the time function can be resolved depends on factors including the type of seismometer used and the size of the earthquake. One important factor is the distance between the source and the receiver, which influences the amount of attenuation. As the pulse travels, the high frequencies that determine the pulse shape are preferentially removed by attenuation, because the amplitude (Eqn 5) decays by $1/e$ in a time $2Q/\omega$, so higher frequencies decay faster for a given Q. Thus the seismogram is smoothed by the effects of both attenuation and the seismometer (Fig. 4.3-9), especially for long-period seismometers, which also suppress high frequencies (e.g., Fig. 4.3-1). As a result, body wave pulses at teleseismic distances can look similar for different source time functions of approximately the same duration (Fig 4.3-10). Conversely, the best resolution for the details of source time functions is

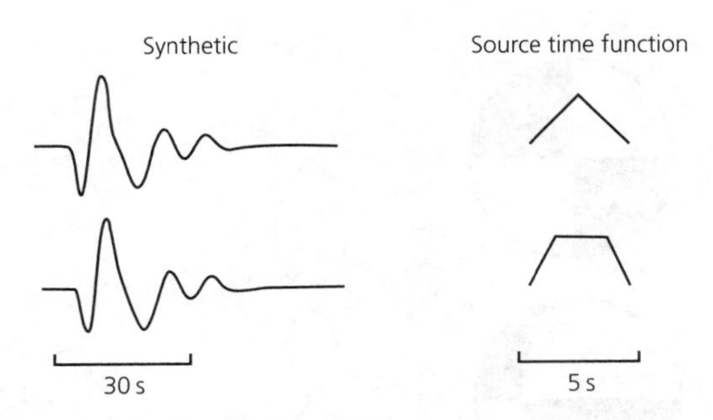

Fig. 4.3-10 Comparison of seismograms synthesized at a teleseismic distance with different source time functions. The effects of the seismometer and attenuation make it difficult to resolve some of the details of the time function. (Stein and Kroeger, 1980. Reproduced with the permission of the American Society of Mechanical Engineers.)

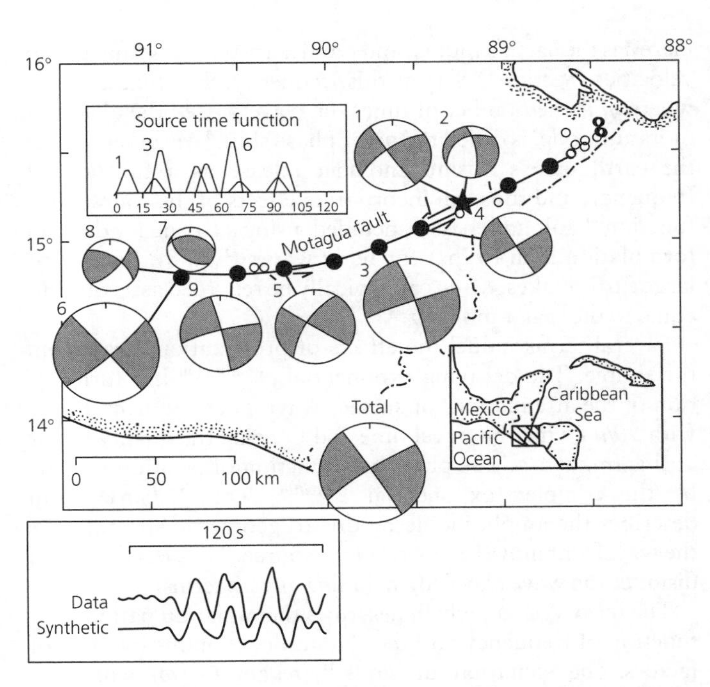

Fig. 4.3-11 Data and synthetic seismogram for the large (M_s 7.5) 1976 Guatemala earthquake. The source is modeled as a series of sub-events along the fault, with positions, timing, relative amplitudes, and mechanisms shown, which gives rise to the complex waveform observed. (After Kikuchi and Kanamori, 1991. © Seismological Society of America. All rights reserved.)

given by strong motion records close to an earthquake and broadband seismometers with uniform response over a wide frequency range.

Larger earthquakes typically occur on longer faults, and thus have longer-duration time functions. As a result, it is often possible to resolve details of the slip process. For example, Fig. 4.3-11 shows complex waveforms from the 1976 Guatemala earthquake.[3] The synthetic seismograms fit the data by assuming that the source consisted of a number of separate sub-events along the fault. Such studies can offer useful insight into the faulting process by showing how the amount and geometry of slip varied along the fault.

A useful way to estimate source time functions is based on the *Green's function*,

$$g(t) = e(t) * q(t), \tag{15}$$

combining the elastic and anelastic effects of propagation from the source to the receiver. The Green's function thus describes the signal that would arrive at the seismometer if the source time function were a delta function. Hence the earthquake's source time function is found by deconvolving the Green's function and the seismometer from the seismogram $u(t)$

$$x(t) = u(t) * [g(t) * i(t)]^{-1}, \quad X(\omega) = \frac{U(\omega)}{G(\omega)I(\omega)}. \tag{16}$$

As we discussed for reflection seismograms (Section 3.3.6), deconvolution can be done in either the time or the frequency domains. Dividing spectra in the frequency domain is easier, but requires care to avoid dividing by small amplitudes which can occur at some frequencies.

Large complex earthquakes can be modeled using Green's functions derived for a simple source in the fault region. The seismogram is treated as the sum of source time functions with different amplitudes, C_j, at different times, τ_j,

$$u(t) = \sum_{j=1}^{K} C_j[x(t-\tau_j) * g(t) * i(t)]. \tag{17}$$

With high-quality data, we will see in Section 4.5.3 that it is possible to go the next step and estimate how the seismic moment release varied on the two-dimensional fault surface as a function of time during the rupture.

4.3.4 *Surface wave focal mechanisms*

Surface waves can be modeled in a conceptually similar way to body waves, and also help resolve earthquake focal mechanisms and depths. In contrast to body wave modeling, which we considered in the time domain using ray theory, we pose surface wave modeling in the frequency domain using a formulation derived from the traveling wave approximation to the earth's normal modes (Section 2.9.6). Thus, for surface waves the contributing factors appear as products of their Fourier transforms (Eqn 4), whereas for body waves (Eqn 12) they appear as convolutions in the time domain (Eqn 3).

We model the transverse component of a Love wave seismogram observed at angular distance θ and azimuth ϕ from an earthquake by its Fourier transform

$$U(\omega, \theta, \phi) = \frac{M(\omega)}{\sqrt{\sin\theta}} e^{-i\pi/4} e^{-i\omega a\theta/c} V(\omega, \phi) e^{-\omega a\theta/2Qu} e^{im\pi/2}$$
$$V(\omega, \phi) = p_L P_L(\omega) + iq_L Q_L(\omega). \tag{18}$$

[3] This M_s 7.5 earthquake, on the Motagua fault which is a transform segment of the boundary between the Caribbean and North American plates (Fig. 5.2-4), caused enormous damage and 22,000 deaths.

Here a is the earth's radius, and c and u are the phase and group velocities (Section 2.8.1) at this frequency. The $(m\pi/2)$ term, where m is the number of times the wave passed the epicenter or its antipode, is called the polar phase shift.[4] $M(\omega)$ represents the earthquake's seismic moment release as a function of frequency, and thus can incorporate effects of the source time function. Fault finiteness is included, using a frequency domain formulation akin to that for body waves (Eqn 8). Except for large earthquakes, $M(\omega)$ can typically be regarded as a constant equal to the scalar moment.

Several terms model the effects of propagation away from the source. The decaying exponential $e^{-\omega a\theta/2Qu}$ is a formulation of the attenuation for surface waves, derived from Eqn 5 with $a\theta/u$ giving the travel time and Q being the quality factor at this frequency. The phase as a function of position is given by the complex exponential $e^{-i\omega a\theta/c}$. The $1/\sqrt{\sin\theta}$ term describes the amplitude decay due to geometric spreading as the wavefront moves away from the source. Thus θ is the actual distance the wave traveled, including any 2π terms.

The term $V(\omega, \phi)$, which describes the radiation pattern as a function of frequency and the azimuth ϕ, contains two sets of factors. The excitation functions $P_L(\omega)$ and $Q_L(\omega)$, which are derived from the radial eigenfunctions for torsional modes with the appropriate frequency, are functions of frequency and the elastic constants at the source depth. These functions weight the *SH*-wave fault geometry factors p_L and q_L (Eqn 14). Because the radiation pattern is a complex number, we can write both amplitude and phase radiation patterns for a given frequency as a function of azimuth

$$|V(\omega, \phi)| = [(p_L P_L(\omega))^2 + (q_L Q_L(\omega))^2]^{1/2},$$
$$\Phi(\omega, \phi) = \tan^{-1}[(q_L Q_L(\omega))/(p_L P_L(\omega))]. \quad (19)$$

Similarly, we can synthesize the vertical component of Rayleigh waves using

$$U(\omega, \theta, \phi) = \frac{M(\omega)}{\sqrt{\sin\theta}} e^{i\pi/4} e^{-i\omega a\theta/c} V(\omega, \phi) e^{-\omega a\theta/2Qu} e^{im\pi/2},$$
$$V(\omega, \phi) = s_R S_R(\omega) + p_R P_R(\omega) + i q_R Q_R(\omega). \quad (20)$$

The radiation pattern $V(\omega, \phi)$ contains excitation functions $S_R(\omega)$, $P_R(\omega)$, and $Q_R(\omega)$, derived from the radial eigenfunctions of spheroidal modes, together with the *P–SV* fault geometry factors s_R, q_R, and p_R (Eqn 14).

Theoretical surface wave spectra can be computed for any fault geometry using the radiation pattern. For example, a vertically dipping dip-slip fault has $s_R = p_R = 0$, $q_R = -\sin(\phi_f - \phi)$, so the only excitation function on which the radiation pattern depends is Q_R. Alternatively, for a vertically dipping strike-slip fault, $s_R = q_R = 0$, $p_R = \sin 2(\phi_f - \phi)$, so the radiation pattern depends on P_R. Thus Rayleigh wave spectral amplitudes for vertically dipping dip-slip and strike-slip faults vary with azimuth as $\sin(\phi_f - \phi)$ and $\sin 2(\phi_f - \phi)$.

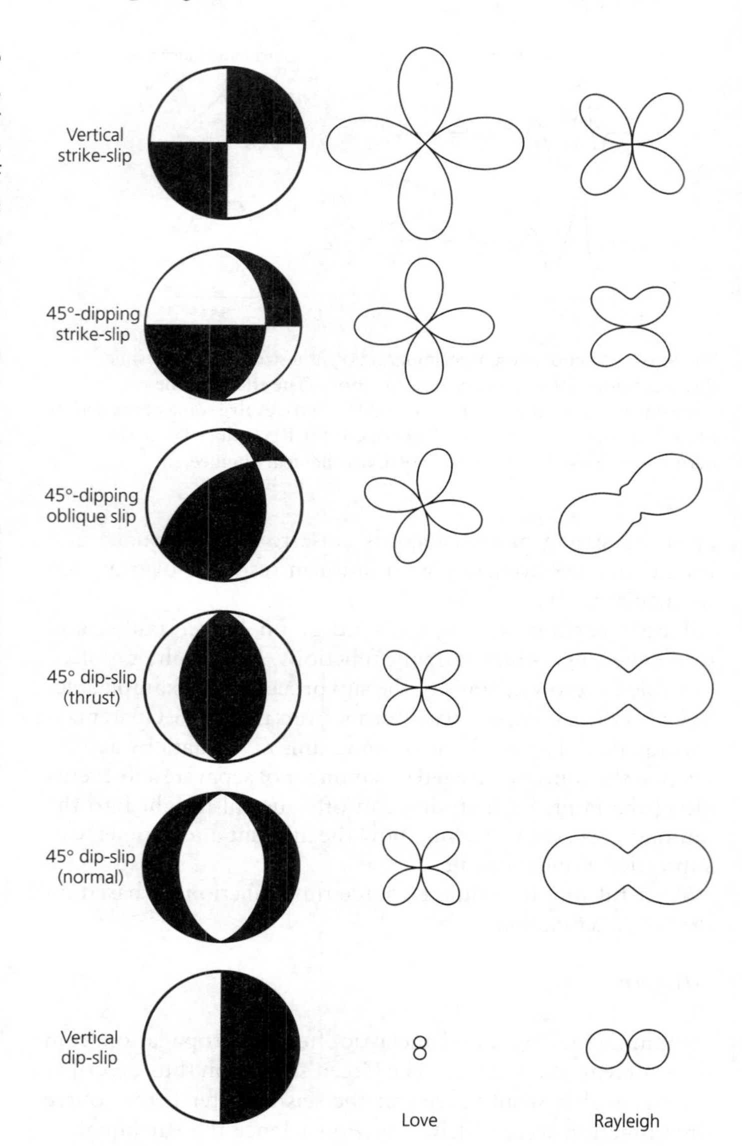

Fig. 4.3-12 Focal mechanisms and surface wave amplitude radiation patterns for six fault geometries. The mechanisms all have one fault plane with a strike of 0°, and the radiation patterns are for a source of constant moment.

Figure 4.3-12 shows theoretical amplitude radiation patterns for Love and Rayleigh waves corresponding to several focal mechanisms, all with a fault plane striking north (0°). The patterns are distinctive: a vertical strike-slip fault has two four-lobed patterns, whereas a 45°-dipping dip-slip fault has a four-lobed Love wave pattern and a two-lobed Rayleigh wave pattern. These radiation patterns are computed for the same seismic moment, and thus show that a vertical strike-slip earthquake is much more efficient at generating Love waves than a vertical dip-slip one. A 45°-dipping oblique-slip mechanism is

[4] This shift arises from the $(l + 1/2)\theta$ in the approximation used to convert normal modes to traveling waves (Eqn 2.9.17) (Brune *et al.*, 1961; Aki and Richards, 1980). For its application to equalization, see Kanamori (1970a).

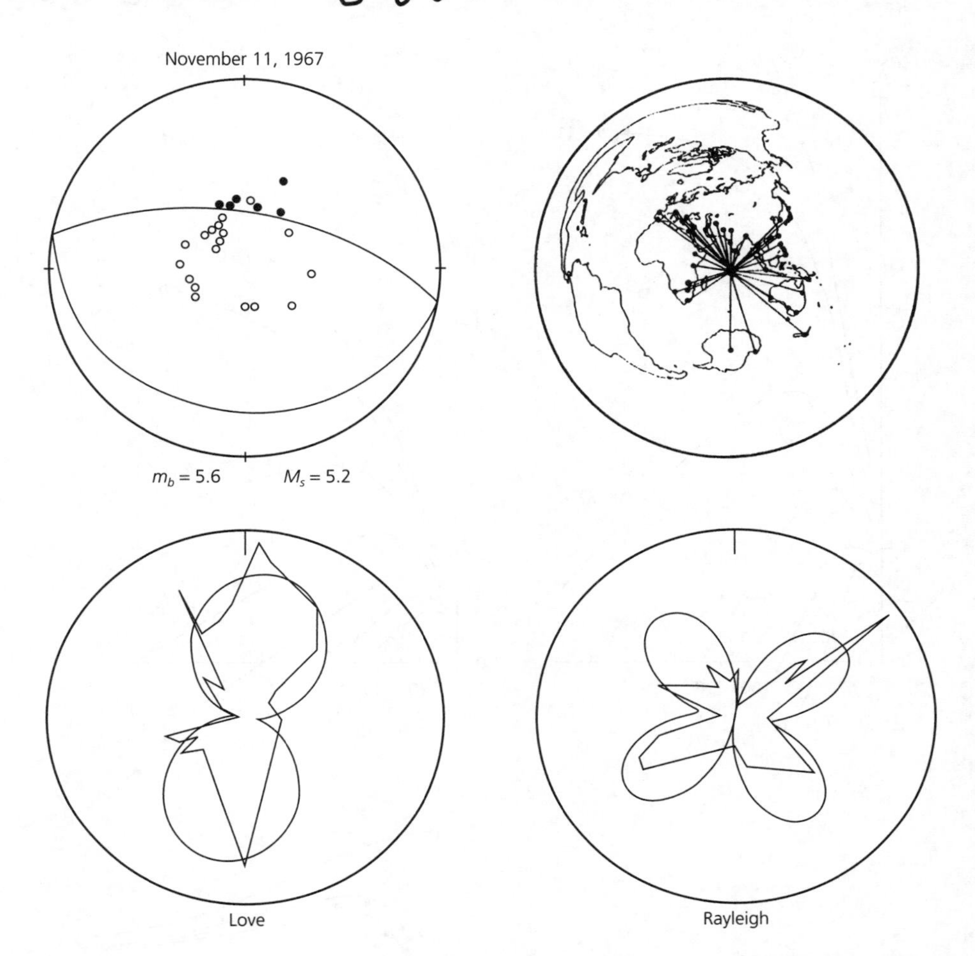

Fig. 4.3-13 Determination of a focal mechanism using surface wave amplitudes. Although *P*-wave first motions cannot constrain both nodal planes, the second plane is constrained by matching the observed Love and Rayleigh wave radiation amplitude patterns. (Stein, 1978.)

intermediate between the 45°-dipping strike-slip and the 45°-dipping thrust mechanisms, and so are the corresponding Love and Rayleigh radiation patterns. Such patterns can be generated for any fault geometry and compared to observations to find the best-fitting source geometry.

To do so, seismograms are Fourier-analyzed to determine the spectral amplitudes at certain frequencies. We can then either model the amplitude at each station, or generate the observed radiation pattern by an equalization correction which simulates a common source-station distance. To do the latter, observations at distance θ, with Fourier transform $U(\omega, \theta, \phi)$, are equalized to a distance θ_0 using

$$U(\omega, \theta_0, \phi) = \left(\frac{\sin\theta}{\sin\theta_0}\right)^{\frac{1}{2}} U(\omega, \theta, \phi) \exp\left[i\left(\frac{\omega a(\theta-\theta_0)}{c} - \frac{m\pi}{2}\right)\right] \exp\left[\frac{\omega a(\theta-\theta_0)}{2Qu}\right]. \tag{21}$$

The $(m\pi/2)$ term, where m is the number of times the path connecting θ and θ_0 goes through the epicenter or its antipode, is the polar phase shift.

Equalization ideally removes all propagation effects, so the spectral amplitude as a function of azimuth should reflect the source's radiation pattern and be comparable to theoretical patterns. Figure 4.3-13 shows an example for a normal faulting earthquake in the diffuse plate boundary zone of the Indian Ocean (Fig. 5.5-5), using Rayleigh and Love waves with the source–receiver paths indicated. Because the first motion data constrained only one E–W striking, north-dipping, nodal plane, the second plane was derived by matching theoretical surface wave amplitude radiation patterns (smooth lines) to the equalized data. Although the observed radiation patterns are somewhat jagged, the fault geometry shown is consistent with the first motions and matches the maximum and minimum amplitude directions of the surface waves.

The equalized data in Fig. 4.3-13 are not as smooth as the theoretical pattern, both because of noise in the data and because the equalization assumes that the attenuation and group velocity are the same for all paths, whereas in reality they vary. As a result, the amplitudes at some stations are higher or lower than predicted. It is possible to reduce this effect by correcting for velocity and Q structure. Even without doing so, such analyses are often valuable for mechanism studies, even for moderate-sized earthquakes like in this example. Phase radiation patterns can also be used, but are generally more sensitive to lateral variations in velocity.

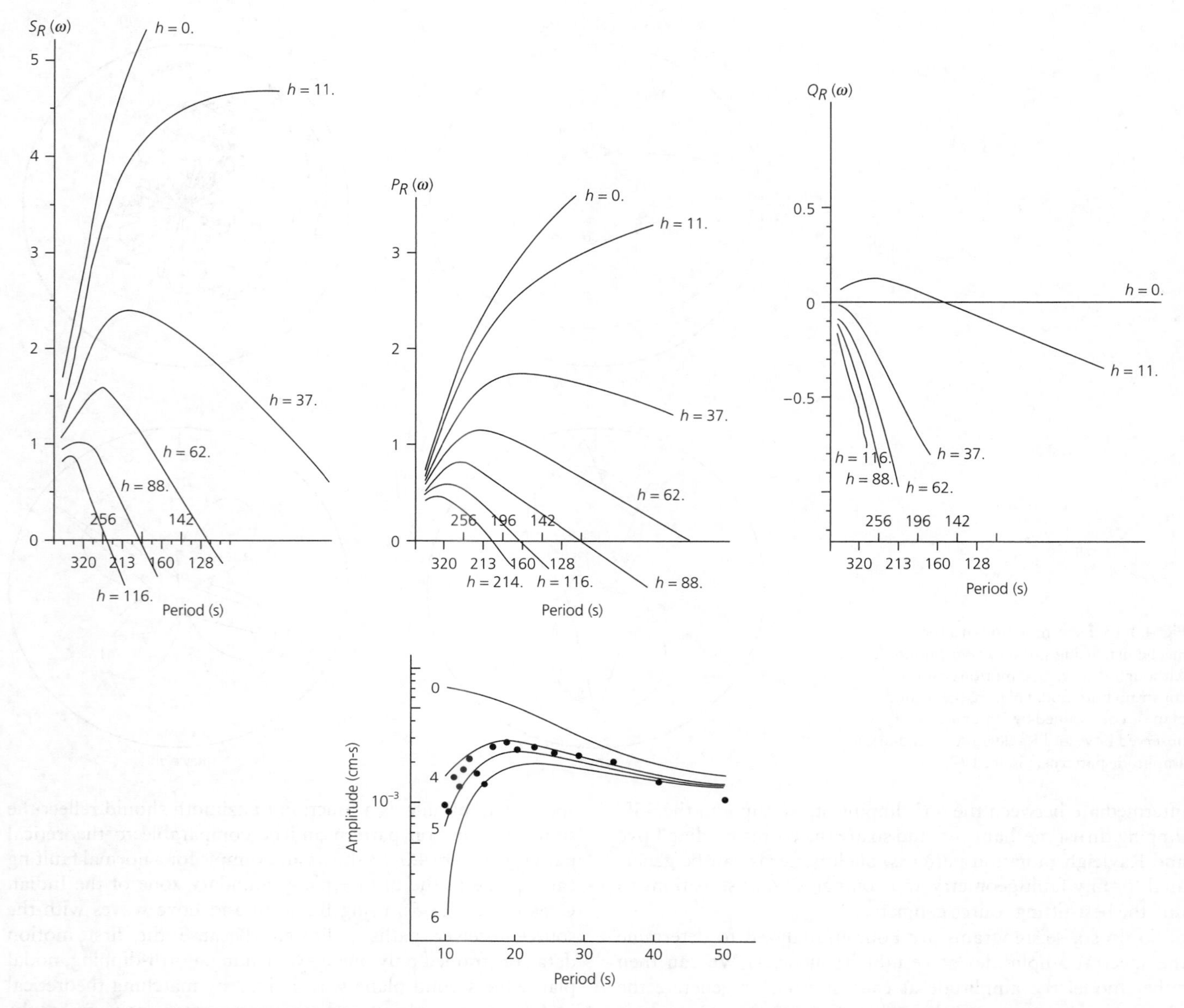

Fig. 4.3-14 Surface wave depth determination uses the variation in Rayleigh wave excitation functions with period and source depth (*top*) (Romanowicz and Guillemant, 1984. © Seismological Society of America. All rights reserved.) For example (*bottom*), the Rayleigh wave spectrum shown is best fit by a 4–5 km focal depth. (Tsai and Aki, 1970. *J. Geophys. Res.*, *75*, 5729–43, copyright by the American Geophysical Union.)

Surface waves can also provide information about earthquake depths because the excitation functions depend on period and source depth, as shown in Fig. 4.3-14 (*top*) for Rayleigh waves. The excitation decreases with source depth, as expected for fundamental mode Rayleigh waves. For a shallow source $Q_R(\omega)$ goes to zero, because this term is proportional to the shear stress generated by the wave, which is zero at the free surface. Figure 4.3-14 (*bottom*) compares an observed surface wave amplitude spectrum to that predicted for various source depths, with the best fit for 4–5 km depth. This process can be formalized by computing the error as a function of assumed source depth and seeking the depth that provides the best fit.

Surface waves can also be used to study fault length and rupture for large earthquakes. Figure 4.3-15 shows an analysis for the great 1964 Alaska earthquake, the second largest ever instrumentally recorded (Fig. 1.2-2). The focal mechanism and geodetic data imply thrust faulting on a roughly NE–SW-striking, shallow NW-dipping fault, due to the subduction of the Pacific plate beneath North America (Fig. 5.2-3). The earthquake was so large that surface waves were unusable until their amplitude had decayed enough, by the fifth station passage (R5 and G5, Fig. 2.7-3). From Fig. 4.3-12, we would expect both the Love and Rayleigh wave amplitude radiation patterns to have minima in the strike direction. However, the observed

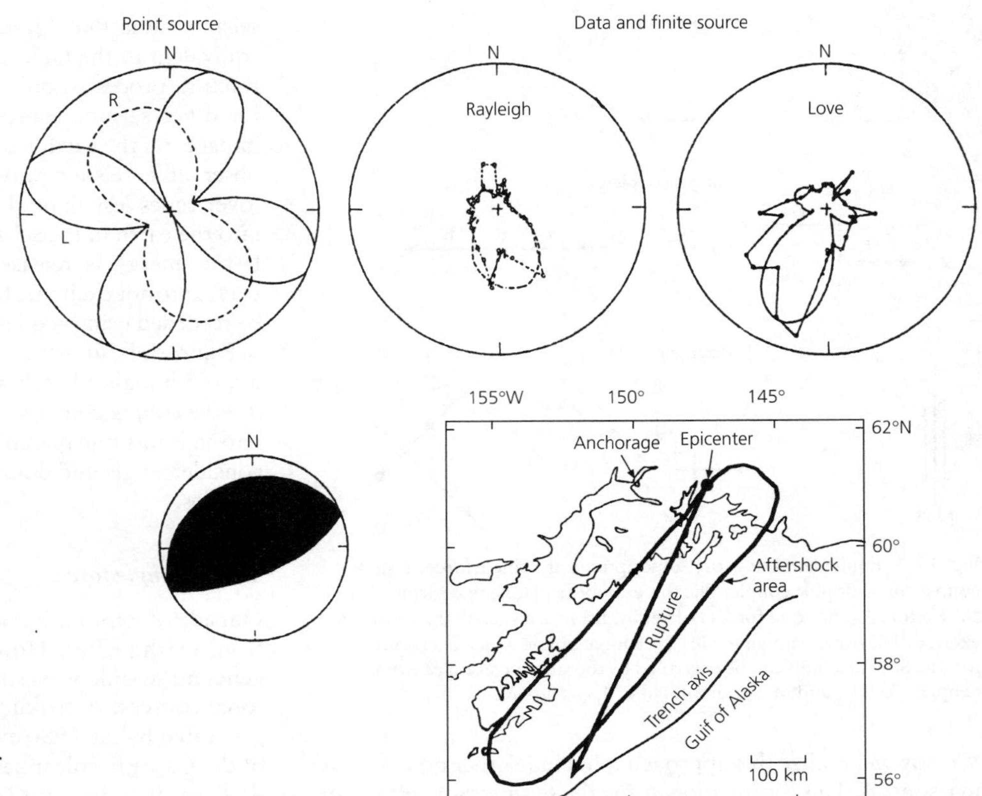

Fig. 4.3-15 Focal mechanism for the great 1964 Alaska earthquake, and the surface wave radiation patterns it predicts if the source is treated as a point (*top left*). Love and Rayleigh waves are shown as solid and dashed lines, respectively. The observed patterns (jagged lines) are quite different, but are reasonably consistent with those predicted by a finite source propagating southwestward along the 600 km-long fault plane, consistent with the large aftershock area (*bottom*). (Kanamori, 1970b. *J. Geophys. Res.*, *75*, 5029–40, copyright by the American Geophysical Union.)

amplitude radiation patterns are quite different, and modeling shows them to be consistent with rupture propagating southwestward along the 600 km-long fault plane. This dimension is consistent with the large aftershock area, and together with the seismic moment (Section 4.6) implies an average fault slip of about 7 meters, bearing out the gigantic nature of the earthquake.[5] In fact, postseismic deformation is still observed with geodetic data (Fig. 4.5-15).

4.3.5 *Once and future earthquakes*

Combining body and surface wave modeling with first motions is often valuable for studying seismograms from older earthquakes. This application arises often in tectonic studies, because in many cases the largest earthquakes occurred prior to the development of global seismic networks in the early 1960s (Section 6.6). Since about 1930, a few stations have reported first motions to the International Seismological Summary. The number of points per earthquake is far less than that available for a modern study, and the data from nonstandardized seismometers are often discordant. However, in some cases body and/or surface wave modeling is useful, especially if the first motions constrain at least one nodal plane. One technique is to use the ratio of Love and Rayleigh wave amplitudes.

This discussion brings out an important difference between first motion and modeling studies. For first motion studies, all we need to know about the seismometer is the polarity, so compressional arrivals are in fact "up" on the seismograms. However, modeling requires knowing the response of the instruments. Fortunately, modern instruments are (at least in theory) standardized, and their calibration can be checked. This is a problem for studies of older earthquakes, because calibrations were often quite poor.

In recent years, modeling approaches have become steadily more powerful. High-quality data from digital broadband seismometers (Section 6.6) have become standard. In addition, laterally homogeneous models for seismic velocity and attenuation have been developed and improved. As a result, inversions of body and surface wave data for many earthquakes, as discussed in the next section, are giving large focal mechanism datasets for tectonic and earthquake source studies.

4.4 Moment tensors

4.4.1 *Equivalent forces*

Our approach so far in this chapter has been to view earthquakes as due to slip on a fault and to estimate their source parameters by forward modeling the radiated seismic waves.

[5] Some of the earthquake damage is shown in Fig. 1.2-11.

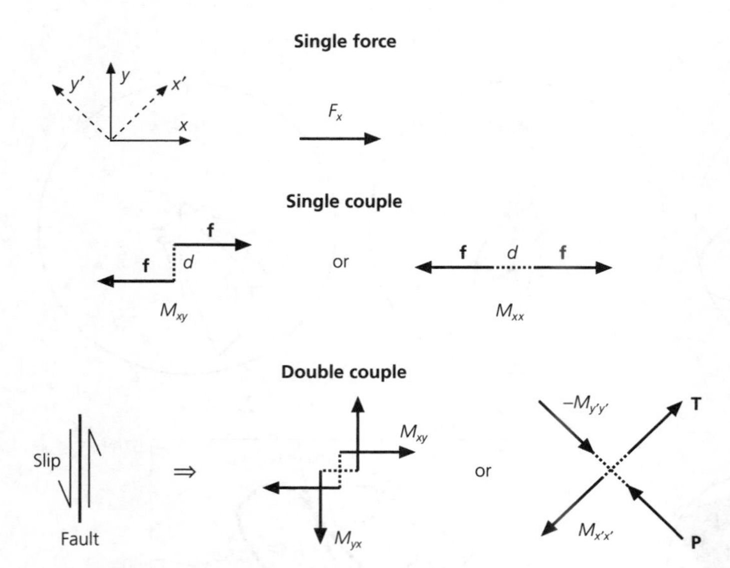

Fig. 4.4-1 Equivalent body force descriptions of a single force, a single couple, and a double couple. The force couple can take two forms. One, shown for M_{xy}, has two forces f offset by distance *d* such that a torque is exerted. The other, shown for M_{xx}, is a force dipole which exerts no torque. Slip on a fault can be described by the superposition of either couples like M_{xy} and M_{yx} or dipoles like $M_{x'x'}$ and $-M_{y'y'}$.

We now generalize this approach to include other types of seismic sources. This formulation, using the *seismic moment tensor*, gives additional insight into the rupture process and greatly simplifies inverting seismograms to estimate source parameters.

We begin by returning to the concept of finding the seismic waves generated by earthquakes due to slip on a fault by solving the equation of motion with the faulting represented by *equivalent body forces* (Section 4.2.3) that yield the same seismic radiation. Although these forces are a seismic source equivalent to the fault motion, they do not describe the actual fracture process. Equivalent body forces can also be derived for other seismic sources, such as explosions, landslides, or impacts on the earth's surface. These phenomena can generate observable seismic waves when they occur rapidly enough (over times less than about an hour) that they release energy into the earth in the seismic wave frequency band (Fig. 2.4-7). If the energy is released more slowly, propagating seismic waves are not excited, although slower crustal deformation can be recorded using geodetic methods (Section 4.5.1).

Figure 4.4-1 illustrates the forces we consider. As noted earlier, earthquakes involving slip upon a fault are modeled as a *double couple* composed of four forces. However, this combination is just one possible combination of forces. Thus we first consider single and double forces, and then work up to double couples.

4.4.2 *Single forces*

Outside of exploration applications, most seismograms result from earthquakes. However, other geophysical phenomena generate seismic waves that are sometimes modeled as single-force sources. A striking example is the large seismic waves generated by the 1980 explosive eruption of Mt St Helens, one of the Cascade volcanoes reflecting the subduction of the Juan de Fuca plate beneath North America (Fig. 5.2-3). The Love and Rayleigh wave radiation patterns (Fig. 4.4-2) are two-lobed, of comparable amplitude, and rotated 90° from each other. Consideration of the patterns for double-couple fault sources shows that such a lobe pattern is expected only for a vertical dip-slip fault (Fig. 4.3-12), and that in this case the Love waves should be much smaller than the Rayleigh waves.

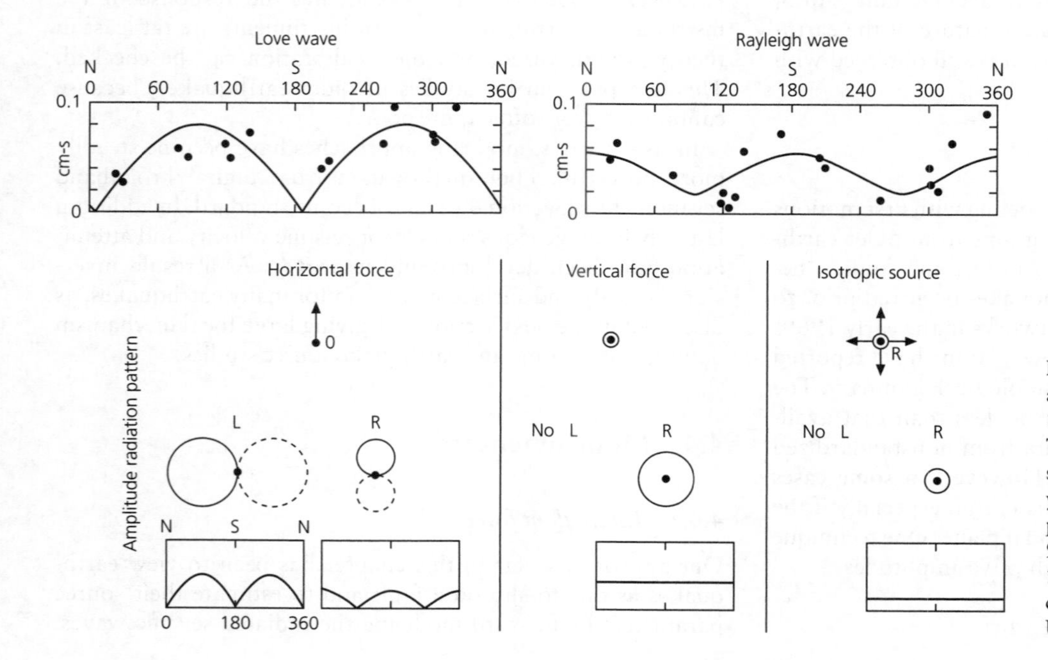

Fig. 4.4-2 *Top*: Observed surface wave amplitude radiation patterns from the May 18, 1980, blast at Mt St Helens. *Bottom*: Theoretical radiation patterns for several seismic sources. Only the horizontal force yields two-lobed Love and Rayleigh wave patterns of comparable amplitude, rotated 90° from each other. (Kanamori and Given, 1982. *J. Geophys. Res.*, *87*, 5422–3, copyright by the American Geophysical Union.)

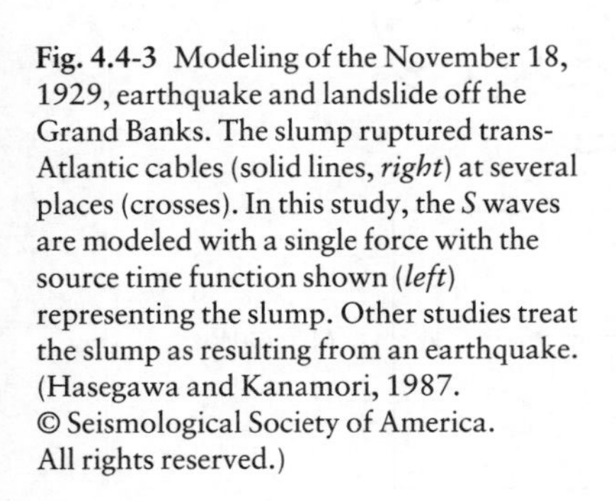

Fig. 4.4-3 Modeling of the November 18, 1929, earthquake and landslide off the Grand Banks. The slump ruptured trans-Atlantic cables (solid lines, *right*) at several places (crosses). In this study, the *S* waves are modeled with a single force with the source time function shown (*left*) representing the slump. Other studies treat the slump as resulting from an earthquake. (Hasegawa and Kanamori, 1987.)

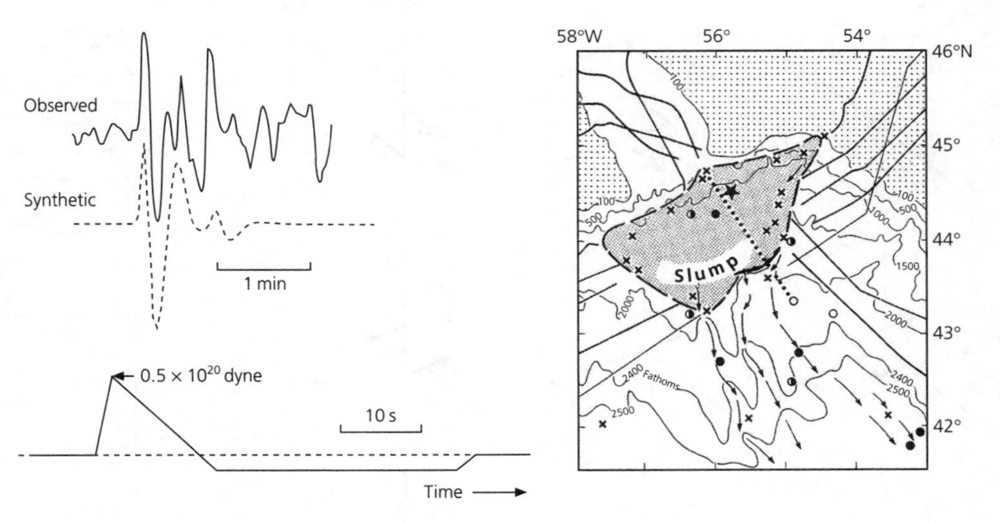

Of the likely non-double-couple sources, both a vertical force and an explosion would produce no Love waves and a circular (rather than lobed) Rayleigh wave radiation pattern. However, a horizontal force can reproduce the observed radiation patterns. The seismic source has thus been modeled with a southward-pointing single force, opposite the direction of the north-directed explosion and northward-flowing landslide. The modeling gives estimates of the force involved in the landslide and explosion, which devastated more than 250 square miles (640 km^2) on the north side of the mountain. This explosion is equivalent to an M_s 5.2 earthquake, significantly bigger than the smaller earthquakes often associated with magma movements within volcanoes.

Landslides have also been modeled by a single force in the direction opposite that of the rock flow. Figure 4.4-3 illustrates this for a large underwater slump (a kind of landslide in which the mass of rock moves as a coherent body) associated with the 1929 M_s 7.2 Grand Banks earthquake. This earthquake, one of the largest in a minor zone of seismicity along the Atlantic continental margin of Canada (Section 5.6.3), was notable because the slump generated powerful sediment flows, known as turbidity currents, which ruptured telephone cables and hence provided important evidence on the speed and force of such currents. As shown, the observed *S* waves are reasonably well modeled by synthetic seismograms for a horizontally oriented single force, implying that the slump itself was the seismic source. However, another study found that the seismograms were well modeled by a double-couple earthquake at about 20 km depth, which triggered the slump. The issue of whether it takes an earthquake to generate such slumps is interesting because such mass movements, which might occur on many heavily sedimented continental margins, can also generate significant tsunamis (Section 1.2.4). The tsunami for this earthquake caused 27 deaths along the Canadian coast, and a slump following an M_s 7.0 earthquake is thought to have caused the devastating 1998 New Guinea tsunami which caused over 2000 deaths.

Meteor impacts should, in principle, generate significant seismic waves. Impacts have been detected seismologically on the moon, but not on earth, where only large meteorites survive passage through the atmosphere. Although it might seem that a meteor impact should be modeled as a vertical force, this would probably not be correct, because the impact's energy would vaporize rock and cause a spherically symmetric explosion similar to an underground nuclear detonation. This idea is supported by the observation that craters produced by meteorites, which are believed to have impacted at very oblique angles, are essentially symmetrical. As we will see, spherically symmetric explosions can be modeled by a set of three orthogonal force couples.

4.4.3 *Force couples*

A force couple consists of two forces acting together. These are similar in concept to electromagnetic dipoles, like that used to model the earth's magnetic field. Two basic couples are shown in Fig. 4.4-1. One consists of a pair of forces offset in a direction normal to the force. The couple M_{xy} consists of two forces of magnitude f, separated by a distance d along the y axis, that act in opposite ($\pm x$) directions. The magnitude of M_{xy} is fd, which in seismology is given in dyn-cm or N-m. To model a couple acting at a point, the limit is taken as d goes to zero such that the product fd stays constant.

The other type of couple, a *vector dipole*, consists of forces offset in the direction of the force. M_{xx} consists of two forces of magnitude f acting in the $\pm x$ directions, separated by d along the x axis. The magnitude is fd, and the limit is taken in the same way. The difference between the two couple types is that the second exerts no torque.

Combining force couples of different orientations into the *seismic moment tensor* **M** (Fig. 4.4-4) gives a general description that can represent various seismic sources. No geophysical processes have been found that are best modeled as single couples, probably because such couples would generate large torques

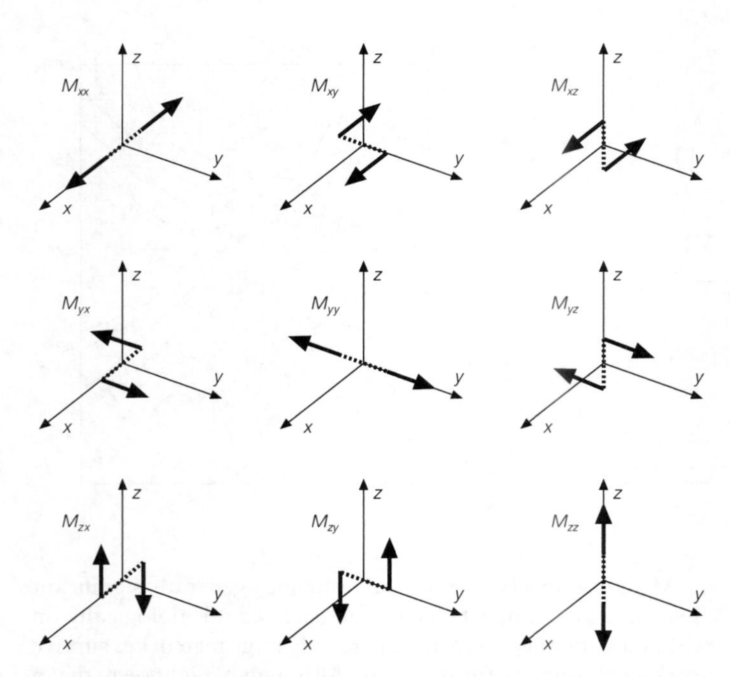

Fig. 4.4-4 The nine force couples which are the components of the seismic moment tensor. Each consists of two opposite forces separated by a distance *d* (dashed line), so the net force is always zero.

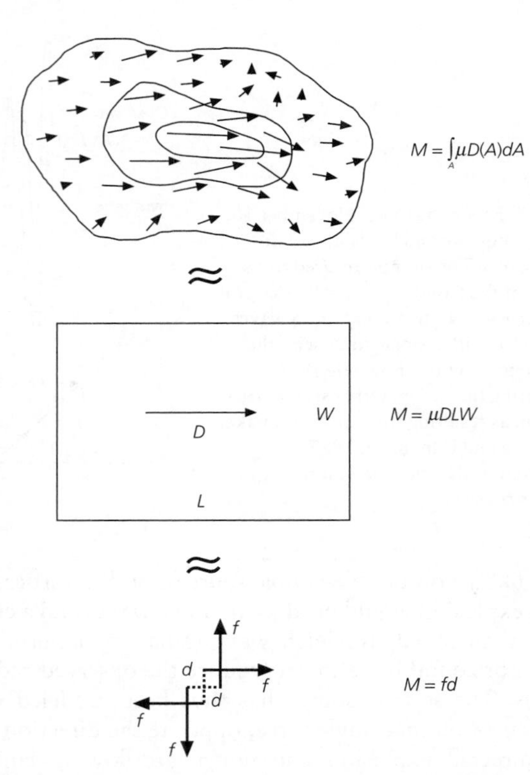

Fig. 4.4-5 Schematic approximations made in modeling the seismic rupture process. *Top*: The rupture process involves a complicated slip function that is variable in space and time. The scalar seismic moment is the integral of this slip process. *Middle*: To infer source parameters, we approximate the rupture as a constant slip $\overline{D}$ on a geometrically simple fault, making the moment a product of the rigidity, average slip, and fault area. *Bottom*: The faulting is further approximated as a double couple of equivalent body forces with moment *fd*.

and thus observable rotations of the earth about different axes. The double and triple sets of couples used to model earthquakes and explosions, respectively, do not generate net torques.[1]

4.4.4 Double couples

Figure 4.4-1 illustrates the relation between an earthquake's fault geometry and the double couple of equivalent body forces. For this example, left-lateral strike-slip in the $\pm y$ directions on a fault in the y–z plane, the equivalent body forces $M_{xy} + M_{yx}$ make up the double-couple source. The M_{yx} couple seems intuitive, because the forces point in the slip directions, but the M_{xy} couple is also needed for reasons including avoiding net torque on the fault.

Because the equivalent body forces are a double couple, they would be the same if the slip were instead right-lateral on a fault in the x–z plane. Thus, as we have noted, seismic waves from a point double-couple source are the same regardless of which plane is the fault plane and which is the perpendicular, auxiliary plane.

The magnitude of the equivalent body forces is M_0, the scalar seismic moment of the earthquake, which has units of dyn-cm, like those of a force couple. Thus if M_{xy} and M_{yx} are couples of unit magnitude, the moment tensor is

$$\mathbf{M} = M_0(M_{xy} + M_{yx}). \qquad (1)$$

[1] Earthquakes can cause measurable changes in the earth's rotation. However, these result not from applied torques, but from vertical redistribution of mass due to static displacements near a fault (Section 4.5).

Hence the moment tensor of an earthquake represents both its fault geometry, via the different components, and its size, via the scalar moment. The moment tensor is a simple mathematical representation that gives the seismic waves produced by a complex rupture involving displacements varying in space and time on a irregular fault (Fig. 4.4-5). In the previous section we approximated the rupture with a constant average displacement $\overline{D}$ over a rectangular fault, and we now approximate it further as a set of force couples. These successive approximations are usually surprisingly successful at matching observed seismograms.

4.4.5 Earthquake moment tensors

As we have seen, the equivalent body forces for seismic sources of different geometries are represented by the seismic moment tensor, $\mathbf{M}$, whose components are the nine force couples

$$\mathbf{M} = \begin{pmatrix} M_{xx} & M_{xy} & M_{xz} \\ M_{yx} & M_{yy} & M_{yz} \\ M_{zx} & M_{zy} & M_{zz} \end{pmatrix}. \qquad (2)$$

In this notation, the earthquake in Fig. 4.4-1 is represented as

$$\mathbf{M} = \begin{pmatrix} 0 & M_0 & 0 \\ M_0 & 0 & 0 \\ 0 & 0 & 0 \end{pmatrix} = M_0 \begin{pmatrix} 0 & 1 & 0 \\ 1 & 0 & 0 \\ 0 & 0 & 0 \end{pmatrix}. \quad (3)$$

We can write the moment tensor in any orthogonal coordinate system because vector and tensor equations are valid regardless of coordinate system. In general, the tensor appears more complicated than Eqn 3 if the fault and slip directions are not oriented neatly relative to the coordinate system. To see this, we write the moment tensor for a double-couple earthquake in an arbitrary coordinate system. The components are given by the scalar moment and the components of $\hat{\mathbf{n}}$, the unit normal vector to the fault plane, and $\hat{\mathbf{d}}$, the unit slip vector,

$$M_{ij} = M_0(n_i d_j + n_j d_i), \quad (4)$$

or

$$\mathbf{M} = M_0 \begin{pmatrix} 2n_x d_x & n_x d_y + n_y d_x & n_x d_z + n_z d_x \\ n_y d_x + n_x d_y & 2n_y d_y & n_y d_z + n_z d_y \\ n_z d_x + n_x d_z & n_z d_y + n_y d_z & 2n_z d_z \end{pmatrix}. \quad (5)$$

This formulation shows two important things. First, the interchangeability of $\hat{\mathbf{n}}$ and $\hat{\mathbf{d}}$ makes the tensor symmetric ($M_{ij} = M_{ji}$). Physically, this shows that slip on either the fault plane or the auxiliary plane yields the same seismic radiation patterns. Second, the trace (sum of diagonal components) of the tensor is zero,[2]

$$\sum_i M_{ii} = M_{ii} = 2M_0 n_i d_i = 2M_0\, \hat{\mathbf{n}} \cdot \hat{\mathbf{d}} = 0, \quad (6)$$

because the slip vector lies in the fault plane and is thus perpendicular to the normal vector. Hence moment tensors corresponding to slip on a fault plane have zero trace. A nonzero trace implies a volume change (explosion or implosion). Such an *isotropic* component does not exist for a pure double-couple source.

Before going further, it is worth briefly considering the tensor properties of M_{ij}. In discussing stress, we noted that a matrix of numbers is a tensor only if it transforms between coordinate systems in a specific way (Eqn 2.13.18). It is easy to prove that the moment tensor for a double couple (Eqn 5) transforms in this manner, because it is a physical entity relating the normal and slip vectors much as the stress tensor relates the normal and traction vectors. At deeper level, M_{ij} is a tensor even for non-double-couple sources because it derives, in a complicated way that we will not discuss, from the change the earthquake causes in stress integrated over the source region. The scalar moment gives the magnitude of the moment tensor $M_0 = (\sum_{ij} M_{ij}^2)^{1/2}/\sqrt{2}$, which is analogous to the magnitude of a vector.

Using the definitions of the normal and slip vectors in terms of fault strike, dip, and slip directions (Section 4.2), we can write the moment tensor for any fault. The reverse process of finding the fault geometry corresponding to a moment tensor is more complicated. However, we need this ability for seismogram inversions that yield the moment tensor. This can be done using some ideas from linear algebra about vector transformations (Section A.5), because the eigenvectors of the moment tensor are parallel to the T, P, and null axes.

To show this, we use the fact (Section 4.2.5) that vectors in these three orthogonal directions $\mathbf{t}$, $\mathbf{p}$, and $\mathbf{b}$ can be written in terms of the fault normal, $\hat{\mathbf{n}}$, and slip vector, $\hat{\mathbf{d}}$, as

$$\begin{aligned} \mathbf{t} &= \hat{\mathbf{n}} + \hat{\mathbf{d}}, \quad t_i = n_i + d_i, \\ \mathbf{p} &= \hat{\mathbf{n}} - \hat{\mathbf{d}}, \quad p_i = n_i - d_i, \\ \mathbf{b} &= \hat{\mathbf{n}} \times \hat{\mathbf{d}}, \quad b_i = \varepsilon_{ijk} n_j d_k. \end{aligned} \quad (7)$$

To prove that these are the eigenvectors and to find the eigenvalues, we begin with $\mathbf{t}$, a vector in the T axis direction, and evaluate

$$\begin{aligned} M_{ij} t_i &= M_0(n_i d_j + n_j d_i)(n_i + d_i) \\ &= M_0(n_i n_i d_j + n_i d_i d_j + n_i n_j d_i + n_j d_i d_i). \end{aligned} \quad (8)$$

Because the normal and slip vectors are perpendicular, ($n_i d_i = 0$) and have unit length ($n_i n_i = d_i d_i = 1$), we see that

$$M_{ij} t_i = M_0(d_j + n_j) = M_0 t_j. \quad (9)$$

Thus the scalar moment M_0 is the eigenvalue associated with $\mathbf{t}$, which is an eigenvector.

Similarly, for the P axis,

$$\begin{aligned} M_{ij} p_i &= M_0(n_i d_j + n_j d_i)(n_i - d_i) \\ &= M_0(n_i n_i d_j + n_i n_j d_i - n_i d_j d_i - n_j d_i d_i) \\ &= M_0(d_j - n_j) = -M_0 p_i, \end{aligned} \quad (10)$$

so $-M_0$ is the eigenvalue associated with $\mathbf{p}$, which is also an eigenvector.

Finally, because M_{ij} is a real symmetric matrix, we know that a third eigenvector is perpendicular to the first two (Section A.5.3). This turns out to be the null axis, $\mathbf{b}$. In Section 4.2.5 we showed that the null axis is perpendicular to the P and T axes:

$$(1/2)(\mathbf{t} \times \mathbf{p}) = -(\hat{\mathbf{n}} \times \hat{\mathbf{d}}) = -\mathbf{b}. \quad (11)$$

To show that $\mathbf{b}$ is an eigenvector, we form

$$\begin{aligned} M_{il} b_l &= M_0(n_i d_l + d_i n_l)(\varepsilon_{ljk} n_j d_k) \\ &= M_0 \varepsilon_{ljk}(n_i d_l n_j d_k + d_i n_l n_j d_k) \\ &= M_0[n_i n_j(\varepsilon_{ljk} d_l d_k) + d_i d_k(\varepsilon_{ljk} n_l n_j)], \end{aligned} \quad (12)$$

[2] Recall the summation convention notation (Section A.3.5) that a repeated index indicates summation.

and recognize that the cross-product of a vector with itself is zero,

$$\varepsilon_{ljk} d_l d_k = \varepsilon_{jkl} d_k d_l = \hat{\mathbf{d}} \times \hat{\mathbf{d}} = 0,$$
$$\varepsilon_{ljk} n_l n_j = \varepsilon_{klj} n_l n_j = \hat{\mathbf{n}} \times \hat{\mathbf{n}} = 0, \tag{13}$$

so the null axis **b** is an eigenvector with associated eigenvalue 0:

$$M_{il} b_l = 0. \tag{14}$$

The fact that the P, T, and null axes are the eigenvectors of the moment tensor lets us simplify it by transforming it into the "natural" coordinate system whose basis vectors are the eigenvectors. Such orthogonal transformations transform a tensor from one orthogonal coordinate system to another, such that its components change, but its physical meaning does not. The transformation matrix with the eigenvectors as columns (Section A.5.3),

$$U = \begin{pmatrix} t_1 & b_1 & p_1 \\ t_2 & b_2 & p_2 \\ t_3 & b_3 & p_3 \end{pmatrix}, \tag{15}$$

gives a diagonal moment tensor for a double couple in the principal axis coordinate system

$$U^{-1}\mathbf{M}U = \begin{pmatrix} M_0 & 0 & 0 \\ 0 & 0 & 0 \\ 0 & 0 & -M_0 \end{pmatrix}. \tag{16}$$

One diagonal element is zero, and the other two are ± the scalar moment. The trace $(M_{xx} + M_{yy} + M_{zz})$, which is not changed by an orthogonal transformation, started as zero in Eqn 6 and so remains zero. Put another way, the isotropic component is an invariant of the moment tensor and does not depend on the coordinate system.

The point of the transformation is that inverting seismograms in a geographic coordinate system yields the moment tensor in that coordinate system. We then find its eigenvectors, the P, T, and null axes, and use Eqn 7 to find the fault normal and slip vectors and hence strike, dip, and slip angles. As part of the same process, the eigenvalues give the scalar moment.

Thus the moment tensor corresponding to a specific faulting geometry can be written in different ways. Figure 4.4-1 shows this in a two-dimensional geometry. The coordinate system oriented along and perpendicular to the fault has the fault normal and slip vectors as basis vectors, and the nonzero moment tensor components are $M_{xy} = M_{yx} = M_0$ (Eqn 3). If we transform the moment tensor to the new (primed) coordinate system with the P and T axes as basis vectors, 45° away from the first set, a two-dimensional version of Eqn 16 gives the moment tensor

Moment tensor	Beachball	Moment tensor	Beachball
$\frac{1}{\sqrt{3}}\begin{pmatrix} 1 & 0 & 0 \\ 0 & 1 & 0 \\ 0 & 0 & 1 \end{pmatrix}$		$-\frac{1}{\sqrt{3}}\begin{pmatrix} 1 & 0 & 0 \\ 0 & 1 & 0 \\ 0 & 0 & 1 \end{pmatrix}$	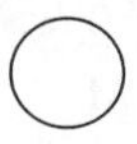
$-\frac{1}{\sqrt{2}}\begin{pmatrix} 0 & 1 & 0 \\ 1 & 0 & 0 \\ 0 & 0 & 0 \end{pmatrix}$		$\frac{1}{\sqrt{2}}\begin{pmatrix} 1 & 0 & 0 \\ 0 & -1 & 0 \\ 0 & 0 & 0 \end{pmatrix}$	
$\frac{1}{\sqrt{2}}\begin{pmatrix} 0 & 0 & -1 \\ 0 & 0 & 0 \\ -1 & 0 & 0 \end{pmatrix}$		$\frac{1}{\sqrt{2}}\begin{pmatrix} 0 & 0 & 0 \\ 0 & 0 & -1 \\ 0 & -1 & 0 \end{pmatrix}$	
$\frac{1}{\sqrt{2}}\begin{pmatrix} -1 & 0 & 0 \\ 0 & 0 & 0 \\ 0 & 0 & 1 \end{pmatrix}$		$\frac{1}{\sqrt{2}}\begin{pmatrix} 0 & 0 & 0 \\ 0 & -1 & 0 \\ 0 & 0 & 1 \end{pmatrix}$	
$\frac{1}{\sqrt{6}}\begin{pmatrix} 1 & 0 & 0 \\ 0 & -2 & 0 \\ 0 & 0 & 1 \end{pmatrix}$		$\frac{1}{\sqrt{6}}\begin{pmatrix} -2 & 0 & 0 \\ 0 & 1 & 0 \\ 0 & 0 & 1 \end{pmatrix}$	
$\frac{1}{\sqrt{6}}\begin{pmatrix} 1 & 0 & 0 \\ 0 & 1 & 0 \\ 0 & 0 & -2 \end{pmatrix}$		$-\frac{1}{\sqrt{6}}\begin{pmatrix} 1 & 0 & 0 \\ 0 & 1 & 0 \\ 0 & 0 & -2 \end{pmatrix}$	

Fig. 4.4-6 A selection of moment tensors and their associated focal mechanisms. The top row shows an explosion (*left*) and an implosion (*right*). The next three rows are for double-couple sources. The bottom two rows show CLVD sources which have a baseball or eyeball/fried-egg appearance. (After Dahlen and Tromp (1998), with moment tensors transformed to the coordinate system with basis vectors pointing north, west, and up.)

$M_{x'x'} = -M_{y'y'} = M_0$. The transformation changes the components, but the physical moment tensor stays the same, so these two different-looking force systems give the same radiated seismic waves. Hence the seismic waves alone provide no way of deciding which is more "real." Given that most earthquakes occur on faults about which we have other knowledge, we generally view earthquakes as slip on a fault rather than dipoles. It is worth recalling that a similar concept appears whenever we transform vector or tensor quantities between coordinate systems. For example, Fig. 2.3-6 showed that a given physical state of stress could be represented either by normal stresses (diagonal terms in the stress tensor) or shear stresses (off-diagonal terms in the stress tensor), depending on the coordinate system.

Figure 4.4-6 shows the diagonalized moment tensor and focal mechanism for some source geometries. The second, third, and fourth rows show end-member double-couple mechanisms. For each, the figure shows a vertical strike-slip (second row), vertical dip-slip (third row), and a 45°-dipping pure thrust fault. The first and last two rows, however, show very different-looking mechanisms, which are discussed next. The moment tensors are given in the coordinate system of Section 4.2.1, with basic vectors pointing north, west, and up. In another coordinate system, such as spherical coordinates, the components of the tensors would differ.

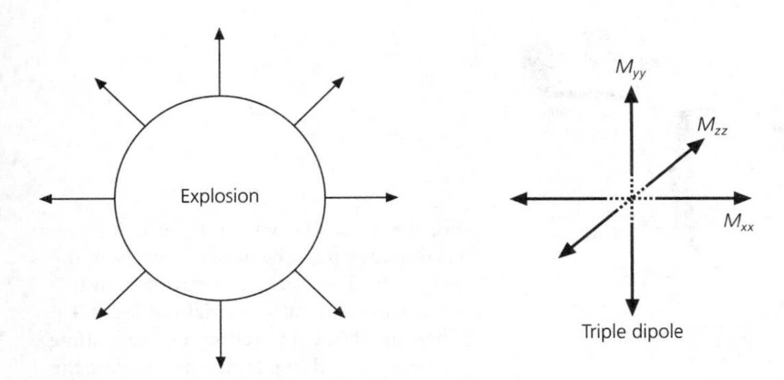

Fig. 4.4-7 An explosive source, which radiates energy equally in all directions, is modeled using a triple dipole as an equivalent body force system.

4.4.6 *Isotropic and CLVD moment tensors*

If all three diagonal terms of the moment tensor are nonzero and equal, the polarity of the first motions (focal mechanism) is the same in all directions. Such a *triple vector dipole* of three equal and orthogonal force couples is the equivalent body force system for an explosion or an implosion (Fig. 4.4-7). The moment tensor looks like

$$\mathbf{M} = \begin{pmatrix} E & 0 & 0 \\ 0 & E & 0 \\ 0 & 0 & E \end{pmatrix}, \tag{17}$$

and has nonzero trace $3E$. A moment tensor with a nonzero isotropic component represents a volume change.

Most explosive sources are man-made mining or nuclear explosions. The ability to identify and locate nuclear explosions seismologically is critical for monitoring nuclear testing (Section 1.2). Natural explosive or implosive sources are rare, but may be associated with fluid and gas migration linked to magmatic processes or with sudden phase transitions of metastable minerals. High-velocity impacts of meteorites could also be modeled with explosive sources.

The physical processes in explosions differ markedly from those for earthquakes. An explosion involves a sudden increase in pressure, which causes nonlinear deformation that can melt and even vaporize rock. As this *shock wave* of pressure expands, its amplitude decreases until the deformations are small enough to occur elastically, yielding a spherical P wave (Section 2.4.3). This propagating wave interacts with interfaces within the earth, including the surface, and generates SV and Rayleigh waves, as seen in the nuclear explosion seismogram in Fig. 1.2-19. Surprisingly, SH waves, including Love waves, are also observed. These would not be expected in a spherically symmetric and isotropic earth, where P–SV and SH waves are decoupled. Several possibilities have been suggested, including *tectonic release* of deviatoric stress near the source, essentially triggering earthquakes, and giving the source both isotropic and double-couple components.

Another class of non-double-couple seismic sources are *compensated linear vector dipoles* (CLVDs). These are sets of three force dipoles that are compensated, with one dipole −2 times the magnitude of the others:

$$\mathbf{M} = \begin{pmatrix} -\lambda & 0 & 0 \\ 0 & \lambda/2 & 0 \\ 0 & 0 & \lambda/2 \end{pmatrix}. \tag{18}$$

The trace of the moment tensor is zero, so there is no isotropic component. CLVDs are illustrated by the strange-looking bottom two rows in Fig. 4.4-6. By contrast with the beachball-looking focal mechanisms of double couples, the first motions for CLVDs look like baseballs (fifth row) or eyeballs (sixth row). Although sources with large CLVD components are rare, they have been identified in several complicated tectonic environments.

Two primary explanations have been offered for CLVD mechanisms. Especially in volcanic areas, it is natural to think of an inflating magma dike, which can be modeled as a crack opening under tension. The moment tensor is for such a crack is[3]

$$\mathbf{M} = \begin{pmatrix} \lambda & 0 & 0 \\ 0 & \lambda & 0 \\ 0 & 0 & \lambda + 2\mu \end{pmatrix}, \tag{19}$$

where λ and μ are the Lamé elastic constants (Eqn 2.3.69). The trace of this tensor is $3\lambda + 2\mu$, which is positive because the crack opened. Thus we can decompose the tensor into two terms:

$$\begin{pmatrix} \lambda & 0 & 0 \\ 0 & \lambda & 0 \\ 0 & 0 & \lambda + 2\mu \end{pmatrix} = \begin{pmatrix} E & 0 & 0 \\ 0 & E & 0 \\ 0 & 0 & E \end{pmatrix} + \begin{pmatrix} -2/3\mu & 0 & 0 \\ 0 & -2/3\mu & 0 \\ 0 & 0 & 4/3\mu \end{pmatrix}. \tag{20}$$

The first term is an isotropic tensor whose diagonal components $E = \lambda + 2/3\mu$ are one-third of the trace, and the second term is a CLVD. Because, as we will see shortly, inversion of moment tensors for shallow earthquakes cannot resolve the isotropic component, the seismic waves from such a crack would look like a CLVD.

An alternative explanation is that CLVDs are due to near-simultaneous earthquakes on nearby faults of different geometries. For example, consider the sum of two double-couple sources with moments M_0 and $2M_0$, expressed in the principal axis coordinate system (Eqn 16):

$$\begin{pmatrix} M_0 & 0 & 0 \\ 0 & 0 & 0 \\ 0 & 0 & -M_0 \end{pmatrix} + \begin{pmatrix} 0 & 0 & 0 \\ 0 & -2M_0 & 0 \\ 0 & 0 & 2M_0 \end{pmatrix} = \begin{pmatrix} M_0 & 0 & 0 \\ 0 & -2M_0 & 0 \\ 0 & 0 & M_0 \end{pmatrix}. \tag{21}$$

[3] Aki and Richards (1980).

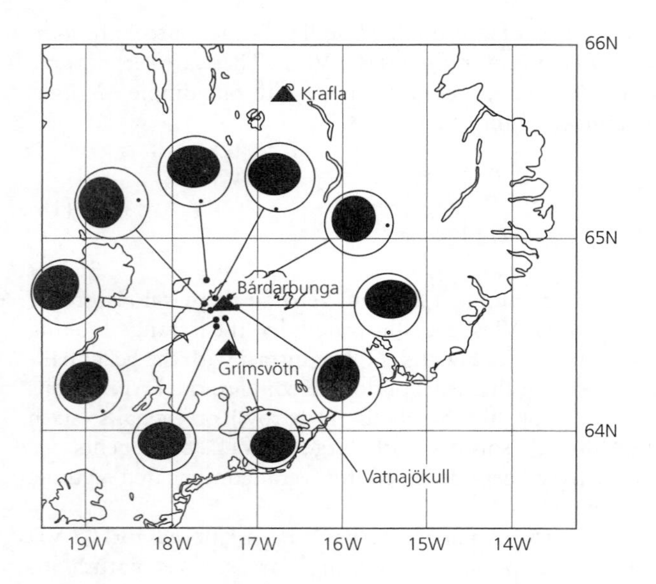

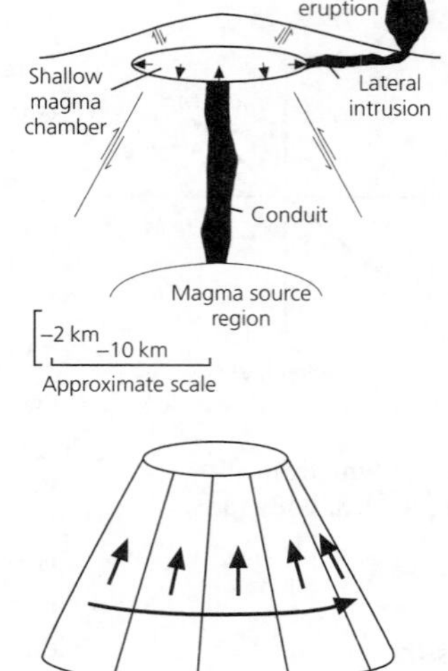

Fig. 4.4-8 CLVD-type focal mechanisms for earthquakes near the Bardarbunga volcano in Iceland. The mechanisms are similar to those shown in the lower right of Fig. 4.4-6. These are thought to reflect reverse faulting on cone-shaped ring faults surrounding the magma chamber. In this model, deflation of the magma chamber increases horizontal compression, so the roof block above the magma chamber subsides with respect to the surrounding rock (*right*). (Nettles and Ekström, 1998. *J. Geophys. Res.*, *103*, 17, 973–83, copyright by the American Geophysical Union.)

Thus, adding these two double couples yields a CLVD. In this example, both double-couple moment tensors are diagonal and so have the same eigenvector directions, but the P, B, and T axes of the first are the T, P, and B axes of the second. Thus, if the first earthquake were strike-slip on a vertical fault, the second would be normal faulting on a 45°-dipping fault (Fig. 4.2-16).

Decomposing a CLVD into double couples bears out the concept that the moment tensors can be decomposed in different ways, with different interpretations. This is because the moment tensor represents the equivalent body force system, so different decompositions reflect the same force system and give the same seismic waves. Hence the seismic waves alone cannot distinguish between alternative decompositions.

Multiple faulting events giving rise to apparent CLVDs have been reported. For example, Fig. 4.4-8 shows CLVD mechanisms at a volcano in Iceland, which have been interpreted as resulting from reverse faulting on cone-shaped ring faults beneath the caldera, triggered by deflation of the magma chamber. Such CLVDs and other non-double-couple seismic sources, like the single force for Mt St Helens (Fig. 4.4-2), occur in volcanic regions where faulting and magmatic processes interact. It is often difficult to distinguish the roles of the two processes, even when geological and other geophysical data are also used. Hence different interpretations of seismic events have been offered in areas including Hawaii and the Long Valley, California, caldera.

4.4.7 *Moment tensor inversion*

In addition to being an elegant representation of the source, the moment tensor has two advantages for source studies. First, it allows us to analyze seismograms without assuming that they result from slip on a fault. In some applications, such as deep earthquakes or volcanic earthquakes, we would like to identify possible isotropic or CLVD components. Second, the moment tensor makes it easier to invert seismograms to find source parameters.

For example, consider the formulation we used to synthesize surface waves (Section 4.3.4). The predicted seismograms depended on fault geometry factors that are complicated products of trigonometric functions of the fault strike, dip, and slip angles. This is not a problem in forward modeling, but makes it hard to invert the seismograms to find the fault angles. The inverse problem is much easier if we write the seismograms as linear functions of components of the moment tensor.

To see this, we represent the source by a vector **m**, containing components of the moment tensor. Although the tensor has nine components, only six are independent, because the tensor is symmetric. We then extend the idea of a Green's function which we previously used to represent the effect on a seismogram of an earthquake with a particular fault geometry (Eqn 4.3.15). Here, we define $G_{ij}(t)$ as the seismogram at the ith seismometer due to the moment tensor component m_j. $G_{ij}(t)$ includes the effects of the seismometer and earth structure along the path from the source to this seismometer, so the ith seismogram is the sum of the Green's functions weighted by the moment tensor components,

$$u_i(t) = \sum_{j=1}^{6} G_{ij}(t) m_j. \tag{22}$$

Because we have many seismograms, we can write this as a vector–matrix equation

$$\mathbf{u} = \mathbf{G}\mathbf{m}, \tag{23}$$

where **u** is a vector composed of the seismograms at n stations and **G** is the Green's function matrix. **G** has as many rows as seismometers and as many columns as moment tensor components, so Eqn 23 looks like

$$\begin{pmatrix} u_1 \\ u_2 \\ \cdot \\ \cdot \\ \cdot \\ \cdot \\ \cdot \\ u_n \end{pmatrix} = \begin{pmatrix} G_{11} & G_{12} & G_{13} & G_{14} & G_{15} & G_{16} \\ G_{21} & G_{22} & G_{23} & G_{24} & G_{25} & G_{26} \\ \cdot & \cdot & \cdot & \cdot & \cdot & \cdot \\ \cdot & \cdot & \cdot & \cdot & \cdot & \cdot \\ \cdot & \cdot & \cdot & \cdot & \cdot & \cdot \\ \cdot & \cdot & \cdot & \cdot & \cdot & \cdot \\ \cdot & \cdot & \cdot & \cdot & \cdot & \cdot \\ G_{n1} & G_{n2} & G_{n3} & G_{n4} & G_{n5} & G_{n6} \end{pmatrix} \begin{pmatrix} m_1 \\ m_2 \\ m_3 \\ m_4 \\ m_5 \\ m_6 \end{pmatrix}. \quad (24)$$

This is an overdetermined system of linear equations with more equations (n) than unknowns (6). We often encounter such systems when we invert large quantities of data to estimate a smaller number of parameters. As we noted in Section 2.8, and will explore in depth in Chapter 7, we cannot invert the matrix **G** because it is not square. Instead, we find the moment tensor that best matches the observed seismograms in a least squares sense, using what is called the *generalized inverse* of **G**,

$$\mathbf{m} = (\mathbf{G}^T\mathbf{G})^{-1}\mathbf{G}^T\mathbf{u}. \quad (25)$$

Thus, because the seismograms are linear functions of the moment tensor components, they can be inverted to find the tensor components.

Although we defer discussing most properties of generalized inverse solutions until Chapter 7, a point worth noting is that how well we can estimate a moment tensor component depends on the Green's function. Equation 22 shows that the seismogram involves products of moment tensor components with their corresponding Green's functions. Thus, if G_{ij} is zero, m_j has no effect on the seismogram, no matter how big it is. Similarly, if G_{ij} is small, m_j has little effect on the seismogram, Conversely, inverting the seismogram to determine m_j essentially involves dividing the seismogram by G_{ij}. Hence, if G_{ij} is small, dividing by it gives a large number, so any small errors or noise in the data produce spuriously large values of m_j. Put another way, we get good estimates of components to which the seismogram is fairly sensitive, but poorer ones for components on which the seismogram depends weakly.[4]

We now consider one inversion approach, a method for surface waves corresponding to the forward modeling in Section 4.3.4. In a coordinate system with the source at the north pole, the vertical component of Rayleigh waves on a seismometer at $\mathbf{r} = (r, \theta, \phi)$ can be written as an inverse Fourier transform:

$$u(\mathbf{r}, t) = \frac{1}{2\pi} \int_{-\infty}^{\infty} U(\omega, \theta, \phi) e^{i\omega t} d\omega. \quad (26)$$

The spectral amplitude $U(\omega, \theta, \phi)$ is a complex number representing the source, the effect of the seismometer, and the elastic and anelastic effects of propagation from the source to the receiver. As in Eqn 4.3.20, we write

$$u(\omega, \theta, \phi) = V(\omega, \phi) H(\omega, \theta),$$

$$H(\omega, \theta) = I(\omega) \frac{e^{i\pi/4}}{\sqrt{\sin\theta}} e^{-i\omega a\theta/c} e^{-\omega a\theta/2Qu} e^{im\pi/2}. \quad (27)$$

$V(\omega, \phi)$ is the radiation pattern term reflecting the effect of source geometry, which we want to find, whereas $H(\omega, \theta)$ represents the effects of the seismometer and of propagation, which we treat as known. $I(\omega)$ is the effect of the seismometer, and the remaining terms are propagation effects, including $e^{-\omega a\theta/2Qu}$, the effect of attenuation as the wave travels a distance θ (including any 2π terms). In these expressions, a is the earth's radius, m is the number of polar or antipolar passages, and c, u, and Q are the phase velocity, group velocity, and attenuation at the frequency ω.

To set up the inversion, we write the radiation pattern, which shows how the amplitude at a given frequency varies with the azimuth (ϕ) of the receiver from the source, in terms of linear combinations of the moment tensor components

$$V(\omega, \phi) = -P_R \left[M_{xy} \sin 2\phi - \frac{1}{2}(M_{yy} - M_{xx}) \cos 2\phi \right] + \frac{1}{3}(S_R + N_R)M_{zz} + \frac{1}{6}(2N_R - S_R)(M_{xx} + M_{yy}) + iQ_R(M_{yz} \sin\phi + M_{xz} \cos\phi). \quad (28)$$

This expression is analogous to the radiation pattern for Rayleigh waves due to slip on a fault (Eqn 4.3.20). The difference is that here the seismic source is written in terms of moment tensor components, rather than as products of trigonometric functions of the fault strike, dip, and slip angles. Thus Eqn 28 represents more general seismic sources than double couples due to slip on a simple fault.

As before, the radiation pattern depends on excitation functions derived from the radial eigenfunctions of spheroidal modes of the appropriate frequency, which describe how a source at a given depth causes displacements as a function of frequency. However, in addition to the excitation functions in (Eqn 4.3.20) (P_R, S_R, Q_R), we have the excitation function N_R that applies to an isotropic source. To see this, recall that for an explosion the moment tensor (Eqn 17) has equal diagonal elements ($M_{xx} = M_{yy} = M_{zz} = M_0$) and zeroes off the diagonal ($M_{xy} = M_{yz} = M_{xz} = 0$). Substituting these into Eqn 28 yields $V(\omega, \phi) = M_0 N_R$, which is a radiation pattern that depends on N_R and is azimuthally symmetric, as expected for an explosion. Conversely, if the source has no isotropic component ($M_{xx} + M_{yy} + M_{zz} = 0$), N_R drops out of Eqn 28.

We can formulate the inverse problem using Eqn 28. At a given frequency, separating $V(\omega, \phi)$ into real and imaginary parts yields the matrix equation

[4] We will formalize this idea using eigenvalues in Section 7.3, but before doing so, we can see intuitively that an estimate of the number of white cats in a dimly lit room will be better than that of black cats.

$$\begin{pmatrix} \mathrm{Re}(V(\omega, \phi)) \\ \mathrm{Im}(V(\omega, \phi)) \end{pmatrix} = B\mathbf{m}, \tag{29}$$

where $\mathbf{m}$ is a vector composed of the moment tensor components that we seek to find,

$$\mathbf{m} = \begin{pmatrix} M_{xy} \\ M_{yy} - M_{xx} \\ M_{zz} \\ M_{xx} + M_{yy} \\ M_{yz} \\ M_{xz} \end{pmatrix}, \tag{30}$$

and B is the known matrix

$$B = \begin{pmatrix} -P_R \sin 2\phi & \frac{P_R}{2}\cos 2\phi & \frac{1}{3}(S_R + N_R) & \frac{1}{6}(2N_R - S_R) & 0 & 0 \\ 0 & 0 & 0 & 0 & Q_R \sin\phi & Q_R \cos\phi \end{pmatrix} \tag{31}$$

containing the excitation functions and azimuthal dependence.

To invert seismograms for the moment tensor, we divide the Fourier transform of the seismogram from the station at $\mathbf{r}_i$ by the propagation and seismometer term $H(\omega, \theta_i)$ (Eqn 27) to find the complex amplitude $V(\omega, \phi_i)$. Data from only one seismic station yields two equations in six unknowns, so we cannot find $\mathbf{m}$. However, with data at three or more stations, all six components of $\mathbf{m}$ can in principle be found. We form a vector $\mathbf{v}$ from the $V(\omega, \phi_i)$ values observed at each of the n stations. We similarly use the values of B for each station, and write a vector–matrix equation equating the observed amplitudes $\mathbf{v}$ to those predicted by the known matrix B and the moment tensor $\mathbf{m}$ that we seek,

$$\mathbf{v} = B\mathbf{m}, \tag{32}$$

where

$$\mathbf{v} = \begin{pmatrix} \mathrm{Re}\,V(\omega, \phi_1) \\ \mathrm{Im}\,V(\omega, \phi_1) \\ \cdot \\ \cdot \\ \cdot \\ \mathrm{Re}\,V(\omega, \phi_n) \\ \mathrm{Im}\,V(\omega, \phi_n) \end{pmatrix} \tag{33}$$

and

$$B = \begin{pmatrix} -P_R \sin 2\phi_1 & \frac{P_R}{2}\cos 2\phi_1 & \frac{1}{3}(S_R + N_R) & \frac{1}{6}(2N_R - S_R) & 0 & 0 \\ 0 & 0 & 0 & 0 & Q_R \sin\phi_1 & Q_R \cos\phi_1 \\ \cdot & \cdot & \cdot & \cdot & \cdot & \cdot \\ \cdot & \cdot & \cdot & \cdot & \cdot & \cdot \\ \cdot & \cdot & \cdot & \cdot & \cdot & \cdot \\ -P_R \sin 2\phi_n & \frac{P_R}{2}\cos 2\phi_n & \frac{1}{3}(S_R + N_R) & \frac{1}{6}(2N_R - S_R) & 0 & 0 \\ 0 & 0 & 0 & 0 & Q_R \sin\phi_n & Q_R \cos\phi_n \end{pmatrix}. \tag{34}$$

With more than three stations, there are more equations than unknowns and Eqn 32 is solved using the least squares solution (Eqn 25), giving

$$\mathbf{m} = (B^T B)^{-1} B^T \mathbf{v}. \tag{35}$$

This solution gives the moment tensor that best predicts the observed spectral amplitudes. It is estimated at a given frequency for a seismic source that is a delta function in time. Time variation in the source can be examined by solving for $\mathbf{m}$ at different frequencies.

An important limitation results from the fact that the middle two columns in matrix B, corresponding to M_{zz} and $M_{xx} + M_{yy}$, do not contain ϕ, and so have no azimuthal variation. Therefore, no matter how many stations we use at a given frequency, we are solving only for the sum

$$\frac{1}{3}(S_R + N_R)M_{zz} + \frac{1}{6}(2N_R - S_R)(M_{xx} + M_{yy}), \tag{36}$$

which is the same at all stations. The inversion thus cannot find $M_{xx} + M_{yy}$ and M_{zz} separately, but only their sum, which is the isotropic portion of the source corresponding to possible volume changes.

One way to deal with this problem is to use data at different frequencies, where the coefficients of M_{zz} and $M_{xx} + M_{yy}$ are quite different. This is often difficult because these coefficients vary slowly with frequency for shallow earthquakes (consider $S_R(\omega)$ for the 11 km-deep earthquake in Fig. 4.3-14). Surface wave moment tensor inversions thus often constrain the source to have no isotropic portion, so that $M_{xx} + M_{yy} = -M_{zz}$. In this case,

$$V_r(\omega, \phi) = -P_R\left[M_{xy}\sin 2\phi - \frac{1}{2}(M_{yy} - M_{xx})\cos 2\phi\right]$$
$$-\frac{1}{2}S_R(M_{yy} + M_{xx}) + iQ_R(M_{yz}\sin\phi + M_{xz}\cos\phi), \tag{37}$$

so the inversion is for a vector with five components. N_R, the excitation function for an isotropic source, no longer enters into the radiation pattern.

We then rewrite the inversion equation (Eqn 32) as

$$\mathbf{v} = A\mathbf{m}, \tag{38}$$

and solve for

$$\mathbf{m} = \begin{pmatrix} M_{xy} \\ M_{yy} - M_{xx} \\ M_{yy} + M_{xx} \\ M_{yz} \\ M_{xz} \end{pmatrix}, \tag{39}$$

given the known matrix

$$A = \begin{pmatrix} -P_R\sin 2\phi_1 & \frac{1}{2}P_R\cos 2\phi_1 & -\frac{1}{2}S_R & 0 & 0 \\ 0 & 0 & 0 & Q_R\sin\phi_1 & Q_R\cos\phi_1 \\ -P_R\sin 2\phi_2 & \frac{1}{2}P_R\cos 2\phi_2 & -\frac{1}{2}S_R & 0 & 0 \\ 0 & 0 & 0 & Q_R\sin\phi_2 & Q_R\cos\phi_2 \\ \cdot & \cdot & \cdot & \cdot & \cdot \\ \cdot & \cdot & \cdot & \cdot & \cdot \\ \cdot & \cdot & \cdot & \cdot & \cdot \\ -P_R\sin 2\phi_n & \frac{1}{2}P_R\cos 2\phi_n & -\frac{1}{2}S_R & 0 & 0 \\ 0 & 0 & 0 & Q_R\sin\phi_n & Q_R\cos\phi_n \end{pmatrix}. \tag{40}$$

The solution gives five moment tensor components, because adding and subtracting m_2 and m_3 yields M_{xx} and M_{yy}. M_{zz} is then found from $-(M_{xx} + M_{yy})$, but is not independent of them.

Another significant difficulty in surface wave moment tensor inversion stems from the fact that the excitation function Q_R is zero at the earth's surface (Fig. 4.3-14) because it is proportional to the shear stress. At shallow depths Q_R is small, so M_{xz} and M_{yz} are poorly determined for shallow earthquakes (< 30 km when inverting at 256 s). This leaves only three tensor components well determined, which are insufficient to determine the fault geometry.

This problem can be addressed in several ways. One is to invert shorter-period waves which have larger amplitudes (Fig. 4.3-14). However, the effects of lateral heterogeneity increase for shorter periods, due to the shorter wavelengths. A second approach is to constrain M_{xz} and M_{yz} to be zero and invert for only the three components M_{xx}, M_{yy}, M_{xy}. This forces one eigenvector to be vertical and makes the major double couple take one of three forms: pure strike-slip on a vertical plane (vertical null axis), thrust faulting on a 45°-dipping plane (vertical T axis), or normal faulting on a 45°-dipping plane (vertical P axis). An interesting way to view this is to note that shallow earthquakes on vertical dip-slip faults, for which the only nonzero fault geometry factor (Eqn 4.3.20) is q_R, have radiation patterns proportional to $Q_R(\omega)$ and so excite surface waves very inefficiently. Hence constraining M_{xz} and M_{yz} to be zero excludes any vertical dip-slip component from the focal mechanism, so a complete solution requires other data, such as first motions or geological knowledge. A third method is to constrain one nodal plane from first motions and then do a linear inversion for the second plane.

We can also use this formulation to invert transverse component Love wave data, using the analogous expressions

$$U(\omega, \theta, \phi) = V(\omega, \phi)I(\omega)\frac{e^{-i\pi/4}}{\sqrt{\sin\theta}}e^{-i\omega a\theta/c}e^{-\omega a\theta/2Qu}e^{im\pi/2}, \tag{41}$$

$$V(\omega, \phi) = P_L\left[\frac{1}{2}(M_{xx} - M_{yy})\sin 2\phi - M_{xy}\cos 2\phi\right]$$
$$+ iQ_L[-M_{xz}\sin\phi + M_{yz}\cos\phi]. \tag{42}$$

4.4.8 *Interpretation of moment tensors*

In general, once a moment tensor has been found by inverting seismograms, it will be more complicated than expected for a double couple. Even if the source were a pure double couple, noise in the data and imperfect knowledge of earth structure would likely produce a tensor that, once diagonalized, would look like

$$\mathbf{M} = \begin{pmatrix} \lambda_1 & 0 & 0 \\ 0 & \lambda_2 & 0 \\ 0 & 0 & \lambda_3 \end{pmatrix} \quad |\lambda_1| \geq |\lambda_2| \geq |\lambda_3|, \tag{43}$$

with eigenvectors $\hat{\mathbf{n}}_1$, $\hat{\mathbf{n}}_2$, and $\hat{\mathbf{n}}_3$.

If $\mathbf{M}$ represents a double couple, then $\lambda_1 = -\lambda_2$, and $\lambda_3 = 0$. However, unless the moment tensor was constrained to satisfy these conditions, it generally will not do so. In most cases, $\lambda_1 \approx -\lambda_2$, and $|\lambda_2| \gg |\lambda_3|$, so M is approximately, but not exactly, a double couple. In this case, we interpret the moment tensor by decomposing it, as we did for the CLVD examples in Section 4.4.6. If there is an isotropic component, we remove it via

$$\begin{pmatrix} \lambda_1 & 0 & 0 \\ 0 & \lambda_2 & 0 \\ 0 & 0 & \lambda_3 \end{pmatrix} = \begin{pmatrix} E & 0 & 0 \\ 0 & E & 0 \\ 0 & 0 & E \end{pmatrix} + \begin{pmatrix} \lambda_1' & 0 & 0 \\ 0 & \lambda_2' & 0 \\ 0 & 0 & \lambda_3' \end{pmatrix}, \tag{44}$$

where $E = (\lambda_1 + \lambda_2 + \lambda_3)/3$. The remaining term is a deviatoric moment tensor, with zero isotropic component and components equal to the deviatoric eigenvalues $\lambda_1' = \lambda_1 - E$, $\lambda_2' = \lambda_2 - E$, and $\lambda_3' = \lambda_3 - E$. If needed, the deviatoric eigenvalues are renumbered so that $|\lambda_1'| \geq |\lambda_2'| \geq |\lambda_3'|$. If the inversion has no isotropic component, the deviatoric moment tensor is the moment tensor resulting from the inversion.

The deviatoric moment tensor can be decomposed in several ways. One is in terms of two double couples, called the major and minor double couples:

$$\begin{pmatrix} \lambda_1' & 0 & 0 \\ 0 & \lambda_2' & 0 \\ 0 & 0 & \lambda_3' \end{pmatrix} = \begin{pmatrix} \lambda_1' & 0 & 0 \\ 0 & -\lambda_1' & 0 \\ 0 & 0 & 0 \end{pmatrix} + \begin{pmatrix} 0 & 0 & 0 \\ 0 & -\lambda_3' & 0 \\ 0 & 0 & \lambda_3' \end{pmatrix}. \tag{45}$$

The first tensor is the major double couple, with scalar moment $|\lambda_1'|$, and the second is the minor double couple, with scalar moment $|\lambda_3'|$. Usually, the magnitude of the major double couple is much larger, and we treat it as the earthquake's source mechanism.

As an example, consider M for an intermediate-depth thrust earthquake in the Kuril subduction zone near Japan.[5] The moment tensor inverted from Rayleigh waves of period 256 s recorded on the IDA network of digital very long-period seismometers was

$$M = \begin{pmatrix} 0.12 & -0.17 & -0.06 \\ -0.17 & -1.54 & -1.44 \\ -0.06 & -1.44 & 1.43 \end{pmatrix}, \tag{46}$$

where the components are in units of 10^{27} dyn-cm. Diagonalizing the matrix yields

$$\begin{pmatrix} -2.14 & 0 & 0 \\ 0 & 2.01 & 0 \\ 0 & 0 & 0.13 \end{pmatrix} = \begin{pmatrix} -2.14 & 0 & 0 \\ 0 & 2.14 & 0 \\ 0 & 0 & 0 \end{pmatrix} + \begin{pmatrix} 0 & 0 & 0 \\ 0 & -0.13 & 0 \\ 0 & 0 & 0.13 \end{pmatrix}, \tag{47}$$

with eigenvectors $\hat{\mathbf{n}}_1 = (0.80, 0.92, 0.37)$, $\hat{\mathbf{n}}_2 = (0.00, -0.38, 0.93)$, and $\hat{\mathbf{n}}_3 = (-0.99, 0.07, 0.03)$. The isotropic component was constrained in the inversion to be zero. Because the minor double couple has a moment only 6% that of the major double couple, we assume that the major double couple represents the earthquake mechanism. $\hat{\mathbf{n}}_1$ is the P axis of the double couple, $\hat{\mathbf{n}}_2$ is the T axis, and $\hat{\mathbf{n}}_3$ is the null axis. Using these axes and our tectonic preconceptions (which are not, of course, always valid), we decide (using a stereonet or a computer) that the earthquake was a thrust on a fault plane striking N189°E and dipping 23°W. The auxiliary plane strikes N3°E and dips 67°E.

In this case, we discarded the minor double couple and assumed that the earthquake was a single double couple. It is likely that the minor double couple often results from lateral heterogeneity in the earth (the velocity and attenuation models used in this inversion were laterally homogeneous), noise in the data, and deviation of the earthquake from a point source. You may recall from the surface wave example in Fig. 4.3-13 that the data were approximately fit by the amplitude radiation pattern predicted by the focal mechanism, but some stations had higher amplitudes, whereas others had lower amplitudes. Similar effects can occur for the amplitude and phase data in a moment tensor inversion. As a result, even if the source were a pure double couple, the inversion fits the deviations in the data from the predictions of the best-fitting double couple, and so yields a moment tensor differing somewhat from the double couple. Thus the better the inversion method reflects the earth's heterogeneity and source complexity, the less the tendency for there to be spurious portions of the moment tensor. In some cases, however, the minor double couple may have physical significance, such as for simultaneous ruptures on nearby faults with different orientations.

The moment tensor can be decomposed in other ways. One is into a double couple and a CLVD:

$$\begin{pmatrix} \lambda_1' & 0 & 0 \\ 0 & \lambda_2' & 0 \\ 0 & 0 & \lambda_3' \end{pmatrix} = \begin{pmatrix} \lambda_1' + \lambda_3'/2 & 0 & 0 \\ 0 & -\lambda_1' - \lambda_3'/2 & 0 \\ 0 & 0 & 0 \end{pmatrix} + \begin{pmatrix} -\lambda_3'/2 & 0 & 0 \\ 0 & -\lambda_3'/2 & 0 \\ 0 & 0 & \lambda_3' \end{pmatrix}. \tag{48}$$

The relative strength of the double couple and CLVD is given by the ratio of the smallest and largest deviatoric eigenvalues, $\varepsilon = \lambda_3'/\lambda_1'$. $\varepsilon = 0$ indicates a pure double couple, and $\varepsilon = \pm 0.5$ shows a pure CLVD source. About 4% percent of the mechanisms in the Harvard global moment tensor catalog, derived from inversions that are not constrained to yield double couples, have $|\varepsilon| \geq 0.3$. Some of these may be artifacts of the inversion process similar to spurious minor double couples, but some appear to be real source effects.

However, as our CLVD example (Section 4.4.6) showed, both moment tensor decompositions and their interpretations are not unique. For example, Eqn 45 showed a decomposition into a major double couple with moment λ_1' and a minor double couple with moment λ_3'. We could also decompose the tensor with the same major double couple but a minor double couple with moment λ_2':

[5] This example was provided by A. Michael.

$$\begin{pmatrix} \lambda_1' & 0 & 0 \\ 0 & \lambda_2' & 0 \\ 0 & 0 & \lambda_3' \end{pmatrix} = \begin{pmatrix} \lambda_1' & 0 & 0 \\ 0 & 0 & 0 \\ 0 & 0 & -\lambda_1' \end{pmatrix} + \begin{pmatrix} 0 & 0 & 0 \\ 0 & \lambda_2' & 0 \\ 0 & 0 & -\lambda_2' \end{pmatrix}. \quad (49)$$

The two decompositions sum to the correct value for each tensor component, which is the equivalent body force, but using tensors of differing scalar moments. This is analogous to the way a vector can be decomposed into various sums of vectors with different magnitudes.

Moment tensor solutions have become an important tool of global seismology. Globally distributed broadband digital seismometers permit reliable focal mechanisms to be generated within minutes after most earthquakes with $M_s \geq 5.5$ and made publicly available through e-mail and the Internet. Several organizations carry out this service, including the Harvard centroid moment tensor (CMT) project. The CMT method inverts two parts of seismograms: long-period ($T > 40$ s) body waves and very long-period ($T > 135$ s) surface waves, called *mantle waves*. The CMT inversion yields both a moment tensor and a centroid time and location. This location often differs from that listed in earthquake bulletins, such as that of the International Seismological Centre (ISC), because the two locations tell different things. Earthquake location bulletins based upon arrival times of body wave phases like P and S give the *hypocenter*: the point in space and time where rupture began. CMT solutions, using full waveforms, give the *centroid*, or average location in space and time, of the seismic energy release. As a result, CMT origin times are almost always later than ISC times. The availability of large numbers of high-quality mechanisms (the Harvard project has produced more than 17,000 solutions since 1976) is of great value in many applications, especially tectonic studies.

4.5 Earthquake geodesy

4.5.1 *Measuring ground deformation*

So far in this chapter we have studied earthquakes using transient displacements due to the propagating seismic waves they generate. However, the large, rapid deformation in an earthquake results from a complex deformation field which extends over a broad region and a long time. Hence, additional information about earthquakes and the processes causing them can be obtained by measuring slow ground deformation using techniques from *geodesy*, the science of the earth's shape. Most such techniques rely on detecting the motion of geodetic monuments,[1] which are markers in the ground.

Until recently, these measurements were typically made by triangulation, which measures the angles between monuments using a theodelite, or trilateration, which measures distances with a laser. Vertical motion was measured by leveling, using a precise level to sight on a distant measuring rod. However, the advent of geodetic methods using signals from space permits all three components of position to be measured to sub-centimeter precision. As a result, geodetic data before and after earthquakes now give coseismic motion to high precision much more easily than was previously possible.

Although the space-based technologies are among the most complex used in the earth sciences, in essence they use electromagnetic waves in ways analogous to those we have discussed for seismic waves. Three of these techniques are used to locate geodetic markers. Very Long Baseline Interferometry (VLBI) uses the difference in the time when radio signals from distant quasars arrive at different points on earth. Satellite Laser Ranging (SLR) uses the time required by light from ground-based lasers to bounce off satellites. The third approach relies on the travel time of radio signals between satellites and ground stations.

Although the various systems provide similar data, the third approach via the Global Positioning System (GPS)[2] is presently the system of choice for most tectonic applications. GPS was developed in the late 1970s by the US Department of Defense for real-time positioning and navigation. A constellation of satellites transmit coded timing signals on a pair of microwave carrier frequencies synchronized to very precise on-board atomic clocks. The timing signals are modulations of the carrier frequencies, analogous to those we discussed in the context of phase and group velocities (Section 2.8.1). By determining the ranges to a minimum of four satellites from the signal delays and the broadcast satellite orbit information, a single GPS receiver can determine its three-dimensional position to a precision of 5 to 100 meters, depending on the level of signal degradation imposed by the military (Fig. 4.5-1).[3] This operation is conceptually the same as locating an earthquake from arrivals at multiple seismometers, which we discuss in Section 7.2. GPS positions are two to three times more precise in the horizontal than in the vertical direction, because radio signals arrive only from above, just as earthquake locations are less precise in depth because waves arrive only from below.

The improvement to cm level or better precision is obtained by using the phase delays of the microwave carriers. Because the carriers have higher frequencies than the modulations,

[1] The most familiar monuments are the metal disks attached to rocks often seen at mountain peaks, but various other designs are also used in hope of minimizing the effects of soil or near-surface motion that mask the tectonic movement. In soft sediment, monuments are often steel rods driven deep into the earth. The popular term for monuments is "benchmarks," although geodesists traditionally reserve this term for monuments used to study vertical motions.

[2] Acronyms abound in space geodesy, given its space and military origins. Alternative meanings have been offered: the large VLBI project teams suggest "Very Large Bunch of Investigators," and the languid pace of GPS surveys prompted "Great Places to Sleep." There are also second-level acronyms involving other acronyms, such as IGS for International GPS Service.

[3] The Department of Defense can degrade GPS positioning via selective availability, which introduces errors in the satellite clocks. This capability, which was discontinued in May 2000, reduced the precision of single receiver positions but had little effect on precise geodetic positions.

Constellation of GPS satellites

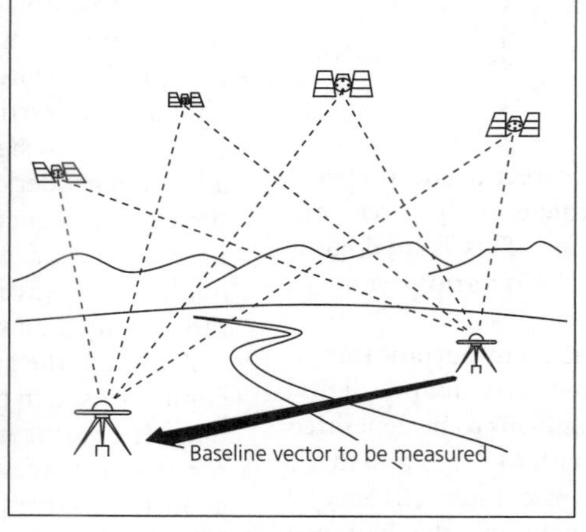

Relative positioning

Fig. 4.5-1 *Left*: The Global Positioning System (GPS) uses a constellation of satellites that transmit timing signals. *Right*: Using precise positions based on signals from multiple satellites recorded at multiple receivers, measurements over time yield relative velocities to precisions of a few mm/yr or better.

their phase can yield more precise locations, much as higher-frequency seismic waves can reveal more detailed velocity structure (Section 3.2.3). The carrier wavelengths are 19 and 24 cm, so precise phase measurements can resolve positions to a fraction of these wavelengths. The use of differential signals from multiple satellites recorded at multiple receivers reduces clock errors. Combining both transmitted frequencies removes the effects of the passage of the GPS radio signals through the ionosphere. Position errors due to signal delays from water vapor in the troposphere can be reduced by estimating the delays using an inversion process similar to solving for seismic velocity structure.

The final element for high-precision surveys is provided by continuously operating global GPS tracking stations and data centers. These provide high-precision satellite orbit and clock information, earth rotation parameters, and a global reference frame. Using this information, GPS studies can achieve positions better than 10 mm, so measurements over time yield relative velocities to precisions of a few mm/yr or better, even for sites thousands of kilometers apart. The uncertainty of the velocity estimate depends on the precision of the estimated positions and the time interval between them.

GPS data are collected in two modes. In survey mode, GPS antennas are set up over monuments for short periods, and the sites are reoccupied later. Alternatively, continuously recording GPS receivers are installed permanently. Continuous GPS can provide significantly more precise data, albeit at higher cost (in the USA, a 25-station network can presently be occupied in survey mode for the cost of a single continuous station).

The biggest limitation of geodetic data for earthquake studies is that the positions of geodetic markers before the earthquake are needed. Thus effort and resources are required to install and survey monuments in advance, in hopes that an earthquake will occur nearby. In active seismic areas that are convenient for study, this condition can sometimes but not always often be met. A way around this difficulty is provided by Synthetic Aperture Radar interferometry (InSAR) from satellites.

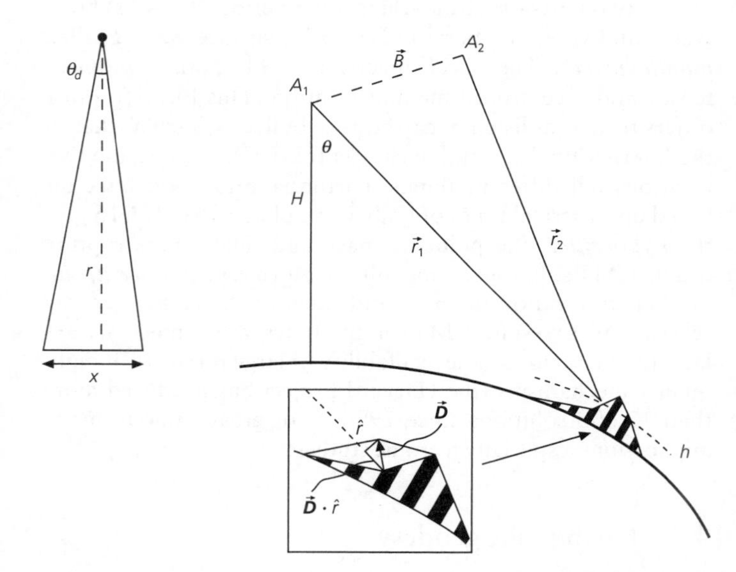

Fig. 4.5-2 *Left*: Geometry of radar imaging from space. A physical antenna's angular resolution is $\theta_d = \lambda / d = x / r$, so x is the resolution on the earth's surface achievable by a radar with antenna length d and wavelength λ operating at altitude r. Synthetic aperture radar dramatically improves the resolution. *Right*: Geometry of the InSAR method. The insert illustrates the relation between the crustal motion D and the resulting range change $\delta r = (D \cdot \hat{\mathbf{r}})$. (After Bürgmann *et al.*, 2000. Reproduced with the permission of Annual Reviews, Inc.)

The synthetic aperture method allows high-resolution radar mapping from spacecraft or aircraft. The resolution of a physical radar can be estimated using the single slit diffraction concept (Fig. 2.5-18), in which the angle θ_d between successive zeros in the diffraction pattern is λ / d, where d is the slit width, and λ is the wavelength. For radar, d is the antenna length, so a radar a distance r above the earth's surface could resolve objects of size x, where (Fig. 4.5-2, *left*)

$$\theta_d = \lambda / d = x / r. \quad (1)$$

Because the radar wavelengths are 10s of centimeters, a radar antenna a few meters long orbiting hundreds of kilometers above the earth can normally resolve topography only on a scale of kilometers. However, SAR uses signal processing to combine information collected by a moving satellite to simulate an antenna much larger than the satellite's real antenna. For example, a real 10 m antenna can be used as a 4 km synthetic antenna. The synthetic antenna can thus resolve both topography and crustal deformation on a "footprint" of tens of meters.

Figure 4.5-2 (*right*) illustrates the technique. The phase difference between radar signals with wavelength λ reflected from the earth's surface and recorded by antennas at position A_1 and A_2 is

$$\phi = (4\pi/\lambda)(r_2 - r_1), \tag{2}$$

where r_i is the range from the antenna at A_i to the reflection point. The antenna baseline separation vector **B** and satellite flight height H are known from the satellite orbits. Because the baseline length $|\mathbf{B}|$ is much shorter than the ranges r_i, an analysis like that used to derive the earthquake rupture time (Fig. 4.3-2) shows that the elevation of the reflecting point is $h = H - r_1 \cos\theta$, so topography can be mapped from space. This method, called interferometry,[4] is used for both earth and planetary mapping, such as the Magellan mission to Venus.

Two such radar images can detect ground motion between successive measurements. If differences in satellite positions between the measurements are removed, a vector surface displacement D causes a phase change

$$\phi \approx (4\pi/\lambda)\delta r, \quad \delta r = (D \cdot \hat{\mathbf{r}}), \tag{3}$$

where δr is the projection (scalar product, Section A.3.3) of the vector displacement along $\hat{\mathbf{r}}$, the look direction connecting the satellite and reflection point. To find the full displacement vector, observations from ascending (moving north) and descending (moving south) tracks of the satellite, or different satellites, can be combined.

The results are shown as a phase difference map, called a differential interferogram. Figure 4.5-3 (*top*) shows such an image of the phase differences resulting from the 1992 Landers (M_w 7.3) and Big Bear (M_w 6.2) earthquakes in the Mojave desert of southern California. A range change δr of $\lambda/2$ causes a phase change of 2π that appears as one fringe (full shading change) in the map. In this case, the C-band radar has a

Fig. 4.5-3 *Top*: SAR interferogram constructed from radar images taken on April 24, 1992, and June 18, 1993, showing the displacements resulting from the 1992 Landers and Big Bear earthquakes. The shaded fringes are interference patterns obtained by comparing the images. Each cycle of shading represents 28 mm of change in the distance between the satellite and the ground, so the static displacement is on the order of tens of centimeters. *Bottom*: Synthetic interferogram computed using a model of the static displacements predicted by the focal mechanisms. The images are 92.2 km across in width. (B. Hernandez, personal communication, 1999, based upon Hernandez *et al.*, 1997. *Geophys. Res. Lett.*, *24*, 1579–82, copyright by the American Geophysical Union.)

[4] Interferometry, using phase differences of traveling waves to make precise distance and time measurements, has many applications. In seismology, the time between arriving waves is measured by cross-correlation (Sections 3.3.6, 6.3.4). GPS and VLBI use the phase differences of radio waves to measure positions. Perhaps the most famous application of interferometry is the Michelson–Morley experiment in the 1880s, which showed that the speed of light was the same in all directions despite the earth's motion through space, and thus played a key role in the birth of the theory of relativity.

frequency of 5.2 GHz, so a fringe corresponds to 28 mm of motion. The observed fringe pattern is coherent over large areas where deformation is resolved. The pattern is reasonably similar to a synthetic interferogram (Fig. 4.5-3, *bottom*) generated for a detailed model of the Landers rupture, which involved several meters of right-lateral strike-slip on a complex set of NW-striking faults extending for about 85 km.

InSAR has several attractive features for earthquake studies. Although radar images before an earthquake are needed, satellites can acquire them over areas far too large for geodetic monuments to have been installed everywhere. In addition, InSAR maps deformation on a spacing of tens of meters, far denser than is practical with geodetic monuments. Moreover, InSAR is especially sensitive to vertical motions, the component for which the GPS is the least precise. InSAR has several limitations. It recovers motion only in the look direction. It cannot be used in some areas of steep topography, where the radar beam cannot penetrate, or where the slope facing the radar is so steep that several points have the same range to the radar. Another limitation is that nontectonic changes between images, such as those due to vegetation growth or weather conditions (which affect radio wave propagation in the atmosphere), can mask the effects of crustal motion. However, when such decorrelation between successive images is not a problem, as in deserts or other bare rock settings, InSAR is a powerful tool. Finally, InSAR provides relative changes within an image that is tens to a hundred kilometers across, but does not provide absolute positions on a plate-wide or global scale. This poses no problems for individual earthquake studies, but means that it alone cannot be used for large-scale applications like plate boundary studies. In many applications, InSAR and GPS are both being combined with seismological data. These techniques are also being applied together with seismology to study ground deformation at volcanoes.

The advent of space-based methods like GPS and InSAR, which make collecting geodetic data faster and easier, have made earthquake geodesy and seismic wave studies common overlapping approaches to earthquake studies. Hence, although seismology and earthquake geodesy were long viewed as very distinct, owing to their different instrumentation, earthquake geodesy is increasingly viewed as very low-frequency seismology (or earthquake seismology as high-frequency geodesy).

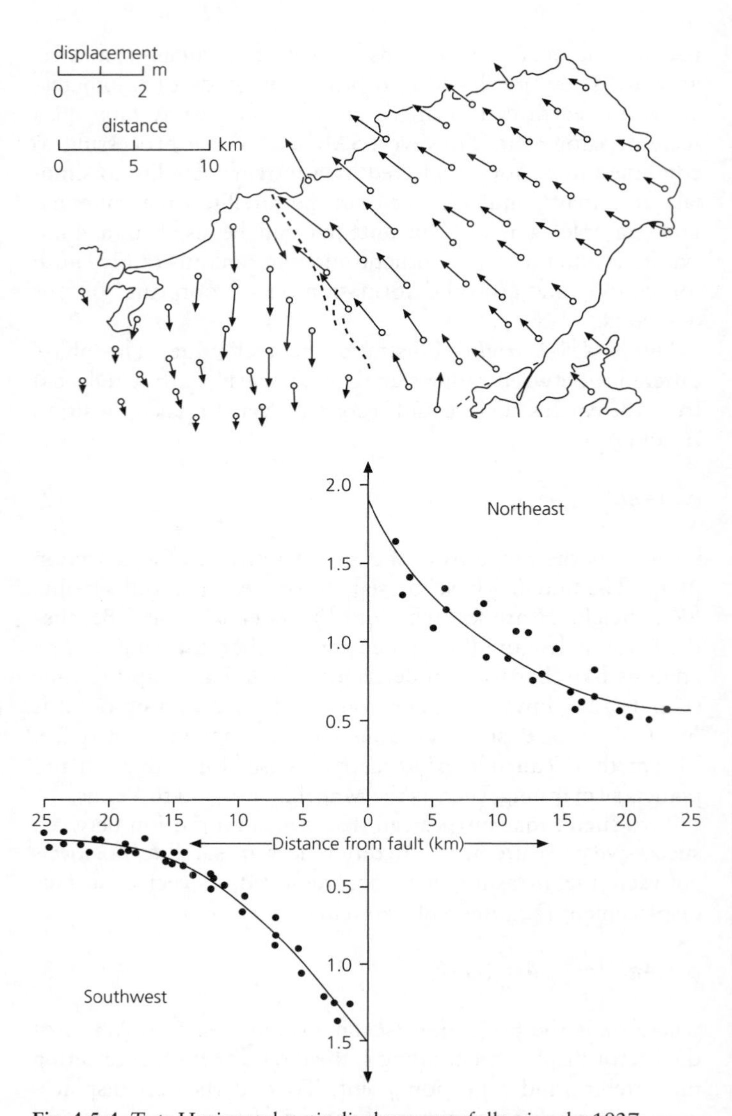

Fig. 4.5-4 *Top*: Horizontal static displacements following the 1927 Tango, Japan, earthquake. The dashed line shows the fault trace. (*Bottom*): Decay of fault-parallel displacements with distance perpendicular to the fault. (After Chinnery, 1961. © Seismological Society of America. All rights reserved.)

4.5.2 *Coseismic deformation*

Seismic source theory shows that the static coseismic displacements produced by earthquakes have radiation patterns analogous to the propagating wave displacements shown in Fig. 4.2-6 and 4.2-7, and so can also provide important information about the fault geometry and slip. An important feature of these displacements is that they contain $1/r^2$ terms, compared to $1/r$ terms for the propagating waves (Eqns 1 and 2). Thus, compared to the propagating waves, the static displacements decay more rapidly with distance from the earthquake. Hence we typically describe the static displacements using Cartesian coordinates near a fault, rather than the spherical coordinates used for teleseismic waves.

A classic example, shown in Fig. 4.5-4, is that of the static displacements following the 1927 M_s 7.5 Tango, Japan, earthquake. The displacements change direction across the fault trace, showing that the earthquake involved primarily left-lateral strike slip. The fault-parallel displacement component decays rapidly with distance from the fault.

Although the full expressions for the static displacements due to slip on a fault are complicated, we can gain considerable insight from the simple case of pure strike-slip faulting on an infinitely long vertically dipping fault. In this case (Fig. 4.5-5,

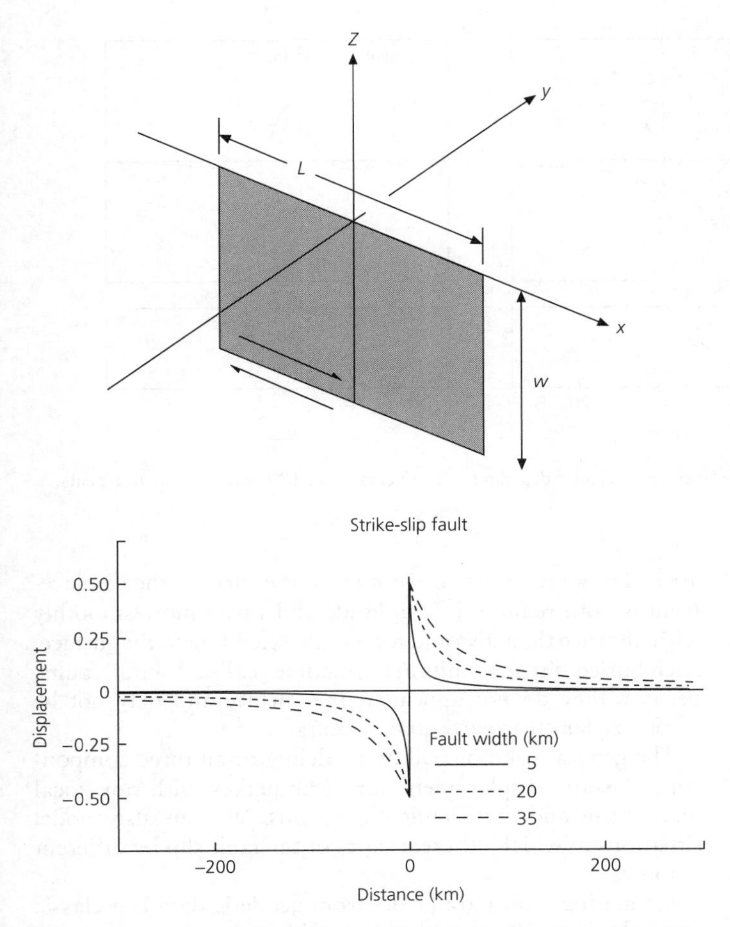

Fig. 4.5-5 *Top*: Geometry of the vertically dipping strike-slip fault model. L and W are fault length and width. *Bottom*: Predicted fault-parallel static displacements, normalized to the maximum offset, for an infinite strike-slip fault, for different fault widths.

top), the fault-parallel displacement in the x direction, $u(y)$, varies with the distance from the fault y as

$$u(y) = \pm D/2 - (D/\pi)\tan^{-1}(y/W), \qquad (4)$$

where D is the slip across the fault, and W is the depth to which faulting extends, called the fault width. The $\pm D$ term is positive for $y > 0$, negative for $y < 0$. This model assumes that the slip is uniform all over the fault plane. Figure 4.5-5 (*bottom*) shows this solution for several different fault widths. Near the fault, $y \to 0$, so the inverse tangent is zero, and $u(0) = \pm D/2$. The displacement decays away from the fault, so by a distance equal to the fault width ($y/W = 1$) the inverse tangent is $\pi/4$ and the displacement is D/4, or half that at the fault. Far from the fault, $y/W \to 0$, and the displacement dies off. Hence the distance over which the displacement extends gives information about the fault width. For example, the data in Fig. 4.5-5 indicate a fault width of about 10 km.

For this infinite fault, fault-parallel displacement extends to infinity along the fault. Calculations for finite-length faults

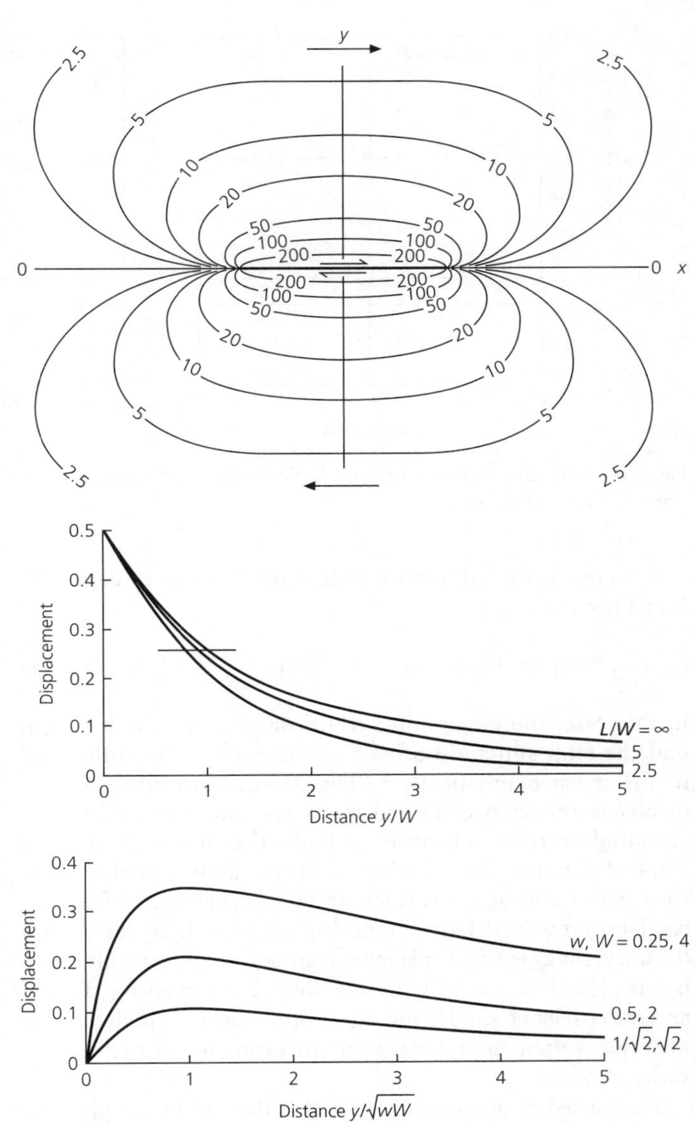

Fig. 4.5-6 *Top*: Fault-parallel static displacements for a finite vertically dipping strike-slip fault. Contours are labeled in units of 10^{-3} times the maximum offset. (Chinnery, 1961.) *Center*: Predicted fault-parallel static displacements, normalized to the maximum offset, for strike-slip faults with different fault widths (W) and lengths (L). The horizontal bar is where displacement has dropped to half its value at the fault. *Bottom*: Predicted fault-parallel static displacements for three buried infinite strike-slip faults extending from depth w to depth W, all with the same slip. (Mavko, 1981. Reproduced with the permission of Annual Reviews Inc.)

show that the displacement tapers off rapidly past the fault ends (Fig. 4.5-6, *top*). In addition, there is some fault-normal (y direction) motion. For finite faults (Fig. 4.5-6, *center*), the decay of fault-parallel displacement perpendicular from the fault (in the y direction) depends somewhat on the ratio of the fault width to fault length, W/L. Thus the fault width estimated from the decay depends on the assumed length.

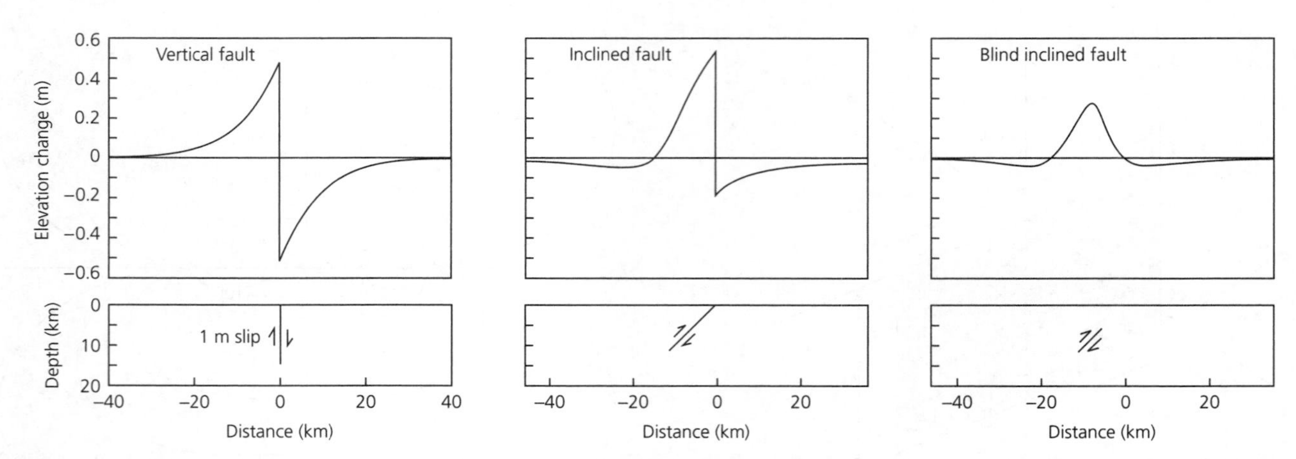

Fig. 4.5-7 Vertical component of static displacement as a function of distance from various pure dip-slip faults. (Yeats *et al.*, 1997; after Stein and Yeats, 1989. Courtesy of H. Iken.)

If a fault is buried and extends from depth w to depth W, Eqn 4 becomes

$$u(y) = (D/\pi)[\tan^{-1}(y/w) - \tan^{-1}(y/W)]. \quad (5)$$

In this case, the maximum surface displacement is less than half the fault slip and occurs a distance from the fault equal to the mean depth $(wW)^{1/2}$ (Fig. 4.5-6, *bottom*). Thus the displacement fields of buried faults are smoother and lower-amplitude versions of those for faults that reach the surface. These differences occur because a buried fault is further away from each point on the surface, and the higher spatial frequencies (shorter wavelengths) in the displacement decay faster with distance, making the displacement smoother. As a result, there is a trade-off between the fault's down-dip dimension $W - w$ and the coseismic slip D, and one is often assumed to determine the other. Often, fault dimensions are estimated from the aftershock zone.

The buried fault solution (Eqn 5) is derived by simply adding to Eqn 4 a fictitious second fault extending from the surface to the fault top w, with the same slip but in the opposite direction. This is an example of the general principle that we can superimpose static solutions for simple geometries to obtain the solution for a complicated geometry. We also do this for the propagating waves from complex faults, as we will see shortly. The solutions can be added because they satisfy linear elasticity.

Solutions are also available for dip-slip faults. Figure 4.5-7 shows solutions for the vertical component of static displacement as a function of distance from various pure dip-slip faults. For vertical dip, the solution looks like the strike-slip solution turned vertically. If the dip is not vertical, the displacement varies in magnitude as well as sign across the fault. The higher amplitudes are above the thrust fault, on the hanging wall block. Interestingly, seismic wave amplitudes for this geometry are also often highest on the hanging wall, and can cause significant damage when such earthquakes occur under populated areas. For a fault that does not reach the surface, the displacement is both reduced in amplitude and varies more smoothly with distance than it would for a fault extending to the surface. Such buried dip-slip faults are sometimes called "blind" faults, because they do not appear at the surface and may not be recognized until an earthquake occurs.

The general solutions allow modeling of all three components of static displacement for earthquakes with any focal mechanism and finite fault dimensions. We can also model situations in which different parts of the fault slip by different amounts.

Estimating fault parameters from geodetic data is a classic example of an inverse problem with a highly non-unique solution, because various combinations of fault parameters predict similar deformation. Figure 4.5-8 shows six solutions that all give reasonable fits to the Tango earthquake data (Fig. 4.5-4). Model I is an infinite fault with uniform slip at depth, model II is an infinite fault with slip tapering to zero at depth, and models III and IV are finite faults with uniform and variable slip, respectively. Model V is the most complicated, in that it assumes that the material near the fault is weaker than that further away.

4.5.3 *Joint geodetic and seismological earthquake studies*

Combining geodetic and seismic wave observations gives more information than either data type alone. The two data types are nicely complementary. For example, although seismic waves have an ambiguity in distinguishing between the fault plane and the auxiliary plane, the geodetic data do not, as shown by the fact that the Tango earthquake data (Fig. 4.5-4) and static displacement models (Fig. 4.5-6, *top*) do not have a nodal plane perpendicular to the fault plane. Both data types can give good constraints on the fault geometry and slip on it, and aftershock locations often provide the best constraint on fault dimensions. However, geodetic data that depend on the difference in position before and after an earthquake provide no information

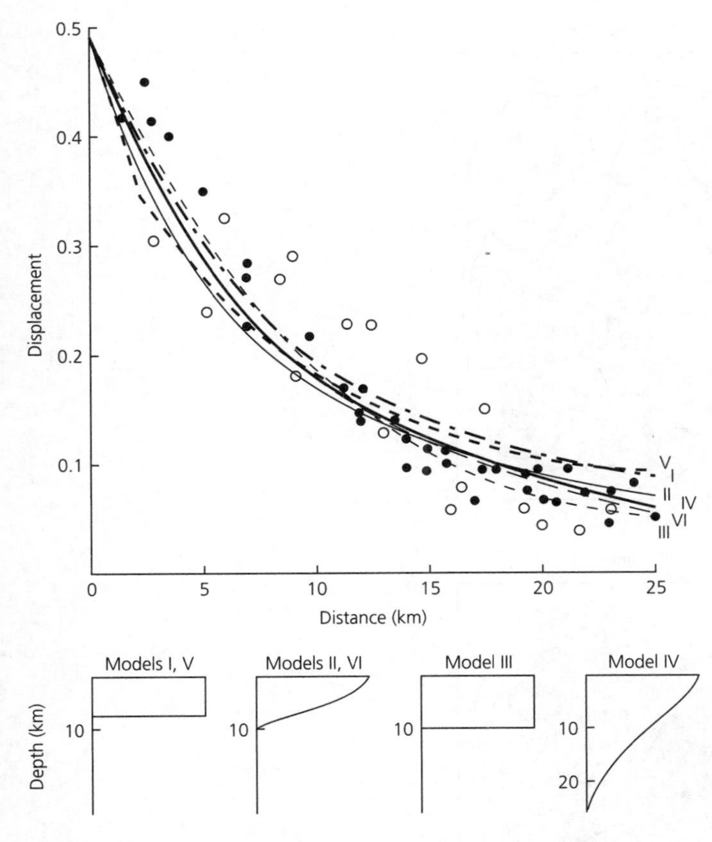

Fig. 4.5-8 Comparison of different fault models that predict coseismic deformation similar to that observed for the Tango earthquake (Fig. 4.5-4). Distance is perpendicular to the fault. The data are normalized by the fault offset, and points from the SW side (closed dots) are multiplied by –1 and plotted with points from the NE side (open dots). (Mavko, 1981. Reproduced with the permission of Annual Reviews, Inc.)

about what happened during the earthquake, whereas seismological data can sometimes show how the rupture evolved.

Figure 4.5-9 illustrates an example of combining geodetic and seismological data for the 1994 M_s 6.7 Northridge earthquake which occurred on a buried thrust fault in the San Fernando Valley, near Los Angeles.[5] The focal mechanism and aftershock distribution indicate thrust faulting on a NW-striking, SW-dipping fault. The geodetic (GPS) data show significant vertical and horizontal motions concentrated above the buried fault. The directions and magnitudes of the static deformation, including the motion of down-dip sites toward the fault and the high amplitudes above the fault, are what we would expect for this geometry (Fig. 4.5-7). These data can be modeled quite well by assuming that about 2.5 m of slip

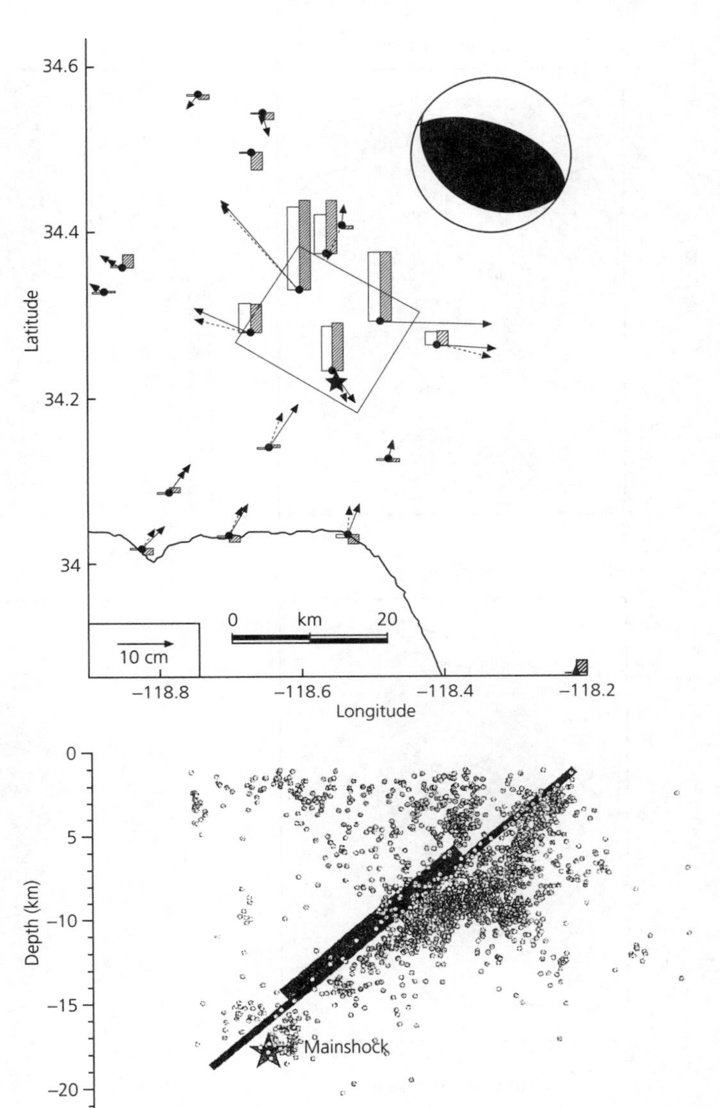

Fig. 4.5-9 Geodetic and seismological results for the 1994 Northridge earthquake. *Top*: The horizontal (solid arrows) and vertical (solid bars) motions observed by GPS are well matched (dashed arrows and open bars) by a fault model derived from these data. Negative uplift is shown by bars below the station locations (dots). *Bottom*: Aftershock locations (dots) and geometry of fault models with uniform slip (thick line) and variable slip on a longer fault (thin line), both of which fit the data. (After Hudnut *et al.*, 1996; Thio and Kanamori, 1996; and Wald *et al.*, 1996. © Seismological Society of America. All rights reserved.)

occurred on a fault plane similar to that which one would infer from the aftershocks. Two geodetic solutions are shown, one with uniform slip and one with variable slip on a larger fault.

Because high-quality geodetic and seismological data are available, considerable detail about the slip distribution has been inferred. Strong motion data from seismometers close to

[5] This earthquake, which is one of the most studied owing to the extensive seismological and geodetic networks in the area, gave rise to some of the highest ground accelerations ever recorded. It illustrates that even a moderate magnitude earthquake can do considerable damage in a populated area. Although the loss of life (58 deaths) was small due to earthquake-resistant construction (Section 1.2.2), the 20 billion dollars in damage makes it the most costly earthquake to date in the USA.

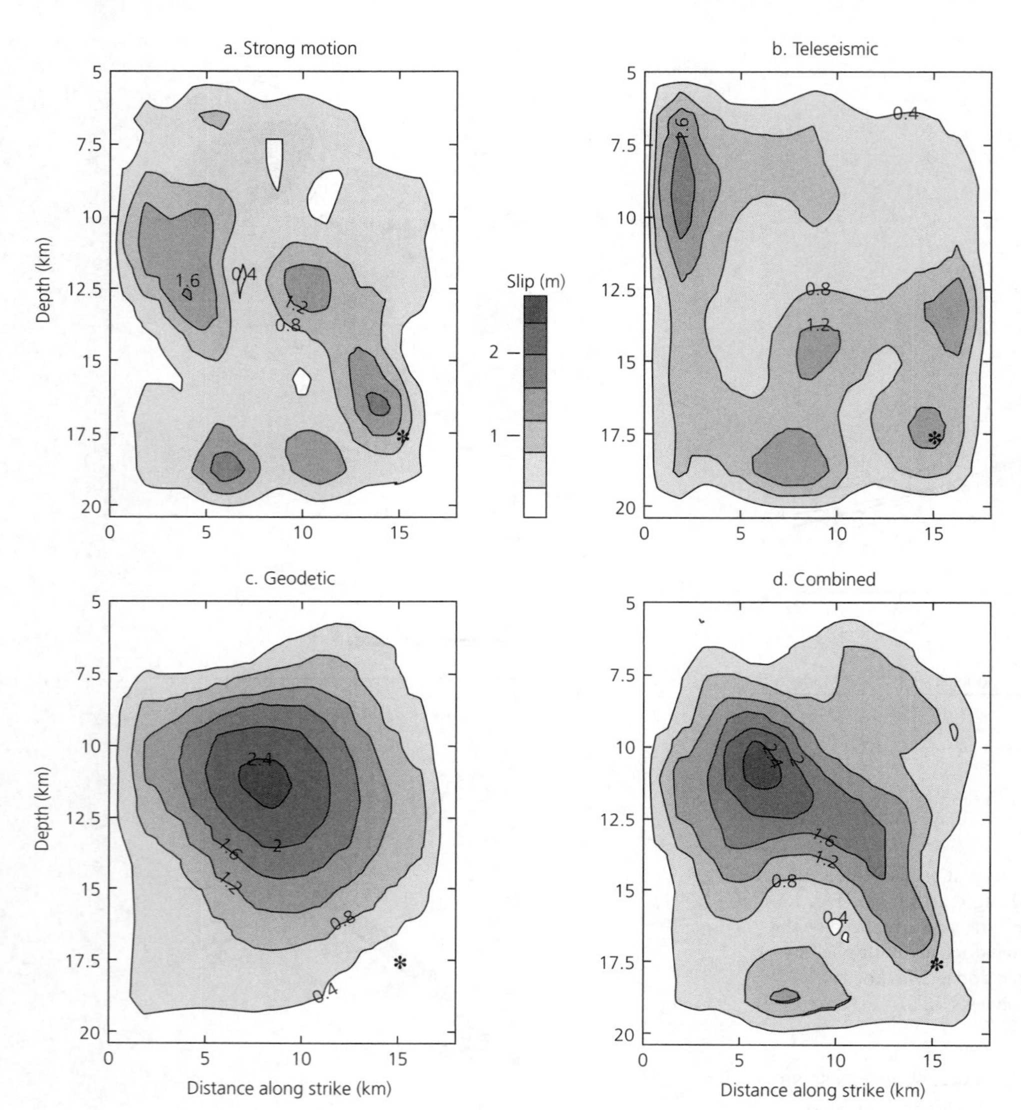

Fig. 4.5-10 Comparison of results of slip inversions for the Northridge earthquake using various datasets. The fault plane is viewed from the southwest and above. The epicenter is marked by a star. (Wald *et al.*, 1996.)

the earthquake are especially valuable because they contain high-frequency details about the source time function, and thus slip process, which can be lost in teleseismic data due to attenuation (Fig. 4.3-10). Figure 4.5-10 shows maps of the slip distribution on the fault plane estimated first by inverting the strong motion, teleseismic, and geodetic data separately, and then by a joint inversion. The seismic inversions extend analysis like that shown in Fig. 4.3-11, which resolved the source time function into sub-events, to locate sub-events on the fault plane. Interestingly, the largest slip is not at the epicenter (star). The results for the different data types differ because each is sensitive to different features of the slip. For example, the geodetic data yield a much smoother image than the seismic data, which can resolve the rupture process, whereas the GPS data sample only its end result. Thus, both waveform datasets yield a high-slip region near the fault's northwest corner. Figure 4.5-11 shows the time evolution of the rupture inferred from the waveforms. Rupture began at the epicenter and then propagated up-dip and northwestward. Such models are giving our best look to date into the rupture process, and are being combined with experimental and theoretical studies of rock fracture (Section 5.7) to explore the complex physics of earthquake faulting.

Geodetic data after earthquakes also sometimes show a phenomenon called afterslip or postseismic slip, in which deformation goes on "silently" (without a seismic signal) for some time after an earthquake and its seismologically observed aftershocks. For plate boundaries, this motion is sometimes thought of as a postseismic portion of the seismic cycle, during which the motion slows from the rapid coseismic motion to the slower steady interseismic motion. However, as discussed in Section 5.7.6, it is often unclear whether the postseismic motion reflects continued slip on the earthquake fault, the response of the lithosphere to the earthquake having a time-varying

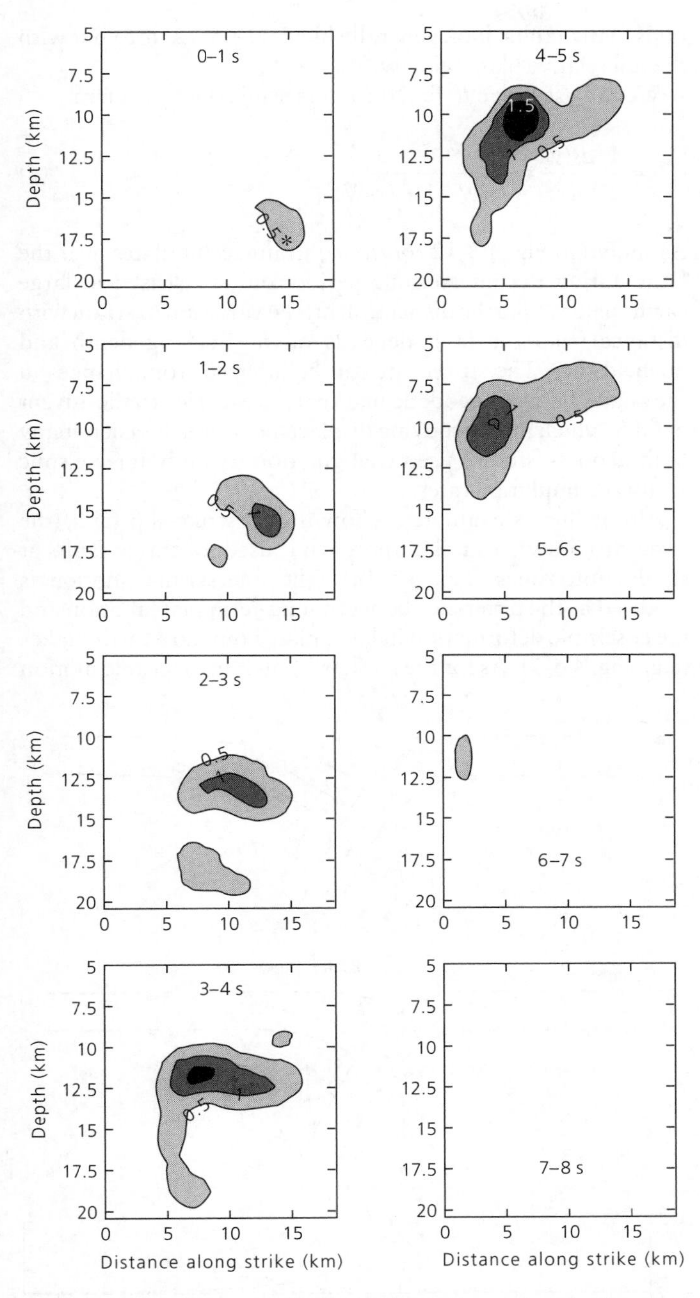

Fig. 4.5-11 Time history results for the Northridge earthquake. Rupture appears to have begun at the epicenter (star) and then propagated up-dip and northwestward. The geometry is the same as in Fig. 4.5-10. (Wald *et al.*, 1996. © Seismological Society of America. All rights reserved.)

viscous component in addition to purely elastic instantaneous deformation, or both.

4.5.4 *Interseismic deformation and the seismic cycle*

Geodesy gives insight into the seismic cycle before, after, and between earthquakes, whereas we can only study the seismic

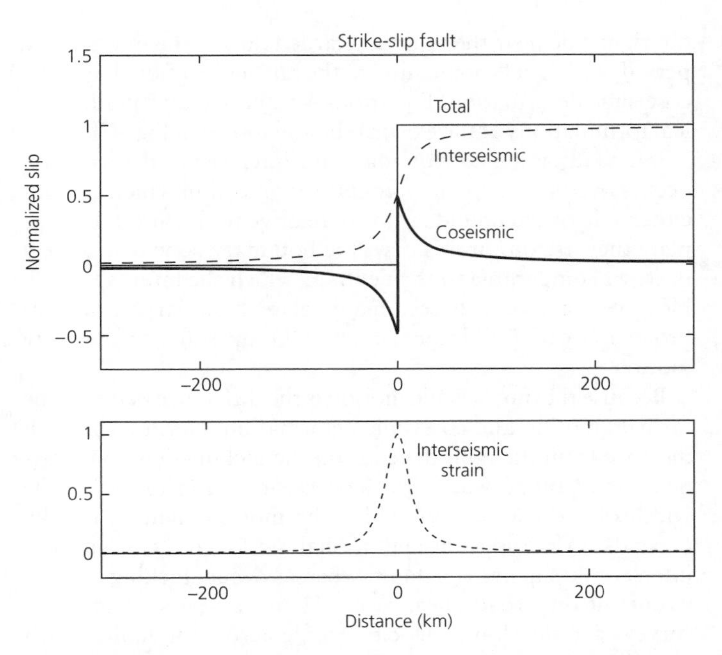

Fig. 4.5-12 *Top*: Coseismic (heavy solid line), interseismic (dashed line), and total or far-field (thin solid line) motions in the fault-parallel (x) direction as functions of fault-perpendicular distance (y) for an elastic rebound model of the seismic cycle on an infinite, vertically dipping, strike-slip fault. *Bottom*: Interseismic strain for this model.

waves once an earthquake occurs. To see this, consider a simple elastic rebound (Fig. 4.1-3) model of an infinite strike-slip fault at a plate boundary, assuming that large earthquakes release all the strain which accumulates between earthquakes. After an earthquake, material on the right ($+y$) side far from the fault moves at the far-field rate v relative to the left ($-y$) side of the fault, and so has moved a distance vt by time t (Fig. 4.5-12, *top*). However, between earthquakes the fault is locked down to depth W, although it slips freely below, so material at the fault does not move between earthquakes. When the next large earthquake occurs, completing the seismic cycle, everything to the right of the fault must have moved a distance vt. The earthquake's coseismic displacement will be given by Eqn 4 with $D = vt$, so the coseismic slip $u(y)$ is less than D except at the fault. This means that points away from the fault already have moved part of the distance D before the earthquake. Similarly, everything on the left side must have had no net motion from the seismic cycle, even though material near the fault moved "backward" (in the $-x$ direction) during the earthquake.

Thus the fault-parallel interseismic motion $s(y)$ is found by subtracting the coseismic slip from the far-field (or net) motion, giving

$$s(y) = D/2 + (D/\pi) \tan^{-1} (y/W). \qquad (6)$$

Hence, as shown in Fig. 4.5-12 (*top*), material on the left side near the locked fault is "dragged along" during the interseismic period, and then rebounds during the earthquake. Material on the right side near the fault is retarded during the interseismic period, and then "catches up" to the far-field motion due to the coseismic deformation. Equations 4 and 6 are thus mathematical formulations of the elastic rebound model in Fig. 4.1-3.

If the fault is a plate boundary, the interseismic deformation occurs over a finite *plate boundary zone* within which sites on either side of the boundary move relative to the interior of the plate they are on. In this case, the boundary zone is relatively narrow, comparable to the depth to which the fault is locked. However, as we will see, many plate boundary zones are broader because additional faults take up some of the plate motion.

Because the interseismic motion is the difference between the far-field motion and coseismic deformation, its variation with distance from the fault depends on the locking depth and far-field rate. Comparison with the coseismic slip shows that the width of the zone across which the motion changes rapidly depends on the locking depth. Shallow locking concentrates interseismic slip near the fault, whereas deeper locking spreads it out into a broad shear zone. Hence a series of geodetic surveys can develop a velocity profile across the fault, which we can interpret by setting $D = vt$ in Eqn 6 and dividing the change in positions between surveys by the time between them. Figure 4.5-13 shows a profile across the much-photographed (Fig. 4.1-1) Carrizo Plain segment of the San Andreas fault. The data are reasonably well fit by a far-field rate of about 35 mm/yr. As we will discuss in the next chapter, this rate is less than the total (approximately 45 mm/yr) motion between the Pacific and North American plates, showing that some of the plate motion occurs away from the San Andreas fault over a broader plate boundary zone. In fact, we will see that space geodetic profiles across the broad boundary zone, which contains many faults, look generally like Fig. 4.5-12 (*top*) but with the full relative plate velocity.

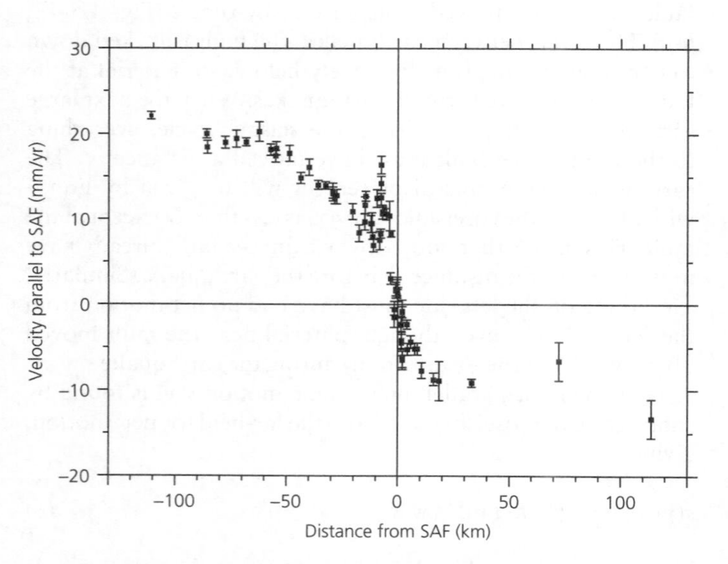

Fig. 4.5-13 GPS data showing fault-parallel horizontal interseismic motion across the Carrizo Plain segment of the San Andreas fault. (Z.-K. Shen, personal communication, 2000.)

We can use Eqn 6 to find the interseismic shear strain rate

$$\dot{e}_{xy} = \frac{1}{2}\frac{ds(y)}{dy} = \frac{v}{2\pi W}\frac{1}{[1 + (y/W)^2]}. \tag{7}$$

As shown in Fig. 4.5-12 (*bottom*), strain accumulates near the fault during the interseismic period and is released in large earthquakes. Like the displacement, the variation of strain with distance from the fault depends on the locking depth and far-field rate. The strain rate can be inferred from changes in the angles between geodetic markers. Thus, prior to the advent of GPS, which made studying displacements much easier, many fault geodesy studies used triangulation to study interseismic strain accumulation rates.

Although this example is shown for a strike-slip fault (the easiest to draw), a similar approach is used for thrust faults at subduction zones (Fig. 4.5-14). The interseismic motion is modeled as the difference between long-term plate motion and the coseismic deformation in large plate boundary earthquakes (e.g., Fig. 4.5-7). As for the strike-slip case, interseismic motion

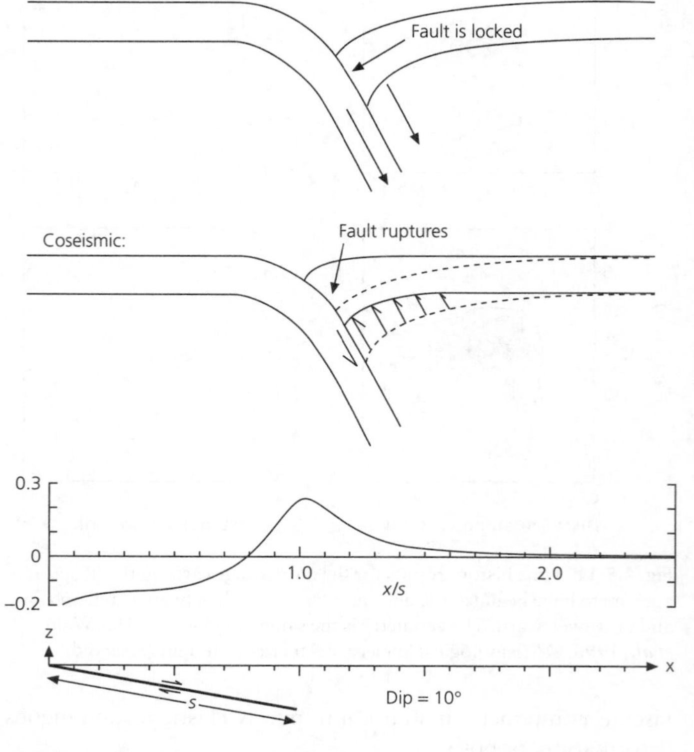

Fig. 4.5-14 *Top*: Two stages in the earthquake cycle at a subduction zone. *Bottom*: Predicted interseismic vertical motion due to a locked fault at a subduction zone. The vertical motion is normalized by the locked plate convergence rate, and the horizontal distance is normalized by the distance between the trench and end of the locked fault. (Savage, 1983. *J. Geophys. Res.*, *88*, 4984–96, copyright by the American Geophysical Union.)

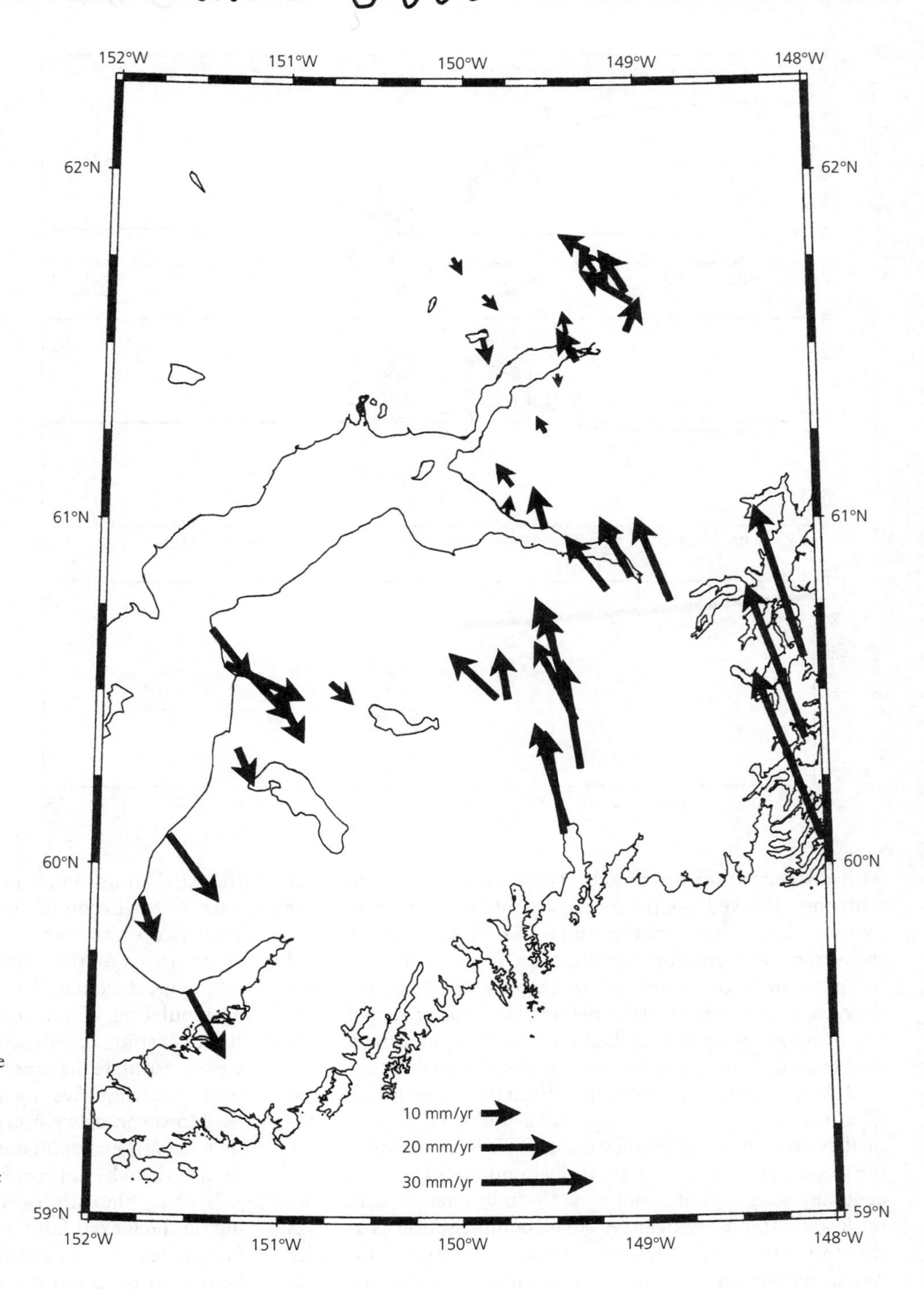

Fig. 4.5-15 GPS velocities relative to North America for some sites near the rupture zone of the great 1964 Alaska earthquake. The eastern sites move in the plate convergence direction, as expected for interseismic motion, whereas sites to the west move in the opposite direction, implying postseismic motion. (Freymueller *et al.*, 2000. *J. Geophys. Res., 105*, 8079–101, copyright by the American Geophysical Union.)

occurs in a boundary zone extending some distance from the fault defining the nominal plate boundary. Modeling predicts interseismic subsidence and landward motion for most sites above the locked fault, and uplift further inland (Fig. 4.5-14, *bottom*). The motion has largely decayed by a distance equal to twice that between the trench and the locked fault end.

Thus geodetic data near trenches can identify the interseismic deformation and provide insight into the mechanics of the subduction interface and future large earthquakes on it. Figure 4.5-15 shows GPS velocities relative to the stable interior of North America for some sites near the rupture zone of the great 1964 Alaska earthquake (Fig. 4.3-15). Sites to the east of the area shown move northwest, in the direction of Pacific plate subduction beneath North America, as we would expect for the interseismic motion of sites on the overriding plate above a locked fault. The motion decays rapidly landward

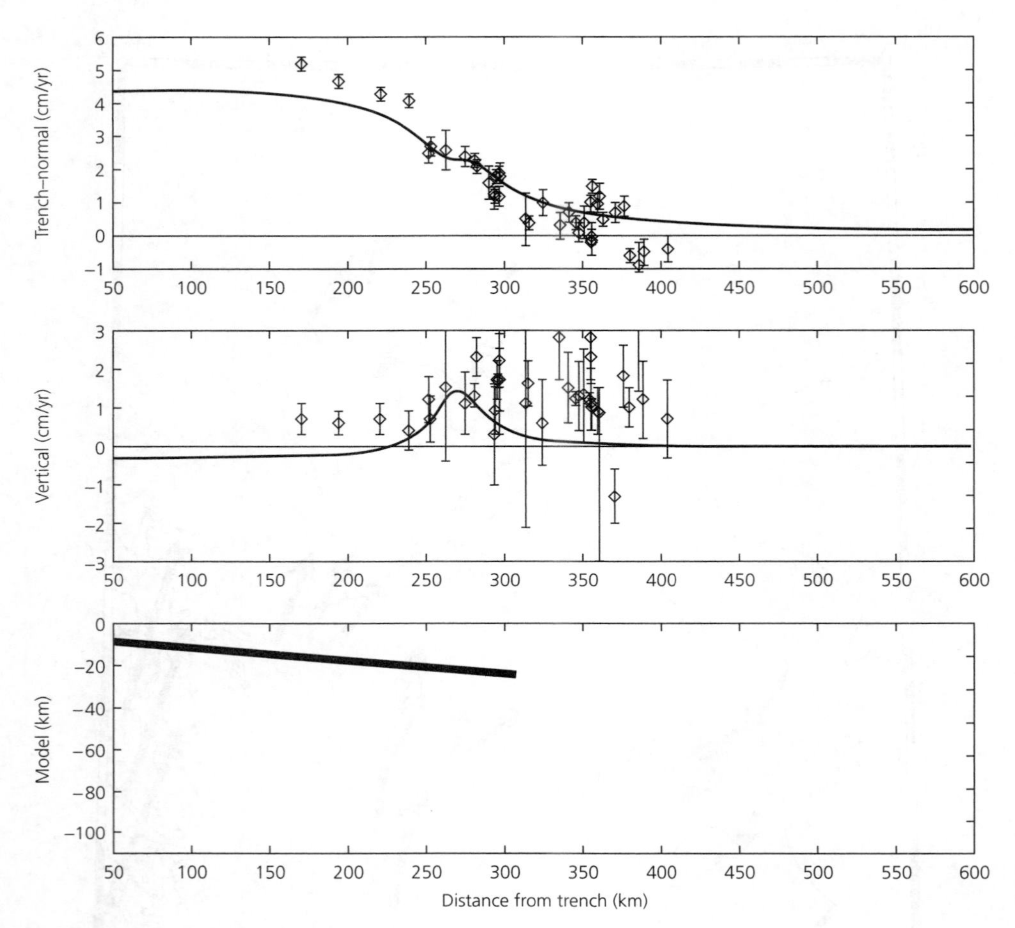

Fig. 4.5-16 Profiles of horizontal (*top*) and vertical (*center*) GPS velocities relative to North America for eastern sites in Fig. 4.5-15. The data are reasonably similar to predictions (solid line) for a locked fault model (*bottom*). Note that uncertainties for the vertical GPS data are larger than for the horizontal data. (Freymueller *et al.*, 2000. *J. Geophys. Res.*, *105*, 8079–101, copyright by the American Geophysical Union.)

with distance from the trench. These observations, together with the observed uplift, are reasonably consistent with the expected interseismic motion (Fig. 4.5-16). However, sites to the west move in the opposite direction, toward the trench, and so appear instead to show continuing postseismic motion. The differences between the two regions may reflect the complex slip history in the great earthquake or long-term differences in the behavior of different parts of the plate interface.

Hence, in general, geodetic data from the interseismic period give insight into the mechanics of a fault and future earthquakes on it, even before they occur. This is gratifying because the seismic cycle is so long, typically hundreds of years, that we generally have to wait a long time to study a major earthquake on a given fault segment. A slight compensation is that, as we wait, estimates of geodetic velocities improve. Consider measuring the rate v of motion of a monument that started at position x_1 and reaches x_2 in time T. If the position uncertainty is given by its standard deviation σ, then the propagation of errors relation (Eqn 6.5.18) discussed in Chapter 6 shows that

$$v = (x_1 - x_2)/T \quad \text{implies} \quad \sigma_v = \sqrt{2}\sigma/T, \tag{8}$$

where σ_v is the uncertainty of the inferred rate. Thus the longer we wait, the smaller the velocity uncertainty becomes, even if the data do not become more precise. Hence, older geodetic data — for example, those taken shortly after the 1906 San Francisco earthquake — can be of great value even if their errors are larger than those of more modern data.

The geodetic data let us see the rate at which locked slip is accumulating, and hence infer the maximum possible slip in a future earthquake, depending on when it occurs. Conversely, we can estimate the time until a future earthquake from records of past earthquakes, by assuming what the coseismic slip will be. However, as we noted in Section 1.2 and will discuss further, the large earthquakes are variable enough that attempts to predict them by approaches like this have not been successful.

In some places, geodetic data imply that slip is accumulating on the locked fault at a rate less than the far-field motion. For the San Andreas example shown, this difference seems to be due to plate motion taken up elsewhere. In other places, the difference is thought to indicate that some of the plate boundary slip occurs by aseismic slip or sliding (perhaps as "silent earthquakes") on the fault, and hence will not appear in future earthquakes. As discussed in the next chapter, the idea that significant portions of the motion on many plate boundaries occurs aseismically is also suggested by earthquake history studies. Such aseismic fault creep has been observed geodetically in some areas.

The area of locking has interesting implications. Because an earthquake's seismic moment is the product of the fault area, coseismic slip, and rigidity, the fault width and rate at which slip accumulates give insight into the maximum seismic moment that the locked fault could release in a future earthquake. The San Andreas data (Fig. 4.5-13) indicate that the vertically dipping fault is locked to a depth of about 20 km, which is similar to the maximum depth of small earthquakes and the inferred lower extent of rupture in large earthquakes along the fault. As discussed in Section 5.7, this depth is generally consistent with studies of rock strength and friction, which imply that rocks deeper than about 20 km are weak and undergo stable sliding rather than accumulate elastic strain for future earthquakes. The Alaska situation is quite different because the plate interface has a shallow dip (Fig. 4.5-16), so there is a large fault area at depths shallow enough to accumulate strain and then rupture. Hence, as we will see in Section 4.6, the largest earthquakes occur at shallow-dipping subduction zones and are much bigger than those for transform boundaries. In either environment, however, it is not clear whether the entire locked region contributes to the seismic slip or whether part of the fault slips rapidly in the earthquake and another part contributes to aseismic afterslip.

To complicate matters even further, it is worth bearing in mind that we still do not have good geodetic data spanning even one full seismic cycle, much less such data combined with detailed studies of earthquakes at either end. Hence we have little insight into the different possible time-variable effects like afterslip or the transient effects due to earthquakes on nearby faults or other segments of the same fault. Thus it may be quite some time before many of these issues are resolved.

An intuitive way to summarize some of these ideas is to think of the seismic cycle as a fault's "slip budget," analogous to personal finances. Given our income (plate motion), we spend some immediately (aseismic slip) and save some (locked slip). The savings are used for major purchases (earthquakes) at a rate depending on the price of individual purchases (coseismic slip), expenses associated with these major purchases (postseismic slip), and our saving rate (locked slip). Thus, although we can estimate roughly when we might make a future large purchase, the actual date depends on unpredictable changes in the price (variable earthquake size) and changes in our savings beyond our steady income and regular expenses, due to gifts or unanticipated expenses (effects of other earthquakes). Thus even in this simple analogy the earthquake cycle is complicated.

4.6 Source parameters

4.6.1 Magnitudes and moment

So far in this chapter we have discussed using seismic waves radiated by earthquakes to study their source geometry and focal depth. While recognizing the limitations on what the seismic waves can tell us about the actual source process, we have seen that for most earthquakes, assuming a simple fault geometry and source model allows us to estimate parameters that are generally consistent with other data and our geological instincts. We thus proceed further in using seismic waves to learn more about the faulting process.

In fact, even before earthquake mechanisms were studied, seismologists' second need after learning to locate earthquakes was to quantify their size, both for scientific purposes and to discuss their effects on society. The first measure introduced was the *magnitude*, which is based on the amplitude of the resulting waves recorded on a seismogram. The concept is that the wave amplitude reflects the earthquake size once the amplitudes are corrected for the decrease with distance due to geometric spreading and attenuation. Magnitude scales thus have the general form

$$M = \log(A/T) + F(h, \Delta) + C, \qquad (1)$$

where A is the amplitude of the signal, T is its dominant period, F is a correction for the variation of amplitude with the earthquake's depth h and distance Δ from the seismometer, and C is a regional scale factor.[1] Magnitude scales are thus logarithmic, so an increase in one unit, as from magnitude "5" to "6," indicates a ten-fold increase in seismic wave amplitude. Measured magnitudes range more than 10 units[2] because the displacements measured by seismometers span more than a factor of 10^{10}.

The earliest magnitude scale, introduced by Charles Richter in 1935 for southern California earthquakes, is the *local magnitude*, M_L, often referred to as the "Richter scale." Figure 4.6-1 shows how M_L is determined from the amplitude measured on a specific seismograph, known as the *Wood–Anderson* seismograph. The magnitude of the largest arrival (often the S wave) is measured and corrected for the distance between the source and the receiver, given by the difference in the arrival times of the P and S waves. The scale

$$M_L = \log A + 2.76 \log \Delta - 2.48, \qquad (2)$$

defined for earthquakes in southern California, is a form of Eqn 1 with the instrument period (0.8 s) and nearly constant (shallow) depth incorporated in the constants, and the distance in km. Richter magnitudes in their original form are no longer used because most earthquakes do not occur in California and Wood–Anderson seismographs are rare. However, local magnitudes are sometimes still reported because many buildings have resonant frequencies near 1 Hz, close to that of a Wood–Anderson seismograph, so M_L is often a good indication of the structural damage an earthquake can cause.

With time, various local and global magnitude scales evolved. For global studies, the primary two were the *body*

[1] We use the notations "log" for $\log_{10}$, and "ln" for the natural $\log_e$.

[2] Magnitudes can be negative for very small displacements; a magnitude −1 earthquake might correspond to a hammer blow.

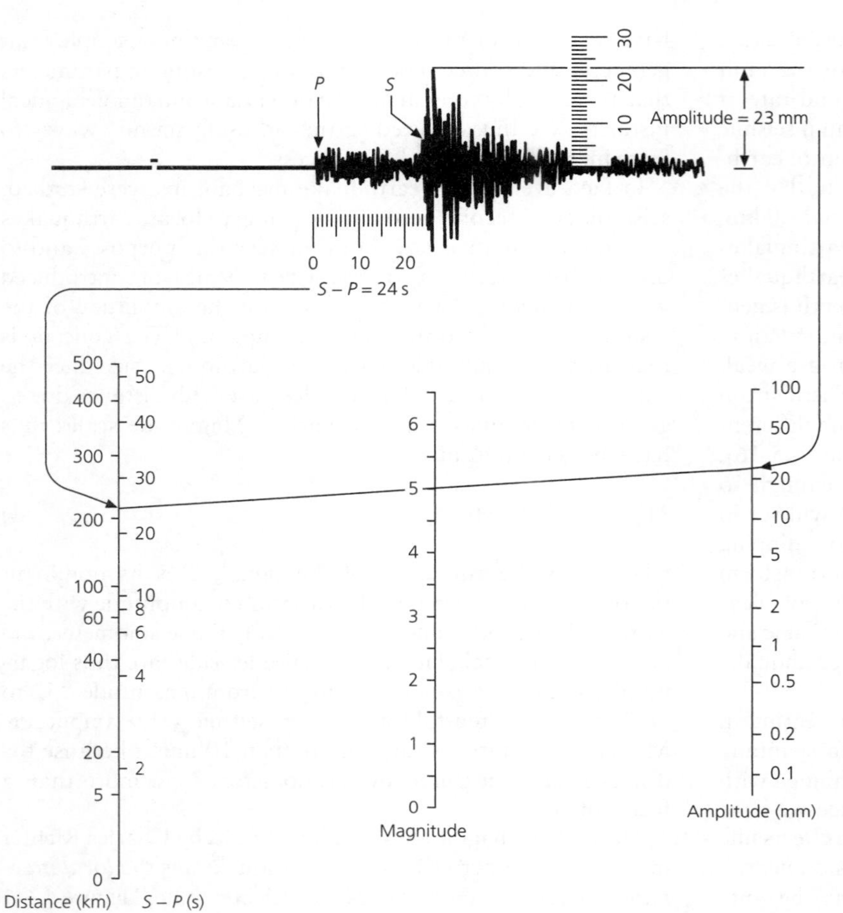

Fig. 4.6-1 The Richter scale for local magnitude, M_L. The magnitude is found from the amplitude of the largest arrival and the $S-P$ travel time difference. In this example, the maximum amplitude is 23 mm and the $S-P$ time is 24 s, making $M_L = 5.0$. (From *Earthquakes* by Bruce A. Bolt © 1978, 1988, 1993 by W. H. Freeman and Co. Used by permission.)

wave magnitude, m_b, and *surface wave magnitude*, M_s. m_b is measured from the early portion of the body wave train, usually the *P* wave, using

$$m_b = \log (A/T) + Q(h, \Delta), \qquad (3)$$

where *A* is the ground motion amplitude in microns after the effects of the seismometer are removed, *T* is the wave period in seconds, and *Q* is an empirical term depending on the distance and focal depth. This function can be derived either as a global average or for a specific region, as shown by Fig. 4.6-2. Measurements of m_b depend on the seismometer used and the portion of the wave train measured. Common US practice uses the first 5 s of the record and periods less than 3 s, usually about 1 s, on instruments with peak response near 1 s. m_b is measured out to 100° distance, beyond which diffraction around the core has a complicated effect on the amplitude.

The surface wave magnitude, M_s, is measured using the largest amplitude (zero to peak) of the surface waves

$$M_s = \log (A/T) + 1.66 \log \Delta + 3.3 \text{ or}$$
$$M_s = \log A_{20} + 1.66 \log \Delta + 2.0, \qquad (4)$$

where the first form is general, and the second uses the amplitudes of Rayleigh waves with a period of 20 s, which often have the largest amplitudes. In these relations, *A* is the ground motion amplitude in microns after the effects of the seismometer are removed, *T* is the wave period in seconds, and the distance Δ is in degrees.

As measures of earthquake size, magnitudes have two major advantages. First, they are directly measured from seismograms without sophisticated signal processing. Second, they yield units of order 1 which are intuitively attractive: magnitude 5 earthquakes are moderate, magnitude 6 are strong, 7 are major, and 8 are great.

However, magnitudes have two related limitations. First, they are totally empirical and thus have no direct connection to the physics of earthquakes. A striking illustration of this is that Eqns 1–4 are not even dimensionally correct—logarithms can be taken only for dimensionless quantities, whereas these expressions involve ratios of displacement to period. A second difficulty is with the numbers that emerge. Magnitude estimates vary noticeably with azimuth, due to the amplitude radiation patterns (Section 4.3), although this difficulty can be reduced by averaging results. The different magnitude scales

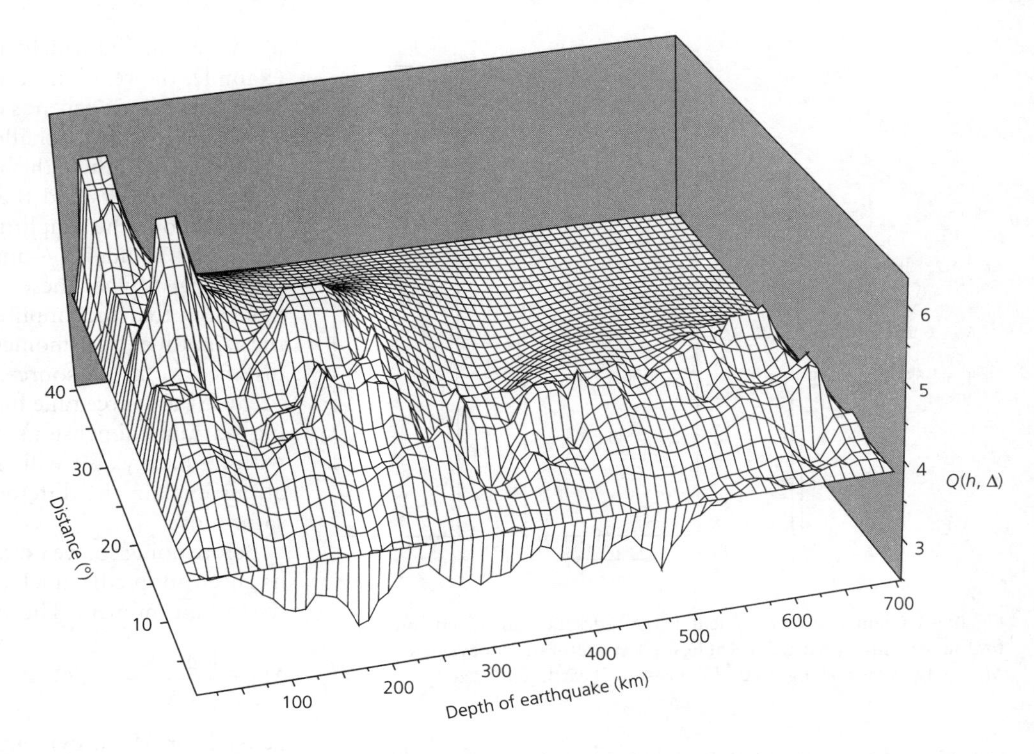

Fig. 4.6-2 Q factor for body wave magnitude m_b derived for P waves from earthquakes in the Tonga region recorded by a temporary deployment of seismometers. Q depends on focal depth and epicentral distance. (Wysession *et al.*, 1996. © Seismological Society of America. All rights reserved).

Table 4.6-1 Source parameters for selected earthquakes.

Earthquake	Body wave magnitude, m_b	Surface wave magnitude, M_s	Fault area (km²) (length × width)	Average dislocation (m)	Moment (dyn-cm), M_0	Moment magnitude, M_w
Truckee, 1966	5.4	5.9	10 × 10	0.3	8.3×10^{24}	5.9
San Fernando, 1971	6.2	6.6	20 × 14	1.4	1.2×10^{26}	6.7
Loma Prieta, 1989	6.2	7.1	40 × 15	1.7	3.0×10^{26}	6.9
San Francisco, 1906		7.8	450 × 10	4	5.4×10^{27}	7.8
Alaska, 1964	6.2	8.4	500 × 300	7	5.2×10^{29}	9.1
Chile, 1960		8.3	800 × 200	21	2.4×10^{30}	9.5

Sources: Values from Geller (1976), Wallace *et al.* (1991), and Wald *et al.* (1993).

yield different values. Moreover, body and surface wave magnitudes do not correctly reflect the size of large earthquakes.

The latter two effects are illustrated in Table 4.6-1, which gives magnitudes for various earthquakes, ordered by increasing scalar moment.[3] As shown, m_b and M_s differ significantly. The earthquakes with moments greater than that of the San Fernando earthquake all have m_b 6.2, even as the moment increases by a factor of 20,000. Similarly, the earthquakes larger than the San Francisco earthquake have M_s about 8.3, even as the moment increases by a factor of 400. This effect, called *magnitude saturation*, is a general phenomenon for m_b above about 6.2 and M_s above about 8.3.

Earthquake source parameter data like those in Table 4.6-1, some of which are shown in Fig. 4.6-3, are used to investigate issues related to earthquake size. Before doing so, it is worth briefly discussing how the tectonic setting affects earthquake size. All of these earthquakes, except for Chile, reflect deformation in the broad boundary zone between the North American and Pacific plates (Fig. 5.2-3). The San Fernando earthquake occurred on a buried thrust fault in the Los Angeles area, similar to the Northridge earthquake (Figs 4.5-9 and 4.5-10). These relatively short faults are part of an oblique trend in the boundary zone, so the fault areas tend to be roughly rectangular. Their down-dip width seems controlled by the fact that rocks deeper than about 20 km are weak and undergo stable sliding rather than accumulating elastic strain for future earthquakes, as discussed in the context of fault locking (Section 4.5.4). The next largest earthquake, Loma Prieta, occurred either close to or on a short segment of the San Andreas fault (Fig. 1.2-16), and hence on a somewhat longer fault of comparable width. The San Francisco earthquake ruptured a long segment of the San Andreas fault with significantly larger slip, but because the fault is vertical, still had a narrow width. Thus the 1906 earthquake illustrates approximately the maximum size of

[3] Seismic moments are reported either in dyn-cm or N-m, with 1 N-m = 10^7 dyn-cm.

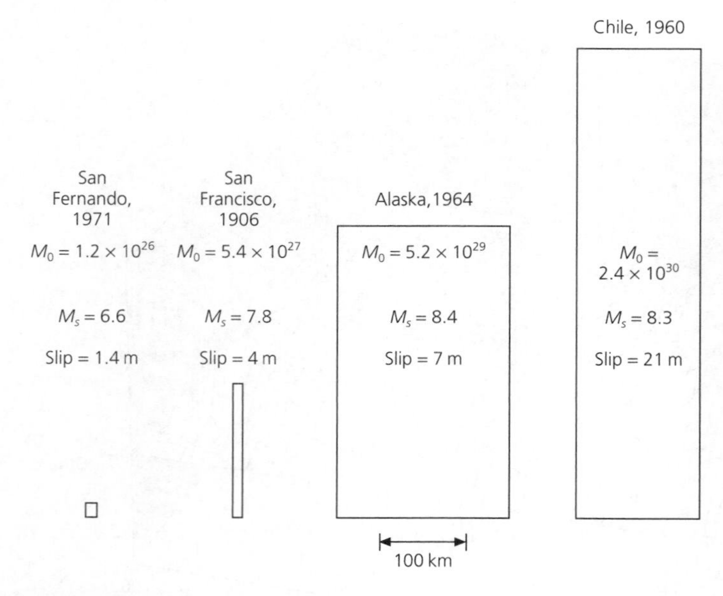

Fig. 4.6-3 Comparison of moment, magnitudes, fault area, and fault slip for four earthquakes listed in Table 4.6-1. M_s saturates for events with $M_w > 8$ and so is no longer a useful measure of earthquake size.

continental transform earthquakes. However, the Alaska and Chilean earthquakes had much larger rupture areas because they occurred on shallow-dipping subduction thrust interfaces. As shown in Fig. 4.5-16, these faults can have widths of hundreds of km on which elastic strain can build up and eventually be released seismically. As will be discussed shortly, the larger fault dimensions give rise to greater slip, so the combined effects of larger fault area and more slip cause the largest earthquakes to occur at subduction zones rather than on transforms.

It is important to realize that values like those in Table 4.6-1 are estimates with considerable uncertainties due to various causes. First, there are uncertainties due to the earth's variability and deviations from the mathematical simplifications used. For example, even with high-quality modern data, seismic moment estimates for the Loma Prieta earthquake vary by about 25%, and M_s values vary by about 0.2 units. Second, the estimation techniques vary. The actual approaches used to compute magnitudes have changed with time (note that the pre-1964 earthquakes do not have m_b values) in various ways. Uncertainties for historic earthquakes are especially large; for example, fault length estimates for the 1906 San Francisco earthquake vary from 300 to 500 km, M_s has been estimated at 8.3 but is now thought to be about 7.8, and the fault width is essentially unknown and inferred from the depths of more recent earthquakes and geodetic data. Third, different techniques (body waves, surface waves, geodesy, geology) can yield different estimates. Fourth, the fault dimensions and dislocations shown are average values for quantities that can vary significantly along the fault (Fig. 4.5-10). As a result, different studies yield varying and sometimes inconsistent values, depending on which parameters are estimated directly from data, which are assumed, and which are inferred by combining others. For example, the relation between the seismic moment, slip, and fault dimensions depends on the rigidity assumed (typically 3–5×10^{11} dyn/cm^2 for shallow earthquakes). Even so, such data are sufficient to show the basic effects of interest.

We can understand these effects given what we have discussed about the amplitudes of body and surface waves in Sections 4.2 and 4.3 — information that was not available to seismologists when these magnitude scales were developed. We have seen that the amplitudes depend on the scalar moment, the azimuth of a seismometer relative to the fault geometry, the distance from the source, and the source depth. Moreover, because the source time function has a finite duration, depending on fault dimensions and rise time, the amplitudes vary with frequency. We will see shortly that these frequency variations explain the differences between magnitudes and their saturation.

Before doing so, we note the simple and elegant solution that has been adopted: namely, defining a magnitude scale based on the seismic moment. The *moment magnitude*,

$$M_w = \frac{\log M_0}{1.5} - 10.73, \qquad (5)$$

defined for M_0 in dyn-cm, has several advantages. It gives a magnitude directly tied to earthquake source processes that does not saturate. Moreover, it preserves the simplicity of the magnitude scale by giving values of order 1 compatible with other magnitude scales. As we will see, M_w is comparable to M_s until M_s saturates at about 8.2, but then increases. The largest seismically recorded earthquake, the 1960 Chile event listed in Table 4.6-1, had M_w 9.5. Moment magnitude has become the common measure of the magnitude of large earthquakes. Estimation of M_0 (and therefore M_w) requires more analysis of seismograms than for m_b or M_s. However, semi-automated programs like the Harvard CMT project or comparable regional analyses now regularly compute moment magnitudes for most earthquakes larger than about M_w 5.

4.6.2 *Source spectra and scaling laws*

The relations between the moment and various magnitudes arise from the spectrum of the radiated seismic waves. We saw in Section 4.3.2 that the radiated waves depend on the product of the scalar moment and the source time function generated by the earthquake. We used a simple model in which the time function was the convolution of two "boxcar" time functions due to the finite length of the fault and the finite rise time of the faulting at any point. The Fourier transform of the resulting time function is the product of the transforms of the boxcars.

The transform of a boxcar of height $1/T$ and length T is

$$F(\omega) = \int_{-T/2}^{T/2} \frac{1}{T} e^{i\omega t} dt = \frac{1}{Ti\omega}(e^{i\omega T/2} - e^{-i\omega T/2}) = \frac{\sin(\omega T/2)}{\omega T/2}. \qquad (6)$$

This function, sometimes written as sinc $x = (\sin x)/x$, appears in many applications in which only part of a signal is selected. In Section 2.5.10 it described the amplitude resulting when part of a plane wave diffracts through a slit. In Section 6.3, we will use it to describe the effect on a time series spectrum from using only part of the series. Here, the sinc function describes the fact that the source pulse has finite duration.

Thus the spectral amplitude of the source signal is the product of the seismic moment and two sinc terms,

$$|A(\omega)| = M_0 \left| \frac{\sin(\omega T_R/2)}{\omega T_R/2} \right| \left| \frac{\sin(\omega T_D/2)}{\omega T_D/2} \right|, \tag{7}$$

where T_R and T_D are the rupture and rise times. Often, we use the logarithm of Eqn 7,

$$\log A(\omega) = \log M_0 + \log [\text{sinc}\ (\omega T_R/2)] + \log [\text{sinc}\ (\omega T_D/2)]. \tag{8}$$

A useful approximation is to treat sinc x as 1 for $x < 1$, and $1/x$ for $x > 1$, as shown in Fig. 4.6-4 (*top*). In this approximation a plot of $\log | A(\omega) |$ versus $\log \omega$ is just three segments, corresponding to different frequency ranges (Fig. 4.6-4, *bottom*). Assuming $T_R > T_D$, we have

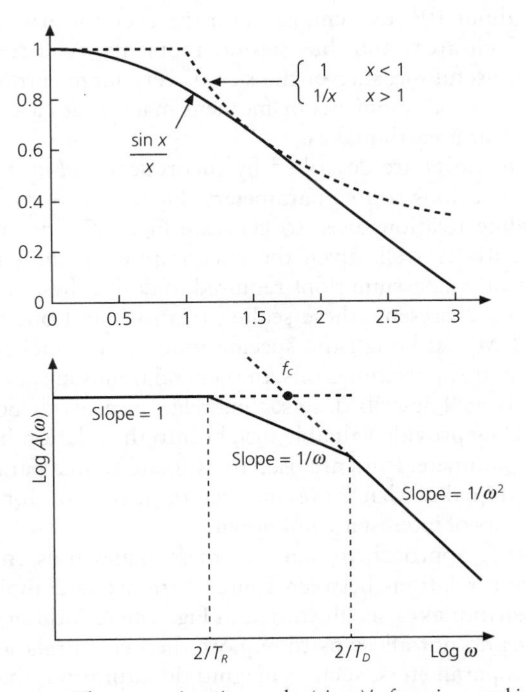

Fig. 4.6-4 *Top*: The approximation to the $(\sin x)/x$ function used in modeling the source spectrum. *Bottom*: Theoretical source spectrum of an earthquake, modeled as three regions with slopes of 1, ω^{-1}, and ω^{-2}, divided by angular frequencies corresponding to the rupture and rise times, T_R and T_D. Another common approximation uses a single corner frequency, f_c, at the intersection of the first and third spectrum segments. The flat segment extending to zero frequency gives M_0.

$$\log | A(\omega) | = \begin{cases} \log M_0 & \omega < 2/T_R \\ \log M_0 - \log (T_R/2) - \log \omega & 2/T_R < \omega < 2/T_D \\ \log M_0 - \log (T_R T_D/4) - 2 \log \omega & 2/T_D < \omega \end{cases} \tag{9}$$

This plot is divided into three regions by the frequencies $2/T_R$ and $2/T_D$, which are called *corner frequencies*. The spectrum is flat for frequencies less than the first corner, goes as ω^{-1} between the corners, and decays as ω^{-2} for the high frequencies. Thus the spectrum is parametrized by three factors: seismic moment, rise time, and rupture time. It is worth noting that other source spectral models have been used. A third corner frequency can be added to this model, representing the effects of fault width and yielding an ω^{-3} segment at high frequency. Other models have a single corner frequency (dashed line in Fig. 4.6-4, *bottom*) that combines the effects of rise and rupture time. As a result, the interpretation of observed earthquake spectra depends somewhat on the source model.

To see how the source spectrum varies with earthquake size, we first note that the seismic moment is the scale factor for the spectral amplitude at low frequencies $\omega \to 0$. This is the reason why it is also called the "static" moment. It is defined (Section 4.2.3) as the product of the rigidity at the source depth, μ, the average slip (or dislocation) on the fault, $\bar{D}$, and the fault area, S. The fault area can be written in terms of a shape factor f and the square of a dimension L, so

$$M_0 = \mu \bar{D} S = \mu \bar{D} f L^2. \tag{10}$$

For large earthquakes, faults are often treated as approximately rectangular, so L is the length, and f is the ratio of width to length. Another common approach uses a circular fault model for which L is the radius and $f = \pi$.

The rupture time (Eqn 4.3.8) needed for the rupture to propagate along the fault is approximately

$$T_R = L/v_R = L/(0.7\beta), \tag{11}$$

if we assume that the rupture velocity is about 0.7 times the shear velocity. The rise time needed for the dislocation to reach its full value at any point on the fault has been predicted to be about

$$T_D = \mu \bar{D}/(\beta \Delta\sigma) = 16 L f^{1/2}/(7\beta\pi^{1.5}), \tag{12}$$

where $\Delta\sigma$ is the stress drop in the earthquake, a quantity that we will discuss shortly. Assuming a shear velocity of about 4 km/s, Eqns 11 and 12 yield approximately

$$T_R = 0.35L, \quad T_D = 0.1 L f^{1/2}. \tag{13}$$

Table 4.6-1 shows that the Truckee and San Fernando earthquakes occurred on approximately square faults ($f = 1$), Loma Prieta and Alaska had $L \approx 2W$, or $f \approx 0.5$, and the San Francisco

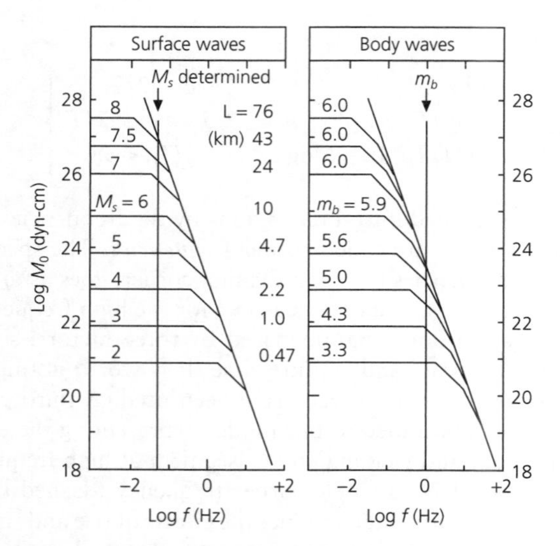

Fig. 4.6-5 Theoretical source spectra of surface and body waves. The two are identical at frequencies below the ω^{-2} corner frequency. This model includes a fault-width corner frequency, and thus an ω^{-3} segment at high frequency. m_b, reflecting the amplitude at 1 s, saturates at about 6 for earthquakes with moment above about 10^{25} dyn-cm. M_s, measured at 20 s, saturates at about 8 for moments greater than about 5×10^{27} dyn-cm. The x axis is in frequency (Hz) rather than angular frequency (ω). (Geller, 1976.)

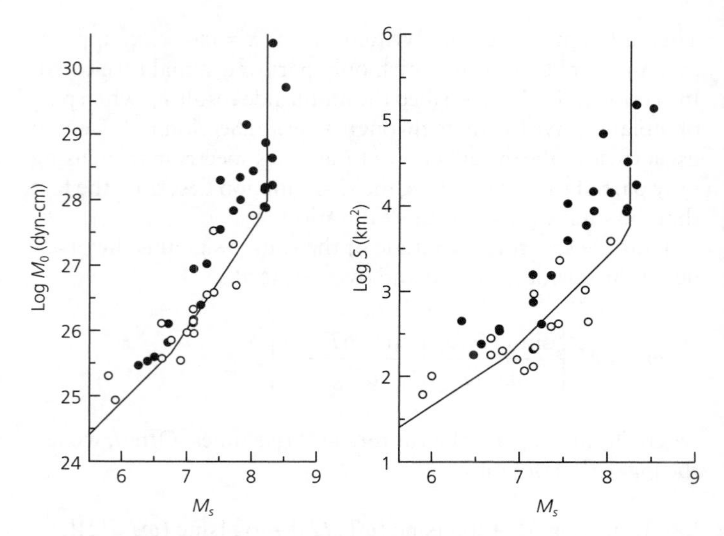

Fig. 4.6-6 Plots of M_s versus log M_0 and M_s versus log of fault area (S) show the saturation of surface wave magnitude. M_s saturates even as the moment and fault areas increase. Lines show the predictions of the scaling relations in Table 4.6-2. Open and closed circles denote intraplate and interplate earthquakes, respectively. (Geller, 1976.)

earthquake occurred on a long narrow fault with $L \gg W$, or $f < 0.1$. Thus, in these cases or for a circular fault, $T_R > T_D$, as drawn in Fig. 4.6-4.

As we will see, earthquake stress drops are approximately independent of seismic moment, implying that the slip is proportional to fault length. Hence, for an assumed stress drop, we can compute theoretical spectra for various moments and fault lengths (Fig. 4.6-5). The results show why m_b and M_s differ, and why both magnitude scales saturate. As the fault length increases, the seismic moment, rupture time, and rise time increase. Thus the corner frequencies move to the left, to lower frequencies. The moment, M_0, determines the zero-frequency level, which rises as the earthquake becomes larger. However, the surface wave magnitude, M_s, is measured at a period of 20 s, and so depends on the spectral amplitude at this period. For earthquakes with moments less than about 10^{26} dyn-cm, a 20 s period corresponds to the flat part of the spectrum, so M_s increases with moment. However, for larger moments, 20 s is to the right of the first corner frequency, so M_s does not increase at the same rate as the moment. Once the moment exceeds about 5×10^{27} dyn-cm, 20 s is to the right of the second corner, on the ω^{-2} portion of the spectrum. Thus M_s saturates at about 8.2, even if the moment increases. A similar effect occurs for body wave magnitude, which depends on the amplitude at a period of 1 s. Because this period is shorter than the 20 s used for M_s, m_b saturates at a lower moment (about 10^{25} dyn-cm), and remains about 6 even for much larger earthquakes. Similar saturation effects occur for other magnitude scales which are measured at specific frequencies.

Another way to view magnitude saturation is shown by the data for various earthquakes in Fig. 4.6-6. For earthquakes above about 10^{28} dyn-cm, M_s saturates even for progressively larger fault areas and thus seismic moments. As a result, M_s is not a useful measure of the size of very large earthquakes. For this reason, moments or moment magnitudes are used to describe large earthquakes.

These effects are described by theoretical *scaling relations* between various source parameters. Figure 4.6-6 shows that the scaling relations used to generate Fig. 4.6-5 describe the data relatively well, given the uncertainties in the data and the simplifying assumptions required to derive these relations. Table 4.6-2 presents these scaling relations and one relating m_b and M_s. Although the specific numerical values in these relations are approximations, the general trends in the data are relatively well described, so scaling relations provide powerful tools. They provide valuable insight into the relation between source parameters and are used to estimate source parameters for earthquakes that have not yet occurred, or for which parameters of interest are unknown.

Another approach to some of these issues uses empirical regression relations between source parameters compiled for many earthquakes, as illustrated in Fig. 4.6-7. Although these relations do not allow us to explore theoretical relationships between parameters, such as magnitude saturation, they offer useful inferences about past and potential earthquakes. For example, these regressions imply that an earthquake on a 100 km-long fault would have an average slip of about 2 m and M_w about 7.4, whereas on a 10 km-long fault we expect about 0.3 m slip and M_w about 6.2. As for the scaling laws, these estimates should be taken as useful averages. For example, we

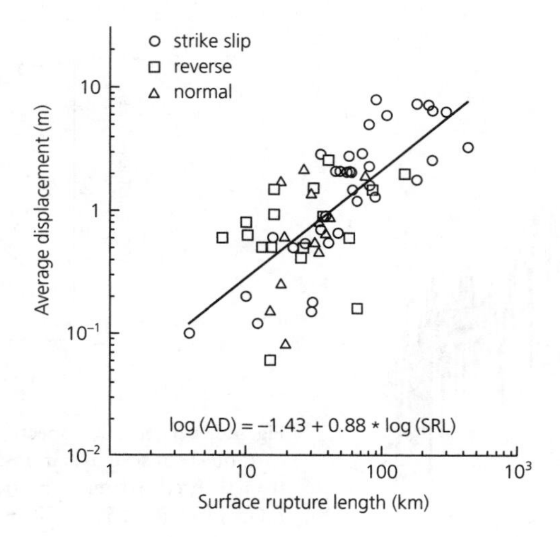

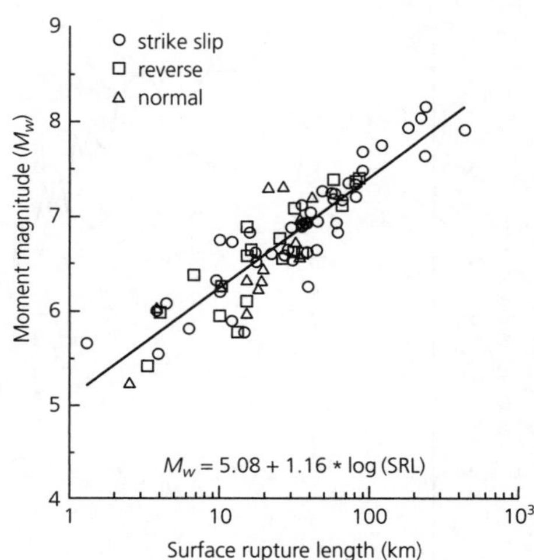

Fig. 4.6-7 Empirical relations showing the average slip, fault length, and moment magnitude for a compilation of earthquakes. (Wells and Coppersmith, 1994. © Seismological Society of America. All rights reserved.)

Table 4.6-2 Earthquake scaling relations.

m_b and M_s are related by	
$m_b = M_s + 1.33$	$M_s < 2.86$
$m_b = 0.67M_s + 2.28$	$2.86 < M_s < 4.90$
$m_b = 0.33M_s + 3.91$	$4.90 < M_s < 6.27$
$m_b = 6.00$	$6.27 < M_s$.
Assuming $L = 2W$, M_s and fault area (in km^2) are related by	
$\log S = 0.67M_s - 2.28$	$M_s < 6.76$
$\log S = M_s - 4.53$	$6.76 < M_s < 8.12$
$\log S = 2M_s - 12.65$	$8.12 < M_s < 8.22$
$M_s = 8.22$	$S > 6080$ km^2.
Assuming a stress drop of 50 bars, $\log M_0$ (in dyn-cm) and M_s are related by	
$\log M_0 = M_s + 18.89$	$M_s < 6.76$
$\log M_0 = 1.5M_s + 15.51$	$6.76 < M_s < 8.12$
$\log M_0 = 3M_s + 3.33$	$8.12 < M_s < 8.22$
$M_s = 8.22$	$\log M_0 > 28$.

Source: Geller (1976).

would be surprised by 10 m of motion on a 100 km-long fault, but not by 1 or 4 m.

4.6.3 *Stress drop and earthquake energy*

The relationship between the slip in an earthquake, its fault dimensions, and its seismic moment is closely tied to the magnitude of the stress released by the earthquake, or *stress drop*. As discussed in Section 4.5.4, the earthquake releases the strain that has accumulated over time near the fault, so the radiated seismic waves are used to estimate the stress change.

To do this, we assume that the earthquake's slip, D, occurs on a fault with characteristic dimension L, and so causes a strain change of approximately

$$\varepsilon_{xx} = \frac{\partial u_x}{\partial x} \approx \frac{\bar{D}}{L}, \tag{14}$$

so the stress drop averaged over the fault is approximately

$$\Delta\sigma \approx \mu\bar{D}/L. \tag{15}$$

From seismological observations alone, the best-constrained quantity is the seismic moment, so we estimate the average slip, $\bar{D}$, from the seismic moment as

$$\bar{D} \approx cM_0/(\mu L^2), \tag{16}$$

where c is a factor depending on the fault's shape. Thus the stress drop is proportional to the moment and inversely proportional to the fault dimension cubed or the 3/2 power of the fault area:

$$\Delta\sigma = cM_0/L^3 = cM_0/S^{3/2}. \tag{17}$$

The specific relation and values of c depend on the fault shape and the rupture direction. For example, the stress drop on a circular fault with a radius R is

$$\Delta\sigma = \frac{7}{16}\frac{M_0}{R^3}, \tag{18}$$

strike-slip on a rectangular fault with length L and width w yields

$$\Delta\sigma = \frac{2}{\pi}\frac{M_0}{w^2 L}, \tag{19}$$

and dip-slip on a rectangular fault gives

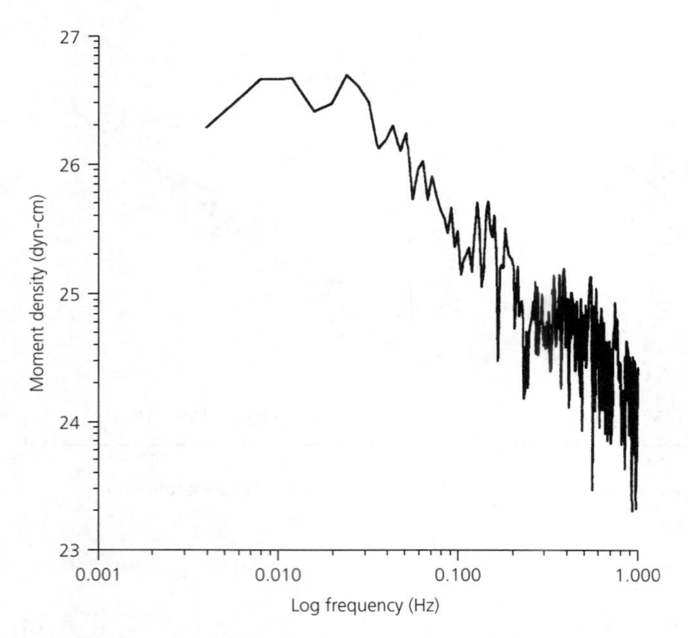

Fig. 4.6-8 Amplitude spectrum averaged from *P* waves recorded at globally distributed broadband seismometers for the October 21, 1995, earthquake near Chiapas, Mexico. (Rebollar *et al.*, 1999. © Seismological Society of America. All rights reserved.)

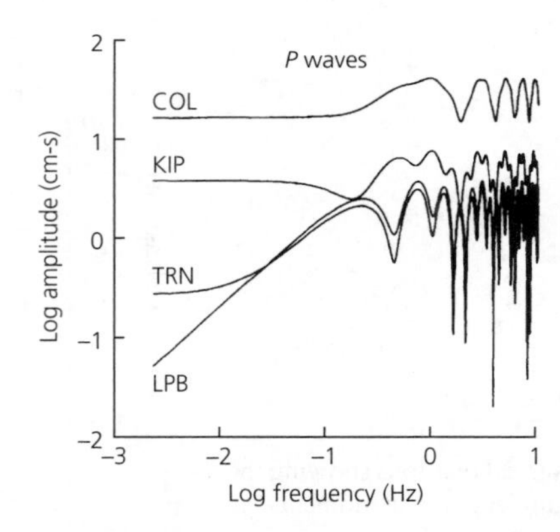

Fig. 4.6-9 Theoretical spectra for a shallow (about 8 km focal depth) earthquake at several stations. Due to free surface effects, the spectra differ from theoretical source spectra like that in Fig. 4.6-4. (Langston, 1978. *J. Geophys. Res.*, *83*, 3422–6, copyright by the American Geophysical Union.)

$$\Delta\sigma = \frac{4(\lambda+\mu)}{\pi(\lambda+2\mu)}\frac{M_0}{w^2L} \approx \frac{8}{3\pi}\frac{M_0}{w^2L}, \tag{20}$$

where the last form assumes $\lambda = \mu$.

These equations let us estimate the stress drop from an observed seismic moment and inferred fault dimensions. If we know the fault dimensions from other observations, this process is straightforward. For example, the fault area of the great 1964 Alaska earthquake can be estimated from the aftershock area, the source finiteness shown by surface waves (Fig. 4.3-15), and geodetic data. Thus using the values in Table 4.6-1 and Eqn 20 with $\lambda = \mu$ yields an average stress drop estimate

$$\Delta\sigma = \frac{8}{3\pi}\frac{5.2\times10^{29}\ \text{dyn-cm}}{9\times10^{14}\ \text{cm}^2\ 5\times10^{7}\ \text{cm}} \approx 10^7\ \text{dyn/cm}^2 = 10\ \text{bars}. \tag{21}$$

However, without independent knowledge of fault dimensions, estimating the stress drop is harder. One approach uses the spectrum to identify corner frequencies and estimate the rupture time and hence fault dimensions. Figure 4.6-8 illustrates this for a M_w 7.1 earthquake occurring at 165 km depth in the subduction zone beneath Mexico. Analysis of the spectrum with a single corner frequency model like that in Fig. 4.6-4, and assuming a circular fault with rupture velocity of 3 km/s, yielded a rupture duration of 22 s and a stress drop of about 65 bars.[4] The low-frequency portion of the spectrum yields a moment of 5.2×10^{26} dyn-cm, in reasonable agreement with other studies, which found 4.6 and 7.1×10^{26} dyn-cm.

In many cases the spectrum is not directly amenable to corner frequency analyses. The earthquake in Fig. 4.6-8 was deep enough that the spectrum of the direct *P* wave could be found without contamination from later-arriving surface reflections. However, for shallow earthquakes, *P*, *pP*, and *sP* often overlap (Fig. 4.3-7), yielding a combined spectrum quite different from the source pulse. Figure 4.6-9 illustrates this effect for a shallow earthquake. As shown, the spectra differ significantly between stations, due to the variation in amplitude between direct and reflected arrivals, and cannot be used to find corner frequencies or the seismic moment. This difficulty can be addressed by modeling the body waves, including the free surface reflections, and estimating the source time function duration by matching the observed waveforms. Given a duration estimate and an assumed fault geometry, the fault length and stress drop are estimated in the same way as in the corner frequency analysis.

These examples illustrate that estimating the fault dimensions and stress drop is challenging, whether it is done in the time domain, by modeling or inverting waveforms, or in the frequency domain. First, the parameter required is estimated only with modest precision, as shown by the issue of choosing the corner frequency even with high-quality data like that in Fig. 4.6-8. The uncertainty is compounded by the fact that inferring a source dimension from the corner frequency or source time function requires assuming the rupture velocity and fault geometry. Moreover, the estimated stress drop depends on $1/L^3$, so uncertainty in the fault dimension causes a large uncertainty in $\Delta\sigma$. Figure 4.6-10 illustrates this issue via synthetic *P* waves for different source time function durations. As shown, the seismogram depends only moderately on the source time function. However, small differences in time function duration correspond to larger differences in stress drop, even for an assumed rupture velocity and fault geometry.

[4] Stress drops are reported either in bars (1 bar = 10^6 dyn/cm^2) or MegaPascals (1 MPa = 10 bars).

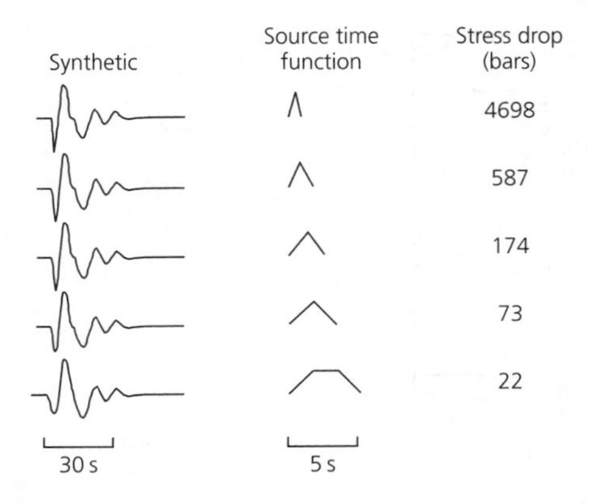

Fig. 4.6-10 Synthetic seismograms for different source time functions and corresponding inferred stress drops. A small uncertainty in source time function duration results in a large uncertainty in stress drop. (Stein and Kroeger, 1980. Reproduced with the permission of the American Society of Mechanical Engineers.)

This brings up the interesting question of how the uncertainty in a quantity like stress drop, which is inferred by combining parameters estimated from data using model assumptions, is related to the uncertainties involved in each step. A common approach, derived in Section 6.5.1, uses the propagation of errors relation (Eqn 6.5.19), which involves the partial derivatives of the parameters going into the final quantity. To use this, we write the stress drop (Eqn 17) with the fault dimension equal to the product of rupture velocity and rupture time,

$$\Delta\sigma = f(c, M_0, v_R, T_R) = cM_0/(v_R T_R)^3. \tag{22}$$

The standard deviation, or uncertainty, in the stress drop is thus approximately

$$\sigma_f^2 = \sigma_c^2\left(\frac{\partial f}{\partial c}\right)^2 + \sigma_{M_0}^2\left(\frac{\partial f}{\partial M_0}\right)^2 + \sigma_{v_R}^2\left(\frac{\partial f}{\partial v_R}\right)^2 + \sigma_{T_R}^2\left(\frac{\partial f}{\partial T_R}\right)^2. \tag{23}$$

We can compute the partial derivatives and use estimates of the parameters and their uncertainties to estimate the resulting uncertainty in the stress drop. For example, we might assume the seismic moment, rupture time, and rupture velocity are uncertain to about 25%. As Eqns 18–20 show, different models give different shape factors, and various methods are used to interpret corner frequencies, so c is uncertain to at least 50% if we have no other knowledge of the fault geometry. Depending on the values used, it seems that the precision of a stress drop estimate is often a factor of 2 or 3. The accuracy — how this value is related to the physical process of faulting — is hard to assess, because the form of the time function and its relation to other source parameters are derived for simple source models, which may or may not describe real faulting very well. Hence the stress drop is more usefully viewed as a characterization of the source spectrum than as giving direct insight into the physics of the source.

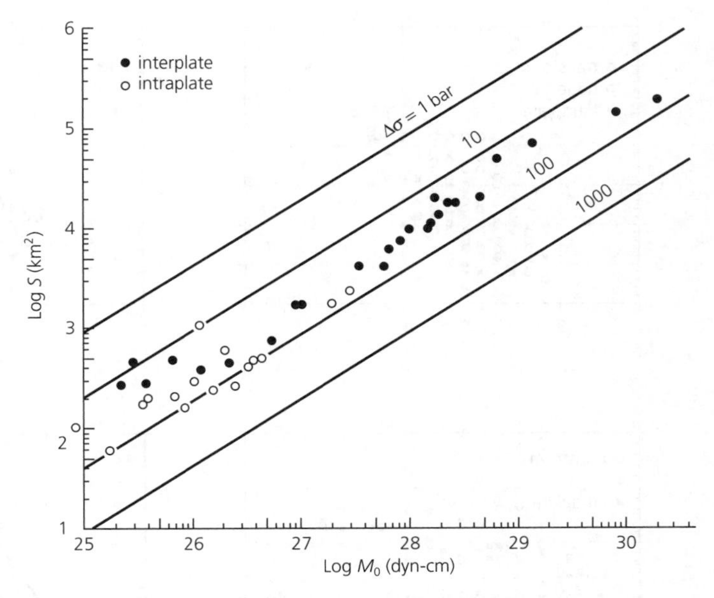

Fig. 4.6-11 Stress drops for interplate (plate boundary) and intraplate (plate interior) earthquakes. The earthquakes are plotted in terms of log fault area and log seismic moment. The lines for constant stress drop values have slopes of 2/3, as shown by Eqn 17. Most earthquakes have $\Delta\sigma = 10 - 100$ bars, with intraplate and interplate events trending toward the higher and lower ends of the range, respectively. (Kanamori and Anderson, 1975. © Seismological Society of America. All rights reserved.)

With all these difficulties, it is encouraging that earthquake stress drop studies typically yield values in the 10–100s of bars, as shown in Fig. 4.6-11. The stress drop is essentially constant over five orders of magnitude in moment, although there appear to be small differences between tectonic environments. Stress drops for interplate events average about 30 bars, whereas intraplate stress drops sometimes exceed 100 bars.

There also seem to be differences among earthquakes at different plate boundary types. Figure 4.6-12 shows M_0/τ^3, the ratio of seismic moment to the observed total time function duration (rise time plus rupture time), for some oceanic ridge, transform, and intraplate earthquakes. This quantity is approximately proportional to stress drop (Eqn 22) and is hopefully less model-dependent than stress drop estimates. For a given M_w, the ratio seems smaller for transform earthquakes than for ridges, perhaps implying lower stress drops.

Another way to use such data is to take the different magnitudes to study how energy release varies with frequency. Compared to ridge earthquakes, transform earthquakes often have large M_s relative to m_b and large M_w relative to M_s, suggesting that seismic wave energy is relatively greater at longer periods. Earthquakes that preferentially radiate at longer periods are called "slow" earthquakes. Slow earthquakes have been noted in various environments. For example, slow earthquakes underwater in the appropriate locations and focal geometry can cause very large tsunamis (Section 1.2.4) that are not predicted by tsunami warning systems based on real-time assessments of m_b or M_s.

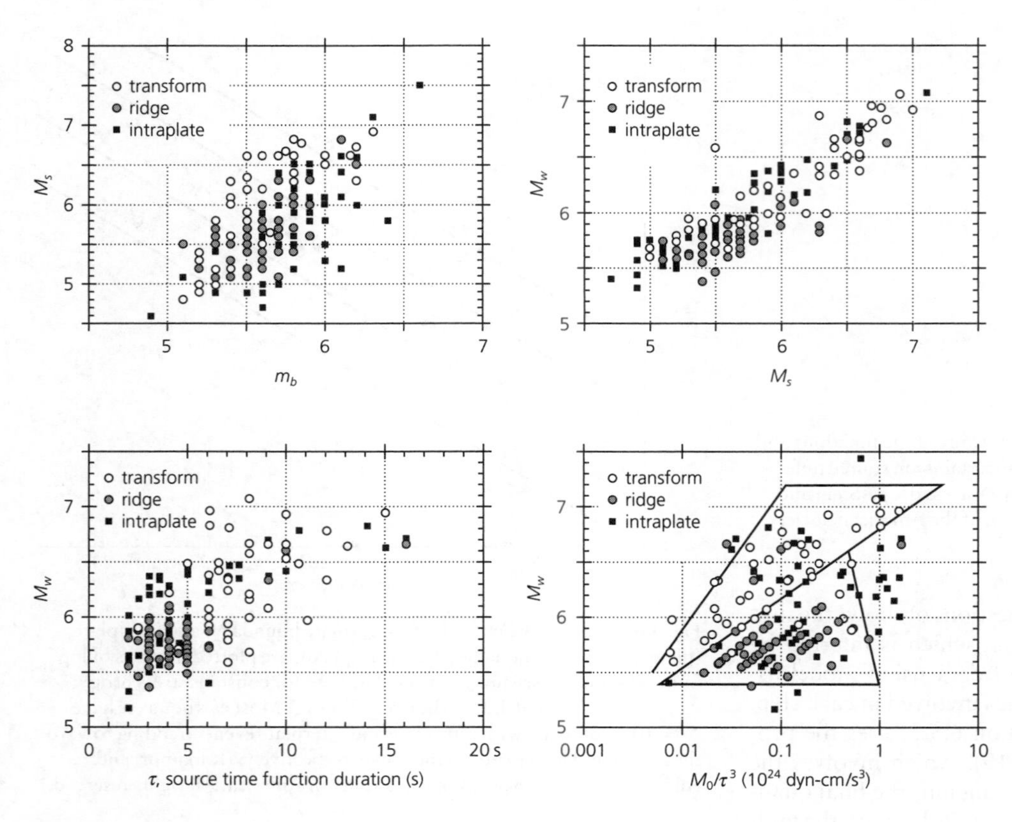

Fig. 4.6-12 Source parameters for some oceanic ridge, transform, and intraplate earthquakes. The transform earthquakes have relatively longer time functions and higher M_s/m_b, M_w/M_s, and M_0/τ^3 ratios, implying that they are "slow" earthquakes, perhaps with lower stress drop. (Stein and Pelayo, 1991. Reproduced with the permission of the Royal Society of London.)

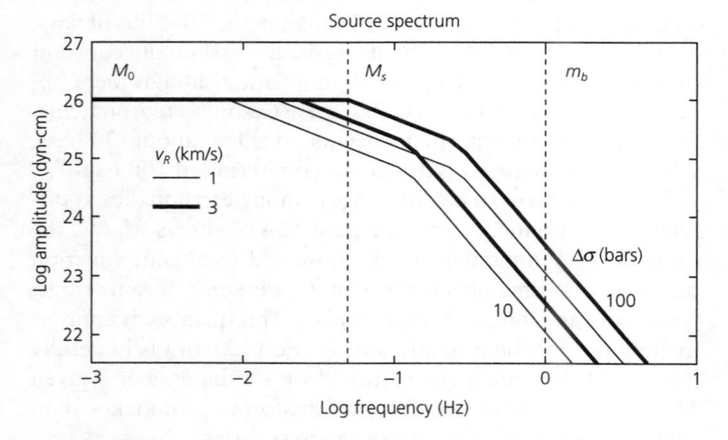

Fig. 4.6-13 Theoretical source spectra for earthquakes with the same seismic moment and fault shape. For each pair of spectra with the same rupture velocity, the left curve for lower stress drop corresponds to larger fault dimensions, and hence longer time functions and smaller corner frequencies. This earthquake would be "slower" with less high-frequency radiation and lower M_s and m_b. Similar effects occur for slower rupture velocity. The *x* axis is in frequency (Hz) rather than angular frequency (ω).

Differences between m_b, M_s, and M_w can reflect differences in stress drop. Figure 4.6-13 illustrates this using theoretical source spectra for earthquakes with the same seismic moment. For a given moment and fault shape, Eqn 17 shows that a lower stress drop corresponds to larger fault dimensions, and hence longer time functions and smaller corner frequencies. Thus, given two earthquakes with the same rupture velocity, the one with lower stress drop will have less high-frequency radiation, and thus lower M_s and m_b. Similar effects can result from a slower rupture velocity, which also gives a longer time function for a given fault dimension. These two possibilities can be distinguished when the rupture velocity can be inferred from the relative time between sub-events, as in Figs 4.3-11 or 4.5-11.

Thus the stress drop both characterizes earthquake source spectra and gives insight into the physics of faulting. From a source spectrum view, earthquake magnitudes saturate because the stress drop is essentially constant as earthquake moment increases, so the ratio of the slip to fault length remains constant. As a result, larger-moment earthquakes have longer faults and hence lower corner frequencies. From a fault mechanics view, the fact that the ratio of the slip to fault length is constant indicates that strain release in earthquakes is roughly constant, at about

$$\varepsilon_{xx} \approx \bar{D}/L \approx \Delta\sigma/\mu \approx 10^{-4}, \tag{24}$$

assuming a stress drop of 50 bars and $\mu = 5 \times 10^{11}$ dyn/cm^2, which are average values for earthquakes in the crust and the upper mantle.

This brings us to the important and unresolved issue, which will be discussed in Section 5.7, that the 10–100 bar stress drops found for earthquakes are much less than the strength of rock found in laboratory friction experiments. One possibility is that the low stress drop reflects the average of highly variable slip over a fault plane, whereas strength is much higher in strong patches (sometimes called asperities) where the largest slip occurs. However, other data, such as the absence of heat flow anomalies at faults, also imply that faults are weaker than expected from the laboratory results. As a result, it is not clear whether earthquakes release most of the stress built up on a fault or only a small fraction of it. It is similarly unclear how to interpret the possible variations in energy release as a function of period, and perhaps stress drop, in different tectonic environments. Intuitively, they might reflect interplate earthquakes occurring more frequently than intraplate events on better-established, and perhaps thus weaker, faults. Similarly, established transforms may be weaker than newly formed near-ridge crust.

This discussion leads naturally to the question of how the seismic wave energy radiated by an earthquake is related to its moment and magnitude. To address this, recall that work equals force times distance, so the strain energy released is the product of the average stress during faulting, $\bar{\sigma}$, the average slip, and the fault area,

$$W = \bar{\sigma}\bar{D}S. \quad (25)$$

If the stresses before and after faulting are σ_0 and σ_1, then $\Delta\sigma = \sigma_0 - \sigma_1$, and $\bar{\sigma} = \sigma_1 + (\Delta\sigma)/2$. Some of this energy, H, is lost to friction, so the radiated seismic energy is

$$E = W - H = \bar{\sigma}\bar{D}S - \sigma_f\bar{D}S, \quad (26)$$

where σ_f is the frictional stress, or

$$E = (\Delta\sigma/2)\bar{D}S + (\sigma_1 - \sigma_f)\bar{D}S = E_0 + (\sigma_1 - \sigma_f)\bar{D}S. \quad (27)$$

Thus the quantity

$$E_0 = (\Delta\sigma/2)\bar{D}S = (\Delta\sigma/2\mu)M_0 \quad (28)$$

is a lower bound on the radiated seismic energy (Fig. 4.6-14). If faulting stops once the final stress equals the frictional stress, $\sigma_1 = \sigma_f$, then $E_0 = E$, the radiated energy. Note that the radiated energy is proportional to the stress drop.

The ratio of the radiated energy to the total strain energy release is called the *seismic efficiency*,

$$\eta = E/W = \Delta\sigma/(2\bar{\sigma}), \quad (29)$$

where the last form assumes that $E_0 = E$. The efficiency depends on the final stress or, equivalently, the ratio of stress drop to the average stress. The case $\Delta\sigma \ll \bar{\sigma}$ is called partial stress drop, whereas $\Delta\sigma \approx 2\bar{\sigma}$ corresponds to near-total stress drop. It is still unresolved which of these cases is appropriate for earthquakes, because, of all the parameters in this model, only the stress drop can be directly estimated from seismological data.

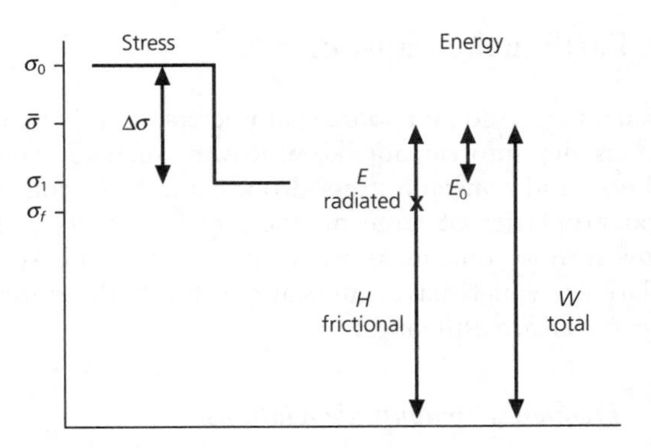

Fig. 4.6-14 Schematic illustration of the relation between the total strain energy released in faulting (W) and its portions radiated seismically (E) and dissipated by friction (H). In the model, these depend on the initial and final stresses (σ_0 and σ_1), their average ($\bar{\sigma}$), the stress drop ($\Delta\sigma$), and the frictional stress (σ_f). If the final stress equals the frictional stress, $E_0 = E$. Of these quantities, only stress drop can be estimated directly from seismological data.

This model of the seismic energy radiated in earthquakes underlies the concept of moment magnitude. Assuming a stress drop of 50 bars and $\mu = 5 \times 10^{11}$ dyn/cm^2, as in Eqn 24, Eqn 28 yields

$$E_0 = M_0/(2 \times 10^4), \quad \log E_0 = \log M_0 - 4.3, \quad (30)$$

where E_0 is in ergs. Inverting the definition of moment magnitude Eqn 5 gives

$$\log M_0 = 1.5M_w + 16.1, \quad (31)$$

so the second part of Eqn 30 becomes

$$\log E_0 = 1.5M_w + 11.8. \quad (32)$$

This relation illustrates that an increase in earthquake magnitude of 1 unit, for example from 5 to 6, increases the radiated energy by a factor of $10^{1.5}$, or about 32. Hence a magnitude 7 earthquake releases 10^3, or, 1000 times more energy than a magnitude 5 event. This ratio is strictly valid only for earthquakes with the same stress drop, but is a good general approximation.

Equation 30 also illustrates the intriguing fact that although the seismic moment has the dimensions of energy (1 erg = 1 dyn-cm), the radiated energy is only $1/(2 \times 10^4)$, or 0.00005, of the seismic moment released. This is because the seismic moment is not an energy, but instead is fundamentally related to the integral of the stress change over the earthquake source region, which gives the moment dimensions of dyn/cm^2 × cm^3, or dyn-cm. We can view Eqn 28 as converting the moment to a strain, and then multiplying by the stress acting during the earthquake to find the strain energy radiated. To illustrate that seismic moment and energy are different, seismic moment is quoted in dyn-cm (or N-m), and seismic energy is given in ergs (or J), even though the units are equivalent.

4.7 Earthquake statistics

In discussing earthquake source parameters, we saw that interesting insights into earthquake processes, such as magnitude saturation and constant stress drop, came from considering general properties of large numbers of earthquakes. Hence we now turn to some ideas about the statistics of earthquake populations, which have implications for both source processes and hazard estimation.[1]

4.7.1 Frequency–magnitude relations

As mentioned in Section 1.2, the number of earthquakes that occur yearly around the world varies with magnitude, with successively smaller earthquakes being more common. This observation was quantified by Gutenberg and Richter[2] in the 1940s via the logarithmic earthquake *frequency–magnitude* relation

$$\log N = a_1 - bM, \tag{1}$$

in which N is the number of earthquakes with magnitude greater than M occurring in a given time. The distribution is described by a linear relation, with constants a_1 and b. It turns out that although the intercept, a_1, depends on the number of earthquakes in the time and region sampled, the slope, b, is generally about 1. This is shown in Fig. 4.7-1 for the nearly 13,000 earthquakes with $M_s \geq 5$ for the 30 years between 1968 and 1997. There is an approximately tenfold increase in the number of earthquakes for successively smaller magnitudes: annually around the world there are about one $M_s = 8$ earthquake, 10 $M_s = 7$ events, 100 $M_s = 6$ events, and so forth.

A striking feature of this relation, sometimes called the *Gutenberg–Richter relation*, is that it also applies in individual seismic areas, with b generally about 1. Thus, although the number of earthquakes depends on how seismically active an area is, the relative frequency (M > 6 earthquakes about 10 times more common than M > 7, etc.) still applies. For example, in the past 1300 years Japan is estimated to have had about 190 earthquakes with M > 7 and 20 with M > 8. Similarly, since 1816 southern California has had about 180 earthquakes with M > 6, 24 with M > 7, and 1 with M > 8; whereas the New Madrid (central USA) seismic zone has had about 16 earthquakes with M > 5 and 2 with M > 6. Although the precise numbers, especially for the rarer large earthquakes, depend on the period chosen and uncertainties in estimating magnitudes prior to the invention of the seismometer (in about 1890), the logarithmic decay still appears.

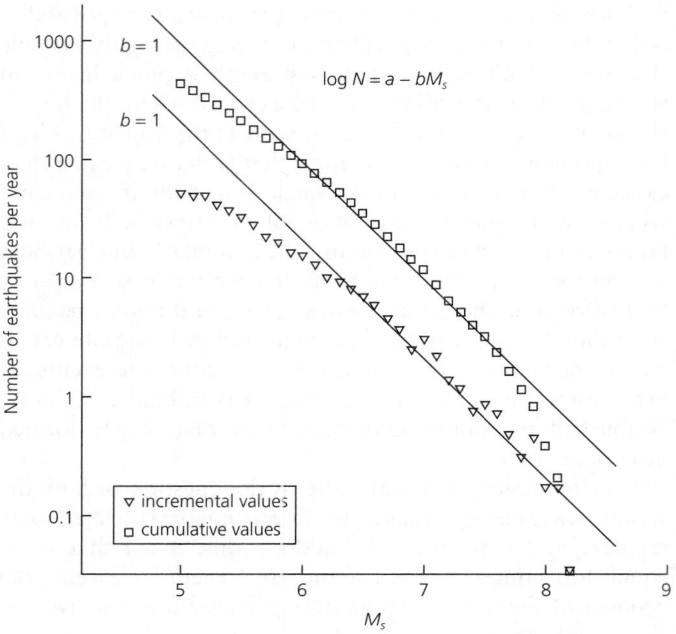

Fig. 4.7-1 Frequency–magnitude plot for all earthquakes with $M_s \geq 5.0$ during 1968–97 listed in the catalog of the National Earthquake Information Center. The logarithm of the numbers of earthquakes as a function of magnitude gives a line with slope (b) about 1. The values are shown both as a cumulative curve for the number of earthquakes per year with magnitude greater than or equal to a certain value and as incremental values in 0.1 magnitude unit bins.

Such a pattern, called fractal scaling, self-similarity, or scale invariance, is common in nature. For instance, a coastline or river drainage pattern looks similar when viewed at scales of 1, 10, 100, or 1000 km. The idea that the distribution of earthquake size is invariant with respect to scale except for the largest earthquakes is part of the rationale for the hypothesis that earthquakes are unpredictable, because there is no way to predict which small earthquakes will grow into large ones (Section 1.2.6).

The frequency–magnitude relation applies not only to the cumulative number N of earthquakes greater than a given magnitude, but to the incremental numbers n in a magnitude range M to $M + dM$. To see this, we write Eqn 1 as

$$N = 10^{a_1 - bM}, \tag{2}$$

differentiate it with respect to M, and take the logarithm, so

$$\log\left(\frac{dN}{dM}\right) = \log n = a_2 - bM \tag{3}$$

where a_2 is a new constant. Thus although the intercept a changes, the slope b stays constant. The data in Fig. 4.7-1 show

[1] Seismologists, like other geoscientists, have an ambivalence toward statistics, finding them valuable but often insufficient to require discarding models that do not rise to statistical significance. Often our attitude recalls the adage that statistics should be used as a drunk uses a lamp post — more for support than illumination. Sometimes this works; when asked whether he had statistically tested his exciting magnetic anomaly results showing symmetric sea floor spreading at mid-ocean ridges, F. Vine said that he never touched statistics but just dealt with facts (Menard, 1986).

[2] The many important contributions to seismology of Beno Gutenberg (1889–1960) and Charles Richter (1900–85) include quantifying global and regional seismicity.

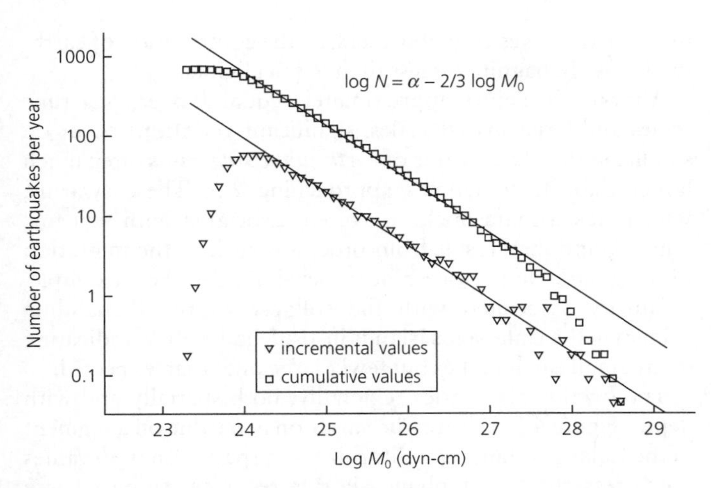

Fig. 4.7-2 Frequency–magnitude plot of all earthquakes during 1976–98 with seismic moments measured by the Harvard CMT project. The slope (β) of this distribution (solid lines) is –2/3, consistent with a b value of 1. The values are shown both as a cumulative curve for the number of earthquakes per year with log M_0 greater than or equal to a certain value, and as incremental values in 0.1 log M_0 bins.

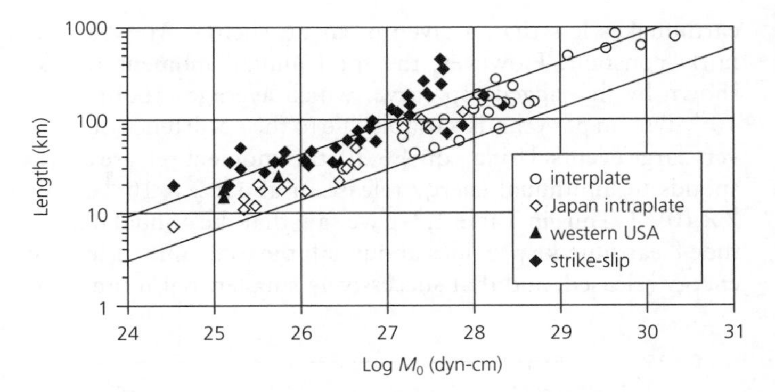

Fig. 4.7-3 Log–log plot of fault length versus seismic moment. Most earthquakes fall between the solid lines with slopes of 1/3, showing M_0 proportional to L^3. However, strike-slip earthquakes (solid diamonds) have moments lower than expected for their fault lengths, because above a certain moment the fault width reaches a maximum, so the fault grows only in length. (Romanowicz, 1992. *Geophys. Res. Lett.*, *19*, 481–4, copyright by the American Geophysical Union.)

this effect, although the linear fit is better for the cumulative values because the numbers are larger, and hence less affected by time sampling. Using more earthquakes by sampling longer intervals and/or larger areas produces better fits. Conversely, the shorter the time or the smaller the area, the more the fit is degraded by the statistics of small numbers, as discussed shortly.

Although the data in Fig. 4.7-1 are generally well described by the linear relation, there are deviations. The data deviate from the $b = 1$ line for very small ($M_s < 3$) magnitudes, because the global earthquake catalog is incomplete, with many small earthquakes not detected. The deviation for large ($M_s > 7.5$) earthquakes is expected, because the surface wave magnitude saturates (Fig. 4.6-6). To address this issue we can use the seismic moment, which better indicates the size of large earthquakes. Using the definition of moment magnitude (Eqn 4.6.5) in Eqn 1 yields

$$\log N = a_1 - b(\log M_0/1.5 - 10.73) = \alpha - \beta \log M_0. \quad (4)$$

This linear relation, with slope $\beta = b/1.5 \approx 2/3$, is shown in Fig. 4.7-2 for global earthquakes. The equation can also be written in an incremental form analogous to Eqn 3.

The data in Fig. 4.7-2 deviate from the linear frequency–moment relation at large and small moments, just as the frequency–magnitude data did. The deviation for small earthquakes is likely in part to be due to the incomplete earthquake catalog, but is also expected from energy considerations, as discussed in problem 20. However, the deviation at large moments is more puzzling, since moments do not saturate. For moments above 10^{27} dyn-cm, the data are more consistent with $\beta \geq 1$ than $\beta = 2/3$. In other words, there are fewer earthquakes than expected for a given moment.

A model for this phenomenon based on the concept of scale invariance assumes that the probability of an earthquake of a given size on a fault is inversely proportional to the area of faulting involved, so the number N of earthquakes with fault area greater than S should obey a frequency–area relation like those for magnitude or moment

$$\log N = c - \log S. \quad (5)$$

We saw in Eqn 4.6.17 that for constant stress drop the moment is proportional to $S^{3/2}$, or the fault dimension L cubed, so we expect

$$\log N = c - 2/3 \log M_0, \quad (6)$$

which is consistent with the observations showing $\beta \approx 2/3$. However, we have seen that for large transform fault earthquakes, which occur on vertical faults, the width (down-dip extent) stays narrow even as fault length increases (Fig. 4.6-3). As a result, the seismic moment for such earthquakes is no longer proportional to L^3, and is smaller than for other earthquakes of comparable fault length, as shown in Fig. 4.7-3. If both the fault slip and the fault width no longer increase with length, then the fault area, moment, and number of earthquakes should be proportional to L, so by Eqns 4 and 5 we find that $\beta = 1$. Such an increase is suggested by the data for the largest earthquakes in Fig. 4.7-2.

The frequency–moment data give insight into earthquake energy release because the radiated energy is proportional to the seismic moment (Eqn 4.6.30). The few largest earthquakes release much more energy than the many smaller earthquakes. In fact, the largest earthquake in a given year often releases more energy than the rest of the year's earthquakes. This effect is illustrated in Fig. 4.7-4, which shows the cumulative seismic moment release since 1976. The annual moment release by earthquakes less than a given moment, such as $M_w < 7.5$, is fairly constant. However, the total annual moment release shown by the jagged top curve, which averages about 3.5×10^{28} dyn-cm per year, is variable due to the occurrence of a few very large events. Using Eqn 4.6.30, this moment release corresponds to an annual energy release of about 2×10^{24} erg or 2×10^{17} J. Thus in Table 1.2-1 we saw that the annual magnitude 8 earthquake provides about half the total annual seismic energy released, and that successively smaller, but more common, earthquakes contribute less, so the contribution of earthquakes with magnitudes less than 6 is negligible.

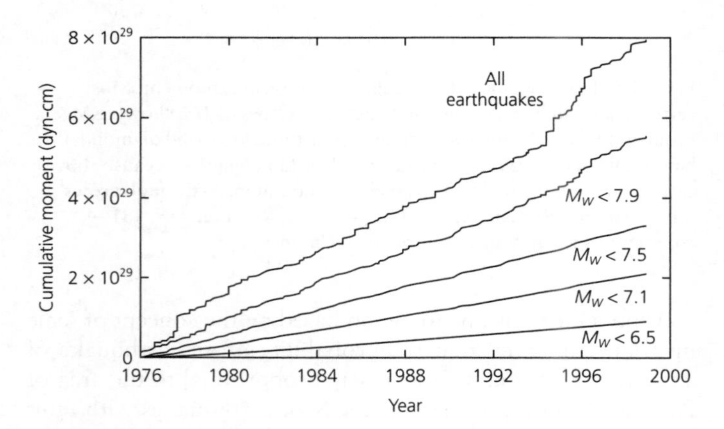

Fig. 4.7-4 Cumulative seismic moment for the earthquakes in Fig. 4.7-2. The total global seismic moment release is dominated by the few largest events. The total moment for 1976–98 is about 1/3 that of the giant 1960 Chilean earthquake.

Although b values approximately equal 1 over long time scales and large spatial scales, significant variations occur on smaller scales. The b value of *earthquake swarms* is often much larger than 1, sometimes approaching 2.5. These swarms, which lack a mainshock, are often associated with volcanic regions, and may result from processes such as the migration of magmatic fluids or caldera development. For example, seismicity associated with the collapse of the Fernandina caldera in the Galapagos Islands in 1968 had $b \approx 1.9$, indicating many small earthquakes but fewer large ones than expected.

The b value also varies regionally, both spatially and with depth. Figure 4.7-5 shows the variation in b value on a segment of the Calavaras fault in California. Some patches have b values much less than 1, implying shorter recurrence time. These patches have been interpreted as possible asperities or stress concentrations, perhaps reflecting variations in frictional properties along the fault, which may control the recurrence of the next large earthquake and have large moment release during it.

Other intriguing possible deviations from $b = 1$ have been reported. Figure 4.7-6 (*left*) shows earthquake magnitudes and frequencies for large earthquakes inferred from geological paleoseismic studies, which deviate from the seismologically determined frequency–magnitude data. These observations have been interpreted as showing large (sometimes termed characteristic) earthquakes more frequent than would be expected from the linear relation derived from the instrumental data,

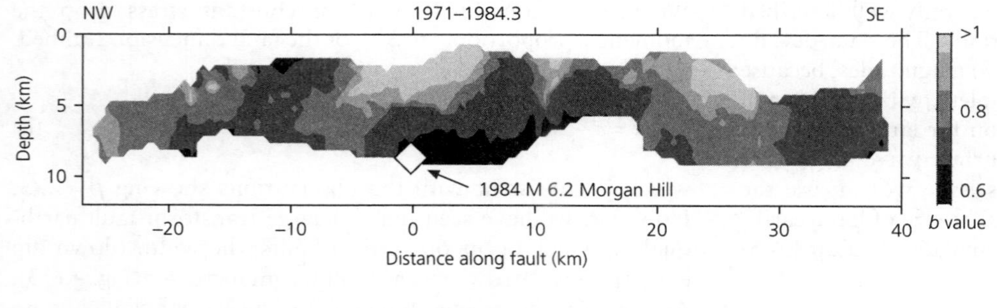

Fig. 4.7-5 Variation of b values for small earthquakes with depth and distance along the Morgan Hill segment of the Calaveras fault during 1971–84. Regions with low b values may have a shorter time until the next large earthquake. The 1984 Morgan Hill earthquake occurred in a region of low b values. (Wiemer and Wyss, 1997. *J. Geophys. Res., 102*, 15, 115–28, copyright by the American Geophysical Union.)

Fig. 4.7-6 Deviations from a linear frequency–magnitude relation. *Left*: Paleoseismic results (box) for the Wasatch fault zone (Utah) showing large earthquakes more frequent than expected from the instrumental seismicity (dots). The solid line is a model for this effect. (Youngs and Coppersmith, 1985. © Seismological Society of America. All rights reserved.) *Right*: Incremental frequency–magnitude data for instrumentally studied earthquakes in continental interiors, showing larger earthquakes less frequent than expected from the smaller earthquakes. (Triep and Sykes, 1997. *J. Geophys. Res., 102*, 9923–48, copyright by the American Geophysical Union.)

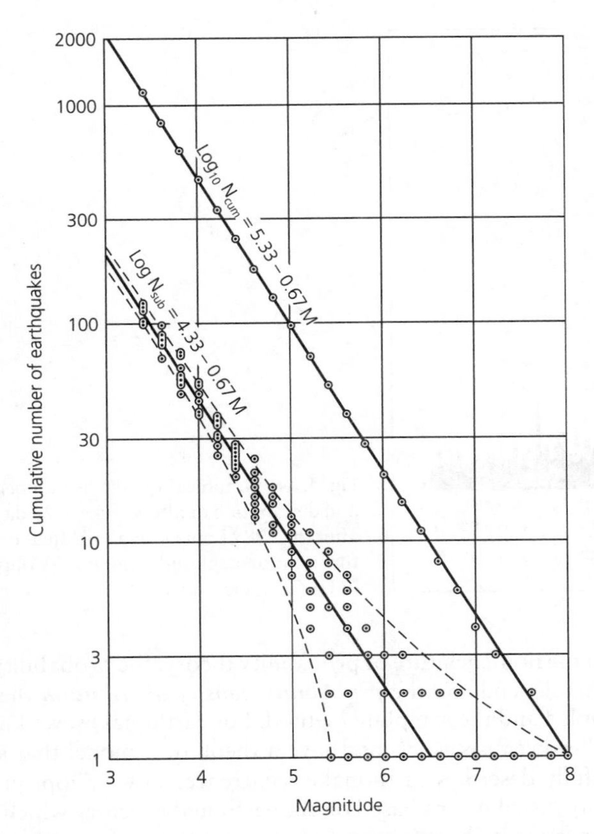

Fig. 4.7-7 Numerical simulation showing apparent deviations from a linear frequency–magnitude relation resulting from small sample size. When data satisfying a linear relation (upper solid line) are divided into subsets, most subsets have largest earthquakes either above or below the ideal linear relation for the subsets (lower solid line). (Howell, 1985.)

although the opposite has also been observed (Fig. 4.7-6, *right*). It is not yet clear whether these effects are real or due to differences in magnitude and frequency estimation between seismological and geological approaches. A study using seismological data for continental interiors finds that the largest earthquakes are less frequent than expected from the smaller earthquakes (Fig. 4.7-6, *right*). These observations are interpreted as showing a possible small deviation toward higher frequency at about M_w 7, followed by a significant decrease as observed in the global data (Fig. 4.7-1), presumably due to finite fault width. The Gutenberg–Richter relation can be modified to describe the different deviations from linearity.

Some deviations of the largest earthquakes in an area from a linear frequency–magnitude relation may reflect small sampling (Section 6.5.2). Figure 4.7-7 illustrates this effect by dividing an earthquake population that follows a Gutenberg–Richter distribution into ten subsets. Because only one subset contains the largest earthquake, and some have a much smaller largest earthquake, the frequency–magnitude relations for the subsets have considerable scatter. The largest earthquakes appear in some cases more and in other cases less frequent than for the total population. Thus the *b* value is reasonably well estimated from the smaller earthquakes, but not the largest ones. This effect may occur if the number of large earthquakes in the study is small. For example, we might expect this for individual faults, even in a region whose overall seismicity obeys a Gutenberg–Richter distribution.

A final point worth noting is that although the Gutenberg–Richter distribution predicts the frequency of arbitrarily large earthquakes, such earthquakes may not actually occur. As we have seen, the area available for faulting limits earthquake size. For example, we will see in Section 5.3.3 that the maximum moment of mid-ocean ridge earthquakes varies inversely with spreading rate. As a result, regional studies often assume the existence of a maximum magnitude, sometimes based on the earthquake history, in the Gutenberg–Richter distribution. This assumption has interesting implications, because on many plate boundaries the motion inferred from earthquake histories seems to be significantly less than the plate motion (Sections 5.3.3, 5.4.3, 5.6.2). Hence, either the missing motion occurs in very large, rare earthquakes, or much of the motion occurs by aseismic processes. Which of these is the case is also of interest for seismic hazard studies.

4.7.2 *Aftershocks*

The smaller *aftershocks* following a *mainshock* have a characteristic distribution in size and time. As previously noted (e.g., Fig. 4.5-9), most aftershocks occur on or near the mainshock's fault plane, so their locations are used to distinguish between the fault and auxiliary planes and to estimate the fault area. The largest aftershock is usually more than a magnitude unit smaller than the mainshock, and the aftershocks have a size distribution with *b* near 1, so the total energy released by the aftershocks is usually less than 10% of that of the mainshock.

Most of the aftershocks occur soon after the mainshock, and the remainder decay with time in a quasi-hyperbolic manner. This decay is described by a relation now called *Omori's law*,[3]

$$n = \frac{C}{(K + t)^P}, \tag{7}$$

where n is the frequency of aftershocks at a time t after the mainshock, with K, C, and P as fault-dependent constants. P is typically about 1. This decay is illustrated by the aftershocks of the 1989 (M_s 7.1) Loma Prieta earthquake (Fig. 4.7-8).

The aftershock decay is thought to reflect stress readjustment following the stress changes due to the main shock. An intriguing exception to Omori's law is that most deep earthquakes have many fewer, and often no, detected aftershocks. This difference may reflect deep earthquakes resulting from phase

[3] Fusakichi Omori (1868–1923), considered the founder of Japanese seismology, participated in the commission that studied the 1906 San Francisco earthquake (Section 4.1) and correctly assured worried citizens that no comparable earthquake would be expected for at least the next 50 years.

Magnitude	Number	Effect
5	2	Damaging
4	20	Strong
3	65	Perceptible
2	384	Not felt
1	1855	Not felt
<1	2434	Not felt
Total	4760	

4760 aftershocks of the Loma Prieta earthquake had been recorded by noon on November 7, 1989. The diminishing number of aftershocks with time is typical for large California earthquakes.

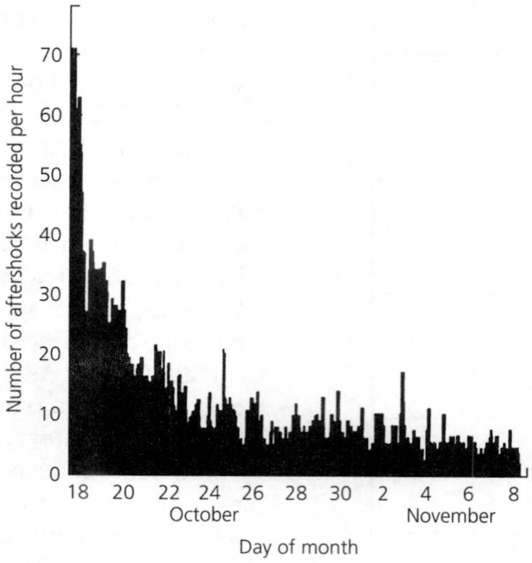

Fig. 4.7-8 Graphic showing the number and distribution of aftershocks in 22 days after the 1989 Loma Prieta earthquake as functions of magnitude and time. (Courtesy of US Geological Survey.)

changes in mantle minerals (Section 5.4.2), which could produce slip only once on a fault surface, in contrast to frictional sliding, which can recur.

4.7.3 *Earthquake probabilities*

A natural use of earthquake statistics is to estimate the probability of future earthquakes. These probabilities are interesting from the standpoint of earthquake physics, and crucial for attempts to forecast the hazards due to large, damaging earthquakes (Section 1.2.5).

The challenge of estimating earthquake probabilities can be illustrated by a simple analogy. Problems in probability are often couched as games of chance, but earthquakes have the special feature that the game's rules are unknown. To see this, consider estimating the probability that particular playing cards will be dealt from a deck. If the game begins with a full deck, there is a 25% (13/52) chance of drawing a spade, an 8% (4/52) chance of an ace, and a 2% (1/52) chance of the ace of spades. These chances are analogous to the prospects of having a magnitude 6, 7, or 8 earthquake in a year. As play continues, there are several possible cases. If the deck is shuffled after every draw, the probabilities do not change. Alternatively, if the deck is not shuffled, the probabilities change depending on the cards that have been drawn. For example, if no aces have yet appeared, the probability of an ace increases with each draw. However, if cards are dealt from under the table, we do not know what cards the deck began with (there may be no aces or eight of them) and whether it is shuffled. We must infer what the deck contains, how it is shuffled, and what cards will appear, with no information except the cards already drawn. Hence if no aces have appeared after a large number of draws, the probability of an ace may be high (because the remaining cards contain several) or low (because the starting deck had few).

In the nomenclature of probability theory, the probability of events depends on the *probability density distribution* that is sampled and the sampling method. For earthquakes, we know neither because we do not have a theoretical model that successfully describes earthquake recurrence, so we adopt probability distributions based on the earthquake history which for most faults is short (only a few recurrences) and complicated. As a result, various distributions grossly consistent with the limited history are used and can produce quite different estimates.

The simplest model describes earthquake occurrence by a *Poisson distribution* often used to describe rare events.[4] We assume that the probability of n large earthquakes in an area or on a fault during time t is

$$p(n, t, \tau) = (t/\tau)^n e^{-t/\tau}/n!, \tag{8}$$

where $1/\tau$ is the number expected in a year from the regional Gutenberg–Richter distribution or some variant, so τ is the mean recurrence time. The probability of one or more earthquakes is found from the probability that none will happen, using the certainty ($p = 1$) that an earthquake either will or will not happen, so

$$p(n \geq 1, t, \tau) = 1 - p(0, t, \tau) = 1 - e^{-t/\tau} \approx t/\tau, \tag{9}$$

where the last step used the Taylor series expansion $e^x \approx 1 - x$, and so is valid for $t \ll \tau$. In this model, the probability that an earthquake will occur in an interval of time t starting from now does not depend on when "now" is, because a Poisson process has no "memory." On average, earthquakes are separated by time τ, but when the last earthquake occurred has no effect.

[4] Examples include volcanic eruptions, radioactive decay, and the number of Prussian soldiers killed by their horses.

The Poisson model is the simplest null hypothesis against which we can compare other models. However, its time-independence in which earthquakes are implicitly random events is not appealing, because almost all of our seismological instincts favor earthquake cycle models, in which strain builds up slowly from one major earthquake to the next.[5] In this case, the probability of a large earthquake should be small immediately after a large earthquake, and then grow with time. This is described by time-dependent models in which the probability of a large earthquake a time t after the past one is given by a probability density distribution $p(t, \tau, \sigma)$ that depends on the average and variability of the recurrence times, described by the mean τ and the standard deviation σ. In other words, p gives the probability that the recurrence time for this earthquake will be t, given an assumed distribution of recurrence times. The cumulative probability that the earthquake will occur by time T since the past earthquake is found by integrating the density function

$$P(T) = \int_0^T p(t, \tau, \sigma)dt. \qquad (10)$$

We seek to estimate how likely an earthquake is between now and some future time. Formally, this is the conditional probability that the earthquake will occur between time T_0 (now) and a future time T, given the condition that it has not yet happened by time T_0. To do this, we use *Bayes's theorem*, which states that $P(A|B)$, the conditional probability of event A given that event B has occurred, is the ratio of the joint probability $P(A, B)$ of both A and B to $P(B)$, the probability of event B:

$$P(A|B) = P(A, B)/P(B). \qquad (11)$$

In this case, the conditional probability $C(T, T_0)$ that the earthquake will occur between T_0 and T is the ratio of the probability that it will occur in that interval to the probability that it has not yet happened by T_0, which is just 1 minus the probability that it has. Hence

$$C(T, T_0) = (P(T) - P(T_0))/(1 - P(T_0)). \qquad (12)$$

The denominator is less than one, so the conditional probability is greater than the joint probability (numerator) because the fact that the earthquake has not happened makes it more likely.

This approach can be used with any assumed probability density function. The simplest is to assume that earthquake recurrence follows the familiar *Gaussian* or *normal* (bell curve) distribution (Section 6.5.1)

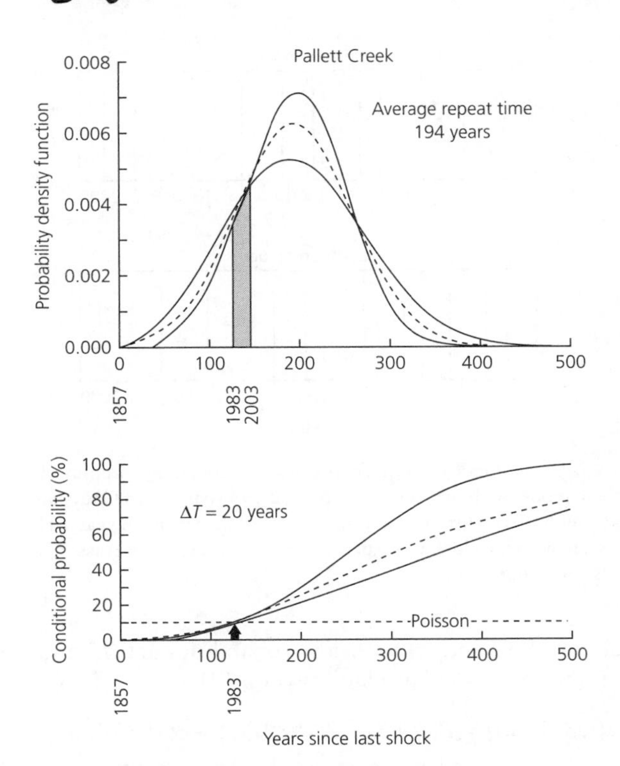

Fig. 4.7-9 Earthquake probability estimate for a segment of the San Andreas fault on which the last major earthquake occurred in 1857. *Top*: Probability density functions, with the interval 1983–2003 shaded. The dashed line is for a Gaussian distribution, with mean and standard deviations of 194 and 58 years, and the solid lines are for an alternative (Weibull) distribution. *Bottom*: Conditional probability that the next large earthquake will occur in the next 20 years, as a function of time since 1857. As of 1983 (arrow), the probabilities for the time-dependent models were comparable to those for a time-independent Poisson model. (Sykes and Nishenko, 1984. *J. Geophys. Res.*, *89*, 5905–27, copyright by the American Geophysical Union.)

$$p(t, \tau, \sigma) = \frac{1}{\sigma\sqrt{2\pi}} \exp\left[\frac{-1}{2}\left(\frac{t-\tau}{\sigma}\right)^2\right]. \qquad (13)$$

This distribution is often described using the normalized variable $z = (t - \tau)/\sigma$ describing how far, in terms of the standard deviation, t is from its mean.

Figure 4.7-9 shows such an analysis for the segment of the San Andreas fault including the Pallett Creek site (Fig. 1.2-15), on which the last major earthquake was the 1857 Fort Tejon earthquake. The analysis uses a Gaussian distribution with a mean and standard deviation of 194 and 58 years, corresponding to the most recent five major earthquakes. The upper panel shows the probability density function for this distribution (dashed line) and two others. These are used to estimate the conditional probability that a major earthquake would occur between 1983 (the study time) and 2003. These times are 126 and 146 years since 1857, and so correspond to normalized

[5] Of course, these instincts favoring determinism may ultimately prove incorrect — Einstein initially rejected quantum mechanics, arguing that "God does not play dice."

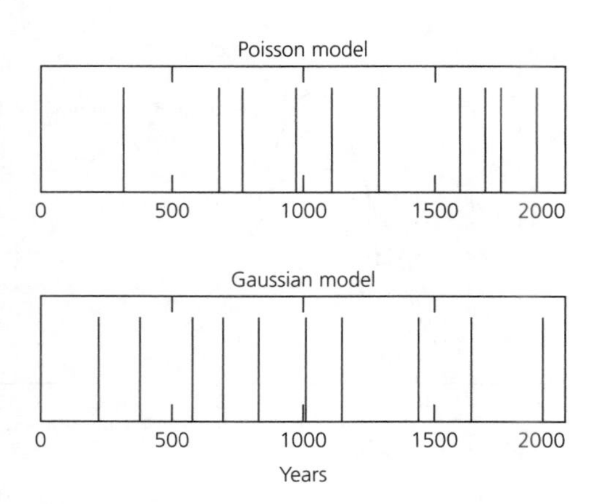

Fig. 4.7-10 Synthetic earthquake histories computed by sampling a Poisson model with the recurrence time of 194 years and a Gaussian model with this recurrence time and standard deviation 58 years. The Gaussian model yields a more periodic series, whereas the Poisson model yields clustering.

times of −1.17 and −0.83, with probabilities of 0.12 and 0.20. Thus the conditional probability (Eqn 12) is

$$C(2003, 1983) = (P(2003) - P(1983))/(1 - P(1983)) = (0.20 - 0.12)/(1.0 - 0.12) = 0.09, \quad (14)$$

or 9%. The probability for successive 20-year intervals increases with time, and so is 29% if the earthquake has not occurred by 2057, and 56% if it has not occurred by 2157.

It is interesting to compare these time-dependent probabilities to those predicted by the time-independent Poisson model. For an assumed mean recurrence time of 194 years, the probability in 20 years is 10%. Thus for times since the previous earthquake less than about 2/3 of the assumed recurrence interval, the Poisson model predicts higher probabilities. At about 2/3 of the interval, in this case about 1986, the models predict comparable probabilities. At later times the Gaussian model predicts progressively greater probabilities. This comparison illustrates the seismic gap concept: a gap exists when it has been long enough since the last major earthquake that time-dependent models predict an earthquake probability much higher than expected from time-independent models.

The differences between the models can be illustrated by comparing the earthquake histories that each predicts. Figure 4.7-10 shows synthetic earthquake histories generated by randomly sampling probability distributions with the parameters used in Fig. 4.7-9. In the simulation, both models yield ten earthquakes after an earthquake at time zero. The earthquakes from the Poisson model have a mean recurrence of 189 years and a standard deviation of 107 years, whereas those for the Gaussian model have a mean and standard deviation of 191 and 58 years, respectively. The difference results from the fact that the Poisson process is time-independent, so there are both shorter and longer intervals between earthquakes than for the Gaussian process, which is more regular. The Poisson process thus shows clustered earthquakes resulting from the random sampling. In the limit of very long histories, the Poisson process has a standard deviation of recurrence intervals equal to its mean. Thus a recurrence history with standard deviation close to the mean favors a Poisson process, whereas a standard deviation significantly smaller than the mean suggests a Gaussian or other time-dependent process. How to interpret the limited earthquake histories available is an interesting question, as illustrated by this simple example with ten recurrences, which is longer than usually available.

These examples bear out that estimates of earthquake probabilities depend significantly on both the probability distribution used and the parameters for that distribution, which are generally not well constrained by observations. For example, the analysis in Fig. 4.7-9 used a Gaussian distribution with a mean and standard deviation of 194 and 58 years, corresponding to the most recent five major earthquakes at Pallett Creek. Alternatively, the past ten earthquakes there yield a recurrence with a mean and standard deviation of 132 and 105 years (Section 1.2.5). Other probability distributions give different probability estimates, as illustrated by the curves in Fig. 4.7-9 corresponding to Poisson and Weibull distributions. Similarly, different estimates would result from using a log-normal distribution in which the natural logarithm of recurrence time is normally distributed, so recurrence intervals longer than the mean are more likely than shorter ones.

Hence earthquake forecasts are easy to make, but hard to test. Because the estimates must be tested using data that were not used to derive them, hundreds or thousands of years

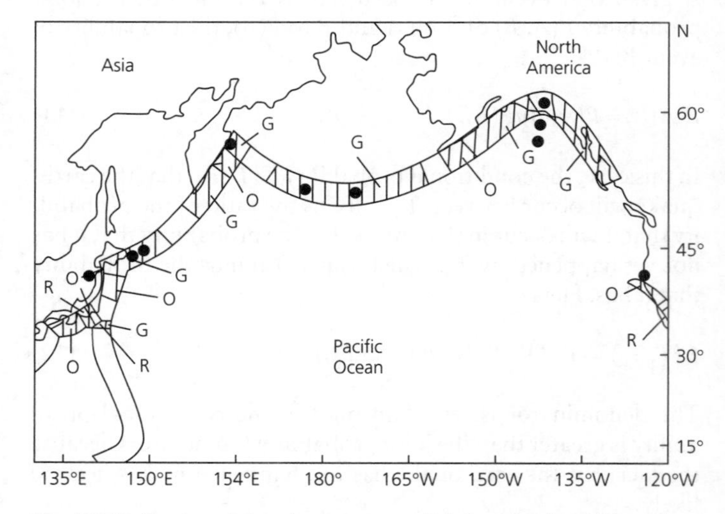

Fig. 4.7-11 Portion of the seismic gap map (McCann *et al.*, 1979) used by Kagan and Jackson (1991) to test the gap hypothesis. The shaded segments of the plate boundaries had been assigned seismic potentials of high (red, R), intermediate (orange, O), and low (green, G). Unshaded segments were regarded as having uncertain potential. During the ten years following the map's publication, ten large (M > 7) earthquakes (dots) occurred in these regions. None were in the high- or intermediate-risk segments, and five were in the low-risk segments. (Stein, 1992. Reproduced with permission from *Nature*.)

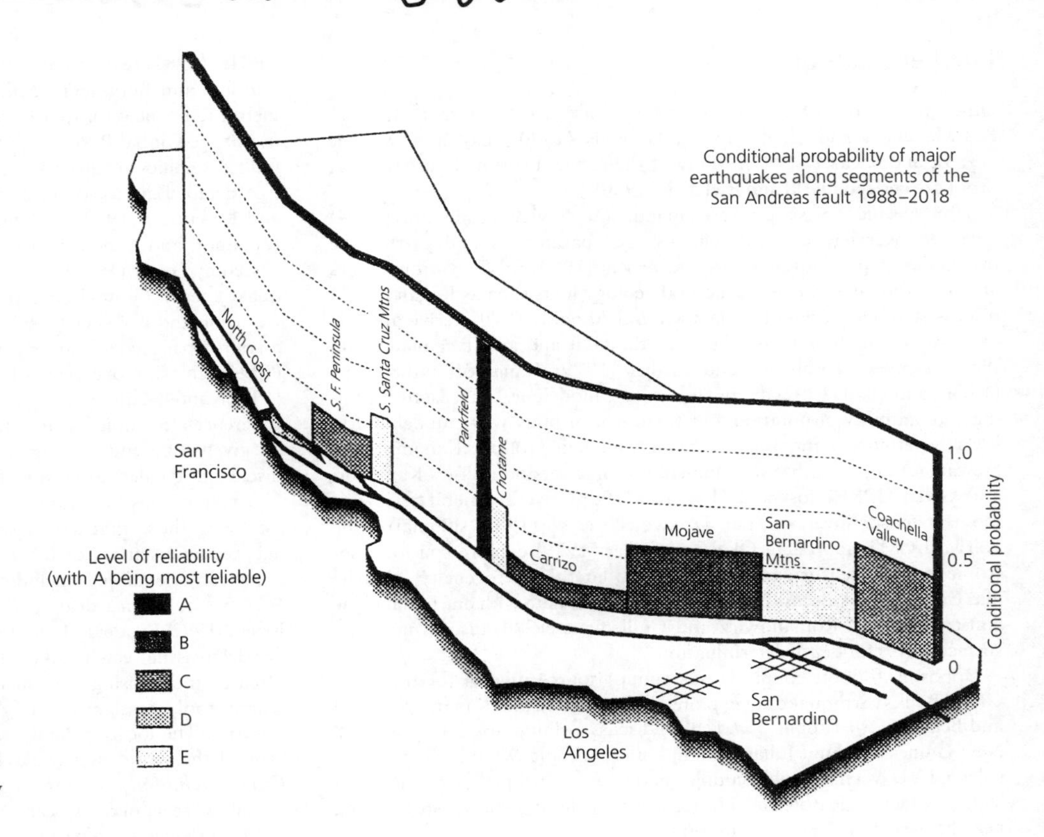

Fig. 4.7-12 Conditional probabilities of major earthquakes estimated for segments of the San Andreas fault for the period 1988–2018. (Agnew *et al.*, 1988. Courtesy of the US Geological Survey.)

(multiple recurrences) will be needed to assess how well various models predict large earthquakes on specific faults or fault segments. The first challenge is to show that a model predicts future earthquakes significantly better than the simple time-independent Poissonian model.

Given human impatience, attempts have been made to conduct alternative tests using smaller earthquakes or many faults over a short time interval. To date, the results are not encouraging. As discussed in Section 1.2.5, the history of relatively small (M 5–6) earthquakes near Parkfield, California, was used in 1985 to predict at 95% confidence level that the next one would occur by 1993, whereas the earthquake has not materialized to date (2002). Presumably the earthquake will occur eventually, although its conditional probability seems to have been overestimated and might even be assumed to be decreasing, because the longer the earthquake is delayed, the longer the mean recurrence interval inferred from the earthquake history becomes.[6] Moreover, a global test of the seismic gap hypothesis, which examined how well a gap map (Fig. 4.7-11) forecast the locations of major earthquakes, found that the map did no better than random guessing. In fact, many more large earthquakes occurred in areas identified as low risk than in the presumed higher-risk gaps. This result, which appears inconsistent with ideas of earthquake cycles and seismic gaps, has led to various interpretations, including that the gap model applies only to the largest events that break major portions of the plate boundary.

Perhaps the most sophisticated large-scale earthquake probability studies have been in California. Figure 4.7-12 shows conditional probabilities estimated along segments of the San Andreas fault. Such models can also include factors such as variable slip in earthquakes and stress changes due to nearby earthquakes (Section 5.7). Testing more complicated models with more adjustable parameters, however, will be even more challenging and take even longer.

Hence, at present, estimates of earthquake probabilities have large uncertainties. For example, using the complex Pallett Creek earthquake series (Fig. 1.2-15), in 1989 the range of probabilities for a major earthquake before 2019 was estimated as about 7–51%.[7] Thus it has been suggested that it is only meaningful to quote probabilities in broad ranges, such as low (<10%), intermediate (10–90%), or high (>90%).[8] However, despite these formidable difficulties, estimation of earthquake probabilities seems certain to remain an active research area. If some probability model is ultimately demonstrated to be reasonably successful, its use could advance efforts to estimate earthquake hazards.

[6] This situation, discussed by Davis *et al.* (1989), has been likened to waiting for a bus — the longer the bus fails to arrive, the less likely its arrival seems. A homework problem illustrates these issues.

[7] Sieh *et al.* (1989).

[8] Savage (1991).

Further reading

Other treatments of earthquake sources are given by texts such as Ben-Menahem and Singh (1981), Gubbins (1990), Lay and Wallace (1995), and Shearer (1999). Many of the results presented here without proof are derived in Aki and Richards (1980).

Some specific topics are covered in individual reviews. Kanamori (1994) gives an overview of earthquake source parameters and earthquake mechanics; papers in Kanamori and Boschi (1983) review various topics about earthquake sources. Structural geology texts such as Ragan (1968) discuss stereonet techniques. Jarosch and Aboodi (1970) derive analytic expressions for the relations between the fault and auxiliary planes and the stress axes. Helmberger and Burdick (1979), Kanamori and Stewart (1976), and Okal (1992) discuss body wave modeling. For reasons including compatibility of notation, our treatment of body wave modeling follows the latter two, that for surface wave modeling follows Kanamori and Stewart (1976), and that for moment tensor inversion follows Kanamori and Given (1981). Jost and Hermann (1989) give a general review of moment tensor inversion, and Dziewonski *et al.* (1981) summarize the Harvard CMT approach. Okal and Geller (1979) explore spurious isotropic moment tensor components due to lateral heterogeneity, Michael and Geller (1984) discuss inverting surface wave data with one nodal plane constrained, and Romanowicz and Guillemant (1984) discuss inverting surface waves for depth determination.

Opposing (double-couple versus slump) source models for the 1929 Grand Banks earthquake are explored by Hasegawa and Kanamori (1987) and Bent (1995); Tappin *et al.* (1999) discuss a slump origin for the 1998 New Guinea tsunami. Julian and Sipkin (1985) and Wallace (1985) consider CLVD versus double-couple models for earthquakes in the Long Valley caldera. Heaton and Hartzell (1988) discuss source study using near-field earthquake ground motions.

General treatments of geodesy include those by Lambeck (1988) and Torge (1991). Geodetic solutions for faults are given by Okada (1985). Mavko (1981) reviews fault models and the use of geodetic data to study faulting, and Burgmann *et al.* (2000) review the use of radar interferometry. References for topics related to the tectonic setting of earthquakes, use of the Global Positioning System, and the relation of earthquakes to fault mechanics are given at the end of Chapter 5.

A detailed discussion of earthquake magnitudes is presented by Geller and Kanamori (1977). Relations between fault parameters are given in Kanamori and Anderson (1975); source spectra and scaling laws are discussed by Geller (1976). Our treatment of moment magnitude and earthquake energy follows Kanamori (1977a) and Hanks and Kanamori (1979). Atkinson and Beresnev (1997) discuss the relation between stress drop as a source parameter and as a tectonic quantity. Okal and Romanowicz (1994) give an overview of frequency–magnitude relations. Turcotte (1992) and Main (1996) review self-similar models for earthquakes. References for topics related to earthquake forecasting and seismic gaps are given at the end of Chapter 1. In particular, Kagan and Jackson (1991) discuss the challenge of testing forecasts.

A voluminous literature deals with studies of individual earthquakes, especially those that are of special interest because of their size, damage, tectonic setting, or location near centers of seismological research. Some recent examples include issues of the *Bulletin of the Seismological Society of America* dealing with the 1989 Loma Prieta (October 1991 issue), 1992 Landers (June 1994 issue), and 1994 Northridge (February 1996 issue) earthquakes. Detailed studies of other earthquakes can often be found using the American Geological Institute's *Georef* WWW search tool, available through many earth science departments and libraries. The locations and focal mechanisms of post-1977 earthquakes around the world are available at *http://www.seismology.harvard.edu/CMTsearch.html*, and information about earthquakes in specific areas, including seismograms, can often be found at WWW sites compiled at *http://www.geophys.washington.edu/seismosurfing.html* or *http://www.iris.edu*.

Problems

1. Using the travel time chart in Fig. 3.5-4 for earthquakes at a depth of 600 km, graph the take-off angle of the *P* wave for stations at distances from 2000 to 10,000 km. Assume that the *P* velocity at 600 km depth is 10 km/s. Use enough points for a smooth graph.
2. Plot the following focal mechanisms on a stereonet by using the relations in Section 4.2.5 to find the second nodal plane. Indicate the compressional and dilatational quadrants, mark the P and T axes, and describe the type of faulting. Use the conventions of Fig. 4.2-2, and remember that dip is defined from the $-x_2$ axis and is less than 90°.
   (a) $\phi=330°, \delta=65°, \lambda=70°$
   (b) $\phi=280°, \delta=60°, \lambda=270°$
   (c) $\phi=280°, \delta=60°, \lambda=90°$
   (d) $\phi=40°, \delta=80°, \lambda=20°$
   (e) $\phi=40°, \delta=80°, \lambda=200°$
3. Figure P4.1 gives a stereonet and first motion data for four earthquakes on stereonets of the same scale. Closed circles show compressions, and open circles show dilatations. To evaluate the focal mechanism for each earthquake:
   (i) Find nodal planes that you consider the best solution. Show these planes on the first motion plots, and measure their strikes and dips.
   (ii) Find two planes bounding the acceptable range for each nodal plane.
   (iii) For each best choice nodal plane, give the motion (right-lateral strike-slip, left-lateral strike-slip, dip-slip – thrust or normal) implied by the focal mechanism for slip on that nodal plane. If the faulting is a combination of the above, give the dominant type.
   (iv) Find the *B*, *P*, and *T* axes and the two possible slip angles (one for each nodal plane) implied by the best choice nodal planes. Check that these are consistent with the answers to part iii.
4. If a *P* wave leaves the focal sphere exactly on a nodal plane, it should theoretically have zero amplitude. Explain why this is not the case in reality.
5. Derive the travel time for *sP* (Eqn 4.3.11) using a geometry similar to that of Fig. 4.3-6 (*bottom*).
6. For a fault plane solution in which one plane has strike ϕ_1 and dip δ_1, and the second plane is striking at ϕ_2, show that $\tan \lambda_1 = \cot(\phi_2 - \phi_1)/\cos \delta_1$. For what angles will this not apply?
7. Compute the Love wave amplitude radiation pattern for an isotropic source.
8. Use the expression for the moment tensor of a double couple (Eqn 4.4.5) to prove that it obeys the tensor transformation law (Eqn 2.3.18).
9. (a) Show how the moment tensor for a vertical dipole can be decomposed into an isotropic source and a CLVD.

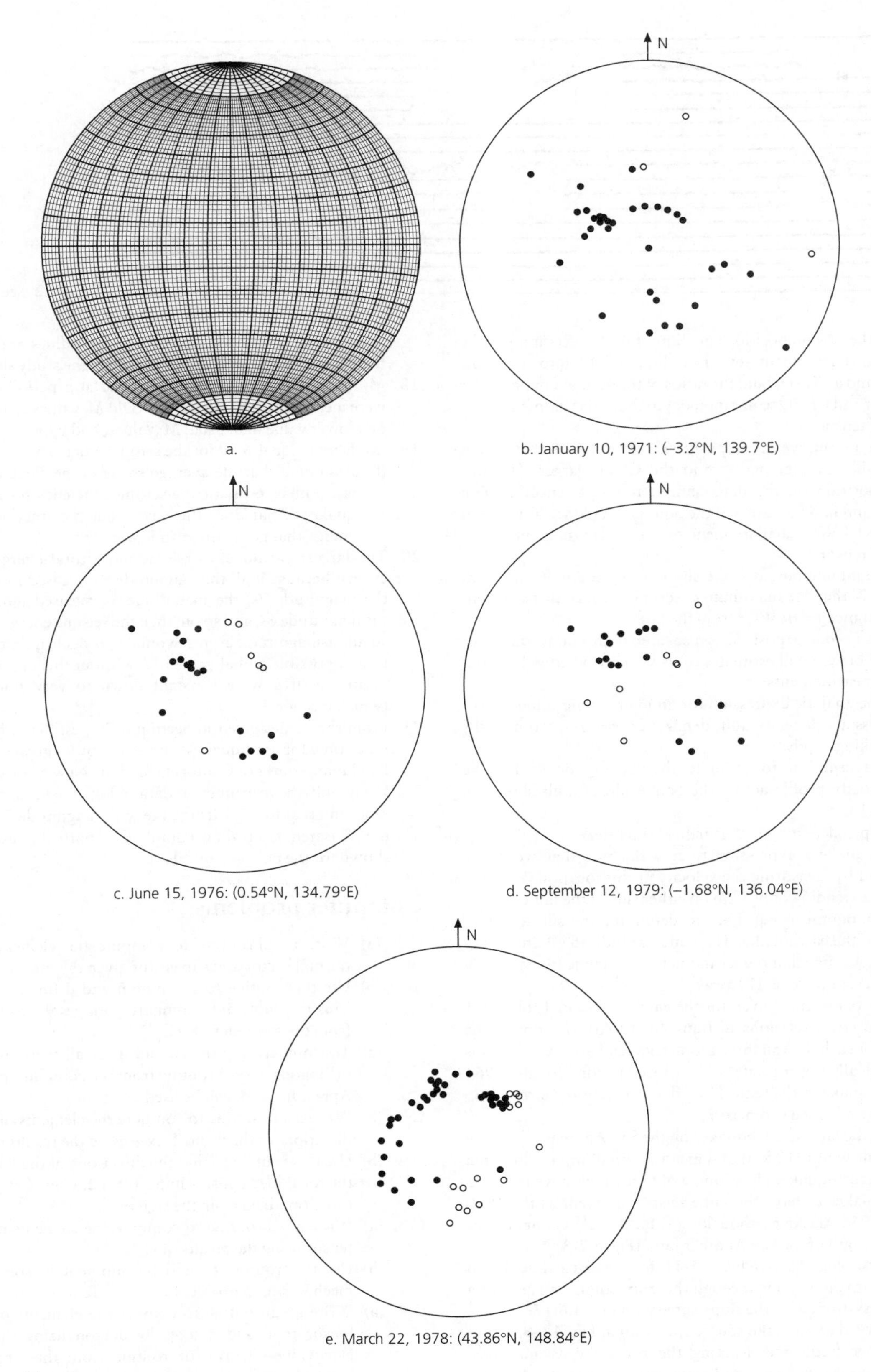

Fig. P4.1 See problem 3.

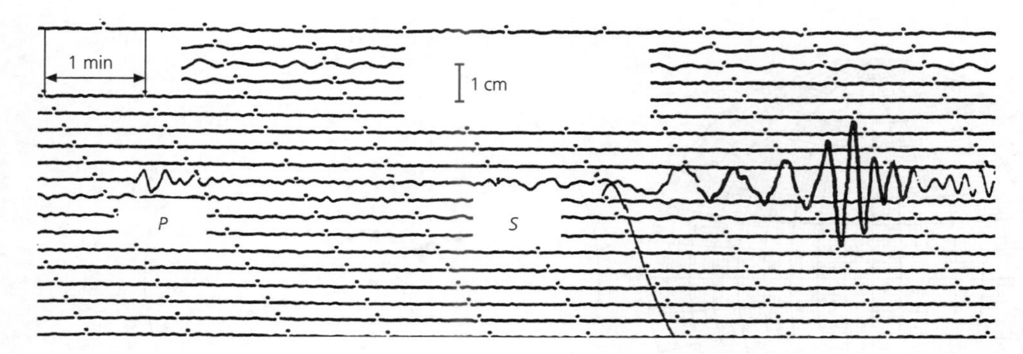

Fig. P4.2 See problem 13.

(b) Using the decomposition in Eqn 4.4.48, decompose the diagonalized moment tensor in Eqn 4.4.47 into a double couple and a CLVD. Find the ratios of the double-couple scalar moment and CLVD scalar moment to the scalar moment of the original tensor.

(c) Give an alternative decomposition to Eqn 4.4.48 that makes the double couple smaller and the CLVD larger. Use this decomposition on the diagonalized moment tensor in Eqn 4.4.47, and find the ratios of the double-couple scalar moment and the CLVD scalar moment to the scalar moment of the original tensor.

10. Show for an infinite buried strike-slip fault extending from depth w to depth W that the maximum coseismic surface displacement occurs at distance $y = (wW)^{1/2}$ from the fault.

11. Assume that a geodetic position is measured with an uncertainty of 3 mm. How precise will estimates of its velocity be after 1, 5, and 10 years of measurements?

12. (a) Using the analytic expression for an interseismic velocity profile across a strike-slip fault, define a criterion to estimate the fault locking depth.

(b) Use this criterion to estimate the locking depth for the GPS velocity profile across the San Andreas fault shown in Fig. 4.5-13.

(c) For this profile, estimate the far-field slip rate.

(d) Use the analytic expression to find the rate that would be estimated by measuring the velocity at this location, but on a baseline extending only 5 km on either side of the fault.

13. Use the seismogram in Fig. P4.2 to determine the surface wave magnitude of the earthquake. The scale bar indicates 1 cm on the seismogram. Assume that the seismometer's magnification is 3000, and that the earthquake is 17° away.

14. Use the fault parameters given for the earthquakes in Table 4.6-1 and the theoretical relations in Eqns 4.6.18–20 to estimate the stress drop for each. Use all three geometries, and note which seems most geologically appropriate. (Part of this is done for the 1964 Alaska earthquake in the text.) How does the inferred stress drop depend on the assumed geometry?

15. Assume that the largest earthquakes on the San Andreas fault have the same fault width (10 km) and average slip (4 m) as estimated for the 1906 earthquake. How long would the fault have to be for these earthquakes to have the same seismic moment as the 1960 Chilean or 1964 Alaska earthquakes (Table 4.6-1)? Compare this value to the length of the San Andreas fault (Fig. 5.2-3).

16. Plot log S versus log M_0, as in Fig. 4.6-11, for the six earthquakes in Table 4.6-1. If you fit a line through these six points and assume a constant stress drop, does the slope agree with Eqn 4.6.17?

17. For the observed earthquake source spectrum in Fig. 4.6-8, estimate the corner frequency. Making the necessary assumptions, estimate a source dimension and stress drop. Given the different assumptions and models possible, your values are likely to differ from the 30 km and 65 bars inferred by the study shown.

18. M_s magnitudes are usually measured at a period of 20 s. If they were measured at 30 s instead, would M_s values saturate at a higher or a lower value than usual M_s values, and why?

19. (a) Derive Eqn 4.6.29 for the seismic efficiency.

(b) Assuming that the average stress in the earth during faulting is 1.5 kbar, estimate the seismic efficiency for a typical earthquake? What does this say about the fraction of the strain energy that goes into seismic waves?

20. The largest earthquakes release more total energy than smaller events, because if all the magnitude 6s released more energy than the magnitude 7s, the magnitude 5s released more energy than the magnitude 6s, and so on, then the seismic energy released by the smallest-magnitude events would approach infinity. What is the largest possible global value of b without this impossible scenario occurring, if b were constant down to very small magnitudes (which it is not)?

21. From the values given in Section 4.7.1, estimate the mean recurrence time for earthquakes with magnitudes greater than 6, 7, and 8 in Japan, southern California, and the New Madrid seismic zone.

22. Using only the instrumental data in Fig. 4.7-6, estimate the recurrence interval for an earthquake with magnitude 7.5 or greater in the Wasatch fault zone (Utah). Compare this estimate to that shown for the paleoseismic data.

Computer problems

C-1. (a) Write a subroutine to compute the elements of a fault's normal vector and slip vector given the three fault angles.

(b) Use this routine to compute $\hat{\mathbf{n}}$ and $\hat{\mathbf{d}}$ for the focal mechanisms in problem 2. Compare your results to those obtained from the stereonet.

(c) Test numerically that $\hat{\mathbf{n}}$ and $\hat{\mathbf{d}}$ for all these mechanisms are orthogonal. A subroutine from the computer problems in the Appendix, C-4, can be used.

C-2. (a) Write a subroutine to compute the elements of vectors in the directions of the P and T axes using the results of C-1.

(b) Use this routine to find the directions of the P and T axes for the focal mechanisms in problem 2. Compare your results to those obtained from the stereonet.

C-3. (a) Write a subroutine to compute the elements of the moment tensor using the results of C-1.

(b) Use this routine to find the moment tensors for the focal mechanisms in problem 2.

C-4. (a) Write a subroutine to convert the elements of the moment tensor to P and T axes by diagonalizing the tensor. The eigenvalue–eigenvector routine from the Appendix, problem C-12, may be useful.

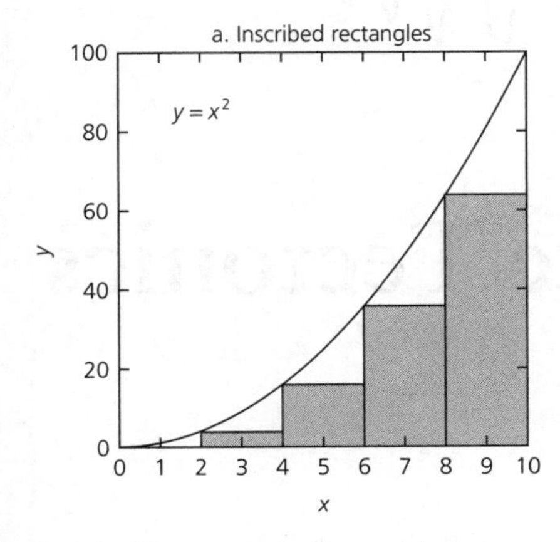

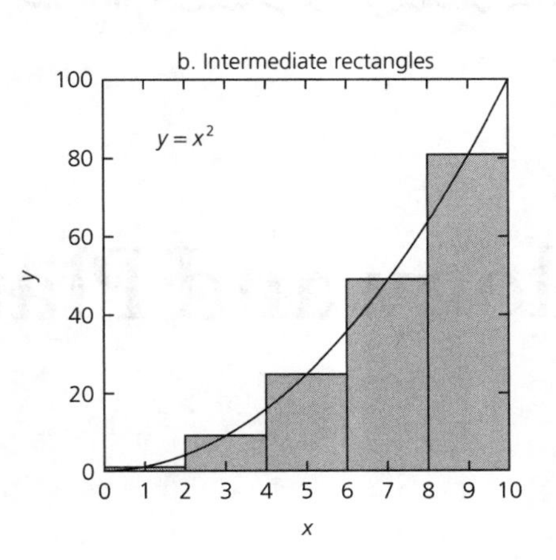

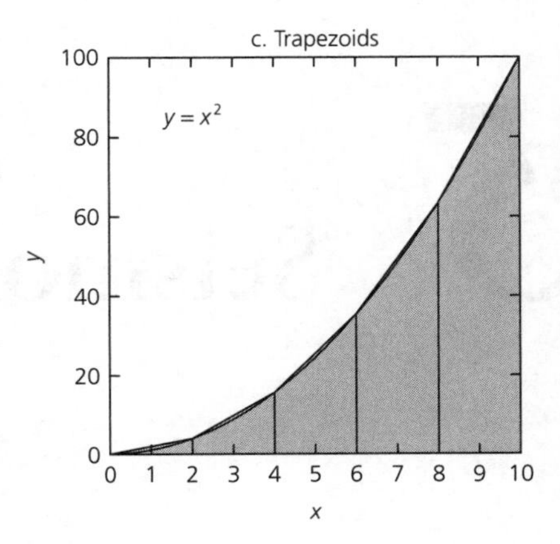

Fig. P4.3 See problem C-6.

(b) Use this routine to find the directions of the P and T axes for the focal mechanisms in problem 2. Compare your results to those obtained in C-2.

C-5. Write subroutines to generate the amplitude radiation patterns for Love and Rayleigh waves. Use these, with values of the excitation functions

$$P_L = -2.75, Q_L = -0.34, \quad \text{and} \quad S_R = 4.0, P_R = 2.7, Q_R = -1.6$$

to replicate the examples of Fig. 4.3-12.

C-6. Figure P4.3 shows three ways to evaluate integrals numerically. To see how these work:

(a) Analytically integrate the function $y = x^2$ over the interval $0 \le x \le 10$.

(b) Write a subroutine to numerically integrate this function using inscribed rectangles as in Fig. P4.3a. Try this with intervals of 2 (as shown) and 0.02. What is the percentage difference between these results and the true value in part (a)?

(c) Repeat (b) using intermediate rectangles, as shown in Fig. P4.3b.

(d) Repeat (b) using trapezoids, as in Fig. P4.3c.

C-7. (a) Write a subroutine that uses one of the methods in C-6 to integrate the Gaussian probability function $p(t, \tau, \sigma)$ (Eqn 4.7.13) over an interval from $-t$ to t.

(b) Use the subroutine to find the integral of $p(t, \tau, \sigma)$ (Eqn 4.7.13) over the interval $-10 \le t \le 10$ with $\tau = 0$ and $\sigma = 5$, and explain the result.

C-8. (a) Write a program to estimate the conditional probability, using Gaussian and Poisson models, that an earthquake will occur in a specified time interval, given the time of the last earthquake and the mean and standard deviations of the recurrence time. The routine in C-7 will be useful for the Gaussian model.

(b) Check the routine using the San Andreas example in Fig. 4.7-9 for 20-year periods beginning in 1983, 2057, and 2157.

(c) Calculate the values for the same periods, but using a mean recurrence of 132 years and a standard deviation of 105 years, which correspond to the full Pallett Creek earthquake series. Explain how and why the results change.

C-9. Use the routine from C-8 to estimate the Poisson and Gaussian conditional probabilities of a major earthquake in the New Madrid seismic zone in the next 20 years, assuming that the past one occurred in 1812. Assume that major earthquakes have:

(a) a mean recurrence time of 500 years with standard deviation 100 years.

(b) a mean recurrence time of 750 years with standard deviation 250 years.

(c) a mean recurrence time of 1000 years with standard deviation 500 years.

C-10. Write a subroutine (or set up a spreadsheet) to compute the mean and standard deviation of series of numbers.

C-11. By combining the results from C-8 and C-10:

(a) Find the mean and standard deviation of recurrence intervals for the series of Parkfield earthquakes that occurred in the years 1857, 1881, 1901, 1922, 1934, and 1966. Compute Poisson and Gaussian conditional probabilities starting in 1985 for an earthquake in the eight-year interval until 1993.

(b) Do the same calculation if the 1934 earthquake had occurred in 1944, as implicitly assumed when the prediction discussed in Section 1.2.5 was made. How do the values change and why?

(c) The awaited earthquake may or may not have occurred by the time you do this problem. In either event, assume that it has not occurred by 2010, and find the mean and standard deviation of the recurrence times from the dates in (a), also including the interval 1966–2010. Calculate the Poisson and Gaussian conditional probabilities that the earthquake will occur in eight years from 2010.

(d) Do the same assuming the earthquake has not occurred by 2020.

(e) Compare the results of (a), (c), and (d) and explain the differences.

5 Seismology and Plate Tectonics

The acceptance of continental drift has transformed the earth sciences from a group of rather unimaginative studies based on pedestrian interpretations of natural phenomena into a unified science that holds the promise of great intellectual and practical advances.

J. Tuzo Wilson, *Continents Adrift and Continental Aground*, 1976

5.1 Introduction

Two of the major advances in the earth sciences since the 1960s have been the growth of global seismology and the development of our understanding of global plate tectonics. The two are closely intertwined because seismological advances provided some of the crucial data that make plate tectonics the conceptual framework used to think about large-scale processes in the solid earth.

The theory of plate tectonics grew out of the earlier theory of continental drift, proposed in its modern form by Alfred Wegener in 1915. The idea that continents drifted apart was an old one, rooted in the remarkable fit of the coasts of South America and Africa. Still, without compelling evidence for motion between continents, the idea that such motions were physically impossible prevented most geologists from accepting Wegener's ideas. By the 1970s the story was very different. Geologists accepted continental drift in large part because paleomagnetic measurements, based on the geometry and history of the earth's magnetic field, showed that continents had in fact moved over millions of years. Combination of these observations with results from seismology and marine geology and geophysics led to the realization that all parts of the earth's outer shell, not just the continents, were moving.

Plate tectonics is conceptually simple: it treats the earth's outer shell as made up of about 15 rigid plates, about 100 km thick, which move relative to each other at speeds of a few cm per year.[1] The plates are rigid in the sense that little (ideally no) deformation occurs within them, so deformation occurs at their boundaries, giving rise to earthquakes, mountain building, volcanism, and other spectacular phenomena. These strong plates form the earth's *lithosphere*, and move over the weaker *asthenosphere* below. The lithosphere and asthenosphere are mechanical units defined by their strength and the way they deform. The lithosphere includes both the crust and part of the upper mantle.

Figure 5.1-1 shows the three basic types of plate boundaries. Warm mantle material upwells at *spreading centers*, also known as mid-ocean ridges, and then cools. Because the strength of rock decreases with temperature (Section 5.7.3), the cooling material forms strong plates of new oceanic lithosphere. The cooling oceanic lithosphere moves away from the ridges, and eventually reaches *subduction zones*, or trenches,[2] where it descends in *downgoing slabs* back into the mantle, reheating as it goes. The direction of the relative motion between two plates at a point on their common boundary determines the nature of the boundary. At spreading centers both plates move away from the boundary, whereas at subduction zones the subducting plate moves toward the boundary. At the third boundary type, *transform faults*, relative plate motion is parallel to the boundary.

As discussed in Section 3.8, seismology shows that the structure of the mantle and the core varies with depth, due to changes in temperature, pressure, mineralogy, and composition. Plate tectonics describes the behavior of the lithosphere, the strong outer shell of the mantle, which is the cold outer boundary layer of the thermal convection system involving the mantle and the core that removes heat from the earth's interior. Although much remains to be learned about this convective system, especially in the lower mantle and the core (Fig. 5.1-2), there is general agreement that at shallow depths the warm,

[1] This is about the speed at which fingernails grow.

[2] Boundaries are described either as mid-ocean ridges and trenches, emphasizing their morphology, or as spreading centers and subduction zones, emphasizing the plate motion there. The latter nomenclature is more precise, because there are elevated features in the ocean basins that are not spreading ridges, and spreading centers like the East African rift exist within continents.

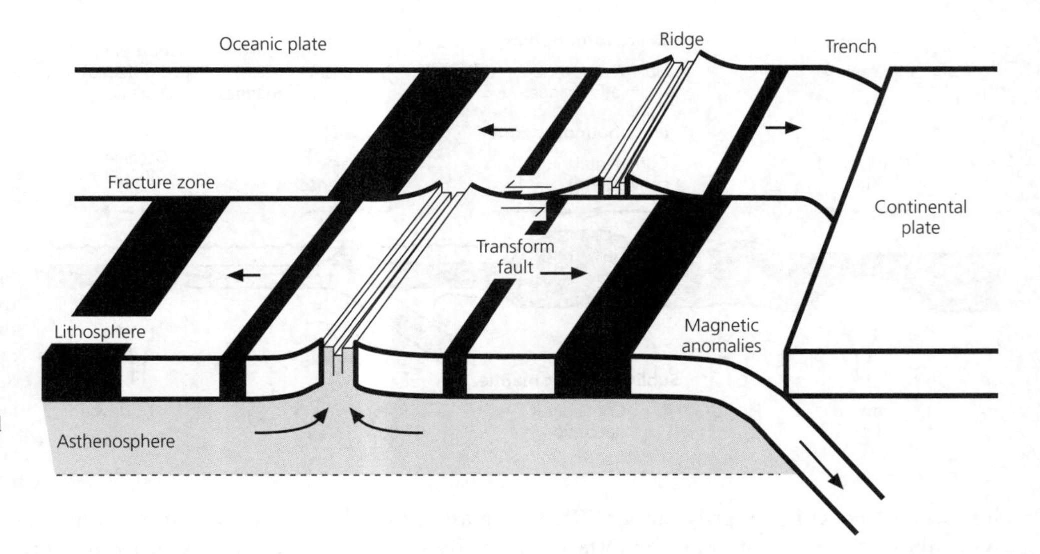

Fig. 5.1-1 Plate tectonics at its simplest. Oceanic lithosphere is formed at ridges and subducted at trenches. At transform faults, plate motion is parallel to the boundaries. Each boundary type has typical earthquakes.

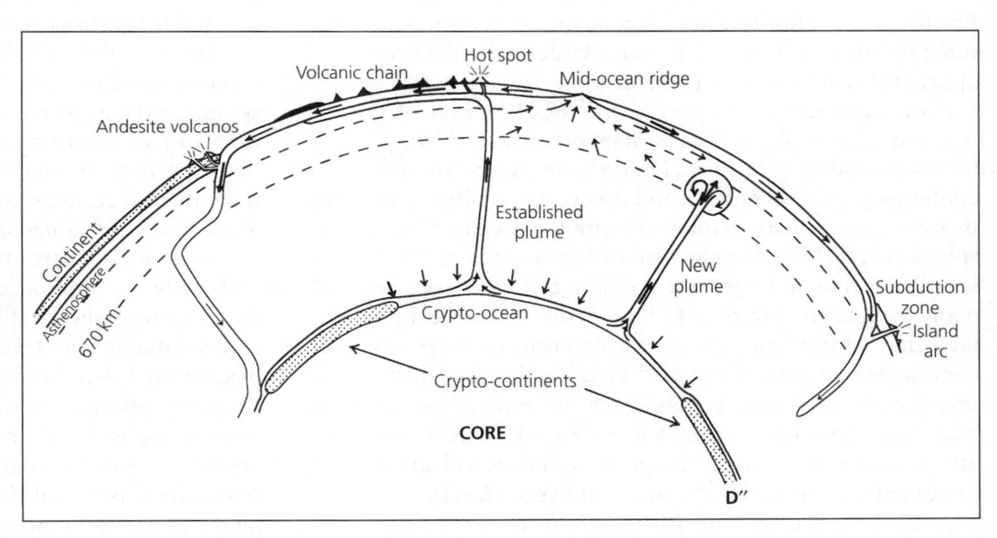

Fig. 5.1-2 Schematic diagram showing ideas about mantle convection. Ridges reflect upper mantle upwelling. Slabs penetrate into the lower mantle, causing heterogeneity there, and in some cases descend to the base of the mantle. Mantle (hot spot) plumes reflect lower mantle upwelling. Many features shown are controversial and subject to change without notice. (Modified from Stacey, 1992.)

and hence less dense, material rising below spreading centers forms upwelling limbs, whereas the relatively cold, and hence dense, subducting slabs form downwelling limbs. Although the lithosphere is a very thin layer compared to the rest of the mantle (100 km is 1/29 of the mantle's radius), it is where the greatest temperature change occurs, from about 1300° to 1400°C at a depth of 100 km to about 0°C at the surface. For this reason, the lithosphere is called a thermal boundary layer. Because of this temperature change, the lithosphere is much stronger than the underlying rock, and so is also a mechanical boundary layer. This strong boundary layer is thought to be a primary reason why plate tectonics is much more complicated than expected from simple convection models. Moreover, the lithosphere, which contains the crust, is also a chemical boundary layer distinct from the remainder of the mantle. Continental lithosphere is especially distinct: although individual plates can contain both oceanic and continental lithosphere, the latter is made of less dense rock than the former (recall the differences between granitic and basaltic rocks discussed in Section 3.2), and so does not subduct. The oceanic lithosphere is continuously subducted and reformed at ridges, and so never gets older than about 200 Myr. The continental lithosphere, however, can be billions of years old.

Put another way, plate tectonics is the primary surface manifestation of the heat engine whose nature and history govern the planet's thermal, mechanical, and chemical evolution.[3] Earth's heat engine is characterized by the balance between three modes of heat transfer from the interior: the plate tectonic cycle involving the cooling of oceanic lithosphere; mantle plumes, which are thought to be a secondary feature of mantle convection; and heat conduction through continents that are not subducted and hence do not participate directly in the oceanic plate tectonic cycle. Based on estimates from sea floor topography and heat flow, discussed shortly, terrestrial heat

[3] It has been said that heat is the geological lifeblood of planets.

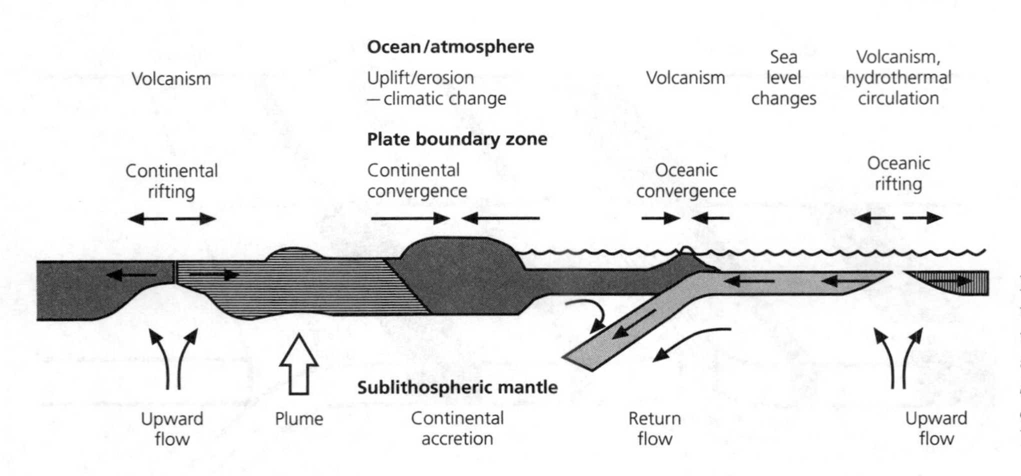

Fig. 5.1-3 Cartoon summarizing some of the primary modes of interaction between the solid earth's interior and the fluid ocean and atmosphere system. (Stein *et al.*, 1995. *Seafloor Hydrothermal Systems*, 425–45, copyright by the American Geophysical Union.)

loss seems to occur primarily (about 70%) via plate tectonics, with about 5% via hot spots (mantle plumes). By contrast, Earth's grossly similar sister planets, Mars and Venus, seem to function quite differently, because large-scale plate tectonics appears absent, at least at present.

Plate tectonics is also crucial for the evolution of Earth's ocean and atmosphere, because it involves many of the primary means (including volcanism, hydrothermal circulation through cooling oceanic lithosphere, and the cycle of uplift and erosion) by which the solid earth interacts with the ocean and the atmosphere (Fig. 5.1-3). The chemistry of the oceans and the atmosphere depends in large part on plate tectonic processes, and many long-term features of climate are influenced by mountains that are uplifted by plate convergence and the positions of continents that control ocean circulation. In fact, the presence of plate tectonics may explain how life evolved on earth (at mid-ocean ridge hot springs) and be crucial for its survival (the atmosphere is maintained by plate boundary volcanism, and plate tectonics raises the continents above sea level).

As a result, plate tectonics is heavily studied by earth scientists. Our goal in this chapter is to introduce some of the ways in which seismology contributes to these studies. Some sources for more general and more detailed treatments of these topics are listed at the end of the chapter.

Seismology plays several key roles in our studies of plate tectonics. The distribution of earthquakes provides strong evidence for the idea of essentially rigid plates, with deformation concentrated on their boundaries. Figure 5.1-4 shows maps of global seismicity covering the time period 1964–97. Such maps did not become available until the early 1960s, when the World Wide Standardized Seismographic Network (WWSSN) allowed accurate locations for earthquakes of magnitude 5 or greater anywhere in the world. The map shows several remarkable patterns.

The mid-ocean ridge system, where the oceanic lithosphere is created, is beautifully outlined by the earthquake locations. For example, the Mid-Atlantic ridge and East Pacific rise can be followed using epicenters for thousands of kilometers. The locations of the trenches, where oceanic lithosphere is subducted, are even more apparent in the lower panel showing earthquakes with focal depths greater than 100 km, because mid-ocean ridge earthquakes are shallow and thus do not appear.

It is especially impressive to plot the locations of earthquakes on cross-sections across trenches (Fig. 5.1-5). Inclined zones of seismicity delineate the subducting oceanic plates, which travel time and attenuation studies show to be colder and stronger than the surrounding mantle. These zones, identified before their plate tectonic significance became clear, are known as *Wadati–Benioff zones* after their discoverers.[4]

The *interplate* earthquakes both delineate plate boundaries and show the motion occurring there. We will see that the direction of faulting reflects the spreading at mid-ocean ridges and subduction at trenches. The earthquake locations and mechanisms also show that plate boundaries in continents are often complicated and diffuse, rather than the simple narrow boundaries assumed in the rigid plate model that are a good approximation to what we see in the oceans. For example, seismicity shows that the collision of the Indian and Eurasian plates creates a deformation zone which includes the Himalayas but extends far into China. Similarly, the northward motion of the Pacific plate with respect to North America creates a broad seismic zone, indicating that the plate boundary zone spans much of the western USA and Canada.

In addition, *intraplate* earthquakes occur within plate interiors, far from boundary zones. For example, Fig. 5.1-4 shows earthquakes in eastern Canada and central Australia. Such earthquakes are much rarer than plate boundary zone earthquakes, but are common enough to indicate that plate interiors are not perfectly rigid. In some cases these earthquakes are associated with intraplate volcanism, as in Hawaii. Intraplate earthquakes are studied to provide data about where and how the plate tectonic model does not fully describe tectonic processes.

[4] Kiyoo Wadati (1902–95) discovered the existence of deep seismicity and its geometry under Japan; Hugo Benioff (1899–1968), also known for important contributions to seismological instrumentation, discussed the global nature of deep earthquakes and their relation to surface features (Fig. 1.1-10).

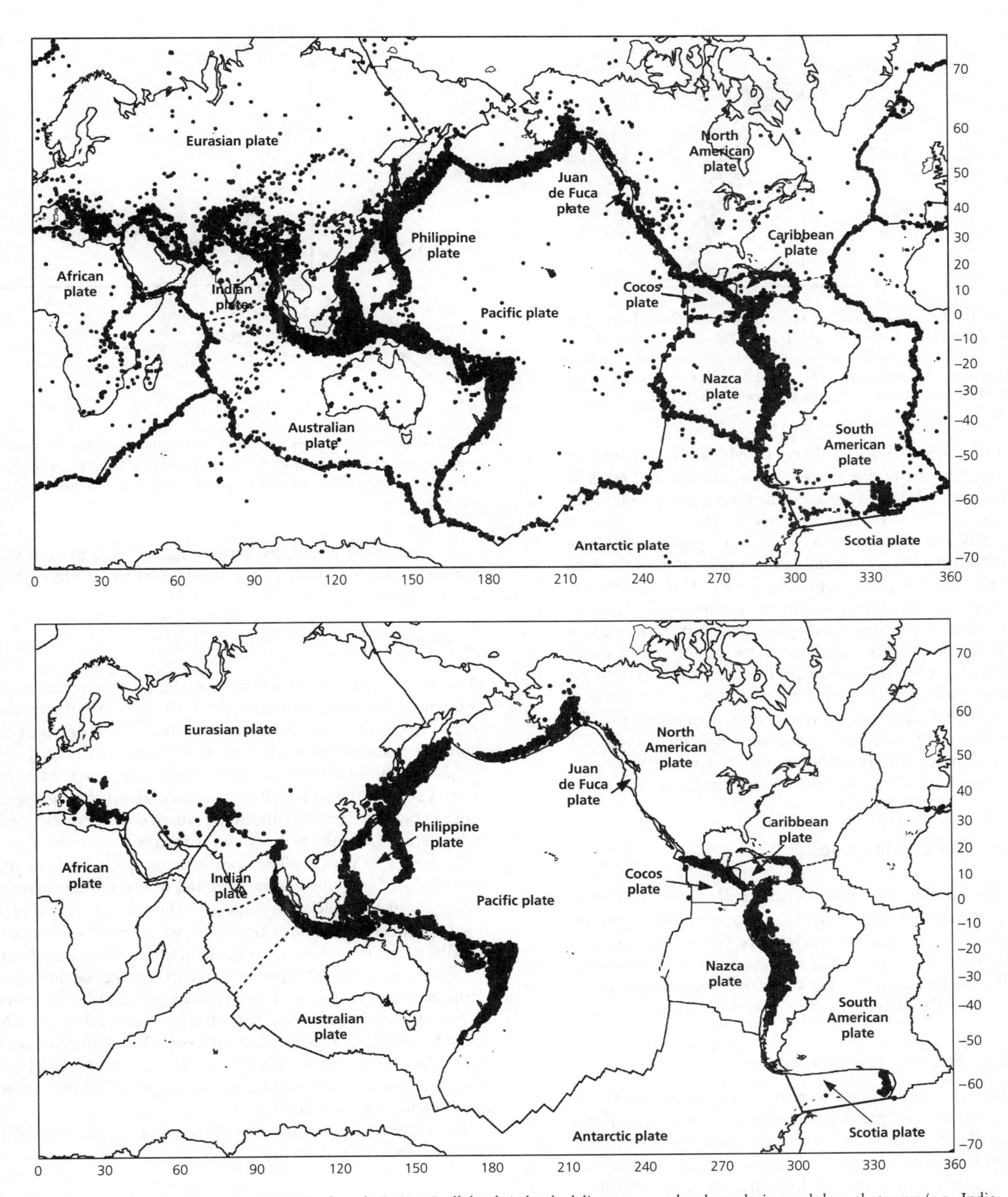

Fig. 5.1-4 Global seismicity (1964–97). *Top*: Earthquakes ($m_b \geq 5$, all depths) clearly delineate most plate boundaries, and show that some (e.g., India-Eurasia) are diffuse. Many intraplate earthquakes show internal plate deformation. *Bottom*: The locations of seismicity (of all magnitudes) below 100 km indicate the subduction zones.

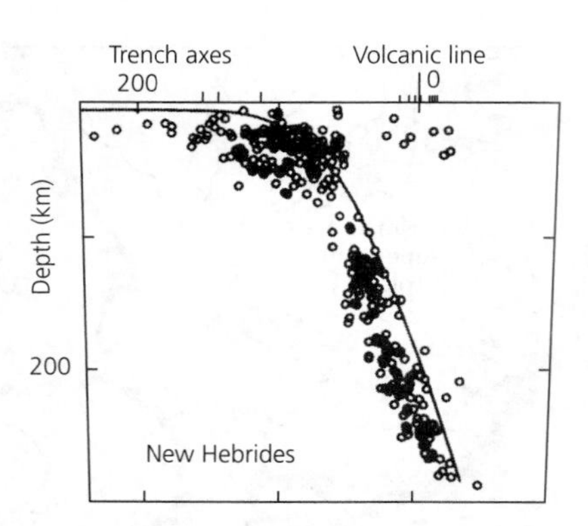

Fig. 5.1-5 Seismicity cross-section perpendicular to the New Hebrides trench showing the Wadati–Benioff zone. This dipping plane of earthquakes indicates the position of the subducting plate. (Isacks and Barazangi, 1977. *Island Arcs, Deep Sea Trenches and Back Arc Basins*, 99–114, copyright by the American Geophysical Union.)

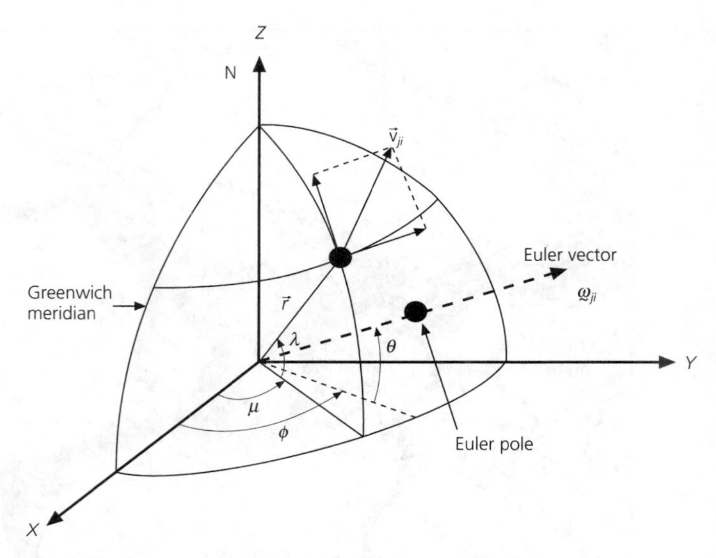

Fig. 5.2-1 Geometry of plate motions. Linear velocity at point **r** is given by $\mathbf{v}_{ji} = \boldsymbol{\omega}_{ji} \times \mathbf{r}$. The Euler pole is the intersection of the Euler vector with the earth's surface. Note that west longitudes and south latitudes are negative.

In summary, seismology provides crucial information about both *plate kinematics*, the directions and rates of plate motions, and *plate dynamics*, the forces causing plate motions. As we will see, seismicity is one of the major tools used to identify and delineate plate boundary zones, and earthquake mechanisms are among the primary data used to determine the motion within plate boundary zones. The mechanisms also provide information about the stresses acting at plate boundaries and within plates, which, together with earthquake depths and seismic velocity structure, are important in developing ideas about the forces involved and the physical processes by which rocks deform and cause earthquakes. Conversely, plate motion data are used to draw inferences about the locations and times of future earthquakes and their societal risks. Thus it is often hard, and sometimes pointless, to decide where seismology ends and plate tectonics begins, or vice versa.

5.2 Plate kinematics

Understanding the distribution and types of earthquakes requires an understanding of the geometry of plate motions, or plate kinematics. In this section we sketch some basic results, of which we assume most readers have some knowledge. As full exploration of this topic is beyond our scope, readers are encouraged to delve into the suggested literature.

5.2.1 *Relative plate motions*

A basic principle of plate tectonics is that the relative motion between any two plates can be described as a rotation about an *Euler pole*[1] (Fig. 5.2-1). This condition controls the types of boundaries and the focal mechanisms of earthquakes resulting from relative motions, as discussed later. Specifically, at any point **r** along the boundary between plate i and plate j, with latitude λ and longitude μ, the *linear velocity* of plate j with respect to plate i is

$$\mathbf{v}_{ji} = \boldsymbol{\omega}_{ji} \times \mathbf{r}. \quad (1)$$

This is the usual formulation for rigid body rotations in mechanics. **r** is the position vector to the point on the boundary, and $\boldsymbol{\omega}_{ji}$ is the angular velocity vector, or *Euler vector*. Both vectors are defined from an origin at the center of the earth.

The direction of relative motion at any point on the boundary is a small circle, a parallel of latitude *about the Euler pole* (not a geographic parallel about the North Pole!). For example, in Fig. 5.2-2 (*top*) the pole shown is for the motion of plate 2 with respect to plate 1. The convention used is that the first named plate ($j = 2$) moves counterclockwise (in a right-handed sense) about the pole with respect to the second named plate ($i = 1$). The segments of the boundary where relative motion is parallel to the boundary are transform faults. Thus transforms are small circles about the pole, and earthquakes occurring on them should have pure strike-slip mechanisms. Other segments have relative motion away from the boundary, and are thus spreading centers. Figure 5.2-2 (*bottom*) shows an alternative case. The pole here is for plate 1 ($j = 1$) with respect to plate 2 ($i = 2$), so plate 1 moves toward some segments of the boundary, which are subduction zones.

The magnitude, or rate, of relative motion increases with distance from the pole because

$$|\mathbf{v}_{ji}| = |\boldsymbol{\omega}_{ji}|\,|\mathbf{r}|\sin\gamma, \quad (2)$$

where γ is the angle between the Euler pole and the site (corresponding to a colatitude about the pole). All points on a plate

[1] This term comes from Euler's theorem, which states that the displacement of any rigid body (in this case, a plate) with one point (in this case, the center of the earth) fixed is a rotation about an axis.

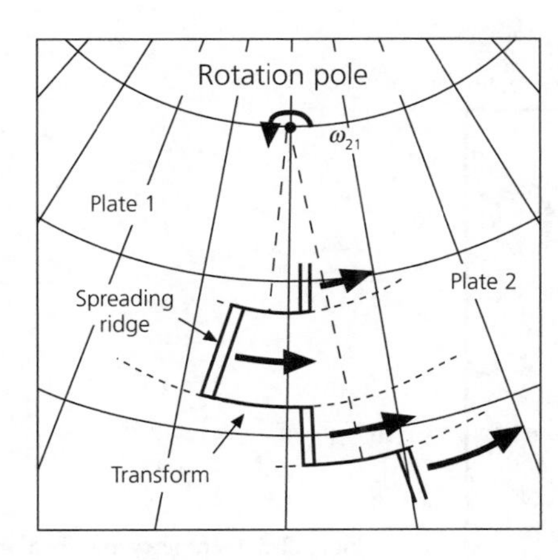

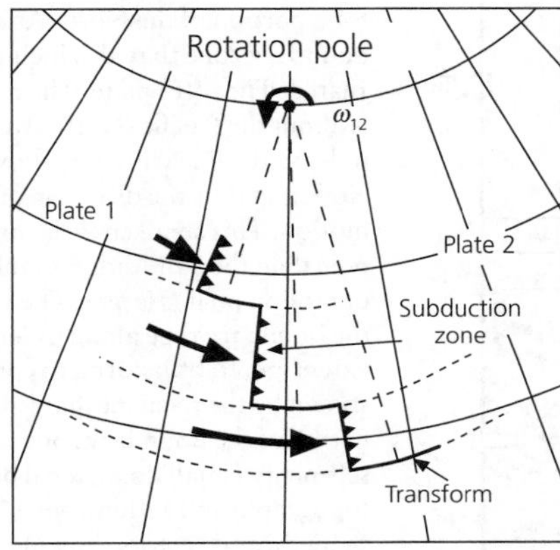

Fig. 5.2-2 Relationship of motions on plate boundaries to the Euler pole. Relative motions occur along small circles about the Euler pole (short dashed lines) at a rate that increases with distance from the pole. Note the difference the sense of rotation makes: $\boldsymbol{\omega}_{ji}$ is the Euler vector corresponding to the rotation of plate *j* counterclockwise with respect to *i*.

boundary have the same angular velocity, but the magnitude of the linear velocity varies from zero at the pole to a maximum 90° away.

The components of the vectors can be written in Cartesian (x, y, z) coordinates (Fig. 5.2-1). The position vector is

$$\mathbf{r} = (a \cos \lambda \cos \mu, a \cos \lambda \sin \mu, a \sin \lambda), \tag{3}$$

where a is the earth's radius. Similarly, if the Euler pole is at latitude θ and longitude ϕ, the Euler vector is written (neglecting the ij subscripts for simplicity) as

$$\boldsymbol{\omega} = (|\boldsymbol{\omega}| \cos \theta \cos \phi, |\boldsymbol{\omega}| \cos \theta \sin \phi, |\boldsymbol{\omega}| \sin \theta), \tag{4}$$

where the magnitude, $|\boldsymbol{\omega}|$, is the scalar angular velocity or rotation rate. To find the Cartesian components of the linear velocity $\mathbf{v}$, we evaluate the cross product (Eqn 1) using its definition (Eqn A.3.28), and find

$$\mathbf{v} = (v_x, v_y, v_z),$$

$$v_x = a|\boldsymbol{\omega}|(\cos \theta \sin \phi \sin \lambda - \sin \theta \cos \lambda \sin \mu)$$

$$v_y = a|\boldsymbol{\omega}|(\sin \theta \cos \lambda \cos \mu - \cos \theta \cos \phi \sin \lambda)$$

$$v_z = a|\boldsymbol{\omega}| \cos \theta \cos \lambda \sin (\mu - \phi). \tag{5}$$

At the point $\mathbf{r}$, the north–south and east–west unit vectors can be written in terms of their Cartesian components using Eqn A.7.4,

$$\hat{\mathbf{e}}^{NS} = (-\sin \lambda \cos \mu, -\sin \lambda \sin \mu, \cos \lambda),$$

$$\hat{\mathbf{e}}^{EW} = (-\sin \mu, \cos \mu, 0), \tag{6}$$

so we find the north–south and east–west components of $\mathbf{v}$ by taking dot products of its Cartesian components (Eqns 5) with the unit vectors (Eqns 6), and obtain

$$v^{NS} = a|\boldsymbol{\omega}| \cos \theta \sin (\mu - \phi),$$

$$v^{EW} = a|\boldsymbol{\omega}|[\sin \theta \cos \lambda - \cos \theta \sin \lambda \cos (\mu - \phi)]. \tag{7}$$

We can then find the rate and direction of plate motion,

$$\text{rate} = |\mathbf{v}| = \sqrt{(v^{NS})^2 + (v^{EW})^2}$$

$$\text{azimuth} = 90° - \tan^{-1}[(v^{NS})/(v^{EW})], \tag{8}$$

such that azimuth is measured in the usual convention, degrees clockwise from North.

In evaluating these expressions, it is important to be careful with dimensions. Although rotation rates are typically reported in degrees per million years, they should be converted to radians per year. The resulting linear velocity will have the same dimensions as Earth's radius. By serendipity, converting radius in km to mm and Myr to years cancel out, so only the degrees to radians ($\times \pi/180°$) conversion actually needs to be done to obtain a linear velocity in mm/yr. Plate motions are often quoted as mm/yr, because a year is a comfortable unit of time for humans and 1 mm/yr corresponds to 1 km/Myr, making it easy to visualize what seemingly slow plate motion accomplishes over geologic time.

To see how this works, consider Fig. 5.2-3, which shows the North America–Pacific boundary zone. The map is drawn in a projection about the Euler pole, so the expected relative motion is parallel to small circles like the one shown. By analogy to Fig. 5.2-2, this geometry predicts NW–SE-oriented spreading along ridge segments in the Gulf of California, which are rifting Baja California away from the rest of Mexico. Further north, the San Andreas fault system is essentially parallel to the relative motion, so is largely a transform fault. In Alaska, the eastern Aleutian arc is perpendicular to the plate motion, so the Pacific plate subducts beneath North America. Thus this plate boundary contains ridge, transform, and trench portions, depending on the geometry of the boundary.[2] In addition, the

[2] A good way to visualize the plate motion is to photocopy Fig. 5.2-3, cut along the boundary of the Pacific plate, and then photocopy the "Pacific" onto another piece of paper. Putting the "Pacific" beneath "North America" and rotating around a thumbtack through the pole shows the ridge, transform, and trench motions both forward and backward in time.

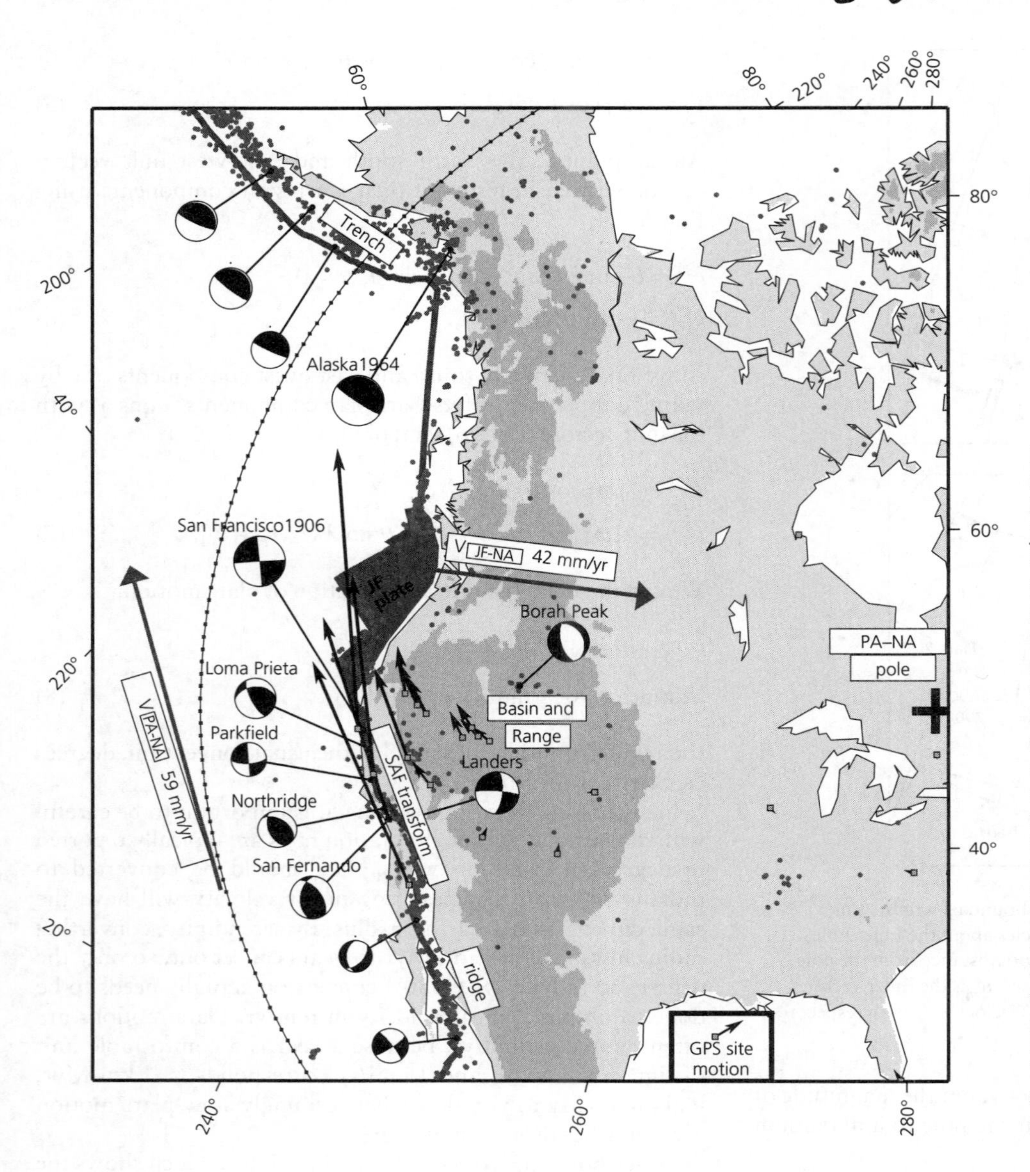

Fig. 5.2-3 Geometry and focal mechanisms for a portion of the North America–Pacific boundary zone that also includes the small Juan de Fuca (JF) plate. The map projection is about the Pacific–North America Euler pole, so the line with dots shows a small circle and thus the direction of plate motion. This small circle is further from the pole than the San Andreas fault, so the rate of motion on it is larger. The variation in the boundary type along its length from extension, to transform, to convergence, is shown by the focal mechanisms. The diffuse nature of the boundary zone is shown by seismicity (small dots), focal mechanisms, topography (elevation above 1000 m is shaded), and vectors showing the motion of GPS and VLBI sites (squares) (Bennett *et al.*, 1999) with respect to the stable interior of North America. The velocity scale is shown by the plate motion arrows; some site motion vectors are too small to be seen. (Stein and Klosko, 2002. From *The Encyclopedia of Physical Science and Technology*, ed. R. A. Meyers, copyright 2002 by Academic Press, reproduced by permission of the publisher.)

boundary zone contains the small Juan de Fuca plate, which subducts beneath the Pacific Northwest at the Cascadia subduction zone.

Equation 8 lets us find how the motion varies. The predicted motion of the Pacific plate with respect to the North American plate at a point on the San Andreas fault (36°N, 239°E) has a rate of 46 mm/yr at an azimuth of N36°W. The predicted direction agrees reasonably well with the average trend of the San Andreas fault, N41°W. Thus, to first order, the San Andreas is a Pacific–North America transform plate boundary with right-lateral motion. However, there are some deviations from pure transform behavior. As we will see, the rate on the San Andreas fault is less than the total plate motion because some of the motion occurs elsewhere within the broad plate boundary zone. In addition, in some places the San Andreas trend differs enough from the plate motion direction that dip-slip faulting occurs. Hence we think of the San Andreas as the primary feature of the essentially strike-slip portion of the plate boundary zone.

Similarly, at a point on the Aleutian trench near the site of the great 1964 Alaska earthquake (Fig. 4.3-15) (62°N, 212°E), we predict Pacific motion of 53 mm/yr at N14°W with respect to North America. This motion is into the trench, which is a Pacific–North America subduction zone. It is worth noting that for a given convergent relative motion either plate can be subducting. However, the relative direction is important, so the plates cannot be interchanged: if N14°W were the direction of motion of North America with respect to the Pacific, the motion would be away from the boundary, which would then be a spreading center with the same rate. As for the San Andreas, the actual boundary zone shown by earthquakes and other deformation is wider and more complicated than the ideal.

Earthquake focal mechanisms within the boundary zone are consistent with the overall plate motions and illustrate some of their complexities. In the Gulf of California we see both strike-slip faulting along oceanic transforms and normal faulting on ridge segments. The San Andreas fault system, composed of the main fault and some others, has both pure strike-slip earthquakes (Parkfield) and earthquakes with some dip-slip motion (Northridge (Section 4.5.3), San Fernando, and Loma Prieta) when it deviates from pure transform behavior. The seismicity also shows that the plate boundary zone is quite broad. Although the San Andreas fault system is the locus of most of the plate motion (Fig. 4.5-13) and hence large earthquakes, seismicity extends as far eastward as the Rocky Mountains. For example, the Landers earthquake shows strike-slip motion east of the San Andreas, and the Borah Peak earthquake illustrates the extensional faulting that occurs in the Basin and Range. These focal mechanisms are consistent with the motions shown by space-based geodetic measurements, discussed shortly, and with geologic studies.

5.2.2 *Global plate motions*

The relative plate motions show how the plate boundary geometry is evolving and has evolved. The Juan de Fuca plate is subducting under North America faster than new lithosphere is being added to it by sea floor spreading at its boundary with the Pacific plate, so this plate was larger in the past and is shrinking. Rotating the Pacific plate backwards with respect to North America shows that 10 million years ago the Gulf of California had not yet begun to open by sea floor spreading. These changes are part of the evolution of the plate boundary in western North America, in which the large oceanic Farallon plate that used to be between the Pacific and North American plates began subducting under North America at about 40 Ma,[3] leaving the Juan de Fuca plate as a remnant and forming the San Andreas fault.

At this point you may be wondering how Euler poles are found. Until recently, this was done by combining three different types of data from different boundaries. The rates of spreading are found from sea floor magnetic anomalies, which form as the hot rock at ridges cools and acquires magnetization parallel to the earth's magnetic field. Because the history of reversals of the earth's magnetic field is known, the anomalies can be dated, so their distance from the ridge where they formed shows how fast the sea floor moved away from the ridge. The directions of motion are found from the orientations of transform faults and the slip vectors of earthquakes on transforms and at subduction zones. Euler vectors are found from the relative motion data, using geometrical conditions we have discussed. The process is easy to visualize. Because slip vectors and transform faults lie on small circles about the pole, the pole must lie on a great circle at right angles to them (Fig. 5.2-2). Similarly, the rate of plate motion increases with the sine of the distance from the pole (Eqn 2). These constraints make it possible to locate the poles. Determination of Euler vectors for all the plates can thus be treated as an overdetermined least squares problem whose solution (Section 7.5) gives a *global relative plate motion model*. Because these models use spreading rates determined from magnetic anomaly data that span several million years, they describe plate motions averaged over the past few million years.[4]

Table 5.2-1 gives such a model, known as NUVEL-1A,[5] which specifies the motions of plates (Fig. 5.2-4) with respect to North America. The vectors follow the convention that each named plate moves counterclockwise relative to North America. Although the table lists only Euler vectors with respect to North America, the motion of plates with respect to other plates is easily found using vector arithmetic. For example,

$$\omega_{ij} = -\omega_{ji}, \tag{9}$$

so we reverse the plate pair using the negative of the Euler vector. The pole for the new plate pair is the antipole, with latitude of opposite sign and longitude increased by 180°. The magnitude (rotation rate) stays the same. We can also reverse the plate pair by keeping the same pole and making the rotation rate negative (clockwise rather than counterclockwise). Although we usually use positive rotation rates, negative ones sometimes help us visualize the motion. For example, the table shows the Pacific–North America pole at about −49°N, 102°E, so the North America–Pacific pole is at about 49°N, (102 + 180 = 282)°E, which is in southeastern Canada. Thus, about this pole, North America rotates counterclockwise with respect to the Pacific, or the Pacific rotates clockwise with respect to North America, as shown in Fig. 5.2-3.

For other plate pairs we assume that the plates are rigid, so all motion occurs at their boundaries. We can then add Euler vectors,

$$\omega_{jk} = \omega_{ji} + \omega_{ik} \tag{10}$$

because the motion of plate j with respect to plate k equals the sum of the motion of plate j with respect to plate i and the motion of plate i with respect to plate k. Thus if we start with a set of vectors all with respect to one plate, e.g., i, we use

$$\omega_{jk} = \omega_{ji} - \omega_{ki} \tag{11}$$

to form any Euler vector needed. These operations are easily done using the Cartesian components (Eqn 4), as shown in this chapter's problems. We can also perform the analogous operations on linear velocity vectors at a specific site.

[3] "Ma" is often used to denote millions of years before the present.

[4] The most recent magnetic reversal occurred about 780,000 years ago, so any plate model based on paleomagnetic data must average at least over that interval.

[5] NUVEL-1 (Northwestern University VELocity) was developed as a new ("nouvelle") model (DeMets *et al.*, 1990). The multiyear development prompted the suggestion that "OLDVEL" might be a better name. Due to changes in the paleomagnetic time scale the model was revised to NUVEL-1A (DeMets *et al.*, 1994). This change caused a slight difference in the rates of relative motion, but not in the poles and hence directions of relative motion.

Table 5.2-1 Euler vectors with respect to North America (NA).

Plate	Pole latitude (°N)	Longitude (°E)	|ω| (°/Myr)
Pacific (PA)	−48.709	101.833	0.7486
Africa (AF)	78.807	38.279	0.2380
Antarctica (AN)	60.511	119.619	0.2540
Arabia (AR)	44.132	25.586	0.5688
Australia (AU)	29.112	49.006	0.7579
Caribbean (CA)	74.346	153.892	0.1031
Cocos (CO)	27.883	−120.679	1.3572
Eurasia (EU)	62.408	135.831	0.2137
India (IN)	43.281	29.570	0.5803
Nazca (NZ)	61.544	−109.781	0.6362
South America (SA)	−16.290	121.876	0.1465
Juan de Fuca (JF)	−22.417	67.203	0.8297
Philippine (PH)	−43.986	−19.814	0.8389
Rivera (RI)	22.821	−109.407	1.8032
Scotia (SC)	−43.459	123.120	0.0925
NNR*	2.429	93.965	0.2064

Source: After DeMets *et al.* 1994.
*No net rotation, defined in Section 5.2.4.

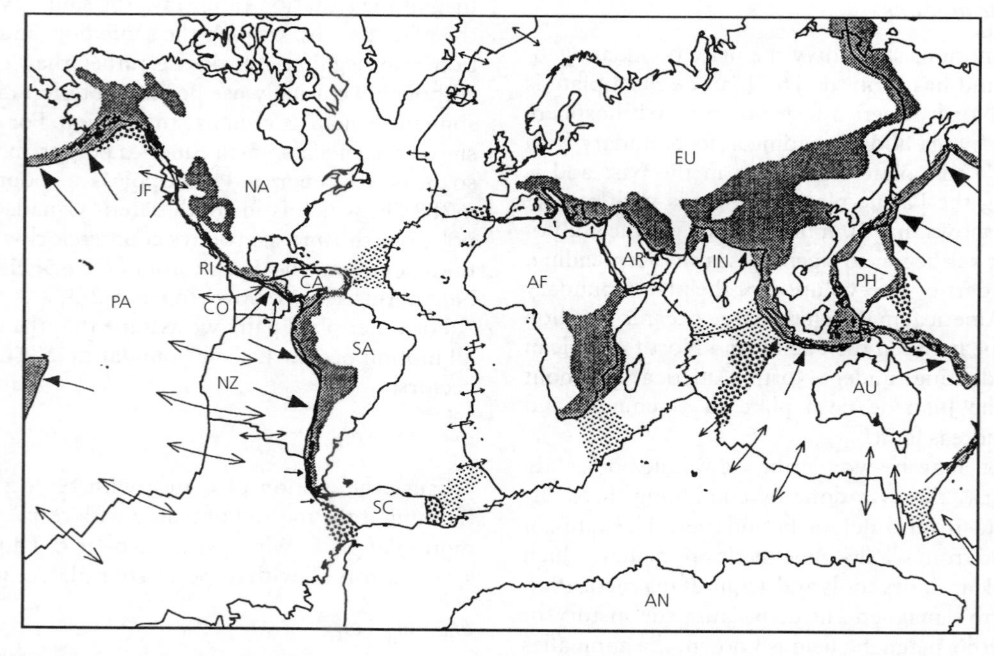

Fig. 5.2-4 Relative plate motions for the NUVEL-1 global plate motion model. Arrow lengths are proportional to the displacement if plates maintain their present relative velocity for 25 Myr. Divergence across mid-ocean ridges is shown by diverging arrows. Convergence is shown by single arrows on the underthrust plate. Plate boundaries are shown as diffuse zones implied by seismicity, topography, or other evidence of faulting. Fine stipple shows mainly subaerial regions where the deformation has been inferred from seismicity, topography, other evidence of faulting, or some combination of these. Medium stipple shows mainly submarine regions where the nonclosure of plate circuits indicates measurable deformation; in most cases these zones are also marked by earthquakes. Coarse stipple shows mainly submarine regions where the deformation is inferred mostly from the presence of earthquakes. The geometry of these zones, and in some cases their existence, is under investigation. (Gordon and Stein, 1992. *Science, 256*, 333–42, copyright 1992 American Association for the Advancement of Science.)

Such vector addition is important because we only have certain types of data for individual boundaries (Fig. 5.2-5). Although spreading centers provide rates from the magnetic anomalies and azimuths from both transform faults and slip vectors, only the direction of motion is directly known at subduction zones. As a result, convergence rates at subduction zones are estimated by global closure, combining data from all plate boundaries (Section 7.5). Thus the predicted rate at which

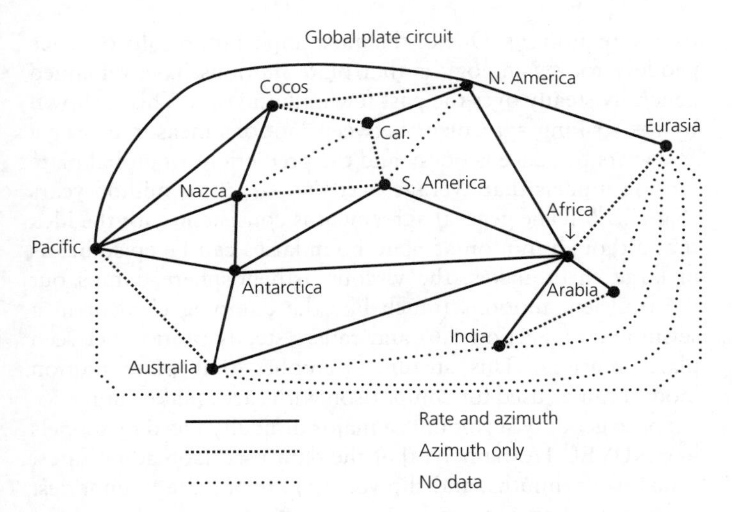

Fig. 5.2-5 Global plate circuit geometry for the NUVEL-1 plate motion model. Relative motion data are used on the boundaries indicated. (De Mets *et al.*, 1990. *Geophys. J. Int.*, *101*, 425–78.)

the Cocos plate subducts beneath North America, causing large earthquakes in Mexico, depends on the measured rates of Cocos–Pacific spreading on the East Pacific rise and Pacific–North America spreading in the Gulf of California. In some cases, such as relative motion between North and South America, no direct data were used because the boundary location and geometry are unclear, so the relative motion is inferred entirely from closure. Not surprisingly, the motions of plate pairs based on both rate and azimuth data appear to be better known.

Figure 5.2-4 shows the predicted relative motions at plate boundaries around the world. As shown for the Pacific–North America boundary in Fig. 5.2-3 and discussed in general terms in later sections, the predicted motions correspond to the earthquake mechanisms. Moreover, we can use the plate motions to make inferences about future earthquakes. For example, even though we do not have seismological observations of large earthquakes along the boundary between the Juan de Fuca and North American plates, the plate motions predict that such earthquakes could result from the subduction of the Juan de Fuca plate beneath North America. Evidence for this subduction is given by the presence of the Cascade volcanoes (such as Mount Saint Helens and Mount Rainer) and paleoseismic records (Section 1.2.5) that are interpreted as evidence of large past earthquakes.

Figure 5.2-4 also illustrates that boundaries between plates are often diffuse. Seismicity, active faulting, and elevated topography often indicate a broad zone of deformation between plate interiors. This effect is evident in continental lithosphere, such as the India–Eurasia collision zone in Asia or the Pacific–North America boundary zone in the western USA, but can also sometimes be seen in oceanic lithosphere, as in the Central Indian Ocean. Plate boundary zones cover about 15% of the earth's surface, and about 40% of the earth's population lives within them.

Earthquakes are among the best tools for investigating plate boundary zones and other deviations from plate rigidity. They provide one of the best indicators of the location of boundary zones, so new earthquakes often change our views. We also use plate motion data, many of which are earthquake slip vectors. For example, Fig. 5.2-4 shows zones of seismicity in the Central Indian Ocean (Section 5.5.2) as boundaries between distinct Indian and Australian plates, rather than as within a single Indo-Australian plate, because spreading rates along the Central Indian Ocean ridge are better fit by a two-plate model. A similar argument justifies the assumption of a small Rivera plate distinct from the Cocos plate. Another approach is to use the global plate circuit closures (Fig. 5.2-5). Recall that forming a Euler vector from two others (Eqn 10) assumes that all three plates are rigid. Hence this assumption can be used to test for deviations from rigidity. To do this, we form a *best-fitting vector* for a plate pair, using only data from that pair of plates' boundary, and a *closure fitting vector* from data elsewhere in the world. If the plates were rigid, the two vectors would be the same. However, a significant difference between the two indicates a deviation from rigidity, or another problem with the plate motion model. For example, such analysis shows systematic deviations along some subduction zones, suggesting that the slip vectors of the trench earthquakes do not exactly reflect plate motions because a sliver of forearc material in the overriding plate moves separately from the remainder of the overriding plate (Section 5.4.3).

A variant of this approach is to examine the Euler vectors for three plates that meet at a *triple junction*, compute best-fitting Euler vectors for each of the three plate pairs, and sum them. For rigid plates, Eqn 10 shows that the sum should be zero. However, when this was done for the junction in the Central Indian Ocean, assuming that it was where the African, Indo-Australian, and Antarctic plates met, the Euler vector sum differed significantly from zero, indicating deviations from plate rigidity. As plate motion data improve, it seems that what was treated as a three-plate system may include as many as six resolvable plates (Antarctica, distinct Nubia (West Africa) and Somalia (East Africa), India, Australia, and Capricorn (between India and Arabia)). Hence models of plate boundaries and motions improve with time (Fig. 1.1-9). For example, although the model in Fig. 5.2-4 has a single African plate, recent models seek to resolve the motion between Nubia and Somalia (Fig. 5.6-2).

5.2.3 *Space-based geodesy*

New plate motion data have become available in recent years due to the rapidly evolving techniques of space-based geodesy. Using space-based measurements to determine plate motions was suggested by Alfred Wegener when he proposed the theory of continental drift in 1915. Wegener realized that proving continents moved apart was a formidable challenge. Although geodesy — the science of measuring the shape of, and distances on, the earth — was well established, standard surveying

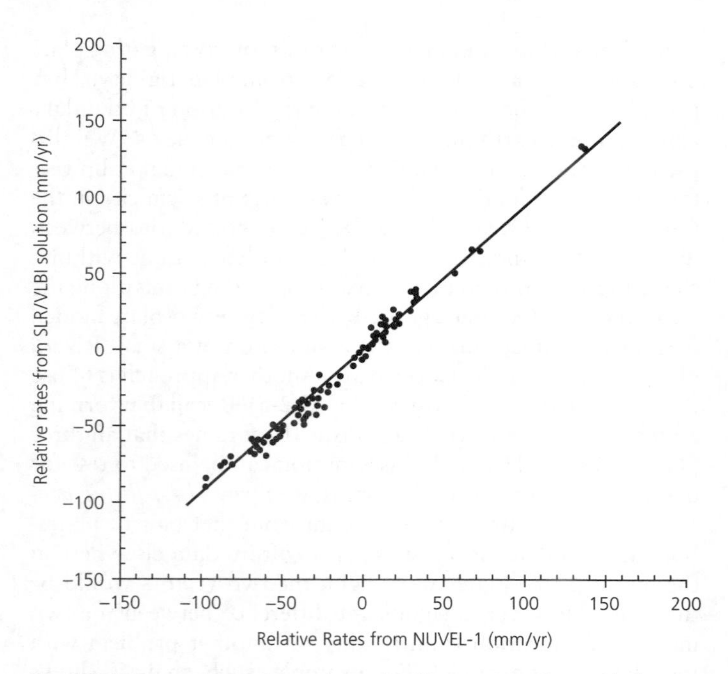

Fig. 5.2-6 Comparison of rates determined by space geodesy with those predicted by the NUVEL-1 global plate motion model. The space geodetic rates are determined from sites located away from plate boundaries to reduce the effects of deformation near the boundaries. The slope of the line is 0.94, indicating that plate motions over a decade are very similar to those predicted by a model averaging over 3 million years. (Robbins *et al.*, 1993. *Contributions of Space Geodesy to Geodynamics*, 21–36, copyright by the American Geophysical Union.)

methods offered no hope of measuring slow motions between continents far apart. Wegener thus decided to measure the distance between continents using astronomical observations.[6] However, because measuring continental drift called for measurement accuracies far greater than ever before to show small changes in positions over a few years, Wegener's attempts failed, and the idea of continental drift was largely rejected.

By the 1970s the story was very different. Geologists accepted continental drift, in large part because paleomagnetic measurements showed that continents had in fact moved over millions of years. It thus seemed natural to see if modern space-based technology could accomplish Wegener's dream of measuring continental motions over a few years. Three basic approaches were attempted. Each faced formidable technical challenges — and all succeeded. Hence, using the techniques discussed in Section 4.5.1, plate motions can now measured to a precision of a few mm/yr or better, using a few years of data from systems including Very Long Baseline Interferometry (VLBI), Satellite Laser Ranging (SLR), and the Global Positioning System (GPS).

Space geodesy measures both the rate and the azimuth of the motions between sites, and can thus be used to compute relative plate motions. One of the most important results of space geodesy for seismology is that plate motions have remained generally steady over the past few million years. This is shown by the striking agreement between motions measured over a few years by space geodesy and the predictions of global plate motion models that average over the past three million years (Fig. 5.2-6). The general agreement is consistent with the idea that although motion at plate boundaries can be episodic, as in large earthquakes, the viscous asthenosphere damps out the transient motions (much like the damping element in a seismometer, Section 6.6) and causes steady motion between plate interiors. This steadiness implies that plate motion models can be used for comparison with earthquake data.

Space geodesy surmounts a major difficulty faced by models like NUVEL-1A: namely, that the data used (spreading rates, transform azimuths, and slip vectors) are at plate boundaries, so the model provides only the net motion across a boundary. By contrast, space geodesy can also measure the motion of sites within plate boundary zones. For example, Fig. 5.2-3 shows the motions of GPS and VLBI sites within the North America–Pacific boundary zone. Sites in eastern North America move so slowly — less than 2 mm/yr — with respect to each other that their motion vectors cannot be seen on this scale. These sites thus define a rigid reference frame for the stable interior of the North American plate. Sites west of the San Andreas fault move at essentially the rate and direction predicted for the Pacific plate by the global plate motion model. The site vectors show that most of the plate motion occurs along the San Andreas fault system, but significant motions occur for some distance eastward. The geodetic motions are consistent with the focal mechanisms and geological data. Thus, as discussed further in Section 5.6, the different data types are used together to study how the seismic and aseismic portions of the deformation vary in space and time in the diffuse deformation zones that characterize many plate boundaries. This is done both on large scales, as shown here, and for studies of smaller areas and individual earthquakes (Section 4.5).

Space geodesy is also used to study the relatively rare, but sometimes large, earthquakes within plates. Global plate motion models give no idea where or how often intraplate earthquakes should occur, beyond the trivial prediction that they should not occur because there is no deformation within ideal rigid plates. Space geodesy is being combined with earthquake locations, focal mechanisms, and other geological and geophysical data to investigate the motions and stresses within plates and how they give rise to intraplate earthquakes (Section 5.6.3).

5.2.4 Absolute plate motions

So far, we have discussed the relative motions between plates, which have traditionally been of greatest interest to seismologists because most earthquakes reflect these motions. However, in some applications it is important to consider *absolute* plate motions, those with respect to the deep mantle.

In general, both plates and plate boundaries move with respect to the deep mantle. To see this, assume that the African

[6] Using an extraterrestrial reference has a long history; in about 230 BC Eratosthenes found the Earth's size from observations of the sun's position at different sites, and navigators have found their positions by observing the sun and stars.

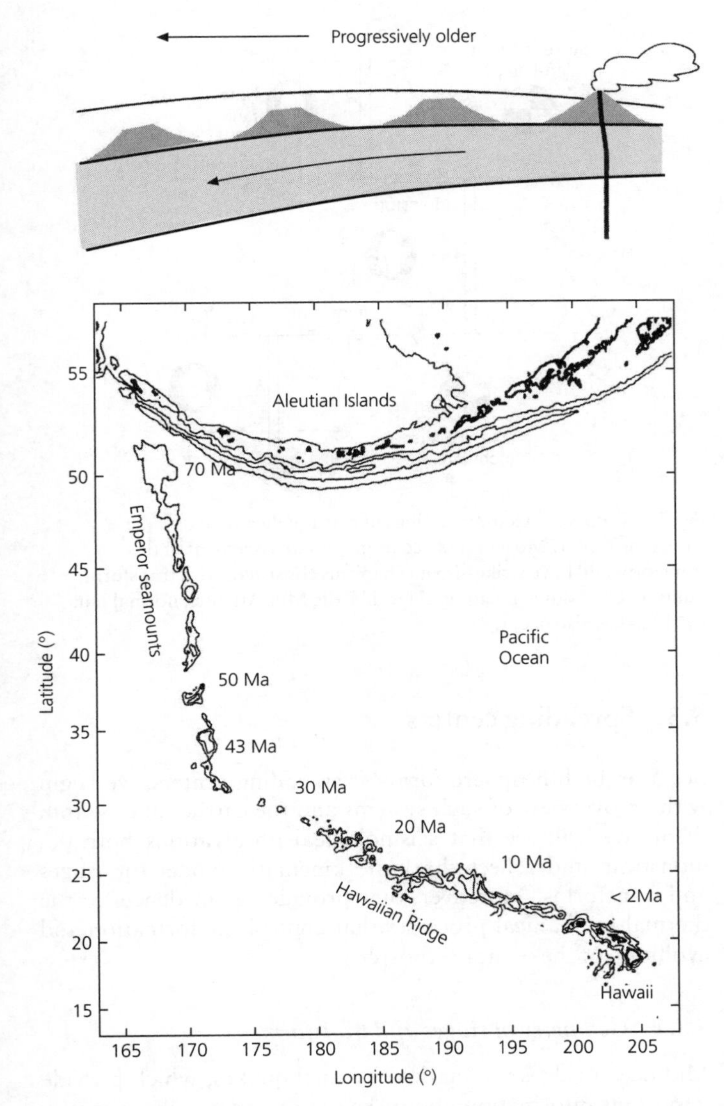

Fig. 5.2-7 *Top*: Illustration of the formation of a volcanic island chain by plate motion over a fixed hot spot. *Bottom*: Ages, in millions of years, of volcanoes in the Hawaiian–Emperor chain.

plate were not moving with respect to the deep mantle. In this case, as lithosphere was added to the plate by sea floor spreading at the Mid-Atlantic ridge (Fig. 5.2-4), both the ridge and the South American plate would move westward with respect to the mantle. Conversely, as the African plate lost area by subduction beneath the Eurasian plate in the Mediterranean, the trench would "roll backward," causing both it and Eurasia to move southward relative to the mantle. Such motions can have important consequences for processes at plate boundaries (e.g. Fig. 5.3-10).

Absolute plate motions cannot be measured directly. Hence we infer these motions in two ways. One uses the *hot spot* hypothesis, in which certain linear volcanic trends result from the motion of a plate over a hot spot, or fixed source of volcanism, which causes melting in the overriding plate (Fig. 5.2-7). If the overriding plate is oceanic, its motion causes a progression from active volcanism that builds the islands, to older islands, to underwater seamounts as the sea floor moves away from the hot spot, cools, and subsides. This process leaves a broad, shallow, topographic swell around the hot spot and a characteristic volcanic age progression away from it, as shown for the Hawaiian–Emperor seamount chain. The ages of volcanism range from present, on the currently active island of Hawaii, to a few million years on the other Hawaiian islands,[7] to about 28 Ma at Midway island, and about 70 Ma where the seamount chain vanishes into the Aleutian trench. Thus the direction and age of the volcanic chain give the motion of the plate with respect to the hot spot. For example, the bend in the Hawaiian–Emperor seamount chain has been interpreted as indicating that the Pacific plate changed direction about 40 million years ago. Hence using hot spot tracks beneath different plates, and assuming that the hot spots are fixed with respect to the deep mantle (or move relative to each other more slowly than plates), yields a hot spot reference frame.

It is often further assumed that hot spots result from plumes of hot material rising from great depth, perhaps even the core-mantle boundary (Fig. 5.1-2). The concepts of hot spots and plumes are attractive and widely used, but the relation between the persistent volcanism and possible deep mantle plumes remains a subject of active investigation because there are many deviations from what would be expected. Some hot spots move significantly, some chains show no clear age progression, evidence for plate motion changes associated with bends like that in Fig. 5.2-7 is weak, and oceanic heat flow data show little or no thermal anomalies at the swells. Seismological studies find low-velocity anomalies, but assessing their depth extent and relation to possible plumes is challenging. However, the hot spot reference frame is similar to one obtained by assuming there is no net rotation (NNR) of the lithosphere as a whole, and hence that the sum of the absolute motion of all plates weighted by their area is zero. Thus despite unresolved questions about the nature and existence of hot spots and plumes, NNR reference frames are often used to infer absolute motions.

To compute absolute motions, we recognize that motions in an absolute reference frame correspond to adding a rotation to all the plates. Thus we use the Euler vector formulation and treat the absolute reference frame as mathematically equivalent to another plate. We define Ω_i as the Euler vector of plate i in an absolute reference frame. For example, Table 5.2-1 gives the NNR Euler vector relative to the North American plate ($\boldsymbol{\omega}_{NNR-NA}$), so its negative ($\boldsymbol{\omega}_{NA-NNR}$) is the absolute Euler vector Ω_{NA} for North America in the NNR reference frame. The linear velocity at a point **r** is found by analogy to Eqn 1:

$$\mathbf{v}_i = \Omega_i \times \mathbf{r}. \qquad (12)$$

Thus we find the motion of North America with respect to the hot spot thought to be producing the volcanism and earthquakes in Yellowstone National Park (44°, –110°) to be

[7] This age progression was recognized by native Hawaiians, who attributed it to the order in which the volcano goddess Pele plucked the islands from the sea.

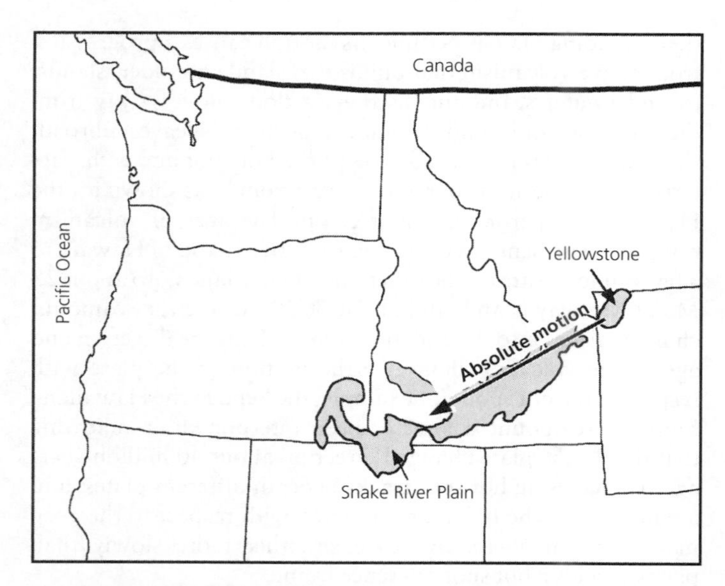

Fig. 5.2-8 Comparison of the predicted absolute motion of North America to the Snake River Plain basalts, which are thought to be the track of a hot spot now producing volcanism in Yellowstone National Park. (After Smith and Braile, 1994. *J. Volcan. Geotherm. Res.*, *61*, 121–87, with permission from Elsevier Science.)

18 mm/yr directed N239°E. This motion is along the trend connecting the present volcanism in Yellowstone to the Snake River Plain basalts (Fig. 5.2-8), which are thought to be its track, a continental analogy to the Hawaiian–Emperor seamount chain.

Relative and absolute Euler vectors are simply related because

$$\boldsymbol{\omega}_{ij} = \boldsymbol{\Omega}_i - \boldsymbol{\Omega}_j, \tag{13}$$

the relative Euler vector for two plates, is the difference between their absolute Euler vectors. Thus, if we know one plate's absolute motion, we can find all the others from the relative motions. For example, the absolute motion of the Pacific plate can be found from Table 5.2-1, which gives its vector relative to North America, using

$$\boldsymbol{\Omega}_{PA} = \boldsymbol{\omega}_{PA-NA} + \boldsymbol{\Omega}_{NA}. \tag{14}$$

Absolute motions are important in several seismological applications. Seismology is used to study hot spots and their effects, including the resulting intraplate earthquakes like those associated with the volcanism in Hawaii. For example, Fig. 2.8-5 illustrated the use of surface wave dispersion to study the velocity structure under the Walvis ridge, which is thought to be the track produced by a hot spot under the Mid-Atlantic ridge. A second application involves seismic anisotropy in the mantle (Section 3.6), which is thought to reflect flow of olivine-rich material in a direction that is often consistent with the predicted absolute plate motions. Thus seismic anisotropy, seismic velocities, and absolute motions are being combined to model mantle flow.

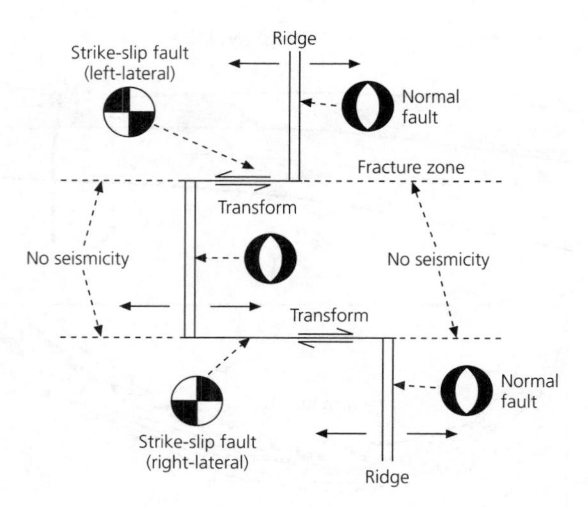

Fig. 5.3-1 Possible tectonic settings of earthquakes at an oceanic spreading center. Most events occur on the active segment of the transform and have strike-slip mechanisms consistent with transform faulting. On a slow-spreading ridge, like the Mid-Atlantic, normal fault earthquakes also occur.

5.3 Spreading centers

Because the lithosphere forms at spreading centers, we begin with an overview of such systems and the earthquakes within them. We will see that seismological observations both demonstrate and reflect the basic kinematic model for ridges and transforms. Moreover, they provide key evidence for the thermal-mechanical processes that control the formation and evolution of the oceanic lithosphere.

5.3.1 *Geometry of ridges and transforms*

Mid-ocean ridges are marked by earthquakes, which provide important information about the sea floor spreading process. Figure 5.3-1 is a schematic diagram of a portion of a spreading ridge offset by transform faults. Because new lithosphere forms at ridges and then moves away, transform faults are segments of the boundaries between plates, across which lithosphere moves in opposite directions. A given pair of plates can have either right- or left-lateral motion, depending on the direction in which a transform offsets the ridge; both reflect the same direction of relative plate motion. This motion across the transform is not what produced the offset of the ridge crest. In fact, in the usual situation such that spreading is approximately symmetric (equal rates on either side), the length of the transform will not change with time. This is a very different geometry from a transcurrent fault, where the offset between ridge segments is produced by motion on the fault and increases with time.

The focal mechanisms illustrate these ideas. Figure 5.3-2 (*top*) shows a portion of the Mid-Atlantic ridge composed of north–south-trending ridge segments that are offset by transform faults such as the Vema transform that trend approxim-

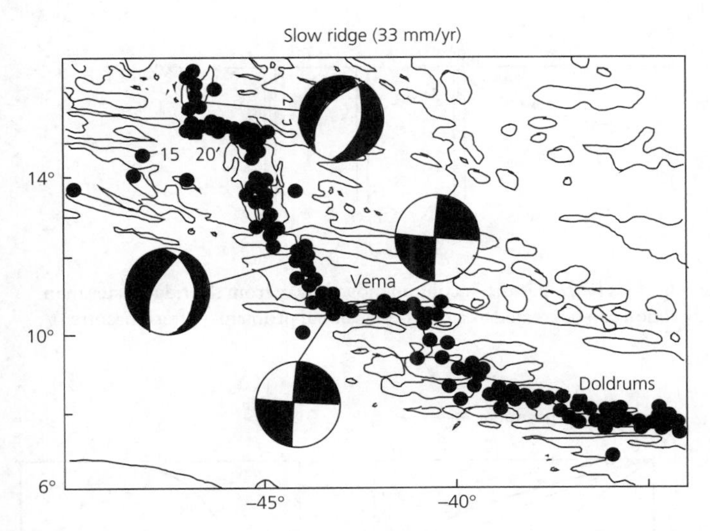

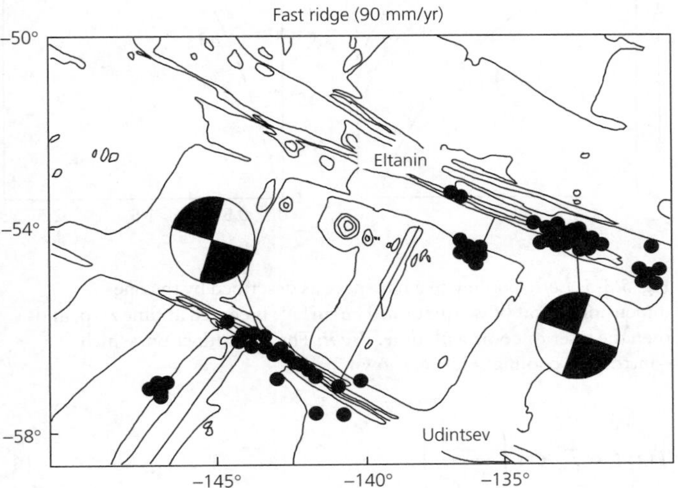

Fig. 5.3-2 Maps contrasting faulting on slow- and fast-spreading centers. *Top*: The slow Mid-Atlantic ridge has earthquakes on both the active transform and the ridge segments. Strike-slip faulting on a plane parallel to the transform azimuth is characteristic. On the ridge segments, normal faulting with nodal planes parallel to the ridge trend is seen. *Bottom*: The fast East Pacific rise has only strike-slip earthquakes on the transforms. (Stein and Woods, 1989.)

ately east–west. Both the ridge crest and the transforms are seismically active. The mechanisms show that the relative motion along the transform is right-lateral. Sea floor spreading must be occurring on the ridge segments to produce the observed relative motion. For this reason, earthquakes occur almost exclusively on the active segment of the transform fault between the two ridge segments, although an inactive extension known as a fracture zone extends to either side. Although no relative plate motion occurs on the fracture zone,[1] it is often marked by a topographic feature due to the contrast in lithospheric ages across it.

[1] Unfortunately, some transform faults named before this distinction became clear are known as "fracture zones" along their entire length.

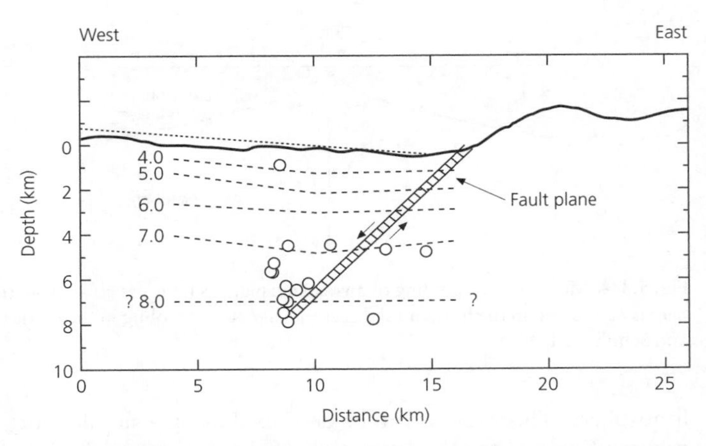

Fig. 5.3-3 Cross-section through the Mid-Atlantic ridge. The fault plane inferred from the focal mechanisms of large earthquakes is consistent with the locations of microearthquakes (dots) determined using ocean bottom seismometers. Dashed lines show P-wave velocity structure. (Toomey *et al*., 1988. *J. Geophys. Res.*, *93*, 9093–112, copyright by the American Geophysical Union.)

Earthquakes also occur on the spreading segments. Their focal mechanisms show normal faulting, with nodal planes trending approximately along the ridge axis. These normal fault earthquakes are thought to be associated with the formation of the axial valley. For example, Fig. 5.3-3 shows a cross-section through the Mid-Atlantic ridge. The fault planes inferred from teleseismic focal mechanisms and the locations of microearthquakes determined using ocean bottom seismometers are consistent with normal faulting along the east side of the valley. Slip on this fault over 10,000 years would be enough to produce the observed geometry, including the eastward tilt of the valley floor.

The seismicity differs along the East Pacific rise. Here (Fig. 5.3-2, *bottom*) earthquakes occur on the transform faults with the expected strike-slip mechanisms, but few earthquakes occur on the ridge crest. This is probably because the East Pacific rise has an axial high, rather than the axial valley that occurs at the Mid-Atlantic ridge.[2] This difference appears to reflect the spreading rates: ridges spreading at less than about 60 mm/yr usually have axial valleys, whereas faster-spreading ridges have axial highs and thus do not have ridge crest normal faulting.

These examples show the spreading process at its simplest, but there can be complexities. Spreading can be asymmetric (one flank faster than the other) or oblique, such that the spreading is not perpendicular to the ridge axis. In addition, the geometry of a ridge system can change with time, as discussed in Section 5.3.3.

5.3.2 *Evolution of the oceanic lithosphere*

To understand the difference between fast- and slow-spreading ridges, and the nature of the earthquakes associated with them, it is important to understand the evolution of the oceanic

[2] This is often shown incorrectly on older maps.

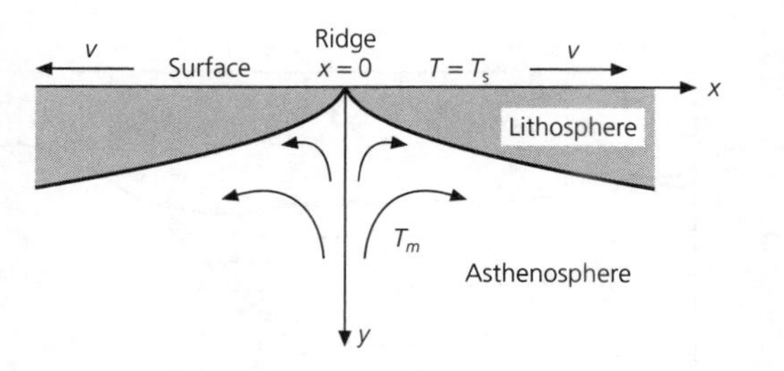

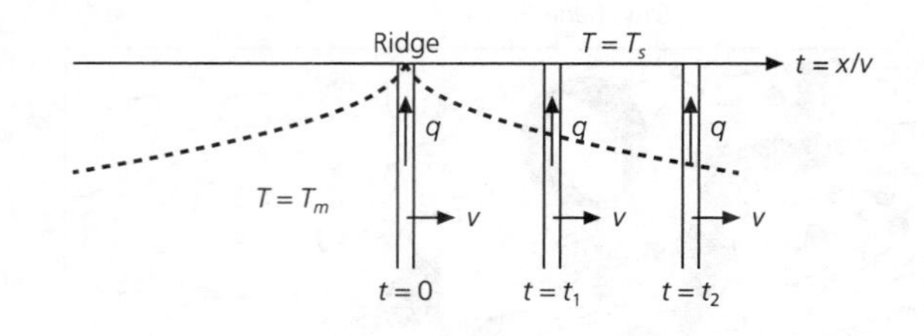

Fig. 5.3-4 Model for the cooling of an oceanic plate as it moves away from the ridge axis (*left*). Because a column moves away from the ridge faster than heat is conducted in the horizontal direction (*right*), the cooling in the vertical direction can be treated as a one-dimensional problem. (After Turcotte and Schubert, 1982.)

lithosphere. This process can be described using a simple, but powerful, model for the formation of the lithosphere by hot material at the ridge, which cools as the plate moves away.

In this model, material at the ridge at a mantle temperature T_m (1300–1400 °C) is brought to the ocean floor, which has a temperature T_s. The material then moves away at a velocity v, while its upper surface remains at T_s (Fig. 5.3-4). Because the plate moves away from the ridge faster than heat is conducted horizontally, we can consider only vertical heat conduction. Mathematically, this is the same as the cooling of a halfspace originally at temperature $T = T_m$, whose surface is suddenly cooled to T_s at time $t = 0$.

The temperature as a function of depth and time is given by the one-dimensional heat flow equation, which relates the temperature change with time in a piece of material to the rate at which heat is conducted out of it,

$$\frac{\partial T(z,t)}{\partial t} = \frac{k}{\rho C_p}\frac{\partial^2 T(z,t)}{\partial z^2} = \kappa\frac{\partial^2 T(z,t)}{\partial z^2}. \quad (1)$$

κ, known as the *thermal diffusivity*, is a property of the material that measures the rate at which heat is conducted. It has units of distance squared divided by time, and is defined as $\kappa = k/\rho C_p$, where k is the thermal conductivity, ρ is the density, and C_p is the specific heat at constant pressure.

The well known solution to Eqn 1 is

$$T(z,t) = T_s + (T_m - T_s)\,\mathrm{erf}\left(\frac{z}{2\sqrt{\kappa t}}\right), \quad (2)$$

where

$$\mathrm{erf}(s) = \frac{2}{\sqrt{\pi}}\int_0^s e^{-\sigma^2}d\sigma \quad (3)$$

is known as the error function. Figure 5.3-5 (*right*) shows how this function varies between erf (0) = 0 and erf (3) ≈ 1. Thus cooling starts at the surface and deepens with time (Fig. 5.3-5, *left*).

Assuming that any column of oceanic lithosphere cools this way, and that the sea floor temperature is $T_s = 0$ °C, then

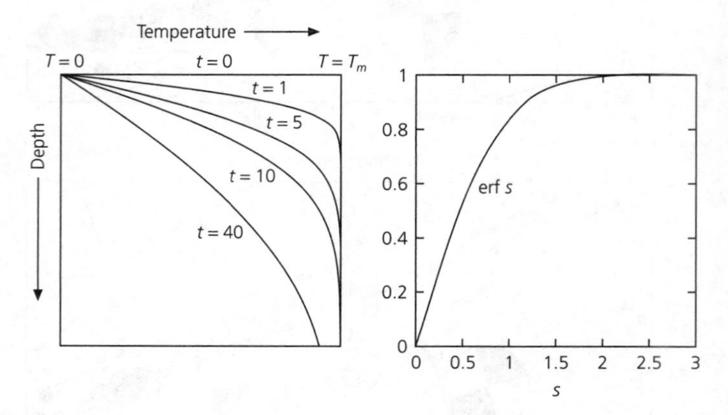

Fig. 5.3-5 *Left*: Cooling of a halfspace as described by the one-dimensional heat flow equation. The surface is cooled at time zero, and then the interior cools with time. *Right*: The error function, which controls the cooling solution shown.

$$T(z,t) = T_m\,\mathrm{erf}\left(\frac{z}{2\sqrt{\kappa t}}\right) \quad (4)$$

gives the temperature at a depth z for material of age t. The lithosphere moves away from the ridge at half the total spreading rate, so the age of the lithosphere is $t = x/v$, its distance from the ridge divided by the half-spreading rate v. Thus the temperature (Eqn 4) as a function of distance and depth is

$$T(x,z) = T_m\,\mathrm{erf}\left(\frac{z}{2\sqrt{\kappa x/v}}\right). \quad (5)$$

It is useful to think of *isotherms*, lines of constant temperature, in the plate. An isotherm is a curve on which the argument of the error function is constant,

$$\frac{z_c}{2\sqrt{\kappa t}} = c, \quad \text{or} \quad z_c = 2c\sqrt{\kappa t}, \quad (6)$$

so that the depth to a given temperature increases as the square root of the lithospheric age.

This is an example of a general feature of heat conduction problems: setting $c = 1$ and examining Fig. 5.3-5 for erf (1)

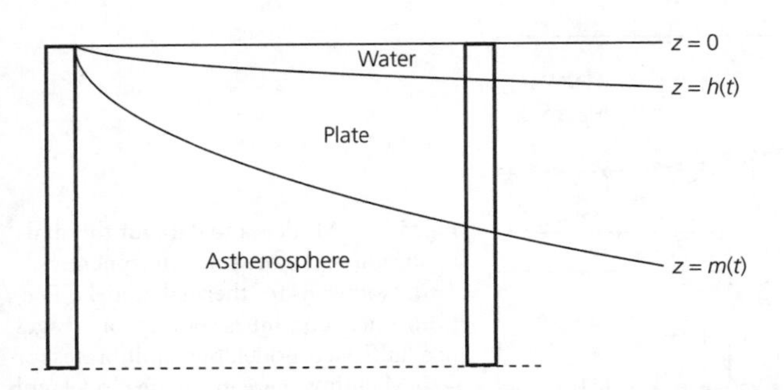

Fig. 5.3-6 The increase in ocean depth with lithospheric age due to the cooling of the lithosphere can be modeled using isostasy, the assumption that the mass in a vertical column is the same for all ages.

shows that most of the temperature change has propagated a distance $2\sqrt{\kappa t}$ in a time t. For example, after a lava flow erupts, it cools as the square root of time. Such square root of time behavior occurs for any process described by a diffusion equation, of which the heat equation is an example.

The concept that the lithosphere cools with time such that isotherms deepen with the square root of age has many observable consequences. The simplest is that ocean depth should vary with age, which makes sense, because spreading centers are ridges precisely because the ocean deepens on either side. To model this effect, we consider the mass in two columns, one at the ridge and one at age t, and invoke the idea of isostasy, which means that the masses in the two columns balance (Fig. 5.3-6).[3]

Assume that the lithosphere, defined by the $T = T_m$ isotherm, has thickness zero at the ridge and $z = m(t)$ at age t, where the water depth is $h(t)$. Similarly, we assume that the asthenosphere is at temperature T_m and has density ρ_m. However, the temperature and thus density in the cooling lithosphere vary, such that at the point (z, t) the temperature is $T(z, t)$ and the corresponding density is

$$\rho(z,t) \approx \rho_m + \frac{\partial \rho}{\partial T}[T(z,t) - T_m] = \rho_m + \rho'(z,t). \tag{7}$$

The change in density due to temperature, at constant pressure, is given by the coefficient of thermal expansion,

$$\alpha = \frac{1}{V}\left(\frac{\partial V}{\partial T}\right)_P = -\frac{1}{\rho}\left(\frac{\partial \rho}{\partial T}\right)_P \tag{8}$$

(the minus sign is because $\partial\rho/\partial T$ is negative). Thus the density perturbation for the halfspace cooling model is

$$\rho'(z,t) = \alpha\rho_m[T_m - T(z,t)] = \alpha\rho_m T_m\left[1 - \text{erf}\left(\frac{z}{2\sqrt{\kappa t}}\right)\right]. \tag{9}$$

If the density of water is ρ_w, equal mass in the two columns requires that

$$\rho_m m(t) = \rho_w h(t) + \int_{h(t)}^{m(t)} [\rho_m + \rho'(z,t)]dz, \tag{10}$$

which gives the isostatic condition for ocean depth,

$$h(t) = \frac{1}{(\rho_m - \rho_w)}\int_{h(t)}^{m(t)} \rho'(z,t)dz. \tag{11}$$

Because temperature and density in the plate are defined for all values of z (the thickness of the plate is defined as some chosen isotherm), let $z' = z - h(t)$ and $m(t) \to \infty$. Then

$$h(t) = \frac{\alpha\rho_m T_m}{(\rho_m - \rho_w)}\int_0^\infty \left[1 - \text{erf}\left(\frac{z'}{2\sqrt{\kappa t}}\right)\right]dz'. \tag{12}$$

To evaluate the integral, substitute $s = z'/2\sqrt{\kappa t}$ and integrate by parts (try it!) to show that

$$\int_0^\infty [1 - \text{erf}(s)]\,ds = 1/\sqrt{\pi}. \tag{13}$$

Thus ocean depth should increase as the square root of plate age,

$$h(t) = 2\sqrt{\frac{\kappa t}{\pi}}\frac{\alpha\rho_m T_m}{(\rho_m - \rho_w)}. \tag{14}$$

The cooling of the lithosphere should also cause heat flow at the sea floor to vary with age. By Fourier's law of heat conduction, the heat flow at the sea floor is the product

$$q = k\frac{dT}{dz} \qquad z = 0 \tag{15}$$

of the temperature gradient at the sea floor and the thermal conductivity k.[4] An easy approximation to see how heat flow varies with age is to consider the T_m isotherm as the base of the lithosphere, so that the thickness of the lithosphere increases

[3] Isostasy is the general idea that topography results from equal masses in different columns. Here we consider thermal isostasy, in which density changes produced by temperature variations cause topographic differences. Another common model, Airy isostasy, is used to explain the relation between crustal thickness variations and topography, such as crustal roots under mountains.

[4] Normally, this equation requires a minus sign because heat flows from hot objects to cold ones. Without this sign, hot objects would get hotter. There is none here because of our customary but inconsistent definitions: heat flow is measured upward whereas depth is measured downward.

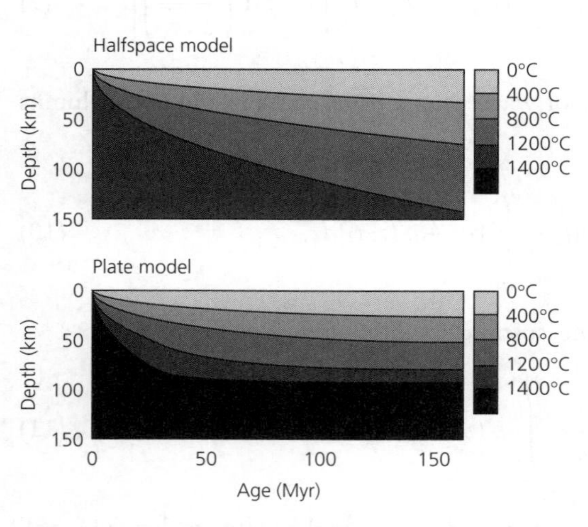

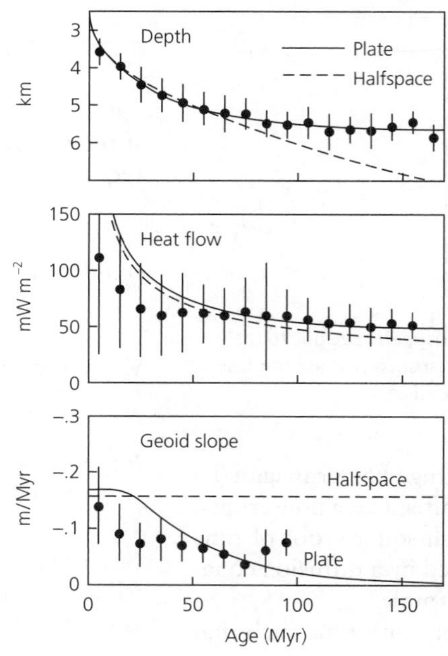

Fig. 5.3-7 Models and data for thermal evolution of the oceanic lithosphere. *Left*: Isotherms for thermal models. The lithosphere continues cooling for all ages in a halfspace model, but equilibrates for ~70 Ma lithosphere in a plate model with a 95 km-thick thermal lithosphere. The plate model shown has a higher basal temperature than the halfspace model. *Right*: Comparison of thermal model predictions to different data. All show a lithospheric cooling signal, and are better (but far from perfectly) fit by the predictions of a plate model (solid lines) than by those of a halfspace model (dashed lines). (Richardson *et al.*, 1995. *Geophys. Res. Lett.*, 22, 1913–16, copyright by the American Geophysical Union.)

with the square root of age. Approximating the gradient at the surface by the average gradient through the lithosphere,

$$q(t) \approx k\frac{\Delta T}{\Delta z} \approx \frac{kT_m}{\sqrt{\kappa t}} \tag{16}$$

predicts that the heat flow decreases as the square root of age. The same result can be obtained by differentiation of the temperature structure (Eqn 4) using

$$\frac{d}{dz}\operatorname{erf}(s) = \frac{d}{dz}\frac{2}{\sqrt{\pi}}\int_0^s e^{-\sigma^2}d\sigma = \frac{2}{\sqrt{\pi}}e^{-s^2}\frac{ds}{dz}, \tag{17}$$

which gives

$$q(t) = k\frac{dT}{dz}\bigg|_{z=0} = k\frac{2T_m}{\sqrt{\pi}}e^{\frac{-z^2}{4\kappa t}}\frac{1}{2\sqrt{\kappa t}}\bigg|_{z=0} = \frac{kT_m}{\sqrt{\pi\kappa t}}. \tag{18}$$

This model, which predicts that lithospheric thickness, heat flow, and ocean depth vary as the square root of age for all ages is called a halfspace model (Fig. 5.3-7, *upper left*). In it, the lithosphere is the upper layer of a halfspace that continues cooling for all time. (In reality, oceanic lithosphere never gets older than 200 million years old because it gets subducted.) The model does a good job of describing the average variation in ocean depth and heat flow with lithospheric age.

However, because ocean depth seems to "flatten" at about 70 Myr, we often use a modification called a plate model (Fig. 5.3-7, *lower left*), which assumes that the lithosphere evolves toward a finite plate thickness L with a fixed basal temperature T_m. In this model,

$$T(x,z) = T_m\left[\frac{z}{L} + \sum_{n=1}^{\infty} c_n \exp\left(-\frac{\beta_n x}{L}\right)\sin\left(\frac{n\pi z}{L}\right)\right], \tag{19}$$

where $c_n = 2/(n\pi)$, $\beta_n = (R^2 + n^2\pi^2)^{1/2} - R$, $R = vL/(2\kappa)$. The constant R, known as the thermal Reynolds number, relates the rates at which heat is transported horizontally by plate motion and conducted vertically. In this model isotherms initially deepen as the square root of age, but eventually level out. The flattening reflects the fact that heat is being added from below, which the model approximates by having old lithosphere reach a steady-state thermal structure that is simply a linear geotherm (Fig. 5.3-8, *top*). As a result, the predicted sea floor depth and heat flow also behave for young ages like in the halfspace model, but evolve asymptotically toward constant values for old ages. Both have simple interpretations: the heat flow is proportional to the geotherm, and thus T_m/L, whereas the depth is proportional to the thermal subsidence and hence heat lost since the plate formed at the ridge, and thus the product T_mL. The model parameters can be estimated by an inverse problem, finding those that best fit a set of depth and heat flow data versus age (Fig. 5.3-8, *bottom*).

Comparison with data shows that the plate thermal model is a good, but not perfect, fit to the average data because processes other than this simple cooling are also occurring. For example, ocean depth is also affected by uplift associated with hot spots (Section 5.2.4). Water flow in the crust transports some of the heat for ages less than about 50 Ma, making the observed heat flow lower than the model's predictions, which assume that all heat is transferred by conduction. Some topographic effects, including the spectacular volcanic oceanic plateaus, result from crustal thickness variations. Because these and other effects vary from place to place, the data vary about their average values for a given age.

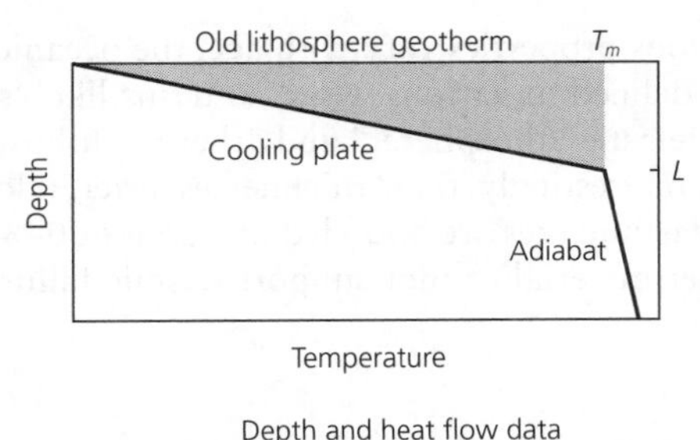

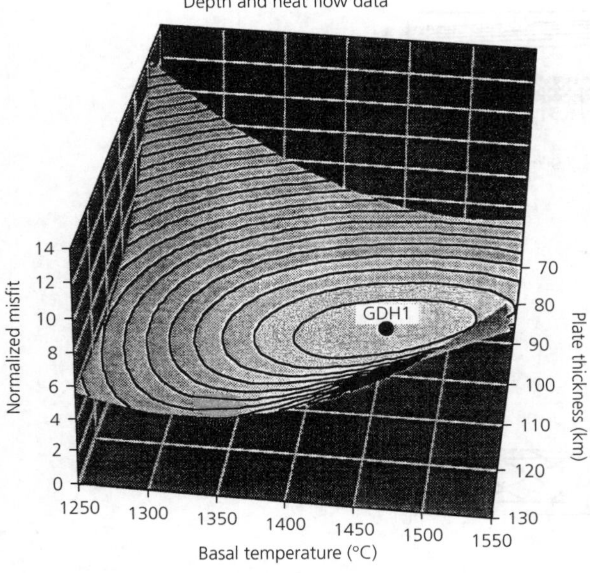

Fig. 5.3-8 *Top*: Asymptotic thermal structure for old lithosphere in a plate model. The sea floor subsidence from the ridge, and thus ocean depth, is proportional to the shaded area between the geotherm and $T = T_m$, whereas heat flow is proportional to the geotherm. A schematic adiabatic temperature gradient (Section 5.4.1) is shown beneath the plate. (Stein and Stein, 1992. Reproduced with permission from *Nature*.) *Bottom*: Fitting process used for thermal model parameters. The misfit to a set of depth and heat flow data has a minimum at the point labeled GDH1, a plate thermal thickness of 95 ± 15 km and basal temperature of 1450 ± 250°C. (Stein and Stein, 1996. *Subduction*, 1–17, copyright by the American Geophysical Union.)

We can view ocean depth, heat flow, and several other properties of the oceanic lithosphere as observable measures of the temperature in the cooling lithosphere. Because the observables depend on different combinations of parameters (Table 5.3-1), they can be used together to constrain individual parameters (a halfspace model corresponds to an infinitely thick plate). The depth depends on the integral of the temperature (Eqn 11), whereas the heat flow depends on its derivative at the sea floor (Eqn 15). Similarly, the slope of the geoid, a function of the gravity field depending on a weighted integral of the density, also varies with age in general agreement with the plate model's prediction (Fig. 5.3-7).

In addition, the elastic thickness of the lithosphere inferred from the deflection caused by loads such as seamounts (Fig. 5.3-9a), the maximum depth of intraplate earthquakes within the oceanic lithosphere (Fig. 5.3-9b), and the depth to

Table 5.3-1 Constraints on thermal models $T(z, t)$.

Observable	Proportional to	Reflects
Young ocean depth	$\int T(z,t)dz$	$k^{1/2}\alpha T_m$
Old ocean depth	$\int T(z,t)dz$	$\alpha T_m L$
Old ocean heat flow	$\left.\frac{\partial T(z,t)}{\partial z}\right\vert_{z=0}$	kT_m/L
Geoid slope	$\frac{\partial}{\partial t}\int zT(z,t)dz$	$k\alpha T_m \exp(-kt/L^2)$

Source: Stein and Stein (1996).

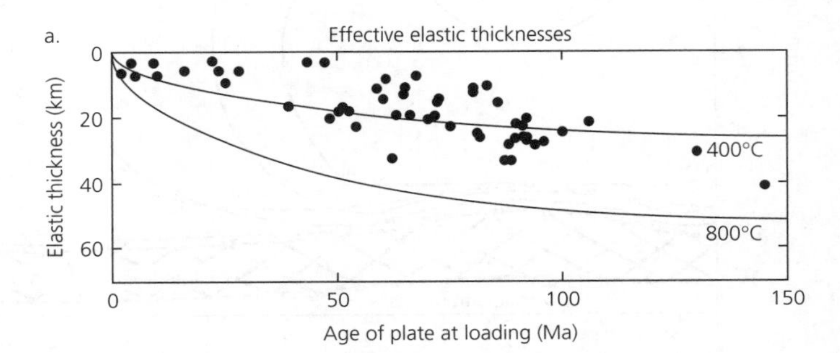

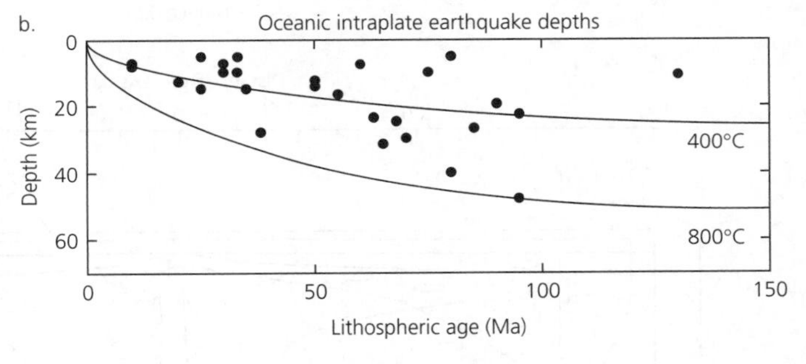

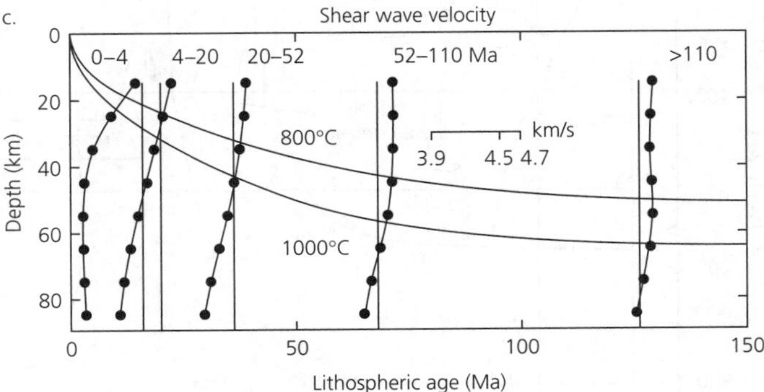

Fig. 5.3-9 Comparison of isotherms as functions of age for a plate model to three datasets whose variation with age is consistent with cooling of the lithosphere. The effective elastic thickness (a), deepest intraplate seismicity (b), and depth to the low-velocity zone, shown by velocity profiles at different ages (c), all increase with age. (After Stein and Stein, 1992. Reproduced with permission from *Nature*.)

the low-velocity zone determined from surface wave dispersion (Figs. 5.3-9c and 2.8-7), all increase with age. Hence the cooling of oceanic lithosphere causes the expected increase in strength and seismic velocity. Moreover, as discussed in Section 5.5, the resulting density increase is thought to provide a major force driving plate motions.

Because various properties vary with age, the oceanic lithosphere can be defined in various ways, so terms like "seismic lithosphere," "elastic lithosphere," and "thermal lithosphere" are often used. Interestingly, these thicknesses differ. It looks as if the deepest earthquakes are bounded by about 600–800 °C, such that hotter material cannot support seismic failure. The

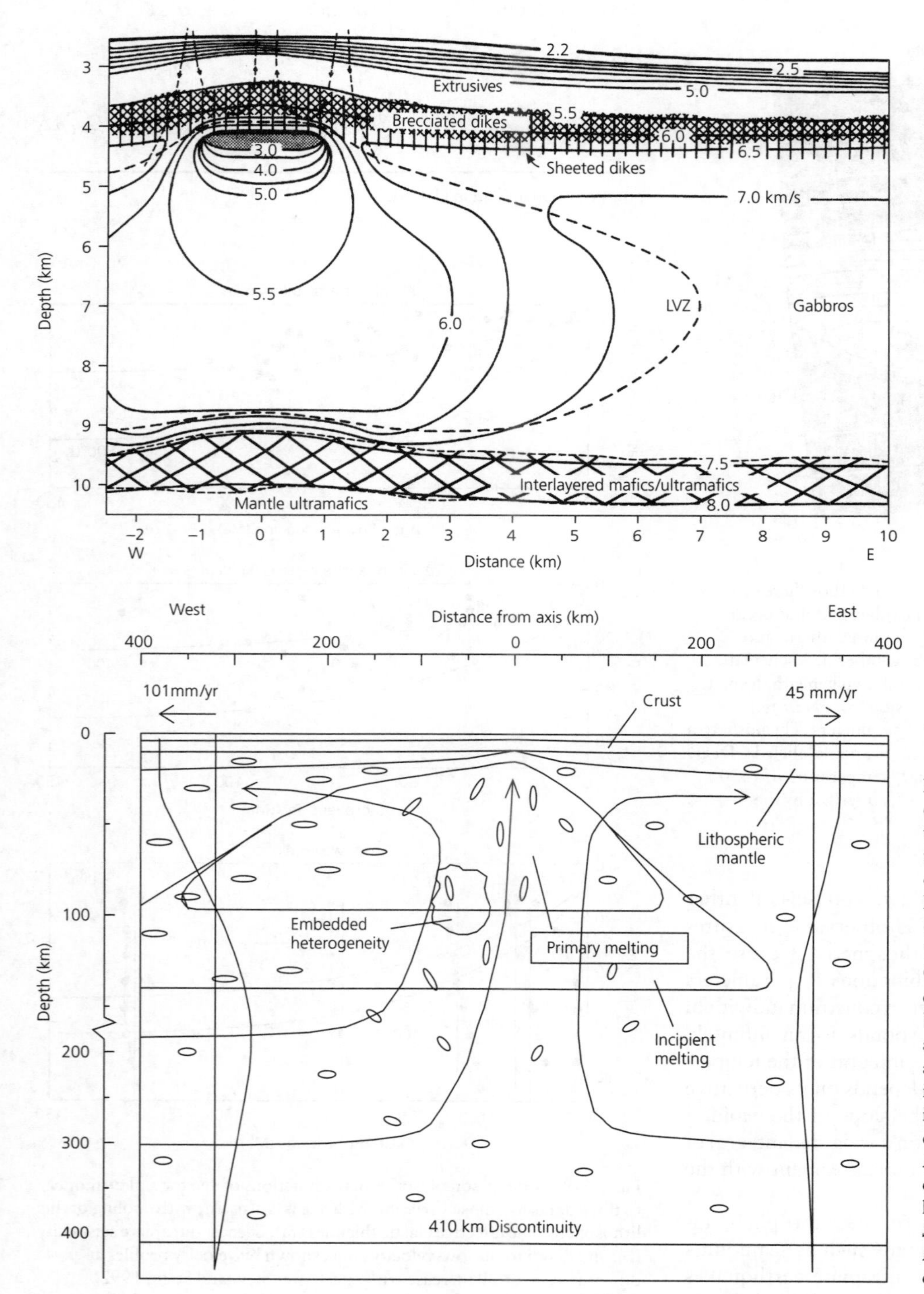

Fig. 5.3-10 *Top*: Geological interpretation of a multichannel seismic velocity study on the East Pacific rise. A low-velocity region under the axis is interpreted as a hot region of melting, capped by a magma lens. Dashed lines are possible paths of water circulation. (Vera *et al.*, 1990. *J. Geophys. Res., 95*, 15,529–56, copyright by the American Geophysical Union.) *Bottom*: Schematic cross-section across the East Pacific rise. The broad region of low velocities is interpreted as the primary melting region. Small ellipses are directions of preferred olivine alignment inferred from anisotropy. Lines with arrows indicate inferred mantle flow, causing the distortion shown of an initially vertical line. Absolute velocities of the two plates (Pacific on left, Nazca on right) are given by small horizontal arrows. (Forsyth *et al.*, 1998. *Science, 280*, 1215–18, copyright 1998 American Association for the Advancement of Science.)

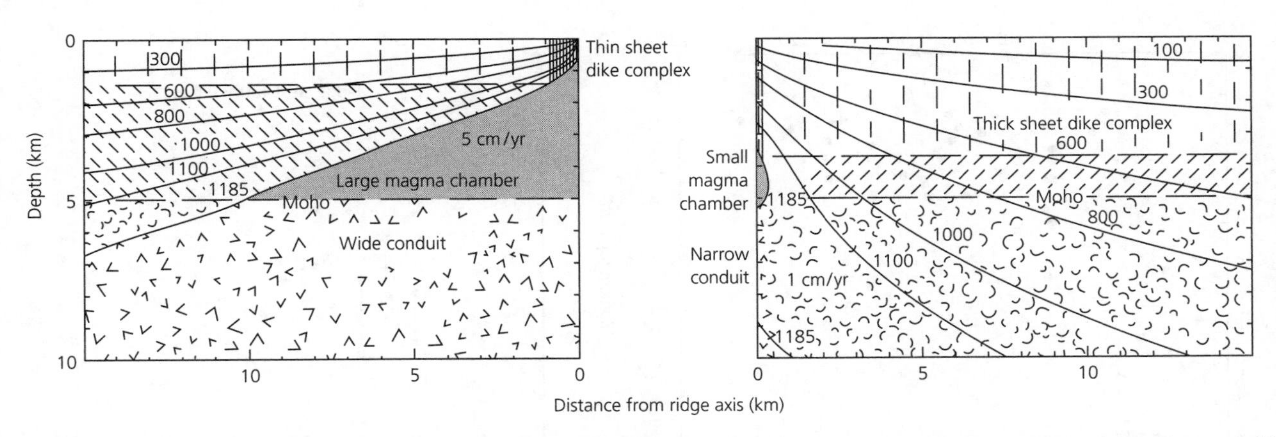

Fig. 5.3-11 Thermal and petrological model for the difference between fast-spreading (*left*) and slow-spreading (*right*) ridges. (Sleep and Rosendahl, 1979. *J. Geophys. Res.*, *84*, 6831–9, copyright by the American Geophysical Union.)

elastic thickness corresponds approximately to the 400 °C isotherm, whereas the low-velocity zone begins approximately below the 1000 °C isotherm (Fig. 5.3-9c). These differences, discussed in Section 5.7, likely result from rock being stronger for more rapid deformation. All of these thicknesses, however, only approximate what we would like to know but cannot directly measure: the depth of the base of the moving plate, which is likely to be a gradational rather than a distinct boundary.

5.3.3 Ridge and transform earthquakes and processes

Seismology makes important contributions to understanding the properties and behavior of spreading centers. Ocean bottom seismometers yield locations of microearthquakes and data for travel time and waveform studies. Larger earthquakes are also studied using teleseismic body and surface waves. The seismological results are being integrated with marine geophysical and petrological data to develop better models. For example, Fig. 5.3-10 (*top*) shows a geological interpretation of a multichannel seismic study (Section 3.3) that used air gun and explosive sources to image velocity structure under the East Pacific rise to a depth of about 10 km. A low-velocity region under the axis is interpreted as a hot melting region capped by a magma lens. Other studies using ocean bottom seismometers and distant earthquake sources map the structure to greater depth, including inferring flow directions under the ridge axis using anisotropy (Fig. 5.3-10, *bottom*). Such studies are finding interesting features of the spreading process. For example, the broad region of low velocity presumed to be the primary melting area extends further west than east of the axis. This asymmetry may occur because the westward absolute motion of the Pacific plate is much faster than the eastward absolute motion of the Nazca plate, causing the ridge to migrate westward relative to the deep mantle. Thus the spreading process, which depends on the relative plate motion (spreading rate), also seems affected by the absolute motion.

Some effects of the spreading rate are illustrated by a model shown in Fig. 5.3-11. At a given distance from the ridge, faster spreading produces younger lithosphere and isotherms closer to the surface than does slow spreading. If the region beneath the 1185 °C isotherm and above the Moho depth of 5 km is considered to be a magma chamber, a fast ridge has a larger magma chamber. Hence crust moving away from a fast-spreading ridge is more easily replaced than that moving away from a slow ridge. Thus, in contrast to the axial valley and normal faulting earthquakes on a slow ridge, a fast ridge has an axial high and an absence of earthquakes. Similarly, both the depths and the maximum seismic moments[5] of ridge crest normal faulting earthquakes decrease with spreading rate (Fig. 5.3-12). These observations are consistent with the fault area decreasing on faster-spreading and hotter ridges, because faulting requires that rock be below a limiting temperature, above which it flows (Section 5.7). The idea that the faulting depends on temperature is also implied by the increase in the maximum depth of oceanic intraplate earthquakes with age (Fig. 5.3-9b).

Transform fault earthquakes also depend on thermal structure. The temperatures along a transform fault should be essentially the average of the expected temperature on the two sides; coolest at the transform midpoint and hottest at either end (Fig. 5.3-13). As expected from the area available for faulting, the maximum seismic moment for transform earthquakes decreases with spreading rate (Fig. 5.3-14), consistent with the idea of faulting limited to a zone bounded by the isotherms.

An interesting question is how the seismic moments of transform earthquakes relate to the plate motion. The average slip rate from earthquakes can be inferred from the total seismic moment released on a transform, assuming that

$$\text{seismic slip rate} = \frac{\text{total seismic moment}}{(\text{fault area})(\text{rigidity})(\text{time period})}. \qquad (20)$$

[5] Recall (Section 4.6) that the seismic moment is the product of the rigidity, the slip in the earthquake, and the fault area.

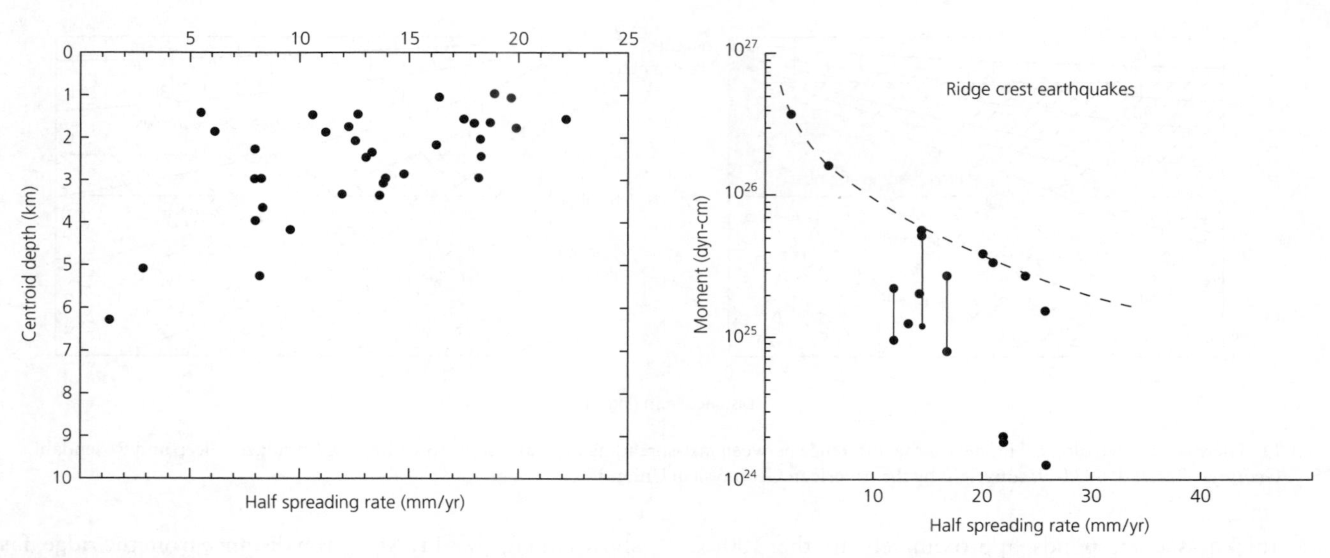

Fig. 5.3-12 *Left*: Shallowing of focal depth for ridge crest normal fault earthquakes with half-spreading rate. (After Huang and Solomon, 1988. *J. Geophys. Res.*, *93*, 13, 445–77, copyright by the American Geophysical Union.) *Right*: Corresponding decrease in maximum seismic moment. (After Solomon and Burr, 1979. *Tectonophysics, 55*, 107–26, with permission from Elsevier Science.)

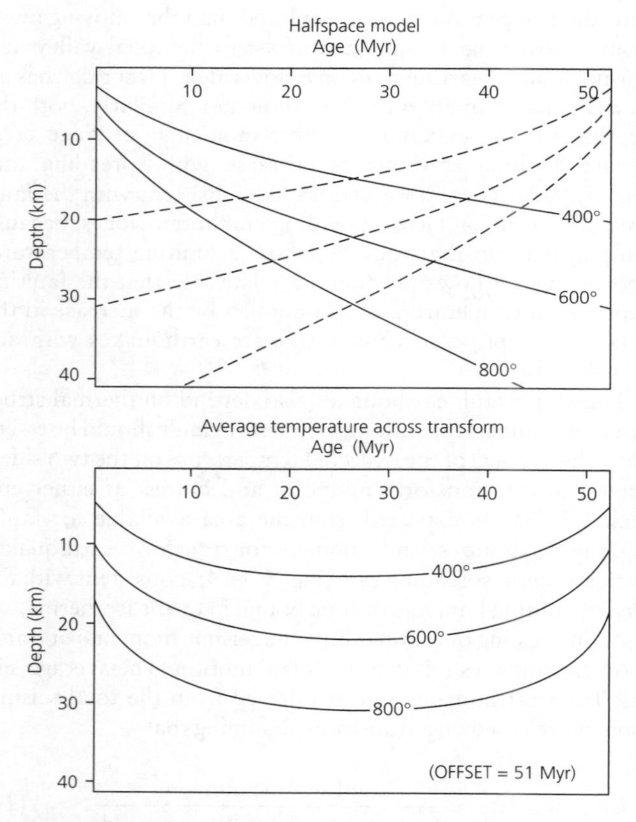

Fig. 5.3-13 Thermal model of the Romanche Transform. *Top*: Temperatures on either flank predicted by the cooling halfspace model. *Bottom*: Average temperature distribution along the transform. (After Engeln *et al.*, 1986. *J. Geophys. Res.*, *91*, 548–77, copyright by the American Geophysical Union.)

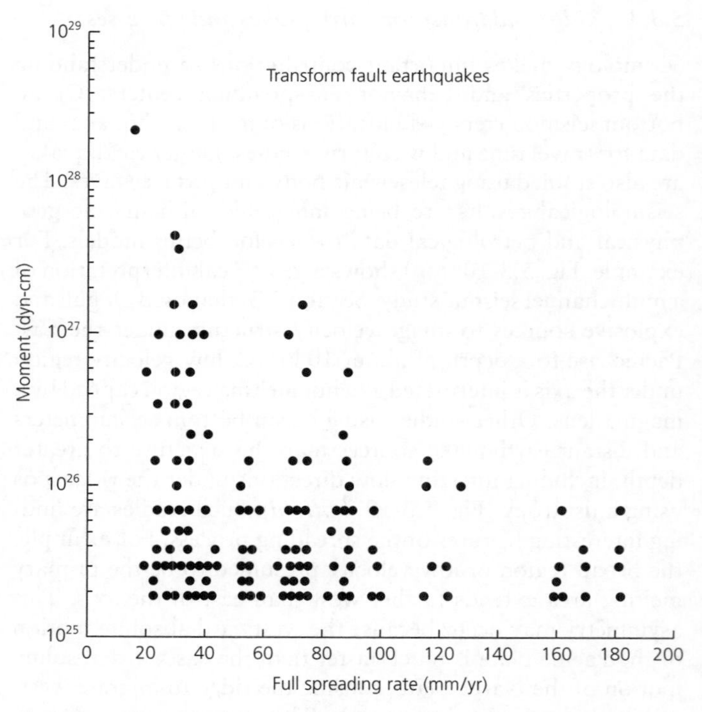

Fig. 5.3-14 Seismic moment versus spreading rate for oceanic transforms. The maximum moment decreases with spreading rate, as expected from thermal considerations. (After Solomon and Burr, 1979. *Tectonophysics, 55*, 107–26, with permission of Elsevier Science.)

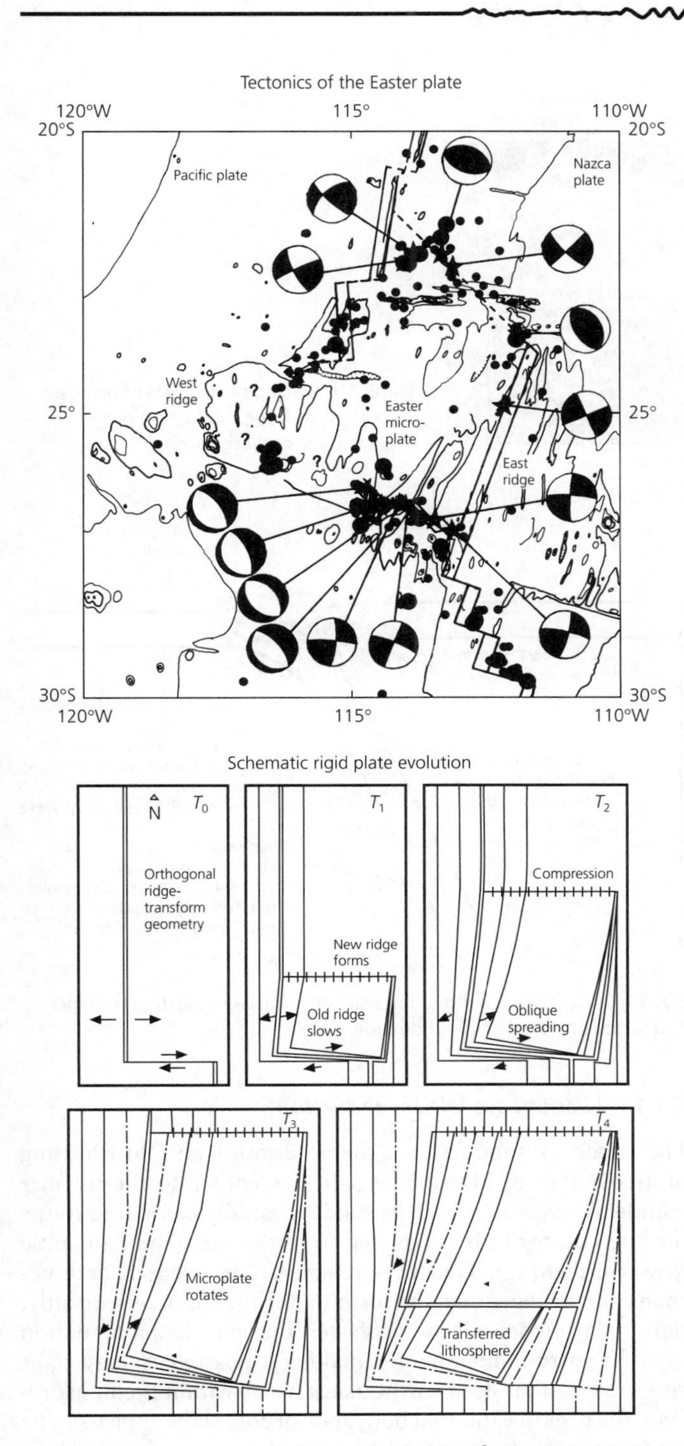

Fig. 5.3-15 The Easter microplate on the East Pacific rise. *Top*: Seismicity (dots) and focal mechanisms in the microplate region. Note the normal faulting on the southern boundary. (After Engeln and Stein, 1984.) *Bottom*: Schematic model for the evolution of a rigid microplate between two major plates by rift propagation. Successive isochrons illustrate the northward propagation of the east ridge, slowing of spreading on the west ridge, the rotation of the microplate, the reorientation of the two ridges, and the conversion of the initial transform into a slow and obliquely spreading ridge. (Engeln *et al.*, 1988. *J. Geophys. Res.*, *93*, 2839–56, copyright by the American Geophysical Union.)

Using this relation requires inferring the fault area, which depends on both the transform length and the depth to which faulting occurs. Assuming the area above the 600–700 °C isotherms fails seismically, the seismic slip rate for major Atlantic transforms is generally less than predicted by the plate motion. Thus, if the time period sampled is long enough to be representative — a major question — some of the plate motion occurs aseismically. The issue of how much slip occurs seismically remains unresolved, as we will see when we discuss subduction zones (Section 5.4.3) and intraplate deformation zones (Section 5.6.2).

In addition, seismology helps study how ridge-transform systems evolve. For example, the East Pacific rise near Easter Island contains two approximately parallel sections (Fig. 5.3-15, *top*). Earthquakes occur on these ridges, but not between them, suggesting that the area in between is an essentially rigid microplate. The normal fault earthquakes on the microplate's southern boundary are surprising because the East Pacific rise here is a very fast-spreading (15 cm/yr) ridge, which should not have normal fault earthquakes (Fig. 5.3-12). Magnetic anomalies show that the east ridge segment is propagating northward and taking over from the old (west) ridge segment. Figure 5.3-15 (*bottom*) shows a simplified model of this process. Because finite time is required for the new ridge to transfer spreading from the old ridge, both ridges are active at the same time, and the spreading rate on the new ridge is very slow at its northern tip and increases southward. As a result, the microplate rotates, causing compression (thrust faulting) and extension (normal faulting) at its north and south boundaries, respectively. Ultimately the old ridge will die, transferring lithosphere originally on the Nazca plate to the Pacific plate, and leaving inactive fossil ridges on the sea floor. Both V-shaped magnetic anomalies characteristic of ridge propagation and fossil ridges are widely found in the ocean basins, showing that this is a common way that ridges reorganize. Even for smaller (a few km) propagating ridge systems, studies of the associated earthquakes can yield useful information about the propagation process.

5.4 Subduction zones

We have seen that earthquakes at spreading centers, which at shallow depths are upwelling limbs of the mantle convection system, reflect the processes forming oceanic lithosphere there. In a similar way, earthquakes at subduction zones, downwelling limbs of the convection system, reflect the processes by which oceanic lithosphere reenters the mantle. Plate convergence takes different forms, depending on the plates involved. Figure 5.4-1 shows the basic model for a situation where oceanic lithosphere of one plate subducts beneath oceanic lithosphere of the overriding plate. Typically, a volcanic island arc forms, and sea floor spreading occurs behind the arc, forming a back-arc basin or marginal sea. Earthquakes occur both at the trench and to great depth, forming a dipping

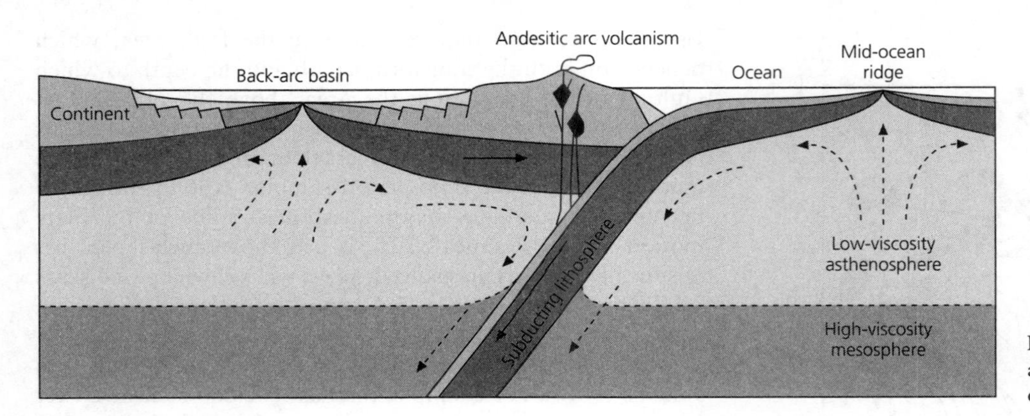

Fig. 5.4-1 Schematic diagram of processes associated with the subduction of one oceanic plate beneath another.

Wadati–Benioff zone. By contrast, when oceanic lithosphere subducts beneath a continent, a mountain chain like the Andes forms on the continent, and the oceanic lithosphere forms a Wadati–Benioff zone. Finally, because continental crust cannot subduct, convergence between two continental plates, as in the Himalayas, causes crustal thickening, mountain building, and shallow earthquakes but does not create a Wadati–Benioff zone.

Subduction zones have a wide variety of earthquakes with different focal mechanisms and depths. There are shallow (less than 70 km deep), intermediate (70–300 km deep), and deep (more than 300 km deep) focus earthquakes.[1] These earthquakes occur in different tectonic environments. The intermediate and deep earthquakes forming the Wadati–Benioff zone occur in the cold interiors of downgoing slabs. The shallow earthquakes are associated with the interaction between the two plates. The largest and most common of these shallow earthquakes occur at the interface between the plates, and release the plate motion that has been locked at the plate interface. In addition, shallow earthquakes can occur within both the overriding and the subducting plates. Figure 5.4-2 shows some features of seismicity observed in subduction zones. Not all features have been observed at all places. For example, the dips and shapes of subduction zones vary substantially. Some show double planes of intermediate or deep seismicity, whereas others do not.

In discussing subduction zones, we follow an approach similar to that used in the last section for ridges. We introduce thermal models for subduction, then use them to gain insight into earthquake and seismic velocity observations. We will see that seismological observations, thermal models, and calculations of the behavior of materials at high temperature and pressure are combined to investigate these complicated regions. In general, the seismological observations are fairly clear, but they can be interpreted in terms of a variety of models. As a result, subduction zone studies remain active, fruitful, and exciting.

[1] Slightly different definitions have been used for these depth ranges; for example, 325 km has also been used as the upper limit for deep earthquakes.

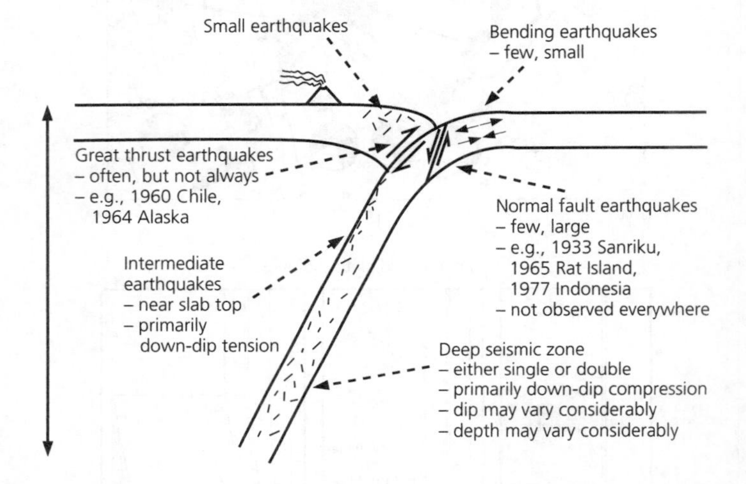

Fig. 5.4-2 Composite subduction zone showing some earthquake types. Not all are observed at all subduction zones.

5.4.1 *Thermal models of subduction*

The essence of subduction is the penetration and slow heating of a cold slab of lithosphere as it descends into the warmer mantle. As we will see, slabs subduct rapidly compared to the time needed for heat conducted from the surrounding mantle to warm them up. Thus they remain colder, denser, and mechanically stronger than the surrounding mantle. Consequently, slabs transmit seismic waves faster and with less attenuation than the surrounding mantle, making it possible to map slabs and to show that deep earthquakes occur within them. Moreover, the negative thermal buoyancy of cold slabs appears to be the primary force driving plate motions and provides a major source of stress within them that causes deep earthquakes.

To explore the thermal evolution of slabs, we use two approaches. First, we discuss a simplified analytic thermal model that allows insights into the physics. We then discuss numerical models that incorporate additional effects in the hope of providing a more realistic description. We highlight some significant points, and more complete information can be found in the references.

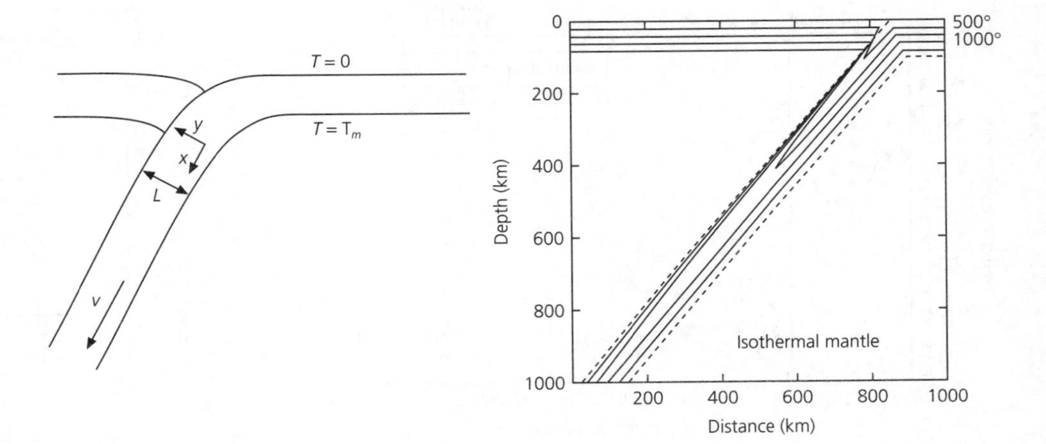

Fig. 5.4-3 An analytic model for temperatures in a subducting plate. *Left*: model geometry. *Right*: Results, showing the cold slab heating up as it descends through the hotter surrounding mantle.

The analytic model (Fig. 5.4-3) considers a semi-infinite slab of thickness L subducting at rate v. The surrounding mantle is at temperature T_m, and the plate enters the trench with a linear temperature gradient from $T = 0$ at its top to T_m at its base. We define the x axis down the dip of the slab, and the y axis across the slab. The evolution of the region is given by a slightly more complicated version of the heat equation (Eqn 5.3.1) used to model the cooling of the lithosphere as it moves away from the ridge. This version,

$$\rho C_p\left(\frac{\partial T}{\partial t} + v\nabla T\right) = \nabla \cdot (k\nabla T) + \varepsilon, \qquad (1)$$

describes the evolution of the temperature field, $T(x, y, t)$, as a function of time and the two space coordinates. In addition to the heat conduction term $\nabla \cdot (k\nabla T)$, Eqn 1 includes a $v\nabla T$ term describing the transfer (or advection) of heat by movement of material, and the ε term representing additional sources or sinks of heat such as radioactivity and phase changes. This form allows key parameters such as the density ρ, specific heat C_p, thermal conductivity k, and heat sources or sinks ε to vary with position. For a simple analytic solution, we assume that the problem is steady state ($\partial T/\partial t = 0$) and neglect heat sources and sinks ($\varepsilon = 0$). We further assume that the physical properties of the material (ρ, C_p, k, and hence the thermal diffusivity $\kappa = k/\rho C_p$) are independent of position.

With these simplifications, Eqn 1 becomes

$$\rho C_p v \frac{\partial T}{\partial x} = k\left(\frac{\partial^2 T}{\partial x^2} + \frac{\partial^2 T}{\partial y^2}\right), \qquad (2)$$

which has a series solution

$$T(x, y) = T_m[1 + 2\sum_{n=1}^{\infty} c_n \exp(-\beta_n x/L) \sin(n\pi y/L)], \qquad (3)$$

with

$$c_n = (-1)^n/(n\pi), \quad \beta_n = (R^2 + n^2\pi^2)^{1/2} - R, \quad R = vL/(2\kappa).$$

R, the dimensionless thermal Reynolds number, is the ratio of the rate at which cold material is subducted to that at which it heats up by conduction. This solution resembles the temperature field in the plate model of cooling lithosphere (Eqn 5.3.19), because both models describe the thermal evolution of a plate of finite thickness with temperature boundary conditions at the top, bottom, and one end. In the previous case the plate cools, whereas in this case it heats up.

To find how far along the slab a given isotherm penetrates, we approximate the series by its first term and use the fact that $R \gg \pi$, so

$$T(x, y) \approx T_m[1 - (2/\pi) \exp(-\pi^2 x/(2RL)) \sin(\pi y/L)]. \qquad (4)$$

Solving for the point where $\partial T/\partial y = 0$ yields $y = L/2$, the middle of the slab. In fact, taking additional terms shows that this point is actually closer to the colder top (Fig. 5.4-3). Using the first-term approximation, a temperature T_0 goes furthest into the subduction zone at

$$T_0(x_0, L/2) = T_m[1 - (2/\pi) \exp(-\pi^2 x_0/(2RL)], \qquad (5)$$

and reaches a maximum down-dip distance

$$x_0 = -vL^2/(\pi^2\kappa) \ln[\pi(T_m - T_0)/(2T_m)]. \qquad (6)$$

To convert this distance to depth in the mantle, we multiply by $\sin\delta$, where δ is the slab dip. This correction converts the subduction rate v to the slab's vertical descent rate $v \sin\delta$. Thus an isotherm's maximum depth is proportional to the subduction rate and the square of the plate thickness, so faster subduction or a thicker slab allows material to go deeper before heating up. If we assume that the square of the plate thickness is proportional to its age, the maximum depth to an isotherm in the downgoing slab is proportional to the vertical descent rate times the age, t, of the subducting lithosphere.

This idea can be tested by assuming, as we did for spreading center earthquakes, that the maximum depth of earthquakes is temperature-controlled, so earthquakes should cease once material reaches a temperature that is too high. To compare

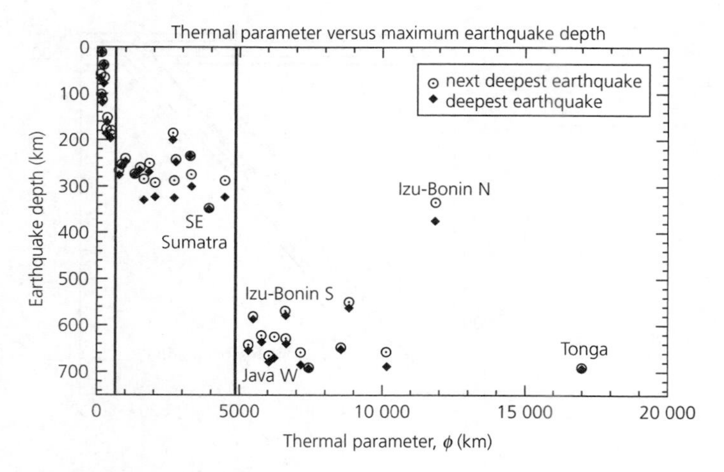

Fig. 5.4-4 Maximum earthquake depths for different subduction zones as a function of thermal parameter, the product of vertical descent rate and lithospheric age. If earthquakes are limited by temperature, this observation is consistent with the simple thermal model's prediction that the maximum depth to an isotherm should vary with the thermal parameter. (After Kirby *et al.*, 1996b. *Rev. Geophys.*, *34*, 261–306, copyright by the American Geophysical Union.)

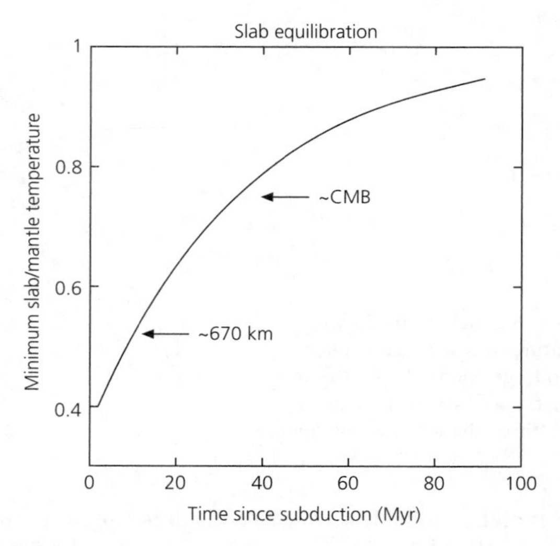

Fig. 5.4-5 Minimum temperature within a slab as a fraction of the mantle temperature, as a function of the time since subduction, computed using the analytic thermal model (Fig. 5.4-3). The coldest portion reaches half the mantle temperature in about 10 Myr, by which time a typical slab is approximately at 670 km depth, and 80% of it in 40 Myr, by which time a slab that continued descending at the same rate would reach the core–mantle boundary. Slabs can thus remain thermally distinct for long periods of time. (Stein and Stein, 1996. *Subduction*, 1–17, copyright by the American Geophysical Union.)

various subduction zones, we examine the maximum depth of earthquakes as a function of their *thermal parameter*

$$\phi = tv \sin \delta. \tag{7}$$

Figure 5.4-4 shows that the maximum depth increases with thermal parameter, and deep earthquakes below 300 km occur only for slabs with a thermal parameter greater than about 5000 km.

However, the fact that the earthquakes stop does not mean that the slab has equilibrated with the surrounding mantle. Figure 5.4-5 shows the predicted minimum temperature within a slab as a function of time since subduction, assuming it maintains its simple planar geometry and does not buckle or thicken. The coldest portion reaches only about half the mantle temperature in about 10 Myr, which is about the time required for the slab to reach 660 km. Thus the restriction of seismicity to depths shallower than 660 km does not indicate that the slab is no longer a discrete thermal and mechanical entity. From a thermal standpoint, there is no reason for slabs not to penetrate into the lower mantle, an issue we discuss shortly. If a slab descended through the lower mantle at the same rate (in fact, it would probably slow down due to the more viscous lower mantle), it would retain a significant thermal anomaly at the core–mantle boundary, consistent with some models of that region (Section 3.8.4).[2]

The thermal model can be improved with simple modifications. Although we assumed that the slab subducts into an isothermal mantle, temperature should increase with depth, as the material is compressed due to increasing pressure from the overlying rock. Because the mantle below the lithosphere is thought to be convecting, it is often assumed that self-compression occurs adiabatically, such that material moving vertically neither loses nor gains heat. In this case, equilibrium thermodynamics requires that the effects of temperature and pressure changes exactly offset each other,

$$dS = \frac{C_p}{T}\, dT - \frac{\alpha}{\rho}\, dP = 0, \tag{8}$$

so that the entropy S does not change. This condition gives the adiabatic temperature gradient, or *adiabat*, as

$$\left(\frac{dT}{dP}\right)_s = \frac{\alpha}{\rho C_p} T, \tag{9}$$

where α is the coefficient of thermal expansion. Because pressure increases with depth as $dP/dz = \rho g$, temperature increases with depth as

$$\left(\frac{dT}{dz}\right)_s = \frac{\alpha g}{C_p} T. \tag{10}$$

We can thus correct the temperatures for the isothermal mantle case to include adiabatic heating. Using the entropies requires using absolute (Kelvin) temperatures, equal to the Celsius temperature plus 273.15°. Thus if the absolute temperature at

[2] The oceanic lithosphere takes about 70 Myr to cool to equilibrium with the mantle below, and so takes about half that time to heat up again from both sides after it subducts.

depth z_0, the base of the plate, is T_0^K, we integrate Eqn 10 to find the absolute temperature at depth z,

$$T^K(z) = T_0^K \exp\,[(\alpha g/C_p)(z - z_0)]. \qquad (11)$$

Another possibly important effect is that of heat sources and sinks. For example, the olivine to spinel transition, which gives rise to the 410 km discontinuity outside the slab, should release heat as it occurs in the slab. Heat might also be generated by friction at the top of the downgoing slab. The heat produced is the product of the subduction rate and the shear stress on the slab interface. The magnitude of this effect is difficult to estimate. It should not be significant unless the shear stress is greater than a few kilobars. As discussed later (Section 5.7.5), the stress on faults is unknown. A further complexity results from the fact that the viscosity of the mantle, which controls the stress, decreases exponentially with temperature. Thus, if frictional heating raises the temperature at the slab interface, viscosity, and hence stress, would decrease, tending to counteract the effect.

To address these complexities, we use numerical models to solve the heat equation at every point in the slab. These models allow parameters such as density to vary with position. In addition, heat sources and sinks such as radioactive heating, phase changes, and frictional heating can be incorporated. The results of such calculations are similar to those of the analytic model and are used to explore how temperatures should vary between subduction zones. For example, Fig. 5.4-6 compares models for a relatively younger and slower-subducting slab (thermal parameter about 2500 km), approximating the Aleutian arc, and an older, faster-subducting slab (thermal parameter approximately 17,000 km), approximating the Tonga arc. As expected, the slab with the higher thermal parameter warms up more slowly, and is thus colder. This prediction is consistent with the observation that Tonga has deep earthquakes, whereas the Aleutians do not (Fig. 5.4-4).

Although we can compute such thermal models, a question is whether they make sense. We test them using two seismological datasets: earthquake locations and seismic velocities. Travel time tomography (Section 7.3) across subduction zones shows high-velocity slabs (Fig. 5.4-7). These results are compared to the velocities predicted using a thermal model of the subducting slab and laboratory values for the variation in velocity with temperature. The model predicts coldest temperatures in the slab interior where the earthquakes occur. Because the tomographic inversion finds the velocity within rectangular cells, the model is converted to that grid and then "blurred" because the seismic rays do not uniformly sample the slab. As shown by the hit count, the number of rays sampling each cell, most rays go down the high-velocity slab, yielding a somewhat distorted image. The fact that this image and the tomographic result are similar suggests that the model is a reasonable description of the actual slab. A similar conclusion emerges from the observation that the tomographic result also resembles parts of the model image that are artifacts, velocity anomalies that are not present in the original model. These artifacts, generally of low amplitude, cause the slab to appear to broaden, shallow in dip, or flatten out. Hence, although slab thermal models are simplifications of complicated real slabs, and many key parameters are not well known, it seems likely that the models are reasonable approximations (perhaps accurate to a few hundred degrees) to the temperatures within actual slabs.

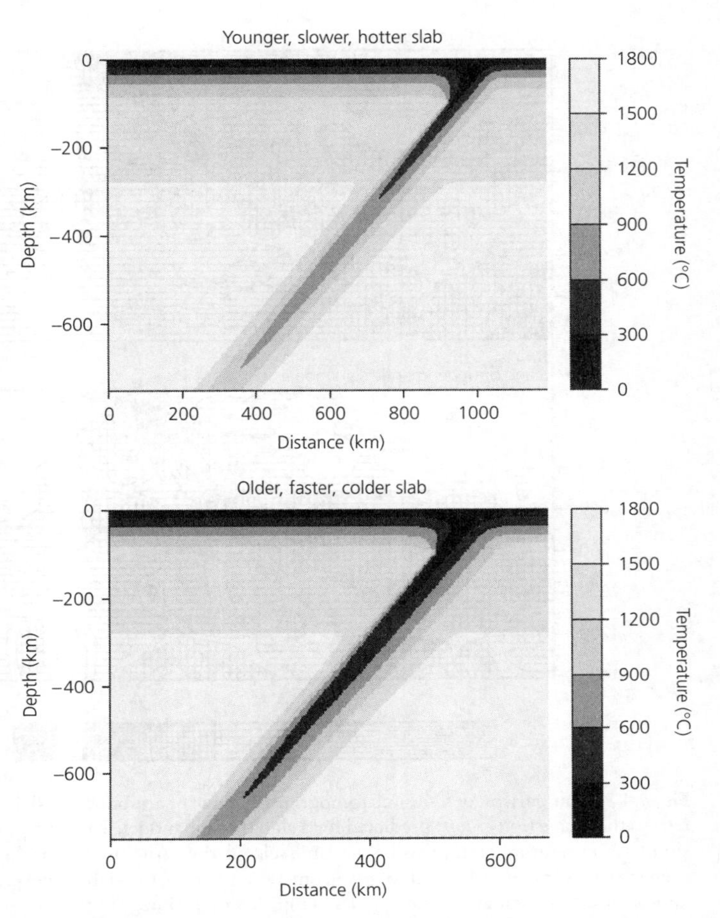

Fig. 5.4-6 Comparison of thermal structure for a relatively younger, slower-subducting slab (50 Myr-old lithosphere subducting at 70 mm/yr; thermal parameter about 2500 km), which approximates the Aleutian arc, and an older, faster-subducting slab (140 Myr-old lithosphere subducting at 140 mm/yr; thermal parameter about 17,000 km) which approximates the Tonga arc. (Stein and Stein, 1996. *Subduction*, 1–17, copyright by the American Geophysical Union.)

Seismology provides other tools to study the contrast between the cold, rigid, downgoing plate and the hotter, less rigid material around it. Figure 3.7-20 showed that a cold slab transmits seismic energy with less attenuation than its surroundings. Figure 5.4-8 shows some of the earliest data for this effect: seismograms from a deep earthquake are contrasted at stations NIU, to which waves travel through the downgoing plate; and VUN, to which waves arrive through the surrounding mantle. The VUN record shows much more long-period energy, especially for S waves, than that at NIU. Thus the

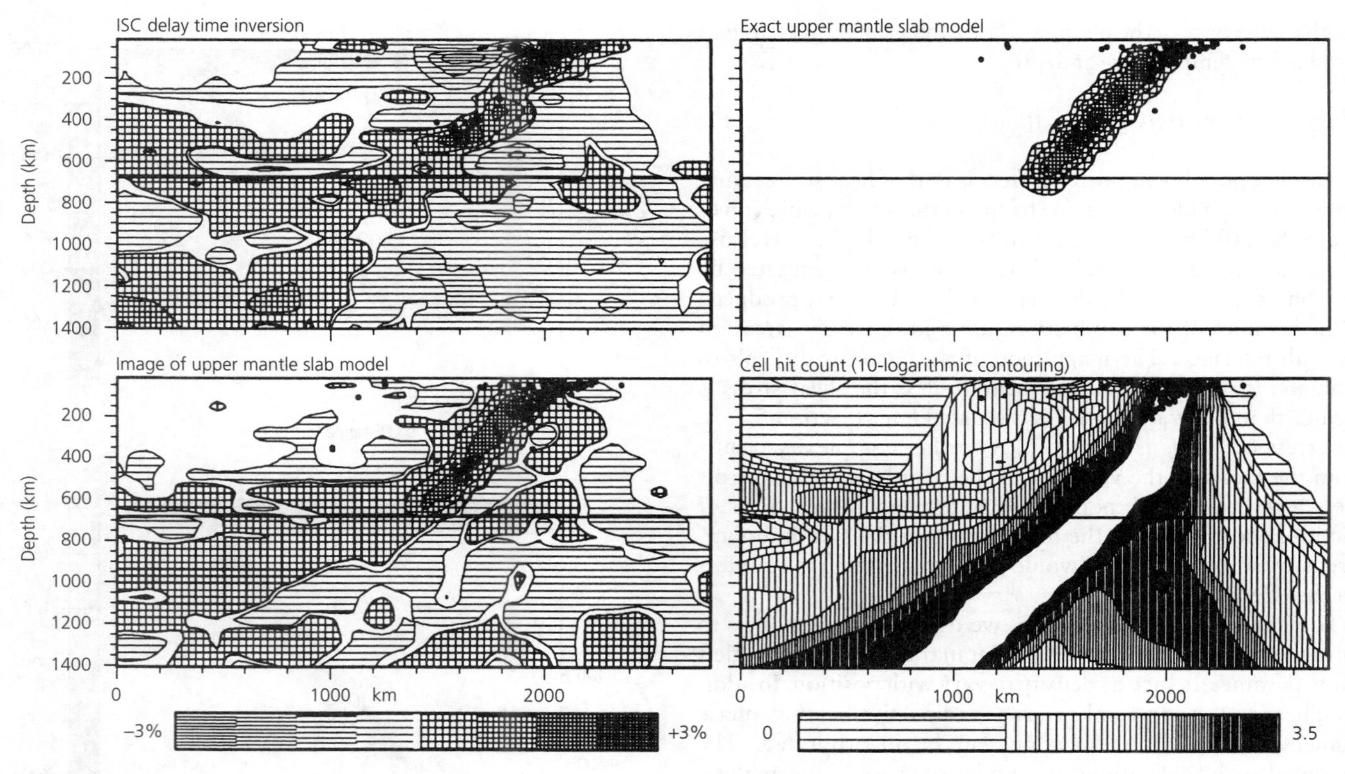

Fig. 5.4-7 Comparison of a seismic tomographic image of a subducting slab, indicated by the velocity anomaly and earthquake hypocenters (dots) (*upper left*) to the image (*lower left*) predicted for a slab thermal model. The seismic velocity anomaly predicted by the thermal model (*upper right*) is imaged by a simulated tomographic study using the same seismic ray path sampling as the data. The hit count (*lower right*) shows the number of rays sampling each cell used in the inversion. As a result of ray geometry and noise, the slab model gives a somewhat distorted image (*lower left*), showing how the model would appear in such a tomographic study. The similarity of the image of the model and the tomographic result suggests that the model generally describes the major features of the actual slab. Left scale bar gives velocity perturbations in percent, with positive values representing fast material. Right scale bar is for hit count, showing values as logarithm to base 10; the white region in the hit count plot is densely sampled and off scale. (Spakman *et al.*, 1989. *Geophys. Res. Lett.*, *16*, 1097–110, copyright by the American Geophysical Union.)

high-frequency components were more absorbed on the path to VUN due to higher attenuation (lower Q) than on the more rigid slab path to NIU. In addition, the sharp contrast in seismic velocity at the top of slabs can be detected using reflected and converted seismic waves (Fig. 2.6-15).

5.4.2 Earthquakes in subducting slabs

The deep and intermediate earthquakes forming the Wadati–Benioff zone extend in some places to depths of almost 700 km (Fig. 5.4-9). These are the deepest earthquakes that occur: away from subduction zones, earthquakes below about 40 km are rare. The Wadati–Benioff zone earthquakes illustrate that material cold enough to fail seismically (rather than flow) is being subducted, and give our best information about the geometry and mechanics of slabs.

The number of earthquakes as a function of depth illustrates why we distinguish intermediate and deep earthquakes; seismicity decreases to a minimum near about 300 km, and then increases again. Deep earthquakes, those below about 300 km, are thus generally treated as distinct from intermediate earthquakes. Deep earthquakes peak at about 600 km, and then decline to a minimum before 700 km. The focal mechanisms also vary with depth; those shallower than 300 km show generally down-dip tension, whereas those below 300 km show generally down-dip compression (Fig. 5.4-10).

Various explanations for this distribution of earthquakes and focal mechanisms are under consideration. One is that near the surface the slab is extended by its own weight, whereas at depth it encounters stronger lower mantle material, causing down-dip compression. Another possible factor may be mineral phase changes that occur at different depths in the cold slab than in the surrounding mantle.

It is generally assumed that the most crucial effect is the negative buoyancy (sinking) of the cold and dense slabs. The thermal model gives the force driving the subduction due to the integrated negative buoyancy of a slab resulting from the density contrast between it and the warmer and less dense material at the same depth outside. Because the slab does not have a discrete lower end in the analytic model, the net force is

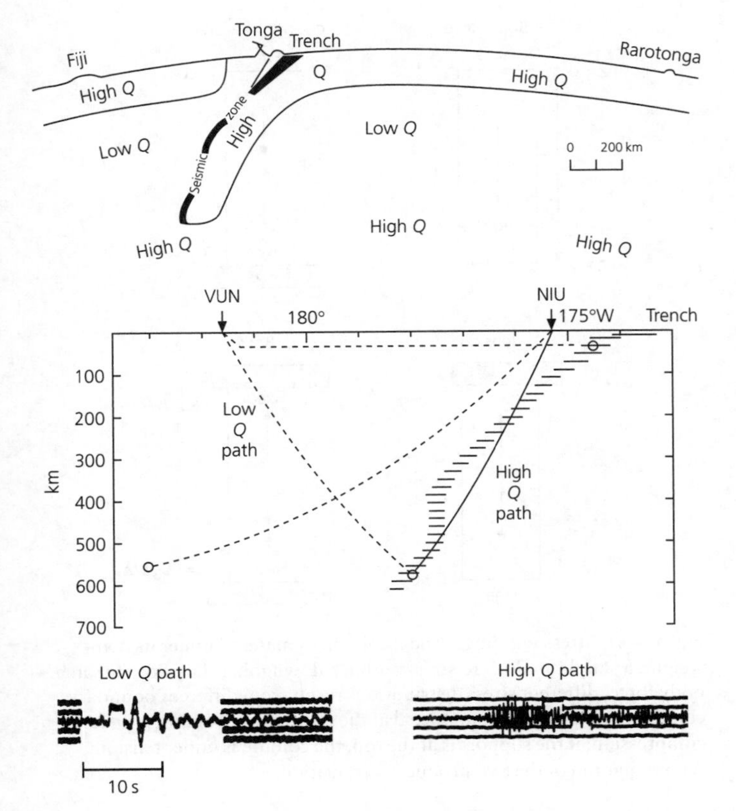

Fig. 5.4-8 Seismological observations showing the difference between the cold slab and hotter ambient mantle. Comparison of the seismograms at NIU and VUN shows that high frequencies are transmitted better by the slab, so the slab is a less attenuating, or higher Q path. (Oliver and Isacks, 1967. *J. Geophys. Res.*, 72, 4259–75, copyright by the American Geophysical Union.)

Fig. 5.4-9 Distribution of seismicity with depth.

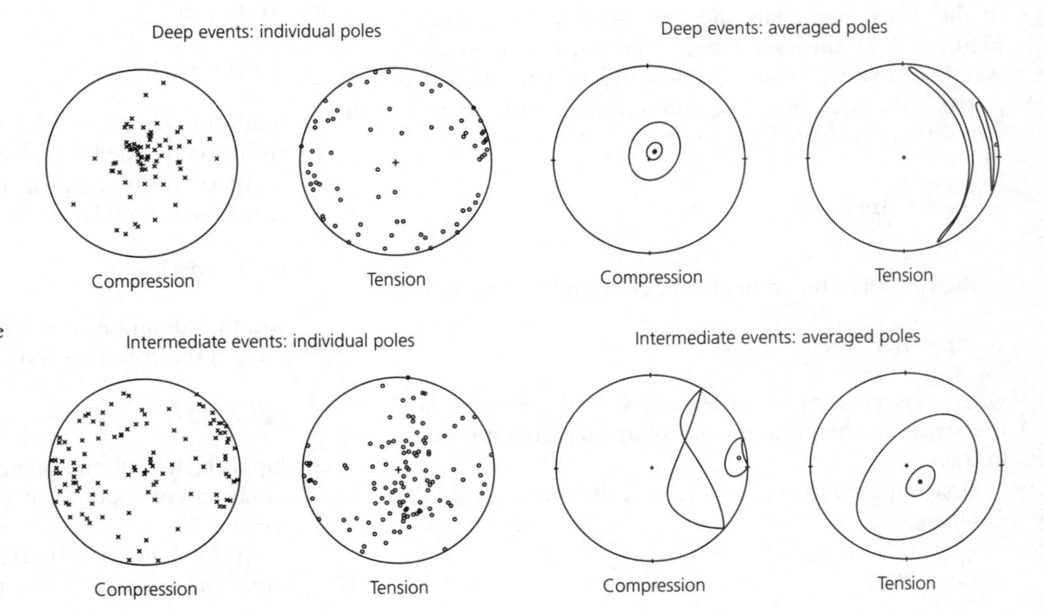

Fig. 5.4-10 Stress orientations inferred from focal mechanisms of subduction zone earthquakes. The P and T axes are rotated so that the down-dip direction is at the center of each plot, and their distributions are contoured. *Top*: Events below 300 km are dominated by down-dip compression. *Bottom*: Events from 70–300 km are dominated by down-dip tension. (After Vassiliou, 1984. *Earth Planet. Sci. Lett.*, 69, 195–201, with permission from Elsevier Science.)

$$F = \int_0^L \int_0^\infty g[\rho(x, y) - \rho_m] dx dy. \tag{12}$$

If material outside the slab is at temperature T_m and density ρ_m, material in the slab at the point (x, y) has density

$$\rho(x, y) \approx \rho_m + \frac{\partial \rho}{\partial T}[T(x, y) - T_m] = \rho_m + \rho'(x, y). \tag{13}$$

As for the cooling plate (Eqn 5.3.9), the density perturbation is

$$\rho'(x, y) = \alpha \rho_m [T_m - T(x, y)], \tag{14}$$

so for the analytic temperature model (Eqn 3) the integral over the slab yields a force

$$F = \frac{g \alpha \rho_m T_m v L^3}{24\kappa}. \tag{15}$$

This force, known as "slab pull," is the plate driving force due to subduction. Specifically, it is the negative buoyancy associated with a cold downgoing limb of the convection pattern. Its significance for stresses in the downgoing plate and for driving plate motions depends on its size relative to the resisting forces at the subduction zone. There are several such forces. As the slab sinks into the viscous mantle, the material displaced causes a force depending on the viscosity of the mantle and the subduction rate. The slab is also subject to drag forces on its sides and to resistance at the interface between the overriding and downgoing plates, which is often manifested as earthquakes.

To gain insight into the relative size of the negative buoyancy ("slab pull") and resistive forces, we consider the stress in the downgoing slab and the resulting focal mechanisms. Figure 5.4-11 shows a simple analogy, the stress due to the weight of a vertical column of length L of material with density ρ. Using the equilibrium equation (Eqn 2.3.49), we equate the stress gradient to the body force,

$$\frac{\partial \sigma_{zz}(z)}{\partial z} = -\rho g, \tag{16}$$

so the stress as a function of depth is found by integration,

$$\sigma_{zz}(z) = -\rho g z + C, \tag{17}$$

where C is a constant of integration. To determine C, and thus the stress in the column, the boundary conditions must be known.

First, suppose the stress is zero at the top, $z = 0$. In this case $C = 0$ and

$$\sigma_{zz}(z) = -\rho g z, \tag{18}$$

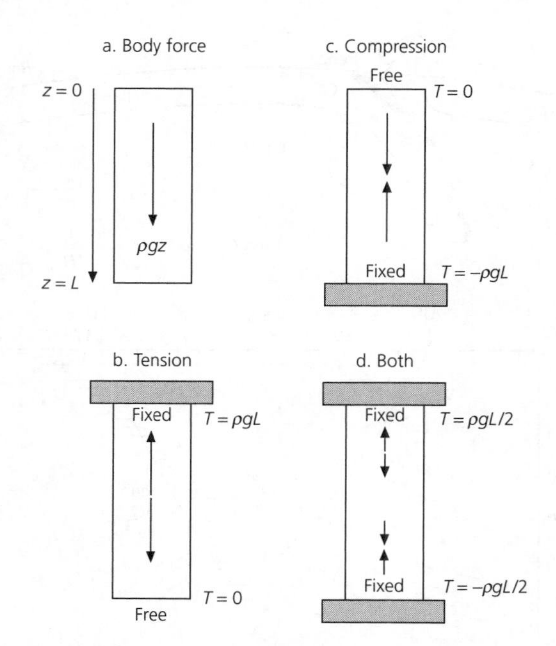

Fig. 5.4-11 Stress within a vertical column of material under its own weight, a simple analogy to stress within a downgoing slab. For the same body force, different stress distributions result from different boundary conditions. If the load is supported at the bottom, the column is under compression; if the support is at the top, the column is under tension. A combination of the two produces a transition.

which is negative, corresponding to compression everywhere. The forces required at the top and the bottom to maintain equilibrium are given by the relation between the traction, stresses, and outward normal vector on a surface (Eqn 2.3.8),

$$T_z = \sigma_{zz} n_z. \tag{19}$$

At the top $T_z(0) = 0$, whereas at the bottom a force

$$T_z(L) = -\rho g L \tag{20}$$

holds the column up. This situation is like a column of material sitting on the earth's surface, under compression everywhere.

Alternatively, suppose the stress is zero at the bottom. In this case the constant is chosen so that

$$\sigma_{zz}(z) = \rho g(L - z) \tag{21}$$

and the column is in extension (σ_{zz} positive) everywhere. The force at the bottom is zero, and the force at the top,

$$T_z(0) = \rho g L, \tag{22}$$

supports the column, because n_z points in the $-z$ direction. This situation corresponds to the material hanging under its own weight.

If the column is supported equally at both ends, the forces at either end are equal, so we find the stress from the condition

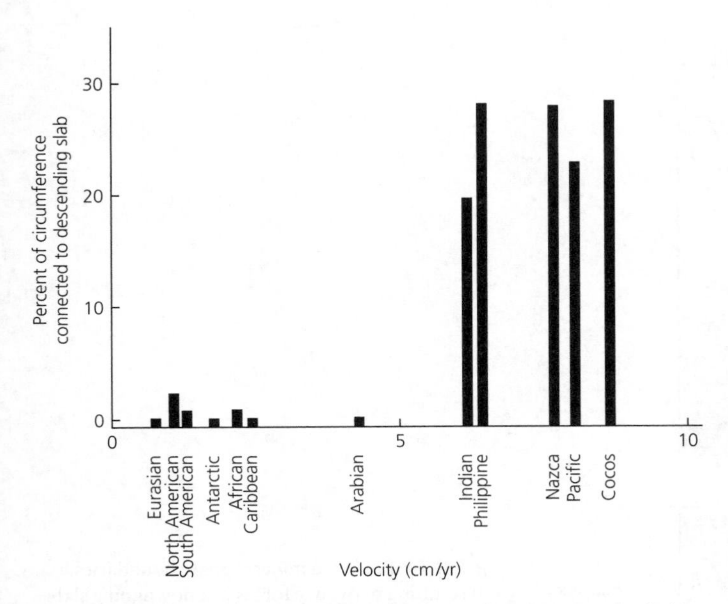

Fig. 5.4-12 The absolute velocity of lithospheric plates increases with the fraction of the plate's boundary formed by subducting slabs, suggesting that slabs provide a major driving force for plate motions. (Forsyth and Uyeda, 1975.)

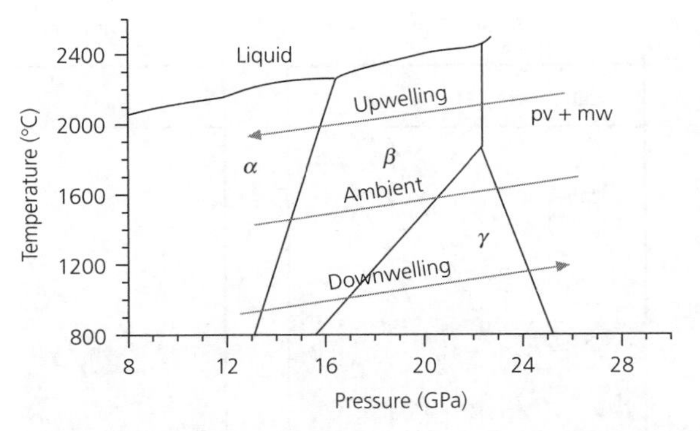

Fig. 5.4-13 Phase diagram for transitions in olivine with increasing depth. The phase boundaries as functions of temperature and pressure are known as Clapeyron curves. The downwelling and upwelling lines contrast conditions in slabs and plumes, respectively, to those in the ambient mantle. A reaction with a positive slope, such as the olivine (α phase) to spinel (*β* phase) change thought to give rise to the 410 km discontinuity outside the slab, is displaced upward (to lower pressure) within the cold slab. By contrast, the *γ* spinel to perovskite plus magnesiowustite (pv + mw) transition has a negative slope, so the 660 km discontinuity should be deeper in slabs than outside. (After Bina and Liu, 1995. *Geophys. Res. Lett.*, *22*, 2565–8, copyright by the American Geophysical Union.)

$$T_z(0) = -T_z(L), \tag{23}$$

which gives

$$\sigma_{zz}(z) = \rho g(L/2 - z). \tag{24}$$

Thus the column is in extension in its upper half, $z < L/2$, and in compression below this point.

The stress in the column shows how the body force due to gravity is balanced by forces on the boundaries. By analogy, if the downgoing slab were in tension, the negative buoyancy force must exceed the resistive forces at the subduction zone, and the slab would be "pulling" on and supported by the remainder of the plate outside the subduction zone. In fact, most earthquakes in the deeper portions of the slab show down-dip compression, whereas the intermediate earthquakes show down-dip tension (Fig. 5.4-10). This situation is like the column supported at both ends.

These ideas about the forces within subduction zones are consistent with two important pieces of data. First, the average absolute velocity of plates increases with the fraction of their area attached to downgoing slabs (Fig. 5.4-12), suggesting that slabs are a major determinant of plate velocities. Second, as discussed in Section 5.5.2, earthquakes in old oceanic lithosphere have thrust mechanisms, demonstrating deviatoric compression. Thus the net effect of the subduction zone on the remainder of the plate is not a "pull," so the term "slab pull" is misleading. Instead, as implied by the slab stress models, the "slab pull" force is balanced by local resistive forces, a combination of the effects of the viscous mantle and the interface between plates. This situation is like an object dropped in a viscous fluid, which is accelerated by its negative buoyancy until it reaches a terminal velocity determined by its density and shape and the viscosity and density of the fluid.

An interesting possible complication is that slabs are not just thermally different from their surroundings; they are probably also mineralogically different. Slabs extend through the mantle transition zone, where mineral phase changes are thought to occur (Section 3.8). However, because a downgoing slab is colder than material at that depth elsewhere, phase changes within the slab are displaced relative to their normal depth. The displacement can be calculated using the thermodynamic relation, known as the Clapeyron equation, for the boundary between two phases as a function of pressure and temperature. If ΔH and ΔV are the heat and volume changes resulting from the phase change, then a change dT in temperature moves the phase change by a pressure dP given by the Clapeyron slope (the reciprocal of Eqn 9),

$$\gamma = \frac{dP}{dT} = \frac{\Delta H}{T\Delta V}. \tag{25}$$

For example, the 410 km discontinuity is attributed to the phase change with increased pressure from olivine to a denser spinel structure (the *β* phase, wadsleyite) described by a phase diagram like that in Fig. 5.4-13. Because the spinel phase is denser, ΔV is less than zero. This reaction is exothermic (gives off heat), so ΔH is also negative, causing a positive Clapeyron slope. If we know the depth (pressure) and temperature at which a phase change occurs in the mantle, the Clapeyron equation gives its position in the slab. The slab is colder than

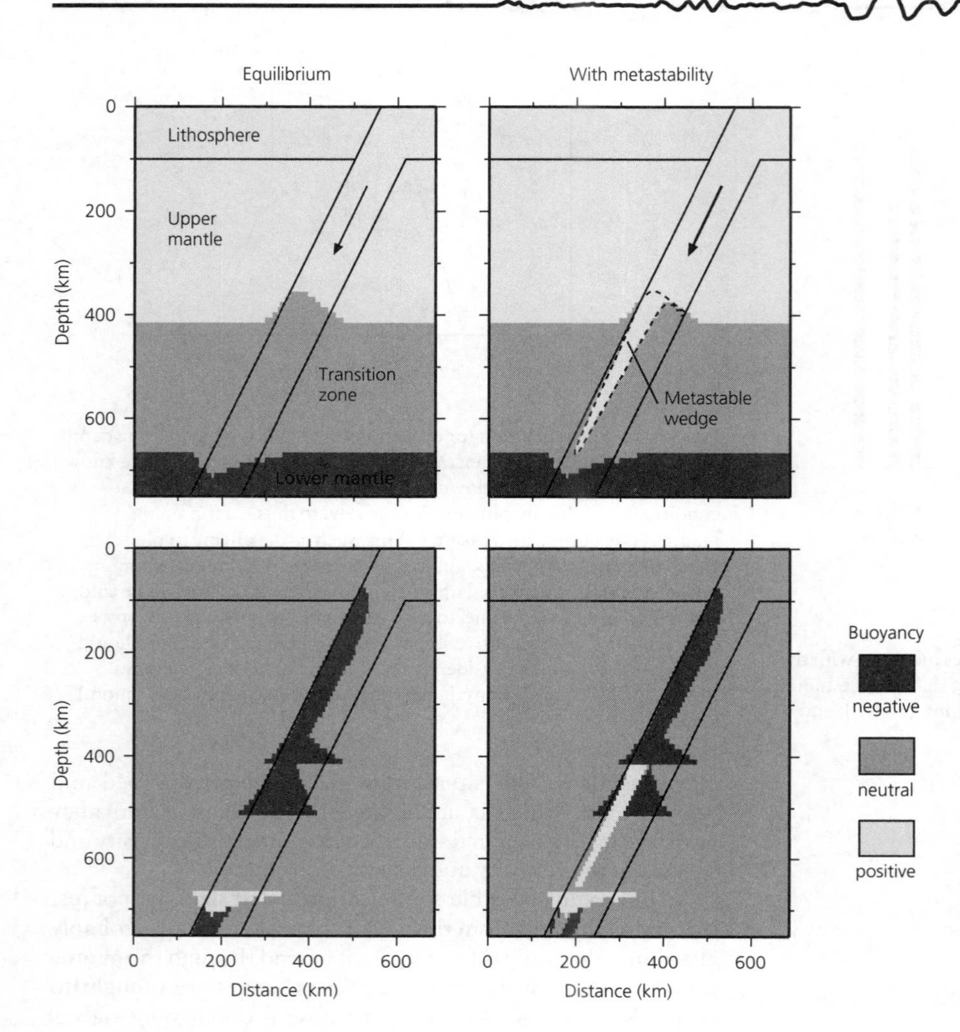

Fig. 5.4-14 Predicted mineral phase boundaries and resulting buoyancy forces in a downgoing slab without (*left panels*) and with (*right panels*) a metastable olivine wedge. Assuming equilibrium mineralogy the cold slab has negative thermal buoyancy, negative compositional buoyancy associated with the elevated 410 km discontinuity, and positive compositional buoyancy associated with the depressed 660 km discontinuity. A metastable wedge gives positive compositional buoyancy and hence decreases the force driving subduction. Negative buoyancy favors subduction, whereas positive buoyancy opposes it. (Stein and Rubie, 1999. *Science, 286*, 909–10, copyright 1999 American Association for the Advancement of Science.)

the ambient mantle ($dT < 0$), so this phase change occurs at a lower pressure ($dP < 0$), corresponding to a shallower depth. Converting the pressure change to depth, the vertical displacement of this phase change is

$$\frac{dz}{dT} = \frac{\gamma}{\rho g}. \tag{26}$$

By contrast, the ringwoodite (γ spinel phase) to perovskite plus magnesiowustite transition, thought to give rise to the 660 km discontinuity, is endothermic (absorbs heat), so ΔH is positive. Because this is a transformation to denser phases (ΔV less than zero), the Clapeyron slope is negative, and the 660 km discontinuity should be deeper in slabs than outside. These opposite effects — upward deflection of the 410 km and downward deflection of the 660 km discontinuities (Fig. 5.4-14) — have been observed in travel time studies. An interesting way to think about these is to note that the negative buoyancy associated with the elevated 410 km discontinuity helps the subduction, whereas the positive buoyancy associated with the depressed 660 km discontinuity opposes the subduction. The reverse effect should not occur at the 660 km discontinuity for upcoming plumes, however, because the phase diagram shows that at these higher temperatures the Clapeyron curve for the perovskite plus magnesiowustite transition is vertical, so the transition is not displaced (Fig. 5.4-13).

The position of the olivine–spinel phase change may be further affected. The Clapeyron slope predicts what happens if a phase change occurs at equilibrium. However, the phase change actually occurs by a process in which grains of the high-pressure phase nucleate on the boundaries between grains of the lower-pressure phase and then grow with time (Fig. 5.4-15). Studies of mineral nucleation and growth rates suggest that in the coldest slabs the phase transformation cannot keep pace with the rate of subduction, causing a wedge of olivine in the cold slab core to persist metastably[3] to greater depths (Fig. 5.4-14).

[3] *Metastability* describes the situation where a mineral phase survives outside its equilibrium stability field in temperature–pressure space. Such metastable persistence is expected because the relatively colder temperatures in slabs should inhibit reaction rates. This effect explains why diamonds, which are unstable at the low pressures of earth's surface, survive metastably rather than transform to graphite. The situation in slabs is similar to that of supercooled water, which persists as a liquid at temperatures below its equilibrium freezing point.

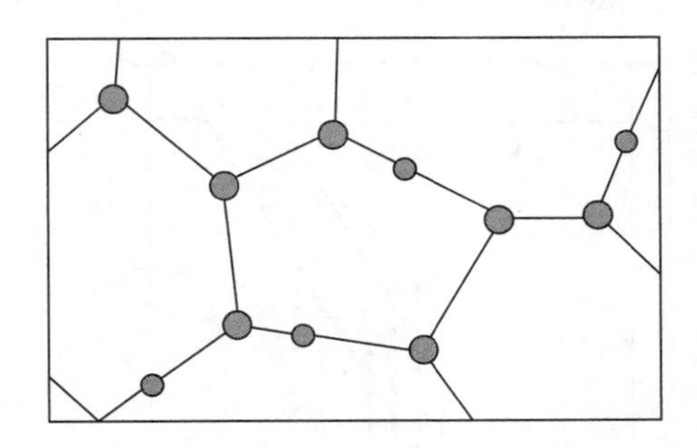

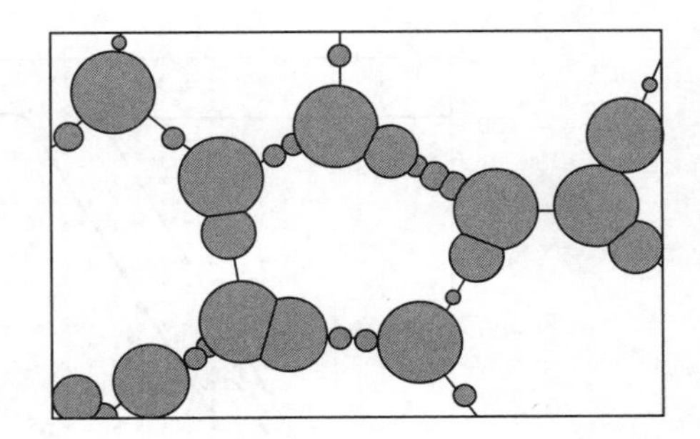

Fig. 5.4-15 Diagram showing the early stages of a phase transformation. Grains of the new phase (shaded) nucleate on grain boundaries and grow by consuming the original phase until none remains. (Kirby *et al.*, 1996b. *Rev. Geophys.*, *34*, 261–306, copyright by the American Geophysical Union.)

The deflections of the phase boundaries have several possible consequences. First, phase changes affect the thermal structure of the slab due to the heat of the phase change. Thus the exothermic olivine–spinel change should add heat to slabs. This effect is simulated in thermal models by increasing the temperature at the phase change. Second, the phase boundaries are probably important for the buoyancy and stresses within slabs. We have already discussed the idea that the cold slabs are denser than their surroundings, causing negative thermal buoyancy, which favors sinking. The phase boundaries cause additional mineralogical buoyancy. For example, if the olivine–spinel boundary is uplifted in the slab, the presence of slab material denser than at that depth outside causes additional negative buoyancy. However, if a wedge of metastable olivine exists, it would be less dense than material at that depth outside and produce positive buoyancy (Fig. 5.4-14) in addition to that caused by the downward deflection of the 660 km discontinuity. Although the net buoyancy must be negative because slabs subduct, the details of the buoyancy can be important. For example, metastable olivine may help regulate subduction rates. Faster subduction would cause a larger wedge of low-density metastable olivine, reducing the driving force and slowing the slab.

A third possibility is that a phase change causes deep earthquakes. Although this idea is a natural consequence of the observation that deep earthquakes occur at transition zone depths, it was not given serious consideration for a long time because deep earthquake focal mechanisms show slip on a fault, rather than isotropic implosions (Section 4.4.6). However, laboratory studies now suggest that an instability

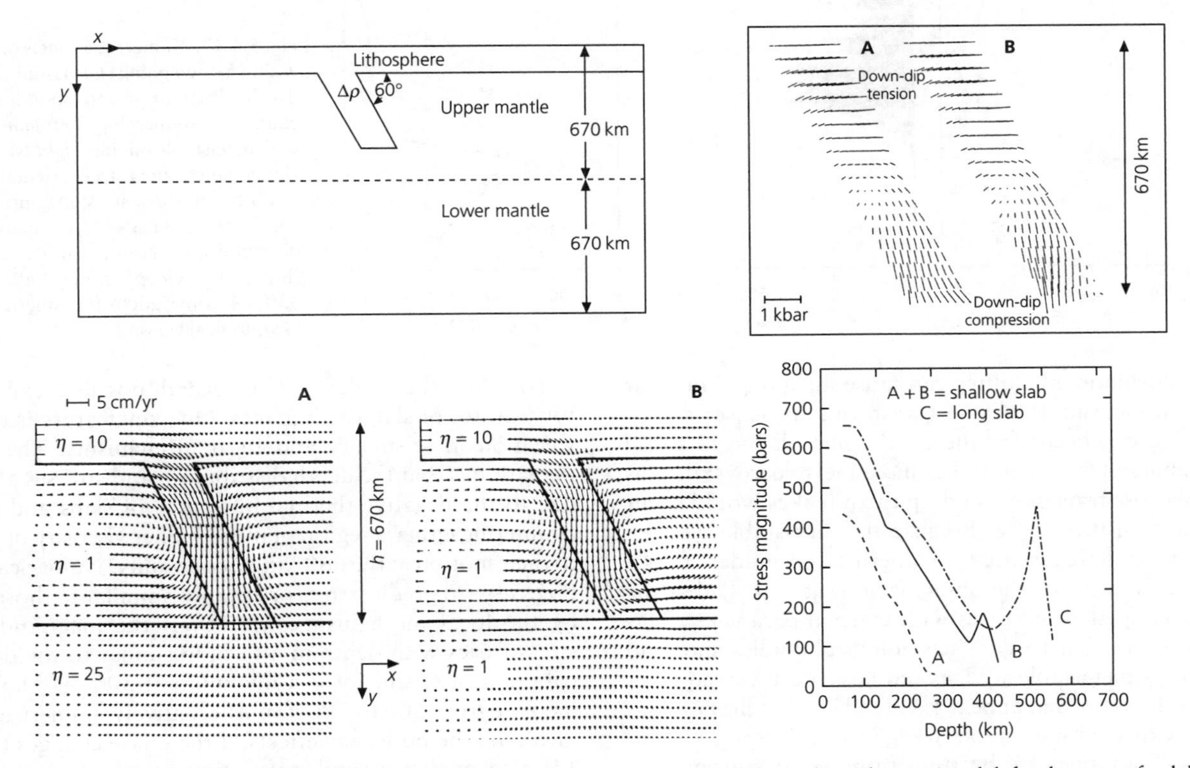

Fig. 5.4-16 Numerical model of mantle flow fields (*lower left*) and resulting stresses (*upper right*) within a downgoing slab for the cases of a slab that (A) encounters higher-viscosity material below 670 km and (B) cannot penetrate below this depth. η values show relative viscosities. Both predict down-dip tension in the upper portion of the slab and down-dip compression in the lower portion. The calculated stresses are highest near the bottom of the slab. (Vassiliou *et al.*, 1984. *J. Geodynam.*, *1*, 11–28, with permission from Elsevier Science.)

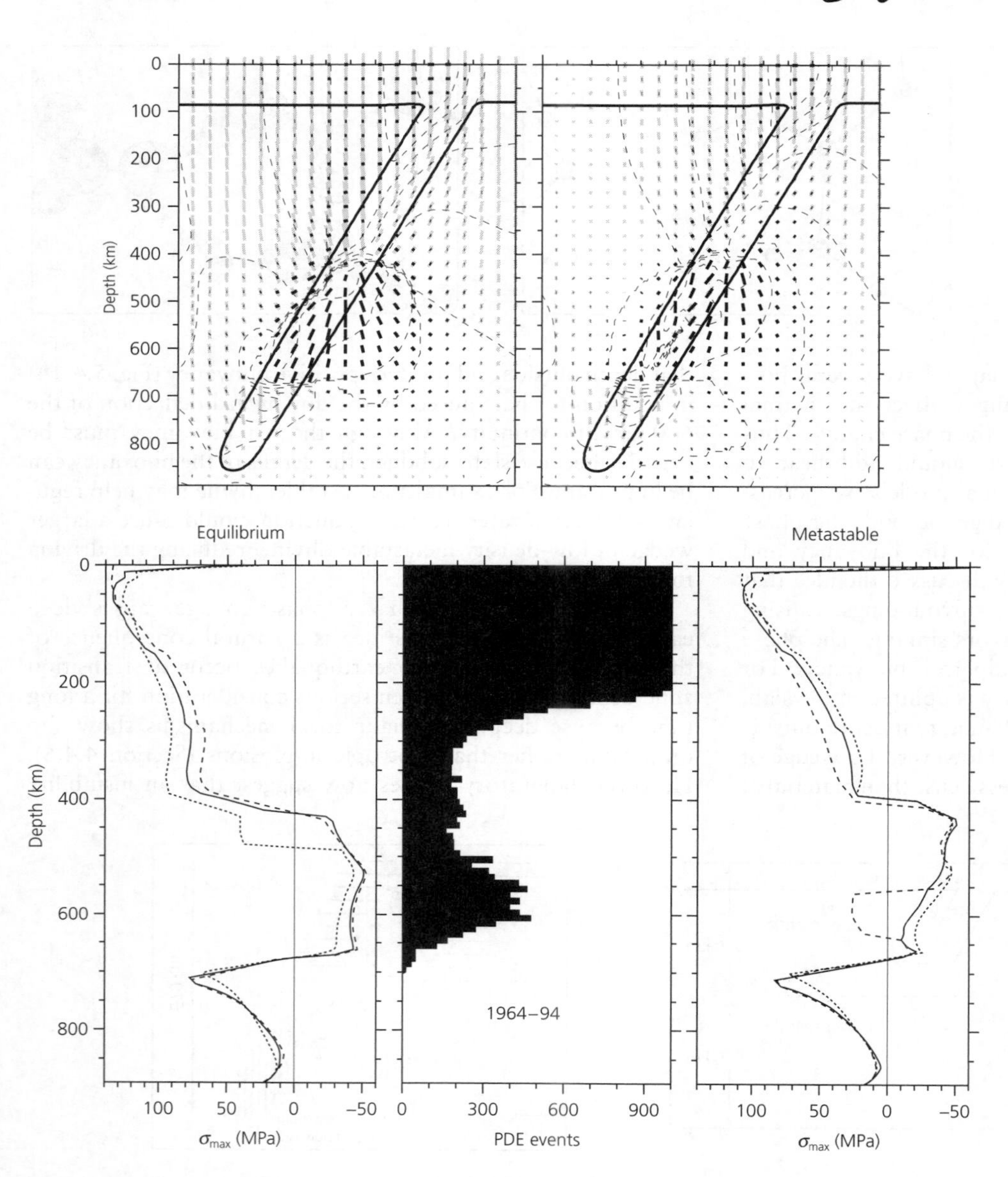

Fig. 5.4-17 Numerical models of stresses within a downgoing slab assuming the density distribution corresponding to equilibrium mineralogy (*left panels*) and with metastable olivine (*right panels*). Upper panels show stress orientations, and lower panels show stress magnitudes, with compression as negative, compared to the distribution of seismicity (*lower center*). (Bina, 1997. *Geophys. Res. Lett.*, *24*, 3301–4, copyright by the American Geophysical Union.)

called transformational faulting can cause slip along thin shear zones where metastable olivine transforms to denser spinel. Such faulting can occur for the exothermic olivine to spinel transition, but not for the endothermic spinel to perovskite plus magnesiowustite transition, so deep earthquakes would occur only in the transition zone. Because the metastable wedge's lower boundaries are essentially isotherms, this model offers a physical mechanism for the observation (Fig. 5.4-4) that the depth of earthquakes increases with thermal parameter. This idea is attractive, but to date seismological studies show no evidence for a metastable wedge, and large deep earthquakes occur on fault planes that appear to extend beyond the boundaries of the expected metastable wedge. If such wedges exist, earthquakes may nucleate by transformational faulting, but then propagate outside the wedge via another failure mechanism.

Together these ideas offer several possible explanations for features of slab earthquakes. One key feature is the depth variation in seismicity and focal mechanisms. The first explanation is that the depth distribution and stresses are largely due to the negative thermal buoyancy of slabs and their encountering either a region of much higher viscosity or a barrier to their motion at the 660 km discontinuity. Numerical models (Fig. 5.4-16) predict stress orientations similar to those implied by the focal mechanisms. Moreover, the magnitude of the stress varies with depth in a fashion similar to the depth distribution of seismicity — a minimum at 300–410 km and an increase from 500 to 700 km. Alternatively, numerical models including the buoyancy effects of the phase changes (Fig. 5.4-14) also predict a similar variation in stress magnitude and orientation with depth (Fig. 5.4-17), without invoking a barrier or higher viscosity in the lower mantle. Thus, in such

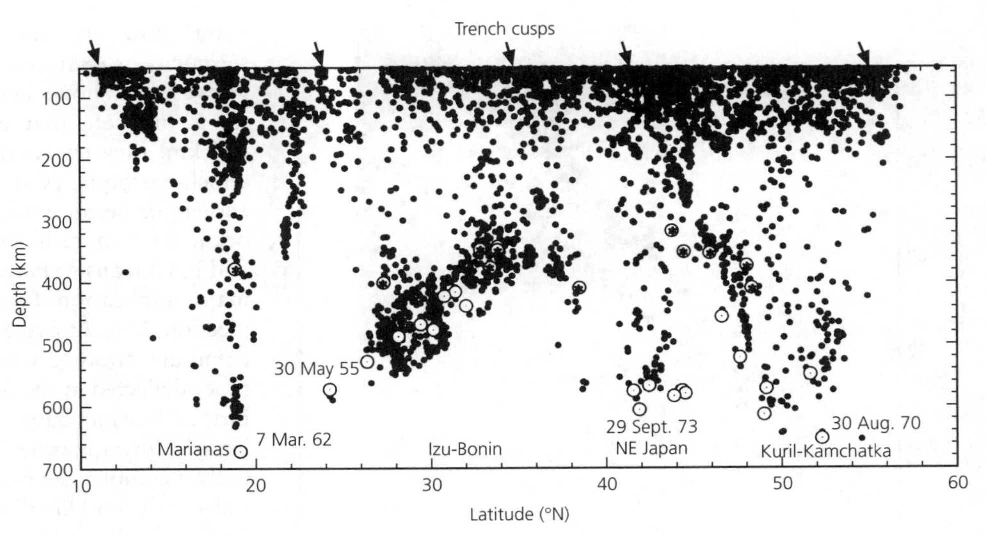

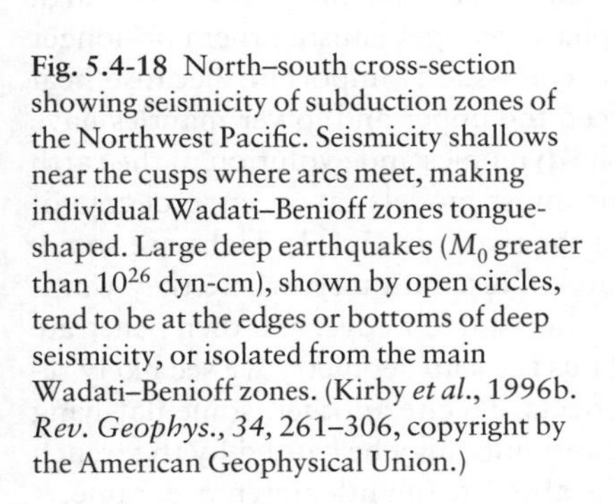
Fig. 5.4-18 North–south cross-section showing seismicity of subduction zones of the Northwest Pacific. Seismicity shallows near the cusps where arcs meet, making individual Wadati–Benioff zones tongue-shaped. Large deep earthquakes (M_0 greater than 10^{26} dyn-cm), shown by open circles, tend to be at the edges or bottoms of deep seismicity, or isolated from the main Wadati–Benioff zones. (Kirby *et al.*, 1996b. *Rev. Geophys.*, *34*, 261–306, copyright by the American Geophysical Union.)

models, deep earthquakes need not be physically different from intermediate ones, because the minimum in seismicity reflects a stress minimum.

A second key issue is how deep earthquakes can occur at all. As discussed in Section, 5.7, the strength of rock that must be exceeded for fracture increases with pressure. The pressures deep in a subducting slab should be high enough to prevent fracture. One possibility is that the slabs become hot enough that water released by decomposition of hydrous minerals lubricates (reduces the effective stress on) faults. Another possibility, mentioned earlier, is transformational faulting in metastable olivine. It is also possible that the earthquakes occur by very rapid creep, possibly associated with weakening due to unusually small spinel grains formed in the coldest slabs.

The different explanations offered by these models all have attractive features and may be true in part. However, although such simple models based on idealized slabs explain some gross features of deep earthquakes, none fully explains the complexity of deep earthquakes. As shown by Fig. 5.4-18, a cross-section along the subduction zones of the Northwest Pacific, deep seismicity is "patchy" and variable. For example, it shallows dramatically at the cusps between the Marianas, Izu-Bonin, NE Japan, and Kuril-Kamchatka arcs. Moreover, the largest earthquakes occur at the edges of the regions of deep seismicity, as especially evident at the northern edge of the Izu-Bonin seismicity. These sites may reflect tears in the down-going lithosphere at the junctions between arcs, where hot mantle material penetrates slabs. A further complexity is that some deep earthquakes occur in unusual locations off the down-dip extension of the main Wadati–Benioff zones and have focal mechanisms differing from those of the deepest earthquakes in the main zone (Fig. 5.4-19). Some other deep earthquakes are isolated from actively subducting slabs. Such unusual earthquakes may occur in slab fragments where metastable olivine survives, and thus have mechanisms related to local stresses rather than those expected for continuous slabs.

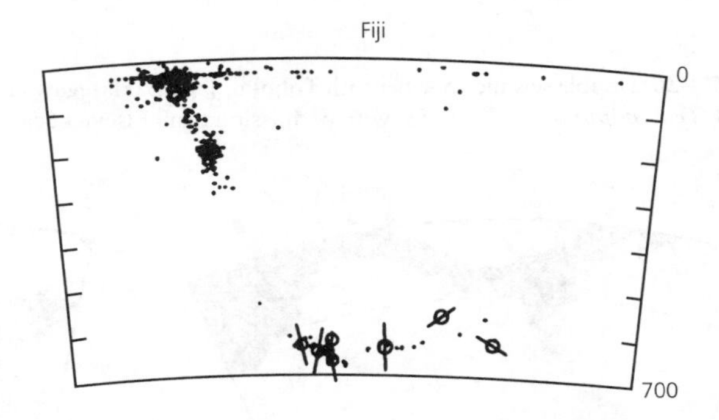

Fig. 5.4-19 Seismicity cross-section for the Fiji subduction zone, showing "outlier" deep earthquakes. Lines through symbols show P axes, which often differ from those for the main Wadati–Benioff zone. (Lundgren and Giardini, 1994. *J. Geophys. Res.*, *99*, 15, 833–42, copyright by the American Geophysical Union.)

Another interesting observation from precise earthquake locations in some subduction zones (Fig. 5.4-20) shows that the Wadati–Benioff zone is made up of two distinct planes, separated by 30–40 km. The upper plane seems to coincide with the conversion plane for *ScSp* (Fig. 2.6-15), a sharp velocity contrast that is presumably near the slab top. Focal mechanisms suggest that the upper plane is in down-dip compression and the lower one in down-dip extension. A variety of models have been proposed. One is that the double plane results from "unbending" of the slab — the release of the bending stresses produced when the slab began to subduct. Another model is that the slab "sags" under its own weight, because at depth it runs into a more viscous mesosphere, while at intermediate depths it encounters a less viscous asthenosphere. Explaining the phenomenon is complicated by the observation that only some subduction zones have double zones.

The nature of deep earthquakes, especially the mechanism restricting them to the transition zone, has implications for

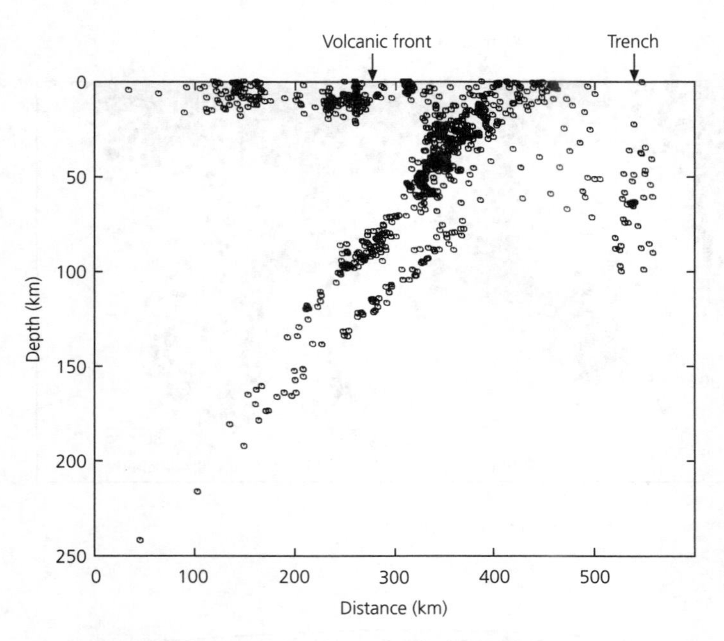

Fig. 5.4-20 Double seismic zone beneath Tohoku, Japan. (Hasegawa *et al.*, 1978. *Tectonophysics*, 47, 43–58, with permission from Elsevier Science.)

mantle flow. The simplest explanation for the cessation of deep seismicity is that slabs cannot penetrate the lower mantle. However, as shown in Fig. 5.4-21, tomographic studies (Chapter 7) indicate that although some slabs are deflected at 660 km, they eventually penetrate deeper. Hence models in which earthquakes stop either because the stress is not high enough or because the phase changes causing them no longer occur seem more likely. The issue is important because heat and mass transfer between the upper and lower mantles have major implications for the dynamics and evolution of the earth (Section 3.8). At present, most models favor some degree of communication between the two (Fig. 5.1-2). Slabs are sometimes deflected at the 660 km discontinuity, where they warm further, lose any buoyant metastable wedge, and then penetrate into the lower mantle. Thus the slab geometry we see likely reflects a complex set of effects. To cite another, some flat-lying slabs at the 660 km discontinuity may be caused by the trench "rolling backward" in the absolute (mantle) reference frame.

There has also been considerable discussion about the nature of intermediate depth earthquakes. Figure 5.4-22 shows a

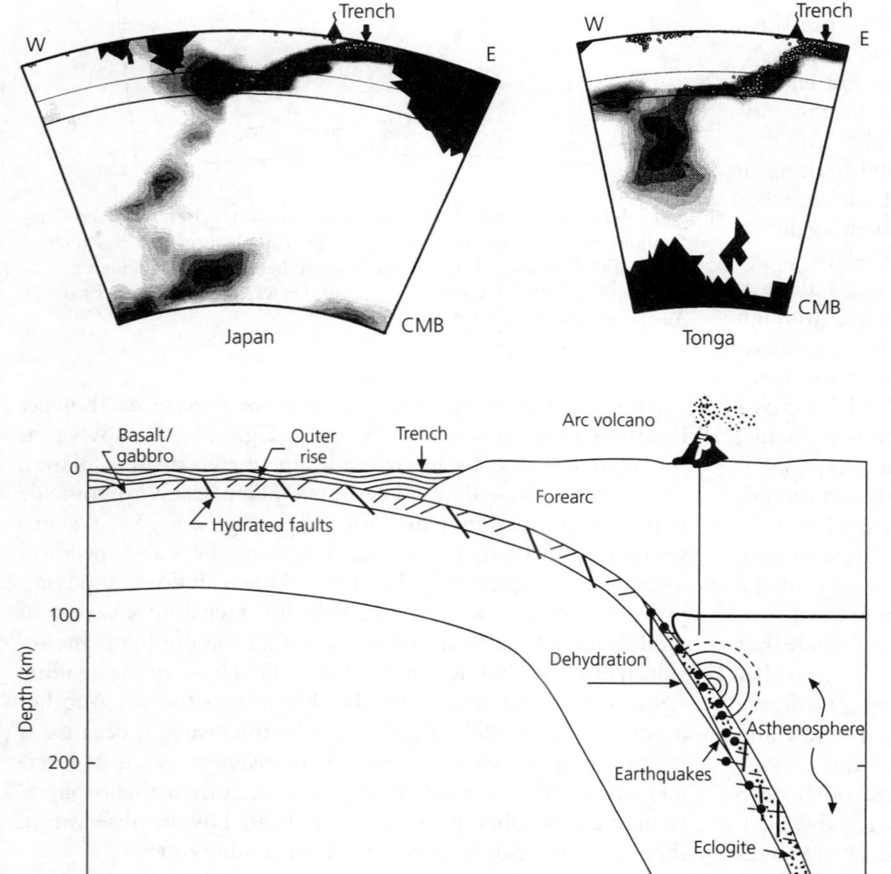

Fig. 5.4-21 Tomographic images across Pacific subduction zones with deep earthquakes. Horizontal lines are at 410 and 660 km depth. White dots are earthquake hypocenters. The Wadati–Benioff zone seismicity generally coincides with the high-velocity anomaly (dark regions) due to the cold subducting slab. Slabs are deflected at the base of the transition zone before penetrating into the lower mantle. (van der Hilst *et al.*, 1998. *The Core–Mantle Boundary Region*, 5–20, copyright by the American Geophysical Union.)

Fig. 5.4-22 Schematic model for intermediate depth earthquakes. Earthquakes are assumed to occur in subducting crust and be associated with the dehydration of mineral phases and the gabbro to eclogite transition. (Kirby *et al.*, 1996a. *Subduction*, 195–214, copyright by the American Geophysical Union.)

schematic model in which the earthquakes are presumed to occur in subducting oceanic crust, rather than throughout the subducting mantle that makes up most of the slabs, because detailed location studies show that the earthquakes are close to the top of the subducting slabs. The crust should undergo two important mineralogical transitions as it subducts. Hydrous (water-bearing) minerals formed at fractures and faults should warm up and dehydrate. Eventually, the gabbro transforms to eclogite, a rock of the same chemical composition composed of denser minerals.[4] Under equilibrium conditions, eclogite should form by the time slab material reaches about 70 km depth. However, travel time studies in some slabs find a low-velocity waveguide interpreted as subducting crust extending to deeper depths. Hence it has been suggested that the eclogite-forming reaction is slowed in cold downgoing slabs, allowing gabbro to persist metastably. Once dehydration occurs, the freed water weakens the faults, favoring earthquakes and promoting the eclogite-forming reactions. In this model the intermediate earthquakes occur by slip on faults, but the phase changes favor faulting. The extensional focal mechanisms may also reflect the phase change, which would produce extension in the subducting crust. Support for this model comes from the fact that the intermediate earthquakes occur below the island arc volcanoes, which are thought to result when water released from the subducting slab causes partial melting in the overlying asthenosphere.

The fact that various explanations are under discussion illustrates the difficulty in understanding the complex thermal structure, mineralogy, rheology, and geometry of real slabs. We can think of the deep subduction process as a chemical reactor that brings cold shallow minerals into the temperature and pressure conditions of the mantle transition zone, where these phases are no longer thermodynamically stable (Fig. 5.4-23). Because we have no direct way of studying what is happening and what comes out, we seek to understand this system by studying earthquakes that somehow reflect what is happening. This is a major challenge, and we have a long way to go.

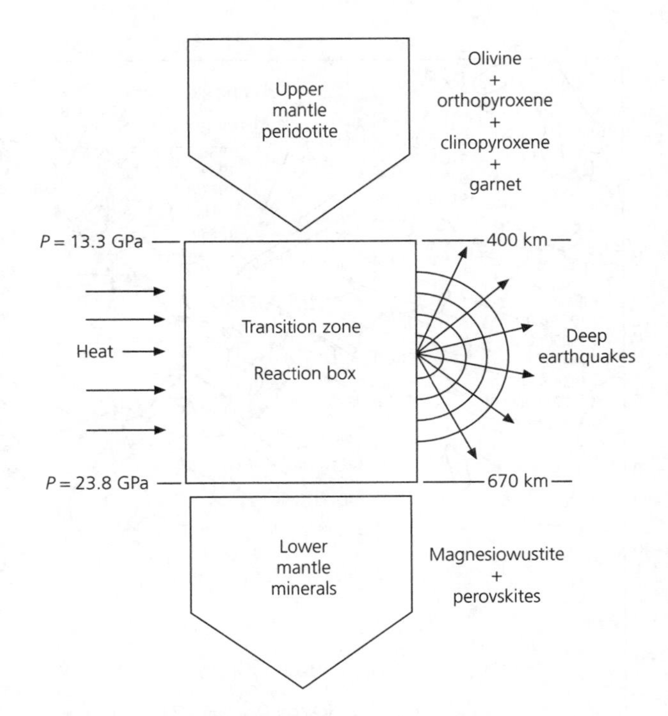

Fig. 5.4-23 Cartoon of subducting slabs in the transition zone as a chemical reactor. (Kirby *et al.*, 1996b. *Rev. Geophys.*, *34*, 261–306, copyright by the American Geophysical Union.)

5.4.3 *Interplate trench earthquakes*

Much of what is known about the geometry and mechanics of the interaction between plates at subduction zones comes from the distribution and focal mechanisms of shallow earthquakes at the interface between the plates. These include the largest earthquakes that occur, as illustrated by Fig. 5.4-24, showing the largest earthquakes (surface wave magnitude greater than 8.0) during 1904–76. Among these are the two largest earthquakes ever recorded seismologically: the 1960 Chilean (M_0 2 $\times 10^{30}$ dyn-cm, M_s 8.3) and 1964 Alaska (M_0 5 $\times 10^{29}$ dyn-cm, M_s 8.4) earthquakes. Figure 5.4-25 shows the geometry of the Chilean earthquake: 21 meters of slip occurred on a fault 800 km long along strike, and 200 km wide down-dip. The mechanism shows thrusting of the South American plate over the subducting oceanic lithosphere of the Nazca plate. The aftershock zone was 800 km long, and the surface deformation was dramatic, reaching 6 meters of uplift in places. Thrust earthquakes of this type, although smaller, make up most of the large, shallow events at subduction zones. Such *interplate* earthquakes release the plate motion that has been locked at the plate interface. As we saw in Section 4.6.1, these can be much bigger than the largest earthquakes at transform fault boundaries like the San Andreas. For example, even the 1906 San Francisco earthquake was tiny (100 times smaller seismic moment) compared to the 1964 Alaska earthquake, although both occurred along different segments of the same plate boundary. The difference reflects the fact that faulting occurs only when rock is cooler than a limiting temperature. Thus a vertically dipping transform like the San Andreas has a much shorter cold down-dip extent than the shallow-dipping thrust interfaces (sometimes called megathrusts) at subduction zones.

Major thrust earthquakes at the interface between subducting and overriding plates directly indicate the nature of subduction. In most cases, their focal mechanisms show slip toward the trench, approximately in the convergence direction predicted by global plate motion models or space-based geodesy (Section 5.2) (Fig. 5.2-3). However, in some cases when the plate motion is oblique to the trench, a forearc sliver moves separately from the overriding plate (Fig. 5.4-26). This effect,

[4] Most of the oceanic crust consists of gabbro, the intrusive version of the extrusive basalt seen at mid-ocean ridges (Section 3.2.5). With increasing pressure, gabbro becomes eclogite as feldspar and pyroxene transform to garnet.

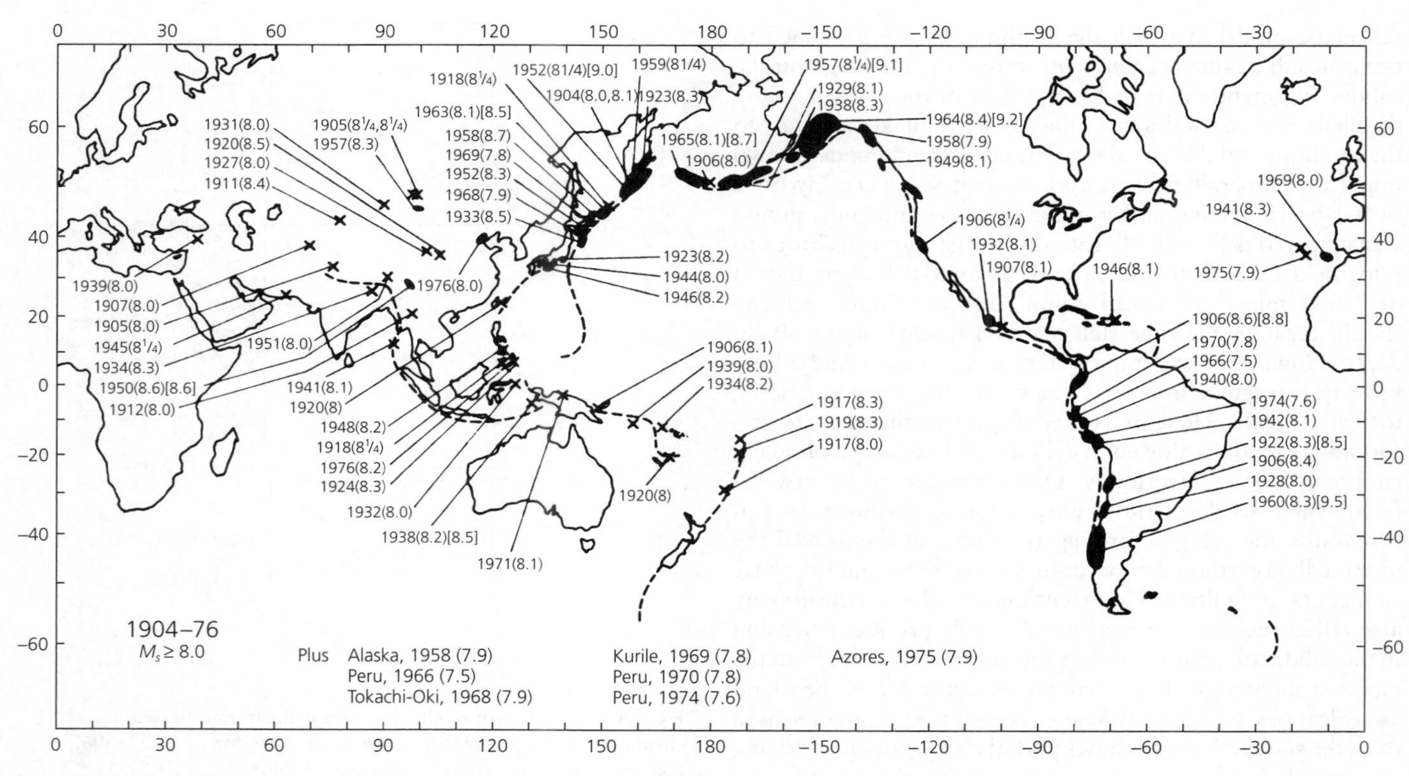

Fig. 5.4-24 Location of the largest earthquakes between 1904 and 1976. M_s values are in parentheses and M_w values in square brackets. Most are at subduction zones and result from thrust faulting at the interface between the two plates. (Kanamori, 1978. Reproduced with permission from *Nature*.)

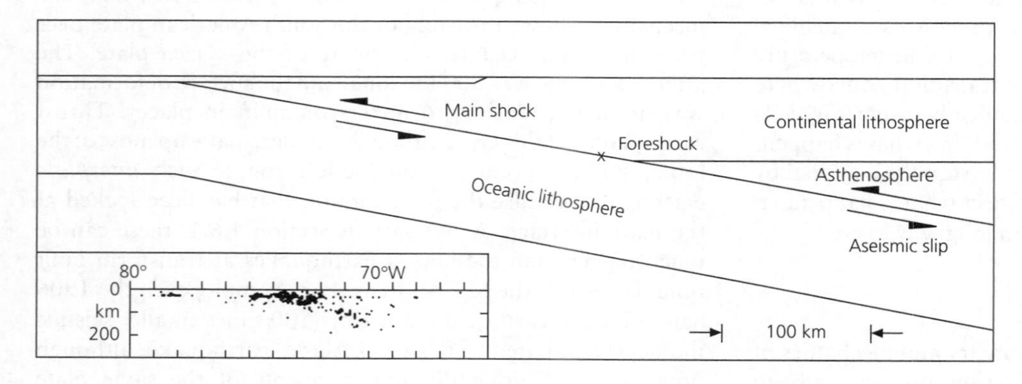

Fig. 5.4-25 Fault geometry and aftershock distribution (*insert*) for the 1960 Chilean earthquake. (Kanamori and Cipar, 1974. *Phys. Earth Planet. Inter.*, 9, 128–36, with permission from Elsevier Science.)

called *slip partitioning*, makes earthquake slip vectors at the trench trend between the trench-normal direction and the predicted convergence direction, and causes strike-slip motion between the forearc and the stable interior of the overriding plate. This effect can be seen in plate motion studies and with GPS data, and can cause misclosure of plate circuits. In the limiting case of pure slip partitioning, pure thrust faulting would occur at the trench, and all the oblique motion would be accommodated by trench-parallel strike-slip.

How the thrust earthquakes release the accumulated plate motion is both interesting scientifically and important for assessing earthquake hazards. In many subduction zones, thrust earthquakes have characteristic patterns in space and time. For example, large earthquakes have occurred in the Nankai trough area of southern Japan approximately every 125 years since 1498 with similar fault areas (Fig. 5.4-27).[5] In some cases the entire region seems to have slipped at once; in others, slip was divided into several events over a few years.

[5] Due to its location where between four and six plates (North America, Pacific, Philippine, Eurasia, and perhaps Okhotsk and Amuria) interact, Japan has a high level of seismicity, which was originally attributed to the motion of the namazu, a giant underground catfish. As a result, Japan has an outstanding tradition of seismology and some of the best data in the world for studying subduction-related earthquakes.

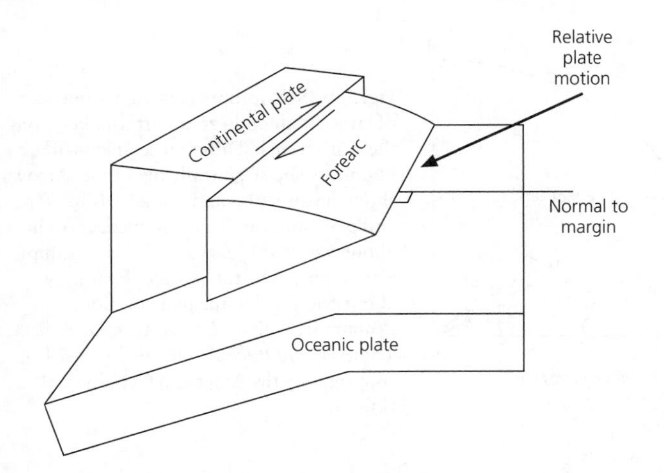

Fig. 5.4-26 Schematic illustration of forearc sliver motion when convergence is oblique. (Courtesy of D. Davis.)

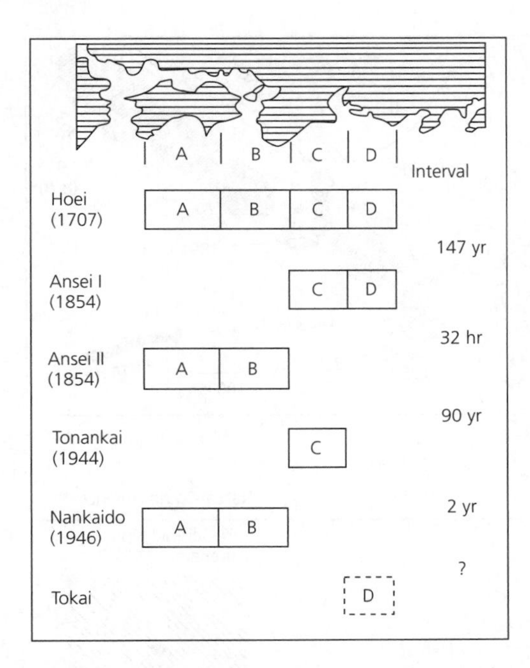

Fig. 5.4-27 Time sequence of large subduction zone earthquakes along the Nankai trough, suggesting both some space and time periodicity and some variability. (Ando, 1975. *Tectonophysics*, *27*, 119–40, with permission from Elsevier Science.)

Given such repeatability, it seems likely that a segment of a subduction zone that has not slipped for some time constitutes a seismic gap and is due for an earthquake. For example, the Tokai area (segment D) may be such as case and is the focus of extensive earthquake prediction studies. However, despite the intuitive appeal of the gap idea, efforts to predict the location of future earthquakes using it have not generally been successful (Sections 1.2.5, 4.7.3).

One difficulty is that not all of the plate motion occurs seismically. Figure 5.4-28 shows that during 1952–73 a large segment of the Kuril trench slipped in a series of six major earthquakes with similar thrust fault mechanism. Seismic moment studies show that the average slip was 2–3 meters. Since the previous major earthquake sequence in the area occurred about 100 years earlier, the average seismic slip rate is 2–3 cm/yr, about one-third of the plate motion predicted from relative motion models. The remaining two-thirds of the slip occurs aseismically, as postseismic or interseismic motion. Similar studies around the world find that the fraction of plate motion that occurs as seismic slip, sometimes called the *seismic coupling* factor, is generally much less than 1, implying that much of the plate motion occurs aseismically if the time interval sampled is adequate.

The Chilean subduction zone shows the other extreme. The seismic slip rate, estimated from the slip in the great 1960 earthquake and historical records indicating that major earthquakes occurred about every 130 years during the past 400 years, exceeds the convergence rate predicted by plate motion models (Fig. 5.4-29). Because the convergence rate is an upper bound on the seismic slip rate, the two estimates are inconsistent. One possibility is that the seismic slip is overestimated: either the earlier earthquakes were significantly smaller than the 1960 event or their frequency in the past 400 years is higher than the long-term average.

More generally, these examples illustrate the difficulty in inferring seismic slip from historical seismicity, owing to problems including the variability of earthquakes on a given plate boundary, the issue of whether the time sample is long enough, and the difficulty in estimating source parameters for earthquakes that pre-dated instrumental seismology. Given the uncertainties in estimating the slip in an earthquake even with seismological data (Section 4.6), doing so without such data is particularly challenging. An alternative approach to estimating plate coupling, discussed in Sections 4.5.4 and 5.6.2, uses GPS geodesy to measure the deflection of the overriding plate, which will be released in future large earthquakes. This deflection depends on the mechanical coupling at the interface, so directly measures what we infer indirectly from the earthquake history. However, the GPS data sample only the present earthquake cycle, which may not be representative of long-term behavior.

Perhaps for similar reasons, efforts to interpret the seismic slip fraction in terms of the physical processes of subduction have not yet been successful. Although the term "seismic coupling" implies a relation between the seismic slip fraction with properties such as the mechanical coupling between the subducting and overriding lithospheres, this has been hard to establish. This relation was originally posed in terms of two end members: coupled Chilean-type zones with large earthquakes and uncoupled Mariana-style zones with largely aseismic subduction. The largest subduction zone earthquakes appear to occur where young lithosphere subducts rapidly (Fig. 5.4-30, *top*), where we might expect the minimum "slab pull" effects and hence the strongest coupling. However,

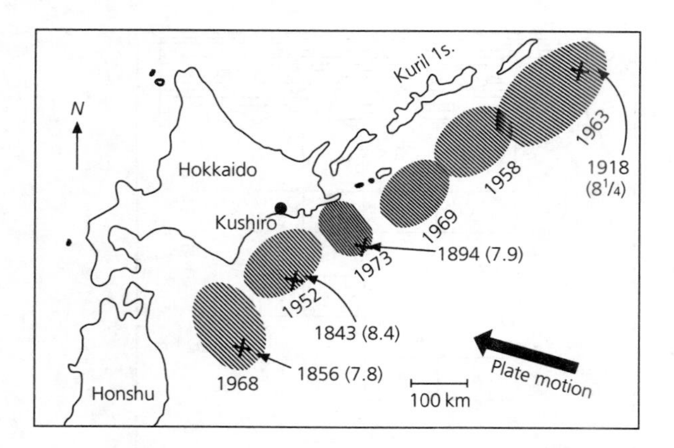

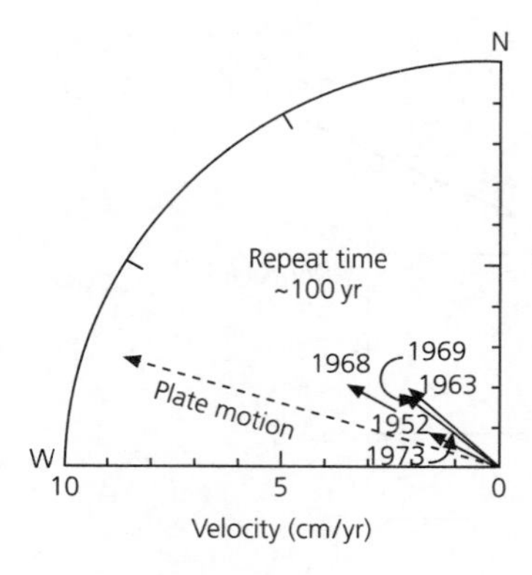

Fig. 5.4-28 Rupture areas for a sequence of large subduction zone earthquakes along the Kuril trench. Different segments of the boundary slip seismically over time. Arrows show the direction and rate of seismic slip and plate motion. If such sequences occur about every 100 years and this time sample is representative, the seismic slip is only about one-third of the plate motion. (Kanamori, 1977b. *Island Arcs, Deep Sea Trenches and Back Arc Basins*, 163–74, copyright by the American Geophysical Union.)

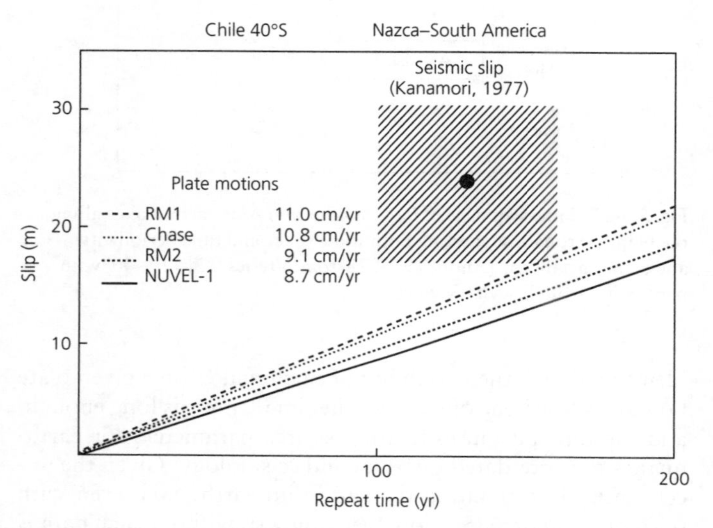

Fig. 5.4-29 Comparison of seismic slip rate and plate motions for the area of the great 1960 Chilean earthquake. Shaded region gives slip rate estimated from slip in the 1960 event and recurrence of large trench earthquakes in the last 400 years. The estimated slip rate exceeds that predicted by any of the four plate motion models shown. (Stein *et al.*, 1986. *Geophys. Res. Lett.*, *13*, 713–16, copyright by the American Geophysical Union.)

efforts to correlate the seismic slip fraction with subduction zone properties such as convergence rate or plate age find no clear pattern (Fig. 5.4-30, *bottom*). It has also been suggested that seismic coupling may be lowest for sedimented trenches and where normal stress on the plate interface is low, although these plausible ideas have yet to be demonstrated. Thus, although seismic coupling can be defined from the seismic slip fraction, its relation to the mechanics of plate coupling is still unclear. It appears that most subduction zones have significant components of aseismic slip, as do oceanic transforms and many continental plate boundaries (Section 5.6.2). Hence, even given the considerable uncertainties in such estimates, it appears common for a significant fraction of plate motion to occur aseismically.

The difficulty in estimating seismic coupling and understanding the process of aseismic plate motion has consequences for estimating the recurrence of earthquakes on a plate boundary and the seismic gap concept. It may be difficult to distinguish between gaps and areas where much of the slip is aseismic. For example, we would not want to say both that areas with recent major seismicity have high seismic hazard and that areas with little recent seismicity are gaps with high seismic hazard.[6] Moreover, as discussed in Sections 1.2 and 4.7.3, the process of earthquake faulting may be sufficiently random that it is hard to use the plate motion rate and seismic history to usefully predict how long it will be until the next large earthquake.

Although most shallow subduction zone seismicity is at the plate interface, some earthquakes occur within either plate. Some appear to result from flexural bending of the downgoing plate as it enters the trench (Fig. 5.4-31). Focal depth studies show a pattern of normal faulting in the upper part of the plate to a depth of 25 km, and thrusting in its lower part, between 40 and 50 km. These observations constrain the position of the neutral surface dividing the mechanically strong lithosphere (Section 5.7.4) into upper extensional and lower compressional zones. In some cases the normal fault earthquakes are so large that they may be "decoupling" events due to "slab pull" that rupture the entire downgoing plate (Fig. 5.4-32). Aftershock distributions and studies of the rupture process indicate that faulting extended through a major portion, and perhaps all, of the lithosphere. Rupture through the entire lithosphere favors the decoupling model. If only a portion of the lithosphere breaks, the interpretation is more complicated. Rupture may have been restricted to one side of the neutral surface (in the flexural model) or reflect the material below being too hot and weak for seismic rupture. In the latter case, the entire lithosphere could have failed, with the deeper rupture being aseismic.

[6] The observation that more recent grizzly bear attacks have occurred in Montana than in Illinois might indicate either a perilous "gap" in Illinois or a greater intrinsic hazard in Montana.

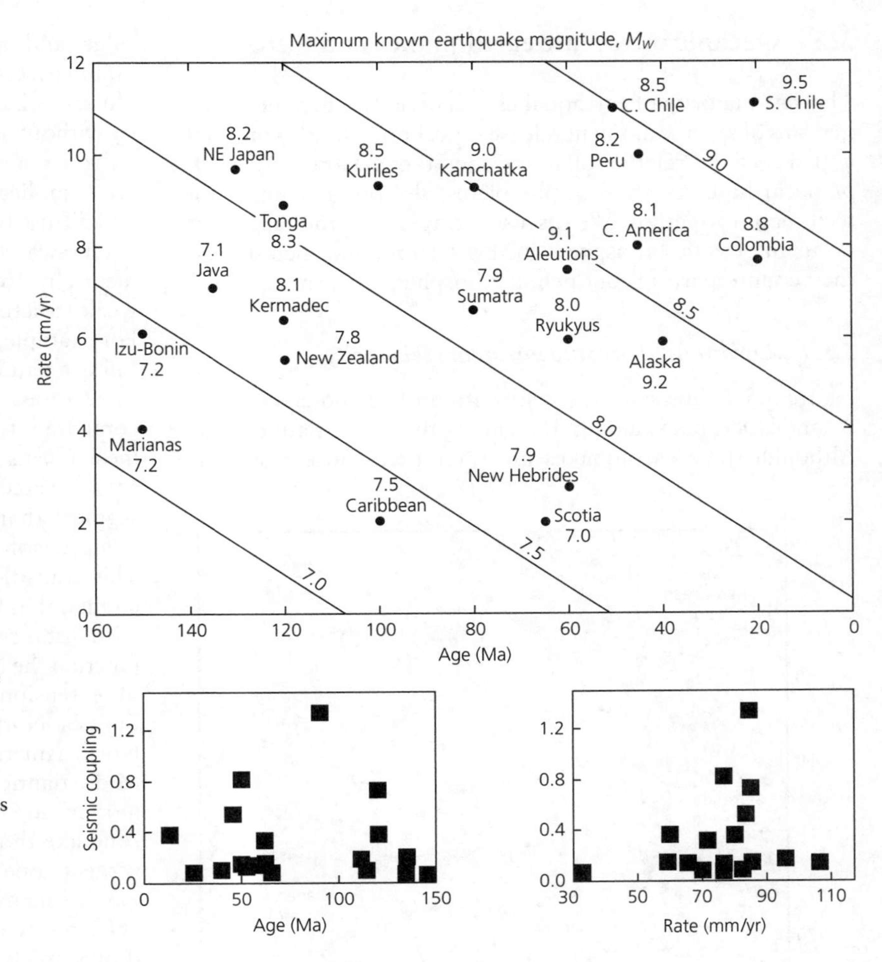

Fig. 5.4-30 *Top*: Variation in the magnitude (M_w) of the largest known subduction thrust fault earthquake between subduction zones as a function of the convergence rate and age of the subducting lithosphere. (Ruff and Kanamori, 1980. *Phys. Earth Planet. Inter.*, *23*, 240–52, with permission from Elsevier Science.) *Bottom*: Seismic coupling fraction estimated from historical seismicity at various subduction zones. Although most subduction zones show considerable aseismic slip, there is no obvious correlation with either age of the subducting lithosphere (*left*) or subduction rate (*right*). (Pacheco *et al.*, 1993. *J. Geophys. Res.*, *98*, 14, 133–59, copyright by the American Geophysical Union.)

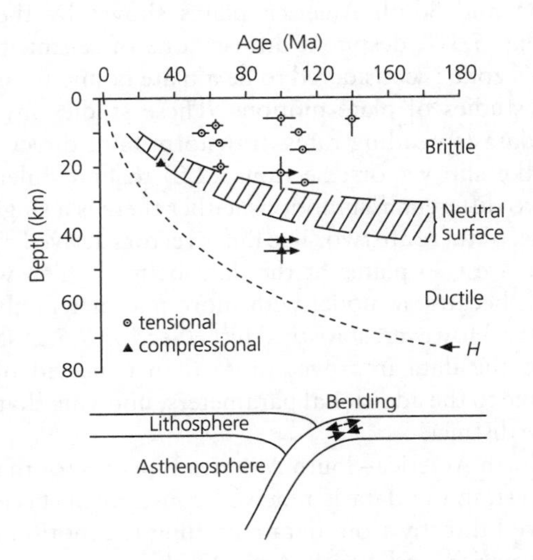

Fig. 5.4-31 Focal depths of flexural earthquakes due to the bending of subducting plates as they enter the trench. Tensional events occur above the neutral surface, and compressional events occur below it. The plate mechanical thickness, H, increases with age, as expected from thermal models. (After Bodine *et al.*, 1981. *J. Geophys. Res.*, *86*, 3695–707, copyright by the American Geophysical Union.)

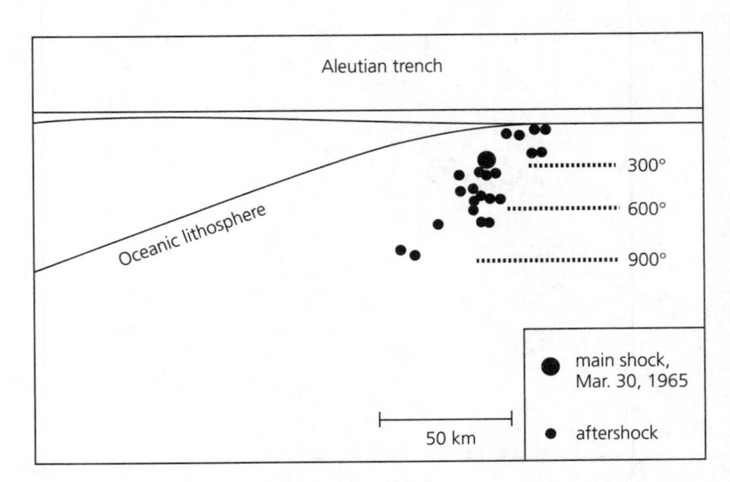

Fig. 5.4-32 Large normal faulting earthquakes at trenches, such as the 1965 M_s 7.5 Rat Island earthquake, may be due to flexure or failure of the lithosphere under its own weight. The extent of aftershocks, which appear not to cut the entire lithosphere, may reflect the extent of rupture or be a temperature effect. (Wiens and Stein, 1985. *Tectonophysics*, *116*, 143–62, with permission from Elsevier Science.)

5.5 Oceanic intraplate earthquakes and tectonics

The vast majority of earthquakes — especially when measured in terms of seismic moment release — occur on plate boundaries and reflect the relative plate motions there. However, *intraplate* earthquakes, those within plates, also provide important tectonic information. We discuss intraplate earthquakes that occur in oceanic lithosphere in this section, and then discuss their counterparts in continental lithosphere in the next.

5.5.1 *Locations of oceanic intraplate seismicity*

Figure 5.5-1 illustrates the distribution of earthquakes in the Atlantic Ocean, excluding those along the Mid-Atlantic ridge. Although these earthquakes are rarer than those along the ridges and transforms making up the Mid-Atlantic ridge plate boundaries, there are enough to justify interest. They nicely illustrate that plates deviate from the ideal case of perfect rigidity without internal deformation, such that all motion occurs at narrow boundaries. Instead, as noted in Section 5.2, real plates are complicated entities that have both internal deformation and diffuse boundary zones.

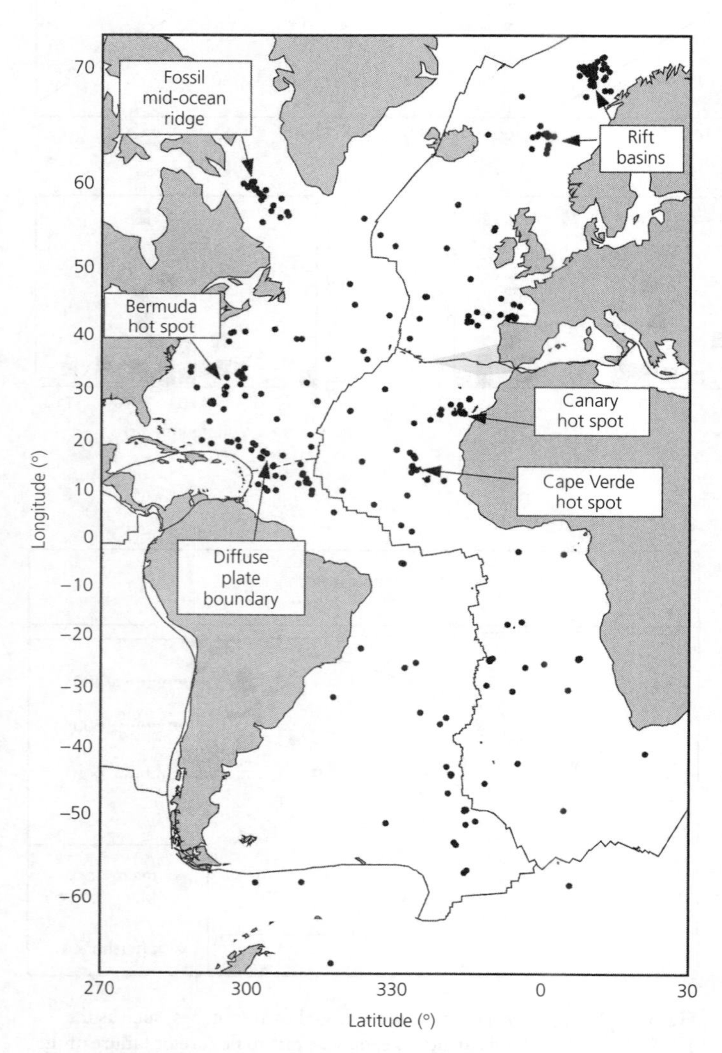

Fig. 5.5-1 Distribution of earthquakes in the Atlantic Ocean other than those on ridge and transform segments of the Mid-Atlantic ridge system. (Wysession *et al.*, 1995. © Seismological Society of America. All rights reserved.)

One way to think about these earthquakes is to consider a hierarchy, from slow-moving plate boundaries, to recognizable weak structures, and then to apparently isolated earthquakes. For example, the Atlantic portion of the boundary between the Eurasian and African plates, which stretches from Gibraltar to the Azores, is poorly defined by topography and seismicity compared to the Mid-Atlantic ridge. However, the focal mechanisms (Fig. 5.5-2, *top*) show a transition from extension at the Terceira Rift near the Azores, to strike-slip along a segment that includes the mapped Gloria transform fault, to compression near Gibraltar, and then into the Mediterranean. This transition reflects the fact that the Euler pole is close enough that the relative motions are small and change rapidly with distance (Fig. 5.5-2, *bottom*). For example, near the triple junction the NUVEL-1A model (Table 5.2-1) predicts 4 mm/yr of extension resulting from the small difference between Eurasia–North America (23 mm/yr at N97°E) and Africa–North America (20 mm/yr at N104°E) spreading across the Mid-Atlantic ridge. Even in the western Mediterranean, the motions are too slow to generate a well-developed subduction zone like those of the Pacific, but instead cause a broad convergent zone indicated by large earthquakes like the 1980 M_s 7.3 El Asnam, Algeria, earthquake.

Even slower motion appears to be why sea floor topography shows no clear evidence for the boundary between the North American and South America plates shown by the dashed line in Fig. 5.5-1, despite a diffuse zone of seismicity in this area. This zone is considered to be a plate boundary, based on detailed studies of plate motions. These studies invert plate motion data (spreading rates, transform fault directions, and earthquake slip vectors; Section 5.2.2) to find Euler vectors under two different assumptions: either there is a single American plate, or there are two. The Euler vectors derived by assuming there are two plates fit the data better, which would be expected, because a model with more parameters always fits data better. However, statistical tests (Section 7.5.2) show that the fit to the data improves more than expected purely by chance due to the additional parameters, implying that the two plates are distinct.

The North America–South America Euler vector that results from inverting the data is not well constrained, because it is not derived directly from data recording the motion between North America and South America, but is estimated from closure of the plate circuit (Fig. 5.2-5). Thus the estimate of motion results from the difference between North America–Africa and South America–Africa motions, which are quite similar (if they were not, the data would clearly show two distinct American plates). The predicted motion along the North

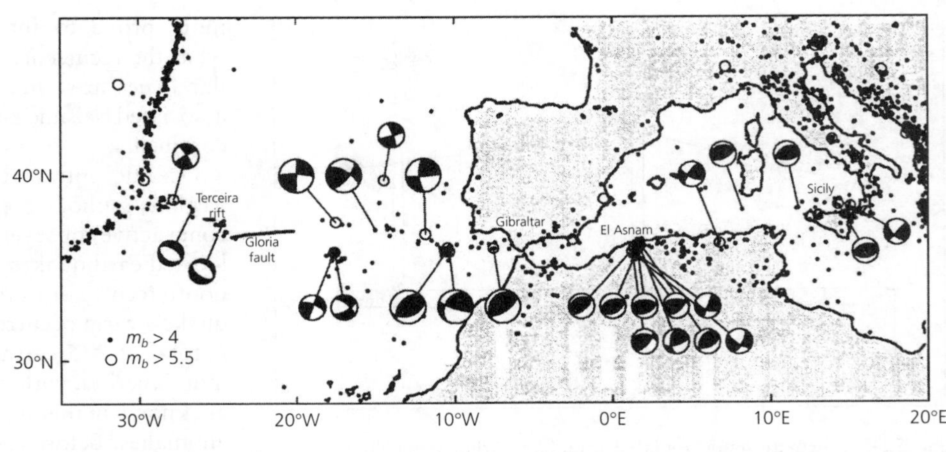

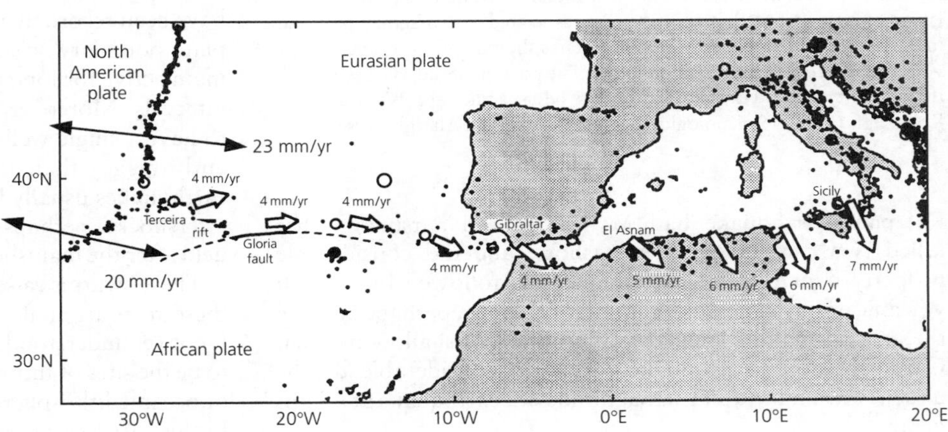

Fig. 5.5-2 *Top*: Focal mechanisms along the western section of the Eurasia–Africa plate boundary. Note the transition from extension near the Azores, to strike-slip (the Gloria fault is a transform), to compression near Gibraltar and into the Mediterranean. *Bottom*: Motions with respect to Africa along the boundary predicted by an Euler pole slightly south of the mapped area, near 20°N, 20°W. The dashed line is a small circle about this pole. (Argus *et al.*, 1989. *J. Geophys. Res.*, *94*, 5585–5602, copyright by the American Geophysical Union.)

America–South America boundary is only about 1 mm/yr — much slower than the approximately 20 mm/yr along the Mid-Atlantic ridge. The North America–South America boundary is thus considered a diffuse, slow-moving boundary zone, although its location and motion are not well constrained. Another reason for treating this as a boundary zone is that paleomagnetic reconstructions find that over the past 70 Myr the two plates have moved relative to each other as the Atlantic Ocean opened.

In general, 1–2 mm/yr is an approximate lower limit for plate boundary deformation. Regions with motions faster than this are generally viewed as plate boundaries, and slower deformation is generally treated as intraplate. However, there is no generally accepted criterion, and evidence from seismicity and topography is also considered. Put another way, in many cases one can regard a region as either a slow-moving plate boundary zone or a zone of intraplate deformation, and "intraplate" earthquakes are often just ones not on an obvious plate boundary.

The Atlantic example (Fig. 5.2-1) shows that in addition to the North America–South America boundary zone, some intraplate seismicity is concentrated in other areas associated with tectonic features. For example, seismicity between Greenland and North America is likely related to the former spreading ridge that opened this part of the Atlantic (the Labrador Sea). Although this spreading stopped about 43 Myr ago, the fossil ridge appears to remain a weak zone along which intraplate stresses cause some motion. Intraplate seismicity is often associated with such fossil structures. Concentrations of seismicity are also associated with the Bermuda (32°N, 65°W), Cape Verde (17°N, 25°W), and Canary (26°N, 17°W) hot spots. Focal mechanism studies are consistent with the earthquakes reflecting heating of the lithosphere by the hot spots.

Hawaii, the most impressive hot spot trace in the oceans (Fig. 5.2-7),[1] provides the best example of intraplate earthquakes associated with hot spot processes (Fig. 5.5-3). Small earthquakes are associated with magma upwelling in the rift zones. Larger earthquakes, which occur on a time scale of tens of years, reflect sliding of the volcanic edifice on subhorizontal faults that are thought to be a layer of weak sediments at the top of the old oceanic crust on which the volcanic island formed. These earthquakes can be quite large — the 1975

[1] Numerical models that infer the amount of upwelling mantle material from how elevated the sea floor is relative to the normal depth–age curves estimate that Hawaii has a buoyancy flux 5–10 times greater than that of Bermuda (Sleep, 1990).

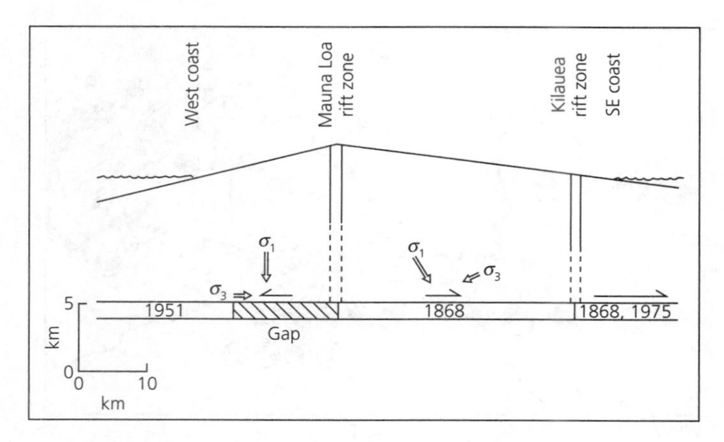

Fig. 5.5-3 Schematic model for large intraplate earthquakes below the island of Hawaii. Small earthquakes are associated with magma upwelling in the rift zones. Larger earthquakes, at dates shown, reflect sliding of the volcanic edifice on subhorizontal faults. The portion of the basal fault that has not ruptured in historic time may be a seismic gap. (Wyss and Koyanagi, 1992. © Seismological Society of America. All rights reserved.)

Kalapana earthquake had M_s 7.2, caused a tsunami that killed two campers on the seashore, and did considerable property damage. The earthquake was followed by a small volcanic eruption near the summit of Kilauea, perhaps because the ground shaking triggered an eruption of shallow magma. Curiously, some earthquakes occur to considerable depths under Hawaii, including a magnitude 6.2 earthquake at 48 km depth.

Although many oceanic intraplate earthquakes are associated with tectonic features, some appear to occur far from plate boundaries, hot spots, or major bathymetric features. Thus the stresses generated by plate driving forces and other sources, including mantle flow near hot spots, appear to reactivate weak zones in the plate resulting from small-scale structure acquired during the lithosphere's evolution.[2]

These earthquakes can be dramatic. For example, the enormous (M_w 8.2) intraplate earthquake that occurred near the Balleny Islands in an oceanic part of the Antarctic plate (63°S, 149°E) in March 1998 was the largest earthquake that had occurred on earth for several years. The fault inferred from waveform modeling (Section 4.3) followed no observable lineaments and cut straight across existing fracture zones. Moreover, in the previous hundred years, no other earthquakes had been located in this region. It is not clear what caused the earthquake or whether this area has any special properties or stress acting there. Although the earthquake occurred south of a puzzling hypothesized deformation zone in the extreme southeast corner of the Australian plate (Fig. 5.2-4), its fault plane solution is inconsistent with its being on the boundaries of a microplate. It is thus unclear whether this area is now any more prone to future earthquakes than other areas, and what the recurrence time of such earthquakes might be. Similar issues arise in considering the intraplate seismicity and associated seismic hazard in the more structurally complex continents.

Oceanic intraplate seismicity often occurs in swarms. Regions without previously known seismicity sometimes become active for several years, with hundreds of teleseismically located earthquakes.[3] The seismicity then dies out, and seems not to recur. For example, during 1981–3, an intraplate earthquake swarm occurred near the Gilbert Islands in Micronesia. A total of 225 earthquakes were detected, mostly over a 15 month period, with 87 above m_b 5. No major tectonic features are known in this area, and a ship survey found no bathymetric anomalies. Before and after the swarm, no other earthquakes have been recorded in this region. The swarms thus differ from plate boundary seismicity, which occurs on features that remain active for long periods even if there are intervening quiet intervals. Moreover, the intraplate swarms often appear not to have a single well-developed fault, and no event is significantly bigger than the others. By contrast, plate boundary earthquakes usually have one or two main ruptures and many aftershocks, perhaps reflecting local adjustments to the stress field after the mainshock has ruptured the entire fault.

These swarms raise an interesting issue. We can assume that these areas are analogous to plate boundaries in having special, if not yet understood, tectonic significance. If so, they are likely to be the sites of future swarms. Alternatively, perhaps all areas of oceanic lithosphere are equally susceptible to such swarms. In this case, over time, swarms will occur in many places, and future swarms are no more likely in one place than another. We will see that similar issues surface in trying to estimate seismic hazards due to intraplate earthquakes within continents.

5.5.2 *Forces and stresses in the oceanic lithosphere*

In addition to using oceanic intraplate seismicity to investigate the specific processes acting at individual sites, we study the seismicity to learn about plate-wide processes. For example, Fig. 5.5-4 shows the variation of mechanism type with lithospheric age. Most of the oceanic lithosphere seems to be in horizontal deviatoric compression, as shown by thrust and strike-slip mechanisms. This compression is in approximately the spreading direction, and is thought to be related to "ridge push": the plate driving force due to lithospheric cooling and subsidence. The major exceptions are the extensional events occurring in the central Indian Ocean. Although originally regarded as intraplate, these earthquakes now appear to be in a diffuse plate boundary zone (Section 5.2.2). In the model shown, the focal mechanisms (Fig. 5.5-5) reflect counterclockwise rotation of Australia with respect to India, causing normal fault earthquakes in the young lithosphere near the Euler pole

[2] This situation is analogous to timbers creaking as a wooden boat rocks in the waves.

[3] There may be many more smaller earthquakes associated with these swarms, but because the swarms often occur in remote regions, only the larger events are detected.

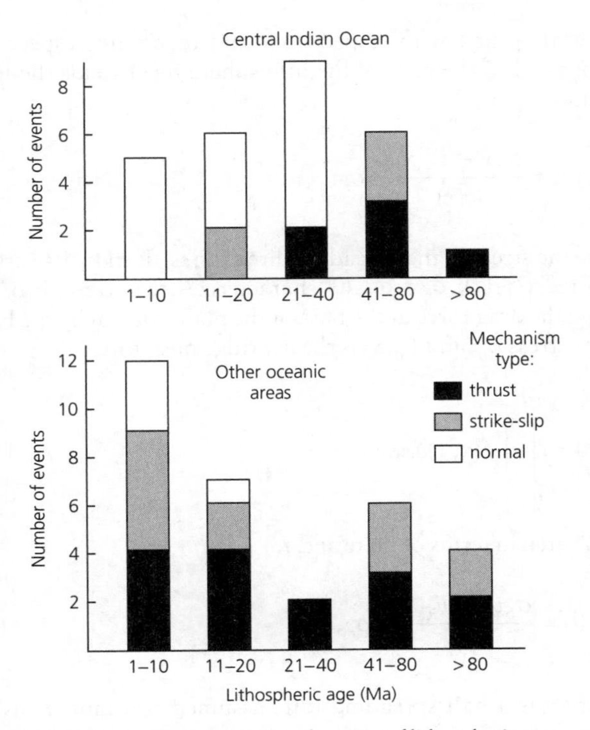

Fig. 5.5-4 Focal mechanism type as a function of lithospheric age for oceanic intraplate earthquakes. Older oceanic lithosphere is in compression, whereas younger lithosphere has both extensional and compressional mechanisms. Extensional events are located primarily in the central Indian Ocean. (Wiens and Stein, 1984. *J. Geophys. Res.*, *89*, 11, 442–64, copyright by the American Geophysical Union.)

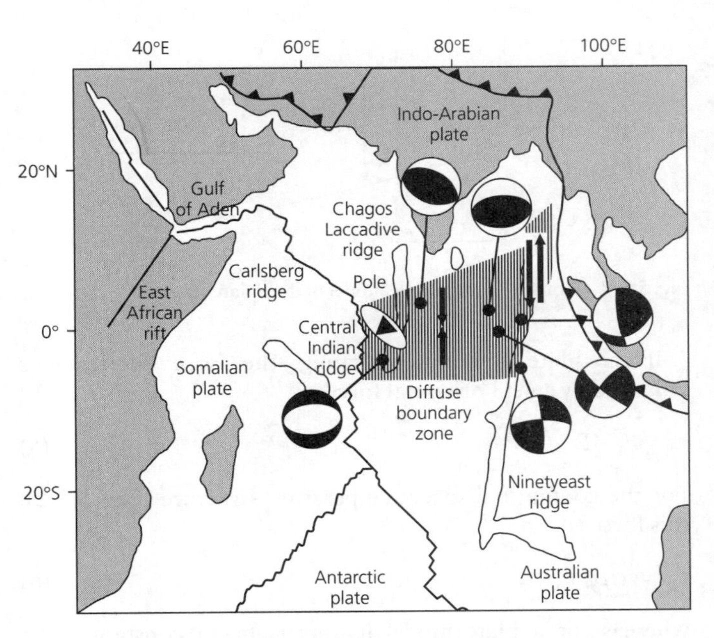

Fig. 5.5-5 Schematic map of earthquake mechanisms in the central Indian Ocean, shown here as a diffuse boundary zone (shaded) between the Indian and Australian plates. Later studies have refined the location and geometry of the boundary zone (Fig. 5.2-4) and pole (triangle) (Wiens *et al.*, 1985. *Geophys. Res. Lett.*, *12*, 429–32, copyright by the American Geophysical Union.)

and thrust and strike-slip earthquakes to the east. These earthquakes reach magnitude 7 on the Ninetyeast ridge.[4]

The general trend of compressive mechanisms in the oceanic plates is consistent with the plate driving force due to the cooling of the oceanic lithosphere. Consider a plate, defined as the area above the $m(t)$ isotherm, out to age t, where the water depth is $h(t)$ (Fig. 5.5-6). The plate is cooler, and thus denser, than material below. The thermal model we used for ocean depth and heat flow also predicts the resulting force.

The total horizontal force on the base of the lithosphere, F_1, equals the integrated horizontal pressure force of the asthenosphere at the ridge, because the material is in hydrostatic equilibrium:

$$F_1 = \int_0^{m(t)} \rho_m gz dz = \rho_m g(m(t))^2/2. \quad (1)$$

Similarly, F_2, the horizontal force due to water pressure on the plate, equals the integrated horizontal pressure force of the water,

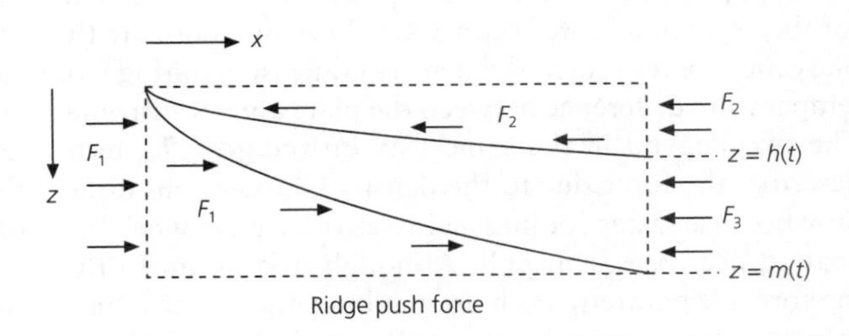

Fig. 5.5-6 Derivation of the "ridge push" force.

$$F_2 = \int_0^{h(t)} \rho_w gz dz = \rho_w g(h(t))^2/2. \quad (2)$$

F_3 is the remaining horizontal force due to lithospheric pressure $P(z, t)$,

$$F_3 = \int_{h(t)}^{m(t)} P(z, t) gz dz, \quad (3)$$

where the pressure depends on the density perturbation due to lithospheric cooling (Eqn 5.3.7),

$$P(z, t) = \rho_w gh(t) + g \int_{h(t)}^{z} [\rho_m + \rho'(z', t)] dz'. \quad (4)$$

[4] Although hot spot tracks like the Ninetyeast and Chagos-Laccadive ridges have been termed "aseismic" ridges, to distinguish them from spreading ridges, these two are more seismically active in terms of moment release than many spreading ridges.

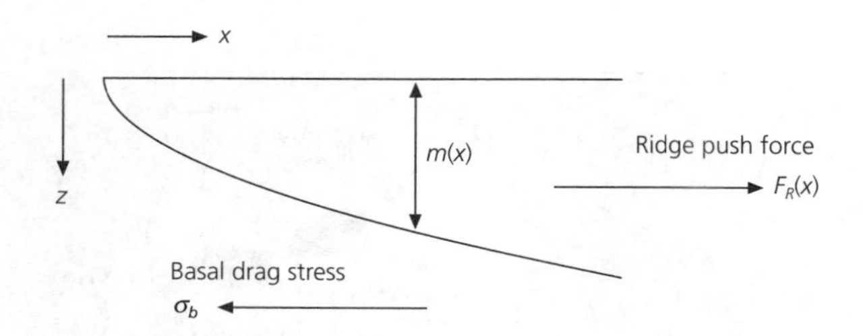

Fig. 5.5-7 Geometry for a simple model of intraplate stresses.

If the plate is not accelerating, the force difference is balanced by a net horizontal force

$$F_R = F_1 - F_2 - F_3. \tag{5}$$

For the cooling halfspace temperature structure (Eqn 5.3.2), this force is

$$F_R = g\alpha\rho_m T_m \kappa t, \tag{6}$$

whereas for a plate model it approaches a constant value for old lithosphere. The convention of calling this force "ridge push" is confusing because it is zero at the ridge and increases linearly with plate age. It results not from force at the ridge but from the total force due to the density anomaly within the cooling plate out to any given age.

The expression for the "ridge push" force is similar to that for the "slab pull" force (Eqn 5.4.15) because both are thermal buoyancy forces due to the density contrast resulting from the temperature difference between the plate and its surroundings. The two depend in the same way on the $g\alpha\rho_m T_m$ term that describes the force due to the density contrast, but differently on κ because faster cooling increases ridge push whereas faster heating decreases slab pull. Although it is useful to think of the forces separately, both are net buoyancy forces due to the mantle convection system of which the plates are a part.[5]

To discuss the stresses within the oceanic lithosphere, we compare the ridge push force to the other forces applied at the boundaries of the plate. These include forces at the plate base and forces at the subduction zone. As for the downgoing slab, earthquake focal mechanisms constrain the relative size of the forces. Here, we use the observation (Fig. 5.5-4) that stress in the spreading direction is typically compressive at all ages.

Consider a simple model of stress in the oceanic lithosphere, using the geometry of Fig. 5.5-7. Using the stress equilibrium equation (Eqn 2.3.49) in the spreading (x) direction, we relate the deviatoric stresses to the body force $f(x, z)$, which is the contribution to ridge push from the material at (x, z),

$$\frac{\partial \sigma_{xx}(x, z)}{\partial x} + \frac{\partial \sigma_{xz}(x, z)}{\partial z} + f(x, z) = 0. \tag{7}$$

[5] Verhoogen (1980) offers the analogy that rain occurs because of the negative buoyancy of the drops relative to the surrounding air, as part of the process by which solar heat evaporates water which rises as vapor due to positive buoyancy and is transported by wind to the point where it cools, condenses into drops, and then falls.

Integrating first with respect to x and then with respect to z from $z = 0$ to the base of the lithosphere $m(x)$ yields the force balance

$$\bar{\sigma}_{xx}(x) = \frac{\sigma_b x - F_R(x)}{m(x)} + \sigma_r. \tag{8}$$

Here the stress in the spreading direction is given by its vertical average $\bar{\sigma}_{xx}(x)$; $\sigma_r = \bar{\sigma}_{xx}(0)$ characterizes the strength of the ridge; the drag force at the base of the plate is given by the basal shear stress σ_b; and $F_R(x)$ is the net ridge push force

$$F_R(x) = \int_0^{m(t)} \int_0^{x} f(x, z)\,dx\,dz. \tag{9}$$

Written in terms of plate age, t,

$$\bar{\sigma}_{xx}(t) = \frac{\sigma_b v t - F_R(t)}{m(t)} + \sigma_r, \tag{10}$$

where v is a half spreading rate, assumed constant. A useful form for comparing different plates comes from the usual assumption that the basal drag force equals the product of absolute velocity u and drag coefficient C ($\sigma_b = Cu$),

$$\bar{\sigma}_{xx}(t) = \frac{Cuvt - F_R(t)}{m(t)} + \sigma_r \tag{11}$$

Thus a drag depending on absolute velocity is applied over an area proportional to the spreading rate. For simplicity, we assume that $v = u$, spreading rate equals absolute velocity (the ridge is fixed with respect to the mantle), so the net drag force is proportional to velocity squared.

A subduction zone would provide a boundary condition on the oldest lithosphere. For example, if focal mechanisms in the lithosphere near trenches were extensional, an extensional condition could be imposed. Because such mechanisms are not seen, it is often assumed that the negative buoyancy of slabs (slab pull) is balanced by local resistive forces (Section 5.4.2). Thus, although the ridge push force is probably smaller than the slab pull forces, the thrust fault mechanisms suggest that it is more crucial for determining stress in oceanic lithosphere.

Although this stress model is schematic and does not describe any individual plate, it lets us use focal mechanism observations to estimate several important quantities. Figure 5.5-8 shows the predicted intraplate stress as a function of plate age and drag coefficient. For zero drag the stress is purely compressive ($\bar{\sigma}_{xx} < 0$) and varies as $\sqrt{t}$, because the force increases linearly with age, whereas the plate thickens as its square root. For larger drag coefficients, $\bar{\sigma}_{xx}$ follows $\sqrt{t}$ curves corresponding to less and less compression, until the lithosphere is in extension for all ages. All lithospheric plates

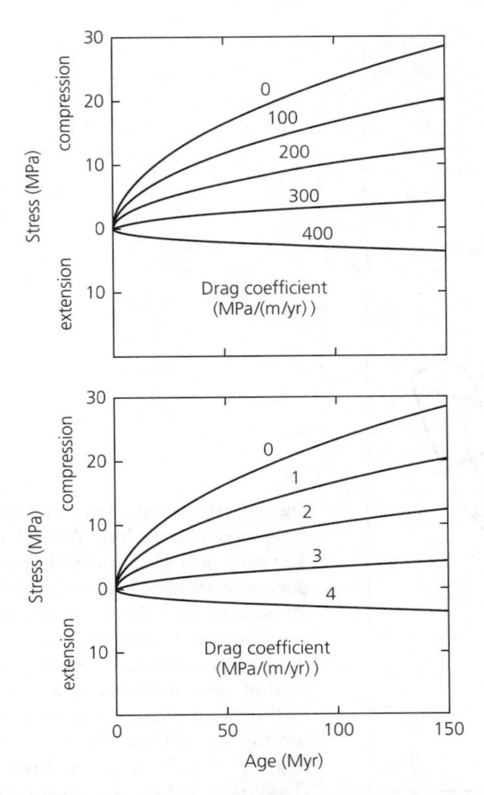

Fig. 5.5-8 Intraplate stress in the spreading direction as a function of lithospheric age and assumed basal drag coefficient for slow-moving (1 cm/yr, *top*) and fast-moving (10 cm/yr, *bottom*) plates. The compressional stresses in old oceanic lithosphere place an upper bound on the drag coefficient of 4 MPa/(m/yr). (Wiens and Stein, 1985. *Tectonophysics, 116*, 143–62, with permission from Elsevier Science.)

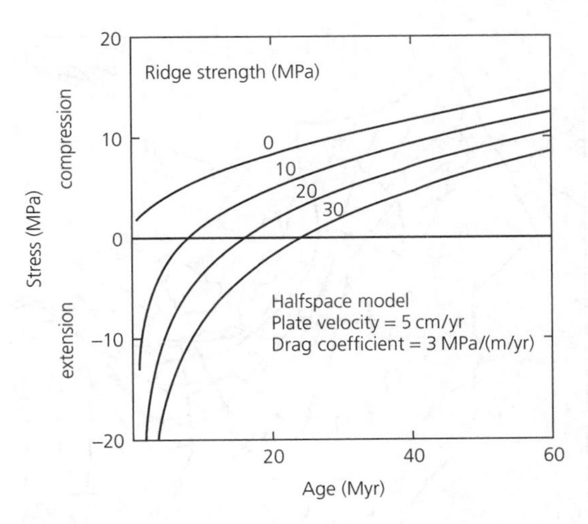

Fig. 5.5-9 Intraplate stress in the spreading direction as a function of lithospheric age computed for several values of ridge strength. The age of the transition from ridge-normal extension to compression increases with the strength of the ridge. (Wiens and Stein, 1984. *J. Geophys. Res., 89*, 11, 442–64, copyright by the American Geophysical Union.)

appear to be in compression, so a rapidly moving plate (such as the Pacific, which moves at about 10 cm/yr) constrains the drag coefficient to less than about 4 MPa/(m/yr). Similar results emerge for a cooling plate model.

This model assumes a zero stress boundary condition at the ridge axis, so the axis has no tensile strength. The predicted stress in young lithosphere, especially the location of a possible transition from compression to extension in the direction of spreading, would be sensitive to the strength of the ridge (Fig. 5.5-9). Models with substantial strength at the axis predict a wide band of extension in the spreading direction. Since such a zone of normal-faulting earthquakes is not observed, the axis seems weak.

Although this simple model describes only a hypothetical average plate, more sophisticated models use realistic plate geometries to calculate the stresses expected from ridge push, slab pull, and basal drag forces. These models' predictions can be compared to earthquake focal mechanisms and other data for specific areas. For example, Fig. 5.5-10 shows stresses predicted for the Indian Ocean region. Although the model was calculated assuming a single Indo-Australian plate, it predicts stresses in the region now considered a diffuse boundary zone (Fig. 5.5-5) that are generally consistent with the focal mechanisms and the folding seen in gravity and seismic reflection data.

5.5.3 *Constraints on mantle viscosity*

The last section's analysis relating earthquake mechanisms to drag at the base of the lithosphere also gives insight into the viscosity of the mantle. The viscosity,[6] the proportionality constant between shear stress and the strain rate (or velocity gradient), controls how the mantle flows in response to applied stress, and is thus crucial for mantle convection. If the drag on the base of a plate is due to motion over the viscous mantle, compressive earthquake mechanisms in old lithosphere constrain the viscosity.

Consider a simple two-dimensional geometry where mass flux due to the moving plate is balanced by a return flow at depth (Fig. 5.5-11, *top*). The drag coefficient is proportional to the viscosity and inversely proportional to the flow depth. Figure 5.5-12 shows that the basal drag constraint from the focal mechanism data, $C \leq 4$ MPa/(m/yr), requires an average mantle viscosity less than 2×10^{20} poise if flow occurs to a depth of 700 km in the upper mantle, or 10^{21} poise if flow occurs in the entire mantle. These values are lower than the $1–5 \times 10^{22}$ poise typically estimated from glacial rebound, earth rotation, and satellite orbits.

This discrepancy can be reconciled by assuming that the plate is underlain by a thin, low-viscosity asthenosphere (Fig. 5.5-11, *bottom*). The low-viscosity layer, in which only a fraction of the return flow occurs, decouples the plates from the underlying

6 Viscosity, defined in Section 5.7, is given in cgs units as poise (dyn-s/cm^2) or in SI units as Pascal-seconds (1 poise = 0.1 Pa-*s*).

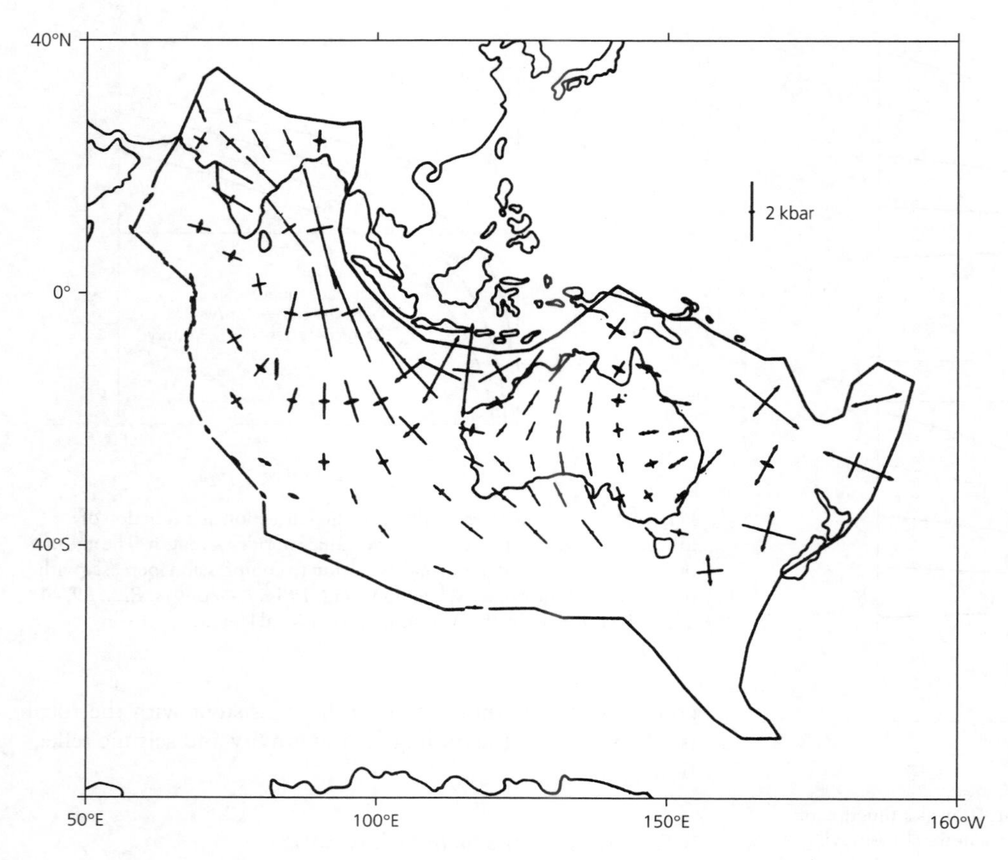

Fig. 5.5-10 Intraplate stress predicted by a force model for the Indo-Australian plate The bars show the principal horizontal deviatoric stresses, with arrowheads marking tension. The location and orientation of the highest stresses, such as the transition between compression and tension, are generally consistent with earthquake mechanisms in the region now regarded as a diffuse plate boundary (Fig. 5.5-5). (Cloetingh and Wortel, 1985. *Geophys. Res. Lett.*, *12*, 77–80, copyright by the American Geophysical Union.)

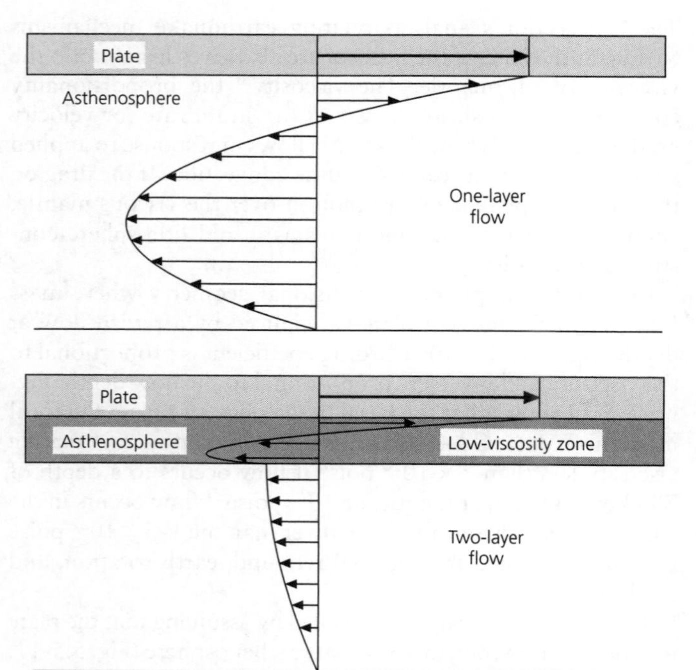

Fig. 5.5-11 *Top*: Velocity profile associated with a return flow of uniform-viscosity asthenosphere that balances the mass flux due to plate motions. *Bottom*: Velocity profile associated with a return flow of two layers of different viscosity. The upper, low-viscosity layer decouples the plates from the underlying mantle. (McKenzie and Richter, 1978.)

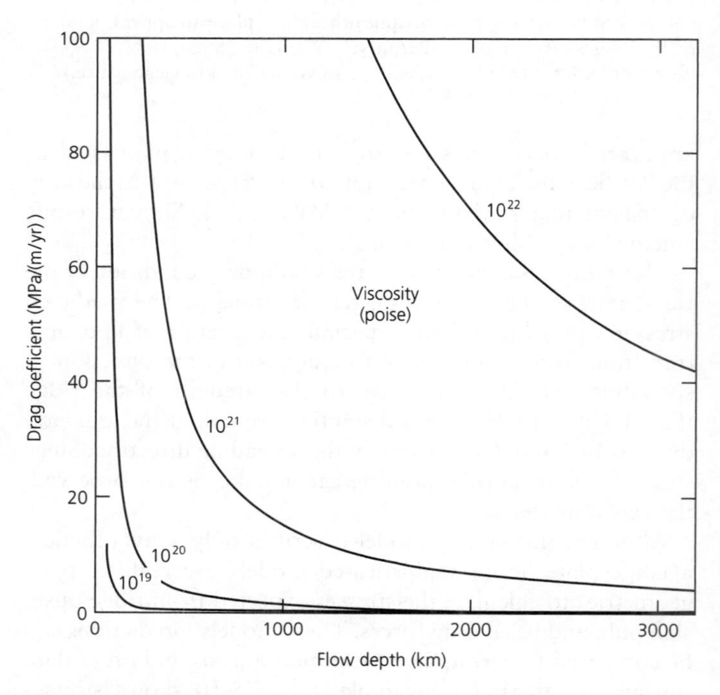

Fig. 5.5-12 Basal drag coefficients as a function of the mantle viscosity and flow depth, assuming single-layer flow. (Wiens and Stein, 1985. *Tectonophysics, 116*, 143–62, with permission from Elsevier Science.)

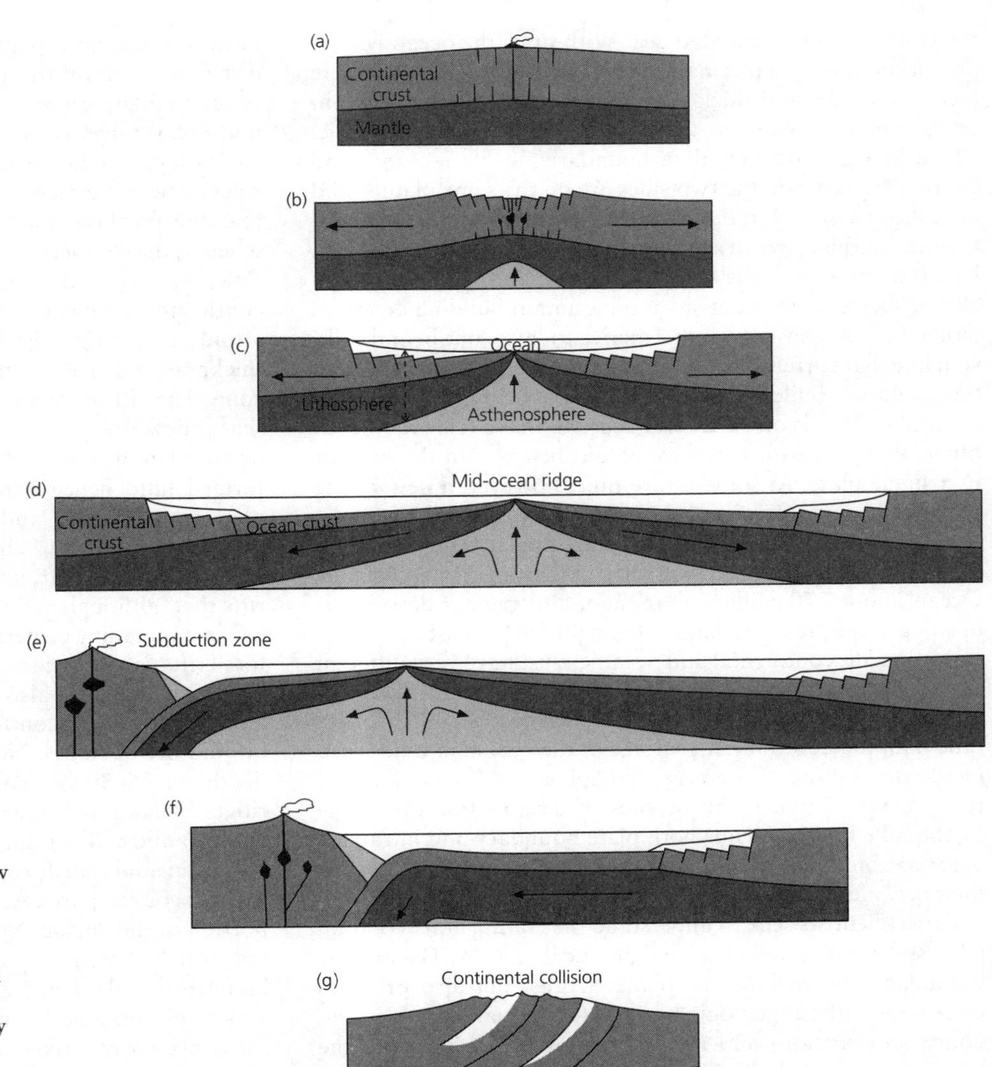

Fig. 5.6-1 Schematic illustration of the Wilson cycle, the fundamental geological process controlling the evolution of the continents. (a)–(b): A continent rifts, such that the crust stretches, faults, and subsides. (c): Sea floor spreading begins, forming a new ocean basin. (d): The ocean widens and is flanked by sedimented passive margins. (e): Subduction of oceanic lithosphere begins on one of the passive margins, closing the ocean basin (f) and starting continental mountain building. (g): The ocean basin is destroyed by a continental collision, which completes the mountain building process. At some later time, continental rifting begins again.

mantle. Viscosity values that satisfy the focal mechanisms are consistent with constraints from gravity and glacial isostasy, and such decoupling is consistent with the lack of correlation between oceanic plate area and absolute velocity (Fig. 5.4-12).

5.6 Continental earthquakes and tectonics

Although the basic relationships between plate boundaries, plate interiors, and earthquakes apply to continental as well as oceanic lithosphere, the continents are more complicated. The continental crust is much thicker, less dense, and has different mechanical properties from the oceanic crust. As a result, plate boundaries in continental lithosphere are generally broader and more complicated than in the oceanic lithosphere (Fig. 5.2-4).

Studies of continental plate boundaries, which rely heavily on seismology, provide important insights into the fundamental geological processes controlling the evolution of the continents. The basic process, known as the *Wilson cycle*,[1] is illustrated in Fig. 5.6-1. A continental region undergoes extension, such that the crust is stretched, faulted, and subsides, yielding a rift valley like the present East African rift. Because the uppermost mantle participates in the stretching, hotter mantle material upwells, causing partial melting and basaltic volcanism. Sometimes the extension stops after only a few tens of kilometers, leaving a failed or fossil rift such as the 1.2 billion-year-old mid-continent rift in the central USA. In other cases the extension continues, so the continental rift evolves into an oceanic spreading center (identifiable from sea floor magnetic anomalies), which forms a new ocean basin like

[1] Named after J. Tuzo Wilson (1908–93), whose key role in developing plate tectonic theory included introducing the ideas of transform faults, hot spots, and that the Atlantic had closed and then reopened.

the Gulf of Aden or the Red Sea. With time, the ocean widens and deepens due to thermal subsidence of oceanic lithosphere (Section 5.3.2), and thick sediments accumulate on the continental margins, such as those on either side of the Atlantic. These margins are not plate boundaries — the oceanic and continental crust on the two sides are on the same plate — and are called *passive margins*, to distinguish them from active continental margins, which are plate boundaries. Subduction often begins along one of the passive margins, and the ocean basin closes, such that magmatism and mountain building occur, as along the west coast of South America today. Continental collision like that currently in the Himalayas occurs eventually, and the mountain building process reaches its climax. If the continental materials on either side cease to move relative to each other, this process leaves a mountain belt within the interior of a single plate. At some future time, however, a new rifting phase can begin, often near the site of the earlier rifting, and a new ocean will start to grow. Thus the Appalachian Mountains record a continental collision that closed an earlier Atlantic Ocean about 270 million years ago, and remain despite the opening of the present Atlantic Ocean during the past 200 Myr.

As a result, continental and oceanic crust have very different life cycles. Because the relatively less dense continental crust is not subducted, the continents have accreted over a much longer time than the 200-million-year age of the oldest oceanic crust. Hence the continents preserve a complex set of geologic structures, many of which can be sites of deformation, including earthquake faulting. Thus both plate boundary and intraplate deformation zones within continents are more complex than their oceanic counterparts.

Earth scientists seek to understand the continental evolution process for both intellectual and practical reasons. The process is fundamental to how the planet works, but also provides information about geologic hazards (earthquakes, volcanism, uplift, and erosion) and mineral resources. In addition, the large mountain belts have major impacts on earth's climate. Seismology contributes to these studies by providing data about earthquakes and velocity structure in regions where different parts of the evolutionary cycle occur today or occurred in the past. These data are combined with other geophysical and geological data to form an integrated picture of the complicated continental evolution processes. Hence, although the processes are not fully understood, important progress continues to be made.

5.6.1 *Continental plate boundary zones*

As for oceanic boundaries, we seek to first describe the motion (kinematics) within boundary zones, and then to combine the kinematics with other data to investigate their mechanics (dynamics). One example is the East African rift (Fig. 5.6-2), a spreading center between the Nubian (West Africa) and Somalian (East Africa) plates. The extension rate is so slow, less than 10 mm/yr, that it is hard to resolve in plate motion models, and the two plates are often treated as one (Fig. 5.2-4). However, the rift topography, normal faulting, and seismicity distribution show the presence of an extensional boundary zone broader, more diffuse, and more complex than at a mid-ocean ridge. For example, the seismicity ends in southern Africa and has no clear connection to the southwest Indian ridge, where the plate boundary must go. A recent estimate is that the northern East Africa rift opens at about 6 mm/yr, whereas the southern part opens at about half that, because the Euler pole is to the south. Some of the complexity of such continental extensional zones results from the fact that, unlike a mid-ocean ridge, the lithosphere starts off with reasonable thickness and then is stretched and thinned in the extending zone. The rifting process can eventually progress far enough that a new oceanic spreading center forms. This has already occurred in the Gulf of Aden and the Red Sea, which are newly formed (and hence narrow) oceans separating the Arabian plate from Somalia and Nubia at rates of about 22 and 16 mm/yr, respectively. Whether the East African rift will evolve this far is still unclear, because the geologic record shows many rifts that, although active for some time, failed to develop into oceanic spreading centers and simply died. As we will see, these fossil rifts can be loci for intraplate earthquakes.

The earthquakes also indicate that the thermal and mechanical structure of continental rifts is more complicated than on mid-ocean ridges. Normal-faulting earthquakes extend to depths of 25–30 km, considerably deeper than at mid-ocean ridges. Hence the lower crust appears to be surprisingly stronger and colder than might be expected in an active rift.

Continental transforms are also more complicated than their oceanic counterparts. As we saw in Section 5.2, the transform portion of the Pacific–North America plate boundary in western North America is an active seismic zone hundreds of kilometers wide (Fig. 5.2-3), in contrast to widths of less than 10 km for oceanic transforms. Thus the focal mechanisms show primarily strike-slip motion on the San Andreas fault itself and demonstrate complexities including thrust faulting for events like the 1971 San Fernando and 1994 Northridge earthquakes and normal faulting due to the regional extension in the Basin and Range province. The earthquakes and space-geodetic data show that although most of the motion occurs along the San Andreas (Fig. 4.5-13) and nearby faults, a reasonable fraction of the motion occurs elsewhere (Figs. 5.6-3 and 5.2-3). The boundary zone is further complicated by volcanism in areas including the Long Valley caldera in eastern California and the Yellowstone hot spot, which also have associated seismicity. Hence, we think of a boundary zone in which the overall steady motion between the plate interiors is distributed in both space and time (Fig. 5.6-4). Although much of the motion occurs in occasional large earthquakes or steady creep on the main boundary segment, some deformation occurs elsewhere in the zone.

The breadth of continental plate boundary zones has important implications for seismic hazards within them. Because ground shaking decays rapidly with distance (Fig. 1.2-5), nearby smaller earthquakes within a boundary zone, but not

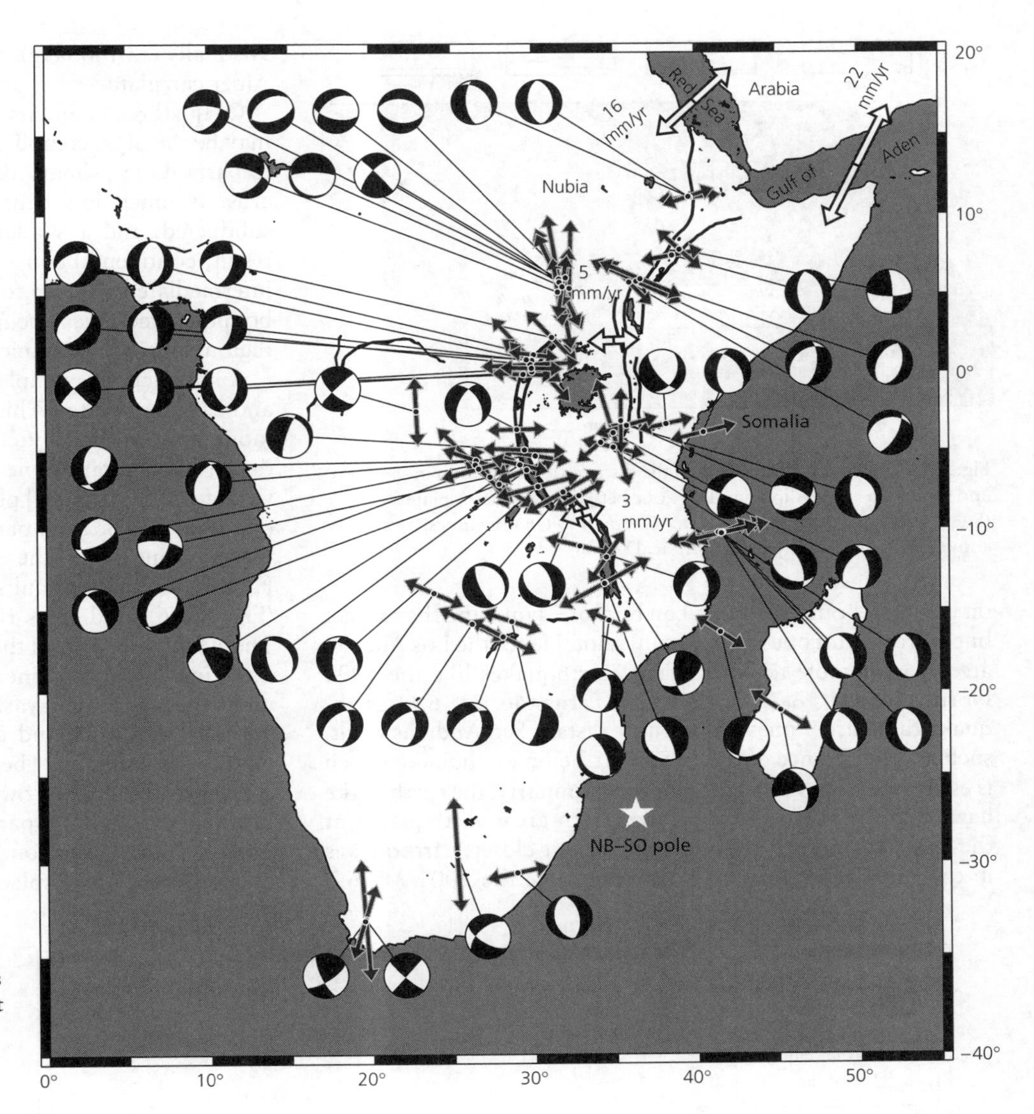

Fig. 5.6-2 Seismicity and focal mechanisms (*T* axes shown by black arrows) for the East African rift system, with relative plate motions (white arrows) from Chu and Gordon (1998, 1999).

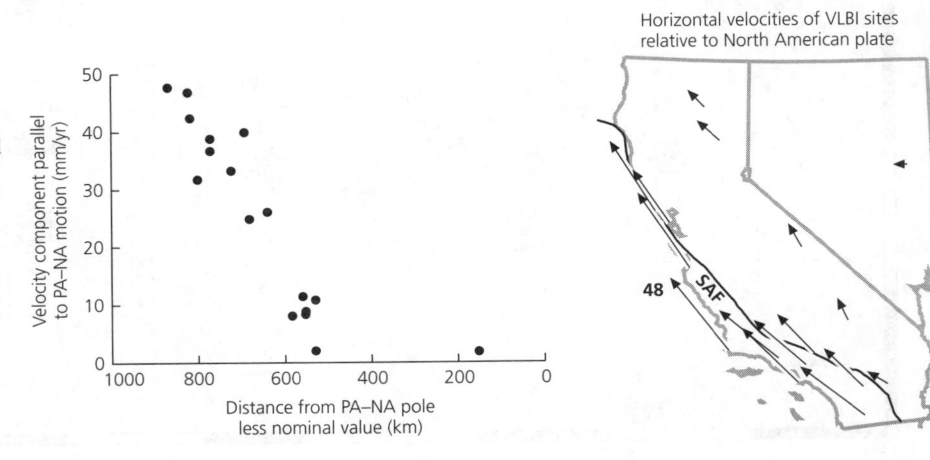

Fig. 5.6-3 Variation in motion of space-geodetic sites across part of the Pacific–North America boundary zone. *Right*: Horizontal velocities of sites in California, Nevada, and Arizona relative to stable North America. The velocity of the southwesternmost site nearly equals the predicted 48 mm/yr velocity of the Pacific plate relative to the North American plate. *Left*: Component of motion tangent to small circles centered on the Pacific–North America Euler pole versus angular distance from that pole. Velocities increase with distance from the Euler pole, with a discontinuity due to the approximately 35 mm/yr of time-averaged slip across the San Andreas fault. (Gordon and Stein, 1992. *Science*, *256*, 333–42, copyright 1992 American Association for the Advancement of Science.)

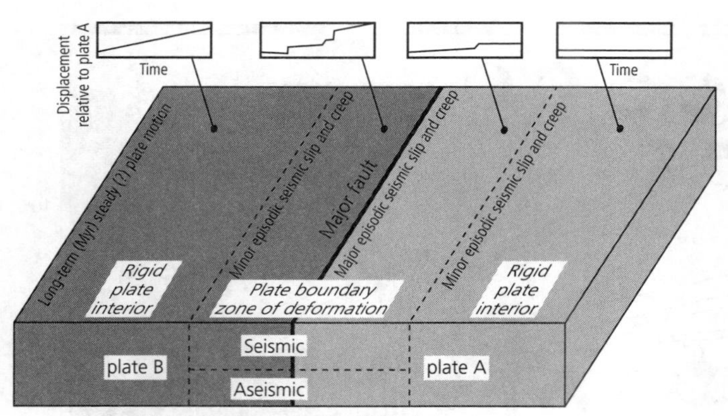

Fig. 5.6-4 Schematic illustration of the distribution of motion in space and time for a strike-slip boundary zone between two major plates. (Stein, 1993. *Contributions of Space Geodesy to Geodynamics*, 5–20, copyright by the American Geophysical Union.)

on the main boundary fault, can be more damaging than larger but more distant ones on the main fault. Hence the Los Angeles area is vulnerable to both nearby earthquakes like the 1994 Northridge (M_w 6.7) or 1971 San Fernando (M_s 6.6) earthquakes and larger ones on the more distant San Andreas Fault, such as a recurrence of the 1857 Fort Tejon earthquake which is estimated to have had M_w about 8. Similarly, the earthquake hazard in the Seattle area involves both great earthquakes at the subduction interface and smaller, but closer, earthquakes in the subducting Juan de Fuca plate (like the 2001 M_w 6.7 Nisqually earthquake) or at shallow depth in the North American plate.

Of the three boundary types, continental convergence zones may be the most complicated compared to their oceanic counterparts. One primary difference is that because continental crust is much less dense than the upper mantle, it is not subducted, and a Wadati–Benioff zone is not formed. As a result, continental convergence zones in general do not have intermediate and deep focus earthquakes. However, the plate boundary tectonics occur over a broader and more complex region than in the oceanic case.

A spectacular example is the collision between the Indian and Eurasian plates. This area is the present type example of mountain building by continental collision, which has produced a boundary zone extending thousands of km northward from the nominal plate boundary at the Himalayan front (Fig. 5.6-5). The total plate convergence is taken up in several ways. About half of the convergence occurs across the locked Himalayan frontal faults such as the Main Central Thrust (Fig. 5.6-6), and gives rise to large destructive earthquakes. These faults are part of the interface associated with the underthrusting Indian continental crust, which thickens the crust under the high Himalayas. However, the earthquakes also show normal faulting behind the convergent zone, in the Tibetan plateau, presumably because the uplifted and thickened crust spreads under its own weight. GPS data (Fig. 5.6-5) show that this extension is part of a large-scale process of crustal "escape," or "extrusion," in which large fragments of continental crust are displaced eastward by the collision along

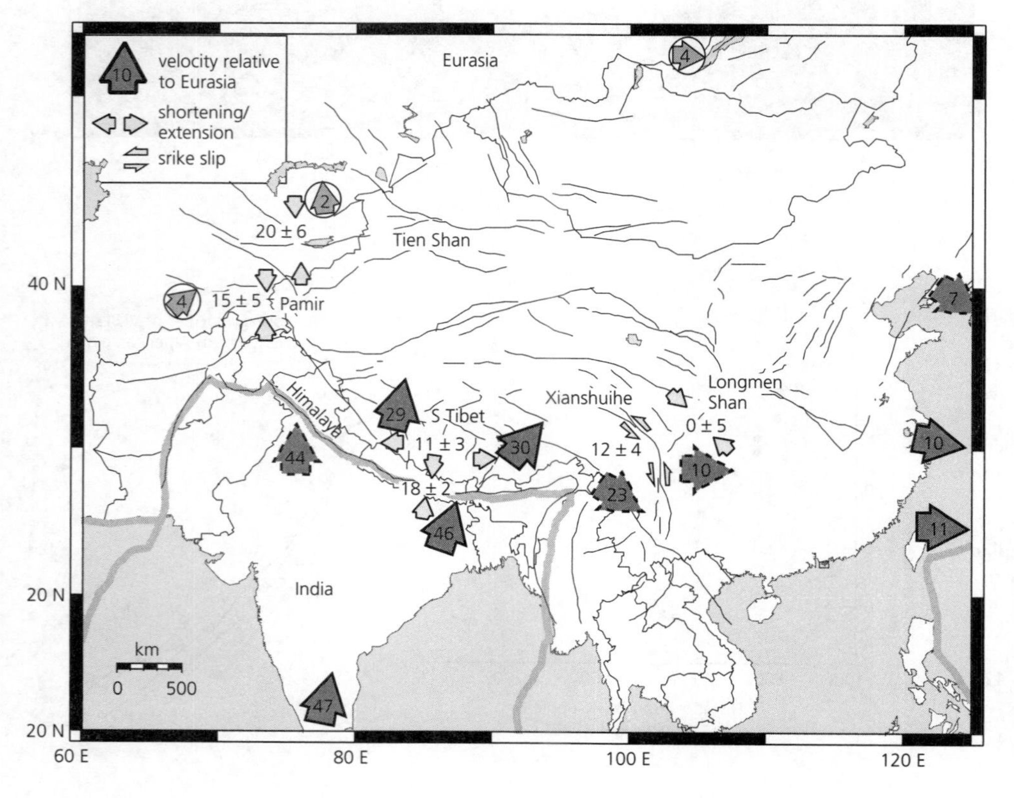

Fig. 5.6-5 Summary of crustal motions determined using space geodesy in the India–Eurasia plate collision zone. Large arrows indicate velocities relative to Eurasia. Arrows in circles show velocities with no significant motion with respect to Eurasia. Small arrows show local relative deformation. (Larson *et al.*, 1999. *J. Geophys. Res.*, *104*, 1077–94, copyright by the American Geophysical Union.)

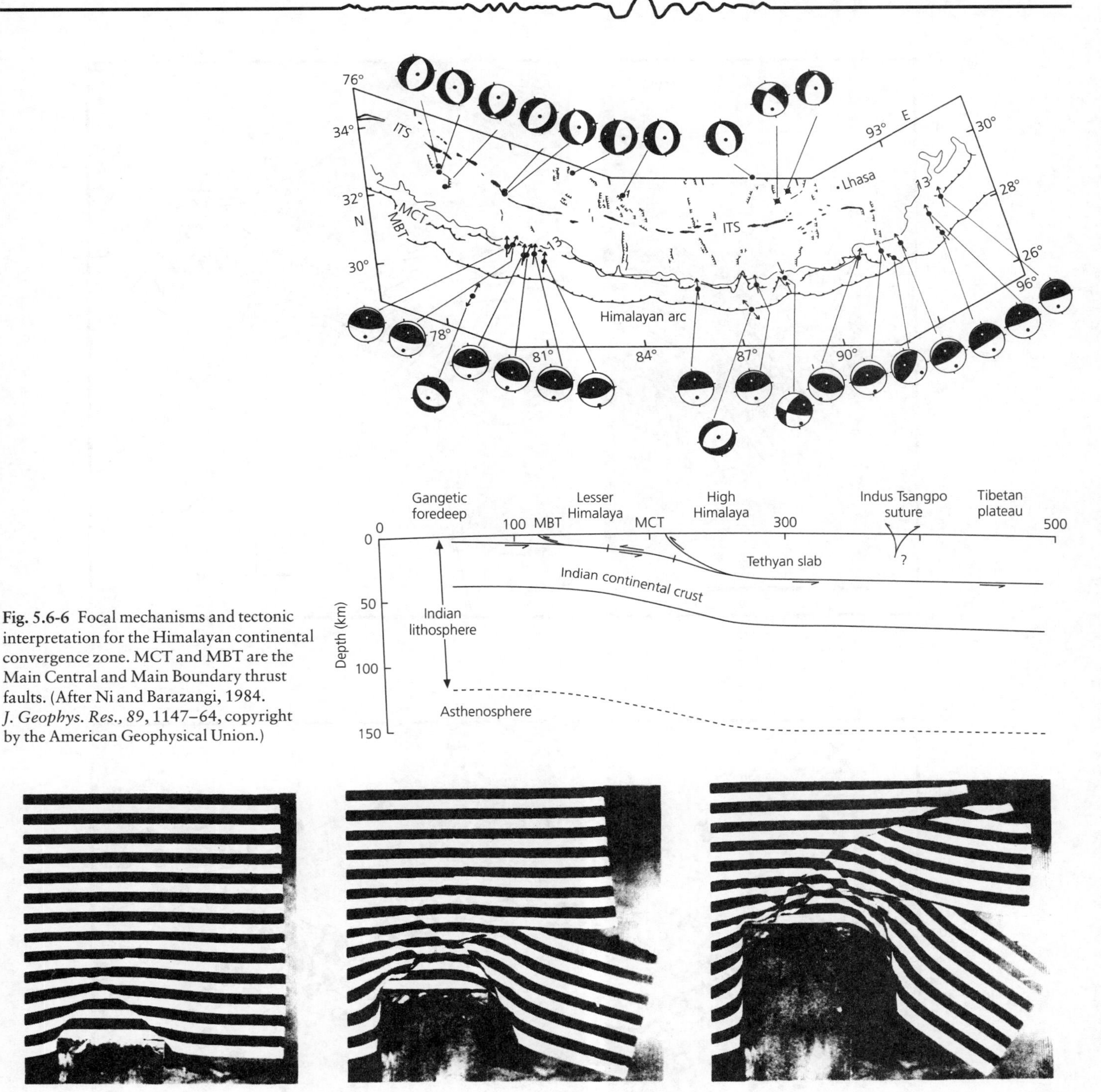

Fig. 5.6-6 Focal mechanisms and tectonic interpretation for the Himalayan continental convergence zone. MCT and MBT are the Main Central and Main Boundary thrust faults. (After Ni and Barazangi, 1984. *J. Geophys. Res., 89*, 1147–64, copyright by the American Geophysical Union.)

Fig. 5.6-7 Demonstration of the deformation of Asia, modeled by a striped block of plasticine, as the result of a collision with a rigid block simulating the Indian subcontinent. The plasticine is constrained on the left side, so the impact forces blocks to be extruded to the right, analogous to the eastward motion of blocks in Indochina and China. (Tapponnier *et al.*, 1982. *Geology, 10*, 611–16, with permission of the publisher, the Geological Society of America, Boulder, Co. © 1982 Geological Society of America.)

major strike-slip faults. This extrusion has been modeled assuming that India acts as a rigid block indenting a semi-infinite plastic medium (Asia), giving rise to a complicated faulting and slip pattern (Fig. 5.6-7). The extent of the collision is illustrated by GPS data and focal mechanisms showing that the Tien Shan intracontinental mountain belt, 1000–2000 km north of the Himalayas, accommodates almost half the net plate convergence in the western part of the zone.

In addition to providing data about a collision region's kinematics, seismological studies provide insight into its mech-

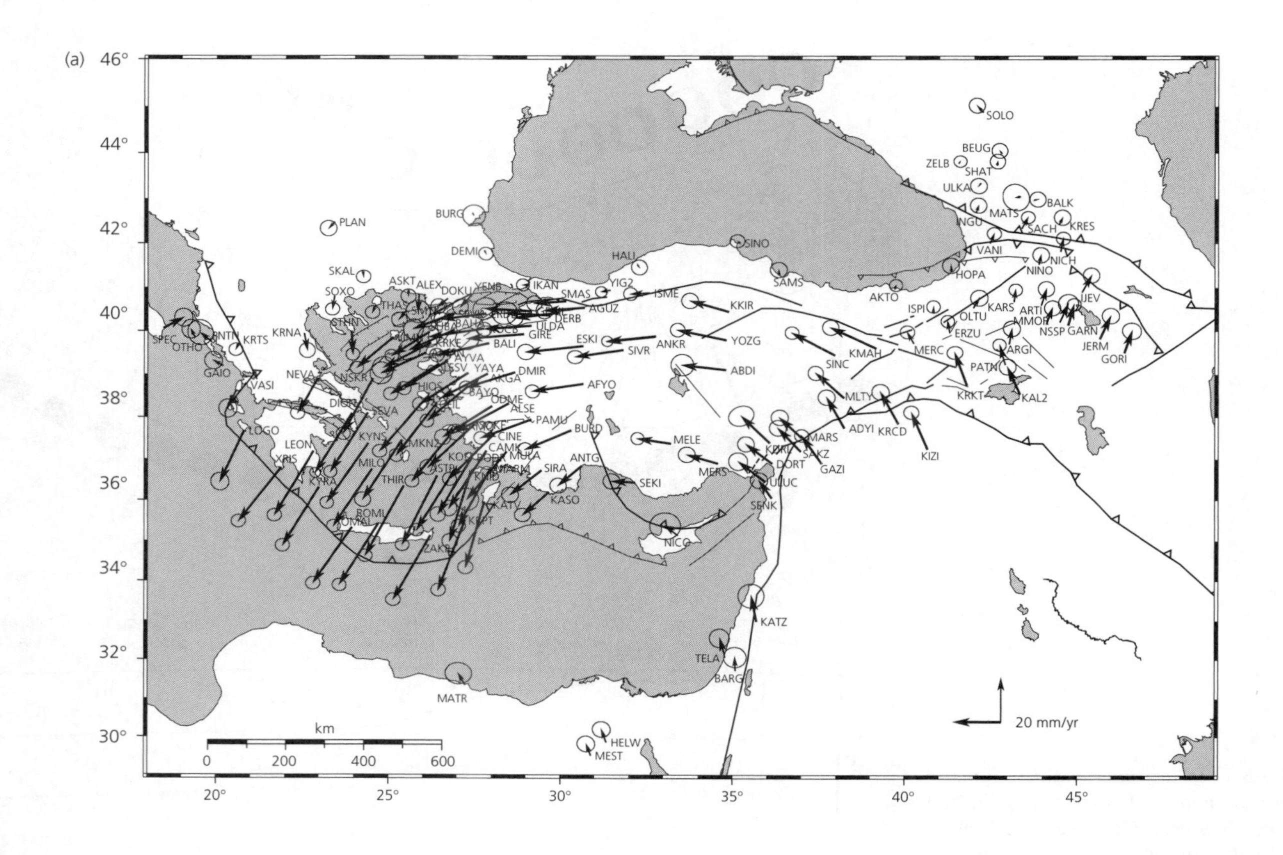
(a)
46°
44°
42°
40°
38°
36°
34°
32°
30°
20°
25°
30°
35°
40°
45°
SOLO
BEUG
ZELB
SHAT
ULKA
BALK
MATS
SACH
KRES
VANI
NICH
NINO
HOPA
PLAN
BURG
DEMI
SINO
SAMS
HALI
AKTO
ISPI
OLTU
KARS
ARTI
JEV
MMOR
NSSP
GARN
JERM
GORI
ERZU
MERC
ARGI
PATN
KRKT
KAL2
KIZI
KRCD
ADYI
MLTY
SINC
KMAH
KKIR
YOZG
ABDI
ANKR
ESKI
SIVR
ISME
YIG2
SMAS
AGUZ
IKAN
YENB
DOKU
ALEX
ASKT
SKAL
SOXO
STHN
THAS
KRNA
SPEC
OTHO
KRTS
GAIO
VASI
NEVA
NSKR
SEVA
LOGO
LEON
XRIS
KYNS
MILO
THIR
ROMI
KASO
SIRA
ANTG
BURD
PAMU
AFYO
DMIR
YAYA
AYVA
BALI
GIRE
ULDA
DERB
ALSE
ODME
CINE
CAMK
MULA
MELE
MERS
SEKI
NICO
KATZ
TELA
BARG
MATR
HELW
MEST
ZAKR
MARS
GAZI
DORT
ULUC
SENK
20 mm/yr
km
0
200
400
600

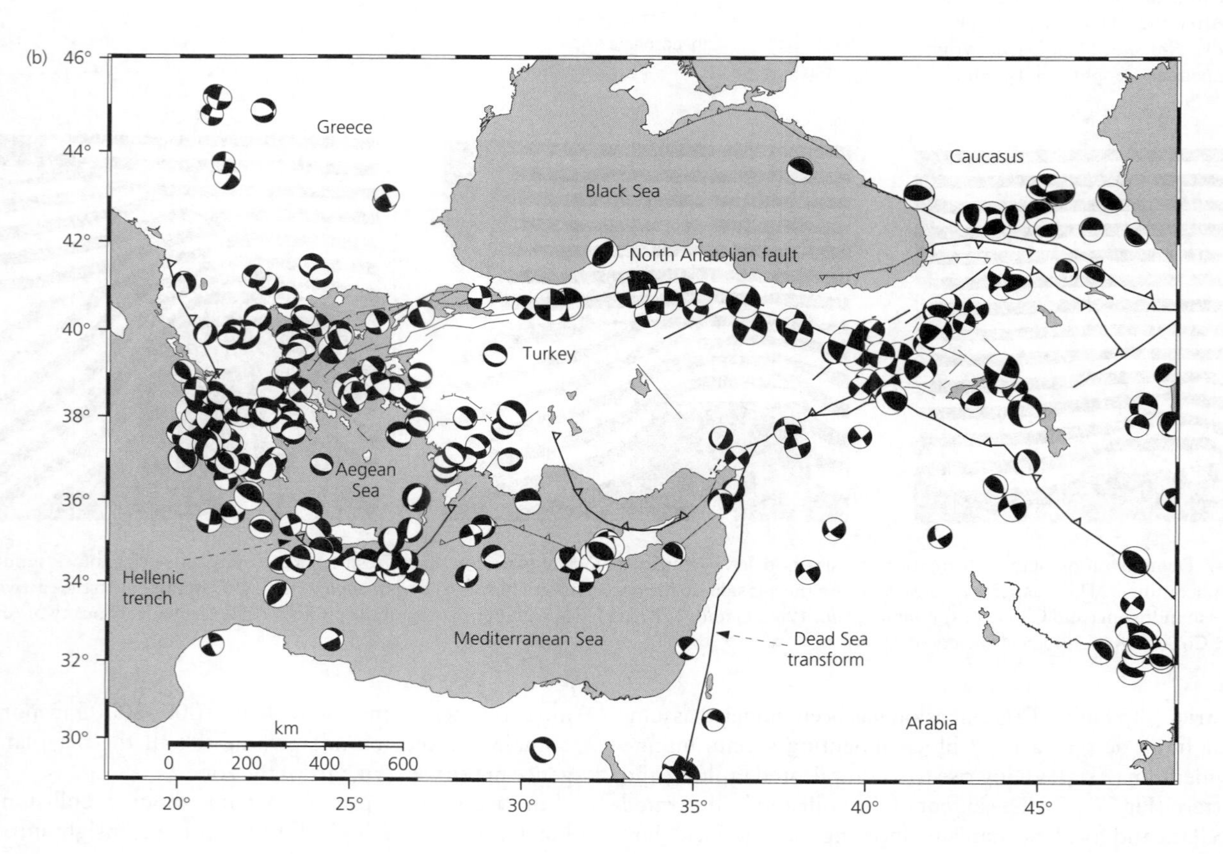
(b)
46°
44°
42°
40°
38°
36°
34°
32°
30°
20°
25°
30°
35°
40°
45°
Greece
Black Sea
Caucasus
North Anatolian fault
Turkey
Aegean Sea
Hellenic trench
Mediterranean Sea
Dead Sea transform
Arabia
km
0
200
400
600

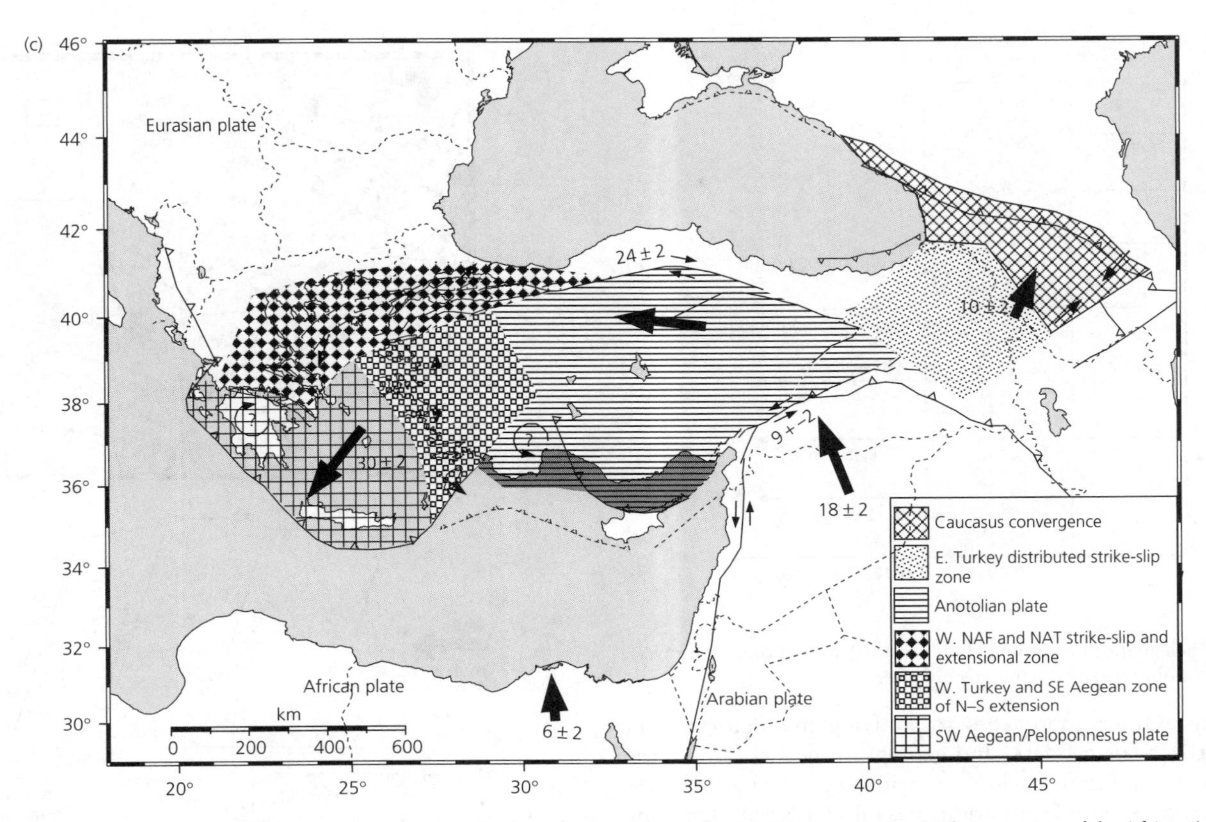

Fig. 5.6-8 GPS observations of motions relative to Eurasia (a), focal mechanisms (b), and tectonic interpretation (c) for a portion of the Africa–Arabia–Eurasia plate collision zone. Note strike-slip along the North Anatolian fault, extension in western Anatolia and the Aegean region, and compression in the Caucasus mountains. Rates are in mm/yr. (McClusky *et al.*, 2000. *J. Geophys. Res.*, *105*, 5695–5719, copyright by the American Geophysical Union.)

anics. The collision process is thought to involve a complex interplay between forces due directly to the collision, gravitational forces due to the resulting uplift and crustal thickening, and forces from the resulting mantle flow. Earthquake depths and studies of seismic velocity, attenuation, and anisotropy are providing data on crustal thicknesses, thermal and mechanical structures, and mantle flow. For example, *P*-wave travel time tomography shows high velocity under the presumably cold Himalayas, which contrasts with low velocity under Tibet. These and other seismological data are consistent with the idea that Tibet deforms easily during the collision.

An equally complicated situation occurs in the eastern Mediterranean collision zone involving the African, Arabian, and Eurasian plates. Combining GPS and focal mechanism data shows the complex motions. Figure 5.6-8 (a) shows the motions of sites in the western Mediterranean relative to Eurasia. Northern portions of Arabia move approximately N40°W, consistent with global plate motion models. Western Turkey rotates as the Anatolian plate about a pole near the Sinai peninsula. Anatolia is thus "squeezed" westward between Eurasia and northward-moving Arabia (Fig. 5.6-8, c).[2] The motion across the North Anatolian fault, about 25 mm/yr, gives rise to large right-lateral strike-slip earthquakes (Fig. 5.6-8, b) such as the 1999 M_s 7.4 Izmit earthquake, which occurred about 100 km east of Istanbul and caused more than 30,000 deaths. To the west, the data show interesting deviations from a rigid Anatolian plate. The increasing velocities toward the Hellenic trench, where the Africa plate subducts below Crete and Greece, show that western Anatolia and the Aegean region are under extension, consistent with the normal fault mechanisms. This region may be being "pulled" toward the arc, perhaps by an extensional process similar to oceanic back-arc spreading, as the trench "rolls back" (Section 5.2.4). By contrast, eastern Turkey is being driven north-ward into Eurasia, causing compression that appears as the thrust fault earthquakes in the Caucasus mountains. The Dead Sea transform separates Arabia from the region to the west, sometimes viewed as the Sinai microplate. Strike-slip motion along this fault gives rise to the earthquakes mentioned in the Bible that repeatedly destroyed famous cities like Jericho.

5.6.2 *Seismic, aseismic, transient, and permanent deformation*

The examples in the previous section illustrate that earthquakes give powerful insights into the crustal deformation shaping the

[2] Consider a melon seed squeezed between a thumb and a forefinger.

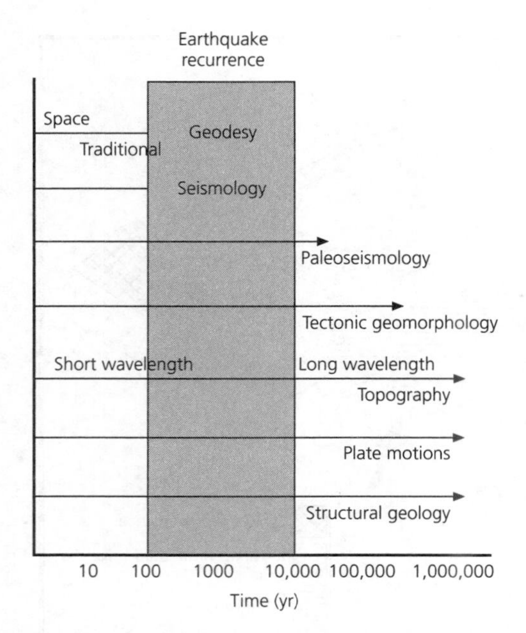

Fig. 5.6-9 Schematic illustration of how crustal deformation on various time scales is observed by different techniques.

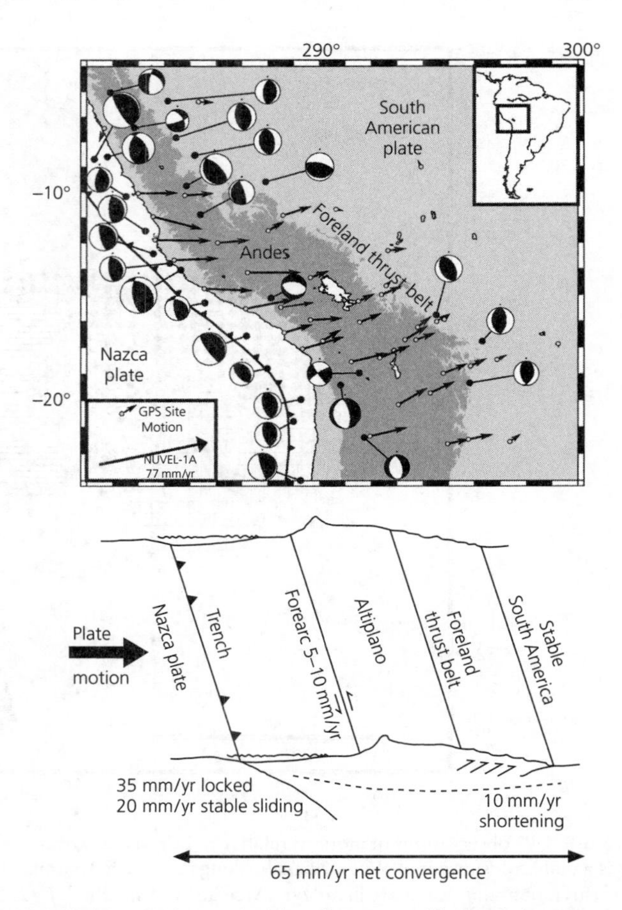

Fig. 5.6-10 *Top*: GPS site velocities relative to stable South America (Norabuena *et al.*, 1998. *Science, 279*, 358–62, copyright 1998 American Association for the Advancement of Science), and selected earthquake mechanisms in the boundary zone. Rate scale is given by the NUVEL-1A vector. *Bottom*: Cross-section showing approximate velocity distribution inferred from GPS data. (Stein and Klosko, 2002. From *The Encyclopedia of Physical Science and Technology*, ed. R. A. Meyers, copyright 2002 by Academic Press, reproduced by permission of the publisher.)

continents. Other approaches to studying this deformation, including various geodetic and geological means, sample the deformation in different ways on various time scales (Fig. 5.6-9). Hence, considerable attention goes into understanding how what we see with these different techniques are related. For example, as discussed earlier (Sections 4.5.4, 5.4.3), in many places only part of the plate motion seems to occur as earthquakes, and the rest takes place as aseismic slip. A related question is how the deformation shown by earthquakes, which has a time scale of a few years, is related to the longer-term deformation that is recorded by topography and the geologic record.

To explore these ideas, consider the distribution of motion within the boundary zone extending from the stable interior of the oceanic Nazca plate, across the Peru–Chile trench to the coastal forearc, across the high Altiplano and foreland thrust belt, and into the stable interior of the South American continent. Figure 5.6-10 shows GPS site velocities relative to stable South America, which would be zero if the South American plate were rigid and all motion occurred at the trench plate boundary. However, the site velocities are highest near the coast and decrease relatively smoothly from the interior of the Nazca plate to the interior of South America.

Figure 5.6-10 (*bottom*) shows an interpretation of these data. In this model, about half of the plate convergence (approximately 35 mm/yr) is locked at the subduction interface, causing elastic strain of the overriding plate that will be released in large interplate thrust earthquakes (Section 4.5.4) like those whose focal mechanisms are shown. Thus the locked fraction of the plate motion corresponds to the seismic slip rate, perhaps via a process in which only a fraction of the interface is locked at any time. Approximately 20 mm/yr of the plate motion occurs by stable sliding at the trench, which does not deform the overriding plate. This portion of the plate motion corresponds to aseismic slip. The rest occurs across the sub-Andean foreland fold-and-thrust belt, causing permanent shortening and mountain building, as shown by the inland thrust fault mechanisms. This portion of the plate motion would be considered aseismic slip if we considered only the fraction of the plate motion that appears in the trench seismic moment release, whereas in reality it occurs as inland deformation. These interpretations come from analyzing the GPS data in the convergence direction relative to the stable interior of South America (Fig. 5.6-11). If all the convergence were locked on the interplate thrust fault, the predicted rates would exceed those observed within about 200 km of the trench. However, if only about half of the predicted convergence goes into locking the fault, the predicted rates near the trench are less, because only the portion of the slip locked at the interface deforms the overriding plate. Similarly, the data farther than about 300 km

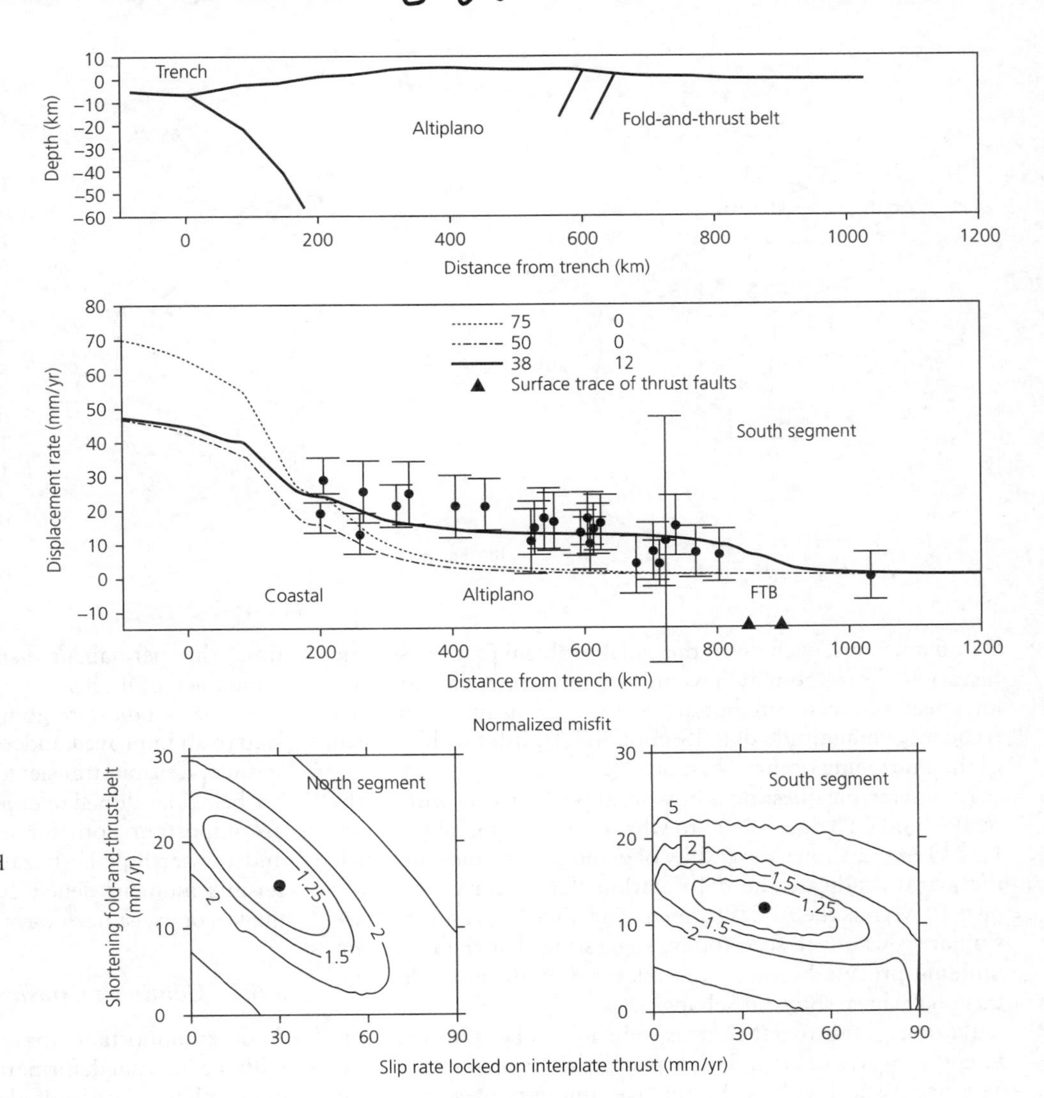

Fig. 5.6-11 Derivation of the model in Fig. 5.6-10 (*bottom*). *Top*: Model geometry, assuming partial slip locked at the plate boundary and shortening in the eastern Andes. *Center*: GPS site velocities in the convergence direction and various models, given by the rates of locked slip and shortening. Solid line shows predictions of best-fitting model, including both partial slip locked at the plate boundary and shortening in the eastern Andes. Short dashed line shows predictions of model with all slip locked on the plate boundary and no shortening. Long-short dashed line shows predictions of model with no shortening and partial slip locked on the plate boundary equal to the sum of best-fitting slip and shortening. *Bottom*: Contour plot showing misfit to the data as a function of the slip rate locked on the plate boundary and shortening rate in the eastern Andes. The best fits (dots) occur for about 30–40 mm/yr of locking and about 10–20 mm/yr shortening. (Norabuena *et al.*, 1998. *Science*, *279*, 358–62, copyright 1998 American Association for the Advancement of Science.)

from the trench are better fit by assuming that about 10 mm/yr motion is locked on thrust faults in the eastern Andes. The locking and shortening rates are the best-fit parameters for this simple model, which does not include other possible complexities such as deformation in the Altiplano.

The idea that about 40% of the plate motion at the trench occurs by aseismic slip seems plausible, because studies using the history of large earthquakes at trenches often estimate that only about half the slip occurs seismically (Fig. 5.4-30). Given the problems of estimating source parameters of earthquakes from historical data, it is encouraging that the geodetic answer seems similar.

The relation between the shortening rate in the thrust belt inferred from GPS data and that implied by the earthquakes can also be studied. Assessing the seismic slip rate is a little more complicated than for transform faults (Section 5.3.3) or subduction zone thrust faulting (Section 5.4.3), because in continental deformation zones earthquakes occur over a distributed volume, rather than on a single fault, and have diverse focal mechanisms. Thus we sum the earthquakes' moment tensors (Section 4.4) to estimate a seismic strain rate tensor[3] using

$$\dot{e}_{ij} = \sum M_{ij}/(2\mu V t), \tag{1}$$

where t is the time interval, and μ is the rigidity. V, the assumed seismic source volume, the product of the length and width of the zone of seismicity and the depth to which seismicity extends. For example, the thrust belt can be assumed to be approximately 2000 km long, 250 km wide, and faulting extends to about 40 km depth. We can then diagonalize the result and consider the eigenvalue associated with the P axis. Scaling this value by the assumed zone width gives an estimate of the shortening rate. The resulting value, less than 2 mm/yr, is significantly less than the approximately 10 mm/yr indicated by the

[3] Strain rates are often written using a dot to indicate the time derivative.

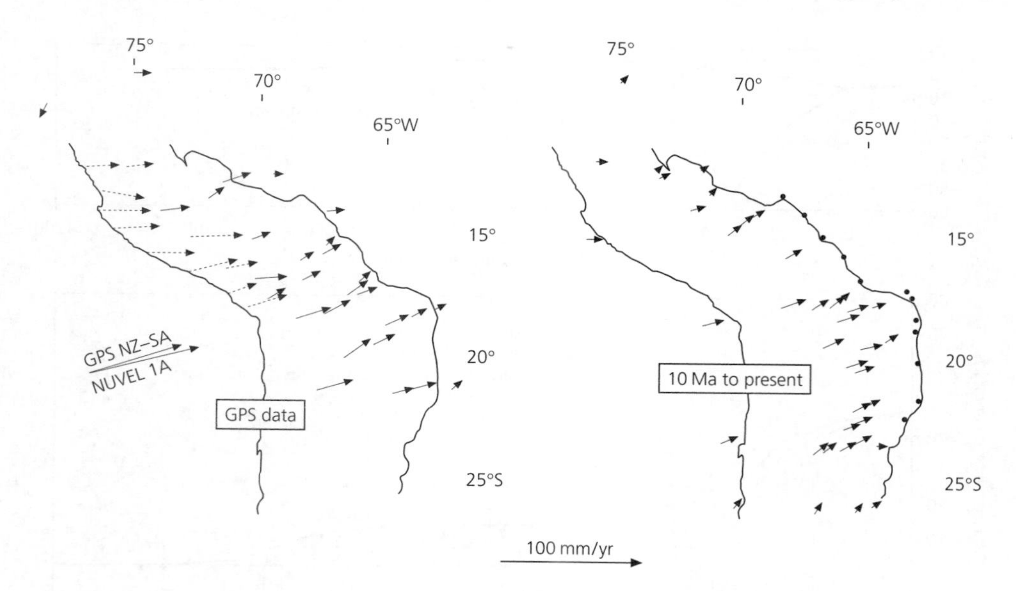

Fig. 5.6-12 Comparison of shortening across the Andes with respect to stable South America from GPS data (*left*) and geological studies (*right*). The dashed GPS vectors reflect elastic strain due to the earthquake cycle at the trench, and are not directly comparable to the permanent shortening in the geological data. Motion decreases toward the eastern extent of the mountain range, shown by the solid line. The geological vectors are largest at about 18°S and decrease to the north and south, showing how the variation in shortening that built the Andes bent them and made them widest about this point. (Hindle *et al.*, 2002.)

GPS data. Thus, even given the usual problem that the seismic history is short and may have missed the largest earthquakes, an effect one can attempt to correct for using earthquake frequency–magnitude data (Section 4.7.1), it looks like much of the shortening occurs aseismically.

An interesting question is how what we see today with earthquakes and GPS data relates to what occurs over geologic time. Figure 5.6-12 shows the results of geological studies, in which the arrows indicate the deformation that occurred over the past 10 Myr as the Andes formed. The directions and rates are similar to what are seen today, suggesting that the mountain building process has occurred relatively uniformly, although there have been some rate changes.[4]

Putting all this together gives some ideas about how the different measures of crustal deformation are related in this area. The first issue involves the relative amounts of seismic and aseismic deformation. It appears that about half of the plate motion at the trench occurs seismically. Similar fractions are also seen in other subduction zones (Fig. 5.4-30), implying that stable sliding at trenches is relatively common. Moreover, only about 10–20% of the shortening in the foreland thrust belt appears to occur seismically. Thus aseismic, and presumably permanent, deformation of rocks in the thrust belt seems like a major phenomenon. Similar results have also been observed for other continental deformation zones (Fig. 5.6-13). The next issue is that of permanent versus transient deformation. In the model of Fig. 5.6-11, the deformation of the South American plate due to the locked slip at the trench is transient, and will be released in the upcoming large trench earthquake. However, it seems likely that the deformation of the foreland thrust belt is permanent, and goes into faulting and folding rocks. Over time, this permanent displacement adds up (Fig. 5.6-12) to build the mountains.

Similar studies are going on around the world, and should lead to an improved understanding of the partitioning between seismic, aseismic, transient, and permanent deformation. Models are being developed to explore these issues (Section 5.7), which are important both for understanding continental evolution and for earthquake hazard assessment, because an apparent seismic moment deficit could indicate either overdue earthquakes or aseismic deformation.

5.6.3 *Continental intraplate earthquakes*

Another important application of earthquake studies deals with the internal deformation of the continental portions of the major plates. Although idealized plates would be purely rigid, intraplate earthquakes reflect the important and poorly understood tectonic processes of intraplate deformation. As in the oceans (Section 5.5.1), there appears to be a hierarchy of places that have such earthquakes. There are areas like the East African rift that can be thought of as either slow-moving plate boundaries or intraplate deformation, less active zones associated with either fossil structures or other processes like hot spots, and then intraplate earthquakes that are not easily correlated with any particular structure or cause.

One example is the New Madrid area in the central USA, which had large earthquakes in 1811–12 and has small earthquakes today. Other continental interiors, including Australia, western Europe, and India, have also had significant intraplate earthquakes. Because motion in these zones is at most a few mm/yr, compared to the generally much more rapid plate boundary motions, seismicity is much lower (Fig. 5.6-14) and thus harder to study. This difficulty is compounded by the fact that, unlike at plate boundaries, where plate motions give insight into why and how often earthquakes occur, we have little

[4] The similarity of the focal mechanism, GPS, and geological data illustrates the principle of *uniformitarianism*, that studying present processes gives insight into the past, a tenet of geology since Lyell and Hutton's seminal work almost two centuries ago.

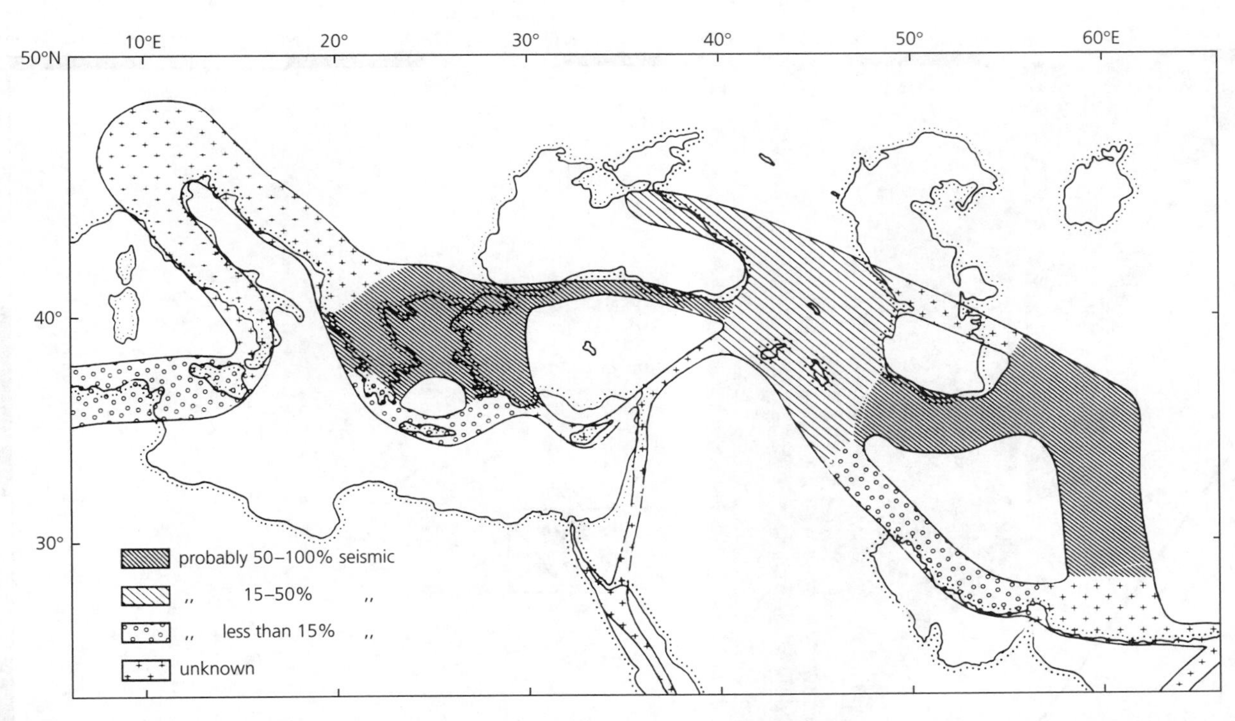

Fig. 5.6-13 Estimates of seismic deformation fractions for areas in the Mediterranean and Middle East. Seismicity appears to account for most or all of the deformation in western Turkey, Iran, and the Aegean, much of the deformation in the Caucasus and eastern Turkey, and little of the deformation in the Zagros and the Hellenic trench. (Jackson and McKenzie, 1988.)

idea of what causes intraplate earthquakes, and no direct way to estimate how often they should occur. As a result, progress in understanding these earthquakes is much slower than for earthquakes on plate boundaries, and key issues may not be resolved for a very long time.

Geodetic data illustrate the challenge. For example, comparison of the absolute velocities of GPS sites in North America east of the Rocky Mountains to velocities predicted by modeling these sites as being on a single rigid plate shows that the interior of the North American plate is rigid at least to the level of the average velocity residual, less than 1 mm/yr (Fig. 5.6-15). Similar results emerge from studies across the New Madrid zone itself and for the interiors of other major plates, showing that plates thought to have been rigid on geological time scales are quite rigid on decadal scales. For example, 1 mm/yr motion spread over 100 or 1000 km distance corresponds to strain rates of 10^{-8} and 10^{-9} yr^{-1} (3×10^{-16} and 3×10^{-17} s^{-1}), respectively. Because the geodetic data include measurement errors due to effects including instabilities of the geodetic markers, it seems likely that the tectonic strains are even smaller. However, over long enough time, even such small motions can accumulate enough slip for large earthquakes to occur.

This idea is consistent with what is known about large intraplate earthquakes. Although there is little seismological data for such events because they are rare, insight can be obtained from combining the seismological data with geodetic, paleoseismological, and other geological and geophysical data. For example, intensities estimated from historical accounts of the 1811–12 New Madrid earthquakes (Fig. 1.2-4) suggest magnitudes in the low 7 range. Paleoseismic studies (Section 1.2) indicate that several previous large earthquakes, presumably comparable to those of 1811–12, occurred 500–800 years apart. Thus, in 500–1000 years (Fig. 5.6-16, *top*) steady strain accumulation less than 2 mm/yr could provide up to 1–2 m of motion available for future earthquakes, suggesting that they would be about magnitude 7. A similar view comes from considering the earthquake history for the area. As discussed in Section 4.7.1, earthquakes of a given magnitude are approximately ten times less frequent than those one magnitude unit smaller. Thus, although the instrumental data contain no earthquakes with magnitude greater than 5, both these and a historical catalog in which magnitudes were estimated from intensity data can be extrapolated to imply that a magnitude 7 earthquake would occur about once every 1400 ± 600 years (Fig. 5.6-16, *bottom*). Hence, as expected, major intracontinental earthquakes occur substantially less frequently than comparable plate boundary events (Fig. 5.6-17). However, because of the lower attenuation in continental interiors (Section 3.7.10), such earthquakes can cause greater shaking than ones of the same magnitude on a plate boundary (Fig. 1.2-5).

Such earthquakes are generally thought to be due to the reactivation of preexisting faults or weak zones in response to either local or intraplate stresses. The New Madrid earthquakes, for example, are thought to occur on faults associated

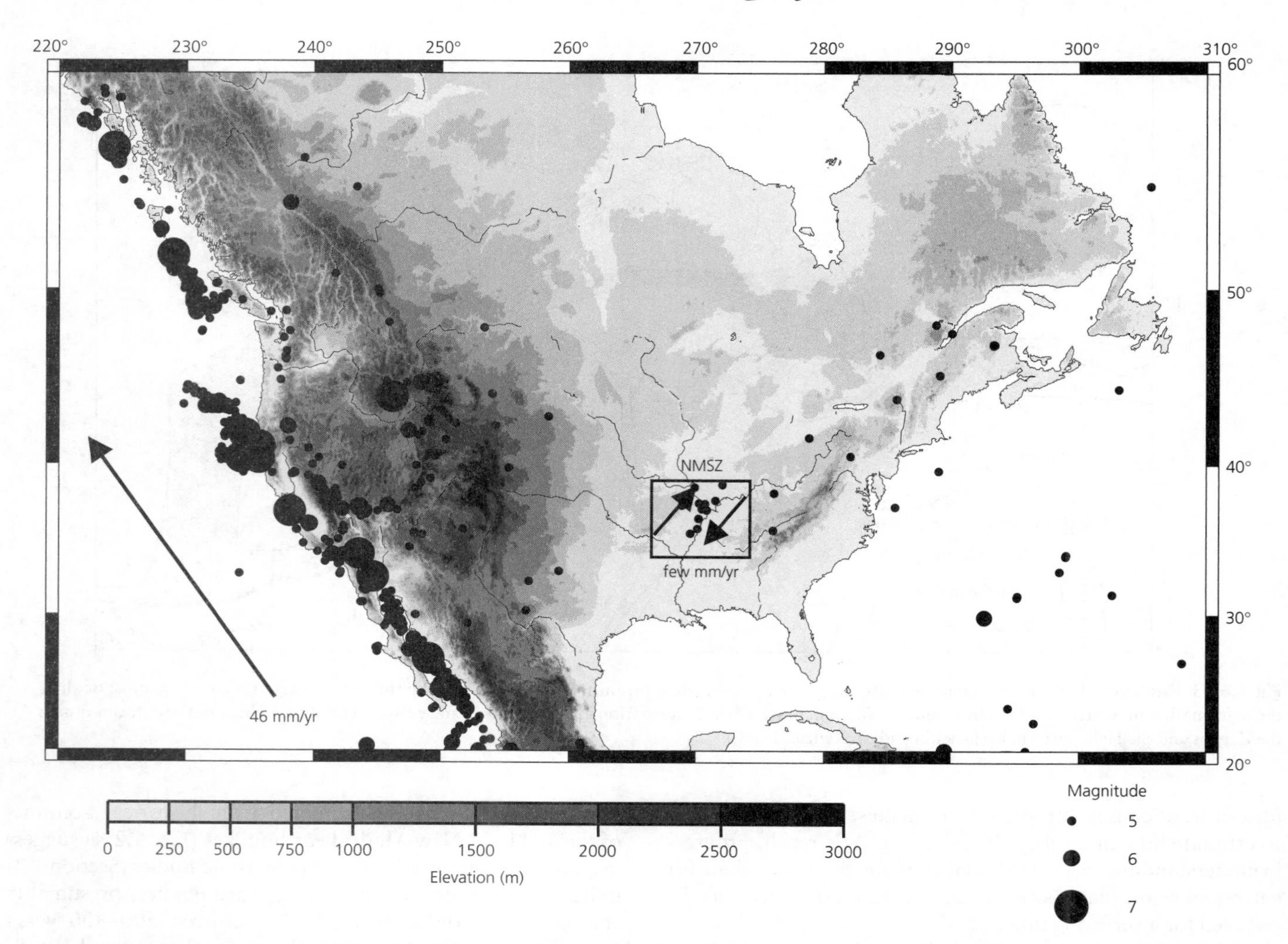

Fig. 5.6-14 Seismicity (magnitude 5 or greater since 1965) of the continental portion of the North American plate and adjacent area. Seismicity and deformation are concentrated along the Pacific–North America plate boundary zone, reflecting the relative plate motion. The remaining eastern portion of the continent, approximately that east of 260°, is much less seismically active. Within this relatively stable portion of the continent, seismicity, and thus presumably deformation, are concentrated in several zones, most notably the New Madrid seismic zone. (Weber *et al.*, 1998. *Tectonics, 17*, 250–66, copyright by the American Geophysical Union.)

with a Paleozoic failed continental rift, now buried beneath thick sediments deposited by the Mississippi river and its ancestors (Fig. 5.6-18). As a result, the faults are not exposed at the surface, so most ideas about them are based on inferences from seismology and other data. The intraplate stress field has been studied by combining focal mechanism and fault orientations with data from drill holes and *in situ* stress measurements (Fig. 5.6-19). In general, the eastern USA shows a maximum horizontal stress oriented NE–SW, consistent with the predictions of the stresses due to plate driving forces. Similar stress maps are being developed for other areas and are being used to investigate both intraplate deformation and plate driving forces. As noted in Section 3.6.5, it appears that seismic anisotropy in the lower continental crust may reflect the stress field that acted during a major tectonic event such as mountain building.

An intriguing question is why intraplate stresses cause earthquakes on particular faults, given that many weak zones could serve this purpose. Geological and paleoseismic data, together with the absence of significant fault-related topography, suggest that individual intraplate seismic zones may be active for only a few thousands of years, so intraplate seismicity migrates. This possibility is akin to that suggested for intermittent oceanic intraplate earthquake swarms. If so, there is nothing special about New Madrid or the other concentrations of intraplate seismicity we observe now — these zones will die off and be replaced by others. Moreover, there are enough tectonic structures available that (typically small) earthquakes will occur almost randomly throughout continental interiors.

A special case of this phenomenon occurs at passive continental margins, where continental and oceanic lithospheres join. Although these areas are in general tectonically inactive,

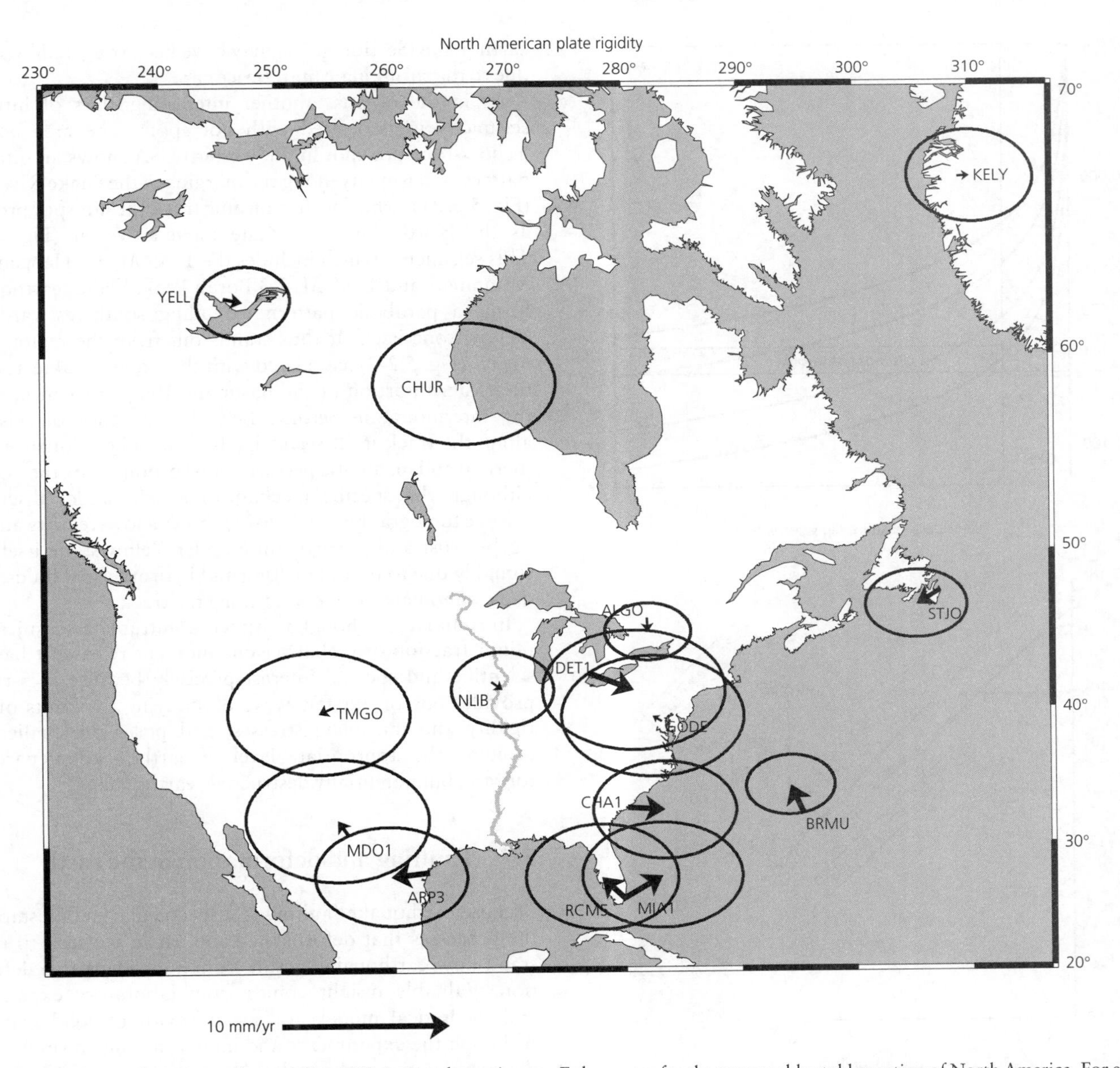

Fig. 5.6-15 Locations of continuously recording GPS sites used to estimate a Euler vector for the presumably stable portion of North America. For each, the misfit between the observed velocity and that predicted for a single plate is shown. The average misfit is less than 1 mm/yr, showing that eastern North America is quite rigid. (Newman *et al.*, 1999. *Science, 284*, 619–21, copyright 1999 American Association for the Advancement of Science.)

magnitude 7 earthquakes can occur, as on the eastern coast of North America (Fig. 5.6-20). Such earthquakes may be associated with stresses, including those due to the removal of glacial loads, which reactivate faults remaining from the original continental rifting (Fig. 5.6-1). Although such earthquakes are observed primarily on previously glaciated margins, they also occur on nonglaciated passive margins, perhaps due to sediment loading. In some cases large sediment slides occur, as was noted for the 1929 M_s 7.2 earthquake on the Grand Banks of Newfoundland, because the slides broke trans-Atlantic telephone cables and generated a tsunami that caused 27 fatalities.[5] An interesting unresolved question is whether tectonic faulting is required for such earthquakes, or whether the slump itself can account for what is seen on seismograms. Some studies find that the seismograms are best fit by a double-couple fault source, whereas others favor a single force consistent with the slump (Fig. 4.4-3). The issue is important because slumps occur in the sedimentary record along many passive

[5] These deaths account for all but one of Canada's known earthquake fatalities to date, although this situation could change after a large Cascadia subduction zone earthquake.

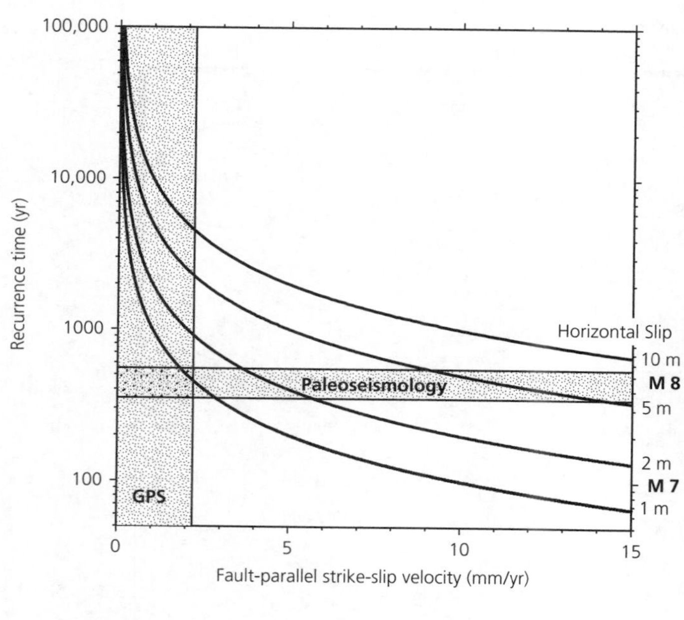

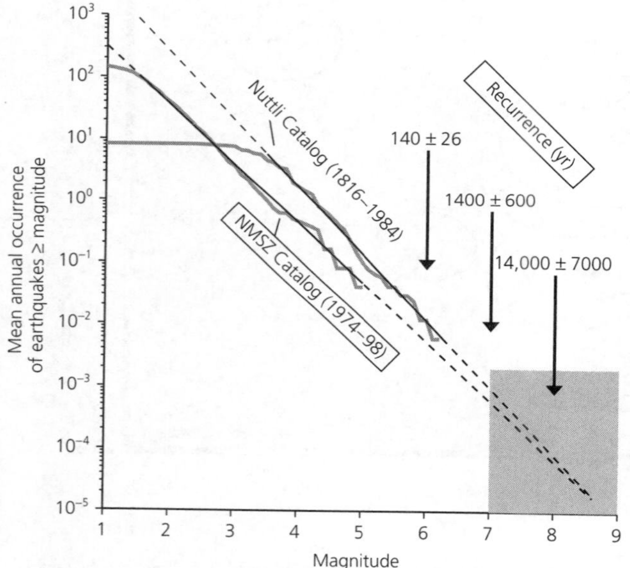

Fig. 5.6-16 *Top*: Relation between interseismic motion and the expected recurrence of large New Madrid earthquakes. The recurrence estimates from paleoseismic studies and geodetic data are jointly consistent with slip in the 1811–12 earthquakes of about 1 m, corresponding to a low magnitude 7 earthquake. *Bottom*: Earthquake frequency–magnitude data for the New Madrid zone. Both the instrumental and historic (1816–1984) data predict a recurrence interval of about 1000 yr for magnitude 7 earthquakes. (Newman *et al.*, 1999. *Science, 284*, 619–21, copyright 1999 American Association for the Advancement of Science.)

margins, even those that have not been recently deglaciated. Stresses associated with the removal of glacial loads may also play a role in causing earthquakes within continental interiors such as the northeastern USA and eastern Canada. It has also been suggested that the huge 1998 Balleny Island intraplate earthquake (Section 5.5.1) may have been triggered by stresses due to the shrinking Antarctic ice cap.

As in the oceans, another interesting class of intraplate seismicity is associated with hot spots. The area near the Yellowstone hot spot in the western USA shows an intriguing pattern of seismicity along the margins of the Snake River plain (Fig. 5.6-21), which is the volcanic track the hot spot produced as the North American plate moved over it (Fig. 5.2-8). This seismicity, which includes the 1959 M_s 7.5 Hebgen Lake, Montana,[6] and 1983 M_s 7.3 Borah Peak, Idaho, earthquakes, forms a parabolic pattern extending southwestward from Yellowstone itself. It thus stands out from the regional seismicity (Fig. 5.2-3) associated with the extensional tectonics of the eastern portion of the Basin and Range province, termed the Intermountain Seismic Belt. The absence of seismicity along the track itself seems likely to be a consequence of the thermal and magmatic perturbations produced by the hot spot, although the specific mechanism is still under discussion. Seismic tomography (Fig. 5.6-21) shows a low-velocity anomaly in the crust and upper mantle under Yellowstone itself, presumably due to partial melting and hydrothermal fluids, and a deeper anomaly that persists along the track.

In summary, although continental intraplate seismicity is a minor fraction of global seismic moment release, it has both scientific and societal interest precisely because it is rare. It provides one of our few ways of studying the limits of plate rigidity and intraplate stresses, and poses the challenge of deciding the appropriate level of earthquake preparedness for rare, but potentially destructive, earthquakes.

5.7 Faulting and deformation in the earth

Because earthquake faulting is a spectacular manifestation of the processes that deform the solid earth, we seek to understand how earthquakes result from and reflect this deformation. Valuable insight comes from laboratory experiments and theoretical models for the behavior of solid materials. Although the experiments and models are much simpler than the complexities of the real earth, they allow us to think about key features. Seismology and geophysics thus exploit research devoted to material behavior by a range of disciplines, including engineering, materials science, and solid state physics. We touch only briefly on some basic ideas, and more information can be found in the references at the end of the chapter.

5.7.1 *Rheology*

Materials can be characterized by their *rheology*, the way they deform. In seismology we typically take a continuum

[6] This earthquake triggered an enormous landslide that buried a campground, causing 28 deaths and dammed the Madison River, forming Quake Lake. These dramatic effects are still visible today and make the site well worth visiting. A visitor center and parking lot are built on the slide.

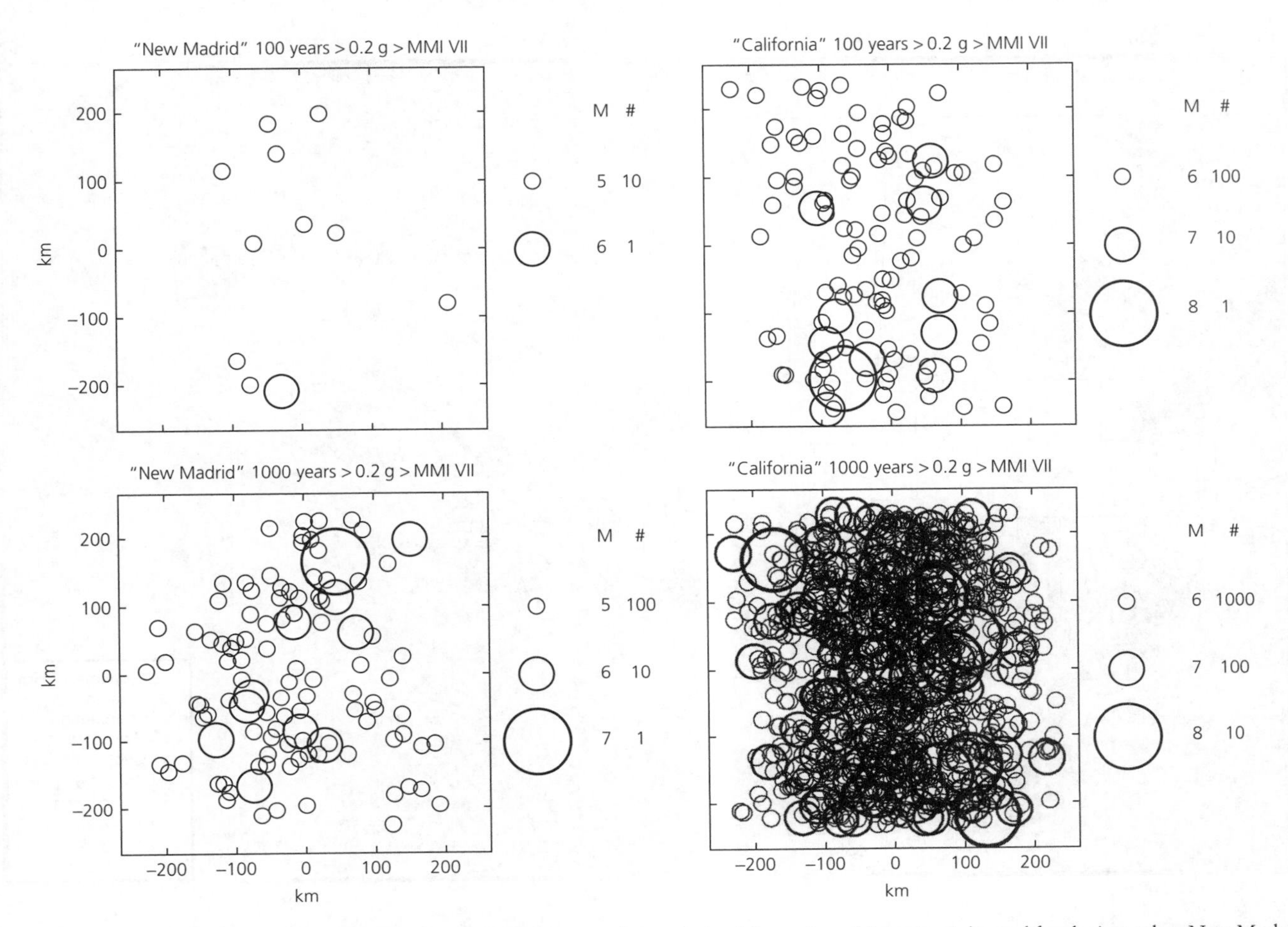

Fig. 5.6-17 Schematic illustration of the relation between the recurrence times of seismicity and resulting seismic hazard for the intraplate New Madrid seismic zone and the southern California plate boundary zone. Seismicity is assumed to be randomly distributed about an N–S line through 0, with California 100 times more active, but New Madrid earthquakes causing potentially serious damage (circles show areas with acceleration 0.2 g or greater, Table 1.2-4) over an area comparable to that for a California earthquake one magnitude unit larger.

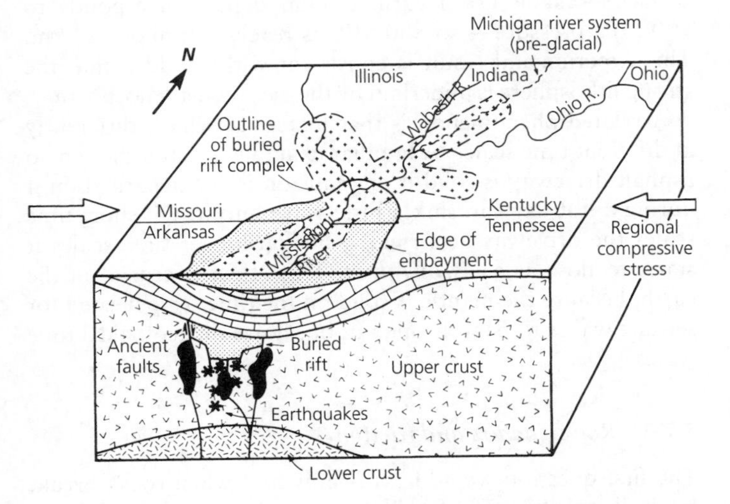

Fig. 5.6-18 Schematic tectonic model for the New Madrid earthquakes. (Braile *et al.*, 1986. *Tectonophysics, 131*, 1–21, with permission from Elsevier Science.)

approach, considering the earth to be a continuous deformable material. This means that we focus on its aggregate behavior (Section 2.3) rather than on how its behavior is determined by what happens at a microscopic scale.

To do this, consider the strain that results from compressing a rock specimen. The simplest case is shown in Fig. 5.7-1a. For small stresses, the resulting strain is proportional to the applied stress, so the material is purely *elastic*. Elastic behavior happens when seismic waves pass through rock, because the strains are small (Section 2.3.8). However, once the applied stress reaches a value σ_f, known as the rock's *fracture strength*, the rock suddenly breaks. Such *brittle* fracture is the simplest model for what happens when an earthquake occurs on a fault. Thus brittle fracture — a deviation from elasticity — generates elastic seismic waves.

Other materials show a change in the stress–strain curve for increasing stresses (Fig. 5.7-1b). For stresses less than the *yield stress*, σ_o, the material acts elastically. Thus, if the stress is released, the strain returns to zero. However, for stresses greater

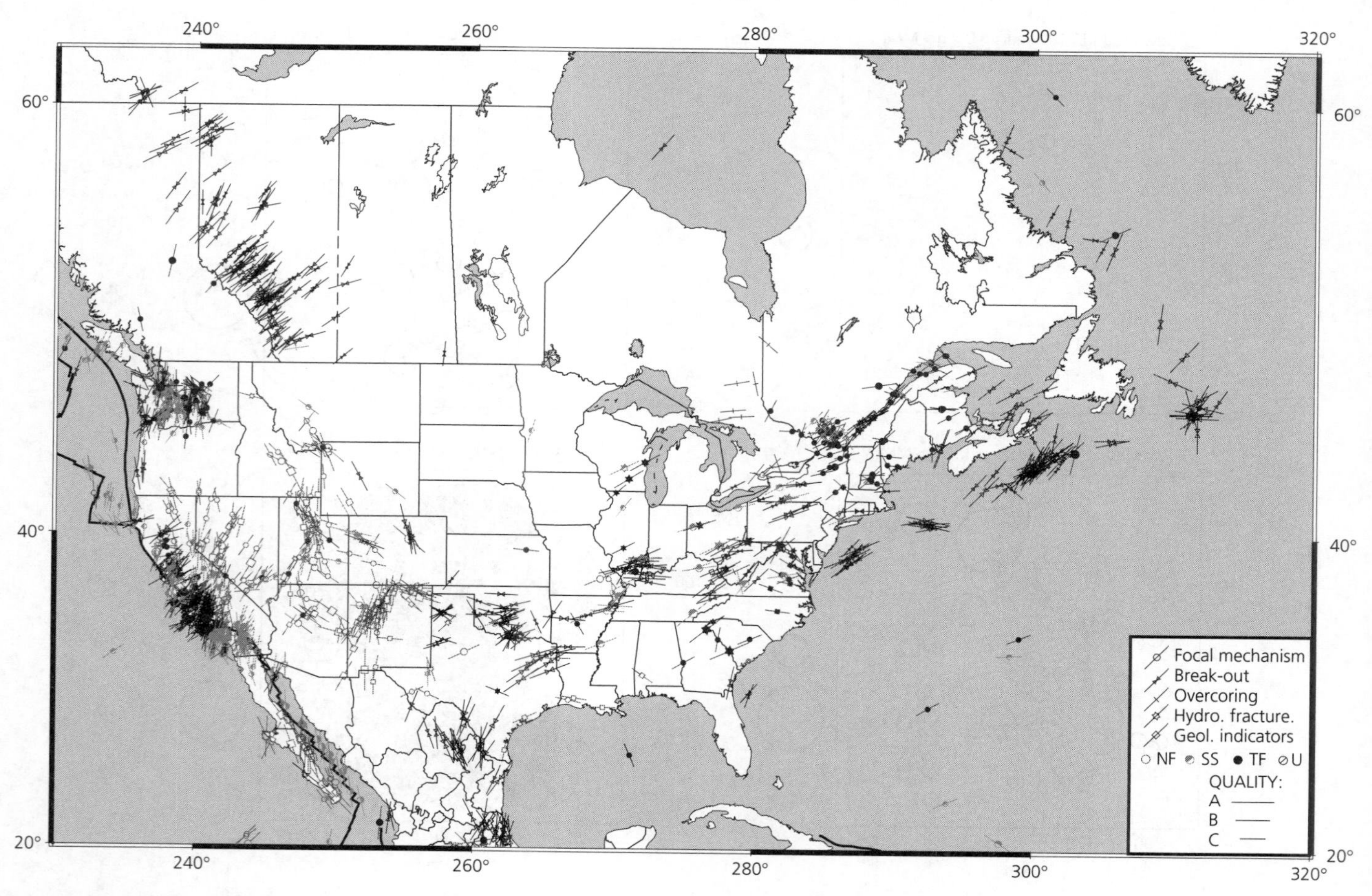

Fig. 5.6-19 Stress map for North America. (World Stress Map project, Muller *et al.*, 2000.)

than the yield stress, releasing the stress relieves the elastic portion of the strain, but leaves a permanent deformation (Fig. 5.7-1c). If the material is restressed, the stress–strain curve now includes the point of the permanent strain. The material behaves as though its elastic properties were unchanged, but the yield strength has increased from σ_o to σ_o'. The portion of the stress–strain curve corresponding to stress above the yield stress is called *plastic* deformation, in contrast to the elastic region where no permanent deformation occurs. Materials showing significant plasticity are called *ductile*. A common approximation is to treat ductile materials as *elastic-perfectly plastic*: stress is proportional to strain below the yield stress and constant for all strains when stress exceeds the yield stress (Fig. 5.7-2).

An important result of laboratory experiments is that at low pressures rocks are brittle, but at high pressures they behave ductilely, or flow. Figure 5.7-3 shows experiments where a rock is subjected to a compressive stress σ_1 that exceeds a confining pressure σ_3. For confining pressures less than about 400 MPa the material behaves brittlely — it reaches the yield strength, then fails. For higher confining pressures the material flows ductilely. These pressures occur not far below the earth's surface — as discussed earlier, 3 km depth corresponds to 100 MPa pressure — so 800 MPa is reached at about 24 km. This experimental result is consistent with the idea that the strong lithosphere is underlain by the weaker asthenosphere.

A related phenomenon is that materials behave differently at different time scales. A familiar example is that although an asphalt driveway is solid if one falls on it, a car parked on it during a hot day can sink a little ways into it. On short time scales the driveway acts rigidly, but on longer time scales it starts to flow as a *viscous fluid*. This effect is crucial in the earth, because the mantle is solid on the time scale needed for seismic waves to pass through it, but flows on geological time scales.

5.7.2 *Rock fracture and friction*

The first question we address is how and when rocks break. In the brittle regime of behavior, the development of faults and the initiation of sliding on preexisting faults depend on the applied stresses.

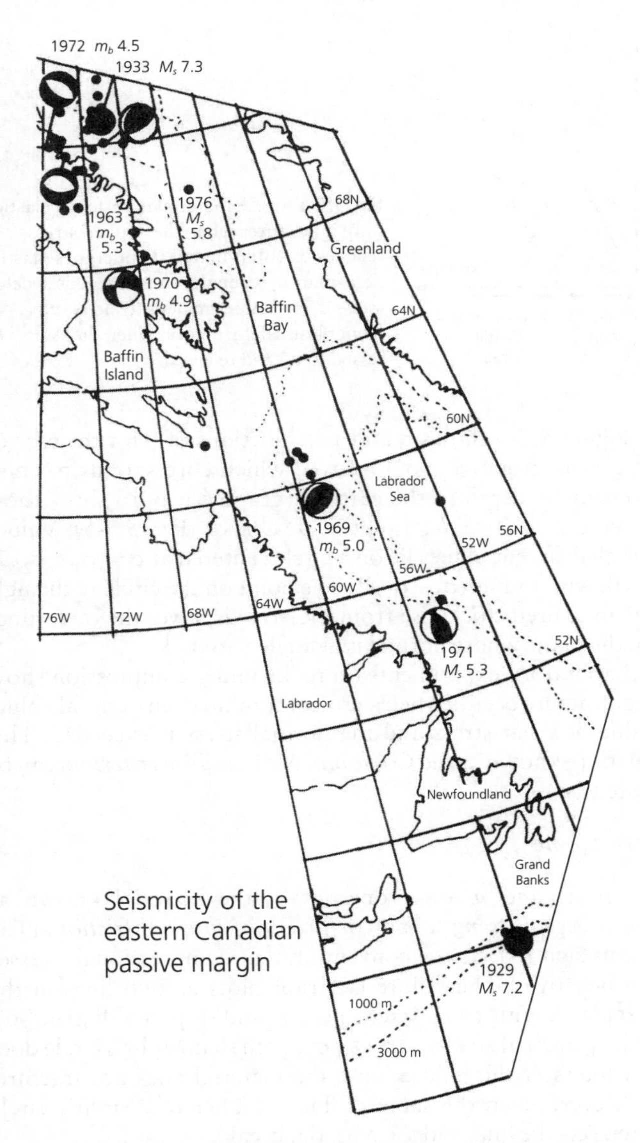

Fig. 5.6-20 Earthquakes along the passive continental margin of eastern Canada. These earthquakes may have occurred on faults remaining from continental rifting. (Stein *et al.*, 1979. *Geophys. Res. Lett.*, *6*, 537–40, copyright by the American Geophysical Union.)

Given a stress field specified by a stress tensor, we use the approach of Section 2.3.3 to find the variation in normal and shear stress on faults of various orientations. For simplicity, we consider the stress in two dimensions. If the coordinate axes $(\hat{e}_1, \hat{e}_2)$ are oriented in the principal stress directions, the stress tensor is diagonal,

$$\sigma_{ij} = \begin{pmatrix} \sigma_1 & 0 \\ 0 & \sigma_2 \end{pmatrix}. \tag{1}$$

To find the stress on a plane whose normal $\hat{e}'_1$ is at an angle of θ from $\hat{e}_1$, the direction of σ_1 (Fig. 5.7-4), we transform the stress

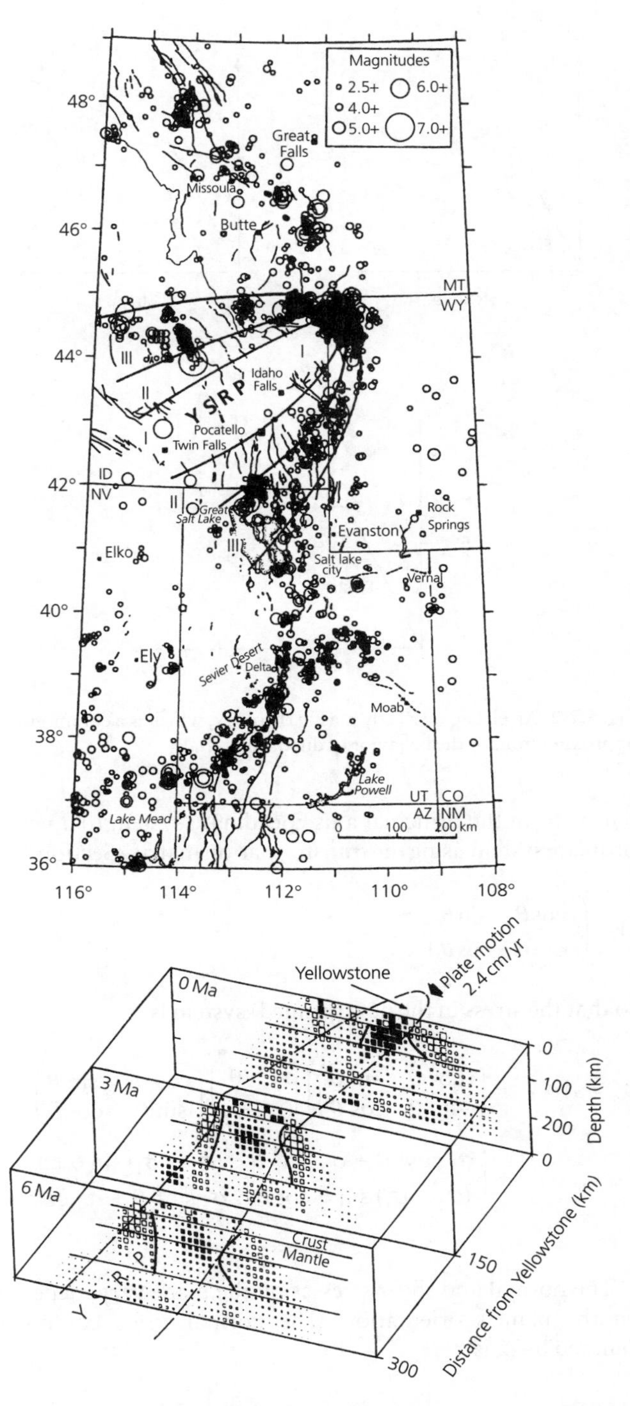

Fig. 5.6-21 *Top*: Seismicity (1900–85) of the Intermountain area of the western USA. Superimposed on the regional seismicity are earthquakes forming a parabola along the margins of the Yellowstone–Snake River plain (YRSP), the volcanic track of the Yellowstone hot spot. *Bottom*: *P*-wave velocities across the hot spot track, shown by squares scaled in size to the differences from a uniform-velocity model. The largest symbols are ±3%, with dark and open symbols showing low and high velocities. (Smith and Braile, 1994. *J. Volcan. Geotherm. Res.*, *61*, 121–87, with permission from Elsevier Science.)

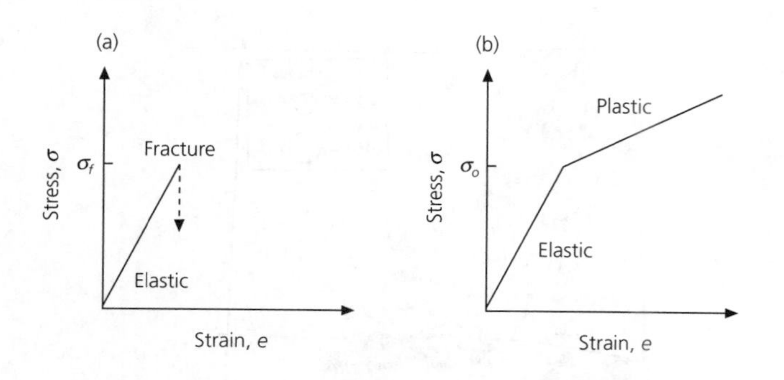

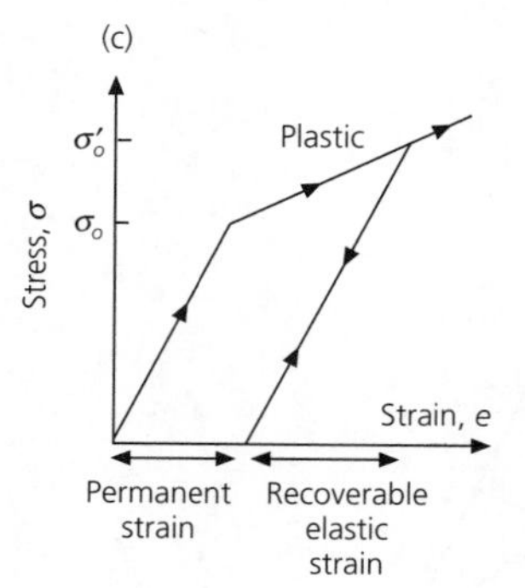

Fig. 5.7-1 (a): A material is perfectly elastic until it fractures when the applied stress reaches σ_f. (b): A material undergoes plastic deformation when the stress exceeds a yield stress σ_o. (c): A permanent strain results from plastic deformation when the stress is raised to σ'_o and released.

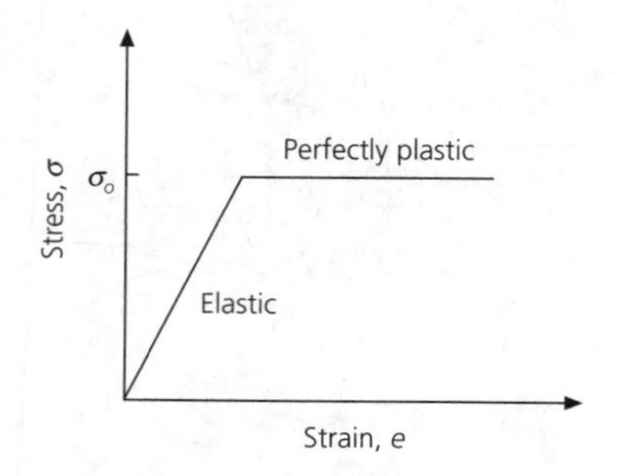

Fig. 5.7-2 An elastic–perfectly plastic rheology, which is a commonly used approximation for the behavior of ductile materials.

tensor from the principal axis coordinate system to a new coordinate system using the transformation matrix (Section 2.3.3)

$$A = \begin{pmatrix} \cos\theta & \sin\theta \\ -\sin\theta & \cos\theta \end{pmatrix} \quad (2)$$

so that the stress in the new (primed) system is

$$\sigma'_{ij} = A\sigma A^T = \begin{pmatrix} \cos\theta & \sin\theta \\ -\sin\theta & \cos\theta \end{pmatrix} \begin{pmatrix} \sigma_1 & 0 \\ 0 & \sigma_2 \end{pmatrix} \begin{pmatrix} \cos\theta & -\sin\theta \\ \sin\theta & \cos\theta \end{pmatrix}$$
$$= \begin{pmatrix} \sigma_1\cos^2\theta + \sigma_2\sin^2\theta & (\sigma_2-\sigma_1)\sin\theta\cos\theta \\ (\sigma_2-\sigma_1)\sin\theta\cos\theta & \sigma_1\sin^2\theta + \sigma_2\cos^2\theta \end{pmatrix}. \quad (3)$$

The normal and shear stresses on the plane vary, depending on the plane's orientation. The normal stress component, denoted by σ, is

$$\sigma = \sigma'_{11} = \sigma_1\cos^2\theta + \sigma_2\sin^2\theta = \frac{(\sigma_1+\sigma_2)}{2} + \frac{(\sigma_1-\sigma_2)}{2}\cos 2\theta, \quad (4a)$$

and the shear component, denoted by τ, is

$$\tau = \sigma'_{12} = (\sigma_2-\sigma_1)\sin\theta\cos\theta = \frac{(\sigma_2-\sigma_1)}{2}\sin 2\theta. \quad (4b)$$

Figure 5.7-4 shows σ and τ as functions of θ for the case of σ_1 and σ_2 negative ($|\sigma_1| > |\sigma_2|$), which corresponds to compression at depth in the earth. A graphic way to show these is with *Mohr's circle*, a plot of σ versus τ (Fig. 5.7-5). Values for all different planes lie on a circle centered at $\sigma = (\sigma_1+\sigma_2)/2$, $\tau = 0$, with radius $(\sigma_2-\sigma_1)/2$. The point on the circle with angle 2θ, measured clockwise from the $-\sigma$ axis, gives the σ, τ values on the plane whose normal is at angle θ to σ_1.[1]

Laboratory experiments on rocks under compression show that fracture occurs when a critical combination of the absolute value of shear stress and the normal stress is exceeded. This relation, known as the *Coulomb–Mohr failure criterion*, can be stated as

$$|\tau| = \tau_o - n\sigma, \quad (5)$$

where τ_o and n are properties of the material known as the *cohesive strength* and *coefficient of internal friction*. (The minus sign reflects the convention that compressional stresses are negative.) The failure criterion plots as two lines in the τ–σ plane, with τ axis intercepts $\pm\tau_o$ and slope $\pm n$ (Fig. 5.7-6). If the principal stresses are σ_1, σ_2, such that Mohr's circle does not intersect the failure lines, the material does not fracture. However, given the same σ_2 but a higher σ'_1, Mohr's circle intersects the line, and the material breaks.

The failure lines show how much shear stress, τ, can be applied to a surface subject to a normal stress σ before failure occurs. The cohesive strength is the minimum (absolute value) shear stress for failure. The coefficient of internal friction indicates the additional shear stress sustainable as the normal stress increases. Thus, deeper in the crust, where the pressure and hence normal stress are higher, rocks are stronger, and higher shear stress is required to break them.

The failure lines and Mohr's circle show on which plane failure occurs for a given stress state. To find θ, the angle between the plane's normal and the maximum compressive stress (σ_1) direction, we write the failure lines as

$$|\tau| = \tau_o - \sigma\tan\phi, \quad (6)$$

[1] Following the seismological convention of compressive stresses being negative, Mohr's circle is shown for $\sigma < 0$. The opposite convention is often used in rock mechanics, e.g. Figs. 5.7-3 and 5.7-10.

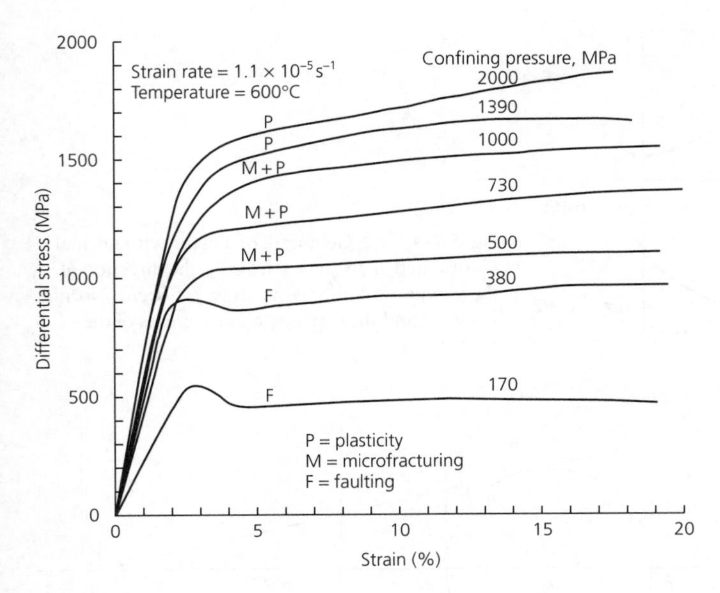

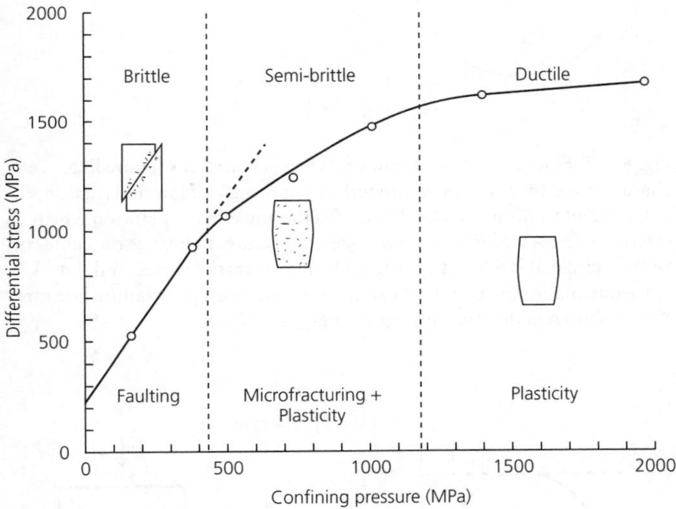

Fig. 5.7-3 Results of an experiment in which rocks are subjected to a compressive stress σ_1 greater than the confining pressure σ_3. *Top*: Differential stress $(\sigma_1 - \sigma_3)$ versus strain (compare to Figs 5.7-1 and 2) curves for various confining pressures. *Bottom*: Ultimate strength ($\sigma_1 - \sigma_3$ at 10% strain rate, from top) for various confining pressures. For low (< 400 MPa) confining pressures, the material fractures, and its strength increases with pressure. For higher pressures, the material is ductile, and its strength increases only slowly with pressure. A semi-brittle transition regime, in which both microfractures and crystal plasticity occur, separates the brittle and ductile regimes. (Kirby, 1980. *J. Geophys. Res.*, *85*, 6353–63, copyright by the American Geophysical Union.)

where $n = \tan \phi$, and ϕ, the *angle of internal friction*, is formed by extending the failure line to the σ axis (Fig. 5.7-7). Fracture occurs at point F, where the failure line is tangent to Mohr's circle. Considering the right triangle AFB, we see that

$$\phi = 2\theta - 90°, \quad \text{so} \quad \theta = \phi/2 + 45°. \tag{7}$$

For example, in introducing the relation between fault plane solutions and crustal stresses in Section 2.3.5, we made the simplest assumption that fracture occurs at 45° to the principal stress axes, corresponding to the case $\phi = 0°$, $n = 0$, $\theta = 45°$. Physically, this means that the normal stress has no effect on the strength of the rock. However, rocks typically have n about 1, so $\phi = 45°$, $\theta = 67.5°$, and the fault plane is closer (22.5°) to the maximum compression (σ_1) direction (Fig. 5.7-8). This idea is important when using P and T axes of focal mechanisms to characterize stress directions.

Figure 5.7-7 also shows how to find the stresses when fracture occurs. Consider the point T on the failure line such that $\overline{T\sigma_2}$ is perpendicular to the σ axis. Because the angle $AT\sigma_2$ is θ (triangles AFT and Aσ_2T are congruent),

$$\overline{T\sigma_2} = A\sigma_2 \cot \theta, \tag{8}$$

or, since $\overline{A\sigma_2} = (\sigma_2 - \sigma_1)/2$,

$$\overline{T\sigma_2} = \frac{(\sigma_2 - \sigma_1)}{2} \cot \theta. \tag{9}$$

Similarly,

$$\overline{T\sigma_2} = \tau_o - \sigma_2 \tan \phi \tag{10}$$

(the minus sign is because σ_2 is negative), so

$$\frac{(\sigma_2 - \sigma_1)}{2} \cot \theta = \tau_o - \sigma_2 \tan \phi. \tag{11}$$

This relation can be written in terms of the angle of the fracture plane, using Eqn 7 and trigonometric identities,

$$\tan \phi = -\cot 2\theta = \frac{-1}{\tan 2\theta} = \frac{\tan^2 \theta - 1}{2 \tan \theta}, \tag{12}$$

yielding

$$\sigma_1 = -2\tau_o \tan \theta + \sigma_2 \tan^2 \theta. \tag{13}$$

We will use this relation between the stresses when fracture occurs to estimate the maximum stresses in the crust.

Similar analyses show when the shear stress is high enough to overcome friction and cause sliding on a previously existing fault. The results are similar to those for a new fracture in unbroken rock, except that at low stress levels the preexisting fault has no cohesive strength. Thus slip on the fault occurs when $|\tau| = -\mu\sigma$, where μ is the *coefficient of sliding friction*, which can be expressed by an *angle of sliding friction*

$$\tan \alpha = \mu. \tag{14}$$

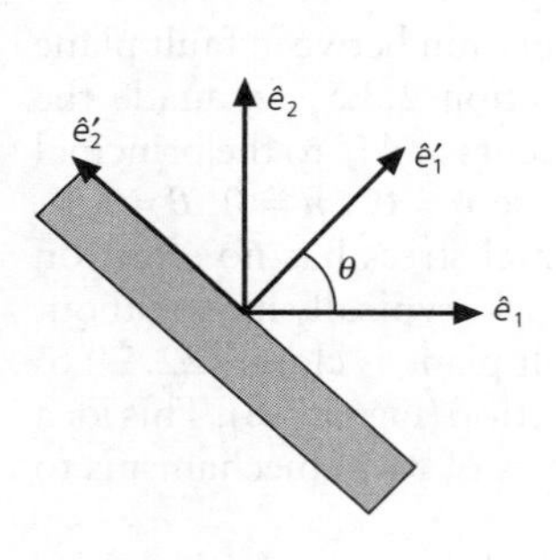

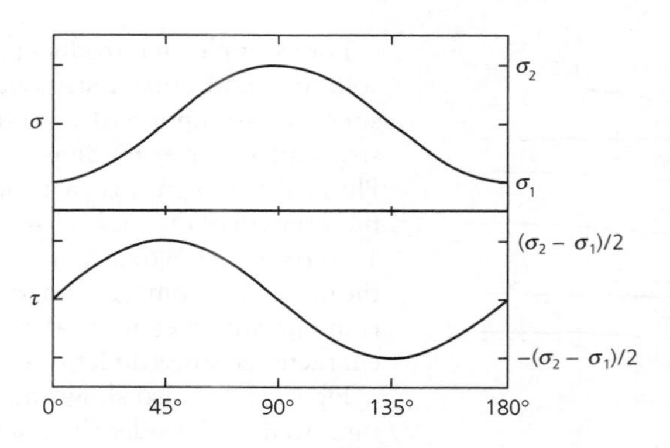

Fig. 5.7-4 *Left*: Geometry of a plane with normal $\hat{e}'_1$, oriented at an angle θ from $\hat{e}_1$, the direction of the maximum compressive stress σ_1. *Right*: Normal stress, σ, and shear stress, τ, as functions of the angle θ.

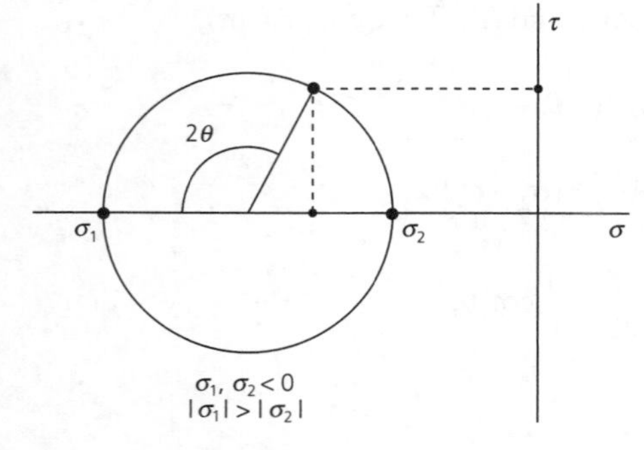

Fig. 5.7-5 Mohr's circle: Given a state of stress described by principal stresses σ_1 and σ_2, the normal stress, σ, and the shear stress, τ, for planes of all orientations lie on a circle with radius $(\sigma_2 - \sigma_1)/2$. The point on the circle with angle 2θ, measured clockwise from the $-\sigma$ axis, gives σ and τ on a plane whose normal is at an angle θ from the direction of σ_1.

Fig. 5.7-7 Fracture occurs at point F, where a material's failure line, characterized by its cohesive strength, τ_o, and angle of internal friction, ϕ, is tangent to Mohr's circle. Hence θ is the angle of the plane on which fracture occurs, and F gives the stresses at fracture. Point A is the center of Mohr's circle, B is where the failure line intersects the σ axis, and $\overline{T\sigma_2}$ is perpendicular to the σ axis. For simplicity, only the upper failure line for $\tau > 0$ is shown in this and subsequent figures.

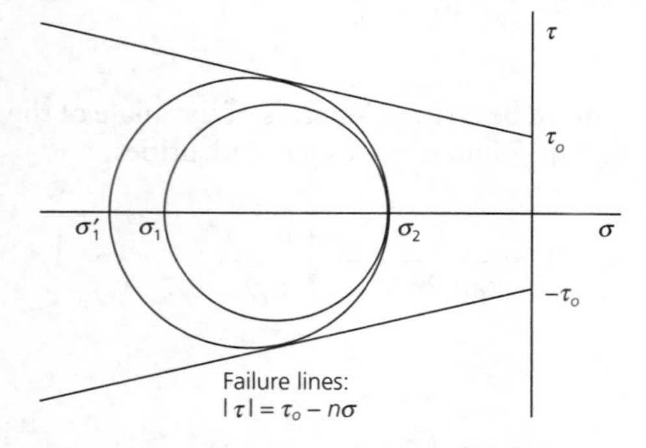

Fig. 5.7-6 The Coulomb–Mohr failure criterion assumes that a material fractures when Mohr's circle intersects the failure line.

Figure 5.7-9 shows the Mohr's circle representation of a rock with preexisting faults. In addition to the failure line, there is a frictional sliding line corresponding to

$$\tau = -\mu\sigma = -\sigma \tan \alpha. \tag{15}$$

Because the sliding line starts at the origin, it is initially below the failure line. Assume that the stresses are large enough that Mohr's circle touches the failure line at the point yielding frac-

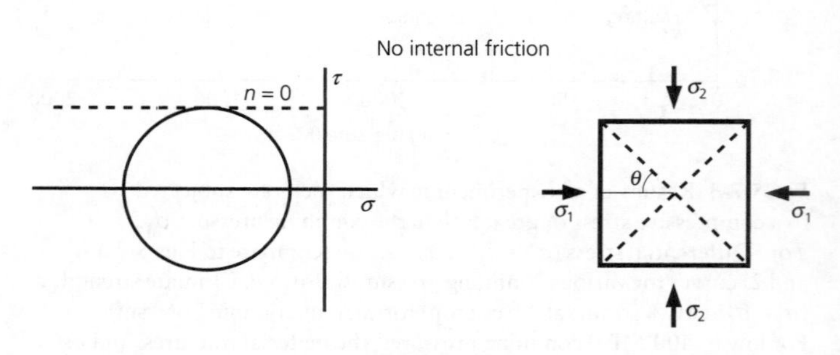

Fig. 5.7-8 With no internal friction, fracture occurs at an angle of 45°. For $n = 1$, the fracture angle is 67.5°, and the fault plane is closer (22.5°) to the maximum compression (σ_1) direction.

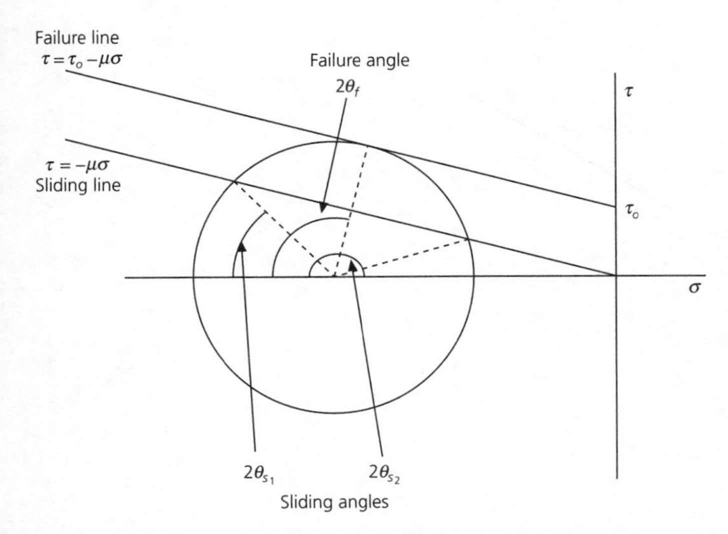

Fig. 5.7-9 Mohr's circle for sliding on a rock's preexisting faults. A new fracture would form at an angle θ_f, given by the failure line. However, slip will occur on a preexisting fault if there are any with angles between θ_{s_1} and θ_{s_2}, given by the intersection of the circle with the frictional sliding line.

ture on a plane corresponding to an angle θ_f. Similarly, the frictional sliding line intercepts the circle at two points, corresponding to angles θ_{s_1} and θ_{s_2}. Thus the rock can fail in several ways. If there are preexisting faults with angles θ_{s_1} or θ_{s_2}, slip on these faults may occur. Alternatively, a new fracture may form on the plane given by θ_f. However, because this fracture occurs at higher shear stress than is needed for frictional sliding on the preexisting faults, sliding is favored over the formation of a new fracture. Thus, if the stress has gradually risen to this level, sliding on preexisting faults would probably have prevented a new fracture from forming.

This effect can have seismological consequences. The simplest way to use focal mechanisms to infer stress orientations is to assume that the earthquakes occurred on newly formed faults. However, if the rock had been initially faulted, the earthquakes may have occurred on preexisting faults. In the representation of Fig. 5.7-9, if faults exist with normals oriented between θ_{s_1} and θ_{s_2} to the maximum compressive stress, slip on these faults will occur rather than the formation of a new fracture. Thus the inferred stress direction will be somewhat inaccurate. For example, the thrust focal mechanisms along the Himalayan front (Fig. 5.6-6) or eastern Andean foreland thrust belt (Fig. 5.6-10) have fault planes that rotate as the trend of the mountains changes, suggesting that the fault planes are controlled by the existing structures, so the P axes only partially reflect the stress field. A similar pattern appears for T axes along the East African rift (Fig. 5.6-2). In general, stress axes inferred from many fault plane solutions in an area seem relatively coherent (Fig. 5.6-19). Thus we assume that the crust contains preexisting faults of all orientations, so the average stress orientation inferred from the focal mechanisms is not seriously biased.

At this point, it is worth noting other complexities. Both the failure and sliding curves may be more complicated than straight lines. These curves, known as Mohr envelopes, can be derived from experiments at various values of stress. Additional complexity comes from the fact that water and other fluids are often present in rocks, especially in the upper crust. The fluid pressure, known as the *pore pressure*, reduces the effect of the normal stress and allows sliding to take place at lower shear stresses. This effect is modeled by replacing the normal stress σ with $\bar{\sigma} = \sigma - P_f$, known as the *effective normal stress*, where P_f is the pore fluid pressure.[2] Because the pore pressure is defined as negative, the effective normal stress is reduced (less compressive). Similarly, effective principal stresses taking into account pore pressure,

$$\bar{\sigma}_1 = \sigma_1 - P_f \quad \text{and} \quad \bar{\sigma}_2 = \sigma_2 - P_f, \tag{16}$$

are used in the fracture theory.

The relations we have discussed can be used to estimate the maximum stresses that the crust can support. Laboratory experiments (Fig. 5.7-10) for sliding on existing faults in a variety of rock types find relations sometimes called *Byerlee's law*:

$$\tau \approx -0.85\bar{\sigma}, \quad |\bar{\sigma}| < 200 \text{ MPa}$$
$$\tau \approx 50 - 0.6\bar{\sigma}, \quad |\bar{\sigma}| > 200 \text{ MPa}. \tag{17}$$

These relations, written in terms of the normal and shear stresses on a fault, can be used to infer the principal stress as a function of depth. To do so, we write the minimum compressive stress as σ_3, because we are in three dimensions. We assume that the crust contains faults of all orientations, and that the stresses cannot exceed the point where Mohr's circle is tangent to the frictional sliding line, or else sliding will occur (Fig. 5.7-11). At shallow depths where $|\bar{\sigma}| < 200$ MPa, Eqn 17 shows that $\tau_0 = 0$. Thus Eqn 13, the relation between the stresses when fracture occurs, yields

$$\bar{\sigma}_1 = \bar{\sigma}_3 \tan^2 \theta_s. \tag{18}$$

Using Eqn 7 for the case of frictional sliding,

$$\theta_s = \alpha/2 + 45°, \tag{19}$$

and the values in Eqn 17 give

$$\mu = \tan \alpha = 0.85, \quad \alpha \approx 41°, \quad \theta_s \approx 66°, \quad \tan^2 66° \approx 5, \tag{20}$$

so the stresses are related by

$$\bar{\sigma}_1 \approx 5\bar{\sigma}_3. \tag{21}$$

At greater depths, where $|\bar{\sigma}| > 200$ MPa, $\alpha \approx 31°$ and $\theta_s = 60.5°$, so the stresses are related by

$$\bar{\sigma}_1 \approx -175 + 3.1\bar{\sigma}_3. \tag{22}$$

[2] The role of pore pressure in making sliding easier can be seen by trying to slide an object across a dry table and then wetting the table.

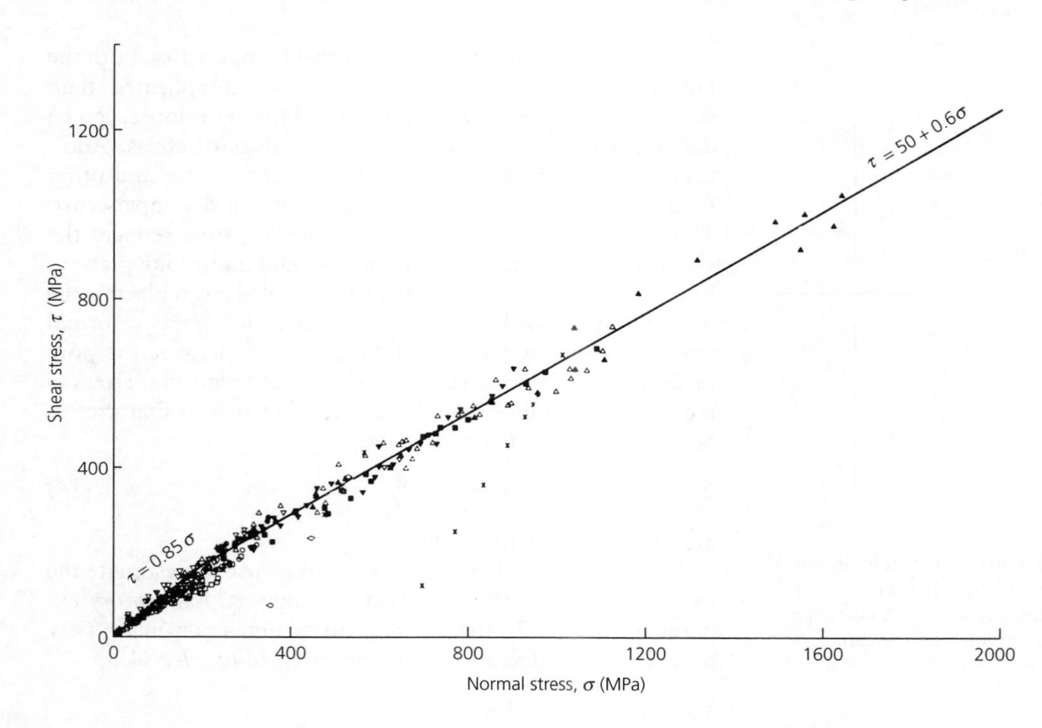

Fig. 5.7-10 Shear stress versus normal stress for frictional sliding, compiled for various rock types. Compressive stress is positive. (Byerlee, 1978. *Pure Appl. Geophys.*, *116*, 615–26, reproduced with the permission of Birkhauser.)

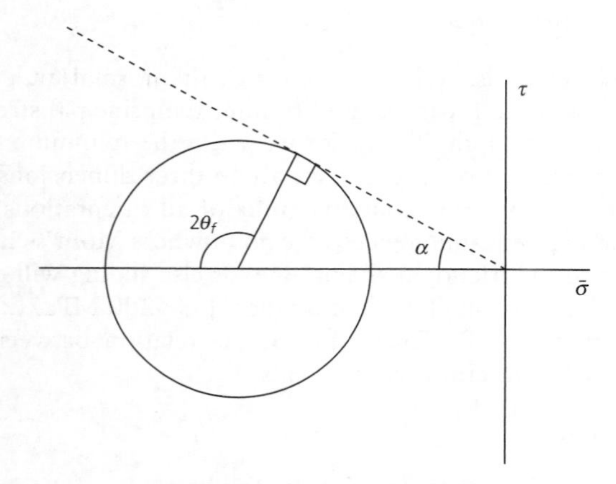

Fig. 5.7-11 Mohr's circle and sliding line for $|\bar{\sigma}| < 200$ MPa. If the lithosphere contains fractures in all directions, the stresses cannot exceed those at the point where Mohr's circle is tangent to the sliding line, because sliding would occur.

We assume that one principal stress, σ_1 or σ_3, is the vertical stress due to the lithostatic pressure as a function of depth (z),

$$\sigma_V = -\rho g z. \tag{23}$$

The other principal stress, which must be horizontal, is denoted σ_H. The pore pressure $P_f(z)$ is unknown. One common assumption is that the rock is dry, so $P_f(z) = 0$. Another is that the pore pressure is *hydrostatic*, which is equivalent to assuming that pores are connected up to the surface, so

$$P_f(z) = -\rho_f g z, \tag{24}$$

where ρ_f is the density of the fluid, which is usually water, with $\rho_f = 1$ g/cm^3. Alternatively, the pore pressure can be assumed to be a fixed fraction of the lithostatic pressure (Section 2.3.6).

We now can find the *strength* of the crust, defined by the maximum difference between the horizontal and vertical stresses that the rock can support. At shallow depths where $|\bar{\sigma}| < 200$ MPa, Eqn 21 shows that $\bar{\sigma}_1 = 5\bar{\sigma}_3$. There are two possibilities, depending on whether the vertical stress is the most ($\bar{\sigma}_1$) or least ($\bar{\sigma}_3$) compressive. If the vertical stress is the most compressive,

$$\sigma_V = \sigma_1, \quad \bar{\sigma}_1 = \sigma_V - P_f = -\rho g z - P_f(z)$$
$$\sigma_H = \sigma_3, \quad \bar{\sigma}_3 = \bar{\sigma}_1/5 = -(\rho g z + P_f(z))/5. \tag{25}$$

Alternatively, if the vertical stress is the least compressive,

$$\sigma_V = \sigma_3, \quad \bar{\sigma}_3 = \sigma_V - P_f = -\rho g z - P_f(z)$$
$$\sigma_H = \sigma_1, \quad \bar{\sigma}_1 = 5\bar{\sigma}_3 = -5(\rho g z + P_f(z)). \tag{26}$$

In the first case,

$$\sigma_H - \sigma_V = \sigma_3 - \sigma_1 = 0.8(\rho g z + P_f(z)), \tag{27}$$

corresponds to an extensional (positive) stress. In the second,

$$\sigma_H - \sigma_V = \sigma_1 - \sigma_3 = -4(\rho g z + P_f(z)) \tag{28}$$

corresponds to a compressive (negative) stress that is much greater in absolute value. Thus, at any depth, the crust can

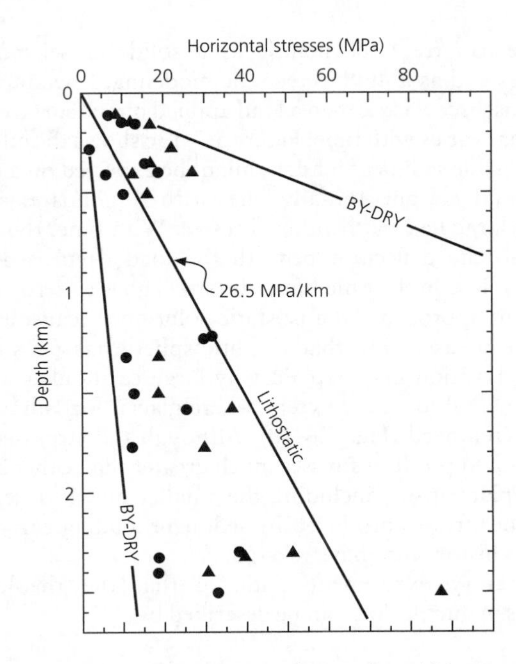

Fig. 5.7-12 Horizontal stresses measured in southern Africa. Dots are for horizontal stresses being the least compressive (σ_3), and triangles are for horizontal stresses being the most compressive (σ_1). The lithostatic stress gradient (26.5 MPa/km) is shown, along with Byerlee's law (BY) for zero pore pressure (DRY). The stronger line is for compression, and the weaker one is for extension. The observed stresses are within the maximum and minimum BY-DRY lines. (Brace and Kohlstedt, 1980. *J. Geophys. Res., 85*, 6248–52, copyright by the American Geophysical Union.)

support greater compressive deviatoric stress than extensional deviatoric stress (Fig. 5.7-12).

5.7.3 *Ductile flow*

When rocks behave brittlely, their behavior is not time-dependent; they either strain elastically or fail. By contrast, the deformation of ductile rock depends on time. A common model for the time-dependent behavior is a *Maxwell viscoelastic material*, which behaves like an elastic solid on short time scales and like a viscous fluid on long time scales. This model can describe the mantle because seismic waves propagate as though the mantle were solid, whereas postglacial rebound and mantle convection occur as though the mantle were fluid.

To see this difference, consider two types of deformation in one dimension. For an elastic solid subjected to elastic strain $e_E = e_{11}$,

$$\sigma = Ee_E, \tag{29}$$

where E is Young's modulus, and σ is σ_{11}. The simplest viscous fluid obeys

$$\sigma = 2\eta \frac{de_F}{dt}, \tag{30}$$

where η is the viscosity, and e_F is the fluid portion of the strain. This equation defines the viscosity, the property that measures a fluid's resistance to shear.[3]

We often think of an elastic material as a spring, which exerts a force proportional to distance. Thus stress and strain are proportional at any instant, and there are no time-dependent effects. By contrast, the viscous material is though of as a dashpot, a fluid damper that exerts a force proportional to velocity. Hence the stress and strain rate are proportional, and the material's response varies with time. These effects are combined in a viscoelastic material, which can be thought of as a spring and dashpot in series (Fig. 5.7-13). The combined elastic and viscous response comes from the combined strain rate

$$\frac{de}{dt} = \frac{de_E}{dt} + \frac{de_F}{dt} = \frac{1}{E}\frac{d\sigma}{dt} + \frac{\sigma}{2\eta}. \tag{31}$$

This differential equation, the rheological law for a Maxwell substance, shows how the stress in the material evolves after a strain e_o is applied at time $t = 0$ and then remains constant. At $t = 0$ the derivative terms dominate, so the material behaves elastically, and has an initial stress

$$\sigma_o = Ee_o. \tag{32}$$

For $t > 0$, $de/dt = 0$, so

$$\frac{d\sigma}{dt} = -\frac{E}{2\eta}\sigma, \tag{33}$$

whose integral is

$$\sigma(t) = \sigma_o \exp\left[-(Et/2\eta)\right]. \tag{34}$$

Thus stress relaxes from its initial value as a function of time (Fig. 5.7-13). A useful parameter is the *Maxwell relaxation time*,[4]

$$\tau_M = \frac{2\eta}{E} \approx \frac{\eta}{\mu}, \tag{35}$$

required for the stress to decay to e^{-1} of its initial value. For times less than τ_M the material can be considered an elastic solid, whereas for longer times it can be considered a viscous fluid.

For example, if the mantle is approximately a Poisson solid with $\mu \approx 10^{12}$ dyn/cm^2 and $\eta \approx 10^{22}$ poise, its Maxwell time is about 10^{10} s or 300 years. Although the viscosity is not that well known, so estimates of the Maxwell time vary, it is clear

[3] In familiar terms, viscosity measures how "gooey" a fluid is. Maple syrup is somewhat more viscous than water, and the earth's mantle is about 10^{24} times more viscous.

[4] Definitions of the Maxwell time vary, but always involve the ratio of the viscosity to an elastic constant.

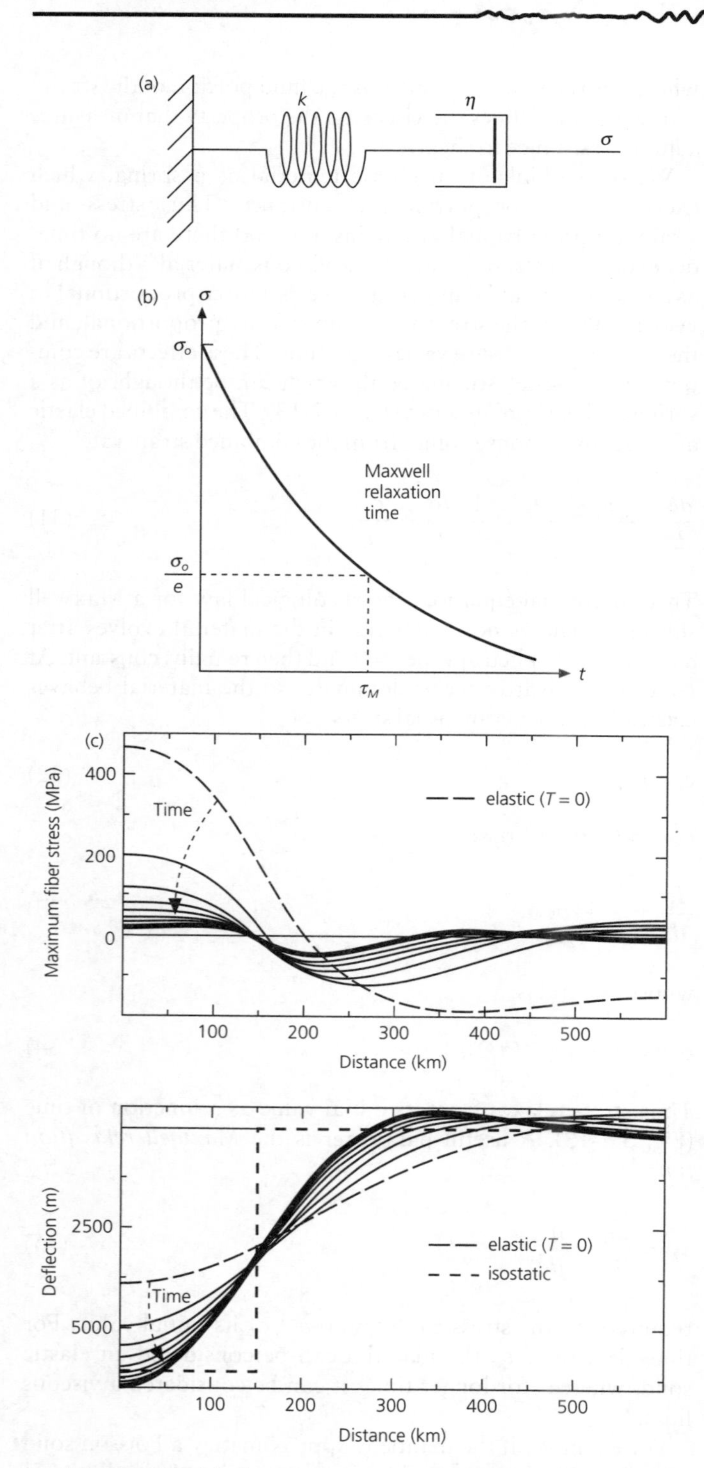

Fig. 5.7-13 (a) Model of a viscoelastic material as an elastic spring and viscous dashpot in series. (b) Stress response of a viscoelastic material to an applied strain. The Maxwell relaxation time, τ_M, is the time the stress takes to decay to e^{-1} of its initial value. (c) Evolution of the deflection and bending stress produced by a sediment load on a viscoelastic earth. At first the earth responds elastically, as shown by the long-dashed line, but with time it flows, so the deflection beneath the load deepens and the stresses relax. (Stein *et al.*, 1989, with kind permission from Kluwer Academic Publishers.)

that we can treat the mantle as a solid for seismological purposes and as a fluid in tectonic modeling. If we model the mantle as viscoelastic, then a load applied on the surface has an effect that varies with time. Figure 5.7-13c shows the effect of a 150 km-wide sediment load, as might be expected on a passive continental margin. Initially, the earth responds elastically, causing large flexural bending stresses. With time, the mantle flows, so the deflection beneath the load deepens and the stresses relax. In the time limit, the stress goes to zero, and the deflection approaches the isostatic solution, because isostasy amounts to assuming that the lithosphere has no strength. Stress relaxation may explain why large earthqukes are rare at continental margins, except where glacial loads have been recently removed (Fig. 5.6-20). Although the large sediment loads should produce stresses much greater than other sources of intraplate stress, including the smaller and less dense ice loads, the stresses produced by sediment loading early in the margin's history may have relaxed.

Laboratory experiments indicate that the rheology of minerals in ductile flow can be described by

$$\frac{de}{dt} = \dot{e} = f(\sigma)\, A \exp\,[-(E^* + PV^*)/RT], \tag{36}$$

where T is temperature, R is the gas constant, and P is pressure. $f(\sigma)$ is a function of the stress difference $|\sigma_1 - \sigma_3|$, and A is a constant. The effects of pressure and temperature are described by the *activation energy* E^* and the *activation volume* V^*. Observed values of $f(\sigma)$ are often fit well by assuming

$$f(\sigma) = |\sigma_1 - \sigma_3|^n$$

$$\dot{e} = |\sigma_1 - \sigma_3|^n A \exp\,[-(E^* + PV^*)/RT]. \tag{37}$$

The rheology of such fluids is characterized by a power law. If $n = 1$, the material is called *Newtonian*, whereas a non-Newtonian fluid with $n = 3$ is often used to represent the mantle. From Eqn 30 we see that for a Newtonian fluid the viscosity depends on both temperature and pressure:

$$\eta = (1/2A) \exp\,[(E^* + PV^*)/RT]. \tag{38}$$

Thus the viscosity decreases exponentially with temperature. This decrease is assumed to give rise to a strong lithosphere overlying a weaker asthenosphere, and the restriction of earthquakes to shallow depths.[5] For a non-Newtonian fluid, Eqn 30 gives the *effective viscosity*, the equivalent viscosity if the fluid were Newtonian.

We think of equations like Eqn 37 as showing the strength, or maximum stress difference $|\sigma_1 - \sigma_3|$, that the viscous material can support. This stress difference depends on temperature, pressure, strain rate, and rock type. The material

[5] Temperature-dependent viscosity is an effect familiar to automobile drivers in cold temperatures, when the engine and the transmission became noticeably sluggish.

is stronger at higher strain rates, and weakens exponentially with high temperatures. At shallow depths, the small pressure effect is often neglected, so the activation volume V^* is treated as zero. For example, a commonly used flow law for dry olivine is[6]

$$\dot{e} = 7 \times 10^4 |\sigma_1 - \sigma_3|^3 \exp\left(\frac{-0.52\ \text{MJ/mol}}{RT}\right) \quad \text{for} \quad |\sigma_1 - \sigma_3| \leq 200\ \text{MPa}$$

$$= 5.7 \times 10^{11} \exp\left[\frac{-0.54\ \text{MJ/mol}}{RT}\left(1 - \frac{[\sigma_1 - \sigma_2]}{8500}\right)^2\right] \quad \text{for} \quad |\sigma_1 - \sigma_3| > 200\ \text{MPa}, \qquad (39)$$

where $\dot{e}$ is in s^{-1}. Similarly, for quartz,

$$\dot{e} = 5 \times 10^6 |\sigma_1 - \sigma_3|^3 \exp\left(\frac{-0.19\ \text{MJ/mol}}{RT}\right) \quad \text{for} \quad |\sigma_1 - \sigma_3| < 1000\ \text{MPa}. \qquad (40)$$

At a given strain rate, quartz is much weaker (can sustain a smaller stress difference) than olivine. Thus the quartz-rich continental crust should be weaker that the olivine-rich oceanic crust, an effect whose tectonic consequences are discussed next.

5.7.4 *Strength of the lithosphere*

The strength of the lithosphere as a function of depth depends upon the deformation mechanism. At shallow depths, rocks fail by either brittle fracture or frictional sliding on preexisting faults. Both processes depend in a similar way on the normal stress, with rock strength increasing with depth. However, at greater depths the ductile flow strength of rocks is less than the brittle or frictional strength, so the strength is given by the flow laws and decreases with depth as the temperatures increase. This temperature-dependent strength is the reason why the cold lithosphere forms the planet's strong outer layer.

To calculate the strength, a strain rate and a geotherm giving temperature as a function of depth are assumed. At shallow depths the strength, the maximum stress difference before frictional sliding occurs, is computed using Eqns 27 and 28. At some depth, the frictional strength exceeds the ductile strength allowed by the flow law, so for deeper depths the maximum strength is given by the flow law. Figure 5.7-14 shows a strength plot, known as a *strength envelope*, for a strain rate of $10^{-15}\ s^{-1}$ and a temperature gradient appropriate for old oceanic lithosphere or stable continental interior. In the frictional region, curves are shown for various values of λ, the ratio of pore pressure to lithostatic pressure. The higher

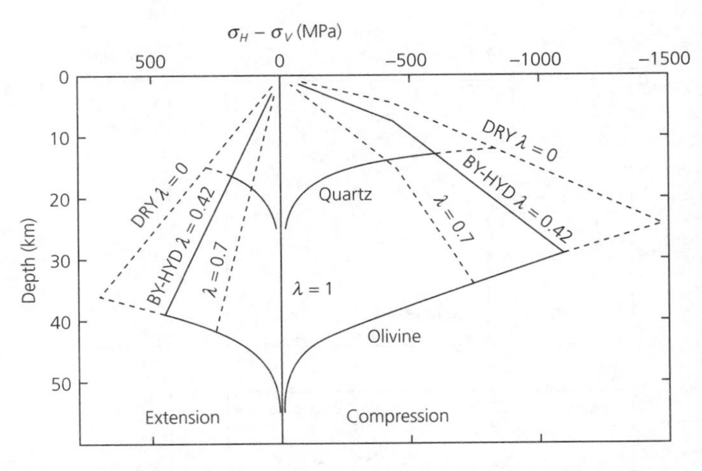

Fig. 5.7-14 Strength envelopes as a function of depth for various values of λ, the ratio of pore pressure to lithostatic pressure. BY-HYD lines are for Byerlee's law with hydrostatic pore pressure. At shallow depths, strength is controlled by brittle fracture; at greater depths ductile flow laws predict rapid weakening. In the ductile flow regime, quartz is weaker than olivine. In the brittle regime, the lithosphere is stronger in compression (*right side*) than in extension (*left side*). (Brace and Kohlstedt, 1980. *J. Geophys. Res.*, *85*, 6248–52, copyright by the American Geophysical Union.)

pore pressures result in lower strengths. Ductile flow laws are shown for quartz and olivine, minerals often used as models for continental and oceanic rheologies. Strength increases with depth in the brittle region, due to the increasing normal stress, and then decreases with depth in the ductile region, due to increasing temperature. Hence strength is highest at the *brittle–ductile transition*. Strength decreases rapidly below this transition, so the lithosphere should have little strength at depths greater than about 25 km in the continents and 50 km in the oceans. The strength envelopes show that the lithosphere is stronger for compression than for tension in the brittle regime, but the two are symmetric in the ductile regime. Strength envelopes are often plotted using the rock mechanics convention of compression positive.

The actual distribution of strength with depth is probably more complicated, because the brittle–ductile transition occurs over a region of semi-brittle behavior that includes both brittle and plastic processes (Fig. 5.7-3). However, this simple model gives insight into various observations. In particular, we have seen that the depths of earthquakes in several tectonic environments seem to be limited by temperature. This makes sense, because for a given strain rate and rheology the exponential dependence on temperature would make a limiting strength for seismicity approximate a limiting temperature.

To see this, consider Fig. 5.7-15, which shows that as oceanic lithosphere ages and cools, the predicted strong region deepens. This result seems plausible because earthquake depths, seismic velocities, and effective elastic thicknesses imply that the strong upper part of the lithosphere thickens with age (Fig. 5.3-9). The strength envelopes are thus consistent with the observation that the maximum depth of earthquakes within

[6] Brace and Kohlstedt (1980).

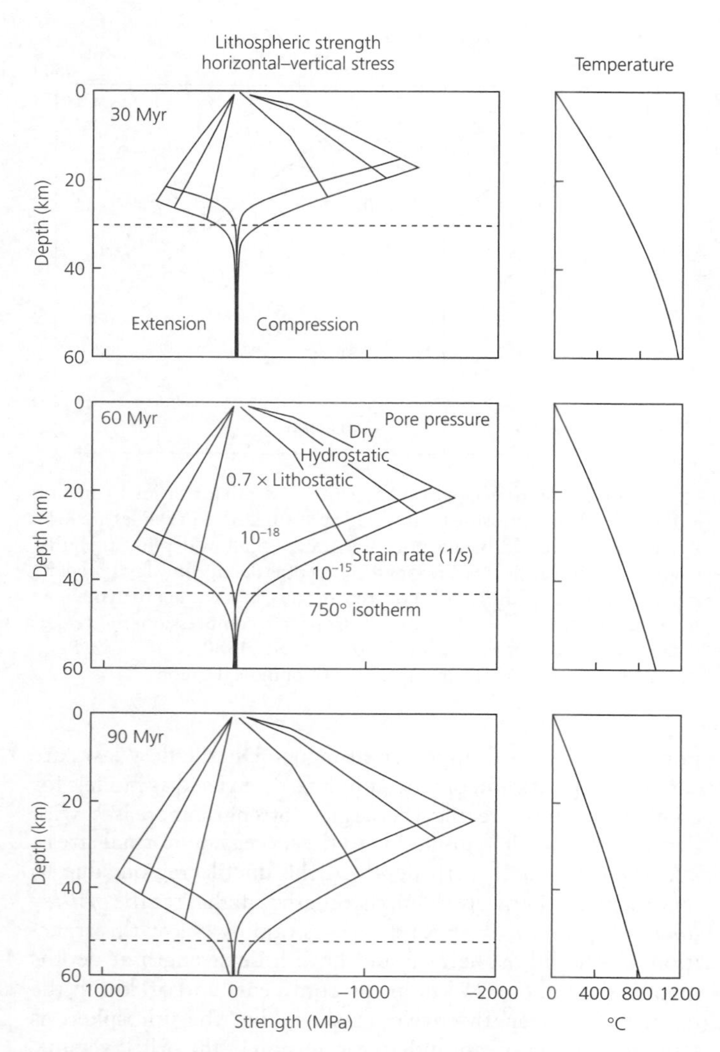

Fig. 5.7-15 Strength envelopes showing maximum stress difference (strength) as a function of depth for an olivine rheology, for geotherms (*right*) corresponding to cooling oceanic lithosphere of different ages. Strength in the brittle regime is reduced by higher pore pressure; strength in the ductile regime is reduced by lower strain rate. The depth range in which the material is strong enough for faulting increases with age. (Wiens and Stein, 1983. *J. Geophys. Res., 88*, 6455–68, copyright by the American Geophysical Union.)

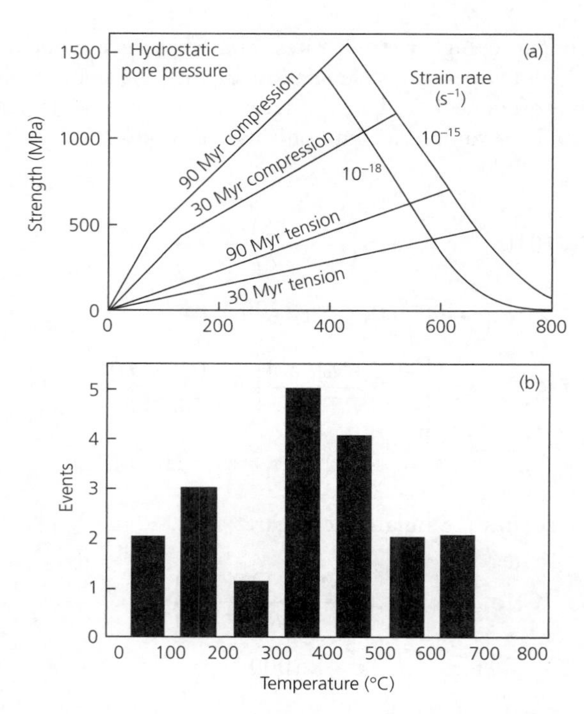

Fig. 5.7-16 Plots of strength and seismicity versus temperature. The strength envelopes explain the observation that intraplate oceanic seismicity occurs only above the 750°C isotherm. (Wiens and Stein, 1985. *Tectonophysics, 116*, 143–62, with permission from Elsevier Science.)

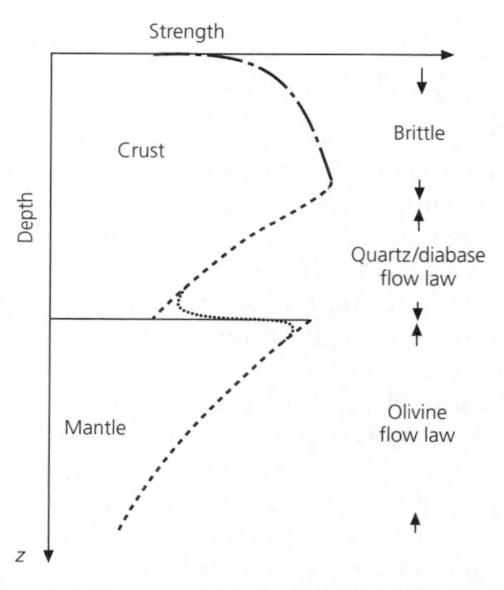

Fig. 5.7-17 Schematic strength envelope for continents. Below the ductile lower crust may be a stronger zone in the olivine-rich mantle. (Chen and Molnar, 1983. *J. Geophys. Res., 88*, 4183–4214, copyright by the American Geophysical Union.)

the oceanic lithosphere is approximately bounded by the 750°C isotherm (Fig. 5.7-16). These envelopes are drawn for strain rates of 10^{-15} and 10^{-18} s^{-1}, appropriate for slow deformation within plates. By contrast, a seismic wave with a period of 1 s, a wavelength of 10 km, and a displacement of 10^{-6} m corresponds to a strain rate of 10^{-10} s^{-1}. The successively greater effective elastic thicknesses, depth of the deepest earthquakes, and depth of the low-velocity zone are thus consistent with strength increasing with strain rate.

The strength envelopes give insight into differences between continental and oceanic lithospheres (Fig. 5.7-17). First, quartz is weaker than olivine at a given temperature (Fig. 5.7-14), consistent with the fact that the limiting temperature for continental seismicity is lower than for oceanic earthquakes (Fig. 5.7-18). Second, the strength profiles differ. The strength of oceanic lithosphere increases with depth and then decreases. However, in continental lithosphere we expect such a profile in

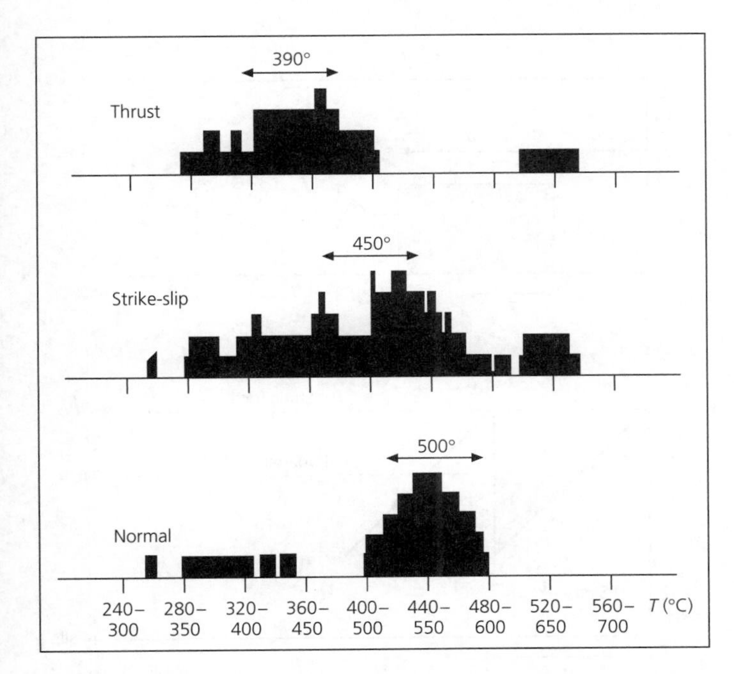

Fig. 5.7-18 Limiting temperatures for continental seismicity. These temperatures are much lower than those for oceanic lithosphere, since the quartz rheology in continents is much weaker than olivine. (Courtesy of J. Strehlau and R. Meissner.)

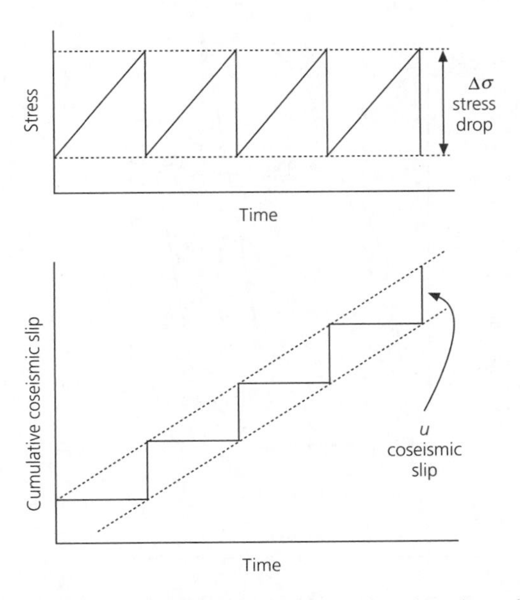

Fig. 5.7-19 Stress and slip history for an idealized earthquake cycle on a plate boundary, in which all earthquakes have the same stress drop and coseismic slip. (Shimazaki and Nakata, 1980. *Geophys. Res. Lett.*, *7*, 279–82, copyright by the American Geophysical Union.)

the quartz-rich crust, but also a second, deeper zone of strength below the Moho, due to the olivine rheology. This "jelly sandwich" profile including a weak zone may be part of the reason why continents deform differently than oceanic lithosphere. For example, some continental mountain building (Fig. 5.6-6) may involve crustal thickening in which slices of upper crust, which are too buoyant to subduct, are instead thrust atop one another. The weaker lower crust may also contribute in other ways to the general phenomenon that continental plate boundaries are broader and more complex than their oceanic counterparts (Fig. 5.2-4).

5.7.5 *Earthquakes and rock friction*

It is natural to assume that earthquakes occur when tectonic stress exceeds the rock strength, so a new fault forms or an existing one slips. Thus steady motion across a plate boundary seems likely to give rise to a cycle of successive earthquakes at regular intervals, with the same slip and stress drop (Fig. 5.7-19). However, we have seen that the earthquake process is more complicated. The time between earthquakes on plate boundaries varies (Fig. 1.2-15), although the plate motion causing the earthquakes is steady. Earthquakes sometimes rupture along the same segments of a boundary as in earlier earthquakes, and other times along a different set (Fig. 5.4-27). Moreover, many large earthquakes show a complicated rupture pattern, with some parts of the fault releasing more seismic energy than others (Fig. 4.5-10). Attempts to understand these complexities often combine two basic themes. Some of the complexity may be due to intrinsic randomness of the failure process, such that some small ruptures cascade into large earthquakes, whereas others do not (Section 1.2.6). Other aspects of the complexity may be due to features of rock friction.

Interesting insight emerges from considering an experiment in which stress is applied until a rock breaks. When the fault forms, some of the stress is released, and then motion stops. If stress is reapplied, another stress drop and motion occur once the stress reaches a certain level. So long as stress is reapplied, this pattern of jerky sliding and stress release continues (Fig. 5.7-20).

This pattern, called *stick-slip*, looks like a laboratory version of what happens in a sequence of earthquakes on a fault. By this analogy, the stress drop in an earthquake relieves only part of the total tectonic stress, and as the fault continues to be loaded by tectonic stress, occasional earthquakes occur. The analogy is strengthened by the fact that at higher temperatures (about 300° for granite), stick-slip does not occur (Fig. 5.7-20). Instead, *stable sliding* occurs on the fault, much as earthquakes do not occur at depths where the temperature exceeds a certain value. Thus, understanding stick-slip in the laboratory seems likely to give insight into the earthquake process.

Stick-slip results from a familiar phenomenon: it is harder to start an object sliding against friction than to keep it going once it is sliding. This is because the *static friction* stopping the object from sliding exceeds the *dynamic friction* that opposes motion once sliding starts.[7] To understand how this difference

[7] This effect is the basis of cross-country skiing, where loading one ski makes it grip the snow, while unloading the other lets it glide.

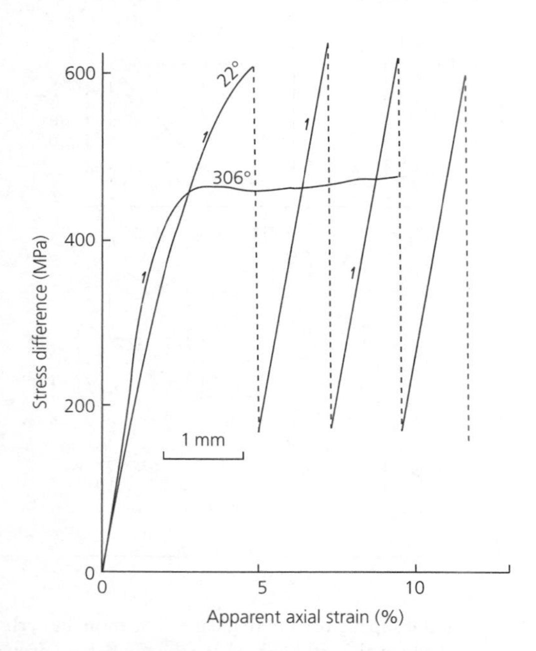

Fig. 5.7-20 Force versus slip history for a rock sample. At low temperature, so long as stress is reapplied, a stick-slip pattern of jerky sliding and stress release continues. By contrast, stable sliding occurs at high temperature. (Brace and Byerlee, 1970. *Science, 168*, 1573–5, copyright 1970 American Association for the Advancement of Science.)

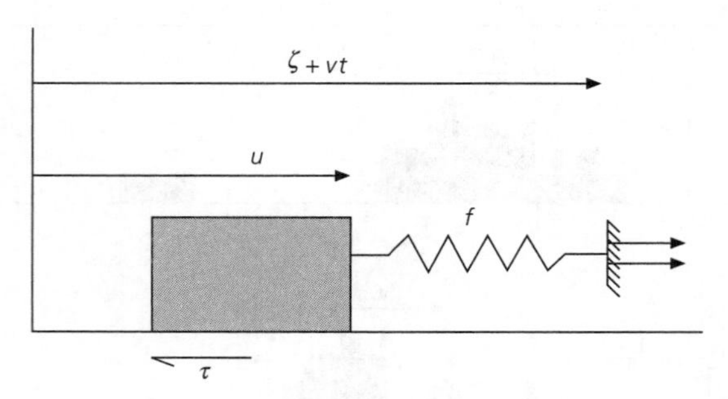

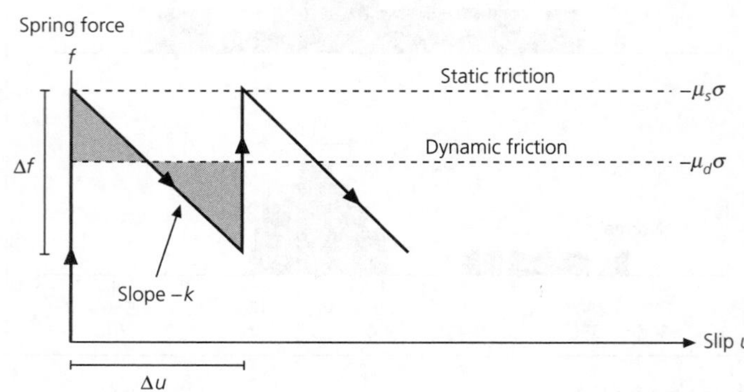

Fig. 5.7-21 A simple spring and slider block analog for stick-slip as a model for earthquakes. The slider is loaded by force f due to the spring end moving at velocity v. Before sliding, the block is retarded by a static friction force $\tau = -\mu_s\sigma$, but once sliding starts, the friction force decreases to $-\mu_d\sigma$. A series of slip events occur, each with slip Δu and force change (stress drop) Δf.

causes stick-slip, and get insight into stick-slip as a model for earthquakes, consider the experiment in Fig. 5.7-21. It turns out that if an object is pulled across a table with a rubber band, jerky stick-slip motion occurs.[8] Thus a steady load, combined with the difference in static and dynamic friction, causes an instability and a sequence of discrete slip events.

We analyze this situation assuming that a block (sometimes called a slider) is loaded by a spring that applies a force f proportional to the spring constant (stiffness) k and the spring extension. If the loading results from the spring's far end moving at a velocity v, the spring force is

$$f = k(\zeta + vt - u), \tag{41}$$

where u is the distance the block slipped, and ζ is the spring extension when sliding starts at $t = 0$. This motion is opposed by a frictional force $|\tau| = -\mu\sigma$ equal to the product of σ, the compressive (negative) normal stress due to the block's weight, and the friction coefficient, μ. By Newton's second law that force equals mass times acceleration,

$$m\frac{d^2u}{dt^2} = f - \tau = k(\zeta + vt - u) + \mu\sigma. \tag{42}$$

However, the block starts sliding only once the spring force exceeds the frictional force, so just before sliding starts at $t = 0$,

$$0 = k\zeta + \mu_s\sigma, \tag{43}$$

where μ_s is the static friction coefficient. For simplicity, assume that at the instant sliding starts, the friction drops to its dynamic value μ_d, and

$$m\frac{d^2u}{dt^2} = k(\zeta - u) + \mu_d\sigma. \tag{44}$$

Subtracting Eqn 43 from Eqn 44 gives

$$m\frac{d^2u}{dt^2} = -ku + (\mu_d - \mu_s)\sigma = -ku + \Delta\mu\sigma, \tag{45}$$

which we can use as the equation of motion for the block's slip history $u(t)$ if the loading rate v is slow enough to ignore during the slip event.

A solution to Eqn 45, with initial conditions $u(0) = 0$ and $\frac{du(0)}{dt} = 0$, is

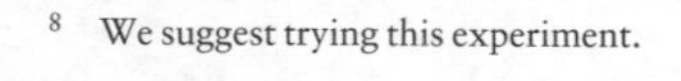

[8] We suggest trying this experiment.

$$u(t) = \frac{\Delta\mu\sigma}{k}(1 - \cos \omega t) \quad \text{(slip)},$$

$$\frac{du(t)}{dt} = \frac{\Delta\mu\sigma}{\sqrt{km}} \sin \omega t \quad \text{(velocity)},$$

$$\frac{du^2(t)}{dt^2} = \frac{\Delta\mu\sigma}{m} \cos \omega t \quad \text{(acceleration)}, \tag{46}$$

where $\omega = \sqrt{k/m}$. As shown, the block starts slipping because the spring force exceeds the friction force. During the slip event, the spring force decreases as the spring shortens, until it becomes less than the friction force and the block slows and eventually stops. The block stops once the shaded area above the spring force line equals that below the line, or when the work done accelerating the block equals that which decelerated it. If the spring end continues to move, loading continues until the spring force again equals the static friction force and another slip event occurs.

It is interesting to think of analogies between this model of slip events and earthquakes. The slip event's duration t_D, analogous to an earthquake rise time (Section 4.3.2), satisfies

$$\frac{du(t_D)}{dt} = 0, \quad t_D = \frac{\pi}{\omega} = \pi\sqrt{m/k}. \tag{47}$$

The total slip during the event is

$$\Delta u = u(t_D) = 2\Delta\mu\sigma/k, \tag{48}$$

and the drop in the spring force, which is analogous to an earthquake stress drop (Section 4.6.3), is

$$\Delta f = 2\Delta\mu\sigma. \tag{49}$$

Thus the rise time depends on the spring constant, but not on the difference between static and dynamic friction. However, the total slip and stress drop depend upon the friction difference. None of these depend upon the loading rate, which is analogous to the rate of plate motion causing earthquakes on a plate boundary. But the loading rate determines the time between successive slip events. Thus, in the plate boundary analogy, the time between large earthquakes depends on the plate motion rate, but their slip and stress drop depend on the frictional properties of the fault and the normal stress. Hence faster-slipping boundaries would have more frequent large earthquakes, but the slip and stress drop in them would not be greater than on a slower boundary with similar frictional properties and normal stress.

Laboratory experiments show that the difference between static and dynamic friction is more complicated than the constant values assumed in this simple model. We can think of the lower dynamic friction as showing either velocity weakening, decreasing as the object moves faster, or slip weakening, decreasing as the object moves further. Frictional models called

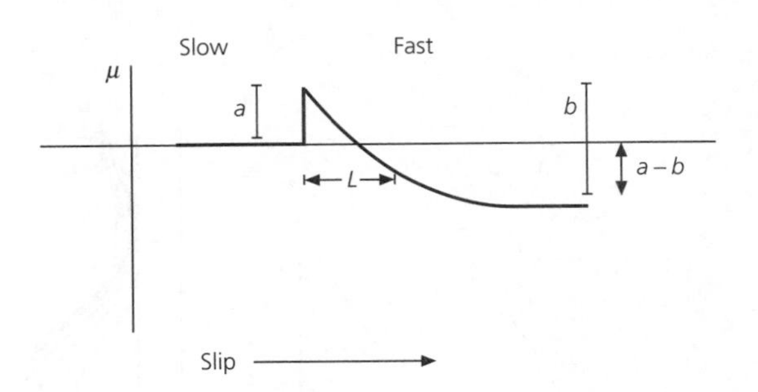

Fig. 5.7-22 Evolution of friction in a simple rate- and state-dependent model. If the slip rate increases by a factor of e, friction increases by a, and then decreases as slip progresses to a steady-state value $a - b$. (After Scholz, 1990. Reprinted with the permission of Cambridge University Press.)

rate- and state-dependent friction with a variable coefficient of sliding friction, μ, are used to describe these effects. In a simple model of this sort,

$$\mu = [\mu_0 + b\psi + a \ln (v/v^*)], \tag{50}$$

where μ_0 is the coefficient of static friction. The friction depends on the slip rate v, normalized by a rate v^*, and a state variable ψ that represents the slip history

$$\frac{d\psi}{dt} = -(v/L)[\psi + \ln (v/v^*)], \tag{51}$$

where L is an experimentally determined characteristic distance. The friction also depends on a and b, which characterize the material.

Figure 5.7-22 illustrates how friction evolves. If the slip rate increases by a factor of e, the friction increases by a, and then decreases as slip progresses, reaching a new steady-state value. With time, ψ reaches a steady-state value given by Eqn 51,

$$0 = -(v/L)[\psi_{ss} + \ln (v/v^*)], \quad \psi_{ss} = -\ln (v/v^*). \tag{52}$$

The steady state friction (Eqn 50) is

$$\mu_{ss} = [\mu_0 + b\psi + a \ln (v/v^*)] = [\mu_0 + (a - b) \ln (v/v^*)], \tag{53}$$

and varies with slip rate as

$$\frac{d\mu_{ss}}{d \ln v} = (a - b), \tag{54}$$

so after the slip velocity change, the net friction change is $(a - b)$. If $(a - b)$ is negative, the material shows velocity weakening, which permits earthquakes to occur by stick-slip. However, for $(a - b)$ positive, the material shows velocity strengthening, and stable sliding is expected. Laboratory results (Fig. 5.7-20) show that $a - b$ for granite changes sign at about 300°, which should be the limiting temperature for earthquakes. Thus the frictional

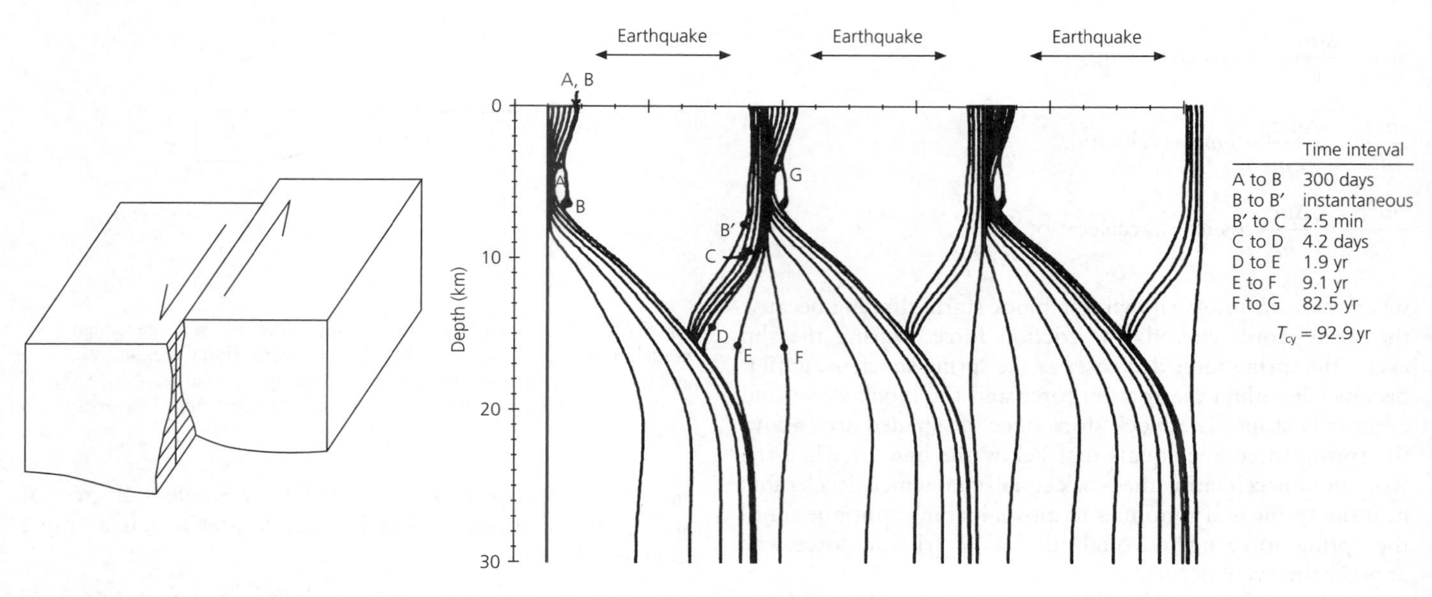

Fig. 5.7-23 Earthquake cycle for a model in which a strike-slip fault with rate- and state-dependent frictional properties is loaded by plate motion. The slip history for three cycles as a function of depth and time is shown by the lines, each of which represents a specific time. Steady motion occurs at depth, and stick-slip occurs above 11 km. (After Tse and Rice, 1986. *J. Geophys. Res.*, *91*, 9452–72, copyright by the American Geophysical Union.)

model predicts a maximum depth for continental earthquakes similar to that predicted by the rock strength arguments.

These results can be used to simulate the earthquake cycle, using fault models analogous to the simple slider model (Fig. 5.7-21). Figure 5.7-23 shows the slip history as a function of depth and time for a model in which a strike-slip fault is loaded by plate motion. The fault is described by rate- and state-dependent frictional properties as a function of depth, such that stick-slip occurs above 11 km. Initially from time A to B, stable sliding occurs at depth, and a little precursory slip occurs near the surface. The earthquake causes 2.5 m of sudden slip at shallow depths, as shown by the curves for times B and B′. As a result, the faulted shallow depths "get ahead" of the material below, loading that material and causing postseismic slip from times B′ to F. Once this is finished, the 93-year cycle starts again with steady stable sliding at depth.

Such models replicate many aspects of the earthquake cycle. An interesting difference, however, is that the models predict earthquakes at regular intervals, whereas earthquake histories are quite variable. Some of the variability may be due to the effects of earthquakes on other faults, or other segments of the same fault. Figure 5.7-24 shows this idea schematically for the slider model in Fig. 5.7-21. Assume that after an earthquake cycle, the compressive normal stress σ on the slider is reduced. This "unclamping" reduces the frictional force resisting sliding, so it takes less time for the spring force to rise again to the level needed for the next slip event. Conversely, increased compression "clamps" the slider more, and so increases the time until the next slip event. In addition, by Eqn 49, the stress drop in the slip event changes when σ changes.

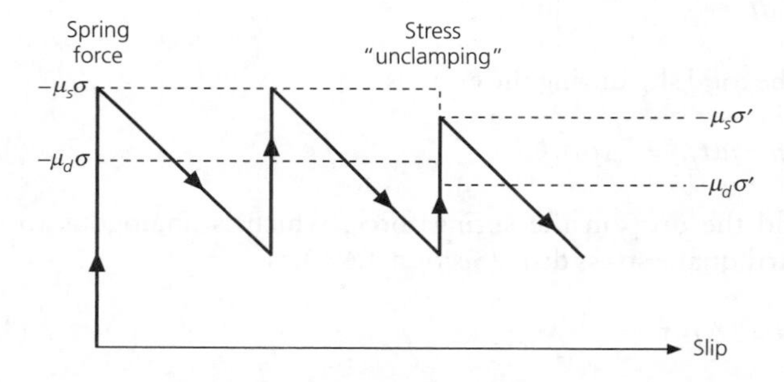

Fig. 5.7-24 Modification of a slider block model (Fig. 5.7-21) to include the effects of changes in normal stress. Reduced normal stress ($|\sigma| < |\sigma'|$) reduces the frictional force, and so "unclamps" the fault and decreases the time until the next slip event.

For earthquakes, the analogy implies that earthquake occurrence on a segment of a fault may reflect changes in the stress on the fault resulting from earthquakes elsewhere. This concept is quantified using the Coulomb–Mohr criterion (Eqn 5) that sliding can occur when the shear stress exceeds that on the sliding line (Fig. 5.7-9), or $\tau > \mu\sigma$. We can thus define the Coulomb failure stress

$$\sigma_f = \tau + \mu\sigma \tag{55}$$

such that failure occurs when σ_f is greater than zero. Whether a nearby earthquake brings a fault closer to or further from failure is shown by the change in Coulomb failure stress due to the earthquake,

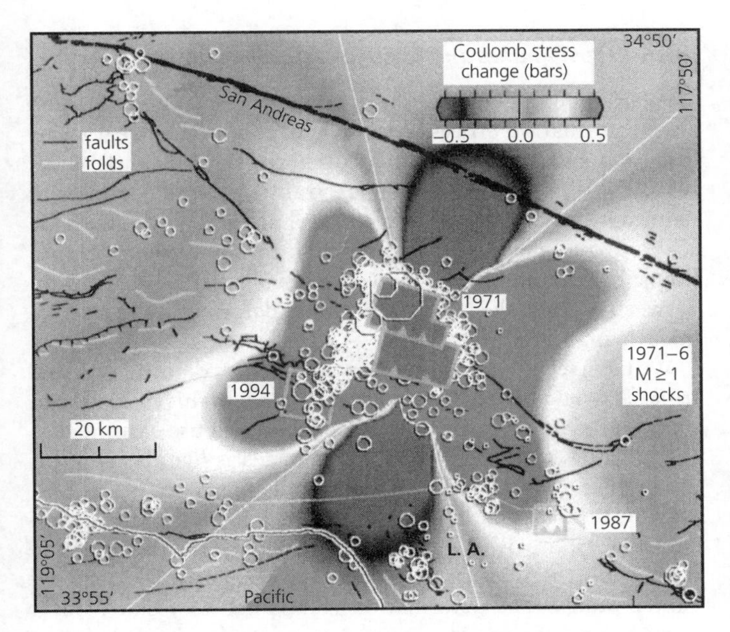

Fig. 5.7-25 Predicted changes in Coulomb failure stress due to the 1971 San Fernando earthquake. The Whittier Narrows and Northridge earthquakes subsequently occurred in regions where the 1971 earthquake increased the failure stress. (Stein *et al*., 1994. *Science, 265*, 1432–5, copyright 1994 American Association for the Advancement of Science.)

$$\Delta\sigma_f = \Delta\tau + \mu\Delta\sigma. \qquad (56)$$

Failure is favored by positive $\Delta\sigma_f$, which can occur either from increased shear stress τ or a reduced normal stress (compression is negative, so $\Delta\sigma > 0$ favors sliding).

Some earthquake observations provide support for this idea. Figure 5.7-25 shows the predicted Coulomb failure stress changes in the Los Angeles region due to the 1971 (M_s 6.6) San Fernando earthquake. The stress change pattern reflects the earthquake's focal mechanism, thrust faulting on a NW–SE-striking fault (Fig. 5.2-3). Two moderate earthquakes, the 1987 Whittier Narrows (M_L 5.9) and 1994 Northridge (M_w 6.7) earthquakes subsequently occurred in regions where the 1971 earthquake increased the failure stress, suggesting that the stress change may have had a role in triggering the earthquakes. A similar pattern has been found after other earthquakes, and some studies have found that aftershocks are concentrated in regions where the mainshock increased the failure stress. Stress triggering may explain why successive earthquakes on a fault sometimes seem to have a coherent pattern. For example, the 1999 M_s 7.4 Izmit earthquake on the North Anatolian fault (Fig. 5.6-8) appears to be part of a sequence of major (M_s 7) earthquakes over the past 60 years, which occurred successively further to the west, and hence closer to the metropolis of Istanbul.

An intriguing feature of such models is that the predicted stress changes are of the order of 1 bar, or only 1–10% of the typical stress drops in earthquakes (Section 4.6.3). Such small stress changes should only trigger an earthquake if the tectonic stress is already close to failure. However, as in the slider model (Fig. 5.7-24), stress changes can affect the time until the tectonic stress is large enough to produce earthquakes. It has been argued that the 1906 San Francisco earthquake reduced the failure stress on other faults in the area, causing a "stress shadow" and increasing the expected time until the next earthquake on these faults. This is consistent with the observation that during the 75 years before the 1906 earthquake, the area had 14 earthquakes with M_w above 6, whereas only one occurred in the subsequent 75 years. Such analyses may help improve estimates of the probability that an earthquake of a certain size will occur on a given fault during some time period. To date, such estimates have large uncertainties (Section 4.7.3), in part because of the large variation in the time intervals between earthquakes. Stress loading models, some of which incorporate rate- and state-dependent friction because simple Coulomb friction does not predict large enough changes in recurrence time, may explain some of the variations and thus reduce these uncertainties.

This discussion brings out the importance of understanding the state of stress on faults. On this issue, the friction models give some insight, but major questions remain. Earthquake stress drops estimated from seismological observations are typically less than a few hundred bars (tens of MPa). Yet, the expected strength of the lithosphere (e.g., Fig. 5.7-14–16) is much higher, in the kilobar (hundreds of MPa) range. The laboratory results (Fig. 5.7-20) and frictional models (Fig. 5.7-21) suggest an explanation for this difference, because in both the stress drop during a slip event is only a fraction of the total stress.

However, the frictional models do not explain an intriguing problem called the "San Andreas" or "fault strength" paradox. As noted in Section 5.4.1, a fault under shear stress τ slipping at rate v should generate fractional heat at a rate equal to τv. Thus, if the shear stresses on faults are as high (kbar or hundreds of MPa) as expected from the strength envelopes, significant heat should be produced. But little if any heat flow anomaly is found across the San Andreas fault (Fig. 5.7-26), suggesting that the fault is much weaker than expected. A similar conclusion emerges from consideration of stress orientation data. Although the Coulomb–Mohr model predicts that the maximum principal stress directions inferred from focal mechanisms, geological data, and boreholes should be about 23° from the San Andreas fault (Fig. 5.7-8), the observed directions are essentially perpendicular to the fault (Fig. 5.6-19), implying that the fault acts almost like a free surface. To date, there is no generally accepted explanation for these observations. The most obvious one is that the effective stress on the fault is reduced by high pore pressure, but there is discussion about whether pressures much higher than hydrostatic pressure could be maintained in the fault zone. An alternative explanation, that the fault zone is filled by low-strength clay-rich fault gouge, faces the difficulty that experiments on such material find that it has normal strength unless pore pressures are high.

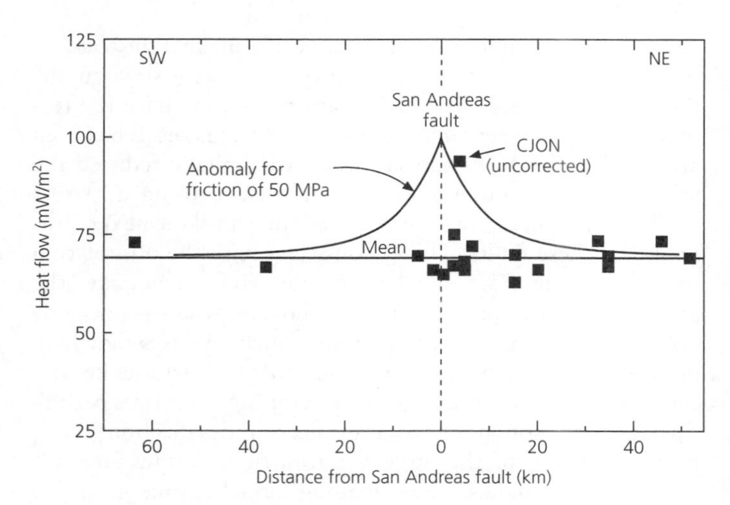

Fig. 5.7-26 Observed (squares) heat flow across the San Andreas fault. The elevated heat flow predicted by shear heating (solid line) is not observed, except for one point (CJON, Cajon pass), where alternative interpretations are possible, implying that the fault is weak. (Lachenbruch and Sass, 1988. *Geophys. Res. Lett., 15*, 981–4, copyright by the American Geophysical Union.)

In summary, ideas based on rock friction are providing important insights into earthquake mechanics. Although many issues remain unresolved, and some attractive notions remain to be fully demonstrated, rock friction seems likely to play a growing role in addressing earthquake issues.

5.7.6 Earthquakes and regional deformation

The large, rapid deformation in earthquakes is often part of a slow deformation process occurring over a broader region. As discussed in Section 5.6.2, there often appear to be differences between the seismic, aseismic, transient, and permanent deformations sampled by different techniques on different time scales. Experimental and theoretical ideas about rheology and lithospheric dynamics are being used to investigate the relation between earthquakes and the regional deformations that produce them.

We have seen that earthquakes often reflect deformation distributed over a broad plate boundary zone. In this case, we can think of the lithosphere as a viscous fluid and use earthquakes as indicators of its deformation. This idea is like the physical model (Fig. 5.6-7) that used deformable plasticine as an analogy for the deformation of Asia resulting from the Himalayan collision. Figure 5.7-27 shows such an analysis for part of the Pacific–North America plate boundary zone in the western United States. The deformation is assumed to result from a combination of forces due to the transform plate boundary and forces due to the potential energy of elevated topography, which tends to spread under its own weight. To test this idea, a continuous velocity field has been interpolated from space-geodetic, fault slip, and plate motion data (Figs 5.2-3 and 5.6-3). The velocity field is treated as being due to the motion of a viscous fluid, and is converted to a strain rate tensor field. This is then compared to the magnitude of the stress tensor inferred

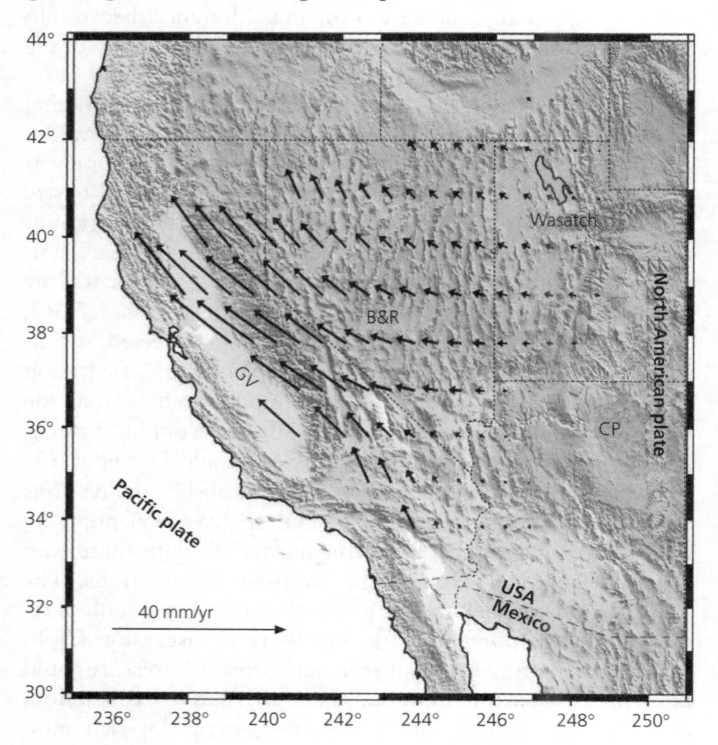

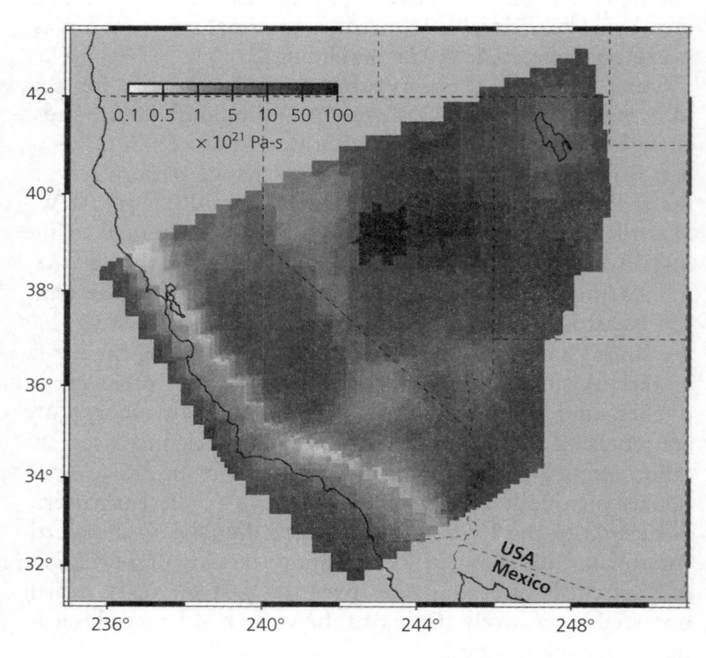

Fig. 5.7-27 *Left*: Estimated velocity field for part of the Pacific–North America plate boundary zone in the western USA. *Right*: Effective viscosity determined by dividing the magnitude of the deviatoric stress tensor by the magnitude of the strain rate tensor. (Flesch *et al.*, 2000. *Science, 287*, 834–6, copyright 2000 American Association for the Advancement of Science.)

from topography and plate boundary forces. The ratio of stress to strain rate at any point, which is the vertically averaged effective viscosity, varies significantly. Low values along the San Andreas fault and western Great Basin show that the strain rates are relatively high for the predicted stress, consistent with a weak lower crust. The Great Valley–Sierra Nevada block has little internal deformation, and thus acts relatively rigidly and appears as a high-viscosity region. Summing seismic moment tensors (Section 5.6.2) yields a seismic strain rate averaging about 60% of the inferred total strain. As discussed earlier, this discrepancy may indicate some aseismic deformation or that the 150 years of historical seismicity is too short for a reliable estimate.

Viscous fluid models can be used to study how the lithosphere deforms on different time scales. For example, as noted in Section 5.6.2, GPS data across the entire Nazca–South America plate boundary zone show faster motion than is inferred from structural geology or topographic modeling. The difference probably occurs because the GPS data record instantaneous velocities that include both permanent deformation and elastic deformation that will be recovered during future earthquakes, whereas the lower geological rates reflect only the permanent deformation. This can be modeled by representing the overriding South American plate using a simple one-dimensional system of a spring, a dashpot, and a pair of frictional plates (Fig. 5.7-28). This system approximates the behavior of the crust: the spring gives the elastic response over short periods, the dashpot gives the viscous response over geological time scales, and the frictional plates simulate the thrust faulting earthquake cycle at the trench. As plate convergence compresses the system, the stress $\sigma(t)$ increases with time until it reaches a yield strength σ_y, when an earthquake occurs, stress drops to σ_b, and the process repeats. Displacement accumulates at a rate v_0 except during earthquakes, when the displacement drops by an amount Δu. The topography and geologic data record the averaged long-term shortening rate v_c shown by the envelope of the sawtooth curve, whereas GPS data record the higher instantaneous velocity v_0. The instantaneous velocity thus results from the portion of the plate motion locked at the trench that deforms the overriding plate elastically (Fig. 4.5-14) and is released as seismic slip in interplate earthquakes. By contrast, the aseismic slip component at the trench has no effect because it does not contribute to locking on the interface and deformation of the overriding plate. Similar models are being explored for other regions where deformation appears to vary on different time scales.

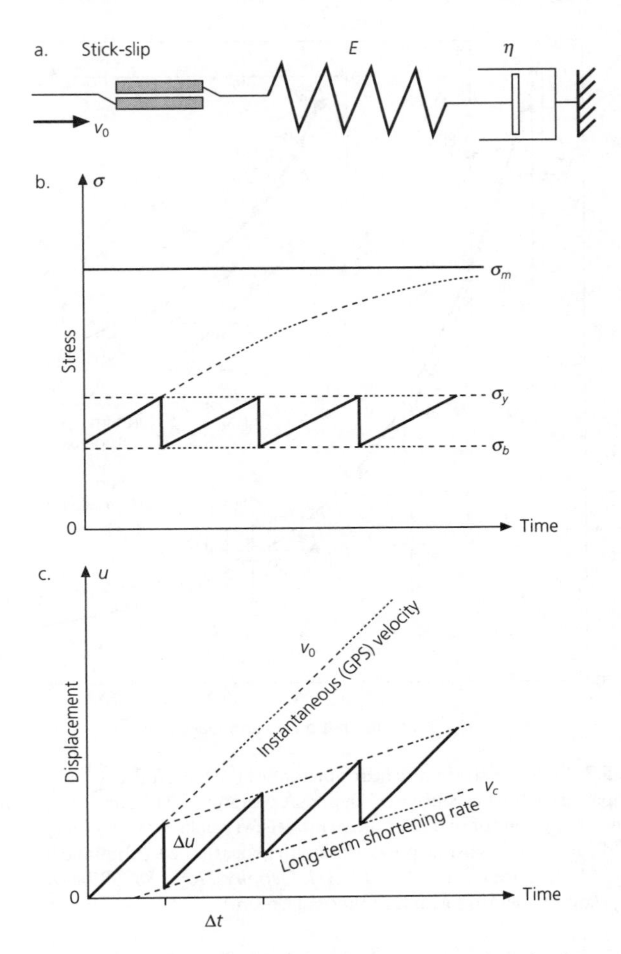

Fig. 5.7-28 a: Model for a viscoelastic–plastic crust to describe the response of the overriding South American plate to the subduction of the Nazca plate. The dashpot represents a viscous body modeling the permanent deformation, the spring represents an elastic body modeling the transient deformation, and the frictional plates represent the earthquake cycle at the trench. b: Stress evolution for the model. c: Displacement history for the model. Displacement accumulates at the instantaneous rate v_0 except during earthquakes, when slip Δu occurs. GPS data record a gradient starting at v_0 from the trench, whereas the envelope of the displacement curve v_c is the long-term shortening rate reflected in geological records and topography. (Liu *et al.*, 2000. *Geophys. Res. Lett.*, *18*, 3005–8, copyright by the American Geophysical Union.)

Viscous fluid models are also used to analyze other aspects of the earthquake cycle. For example, Fig. 5.7-29 shows the strain rate near portions of the San Andreas fault compared to the time since the last great earthquake on that portion of the fault. Postseismic motion seems to continue for a period of years after an earthquake and then slowly decays, presumably due to the steady interseismic motion. A similar picture emerges from GPS and other geodetic results following large trench thrust faulting earthquakes. For a number of years, sites near the trench on the overriding plate move seaward, showing postseismic motion consistent with the earthquake focal mechanism (Fig. 4.5-15). Eventually, however, the sites resume the landward interseismic motion usually seen at trenches (Fig. 5.6-10). Such observations are challenging to interpret, because postseismic afterslip on or near a fault can have effects at the surface similar to viscoelastic flow of the asthenosphere (Fig. 5.7-29), but offer the prospect of improving our understanding of both earthquake processes and the rheology of the lithosphere and the asthenosphere. A tantalizing possibility is that the viscous asthenosphere permits stress waves generated by large earthquakes to travel slowly for large distances and contribute to earthquake triggering.

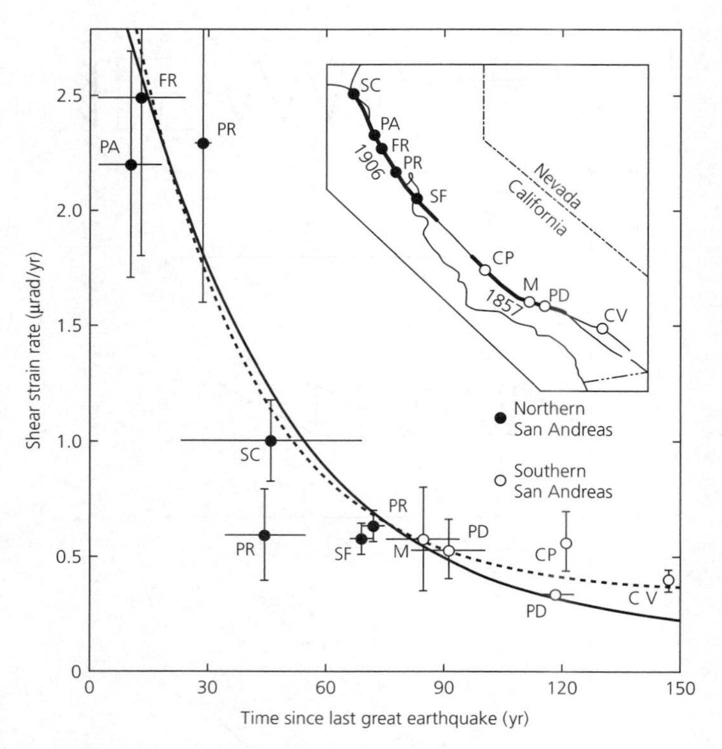

Fig. 5.7-29 Shear strain rate near portions of the San Andreas fault compared to the time since the last great earthquake. The data are similar to the predictions of two alternative models: viscoelastic stress relaxation (solid curve) and aseismic postseismic slip beneath the earthquake fault plane (dashed line). (Thatcher, 1983. *J. Geophys. Res., 88*, 5893–902, copyright by the American Geophysical Union.)

Further reading

Given the comparatively recent discovery of plate tectonics, its importance for most aspects of geology, and its crucial role in the earthquake process, many excellent sources, a few of which are listed here, offer more information about this chapter's topics.

The dramatic development of plate tectonics is discussed from the view of participants by Menard (1986) and in Cox's (1973) collection of classic papers. Basic ideas in plate tectonics are treated in most introductory and structural geology texts. More detailed treatments include Uyeda (1978), Fowler (1990), Kearey and Vine (1990), and Moores and Twiss (1995). Cox and Hart (1986) present the basic kinematic concepts, and global plate motion models are discussed by Chase (1978), Minster and Jordan (1978), and DeMets *et al.* (1990).

Thermal and mechanical aspects of plate tectonics are discussed by Turcotte and Schubert (1982) and Sleep and Fujita (1997). Mid-ocean ridge tectonics and structure are discussed by Solomon and Toomey (1992) and Nicolas (1995). The thermal evolution of oceanic lithosphere is discussed by Parsons and Sclater (1977) and Stein and Stein (1992); McKenzie (1969) presents the subduction zone thermal model we follow. Papers in Bebout *et al.* (1996) cover many aspects of subduction, and Kanamori (1986) reviews subduction zone thrust earthquakes. Lay (1994) treats the nature and fate of subducting slabs, and deep earthquakes are reviewed by Frohlich (1989), Green and Houston (1995), and Kirby *et al.* (1996b). For a derivation of the ridge push force see Parsons and Richter (1980); Wiens and Stein (1985) discuss its application to oceanic intraplate stresses. Yeats *et al.* (1997) cover a wide variety of topics about the relation of earthquakes to regional geology. Rosendahl (1987) reviews continental rifting. Papers in Gregersen and Basham (1989) treat aspects of passive margin and continental interior earthquakes with emphasis on postglacial effects.

Concepts in continental deformation are treated by Molnar (1988) and England and Jackson (1989); Gordon (1998) gives an overview of plate rigidity and diffuse plate boundaries. Applications of space geodesy to tectonics are reviewed by papers in Smith and Turcotte (1993) and by Dixon (1991), Gordon and Stein (1992), and Segall and Davis (1997). Many GPS data and results, including an overview brochure, can be found on the University NAVSTAR Consortium WWW site http://www.unavco.org. Stress maps and their interpretations are discussed by Zoback (1992) and other papers in the same journal issue; stress maps are available at the World Stress Map project WWW site http://www-wsm.physik.uni-karlsruhe.de.

Mantle plumes in general are reviewed by Sleep (1992); Nataf (2000) and Foulger *et al.* (2001) discuss seismic imaging of plumes; Smith and Braile (1994) discuss the Yellowstone hot spot; and Stein and Stein (1993) discuss oceanic hot spot swells. Papers in Peltier (1989) treat many aspects of mantle convection; Silver *et al.* (1988) explore the relationship between subduction, convection, and mantle structure; and Christensen (1995) reviews the effects of phase transitions on mantle convection. The heat engine perspective on global tectonics is discussed by Stacey (1992), and Ward and Brownlee (2000) summarize the arguments advocating a crucial role for plate tectonics in the origin and survival of life on Earth.

Topics involving rock mechanics, flow, and their tectonic applications are discussed by Jaeger (1970), Weertman and Weertman (1975), Jaeger and Cook (1976), Turcotte and Schubert (1982), Kirby (1983), Kirby and Kronenberg (1987), and Ranalli (1987). Scholz (1990) and Marone (1998) cover topics dealing with the relation of rock mechanics to earthquakes, with special emphasis on rock friction. Our treatment of the slider model for faulting follows Scholz (1990). Related topics, including issues of continental deformation and fault strength, are also treated by papers in Evans and Wong (1992). Stein (1999) summarizes the concept of stress triggering of earthquakes.

Problems

1. Assume that Pacific–North America plate motion along the San Andreas fault occurs at 35 mm/yr.
   (a) If all this motion occurs seismically in earthquakes about 22 years apart, which is a typical recurrence interval for the Parkfield fault segment, how much slip would you expect in the earthquakes? From Fig. 4.6-7, estimate likely fault lengths and magnitudes for such earthquakes.
   (b) Give similar estimates if the earthquakes occur about 132 years apart, as at Pallett Creek.
2. Assume that all the earthquakes in the Pallett Creek sequence (Fig. 1.2-15) involved 4 m of seismic slip. Using the time interval from the present to the 1857 earthquake, calculate the seismic slip rate on this portion of the San Andreas fault. Next, do so by averaging the recurrence intervals for the past two earthquakes (1857 and 1812), the past three, and so on for the entire earthquake history. What are the implications of this simple experiment for seismic slip estimates? What other sources of uncertainty should also be considered, and how might they affect this estimate?
3. (a) Use Table 5.2-1 to find the rate that the Juan de Fuca plate subducts beneath North America at 46°N, 125°W.
   (b) If all this motion occurs in large earthquakes, how often would you expect an earthquake if the slip in each were 5 m? How would this estimate change if the slip were 10 or 20 m?
   (c) How would the answers to (b) change if only 25% or 50% of the plate motion occurred by seismic slip?
   (d) Paleoseismic observations and historic records of a tsunami imply that this subduction zone has had very large earthquakes approximately 500 years apart. Suggest some possibilities in view of parts (a)–(c). How might you attempt to distinguish between them?
   (e) The crust subducting at this trench is about 10 million years old. Given the convergence rate and the observations from other trenches in Fig. 5.4-30, what might you infer about the moment magnitude of the largest earthquake expected here? Find the corresponding seismic moment and suggest a plausible fault geometry and amount of slip that would also be consistent with the paleoseismic and plate motion observations.
4. For rigid plates, Eqn 5.2.10 shows that we can find the *angular* velocity vector of one plate from the sum of two others. Show that at a point we can also do this for the *linear* velocity vectors.
5. The news media sometimes ask "How large would the largest possible earthquake be?" Estimate the seismic moment and moment magnitude by assuming that all the trenches in the world (48,000 km) slip at the same time, that 10 m of slip occurs, and the fault width is 250 km.
6. Estimate the thermal Reynolds number R defined in Eqns 5.3.19 and 5.4.3, assuming that $\kappa = 10^{-6}\ m^2s^{-1}$. What does this estimate imply about the processes of plate cooling and subduction?
7. Assume that oceanic lithosphere has a thermal conductivity of 3.1 $Wm^{-1}{}^{\circ}C^{-1}$.
   (a) Find the heat flow for old oceanic lithosphere, assuming a linear temperature gradient (Fig. 5.3-8), a basal temperature of 1450°C, and a plate thickness of 95 km.
   (b) How would this value change for a basal temperature of 1350°C and plate thickness 125 km?
   (c) If the lithosphere under a midplate region were thinned to 50 km while the basal temperature remained 1350°, what would the heat flow be, assuming a linear temperature gradient?
8. A way to get insight into the physics of subduction is to use a classic result from fluid mechanics, called Stokes' problem, which describes the terminal velocity v at which a sphere of radius a and density ρ sinks due to gravity in a fluid with viscosity η and lower density ρ'. The result is $v = 2ga^2(\rho - \rho')/9\eta$. Estimate the subduction velocity of a slab assuming the slab is a sphere with radius equal to half its thickness. To do this, estimate the density contrast from the thermal model (Eqn 5.4.14) and a coefficient of thermal expansion $\alpha = 3 \times 10^{-5}\ {}^{\circ}C^{-1}$. Use a mantle viscosity from Section 5.5.3. Because this is a back-of-the-envelope calculation, there is no correct answer, but you should be able to come up with something reasonable (within an order of magnitude or two of reality).
9. The result that a subducting slab that reaches the core should still be thermally distinct (Fig. 5.4-5) may seem surprising. For another estimate, use the one-dimensional cooling equation in Section 5.3.2 to estimate how long a slab should need to warm up to 90% of the ambient lowermost mantle temperatures, assuming that it were immediately transported to the base of the mantle and that $\kappa = 10^{-6}\ m^2s^{-1}$.
10. Using the definition of the slab pull force (Eqn 5.4.15):
    (a) Write the force in terms of the age of the subducting plate.
    (b) Explain whether this force would be greater or smaller, and why, for increased values of subducting plate age, coefficient of thermal expansion, and thermal diffusivity.
11. Assume that in a subducted slab the depth of the spinel–perovskite phase transition deepens from its usual 660 km outside the slab to 700 km, and that the core of the slab is 800° colder than the surrounding mantle. What is the Clapeyron slope of the phase change?
12. The surface of Venus is much hotter (450°C) than that of Earth. If Venus had plate tectonics and the rocks were similar, so that the temperature gradient in old lithosphere there were the same as on Earth, how would the thickness of the "oceanic" lithosphere differ? How would the slab pull and ridge push forces differ? What other differences might you expect?
13. Express the ratio of the slab pull (Eqn 5.4.15) and ridge push (Eqn 5.5.6) forces. Explain why this ratio depends on thermal diffusivity. Estimate this ratio near a trench where old oceanic lithosphere is subducting, assuming that $\kappa = 10^{-6}\ m^2s^{-1}$.
14. To see if momentum can be responsible for the Indian plate's northward motion long after its collision with Asia began, estimate the momentum of the Indian plate and that of an ocean liner, and compare the two.
15. Use Mohr's circle to show why
    (a) Rocks at depth do not fracture under lithostatic pressure alone.
    (b) The deviatoric stress needed for fracture increases at greater depth.
16. Suppose that a rock is stressed close to its brittle limit. Show graphically which will make the rock fracture sooner: (a) increasing σ_1 or (b) decreasing σ_2 by the same amount (assume a two-dimensional case where σ_1 and σ_2 are both negative, and internal friction exists).
17. Suppose that the fracture line for a particular rock is $\tau = 80 - 0.5\sigma$, where stresses are in MPa. What angle would the normal to a fracture plane make with σ_1? If σ_1 is 400 MPa at failure, what is σ_2?
18. For the slider block earthquake model in Section 5.7.5:
    (a) Derive an expression for the time between successive slip events.

(b) Sketch the force–slip diagram for two different spring constants, and use the sketch to explain how the slip and force drop in a slip event change and why.

(c) For the slider block model, formulate a quantity analogous to an earthquake's seismic moment, and explain why it depends on each term. What is the major difference between this quantity and the seismic moment?

(d) Recall the observation (Fig. 4.6-11) that earthquake stress drops are similar for a wide range of earthquakes. If the slider block model is relevant, what does this imply?

(e) What conditions might correspond to aseismic slip, which could be viewed as the limit of a continuous series of very small slip events?

Computer problems

C-1. (a) Write a subroutine to compute the rate and azimuth of plate motion at a point, given the location and an Euler vector in the form (pole latitude, longitude, magnitude).

(b) Use the Euler vector in Table 5.2-1 to test your program on the San Andreas and Aleutian site examples in Section 5.2.1.

C-2. (a) Find the rate and azimuth of Cocos–North America plate motion at 18.3°N, 102.5°W.

(b) This location is the epicenter of a large 1985 Mexican earthquake, whose mechanism had nodal planes whose strike and dip are (127°, 81°) and (288°, 9°). Infer from the tectonics of the Middle American trench which plane was the fault plane. Using the methods of Section 4.2, determine the azimuth of slip during the earthquake. How does this compare to your predicted azimuth?

C-3. (a) Write a subroutine to add and subtract two Euler vectors given in the form (pole latitude, longitude, magnitude). The output should be a Euler vector in the same form.

(b) Use your program to determine the absolute Euler vector for the Pacific plate using Table 5.2-1.

(c) Determine the rate and azimuth of absolute plate motion at Hawaii (Fig. 5.2-7). Compare the direction to the Hawaiian–Emperor seamount chain.

C-4. Write a program to plot the temperature distribution in the oceanic lithosphere as a function of age using the cooling half-space thermal model (Eqn 5.3.4). Compute erf(s) (Eqn 5.3.3) using either available software or numerical integration as discussed in problem 4C-6.

C-5. (a) Write a program to plot the temperature distribution in a subducting slab using the analytic thermal model (Eqn 5.4.3). Compute it for a plate subducting at 80 mm/yr at an angle of 45°. Make assumptions that seem reasonable and justify them.

(b) Change the program to make the age of the subducting plate a parameter and generate temperature fields for different slabs, as in Fig. 5.4-6.

(c) Using the results of (b) and Fig. 5.4-4, estimate a temperature above which deep earthquakes are not observed.

6 Seismograms as Signals

We shall introduce the concepts of signal and noise. We define the signal as the desired part of the data and the noise as the unwanted part. Our definition of signal and noise is subjective in the sense that a given part of the data is "signal" for those who know how to analyze and interpret the data, but it is "noise" for those who do not. For example, for many years the times of the first arrivals of P- and S-waves were the only signals conveyed by an earthquake, and the rest of the seismogram, such as surface waves and coda waves, had to be considered as useless until appropriate methods of interpretations were found.

Thus, through the application of a new technique to old data, an analyst can experience a moment of discovery as joyful as a data gatherer does using a new observational device.

Aki and Richards, *Quantitative Seismology*, 1980

6.1 Introduction

Seismology uses various techniques to study the displacement field as a function of position and time associated with elastic waves in the earth, and to draw inferences from it about the nature of seismic sources and the earth. Although some techniques depend on specific aspects of seismic waves in the earth, others rely on general properties of functions of space and time.

We thus often use a class of techniques known as *signal processing* or *time series analysis*. Signal processing considers functions of time or space, also called series or signals, in general terms without regard to the specific physics involved. As a result, many wave propagation subjects, including seismology, radar, sonar, and optics, can be treated in similar ways. The signals can have different forms. For example, in seismology, we can treat either a continuous (*analog*) record of ground motion or the *digital* data that result from representing the ground motion as being *sampled* at discrete intervals, providing numbers that can be manipulated using a computer.

In general terms, we can think of *filtering* a signal, or applying some operation that modifies the signal. We have already discussed several examples. A seismometer is a filter, in that it yields a record of ground motion that differs from the actual ground motion. Similarly, processes in the earth such as dispersion or attenuation have effects that can be described as a filter acting on the wave field. We can also consciously apply filters to enhance parts of a seismogram or seismic wave field and suppress others. In this chapter we extend these ideas by considering mathematical approaches that are common to such applications and then seeing how these approaches give additional insight into the physical processes. We discuss some basic concepts and provide references at the end of the chapter for more extensive treatments.

6.2 Fourier analysis

6.2.1 *Fourier series*

In many applications, we use an approach based on the idea that any time series can be decomposed into the sum or integral of harmonic waves of different frequencies, using methods known as *Fourier analysis*. We derived the properties of seismic waves using a harmonic wave, a sinusoid of a single frequency, and noted that any wave could be treated as the sum of harmonic waves. Thus we showed that waves on a string could be viewed as the sum of the string's normal modes, or standing waves (Section 2.2.5), and that waves in a spherical earth can be written as the sum of the earth's normal modes (Section 2.9). This concept is especially useful when the components with various frequencies behave differently. For example, surface waves of different frequencies have different apparent velocities (Section 2.8) and seismic wave attenuation varies with frequency (Section 3.7). Similarly, we will see shortly that seismometers respond differently to ground motion of different frequencies. Fourier analysis lets us decompose the signal into harmonic waves, consider each harmonic wave separately, and then recombine the harmonic waves. Thus we use this approach to analyze situations where the effect of the earth or a seismometer can be described by a filter. We also use Fourier

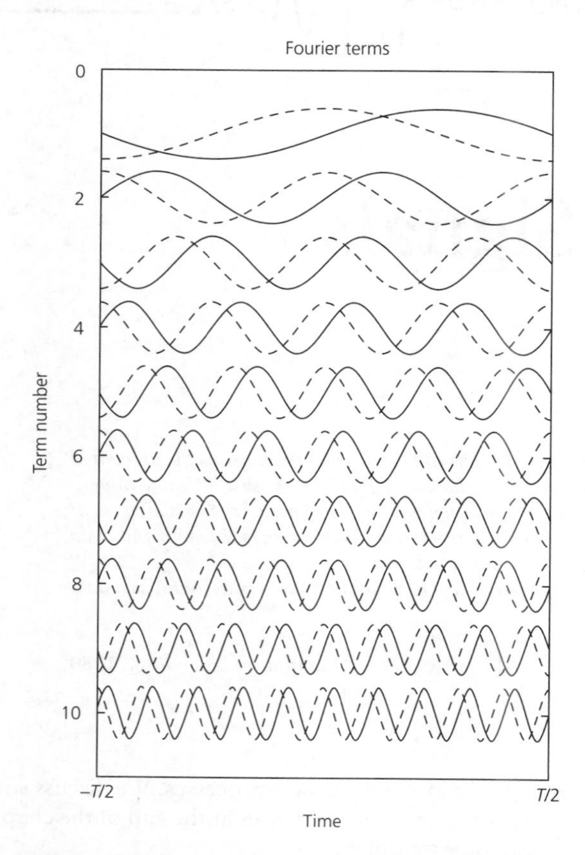

Fig. 6.2-1 Successive terms of a Fourier series. Solid lines are sin $(2n\pi t/T)$; dashed lines are cos $(2n\pi t/T)$.

analysis to filter a signal when the part that interests us overlaps with a part that does not in the time or space domains, but the two can be separated in the frequency or wavenumber domains.

We first consider the decomposition of a signal with a finite duration into a *Fourier series*, or sum of harmonic components with different frequencies. We will see later that as the duration of the signal becomes infinite, the Fourier series becomes the *Fourier transform* integral.

The Fourier series for an arbitrary function of time $f(t)$ defined over the interval $-T/2 < t < T/2$ is

$$f(t) = a_0 + \sum_{n=1}^{\infty} a_n \cos\left(\frac{2n\pi t}{T}\right) + \sum_{n=1}^{\infty} b_n \sin\left(\frac{2n\pi t}{T}\right). \tag{1}$$

This series decomposes $f(t)$ into a sum of Fourier terms that are sine and cosine functions with different periods, because sin $(2n\pi t/T)$ and cos $(2n\pi t/T)$ are periodic with period T/n, or frequency n/T (Fig. 6.2-1). Larger values of n correspond to shorter periods, or higher frequencies. For $n = 0$, the cosine term equals 1 for all values of t, and there is no sine term, because it would be zero.

The sine and cosine Fourier terms are a set of *orthogonal functions*, which means that the integral of the product of two different ones over the interval from $-T/2$ to $T/2$ is always zero:

$$\int_{-T/2}^{T/2} \sin\left(\frac{2m\pi t}{T}\right) \sin\left(\frac{2n\pi t}{T}\right) dt = \frac{T}{2}\,\delta_{mn}(1 - \delta_{m0}), \tag{2}$$

$$\int_{-T/2}^{T/2} \cos\left(\frac{2m\pi t}{T}\right) \cos\left(\frac{2n\pi t}{T}\right) dt = \frac{T}{2}\,\delta_{mn}(1 + \delta_{m0}), \tag{3}$$

$$\int_{-T/2}^{T/2} \cos\left(\frac{2n\pi t}{T}\right) \sin\left(\frac{2m\pi t}{T}\right) dt = 0 \quad \text{for all } m, n, \tag{4}$$

where the Kronecker delta, δ_{mn}, equals 1 for $m = n$ and 0 otherwise (Eqn A.3.37). For the special case $m = n = 0$, the integral in Eqn 2 is zero, and the integral in Eqn 3 is twice the value for any other $m = n$.[1]

To express the Fourier series for a given function, we solve for the coefficients a_n and b_n by multiplying both sides of Eqn 1 by the appropriate sine or cosine term and integrating from $-T/2$ to $T/2$. For example, to find the coefficient a_k, where k is some particular integer, we multiply by cos $(2k\pi t/T)$ and integrate to get

$$\int_{-T/2}^{T/2} \cos\left(\frac{2k\pi t}{T}\right) f(t)\, dt =$$

$$\int_{-T/2}^{T/2} \cos\left(\frac{2k\pi t}{T}\right) \left[a_0 + \sum_{n=1}^{\infty} a_n \cos\left(\frac{2n\pi t}{T}\right) + \sum_{n=1}^{\infty} b_n \sin\left(\frac{2n\pi t}{T}\right)\right] dt. \tag{5}$$

By the orthogonality relations (Eqns 2–4), the only term in the sums on the right-hand side whose contribution to the integral is nonzero is cos $(2\pi kt/T)$, so the equation simplifies to

$$\int_{-T/2}^{T/2} \cos\left(\frac{2k\pi t}{T}\right) f(t)\, dt = a_k \int_{-T/2}^{T/2} \cos^2\left(\frac{2k\pi t}{T}\right) dt = \frac{T}{2} a_k (1 + \delta_{k0}), \tag{6}$$

which shows that the coefficient a_k is

$$a_k = \frac{2 - \delta_{k0}}{T} \int_{-T/2}^{T/2} \cos\left(\frac{2k\pi t}{T}\right) f(t)\, dt. \tag{7}$$

[1] The proofs of Eqns 2–4 are left for the problems.

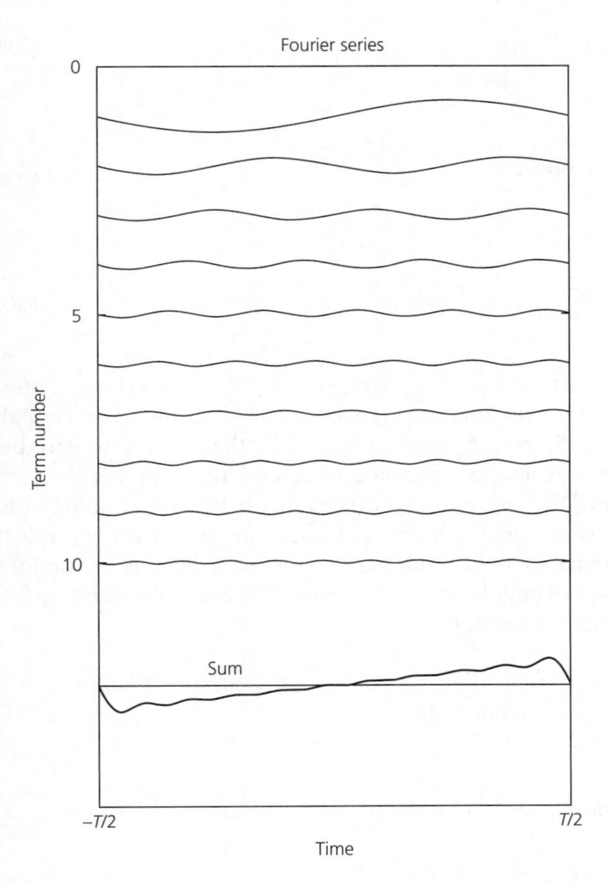

Fig. 6.2-2 The first ten terms of the Fourier series for a ramp function. The terms are weighted by their coefficients and then summed. The first ten terms give a reasonably good representation of the time function, but more terms would do better.

The a_0 term is simply

$$a_0 = \frac{1}{T} \int_{-T/2}^{T/2} f(t)\, dt, \tag{8}$$

which corresponds to the average value of the function. The coefficients of the sine terms are found similarly by

$$b_k = \frac{2}{T} \int_{-T/2}^{T/2} \sin\left(\frac{2k\pi t}{T}\right) f(t)\, dt. \tag{9}$$

Mathematically, what we have done is to consider the function $f(t)$ as being in a vector space whose basis vectors (Section A.3.6) are the sine and cosine Fourier terms. The coefficients a_k and b_k are the components that describe the particular vector $f(t)$. Thus, multiplying each basis function by the appropriate coefficient and then summing yields the function. Similarly, the operation of finding the coefficients using the integrals in Eqns 7–9 corresponds to finding each component of a vector by taking the scalar product with the appropriate unit basis vector (Eqn A.3.27).

Figure 6.2-2 illustrates this idea for a ramp function $f(t) = t/T$. Performing the integrations in Eqns 7–9 gives $a_k = 0$ and $b_k = (-1)^{k+1}/k\pi$. The cosine terms are zero, because the function is odd ($f(t) = -f(-t)$), whereas cosine is an even function ($f(t) = f(-t)$). Conversely, if the function were even, the Fourier series would include only cosine terms. Adding up the first ten sine terms reproduces the ramp reasonably well. If more terms were used, the ramp would be reproduced even better. The terms with small k are longer-period functions, and so describe the long-period features of the time series, whereas those with larger k reproduce the shorter-period features.

We used the Fourier series to express waves on a string as the sum of the string's normal modes (Section 2.2.5). Each normal mode has a spatial eigenfunction, which is a Fourier term, and an eigenfrequency. The amplitude of each Fourier term depends on the source that generated the waves, so different waves are represented by differently weighted sums of the Fourier terms. For the string the Fourier series described the variation of a function in space along a finite string, whereas here we use it to describe the variation of a function of time over a finite period. Because waves are functions of both time and space, Fourier analysis can be used for either variable or both. Fourier series are also used in other geophysical applications to represent functions that vary in space or time over finite domains. For example, we used Fourier series to describe the temperature fields in cooling oceanic lithosphere (Eqn 5.3.19) and in subducting plates (Eqn 5.4.3).

6.2.2 *Complex Fourier series*

The Fourier series (Eqn 1) can be written in a simpler form. First, we use the angular frequencies $\omega_n = 2n\pi/T$, expand the sine and cosine functions into complex exponentials, and regroup terms as

$$f(t) = a_0 + \frac{1}{2} \sum_{n=1}^{\infty} [(a_n - ib_n)e^{i\omega_n t} + (a_n + ib_n)e^{-i\omega_n t}]. \tag{10}$$

Then we use the definitions of the coefficients in Eqns 7–9, again expanding the sine and cosine functions into complex exponentials:

$$(a_n - ib_n)/2 = \frac{1}{T} \int_{-T/2}^{T/2} [\cos \omega_n t - i \sin \omega_n t] f(t)\, dt$$

$$= \frac{1}{T} \int_{-T/2}^{T/2} e^{-i\omega_n t} f(t)\, dt$$

$$(a_n + ib_n)/2 = \frac{1}{T}\int_{-T/2}^{T/2} [\cos \omega_n t + i \sin \omega_n t] f(t)\, dt$$

$$= \frac{1}{T}\int_{-T/2}^{T/2} e^{i\omega_n t} f(t)\, dt. \quad (11)$$

Next, we define

$$F_n = (a_n - ib_n)/2, \quad F_0 = a_0, \quad \text{and} \quad F_{-n} = (a_n + ib_n)/2, \quad (12)$$

so that the Fourier series becomes

$$f(t) = F_0 + \sum_{n=1}^{\infty} F_n e^{i\omega_n t} + \sum_{n=1}^{\infty} F_{-n} e^{-i\omega_n t}. \quad (13)$$

Because $-\omega_n = -2n\pi/T = \omega_{-n}$ and F_{-n} is the complex conjugate of F_n, $(F_{-n} = F_n^*)$, the negative exponentials can be written

$$\sum_{n=1}^{\infty} F_{-n} e^{-i\omega_n t} = \sum_{n=-1}^{-\infty} F_n e^{i\omega_n t}. \quad (14)$$

Making these substitutions in Eqn 10 yields the Fourier series in complex number form:

$$f(t) = \sum_{n=-\infty}^{\infty} F_n e^{i\omega_n t}, \quad (15)$$

with components

$$F_n = \frac{1}{T}\int_{-T/2}^{T/2} f(t)\, e^{-i\omega_n t}\, dt. \quad (16)$$

6.2.3 *Fourier transforms*

The complex Fourier series, which represents a function of time in terms of a sum over discrete angular frequencies ω_n, can be extended into the *Fourier transform* that represents the function as an integral over a continuous range of angular frequencies. Thus, although we used the Fourier series to describe the discrete normal modes of a finite string and the earth, we use the Fourier transform in most seismological applications, because we regard the waves as continuous functions of angular frequency.

To do this, we write Eqn 15 as

$$f(t) = \sum_{n=-\infty}^{\infty} F_n e^{i\omega_n t} \Delta n \quad (17)$$

(because $\Delta n = 1$), and define the difference between the successive angular frequencies

$$\Delta\omega = (2\pi/T)\Delta n \quad (18)$$

so that

$$\Delta n = (T\Delta\omega)/(2\pi) \quad (19)$$

and

$$f(t) = \sum_{n=-\infty}^{\infty} F_n (T/2\pi) e^{i\omega_n t} \Delta\omega. \quad (20)$$

Next, we let the period T over which $f(t)$ is defined go to infinity, so that the angular frequencies ω_n become close enough that the discrete ω_n can be replaced by the continuous variable ω. As a result, $\Delta\omega$ becomes $d\omega$, and the sum becomes an integral. We assert (note the difference between seismology and mathematics texts) that this can be done such that the product TF_n remains finite and can be replaced by the continuous function of angular frequency $F(\omega)$. The Fourier series (Eqn 20) becomes the integral

$$f(t) = \frac{1}{2\pi}\int_{-\infty}^{\infty} F(\omega) e^{i\omega t} d\omega, \quad (21)$$

and the expression for the coefficients (Eqn 16) becomes

$$F(\omega) = \int_{-\infty}^{\infty} f(t) e^{-i\omega t} dt. \quad (22)$$

Equation 22 is called the *Fourier transform*, and Eqn 21 is the *inverse Fourier transform*. These can be defined in alternate ways by interchanging the signs on the exponentials and placing the $1/2\pi$ before either integral.

It may seem strange that by starting with a real function of time $f(t)$ we obtain the transform $F(\omega)$, which is a complex function of angular frequency. The idea of negative angular frequencies may also seem disturbing. In a sense the two offset each other — we obtain a real time function by integrating a complex transform over both positive and negative angular frequencies.

An important feature of the transform and inverse transform is that their dimensions are different. For example, if $f(t)$ is a seismogram that has the dimensions of displacement, its transform $F(\omega)$ has the dimensions of displacement multiplied by time (from the dt term). Thus, if $f(t)$ gives ground motion in centimeters, $F(\omega)$ gives the transform of ground motion in centimeter-seconds.

The Fourier transform, a complex-valued function of angular frequency, can be written in terms of two real-valued functions of angular frequency:

$$F(\omega) = |F(\omega)|\, e^{i\phi(\omega)}, \quad (23)$$

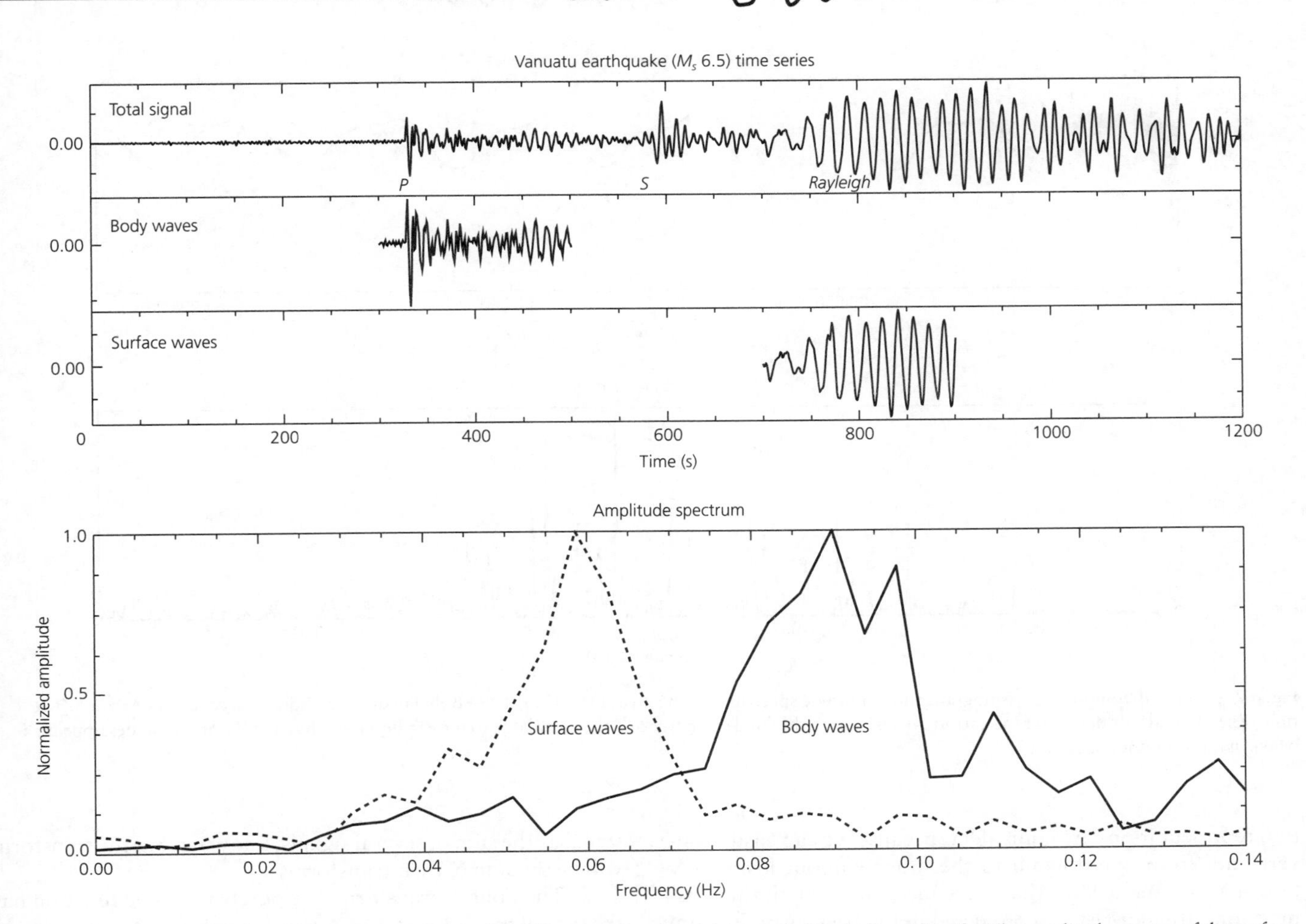

Fig. 6.2-3 Vertical-component seismogram for a moderate-sized ($M_s = 6.5$) earthquake recorded in the South Pacific. The amplitude spectra of the surface waves and a portion of the body waves, obtained by transforming different portions of the seismogram into the frequency domain, show that the surface waves contain longer-period energy than the body waves.

where

$$|F(\omega)| = [F(\omega)F^*(\omega)]^{1/2} = [\mathrm{Re}^2\,(F(\omega)) + \mathrm{Im}^2\,(F(\omega))]^{1/2} \tag{24}$$

is called the *amplitude spectrum*, and

$$\phi(\omega) = \tan^{-1}(\mathrm{Im}\,(F(\omega))/\mathrm{Re}\,(F(\omega))) \tag{25}$$

is the *phase spectrum*.[2]

Both the amplitude and the phase spectra are needed to fully represent the transform, which is also called the complex spectrum. In many applications only the amplitude spectrum is shown, because it indicates how the energy (the square of the amplitude) in the time series depends on frequency. Figure 6.2-3 shows a seismogram for a moderate-size earthquake, together with amplitude spectra for the body and surface wave portions of the seismogram. Looking at the seismogram, we see that the surface waves contain longer-period energy than the body waves. The spectra demonstrate this: the body wave is dominated by energy with frequencies between 0.1 and 0.08 Hz (periods of 10–12 s), whereas the surface wave is dominated by energy with frequencies between 0.07 and 0.05 Hz (periods of 14–20 s). For comparison, Fig. 6.2-4 shows data for a much larger earthquake. The seismogram, from an instrument designed to record at long periods, covers seven days after the earthquake. The large oscillations with periods of about 90,000 s are tides within the solid earth. Superimposed on these is the signal due to the earthquake. The portion of the amplitude spectrum shown indicates the presence of energy at long periods (0.002 Hz corresponds to 500 s period). The energy is concentrated at discrete peaks, corresponding to the earth's normal modes.

The Fourier transform $F(\omega)$ is another way of representing the time series $f(t)$. We speak of $f(t)$ as being in the "time domain," and $F(\omega)$ as being in the "frequency domain." The

[2] The notations Re and Im indicate the real and imaginary portions of a complex number (Section A.2).

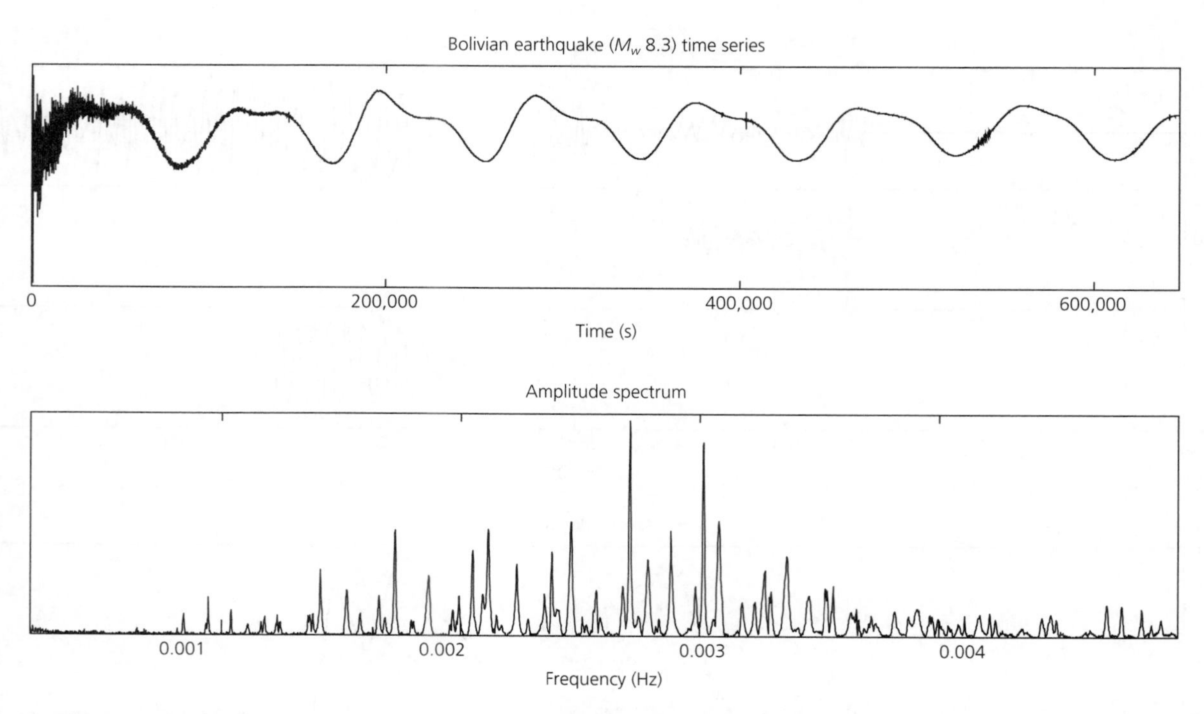

Fig. 6.2-4 Vertical-component seismogram and amplitude spectrum for the great ($M_w = 8.3$) 1994 Bolivian deep earthquake recorded in Arizona. The time series extends for days after the earthquake, showing the solid earth tide and the signal due to the earthquake. The earth's normal modes appear as peaks in the amplitude spectrum.

two representations are equivalent, because we can easily convert data from one domain to the other without losing any information. We will see that some methods of analyzing seismograms are more easily conducted in the frequency domain, and that there is a relation between time and frequency domain operations.

The Fourier transform and inverse transform relate a function of time $f(t)$ and its transform $F(\omega)$, a function of angular frequency. Similar relations apply between other pairs of variables. In seismology, the other commonly used pair is distance and wavenumber. Because the wavenumber is the spatial frequency (Section 2.2.2), it is related to distance in the same way that angular frequency is related to time. Hence, there are applications in which a double Fourier transform is taken to convert a set of seismograms, which describe displacement as a function of space and time, into a function of wavenumber and frequency (Section 3.3.5). A triple Fourier transform can similarly be taken for data in two space dimensions and time.

6.2.4 *Properties of Fourier transforms*

The Fourier transform has a number of interesting properties that we often use, whose proofs are left for the problems.

(1) The Fourier transform is linear: if $F(\omega)$ and $G(\omega)$ are the transforms of $f(t)$ and $g(t)$, then $(aF(\omega) + bG(\omega))$ is the transform of $(af(t) + bg(t))$. This property makes the Fourier transform useful in filtering, because it permits us to treat a signal as the sum of several signals, knowing that the transform will be the sum of their transforms.

(2) The Fourier transform of a purely real time function has the symmetry

$$F(-\omega) = F^*(\omega). \tag{26}$$

Thus for seismograms (which are real because the motion of the ground is purely real), the values of the transform for the negative frequencies can be found from those for positive frequencies. Hence, in filtering seismograms, we can operate on only the positive frequencies and compute the value of the transform at the negative frequencies by taking the conjugate, thus saving computer time and storage space.

(3) The Fourier transform of a time series shifted in time is found by changing the phase of the transform: if the transform of $f(t)$ is $F(\omega)$, the transform of $f(t-a)$ is $e^{-i\omega a}F(\omega)$. In analyzing seismograms it is arbitrary what time we choose as the origin; the amplitude spectrum stays the same, and the phase changes in a simple way. This makes sense, because in the absence of attenuation a wave keeps its shape but changes in phase as it propagates. Similarly, shifting a Fourier transform in frequency causes a phase change in the corresponding time series: the inverse transform of $F(\omega - a)$ is $e^{iat}f(t)$. These relations are sometimes called *shift theorems*.

(4) The Fourier transform of the derivative of a time function is found by multiplication: $(i\omega)F(\omega)$ is the transform of

$df(t)/dt$. Similarly, $(i\omega)^n F(\omega)$ is the transform of $d^n f(t)/dt^n$. This makes differentiation easy on a computer, and is an easy way to change a displacement record into velocity, or velocity into acceleration. This property also makes it easy to solve differential equations (e.g., Eqn 3.7.8) using the Fourier transform, an approach that is often posed as using a sinusoidal trial solution. Hence we sometimes write and operate on the wave equation using the Fourier transform of the wave field (Eqns 2.2.34, 3.3.74).

(5) The total energy in a Fourier transform is the same as that in the time series:

$$\int_{-\infty}^{\infty} |f(t)|^2 dt = \frac{1}{2\pi} \int_{-\infty}^{\infty} |F(\omega)|^2 d\omega, \tag{27}$$

a relation known as *Parseval's theorem*. This relation arises because the time series and its Fourier transform are equivalent representations.

6.2.5 *Delta functions*

In using Fourier transforms, we often need to describe a signal that is concentrated at a single time or frequency. This is done using the Dirac delta function, an entity that is not truly a function, but rather a generalized function that is the limit of a sequence of continuous functions. The delta function can be defined in several ways, each of which offers a different insight into its nature.

A delta function at $t = t_0$, written $\delta(t - t_0)$, is defined as the limit of a Gaussian function that keeps the area constant (= 1) as the width (σ) narrows and the height, $1/\sigma\sqrt{2\pi}$, increases (Fig. 6.2-5):

$$\delta(t - t_0) = \lim_{\sigma \to 0} \frac{1}{\sigma\sqrt{2\pi}} \exp\left[\frac{-1}{2}\left(\frac{t - t_0}{\sigma}\right)^2\right]. \tag{28}$$

Thus the Dirac delta function is a continuous function analogous to the Kronecker delta symbol, δ_{ij} (Eqn A.3.37) which is a function of two discrete variables, i and j. An alternative definition comes from defining the delta function by how it behaves when integrated, a property called "sifting." This is defined as

$$f(t_0) = \int_{-\infty}^{\infty} f(t)\delta(t - t_0)\, dt. \tag{29}$$

Thus the delta function at $t = t_0$ "sifts out" the value of a function at time t_0 if it is multiplied by the function and integrated over all time.

A third definition comes from considering a step, or Heaviside, function $H(t - t_0)$ that is 0 for time before $t = t_0$ and equal to 1 afterwards (Fig. 6.2-5). The delta function $\delta(t - t_0)$ is the derivative of the step, because it is zero except at t_0, when it goes to infinity. Because the delta function is located where its argument is zero, $\delta(t_0 - t)$ is at time t_0, whereas $\delta(t + t_0)$ is at time $-t_0$.

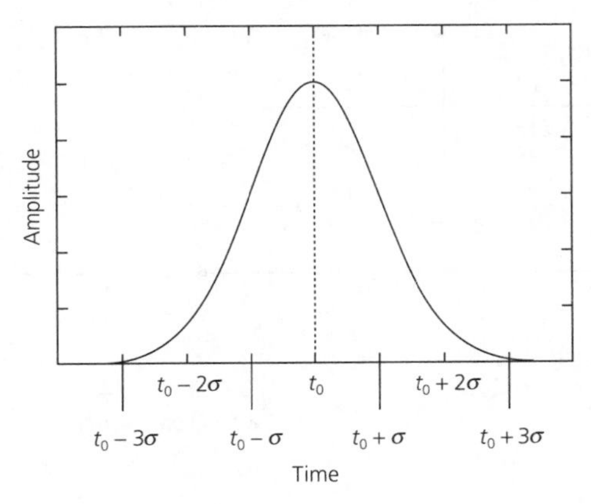

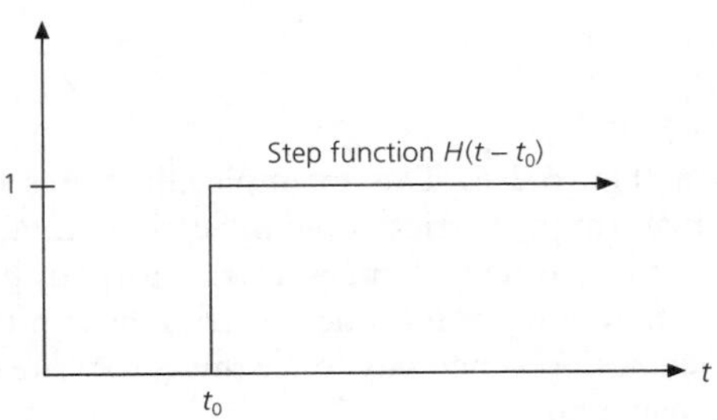

Fig. 6.2-5 Definitions of a delta function at $t = t_0$. *Top*: $\delta(t - t_0)$ is the limit of a Gaussian function with width σ. The area stays equal to 1 as the width narrows and the height increases. *Bottom*: $\delta(t - t_0)$ is the derivative of a step function $H(t - t_0)$ at time $t = t_0$, which is zero at all times except near t_0, when it goes to infinity.

To find the Fourier transform of the delta function, we use the definition of the transform (Eqn 22) with $f(t) = \delta(t - t_0)$,

$$F(\omega) = \int_{-\infty}^{\infty} \delta(t - t_0) e^{-i\omega t} dt = e^{-i\omega t_0}, \tag{30}$$

and evaluate the integral by the sifting property (Eqn 29). If the delta function is at time zero,

$$F(\omega) = \int_{-\infty}^{\infty} \delta(t) e^{-i\omega t} dt = 1. \tag{31}$$

Similarly, for a delta function at $t = t_0$, the amplitude spectrum (Eqn 24) is also

$$|F(\omega)| = (e^{-i\omega t_0} e^{i\omega t_0})^{1/2} = 1, \tag{32}$$

but the phase spectrum (Eqn 25) is

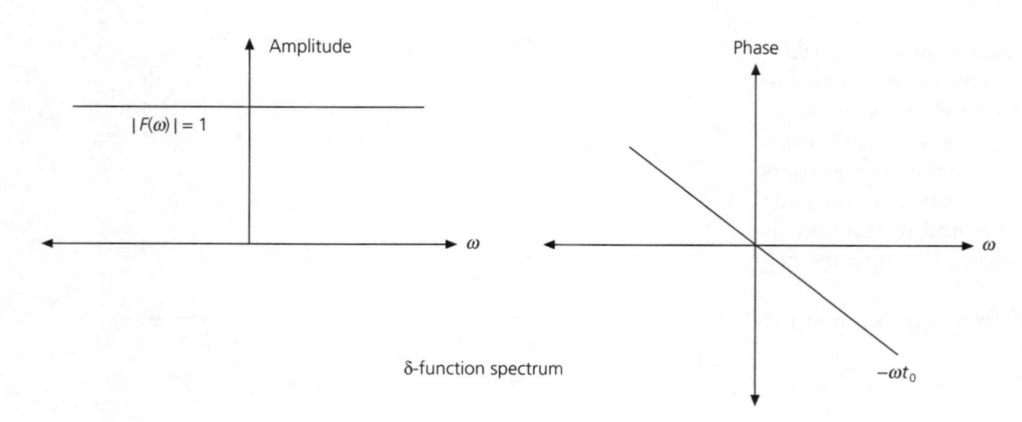

Fig. 6.2-6 The Fourier transform of a delta function, $\delta(t-t_0)$, is $e^{-i\omega t_0}$. Its amplitude spectrum has unit amplitude at all frequencies, and its phase spectrum has a slope of $-t_0$.

$$\phi(\omega) = -\omega t_0, \tag{33}$$

as shown in Fig. 6.2-6. This example illustrates one of the Fourier transform properties noted in Section 6.2.4, that shifting a function by a time t_0 changes its transform by $e^{-i\omega t_0}$.

The delta function's amplitude spectrum has unit amplitude at all frequencies. Another way to see this is to write the inverse transform, using Eqn 21,

$$f(t) = \frac{1}{2\pi}\int_{-\infty}^{\infty} e^{-i\omega t_0} e^{i\omega t} d\omega = \frac{1}{2\pi}\int_{-\infty}^{\infty} e^{i\omega(t-t_0)} d\omega = \delta(t-t_0), \tag{34}$$

which shows that the delta function is an integral or sum of sinusoids of all frequencies. These are in phase only at time t_0, giving a large amplitude, and are out of phase at all other times, giving a zero amplitude (Fig. 6.2-7).

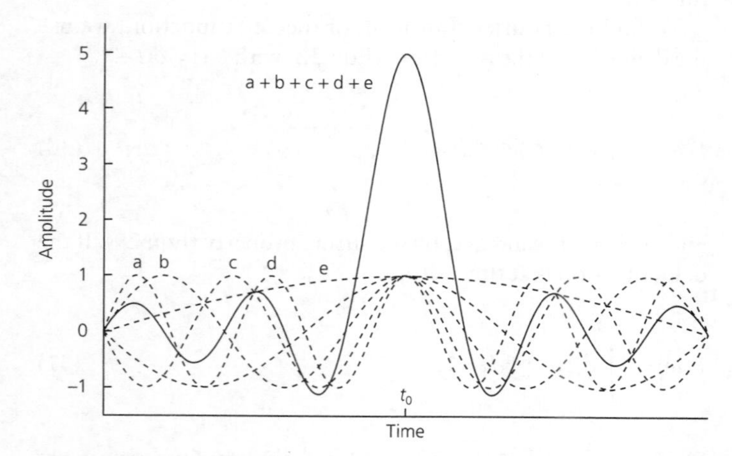

Fig. 6.2-7 Because the Fourier transform of a delta function has unit amplitude at all frequencies, it corresponds to the sum of sinusoids of all frequencies. These are in phase only at time t_0, giving a large amplitude, and are out of phase at all other times, giving zero amplitude. In this example, five sinusoids (dashed lines *a–e*) with unit amplitude (cos $[(2n+1)(t-t_0)]$) are summed (solid line), giving a peak of amplitude 5 at t_0.

Although so far we have discussed delta functions only in the time domain, they are also useful in the frequency domain. The properties of the frequency domain delta functions are analogous to those in the time domain. A delta function at angular frequency ω_0, $\delta(\omega-\omega_0)$, has an inverse transform of

$$f(t) = \frac{1}{2\pi}\int_{-\infty}^{\infty} \delta(\omega-\omega_0) e^{i\omega t} d\omega = \frac{1}{2\pi} e^{i\omega_0 t}. \tag{35}$$

Thus we can express the delta function in terms of its Fourier transform,

$$\delta(\omega-\omega_0) = \frac{1}{2\pi}\int_{-\infty}^{\infty} e^{i\omega_0 t} e^{-i\omega t} dt = \frac{1}{2\pi}\int_{-\infty}^{\infty} e^{i(\omega_0-\omega)t} dt, \tag{36}$$

showing that it is the integral, or sum, of sinusoids that are in phase only at frequency ω_0.

Delta functions in angular frequency give the spectra of sinusoids with a single frequency. For example, a cosine with frequency ω_0, given by

$$f(t) = \cos \omega_0 t = (e^{i\omega_0 t} + e^{-i\omega_0 t})/2, \tag{37}$$

has a Fourier transform of

$$F(\omega) = \frac{1}{2}\int_{-\infty}^{\infty} [e^{i\omega_0 t} + e^{-i\omega_0 t}] e^{-i\omega t} dt = \frac{1}{2}\int_{-\infty}^{\infty} [e^{i(\omega_0-\omega)t} + e^{-i(\omega_0+\omega)t}] dt. \tag{38}$$

By Eqn 36, this is the sum of two delta functions in the frequency domain,

$$F(\omega) = \pi[\delta(\omega-\omega_0) + \delta(\omega+\omega_0)]. \tag{39}$$

Thus the amplitude spectrum of the cosine time function in Eqn 37 consists of two delta functions, one at ω_0 and one at

$-\omega_0$. If the time function were a sine rather than a cosine, the amplitude spectrum would be the same, but the phase spectrum would be different. Given the relation between the transforms of functions shifted in time discussed in the previous section, this makes sense, because a sine function is a time-shifted cosine, and vice versa.

This example illustrates one of the reasons for using Fourier transforms. The frequency domain description of the function is simpler, because a large number of points are needed to accurately describe the cosine as a function of time, but only two complex numbers, the values of the transforms at $\pm\omega_0$, are needed to describe it as a function of frequency. Time series more complicated than a pure cosine are often more easily described in the frequency domain, and processes that act on the time series are also often more easily represented in the frequency domain. In such cases, it is common to work in the frequency domain and then use the inverse transform to generate the final time series.

6.3 Linear systems

Among the uses of Fourier analysis in seismology is modeling different factors affecting a seismogram. First, a seismogram is a record of ground motion that includes the effect of the seismometer. Furthermore, the ground motion combines the effects of the seismic source and the elastic and anelastic earth structure along the propagation path (Section 4.3). To characterize the combined effects of these different factors, we use the idea of a *linear system*, a general representation of any device or process that takes an input signal and modifies it. This representation treats these processes as mathematical operators transforming an input signal into an output signal.

6.3.1 *Basic model*

A linear system is one in which if input signals $x_1(t)$ and $x_2(t)$ produce output signals $y_1(t)$ and $y_2(t)$, the combined input $(Ax_1(t) + Bx_2(t))$ yields $(Ay_1(t) + By_2(t))$ (Fig. 6.3-1). We have previously referred to this feature as the principle of *superposition*. Fortunately, the earth generally behaves this way in transmitting seismic waves. As a result, linear system models are used in a wide variety of seismological applications. Fourier analysis is a natural tool for studying linear systems because the Fourier transform has these same linear properties (Section 6.2.4).

We characterize a linear system by its response to an impulsive delta function in time (Fig. 6.3-2). This *impulse response* $f(t)$ can be used to find the response of the system to an arbitrary input signal. Viewed in the frequency domain, the *impulse*, whose spectral amplitude is equal to 1 at all frequencies, gives rise to an output $F(\omega)$, which is the transform of the impulse response, sometimes called the *transfer function.* Thus, if the input signal is an arbitrary signal $x(t)$, with transform $X(\omega)$, the resulting output spectrum is just the input spectrum times the spectrum of the impulse response,

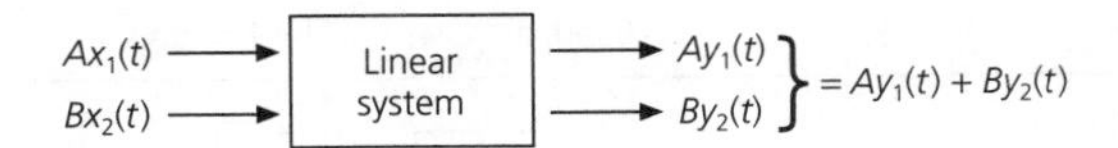

Fig. 6.3-1 Definition of a linear system.

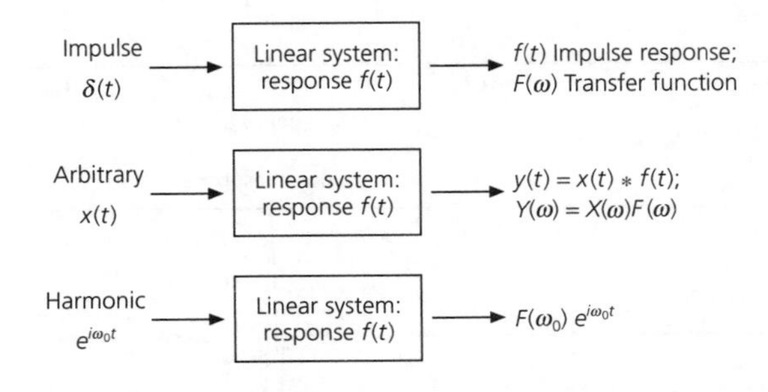

Fig. 6.3-2 Characterization of a linear system by its impulse response $f(t)$ and transfer function $F(\omega)$.

$$Y(\omega) = X(\omega)F(\omega). \tag{1}$$

Because the transforms are generally complex numbers, the phase as well as the amplitude of the input signal is usually modified.

The output in the time domain $y(t)$ can be found by inverting the transform,

$$y(t) = \frac{1}{2\pi}\int_{-\infty}^{\infty} X(\omega)F(\omega)e^{i\omega t}d\omega. \tag{2}$$

To see that this works, note that for the impulse $x(t) = \delta(t)$, $X(\omega) = 1$, and $y(t) = f(t)$. This equation gives another way to think of the impulse response. For a harmonic input signal of unit amplitude $e^{i\omega_0 t}$, whose transform is the delta function in frequency

$$X(\omega) = 2\pi\delta(\omega - \omega_0), \tag{3}$$

the output is

$$y(t) = \frac{1}{2\pi}\int_{-\infty}^{\infty} 2\pi\delta(\omega - \omega_0)F(\omega)e^{i\omega t}d\omega = F(\omega_0)e^{i\omega_0 t}, \tag{4}$$

a harmonic signal of the same frequency with the amplitude of the transfer function at that frequency.

It is interesting to consider the relation between the input time function, the impulse response, and the output time function. To do this, we expand Eqn 2 by writing out the transforms of $X(\omega)$ and $F(\omega)$,

$$y(t) = \frac{1}{2\pi}\int_{-\infty}^{\infty}\left[\int_{-\infty}^{\infty} x(\tau)\,e^{-i\omega\tau}d\tau\right]\left[\int_{-\infty}^{\infty} f(\tau')e^{-i\omega\tau'}d\tau'\right]e^{i\omega t}d\omega, \tag{5}$$

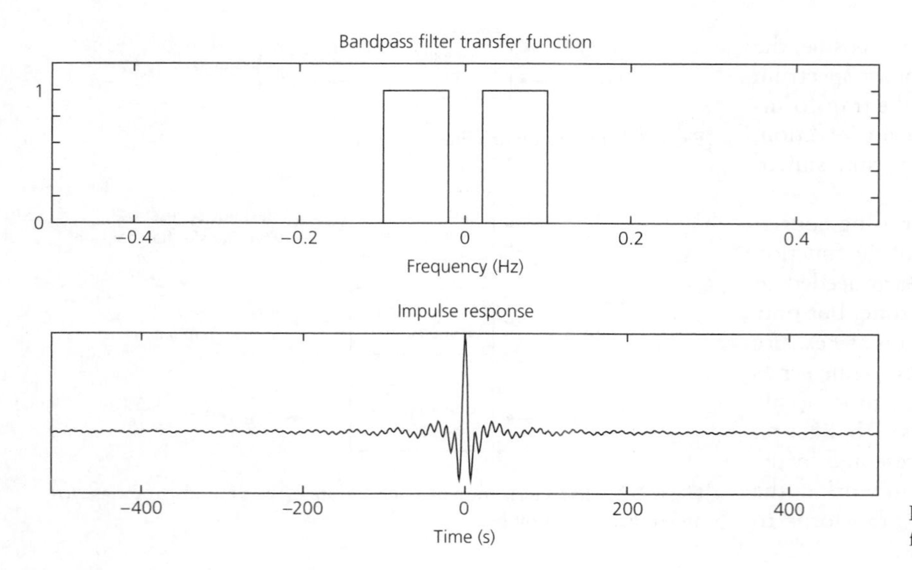

Fig. 6.3-3 A simple bandpass filter specified in the frequency (*top*) and time (*bottom*) domains.

and regrouping terms,

$$y(t) = \int_{-\infty}^{\infty}\int_{-\infty}^{\infty} x(\tau)f(\tau')\left[\frac{1}{2\pi}\int_{-\infty}^{\infty} e^{i\omega(t-\tau'-\tau)}d\omega\right]d\tau d\tau'. \qquad (6)$$

Using the inverse transform of the delta function (Eqn 6.2.34),

$$\frac{1}{2\pi}\int_{-\infty}^{\infty} e^{i\omega(t-\tau'-\tau)}\, d\omega = \delta(t-\tau'-\tau), \qquad (7)$$

we eliminate the frequency integral and obtain

$$y(t) = \int_{-\infty}^{\infty} x(\tau)\left[\int_{-\infty}^{\infty} f(\tau')\delta(t-\tau'-\tau)d\tau'\right]d\tau. \qquad (8)$$

Finally, carrying out the inner integration using the sifting property of the delta function (Eqn 6.2.29) yields

$$y(t) = \int_{-\infty}^{\infty} x(\tau)f(t-\tau)d\tau. \qquad (9)$$

This integral operation, known as the *convolution* of the functions $x(t)$ and $f(t)$, is often written as

$$y(t) = x(t) * f(t). \qquad (10)$$

The output of a linear system is thus the convolution of the input signal and the impulse response. Comparison of Eqns 10 and 1 shows the relation between operations in the two domains: convolution in the time domain corresponds to multiplication in the frequency domain. The reverse is also true: frequency domain convolution corresponds to time domain multiplication.

We thus have two different ways of implementing any operation that can be characterized by a linear system. The effect that the system has on an input signal is specified either by the impulse response in the time domain or by its transform, the transfer function in the frequency domain. For example, to filter a seismogram so that only a certain range of frequencies remains, we can filter in either the frequency or time domains. To do this in the frequency domain, we can define a simple *bandpass filter*, a function which is 1 in the frequency range of interest and 0 for all other frequencies. Figure 6.3-3 (*top*) shows the amplitude spectrum of the filter, whose phase spectrum is defined as zero for all frequencies. To perform the filtering, we multiply this function by the Fourier transform of the seismogram, point by point for all frequencies, and take the inverse transform of the result. The resulting filtered seismogram has only the desired frequencies. Alternatively, however, we could find the impulse response of the bandpass filter by taking the inverse Fourier transform of the amplitude spectrum in the top of Fig. 6.3-3, and filter the data by convolving this impulse response (Fig. 6.3-3, *bottom*) with the seismogram in the time domain.

A few points about this simple filter are worth noting. First, although it is typical to plot the transfer function only for the positive frequencies, the filter is also defined for negative frequencies, to ensure that the resulting signal is real (Section 6.2.4). Second, the peculiar appearance of the impulse response makes sense when we recall that the impulse response describes what comes out of the filter when a delta function comes in (Fig. 6.3-2). The delta function's amplitude spectrum is constant for all frequencies, but only some of these frequencies are transmitted through the filter. The lack of high frequencies is particularly noticeable, and results in the noncausal impulse response beginning before time zero. We noted a similar phenomenon in Section 3.7.8, where anelasticity acted as a

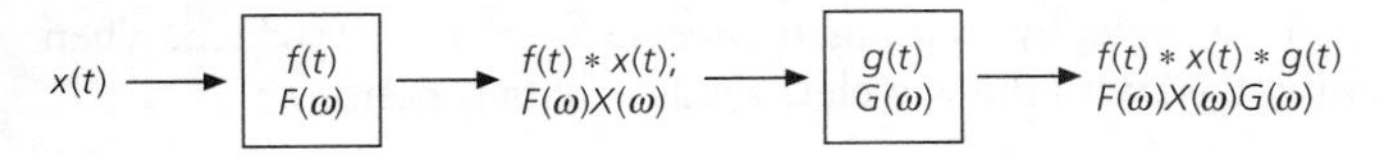

Fig. 6.3-4 When a signal goes through two linear systems in succession, the net output is the convolution of the impulse responses in the time domain, or the product of the transfer functions in the frequency domain.

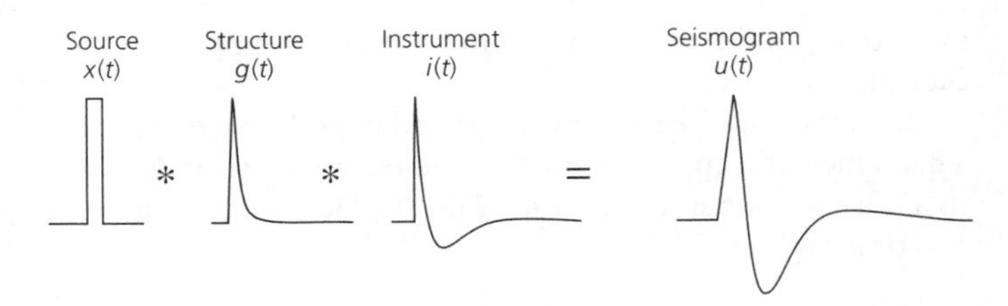

Fig. 6.3-5 A seismogram can be modeled as the convolution of the source signal with operators representing the effects of earth structure and the seismometer. This can be done in the time domain as a set of convolutions, $u(t) = x(t) * g(t) * i(t)$, or in the frequency domain as a set of multiplications, $U(\omega) = X(\omega)G(\omega)I(\omega)$. (After Chung and Kanamori, 1980. *Phys. Earth Planet. Inter.*, *23*, 134–59, with permission from Elsevier Science.)

filter, removing high frequencies and thus making the waveforms noncausal unless the effects of physical dispersion were included. Third, this filter has sharp "corners" at the edges of the passband, although in real applications the corners are smoothed for reasons we discuss shortly.

Because the same effect can be achieved by either time domain or frequency domain filtering, the choice of domain can be made for convenience. Surprisingly, the operations of taking transforms and inverse transforms are sufficiently fast in computation that it generally makes sense to filter in the frequency domain. An attraction of this method is that filters are usually easier to specify in the frequency domain, because it is clear which are the desired and undesired parts of the signal. For example, in Fig. 6.3-3 (*bottom*), the corresponding time domain filter is difficult to visualize intuitively. Similarly, the transfer function, or instrument response, of a seismometer is more easily specified in the frequency domain, as we will discuss in Section 6.6.

6.3.2 *Convolution and deconvolution modeling*

Linear system ideas are so pervasive in seismology that we discussed them in applications such as reflection seismology (Section 3.3.6) and earthquake source studies (Section 4.3) before we justified them mathematically. One reason why these models are so useful is that they are easily generalized to multiple linear systems, so quite complicated physical effects can be described. Specifically, if a signal $x(t)$ goes through two linear systems in succession (Fig. 6.3-4), with impulse responses $f(t)$ and $g(t)$, the net output is either a convolution in the time domain,

$$y(t) = x(t) * f(t) * g(t), \tag{11}$$

or the product of the transfer functions in the frequency domain

$$Y(\omega) = X(\omega)F(\omega)G(\omega). \tag{12}$$

We can extend this to an arbitrary number of linear systems.

A common application is to think of a seismogram as the output resulting from sending a source signal through a set of linear systems. In the simplest case, the seismogram $u(t)$ can be written in terms of three basic effects,

$$u(t) = x(t) * g(t) * i(t), \tag{13}$$

where $x(t)$ is the source signal, $g(t)$ is the response of an operator representing the effects of earth structure along the

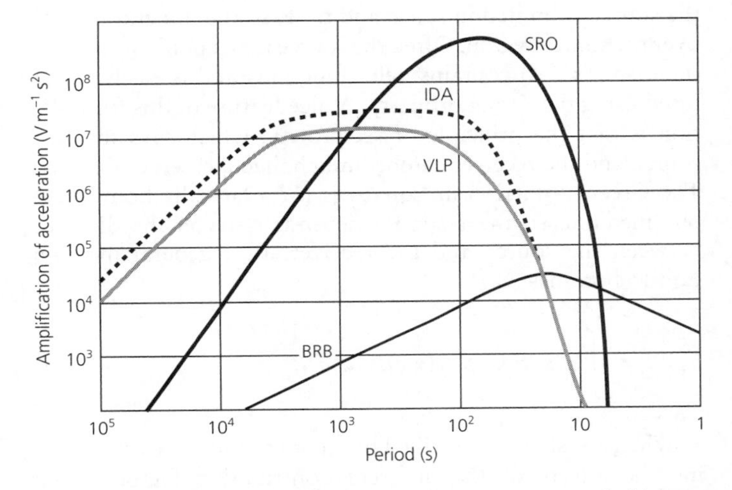

Fig. 6.3-6 Transfer functions for various seismometers, some of which are discussed in Section 6.6. SRO is the Seismic Research Observatory, IDA is International Deployment of Accelerometers, VLP is Very Long Period, and BRB is Broadband. Transfer functions are the frequency domain equivalents of the time domain instrument response shown in Fig. 6.3-5 as $i(t)$.

path of the seismic waves, and $i(t)$ is the impulse response of the seismometer.

Figure 6.3-5 shows a simple example: a seismogram resulting from the convolution of a trapezoidal source function representing the signal emitted by an earthquake with operators giving the effects of earth structure and the seismometer. Each operator can be specified in either domain. For example, the time domain impulse response of a seismometer reflects the fact that its transfer function depends on frequency (Fig. 6.3-6). Once the different effects are characterized by their response in the time or frequency domain, the seismogram due to their combined effects can be obtained.

Convolution can be used to describe the response of a system in space as well as time. For example, probabilistic earthquake hazard maps like Fig. 1.2-3 can be viewed as two-dimensional convolutions in space of an assumed distribution of earthquake sources with an impulse response like Fig. 1.2-5 giving the expected ground motion as a function of earthquake magnitude and distance.

Often the impulse response is defined in both space and time. This is the basic approach used to find the response of the earth to a seismic source (Chapter 4). The displacement at a point $\mathbf{x}$ and time t is

$$u(\mathbf{x}, t) = \iint G(\mathbf{x}-\mathbf{x}'; t-t') f(\mathbf{x}', t') dt' dV', \tag{14}$$

where $G(\mathbf{x} - \mathbf{x}'; t - t')$ is the *Green's function*,[1] the impulse response to a source at position $\mathbf{x}'$ and time t', and $f(\mathbf{x}', t')$ is the distribution of seismic sources. Thus the integral gives the total response due to the distribution of sources. In most cases the source is limited in space and time, so the integral is done over the source region. Often the source is at a point in space or time, so $f(\mathbf{x}', t')$ contains delta functions and is easily integrated using the sifting property. A nice feature of this formulation is that the principle of reciprocity, which says that the source and the receiver can be interchanged, emerges directly. The Green's function in Eqn 14 is for a laterally homogeneous medium, so the response depends only on the distance between the source and the receiver. In a general medium Eqn 14 becomes

$$u(\mathbf{x}, t) = \iint G(\mathbf{x}, t; \mathbf{x}', t') f(\mathbf{x}', t')\, dt' dV'. \tag{15}$$

When a system is described by a convolution, we can examine the effects of the different contributing factors using *deconvolution*. We start with the output and one of the time series that were convolved to form it, and then find the other. For example, in Section 3.3.6 we discussed using seismic reflection data to obtain the sharpest resolution of reflectors in the earth. We assumed that a seismogram $s(t)$ results from convolution of a source pulse, or wavelet, $w(t)$, and an earth structure operator, $r(t)$. $r(t)$, known as a reflector series, is presumed to be a set of delta functions with positions corresponding to the travel time for a reflection from an interface and amplitudes corresponding to the amplitude of the reflected arrival. Thus

$$s(t) = w(t) * r(t) \quad \text{and} \quad S(\omega) = W(\omega)R(\omega). \tag{16}$$

If the travel time differences between the arrivals corresponding to individual reflectors are shorter than the duration of the wavelet, interference can occur, giving a complicated signal. Hence it would be desirable to have a delta function source wavelet whose Fourier transform is simply 1, so that the seismogram would equal the reflector series. Although a physical source wavelet is not a delta function, we simulate such a wavelet by creating an *inverse filter*[2] $w^{-1}(t)$, which, when convolved with the wavelet, yields a delta function:

$$w^{-1}(t) * w(t) = \delta(t). \tag{17}$$

As we saw in Section 3.3.6, the Fourier transform of the inverse filter is just $1/W(\omega)$, so deconvolution can be done by dividing the Fourier transforms

$$S(\omega)/W(\omega) = R(\omega). \tag{18}$$

This sometimes works well, but can be problematic at frequencies where the source wavelet spectrum $W(\omega)$ is small (causing $R(\omega)$ to go to infinity), so a minimum amplitude threshold can be set.

As an alternative, inverse filters can be designed in the time domain to compress the source wavelet into a function as close to a delta function as possible. This approach is a special case of the general problem of finding a shaping filter that converts a given input into a given output. We will shortly discuss another approach, which relies not on the convolution, but on the related cross-correlation operator.

Deconvolution is also used in other applications. A conceptually similar one is modeling seismograms from a distant earthquake as a sum of secondary arrivals generated when the upcoming wave encounters interfaces below the receiver (Fig. 6.3-7). The vertical component is assumed to represent the direct arrival, and is used as a Green's function that is deconvolved from a horizontal component to find a *receiver function* characterizing the structure. The receiver function corresponds to the reflector series in this geometry. Another application of deconvolution is to take seismograms and deconvolve the effects of the seismometer to find the true ground motion, or deconvolve a seismogram to try to find the source pulse due to an earthquake (Section 4.3.3).

6.3.3 *Finite length signals*

We have seen that the Fourier transform describes a signal as the sum of harmonic signals with different frequencies. One important limitation is that the Fourier transform requires integration over all time. In reality, we only have data over a finite interval of time.

To see how this affects our results, consider a *window* function $b(t)$ which selects part of the data. Its effect on the data $f(t)$ is represented by multiplying $f(t)$ by $b(t)$. We then ask how the Fourier transform of the function, including the effect of the window

$$G(\omega) = \int_{-\infty}^{\infty} b(t) f(t) e^{-i\omega t} dt, \tag{19}$$

is related to the transform of the original function, $F(\omega)$.

[1] The same entity is commonly termed a Green's function in physical problems and an impulse response in time series analysis. In seismology the terms are used essentially interchangeably.

[2] The notation $w^{-1}(t)$ does not mean $1/w(t)$.

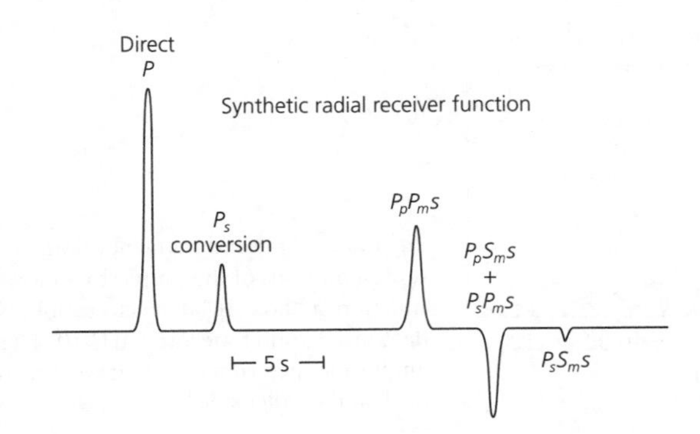

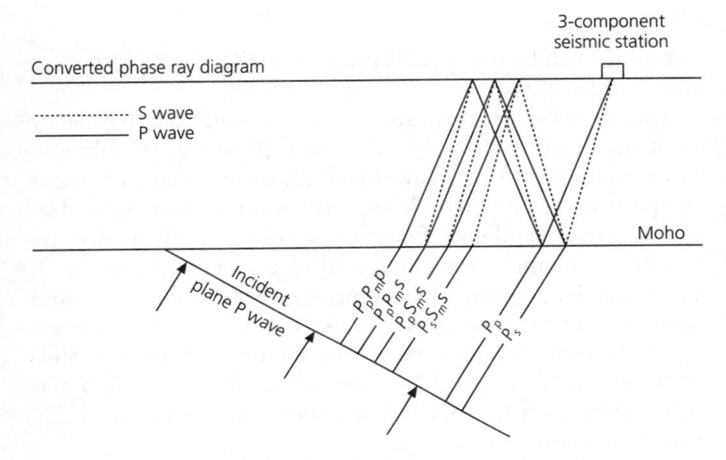

Fig. 6.3-7 Schematic diagram of the receiver function approach. The receiver function, derived by deconvolving the vertical component from a horizontal component, should have arrivals corresponding to the times of seismic wave phases generated when the upcoming wave encounters interfaces below the receiver and amplitudes reflecting the amplitudes of these waves. The receiver function can be used to study the depths of the interfaces and the velocity contrast there. Because a horizontal component is used, the phases predicted involve *P*-to-*S* conversions and their reverberations, as described by the nomenclature used to identify phases (e.g., *PpPms*). Owens *et al.*, 1987. © Seismological Society of America. All rights reserved.)

This question can be answered by writing $b(t)$ and $f(t)$ using their inverse transforms,

$$G(\omega) = \int_{-\infty}^{\infty} \left[\frac{1}{2\pi} \int_{-\infty}^{\infty} B(\omega')e^{i\omega' t} d\omega' \right] \left[\frac{1}{2\pi} \int_{-\infty}^{\infty} F(\omega'')e^{i\omega'' t} d\omega'' \right] e^{-i\omega t} dt$$

$$= \frac{1}{2\pi} \int_{-\infty}^{\infty} B(\omega') \left[\int_{-\infty}^{\infty} F(\omega'') \left(\frac{1}{2\pi} \int_{-\infty}^{\infty} e^{-i\omega t + i\omega' t + i\omega'' t} dt \right) d\omega'' \right] d\omega', \tag{20}$$

recognizing that the inner integral is the Fourier transform of a delta function in frequency (Eqn 6.2.36),

$$G(\omega) = \frac{1}{2\pi} \int_{-\infty}^{\infty} B(\omega') \left[\int_{-\infty}^{\infty} F(\omega'') \delta(\omega - \omega' - \omega'') d\omega'' \right] d\omega', \tag{21}$$

and using the sifting property (Eqn 6.2.29) to obtain

$$G(\omega) = \frac{1}{2\pi} \int_{-\infty}^{\infty} B(\omega')F(\omega - \omega')d\omega' = \frac{1}{2\pi} B(\omega) * F(\omega). \tag{22}$$

Thus the effect of multiplying a time series by a window function is that the spectrum of the time series is convolved with the spectrum of the window function. This is an example of the fact that just as convolution in the time domain corresponds to multiplication in the frequency domain, so multiplication in the time domain corresponds to convolution in the frequency domain.

To see the effect of windowing on the spectrum, consider the simplest window function, a "boxcar" which describes taking only the data in a certain time interval (Fig. 6.3-8),

$$b(t) = 1 \quad \text{for} -T < t < T,$$
$$= 0 \quad \text{otherwise.} \tag{23}$$

Its Fourier transform is

$$B(\omega) = \int_{-T}^{T} e^{-i\omega t} dt = -\frac{e^{-i\omega t}}{i\omega}\bigg|_{-T}^{T} = \frac{2\sin\omega T}{\omega} = \frac{2T\sin\omega T}{\omega T}, \tag{24}$$

whose amplitude spectrum $|B(\omega)|$ has a characteristic shape with a central lobe and smaller side lobes, and equals zero where $x = \omega T = 2n\pi$. The width of the central lobe is $2\pi/T$. This $|(\sin x)/x|$ curve, sometimes called a *sinc* function, is convolved with, and thus modifies, the spectrum $|F(\omega)|$.

For example, if $f(t)$ is a sine wave (Fig. 6.3-9a) whose amplitude spectrum is described by two delta functions, convolution with $B(\omega)$ yields the spectrum of a finite length sine wave, two sinc functions. Thus, taking a finite length of record "smears" the delta functions of the infinite length record's spectrum into broader peaks with side lobes (Figs 6.3-9b). Taking longer records (increasing T) yields sharper spectra (more like the delta function), because the width of the central lobe of the sinc function is proportional to $1/T$.

This effect has an important consequence for analyzing signals containing different frequencies, as shown in Fig. 6.3-9c for a time series with two frequencies. For shorter record lengths (Figs 6.3-9d and e), the spectral peaks broaden until they start to overlap and cannot be resolved separately. Once the width in frequency of the central lobe of the sinc function exceeds the separation between the two spectral peaks (Figs 6.3-9e), they cannot be resolved. Thus the frequency resolution, the minimum separation in frequency for which

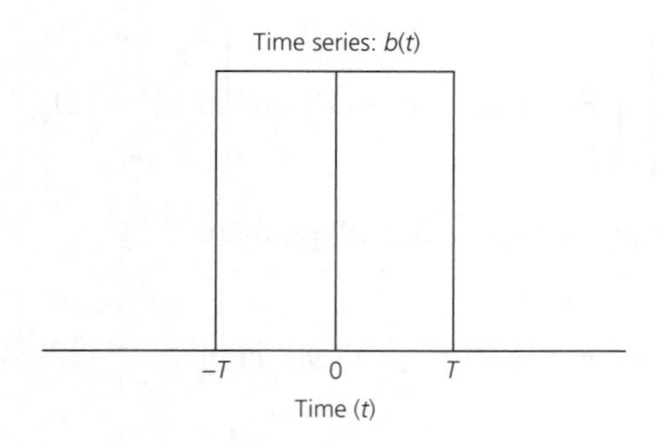

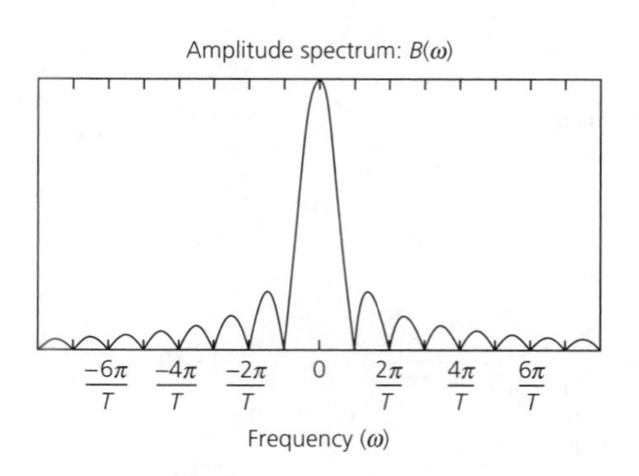

Fig. 6.3-8 Time and frequency domain representations of the simplest window function, a "boxcar" that selects only the data in a certain time interval (*left*). The amplitude spectrum (*right*) has a central peak and smaller side lobes.

Data length and frequency resolution

Time series | Amplitude spectrum

(a) Sine function with period of 10 s |20 s| — −2 0 2 Frequency (Hz)

(b) (a) sampled for a total of 48 s |20 s| — −2 0 2 Frequency (Hz)

(c) Sum of 2 sine functions with 10 s and 20 s periods |20 s| — −2 0 2 Frequency (Hz)

(d) (c) sampled for a total of 36 s |20 s| — −2 0 2 Frequency (Hz)

(e) (c) sampled for a total of 24 s |20 s| — −2 0 2 Frequency (Hz)

Fig. 6.3-9 Effects of finite data length on the spectrum. The spectrum of the sine wave in (a) is "smeared" by taking a short data window (b). For a time series with two frequencies (c), shorter record lengths cause the spectral peaks to broaden (d) until they start to overlap and cannot be resolved separately (e).

two peaks can be resolved, is proportional to the reciprocal of the record length.

This relation between signals in the time and frequency domains demonstrates a fundamental principle. By taking a finite length portion of a time function, we broaden and distort its spectrum in a predictable way. The reverse occurs in the frequency domain; taking a finite portion of the spectrum distorts the time function, as we discussed in considering Fig. 6.3-3. For example, because a seismometer only responds to ground motion in a certain frequency range, the resulting seismogram is a somewhat distorted record of the ground motion. Similarly, physical processes like anelasticity (Section 3.7.8) and diffraction (Section 2.5.10) that remove high frequencies distort the resulting waveforms.

Thus we have an "uncertainty principle" that the product of the "widths" in the two domains is constant; for a time domain record with duration T, the resolution in the frequency domain is proportional to $1/T$. Perfect resolution in frequency requires infinite record length in time, and infinite bandwidth in frequency is needed to represent a time function exactly. These properties are general features of Fourier transform pairs, so also apply to distance and wavenumber.[3]

The sinc or $|\sin x/x|$ function, which we used to represent taking a finite portion of a time series, appears in other similar applications. We saw that diffraction through a slit, in which only part of a wave front is transmitted, is described by a sinc function (Fig. 2.5-18). The sinc function also describes the spectrum of waves radiated from a finite fault (Section 4.6.2).

In real cases, we do not have infinite lengths of data. Moreover, it is not always desirable to take more data. For example, the signal of interest on a seismogram eventually decays into the noise due to attenuation, or is interferred with by a different signal. We seek the best resolution of the spectrum of the signal of interest, but as the record length increases, the noise has a greater effect and increasingly contaminates the spectrum. We thus select a compromise record length and try to obtain the best spectrum. This issue arises in estimating seismic attenuation, which broadens spectral peaks (Section 3.7.7) in a way

[3] The uncertainty principle also appears in quantum physics, where the position and momentum of a particle form a Fourier transform pair. Thus, the better we know a particle's position, the less we know about its momentum, and vice versa.

similar to that of finite record length. Longer records broaden the peaks less, and so give better estimates of attenuation up to the point where the effects of noise degrade the estimates.

Though we can never get around the problem of finite record length, it can be ameliorated by using a different window function than a boxcar. A window function whose "corners" are less "sharp," known as a *taper*, reduces the size of the side lobes and thus the distortion. One simple such function, a cosine taper, is a boxcar with smoother ends:

$$\begin{aligned} W(t) &= \frac{1}{2}\left[1+\cos\frac{\pi(t+T-T_1)}{T_1}\right] && \text{for } -T<t<-T+T_1 \\ &= 1 && \text{for } -T+T_1 \le t \le T-T_1 \\ &= \frac{1}{2}\left[1+\cos\frac{\pi(t-T+T_1)}{T_1}\right] && \text{for } T-T_1<t<T \\ &= 0 && \text{for other times.} \end{aligned} \quad (25)$$

The parameter T_1 is the tapered fraction of the half-length T. Figure 6.3-10 illustrates the effect of tapering data, by comparing the spectra of two windows of the same length. The side lobes for the tapered window are reduced.

Such a taper is often applied in the time domain to data, with $T_1/T \approx 0.1$, before taking spectra. Similarly, bandpass filters are often tapered in the frequency domain. In the frequency domain, a pure bandpass filter is two boxcar functions for the positive and negative frequencies in the passband (Fig. 6.3-3). The corresponding inverse transform thus looks like a sinc function, and causes "ringing," analogous to the side lobes, in the time domain. The ringing can be reduced by tapering the response at the edges of the passbands. For the same reason, the spectrum of a theoretical (synthetic) seismogram computed in the frequency domain is tapered before the inverse Fourier transform is used to produce a synthetic seismogram in the time domain.

This example brings out the general point that, in filtering data, we make certain choices depending on our goals and accept the consequences. There are no absolute criteria for what is best. For example, tapering a filter in the frequency domain reduces the ringing that can produce spurious noncausal arrivals, at the price of distorting the spectrum and waveform. We will see in Section 6.6.5 that this issue appears in designing digital seismometers.

6.3.4 *Correlation*

Often we want to measure how similar two signals are. A common application is identifying a reflected arrival by finding the portion of a seismogram that most resembles a direct arrival or a function that we believe represents the source. To do this, we define the part of the signal we seek to identify as $f(t)$, the remaining portion of the seismogram as $x(t)$, and form the integral

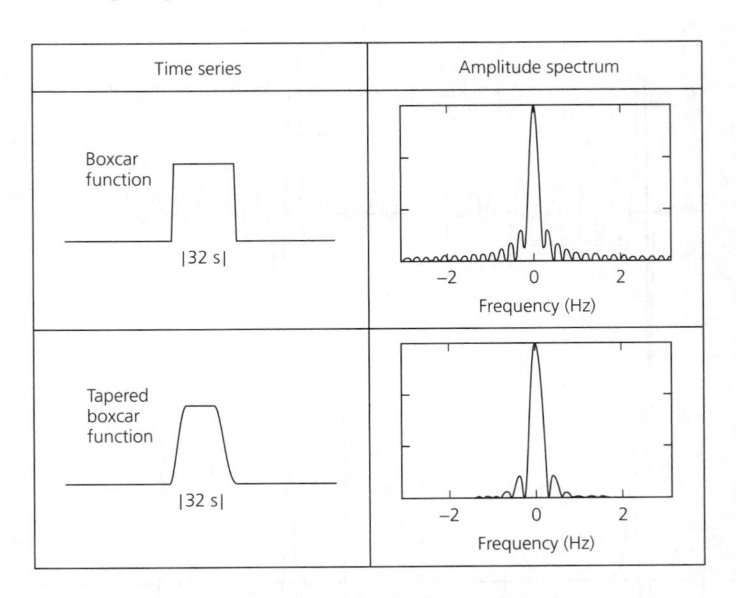

Fig. 6.3-10 Comparison of the spectra of two windows of the same length. The side lobes for the tapered window are reduced, but the central peak is less sharp.

$$C(L) = \lim_{T\to\infty} \frac{1}{T} \int_{-T/2}^{T/2} x(t) f(t+L) dt. \quad (26)$$

$C(L)$, the *cross-correlation* of $x(t)$ and $f(t)$, measures the similarity between $f(t)$ and later portions of $x(t)$ by shifting $f(t)$ by different *lag times*, L, and evaluating the integral of the product as a function of L. The lag for which $C(L)$ is maximum is the time shift that makes the two functions most similar. Although T formally goes to infinity, we set T to an appropriate value, because the data exist only in a finite time range. Thus the $1/T$ factor is a normalization, which is often neglected. Cross-correlation and convolution are similar operations, the major difference being the sign of the time shift.

Figure 6.3-11 shows an example of applying cross-correlation to determine the travel time difference between direct S and SS phases. The SS phase should be similar to S, once S is corrected to include the effects of the additional attenuation on the longer ray path and the $\pi/2$ phase shift due to the surface reflection (Section 3.5.1). Direct S is selected on the seismogram, corrected, and then cross-correlated with the rest of the seismogram. The peak in the cross-correlation gives the lag that measures the arrival time difference between the two phases. Another application of cross-correlation is in exploration seismology, where an assumed Vibroseis source signal is cross-correlated with seismograms, giving peaks at times when reflections occur (Section 3.3.6). In these applications, the cross-correlation is being used to identify reflections, much as could be done by deconvolution, because the cross-correlation is similar to the convolution.

A special case of the cross-correlation is the *auto-correlation*, the cross-correlation of a time series with itself

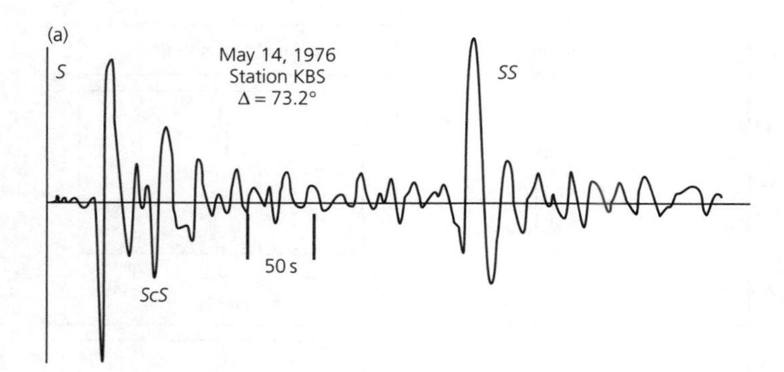

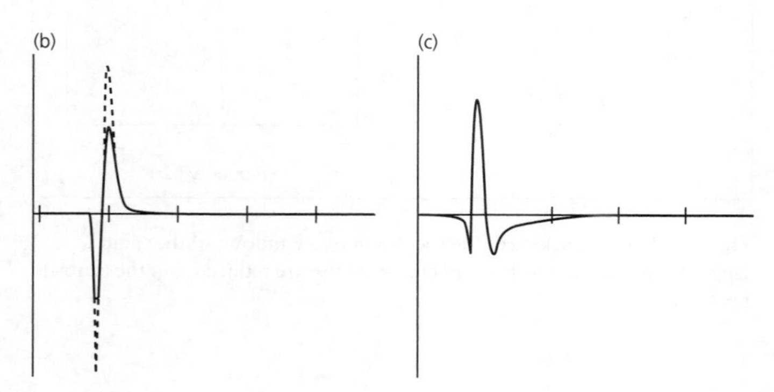

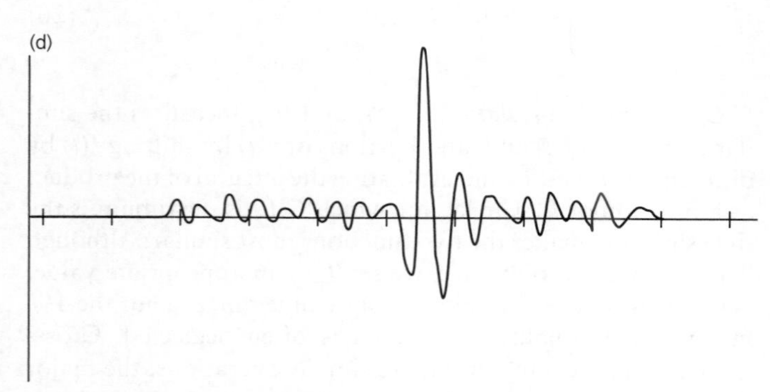

Fig. 6.3-11 Application of the cross-correlation to determine the travel time difference between direct *S* and reflected *SS* phases on a seismogram (a). The direct *S* phase (dashed line in (b)) is corrected for attenuation (solid line in (b)), phase-shifted (c), and then cross-correlated with the rest of the seismogram (d). The peak in the cross-correlation gives the lag that measures the arrival time difference between the two phases. (Kuo *et al.*, 1987. *J. Geophys. Res.*, 92, 6421–36, copyright by the American Geophysical Union.)

$$R(L) = \lim_{T\to\infty} \frac{1}{T} \int_{-T/2}^{T/2} f(t)f(t+L)dt. \tag{27}$$

Not surprisingly, the auto-correlation is maximum at zero lag and is an even function of the lag (Figs 6.3-12 and 3.3-30). When the cross-correlation is used to identify reflections (Figs 6.3-11 and 3.3-31), it makes the seismogram look like the auto-correlation of the signal near the reflection.

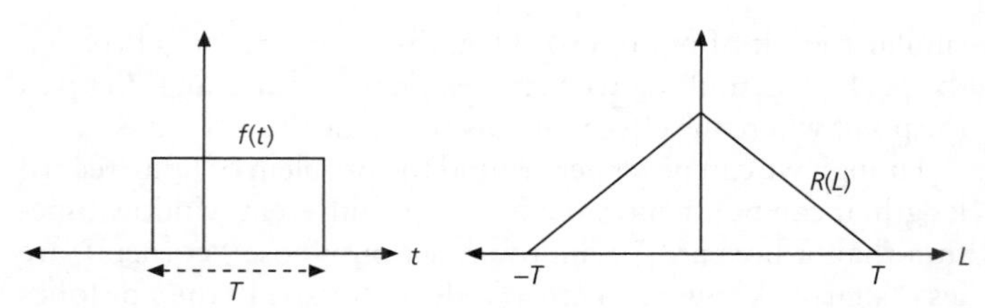

Fig. 6.3-12 Illustration, for a boxcar function, that the auto-correlation is maximum at zero lag and is an even function of the lag.

The auto-correlation is significant in the theory of filtering because it is related to the amplitude spectrum. To see this, consider a function $f(t)$ that is zero except between $-T/2$ and $T/2$. The auto-correlation

$$R(L) = \lim_{T\to\infty} \frac{1}{T} \int_{-T/2}^{T/2} f(t)f(t+L)dt \tag{28}$$

can be expanded using the inverse Fourier transform and using the time shift theorem (Section 6.2.4),

$$R(L) = \lim_{T\to\infty} \frac{1}{2\pi T} \int_{-T/2}^{T/2} f(t) \left[\int_{-\infty}^{\infty} F(\omega)e^{i\omega(t+L)}d\omega \right] dt$$

$$= \lim_{T\to\infty} \frac{1}{2\pi T} \int_{-\infty}^{\infty} F(\omega)e^{i\omega L} \left[\int_{-T/2}^{T/2} f(t)e^{i\omega t}dt \right] d\omega$$

$$= \lim_{T\to\infty} \frac{1}{2\pi T} \int_{-\infty}^{\infty} F(\omega)F(-\omega)e^{i\omega L}d\omega$$

$$= \lim_{T\to\infty} \frac{1}{2\pi T} \int_{-\infty}^{\infty} |F(\omega)|^2 e^{i\omega L}d\omega, \tag{29}$$

where the last step uses the fact that $F(-\omega) = F^*(\omega)$. Thus, if we define the *power spectrum*, a normalized version of the amplitude spectrum,

$$P(\omega) = \lim_{T\to\infty} \frac{1}{T} |F(\omega)|^2, \tag{30}$$

we see that the auto-correlation is the inverse Fourier transform of the power spectrum:

$$R(L) = \frac{1}{2\pi} \int_{-\infty}^{\infty} |P(\omega)| e^{i\omega L}d\omega. \tag{31}$$

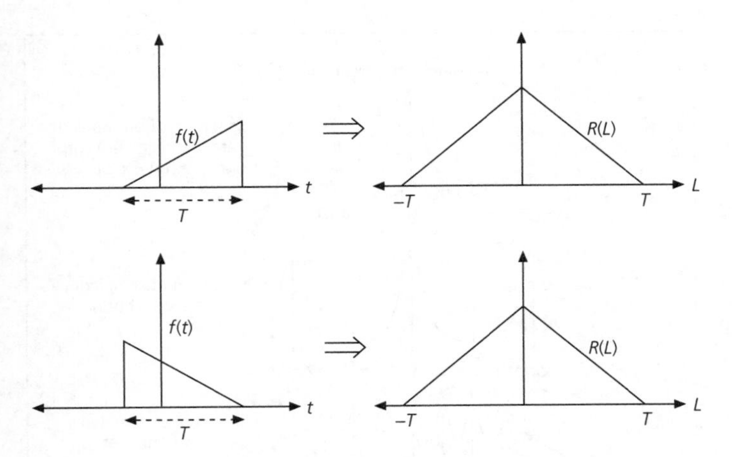

Fig. 6.3-13 Illustration showing that a function has the same auto-correlation if it is reversed in time.

As a result, the auto-correlation of a function contains information only about its amplitude spectrum, but not about its phase. Functions with the same amplitude spectrum but different phase spectra have the same auto-correlation. For example, a function has the same auto-correlation if it is reversed in time (Fig. 6.3-13).

6.4 Discrete time series and transforms

The analysis of seismic data using Fourier transforms requires computers. Thus the ground motion, a continuous function of time, is represented by a signal consisting of the ground motion measured, or *sampled*, at discrete points in time. Early seismometers, which recorded on paper wrapped around a rotating drum, yielded continuous analog seismograms which were digitized to create a discretized seismogram. Modern seismometers typically record the ground motion as a set of amplitude values measured repeatedly over a constant interval, such as 40 times per second (40 sps, "samples per second"). To work with digitized seismograms, the transforms and other mathematical operations that we formulated in Section 6.3 as continuous functions of time are replaced by discretized versions. Working with the discretized data is the subject of *digital* signal processing, whose basic ideas we discuss next.

6.4.1 *Sampling of continuous data*

The operation of sampling a signal at intervals Δt can be represented by multiplying the signal by a series of delta functions (Section 6.2.5) in time spaced Δt apart, called a *Dirac comb* or *Shah* function (Fig. 6.4-1):

$$\nabla(t; \Delta t) \equiv \sum_{n=-\infty}^{\infty} \delta(t - n\Delta t). \tag{1}$$

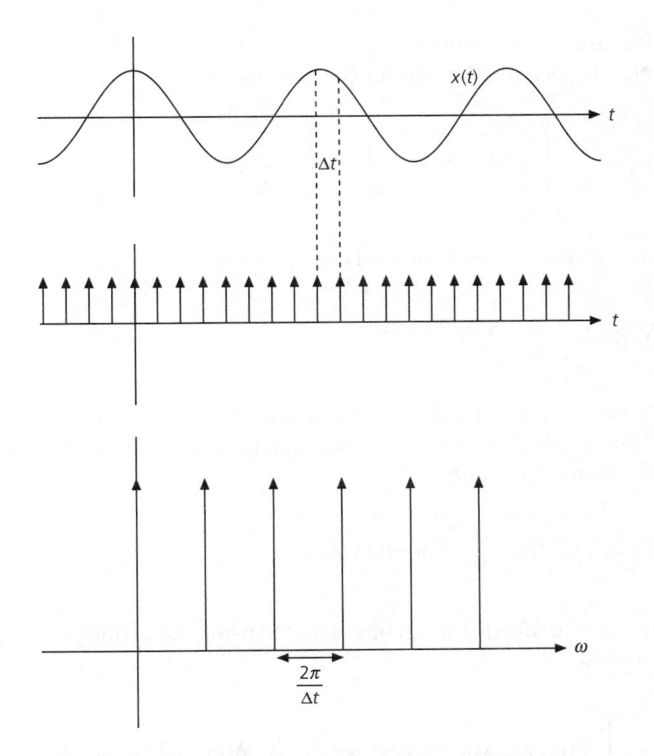

Fig. 6.4-1 Sampling a signal at intervals Δt (*top*) is described by multiplying the signal by a series of delta functions that are spaced Δt apart in time (*center*), called a Dirac comb. The transform of a Dirac comb spaced at Δt in time is a comb spaced $2\pi/\Delta t$ in angular frequency (*bottom*).

To see what this does to the spectrum of the signal being sampled, consider the Fourier transform of the Dirac comb,

$$\int_{-\infty}^{\infty} \nabla(t; \Delta t) e^{-i\omega t} dt = \int_{-\infty}^{\infty} \sum_{n=-\infty}^{\infty} \delta(t - n\Delta t) e^{-i\omega t} dt = \sum_{n=-\infty}^{\infty} e^{-i\omega n\Delta t}, \tag{2}$$

which was evaluated using the sifting property of the delta function (Eqn 6.2.29). It turns out that although the Fourier transform of a single delta function is a complex exponential, the transform of a Dirac comb is another Dirac comb. To see this, note that because $\nabla(t; \Delta t)$ is periodic with period Δt, it can be expanded in a complex Fourier series (Section 6.2.2),

$$\nabla(t; \Delta t) = \sum_{m=-\infty}^{\infty} F_m e^{i\omega_m t} \quad \text{for } \omega_m = 2m\pi/\Delta t, \tag{3}$$

whose coefficients are given by

$$F_m = \frac{1}{\Delta t} \int_{-\Delta t/2}^{\Delta t/2} \nabla(t; \Delta t) e^{-i\omega_m t} dt = \frac{1}{\Delta t} \int_{-\Delta t/2}^{\Delta t/2} \sum_{n=-\infty}^{\infty} \delta(t - n\Delta t) e^{-i\omega_m t} dt. \tag{4}$$

Because in the interval $(-\Delta t/2, \Delta t/2)$ only one delta function, $\delta(t-0)$, occurs, the Fourier coefficients are

$$F_m = \frac{1}{\Delta t}\int_{-\Delta t/2}^{\Delta t/2} \delta(t) e^{-i\omega_m t} dt = \frac{1}{\Delta t} e^{i\omega_m 0} = \frac{1}{\Delta t}; \tag{5}$$

so the Fourier series for the Dirac comb is

$$\nabla(t; \Delta t) = \frac{1}{\Delta t}\sum_{m=-\infty}^{\infty} e^{i2m\pi t/\Delta t}. \tag{6}$$

Now, consider a Dirac comb in the *frequency* domain, $\nabla(\omega; 2\pi/\Delta t)$, which consists of delta functions spaced $2\pi/\Delta t$ apart in angular frequency,

$$\nabla(\omega; 2\pi/\Delta t) \equiv \sum_{n=-\infty}^{\infty} \delta(\omega - n2\pi/\Delta t). \tag{7}$$

Its inverse transform can be evaluated using the sifting property to yield

$$\frac{1}{2\pi}\int_{-\infty}^{\infty} \nabla(\omega; 2\pi/\Delta t) e^{i\omega t} d\omega = \frac{1}{2\pi}\int_{-\infty}^{\infty} \sum_{n=-\infty}^{\infty} \delta(\omega - n2\pi/\Delta t) e^{i\omega t} d\omega$$

$$= \frac{1}{2\pi}\sum_{n=-\infty}^{\infty} e^{i2n\pi t/\Delta t}, \tag{8}$$

which is just $\Delta t/2\pi$ times the Fourier series for $\nabla(t; \Delta t)$ (Eqn 6). Thus the transform of a Dirac comb spaced at Δt in time is $(2\pi/\Delta t)\nabla(\omega; 2\pi/\Delta t)$, a comb spaced $2\pi/\Delta t$ in angular frequency with an amplitude of $2\pi/\Delta t$ (Fig. 6.4-1).

The effects of sampling the signal $x(t)$ at times Δt can be found by writing the sampled signal $\underline{x}(t)$ as the product of the signal and the Dirac comb in time,

$$\underline{x}(t) = x(t)\nabla(t; \Delta t). \tag{9}$$

Because multiplication in the time domain corresponds to convolution in the frequency domain, the transform of the sampled signal, $\underline{X}(\omega)$, can be written as

$$\underline{X}(\omega) = X(\omega) * (2\pi/\Delta t)\,\nabla(\omega; 2\pi/\Delta t). \tag{10}$$

Hence $X(\omega)$ is convolved with the Dirac comb, causing the spectrum of the sampled signal $\underline{X}(\omega)$ to be *periodic* in angular frequency with period $(2\pi/\Delta t)$.

To see what this does, suppose that the signal $x(t)$ is *band limited* such that its spectrum $X(\omega)$ is zero outside the principal angular frequency band $-\pi/\Delta t < \omega < \pi/\Delta t$, the range between the first delta functions on either side of the origin (Fig. 6.4-2a). Thus, after sampling, the adjacent $X(\omega)$ do not overlap (Fig. 6.4-2b), and the spectrum of the sampled time series is the same as that of the original time series in the principal frequency range.

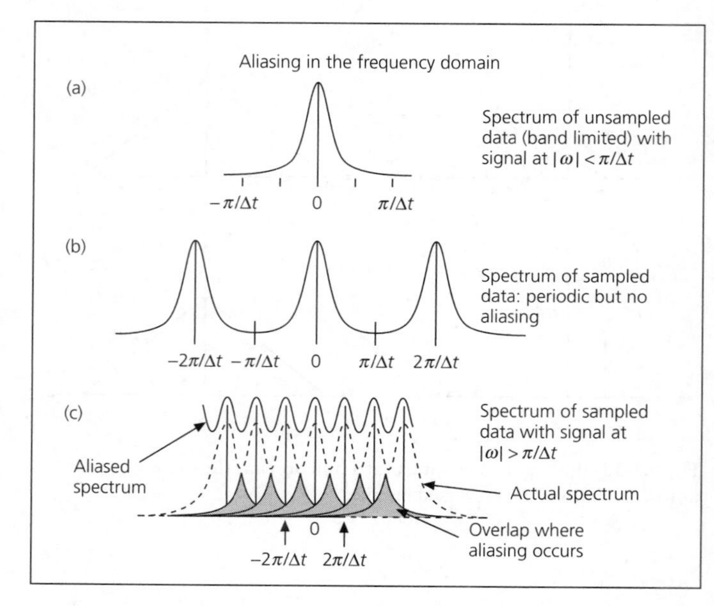

Fig. 6.4-2 Effect of sampling on the frequency amplitude spectrum. The spectrum of the unsampled signal (a) is convolved with a Dirac comb, making the spectrum of the sampled signal periodic in angular frequency with period $(2\pi/\Delta t)$. If the spectrum of the unsampled signal is zero outside the principal angular frequency band $-\pi/\Delta t < \omega < \pi/\Delta t$, the range between the first delta functions on either side of the origin, the spectrum of the sampled signal is the same as that of the original signal in this frequency range (b). Otherwise the spectra overlap after convolution (c), a phenomenon called aliasing that makes the sampled spectrum inaccurate.

On the other hand, if $X(\omega)$ is not limited to this range, the spectra overlap after sampling, so that two adjacent spectra both contribute at these frequencies (Fig. 6.4-2c). The effect of the periodicity is that for angular frequencies $|\omega| > \pi/\Delta t$, or frequencies $|f| > 1/(2\Delta t)$, the spectrum is inaccurate, because the overlap area is *folded* into the principal frequency range. This phenomenon, called *aliasing*, can be avoided by sampling the signal sufficiently densely that the spectra do not overlap. This requires that the sampling interval Δt be such that the corresponding frequency, known as the *Nyquist frequency*,

$$f_N = 1/(2\Delta t) \quad \text{or} \quad \omega_N = \pi/\Delta t, \tag{11}$$

is higher than the highest-frequency component of the signal, so that the spectrum is correctly resolved. The shorter the sampling interval, the higher the Nyquist frequency, the larger the interval over which the spectrum is periodic, and thus the higher the frequency below which the spectrum is correctly resolved. In practice, it is desirable to sample even more densely, perhaps four or more times, than the Nyquist criterion. As we sample more densely, the sampled signal becomes a better representation of the signal, and its spectrum becomes a better representation of the true spectrum.

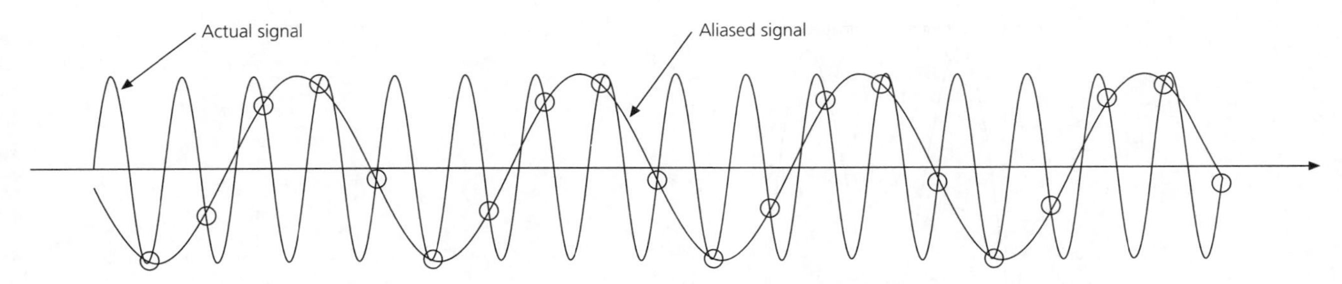

Fig. 6.4-3 In the time domain, aliasing can be viewed by noting that at least two samples per wavelength are needed to reconstruct a sinusoid accurately. Any higher frequencies are aliased into lower ones. In this case, sampling a sine wave at a sampling interval of four-fifths of the period of the wave results in an aliased signal with a period that is four times greater.

Another way to see these ideas is to note that at least two samples per wavelength are needed to reconstruct a sinusoid accurately. Any higher frequencies are aliased into lower ones (Fig. 6.4-3).[1] Aliasing occurs when the data are sampled, and once this occurs, the data cannot be "unaliased." As a result, seismic data are filtered with an analog *anti-aliasing* filter to remove frequencies above the Nyquist frequency before sampling to produce a digital seismogram.

6.4.2 *The discrete Fourier transform*

We now consider the Fourier transform of a sampled time series. If the function $f(t)$ is sampled at N time points that are Δt apart, the function can be represented as

$$f(t) = f(n\Delta t) \quad \text{for } n = 0, 1, \ldots, N-1. \tag{12}$$

To make subsequent derivations easier, we require N to be an even number. The Fourier transform integral,

$$F(\omega) = \int_{-\infty}^{\infty} f(t) e^{-i\omega t} dt, \tag{13}$$

can be written as a summation:

$$F(\omega) = \Delta t \sum_{n=0}^{N-1} f(n\Delta t) e^{-i\omega n\Delta t}. \tag{14}$$

This transform is a continuous function of ω that we approximate using its values at discrete frequency points. Because sampling produces a spectrum that is periodic in angular frequency with period $2\pi/\Delta t$, or twice the Nyquist angular frequency ω_N, we divide this interval into N points as

$$F(\omega) = F(k\Delta\omega) \quad \text{for } k = 0, 1, \ldots, N-1, \tag{15}$$

with

$$\Delta\omega = 2\omega_N/N = 2\pi/N\Delta t = 2\pi/T, \tag{16}$$

where $T = N\Delta t$ is the total length of the data in time, sometimes called the record length. This sampled Fourier transform of a sampled time series is called the *Discrete Fourier Transform* (DFT):

$$F(k\Delta\omega) = \Delta t \sum_{n=0}^{N-1} f(n\Delta t) e^{-ik\Delta\omega n\Delta t} = \Delta t \sum_{n=0}^{N-1} f(n\Delta t) e^{-ikn2\pi/N}. \tag{17}$$

The DFT gives values at angular frequencies

$$0, \Delta\omega, 2\Delta\omega, \ldots (N/2)\Delta\omega, \ldots (N-1)\Delta\omega. \tag{18}$$

The second half of the values represent angular frequencies greater than $(N/2)\Delta\omega$, which equals the Nyquist angular frequency. These points correspond to the negative angular frequencies, wrapped around to follow the positive angular frequencies. For example, the first point after the Nyquist angular frequency occurs for angular frequency

$$\begin{aligned}(N/2+1)\Delta\omega &= (N/2)\Delta\omega + \Delta\omega = \omega_N + \Delta\omega \\ &= -\omega_N + \Delta\omega = -\left(\frac{N}{2} - 1\right)\Delta\omega,\end{aligned} \tag{19}$$

where we use the fact that the spectrum is periodic with period $2\omega_N$. Each successive point corresponds to an increment of $-\Delta\omega$. Thus, we can consider the DFT to give values at angular frequencies

$$0, \Delta\omega, 2\Delta\omega, \ldots \left(\frac{N}{2} - 1\right)\Delta\omega, \omega_N, -\left(\frac{N}{2} - 1\right)\Delta\omega, \ldots, -2\Delta\omega, -\Delta\omega. \tag{20}$$

Graphically, we can think of folding the second half of the DFT about zero frequency to give the values of the spectrum at the negative frequencies (Fig. 6.4-4).

[1] An illustration of sampling issues is that in Western films, wagon wheels sometimes appear to rotate backwards, stop, or rotate only slowly forward. These effects result from differences between the wheels' rotation rate and the movie cameras' sampling rate, typically 24 frames per second.

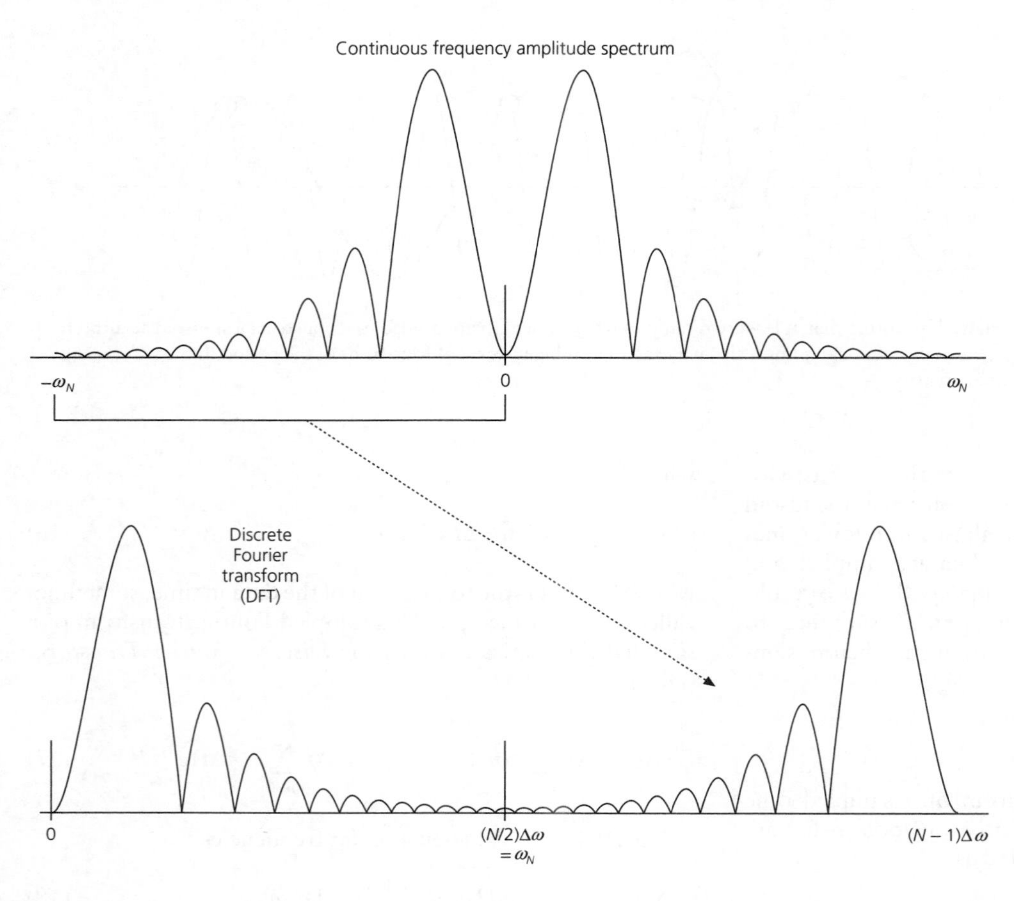

Fig. 6.4-4 Due to the periodicity of the discrete Fourier transform, the second half of the values of the frequency amplitude spectrum, at angular frequencies greater than the Nyquist angular frequency $(N/2)\Delta\omega$, represents the negative angular frequencies.

The fact that the DFT is the sampled spectrum of a sampled time series has two interesting consequences. The *highest* angular frequency that can be resolved is the Nyquist, which depends inversely on the sampling rate in time, because $\omega_N = \pi/(\Delta t)$. On the other hand, the *resolution* in frequency, given by the spacing between successive angular frequency points, $\Delta\omega = 2\pi/(N\Delta t)$, depends inversely on $T = N\Delta t$, the total record length.

For example, to resolve the singlets making up the normal mode multiplet ${}_0S_2$ (Fig. 2.9-16), we would like a frequency resolution of at least 0.0001 cycles/minute, or 1.7×10^{-6} s^{-1}. This requires data extending for $1/1.7 \times 10^{-6}$ s, or more than 160 hours, after the earthquake. However, because the mode's period is 54 minutes, a seismogram sampled every few minutes would be adequate and give a manageable number of data points. We need, however, to prevent aliasing due to surface and body waves that have periods of tens to hundreds of seconds. An easy way to do this would be to start the analysis a day or so after the earthquake, when the shorter-period waves have decayed due to attenuation. This approach uses the earth's anelasticity as a natural anti-aliasing filter. By contrast, reflection seismology requires high temporal resolution to resolve closely spaced interfaces, so reflection data are sampled at high rates such as 250 times per second after an anti-aliasing filter is applied.

By analogy to the DFT, we write the inverse DFT (IDFT) by approximating the inverse Fourier transform integral

$$f(t) = \frac{1}{2\pi}\int_{-\infty}^{\infty} F(\omega)e^{i\omega t}d\omega \tag{21}$$

in the same way, which gives

$$\begin{aligned} f(n\Delta t) &= \frac{1}{2\pi}\sum_{k=0}^{N-1} F(k\Delta\omega)e^{i(k\Delta\omega)(n\Delta t)}\Delta\omega \\ &= \frac{\Delta\omega}{2\pi}\sum_{k=0}^{N-1} F(k\Delta\omega)e^{ikn2\pi/N} \\ &= \frac{1}{N\Delta t}\sum_{k=0}^{N-1} F(k\Delta\omega)e^{ikn2\pi/N}. \end{aligned} \tag{22}$$

An interesting feature of the IDFT comes from the fact that it samples the spectrum at discrete frequencies $\Delta\omega$. Sampling the time series at Δt causes the phenomenon of aliasing, because the spectrum is periodic in angular frequency with period $2\pi/(\Delta t)$. By analogy, sampling the frequency spectrum at $\Delta\omega$ makes the time series periodic with a period of

$$\frac{2\pi}{\Delta\omega} = \frac{2\pi}{2\pi/(N\Delta t)} = (N\Delta t) = T, \tag{23}$$

which is equal to the original record length.[2] This *wraparound* phenomenon can be important, as we shall see when discussing the use of DFTs to carry out convolutions.

6.4.3 Properties of DFTs

For simplicity, we write the DFT and the inverse DFT implicitly assuming a unit sampling interval, $\Delta t = 1$, and define

$$F(k) \equiv F(k\Delta\omega) = \sum_{n=0}^{N-1} f(n)e^{-2\pi ikn/N} \quad \text{for } k \text{ and } n = 0, 1, \ldots, N-1 \tag{24}$$

$$f(n) \equiv f(n\Delta t) = \frac{1}{N}\sum_{k=0}^{N-1} F(k)e^{2\pi ikn/N} \quad \text{for } k \text{ and } n = 0, 1, \ldots, N-1. \tag{25}$$

The two equations are very similar in form and are easy to evaluate – the forward and inverse transforms differ only in the sign of the exponential and the $1/N$ normalization. This is especially clear if we define the complex exponential as $W = e^{-2\pi i/N}$, so the definitions of the DFT and IDFT become

$$F(k) = \sum_{n=0}^{N-1} f(n)W^{kn} \quad \text{and} \quad f(n) = \frac{1}{N}\sum_{k=0}^{N-1} F(k)W^{-kn}. \tag{26}$$

The terms with the complex exponential are periodic in N,

$$W^{kn} = W^{(N+k)n} = W^{k(N+n)}, \tag{27}$$

so the DFT and IDFT can be defined for all integers k, n, j as

$$f(n) = f(jN+n), \quad F(k) = F(jN+k). \tag{28}$$

A formal statement of the relation between the negative and positive frequencies can also be given as

$$f(-n) = f(N-n), \quad F(-k) = F(N-k). \tag{29}$$

We used this relation when we explained how the second half of the DFT corresponds to negative frequencies (Fig. 6.4-4).

Using these definitions, we can show that the discrete transforms have properties that we discussed for the continuous transforms in Section 6.2.4:[3]

(1) The DFT and IDFT are linear: if $A(k)$ and $B(k)$ are the transforms of time series $a(n)$ and $b(n)$, then $\alpha A(k) + \beta B(k)$ is the transform of $\alpha a(n) + \beta b(n)$. Thus we can use the discrete transforms to model linear systems.

(2) The DFT of a real time series (i.e., one for which $f(n) = f^*(n)$) has the symmetry

$$F(-k) = F(N-k) = F^*(k). \tag{30}$$

Thus, as with the continuous transform, the values for the negative frequencies are the conjugates of those for the positive frequencies.

(3) Shifting a time series in time simply changes the phase of the DFT: if the transform of $f(n)$ is $F(k)$, the DFT of $f(n-j)$ is $W^{kj}F(k)$. Similarly, shifting a Fourier transform in frequency changes the phase of the IDFT: the inverse transform of $F(k-m)$ is $W^{-mn}f(n)$.

6.4.4 The fast Fourier transform (FFT)

For these concepts to be useful, the transforms and inverse transforms must be evaluated on a computer. Moreover, it only makes sense to carry out filtering using Fourier transforms if the transform and inverse transform operations are relatively quick. It turns out that an elegant algorithm known as the Fast Fourier Transform (FFT) provides a fast way of carrying out the DFT and IDFT.

The time a computer needs to carry out an algorithm depends on how many arithmetic operations are needed. We would expect that evaluating all N points in the DFT, each of which is the sum of the N terms in the series, would require approximately N^2 operations. The FFT algorithm, however, requires a much smaller number of operations, approximately $N \log_2 N$. The difference is substantial; for $N = 4096$, $N^2 = 16{,}777{,}216$, but $N \log_2 N = 49{,}152$ – about 340 times fewer! As a consequence, the introduction of the FFT made digital signal processing common in seismology and many other disciplines.

Entire books have been written about the FFT, so we only briefly sketch the approach here. The underlying idea is that a simple method can be used to compute the transform of a series of points by splitting it in half. We take a series with N points,

$$f(n) \quad \text{for } n = 0, 1, \ldots, N-1 \tag{31}$$

and form two subseries, one with the odd-numbered points and one with the even-numbered points:

$$a(n) = (f(0), f(2), f(4), \ldots) = f(2n) \quad \text{for } n = 0, 1, \ldots, N/2-1,$$
$$b(n) = (f(1), f(3), f(5), \ldots) = f(2n+1). \tag{32}$$

The DFTs of the two subseries are

$$A(k) = \sum_{n=0}^{N/2-1} a(n)e^{-4\pi ikn/N} \quad \text{and}$$

[2] Because of this periodicity, the record length is considered to be $N\Delta t$ rather than $(N-1)\,\Delta t$.

[3] As for the continuous transforms, the proofs are left for the problems.

$$B(k) = \sum_{n=0}^{N/2-1} b(n)e^{-4\pi ikn/N}, \qquad (33)$$

where k goes from 0 to $N/2 - 1$, and the factor of 4 comes from the fact that the subseries lengths are $N/2$.

The DFT of the original series can be written in terms of the DFTs of the subseries,

$$F(k) = \sum_{n=0}^{N-1} f(n)e^{-2\pi ikn/N}$$
$$= \sum_{n=0}^{N/2-1} [a(n)e^{-2\pi ik(2n)/N} + b(n)e^{-2\pi ik(2n+1)/N}]$$
$$= A(k) + e^{-2\pi ik/N}B(k) \quad \text{for } k = 0, 1, \ldots, N/2 - 1, \qquad (34)$$

giving the first $N/2$ points of $F(k)$. The second $N/2$ points come from replacing k by $k + N/2$,

$$F(k + N/2) = A(k + N/2) + e^{-2\pi i(k+N/2)/N}\, B(k + N/2), \qquad (35)$$

and noting that, because the DFTs of the subseries are periodic with a period equal to their length, $N/2$,

$$A(k + N/2) = A(k) \quad \text{and} \quad B(k + N/2) = B(k). \qquad (36)$$

Because the exponential can be written as

$$e^{-2\pi i(k+N/2)/N} = e^{-\pi i}\, e^{-2\pi ik/N} = -e^{-2\pi ik/N}, \qquad (37)$$

the second half of the transform can be found from the first, using

$$F(k + N/2) = A(k) - e^{-2\pi ik/N}B(k). \qquad (38)$$

In terms of $W = e^{-2\pi i/N}$, the expressions for the two parts of the transform (Eqns 34 and 38) have the simple form of

$$F(k) = A(k) + W^kB(k) \quad \text{and} \quad F(k + N/2) = A(k) - W^kB(k). \qquad (39)$$

This method is called *doubling* — finding the transform of an N-point series from the transforms of its two $N/2$-point subseries. Doubling can be applied recursively, because we can find the transform of each $N/2$-point series from that of two $N/4$-point series, etc. Ultimately, a series of length $N = 2^n$ can be evaluated via $n = \log_2 N$ such stages. In the final stage, the transform of each 2-point series is found from two 1-point series, but the transform of a 1-point series is itself. Various methods can be used to further speed up operations.

Thus, to obtain the FFT of a time series, we treat the data points as N 1-point series, use doubling to form $(N/2)$ 2-point series, and so on until the final N-point transform. The same FFT algorithm can also be used to take the inverse transform. Commonly, the same computer program is used for both forward and inverse FFTs, except that the sign of the exponential must be changed and the $1/N$ normalization remembered (the last being a traditional bane of students).

In using the FFT to transform data as part of a filtering operation, the factor of $1/N$ may be included at any step in the process. Often, however, we use the FFT to obtain the Fourier transform of a time series, and compare this to a result derived in the frequency domain, such as an analytic expression for a synthetic seismogram as a function of ω. In this case, we have to consider the units of both the forward and the inverse DFT. The forward DFT is an approximate way of evaluating the Fourier transform integral (Eqn 13), in which the differential dt is replaced by the difference Δt. Thus, the FFT results are multiplied by Δt. Similarly, the IDFT approximates the inverse transform integral (Eqn 21), with the differential $d\omega$ replaced by the difference $\Delta\omega$. Hence the results from inverting the FFT are multiplied by $\Delta\omega/(2\pi)$. The product of these two factors is $\Delta\omega\Delta t/2\pi = 1/N$, as expected.

This discussion assumes that the series length N is a power of 2. If this is not the case, a number of zeroes necessary to obtain a power of 2 can be added to the end of the time series. Such *zero padding* has the effect of sampling the spectrum more densely, because the sample interval is unchanged, but the frequency interval $\Delta\omega = 2\pi/(N\Delta t)$ decreases. Despite the denser sampling, the real resolution in frequency is not increased beyond that resulting from the real (nonzero) data length. Instead, smooth interpolation is done within the range of actual resolution $\Delta\omega_{real} = 2\pi/T_{nonzero}$.

Finally, it is worth distinguishing between the DFT and the FFT. The DFT is the discrete approximation to the Fourier transform which has the periodic properties we have discussed. The FFT is a clever method for computing the DFT with many fewer operations.

6.4.5 *Digital convolution*

As discussed in Section 6.3.2, the convolution is used in many seismological applications. This operation has some special features when carried out with discretized time series and their transforms.

Given two discrete time series with unit sample period, $x(m)$ with M points $x(0), x(1), \ldots, x(M-1)$ and $f(n)$ with N points $f(0), f(1), \ldots, f(N-1)$, the convolution in the time domain is written, by analogy to the integral definition, as

$$y(t) = x(t) * f(t) = \sum_{m=0}^{M-1} x(m)f(t-m). \qquad (40)$$

We evaluate the summation for each value of t that yields a nonzero value. Because $f(n)$ is zero for n outside the range $(0, N-1)$ and $x(m)$ is zero for m outside the range $(0, M-1)$, there are $N + M - 1$ terms in the convolution, and $y(t)$ is defined for $t = 0, 1, \ldots, N + M - 2$. For example, if $N = 3$ and $M = 4$, the $3 + 4 - 1 = 6$ terms are

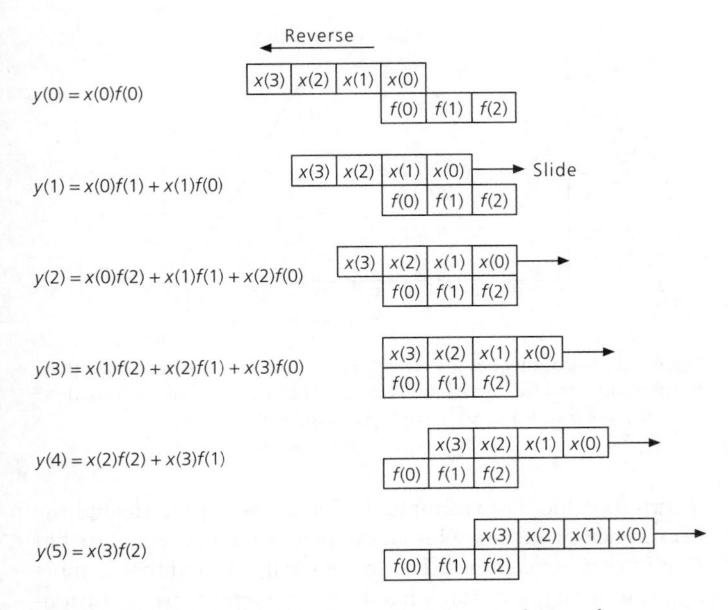

Fig. 6.4-5 Schematic diagram of a time domain convolution of two sampled time series as a reverse, multiply, and slide operation.

$$\begin{aligned}
y(0) &= x(0)f(0) \\
y(1) &= x(0)f(1) + x(1)f(0) \\
y(2) &= x(0)f(2) + x(1)f(1) + x(2)f(0) \\
y(3) &= x(1)f(2) + x(2)f(1) + x(3)f(0) \\
y(4) &= x(2)f(2) + x(3)f(1) \\
y(5) &= x(3)f(2).
\end{aligned} \tag{41}$$

We can think of this operation as reversing the order of $x(m)$ and sliding it past $f(n)$, while conducting all nonzero multiplications (Fig. 6.4-5).

These formulations show that the convolution has more terms than either of the time series being convolved. This has some interesting consequences if we do the convolution in the frequency domain. Because the data are sampled at discrete intervals, convolution in the frequency domain requires taking two discrete Fourier transforms, multiplying them, and then taking the inverse discrete Fourier transform. If $Y(k)$, $X(k)$, and $F(k)$ are the DFTs of $y(t)$, $x(m)$, and $f(k)$, then

$$Y(k) = X(k)F(k) \tag{42}$$

gives the complex spectrum at each angular frequency. This brings out an important point; all the DFTs must be defined at the same frequencies. For a time series of length N with unit sample period ($\Delta t = 1$), the angular frequencies in the DFT are

$$k\Delta\omega = k2\pi/N \quad \text{for } k = 0, 1, \ldots, N-1. \tag{43}$$

Because $x(m)$ and $f(n)$ have different lengths, the points in the two transforms would correspond to different angular frequencies. To avoid this, the two time series are extended with zeroes at their ends, so that their lengths equal the same power of 2.

A further point to bear in mind is that the time series corresponding to the convolution is longer than either of the two series that are convolved. If the number of points in the DFT is less than this length, a wraparound phenomenon similar to aliasing occurs when we invert the transform, due to the periodicity resulting from the sampled transform. The two time series thus need to be extended to a length at least that of their convolution before their DFTs are taken.

6.5 Stacking

Seismology uses data to estimate quantities that describe the earth and seismic sources. Ideally these estimates are both *accurate* and *precise*. Accuracy measures the deviation of the estimate from its true value, whereas precision measures the repeatability of individual estimates. Hence the accuracy depends on systematic errors that bias groups of estimates, whereas the precision depends on random errors that affect individual estimates. Estimates can be precise but inaccurate, or accurate but imprecise. For example, an estimate of an earthquake's location depends on the quality of the travel time data used and the accuracy of the velocity model. High-quality travel time data, together with an incorrect velocity model, can yield a location that is precise in that the data are well fit and so imply small uncertainty, but inaccurate in that the resulting location is not where the earthquake occurred. In such a case the true uncertainty exceeds the formal uncertainty inferred from how well the model fits the data. Conversely, an accurate velocity model and poor travel time data can give a location that is accurate in that it is close to where the earthquake occurred, but imprecise in that the location has a large uncertainty and there are large misfits to the data.

Approaches to improving the accuracy and precision of estimates are often couched in terms of measuring a quantity like the length of a table. Accuracy is improved by using different measuring tools, ideally calibrated against each other. Precision is improved by making multiple measurements, ideally by different people. We follow such approaches for the earth when possible, but face additional complexities. For example, an earthquake is a nonrepeatable experiment, so we cannot make additional measurements. We can use different techniques, but still face difficulties. A case in point is that estimates of an earthquake's depth from travel times and waveform modeling are only partially independent. Both can be biased similarly by incorrect assumptions about the near-source velocity, but the travel times are independent of the assumed source mechanism, and the waveform modeling (which depends on relative arrival times) would not be biased by an error in the absolute timing of individual seismograms.

A further complexity is that different methods can measure related but not identical entities: the earthquake depth ranges inferred from travel times, waveform modeling, aftershock locations, and geodesy differ somewhat, because each measures related but not identical quantities.

Most discussions of these issues focus on random errors because they are easy to estimate from the scatter of measurements. However, it is worth bearing in mind that systematic errors not included in these error estimates can be more significant, as discussed in Section 1.1.2. Systematic errors can come about in surprising ways and have subtle and crucial effects. For example, we have noted that velocity heterogeneities can perturb ray paths and thus bias earthquake focal mechanisms (Section 3.7.3); attenuation variations can bias estimates of the yields of nuclear explosions (Section 1.2.8); errors in the paleomagnetic time scale can bias estimates of plate motions (Section 5.2.2); and effects including an undetected earthquake can change estimates of earthquake recurrence from paleoseismology (Section 1.2.5). Systematic biases are difficult to detect, but sometimes are identified from discrepancies between different approaches. For example, the discrepancy between earth models derived from body waves and those from normal modes suggests physical dispersion due to anelasticity (Section 3.7.8), and the discrepancy between oceanic Love and Rayleigh wave velocities points toward anisotropy (Section 3.6.5). Hence, when data are discordant, as in the differences in earthquake frequency–magnitude relations derived from seismological and paleoseismic data (Section 4.7.1), systematic bias is one possible cause.

In this section, we develop some general ideas about errors and consider some examples. Our focus is one of the most useful methods for improving estimates from seismological data: *stacking*, or taking multiple measurements and averaging them. We do this either by averaging measurements such as travel times from different seismograms, or by adding many seismograms and then estimating parameters. This process has two effects. First, it improves precision by reducing the effects of random noise in the data. Second, if the data are averaged in specific ways, the precision, and perhaps accuracy, can be improved by suppressing some features of the data and thus enhancing desired features.

6.5.1 *Random errors*

We seek to estimate a quantity x from multiple measurements, each of which gives a value x_i due to noise and the limitations of the measurements. With enough measurements, a pattern generally emerges in which the values x_i are distributed about a value x'. If we neglect systematic errors of measurement, we can estimate the value of x from the measured values x_i and say something about how this estimate is related to the unknown true value of x.

For this purpose we view the measured values x_i as random samples from a parent distribution described by the probability density function $p(x)$ that gives the probability of observing

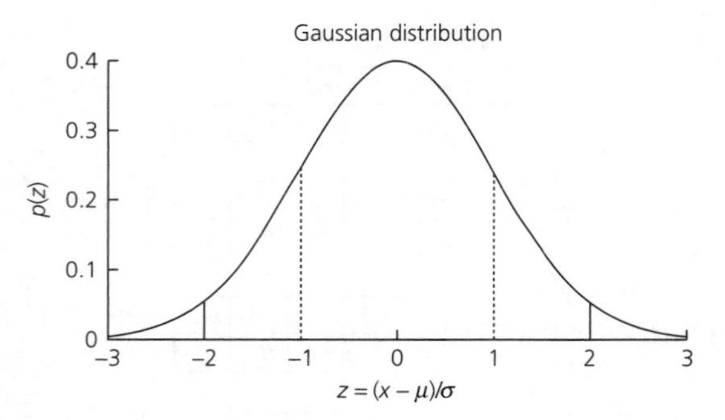

Fig. 6.5-1 Probability density function for a Gaussian distribution with mean μ and standard deviation σ. Ranges within one and two standard deviations of the mean are shown by vertical lines.

a certain value. For example, in Section 4.7.3 we treated the occurrence of earthquakes as samples from a parent distribution of recurrence times. That example illustrated that in most applications it is not clear what the most suitable parent distribution is. It is common to assume that the parent distribution is a Gaussian distribution, also called the "normal distribution," because it often describes the frequencies at which very different phenomena occur. A famous result called the *central limit theorem* shows that this is because a sum of random numbers approaches a Gaussian distribution even if the random numbers are derived from other probability distributions.

For a Gaussian distribution, the probability that the ith measurement would yield a value in the interval $x_i \pm dx$, in the limit as $dx \to 0$, is

$$p(x_i) = \frac{1}{\sigma\sqrt{2\pi}} \exp\left[-\frac{1}{2}\left(\frac{x_i - \mu}{\sigma}\right)^2\right]. \tag{1}$$

The distribution is thus characterized by two parameters: the mean, μ, and the standard deviation, σ. The most probable measurement is the mean value, and values on either side of it are less likely the further from the mean they are. The distribution is often written as a function of the normalized variable $z = (x - \mu)/\sigma$,

$$p(z) = \frac{1}{\sqrt{2\pi}} \exp\,[-z^2/2]. \tag{2}$$

Figure 6.5-1 shows the familiar "bell curve" that results.

A common application is to estimate how likely a measurement is to be within a range z from the mean. To do this, we integrate the probability density function to find the cumulative probability

$$A(z) = \int_{-z}^{z} p(y)dy = \frac{1}{\sqrt{2\pi}} \int_{-z}^{z} \exp\,[-y^2/2]dy. \tag{3}$$

For $z = 1$, we get $A(z) = 0.68$, indicating that there is a 68% probability that a measurement will be within one standard deviation of the mean. Similarly, $A(2) = 0.95$ and $A(3) = 0.997$, indicating a 95% probability that a measurement will be within two standard deviations of the mean, and a greater than 99% probability that it will be within three standard deviations. We used such ideas in estimating earthquake probabilities (Section 4.7.3).

We expect that if we made an infinite number of measurements (samples) without any systematic biases, a histogram of the measurements would look like the parent distribution. The mean of the observed values will be the mean of the distribution

$$\mu = \lim_{N\to\infty} \left[\frac{1}{N} \sum_{i=1}^{N} x_i \right], \tag{4}$$

and the *spread* of the measurements is the *variance* (standard deviation squared) of the distribution,

$$\sigma^2 = \lim_{N\to\infty} \left[\frac{1}{N} \sum_{i=1}^{N} (x_i - \mu)^2 \right]. \tag{5}$$

Thus, if the assumptions we have made are valid, the mean of a large number of measurements, μ, would be the value that we seek.

The difficulty in reality is that only a limited number of measurements are available to estimate μ. As a result, the actual mean μ' is not necessarily equal to μ. We thus ask what method of deriving μ' from the measurements gives the maximum likelihood that μ' is actually the mean of the parent distribution.

To find this, we assume that the parent distribution had mean μ' and standard deviation σ, so the probability that the ith measurement would yield a value in the interval $x_i \pm dx$ in the limit as $dx \to 0$ is

$$p_i(\mu') = \frac{1}{\sigma\sqrt{2\pi}} \exp\left[-\frac{1}{2}\left(\frac{x_i - \mu'}{\sigma} \right)^2 \right]. \tag{6}$$

For N observations, the probability of observing a particular set of values x_i is the product of the probabilities that each individual measurement would have that particular value,

$$p(\mu') = \prod_{i=1}^{N} P_i(\mu') = \left[\frac{1}{\sigma\sqrt{2\pi}} \right]^N \exp\left[-\frac{1}{2} \sum_{i=1}^{N} \left(\frac{x_i - \mu'}{\sigma} \right)^2 \right]. \tag{7}$$

The most probable value of μ' is the one that maximizes $p(\mu')$, the probability of obtaining the set of measurements actually found. To find this value, we set the derivative of the argument of the exponential equal to zero,

$$0 = \frac{d}{d\mu'} \left[-\frac{1}{2} \sum_{i=1}^{N} \left(\frac{x_i - \mu'}{\sigma} \right)^2 \right] = -\frac{1}{2} \sum_{i=1}^{N} \frac{d}{d\mu'} \left[\frac{x_i - \mu'}{\sigma} \right]^2, \tag{8}$$

which occurs for

$$\sum_{i=1}^{N} [x_i - \mu'] = 0, \tag{9}$$

or

$$\mu' = \frac{1}{N} \sum_{i=1}^{N} x_i. \tag{10}$$

This is not surprising — the average value of x_i is the best estimate of the mean. An interesting question is what is the standard deviation σ_N of this estimate of μ'? Specifically, how does the uncertainty associated with this estimate compare to the uncertainty of each individual measurement?

To answer this, we use the *propagation of errors*, a general method for finding the relation between the uncertainty in a function and the uncertainty in the variables that it depends on. If z is a function of multiple variables, then

$$z = f(u, v, \ldots), \tag{11}$$

and we have N measurements of $(u, v, \ldots)$. The mean value of the function is its value for the mean of the arguments,

$$\bar{z} = f(\bar{u}, \bar{v}, \ldots), \tag{12}$$

and its variance is

$$\sigma_z^2 = \lim_{N\to\infty} \frac{1}{N} \sum_{i=1}^{N} (z_i - \bar{z})^2. \tag{13}$$

If we expand z in a Taylor series about its mean value,

$$z_i - \bar{z} = (u_i - \bar{u}) \frac{\partial z}{\partial u} + (v_i - \bar{v}) \frac{\partial z}{\partial v} + \ldots, \tag{14}$$

so

$$\begin{aligned} \sigma_z^2 &= \lim_{N\to\infty} \frac{1}{N} \sum_{i=1}^{N} \left[(u_i - \bar{u}) \frac{\partial z}{\partial u} + (v_i - \bar{v}) \frac{\partial z}{\partial v} + \ldots \right]^2 \\ &= \lim_{N\to\infty} \frac{1}{N} \sum_{i=1}^{N} \left[(u_i - \bar{u})^2 \left(\frac{\partial z}{\partial u} \right)^2 + (v_i - \bar{v})^2 \left(\frac{\partial z}{\partial v} \right)^2 + \ldots \right. \\ &\quad \left. + 2(u_i - \bar{u}) \frac{\partial z}{\partial u} (v_i - \bar{v}) \frac{\partial z}{\partial v} + \ldots \right]. \end{aligned} \tag{15}$$

To simplify this expression, we use the variances of each variable about its mean

$$\sigma_u^2 = \lim_{N\to\infty} \frac{1}{N} \sum_{i=1}^{N} (u_i - \bar{u})^2 \quad \text{and} \quad \sigma_v^2 = \lim_{N\to\infty} \frac{1}{N} \sum_{i=1}^{N} (v_i - \bar{v})^2 \qquad (16)$$

and the *covariances* that describe how fluctuations between variables are correlated:

$$\sigma_{uv}^2 = \lim_{N\to\infty} \frac{1}{N} \sum_{i=1}^{N} (u_i - \bar{u})(v_i - \bar{v}). \qquad (17)$$

Substituting Eqns 16 and 17 into Eqn 15 gives

$$\sigma_z^2 = \sigma_u^2 \left(\frac{\partial z}{\partial u}\right)^2 + \sigma_v^2 \left(\frac{\partial z}{\partial v}\right)^2 + \ldots + 2\sigma_{uv}^2 \left(\frac{\partial z}{\partial u}\right)\left(\frac{\partial z}{\partial v}\right) + \ldots \qquad (18)$$

This relation, called the propagation of errors equation, illustrates that the extent to which the uncertainty in each variable contributes to the uncertainty in a function depends on the partial derivative of the function with respect to that variable. We often assume that the variations in the different variables are uncorrelated (which is not always the case), so we set the covariances equal to zero, and simplify the variance of z to

$$\sigma_z^2 = \sigma_u^2 \left(\frac{\partial z}{\partial u}\right)^2 + \sigma_v^2 \left(\frac{\partial z}{\partial v}\right)^2 + \ldots \qquad (19)$$

This result is a general one that we have already mentioned in the context of estimating the uncertainty of geodetic rates (Eqn 4.5.8) and earthquake source parameters (Eqn 4.6.23).

In the specific application here, we consider the mean to be a function of the observations,

$$z = \mu' = \frac{1}{N} \sum_{i=1}^{N} x_i, \qquad (20)$$

so the error propagation equation can be used with $(u, v, \ldots) = x_i$. Assuming that the variables are independent, so their errors are uncorrelated, we get

$$\sigma_{\mu'}^2 = \sum_{i=1}^{N} \sigma_{x_i}^2 \left(\frac{\partial \mu'}{\partial x_i}\right)^2 = \sum_{i=1}^{N} \sigma_{x_i}^2 \left(\frac{1}{N} \frac{\partial}{\partial x_i} \sum_{i=1}^{N} x_i\right)^2 = \frac{1}{N^2} \sum_{i=1}^{N} \sigma_{x_i}^2. \qquad (21)$$

If all the observations have equal uncertainties ($\sigma_{x_i}^2 = \sigma^2$), then

$$\sigma_{\mu'}^2 = \sigma^2 / N. \qquad (22)$$

Thus the variance of the mean is $1/N$ times the variance of the individual measurements. Hence making N measurements reduces the standard deviation of the mean by $1/\sqrt{N}$. This is the basic idea behind stacking; averaging multiple measurements of some quantity yields an estimate that has a smaller uncertainty than the individual measurements.

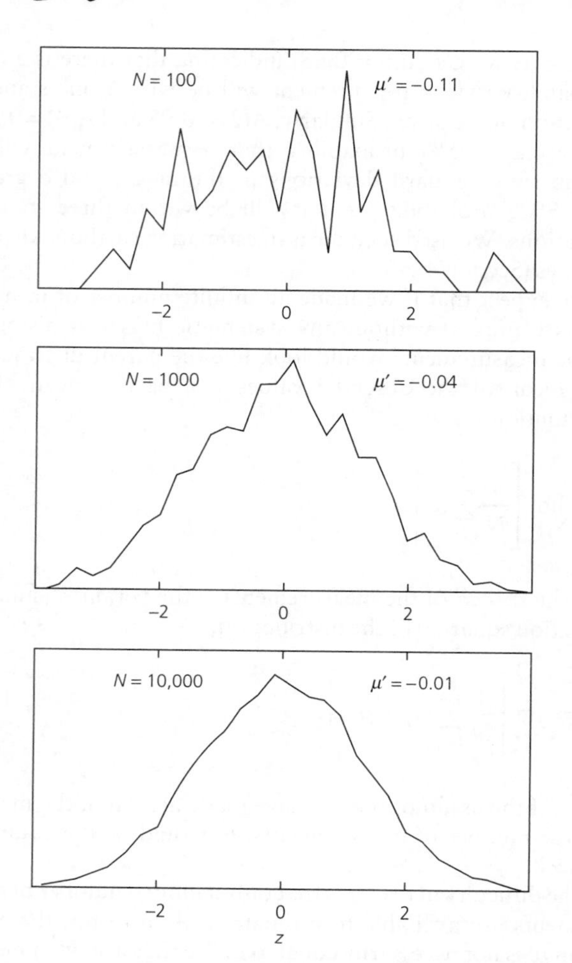

Fig. 6.5-2 Results of drawing N samples from a Gaussian parent distribution with mean zero and a unit standard deviation. For small numbers of samples, the observed distribution can look quite different from the parent distribution, and the sample mean μ' differs from that of the parent distribution. As the number of samples increases, the observed distribution looks increasingly like the parent distribution.

Figure 6.5-2 illustrates this idea. We assume that measurements of some quantity are described by a Gaussian parent distribution with a mean of zero, and we try to estimate this quantity with different numbers of samples. As the number of samples increases, the distribution of samples looks increasingly like the parent distribution, and the sample mean approaches the mean of the parent distribution. However, for a small number of samples, the observed distribution can look quite different from the parent distribution. This issue arises in studying earthquake recurrence, where the few samples available make it difficult to assess whether apparent differences in earthquake history (Section 4.7.1) are significant and what parent distributions and parameters should be used to estimate earthquake probabilities (Section 4.7.3).

This simple Gaussian model is widely used in analyzing data. We assume that each measurement includes the quantity of interest and some *noise*, defined as the portion of the signal that is not of interest. The noise thus reflects both true errors of measurement and processes not under consideration, all of which are assumed to be uncorrelated between measurements. To the extent that these assumptions are valid, stacking data will improve the signal. The random, uncorrelated noise idea often seems to be a good approximation. However, if noise is correlated between measurements, as can occur if the measurement equipment is biased or an "error" source is otherwise common to the measurements, the desired noise reduction will be less. For instance, the structure under a seismometer is studied by means of receiver functions that are derived using the radial and vertical components (Fig. 6.3-7), assuming that the noise on each is uncorrelated. However, noise due to microseismic activity (Section 6.6.3) will be correlated between components and hence can yield spurious layering.

6.5.2 *Stacking examples*

A simple stacking approach is to add seismograms at nearby stations, assuming that they contain a common signal of interest plus "noise" that differs between stations. The noise includes differences in the response of the seismometers and differences in the seismograms generated by the interaction between the upcoming waves and the crustal structure under each seismometer. If the seismometers and crustal structure are similar enough, stacking seismograms should reduce the noise and yield a better representation of the signal of interest than the individual seismograms.

An extension of this idea is used for seismograms at different places or times. If we know theoretically how the signal of interest varies as a function of position or time, we can correct the data to a common position or time and stack them. For example, in CMP stacking of reflection seismic data, traces with a common midpoint are shifted by a time corresponding to the travel time curve of a reflection and then stacked (Section 3.3.4). The reflected arrivals are in phase and thus enhanced, whereas other arrivals with different travel time curves are out of phase and thus suppressed. Although the undesired arrivals are not random noise, they are reduced relative to the reflected arrivals. Random noise in the data is also reduced.

This approach is also useful in observing deeper earth structures, such as mantle discontinuities (Section 3.5.3). Figure 6.5-3 shows an example of stacking large numbers of long-period transverse-component seismograms to enhance precursors to the *SS* arrivals. The precursors, $S_{410}S$, $S_{520}S$, and $S_{660}S$, are underside reflections from the discontinuities at 410, 520, and 660 km depths. However, these phases are weak and are not easily observed above the noise on individual seismograms. Stacking many records enhances these arrivals, allowing the depths of the discontinuities to be studied. Moreover, after removal of the theoretical signals of $S_{410}S$ and $S_{660}S$ (Fig. 6.5-3, *middle*), the stacked record shows the $S_{520}S$ arrival

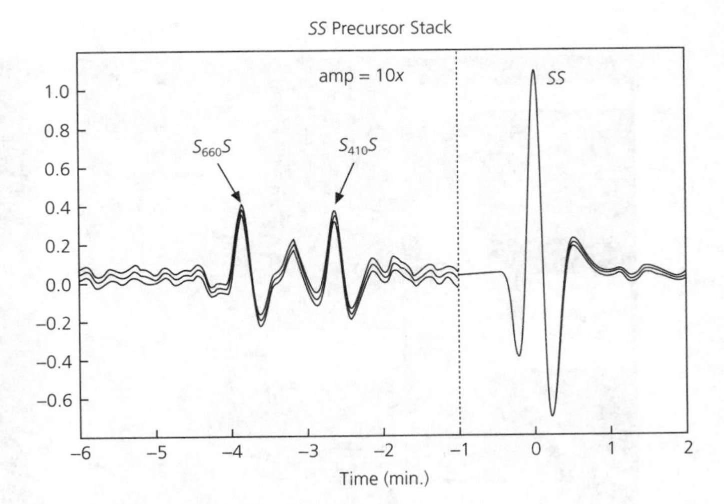

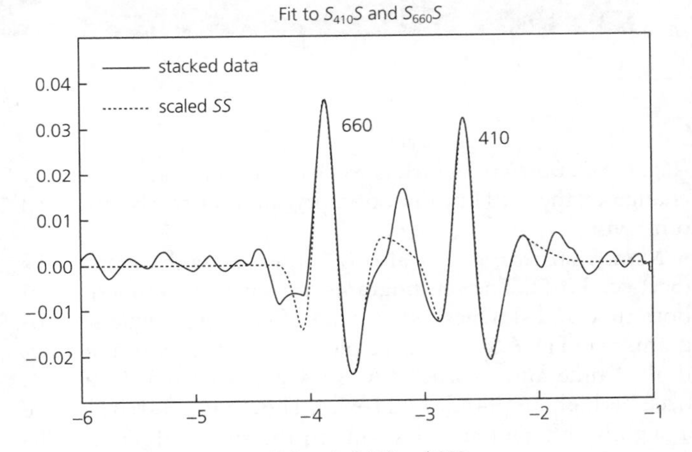

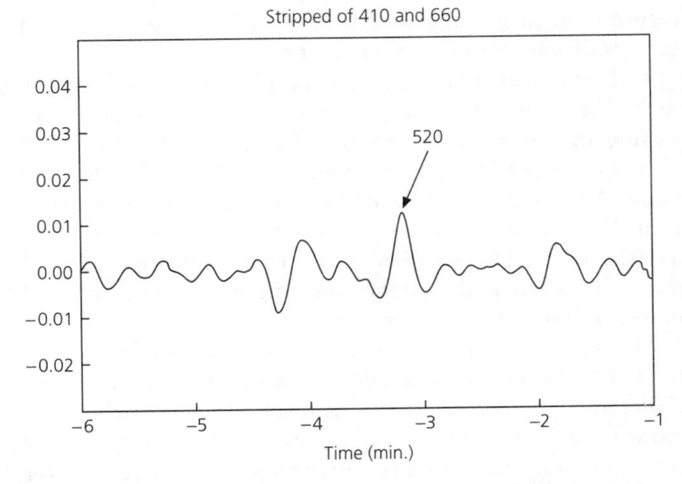

Fig. 6.5-3 Stacking long-period seismograms to identify the depth of mantle discontinuities by enhancing precursors to *SS*. The initial stack (*top*) shows the $S_{410}S$ and $S_{660}S$ underside reflections off the 410 km and 660 km discontinuities, magnified by a factor of 10. A theoretical signal generated from the *SS* wave (*center*) is subtracted from the observed stack to reveal the reflection from the 520 km discontinuity (*bottom*). (Shearer, 1996. *J. Geophys. Res., 101*, 3053–66, copyright by the American Geophysical Union.)

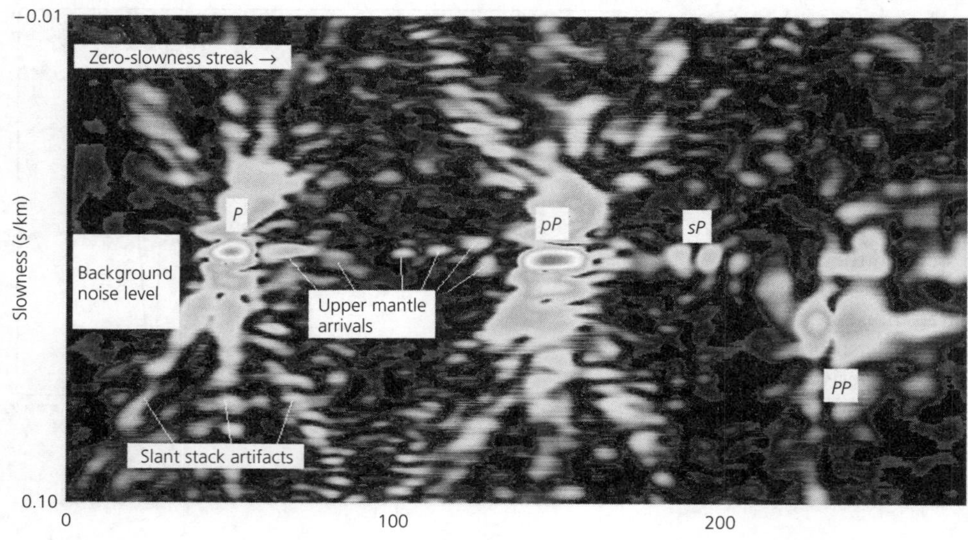

Fig. 6.5-4 Slant stack of seismograms at 279 stations for a deep (476 km) earthquake. The bull's-eyes are concentrations of seismic energy for particular arrivals. (Vidale and Benz, 1992. Reproduced with permission from *Nature*.)

(Fig. 6.5-3, *bottom*), which is weak due to the gradual velocity change at the 520 km discontinuity, and so rarely observed otherwise.

Mantle structures can also be observed with slant stacks (Section 3.3.5). The seismograms are stacked as functions of both time and slowness, so instead of getting a single seismogram, as in Fig. 6.5-3, we get a plot of seismic energy as a function of time and slowness. As shown in Fig. 6.5-4, arrivals occur as high-amplitude bull's-eyes. The *P* and *pP* arrivals have a slightly different slowness due to the small (about 1°) difference in incidence angles. The large arrivals create smeared features that are artifacts of the slant stacking.

Stacking is also used to enhance specific normal modes of the earth. The amplitudes of normal modes vary between stations, because they depend on spherical harmonics that are functions of latitude and longitude, which differ between individual modes (Section 2.9.3). Although simply stacking seismograms from different sites does not make spectral peaks stand out better, correcting for the theoretical variation in amplitude and phase for a given mode and then stacking enhances the mode of interest and suppresses others (Fig. 6.5-5).

Stacking can be applied to very large volumes of data. Figure 6.5-6 shows record sections generated with thousands of digitally recorded seismograms from different earthquakes and seismometers. The seismograms were rotated into vertical, radial, and transverse components, grouped by source–receiver distance, and then those within half-degree intervals were normalized to a common amplitude and stacked. The strong arrivals in the stacked record sections correspond to the major phases shown in the travel time curves. It is interesting to compare this analysis of global seismic data spanning large distance ranges with reflection seismic data analysis (Section 3.3.4). For reflection data, CMP stacking involves forming common

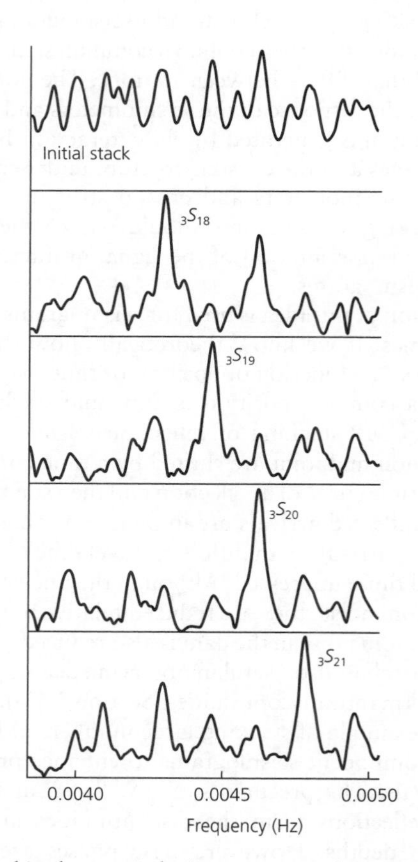

Fig. 6.5-5 Stacking long-period seismograms to enhance specific normal modes of the earth. Although a given mode multiplet is not enhanced by simply stacking seismograms from different sites (*top*), stacking using its predicted variation between sites enhances the multiplet and suppresses others (*lower panels*). (Mendiguren, 1973. *Science, 179*, 179–80, copyright 1973 American Association for the Advancement of Science.)

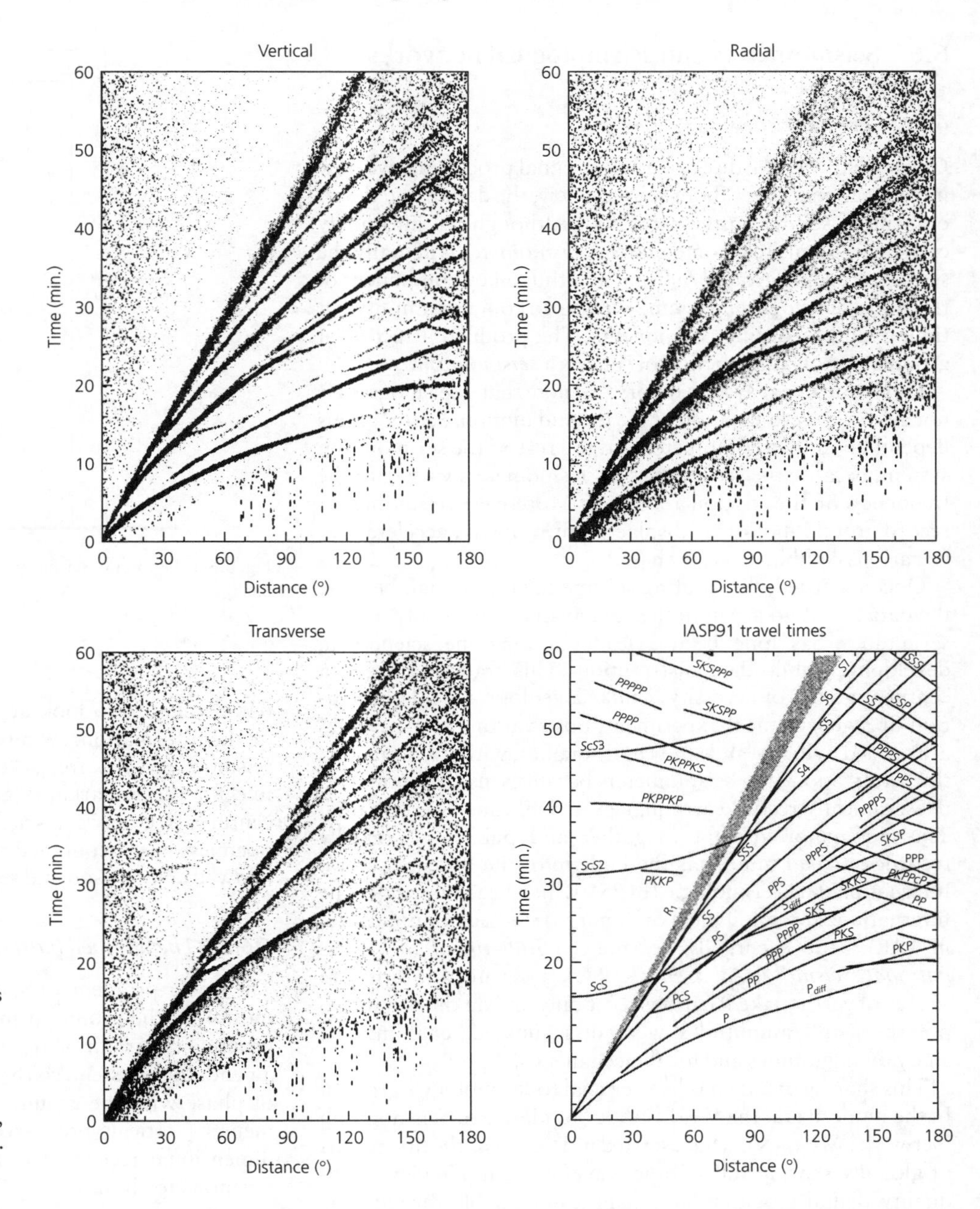

Fig. 6.5-6 Stacking of global seismograms to produce record sections. The three stacks, each for a different component, show distinct arrivals that can be compared to those predicted by the travel time curve for an earth model. (Astiz *et al.*, 1996.)

midpoint gathers and stacking them over all source–receiver distances (offsets) (Fig. 3.3-18), to produce synthetic zero-offset traces on which reflected arrivals are enhanced. These traces are then shown together to produce a common midpoint section, a function of midpoint and time. By contrast, the global data are gathered by common offset, stacked for that offset, and then displayed as a function of offset and time. This operation only reduces noise, rather than enhancing specific arrivals, and so shows various arrivals (direct waves, reflections, surface waves, etc.). Another example was shown in Fig. 2.7.4, where many long-period seismograms were stacked to demonstrate the group and phase velocities of surface waves.

In these or other stacking operations, one possible source of systematic error is incorrect transformation of the data between different times or positions. Interestingly, in the very different cases just discussed, a common difficulty is lateral variation in structure. In the reflection example, structures may dip rather than be flat-lying, causing traces with common midpoints not to sample the same point on a reflector (Fig. 3.3-19). In the global travel time analysis, seismograms for the same source–receiver distance differ when the structure between the source and the receiver differs. An analogous effect occurs for normal modes due to deviations of the structure from spherical symmetry. Nevertheless, because in most cases structure varies primarily with depth, these stacking operations generally work well.

6.6 Seismometers and seismological networks

6.6.1 *Introduction*

Given what we have discussed about signal processing, we now introduce some ideas about *seismometry*, the design and development of seismic instrumentation. Although we informally call such systems seismometers, the *seismometer* is actually the sensor recording ground motion, and thus a key component of the entire *seismograph* system, which also contains amplifying, timing, and recording components. The product, a record of ground motion as a function of time, is a *seismogram*.

Following linear system theory, we note that a seismogram is not an exact representation of the ground motion. Seismograms depend upon the seismometer and the rest of the seismograph system, because the sensitivities of seismometers vary with the frequency of the motion recorded. Moreover, seismometers record ground motion as displacement, velocity, acceleration, or various combinations of these.[1]

Once recorded, distributing seismic data is crucial, because the data are of no use until they are available for study. Hence seismology has long been a leader among the sciences in developing public data distribution. This tradition began a century ago out of necessity. Unlike a geological field observation or a geochemical experiment, observations at many sites are needed to locate and study earthquakes, with the more data the better. Soon after seismometers became sensitive enough to teleseismically record earthquakes, arrival times were shared. The first major attempt to gather and publish seismically recorded arrival times was the bulletin of the Bureau Central International de Séismologie (BCIS), which began in 1904. The International Seismological Summary (ISS) began publication in 1913,[2] and eventually became the *Bulletin of the International Seismological Centre* (ISC), now an authoritative source of earthquake locations. Not only arrival times but also polarities and amplitudes were disseminated, enabling the study of magnitudes and focal mechanisms.

This sharing of data has been crucial to seismology's growth. In the modern era, the World Wide Standardized Seismograph Network (WWSSN), which started in 1962, was the first means of globally sharing full seismic waveform data. Today, high-quality digital global seismic data are available through the Federation of Digital Broad-Band Seismographic Networks (FDSN), of which the stations of the US-sponsored Incorporated Research Institutions for Seismology (IRIS) are a part. Data and results such as earthquake locations are also provided by national and regional data centers. Seismologists anywhere in the world need only a computer and access to the Internet to freely and conveniently obtain terabytes[3] of digital seismic data, software to look at it, and a great deal of other earthquake information. As much as any development in theory or seismometry, this free access to data and software is responsible for the remarkable growth of the field within the past century. Not only can scientists work more efficiently, but this openness has encouraged the sharing of data and models, and allowed comparison and testing of results.

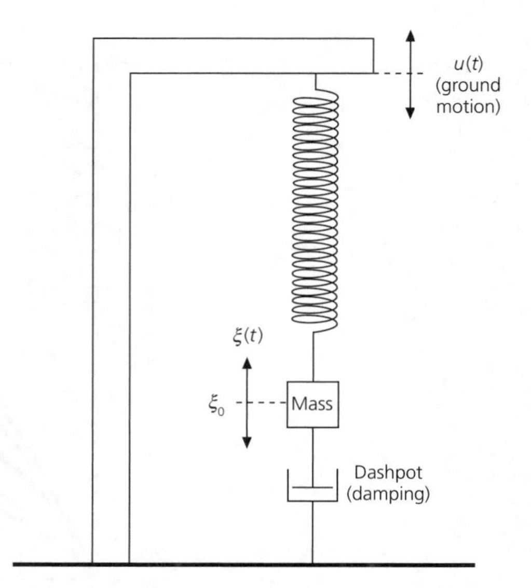

Fig. 6.6-1 Pendulum seismograph consisting of a mass, a spring, and a dashpot.

6.6.2 *The damped harmonic oscillator*

The basic problem of seismometry is how to measure the motion of the ground using an instrument that is also on the ground. The traditional solution is to use an inertial, known as a pendulum, system, so that the motion of the pendulum is out of phase with the ground motion. Three orthogonal seismometers (vertical, north–south, east–west) can give a three-dimensional record of ground motion. A schematic vertical seismometer is shown in Fig. 6.6-1. The key elements of the system are the mass, the spring, and a dashpot, or damping device. We consider such a system in general, without concern for the mechanics of how it is actually implemented.

This mechanical seismometer system is a damped simple harmonic oscillator. If the spring equilibrium length in the absence of ground motion is ξ_0, the spring exerts a force proportional to its extension from equilibrium as a function of time, $\xi(t) - \xi_0$, times a spring constant k. The dashpot, with damping constant d, exerts a force proportional to the velocity between the mass (m) and the earth. So, for a ground motion $u(t)$,

$$m\frac{d^2}{dt^2}[\xi(t) + u(t)] + d\frac{d\xi(t)}{dt} + k[\xi(t) - \xi_0] = 0. \quad (1)$$

[1] This is analogous to the way animals see differently; the electromagnetic radiation is the same, but human eyes respond slightly differently than those of bears (which are very nearsighted), and entirely differently from the hexagonally tiled eyes of flies.

[2] Its original name was the *Monthly Bulletin of the Seismological Committee of the British Association for the Advancement of Science*.

[3] One terabyte (Tbyte) equals 10^{12} bytes.

If we define $\xi(t) - \xi_0$ as $\xi(t)$, the displacement relative to the equilibrium position, Eqn 1 becomes

$$m\ddot{\xi} + d\dot{\xi} + k\xi = -m\ddot{u}, \tag{2}$$

or

$$\ddot{\xi} + 2\varepsilon\dot{\xi} + \omega_0^2\xi = -\ddot{u}, \tag{3}$$

where the single and double dots denote the first and second time derivatives, $\omega_0 = \sqrt{k/m}$ is the natural frequency of the undamped system, and the damping is described by $\varepsilon = d/(2m)$. This is a linear differential equation with constant coefficients that we encountered when we used a damped harmonic oscillator as a model for anelasticity (Section 3.7.5). Thus Eqn 3 is the inhomogeneous (forcing term) version of Eqn 3.7.8, where the damping term ε appeared as $\omega_0/2Q$. To solve it, we assume that

$$u(t) = e^{-i\omega t} \quad \text{and} \quad \xi(t) = X(\omega)e^{-i\omega t} \tag{4}$$

and substitute Eqn 4 into Eqn 3 to yield

$$X(\omega)(-\omega^2 - 2\varepsilon i\omega + \omega_0^2)e^{-i\omega t} = \omega^2 e^{-i\omega t}, \tag{5}$$

or

$$X(\omega) = -\omega^2/(\omega^2 - \omega_0^2 + 2\varepsilon i\omega), \tag{6}$$

which is the instrument response produced by a ground motion $e^{i\omega t}$.

$X(\omega)$ is complex and can be written in terms of the amplitude and phase responses

$$X(\omega) = |X(\omega)|e^{i\phi(\omega)}, \tag{7}$$

where

$$|X(\omega)| = \omega^2/[(\omega^2 - \omega_0^2)^2 + 4\varepsilon^2\omega^2]^{1/2}, \tag{8}$$

$$\phi(\omega) = -\tan^{-1}\frac{2\varepsilon\omega}{\omega^2 - \omega_0^2} + \pi. \tag{9}$$

As shown in Fig. 6.6-2, these functions have several interesting features. First, as the angular frequency of the ground motion, ω, approaches the natural frequency of the pendulum, ω_0, the amplitude response is large. This effect, called *resonance*, is like "pumping" a playground swing at its natural period. Thus the seismometer responds best to ground motion near its natural period.

For frequencies much greater than the natural frequency, $\omega \gg \omega_0$, $|X(\omega)| \to 1$, and $\phi(\omega) \to \pi$, so the seismometer records the ground motion, but with the sign reversed.[4] To see why this

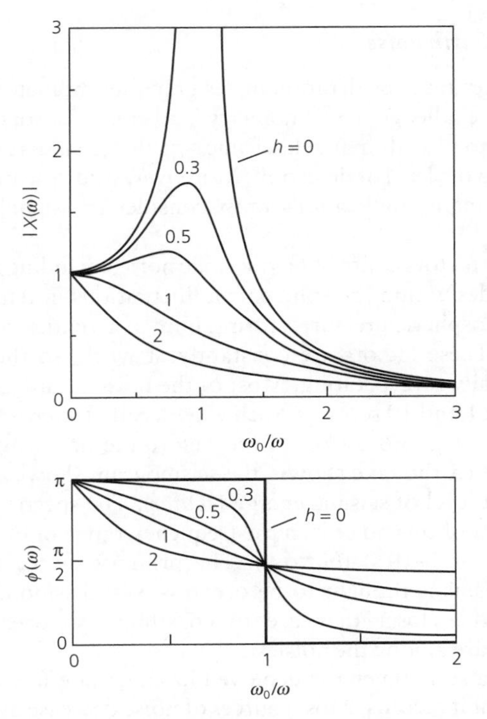

Fig. 6.6-2 Amplitude response $|X(\omega)|$ and phase delay $\phi(\omega)$ for a pendulum seismometer such as that shown in Fig. 6.6-1.

occurs, consider Eqn 3. For $\omega \gg \omega_0$, the $\ddot{\xi}$ term is the largest term on the left-hand side, so $\ddot{\xi}$ approximately equals $\ddot{u}$. Thus the seismometer responds to the ground *displacement*. On the other hand, for frequencies much less than the natural frequency, $\omega \ll \omega_0$, $|X(\omega)| \to \omega^2/\omega_0^2$, and $\phi(\omega) \to 0$. Hence, in this case the seismometer responds to *acceleration*, as can be seen from Eqn 3, because the $\omega_0^2\xi$ term is dominant, so ξ is proportional to $\ddot{u}$. The shape of the instrument response depends on the damping factor $h = \varepsilon/\omega_0$. For $h = 0$, the system is undamped, and the amplitude response is peaked around the resonant frequency, $\omega = \omega_0$. The seismometer amplifies ground motion with periods near its natural period. As damping is increased, the curve is smeared out. Thus the natural period and damping are used to design a seismometer to record ground motion in a particular period range.

Figure 6.6-2 bears a strong resemblance to Fig. 3.7-13, which showed the frequency response for a damped harmonic oscillator as a function of Q. The plots are slightly different, in that Fig. 3.7-13 is plotted as a function of ω, and Fig. 6.6-2 is plotted as ω_0/ω. In addition, Fig. 6.6-2 is normalized to the value at $\omega_0/\omega = 0$. However, the curves convey the same information because h and Q are related as $h = 1/2Q$. The Q values in Fig. 3.7-13 of 5, 15, and 100 correspond to h values of 0.1, 0.03, and 0.005, all of which would plot close to the curve for $h = 0$ in Fig. 6.6-2.

[4] To see this, quickly jiggle an object hanging by a rubber band and note that its motion is out of phase with your hand.

6.6.3 *Earth noise*

An important consideration in designing seismometers is earth noise. A challenge of seismometry is to create sensors sensitive enough to record small teleseismic signals, given that noise sets a limit to the level of detection. Moreover, studies using seismic data in many applications must consider the signal-to-noise ratio.

Many factors contribute to seismic noise, including solar and lunar tides within the solid earth, fluctuations in temperature and atmospheric pressure, storms, human activities, and ocean waves. These factors are constantly at work, so the crust is continually reverberating. Most of the noise occurs at periods between 1 and 10 seconds. Such waves, called *microseisms*, are shown in Fig. 6.6-3 (*top*). Even before the first waves arrive from the earthquake shown, the seismogram shows a roughly constant level of seismic energy (*center*). The spectrum shows that most of this noise is in the frequency range of 0.1–0.2 Hz (periods of 5–10 s) (*bottom*). The primary source for these microseisms is thought to be ocean waves. Seismometers are noisier the closer they are to coastlines, so ocean island stations are among the noisiest.

How a seismometer is deployed has a great effect upon the noise that it records. Most sources of noise decrease away from the surface, so permanent seismometer installations are often in boreholes. For portable seismometers, burying them even half a meter beneath the surface greatly reduces noise from daily temperature fluctuations. Rain generates high frequency noise, and wind, coupled to the ground through the roots of swaying trees, can generate severe long-period noise. Human activity (trucks, trains, machinery, etc.) causes significant ground noise, so seismologists deploying temporary stations face a trade-off between the convenience (continuous power, security, constant temperature, no flooding) of building basements and the lower noise of remote sites.

6.6.4 *Seismometers and seismographs*

Seismometers record ground motions ranging from large high-frequency accelerations near an earthquake to small ultra-long-period normal mode signals. Because no single seismograph can do this, different instruments have evolved to handle the different *dynamic ranges* and *frequency ranges* of seismic waves.

Dynamic range is measured in decibels (dB), which increase by 20 for each order of magnitude increase in amplitude. Thus, if signal A_1 is five orders of magnitude larger than signal A_2, $A_1/A_2 = 10^5$, and the dynamic range is 100 dB. The displacements associated with a magnitude 2 earthquake may be as low as 10^{-10} m, whereas teleseismic displacements from a magnitude 8 earthquake may be on the order of 10^{-1} m, and displacements near a large earthquake can be much greater. Thus the dynamic range of seismometry is at least 180 dB. Similarly, the frequency range of seismometers spans seven orders of magnitude from Earth tides (0.000023 Hz) to ultra-high frequencies of greater than 200 Hz for very shallow structure investigations.

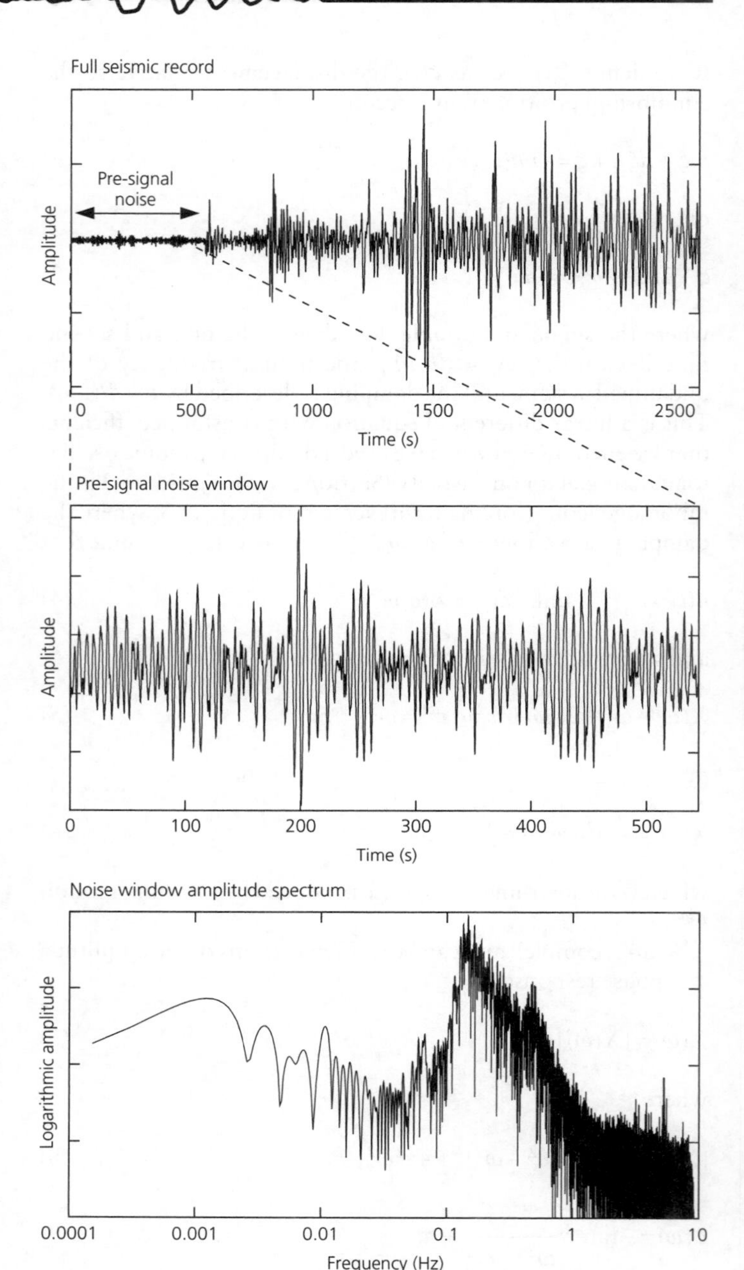

Fig. 6.6-3 Demonstration of seismic noise on a broadband seismogram in Hudson, New York, from an April 7, 1995, Tonga earthquake. *Top*: Seismic noise appears before the first arrival, which is P_{diff}. *Center*: Visual examination of the noise shows waves with a dominant period of about 5–6 s, called *microseisms*. *Bottom*: The spectrum of the noise has largest amplitude in the 5–10 s period range.

The earliest attempts to record the motions of earthquakes used *seismoscopes*, which differ from seismographs in that they record ground motion without time information. The first known seismoscope, built by the Chinese astronomer Chang

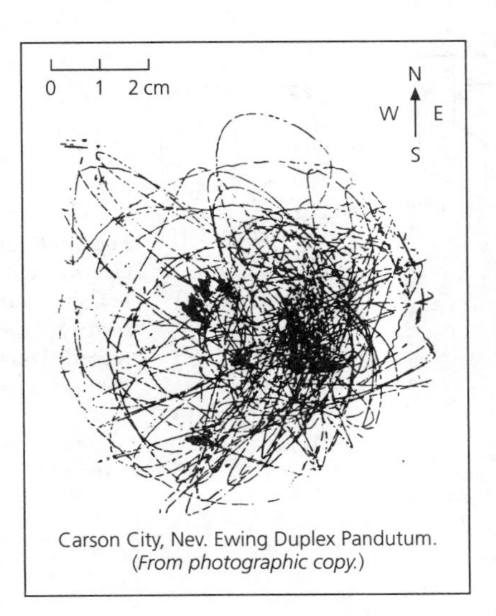

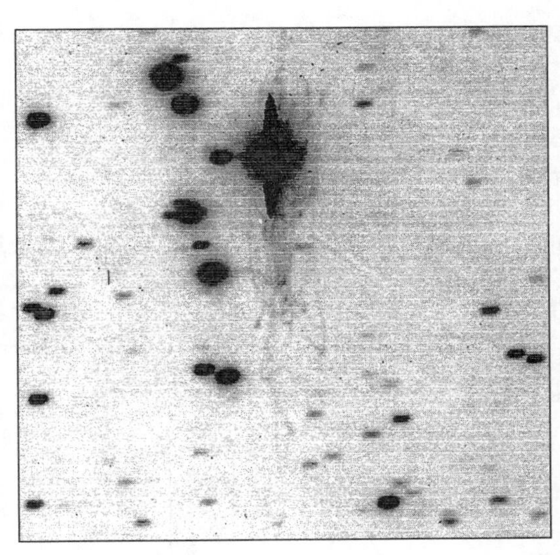

Fig. 6.6-4 Two examples of seismoscope recordings, which show the amplitudes of motions without a record of time. *Left*: Seismogram of the great 1906 San Francisco earthquake, recorded by the Ewing duplex pendulum seismoscope in Carson City, Nevada. (Kanamori, 1988. Importance of historical seismograms for geophysical research, in *Historical Seismograms and Earthquakes of the world*, ed. W.H.K. Lee, H. Myers and K. Shimizaki, copyright 1988 by Academic Press, reproduced by permission of the publisher.) *Right*: Seismogram of a $m_b = 4.3$ earthquake in Hawaii, recorded as a telescope image at the Hawaii Telescope Observatory. The dark images are stars, and the lines emanating from the large star at the upper center of the image result from tilting of the telescope during the earthquake. (Courtesy of L. Meech.)

Heng in about AD 132, consisted of a pendulum inside a 6 ft-diameter jar. Eight dragons' heads with metal balls in their mouths were placed around the rim of the jar, so the balls would drop in the direction from which seismic waves arrived. Later seismoscopes included a pendulum etching a path on a bed of sand (A. Bina, 1751), a collection system for a bowl filled to the brim with mercury (A. Cavalli, 1784), and optical reflection off a basin of mercury (R. Mallet, 1851). Two very different seismoscope recordings are shown in Fig. 6.6-4.

Early seismometers, incorporating a record of the time-dependence of the ground motion, were purely mechanical instruments like that outlined in Section 6.6.2. Seismometry began with the designs of F. Cecchi around 1875, and developed rapidly through the work of seismologists like J. Milne, J. Ewing, and T. Gray. The first teleseismic recording was by a seismograph in Potsdam of a Japanese earthquake in 1889. By the start of the twentieth century a global network of more than 40 seismographs was in operation. Such instruments often produced excellent data but responded best to very large earthquakes because their magnifications were low, only about 100 times the actual ground motion.

Higher magnifications are achieved by using electromagnetic instruments, based on a design introduced by Galitzin in 1914 that is now common. The motion of the pendulum relative to the frame is measured by moving a coil attached to the mass through the magnetic field produced by a magnet fixed to the seismometer frame. The voltage produced in the coil is proportional to the time rate of change of the magnetic field, and thus to the velocity of the mass relative to the frame (Fig. 6.6-5). The sensitivity can be increased by feeding the output from this sensor into a galvanometer, a wire suspended by a thin fiber such that it is deflected by the current produced by the sensor (Fig. 6.6-6). A mirror is attached so that ground motion deflects the mirror and thus changes the position of a beam

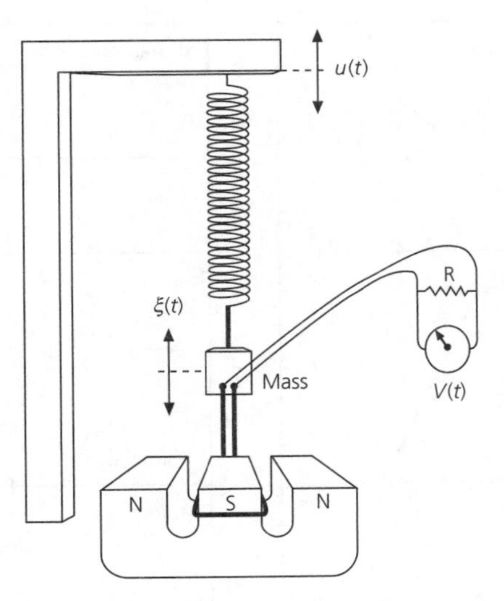

Fig. 6.6-5 Schematic illustration of an electromagnetic seismograph, in which the mass is coupled to an electromagnetic transducer. Motions of the mass move the coil through the magnetic field, generating an electric current. The voltage across the coil is proportional to the relative velocity between the mass and the magnet.

of light hitting a piece of photographic paper. The paper is mounted on a helical drum which turns once per hour.

Thus the response of an electromagnetic analog seismometer system is a combination of the pendulum, transducer (electromagnetic velocity sensor), and galvanometer responses. These are shown as log–log plots in Fig. 6.6-7. The pendulum response (Fig. 6.6-7a, b) is proportional to ω^2 for $\omega < \omega_s$, the pendulum frequency. The transducer response (Fig. 6.6-7c, d) is proportional to ω because it responds to the velocity, the derivative of displacement. The galvanometer response

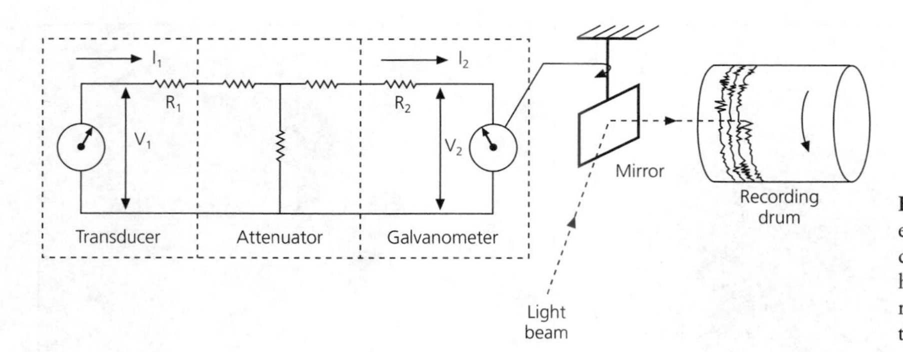

Fig. 6.6-6 Coupling of the transducer of an electromagnetic seismograph to a galvanometer, which deflects a mirror and thus a light beam, causing a time history of the voltage and thus the mass movements to be recorded on photographic paper. Timing pulses deflect the mirror to make minute and hour marks.

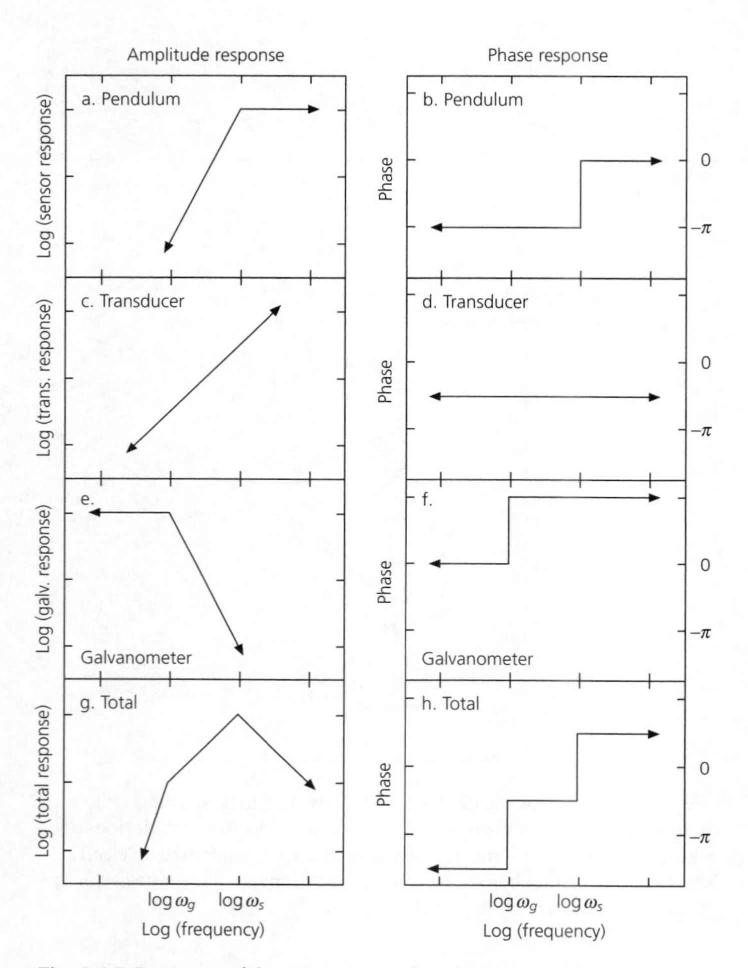

Fig. 6.6-7 Response of the components of an electromagnetic seismograph system. Left panels show the amplitude responses, and right panels show the phase responses. ω_s and ω_g are the pendulum and galvanometer frequencies.

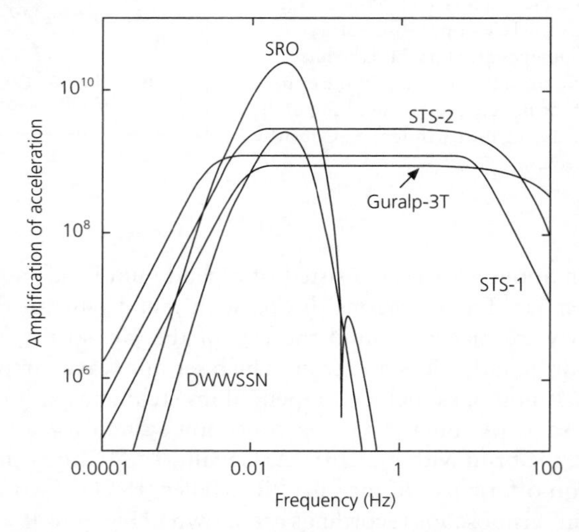

Fig. 6.6-8 Frequency domain instrument responses for several types of seismometers. The SRO and DWSSN sensors have responses peaked at long periods and so do not record high-frequency signals. The STS-1, STS-2, and Guralp-3T sensors are broadband seismometers with a flat response over a wide range of frequencies.

(Fig. 6.6-7e, f) falls off as ω^{-2} for $\omega > \omega_g$, the galvanometer frequency. The combined effect is shown in Fig. 6.6-7g, h. Thus, the response of an electromagnetic seismometer can be "shaped" by choosing the pendulum and galvanometer periods.

Two classic electromagnetic instruments used heavily for years were the World Wide Standardized Seismograph Network (WWSSN) long- and short-period instruments. The long-period (LP) instrument had a pendulum period of 15 s (30 s in some early versions) and a galvanometer period of 100 s. The short-period instrument had a 1 s pendulum and a 0.75 s galvanometer. Each WWSSN station had three LP and three SP instruments oriented to record ground motion in the vertical, east–west, and north-south directions. The resulting response curve of the LP instrument (labeled "DWWSSN" from when some of the WWSSN seismometers were converted to record digitally) is shown in Fig. 6.6-8. Instruments ran at several possible magnifications (gains). The two different instruments were designed to reduce the effects of seismic noise. The LP sensors had peak sensitivity in the 10–40 s range, making them ideal for long-period teleseismic studies. The SP sensors were peaked at around 1 s, a good period with which to pick the travel times of *P* waves.

A sample of the data is shown in Fig. 6.6-9. The record, covering 24 hours, has calibration pulses at the beginning, which can be used to check the amplitude and phase

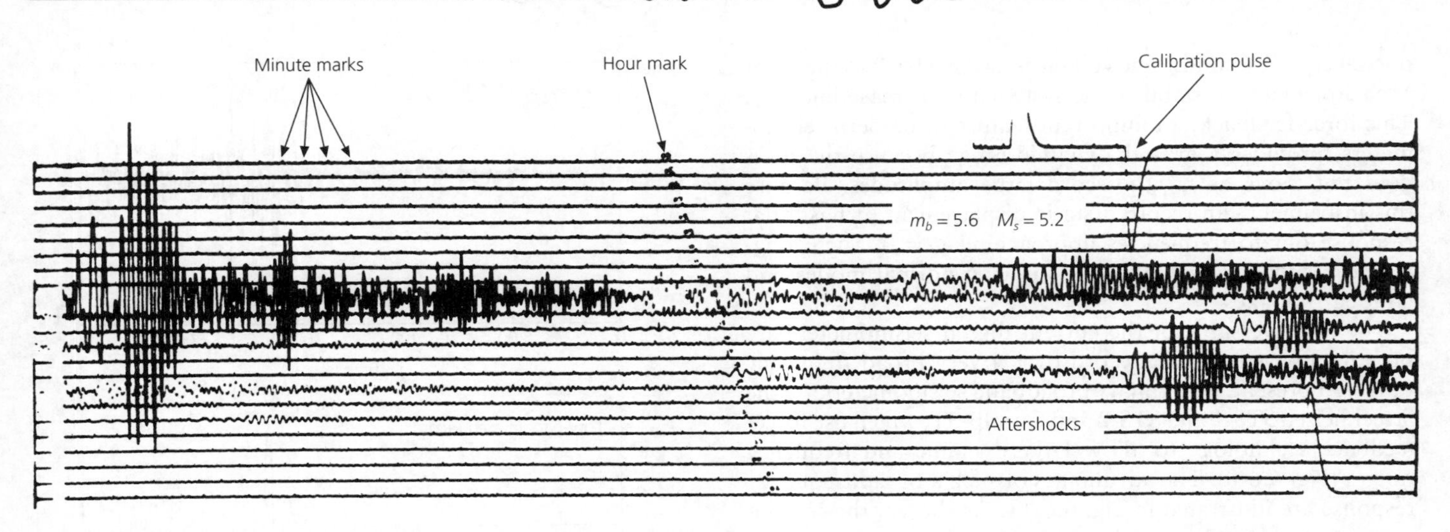

Fig. 6.6-9 Sample WWSSN seismogram, showing the long-period vertical component from an earthquake in the Indian Ocean, recorded 36° away in Pakistan.

calibration. Timing marks, generated by crystal clocks accurate to 1 part in 10^7 are placed at each minute (short mark) and each hour (longer mark). Every sixth hour has no hour mark. This timing allowed arrival times to be read accurately, and the calibration allowed studies using true amplitudes. The seismograms were microfilmed and made available to the seismological community.

Although many results discussed in this text were derived from such data, using WWSSN data was cumbersome. Microfiche records had be acquired, examined in a microfiche reader, copied, and refiled. The traces were then digitized by taping them to a special table that contained a grid of electromagnetic wires and then tracing the seismogram with a cursor. After digitization, the seismogram was interpolated to a desired sampling rate. The hand digitization added a source of error, as it was not always easy to follow the trace of interest, especially for large earthquakes where the surface waves could wrap around the seismic record for several hours. Because of the effort involved, entire Ph.D. dissertations might involve the analysis of only tens or hundreds of seismograms, a task that is now done in minutes to days.

The replacement of analog seismographs by digital broadband instruments has important advantages. The newer seismometers provide better data over a broader frequency band, and the digital data are available via magnetic tape, compact disk, or the Internet, making computer analysis much easier. Routine processing, such as rotating into radial and transverse components and making record sections, has become nearly trivial. Large volumes of data are available and can be processed easily. For example, as of 2000 the IRIS Data Management Center had over 7 Tbytes of digital data available over the Internet either immediately or with only the short delay needed for it to be read from mass storage systems.

Some of the technology involved in more recent seismograph systems is illustrated by one of the first digital seismological

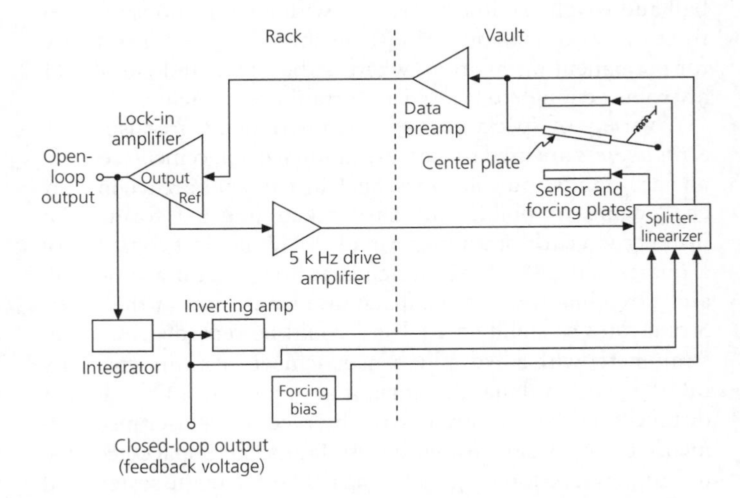

Fig. 6.6-10 Block diagram of the sensing and feedback electronics of an IDA gravimeter recording system. (Agnew *et al.*, 1976. *Eos Trans. Am. Geophys. Un.*, *57*, 180–8, copyright by the American Geophysical Union.)

systems, the instrument used by the International Deployment of Accelerometers (IDA) shown schematically in Fig. 6.6-10. The sensor is a force-feedback gravimeter that detects vertical ground motion by the resulting change in gravity. The gravimeter mass is connected to the center plate of a capacitor whose outer two plates are fixed. As the mass moves, the voltage between the center plate and the outer plates is proportional to the displacement. A 5 kilohertz alternating voltage applied to the outer plates is amplitude-modulated (Section 2.8.1) by the lower-frequency seismic signal. The modulated signal is fed to an amplifier that generates a voltage proportional to the displacement of the mass. This signal then goes to an integrator circuit whose output is proportional to the acceleration of the mass. This is the seismic system's output, which is sampled once every ten seconds. The voltage is also fed back to the outer capacitor plates to stabilize the system and increase linearity. This force-feedback, an important feature of modern seismometers, provides a greater dynamic range because the mass does not move as far to record large amplitudes. Because this instrument can record a static displacement, it has a flat response out to frequencies approaching $\omega = 0$. Such long-period response is valuable for studying normal modes and large earthquakes.

The most versatile of the current digital seismometers are broadband systems that record over a very broad frequency range. At present, the primary broadband seismometers are the Streckheisen STS-1 and STS-2 and Guralp-3T, which use force-feedback technology to allow large dynamic and frequency ranges (Fig. 6.6-8). The advantages of such a broad frequency response are illustrated in Fig. 6.6-11. As shown, the seismogram can be filtered to isolate and give excellent records of two very different overlapping signals. These seismometers are very compact (the three-component STS-2 is the size of a bowling ball and weighs 20 lb)[5] but record with a flat response at over three orders of magnitude in frequency. The STS-1 is designed for permanent installation, whereas the STS-2 and Guralp-3T are robust enough to be used as portable instruments.

A variety of specialized seismic instruments are also used. *Strainmeters* are used to measure gradual displacements, especially near faults and volcanoes. Such instruments are technically challenging to build, and have taken unusual forms. For instance, an early strainmeter made by H. Benioff consisted of a quartz rod 24 m long, attached to the ground at one end, and extending through a capacitance transducer at the other. Strain rates as small as 10^{-15} s^{-1} could be recorded. A recent strainmeter with a hydraulic sensor achieves a strain sensitivity of 10^{-12} with a dynamic range of about 130 dB. Over longer distances, horizontal strains are observed using laser measurements between sites (often across faults) and space-geodetic techniques (Section 4.5), including the GPS satellite system and very long baseline radio interferometry.

At the other end of the spectrum of seismic instrumentation are strong-motion sensors that record strong shaking near an earthquake. Whereas strainmeters record minute displacements, strong-motion sensors, also called accelerometers, record accelerations up to 2 g without breaking or going off scale. For example, horizontal accelerations of 1.25 g were recorded 3 km from the 1971 San Fernando Valley earthquake, and vertical accelerations of 1.74 g were recorded 1 km from the 1979 Imperial Valley earthquake. Thus the seismometer pendulum frequency ω_0 is chosen to exceed the highest frequency of interest (about 20 Hz). These instruments are stable because the small pendulums make the accelerometers less susceptible to tilt and drift than longer-period instruments. A damping parameter (often 0.7 of the critical value) is chosen to

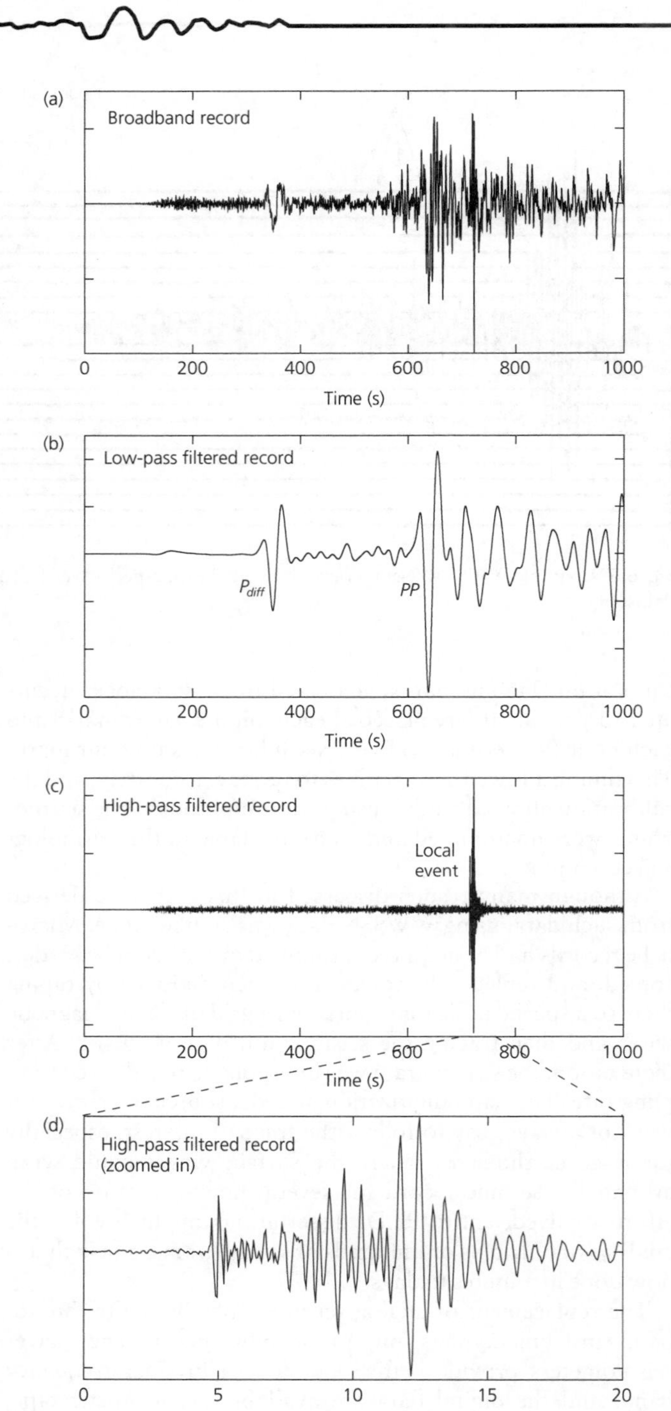

Fig. 6.6-11 STS-2 broadband seismogram recorded in Slippery Rock, PA, from a July 3, 1995, Tonga earthquake. Because the seismometer records a wide range of frequencies, the same seismogram can be used to study both local and teleseismic events. (a): The original broadband record. (b): The same record, low-pass filtered at a frequency of 0.03 Hz, showing the long-period teleseismic signals from the Tonga event. (c): The record high-pass filtered at 0.5 Hz, showing the high-frequency signals from a local event. (d): A zoom-in of the high-pass filtered record shows the full waveform of the local event. The $S-P$ time suggests that the event was 20 km away from the station, probably a local quarry blast.

[5] Before such technology, some mechanical seismometers built in the first half of the twentieth century weighed more than 20 tons because the large mass gave higher long-period magnification, as shown by Eqn 6.

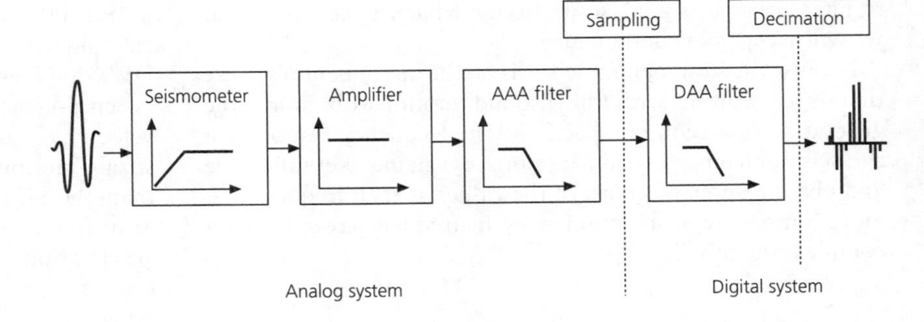

Fig. 6.6-12 Diagram showing the analog-to-digital (ADC) process. The analog part of the system consists of the generation of a seismic signal by the seismometer, its amplification, and analog anti-aliasing (AAA) filtering. The digital part of the system consists of sampling the AAA-filtered signal, filtering the signal further with a digital anti-aliasing (DAA) filter, and then decimating the signal to achieve the desired sampling rate. (Scherbaum, 1996, with kind permission from Kluwer Academic Publishers.)

give a response curve that is flat and directly proportional to ground acceleration from periods of zero to the natural period of the seismometer.

A major advance in seismometry has been in timing, which has long been a difficulty. In the early days of seismology, timing errors played a large part in the mislocation of earthquakes. However, seismometers now receive time signals from GPS satellites, whose atomic clocks are accurate to a billionth of a second. Similarly, although ocean bottom seismometers cannot receive GPS signals, accurate clocks for them are now available.

6.6.5 *Digital recording*

Although digital seismic data are easier to use than analog data, the conversion of continuous ground motion into a digital seismogram is not a trivial matter. Figure 6.6-12 shows how this is done. Ground motion, represented by the waveform at the left, is detected by the seismometer through the motion of the mass. This motion is converted into an analog electrical signal and then amplified. To avoid a spurious signal due to aliasing (Fig. 6.4-3), a combination of anti-aliasing filters is used. Many seismometers use an initial frequency domain low-pass filter as an analog anti-aliasing (AAA) filter. The filtered signal is then oversampled at a rate that is at least twice the frequency of the AAA filter in order to avoid aliasing. This signal is then convolved with a digital anti-aliasing (DAA) filter, often called a finite impulse response (FIR) filter, and finally resampled at twice the desired Nyquist frequency.

An example of a FIR filter is shown in Fig. 6.6-13a, with the resulting signal shown in Fig. 6.6-13c. The FIR filter maintains the shape of the pre-filtered signal, but introduces spurious noncausal arrivals that might be mistaken for early stages of earthquake rupture. These precursory signals result because the FIR filter's impulse response is an emergent signal. This effect can be removed by correcting the phase of the FIR filter to make it causal (Fig. 6.6-13b). This filter does not cause precursory signals (Fig. 6.6-13d), but the shapes of the waveforms are changed. We noted a similar phenomenon in Section 3.7.8, where anelasticity acted as a filter, removing high frequencies and making the waveforms noncausal unless the phase was changed. As discussed in Section 6.3.3, there is no perfect way

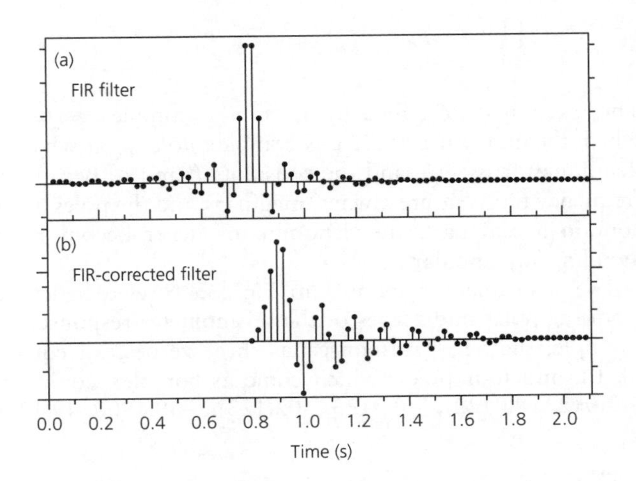

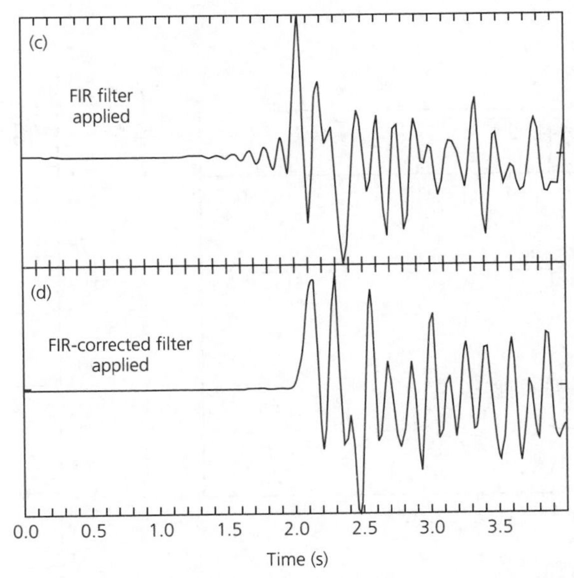

Fig. 6.6-13 Example of a FIR filter, a type of DAA filter, and its effects. When the FIR filter (a) is used for the digital anti-aliasing, the resulting signal (c) retains the wave shape of the original signal, but is preceded by high-frequency artifacts. When a phase-corrected FIR filter (b) is applied instead, the precursory effects vanish (d), but the seismic signal is phase-shifted from the original. (After Scherbaum, 1996, with kind permission from Kluwer Academic Publishers.)

to filter a seismic signal, so we decide what we seek and what we will accept as a consequence.

Because the seismogram depends on the instrument response that is convolved with the ground motion, obtaining the ground motion requires specifying the frequency response of the seismometer. This can be done by giving the amplitude and phase response as a list of the values at each frequency. A more compact representation gives the frequency response as a complex fraction like

$$T(i\omega) = \frac{\beta \prod_{j=1}^{L} (i\omega - z_j)}{\alpha \prod_{k=1}^{N} (i\omega - p_k)}. \qquad (10)$$

The fraction is described by a set of L complex *zeros* z_j at which the numerator is zero, N complex *poles* p_k at which the denominator is zero, and the constants β and α. Because the frequency terms $i\omega$ are always imaginary and the poles always contain a real part, the denominator never becomes zero, avoiding any singular values.

The instrument responses in Fig. 6.6-8 were calculated from the poles and zeroes of the seismometer responses. For example, the STS-1 response has three zeroes, all equal to (0, 0), and four poles, which come as complex conjugates: (−0.0123, 0.0123), (−0.0123, −0.0123), (−39.1800, 49.1200), (−39.1800, −49.1200). These poles provide the corner frequencies and determine the sharpness of the corners. Similarly, the DWWSSN response has five zeroes and 11 poles.

Seismometers record combinations of ground displacement, velocity, or acceleration, depending upon the application. In a strong-motion seismometer, the displacements may be greater than the size of the instrument itself, so accelerations are measured to keep signals on scale. This makes sense because accelerations are primarily responsible for damage to structures and so are considered in strong-motion studies. At the other end of the frequency spectrum, strainmeters are used to study slow tectonic displacements. In fact, if they measured accelerations, the signals would be so small as to be unusable. Most other branches of earthquake seismology fall in between, using the waves from distant earthquakes, and so use seismometers that record ground velocity.

Although different instruments record displacement, velocity, or acceleration, it is simple to convert between them. For instance, given a velocity record, the acceleration is found by taking the derivative of the seismogram, and the displacement record is found by integrating. This is easily done in the frequency domain, because if $F(\omega)$ is the Fourier transform of $f(t)$, then $i\omega F(\omega)$ is the transform of $df(t)/dt$, and $-\omega^2 F(\omega)$ is the transform of $d^2f(t)/dt^2$ (Section 6.2.4). Thus, a velocity seismogram can be converted to acceleration by multiplying the complex value of its transform at each frequency by $i\omega$, or to displacement by dividing by $i\omega$. Of the three, the displacement

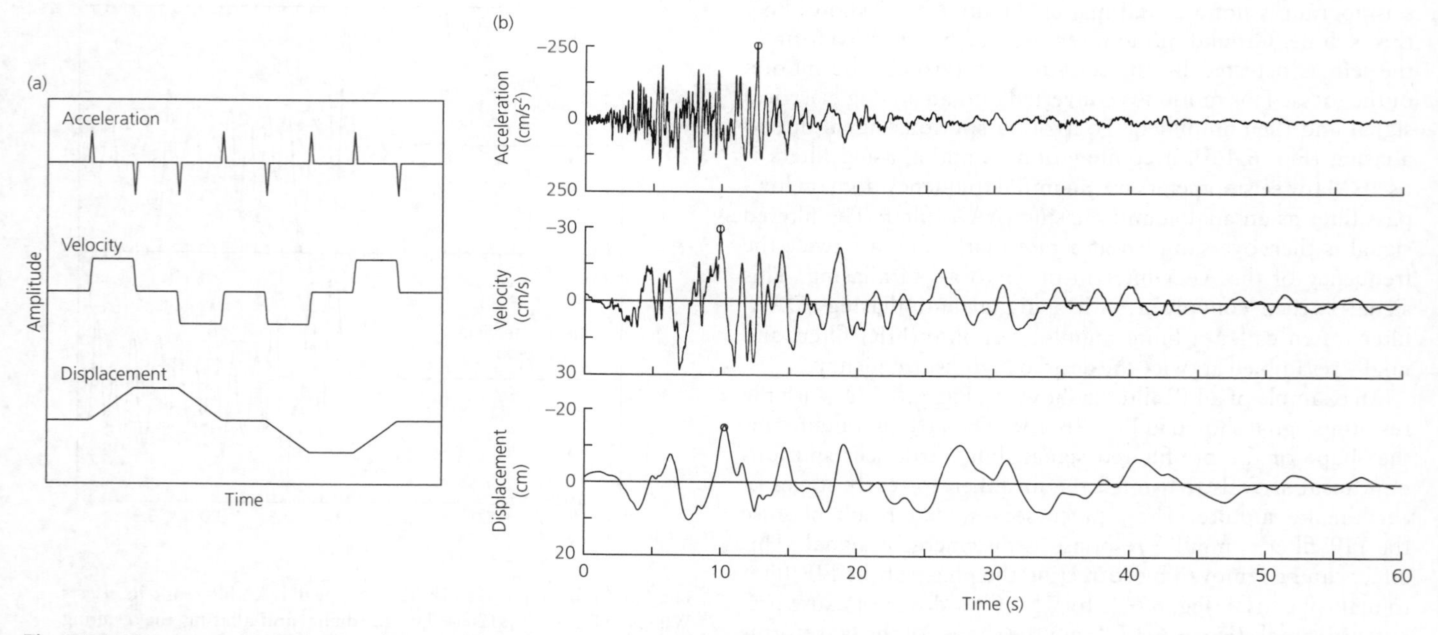

Fig. 6.6-14 Demonstration in the time domain of the relation between displacement, velocity, and acceleration. (a): A synthetic example, consisting of delta function-like acceleration pulses. The velocity and displacement signals are obtained through successive integrations of the accelerogram. (b): A real example, with an accelerogram recorded on the first floor of a building in Los Angeles during the 1971 San Fernando earthquake. The velocity and displacement records were obtained through successive integrations of the accelerogram. (Krinitzsky *et al.*, 1993. *Fundamentals of Earthquake Resistant Construction*. Copyright © 1993. Reprinted by permission of John Wiley & Sons, Inc.)

seismogram has the greatest power at low frequencies, and the acceleration seismogram has the greatest power at high frequencies. In general, displacements have lower frequencies than velocities, and velocities have lower frequencies than accelerations, because integration "smoothes" a signal, whereas differentiation makes it "rougher."[6]

Figure 6.6-14a illustrates this relation with three different versions of the same seismogram. If an accelerogram consists of high-frequency spikes (*top*), then smoother lower-frequency velocity (*center*) and displacement (*bottom*) traces result from integrating once and twice. Figure 6.6-14b shows this effect for a strong-motion seismogram of the 1971 San Fernando earthquake, where the velocity and acceleration records have higher frequencies than the displacement. It is common in earthquake engineering to show the response of a structure to ground motions using a plot that shows the displacement, velocity, and acceleration. Figure 6.6-15 shows this formulation for the data in Fig. 6.6-14b. This representation uses the relation between the Fourier transforms expressed above, so the velocity scale is vertical, whereas the acceleration and displacement scales have opposite slopes as a function of frequency.

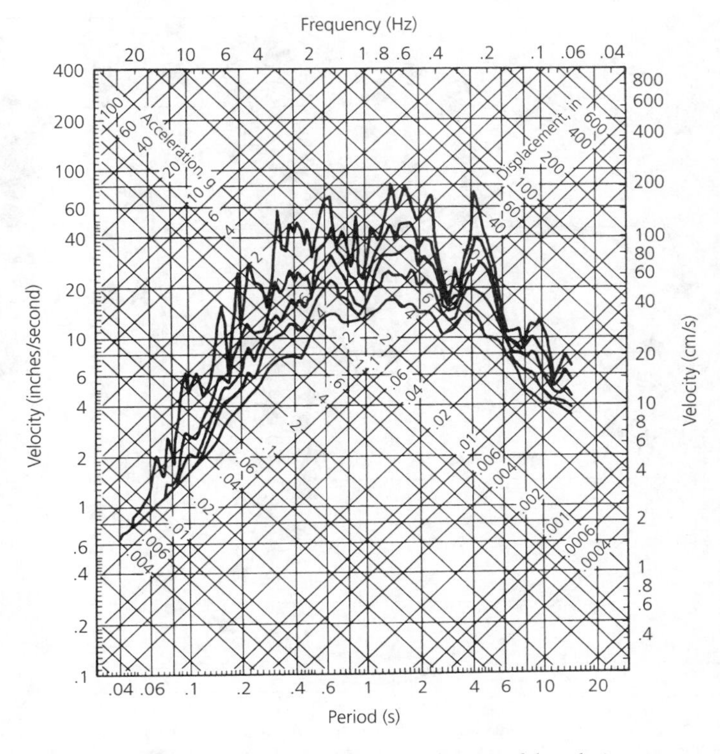

Fig. 6.6-15 Demonstration in the frequency domain of the relation between displacement, velocity, and acceleration. In this example, taken from the accelerogram in Fig. 6.6-14b, a site response spectrum of the building housing the strong-motion seismometer is given as displacement, velocity, and acceleration. The multiple curves show the amplitude of the building response at various levels of damping, with the undamped curve at the top, and successive levels of damping at 2%, 5%, 10%, and 20% of critical damping. (Krinitzsky *et al.*, 1993. *Fundamentals of Earthquake Resistant Construction*. Copyright © 1993. Reprinted by permission of John Wiley & Sons, Inc.)

6.6.6 *Types of networks*

Most seismic experiments require multiple seismometers that are deployed in networks or arrays. Different applications, such as studying regional and global earth structure, resource exploration, seismicity monitoring, or identifying nuclear tests, lead to different deployment geometries. In some cases a unique network of stations is used for a particular application, but often an existing network has a geometry that is a compromise for different objectives.

Although the division is somewhat artificial, deployments of seismometers are often divided into global networks, regional networks, and arrays. Global networks are used to study global patterns of seismicity, plate tectonics, mantle convection, and earth structure. For these purposes seismometers should ideally be spread evenly around the world. This means, however, that the station spacing is too sparse to resolve the entire wave field.[7] Instead, individual measurements at separate stations are combined for applications including locating earthquakes, 3-D tomography, and waveform analyses.

The antithesis of a global network is a local array, where a set of seismometers is deployed with a geometry chosen for a particular goal. Array data are often analyzed as a single entity, as in refraction and reflection studies (Sections 3.2 and 3.3). Other examples are arrays used to locate distant nuclear tests. Data from the array stations are stacked to track the propagation of the wave field across the array, so the wave vector shows the direction the waves came from and the distance they have traveled. One of several exceptions to this division between global networks and arrays is normal mode seismology, where all the stations of a global network are sometimes used as a single array.

Between global networks and arrays are regional networks, which usually focus on the seismicity or structure of a particular region. The data are sometimes analyzed with array techniques, but are more often combined as individual measurements (such as arrival times or amplitudes) in the same way as global network data.

6.6.7 *Global networks*

The global network of seismometers has a rich history. At the start of the twentieth century there were already seismometers

[6] An analogy might be to compare displacement and velocity to the topography and gradient of a mountain. A kilometer of topography over a horizontal wavelength of a meter would be very unusual, but a kilometer of topography over a longer wavelength of 5–10 km would be a normal mountain. Similarly, large vertical gradients are rare at the scale of mountains (El Capitan in Yosemite and the Jungfrau in Switzerland are exceptions), but common at the higher spatial frequency scale of meters, as where a path goes over a boulder.

[7] By analogy to time series, such undersampling is termed *spatial aliasing*.

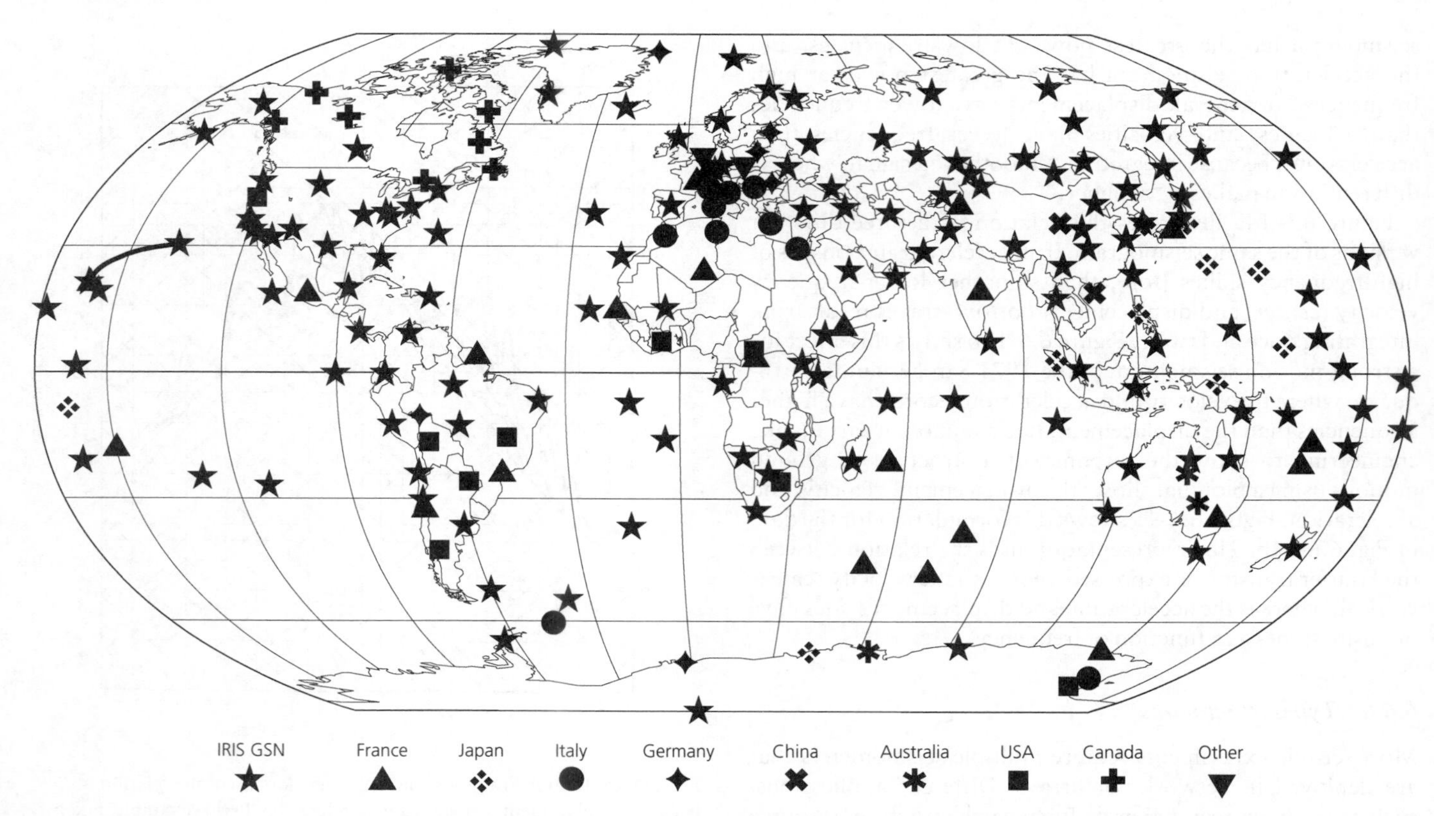

Fig. 6.6-16 Station map of the Federation of Digital Broad-Band Seismographic Networks (FDSN) as of 1999. (Courtesy of the Incorporated Research Institutions for Seismology.)

in locations around the world, operated by groups including many Jesuit institutions. Devastating earthquakes such as the 1906 San Francisco and 1923 Tokyo events spurred the installation of seismometers and the interchange of data. Bulletins of earthquake locations were published by several agencies, the most notable being the ISS/ISC bulletin (Section 6.6.1). By mid-century, the ISS received arrival times from several hundred stations for very large earthquakes. However, there were problems due to a lack of standardization. Different types of seismometers were used, with a wide range in the quality of the response, timing, and station operation practices. As a result, earthquake locations were often poor, and focal mechanisms, which require accurate information about polarities, were rarely derived.

These problems were largely solved with the creation of the World Wide Standardized Seismographic Network. WWSSN seismometers were standardized and had known responses. The network was installed, starting in 1961, to monitor nuclear testing within Eurasia, and had a high density of stations around the borders of the Soviet Union, China, and Eastern Europe. The WWSSN, which reached its peak of about 120 stations in the late 1960s, gave a great boost to geophysics. Several great earthquakes in the 1960s, such as the 1964 Alaska earthquake, provided excellent sources for seismic investigations. WWSSN data were crucial for advances in plate tectonics, earthquake source studies, and global velocity structure.

The first digital stations began to be deployed in the 1970s. Over the next two decades, the number of permanent digital seismometers increased gradually. Following the phase-out of the WWSSN, these became part of the Global Digital Seismic Network, the primary means of global broadband data collection between 1977 and 1986. The GDSN was enhanced by the network of IDA gravimeters, beginning in 1977, and by the French GEOSCOPE network, which has deployed broadband seismometers since 1982.

In 1986, the GDSN gave way to the IRIS Global Seismographic Network (GSN) program, which incorporates many borehole seismometers with an aim toward global coverage, with 128 stations spaced about 2000 km apart. These are extremely quiet, permanent broadband seismic stations of the highest quality. The GSN is part of a larger Federation of Digital Broad-Band Seismographic Networks (FDSN) that also includes the US National Seismographic Network (NSN) and networks from other countries including Canada (CNSN), China (CDSN), France (GEOSCOPE), Germany (GEOFON), Italy (MEDNET), Japan (Pacific 21), and Taiwan (BATS). FDSN station locations are shown in Fig. 6.6-16. Some FDSN stations are also part of the International Monitoring

System (IMS) network used to monitor nuclear testing (Section 1.2.8).

Although the present global network of broadband seismometers relies on land sites, it is hoped that the global network will soon include permanent ocean bottom seismometers (OBS), especially in the Southern Hemisphere, where there is much less land, and coverage is currently very uneven. Although OBS instruments are currently used mostly for temporary deployments, the technology is evolving to the point where permanent sites are practical.

An important aspect of the different networks of high-quality broadband seismometers is considerable standardization in data processing and formatting. All 7 terabytes of seismic data archived by the IRIS DMC[8] as of 2000 are available in a format called SEED (Standard for the Exchange of Earthquake Data), which is the standard for the FDSN. SEED data can be converted into whatever format an investigator requires.

It was not until the mid-1990s, more than 30 years after the start of the WWSSN, that the global number of permanent digital broadband seismometers surpassed the number of WWSSN stations at its heyday. However, digital data from all parts of the FDSN can be retrieved as if it were a single array, making it more powerful than the WWSSN for seismic analyses. Many stations now report in *real time* through satellite telemetry, so seismic signals arrive at data centers a fraction of a second after they occur, allowing better quality control. Efforts are being made to eventually have all GSN stations report in real time, which will be important for applications like tsunami warning. Software has been developed to take real-time data from different networks and display it on the Internet as if it were from a single array. Hence, anyone with a computer and access to the Internet will soon be able to examine global seismic data within seconds of them being recorded.

6.6.8 *Arrays*

For global networks, the precise configuration of individual stations is less important than the total coverage. However, the geometries of seismic arrays are optimized for certain investigations. Arrays can be linear, two-dimensional, and even three-dimensional, incorporating borehole seismometers (Fig. 7.3-8).

There is always a trade-off between the benefits of linear versus two-dimensional arrays. The same number of stations, and therefore cost and time for installation, provides greater resolution if deployed in a linear manner, but the resulting two-dimensional "slice" into the earth does not image the third dimension. Linear arrays have long been the mainstay of *active source* reflection and refraction experiments.[9] A marine linear array is easily deployed by towing hydrophones behind a ship, and similar linear deployments are used for land-based studies. These data are analyzed using techniques discussed in Sections 3.2 and 3.3.

Linear arrays are most useful if the structure being investigated varies most in one direction, as is often the case at plate boundaries. For instance, Fig. 5.3-10 (*bottom*) showed the seismic structure of the East Pacific rise obtained from an array of OBSs. Because the structure of the lithosphere changes much more significantly perpendicular to the ridge than parallel to it, most of the OBSs were deployed in a line crossing the ridge. Most of the remaining seismometers were placed in a second line, parallel to the first. Both lines were aligned along a great circle path to the seismogenic zones of Tonga and South America, so as to maximize the chance of obtaining good signals from distant earthquakes. Similarly, at subduction zones and transform faults structure varies more significantly across the plate boundary than along it, so refraction lines are often placed perpendicular to the boundary. For example, Fig. 3.2-17 showed a cross-section of the western US lithosphere perpendicular to the San Andreas fault that was derived from refraction surveys.

Two-dimensional arrays can create a three-dimensional image of a small region. As a result, two-dimensional arrays have been deployed around hot spots, rifts, plateaus, transform faults, and subduction zones to study their structure and tectonics. Reflection data are also now commonly gathered by two-dimensional surface deployments. An important contributor to this development has been advances in computers and graphics software that make it possible to analyze and model such data and display the resulting earth structure in a comprehensible fashion. Such three-dimensional images are of great importance in exploring for oil and gas and managing existing oil and gas fields.

Special two-dimensional arrays, often consisting of short-period vertical seismometers, have been used to monitor the locations and magnitudes of underground nuclear tests. The most ambitious such array was the circular Large Aperture Seismic Array (LASA), which operated in Montana from the mid-1960s until 1978. LASA was an array of arrays totaling 525 high-frequency vertical seismometers. Twenty-one clusters of 25 seismometers, each covering 7 km^2, were deployed with a total array diameter of 200 km (Fig. 6.6-17). A similar array is the Norwegian Seismic Array (NORSAR), built in 1971, with 22 sub-arrays spanning an area of 100 km^2. Part of NORSAR, the NORESS array, has 24 seismometers distributed within a 3 km-diameter circle. It has counterparts in northern Norway, Finland, and Germany. As with the WWSSN, arrays designed for nuclear monitoring have also been important for studies of earth structure. Array data can be stacked (Section 6.5), allowing small seismic signals to be extracted from noise. The characteristics of the inner core boundary were first quantified using stacked array data for *PKiKP* waves, which reflect at the boundary but are rarely identified on individual seismograms due to their small amplitudes.

[8] Because all data are duplicated in a sort order, and also stored off site, the computer storage needed is four times greater, or 28 Tbytes.

[9] *Active* experiments include their own seismic sources, as opposed to *passive* experiments using earthquake sources.

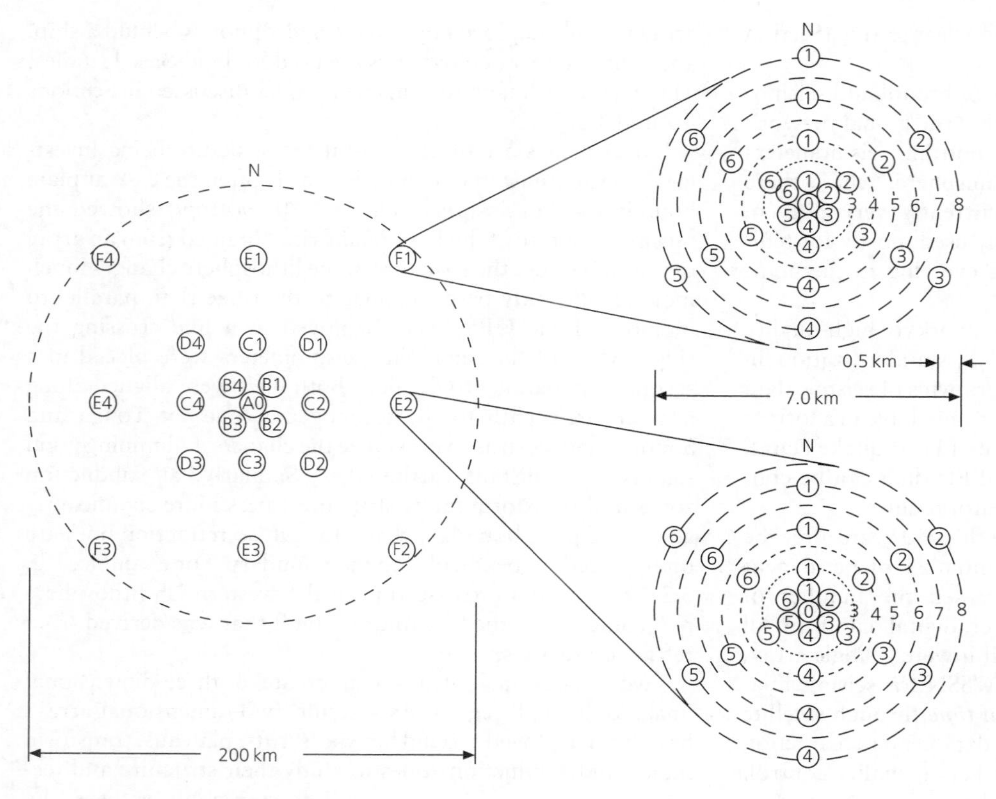

Fig. 6.6-17 Seismometer geometry of the Large Aperture Seismic Array (LASA). (Capon, 1969. *J. Geophys. Res.*, *74*, 3182–94, copyright by the American Geophysical Union.)

6.6.9 *Regional networks*

Regional networks, intermediate between global networks and arrays, are usually constructed to monitor local seismicity or volcanism. Including Alaska, Hawaii, and Puerto Rico, over 3200 seismic stations are part of more than 40 separate US networks (Fig. 6.6-18). Some have only a few stations, and some have hundreds. Many use short-period vertical sensors, but some use accelerometers. For example, the California Strong-Motion Instrumentation Program operates more than 400 accelerometers to provide data for earthquake engineers. Strong-motion data also provide excellent information on source properties because much of the seismic signal is severely attenuated at teleseismic distances. Some networks also incorporate broadband seismometers. For instance, as of 2000, the Southern California Seismographic Network operated 79 broadband stations in addition to its 163 short-period instruments. Regional network stations can also be valuable for earth structure studies, as shown in Fig. 6.6-19.

Many countries have regional networks. For instance, as of 1999, Japan had about 560 stations in operation. These stations have provided valuable data about the subduction process there, including the double seismic zones (Fig. 5.4-20) and *ScS*-to-*P* conversions at the slab top (Fig. 2.6-15).

Regional networks, like global networks, are continually being upgraded. In the USA there are efforts under way, as part of the Advanced National Seismic System (ANSS), to install more broadband and short-period seismometers, and to add about 6000 strong-motion sensors in urban areas at risk from damaging earthquakes. A very ambitious network planned is the USArray, which would have three different components operating simultaneously. First, the number of permanent broadband stations would be increased (Fig. 6.6-20, *left*). Second, 400 portable broadband seismometers would travel around the country. Over eight years, this "bigfoot" array would visit about 2000 sites in the continental USA, with an average station spacing of about 70 km, before going to Alaska and Hawaii (Fig. 6.6-20, *right*). Third, about 2400 seismometers (a mix of broadband, short-period, and high-frequency sensors) would be used as flexible arrays to accompany the moving array. As planned, USArray will be an array at the scale of a regional network. Data from the moving array will be available in near-real time, and can be processed using migration techniques to attain high-resolution imaging deep into the mantle.

Interestingly, because there is an increasing trend toward real-time telemetry for transmitting data from the sensors, seismology is moving toward a situation where data from global networks, regional networks, and many local arrays can be easily combined, largely eliminating the distinctions between networks. This development offers great scientific opportunities.

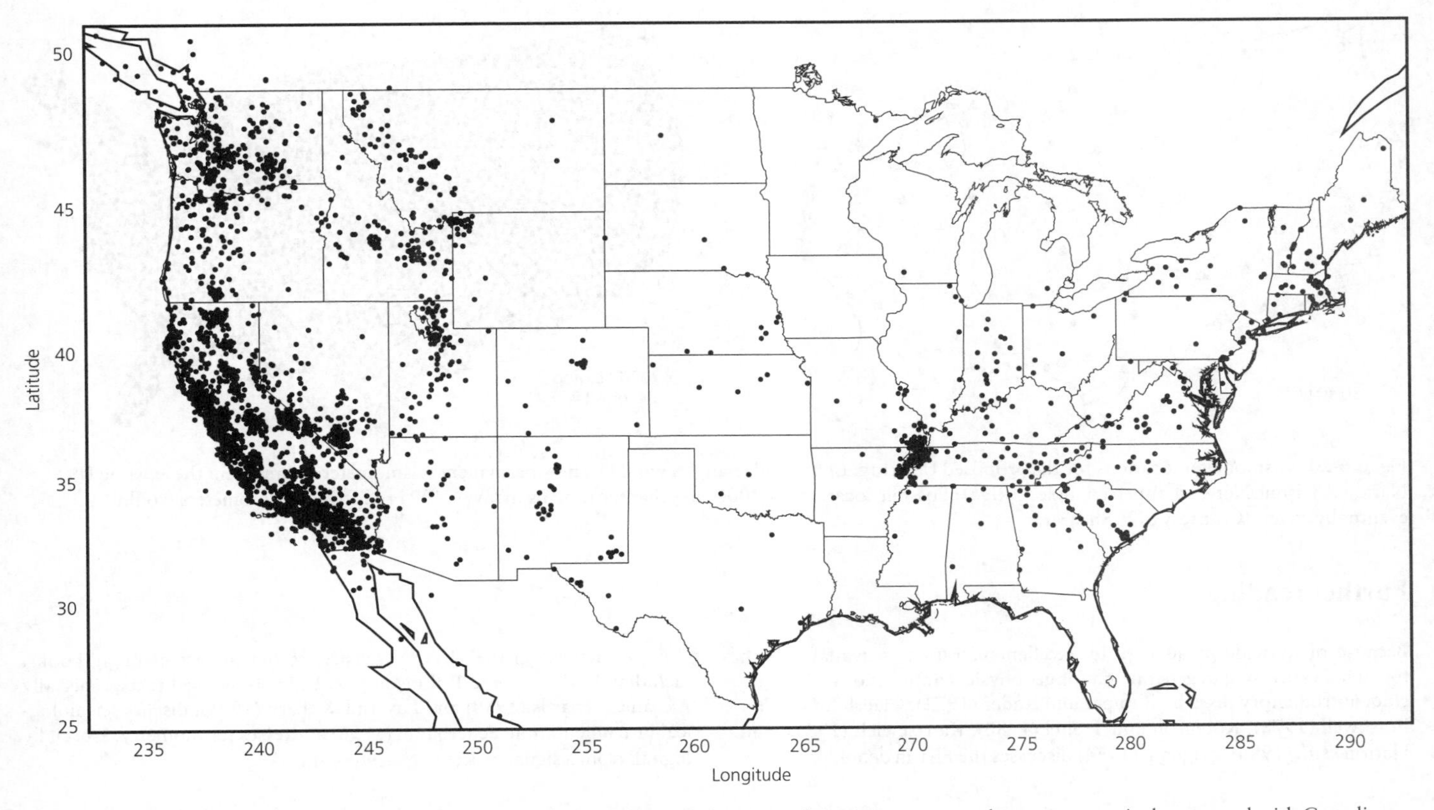

Fig. 6.6-18 Map of regional network seismometer stations in the continental USA as of 1999. Some networks are cooperatively operated with Canadian and Mexican institutions.

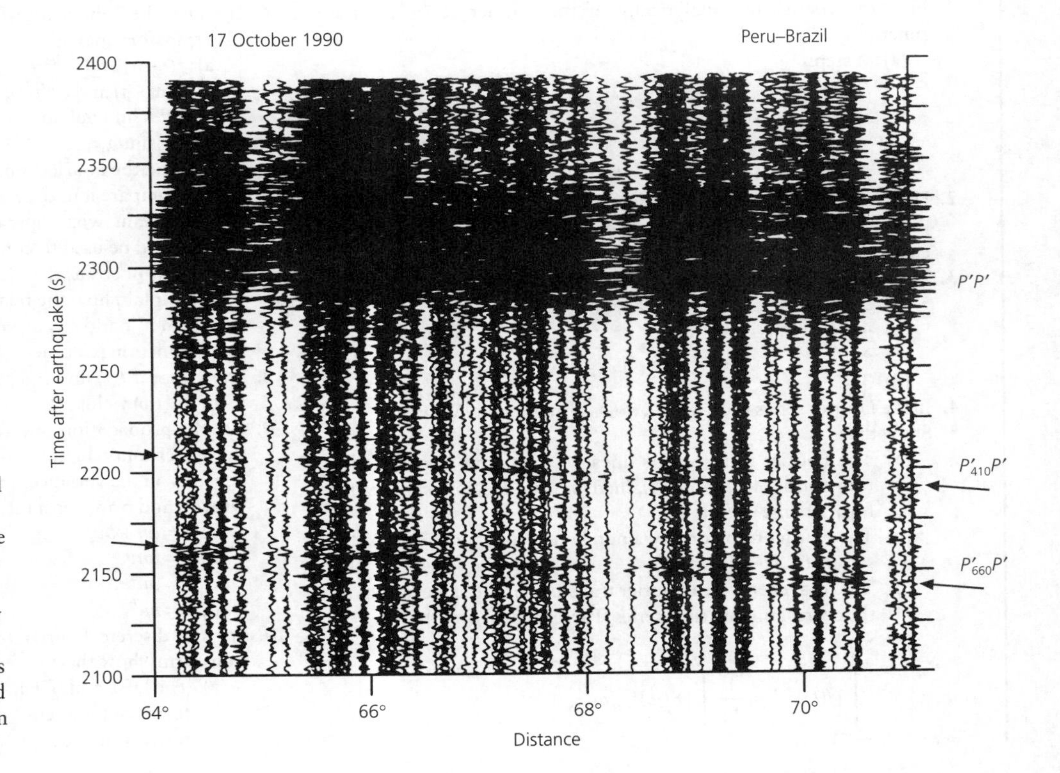

Fig. 6.6-19 Records from the short-period seismometers of California regional networks for an Oct. 17, 1990, earthquake in South America. The data reveal distinct reflections off the sharp 410 km and 660 km mantle discontinuities. The ability to examine large amount of data over a small geographical region greatly increases the resolution of earth structure. (Benz and Vidale, 1993. Reproduced with permission from *Nature*.)

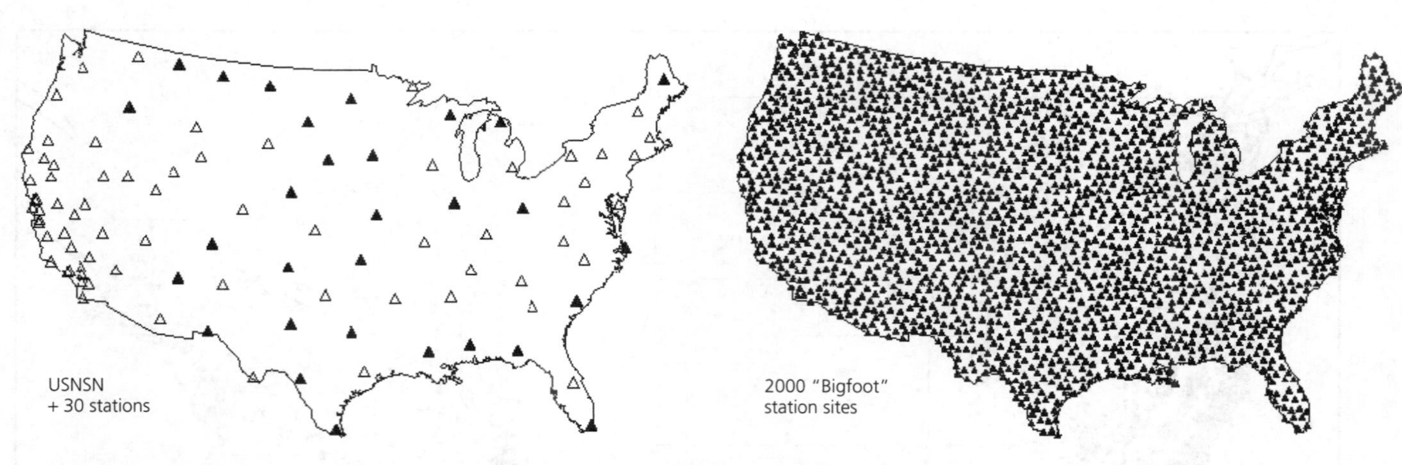

Fig. 6.6-20 Seismometer locations for the proposed USArray. *Left*: Solid triangles would be new permanent seismometers to augment the existing US National Seismic Network (open triangles). *Right*: Possible locations of 2000 sites that the moving array of 400 broadband seismometers would eventually cover. (Courtesy of P. Shearer.)

Further reading

Because of its widespread use, an excellent literature is available both for signal processing in general and for geophysical applications. These include introductory texts by Rabiner and Rader (1972), Claerbout (1976), Bracewell (1978), Robinson and Treitel (1980), Kanasewich (1981), and Hatton *et al.* (1986). Brigham (1974) discusses the FFT in detail.

Error analysis in the physical sciences is the subject of many books, including Bevington and Robinson (1992). Seismological texts, especially Aki and Richards (1980) and Lay and Wallace (1995), discuss seismological instrumentation. Scherbaum (1996) addresses seismometry, especially digital, from a signal processing viewpoint.

Problems

1. Find the coefficients analytically of the Fourier series for the functions
   (a) A step:
   $f(t) = 1 \quad 0 < t < 1/2$
   $\quad -1 \quad -1/2 < t < 0.$
   (b) A ramp: $f(t) = t \quad$ for $-1/2 < t < 1/2$.
2. Use the formulae for the product of sine and cosine functions (Section A.2) to prove the orthogonality relations for the sine and cosine functions (Eqns 6.2.2–4).
3. Express the following complex numbers in $a + ib$ form:
   (a) $e^{i\pi}$
   (b) $4e^{i\pi/2}$
   (c) $e^{-i\pi/2}$
   (d) $3e^{i\pi/3}$
4. In the Fourier series (Eqn 6.2.1), no b_0 term is given. Why?
5. Show that
   (a) The Fourier transform is linear: if $F(\omega)$ and $G(\omega)$ are the transforms of $f(t)$ and $g(t)$, then $(aF(\omega) + bG(\omega))$ is the transform of $(af(t) + bg(t))$.
   (b) The Fourier transform of a purely real-time function has the symmetry $F(-\omega) = F^*(\omega)$.
   (c) The total energy in a Fourier transform is the same as that in the corresponding time series (Parseval's theorem):

$$\int_{-\infty}^{\infty} |f(t)|^2 dt = \frac{1}{2\pi} \int_{-\infty}^{\infty} |F(\omega)|^2 d\omega.$$

6. If $F(\omega)$ is the Fourier transform of $f(t)$, show that the following are also transform pairs:
   (a) $f(t-a)$ and $e^{-i\omega a}F(\omega)$,
   (b) $F(\omega - a)$ and $e^{iat}f(t)$,
   (c) df/dt and $i\omega F(\omega)$.
7. For $f(t) = \sin \omega_0 t$,
   (a) Find the Fourier transform.
   (b) Compare it to the Fourier transform of $f(t) = \cos \omega_0 t$.
   (c) Explain what operation (filter) in the frequency domain could be used to convert the Fourier transform of $\sin \omega_0 t$ to that of $\cos \omega_0 t$.
   (d) Explain how the relation between the Fourier transforms of $\sin \omega_0 t$ and $\cos \omega_0 t$ could be derived using the fact that one function is a time-shifted version of the other.
8. Show that if $f(t)$ and $F(\omega)$ are a transform pair, the inverse transform of $F(\omega)$ yields $f(t)$.
9. Use the propagation of errors relation (Eqn 6.5.18) to show how the uncertainty in the following functions of several variables depends on the variances and covariances of the variables u and v, where a and b are constants:
   (a) $z = au + bv$,
   (b) $z = auv$,
   (c) $z = au/v$,
   (d) $z = au^b$.
10. For the discrete Fourier transform and inverse discrete Fourier transform, show that:
    (a) The DFT and IDFT are linear: if $A(k)$ and $B(k)$ are the transforms of time series $a(n)$ and $b(n)$, then $\alpha A(k) + \beta B(k)$ is the transform of $\alpha a(n) + \beta b(n)$.

(b) The DFT of a real-time series has the symmetry $F(-k) = F(N-k) = F^*(k)$.

(c) If the DFT of $f(n)$ is $F(k)$, the DFT of $f(n-j)$ is $W^{kj}F(k)$, and the IDFT of $F(k-m)$ is $W^{-mn}f(n)$, where $W = e^{-2\pi i/N}$.

11. As derived in Eqn 4.3.10, the depth h of an earthquake can be estimated from the difference in arrival times δt between the direct P wave and pP, the P wave reflected from the surface, using $\delta t = (2h \cos i)/v$ where i and v are the incidence angle and velocity.

(a) Express the depth as a function of the parameters $\delta t, v, i$.

(b) Find the depth for a measured time difference of 2.7 s and assumed velocity of 6.8 km/s and incidence angle of 24°.

(c) Use the propagation of errors relation to show how the uncertainty in depth depends on the uncertainties of the three parameters.

(d) Use the results of (c) to find the uncertainty in depth corresponding to uncertainties (one standard deviation) of 0.5 s in time difference, 0.5 km/s in velocity, and 3° in incidence angle. (Remember to convert to radians.)

Computer problems

C-1. Using the Fourier series coefficients for the step function, derived in problem 1a, plot the first ten terms of the series and their sum. Also plot the sum of the first 20 and 30 terms.

C-2. Write a subroutine to prepare a time series for taking the fast Fourier transform and take it. The subroutine should call a set of separate subroutines that extend the time series to a power of 2 as required, allow for a taper of a length which you input, take the FFT using the subroutine (COOLB) provided (Box 6C-2) or another, and plot the amplitude spectrum. The subroutine should have the option to list the real and imaginary parts of the spectrum, and the amplitude and phase spectra, at each frequency.

C-3. (a) Write a subroutine to generate values of the function $\sin \frac{2\pi t}{T}$ from $t = 0$ to $t = T_{max}$, where the time step Δt, the period T, and the total data length T_{max} are inputs.

(b) Plot this function for $\Delta t = 0.25$, T = 5, $T_{max} = 20$.

(c) Use the results of C-2 to find the amplitude spectrum, with no tapering and with 10% and 20% tapering.

(d) Do parts (b) and (c) for $\Delta t = 0.25$, T = 8, $T_{max} = 50$.

(e) Do parts (b) and (c) for the function

$$\sin \frac{2\pi t}{5} + (0.5) \sin \frac{2\pi t}{8},$$

with $\Delta t = 0.25$, $T_{max} = 256$.

Box 6C-2 COOLB subroutine.[1]

```
      SUBROUTINE COOLB(NN,DATAI, SIGNI)
C CLASSIC - BUT USABLE - FFT PROGRAM
C DATAI IS DATA ARRAY, 2*NP REAL NUMBERS REPRESENTING
C NP COMPLEX POINTS, SO EACH PAIR OF POINTS ARE THE
C (REAL, IMAGINARY) PARTS OF A COMPLEX NUMBER.
C NN IS POWER OF TWO, CAN BE FOUND BY
C NN=(ALOG10(FLOAT(NP))/ALOG10(2.))+.99
C TRANSFORM DIRECTION CONTROLLED BY REAL VARIABLE
C SIGNI (SIGN OF EXPONENTIAL):-1. FORWARD, 1. TO
C INVERT.
C DIMENSIONS: IF TIME SERIES HAS TIME INCREMENT DT,
C TRANSFORM HAS DELTA FREQ=1/(2**NN*DT)
C NOTE: AFTER TAKING INVERSE FFT DIVIDE OUTPUT BY 2**NN
      INTEGER NN
      REAL SIGNI
      DIMENSION DATAI(1)
      N=2**(NN+1)
      J=1
      DO 5 I=1,N,2
      IF(I-J)1,2,2
1     TEMPR=DATAI(J)
      TEMPI=DATAI(J+1)
      DATAI(J)=DATAI(I)
      DATAI(J+1)=DATAI(I+1)
      DATAI(I)=TEMPR
      DATAI(I+1)=TEMPI
2     M=N/2
3     IF(J-M)5,5,4
4     J=J-M
      M=M/2
      IF(M-2)5,3,3
5     J=J+M
      MMAX=2
6     IF(MMAX-N)7,10,10
7     ISTEP=2*MMAX
      THETA=SIGNI*6.2831831/FLOAT(MMAX)
      SINTH=SIN(THETA/2.)
      WSTPR=-2.0 *SINTH*SINTH
      WSTPI=SIN(THETA)
      WR=1.
      WI=0.
      DO 9 M=1,MMAX,2
      DO 8 I=M,N,ISTEP
      J=I+MMAX
      TEMPR=WR*DATAI(J)-WI*DATAI(J+1)
      TEMPI=WR*DATAI(J+1)+WI*DATAI(J)
      DATAI(J)=DATAI(I)-TEMPR
      DATAI(J+1)=DATAI(I+1)-TEMPI
      DATAI(I)=DATAI(I)+TEMPR
8     DATAI(I+1)=DATAI(I+1)+TEMPI
      TEMPR=WR
      WR=WR*WSTPR-WI*WSTPI+WR
9     WI=WI*WSTPR+TEMPR*WSTPI+WI
      MMAX=ISTEP
      GO TO 6
10    RETURN
      END
```

[1] COOLB, written in 1960s vintage Fortran (note the arithmetic IF statements), has been left in original form to illustrate both the persistence of programs that work and the advantages of subsequent developments in programming practice and documentation (Section A.8.2).

C-4. (a) Write a subroutine, using the results of C-2, to use the fast Fourier transform to take a time series, filter it in the frequency domain over a specified passband, and invert the FFT, yielding a filtered time series. The subroutine should have the capability to taper in the frequency domain. This subroutine is best written as a set of subroutines.

(b) Use this routine to filter the time series in C-3e to isolate the two different frequency components.

C-5. (a) Write a subroutine, using the results of C-2 and C-4, to use the fast Fourier transform to convolve two time series.

(b) Use it on two boxcar functions of unit amplitude, one 6 s long and one 3 s long.

C-6. (a) Write a subroutine to do time domain convolution of two functions of different lengths, both sampled at a time step Δt.

(b) Use it on two boxcar functions of unit amplitude, one 6 s long and one 3 s long. Compare the results to those of C-5b.

7 Inverse Problems

Most people, if you describe a train of events to them, will tell you what the result would be. There are few people, however, who, if you told them a result, would be able to evolve from their own inner consciousness what the steps were which led up to that result. This power is what I mean when I talk of reasoning backwards.

Sherlock Holmes, in *A Study in Scarlet* by Arthur Conan Doyle

7.1 Introduction

Throughout this book we have noted that seismology is largely directed at solving inverse problems dealing with earthquake sources and earth structure. We start with the end result, seismograms, and work backwards to characterize the earthquakes that generated the seismic waves and the medium through which the waves passed. To do this, we first addressed the forward problems of how features of seismic waves that are observable from seismograms, such as travel times, amplitudes, waveforms, eigenfrequencies, dispersion, and attenuation, depend on the seismic source and the medium. We have also discussed how the properties of the medium and the source, such as velocity structure and earthquake mechanisms, reflect tectonic processes within the earth. These are specific examples of the fundamental question of what we can say about the earth from seismological and other observations at its surface.

We now end our discussions by addressing some issues in solving inverse problems. Inverse problems can be posed by assuming that we understand the physics of a process which, for a set of model parameters described by a vector **m**, gives rise to a set of observed data described by the vector **d**. The data can thus be considered the result of a function, or operator, A acting on the model parameters,

$$\mathbf{d} = A(\mathbf{m}). \tag{1}$$

The forward problem, predicting the data **d** that would result from a given model described by **m**, is tractable if we understand the process. The corresponding inverse problem, finding what gave rise to a specific set of observed data, is more difficult. We assume that some physical model describes the process, and then use the data to estimate a set of model parameters that are consistent with the data. We solve the inverse problem using either mathematical inverse techniques to find **m** directly from **d**, or trial-and-error techniques that solve the forward problem repeatedly and look for the best solution. Each approach has advantages in some applications.

We have already mentioned solving inverse problems in contexts including studying the cooling of oceanic lithosphere using surface wave dispersion (Section 2.8.3), inverting travel time and amplitude data to find earth structure (Chapter 3), inverting polarity, waveform, and geodetic data to study earthquake mechanisms (Chapter 4), and using earthquake mechanisms to study plate motions and regional tectonics (Chapter 5). We have noted (Section 1.1.2) that although forward problems typically can be solved in a straightforward way, giving a unique solution, inverse problems often have no unique, exact, or "correct" solutions. Because the data are generally somewhat inconsistent due to errors, and our models simplify complex reality, no model exactly describes the data. Similarly, a range of parameters can describe the data equally well for a given model, and we have various models to choose from based on various criteria and preconceptions. Moreover, the data are often insufficient to resolve aspects of the model. We can thus only recognize and accept these limitations on the solutions.[1]

A consequence of these limitations is a trade-off between the model's *resolution*, how detailed it is, and its *stability*, or robustness. For example, inverting travel times with simple earthquake location algorithms using a laterally homogeneous velocity model shows the Wadati–Benioff zones of dipping seismicity. These results are stable, in that they do not depend

[1] This situation is summarized by the title of a paper "Interpretation of inaccurate, insufficient, and inconsistent data" (Jackson, 1972).

Table 7.1-1 Some large-scale reference models.

Model for	Observables inverted and predicted	Parameters estimated	Misfits ("anomalies") indicate
Laterally homogeneous earth structure	Travel times, eigenfrequencies	Average velocity and density versus depth	Lateral velocity variation (subduction zones, continental–ocean differences, etc.)
Relative plate motions	Rates and azimuths of plate motion	Euler vectors	Nonrigid plate behavior (plate interiors and boundary zones)
Thermal evolution of oceanic lithosphere	Variation with age in depth, heat flow, and geoid	Plate thickness, asthenospheric temperature, physical properties (e.g., α, κ, k)	Lateral thermomechanical variations (swells, etc.)

significantly on the details of the location algorithm and velocity model, but have only limited resolution for where in the slab the earthquakes occur. More detailed locations, which are more useful for relating the earthquakes to the physics of subduction, can be derived from sophisticated location algorithms using a laterally variable velocity model that better represents the slab. However, the improved resolution comes at the price of stability, in that it depends on the specific velocity model used.

The results of inverse studies can be viewed in terms of two end members. In one, we use an individual set of data to characterize a specific phenomenon, such as the location of an earthquake or the velocity structure in a specific area. In others, we describe a set of data averaged over a region or the whole earth with a simple physical model characterized by a relatively small, or *sparse*, set of parameters. Such *reference models* — the physical model with a specific set of parameters — are used to characterize large sets of data in a simple way, predict data where no observations exist, and thus identify misfits, or "anomalies," where the data deviate from the model predictions and hence the global average. We then use reference models to draw inferences about the processes that give rise to both the average situation and deviations from it. For example, body wave, surface wave, and normal mode data give average global velocity structure. This structure is used to constrain models of the average radial variations in composition and temperature, and as a reference against which velocity perturbations due to subducting slabs, continental roots, hot spots, ridges, etc. can be identified and analyzed in terms of local processes that perturb the global model. As shown in Table 7.1-1, we can view other reference models in a similar way. For example, the Euler vectors describing a plate's motion are a simple description of its behavior, and places where earthquake mechanisms differ from these predictions indicate deviations from rigid plate behavior. Similarly, simple cooling models of the oceanic lithosphere describe the average variations in depth, heat flow, and the geoid, and so give a reference model for the temperature against which other effects can be identified and modeled.

As illustrated in Fig. 1.1-8, the models are refined over time using new data and model parameterizations. Eventually, the reference model does not improve significantly. When this occurs, we are probably doing about as well as possible with this type of model. For example, as discussed in Section 3.5, laterally homogeneous global seismic velocity models have become sufficiently accurate that more attention is now directed toward the lateral variations.

In this chapter, we discuss several inverse problems to introduce some of the methods used. Because such inverse problems are crucial to seismology and the earth sciences, and also appear in other sciences, considerable attention has been directed toward them. It turns out that physically quite different problems are often described in mathematically similar ways. Our goal is to identity some common themes and approaches, rather than discuss the details. Some more sophisticated treatments are listed in the suggested reading.

7.2 Earthquake location

We first consider the classic inverse problem of locating an earthquake and finding its origin time using the arrival times of seismic waves at various stations. The velocity structure, which determines the ray paths and hence travel times, is crucial. We first regard the velocity structure as known, and then explore how it can also be estimated from the travel times.

7.2.1 *Theory*

Assume that an earthquake occurred at an unknown time t, at an unknown position $\mathbf{x} = (x, y, z)$, known as the *hypocenter*, or *focus* (Fig. 7.2-1). The point (x, y) on the surface above the focus is called the *epicenter*. n seismic stations at locations $\mathbf{x}_i = (x_i, y_i, z_i)$ detect the earthquake at arrival times d'_i, which depend on the origin time t and the travel time between the source and the station $T(\mathbf{x}, \mathbf{x}_i)$:

$$d'_i = T(\mathbf{x}, \mathbf{x}_i) + t. \quad (1)$$

If the velocity structure is known, the forward problem can be written using the formulation

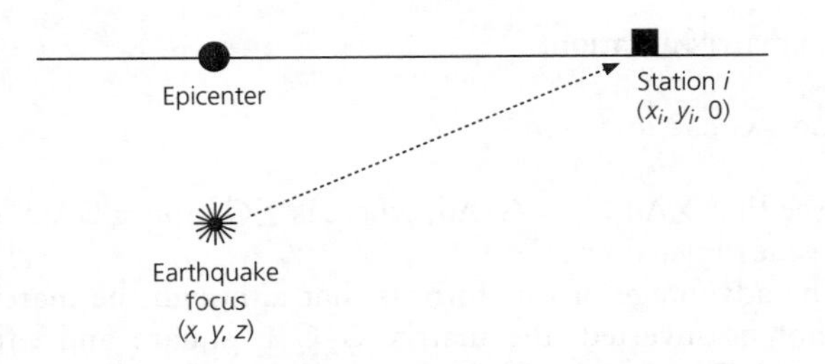

Fig. 7.2-1 Geometry for earthquake location in a homogeneous (uniform velocity) halfspace.

$$\mathbf{d} = A(\mathbf{m}), \quad \text{or} \quad d_i = A(m_j), \tag{2}$$

showing how the data vector, containing the arrival times at the stations, can be computed from an assumed model vector composed of the source location and origin time,

$$\mathbf{m} = (x, y, z, t). \tag{3}$$

The model vector consists of physically different quantities: three space coordinates and an origin time. Because the data and model are vectors, relations between them can be written in terms of either vectors ($\mathbf{d} = A(\mathbf{m})$) or their components ($d_i = A(m_j)$).

The inverse problem can be stated as: given the observed arrival times, find a model that fits them. To do this, we begin with a *starting model* $\mathbf{m}°$, which is an estimate of (or guess at) a model that we hope is close to the solution we seek. The starting model predicts that we would have observed data $d_i° = A(m_j°)$. Unless we are lucky, these predicted data are not what were actually observed. Hence we seek changes Δm_j in the starting model

$$m_j = m_j° + \Delta m_j \tag{4}$$

that will make the predicted data closer to those observed. In general, the data do not depend linearly on the model parameters, so we *linearize* the problem by expanding the data in a Taylor series about the starting model $\mathbf{m}°$ and keeping only the linear term,

$$d_i \approx d_i° + \sum_j \left.\frac{\partial d_i}{\partial m_j}\right|_{\mathbf{m}°} \Delta m_j. \tag{5}$$

This equation can be written in terms of the difference between the observed data and those predicted,

$$\Delta d_i° \equiv d_i' - d_i° \approx \sum_j \left.\frac{\partial d_i}{\partial m_j}\right|_{\mathbf{m}°} \Delta m_j°. \tag{6}$$

Such relations are common in inverse problems. For simplicity, we omit the superscripts and define the partial derivative matrix as

$$G_{ij} = \frac{\partial d_i}{\partial m_j}, \tag{7}$$

so the equation becomes

$$\Delta \mathbf{d} = G\Delta\mathbf{m}, \quad \text{or} \quad \Delta d_i = \sum_j G_{ij}\Delta m_j. \tag{8}$$

Often the Δs are also suppressed, and the equation is written as $\mathbf{d} = G\mathbf{m}$. This makes the notation simpler, but can be confusing at first. In this derivation, we retain the Δs to explicitly indicate changes.

Equation 8 is a vector–matrix equation representing a system of simultaneous linear equations. To solve it, we seek a change in the model $\Delta\mathbf{m}$ that, when multiplied by the known partial derivative matrix G, gives the required change in the data $\Delta\mathbf{d}$. This is an inverse problem, in contrast to the forward problem of finding the change in the data $\Delta\mathbf{d}$ predicted by an assumed change $\Delta\mathbf{m}$ in the model. Many aspects of inverse theory deal with solving such equations under various circumstances. The earthquake location problem considered here is a simple case.

A common complexity is that we generally have arrival time observations at many (often several hundred) seismic stations, and are solving for only four model parameters. In the notation of Eqn 8, j ranges from 1 to 4, and i ranges from 1 to n, where n is much greater than 4. Because each arrival time corresponds to one equation, and each model parameter provides one unknown, G has a number of rows equal to the number of arrival time observations, and a number of columns equal to the number of model parameters. Because there are more (n) equations than unknowns (4), G has more rows than columns, so Eqn 8 looks like

$$\begin{pmatrix} \Delta d_1 \\ \Delta d_2 \\ \cdot \\ \cdot \\ \cdot \\ \cdot \\ \cdot \\ \Delta d_n \end{pmatrix} = \begin{pmatrix} G_{11} & G_{12} & G_{13} & G_{14} \\ G_{21} & G_{22} & G_{23} & G_{24} \\ \cdot & \cdot & \cdot & \cdot \\ \cdot & \cdot & \cdot & \cdot \\ \cdot & \cdot & \cdot & \cdot \\ \cdot & \cdot & \cdot & \cdot \\ \cdot & \cdot & \cdot & \cdot \\ G_{n1} & G_{n2} & G_{n3} & G_{n4} \end{pmatrix} \begin{pmatrix} \Delta m_1 \\ \Delta m_2 \\ \Delta m_3 \\ \Delta m_4 \end{pmatrix}. \tag{9}$$

Such *overdetermined* problems can pose difficulties. One way to see this is to recall that if n were equal to 4 the matrix G would be square (have the same number of rows and columns), so Eqn 8 could be solved by multiplication by the inverse matrix,

$$G^{-1}\Delta\mathbf{d} = G^{-1}G\Delta\mathbf{m} = \Delta\mathbf{m}, \quad \text{or}$$

$$\sum_i G_{ki}^{-1}\Delta d_i = \sum_i G_{ki}^{-1}\left(\sum_j G_{ij}\Delta m_j\right) = \Delta m_k. \tag{10}$$

If the number of arrival time observations exceeds four, this method cannot be used, because G is not square and thus does not have an inverse.[1] Our first instinct might be to use only arrival times at four stations, which would give an exact solution, and assume that the arrival times at the other stations give only extra, redundant information. In an ideal world this would be the case. In reality, the arrival times contain errors due to a variety of possible effects, including reading errors, inaccuracies in the clocks at the stations, and misidentification of the first arrivals. In addition to these errors of measurement, there are systematic errors due to the fact that the velocity structure is not perfectly known and is laterally variable. As a result, the equations are *inconsistent*: no one model can solve them exactly. Moreover, choosing four arrival times might mean selecting data poorer than those discarded. The approach taken instead is to seek the origin time and source location that "best" solve the overdetermined, inconsistent equations.

To do this, we regard the observations d_i' as having errors described by their standard deviations σ_i and find the model that minimizes the misfit,

$$\chi^2 = \sum_i \frac{1}{\sigma_i^2}\left(\Delta d_i - \sum_j G_{ij}\,\Delta m_j\right)^2, \tag{11}$$

which is the prediction error, the normalized sum of the squares of the difference between the observed arrival times and those predicted by the model. x^2, the fitting function to be minimized, weights the data by the reciprocal of their variances so that the most uncertain have the least effect. To find the best fit, we set partial derivatives of the misfit with respect to the change in model parameters Δm_k equal to zero, and use the fact that the model elements are independent, so the partial derivative of the change in one with respect to those in the others is zero,

$$\frac{\partial \Delta m_j}{\partial \Delta m_k} = \delta_{jk}. \tag{12}$$

The partial derivatives of the misfit are

$$\frac{\partial \chi^2}{\partial \Delta m_k} = 0 = 2\sum_i \frac{1}{\sigma_i^2}\left(\Delta d_i - \sum_j G_{ij}\Delta m_j\right)G_{ik}, \tag{13}$$

or

$$\sum_i \frac{1}{\sigma_i^2}\Delta d_i G_{ik} = \sum_i \frac{1}{\sigma_i^2}\left(\sum_j G_{ij}\Delta m_j\right)G_{ik}. \tag{14}$$

If the variances of the data are equal ($\sigma_i^2 = \sigma^2$), that term can be factored out, and

$$\sum_i \Delta d_i G_{ik} = \sum_i \left(\sum_j G_{ij}\Delta m_j\right)G_{ik}, \tag{15}$$

or, in matrix notation,

$$G^T\Delta\mathbf{d} = G^T G\Delta\mathbf{m}. \tag{16}$$

To see that $\sum_i \Delta d_i G_{ik} = G^T\Delta\mathbf{d}$, whereas $\sum_j G_{ij}\Delta m_j = G\Delta\mathbf{m}$, consider the dimensions.

The advantage of this form is that although the matrix G cannot be inverted, the matrix G^TG is square and can be inverted. Equation 16 thus gives $\Delta\mathbf{m}$, the standard least squares solution to a set of equations that cannot be solved exactly, because

$$\Delta\mathbf{m} = (G^TG)^{-1}G^T\Delta\mathbf{d} = G^{-g}\Delta\mathbf{d}, \quad \text{or} \quad \Delta m_j = \sum_i G_{ji}^{-g}\Delta d_i. \tag{17}$$

The operator $(G^TG)^{-1}G^T$, which acts on the data to yield the model, is called the *generalized inverse* of G, and is written as G^{-g}. It provides the "best" solution in a least squares sense, because it gives the smallest squared misfit. The generalized inverse is the analog of the inverse, but for a matrix that is not square, and hence does not have a conventional inverse. If G is square and has an inverse, then $G^{-1} = G^{-g}$. If the data errors are not equal, the least squares solution is weighted by the errors, as shown in problem 5 at the end of this chapter.

To use this method, we begin with a starting model (source location and origin time) $\mathbf{m}^\circ$ and predict the values expected for the data, $\mathbf{d}^\circ = A(\mathbf{m}^\circ)$. We then form the residual vector giving the misfit to the data, $\Delta\mathbf{d}^\circ \equiv \mathbf{d}' - \mathbf{d}^\circ$, evaluate the matrix of partial derivatives about the starting model,

$$G_{ij} = \left.\frac{\partial d_i}{\partial m_j}\right|_{\mathbf{m}^\circ}, \tag{18}$$

and use the generalized inverse (Eqn 17) to find $\Delta\mathbf{m}^\circ$, the change in the starting model that gives a better fit to the data. Thus the new model

$$\mathbf{m}^1 = \mathbf{m}^\circ + \Delta\mathbf{m}^\circ \tag{19}$$

predicts values of the data

$$\mathbf{d}^1 = A(\mathbf{m}^1) \tag{20}$$

that should be closer to the observations than the predictions of the starting model. This can be tested by computing the difference between the observations and the predicted data for the new model $\Delta\mathbf{d}^1 \equiv \mathbf{d}' - \mathbf{d}^1$, and examining the total squared misfit $\Sigma(\Delta d_i^1)^2 = \Sigma(d_i' - d_i^1)^2$. This should be less than the corresponding misfit for the starting model $\Sigma(\Delta d_i^\circ)^2$. The total squared misfit is more useful than the total misfit $\Sigma\Delta d_i$, because the latter could be small for large misfits of opposite signs.

We can often do even better. Remember that the G matrix of partial derivatives was found by expanding the function that predicts the data (travel times) about the starting model in a Taylor series, and taking the linear terms. This expansion

[1] The definition of the inverse (Section A.4.3) requires that both pre- and post-multiplication yield the identify; i.e., $A^{-1}A = AA^{-1} = I$.

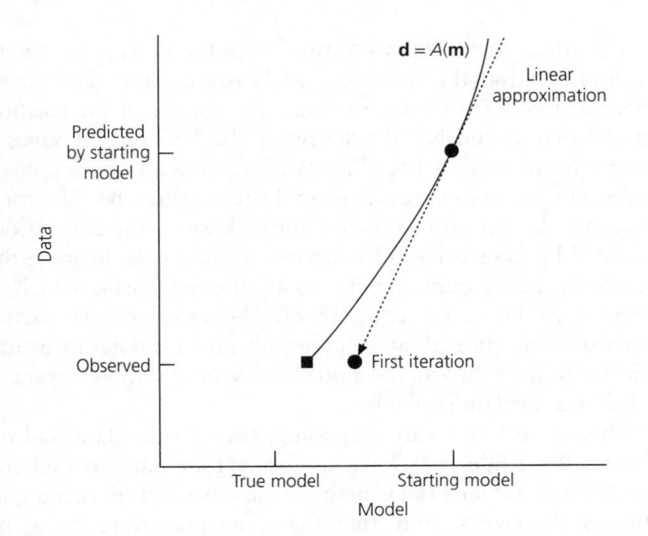

Fig. 7.2-2 Schematic illustration of the effect of linearizing about a starting model in an inverse problem. The new model is found from the difference between the observed data and that predicted for the starting model. The worse the linear approximation is, the more iterations will be needed to reach the true model.

works well if the starting model is "close" to the actual model. If this is not the case, the linear approximation may not be a good one. Figure 7.2-2 illustrates this idea schematically. The actual situation is hard to draw, because each model vector is an element in a four-dimensional (three space and one time) vector space.

As a result, the method can be iterated. Once the model has been changed, a new partial derivative matrix

$$G_{ij} = \left.\frac{\partial d_i}{\partial m_j}\right|_{\mathbf{m}^1} \tag{21}$$

is found by expanding the function that predicts the data about the *new* model. The generalized inverse method is then used to solve

$$\Delta\mathbf{d}^1 = G\Delta\mathbf{m}^1 \tag{22}$$

for a further change in the model $\Delta\mathbf{m}^1$ that reduces the remaining misfit. This process is repeated until successive iterations produce only small changes in the model, and hence in the total misfit to the data (Fig. 7.2-3).

7.2.2 *Earthquake location for a homogeneous medium*

To make these ideas less abstract, we consider the simple case of locating an earthquake in a medium of uniform velocity v. In this case the ray paths connecting an earthquake and seismic stations are straight lines. This geometry approximates a situation where the receivers are close enough to a source that the first arrivals are direct waves in a medium whose velocity does not vary significantly. Seismic waves from an earthquake that

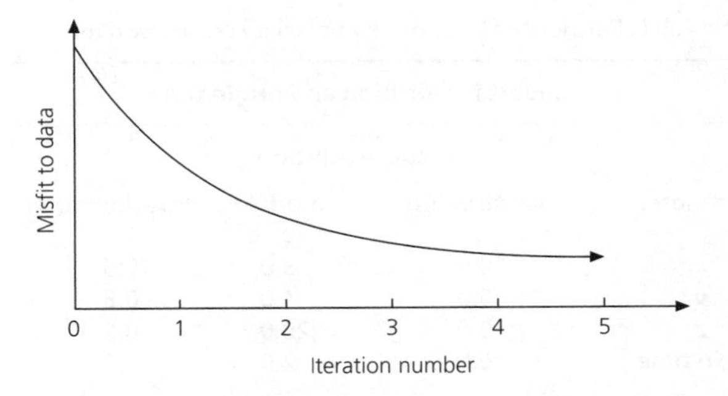

Fig. 7.2-3 Schematic illustration of the variation in misfit to the data as a function of iteration number for an inverse problem.

occurred at time t at location $\mathbf{x} = (x, y, z)$ are recorded by seismic stations at positions $\mathbf{x}_i = (x_i, y_i, z_i)$ with arrival times

$$d_i = T(\mathbf{x}, \mathbf{x}_i) + t = \frac{1}{v}[(x - x_i)^2 + (y - y_i)^2 + (z - z_i)^2]^{1/2} + t. \tag{23}$$

Although the earthquake can occur below the surface, the stations are at the surface $z_i = 0$. The travel times depend only on the distance between source and receiver, $|\mathbf{x} - \mathbf{x}_i|$.

To solve the inverse problem, we form the matrix G_{ij}. Its elements, the partial derivatives of the elements of the data vector d_i (the arrival times at each station) with respect to the model parameters m_j (the location coordinates and origin time of the earthquake) are easily found. Differentiation of the ith element of the data vector is done with respect to the first element of the model vector, which is the x coordinate of the location

$$G_{i1} = \frac{\partial d_i}{\partial m_1} = \frac{\partial d_i}{\partial x} = \frac{\partial T(\mathbf{x}, \mathbf{x}_i)}{\partial x}$$
$$= \frac{(x - x_i)}{v}[(x - x_i)^2 + (y - y_i)^2 + z^2]^{-1/2}. \tag{24}$$

Similar expressions give the partial derivatives with respect to the other two space coordinates of the location. Note that these partial derivatives are functions of the spatial model parameters (x, y, z). The final partial derivative, with respect to origin time, is just

$$G_{i4} = \frac{\partial d_i}{\partial m_4} = \frac{\partial d_i}{\partial t} = 1. \tag{25}$$

Given the G matrix, the earthquake is located by choosing a starting model, forming the difference $\Delta\mathbf{d}$ between the model predictions and the observations, and solving for the change in the model $\Delta\mathbf{m}$ using the procedure in the last section.

Table 7.2-1 (*top*) illustrates a hypothetical example of locating an earthquake with ten stations located within a 100 km square. The earthquake is assumed to have occurred at time 0 seconds at the point (0, 0, 10) km. We then try to locate the

Table 7.2-1 Earthquake location example with error-free data.

Invert for location and origin time				
		model evolution		
parameter	actual value	model for iteration number		
		0	1	2
x	0.0	3.0	−0.5	0.0
y	0.0	4.0	−0.6	0.0
z	10.0	20.0	10.1	10.0
origin time	0.0	2.0	0.2	0.0
station location		residual for iteration number		
		0	1	2
35.0	9.0	−2.1	−0.4	0.0
−44.0	10.0	−3.0	−0.2	0.0
−11.0	−25.0	−3.8	−0.1	0.0
23.0	−39.0	−3.0	−0.2	0.0
42.0	−27.0	−2.6	−0.3	0.0
−12.0	50.0	−2.0	−0.3	0.0
−45.0	16.0	−2.9	−0.2	0.0
5.0	−19.0	−3.7	−0.2	0.0
−1.0	−11.0	−4.1	−0.2	0.0
20.0	11.0	−2.4	−0.4	0.0
error		92.4	0.6	0.0

Invert for location, origin time, and velocity				
		model evolution		
parameter	actual value	model for iteration number		
		0	1	2
x	0.0	3.0	0.2	0.0
y	0.0	4.0	0.3	0.0
z	10.0	20.0	10.2	10.0
origin time	0.0	2.0	0.7	0.0
velocity	5.0	4.0	4.9	5.0
station location		residual for iteration number		
		0	1	2
35.0	9.0	−4.0	−0.9	0.0
−44.0	10.0	−5.6	−1.0	0.0
−11.0	−25.0	−5.7	−0.9	0.0
23.0	−39.0	−5.6	−1.0	0.0
42.0	−27.0	−5.2	−1.0	0.0
−12.0	50.0	−4.6	−0.9	0.0
−45.0	16.0	−5.6	−1.0	0.0
5.0	−19.0	−5.2	−0.9	0.0
−1.0	−11.0	−5.3	−0.9	0.0
20.0	11.0	−3.8	−0.8	0.0
error		261.3	8.3	0.0

earthquake using the computed arrival times at the ten stations as "data." For a starting model, we assume the earthquake occurred at time 2 seconds at (3, 4, 20) km. As discussed in the previous section, we compute the arrival times expected at each station for a source located at the initial estimated position and time, and then form the residual, the difference between the "data" and this prediction (Eqn 6). For the starting model, the total squared misfit is 92.4 s^2.

To reduce the misfit, we form the partial derivative matrix G_{ij} evaluated at the starting model, and use the generalized inverse (Eqn 17) to solve for $\Delta\mathbf{m}^\circ$, the change in the starting model that would best fit the residuals. This change gives a source location of (−0.5, −0.6, 10.1) km and an origin time of 0.2 s. This new estimate is close to the true values. Because for a real case the true model would not be known, the new model is tested by calculating the expected arrival times, forming the residuals, and examining the total squared misfit, which is reduced to 0.6 s^2. To reduce this further, we form the partial derivative matrix evaluated at the new model and iterate again. The resulting change in the model yields the true model exactly, which fits the data perfectly.

This success is hardly surprising, because the data had no errors. We could thus have used any four data to find the model, and avoided the generalized inverse. Before turning to discuss the errors, note that the same procedure could be used to find the velocity. To do so, we regard the velocity as a fifth model parameter, and invert the data for a model vector $m = (x, y, z, t, v)$. The additional partial derivatives are

$$\frac{\partial d_i}{\partial m_5} = \frac{\partial d_i}{\partial v} = -\frac{1}{v^2}[(x-x_i)^2 + (y-y_i)^2 + z^2]^{1/2}. \tag{26}$$

We thus assume a velocity as part of the starting model, find the partial derivative matrix (which now has five columns), and use the generalized inverse to find the changes in the starting model. Table 7.2-1 (*bottom*) illustrates this process for the same example as before, except that we also invert for velocity.

7.2.3 *Errors*

Because earthquakes are located using arrival time data that have errors, the resulting locations and origin times have uncertainties. To assess these uncertainties, we examine how errors in the data affect the generalized inverse solution.

We characterize the errors in the data at the ith station, d_i, by viewing the specific values measured as samples from a parent distribution that includes all possible $d_i^{(k)}$, $k = 1, \ldots \infty$, such that an infinite number of measurements would yield the parent distribution. In this notation, $d_i^{(k)}$ is the kth sample of d_i, the arrival time at station i. Because in real applications the parent distribution for d_i is unknown, it is common to assume a Gaussian distribution with mean $\bar{d}_i$ and standard deviation σ_i, as discussed in Section 6.5. For a large number of measurements (samples) from this distribution, the mean is the average

$$\bar{d}_i = \lim_{K\to\infty} \frac{1}{K}\sum_{k=1}^{K} d_i^{(k)}, \tag{27}$$

and the "spread" of the measurements is the variance

$$\sigma_i^2 = \lim_{K\to\infty}\left(\frac{1}{K}\sum_{k=1}^{K}(d_i^{(k)} - \bar{d}_i)^2\right). \tag{28}$$

If the Gaussian parent distribution is an appropriate choice, there is a 68% probability that any sample will fall in the range $\bar{d}_i \pm \sigma_i$, and a 95% probability that any sample will fall in the range $\bar{d}_i \pm 2\sigma_i$ (Fig. 6.5-1).

The errors at different stations are described by the variance–covariance matrix of the data

$$\sigma_{\mathrm{d}}^2 = \sigma_{d_{ij}}^2 = \lim_{K\to\infty} \frac{1}{K} \sum_{k=1}^{K} (d_i^{(k)} - \bar{d}_i)(d_j^{(k)} - \bar{d}_j). \tag{29}$$

The diagonal ($i = j$) terms are the variances for data at individual stations. The off-diagonal terms ($i \neq j$) are the covariances that describe the relation between errors at pairs of stations. If the errors are uncorrelated between two stations — for example, those due to a station clock — then how a measurement at one station differs from the mean there is unrelated to what occurs at another station, so their covariance is ideally zero. Given a finite number of real data, we expect the covariance to be small. By contrast, if the errors are correlated (for example, if one person were reading seismograms from different stations with a consistent bias), then similar deviations from the mean occur between these stations, and their covariances would be larger. Errors can also be anti-correlated, such that deviations at a station tend to occur in one direction, whereas those at another station tend the other way, yielding negative covariances. Although errors of measurement are likely to be uncorrelated, systematic errors are often correlated. For example, variations in velocity can cause systematic biases that are either correlated or anti-correlated between different stations.

The data are inverted using the generalized inverse solution

$$m_j = \sum_i G_{ji}^{-g} d_i \tag{30}$$

(here the Δs are not written). As a result, the uncertainty in a model parameter can reflect errors in *all* of the data. Thus, even if the errors in the data are uncorrelated, the resulting uncertainties in model parameters can be correlated. To see this, we write the covariances of the model parameters in terms of those for the data

$$\begin{aligned} \sigma_{\mathrm{m}}^2 = \sigma_{m_{ji}}^2 &= \lim_{K\to\infty} \frac{1}{K} \sum_{k=1}^{K} (m_j^{(k)} - \bar{m}_j)(m_i^{(k)} - \bar{m}_i) \\ &= \lim_{K\to\infty} \frac{1}{K} \sum_{k=1}^{K} \left(\sum_p G_{jp}^{-g} (d_p^{(k)} - \bar{d}_p) \right) \left(\sum_s G_{is}^{-g} (d_s^{(k)} - \bar{d}_s) \right) \\ &= \sum_p G_{jp}^{-g} \sum_s G_{is}^{-g} \left(\lim_{K\to\infty} \frac{1}{K} \sum_{k=1}^{K} (d_p^{(k)} - \bar{d}_p)(d_s^{(k)} - \bar{d}_s) \right) \\ &= \sum_p G_{jp}^{-g} \sum_s G_{is}^{-g} \sigma_{d_{ps}}^2 . \end{aligned} \tag{31}$$

This relation can be written in matrix form in terms of σ_{d}^2 and σ_{m}^2, the variance–covariance matrices for the data and model:

$$\sigma_{\mathrm{m}}^2 = G^{-g} \sigma_{\mathrm{d}}^2 (G^{-g})^T. \tag{32}$$

We often assume that the data errors are uncorrelated and equal, so that the data variance–covariance matrix is a constant times the identity matrix,

$$\sigma_{\mathrm{d}}^2 = \sigma^2 \delta_{ij}, \tag{33}$$

and the model variance–covariance matrix is

$$\sigma_{\mathrm{m}}^2 = \sigma^2 (G^T G)^{-1}, \tag{34}$$

as proved in problem 4.

Table 7.2-2 illustrates these ideas for the location example in the previous section. In this case, Gaussian errors with mean zero and standard deviation 0.1 s were added to the arrival times. As a result, the data are inconsistent and cannot be fit exactly by any model. The inversion thus changes the model until a good, but not perfect, fit to the data is achieved. This final model, which is no longer changing much after three

Table 7.2-2 Earthquake location example with errors.

Invert for location and origin time

model evolution

parameter	actual value	model for iteration number 0	1	2	3
x	0.0	3.0	−0.2	0.2	0.2
y	0.0	4.0	−0.9	−0.4	−0.4
z	10.0	20.0	12.2	12.2	12.2
origin time	0.0	2.0	0.0	−0.2	−0.2

station location		residual for iteration number 0	1	2	3
35.0	9.0	−2.0	−0.1	0.1	0.1
−44.0	10.0	−3.0	−0.1	0.0	0.0
−11.0	−25.0	−3.8	0.0	0.1	0.1
23.0	−39.0	−3.2	−0.1	0.0	0.0
42.0	−27.0	−2.8	−0.2	−0.1	−0.1
−12.0	50.0	−2.1	−0.3	−0.1	−0.1
−45.0	16.0	−2.9	−0.1	0.0	0.0
5.0	−19.0	−3.7	−0.1	0.0	0.0
−1.0	−11.0	−4.0	−0.1	0.0	0.0
20.0	11.0	−2.5	−0.3	0.0	0.0
error		93.74	0.33	0.04	0.04

data standard deviation 0.10

model variance–covariance matrix

0.06	0.01	0.01	0.00
0.01	0.08	−0.13	0.01
0.01	−0.13	1.16	−0.08
0.00	0.01	−0.08	0.01

model standard deviation

x	y	z	origin time
0.25	0.28	1.08	0.10

iterations, is close to, but not exactly, the model used to generate the data. This simple example thus has some features of real situations.

The uncertainties in the final model are shown by the model variance–covariance matrix

$$\sigma_{\mathrm{m}}^2 = \begin{pmatrix} \sigma_{xx}^2 & \sigma_{xy}^2 & \sigma_{xz}^2 & \sigma_{xt}^2 \\ \sigma_{yx}^2 & \sigma_{yy}^2 & \sigma_{yz}^2 & \sigma_{yt}^2 \\ \sigma_{zx}^2 & \sigma_{zy}^2 & \sigma_{zz}^2 & \sigma_{zt}^2 \\ \sigma_{tx}^2 & \sigma_{ty}^2 & \sigma_{tz}^2 & \sigma_{tt}^2 \end{pmatrix}. \quad (35)$$

To see that the results seem reasonable, we compare the final inversion model, taking into account its uncertainty, to the true model. The standard deviations of each parameter are given by the square roots of the diagonal terms of the model variance–covariance matrix, so the final model ($x = 0.2 \pm 0.25$ km, $y = -0.4 \pm 0.28$ km, $z = 12.2 \pm 1.08$ km, $t = -0.2 \pm 0.10$ s) is an acceptable representation of the true model.

The model variance–covariance matrix shows some interesting features. The variance of the depth estimate, σ_{zz}^2 is larger than the corresponding terms σ_{xx}^2 and σ_{yy}^2, indicating that the depth is less well constrained than the epicenter. This situation is common, and arises because all the seismometers are at the surface.[2] In some cases when the depth is poorly constrained, it is regarded as fixed, and only the epicenter and the origin are inverted for. The results of multiple inversions, each with the depth fixed at a different value, are compared to see which best fits the data. It is also possible to determine the depth from other criteria, such as the times of surface reflections (Section 4.3), and then invert with the depth fixed.

The uncertainties in the model parameter estimates are correlated, because the off-diagonal elements of the model variance–covariance matrix are nonzero. σ_{zt}^2, the covariance of the depth and origin time uncertainties, is negative, indicating a trade-off between the focal depth and the origin time. At any station, similar arrival times result if the earthquake occurred earlier (t smaller) but deeper (z larger). Similarly, σ_{xy}^2, the covariance of the x and y location uncertainties, is nonzero, so the uncertainties in these two parameters are correlated. A method often used to illustrate this is to extract the 2×2 submatrix

$$\begin{pmatrix} \sigma_{xx}^2 & \sigma_{xy}^2 \\ \sigma_{yx}^2 & \sigma_{yy}^2 \end{pmatrix} \quad (36)$$

and diagonalize it by finding the eigenvalues $\lambda^{(1)}$ and $\lambda^{(2)}$, and the associated eigenvectors $(x_1^{(1)}, x_2^{(1)})$ and $(x_1^{(2)}, x_2^{(2)})$. The uncertainty in the epicenter can then be thought of as an ellipse with semi-major and semi-minor axes $\lambda^{(1)1/2}$ and $\lambda^{(2)1/2}$, oriented in a direction given by $\tan^{-1}(x_1^{(1)}/x_2^{(1)})$. In this case, the semi-major and semi-minor axes have lengths of 0.29 and 0.24 km, and the semi-major axis trends N22°E. An interesting feature of the error ellipse is that its shape and orientation depend on the $(G^TG)^{-1}$ matrix, whereas the variance of the data, σ_{d}^2, controls the size of the ellipse. Because the shape of the error ellipse depends on the geometry of the receivers, it can be examined without reference to specific data. As written, the ellipse is for a confidence level of 1σ (68%), but ellipses are sometimes also given for 2σ (95%), or 3σ (99%).

We have shown that the model variance–covariance matrix depends on the variance–covariance matrix of the data. In the example, we knew the standard deviation of the data and that the errors were uncorrelated. This information would not be available for a real experiment. However, we could estimate the standard deviation of the data from the misfit between the data and the best-fitting model, given by the sample variance s^2,

$$\sigma^2 \approx s^2 = \frac{1}{n-k}\sum_{i=1}^{n}(d_i' - d_i)^2. \quad (37)$$

Here, d_i' are the observations, d_i are the values of the data predicted by the best-fitting model, and k is the number of model parameters determined from the data. Division by $n - k$, the number of degrees of freedom, rather than by n, the number of data, compensates for the improvement in fit resulting from the use of model parameters determined from the data. Thus, for our example, the final squared misfit is 0.4 s^2, and four parameters were determined from the data, so the sample standard deviation is $s = (0.4/(10-4))^{1/2} = 0.08$ s, a value close to the true σ, 0.1 s.

7.2.4 *Earthquake location for more complex geometries*

This formulation is not restricted to locating earthquakes in a homogeneous halfspace. Velocity variations can be incorporated in the function relating the arrival time at the ith station to the origin time t and travel time $T(\mathbf{x}, \mathbf{x}_i)$,

$$d_i' = T(\mathbf{x}, \mathbf{x}_i) + t. \quad (1)$$

For example, a model for locating local earthquakes could have a series of layers. As a result, even for a source at the surface, the travel time curve is a more complicated function of distance (Section 3.2). At close distances, the first arrival is the direct wave. At greater distances, the first arrival becomes a head wave from an interface at depth, with the relevant interface being deeper as the distance increases. The situation is similar, but more complicated, for a source at depth, because at zero distance the travel time is nonzero.

The travel time curve can be found either analytically or by tracing rays. If the receivers are on the surface at (x_i, y_i), the travel time curve $T(r, z)$ depends on the *horizontal* distance between source and receiver,

$$r_i = [(x - x_i)^2 + (y - y_i)^2]^{1/2}, \quad (38)$$

[2] Similarly, vertical positions determined using the GPS (Section 4.5.1) by a process analogous to earthquake location are less precise than the horizontal positions.

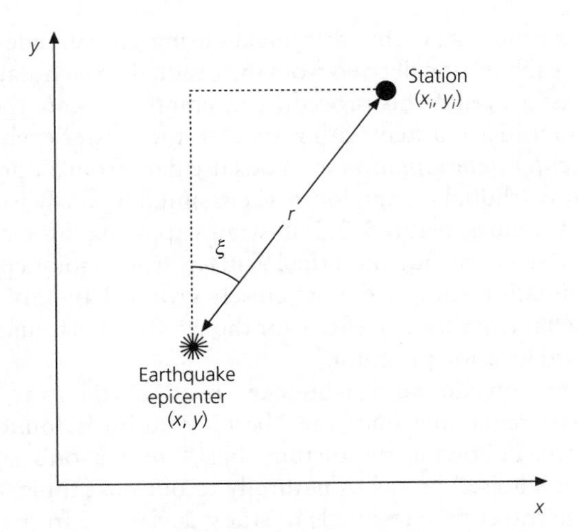

Fig. 7.2-4 Map view of the relation between an earthquake epicenter and a seismic station in Cartesian coordinates.

and the source depth z, so the arrival times are

$$d'_i = T(r_i, z) + t. \tag{39}$$

In this case, the x derivatives are found by

$$\frac{\partial d_i}{\partial x} = \frac{\partial T(r_i, z)}{\partial x} = \frac{\partial T(r_i, z)}{\partial r}\frac{\partial r_i}{\partial x} = \frac{\partial T(r_i, z)}{\partial r}\frac{(x - x_i)}{r_i}, \tag{40}$$

and similarly for the y derivatives. If ζ is the azimuth from the source to the receiver (Fig. 7.2-4),

$$(x - x_i)/r_i = -\sin \zeta_i \quad \text{and} \quad (y - y_i)/r_i = -\cos \zeta_i. \tag{41}$$

If the travel time curve is found numerically, then $T(r_i, z)$ is a set of values for various points (r, z) rather than an explicit function. The procedure for location is still the same, except that the x, y, and z partial derivatives are computed numerically. For example, if we begin by assuming that the source is at $(x^\circ, y^\circ, z^\circ)$, then the partial derivative with respect to r about

$$r_i^\circ = [(x^\circ - x_i)^2 + (y^\circ - y_i)^2]^{1/2} \tag{42}$$

is found using the tabulated travel times for points $(r_i^\circ + \delta/2, z^\circ)$ and $(r_i^\circ - \delta/2, z^\circ)$. Thus the x derivatives are found by approximating the derivative by a difference

$$\frac{\partial T(r_i, z^\circ)}{\partial x} = \frac{\partial T(r_i, z^\circ)}{\partial r}\frac{\partial r_i}{\partial x}$$
$$= \frac{T(r_i^\circ + \delta/2, z^\circ) - T(r_i^\circ - \delta/2, z^\circ)}{\delta}\frac{(x^\circ - x_i)}{r_i^\circ}, \tag{43}$$

and the y derivatives are found similarly. The z derivatives are found numerically by forming the difference between two depths. The inversion is then done as before.

The location of earthquakes for a spherical earth is similar. As before, we assume that velocity varies only with depth. In this case, for an earthquake at colatitude θ, longitude ϕ, focal depth z, and origin time t, we seek to estimate the model vector $\mathbf{m} = (\theta, \phi, z, t)$ from the data.

The travel time to receivers on the surface at colatitudes θ_i and longitude ϕ_i depends on the focal depth and the angular distance from the epicenter (Eqn A.7.7),

$$\cos \Delta_i = \cos \theta \cos \theta_i + \sin \theta \sin \theta_i \cos (\phi_i - \phi). \tag{44}$$

For a travel time curve $T(\Delta, z)$ the arrival times are

$$d_i = T(\Delta_i, z) + t. \tag{45}$$

Several average global travel time curves are available, as in Fig. 3.5-4. In addition, a travel time curve for a specific velocity model can be found numerically by tracing rays.

In this case, the θ derivatives are found using

$$\frac{\partial d_i}{\partial \theta} = \frac{\partial T(\Delta_i, z)}{\partial \theta} = \left.\frac{\partial T(\Delta_i, z)}{\partial \Delta}\right|_{\Delta_i} \frac{\partial \Delta_i}{\partial \theta}. \tag{46}$$

To find the last term, note that

$$\frac{\partial(\cos \Delta_i)}{\partial \theta} = \frac{\partial(\cos \Delta_i)}{\partial \Delta}\frac{\partial(\Delta_i)}{\partial \theta}, \tag{47}$$

so

$$\frac{\partial(\Delta_i)}{\partial \theta} = \left(\frac{\partial(\cos \Delta_i)}{\partial \theta}\right) \Big/ \left(\frac{\partial(\cos \Delta_i)}{\partial \Delta}\right)$$
$$= \frac{1}{\sin \Delta_i}(\sin \theta \cos \theta_i - \cos \theta \sin \theta_i \cos (\phi_i - \phi))$$
$$= \cos \zeta_i, \tag{48}$$

where ζ_i is the azimuth of the ith station with respect to the earthquake (Eqn. A.7.10). Thus the partial derivatives with respect to source colatitude are

$$\frac{\partial d_i}{\partial \theta} = \frac{\partial T(\Delta_i, z)}{\partial \Delta} \cos \zeta_i. \tag{49}$$

Similarly, because by the same method

$$\frac{\partial(\Delta_i)}{\partial \phi} = \left(\frac{\partial(\cos \Delta_i)}{\partial \phi}\right) \Big/ \left(\frac{\partial(\cos \Delta_i)}{\partial \Delta}\right)$$
$$= \frac{1}{\sin \Delta_i}(-\sin \theta \sin \theta_i \sin (\phi_i - \phi))$$
$$= -\sin \theta \sin \zeta_i, \tag{50}$$

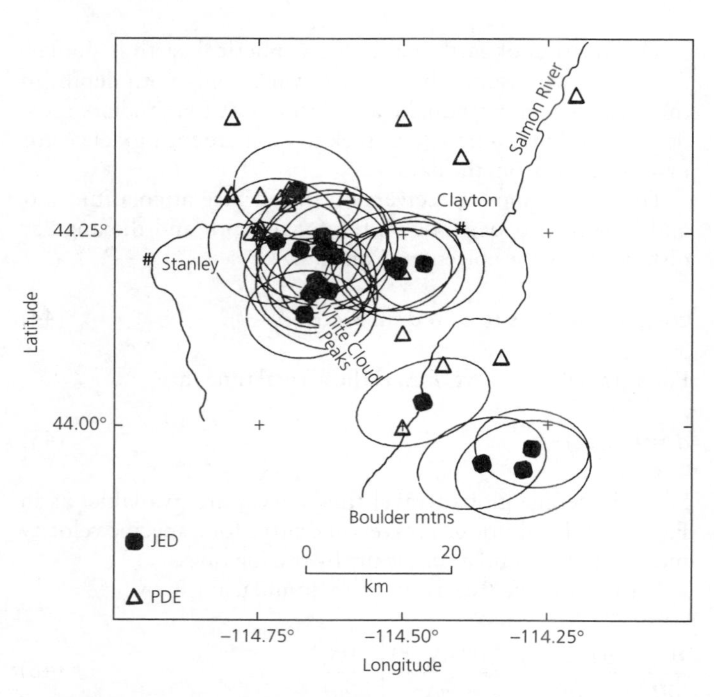

Fig. 7.2-5 Comparison of epicenters for earthquakes in central Idaho derived by a standard location program (PDE, open triangles) and from a joint epicenter determination study (JED, closed symbols). Error ellipses are shown for JED locations. The JED epicenters suggest a narrower source region than the PDE epicenters. (Dewey, 1987. © Seismological Society of America. All rights reserved.)

the partial derivatives with respect to source longitude are

$$\frac{\partial d_i}{\partial \phi} = -\frac{\partial T(\Delta_i, z)}{\partial \Delta} \sin\theta \sin\zeta_i. \tag{51}$$

The two derivatives required from the travel time table, $\partial T(\Delta_i, z)/\partial\Delta$ and $\partial T(\Delta_i, z)/\partial z$, can be approximated by forming differences between tabulated values. This approach is used to locate earthquakes all over the world using teleseismic data, often from hundreds of stations.

We can also locate earthquakes in a laterally varying structure using a numerical representation of the travel time curve. In this case, the travel times, and hence partial derivatives, depend on the actual positions of the source and the receiver, not just on the distance between them. The techniques discussed so far will work, with the modification that the travel times, and hence partial derivatives, must be computed, by tracing rays or otherwise, for each source–receiver pair. The computational effort involved is large enough that laterally homogeneous models are used whenever possible.

A number of methods are sometimes applied to improve locations derived using a laterally homogeneous model. Some treat residuals at individual stations as *station corrections* to be removed. *Master event methods* consider a particular (often the largest) earthquake in a group as the best located, and then locate a group of nearby earthquakes using a travel time correction at each station derived from the residual at each station for the master event. This procedure attempts to locate the other events more accurately with respect to the master event. *Joint hypocenter determination* methods use data from a number of nearby earthquakes, and locate them simultaneously to best fit the travel times. Figure 7.2-5 illustrates applying this technique to a group of earthquakes: the locations from a joint epicenter determination study are more closely grouped and are shifted somewhat from the epicenters for the same events found by the standard location program.

When considering earthquake location, the travel time residuals remaining once the "best" location is found are a nuisance. Following the dictum that "one person's signal is another's noise" brings us naturally to our next topic, the use of these travel time residuals to study deviations from a laterally homogeneous earth model.

7.3 Travel time tomography

In the last section we noted that travel time observations contain information about both the location and the origin time of the seismic source and the velocity structure in the region between the source and receivers. Thus, for the simple halfspace example shown, we also inverted the travel time residuals to find the best velocity. This is analogous to the way in Chapter 3 that we discussed techniques to develop layered models in which velocity varied only with depth. However, we have seen that many of the earth's most interesting processes, such as subduction, cause deviations from a laterally homogeneous velocity model. Methods have thus been developed to use seismological data to investigate laterally heterogeneous structure. For example, we have discussed using lateral variations in surface wave velocities to investigate the cooling of oceanic lithosphere (Section 2.8.3) and migration of seismic reflection data to image variable structure at depth (Section 3.3.7). In this section we introduce the concepts of *travel time tomography*, some of whose results we have seen in Sections 3.7 and 5.4. This discussion illustrates both some further general aspects of inverse problems and some specific features of inverting for earth structure.

7.3.1 *Theory*

Consider the path s of a seismic ray through a medium whose velocity v varies with position. The travel time, T, is

$$T = \int 1/v(s)\,ds = \int u(s)\,ds, \tag{1}$$

the integral of 1/velocity, the slowness, along the ray path. The ray path, in turn, is determined by the velocity distribution. Suppose now that the slowness at various points along the path

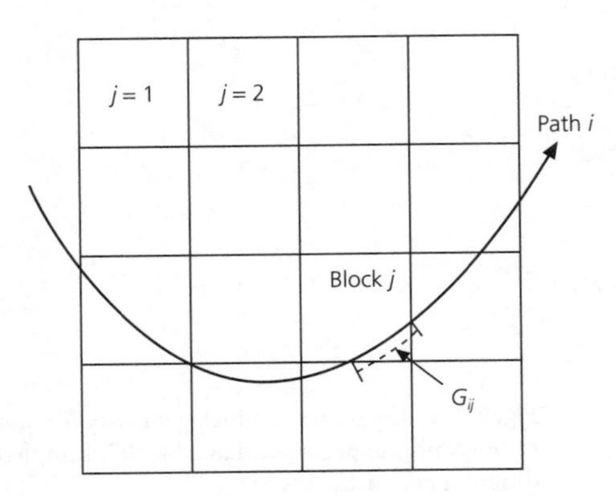

Fig. 7.3-1 Geometry of a region being studied using travel time tomography. The region is divided into blocks *j*, whose perturbations in velocity are to be found from the travel time along ray paths *i*. The velocity outside the blocks is assumed to be laterally homogeneous, so travel time perturbations with respect to the reference model are used to find the velocity perturbations within the blocks.

is perturbed by an amount $\delta u(s)$ small enough that the ray path is essentially unchanged, but the travel time changes by

$$\delta T = \int \delta u(s) ds. \tag{2}$$

We can then use the changes in travel time to study the velocity changes that caused them.

Because the travel time perturbation reflects the slowness perturbation integrated along the ray path, a single observation does not indicate how the perturbation is distributed along the path. A large localized perturbation and a smaller, but more widely distributed, one could give the same effect. To improve resolution, data from ray paths that sample the medium differently can be combined (Fig. 7.3-1). The simplest spatial distribution of the slowness perturbation divides the medium into a number of homogeneous subregions termed blocks, or cells. Thus the integral (Eqn 2) giving the travel time perturbation along the ith ray path is written in discrete form

$$\Delta T_i = \sum_{j=1} G_{ij} \Delta u_j, \tag{3}$$

where G_{ij} is the distance the ith ray travels in the jth block, and Δu_j is the slowness perturbation in the block.

Our goal is to use the observed travel times along a number of paths through the medium to recover the slowness perturbation. Problems of this type, in which observations of properties integrated along a number of paths through the medium are used to infer the two- or three-dimensional distribution of the physical property within a medium, occur in many branches of science and are known collectively as *tomography*.[1] The two- or three-dimensional perturbation can be thought of as an image, which we seek to reconstruct from observations. The observations, one-dimensional integrals through the perturbation, are known as projections.

In travel time tomography, the inverse problem of estimating the slowness perturbation from the observed travel time perturbation has the form discussed in the last section

$$\mathbf{d} = G\mathbf{m}, \quad \text{or} \quad d_i = \sum_j G_{ij} m_j. \tag{4}$$

As before, we do not explicitly write the Δs, so the model vector **m** is the perturbation in slowness from a starting model, and the data vector **d** is the difference between the observed travel times and those predicted by the starting model. The elements of the partial derivative matrix

$$G = \frac{\partial d_i}{\partial m_j} = \frac{\partial T_i}{\partial u_j} \tag{5}$$

equal the distance the ith ray travels in the jth block, which is the partial derivative of the ray's travel time with respect to the slowness in the block.

The matrix G is an operator that relates model vectors and data vectors. As in the location problem, these vectors are physically different quantities with different dimensions. The model vectors have as many elements as there are blocks in the model, whereas the data vectors have a number of elements equal to the number of ray paths. Mathematically, this means that if there are r blocks in the model, any model vector is a vector in an r-dimensional model space. Similarly, if there are n travel times and thus n ray paths, any data vector is a vector in an n-dimensional data space. Because there are generally many more equations (ray paths) than unknowns (model parameters), the system of equations is *overdetermined*. Because the data contain noise, the system of equations is generally also *inconsistent*.

The inverse problem is solved by a procedure like that discussed for the location problem. For the different ray paths, the travel times and the distances traveled in each block are predicted using a starting or reference model. The starting model is generally laterally homogeneous, so the travel times are easily calculated. Travel time residuals are then computed for each ray path by subtracting the times predicted by the starting model from those observed. These travel time residuals form the data vector that is inverted using the generalized inverse to find slowness changes that predict the travel time residuals as well as possible.

To illustrate these ideas, consider a schematic experiment in which a region under a seismic array is divided into four square blocks of unit length (Fig. 7.3-2). Travel time residuals from six ray paths form the data. Four paths (1–4), which can be thought of as due to distant (teleseismic) earthquakes, traverse the model vertically. Two paths (5, 6), which can be thought of as due to local earthquakes, traverse the model horizontally.

[1] This term is Greek for "slice picture."

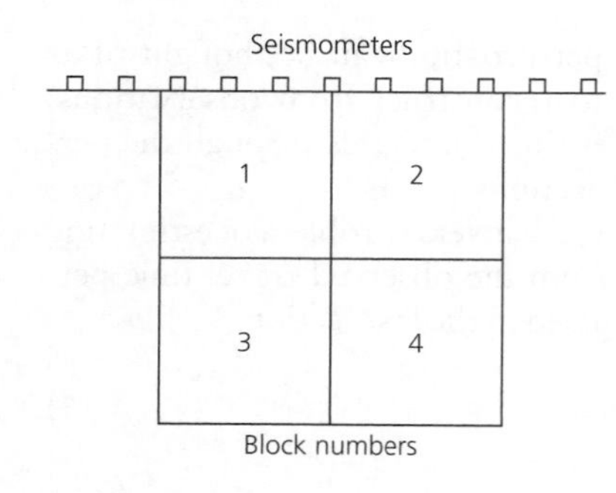

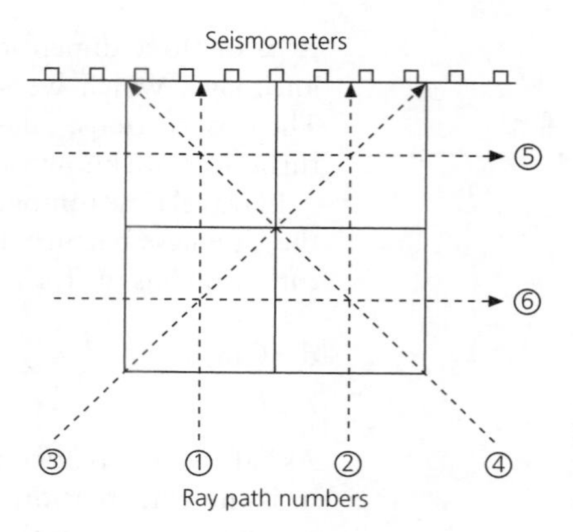

Fig. 7.3-2 Ray path and block geometry for an idealized tomographic experiment. Each block is sampled by three different ray paths.

The reference slowness model is assumed to be appropriate outside the blocks, so the entire travel time residual for each path is attributed to slowness perturbations in the blocks. Thus the problem looks like

$$\begin{pmatrix} 1 & 0 & 1 & 0 \\ 0 & 1 & 0 & 1 \\ 0 & \sqrt{2} & \sqrt{2} & 0 \\ \sqrt{2} & 0 & 0 & \sqrt{2} \\ 1 & 1 & 0 & 0 \\ 0 & 0 & 1 & 1 \end{pmatrix} \begin{pmatrix} m_1 \\ m_2 \\ m_3 \\ m_4 \end{pmatrix} = \begin{pmatrix} d_1 \\ d_2 \\ d_3 \\ d_4 \\ d_5 \\ d_6 \end{pmatrix} \tag{6}$$

We encountered this problem, solving a vector–matrix equation where the matrix is not square, in the last section. As in that case, we form

$$G^T G \mathbf{m} = G^T \mathbf{d} \tag{7}$$

and invert the square matrix G^TG to form the generalized inverse solution

$$\mathbf{m}_g = (G^TG)^{-1} G^T \mathbf{d} = G^{-g}\mathbf{d}. \tag{8}$$

We next ask how $\mathbf{m}_g$, the model found by the inversion, compares to the actual slowness model that gave rise to the travel time data. To compare the two, we substitute $G\mathbf{m}$ for $\mathbf{d}$ in Eqn 8, and find that in this case

$$\mathbf{m}_g = (G^TG)^{-1} G^T G \mathbf{m} = \mathbf{m}, \tag{9}$$

so the inversion correctly resolves the true model. Naturally, if errors are present in the data, these errors propagate into the results of the inversion, as discussed previously.

7.3.2 *Generalized inverse*

An interesting situation occurs in this example if only the four teleseismic ray paths (1–4) are available. The inverse problem becomes finding the four elements of $\mathbf{m}$ from

$$\begin{pmatrix} 1 & 0 & 1 & 0 \\ 0 & 1 & 0 & 1 \\ 0 & \sqrt{2} & \sqrt{2} & 0 \\ \sqrt{2} & 0 & 0 & \sqrt{2} \end{pmatrix} \begin{pmatrix} m_1 \\ m_2 \\ m_3 \\ m_4 \end{pmatrix} = \begin{pmatrix} d_1 \\ d_2 \\ d_3 \\ d_4 \end{pmatrix}. \tag{10}$$

After multiplying by G^T, we attempt to solve this system as before, but find that the matrix G^TG has a zero determinant, so it cannot be inverted. Thus, although the system of equations (7) has four equations for four unknowns, it does not have a unique solution (Section A.4.4). It turns out that this is because the rows of G are not linearly independent. Thus the ray geometry is not adequate to fully resolve the slowness perturbations in the four blocks.

Because this situation occurs frequently in solving inverse problems, methods for dealing with it have been developed. Although a full treatment is beyond our scope, we summarize some key ideas without proof.

In the general case when G is an $n \times r$ matrix, G^TG is an $r \times r$ symmetric matrix that can be decomposed using its eigenvectors and eigenvalues (Section A.5.3)

$$G^TG = V \Lambda V^T, \tag{11}$$

where the columns of matrix V are the r eigenvectors of G^TG

$$V = \begin{pmatrix} v_1^{(1)} & \cdot & \cdot & \cdot & v_1^{(r)} \\ \cdot & \cdot & \cdot & \cdot & \cdot \\ \cdot & \cdot & \cdot & \cdot & \cdot \\ v_r^{(1)} & \cdot & \cdot & \cdot & v_r^{(r)} \end{pmatrix} \tag{12}$$

and Λ is a diagonal matrix with eigenvalues on the diagonal and zeroes elsewhere

$$\Lambda = \begin{pmatrix} \lambda_1 & 0 & \cdot & \cdot & 0 \\ 0 & \lambda_2 & \cdot & \cdot & 0 \\ \cdot & \cdot & \cdot & \cdot & \cdot \\ \cdot & \cdot & \cdot & \cdot & \cdot \\ 0 & 0 & \cdot & \cdot & \lambda_r \end{pmatrix}. \tag{13}$$

Because the eigenvectors are orthogonal,

$$VV^T = V^TV = I, \quad \text{so} \quad V^T = V^{-1}. \tag{14}$$

If G^TG has an inverse,

$$(G^TG)^{-1} = (V\Lambda V^T)^{-1} = V\Lambda^{-1}V^T, \tag{15}$$

where

$$\Lambda^{-1} = \begin{pmatrix} 1/\lambda_1 & 0 & \cdot & \cdot & 0 \\ 0 & 1/\lambda_2 & \cdot & \cdot & 0 \\ \cdot & \cdot & \cdot & \cdot & \cdot \\ \cdot & \cdot & \cdot & \cdot & \cdot \\ 0 & 0 & \cdot & \cdot & 1/\lambda_r \end{pmatrix}. \tag{16}$$

This expression shows that G^TG is singular if at least one eigenvalue is zero. In this case, the p nonzero eigenvalues are used to form the $p \times p$ diagonal matrix

$$\Lambda_p = \begin{pmatrix} \lambda_1 & 0 & \cdot & \cdot & 0 \\ 0 & \lambda_2 & \cdot & \cdot & 0 \\ \cdot & \cdot & \cdot & \cdot & \cdot \\ \cdot & \cdot & \cdot & \cdot & \cdot \\ 0 & 0 & \cdot & \cdot & \lambda_p \end{pmatrix}, \tag{17}$$

and the associated eigenvectors are divided into two matrices:

$$V_p = \begin{pmatrix} v_1^{(1)} & \cdot & \cdot & \cdot & v_1^{(p)} \\ \cdot & \cdot & \cdot & \cdot & \cdot \\ \cdot & \cdot & \cdot & \cdot & \cdot \\ \cdot & \cdot & \cdot & \cdot & \cdot \\ v_r^{(1)} & \cdot & \cdot & \cdot & v_r^{(p)} \end{pmatrix} \quad \text{and} \quad V_0 = \begin{pmatrix} v_1^{(p+1)} & \cdot & \cdot & \cdot & v_1^{(r)} \\ \cdot & \cdot & \cdot & \cdot & \cdot \\ \cdot & \cdot & \cdot & \cdot & \cdot \\ \cdot & \cdot & \cdot & \cdot & \cdot \\ v_r^{(p+1)} & \cdot & \cdot & \cdot & v_r^{(r)} \end{pmatrix}. \tag{18}$$

V_p is the $r \times p$ matrix of the eigenvectors with nonzero eigenvalues, and V_0 is the $r \times (r - p)$ matrix of the eigenvectors with zero eigenvalues.

Similarly, the $n \times n$ matrix GG^T can be decomposed as

$$GG^T = U\Lambda U^T, \tag{19}$$

using its eigenvector matrix U. GG^T has the same p nonzero eigenvalues as G^TG, so the U matrix can be divided into U_p, the $n \times p$ matrix of the eigenvectors with nonzero eigenvalues, and U_0, the $n \times (n - p)$ matrix of the eigenvectors with zero eigenvalues. Although we do not prove it here, it is possible to decompose the matrix G using only the eigenvectors with nonzero eigenvalues:

$$G = U\Lambda V^T = U_p\Lambda_pV_p^T. \tag{20}$$

This decomposition, known as the *Lanczos decomposition*, is important, because a generalized inverse

$$G^{-p} = V_p\Lambda_p^{-1}U_p^T \tag{21}$$

that involves only the eigenvectors with nonzero eigenvalues gives an optimal solution to the inverse problem. This solution provides the best fit to the data while minimizing **m**, the change from the starting model. This is a desirable feature: for example, in the tomographic problem, we start with a laterally homogeneous model, so the best solution is that with least lateral velocity variation consistent with the data.

7.3.3 *Properties of the generalized inverse solution*

The relation between the solution to the inverse problem, the model derived from the data using

$$\mathbf{m}_p = G^{-p}\mathbf{d}, \tag{22}$$

and the "true" (although unknown) model **m**, can be found because the data are related to the "true" model by the forward problem (Eqn 4), so

$$\mathbf{m}_p = G^{-p}G\mathbf{m} = V_p\Lambda_p^{-1}U_p^TU_p\Lambda_pV_p^T\mathbf{m} = V_pV_p^T\mathbf{m}. \tag{23}$$

Thus the matrix $G^{-p}G = V_pV_p^T$ is known as the *model resolution matrix*.

The derivation used the fact that $U_p^TU_p = I$, because the columns of U_p and hence the rows of U_p^T are orthonormal eigenvectors. Similarly, $V_p^TV_p = I$. By contrast, if there are some zero eigenvalues, then $p \neq n$, $U_pU_p^T \neq I$ and $p \neq r$, $V_pV_p^T \neq I$, because the rows of U_p and V_p are no longer orthonormal eigenvectors (because the columns corresponding to the zero eigenvalues have been removed to form the V_0 and U_0 matrices).

To illustrate these ideas, consider the example in Eqn 10. The G matrix yields

$$G^TG = \begin{pmatrix} 3 & 0 & 1 & 2 \\ 0 & 3 & 2 & 1 \\ 1 & 2 & 3 & 0 \\ 2 & 1 & 0 & 3 \end{pmatrix}, \tag{24}$$

which has eigenvalues 0, 2, 4, 6, and hence is singular. The eigenvector matrices are

$$V_p = \begin{pmatrix} -0.5 & -0.5 & 0.5 \\ 0.5 & 0.5 & 0.5 \\ -0.5 & 0.5 & 0.5 \\ 0.5 & -0.5 & 0.5 \end{pmatrix} \quad \text{and} \quad V_0 = \begin{pmatrix} 0.5 \\ 0.5 \\ -0.5 \\ -0.5 \end{pmatrix}. \tag{25}$$

The model resulting from the inversion $\mathbf{m}_p$ is then related to the "true" (although unknown) model **m** by the model resolution matrix

$$\mathbf{m}_p = V_pV_p^T\mathbf{m} = \begin{pmatrix} 0.75 & -0.25 & 0.25 & 0.25 \\ -0.25 & 0.75 & 0.25 & 0.25 \\ 0.25 & 0.25 & 0.75 & -0.25 \\ 0.25 & 0.25 & -0.25 & 0.75 \end{pmatrix}\mathbf{m}. \tag{26}$$

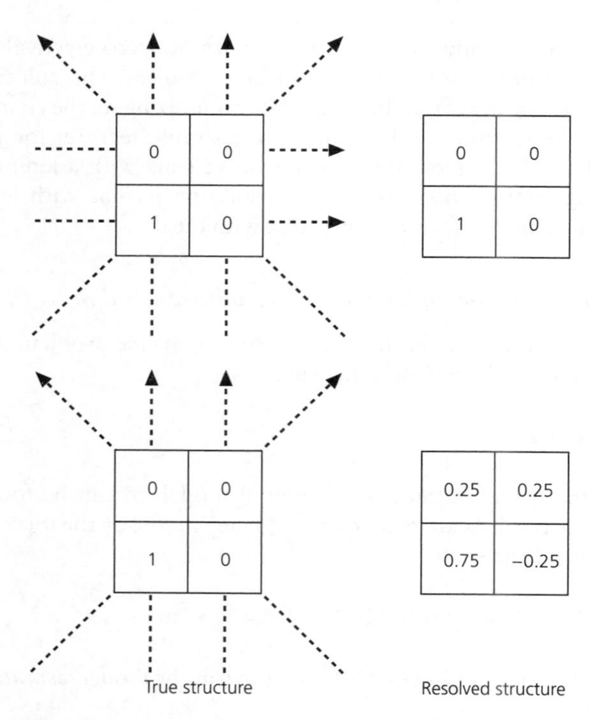

Fig. 7.3-3 Illustration of the "blurring" resulting from the tomographic experiment of Fig. 7.3-2, with incomplete ray coverage. When coverage is adequate, the true slowness perturbation (*top left*) is recovered (*top right*). When coverage is inadequate, the true slowness perturbation (*lower left*) is blurred (*lower right*), although the resulting slowness perturbations yield the correct travel time perturbation for each ray path.

The ith column of the model resolution matrix shows how a unit perturbation in the ith element of the true model maps into various elements of $\mathbf{m}_p$. The true model is thus "blurred" by the inversion. For example (Fig. 7.3-3), inversion of travel time data resulting from a 1% slowness perturbation in block 3 yields a model with 0.25% perturbations in blocks 1 and 2, a 0.75% perturbation in block 3, and a −0.25% perturbation in block 4. These slowness perturbations yield the correct travel time perturbations for the four paths, but because there are no horizontal paths, the solution is not exactly correct. However, most of the perturbation is correctly placed. Note that the resolved structure has a smaller maximum slowness perturbation than the true structure.

The relation between the resolution matrix and the model covariance matrix (Eqn 7.2.32) is interesting. The blurring illustrated by the resolution matrix results from the ray geometry and would occur even if the data contained no errors. In other words, the resolution matrix illustrates how well the inverse problem could be solved for perfect data. Because the data usually contain errors, the uncertainty in the model, given by the model covariance matrix, reflects errors induced in the model by both the ray geometry and the data errors.

Because the resolution matrix shows how a perturbation in any block is resolved by the inversion, it can be used to find how well the inversion can recover an arbitrary slowness anomaly. Thus the ray geometry, which gives the G and hence V matrices their form, controls the resolution. Note that in the first example, in which all six ray paths are used, Eqn 9 shows that the model from the inversion was the true model. In this case the resolution matrix is the identity matrix.

To see how the lack of resolution in the four-ray case arises, consider what would occur if G^TG had no zero eigenvalues and could be inverted. Then, by Eqns 21 and 22, the model derived from the data would be

$$\mathbf{m}_p = V\Lambda_p^{-1}U_p^T\mathbf{d}, \tag{27}$$

because $V_p = V$. The model is thus a linear combination of the columns of V, or the eigenvectors of G^TG. Because there are r (in this case four) linearly independent eigenvectors, and the model vector has r elements, the eigenvectors span the r-dimensional model space. Thus any vector in the model space is a possible model.

If instead, as in this case, some of the eigenvalues are zero, the eigenvectors associated with them are excluded from the V_p matrix. The model

$$\mathbf{m}_p = V_p\Lambda_p^{-1}U_p^T\mathbf{d} \tag{28}$$

is then a linear combination of only the columns of V_p, the eigenvectors associated with the nonzero eigenvalues. In this case, there are $r-p$ (here three) rather than r linearly independent eigenvectors. Hence not all possible vectors in the model space can be constructed. The model resulting from the inversion contains no linear combinations of the eigenvectors associated with the zero eigenvalues.

To illustrate this idea, consider the four-ray case where the eigenvector associated with the zero eigenvalue is (from Eqn 25)

$$\mathbf{v} = (0.5, 0.5, -0.5, -0.5)^T. \tag{29}$$

This vector corresponds to equal slowness perturbations in blocks 1 and 2 and equal perturbations of opposite sign in blocks 3 and 4. Physically, this means changing the slowness everywhere in the upper layer by some amount, and making the opposite change in the lower layer. Because all four teleseismic rays have equal path lengths in the upper and lower layers, their travel times are unaffected, so travel time data cannot resolve any such change.

Another way to see this is to consider Eqn 7 and note that if $\mathbf{v}$ is an eigenvector whose eigenvalue is zero,

$$(G^TG)\mathbf{v} = 0, \tag{30}$$

so that even if the model contains a linear combination of such eigenvectors, they have no effect on the problem. The zero eigenvectors thus limit the resolution of the model. Because any linear combination of these eigenvectors has no effect, the model resulting from the inversion is not unique. It is possible to prove that the generalized inverse G^{-p} finds a "best" model with no contribution from these eigenvectors. Mathematically, the resulting model is restricted to the V_p space and has no components in the V_0 space. As a result, this model is the minimum possible solution consistent with the data. In this application, the minimum model gives the least lateral perturbation in slowness consistent with the travel time data. Philosophically, this is an attractive approach.

The six-ray case, by contrast, had no zero eigenvalues. Because one ray traveled only in the upper layer and another traveled only in the lower layer, a change in the slowness in either layer would affect the travel times. This ray geometry avoids the ambiguity of the four-ray case, so the model is fully resolved. There is no V_0 space, so $V = V_p$, G^TG can be inverted, and the solution is found using the generalized inverse G^{-g} (Eqn 8). To see how this is related to the generalized inverse G^{-p}, we use the Lanczos decomposition (Eqn 20) to expand G:

$$G^TG = (V\Lambda_p U_p^T)(U_p\Lambda_p V^T) = V\Lambda_p^2 V^T, \tag{31}$$

$$(G^TG)^{-1} = V\Lambda_p^{-2}V^T, \tag{32}$$

where the matrix products $\Lambda_p^2 = \Lambda_p\Lambda_p$ and $\Lambda_p^{-2} = \Lambda_p^{-1}\Lambda_p^{-1}$. Thus, if G^TG can be inverted, the generalized inverse

$$G^{-g} = (G^TG)^{-1}G^T = (V\Lambda_p^{-2}V^T)(V\Lambda_p U_p^T) = V\Lambda_p^{-1}U_p^T = G^{-p}. \tag{33}$$

Hence G^{-p} is the general form of the generalized inverse, and G^{-g} is the special form that applies if G^TG can be inverted. The later form, G^{-g}, is easier to compute because it does not require the eigenvector decomposition. Fortunately, it can often be used in applications such as earthquake location.

The eigenvector decomposition also divides the data space into two portions, U_p and U_0, reflecting the nonzero and zero eigenvalues. Data vectors in the U_0 space, linear combinations of the eigenvectors whose eigenvalues are zero, cannot be generated by the operator G for any model. For example, in the six-ray case there cannot be six linearly independent observations because the model has only four parameters. Thus two of the six eigenvectors of the 6×6 matrix GG^T must have zero eigenvalues. These eigenvectors represent travel time observations that should be impossible, given the geometry of the experiment. If the data contained some linear combinations of these eigenvectors, perhaps due to noise in the data, the inversion process could never generate a model capable of matching them.

Figure 7.3-4 summarizes these ideas: the operator G and its generalized inverse G^{-p} relate the model and data spaces. Portions of these spaces are not "illuminated." Any part of the model in the V_0 portion of the model space has no effect on the data, and thus cannot be detected. Thus, if V_0 space exists, the model found by solving the inverse problem is not unique. This situation can only be improved by additional types of data, such as a new set of ray paths in the tomographic example (Fig. 7.3-3).[2] Similarly, any part of the data in the U_0 portion of the data space cannot be described by any possible model.

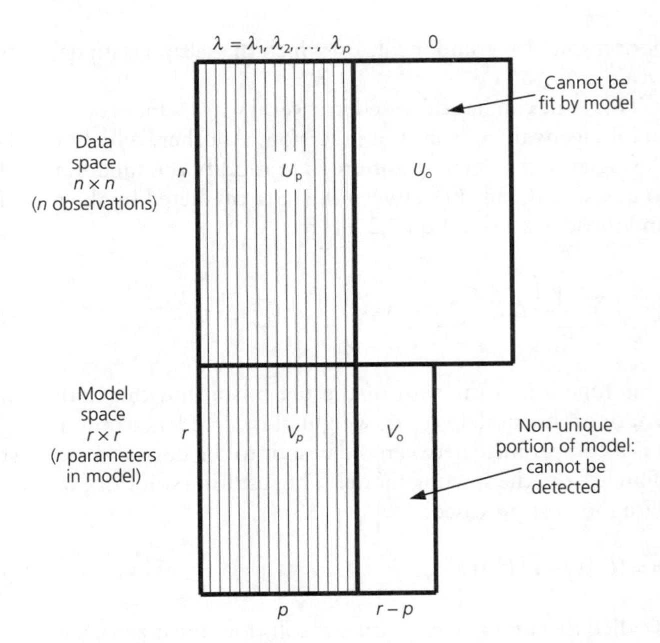

Fig. 7.3-4 Schematic illustration of the relation between the model and data spaces for the inverse problem $\mathbf{d} = G\mathbf{m}$. The observed data $\mathbf{d}$ form a vector in the n-dimensional data space, the model $\mathbf{m}$ sought is a vector in the r-dimensional model space, and the known partial derivative matrix G has dimensions $n \times r$. Matrix U, whose columns are the eigenvectors of the matrix GG^T, can be decomposed into U_p, the matrix of the p eigenvectors with nonzero eigenvalues $\lambda_1, \lambda_2, \ldots, \lambda_p$, and U_0, the matrix of the eigenvectors with zero eigenvalues. Similarly, the matrix V, whose columns are the eigenvectors of the matrix G^TG, can be decomposed into V_p, the matrix of the eigenvectors with nonzero eigenvalues, and V_0, the matrix of the eigenvectors with zero eigenvalues. (After Lanczos, 1961.)

Thus, if a U_0 space exists, the model found by solving the inverse problem is not an exact solution.

7.3.4 *Variants of the solution*

A number of variants of the least squares solution that we have developed using earthquake location and tomography are also used in these and other inverse problems.

One variant arises from the fact that although the eigenvector decomposition gives insights, it may not be the best approach in some real applications. First, it involves significant computations when the matrices are large. Second, it associates difficulties with the eigenvalues that are zero, whereas in real problems complications and noisy data are more likely to yield small, but nonzero, eigenvalues. These small eigenvalues cause the sort of difficulties that occur formally for zero eigenvalues. To see this, note that in Eqn 27 the model is derived by multiplying the data by the matrix Λ^{-1}, which contains the reciprocals of the eigenvalues. Thus the small eigenvalues, representing the worst-constrained features of the data and model spaces, can have large effects on the solution. For example, we noted in Section 4.4.7 that using the generalized inverse to estimate the moment tensor gives good estimates of components on which seismograms depend strongly, but

[2] As Sherlock Holmes says in *The Copper Beeches*, "I have devised seven separate explanations, each of which would cover the facts so far as we know them. But which of these is correct can only be determined by fresh information."

poorer ones for components on which the seismogram depends weakly.

This issue can be addressed in several ways. One is to exclude small eigenvalues from the inversion. Another, which avoids the eigenvector decomposition, is to modify the function used to measure the misfit between the data predicted by the model and those observed (Eqn 7.2.11) to

$$\chi^2 = \sum_i \frac{1}{\sigma_i^2}\left(\Delta d_i - \sum_j G_{ij}\Delta m_j\right)^2 + \varepsilon^2 \sum_j (\Delta m_j)^2. \qquad (34)$$

This function is the sum of the net misfit and the change in length of the model vector, weighted by ε^2. Hence minimizing it is a compromise between the best fit to the data and the least change from the starting model. The resulting solution, written with the Δs suppressed,

$$\mathbf{m} = (G^T G + \varepsilon^2 I)^{-1} G^T \mathbf{d}, \qquad (35)$$

is called the *damped least squares* solution. If ε is zero, we have the best-fit solution (Eqn 7.2.17), whereas larger values of ε reduce or damp the change in the starting model by accepting a poorer fit to the data. The damping parameter ε is chosen empirically to yield a solution that seems plausible, and thus of necessity reflects our ideas about the solution sought, because damping the poorly constrained and undesired changes in the model also damps the better constrained and desired changes.

Another common situation is that we want some data to have greater effect on the solution, usually because we consider them to be better known. We thus incorporate a data-weighting matrix W_d into the solution. The simplest is to weight by $W_d = (\sigma_d^2)^{-1}$, the inverse of the variance–covariance matrix of the data, so the data with the smallest uncertainties have the greatest effect. Problem 5 shows that this *weighted least squares* solution is

$$\mathbf{m} = (G^T W_d G)^{-1} G^T W_d \mathbf{d}. \qquad (36)$$

We may also want to have the model change smoothly, such that each element varies only slightly with respect to its neighbors. For instance, if the model were a continuous function of one variable, we measure the smoothness, or *flatness*, f, of the changes by forming

$$\mathbf{f} = \begin{pmatrix} -1 & 1 & 0 & \cdot & \cdot & 0 \\ 0 & -1 & 1 & \cdot & \cdot & 0 \\ 0 & 0 & 0 & \cdot & \cdot & 0 \\ \cdot & \cdot & \cdot & \cdot & \cdot & 0 \\ \cdot & \cdot & \cdot & \cdot & \cdot & 0 \\ 0 & 0 & 0 & 0 & -1 & 1 \end{pmatrix} \begin{pmatrix} m_1 \\ m_2 \\ m_3 \\ \cdot \\ \cdot \\ m_r \end{pmatrix} = F\mathbf{m}, \qquad (37)$$

where F is the *flatness matrix*, which is a numerical approximation to the derivative at the edges of each element. The overall flatness of the solution is then

$$\mathbf{f}^T\mathbf{f} = \mathbf{m}^T F^T F \mathbf{m} = \mathbf{m}^T W_m \mathbf{m}, \qquad (38)$$

so the matrix $W_m = F^T F$ is a weighting matrix for the model. For more complicated model geometries, F is changed appropriately.

We can combine the model and data weighting in a *weighted damped least squares* inversion, which yields the solution

$$\mathbf{m} = (G^T W_d G + \varepsilon^2 W_m)^{-1} G^T W_d \mathbf{d}. \qquad (39)$$

As noted earlier, the damping parameter ε is chosen empirically. If we do not weight the data and model, the weighting matrices W_m and W_d are identity matrices, and Eqn 39 is just the simple damped least squares solution (Eqn 35).

An example of such an inversion was shown for *P*-wave velocities at the base of the mantle in Fig. 3.5-17. A grid of 660 nodes that were roughly equally spaced were used to represent the base of the mantle. The damping factor, $\varepsilon = 1.2$, was a compromise between the best fit, which minimizes the prediction error, and minimizing the undetermined part of the solution. Because each node is surrounded by 5 or 6 nodes that are roughly equidistant, the rows of the model flatness matrix F were chosen with the diagonal term equal to -1 and the terms of the nearest N neighbors equal to $1/N$ (with $N = 5$ or 6). The data were weighted empirically so that the diagonal elements of the W_d matrix ranked the quality of the observations from 9 (excellent) through 4 (good) to 1 (poor). These choices again bear out that we have various ways of solving inverse problems, so the solution we develop depends on choices about the data we use and the model we seek, based on our ideas about what seems reasonable. Hence our solutions are in part objective and in part subjective, and different approaches yield different solutions.

7.3.5 *Examples*

Studies using travel time tomography yield interesting results for various areas. For example, Fig. 7.3-5 (*top*) shows the model geometry used in a study of the upper mantle in the region including Central Europe, the Mediterranean, and the Middle East. The model contains nine layers, each divided into 1040 1° by 1° blocks. The layer thickness increases with depth from 33 km at the top to 130 km at a depth of 670 km. The data consist of approximately half a million travel times from about 25,000 earthquakes, recorded at stations both within the model region and at distances to 90°.

The data used are travel time anomalies relative to the Jeffreys–Bullen values, which can result from earthquake mislocations as well as variations in seismic velocities. The location and origin time of the earthquakes were thus also inverted for, so the number of unknowns reflects both the number of blocks (9360) and four times the number of earthquakes used. To reduce these numbers, procedures were used to combine data from nearby earthquakes and from stations close to each other. The problem to be solved thus involves approximately 300,000 equations for 20,000 unknowns.

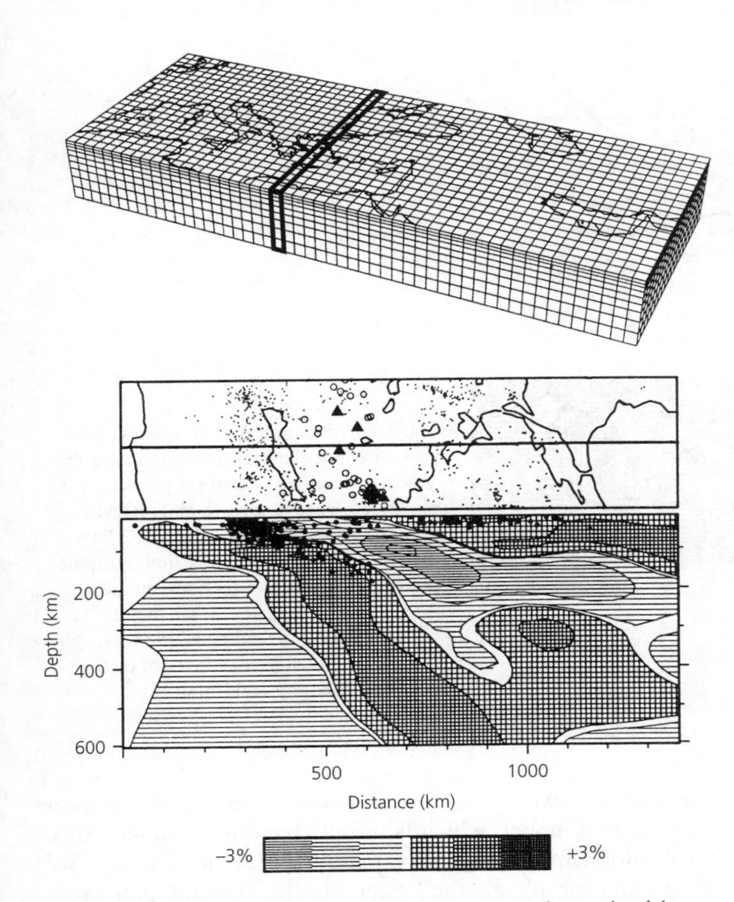

Fig. 7.3-5 *Top*: Block model for a travel time tomographic study of the upper mantle in the region including Central Europe, the Mediterranean, and the Middle East. The heavy line indicates the location of the cross-section shown below. (Spakman and Nolet, 1988, with kind permission from Kluwer Academic Publishers.) *Bottom*: Cross-section through the block model across the Hellenic trench region, showing P-wave velocity perturbations with respect to the JB model. (Spakman *et al.*, 1988. *Geophys. Res. Lett.*, *15*, 60–3, copyright by the American Geophysical Union.)

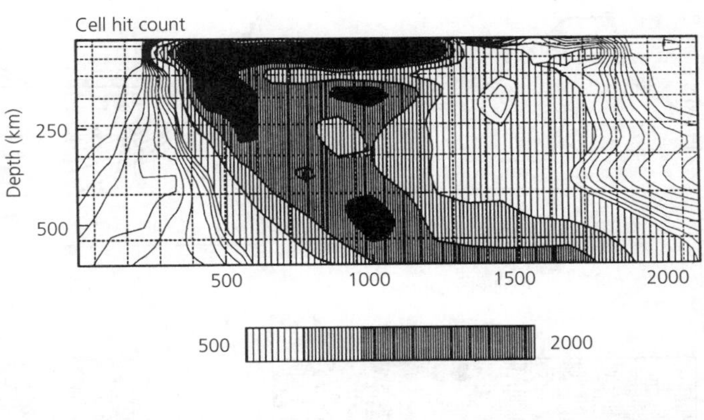

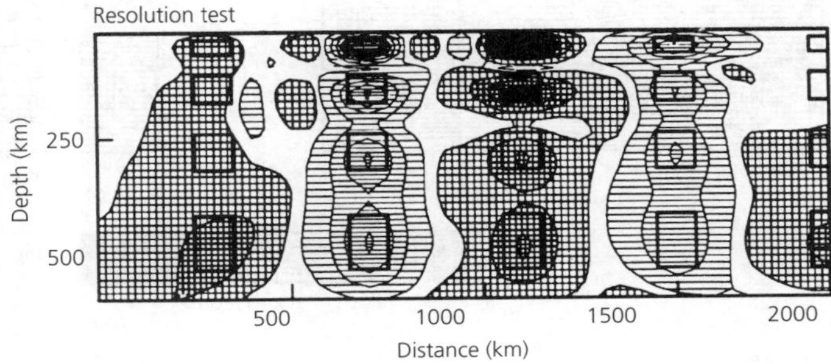

Fig. 7.3-6 Analysis of the tomographic image in Fig. 7.3-5 (*bottom*). *Top*: Hit count plot, showing the number of times each block is sampled. Black regions indicate the best-sampled blocks (hit counts in excess of 2000). *Bottom*: Resolution test using synthetic velocity anomalies. Travel times are generated for a model with 5% velocity perturbations, of alternating sign, in each of the blocks marked by heavy lines. How well the perturbations are recovered illustrates how much the image is blurred. (Spakman and Nolet, 1988, with kind permission from Kluwer Academic Publishers.)

Solving matrix equations of this size poses major difficulties. The matrices are so large (in this case 6×10^9 elements) that they are difficult to store in a computer and operate on. As a result, numerical methods are used, some of which allow only a single row of the matrix to be manipulated at any time. The properties of these algorithms and methods of improving the resulting image form an active research area.

The resulting three-dimensional velocity model can be shown as either cross-sections or map views at various depths. Figure 7.3-5 (*bottom*) shows a cross-section across the Hellenic trench region, where the African plate subducts beneath Crete and the Aegean basin (Fig. 5.6-8). The tomographic image shows velocity anomalies in percent of the velocity predicted for that depth by the JB model. A planar high-velocity (positive) anomaly, presumably the cold downgoing slab, dips NW from the trench and extends to depths well below the deepest earthquakes (dots). Above the slab, a low-velocity (negative) region occurs, presumably due to flow behind the arc. Such observations are valuable for modeling the subduction history and dynamics.

Because tomographic images are solutions to an inverse problem, they are neither unique nor exact. Hence it is important to assess which features in the image are likely to be geologically real, and which are more likely to be artifacts of the inversion. As we have seen, an important factor is how well parts of the model are sampled by the ray paths. Figure 7.3-6 (*top*) shows a *hit count* plot for the section of Fig. 7.3-5 (*bottom*), showing the number of ray paths that sample each block. The better-illuminated regions should be better resolved than poorly sampled regions. Additional insight comes from analyzing how a perturbation in one model block is blurred by the inversion into nearby blocks. This information, given by the resolution matrix (Eqn 23), can also be found by placing a perturbation in one block, computing the forward problem, and inverting the result. Because this would be time consuming for such a large model, perturbations were placed in various blocks, and the combined resolution was estimated by computing synthetic travel time data and inverting it. Figure 7.3-6

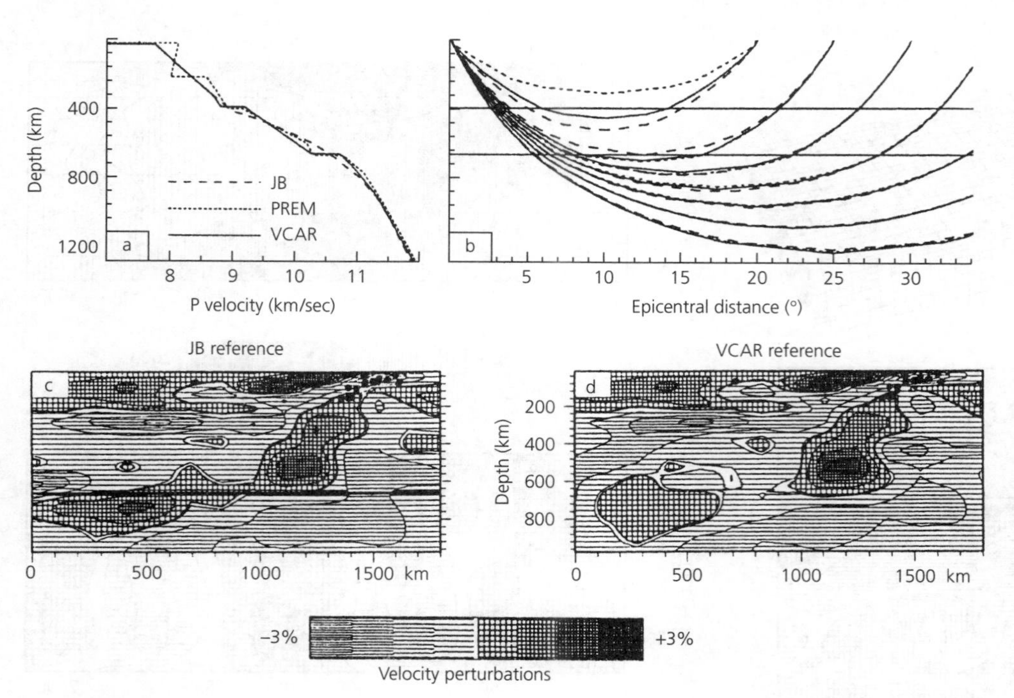

Fig. 7.3-7 Illustration of the effects of the reference model in travel time tomography. Velocity structure (a) and ray paths (b) for global reference models JB and PREM and a local reference model VCAR. The differences (c, d) between the tomographic images reflect differences between the reference models near 600 km depth. (van der Hilst and Spakman, 1989. *Geophs. Res. Lett.*, *16*, 1093–6, copyright by the American Geophysical Union.)

(*bottom*) illustrates this method for a 5% velocity contrast whose sign alternates between columns. If resolution were perfect, the image would be reconstructed exactly: each anomaly would be confined to the original block (heavy line). Due to the ray geometry, the anomalies "blur", but are still concentrated in the correct locations. Comparison with the hit counts shows that better-sampled regions, such as the second column from the left, are better resolved than poorly sampled regions like the lower left column. The reconstructed image is further degraded when the effects of noise in the data are simulated. Even in this case, the inversion results locate the perturbed blocks reasonably well and retrieve the sign of the perturbation. These tests suggest strongly that the high-velocity slab in the image is real.

Typically, the major features of tomographic inversions seem likely to be real, but assessing how much of the detailed structure is real is more difficult. For example, Fig. 5.4-7 showed the results of a numerical experiment to see how well a tomographic study would reconstruct the image of a theoretical subducting slab. It turned out that the general shape of the slab was resolved, but was blurred by artifacts implying velocity anomalies that are not present in the original model. In this case these artifacts, generally of low amplitude, caused the slab to appear to broaden, shallow in dip, or flatten out. The extent to which these artifacts appear depends on ray geometry, so the image could be improved by using upgoing as well as downgoing rays.

Another important factor in tomographic images is the reference model with respect to which the velocity anomalies are shown. In examining images, it is natural to focus on the lateral variations. However, because these variations are with respect to a starting model, which is usually laterally homogeneous, the resulting images depend on the starting model. Figure 7.3-7 shows an example for the Lesser Antilles. The ray paths predicted by the global JB and PREM reference models differ somewhat from those predicted by a model VCAR developed for this region. As a result, tomographic images relative to the JB and VCAR models differ. Although both show the high-velocity North American plate subducting westward beneath the Caribbean, the JB image implies that the slab flattens at the 660 km discontinuity, whereas this suggestion is much less in the VCAR image. The flattening in the JB image results from the fact that the inversion yields "streaks" of velocities relative to JB that are lower than those observed above 660 km, and higher than those observed below 660 km. This effect arises because, compared to VCAR, the JB model predicts higher velocity above 660 km, and lower velocity below. Thus a bias in the reference model can produce spurious lateral heterogeneity. Similar reference model artifacts, in which a common state seems abnormal due to the standard used, appear in various inverse problems and other situations.[3] However, the choice of reference model is subjective, so making a choice requires recognizing its consequences. For example, a global velocity reference model that excluded subducting slabs would be slower than the actual global average, whereas one including slabs would predict slow anomalies elsewhere.

[3] 90% of motorists are said to consider themselves above-average drivers, and "all children are above average" in the mythical town of Lake Wobegon in the radio show *Prairie Home Companion*.

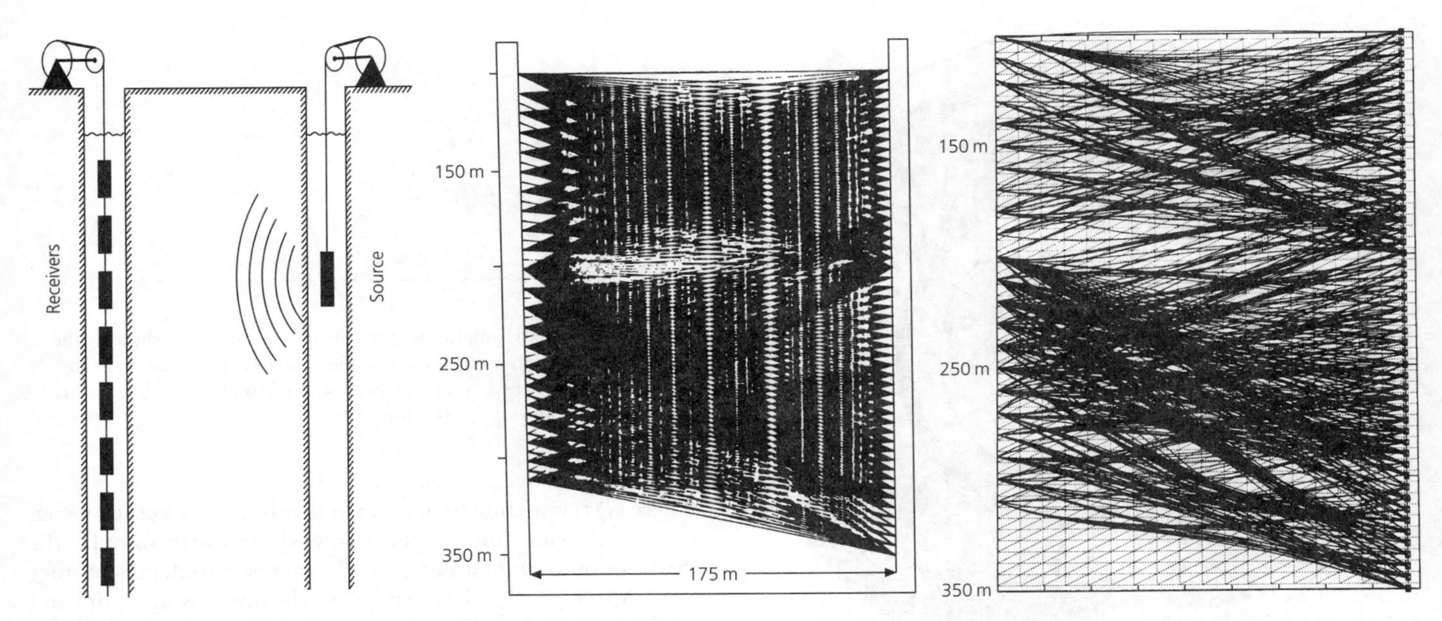

Fig. 7.3-8 An example of cross-borehole tomography in Manitoba, Canada. *Left*: Travel times are recorded from a source at different depths in one borehole to receivers in the other. The experiment is then reversed, yielding dense ray path coverage. *Center*: Straight ray paths computed for the laterally homogeneous starting model. *Right*: Ray paths for the laterally varying model found from the inversion. (Wong *et al.*, 1987.)

In addition to ray geometry and reference model artifacts, it is worth noting that tomographic images can also be affected by something as simple as the contouring scheme used. Sometimes when features are not robust aspects of the image, their tectonic interpretation depends in part on preconceptions, much like the ink-blot tests used by psychologists. Thus, despite the power and value of tomographic images, it is important to bear their limitations in mind.

Tomography is also used in other seismological applications. One important use, providing detailed near-surface images, is illustrated by Fig. 7.3-8 showing tomography between two boreholes. The source and receivers were moved to generate dense coverage with many crossing ray paths. The travel time observations were then inverted for velocity structure. In this experiment, the ray paths were recomputed for the perturbed model and used to compute travel times for later iterations. The differences between the initial and perturbed ray paths show the advantages of recomputing the ray paths for each successive model, a process called *nonlinear tomography*. This updating ensures that the ray paths, and hence predicted travel time anomalies, are consistent with the velocity structure being found. However, for practical reasons it is common to conduct linearized tomography using ray paths from the starting model even as the model is perturbed, and to assume that the resulting errors are small.

It is interesting to compare travel time tomography to the surface wave tomography discussed in Section 2.8.3, where the average surface wave velocity along multiple paths through oceanic lithosphere of various ages is used to infer the velocity structure for each age range. The approach is to find the phase or group wave velocity as a function of frequency for each age range, and then infer the variation in the medium velocity with depth from the dispersion curve giving the variation in apparent velocity as a function of frequency. Hence this is tomography in the lateral direction, and dispersion analysis vertically. We will see in the next section that dispersion analysis is an example of methods that infer earth structure using functions that sample the structure at depth in different ways.

Tomographic methods can be used for waveforms as well as travel times. As noted earlier — for example, in Fig. 3.7-7 — waveforms sample earth structure over broader regions than travel times, which, in the limit, correspond to sampling along narrow geometric rays. Figure 7.3-9 shows some results from global tomography in which velocity perturbations were inferred by fitting both waveforms from 27,000 long-period seismograms and 14,000 travel times. The seismograms include body wave records (from the *P* or *PKP* arrival to the start of the surface waves) and "mantle wave" records, which are low-pass filtered seismograms about 4.5 hours in length. The travel time data include both absolute shear wave arrival times and differential (*SS–S* and *ScS–S*) times. Rather than inverting for the velocity perturbations in blocks, the velocity perturbation was described by a series of orthogonal functions, and the inversion was for the coefficients of the functions. The lateral structure was described by spherical harmonics (Section 2.9.3), and the vertical structure was modeled using Chebyshev polynomials.

In addition, we saw in Section 3.7 that amplitude tomography can infer attenuation variations along the ray paths. Amplitude tomography is similar to medical tomography,[4] in which the image indicates the degree to which X-rays are

[4] The medical term "CAT scan" is for computed axial tomography.

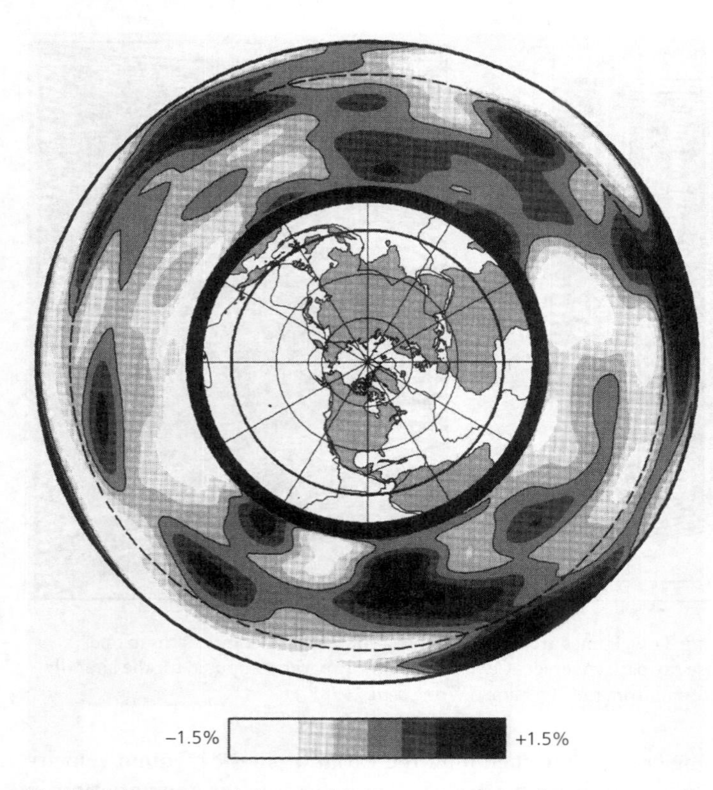

Fig. 7.3-9 Tomographic image of shear wave velocities along a great circle slice through the Equator, obtained by inversion of both waveforms and travel times. (Su *et al.*, 1994. *J. Geophys. Res.*, *99*, 6945–80, copyright by the American Geophysical Union.)

absorbed in different portions of the subject. Medical tomography has the advantages that the subject can be uniformly illuminated from all sides, and that the internal structure is both well understood and subject to later, direct observation.

7.4 Stratified earth structure

Quantities that can be determined using seismological data are often the integrals of a physical property of the earth. For example, the travel time is the integral of slowness along a ray path. As discussed in the last section, although a single travel time gives only the average slowness along the ray path, travel times for different ray paths can be combined to find the spatial distribution of slowness.

A common such problem is finding earth structure for laterally homogeneous or stratified earth models, in which physical properties are assumed to vary only with depth. Frequently, an observable quantity d_i can be expressed as the integral over the radius of a physical property $m(r)$,

$$d_i = \int_0^a G_i(r)m(r)dr, \tag{1}$$

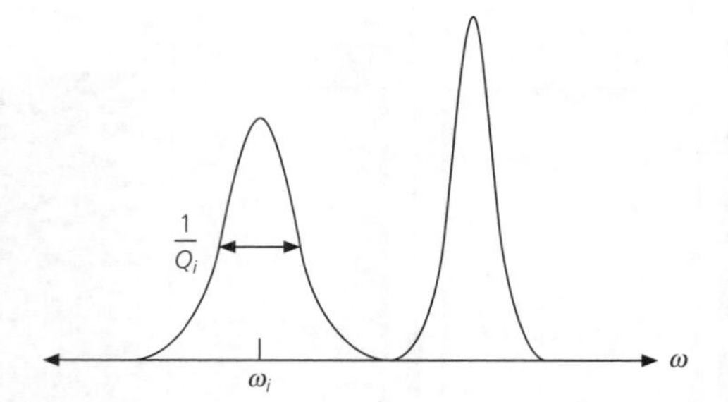

Fig. 7.4-1 Schematic amplitude spectrum of a seismogram, showing the observations used to invert normal mode data for eath structure. Each mode peak is described by a width proportional to Q_i^{-1}, which describes its attenuation, and an eigenfrequency ω_i.

where $G_i(r)$ is a known function of depth called a kernel. Given a set of d_i with different kernels, each of which samples the distribution of $m(r)$ differently, the inverse problem is to infer $m(r)$. Although the relation between the observed quantity and earth structure is sometimes less intuitive than for travel time and slowness, the problems can be formulated in a similar way.

We encountered this idea in discussing Love wave dispersion in Section 2.7.4. The apparent phase velocity along the free surface varies as a function of period, because waves of different period sample the velocity at depth differently. Hence this variation can be used to study the velocity at depth.

7.4.1 *Earth structure from normal modes*

The concepts of inverting observations for the structure of a stratified medium can be illustrated using normal modes (Sections 2.9 and 3.7). The displacement field of the ith mode excited by an earthquake can be written

$$\mathbf{u}_i(t) = \mathbf{C}_i(t) \exp(-\omega_i t/2Q_i). \tag{2}$$

The mode's eigenfrequency ω_i and quality factor Q_i, which describes the attenuation, and thus the width of the peak, can be found from the Fourier transform of the seismogram (Fig. 7.4-1). Because ω_i and Q_i depend on the variation with depth of the seismic velocities, density, and attenuation, these observations can be used to study earth structure.

To do this, we begin with an earth model described by $\alpha(r)$, $\beta(r)$, and $\rho(r)$ and find the eigenfrequencies of the different modes, ω_i. This calculation also gives the partial derivative functions

$$\frac{\partial \omega_i}{\partial \alpha}(r), \quad \frac{\partial \omega_i}{\partial \beta}(r), \quad \frac{\partial \omega_i}{\partial \rho}(r), \tag{3}$$

showing how a mode's eigenfrequency changes if the velocity or density at a given depth is perturbed. The total change in the eigenfrequency is the integral over the radius of the perturbations in the earth model:

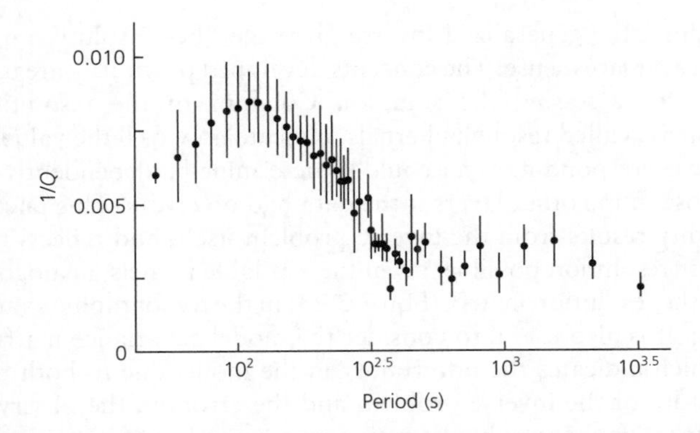

Fig. 7.4-2 Observed attenuation for fundamental spheroidal modes ${}_0S_2$–${}_0S_{191}$. The variation in Q^{-1} with period reflects the depth variation of $q^{-1}(r)$. (Stein *et al.*, 1981. *Anelasticity in the Earth*, 39–53, copyright by the American Geophysical Union.)

$$\Delta\omega_i = \int_0^a \left[\frac{\partial \omega_i}{\partial \alpha}(r)\Delta\alpha(r) + \frac{\partial \omega_i}{\partial \beta}(r)\Delta\beta(r) + \frac{\partial \omega_i}{\partial \rho}(r)\Delta\rho(r) \right] dr. \quad (4)$$

Thus the difference between a measured eigenfrequency and that predicted by an earth model can be inverted to find the perturbation in the model required to fit the data. Although a single mode observation gives only the average over depth of the required perturbation, a set of modes gives more information, because the partial derivatives vary between modes.

We illustrate the method using the corresponding inverse problem for attenuation, which has a simple linear form. If attenuation within the earth is described by the function $q(r)$, the quality factor for the ith mode is

$$Q_i^{-1} = \int_0^a G_i(r) q^{-1}(r) dr, \quad (5)$$

where the kernels $G_i(r)$ are derived from the partial derivatives (Eqn 4), using the formulation of the quality factor as an imaginary part of the frequency that is related to an imaginary part of the velocity (Section 3.7.6). Although the symbol Q is commonly used for both the modes' quality factor and the attenuation as a function of depth, using $q(r)$ for the latter emphasizes the distinction. The problem is written using the reciprocals $q^{-1}(r)$ and Q_i^{-1}, so higher attenuation (larger loss of seismic energy) corresponds to larger values.

Figure 7.4-2 shows measured values of the attenuation of fundamental spheroidal modes, which for periods less than a few hundred seconds correspond to fundamental mode Rayleigh waves. The attenuation is low for the longest-period modes, rises to its highest values at periods slightly above 100 seconds, and then decreases again for the shortest periods (about 50 seconds) shown. This variation occurs because the kernels differ between modes (Fig. 7.4-3). Because Q^{-1} for a mode is the integral of the attenuation weighted by the kernel, the shape of the kernel with depth illustrates a mode's sensitivity to attenuation at various depths. Long-period modes are most sensitive in the lower mantle, periods near 100 seconds sample the low-velocity zone heavily, and periods near 50 seconds are most sensitive to structure in the "lid" region above the low-velocity

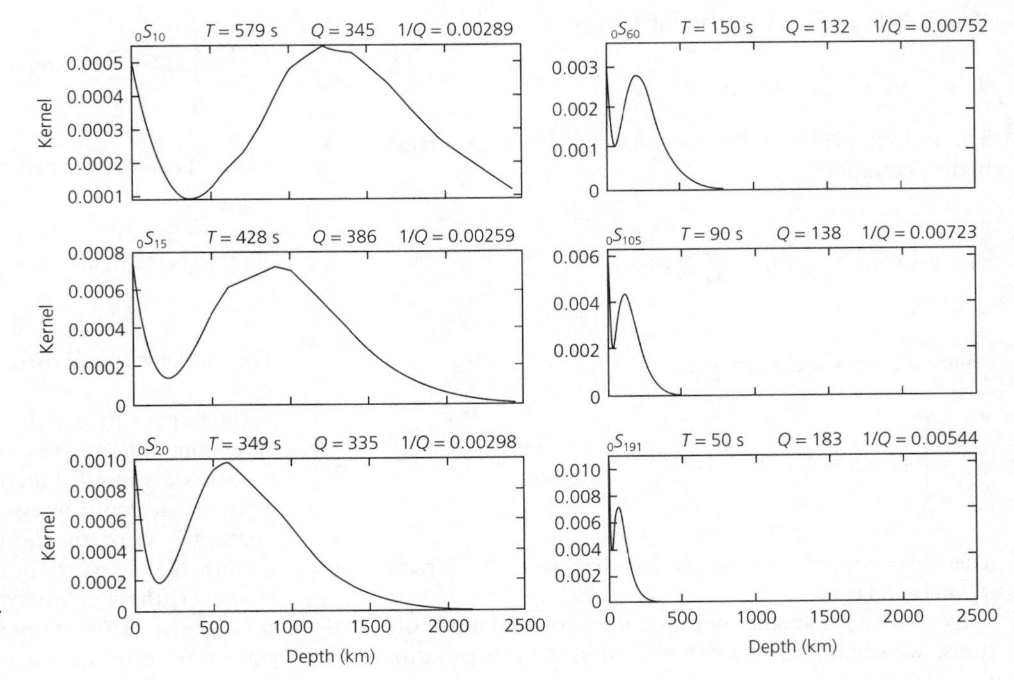

Fig. 7.4-3 Attenuation kernels for various modes, illustrating the different depth sampling. Attenuation values are for the third model in Fig. 7.4-5. (Stein *et al.*, 1981. *Anelasticity in the Earth*, 39–53, copyright by the American Geophysical Union.)

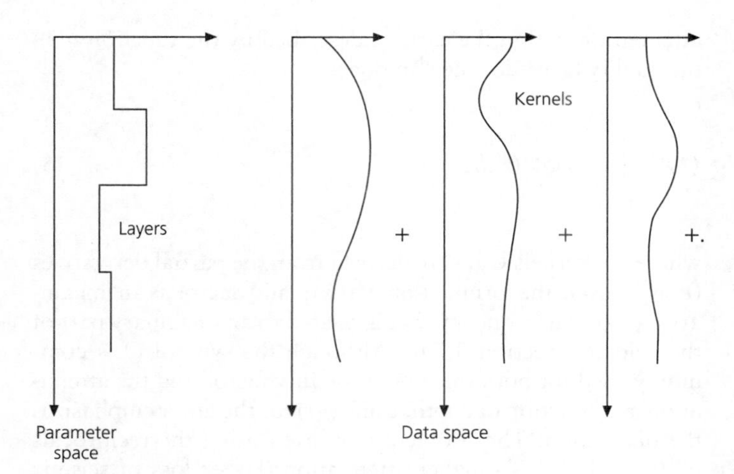

Fig. 7.4-4 Schematic illustration of the model parameterizations for two types of inversion methods. In parameter space inversions, the model is divided into layers; in data space inversions the model is treated as a weighted sum of the kernels.

zone. Q^{-1} is a smooth function of the period, because the kernels of fundamental modes with similar periods are similar.

The inverse problem is to use the observed mode attenuation Q_i^{-1} and the known kernels $G_i(r)$ to infer the function $q^{-1}(r)$ describing the variation of attenuation with depth in the earth that best fits the data. This problem can be approached in several ways, two of which we discuss briefly.

7.4.2 *Parameter and data space inversions*

The most direct approach, *parameter space inversion*, is to regard the unknown model $q^{-1}(r)$ as constant in a set of layers (Fig. 7.4-4, *left*), such that in the jth layer

$$q^{-1}(r) = q_j^{-1}, \quad r_j \leq r \leq r_{j+1}. \tag{6}$$

The inverse problem is then converted from an integral to a matrix equation

$$Q_i^{-1} = \int_0^a G_i(r) \sum_j q_j^{-1} dr = \sum_j A_{ij} q_j^{-1}, \tag{7}$$

where the matrix elements are

$$A_{ij} = \int_{r_j}^{r_{j+1}} G_i(r) dr. \tag{8}$$

The observations are inverted for the value of the parameter q_j^{-1} in each layer.

By choosing a smaller number of layers than mode observations, we obtain an overdetermined system of equations. As before, the generalized inverse gives the "best" solution in a least squares sense. The concepts developed previously are useful for assessing the solution. Columns of the resolution matrix, called resolving kernels, indicate how well the value in the corresponding layer could be determined independently of those in the other layers if the data had no errors. This uncertainty results from the inverse problem itself, and reflects the best resolution possible, given the available kernels, analogous to the resolution matrix (Eqn 7.3.23) in the tomographic example. It is also useful to consider the model covariance matrix, which indicates the uncertainty in the model due to both the nature of the inverse problem and the errors in the observations. Often a weighted average over a number of layers is the best resolution obtainable, analogous to the blurring in travel time tomography.

Parameter space inversion has a few unattractive features. First, the layers in which attenuation is treated as constant must be chosen in advance. This choice might not be a meaningful one. Second, parametrizing the model as constant in these layers yields a model with "steps" at layer boundaries. These steps may be quite unphysical; in many cases our intuition (admittedly sometimes a poor guide) suggests that physical properties should vary smoothly with depth.

In an alternative formulation, *data space inversion*, the unknown model describing attenuation as a function of depth is expanded not into constant layers, but as a weighted sum of the kernels themselves (Fig. 7.4-4, *right*),

$$q^{-1}(r) = \sum_j v_j G_j(r). \tag{9}$$

The inverse problem is then

$$Q_i^{-1} = \int_0^a G_i(r) \sum_j v_j G_j(r) dr = \sum_j A_{ij} v_j, \tag{10}$$

where the matrix elements are

$$A_{ij} = \int_0^a G_i(r) G_j(r) dr. \tag{11}$$

The model is found by inverting for the expansion coefficients v_j.

Data space inversion is less intuitive than parameter space inversion, but has the attractive features that the resulting model is a smooth function of depth, and need not be parametrized in depth in advance. Moreover, it is in some sense "natural" to use the kernels as basis functions for the model, because the observations sample the model along these kernels. However, these solutions often seem too smooth for our instincts, just as the parameter space solutions often seem too jagged. We often both expect changes in properties near certain

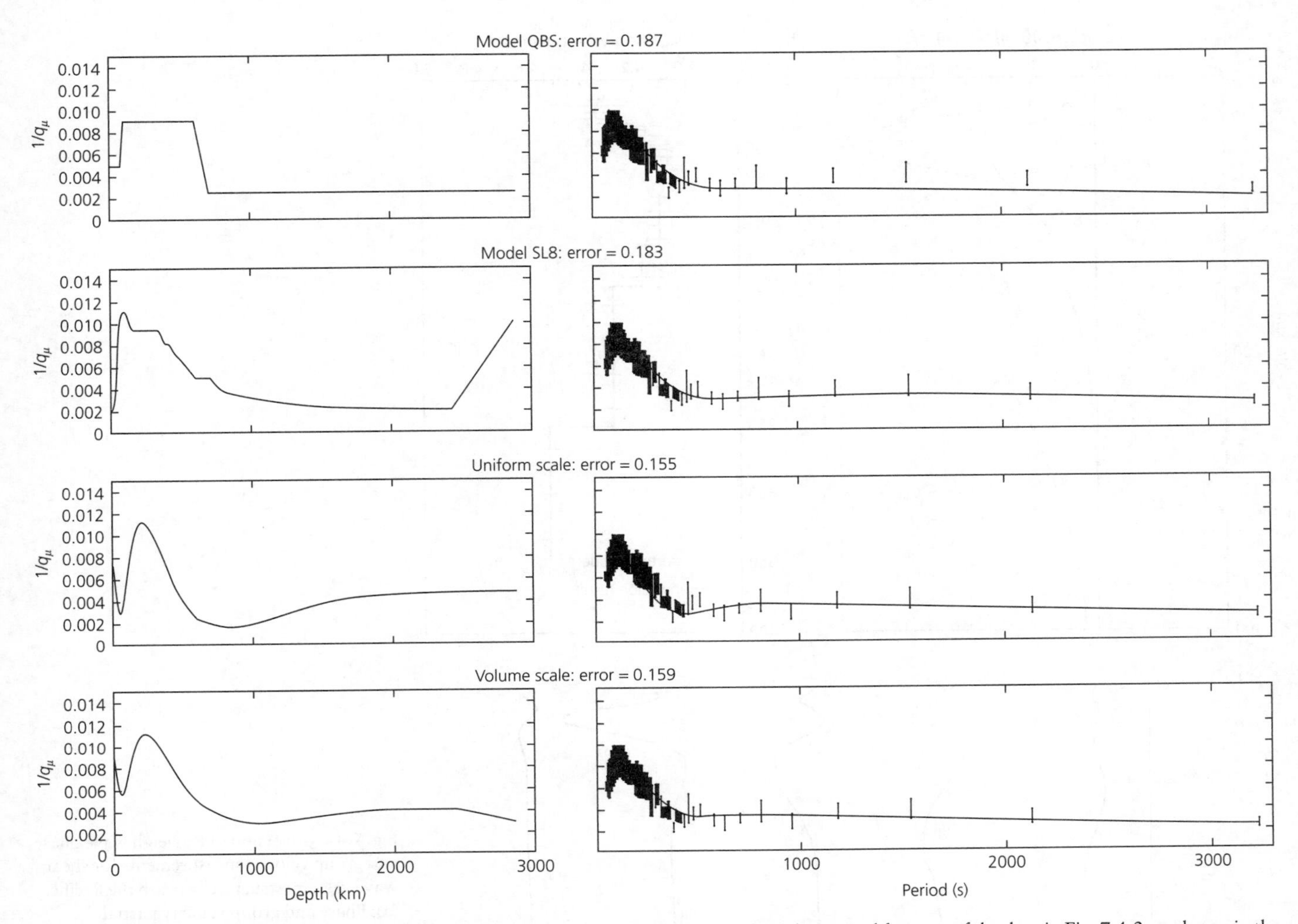

Fig. 7.4-5 Comparison of various attenuation models. Despite the differences, all reproduce the general features of the data in Fig. 7.4-2, as shown in the right hand panels. (Stein *et al.*, 1981. *Anelasticity in the Earth*, 39–53, copyright by the American Geophysical Union.)

depths and are reluctant to force them into the solution. This dilemma is an example of the general issue of deciding how much we want the inversion solution to reflect our preconceptions, some of which may be correct, especially when derived from other data, and some of which may be incorrect. We can choose to focus on what the data require, what the data permit, or a combination of the two.

These issues are illustrated in Fig. 7.4-5, which shows several models for attenuation as a function of depth, all generally consistent with the data in Fig. 7.4-2. Model SL8 was derived by parameter space inversion, whereas the others were derived from data space inversion. The lower two models were derived by inverting the data in Fig. 7.4-2 with different misfit functions, whereas the upper two were derived from different data. Although the models differ, all have low attenuation in the lower mantle, high attenuation in the upper mantle associated with the low-velocity zone, and moderate attenuation in the "lid" above the low-velocity zone. The models illustrate the range of acceptable solutions. For example, the high attenuation zone at the base of the mantle in model SL8 is permissible, and thus survives if included in the starting model, but is not required by the data. This ambiguity results from the fact that the data have little resolution for structure at this depth, as shown by the kernels in Fig. 7.4-3.

7.4.3 *Features of the solutions*

The inverse problem for attenuation (Eqn 5) has a simple form, because each mode's quality factor depends linearly on $q^{-1}(r)$, so the observations can be inverted directly for the attenuation structure. If this is not the case, we linearize about a starting model (Section 7.2.1), so the change in a datum depends linearly on the change in model parameters

$$\Delta d_i = \int_0^a G_i(r)\Delta m(r)dr. \tag{12}$$

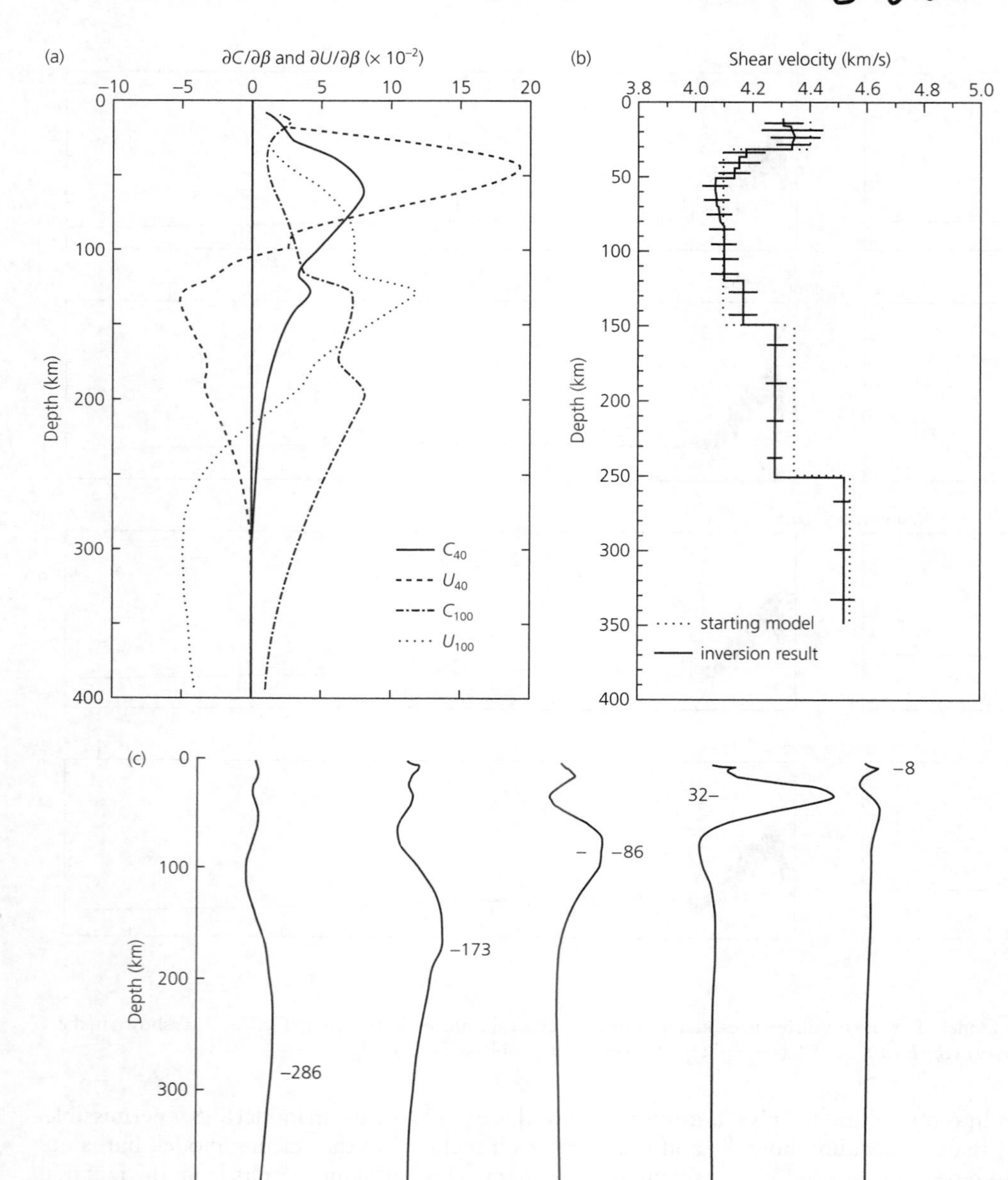

Fig. 7.4-6 Inversion of Rayleigh wave phase and group velocity measurements for shear wave velocity structure beneath the Pacific. (a): Phase and group velocity partial derivatives at 40 and 100 s periods. (b): Starting (dotted line) model and final model derived by parameter space inversion. Horizontal lines indicate the model standard deviation in each layer. (c): Resolving kernels for various depths. The number and horizontal line indicate the depth for each kernel. (Yu and Mitchell, 1979.)

Figure 7.4-6 illustrates a parameter space inversion for vertical shear velocity structure from Rayleigh waves. Using the partial derivatives

$$\frac{\partial C(T)}{\partial \beta}(r), \quad \frac{\partial U(T)}{\partial \beta}(r), \tag{13}$$

which show how the phase and group velocities at a particular period change in response to a shear velocity perturbation at each depth, the starting model is modified to fit the observed dispersion. The resolving kernels that illustrate the vertical "smearing" are largest at the depth for which they are computed, but have nonzero amplitudes at other depths. The best resolution occurs when the kernel is sharply peaked at the desired depth.

As we noted earlier, the generalized inverse solution yields the minimum change in the model that best produces a desired change in the data. Hence the final model is as close to the starting model as possible. Features of a model derived by linearized inversion can thus depend on the starting model. For example, in a parameter space inversion, a layer whose value in the starting model is assumed to differ significantly from adjacent layers will often retain this feature in the solution. One way to avoid this is to start off with a model whose properties are uniform with depth. In other cases, data not included in the inversion can be used to find a starting model more appropriate than a uniform one. Another approach is to do inversions with different starting models and compare the resulting solutions. If the solutions differ, they are likely local minima of the misfit function (Eqn 7.2.11) that the inversion minimized, whereas if the different starting models yield the same solution, it is more likely to be the global minimum that we seek. Yet another approach is to search numerically for the minimum in the model space by varying the model parameters. Such "brute force" approaches, in which we solve the inverse problem by solving the forward problem many times, are attractive when the number of model parameters is small, because they avoid the issue of linearizing about the starting model and show the trade-offs between various parameters. For example, Fig. 5.3-8 showed the trade-off between plate thickness and basal temperature in inverting oceanic depth and heat flow data for thermal structure.

Parameter space and data space inversions can be carried out using more sophisticated variations. For example, parameter space inversion can be smoothed to reduce the jumps at layer boundaries. Data space inversion can be formulated in terms of a set of orthogonal kernels, rather than the actual kernels, which are often quite similar to each other. This approach expands the model in the simplest possible way with the minimum number of parameters. In addition, the model can be constrained to fit the data only within the error bars, rather than attempt to fit the mean value of each datum.

Due to the structure of inverse problems and the range of possible techniques available, various solutions can generally be derived for a set of seismological observations. As a result, inverse problems remain an important research area. The choices, ambiguities, and trade-offs in the solutions of these problems are sometimes key features of the solution. Attempts to explain these issues can be frustrating to nonseismologists, as illustrated by the joke that in response to the question "How much is 2 + 2," an engineer replies "3.9999," a geologist replies, "Somewhere in the mid-single digits," and a geophysicist replies, "How much do you want it to be?"

7.5 Inverting for plate motions

We end our discussion of inverse problems with the issue of determining the Euler vectors that describe relative plate motions. As we have noted, these Euler vectors are derived in part from earthquake focal mechanisms, and are then used as a reference model to predict the directions and rates of plate motions for applications including estimating earthquake recurrence, slip partitioning, and the fractions of seismic and aseismic slip at plate boundaries.

7.5.1 *Method*

The forward problem (Section 5.2.1) is that at any point $\mathbf{r}$ along their boundary, the linear velocity of plate j with respect to plate i is

$$\mathbf{v}_{ji} = \omega_{ji} \times \mathbf{r}, \tag{1}$$

where ω_{ji} is the relative angular velocity, or Euler vector. Hence the rate and direction of plate motion are given by the north–south and east–west components of $\mathbf{v}$,

$$\text{rate} = |\mathbf{v}| = \sqrt{(v^{NS})^2 + (v^{EW})^2},$$
$$\text{azimuth} = 90° - \tan^{-1}[(v^{NS})/(v^{EW})]. \tag{2}$$

The corresponding inverse problem is to find a model, or set of Euler vectors, that best predicts the observed motions. Because Euler vectors can be added, assuming that the plates are rigid, m plates are specified by $m-1$ Euler vectors, and thus their $3(m-1)$ components. Hence we use a data vector $\mathbf{d}$ composed of rates and azimuths to estimate the model vector $\mathbf{m}$ composed of the Euler vector components. Both the model and data vectors consist of physically different quantities: the model vector is made up of Euler pole latitudes, longitudes, and rotation rates

$$\mathbf{m} = (\theta_1, \theta_2, \ldots \theta_{m-1}, \phi_1, \phi_2, \ldots \phi_{m-1}, |\omega_1|, |\omega_2|, \ldots |\omega_{m-1}|), \tag{3}$$

whereas the data vector contains rates and azimuths

$$\mathbf{d} = (r_1, r_2, \ldots r_k, az_1, az_2, \ldots az_{n-k}). \tag{4}$$

As written, the inverse problem is not linear because the data are complicated functions of the model parameters. Thus, as in the previous examples, we linearize about a starting model by forming the partial derivative matrix

$$G_{ij} = \frac{\partial d_i}{\partial m_j}, \tag{5}$$

showing how a change in the jth model parameter affects the prediction of the ith datum. The derivatives are found by differentiating the expressions for v^{NS} and v^{EW} (Eqn 5.2.7). We then have the usual equation

$$\Delta\mathbf{d} = G\Delta\mathbf{m}, \quad \text{or} \quad \Delta d_i = \sum_j G_{ij}\Delta m_j, \tag{6}$$

relating the changes in the data and the model. The system is usually overdetermined, because we generally have data at many sites and solve for only a few plate model parameters. For example, the NUVEL-1 model has 12 plates whose motions were estimated from 1122 data (Fig. 1.1-9). We thus use the weighted least squares solution

$$\Delta\mathbf{m} = (G^T W_d G)^{-1} G^T W_d \Delta\mathbf{d}, \tag{7}$$

where the variance–covariance matrix of the data, $W_d = (\sigma_d^2)^{-1}$, contains our estimates of the uncertainty in rates from magnetic anomalies and the uncertainties in directions associated with estimating transform azimuths and determining earthquake slip vectors. The weighted solution is needed because the uncertainties have different dimensions and vary between data points.

Thus uncertainties in the estimated Euler vectors are given by the model variance–covariance matrix

$$\sigma_{\mathbf{m}}^2 = (G^T W_d G)^{-1}. \tag{8}$$

Uncertainties associated with the Euler poles are often shown by error ellipses analogous to those for earthquake locations, whereas those for the rates are quoted separately. Alternatively, we can view the pole and rate uncertainties as forming a three-dimensional ellipsoid. Hence two Euler vectors are distinct if their error ellipsoids do not overlap. As we have seen, conventional global plate motion studies using magnetic anomalies, transforms, and earthquake slip vectors yield solutions similar to those obtained by using the same formulation to invert the rates and azimuths of plate motions determined by space-based geodesy (Section 5.2.3). This agreement is gratifying, given that the conventional solutions combine data from magnetic anomalies averaged over millions of years, the azimuths of transform faults that formed over long times, and the slip vectors of earthquakes, whereas the space-geodetic solutions based on data spanning only a few years have different uncertainties.

7.5.2 *Testing the results with χ^2 and F-ratio tests*

Given a model derived by inversion, the natural question is, how good is it? This issue is a specific case of the general one of testing how well a model fits data, which is discussed in statistics texts. For our purposes we focus on two issues and note some results without proof.

One common way to test how well a model fits data uses the misfit function χ^2 that we minimized to derive the least squares solution (Eqn 7.2.11). We write it as

$$\chi^2 = \sum_i \frac{(d_i - d_i^m)^2}{\sigma_i^2}, \tag{9}$$

where d_i^m are the data predicted by the model, d_i are the data observed, and σ_i are their uncertainties. Lower values of χ^2 correspond to better fits. However, because a model derived from these data is bound to fit better than one derived without them, we examine the *reduced chi square*

$$\chi_\nu^2 = \chi^2/\nu \tag{10}$$

where the parameter ν, known as the number of *degrees of freedom*, equals $n - p$ where n is the number of data and p is the number of model parameters estimated in the inversion.

If the model is a good fit to the data and our estimates of the uncertainties are reasonable, then we expect χ_ν^2 to be around 1.

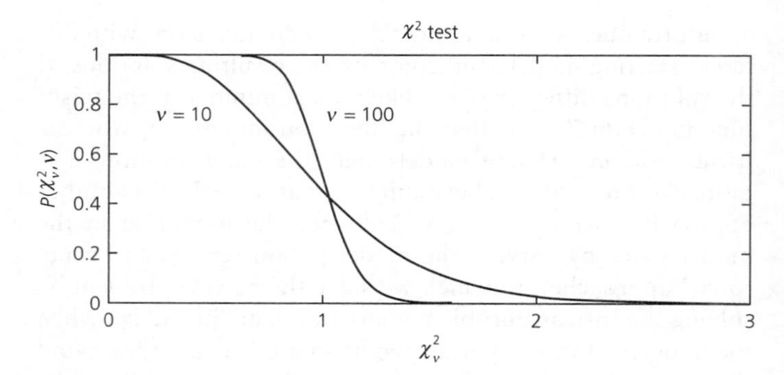

Fig. 7.5-1 Cumulative probability distribution $P(\chi_\nu^2, \nu)$, giving the probability of observing χ_ν^2 above a certain value, plotted for 10 and 100 degrees of freedom. The more the degrees of freedom, the more likely χ_ν^2 is to be near 1, and the less likely much higher or lower values are.

Statistically, this means that there is a reasonable possibility that the observed data are samples from a parent distribution described by the model, given the random uncertainties of measurement. However, if χ_ν^2 is much larger than 1, it is unlikely that the data are samples from this distribution. This issue is addressed using the cumulative probability distribution $P(\chi_\nu^2, \nu)$ given by statistical tables or mathematical software that gives the probability of observing χ_ν^2 above a certain value (Fig. 7.5-1). In other words, this test asks what the probability is that such a high value would be observed purely by chance due to the uncertainties of measurement. The more the degrees of freedom, the less likely a high value is. For example, the chance of observing χ_ν^2 greater than 1.5 is about 13% for $\nu = 10$, but less than 1% for $\nu = 100$. Thus, the more data we have, the more the degrees of freedom, and closer to 1 we expect χ^2 to be. This test does not tell specifically whether the data observed are samples from the distribution predicted by the model, but gives instead some insight into the probability. If χ_ν^2 is too large, there is likely to be something wrong.

One possibility is that the model does not include some crucial factors. For example, a plate motion model may not include an important plate boundary, and so does not describe the data well. In this case, the misfit is greater than expected from considering only random uncertainties of measurement, because systematic errors are also present. Similarly, the misfit to travel time in an earthquake location includes both errors of measurement and the effects of velocity structure like lateral heterogeneity. We sometimes rescale the uncertainties to make $\chi_\nu^2 = 1$, which lets us assign confidence limits using χ_ν^2. This rescaling does not address the causes of the misfit, but implicitly lumps the systematic errors in with the errors of measurement. To do better requires improving the model.

Conversely, if χ_ν^2 is too small, Fig. 7.5-1 indicates that something is also likely to be wrong. For example, for $\nu = 10$, there is only about a 2% chance of observing χ_ν^2 less than 0.3, and the probability is less for more degrees of freedom. This is because the data are unlikely to be fit that well, given errors of measurement. About one-third (100 – 68%) of the data should be misfit by at least 1σ, and about 5% should be outside the 2σ range. Hence a low χ^2_ν value, which we might view as showing an excellent fit, is more likely to imply that the uncertainties in the data have been overestimated, and have thus made χ^2_ν appear too small. For example, χ^2_ν for the NUVEL-1 model is 0.24, whereas it is expected to lie with 95% probability between 0.93 and 1.07. This effect is also seen for other plate motion models, suggesting that the assigned data uncertainties are more like 95% (2σ) confidence limits than one standard deviation. If so, the uncertainties in the model are correspondingly less than implied by the model variance–covariance matrix. Thus the χ^2 test formalizes the adage that if something seems too good to be true, it probably is.[1]

A second issue is whether the number of model parameters is appropriate. As discussed in Chapter 5, there are often several possible plate boundary geometries for an area. Naturally, more plates can describe plate motions in an area better because the model has more parameters. Thus we ask whether the improved fit shown by a lower value of χ^2_ν is more than expected purely by chance due to the additional parameters. For example, a set of data in the x–y plane are always better fit by a higher-order polynomial, such as a quadratic versus a straight line.

This issue can be addressed using the *F-ratio* test, which gives insight into whether a set of data are significantly better fit by a model with more parameters. The idea is that if a set of n data are fit by two models, one with r parameters ($n - r$ degrees of freedom) and a second with p parameters ($n - p$ degrees of freedom) with p greater than r, the second model should fit the data better, and $\chi^2(p)$ should be less than $\chi^2(r)$. To test if the reduction in χ^2 is greater than would be expected simply because additional model parameters are added, we form the statistic

$$F = \frac{[\chi^2(r) - \chi^2(p)]/(p - r)}{\chi^2(p)/(n - p)}. \tag{11}$$

Statistical tables or mathematical software give the probability $P_F(F, \nu_1, \nu_2)$ of observing an F value greater than that observed for a random sample with $\nu_1 = (p - r)$ and $\nu_2 = (n - p)$. Thus, for example, if P_F is 0.01, there is only a 1% chance that the improved fit of the model with more parameters is due purely to chance. Because this test depends on the ratio of χ^2, it is not affected if the uncertainties are consistently over- or under-estimated.

We can use F to test whether the fit to n relative motion data of a model with $p + 1$ plates is significantly better than that of one with p plates. The p plate model has $3(p - 1)$ parameters ($n - 3p + 3$ degrees of freedom), whereas the $p + 1$ plate model has $3p$ parameters ($n - 3p$ degrees of freedom). Thus

$$F = \frac{[\chi^2(p \text{ plates}) - \chi^2(p + 1 \text{ plates})]/3}{\chi^2(p + 1 \text{ plates})/(n - 3p)} \tag{12}$$

is tested using $P_F(F, \nu_1, \nu_2)$ with $\nu_1 = 3$ and $\nu_2 = (n - 3p)$. If the risk that the improved fit would occur by chance is small, perhaps less than 1%, then we treat the additional plate as distinct. Conversely, if the improved fit is likely to result simply from the additional parameters, the data do not strongly indicate the presence of an additional plate. For example, such tests show that although the boundary between them is indistinct, North and South America should be treated as separate plates. This approach is used to investigate complicated regions where the plate geometry is unclear, such as near Japan and in the Indian Ocean. Similarly, we can investigate regions of intraplate deformation to see whether there is resolvable motion.

In many applications these or other statistical tests can be used to examine how well a model fits the data and to gain insight into whether the model is too simple (underparametrized) to explain the data or more complicated (overparametrized) than is required by the data. For example, we can examine cases when adding more layers to a velocity model significantly improves the fit to travel time data, when a more complex earthquake source model fits seismograms significantly better, or when a more complex model of earthquake recurrence describes an earthquake history better. In these applications the statistical tests address only the data used, so a more complex model may be justified based on other data, even if it is not required by the data tested. Moreover, we often suspect that the earth is more complicated than we would like when using simple statistical models. In particular, we often have little a priori knowledge of how to estimate the random and systematic errors. Even so, it is worth subjecting models to tests and seeing how well the data support our beliefs. This testing is a key part of the cycle (Fig. 1.1-8) by which models are refined using new data and model parameterizations.

Further reading

Many discussions of inverse theory, including ours, are based on Lanczos (1961). Applications in the earth sciences, especially seismology, are discussed in texts and reviews including Parker (1977), Aki and Richards (1980), and Menke (1984). Treatments of tomographic methods in seismology are given by Nolet (1987), Thurber and Aki (1987), Spakman and Nolet (1988), Humphreys and Clayton (1988), and Romanowicz (1991). Inversion for the properties of stratified media is reviewed by Wiggins (1972).

Tests for goodness of fit are discussed in statistical texts such as Bevington and Robinson (1992) and Freedman *et al.* (1991); the latter treats the issue of Mendel's results. Chase (1972) and Minster *et al.* (1974) present the inverse problem for plate motions; the latter gives the partial derivatives. Stein and Gordon (1984) and DeMets *et al.* (1990) discuss applications of the *F*-ratio test to plate motions and intraplate deformation.

[1] This approach has been used to argue that Mendel's famous results in 1865 that established the science of genetics are so good — the probability of observing them is 0.004% — that they are suspect. Similarly, instructors have used χ^2 tests to show that students' results reported in laboratory classes are so good that they are unlikely to have actually been obtained.

Problems

1. Show the following matrix identities:
   (a) For an arbitrary (not square) matrix A, the matrices A^TA and AA^T are symmetric.
   (b) For an arbitrary (not square) matrix B and a symmetric matrix A, $(B^T AB)^T = B^TAB$.
   (c) For square matrices A and B such that $(AB)^{-1}$ exists, $(AB)^{-1} = B^{-1}A^{-1}$.
2. Show that if a square matrix G has an inverse, the inverse and generalized inverse are identical.
3. Show that if the variance–covariance matrix of the data is diagonal, $\sigma_{\mathrm{d}}^2 = \sigma_{ij}^2\delta_{ij}$ (with no summation implied), its inverse is another diagonal matrix $W_d = \delta_{ij}/\sigma_{ij}^2$. (Also with no summation implied.)
4. Show that the model variance–covariance matrix (Eqn 7.2.32) $\sigma_{\mathrm{m}}^2 = G^{-g}\sigma_{\mathrm{d}}^2(G^{-g})^T$ reduces to $\sigma_{\mathrm{m}}^2 = \sigma^2(G^TG)^{-1}$ when the data errors are uncorrelated and equal, so the data variance–covariance matrix is a constant times the identity matrix, $\sigma_{\mathrm{d}}^2 = \sigma^2\delta_{ij}$.
5. Show that if the data errors are uncorrelated but not equal, such that the data variance–covariance matrix of the data is the diagonal matrix $\sigma_{\mathrm{d}}^2 = \sigma_{ij}^2\delta_{ij}$ with inverse W_{d} (problem 3):
   (a) The least squares criterion (Eqn 7.2.14) for the inverse problem gives rise to the weighted least squares solution $\Delta\mathbf{m} = (G^TW_dG)^{-1}G^TW_d\Delta\mathbf{d}$.
   (b) The model variance–covariance matrix is $\sigma_{\mathrm{m}}^2 = (G^TW_dG)^{-1}$.
6. For a halfspace with uniform (and known) velocities α and β:
   (a) Show how the location problem can be formulated to use both *P*-wave and *S*-wave first arrival times as data. Write the data vector, model vector, and partial derivatives. How do these differ from the case for *P* waves alone?
   (b) Show how the location problem can be formulated to use only the *difference* between *P*-wave and *S*-wave first arrival times as data. Write the data vector, model vector, and partial derivatives. How do these differ from the case for *P* waves alone? How might you apply this method if only the *P* velocity were known? Under what conditions might this method be useful?
7. For the idealized tomographic experiment in Figure 7.3-2:
   (a) Show how one row of the G matrix in Eqn 7.3.10 can be derived from the others, such that the four teleseismic ray paths are not linearly independent. Give a physical interpretation of this result.
   (b) Find four rows of the G matrix in Eqn 7.3.6 that are linearly independent, and give a physical interpretation of this result.

Computer problems

C-1. Write a subroutine to find the generalized inverse $G^{-g} = (G^TG)^{-1}G^T$ of an $(n \times r)$ matrix G, using a matrix inversion subroutine. As a test, check that the solution satisfies the criterion that for a square matrix G that has an inverse, the inverse and generalized inverse are identical.

C-2. For a homogeneous halfspace with *P*-wave velocity α:
   (a) Write a subroutine to compute the distance and travel time between two points (x, y, z) and (x_i, y_i, z_i). Test this for some simple cases.
   (b) Use the result of (a) to write a program that reads an earthquake location, origin time, and medium velocity and the locations of n seismic stations, and finds the first arrival time at each station.
   (c) Write a subroutine using the result of (a) to compute the partial derivatives of the first arrival time at a station with respect to changes in the model parameters (location, origin time, and medium velocity).
   (d) Modify the result of (b) to compute arrival times for a starting model (assumed location, origin time, and medium velocity), and then locate the earthquake by inverting these synthetic data to find the best-fitting model. The result of C-1 should be useful. Have the program iterate until the model change between iterations is less than a parameter you set. The program should have the option to invert for velocity or hold velocity fixed at an assumed starting value.

C-3. Test the location program with a set of station locations, a "real" origin time and location, and an incorrect starting model. The program should retrieve the "real" model. Once this works for error-free data, add some errors to the travel times, either by using your computer's random number function or by simply choosing some numbers. Invert for the best-fitting model, and see how the result of the inversion changes as the errors become a larger fraction of the travel times. How do the results depend on whether the velocity is held fixed or inverted for?

C-4. Compute and compare χ^2 and χ_ν^2 for C-3 for cases in which you inverted for velocity and in which the velocity is fixed at an incorrect value. Using the *F*-ratio test, does the improved fit due to inverting for velocity seem significant?

Appendix: Mathematical and Computational Background

If you wish to learn about nature, to appreciate nature, it is necessary to understand the language she speaks in. She offers her information only in one form; we are not so unhumble as to demand that she change before we pay attention.

Richard Feynman, *The Character of Physical Law* (1982)

A.1 Introduction

The study of seismology follows a pattern characteristic of many scientific disciplines. We first identify phenomena that we seek to understand, such as the propagation of seismic waves through the solid earth. We then consider the physics of the simplest relevant case, such as the propagation of a wave of a single frequency through a uniform material, formulate the problem mathematically, and derive a solution. From this solution, we build up mathematical solutions to more complex problems, each of which is ideally a better approximation to the complexities of the real earth. Although the simpler problems can be solved analytically, eventually the complexities require numerical techniques.

We thus rely on a set of mathematical techniques often used in physical problems. Experience suggests that although many readers are familiar with most of the mathematics required in this book, a review is often helpful. This appendix briefly summarizes a broad range of material. The first sections treat a variety of mathematical topics. The final section reviews some concepts relevant to the use of computers for scientific calculations.

In using these mathematical techniques, it is worth bearing in mind that we are invoking the special power of mathematics to deal with physical problems. This power is that if a physical problem is posed correctly in mathematical terms, then applying mathematical techniques to this formulation yields quite different, and often apparently unrelated, statements that also correctly describe the physical world. For example, in Section 2.4 we used the equations of elasticity and applied vector calculus to derive the properties of seismic waves that we observe. Similarly, in Section 2.5 we derived an observed physical relation, Snell's law, starting from three different physical formulations. Conversely, we have seen that different physical phenomena can be described using similar mathematical approaches and so have some deep similarities. Although in hindsight such successes may not seem surprising, because many of the mathematical methods we use were developed to solve such physical problems, they illustrate the intimate connection between sciences like seismology and mathematics.[1]

A.2 Complex numbers

In several of our applications, notably in describing propagating waves and their frequency content, complex numbers are helpful. We thus briefly review some of their properties.

The complex number $z = a + ib$, where $i = \sqrt{-1}$, has a real part, a, and an imaginary part, b. These relations are sometimes written $a = \text{Re}(z)$ and $b = \text{Im}(z)$. Complex numbers are typically plotted in the complex plane with their real parts on the x_1 axis and their imaginary parts on the x_2 axis (Fig. A.2-1). Alternatively, a complex number can be written in *polar coordinate* form as

$$z = a + ib = re^{i\theta} = r(\cos\theta + i\sin\theta). \tag{1}$$

[1] Most seismologists are more conservative than Paul Dirac, a leader in the development of quantum physics, who invented the delta function. Dirac regarded mathematical beauty as a guiding principle, stating that "it is more important to have beauty in one's equations than to have them fit experiment."

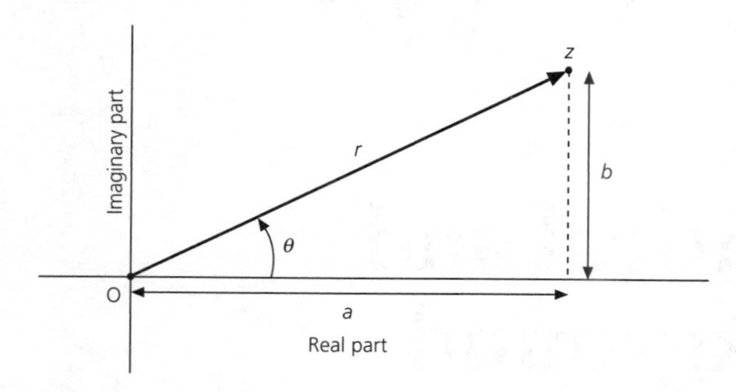

Fig. A.2-1 A number in the complex plane can be represented in terms of its real and imaginary parts, $z = a + ib$, or in polar form $z = re^{i\theta}$.

The *polar coordinates*, the magnitude r and the phase angle θ, can be expressed in terms of the real and imaginary parts as

$$r = \sqrt{a^2 + b^2}, \quad \theta = \tan^{-1}(b/a). \tag{2}$$

and, conversely,

$$a = r\cos\theta, \quad b = r\sin\theta. \tag{3}$$

To describe complex numbers in all four quadrants of the complex plane, θ ranges from 0 to 2π. Because the inverse tangent is periodic with period π, the signs of the real and imaginary parts are used to obtain the correct phase.

Complex numbers are equal when they have the same real and imaginary parts. Two complex numbers in $(a + ib)$ form are added by adding the real parts and the imaginary parts:

$$(a_1 + ib_1) + (a_2 + ib_2) = (a_1 + a_2) + i(b_1 + b_2). \tag{4}$$

Complex numbers can be multiplied either in the $(a + ib)$ form:

$$(a_1 + ib_1)(a_2 + ib_2) = (a_1a_2 - b_1b_2) + i(a_1b_2 + b_1a_2), \tag{5}$$

or in the magnitude and phase form:

$$r_1e^{i\theta_1}r_2e^{i\theta_2} = r_1r_2e^{i(\theta_1+\theta_2)}. \tag{6}$$

The conjugate of a complex number z, z^*, has the same real part and an imaginary part of opposite sign. Because

$$\begin{aligned} z^* &= a - ib = r\cos\theta - ir\sin\theta \\ &= r\cos(-\theta) + ir\sin(-\theta) = re^{-i\theta}, \end{aligned} \tag{7}$$

the conjugate has the same magnitude but the opposite phase. Hence the square of the magnitude of a complex number can be found by multiplication by the complex conjugate,

$$|z|^2 = zz^* = (a + ib)(a - ib) = (a^2 + b^2) = re^{i\theta}re^{-i\theta} = r^2. \tag{8}$$

By combining

$$e^{i\theta} = \cos\theta + i\sin\theta \quad \text{and} \quad e^{-i\theta} = \cos\theta - i\sin\theta \tag{9}$$

we obtain the definitions of the sine and cosine functions in terms of complex exponentials

$$\cos\theta = (e^{i\theta} + e^{-i\theta})/2 \quad \text{and} \quad \sin\theta = (e^{i\theta} - e^{-i\theta})/2i. \tag{10}$$

These relations yield formulae for the trigonometric functions of the sum of the angles because

$$e^{i(\theta_1+\theta_2)} = \cos(\theta_1 + \theta_2) + i\sin(\theta_1 + \theta_2) \tag{11}$$

and, by Eqn 6,

$$\begin{aligned} e^{i(\theta_1+\theta_2)} = e^{i\theta_1}e^{i\theta_2} &= (\cos\theta_1 + i\sin\theta_1)(\cos\theta_2 + i\sin\theta_2) \\ &= (\cos\theta_1\cos\theta_2 - \sin\theta_1\sin\theta_2) \\ &\quad + i(\sin\theta_1\cos\theta_2 + \cos\theta_1\sin\theta_2), \end{aligned} \tag{12}$$

so we can equate the real and imaginary parts and find

$$\cos(\theta_1 + \theta_2) = \cos\theta_1\cos\theta_2 - \sin\theta_1\sin\theta_2 \tag{13}$$

and

$$\sin(\theta_1 + \theta_2) = \sin\theta_1\cos\theta_2 + \cos\theta_1\sin\theta_2. \tag{14}$$

These expressions are symmetric in θ_1 and θ_2, as expected. The corresponding relations for the trigonometric functions of the difference of two angles are found by making θ_2 negative. Setting $\theta_1 = \theta_2$ gives expressions for $\cos(2\theta)$ and $\sin(2\theta)$.

The relations for the product of trigonometric functions of two angles can also be found using complex exponentials

$$\begin{aligned} \cos\theta_1\cos\theta_2 &= \frac{(e^{i\theta_1} + e^{-i\theta_1})}{2}\frac{(e^{i\theta_2} + e^{-i\theta_2})}{2} \\ &= \frac{1}{4}[(e^{i(\theta_1+\theta_2)} + e^{-i(\theta_1+\theta_2)}) + (e^{i(\theta_1-\theta_2)} + e^{-i(\theta_1-\theta_2)})] \\ &= \frac{1}{2}[\cos(\theta_1 + \theta_2) + \cos(\theta_1 - \theta_2)] \end{aligned} \tag{15}$$

and, similarly,

$$\begin{aligned} \sin\theta_1\sin\theta_2 &= \frac{(e^{i\theta_1} - e^{-i\theta_1})}{2i}\frac{(e^{i\theta_2} - e^{-i\theta_2})}{2i} \\ &= -\frac{1}{4}[(e^{i(\theta_1-\theta_2)} + e^{-i(\theta_1-\theta_2)}) - (e^{i(\theta_1+\theta_2)} + e^{-i(\theta_1+\theta_2)})] \\ &= \frac{1}{2}[\cos(\theta_1 - \theta_2) - \cos(\theta_1 + \theta_2)]. \end{aligned} \tag{16}$$

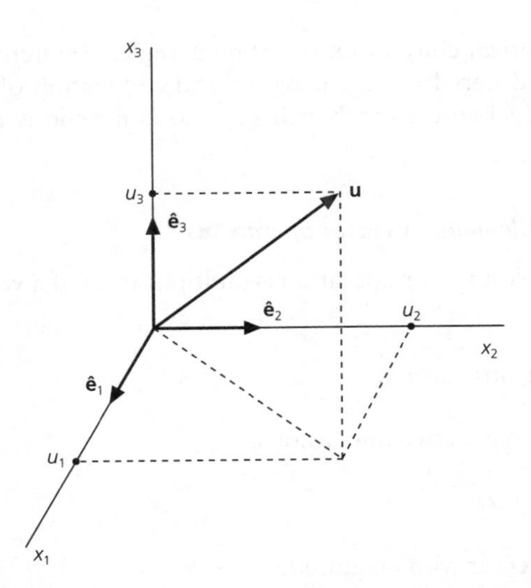

Fig. A.3-1 A vector **u** is expressed by the Cartesian unit basis vectors and its components: $\mathbf{u} = u_1\hat{e}_1 + u_2\hat{e}_2 + u_3\hat{e}_3$.

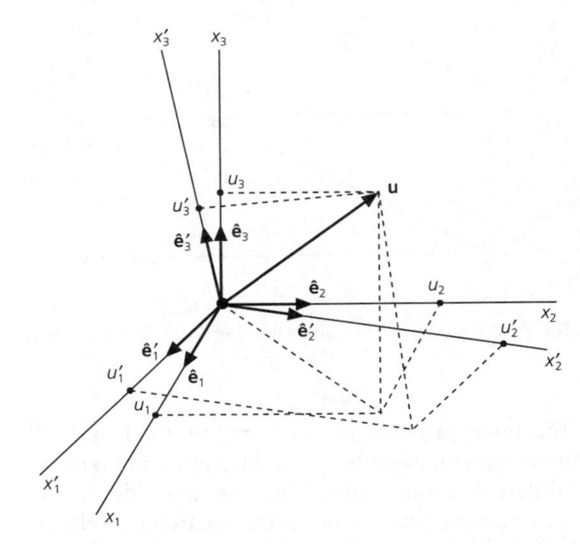

Fig. A.3-2 A vector **u** is described in each of two orthogonal coordinate systems by the Cartesian unit basis vectors of the coordinate system and the components of the vector in the coordinate system: $\mathbf{u} = u_1\hat{e}_1 + u_2\hat{e}_2 + u_3\hat{e}_3 = u'_1\hat{e}'_1 + u'_2\hat{e}'_2 + u'_3\hat{e}'_3$. Although the components differ between coordinate systems, the vector remains the same.

A.3 Scalars and vectors

A.3.1 *Definitions*

In seismology, we deal with several types of physical quantities. The simplest, *scalars*, are numbers describing a physical property at a given point that is independent of the coordinate system used to identify the point. Temperature, pressure, mass, and density are familiar examples. Mathematically, if a point is described in one coordinate system by (x_1, x_2, x_3) and in a second by (x'_1, x'_2, x'_3), the value of a scalar function ϕ in the first coordinate system equals that of the corresponding scalar function in the second

$$\phi(x_1, x_2, x_3) = \phi'(x'_1, x'_2, x'_3). \tag{1}$$

The distance between two points is a scalar because although the coordinates of the points depend on the coordinate system, the distance does not.

Vectors are more complicated entities that have magnitude and direction. In seismology, the most common vector is the motion, or *displacement*, of a piece of material within the earth due to the passage of a seismic wave. Vectors transform between different coordinate systems in a specific way. Thus, if the horizontal ground motion is recorded with seismometers oriented northeast–southwest and northwest–southeast, the north–south and east–west components of the displacement can be found using the properties of vectors. We will see that although the components depend on the coordinate system, the magnitude and direction of the vector remain the same.

Consider the familiar Cartesian coordinate system (Fig. A.3-1) with three mutually perpendicular (orthogonal) coordinate axes. There are two standard notations for these coordinates and axes: either the x_1, x_2, and x_3, or the x, y, and z axes. Each notation has advantages. The x_1, x_2, x_3 notation is more convenient for some derivations, and the x, y, z notation is sometimes clearer in physical problems. We use the x_1, x_2, and x_3 notation in this appendix, and use whichever notation seems more convenient in other discussions.

A point in this coordinate system is described by its x_1, x_2, and x_3 coordinates. Because a vector can be defined by a line from the origin (0, 0, 0) to the point (u_1, u_2, u_3), the three numbers u_1, u_2, and u_3 are the *components* of the vector **u**. A vector is denoted either by boldface type or by a set of its components

$$\mathbf{u} = (u_1, u_2, u_3) = (u_x, u_y, u_z). \tag{2}$$

A Cartesian coordinate system is described by three orthogonal unit basis vectors, $\hat{e}_1$, $\hat{e}_2$, and $\hat{e}_3$, along the x_1, x_2, and x_3 coordinate axes:

$$\hat{e}_1 = (1, 0, 0) \quad \hat{e}_2 = (0, 1, 0) \quad \hat{e}_3 = (0, 0, 1). \tag{3}$$

The caret, or "hat" superscript, indicates a *unit vector*, whose length is 1. The vector **u** is formed from its components and the basis vectors

$$\mathbf{u} = u_1\hat{e}_1 + u_2\hat{e}_2 + u_3\hat{e}_3 = (u_1, u_2, u_3). \tag{4}$$

Now, consider a second Cartesian coordinate system with the same origin and different axes x'_1, x'_2, and x'_3, along which unit basis vectors $\hat{e}'_1$, $\hat{e}'_2$, and $\hat{e}'_3$ are defined (Fig. A.3-2). In this coordinate system the components of **u** are different,

$$\mathbf{u} = u'_1\hat{e}'_1 + u'_2\hat{e}'_2 + u'_3\hat{e}'_3 = (u'_1, u'_2, u'_3). \tag{5}$$

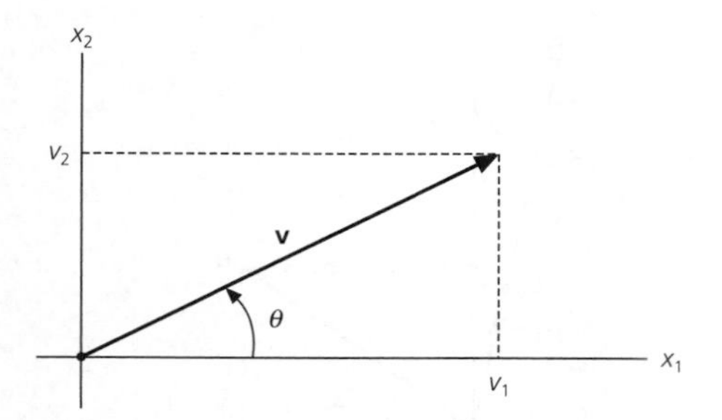

Fig. A.3-3 A vector in two dimensions making an angle θ with the x_1 axis.

Thus the *same* physical vector is represented in a different coordinate system, described by a different set of basis vectors, using different components. The essential idea is that the vector remains the same, or invariant, regardless of the coordinate system, although the numerical values of its components change. Physical laws, like Newton's law stating that the force vector equals the product of the mass and the acceleration vector (the second derivative with respect to time of the displacement vector), are written in vector form because the physical phenomenon does not depend on the coordinate system used to describe it.

The length or *magnitude* of a vector, $|\mathbf{u}|$, is a scalar, and thus the same in different coordinate systems. By the Pythagorean theorem, the length is

$$|\mathbf{u}| = (u_1^2 + u_2^2 + u_3^2)^{1/2} = (u_1'^2 + u_2'^2 + u_3'^2)^{1/2}. \tag{6}$$

The zero vector, **0**, all of whose components are zero in any coordinate system, has zero magnitude.

A vector is specified in either Cartesian coordinates by its components or in polar coordinates by its magnitude and direction. For example, in a two-dimensional (x_1, x_2) coordinate system (Fig. A.3-3), the vector **v** can be written in terms of its components

$$\mathbf{v} = (v_1, v_2) \tag{7}$$

or its magnitude

$$|\mathbf{v}| = (v_1^2 + v_2^2)^{1/2} \tag{8}$$

and direction, given by the angle θ that **v** makes with the x_1 direction

$$\theta = \tan^{-1}(v_2/v_1). \tag{9}$$

Just as $|\mathbf{v}|$ and θ are given by the components, so the components are given by $|\mathbf{v}|$ and θ

$$v_1 = |\mathbf{v}| \cos\theta \quad \text{and} \quad v_2 = |\mathbf{v}| \sin\theta. \tag{10}$$

By analogy, a vector in three dimensions is specified by either its three components or its magnitude and the angles it forms with two of the coordinate axes. It is worth noting that the mathematical convention of defining angles counterclockwise from x_1 differs from the geographical convention of defining angles clockwise from North (x_2), so conversions are often needed.

A.3.2 *Elementary vector operations*

The simplest vector operation is multiplication of a vector by a scalar

$$\alpha\mathbf{u} = (\alpha u_1, \alpha u_2, \alpha u_3). \tag{11}$$

For example, in two dimensions,

$$\alpha\mathbf{v} = (\alpha v_1, \alpha v_2) \tag{12}$$

yields a vector with magnitude

$$((\alpha v_1)^2 + (\alpha v_2)^2)^{1/2} = |\alpha| (v_1^2 + v_2^2)^{1/2} = |\alpha| |\mathbf{v}| \tag{13}$$

whose direction is given by

$$\tan\theta = \alpha v_2/\alpha v_1 = v_2/v_1. \tag{14}$$

Multiplication by a positive scalar thus changes the magnitude of a vector but preserves its direction. Similarly, multiplication by a negative scalar changes the magnitude of a vector and reverses its direction. $\hat{\mathbf{u}}$, a unit vector in the direction of **u** is formed by dividing **u** by its magnitude

$$\hat{\mathbf{u}} = \mathbf{u}/|\mathbf{u}|. \tag{15}$$

The sum of two vectors is another vector whose components are the sums of the corresponding components, so if

$$\mathbf{a} = a_1\hat{\mathbf{e}}_1 + a_2\hat{\mathbf{e}}_2 + a_3\hat{\mathbf{e}}_3 \quad \text{and} \quad \mathbf{b} = b_1\hat{\mathbf{e}}_1 + b_2\hat{\mathbf{e}}_2 + b_3\hat{\mathbf{e}}_3,$$
$$\mathbf{a} + \mathbf{b} = (a_1 + b_1)\hat{\mathbf{e}}_1 + (a_2 + b_2)\hat{\mathbf{e}}_2 + (a_3 + b_3)\hat{\mathbf{e}}_3 = \mathbf{b} + \mathbf{a}. \tag{16}$$

Addition can be done graphically (Fig. A.3-4) by shifting one vector, while preserving its orientation, so that its "tail" is at the "head" of the other, and forming the vector sum. For example, the total force vector acting on an object is the vector sum of the individual force vectors. Equation 16 and Fig. A.3-4 show that vector addition is commutative; it does not matter in which order the vectors are added.

A.3.3 *Scalar products*

There are two methods of multiplying vectors. The first, the *scalar product* (also called the dot product or inner product), yields a scalar:

$$\mathbf{a} \cdot \mathbf{b} = a_1b_1 + a_2b_2 + a_3b_3 = |\mathbf{a}| |\mathbf{b}| \cos\theta, \tag{17}$$

where θ is the angle between two vectors. To see that the two definitions of the scalar product are equivalent, consider a two-

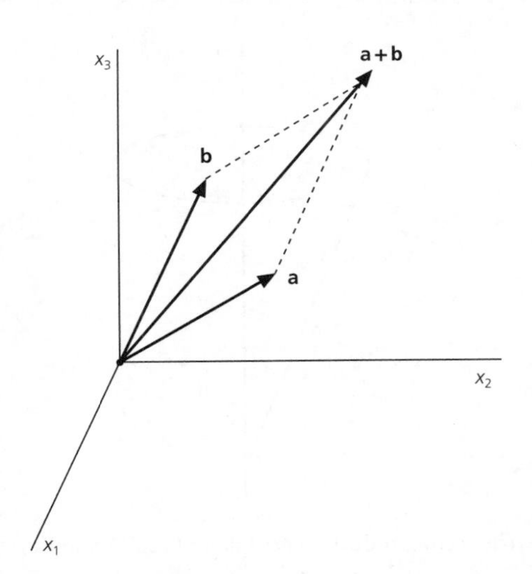

Fig. A.3-4 Addition of vectors **a** and **b**. The addition can be done analytically, by adding components, or graphically. Vector addition is commutative, as the order of addition is irrelevant.

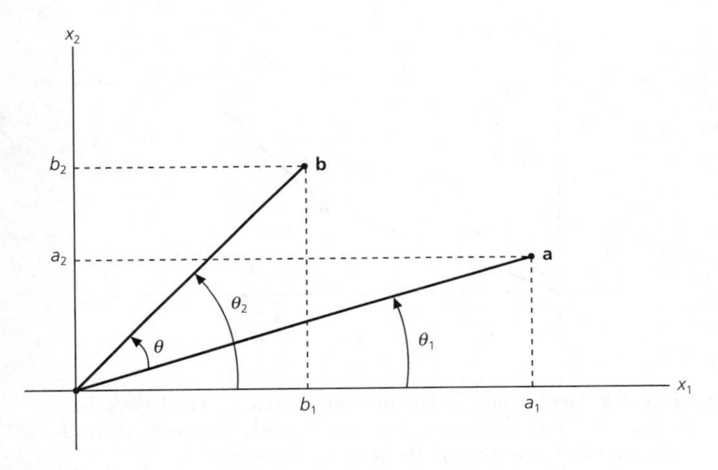

Fig. A.3-5 Derivation of alternative definitions of the scalar product **a** · **b** in two dimensions.

dimensional case (Fig. A.3-5) with $\mathbf{a} = (a_1, a_2)$ and $\mathbf{b} = (b_1, b_2)$. If **a** and **b** make angles θ_1 and θ_2 with the $\hat{e}_1$ axis, then

$$\mathbf{a} \cdot \mathbf{b} = |\mathbf{a}||\mathbf{b}| \cos \theta = |\mathbf{a}||\mathbf{b}| \cos (\theta_2 - \theta_1). \tag{18}$$

Using a trigonometric identity (Eqn A.2.13) we expand

$$\cos \theta = \cos (\theta_2 - \theta_1) = \cos \theta_2 \cos \theta_1 + \sin \theta_2 \sin \theta_1. \tag{19}$$

Because

$$\cos \theta_1 = a_1/(a_1^2 + a_2^2)^{1/2} \quad \text{and} \quad \sin \theta_1 = a_2/(a_1^2 + a_2^2)^{1/2}, \tag{20}$$

and similar definitions hold for θ_2 and **b**, substitutions for the angles in Eqn 18 show that

$$|\mathbf{a}||\mathbf{b}| \cos \theta = \frac{|\mathbf{a}||\mathbf{b}|(a_1 b_1 + a_2 b_2)}{(a_1^2 + a_2^2)^{1/2} (b_1^2 + b_2^2)^{1/2}} = a_1 b_1 + a_2 b_2. \tag{21}$$

Equation 17 shows several features of the scalar product:

- The scalar product commutes: $\mathbf{a} \cdot \mathbf{b} = \mathbf{b} \cdot \mathbf{a}$.
- The scalar product of two perpendicular vectors is zero, because $\cos 90° = 0$.
- The scalar product of a vector with itself is its magnitude squared:

$$\mathbf{a} \cdot \mathbf{a} = a_1 a_1 + a_2 a_2 + a_3 a_3 = |\mathbf{a}|^2. \tag{22}$$

The definition of the scalar product is generalized for vectors with complex components. To see why, note that for a vector $\mathbf{a} = (i, 1, 0)$, where $i = \sqrt{-1}$, Eqn 22 would give a squared magnitude of zero. Because we would like only the zero vector, all of whose elements are zero, to have zero magnitude, Eqn 17 is generalized to

$$\mathbf{a} \cdot \mathbf{b} = a_1^* b_1 + a_2^* b_2 + a_3^* b_3 \tag{23}$$

where * indicates the complex conjugate. Thus the definition of the squared magnitude (Eqn 22) becomes

$$\mathbf{a} \cdot \mathbf{a} = a_1^* a_1 + a_2^* a_2 + a_3^* a_3 = |\mathbf{a}|^2. \tag{24}$$

For example, the squared magnitude of $|(i, 1, 0)|^2 = (i)(-i) + (1)(1) = 2$. These complex definitions reduce to the familiar cases, (Eqns 17 and 22), for vectors with real components.

The relations between the unit basis vectors for a Cartesian coordinate system, $\hat{e}_1$, $\hat{e}_2$, and $\hat{e}_3$, are easily stated using their scalar products. Because each is perpendicular to the other two, the scalar product of any two different ones is zero,

$$\hat{e}_1 \cdot \hat{e}_2 = \hat{e}_1 \cdot \hat{e}_3 = \hat{e}_2 \cdot \hat{e}_3 = 0, \tag{25}$$

and the scalar product of each with itself is its squared magnitude

$$\hat{e}_1 \cdot \hat{e}_1 = \hat{e}_2 \cdot \hat{e}_2 = \hat{e}_3 \cdot \hat{e}_3 = 1. \tag{26}$$

The unit basis set of vectors is *orthonormal*; each is orthogonal (perpendicular) to the others and normalized to unit magnitude.

The *projection*, or component of a vector in a direction given by a unit vector, is the scalar product of a vector with the unit vector. Using this idea, a component of a vector can be found from its projection on the unit basis vector along the corresponding axis. Thus the x_1 component of **u** is

$$\mathbf{u} \cdot \hat{e}_1 = (u_1 \hat{e}_1 + u_2 \hat{e}_2 + u_3 \hat{e}_3) \cdot \hat{e}_1 = u_1, \tag{27}$$

with the other components defined similarly.

A.3.4 *Vector products*

A second form of multiplication, the *vector* or *cross* product, forms a third vector from two vectors by

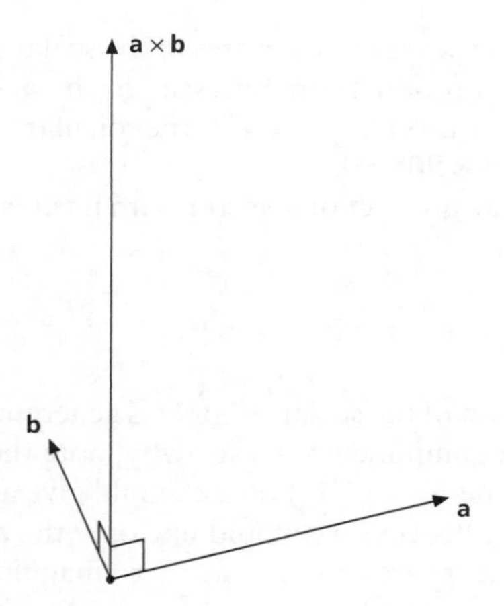

Fig. A.3-6 Illustration of the right-hand rule giving the orientation of the vector product $\mathbf{a}\times\mathbf{b}$.

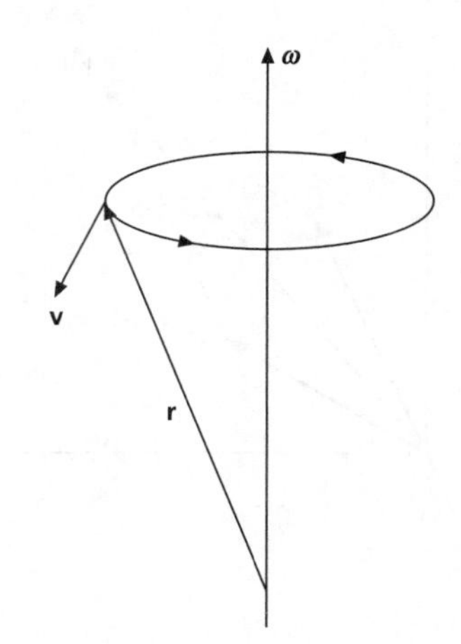

Fig. A.3-7 The vector product $\mathbf{v}=\boldsymbol{\omega}\times\mathbf{r}$ describes a rotation.

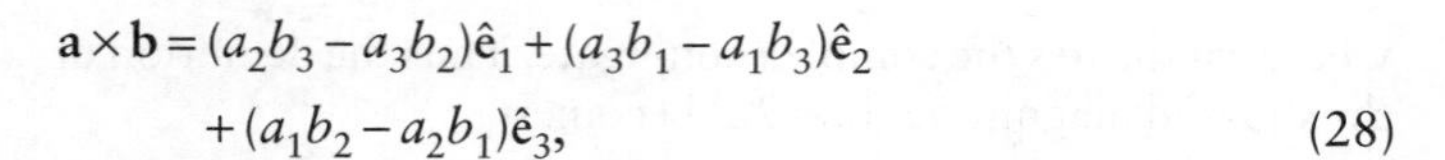

$$\mathbf{a}\times\mathbf{b}=(a_2b_3-a_3b_2)\hat{\mathrm{e}}_1+(a_3b_1-a_1b_3)\hat{\mathrm{e}}_2 +(a_1b_2-a_2b_1)\hat{\mathrm{e}}_3, \quad (28)$$

which can be written as the determinant

$$\mathbf{a}\times\mathbf{b}=\begin{vmatrix} \hat{\mathrm{e}}_1 & \hat{\mathrm{e}}_2 & \hat{\mathrm{e}}_3 \\ a_1 & a_2 & a_3 \\ b_1 & b_2 & b_3 \end{vmatrix}. \quad (29)$$

The vector product of two vectors is perpendicular to both vectors. For example, if **a** and **b** are in the x_1–x_2 plane, $a_3=b_3=0$, and by Eqn 28, the vector product has only an $\hat{\mathrm{e}}_3$ component. This can be shown in general by evaluating $\mathbf{a}\cdot(\mathbf{a}\times\mathbf{b})=\mathbf{b}\cdot(\mathbf{a}\times\mathbf{b})=0$. Geometrically, the direction of the vector product is found by a "right-hand rule" (Fig. A.3-6): if the fingers of a right hand rotate from **a** to **b**, the thumb points in the direction $\mathbf{a}\times\mathbf{b}$. The magnitude of the cross product is

$$|\mathbf{a}\times\mathbf{b}|=|\mathbf{a}|\,|\mathbf{b}|\sin\theta, \quad (30)$$

where θ is the angle between the two vectors. The cross product is zero for parallel vectors because $\sin 0°=0$, so the cross product of a vector with itself is zero.

The vector product often appears in connection with rotations, such as those used to describe the motion of lithospheric plates (Section 5.2). For example, if an object located at a position **r** undergoes a rotation, its linear velocity **v** is given by

$$\mathbf{v}=\boldsymbol{\omega}\times\mathbf{r}, \quad (31)$$

where ω is the rotation vector, which is oriented along the axis of rotation, with a magnitude $|\omega|$ that is the angular velocity (Fig. A.3-7). Similarly, the vector product is used to define the torque, which gives the rate of change of angular momentum. A force **F**, acting at a point **r**, gives a torque

$$\tau=\mathbf{r}\times\mathbf{F}. \quad (32)$$

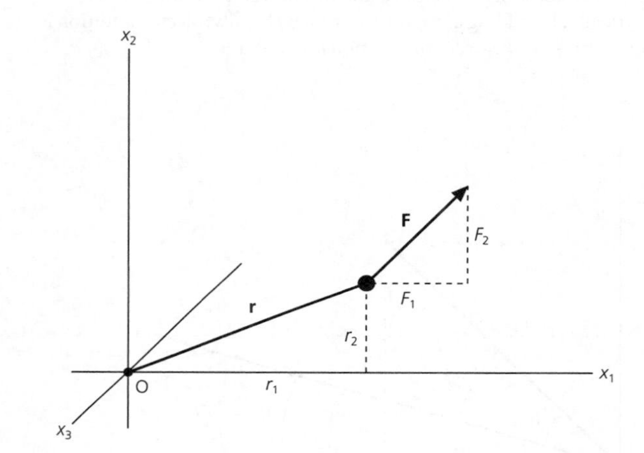

Fig. A.3-8 The x_3 component of the vector product $\tau=\mathbf{r}\times\mathbf{F}$ gives the torque, $r_1F_2-r_2F_1$ about the x_3 axis. In this case r_1F_2 is greater than r_2F_1, so counterclockwise rotation about the x_3 axis occurs.

For example, the torque about the x_3 axis is $\tau_3=(r_1F_2-r_2F_1)$, so each component of the force contributes a counterclockwise torque equal to the component times its lever arm, the perpendicular distance of the point from that axis (Fig. A.3-8).

Some useful identities, whose proofs are left as problems, are

$$\mathbf{a}\cdot(\mathbf{b}+\mathbf{c})=\mathbf{a}\cdot\mathbf{b}+\mathbf{a}\cdot\mathbf{c}$$
$$\mathbf{a}\times(\mathbf{b}+\mathbf{c})=\mathbf{a}\times\mathbf{b}+\mathbf{a}\times\mathbf{c}$$
$$\mathbf{a}\cdot(\mathbf{b}\times\mathbf{c})=\mathbf{b}\cdot(\mathbf{c}\times\mathbf{a})=\mathbf{c}\cdot(\mathbf{a}\times\mathbf{b})$$
$$\mathbf{a}\times(\mathbf{b}\times\mathbf{c})=\mathbf{b}(\mathbf{a}\cdot\mathbf{c})-\mathbf{c}(\mathbf{a}\cdot\mathbf{b}). \quad (33)$$

A.3.5 Index notation

Vector equations, such as the definition of the cross product, can be cumbersome when written in terms of the components. Simplification can be obtained using *index notation*, whereby an index assuming all possible values replaces the subscripts indicating coordinate axes. For example, the vector $\mathbf{u} = (u_1, u_2, u_3)$ is written u_i, where i can be 1, 2, or 3. In this notation, the scalar product is

$$\mathbf{a} \cdot \mathbf{b} = a_1 b_1 + a_2 b_2 + a_3 b_3 = \sum_{i=1}^{3} a_i b_i . \tag{34}$$

Because the sum over all coordinates appears frequently, the *Einstein summation convention* is often used, whereby an index repeated twice implies a summation over that index, and the summation sign is not explicitly written. Hence the scalar product of two real vectors is written

$$\mathbf{a} \cdot \mathbf{b} = a_i b_i, \tag{35}$$

using implied summation over the repeated index i. Similarly, the square of the magnitude of a real vector is

$$|\mathbf{u}|^2 = u_i u_i . \tag{36}$$

A repeated index is called a "dummy" index, like a dummy variable of integration, because it is used only within the summation. The form of the expression indicates that $u_i u_i$ is a scalar; because the repeated index is summed, no index remains "free." By contrast, u_i is a vector, because there is a free index.

Index notation is further simplified by introducing two symbols, δ_{ij} and ε_{ijk}. The *Kronecker delta*, δ_{ij}, is defined

$$\delta_{ij} = 0 \quad \text{if } i \neq j, \qquad = 1 \quad \text{if } i = j. \tag{37}$$

So, for example, $\delta_{11} = 1$, but $\delta_{12} = 0$. Using the Kronecker delta symbol, the relations between the Cartesian basis vectors (Eqns 25, 26) can be written compactly as

$$\hat{\mathbf{e}}_i \cdot \hat{\mathbf{e}}_j = \delta_{ij}. \tag{38}$$

The Kronecker delta, a function of two discrete variables i and j, is analogous to the Dirac delta function which is a function of a continuous variable (Section 6.2.5).

The *permutation symbol*, ε_{ijk}, is defined as

$$\begin{aligned} \varepsilon_{ijk} &= 0 && \text{if any of the indices are the same,} \\ &= 1 && \text{if } i, j, k \text{ are in order, i.e., } (1, 2, 3), (2, 3, 1), \text{ or } (3, 1, 2) \\ &= -1 && \text{if } i, j, k \text{ are out of order, i.e., } (2, 1, 3), (3, 2, 1), (1, 3, 2). \end{aligned} \tag{39}$$

Cases where the indices are in order are known as even, or cyclic, permutations of the indices; those in which the indices are out of order are known as odd permutations. Because of the symmetries in the definition, $\varepsilon_{ijk} = \varepsilon_{jki} = \varepsilon_{kij}$. A useful relation, whose proof is left for the problems, is

$$\varepsilon_{ijk} \varepsilon_{ist} = \delta_{js} \delta_{kt} - \delta_{jt} \delta_{ks} . \tag{40}$$

Using index notation, the definition of the vector product (Eqn 28) becomes

$$(\mathbf{a} \times \mathbf{b})_i = \sum_{j=1}^{3} \sum_{k=1}^{3} \varepsilon_{ijk} a_j b_k = \varepsilon_{ijk} a_j b_k, \tag{41}$$

where the last form uses the summation convention. The notation shows that the cross product yields a vector because only one index, i, remains free after the repeated indices j and k are summed. To see that the index notation gives the correct definition, we expand the $i = 2$ component as

$$\begin{aligned} (\mathbf{a} \times \mathbf{b})_2 &= \varepsilon_{211} a_1 b_1 + \varepsilon_{212} a_1 b_2 + \varepsilon_{213} a_1 b_3 + \varepsilon_{221} a_2 b_1 + \varepsilon_{222} a_2 b_2 \\ &\quad + \varepsilon_{223} a_2 b_3 + \varepsilon_{231} a_3 b_1 + \varepsilon_{232} a_3 b_2 + \varepsilon_{233} a_3 b_3 \\ &= (a_3 b_1 - a_1 b_3), \end{aligned} \tag{42}$$

because the only nonzero ε_{ijk} terms are $\varepsilon_{213} = -1$ and $\varepsilon_{231} = 1$.

Index notation points out an interesting feature of the vector product. Because $a_i b_i = b_i a_i$, the scalar product commutes. By contrast, the properties of the permutation symbol show that

$$\mathbf{a} \times \mathbf{b} = \varepsilon_{ijk} a_j b_k = -\varepsilon_{ijk} b_j a_k = -\mathbf{b} \times \mathbf{a}, \tag{43}$$

so the order matters for the vector product.

Although index notation seems unnatural at first, it does more than simply shorten expressions. The notation explicitly indicates what operations must be performed, and thus makes them easier to evaluate. For example, suppose we seek to show that the cross product of a vector with itself is zero. In contrast to $(\mathbf{a} \times \mathbf{a})$, the notation $\varepsilon_{ijk} a_j a_k$ shows how the cross product should be evaluated. Because $a_j a_k$ is symmetric in the indices j and k, the permutation symbol makes the terms involving any pair of j and k sum to zero. We will see that index notation makes the complicated expressions that we encounter in studying stress and strain easier to evaluate.

A.3.6 Vector spaces

These concepts for vectors can be generalized in several ways. In three dimensions any vector is a weighted combination of three basis vectors. The usual choice of basis vectors along coordinate axes is for simplicity. We could choose any three mutually orthogonal vectors, which need not be of unit length, to be the basis vectors. To see this, remember that a physical vector does not depend on the coordinate system.

Moreover, the idea of vectors in two- or three-dimensional space can be generalized to spaces with a larger number of dimensions. For example, given unit vectors

$$\begin{aligned} &\hat{\mathbf{e}}_1 = (1, 0, 0, 0, 0), \quad \hat{\mathbf{e}}_2 = (0, 1, 0, 0, 0), \quad \hat{\mathbf{e}}_3 = (0, 0, 1, 0, 0), \\ &\hat{\mathbf{e}}_4 = (0, 0, 0, 1, 0), \quad \hat{\mathbf{e}}_5 = (0, 0, 0, 0, 1), \end{aligned} \tag{44}$$

a vector **u** can be formed from the basis vectors and components

$$\mathbf{u} = u_1\hat{\mathbf{e}}_1 + u_2\hat{\mathbf{e}}_2 + u_3\hat{\mathbf{e}}_3 + u_4\hat{\mathbf{e}}_4 + u_5\hat{\mathbf{e}}_5 = (u_1, u_2, u_3, u_4, u_5). \quad (45)$$

This vector is defined in a five-dimensional space, with five axes each orthogonal to the others, because their scalar products are zero. Although this is difficult to visualize (or draw), the mathematics carries through directly from the three-dimensional case. N mutually orthogonal vectors thus provide a basis for an N-dimensional space.

These ideas are formalized in terms of vectors in a general *linear vector space*. For our purposes, a vector space is a collection of vectors **x**, **y**, **z**, satisfying several criteria:

- The sum of any two vectors in the space is also in the space.
- Vector addition commutes: $\mathbf{x}+\mathbf{y}=\mathbf{y}+\mathbf{x}$.
- Vector addition is associative: $(\mathbf{x}+\mathbf{y})+\mathbf{z}=\mathbf{x}+(\mathbf{y}+\mathbf{z})$.
- There exists a unique vector $\mathbf{0}$ such that for all $\mathbf{x}$, $\mathbf{x}=\mathbf{x}+\mathbf{0}$.
- There exists a unique vector $-\mathbf{x}$ such that for all $\mathbf{x}$, $\mathbf{x}+(-\mathbf{x}) = 0$.
- Scalar multiplication is associative: $\alpha(\beta\mathbf{x})=(\alpha\beta)\mathbf{x}$.
- Scalar multiplication is distributive: $\alpha(\mathbf{x}+\mathbf{y}) = \alpha\mathbf{x}+\alpha\mathbf{y}$ and $(\alpha+\beta)\mathbf{x}=(\alpha\mathbf{x}+\beta\mathbf{x})$.

A point worth considering is the number of independent vectors in a vector space. Given N vectors $\mathbf{x}^1, \mathbf{x}^2, \ldots, \mathbf{x}^N$ in a linear vector space, a weighted sum $\sum\alpha_i\mathbf{x}^i$ is called a *linear combination*. The N vectors are *linearly independent* if

$$\sum_{i=1}^{N} \alpha_i\mathbf{x}^i = 0 \text{ only when all } \alpha_i = 0, \quad (46)$$

so that no vector can be expressed as a combination of the others. Otherwise, the vectors are *linearly dependent*, and one can be expressed as a linear combination of the others.

This idea corresponds to that of basis vectors. If N basis vectors are mutually orthogonal, they are linearly independent. Because any vector in an N-dimensional space is a linear combination of N linearly independent basis vectors, the basis vectors *span* the space. Thus the dimension of a vector space is the number of linearly independent vectors within it. For example, we cannot find four linearly independent vectors in three dimensions.

Though vector spaces sound abstract, they are useful in seismology. For example, in Chapter 2 we represent travelling waves by normal modes, which are orthogonal basis vectors in a vector space, so any wave is a weighted sum of them. The modes of a string (Section 2.2.5) form a Fourier series (Chapter 6), in which a function is expanded into sine and cosine functions that are the basis vectors of a vector space. A similar approach is also used for the modes of the spherical earth (Section 2.9). Vector space ideas are also used in inverting seismological observations to study earth structure (Chapter 7).

A.4 Matrix algebra

A.4.1 Definitions

Matrix algebra is a powerful tool often used to study systems of equations. As a result, it appears in seismological applications, including stresses and strains, locating earthquakes, and seismic tomography. We thus review some basic ideas, often stating results without proof and leaving proofs for the problems. Further discussion of these topics can be found in linear algebra texts.

Given a matrix A with m rows and n columns, called an $m \times n$ matrix,

$$A = \begin{pmatrix} a_{11} & a_{12} & \cdots & a_{1n} \\ a_{21} & a_{22} & \cdots & a_{2n} \\ \cdot & \cdot & \cdots & \cdot \\ \cdot & \cdot & \cdots & \cdot \\ a_{m1} & a_{m2} & \cdots & a_{mn} \end{pmatrix} \quad (1)$$

and a second matrix B, also with m rows and n columns, matrix addition is defined by

$$A + B = \begin{pmatrix} a_{11}+b_{11} & a_{12}+b_{12} & \cdots & a_{1n}+b_{1n} \\ a_{21}+b_{21} & a_{22}+b_{22} & \cdots & a_{2n}+b_{2n} \\ \cdot & \cdot & \cdots & \cdot \\ \cdot & \cdot & \cdots & \cdot \\ a_{m1}+b_{m1} & a_{m2}+b_{m2} & \cdots & a_{mn}+b_{mn} \end{pmatrix}. \quad (2)$$

The usual convention is to indicate matrices with capital letters and their elements with lower-case ones.

Matrix multiplication is defined such that for a matrix A that is $m \times n$ and a matrix B that is $n \times r$, the ij^{th} element of the $m \times r$ product matrix $C = AB$ is defined by

$$c_{ij} = \sum_{k=1}^{n} a_{ik}b_{kj} = a_{ik}b_{kj}. \quad (3)$$

The ij^{th} element of C is the scalar product of the ith row of A and the jth column of B. As a result, for matrix multiplication the two matrices need not have the same number of rows and columns, but must have the number of columns in the first matrix equal to the number of rows in the second. Often the numbers of rows and columns in the two matrices allow multiplication in only one order. Thus, in the example above, A "premultiplies" B, or B "postmultiplies" A. A convenient way to remember this is that the number of columns in the first matrix must equal the number of rows in the second, but this dimension does not appear in the product. In the case of $AB = C$, written schematically, we have $[m \times n][n \times r] = [m \times r]$. Hence, in the final form in Eqn 3, the summation convention shows that k is summed out, leaving i and j as free indices, so c_{ij} is a matrix element. Furthermore, even if both AB and BA are allowed, the two products are generally not equal, so matrix multiplication is not commutative.

The *identity matrix*, I, is a square matrix (one with the same number of rows and columns) whose diagonal elements are equal to 1 while all other elements are 0:

$$I = \begin{pmatrix} 1 & 0 & \dots & 0 & 0 \\ 0 & 1 & \dots & 0 & 0 \\ \cdot & \cdot & \dots & \cdot & \cdot \\ \cdot & \cdot & \dots & \cdot & \cdot \\ 0 & 0 & \dots & 1 & 0 \\ 0 & 0 & \dots & 0 & 1 \end{pmatrix}. \tag{4}$$

The identity matrix has the property that for any square matrix A,

$$AI = IA = A. \tag{5}$$

The *transpose* of a matrix A, A^T, is derived by placing the rows of A into the columns of A^T, so for $C = A^T$,

$$c_{ij} = a_{ji}. \tag{6}$$

The transpose has the properties that for matrices A and B,

$$(A+B)^T = A^T + B^T \quad \text{and} \quad (AB)^T = B^T A^T. \tag{7}$$

With these definitions, vector operations can be expressed using matrix algebra, by treating vectors as matrices with one column. For example, premultiplication of a vector by a matrix yields another vector, $\mathbf{y} = A\mathbf{x}$, such that

$$y_i = \sum_j a_{ij} x_j \quad \text{or} \quad y_i = a_{ij} x_j, \tag{8}$$

where the second form uses the summation convention. Each component y_i is the scalar product of the ith row of A with $\mathbf{x}$. Similarly, the scalar product of two vectors is given by the matrix product

$$\mathbf{a} \cdot \mathbf{b} = \mathbf{a}^T \mathbf{b} = \sum_i a_i b_i = a_i b_i. \tag{9}$$

Thus the scalar product of two vectors yields a scalar, because a $1 \times m$ matrix times an $m \times 1$ matrix is a 1×1 matrix, or single value. The squared magnitude of a real vector can be written as

$$|\mathbf{u}|^2 = \mathbf{u} \cdot \mathbf{u} = \mathbf{u}^T \mathbf{u} = \sum_i u_i u_i = u_i u_i. \tag{10}$$

For vectors with complex components, the scalar product (Eqn A.3.23) is

$$\mathbf{a} \cdot \mathbf{b} = \mathbf{a}^{*T} \mathbf{b} = \sum_i a_i^* b_i = a_i^* b_i. \tag{11}$$

This brings us to a minor point of notation. In linear algebra, as in the last few equations, it is common to treat vectors as column vectors represented by $n \times 1$ matrices with n rows and one column

$$\mathbf{u} = \begin{pmatrix} u_1 \\ u_2 \\ \cdot \\ \cdot \\ u_n \end{pmatrix}, \tag{12}$$

whose transposes are row vectors (one row, n columns) like

$$\mathbf{u}^T = (u_1, u_2, \dots u_n). \tag{13}$$

Nonetheless, to save space, we sometimes write

$$\mathbf{u} = (u_1, u_2, \dots u_n), \tag{14}$$

while treating $\mathbf{u}$ as a column vector when required. Strictly speaking, we should call the row vector $\mathbf{u}^T$.

We often encounter matrices that are *symmetric*, or equal their transposes,

$$A = A^T, \quad a_{ij} = a_{ji}. \tag{15}$$

For a matrix A with complex elements, the conjugate matrix A^* is formed by taking the conjugate of each element, and the transpose is generalized to the *adjoint* matrix $A^+ = A^{*T}$, which is the complex conjugate of A^T. Note that if the elements of A are real, $A^+ = A^T$. A matrix A is *Hermitian* if it equals its adjoint

$$A = A^+, \quad a_{ij} = a_{ji}^*. \tag{16}$$

If A is real, "Hermitian" and "symmetric" are equivalent.

A.4.2 *Determinant*

A useful entity is the *determinant* of a matrix, written det A, or $|A|$. For an $n \times n$ matrix,

$$\det A = \sum_{j_1=1}^{n} \sum_{j_2=1}^{n} \dots \sum_{j_n=1}^{n} s(j_1, j_2, \dots j_n) a_{1j_1} a_{2j_2} \dots a_{nj_n}. \tag{17}$$

This complicated sum over n indices, $j_1, j_2, \dots j_n$, uses a generalized form of the permutation symbol

$$s(j_1, j_2, \dots j_n) = \operatorname{sgn} \prod_{1 \le p < q \le n} (j_q - j_p). \tag{18}$$

The sgn function is one times the sign of its argument, so that it equals 1 if its argument is positive, −1 if its argument is negative, and 0 if its argument is zero. For $n = 3$,

$$s(j_1, j_2, j_3) = \operatorname{sgn}[(j_2 - j_1)(j_3 - j_1)(j_3 - j_2)], \tag{19}$$

so that, for example,

$$s(1,2,3)=1, \quad s(2,1,3)=-1, \quad s(1,1,3)=0. \tag{20}$$

Because $s(j_1, j_2, j_3)$ suppresses terms with two equal indices, and assigns others a sign depending on the order of the indices, it is the same as the permutation symbol, $\varepsilon_{j_1 j_2 j_3}$ (Eqn A.3.39).

The definition of the determinant gives the familiar result for $n=2$:

$$\begin{aligned} |A| = \det\begin{pmatrix} a_{11} & a_{12} \\ a_{21} & a_{22} \end{pmatrix} &= \sum_{j_1=1}^{2}\sum_{j_2=1}^{2} s(j_1, j_2) a_{1j_1} a_{2j_2} \\ &= s(1,1)a_{11}a_{21} + s(1,2)a_{11}a_{22} + s(2,1)a_{12}a_{21} + s(2,2)a_{12}a_{22} \\ &= a_{11}a_{22} - a_{12}a_{21}, \end{aligned} \tag{21}$$

because $s(1, 1) = s(2, 2) = 0$, $s(1, 2) = 1$, and $s(2, 1) = -1$. For a matrix with only one element, the determinant equals the matrix element.

Among the properties of determinants that we will find useful in solving systems of equations are:

- The determinant of a matrix equals that of its transpose, $|A| = |A^T|$.
- If two rows or columns of a matrix are interchanged, the determinant has the same absolute value but changes sign.
- If one row (or column) is multiplied by a constant, the determinant is multiplied by that constant.
- If a multiple of one row (or column) is added to another row (or column), the determinant is unchanged.
- If two rows or columns of a matrix are the same, the determinant is zero.

Proving these properties is left for the problems.

A.4.3 *Inverse*

For an $n \times n$ square matrix A, the *inverse* matrix A^{-1} is defined such that multiplication by the inverse gives the identity matrix

$$A^{-1}A = AA^{-1} = I. \tag{22}$$

A^{-1} can be written in terms of the *cofactor matrix*, C, whose elements

$$c_{ij} = (-1)^{i+j} |A_{ij}| \tag{23}$$

are formed from the determinants of A_{ij}, an $(n-1) \times (n-1)$ square matrix formed by deleting the ith row and j^{th} column from A. If $|A|$ is not zero,

$$A^{-1} = C^T / |A|. \tag{24}$$

For the familiar $n=2$ case, see problem 7.

A matrix whose determinant is zero does not have an inverse, and is called *singular*. Because the determinant of a matrix with two equal rows or columns is zero, such a matrix is singular. More generally, a matrix is singular if a row or column is a linear combination of the others.

The inverse of the matrix product AB, if AB is nonsingular, obeys

$$(AB)^{-1} = B^{-1}A^{-1}. \tag{25}$$

A matrix A whose transpose equals its inverse,

$$A^{-1} = A^T, \tag{26}$$

is called *orthogonal*. By extension, a matrix A with complex elements is *unitary* if its adjoint and inverse are equal

$$A^{-1} = A^{+}. \tag{27}$$

A.4.4 *Systems of linear equations*

A vector–matrix representation is often used for systems of linear equations. In this formulation, a system of m equations for n unknown variables x_i,

$$\begin{aligned} &a_{11}x_1 + a_{12}x_2 \ldots + a_{1n}x_n = b_1 \\ &a_{21}x_1 + a_{22}x_2 \ldots + a_{2n}x_n = b_2 \\ &\qquad \ldots \\ &a_{m1}x_1 + a_{m2}x_2 \ldots + a_{mn}x_n = b_m \end{aligned} \tag{28}$$

is written in the form

$$\sum_{j=1}^{n} a_{ij}x_j = b_i \quad \text{or} \quad A\mathbf{x} = \mathbf{b}, \tag{29}$$

by defining the matrix of coefficients and column vectors for the unknowns and right-hand side,

$$A = \begin{pmatrix} a_{11} & a_{12} & \cdots & a_{1n} \\ a_{21} & a_{22} & \cdots & a_{2n} \\ \cdot & \cdot & \cdots & \cdot \\ \cdot & \cdot & \cdots & \cdot \\ \cdot & \cdot & \cdots & \cdot \\ a_{m1} & a_{m2} & \cdots & a_{mn} \end{pmatrix} \quad \mathbf{x} = \begin{pmatrix} x_1 \\ x_2 \\ \cdot \\ \cdot \\ x_n \end{pmatrix} \quad \mathbf{b} = \begin{pmatrix} b_1 \\ b_2 \\ \cdot \\ \cdot \\ \cdot \\ b_m \end{pmatrix}. \tag{30}$$

The coefficient matrix A is $m \times n$, because there is one row for each equation, and one column for each unknown.

The $A\mathbf{x} = \mathbf{b}$ form illustrates that whether a system of equations can be solved depends on the matrix A. A system of equations is called *homogeneous* in the special case that $\mathbf{b} = \mathbf{0}$, and *inhomogeneous* for all other cases in which $\mathbf{b} \neq \mathbf{0}$. We consider here only systems where the number of unknowns and equations are equal, so the coefficient matrix A is square. If A possesses an inverse, both sides can be premultiplied by A^{-1}, and

$$A^{-1}A\mathbf{x} = A^{-1}\mathbf{b} = I\mathbf{x} = \mathbf{x} \tag{31}$$

yields a unique solution vector $\mathbf{x}$. For inhomogeneous systems, computing A^{-1} provides a straightforward manner of solving for the unknown variables x_i. For homogeneous systems of equations, the equation shows that $\mathbf{x} = 0$ if A^{-1} exists. Thus, for a homogeneous system to have a nonzero or *nontrivial* solution, A must be singular. This occurs if the determinant of A is zero, implying that some of the rows (or columns) of A are not linearly independent. If a nontrivial solution of the homogeneous system exists, any constant times that solution is also a solution.

If the coefficient matrix is singular, the corresponding inhomogeneous system of equations does not have unique solutions, and may have none. The existence of A^{-1} and the solvability of the equations thus depend on whether the rows and columns of A are linearly independent. For example, if the rows are linearly dependent, there are fewer independent equations than unknowns and difficulties result, as discussed in the context of inverse problems (Chapter 7).

A.4.5 *Solving systems of equations on a computer*

Standard methods exist to solve linear equations on a computer. Consider the basic problem

$$A\mathbf{x} = \mathbf{b} \qquad (32)$$

$$\begin{pmatrix} a_{11} & a_{12} & a_{13} \\ a_{21} & a_{22} & a_{23} \\ a_{31} & a_{32} & a_{33} \end{pmatrix} \begin{pmatrix} x_1 \\ x_2 \\ x_3 \end{pmatrix} = \begin{pmatrix} b_1 \\ b_2 \\ b_3 \end{pmatrix}$$

in which we solve for $\mathbf{x}$, given A and $\mathbf{b}$. If A were a *triangular* matrix T, with zeroes below the diagonal, it would be easy to solve the system

$$T\mathbf{x} = \mathbf{d} \qquad (33)$$

$$\begin{pmatrix} t_{11} & t_{12} & t_{13} \\ 0 & t_{22} & t_{23} \\ 0 & 0 & t_{33} \end{pmatrix} \begin{pmatrix} x_1 \\ x_2 \\ x_3 \end{pmatrix} = \begin{pmatrix} d_1 \\ d_2 \\ d_3 \end{pmatrix}$$

by starting with the simplest (bottom) equation, solving for x_3, and solving the other equations in succession to find x_2 and then x_1. In other words, the solution

$$x_3 = d_3/t_{33} \qquad (34)$$

can be substituted into the middle equation to find

$$x_2 = (d_2 - t_{23}x_3)/t_{22}. \qquad (35)$$

Then, by substituting x_3 and x_2 into the first equation,

$$x_1 = (d_1 - t_{13}x_3 - t_{12}x_2)/t_{11}. \qquad (36)$$

The importance of this idea is that an arbitrary matrix can be triangularized. Consider that the solution of the system of equations is not changed by any of the following *elementary row operations*:

(i) Rearranging the equations, which corresponds to interchanging rows in the $\mathbf{b}$ vector and matrix, i.e.,

$$\begin{pmatrix} a_{11} & a_{12} & a_{13} \\ a_{31} & a_{32} & a_{33} \\ a_{21} & a_{22} & a_{23} \end{pmatrix} \begin{pmatrix} x_1 \\ x_2 \\ x_3 \end{pmatrix} = \begin{pmatrix} b_1 \\ b_3 \\ b_2 \end{pmatrix}. \qquad (37)$$

The solution is unchanged because the order of the equations is arbitrary.

(ii) Multiplying an equation by a constant c, which corresponds to multiplying a row of A and the corresponding element of $\mathbf{b}$ by a constant, i.e.,

$$\begin{pmatrix} ca_{11} & ca_{12} & ca_{13} \\ a_{21} & a_{22} & a_{23} \\ a_{31} & a_{32} & a_{33} \end{pmatrix} \begin{pmatrix} x_1 \\ x_2 \\ x_3 \end{pmatrix} = \begin{pmatrix} cb_1 \\ b_2 \\ b_3 \end{pmatrix}. \qquad (38)$$

(iii) Adding two equations, which corresponds to adding a multiple of one row to another, i.e.,

$$\begin{pmatrix} ca_{11} + a_{21} & ca_{12} + a_{22} & ca_{13} + a_{23} \\ a_{21} & a_{22} & a_{23} \\ a_{31} & a_{32} & a_{33} \end{pmatrix} \begin{pmatrix} x_1 \\ x_2 \\ x_3 \end{pmatrix} = \begin{pmatrix} cb_1 + b_2 \\ b_2 \\ b_3 \end{pmatrix}. \qquad (39)$$

Thus if the system $A\mathbf{x} = \mathbf{b}$ is transformed into $T\mathbf{x} = \mathbf{d}$ using elementary row operations, the two systems of equations have the same solutions $\mathbf{x}$. This provides a fast method of solving the system: combine A and $\mathbf{b}$ into a single *augmented matrix*

$$(A, \mathbf{b}) = \begin{pmatrix} a_{11} & a_{12} & a_{13} & b_1 \\ a_{21} & a_{22} & a_{23} & b_2 \\ a_{31} & a_{32} & a_{33} & b_3 \end{pmatrix} \qquad (40)$$

and triangularize the augmented matrix to obtain

$$(T, \mathbf{d}) = \begin{pmatrix} t_{11} & t_{12} & t_{13} & d_1 \\ 0 & t_{22} & t_{23} & d_2 \\ 0 & 0 & t_{33} & d_3 \end{pmatrix}, \qquad (41)$$

which represents a set of equations easily solved for $\mathbf{x}$ by the method in Eqns 34–6.

The matrix is triangularized using the following method column by column:

- Find the element of maximum absolute value in the column on or below the diagonal.
- If this "pivot" element is below the diagonal, interchange rows to get it on the diagonal.

- Subtract multiples of the pivot row from rows below it to get zeroes below the diagonal.

The pivoting, though not absolutely necessary, avoids possible numerical difficulties. Note that once a column is zeroed below the diagonal, we do not have to think about it any more.

For an illustration of this method, called *Gaussian elimination with partial pivoting*, consider solving the system of equations

$$\begin{aligned} x_1+x_2&=5,\\ 4x_1+x_2+x_3&=4,\\ 2x_1+2x_2+2x_3&=3. \end{aligned} \tag{42}$$

This can be expressed in matrix form as

$$\begin{pmatrix} 1 & 1 & 0 \\ 4 & 1 & 1 \\ 2 & 2 & 2 \end{pmatrix}\begin{pmatrix} x_1 \\ x_2 \\ x_3 \end{pmatrix}=\begin{pmatrix} 5 \\ 4 \\ 3 \end{pmatrix}, \tag{43}$$

and solved by triangularizing the augmented matrix

$$\begin{pmatrix} 1 & 1 & 0 & 5 \\ 4 & 1 & 1 & 4 \\ 2 & 2 & 2 & 3 \end{pmatrix}. \tag{44}$$

To get zeroes below the diagonal in the first column, we first move 4, the element with the largest absolute value in the first column, to the diagonal by interchanging rows

$$\begin{pmatrix} 4 & 1 & 1 & 4 \\ 1 & 1 & 0 & 5 \\ 2 & 2 & 2 & 3 \end{pmatrix}. \tag{45}$$

We then subtract 1/4 times the first row from second, and 1/2 times the first row from third, leaving

$$\begin{pmatrix} 4 & 1 & 1 & 4 \\ 0 & 0.75 & -0.25 & 4 \\ 0 & 1.5 & 1.5 & 1 \end{pmatrix}. \tag{46}$$

Next, to zero the elements below the diagonal in the second column, we interchange rows to get the pivot for this column, 1.5, on the diagonal:

$$\begin{pmatrix} 4 & 1 & 1 & 4 \\ 0 & 1.5 & 1.5 & 1 \\ 0 & 0.75 & -0.25 & 4 \end{pmatrix} \tag{47}$$

and subtract $0.75/1.5=0.5$ times the second row from the third

$$\begin{pmatrix} 4 & 1 & 1 & 4 \\ 0 & 1.5 & 1.5 & 1 \\ 0 & 0 & -1 & 3.5 \end{pmatrix} \tag{48}$$

to complete the triangularization. We then solve the equations for $\mathbf{x}$, beginning with the bottom one, as in Eqns 34–6.

A similar procedure can be used to invert a matrix. This method uses the idea that two vector–matrix equations

$$A\mathbf{x}=\mathbf{b} \quad \text{and} \quad A\mathbf{y}=\mathbf{c} \tag{49}$$

can be combined into one by forming an augmented matrix from each pair of vectors,

$$X=(\mathbf{x},\mathbf{y}), \quad B=(\mathbf{b},\mathbf{c}), \tag{50}$$

and writing the matrix equation

$$AX=B. \tag{51}$$

Because $\mathbf{x}$, the solution to $A\mathbf{x}=\mathbf{b}$, is not changed by elementary row operations on the augmented matrix $(A, \mathbf{b})$, the corresponding solution to $AX = B$ is unaffected by elementary row operations on the augmented matrix (A, B).

To apply this to matrix inversion, consider a special case

$$AX=I, \tag{52}$$

whose solution $X = A^{-1}$ is the inverse of the $n\times n$ matrix A. X is unaffected by elementary row operations that convert the augmented matrix

$$(A, I)=\begin{pmatrix} a_{11} & \cdot & \cdot & a_{1n} & 1 & \cdot & \cdot & 0 \\ \cdot & \cdot & \cdot & \cdot & \cdot & \cdot & \cdot & \cdot \\ \cdot & \cdot & \cdot & \cdot & \cdot & \cdot & \cdot & \cdot \\ a_{n1} & \cdot & \cdot & a_{nn} & 0 & \cdot & \cdot & 1 \end{pmatrix} \tag{53}$$

to one whose left side is the identity

$$(I, B)=\begin{pmatrix} 1 & \cdot & \cdot & 0 & b_{11} & \cdot & \cdot & b_{1n} \\ \cdot & \cdot & \cdot & \cdot & \cdot & \cdot & \cdot & \cdot \\ \cdot & \cdot & \cdot & \cdot & \cdot & \cdot & \cdot & \cdot \\ 0 & \cdot & \cdot & 1 & b_{n1} & \cdot & \cdot & b_{nn} \end{pmatrix}, \tag{54}$$

so the corresponding equation

$$IX=B \tag{55}$$

shows that the right side of the matrix gives $B=X=A^{-1}$, the inverse of A. The sequence of operations used to diagonalize the left (A) side of the augmented matrix (A, I) are similar to those that triangularize a matrix.

A.5 Vector transformations

In seismology, we often apply two types of transformations to vectors. In the first, the same vector is expressed in two

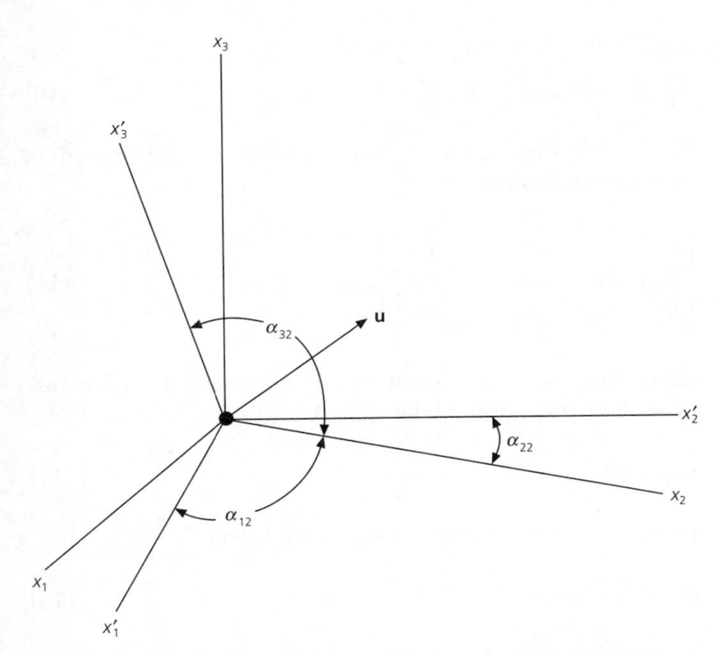

Fig. A.5-1 The relation between two orthogonal coordinate systems with the same origin is described by the angles α_{ij} between the two sets of axes.

different coordinate systems. In the second, some operation converts a vector to another vector expressed in the same coordinate system. In this section we summarize these transformations and their differences.

A.5.1 *Coordinate transformations*

We have seen that vectors remain the same regardless of the coordinate system in which they are defined, although their components differ between coordinate systems. Thus vectors can be defined in one coordinate system (for example, one oriented along an earthquake fault plane) and reexpressed in another (such as a geographic coordinate system). This property is very useful for solving problems and gives valuable insight into the nature of vectors.

To define the relation between vector components and coordinate systems, consider two orthogonal Cartesian coordinate systems (Fig. A.5-1). Because the origins are the same, one coordinate system can be obtained by rotating the other through three angles. The relation between the two sets of unit basis vectors, $\hat{e}_1, \hat{e}_2, \hat{e}_3$ and $\hat{e}'_1, \hat{e}'_2, \hat{e}'_3$, is given by their scalar products, called *direction cosines*,

$$\hat{e}'_i \cdot \hat{e}_j = \cos \alpha_{ij} = a_{ij}, \tag{1}$$

where the angles α_{ij} are the angles between the two sets of axes.

A vector can be expressed in terms of its components in the two coordinate systems

$$\mathbf{u} = u_1\hat{e}_1 + u_2\hat{e}_2 + u_3\hat{e}_3 = u'_1\hat{e}'_1 + u'_2\hat{e}'_2 + u'_3\hat{e}'_3. \tag{2}$$

Given the components u_i in the unprimed system, the components u'_i in the primed system are found by taking the scalar products of the vector with the basis vectors of the primed system:

$$\begin{aligned} u'_1 &= \hat{e}'_1 \cdot \mathbf{u} = (\hat{e}'_1 \cdot \hat{e}_1)u_1 + (\hat{e}'_1 \cdot \hat{e}_2)u_2 + (\hat{e}'_1 \cdot \hat{e}_3)u_3 \\ &= a_{11}u_1 + a_{12}u_2 + a_{13}u_3, \\ u'_2 &= \hat{e}'_2 \cdot \mathbf{u} = a_{21}u_1 + a_{22}u_2 + a_{23}u_3, \\ u'_3 &= \hat{e}'_3 \cdot \mathbf{u} = a_{31}u_1 + a_{32}u_2 + a_{33}u_3. \end{aligned} \tag{3}$$

These can be written as a matrix equation

$$\mathbf{u}' = A\mathbf{u}, \quad \text{or} \quad \begin{pmatrix} u'_1 \\ u'_2 \\ u'_3 \end{pmatrix} = \begin{pmatrix} a_{11} & a_{12} & a_{13} \\ a_{21} & a_{22} & a_{23} \\ a_{31} & a_{32} & a_{33} \end{pmatrix} \begin{pmatrix} u_1 \\ u_2 \\ u_3 \end{pmatrix}, \tag{4}$$

where A is the matrix that transforms a vector from the unprimed to the primed system. Note that this is not a relation between two different vectors $\mathbf{u}$ and $\mathbf{u}'$ — it is a relationship between the *components* of the *same* vector in two coordinate systems. It turns out that the matrix A uniquely describes the transformation between these coordinate systems.

For example, a unit basis vector for the unprimed system

$$\hat{e}_1 = 1\hat{e}_1 + 0\hat{e}_2 + 0\hat{e}_3 = (1, 0, 0) \tag{5}$$

has components in the primed system given by

$$\begin{pmatrix} a_{11} \\ a_{21} \\ a_{31} \end{pmatrix} = \begin{pmatrix} a_{11} & a_{12} & a_{13} \\ a_{21} & a_{22} & a_{23} \\ a_{31} & a_{32} & a_{33} \end{pmatrix} \begin{pmatrix} 1 \\ 0 \\ 0 \end{pmatrix}, \tag{6}$$

and so is written

$$a_{11}\hat{e}'_1 + a_{21}\hat{e}'_2 + a_{31}\hat{e}'_3 = (a_{11}, a_{21}, a_{31}) \tag{7}$$

in the primed system. The last expression is just the first column of A. Similarly, the components of $\hat{e}_2$ and $\hat{e}_3$ in the primed system are the second and third columns of A, respectively. Thus the columns of the transformation matrix A are the basis vectors of the unprimed system written in terms of their components in the primed system.

For example, consider rotating a Cartesian coordinate system by θ counterclockwise about the $\hat{e}_3$ axis, so that the only rotation occurs in the $\hat{e}_1$–$\hat{e}_2$ plane. The $\hat{e}_3$ axis is also the $\hat{e}'_3$ axis (Fig. A.5-2). The elements of the transformation matrix are found by evaluating the scalar products of the basis vectors $a_{ij} = \hat{e}'_i \cdot \hat{e}_j$, so

$$\begin{aligned} &a_{11} = \hat{e}'_1 \cdot \hat{e}_1 = \cos\theta, && a_{12} = \hat{e}'_1 \cdot \hat{e}_2 = \cos(90° - \theta) = \sin\theta, \\ &a_{22} = \hat{e}'_2 \cdot \hat{e}_2 = \cos\theta, && a_{21} = \hat{e}'_2 \cdot \hat{e}_1 = \cos(90° + \theta) = -\sin\theta, \\ &a_{33} = \hat{e}'_3 \cdot \hat{e}_3 = 1, && a_{13} = a_{23} = a_{31} = a_{32} = 0, \end{aligned} \tag{8}$$

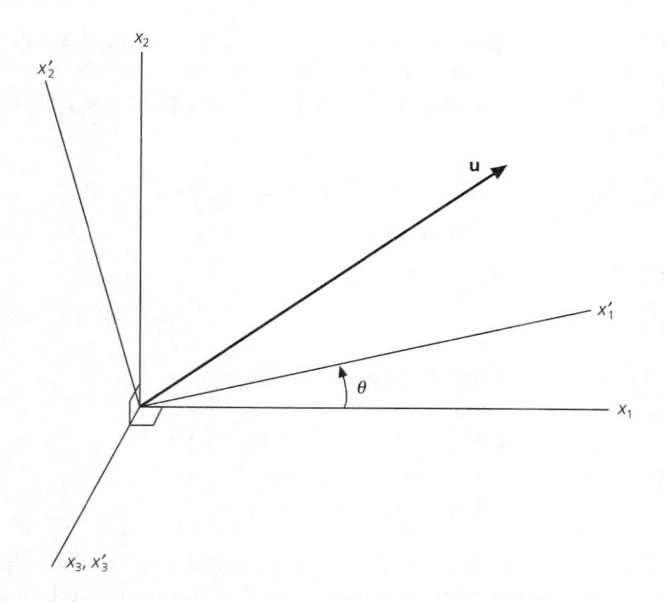

Fig. A.5-2 The relation between the axes of two orthogonal coordinate systems differing by a rotation θ in the x_1–x_2 plane.

and the components of a vector in the two systems are related by

$$\begin{pmatrix} u'_1 \\ u'_2 \\ u'_3 \end{pmatrix} = \begin{pmatrix} \cos\theta & \sin\theta & 0 \\ -\sin\theta & \cos\theta & 0 \\ 0 & 0 & 1 \end{pmatrix} \begin{pmatrix} u_1 \\ u_2 \\ u_3 \end{pmatrix}. \qquad (9)$$

Thus the $\hat{e}_1$ and $\hat{e}'_1$, and the $\hat{e}_2$ and $\hat{e}'_2$ components differ, whereas the $\hat{e}_3$ and $\hat{e}'_3$ components are the same. To check this, consider the case where $\theta = 90°$. As expected, (1, 0, 0) in the unprimed system becomes (0, −1, 0) in the primed system, and (0, 1, 0) in the unprimed system becomes (1, 0, 0) in the primed system, while (0, 0, 1) in the unprimed system remains (0, 0, 1) in the primed system.

Seismologists often use such a geometry. Because the ground motion is a vector, seismometers are generally oriented to record its components in the east–west, north–south, and up–down directions. This decomposition is less useful than decomposing ground motion into its *radial* and *transverse* components, those along and perpendicular to the great circle connecting the earthquake and seismometer. The vertical component is useful as is, so a rotation about the vertical by the angle between East and the great circle connecting the earthquake and seismometer converts the E–W and N–S components into the new representation. The relevant angle, the *back azimuth* to the source from the receiver, is discussed in Section A.7.2.

We can also reverse the transformation. By analogy to Eqn 3, the components in the unprimed system can be found from those in the primed system as

$$u_1 = \hat{e}_1 \cdot \mathbf{u}' = (\hat{e}_1 \cdot \hat{e}'_1)u'_1 + (\hat{e}_1 \cdot \hat{e}'_2)u'_2 + (\hat{e}_1 \cdot \hat{e}'_3)u'_3$$
$$= a_{11}u'_1 + a_{21}u'_2 + a_{31}u'_3,$$

$$u_2 = \hat{e}_2 \cdot \mathbf{u}' = a_{12}u'_1 + a_{22}u'_2 + a_{32}u'_3,$$
$$u_3 = \hat{e}_3 \cdot \mathbf{u}' = a_{13}u'_1 + a_{23}u'_2 + a_{33}u'_3. \qquad (10)$$

Combining these to express the reverse transformation in vector–matrix form,

$$\begin{pmatrix} u_1 \\ u_2 \\ u_3 \end{pmatrix} = \begin{pmatrix} a_{11} & a_{21} & a_{31} \\ a_{12} & a_{22} & a_{32} \\ a_{13} & a_{23} & a_{33} \end{pmatrix} \begin{pmatrix} u'_1 \\ u'_2 \\ u'_3 \end{pmatrix}, \qquad (11)$$

shows that the reverse transformation matrix is just the transpose of the transformation matrix A

$$\mathbf{u} = A^T\mathbf{u}'. \qquad (12)$$

Hence a unit basis vector in the primed system

$$\hat{e}'_1 = 1\hat{e}'_1 + 0\hat{e}'_2 + 0\hat{e}'_3 \qquad (13)$$

becomes, by the matrix transformation,

$$a_{11}\hat{e}_1 + a_{12}\hat{e}_2 + a_{13}\hat{e}_3 \qquad (14)$$

in the unprimed system. This is the first row of A, so the rows of A are the primed basis vectors expressed in the unprimed coordinates. This is natural because the transformations are related by the matrix transpose.

Alternatively, the reverse transformation can be found directly by starting with $\mathbf{u}' = A\mathbf{u}$ and multiplying both sides by the inverse matrix

$$A^{-1}\mathbf{u}' = A^{-1}A\mathbf{u} = \mathbf{I}\mathbf{u} = \mathbf{u}. \qquad (15)$$

Comparison with Eqn 12 shows that the inverse of the transformation matrix equals its transpose, so the transformation matrix is an orthogonal matrix. This seems reasonable because the columns of A, which represent orthogonal basis vectors, are orthogonal. Similarly, the rows of A are orthogonal. As a result, such coordinate transformations are called *orthogonal transformations*. An important feature of orthogonal transformations, whose proof is left as a homework problem, is that they preserve the length of vectors.

The transformation relations, Eqns 4 and 12, provide a mathematical definition of a vector. Any vector must transform between coordinate systems in this way. A set of three entities defined at points in space (for example, temperature, pressure, and density) that does not obey the transformation equations is not a vector.

A.5.2 *Eigenvalues and eigenvectors*

The product of an arbitrary $n \times n$ matrix A and an arbitrary n-component vector $\mathbf{x}$

$$\mathbf{y} = A\mathbf{x} \tag{16}$$

is also a vector in n dimensions. This is not the same as coordinate transformation; the vector $\mathbf{x}$ is transformed into another distinct vector, with both vectors expressed in the same coordinate system.

A physically important class of transformations convert a vector into one parallel to the original vector, so that

$$A\mathbf{x} = \lambda\mathbf{x}, \tag{17}$$

where A is a matrix, and λ is a scalar. The only effect of the transformation is that the length of $\mathbf{x}$ changes by a factor of λ. For a given A, it is useful to know which vectors $\mathbf{x}$ and scalars λ satisfy this equation.

In three dimensions, the case most commonly encountered, Eqn 17 can be written

$$(A - \lambda I)\mathbf{x} = 0$$

$$\begin{pmatrix} a_{11} - \lambda & a_{12} & a_{13} \\ a_{21} & a_{22} - \lambda & a_{23} \\ a_{31} & a_{32} & a_{33} - \lambda \end{pmatrix} \begin{pmatrix} x_1 \\ x_2 \\ x_3 \end{pmatrix} = \begin{pmatrix} 0 \\ 0 \\ 0 \end{pmatrix}. \tag{18}$$

This is a homogeneous system of linear equations, so nontrivial solutions exist only if the matrix $(A - \lambda I)$ is singular. We thus seek values of λ such that the determinant

$$|(A - \lambda I)| = \det \begin{pmatrix} a_{11} - \lambda & a_{12} & a_{13} \\ a_{21} & a_{22} - \lambda & a_{23} \\ a_{31} & a_{32} & a_{33} - \lambda \end{pmatrix} = 0. \tag{19}$$

Evaluating the determinant gives the *characteristic polynomial*

$$\lambda^3 - I_1\lambda^2 + I_2\lambda - I_3 = 0, \tag{20}$$

which depends on three constants called the *invariants* of A:

$$I_1 = a_{11} + a_{22} + a_{33},$$

$$I_2 = \det\begin{pmatrix} a_{11} & a_{12} \\ a_{21} & a_{22} \end{pmatrix} + \det\begin{pmatrix} a_{22} & a_{23} \\ a_{32} & a_{33} \end{pmatrix} + \det\begin{pmatrix} a_{11} & a_{13} \\ a_{31} & a_{33} \end{pmatrix},$$

$$I_3 = \det A. \tag{21}$$

I_1, the first invariant, or *trace*, of A, is the sum of the diagonal elements of A. The invariants of a matrix have significance for stresses, strains, and earthquake moment tensors, because they are not changed by orthogonal transformations.

The characteristic polynomial is a cubic equation in λ with three roots, or *eigenvalues*, λ_m for which the determinant $|A - \lambda I|$ is zero. For each eigenvalue there is an associated nontrivial *eigenvector*, $\mathbf{x}^{(m)}$, satisfying

$$A\mathbf{x}^{(m)} = \lambda_m \mathbf{x}^{(m)}. \tag{22}$$

The components of the eigenvector, $x_1^{(m)}$, $x_2^{(m)}$, $x_3^{(m)}$, are found by solving

$$\begin{pmatrix} a_{11} - \lambda_m & a_{12} & a_{13} \\ a_{21} & a_{22} - \lambda_m & a_{23} \\ a_{31} & a_{32} & a_{33} - \lambda_m \end{pmatrix} \begin{pmatrix} x_1^{(m)} \\ x_2^{(m)} \\ x_3^{(m)} \end{pmatrix} = \begin{pmatrix} 0 \\ 0 \\ 0 \end{pmatrix}. \tag{23}$$

Each eigenvalue and its associated eigenvector form a pair satisfying Eqn 22. In general, an eigenvalue and the eigenvector associated with a different eigenvalue will not satisfy the equation.

For example, the eigenvalues of

$$A = \begin{pmatrix} 3 & -1 & 0 \\ -1 & 2 & -1 \\ 0 & -1 & 3 \end{pmatrix} \tag{24}$$

are found by solving the characteristic polynomial

$$\lambda^3 - 8\lambda^2 + 19\lambda - 12 = 0, \tag{25}$$

whose roots are $\lambda_1 = 4$, $\lambda_2 = 3$, $\lambda_3 = 1$. Next, the equations

$$\begin{pmatrix} 3 - \lambda_m & -1 & 0 \\ -1 & 2 - \lambda_m & -1 \\ 0 & -1 & 3 - \lambda_m \end{pmatrix} \begin{pmatrix} x_1^{(m)} \\ x_2^{(m)} \\ x_3^{(m)} \end{pmatrix} = \begin{pmatrix} 0 \\ 0 \\ 0 \end{pmatrix} \tag{26}$$

are solved for each eigenvalue to yield the associated eigenvector. Thus for $\lambda_3 = 1$,

$$\begin{aligned} 2x_1^{(3)} - x_2^{(3)} &= 0, \\ -x_1^{(3)} + x_2^{(3)} - x_3^{(3)} &= 0, \\ -x_2^{(3)} + 2x_3^{(3)} &= 0. \end{aligned} \tag{27}$$

All three unknowns cannot be found uniquely, because these are homogeneous equations. We thus set $x_1^{(3)}$ equal to 1 and find the other two unknowns, $x_2^{(3)} = 2$, $x_3^{(3)} = 1$. Similarly, the other eigenvectors are found by substituting λ_2 and λ_1 in Eqn 26, so

$$\mathbf{x}^{(3)} = (1, 2, 1), \quad \mathbf{x}^{(2)} = (1, 0, -1), \quad \mathbf{x}^{(1)} = (1, -1, 1). \tag{28}$$

Because the eigenvectors are solutions to a set of homogeneous equations, any multiple of an eigenvector is also an eigenvector. The eigenvectors thus determine a direction in space, but the magnitude of the vector is arbitrary. Often the eigenvectors are normalized to unit magnitude. The set we have found can be written as

$$\begin{aligned} &\mathbf{x}^{(1)} = (1/\sqrt{3}, -1/\sqrt{3}, 1/\sqrt{3}), \quad \mathbf{x}^{(2)} = (1/\sqrt{2}, 0, -1/\sqrt{2}), \\ &\mathbf{x}^{(3)} = (1/\sqrt{6}, 2/\sqrt{6}, 1/\sqrt{6}). \end{aligned} \tag{29}$$

Sometimes complications arise, as for the matrix

$$A = \begin{pmatrix} 1 & 0 & 0 \\ 0 & 0 & 0 \\ 0 & 0 & 1 \end{pmatrix} \tag{30}$$

with eigenvalues 1, 1, and 0. Using the method given above to find the eigenvector for $\lambda_3 = 0$ by setting $x_1^{(3)} = 1$ yields no solution. Setting $x_2^{(3)} = 1$, however, yields a correct solution for the eigenvector, (0, 1, 0). Because this has no $\hat{\mathbf{e}}_1$ component, we could not have set $x_1^{(3)} = 1$ and found the other components.

This example illustrates a complication that arises for a *degenerate*, or repeated, eigenvalue: e.g., $\lambda_1 = \lambda_2 = 1$. In this case, the eigenvalue corresponds not to an eigenvector but to an entire plane, and any vector contained within it is an eigenvector. Two eigenvectors spanning this plane can be found by finding the eigenvector of the nondegenerate eigenvalue, and then choosing two independent vectors orthogonal to it. Because the eigenvector for the nondegenerate eigenvalue is (0, 1, 0), two possible orthogonal eigenvectors for the degenerate eigenvalue are (1, 0, 0) and (0, 0, 1).

A.5.3 *Symmetric matrix eigenvalues, eigenvectors, diagonalization, and decomposition*

The eigenvalues and eigenvectors of a symmetric matrix have interesting properties. An $n \times n$ matrix H has a characteristic polynomial of degree n, each of whose n roots is an eigenvalue. Consider two eigenvalues and their associated eigenvectors

$$H\mathbf{x}^{(i)} = \lambda_i \mathbf{x}^{(i)}, \quad H\mathbf{x}^{(j)} = \lambda_j \mathbf{x}^{(j)}. \tag{31}$$

Multiplication of the first equation by $\mathbf{x}^{(j)T}$ (the transpose of $\mathbf{x}^{(j)}$) and the second equation by $\mathbf{x}^{(i)T}$ yields

$$\mathbf{x}^{(j)T} H\mathbf{x}^{(i)} = \lambda_i \mathbf{x}^{(j)T}\mathbf{x}^{(i)}, \quad \mathbf{x}^{(i)T} H\mathbf{x}^{(j)} = \lambda_j \mathbf{x}^{(i)T}\mathbf{x}^{(j)}. \tag{32}$$

Transposing both sides of the second part of Eqn 32 and subtracting it from the first gives

$$\mathbf{x}^{(j)T} H\mathbf{x}^{(i)} - \mathbf{x}^{(j)T} H^T\mathbf{x}^{(i)} = (\lambda_i - \lambda_j)\mathbf{x}^{(j)T}\mathbf{x}^{(i)}. \tag{33}$$

Because H is symmetric, it equals its transpose, $H = H^T$, so the left-hand side is zero

$$0 = (\lambda_i - \lambda_j)\mathbf{x}^{(j)T}\mathbf{x}^{(i)}. \tag{34}$$

Thus, if $i \neq j$ and the two eigenvalues are different, their associated eigenvectors must be orthogonal so that their scalar product $\mathbf{x}^{(j)T}\mathbf{x}^{(i)}$ is zero. Thus, for a symmetric matrix, eigenvectors associated with distinct eigenvalues are orthogonal.

This result lets us diagonalize a symmetric matrix. To illustrate this for a 3×3 case, consider a matrix U whose columns are the eigenvectors of the symmetric matrix H

$$U = \begin{pmatrix} x_1^{(1)} & x_1^{(2)} & x_1^{(3)} \\ x_2^{(1)} & x_2^{(2)} & x_2^{(3)} \\ x_3^{(1)} & x_3^{(2)} & x_3^{(3)} \end{pmatrix}. \tag{35}$$

If the eigenvalues of H are distinct, the eigenvectors of H, and hence the columns of the eigenvector matrix, are orthogonal, so U is an orthogonal matrix satisfying $U^{-1} = U^T$.

The entire set of eigenvalue–eigenvector pairs, each of which satisfy $H\mathbf{x}^{(i)} = \lambda_i \mathbf{x}^{(i)}$, can be written as the matrix equation

$$HU = U\Lambda, \tag{36}$$

where Λ is the diagonal matrix with eigenvalues on the diagonal

$$\Lambda = \begin{pmatrix} \lambda_1 & 0 & 0 \\ 0 & \lambda_2 & 0 \\ 0 & 0 & \lambda_3 \end{pmatrix}. \tag{37}$$

Premultiplying both sides of Eqn 36 by the inverse of the eigenvector matrix yields

$$U^{-1}HU = U^THU = \Lambda, \tag{38}$$

which shows how the eigenvector matrix can be used to diagonalize a symmetric matrix. This result can also be stated as

$$H = U\Lambda U^T, \tag{39}$$

which illustrates how a symmetric matrix can be decomposed into a diagonal eigenvalue matrix and the orthogonal eigenvector matrix. Similar results apply for complex Hermitian matrices.

We will see that if a matrix contains the components of vectors expressed in a coordinate system, the physical problem under discussion can be simplified by diagonalizing the matrix. This corresponds to rewriting the problem in its "natural" coordinate system, whose basis set is the eigenvectors, an idea used in discussing stresses in the earth (Section 2.3.4) and the seismic moment tensor (Section 4.4.5).

A.6 Vector calculus

A.6.1 *Scalar and vector fields*

Many phenomena in seismology depend on how physical quantities vary in space. Some, like density or temperature, are *scalar fields*, scalar valued functions of the position vector $\mathbf{x}$ denoted by expressions like $\phi(\mathbf{x})$ or $\phi(x_1, x_2, x_3)$. Similarly, a vector that varies in space is described by a *vector field*. For example, seismic waves are described by the variation in the displacement vector

$$\mathbf{u}(\mathbf{x}) = \mathbf{u}(x_1, x_2, x_3)$$
$$= u_1(x_1, x_2, x_3)\hat{\mathbf{e}}_1 + u_2(x_1, x_2, x_3)\hat{\mathbf{e}}_2 + u_3(x_1, x_2, x_3)\hat{\mathbf{e}}_3 \quad (1)$$

as a function of position, and result in turn from forces derived from spatial derivatives of the stress tensor.

Spatial variations of scalar, vector, or tensor fields are described using the vector differential operator "del", $\mathbf{\nabla}$,

$$\mathbf{\nabla} = \left(\hat{\mathbf{e}}_1 \frac{\partial}{\partial x_1}, \hat{\mathbf{e}}_2 \frac{\partial}{\partial x_2}, \hat{\mathbf{e}}_3 \frac{\partial}{\partial x_3} \right). \quad (2)$$

This operator has the form of a vector, but has meaning only when applied to a scalar, vector, or tensor field. We first review uses of the $\mathbf{\nabla}$ operator in Cartesian coordinates, and in the next section discuss the more complicated forms for spherical coordinates.

A.6.2 *Gradient*

The simplest application of the $\mathbf{\nabla}$ operator is the *gradient*, a vector field formed from the spatial derivatives of a scalar field. If $\phi(\mathbf{x})$ is a scalar function of position, the gradient is defined by

$$\text{grad } \phi(\mathbf{x}) = \mathbf{\nabla}\phi(\mathbf{x}) = \frac{\partial \phi(\mathbf{x})}{\partial x_1}\hat{\mathbf{e}}_1 + \frac{\partial \phi(\mathbf{x})}{\partial x_2}\hat{\mathbf{e}}_2 + \frac{\partial \phi(\mathbf{x})}{\partial x_3}\hat{\mathbf{e}}_3, \quad (3)$$

where $\partial \phi(\mathbf{x})/\partial x_1$ is the partial derivative of $\phi(x_1, x_2, x_3)$ with respect to x_1, for x_2 and x_3 held constant. The gradient is a vector field whose components equal the partial derivative with respect to the corresponding coordinate.

Expressions like Eqns 1 and 3 can be written more compactly if the dependences on position are not written explicitly, i.e.,

$$\mathbf{\nabla}\phi = \frac{\partial \phi}{\partial x_1}\hat{\mathbf{e}}_1 + \frac{\partial \phi}{\partial x_2}\hat{\mathbf{e}}_2 + \frac{\partial \phi}{\partial x_3}\hat{\mathbf{e}}_3. \quad (4)$$

In this notation, it is implicit that ϕ, its derivatives, and hence the gradient, vary with position.

For example, the elevation $\phi(x_1, x_2)$ is a scalar field describing the topography as a function of position in a two-dimensional region. This is often plotted using topographic contours (Fig. A.6-1), curves along which ϕ is constant. At any point, $\partial\phi/\partial x_1$ is the slope in the x_1 direction, and $\partial\phi/\partial x_2$ is the slope in the x_2 direction.

The gradient can be used to find the slope in any direction. The projection of a vector in a given direction is the scalar product of the vector and the unit normal vector in that direction, $\hat{\mathbf{n}} = (n_1, n_2)$. Thus the scalar product of the gradient with the normal vector,

$$\hat{\mathbf{n}} \cdot \mathbf{\nabla}\phi = n_1 \frac{\partial \phi}{\partial x_1} + n_2 \frac{\partial \phi}{\partial x_2}, \quad (5)$$

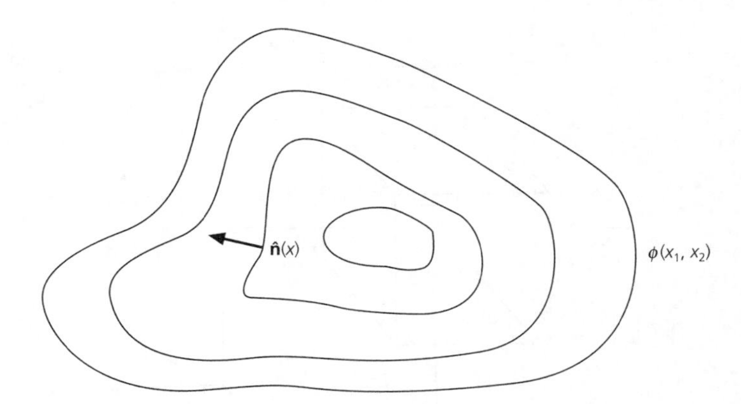

Fig. A.6-1 A scalar field demonstrating the concept of a gradient. If $\phi(x_1, x_2)$ gives the elevation, the gradient can be used to find the slope in the $\hat{\mathbf{n}}$ direction at a point (x_1, x_2).

gives the *directional derivative* in the $\hat{\mathbf{n}}$ direction. Because both $\hat{\mathbf{n}}$ and $\mathbf{\nabla}\phi$ are functions of position, the directional derivative varies in space. At any point, the maximum value of the scalar product occurs for $\hat{\mathbf{n}}$ parallel to the gradient, so the gradient points in the direction of the steepest slope along which ϕ changes most rapidly. The scalar product is zero when $\hat{\mathbf{n}}$ is perpendicular to the gradient, so the gradient is perpendicular to curves of constant ϕ. These concepts are also used in three dimensions.

In index notation, the gradient is written as

$$(\mathbf{\nabla}\phi)_i = \frac{\partial \phi}{\partial x_i} = \phi_{,i}, \quad (6)$$

where the last form uses a common (if sometimes confusing) notation in which differentiation is indicated by a comma. The notation, with one free index, shows that the gradient is a vector. By contrast, the directional derivative, written as

$$\hat{\mathbf{n}} \cdot \mathbf{\nabla}\phi = n_i \frac{\partial \phi}{\partial x_i} = n_i \phi_{,i}, \quad (7)$$

has an implied sum over i and is a scalar.

Often, the gradients of quantities are important physically because an effect depends on spatial variations of a field. For example, the flow of heat depends on the gradient of the temperature field (Sections 5.3.2, 5.4.1), and the gradient of the pressure field in the atmosphere is important for the weather.

A.6.3 *Divergence*

A related operation that describes the spatial variation of a vector field is the *divergence*. The divergence of a vector field $\mathbf{u}(\mathbf{x})$ is given by the scalar product of the $\mathbf{\nabla}$ operator with $\mathbf{u}(\mathbf{x})$ as

$$\text{div } \mathbf{u} = \mathbf{\nabla} \cdot \mathbf{u} = \frac{\partial u_1}{\partial x_1} + \frac{\partial u_2}{\partial x_2} + \frac{\partial u_3}{\partial x_3}, \quad (8)$$

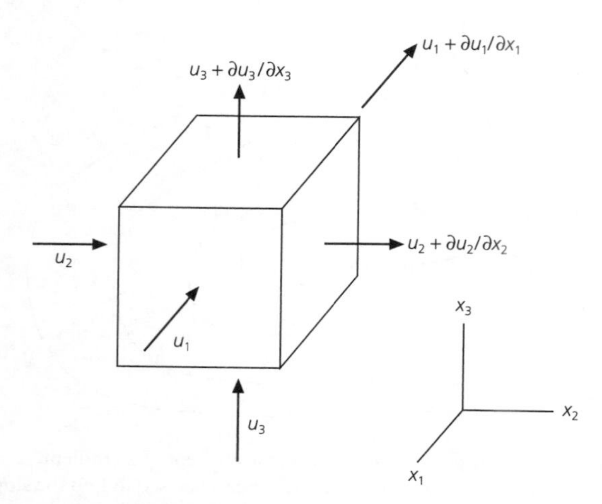

Fig. A.6-2 The divergence, formed from the differences between the flow into one face of a volume and the flow out of the opposite face, gives the net flow through a unit volume.

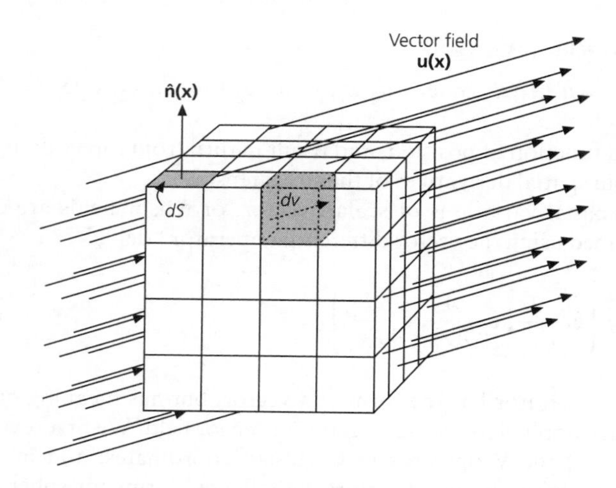

Fig. A.6-3 Geometry for the divergence theorem: $\hat{\mathbf{n}}(\mathbf{x})$ is a unit vector pointing outward at the point $\mathbf{x}$ from an element dS of the surface S that encloses a volume dV.

which yields a scalar field because the vector components and their derivatives are functions of position.

The divergence frequently arises in conservation equations. For example, if $\mathbf{u}(\mathbf{x})$ is the velocity as a function of position in a fluid, $\nabla \cdot \mathbf{u}(\mathbf{x})$ gives the net outflow of material per unit time from a unit volume at position $\mathbf{x}$ (Fig. A.6-2). To see this, note that, to first order, the net flow in the x_2 direction is the difference between the flow out the far side, $u_2 + \partial u_2/\partial x_2$, and that into the near side, u_2, given as

$$u_2 + \frac{\partial u_2}{\partial x_2} - u_2 = \frac{\partial u_2}{\partial x_2}. \tag{9}$$

Adding similar terms for the net flow in the x_1 and x_3 directions gives the divergence (Eqn 8). If the divergence is positive, there is a net outward flow, whereas a negative divergence indicates a net inflow.

This idea can be applied to any vector field $\mathbf{u}(\mathbf{x})$. Consider the problem of finding the net outflow from a region with volume V and surface S. If $\hat{\mathbf{n}}(\mathbf{x})$ is the unit normal vector pointing outward at a point $\mathbf{x}$ on the surface (Fig. A.6-3), the scalar product $\hat{\mathbf{n}}(\mathbf{x}) \cdot \mathbf{u}(\mathbf{x})$ gives the outward *flux* per unit area at that point. Integrating the flux over the surface then gives the total flux. Another way to compute the total flux is to integrate the divergence over the volume. These two methods give the same flux, so

$$\int_S \hat{\mathbf{n}} \cdot \mathbf{u} dS = \int_V \nabla \cdot \mathbf{u} dV. \tag{10}$$

This relation, *Gauss's theorem*, or the *divergence theorem*, says that what accumulates inside a volume is determined by the integral over its surface of what goes out. If we think of the volume as many adjacent cells, the flow out of one cell is the flow into an adjacent cell, which cancels to zero. Only flow in or out of the volume's surface is not canceled out in this way. Written in full, $\int dV$ is a triple integral over the volume, and $\int dS$ is a double integral over the surface.

In index notation, using the summation convention, the divergence is written

$$\nabla \cdot \mathbf{u} = \frac{\partial u_i}{\partial x_i} = u_{i,i}, \tag{11}$$

which is a scalar because no free index remains. Gauss's theorem is written

$$\int_S u_i n_i dS = \int_V \frac{\partial u_i}{\partial x_i} dV, \tag{12}$$

or, using the comma notation for derivatives,

$$\int_S u_i n_i dS = \int_V u_{i,i} dV. \tag{13}$$

As before, it is implicit in the notation that the field $\mathbf{u}$, its derivatives, and the normal vector $\hat{\mathbf{n}}$ vary with position.

A.6.4 Curl

The *curl* operator, the cross product of the ∇ operator with a vector field, yields another vector field

$$\nabla \times \mathbf{u} = \hat{\mathbf{e}}_1\left(\frac{\partial u_3}{\partial x_2} - \frac{\partial u_2}{\partial x_3}\right) + \hat{\mathbf{e}}_2\left(\frac{\partial u_1}{\partial x_3} - \frac{\partial u_3}{\partial x_1}\right) + \hat{\mathbf{e}}_3\left(\frac{\partial u_2}{\partial x_1} - \frac{\partial u_1}{\partial x_2}\right). \tag{14}$$

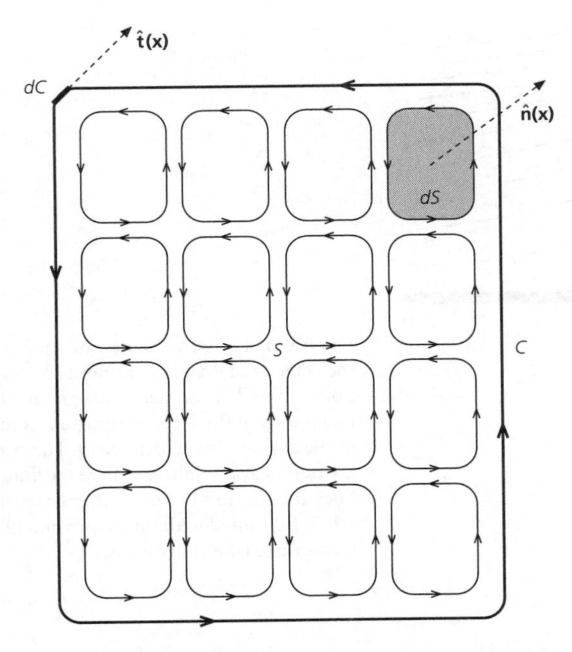

Fig. A.6-4 Geometry for Stokes' theorem: $\hat{\mathbf{n}}(\mathbf{x})$ is a unit vector pointing outward at the point $\mathbf{x}$ from an element dS of the surface S. dC is an element of the curve C bounding S, with tangent $\hat{\mathbf{t}}(\mathbf{x})$.

This can be written as a determinant

$$\nabla \times \mathbf{u} = \det \begin{pmatrix} \hat{\mathbf{e}}_1 & \hat{\mathbf{e}}_2 & \hat{\mathbf{e}}_3 \\ \dfrac{\partial}{\partial x_1} & \dfrac{\partial}{\partial x_2} & \dfrac{\partial}{\partial x_3} \\ u_1 & u_2 & u_3 \end{pmatrix}, \tag{15}$$

or, using index notation, in a compact form as

$$\nabla \times \mathbf{u} = \varepsilon_{ijk} \frac{\partial u_k}{\partial x_j} = \varepsilon_{ijk} u_{k,j}. \tag{16}$$

Some physical insight into the curl comes from *Stokes' theorem*, which relates the integral of the curl of a vector field over a surface S to the line integral around a curve C bounding S (Fig. A.6-4) as

$$\int_C \mathbf{u} \cdot \hat{\mathbf{t}} \, dC = \int_S (\nabla \times \mathbf{u}) \cdot \hat{\mathbf{n}} \, dS. \tag{17}$$

Here dS is an element of surface area with normal $\hat{\mathbf{n}}(\mathbf{x})$, and dC is an element of the curve with tangent $\hat{\mathbf{t}}(\mathbf{x})$. Analogous to the case of Gauss's theorem applied to a volume, we can think of the surface as composed of infinitesimal tiles, each with a line integral of $\mathbf{u} \cdot \hat{\mathbf{t}}$ around it. The border of each tile is shared with another tile, but, because the line integral, or *circulation*, is computed in a counterclockwise manner, the integrals along this border are the same but of opposite sign for the two tiles, and therefore cancel. The segments of the line integrals cancel between all the tiles except those on the outer border that have no adjacent circulation to cancel them.

If the line integral is nonzero, the vector field has a net rotation along the curve, so the integral of its curl over the surface is nonzero. The curl of a vector field shows where rotations arise. A common application is describing the velocity field of a moving fluid. The upper portion of Fig. A.6-5 shows streamlines, lines parallel to the velocity vector at any point, for a viscous fluid flowing past a circular object. The velocity is zero at the object, and increases with distance away from it. The flow is symmetric on the bottom of the object. The lower portion of the figure shows contours of the curl of the velocity field with larger values, indicating greater rotations, close to the object.

Two useful identities, whose proofs are left for the problems, are that the curl of a gradient and the divergence of a curl are zero:

$$\nabla \cdot (\nabla \times \mathbf{u}) = 0 \tag{18}$$

$$\nabla \times (\nabla \phi) = 0. \tag{19}$$

Equation 19 can be used with Stokes' theorem to show that for a vector field written as the gradient of a scalar, the curl, and hence circulation around an arbitrary curve, are zero. This idea is used in mechanics to prove that a conservative force (one that can be written as the gradient of a potential) has a line integral that is independent of path, because its circulation around any path is zero. These relations give insight into seismic waves, because P waves have no curl and S waves have no divergence (Section 2.4.1).

A.6.5 Laplacian

The *Laplacian* operator is formed by taking the divergence of the gradient of a scalar field, which yields a scalar field

$$\nabla^2 \phi = \nabla \cdot \nabla \phi = \frac{\partial^2 \phi}{\partial x_1^2} + \frac{\partial^2 \phi}{\partial x_2^2} + \frac{\partial^2 \phi}{\partial x_3^2} = \phi_{,,i}, \tag{20}$$

where the last form uses index notation and the summation convention. By analogy, the Laplacian of a vector field is a vector field whose components in Cartesian coordinates are the Laplacians of the original vector components,

$$\nabla^2 \mathbf{u} = (\nabla^2 u_1, \nabla^2 u_2, \nabla^2 u_3). \tag{21}$$

For example, the $\hat{\mathbf{e}}_1$ component of $\nabla^2 \mathbf{u}$ is

$$\frac{\partial^2 u_1}{\partial x_1^2} + \frac{\partial^2 u_1}{\partial x_2^2} + \frac{\partial^2 u_1}{\partial x_3^2}. \tag{22}$$

In Cartesian coordinates, the Laplacian of a vector satisfies

$$\nabla^2 \mathbf{u} = \nabla(\nabla \cdot \mathbf{u}) - \nabla \times (\nabla \times \mathbf{u}), \tag{23}$$

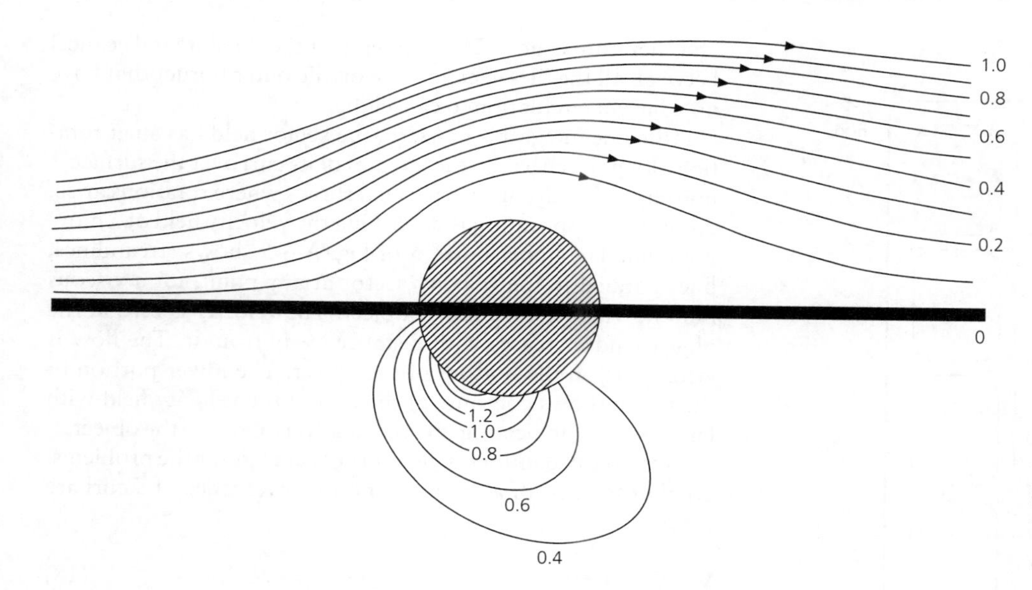

Fig. A.6-5 *Top*: streamlines showing the velocity of fluid flow around an object. Numbers on streamlines show the magnitude of the velocity. *Bottom*: contours of the curl for this velocity field. The curl is greatest near the sphere, where the fluid flow lines are the most curved. (After Batchelor, 1967. Reprinted with the permission of Cambridge University Press.)

an obscure-looking relation that is useful in deriving the existence of *P* and *S* waves.

A.7 Spherical coordinates

The vector operations discussed so far were performed in Cartesian coordinates, in which the unit basis vectors $(\hat{\mathbf{e}}_1, \hat{\mathbf{e}}_2, \hat{\mathbf{e}}_3)$ point in the same direction everywhere. There are, however, situations in which non-Cartesian coordinate systems without these nice properties are useful. In particular, *spherical* coordinates often simplify the solution of problems with a high degree of symmetry about a point.

A.7.1 *The spherical coordinate system*

In a spherical coordinate system, a point defined by a position vector $\mathbf{x}$ is described by its radial distance from the origin, $r = |\mathbf{x}|$, and two angles. θ is the *colatitude*, or angle between $\mathbf{x}$ and the x_3 axis, and ϕ, the *longitude*, is measured in the x_1–x_2 plane. Often the *latitude*, $90° - \theta$, is used instead of the colatitude. Spherical coordinates are often used in seismology because the earth is approximately spherically symmetric, varying with depth much more than laterally. Thus properties like velocity and density are often approximated as functions only of r, independent of θ and ϕ.

Figure A.7-1 shows the relations between rectangular and spherical coordinates. If the vector $\mathbf{x}$ is written as

$$\mathbf{x} = x_1\hat{\mathbf{e}}_1 + x_2\hat{\mathbf{e}}_2 + x_3\hat{\mathbf{e}}_3, \tag{1}$$

then its components in rectangular coordinates (x_1, x_2, x_3) are described by spherical coordinates as

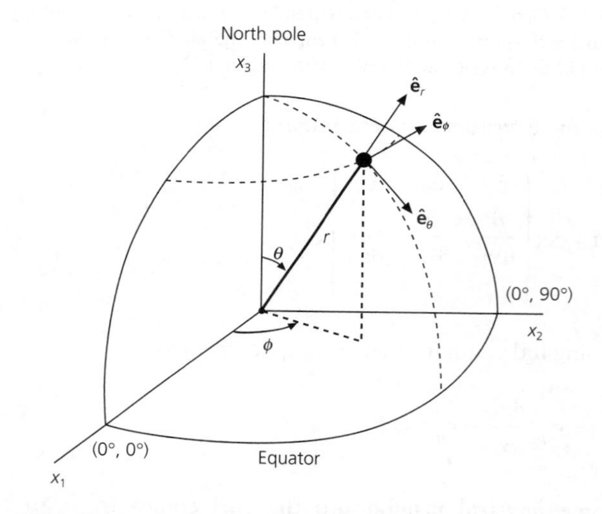

Fig. A.7-1 Relations between spherical (r, θ, ϕ) and Cartesian coordinates (x_1, x_2, x_3). (After Marion, 1970. From *Classical Dynamics of Particles and Systems*, 2nd edn, copyright 1970 by Academic Press, reproduced by permission of the publisher.)

$$\mathbf{x} = \begin{pmatrix} x_1 \\ x_2 \\ x_3 \end{pmatrix} = \begin{pmatrix} r\sin\theta\cos\phi \\ r\sin\theta\sin\phi \\ r\cos\theta \end{pmatrix}. \tag{2}$$

Conversely, the spherical coordinates r, θ, and ϕ can be written as

$$r = (x_1^2 + x_2^2 + x_3^2)^{1/2}, \quad \theta = \cos^{-1}(x_3/r), \quad \phi = \tan^{-1}(x_2/x_1). \tag{3}$$

In the equatorial $(x_1$–$x_2)$ plane, $\theta = 90°$, $\cos\theta = 0$, $\sin\theta = 1$, so $x_1 = r\cos\phi$, $x_2 = r\sin\phi$, and $x_3 = 0$. This is the same as the polar

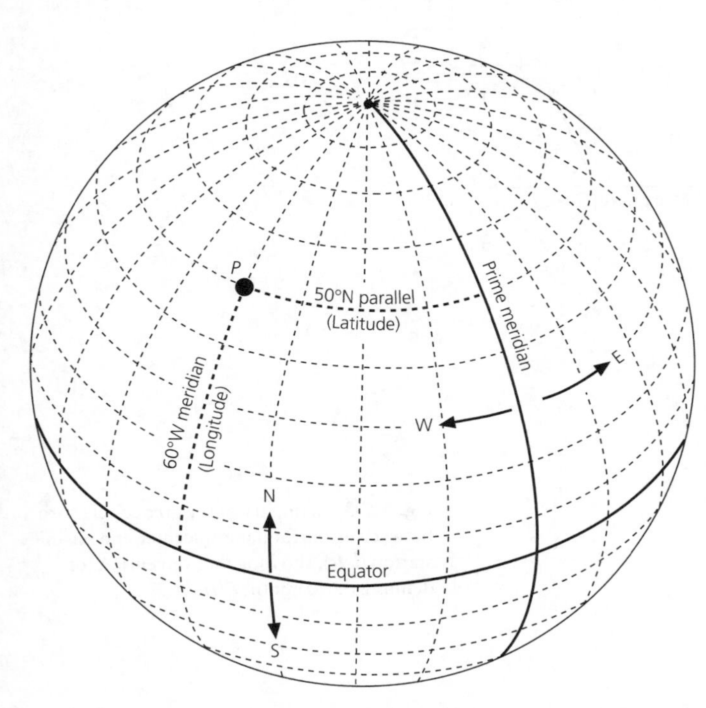

Fig. A.7-2 Geometry of the latitude and longitude system used to locate points on the earth's surface. A point P at 50°N, 60°W ($\theta = 40°$, $\phi = -60°$) is shown. (After Strahler, 1969.)

coordinate system described in Section A.3.1. Along the x_3 axis we have $\theta = 0°$, so $x_1 = x_2 = 0$, and $x_3 = r$. Any of these expressions written in terms of colatitude θ can be converted to latitude $\lambda = 90° - \theta$, using $\cos\theta = \sin\lambda$ and $\sin\theta = \cos\lambda$.

This coordinate system is the familiar one (Fig. A.7-2) used to locate points within the earth or on its surface, $r = a$. For this purpose, the origin is placed at the center of the earth, and the x_3 axis is defined by a line from the center of the earth through the north pole. The intersections of planes containing the x_3 axis with the earth's surface define *meridians*, lines of constant longitude. The x_1 axis intersects the equator at the *prime meridian*, on which ϕ is defined as zero, which has been chosen to run through Greenwich, England. The intersection of planes perpendicular to the x_3 axis with the earth's surface define *parallels*, lines of constant colatitude or latitude. Meridians are a special case of *great circles*, lines on the surface defined by the intersection of a plane through the origin with the surface of the spherical earth. Parallels are a special case of *small circles*, which are lines on the surface defined by the intersection of the surface of the spherical earth with a plane normal to a radius vector.

These conventions allow the colatitude θ ($0° \le \theta < 180°$) and longitude ϕ ($0° \le \phi < 360°$) to define a unique point on the earth's surface. Often locations are described in terms of latitudes north and south of the equator, and longitudes east and west of Greenwich. North and south latitudes correspond, respectively, to colatitudes less than or greater than 90°. Because ϕ measures longitude east of the prime meridian, west longitudes correspond to values of ϕ less than 0° or greater than 180°. Thus a point at (10°S, 110°W) has $\theta = 90° + 10° = 100°$, and $\phi = -110° = 360° - 110° = 250°$.

At any point, unit spherical basis vectors ($\hat{e}_r$, $\hat{e}_\theta$, $\hat{e}_\phi$) can be defined in the direction of increasing r, θ, and ϕ. $\hat{e}_r$ points away from the origin, and gives the upward vertical direction. $\hat{e}_\theta$ points south, and $\hat{e}_\phi$ points east. These two are sometimes written in terms of north- and east-pointing unit vectors, $\hat{e}_{NS} = -\hat{e}_\theta$ and $\hat{e}_{EW} = \hat{e}_\phi$.

An important feature of the unit spherical basis vectors is that at different points they are oriented differently with respect to the Cartesian axes. The Cartesian unit basis vectors ($\hat{e}_1$, $\hat{e}_2$, $\hat{e}_3$) point in the same direction everywhere. By contrast, for example, $\hat{e}_r$ points in the $\hat{e}_3$ direction at the north pole, and in the $-\hat{e}_3$ direction at the south pole. This effect is described by the Cartesian ($\hat{e}_1$, $\hat{e}_2$, $\hat{e}_3$) components of the unit spherical basis vectors, at a point with colatitude θ and longitude ϕ:

$$\hat{e}_\phi = \begin{pmatrix} -\sin\phi \\ \cos\phi \\ 0 \end{pmatrix}, \quad \hat{e}_\theta = \begin{pmatrix} \cos\theta\cos\phi \\ \cos\theta\sin\phi \\ -\sin\theta \end{pmatrix}, \quad \hat{e}_r = \begin{pmatrix} \sin\theta\cos\phi \\ \sin\theta\sin\phi \\ \cos\theta \end{pmatrix}. \quad (4)$$

The dependence on the colatitude and longitude describes how the orientation with respect to the Cartesian axes changes.

At any point, the spherical basis vectors ($\hat{e}_r$, $\hat{e}_\theta$, $\hat{e}_\phi$) form an orthonormal set. For problems whose spatial extent is small enough that the curvature of the earth can be ignored, these basis vectors provide a useful local coordinate system.

A.7.2 *Distance and azimuth*

Spherical coordinates are especially useful in describing the geographic relation between two points on the earth's surface. A common application is to find the distance between points and the direction of the great circle arc joining them. A great circle arc is the shortest path between points on a sphere, so if seismic velocity varies only with depth, the fastest path along the surface is the great circle arc, and the fastest paths through the interior are in the plane of the great circle and the center of the earth. Because velocities vary laterally by only a few percent throughout most of the earth (and imperceptibly in the liquid outer core), this is a good approximation for most seismic applications. The source-to-receiver distance is often given in terms of the angle Δ subtended at the center of the earth by the great circle arc between the two points (Fig. A.7-3). If Δ is expressed in radians, then the length s (in km) of the arc along the earth's surface is $R\Delta$, where R is the earth's radius ($\approx$ 6371 km). If Δ is expressed in degrees, $s = R\Delta\pi/180$, so one degree of arc equals 111.2 km.

Consider the great circle arc connecting an earthquake whose epicenter is at (θ_E, ϕ_E) and a seismic station at (θ_S, ϕ_S). Seismic waves that traveled along the great circle arc (or in the plane of this arc and the center of the earth) left the earthquake in a direction given by the *azimuth* angle ζ measured clockwise from the local direction of north at the epicenter to the great

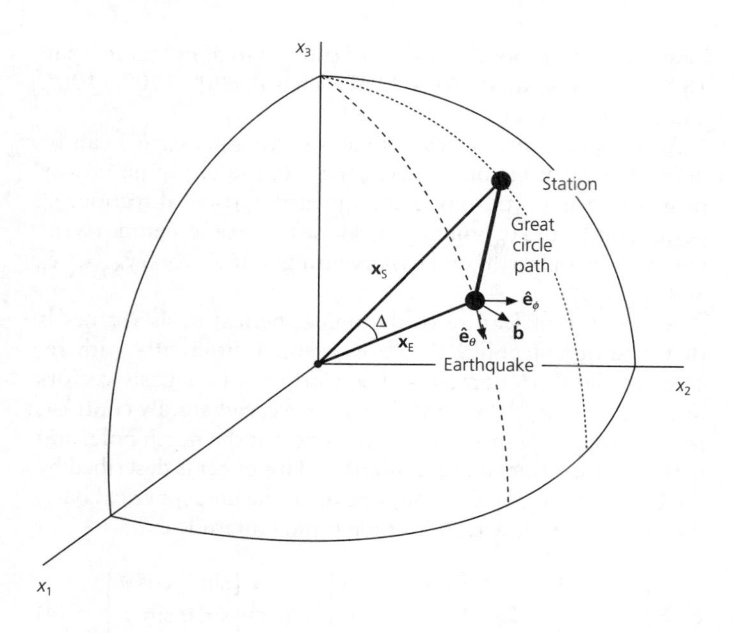

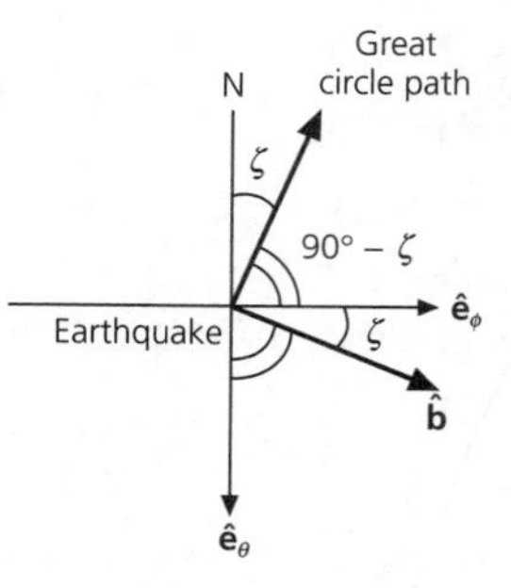

Fig. A.7-3 Geometry of the great circle path between an earthquake epicenter and seismic station (*left*), showing the convention for defining the azimuth, ζ (*right*).

circle arc. These waves arrive at the seismometer from a direction described by the *back azimuth* angle ζ' measured clockwise from the local direction of north at the seismometer to the great circle arc. To find these quantities, the Cartesian components of the position vectors for the earthquake and the station are written, using Eqn 2:

$$\mathbf{x}_E = \begin{pmatrix} R \sin\theta_E \cos\phi_E \\ R \sin\theta_E \sin\phi_E \\ R \cos\theta_E \end{pmatrix} \quad \mathbf{x}_S = \begin{pmatrix} R \sin\theta_S \cos\phi_S \\ R \sin\theta_S \sin\phi_S \\ R \cos\theta_S \end{pmatrix}. \tag{5}$$

The distance Δ, the angle between $\mathbf{x}_S$ and $\mathbf{x}_E$, is given by the scalar product

$$\mathbf{x}_S \cdot \mathbf{x}_E = R^2 \cos\Delta, \tag{6}$$

so

$$\Delta = \cos^{-1}\left[\cos\theta_E \cos\theta_S + \sin\theta_E \sin\theta_S \cos(\phi_S - \phi_E)\right]. \tag{7}$$

This formula defines Δ uniquely between 0 and 180°. This shorter portion of the great circle is called the *minor arc* connecting the two points; the longer portion, known as the *major arc*, is (360° – Δ) degrees long.

To compute the azimuth from the earthquake to the station, consider $\hat{\mathbf{b}}$, a unit vector normal to the great circle in the local horizontal plane at $\mathbf{x}_E$, which is written using the vector product of the position vectors

$$\mathbf{x}_S \times \mathbf{x}_E = \hat{\mathbf{b}} R^2 \sin\Delta. \tag{8}$$

Evaluation of the vector product gives

$$\hat{\mathbf{b}} = \frac{1}{\sin\Delta} \begin{pmatrix} \sin\theta_S \cos\theta_E \sin\phi_S - \sin\theta_E \cos\theta_S \sin\phi_E \\ \cos\theta_S \sin\theta_E \cos\phi_E - \cos\theta_E \sin\theta_S \cos\phi_S \\ \sin\theta_S \sin\theta_E \sin(\phi_E - \phi_S) \end{pmatrix}. \tag{9}$$

The azimuth angle ζ, measured clockwise from north, is then given (Fig. A.7-3) by

$$\cos\zeta = \hat{\mathbf{b}} \cdot \hat{\mathbf{e}}_\phi = \frac{1}{\sin\Delta}\left(\cos\theta_S \sin\theta_E - \sin\theta_S \cos\theta_E \cos(\phi_S - \phi_E)\right) \tag{10}$$

and

$$\sin\zeta = \hat{\mathbf{b}} \cdot \hat{\mathbf{e}}_\theta = \frac{1}{\sin\Delta} \sin\theta_S \sin(\phi_S - \phi_E). \tag{11}$$

Use of both $\sin\zeta$ and $\cos\zeta$ makes the angle ζ unambiguous ($0° \le \zeta < 360°$). The azimuth from an earthquake to a receiver is useful, because earthquakes radiate more energy in some directions than in others (Chapter 4), so measurements at different azimuths yield information about the source.

The back azimuth ζ', obtained by reversing the indices E and S in Eqns 10 and 11, shows the direction from which seismic energy arrives at a seismometer. Seismometers typically record the north–south and east–west components of horizontal ground motion. Using the back azimuth, these observations can be converted into *radial* (along the great circle path) and *transverse* (perpendicular to the great circle path) components by a vector transformation (Eqn A.5.9). This distinction is made because waves appearing on these components propagated differently (Section 2.4). The azimuth and back azimuth

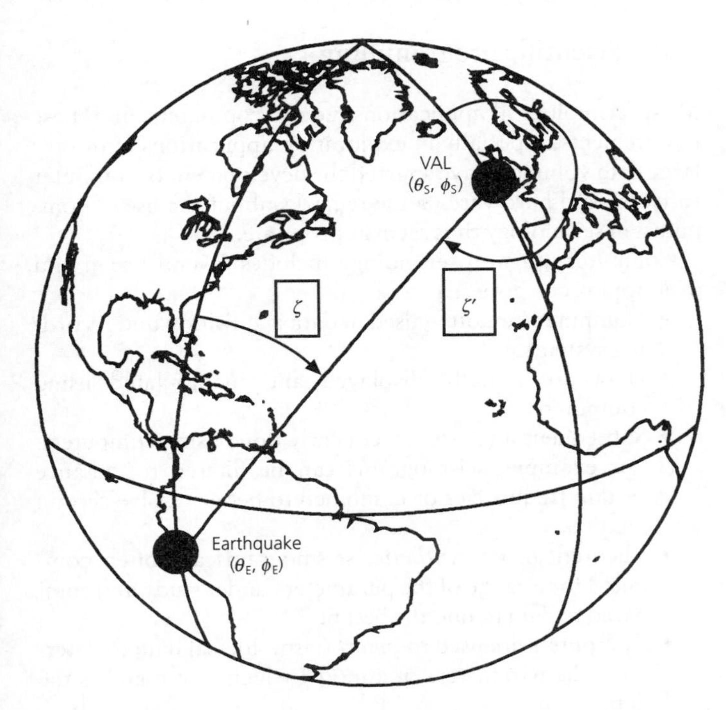

Fig. A.7-4 Geometry of the great circle path for an earthquake in the Peru trench recorded at station VAL (Valentia, Ireland). The azimuth, ζ, and back azimuth, ζ', are not simply related, due to the sphericity of the earth.

angles are measured clockwise from north, a geographic convention which contrasts with the mathematical one of measuring angles counterclockwise from the x_1 direction. Figure A.7-4 illustrates this geometry for an earthquake in the Peru trench ($\theta_E = 102°$, $\phi_E = -78°$) recorded at station VAL (Valentia, Ireland; $\theta_S = 38°$, $\phi_S = -10.25°$). The resulting distances and azimuths are $\Delta = 86°$, $\zeta = 35°$, $\zeta' = 245°$.[1]

This analysis assumes that the earth is perfectly spherical. In fact, the earth is flattened by its rotation into a shape close to an oblate ellipsoid, so the radius varies with colatitude approximately as

$$r(\theta) = R_e(1 - f\cos^2\theta), \tag{12}$$

where R_e is the equatorial radius, 6378 km. The flattening factor f is approximately 3.35×10^{-3}, or about 1/298, so the polar radius R_p is 6357 km. An average radius can be defined as the radius of a sphere with the same volume as the earth, if it were a perfect ellipsoid. Because the volume of an ellipsoidal earth would be $(4/3)\pi R_e^2 R_p$, and a sphere of radius R has volume $(4/3)\pi R^3$, the average radius is 6371 km. For certain applications the ellipticity is included in precise distance calculations.

[1] These distance–azimuth equations also have nonseismological applications because ships and aircraft follow the shortest (great circle) paths between two points when possible.

A.7.3 *Choice of axes*

Spherical coordinates are also used with axes different from the geographic ones. Because the physics of a problem does not depend on the choice of coordinates, a set of coordinates that simplifies the relevant expressions is used. For example, in earthquake source studies, the x_3 axis can be chosen to go from the center of the earth to the location of the earthquake. The prime meridian, and hence x_1, axis can be selected so that the fault is oriented in the direction $\phi = 0$. These axes simplify the description of the seismic waves radiated by an earthquake, because the distance Δ from the source is now the colatitude. Moreover, the radiation pattern generally has a high degree of symmetry about the fault, so simple functions of ϕ appear. By contrast, the radiation pattern need have no symmetry about the North pole and Greenwich meridian, so a description in those coordinates would usually be more complicated.

Fortunately, a coordinate system referred to the earthquake location does not make describing the propagation of waves from the source any more difficult. Because earth structure varies primarily with depth, the spherical symmetry about the center of the earth is independent of the axis orientation chosen. The geographical convention in which the earth rotates about the x_3 axis is helpful for navigation. In most seismological applications, however, the north direction has no particular significance because the propagation of seismic waves is essentially unaffected by the earth's rotation. The choice of a prime meridian is arbitrary; in the early nineteenth century some American maps had it through Washington DC, and some French maps had it through Paris.

A.7.4 *Vector operators in spherical coordinates*

Because at a point on the sphere the unit spherical basis vectors are oriented up, south, and east, the basis vectors at different locations are generally not parallel. This makes the vector differential operators more complicated, because these operators involve taking spatial derivatives of vectors. In Cartesian coordinates the unit basis vectors are not affected by this differentiation because they do not change orientation, so only derivatives of the components need be taken. In spherical coordinates, because a vector **u** is

$$\mathbf{u} = u_r \hat{\mathbf{e}}_r + u_\theta \hat{\mathbf{e}}_\theta + u_\phi \hat{\mathbf{e}}_\phi, \tag{13}$$

differential operators acting on **u** must incorporate the derivatives of the basis vectors. Thus, in spherical coordinates, for a scalar field ψ and a vector field **u**:

$$\text{grad}\,\psi = \hat{\mathbf{e}}_r \frac{\partial \psi}{\partial r} + \hat{\mathbf{e}}_\theta \frac{1}{r}\frac{\partial \psi}{\partial \theta} + \hat{\mathbf{e}}_\phi \frac{1}{r\sin\theta}\frac{\partial \psi}{\partial \phi} \tag{14}$$

$$\text{div}\,\mathbf{u} = \frac{1}{r^2}\frac{\partial}{\partial r}(r^2 u_r) + \frac{1}{r\sin\theta}\frac{\partial}{\partial \theta}(\sin\theta\, u_\theta) + \frac{1}{r\sin\theta}\frac{\partial u_\phi}{\partial \phi} \tag{15}$$

$$\text{curl}\,\mathbf{u} = \hat{\mathbf{e}}_r \frac{1}{r\sin\theta}\left(\frac{\partial}{\partial\theta}(\sin\theta\, u_\phi) - \frac{\partial u_\theta}{\partial\phi}\right)$$
$$+\, \hat{\mathbf{e}}_\theta \frac{1}{r\sin\theta}\left(\frac{\partial u_r}{\partial\phi} - \sin\theta\frac{\partial}{\partial r}(ru_\phi)\right)$$
$$+\, \hat{\mathbf{e}}_\phi \frac{1}{r}\left(\frac{\partial}{\partial r}(ru_\theta) - \frac{\partial u_r}{\partial\theta}\right) \tag{16}$$

$$\nabla^2\psi = \frac{1}{r^2}\frac{\partial}{\partial r}\left(r^2\frac{\partial\psi}{\partial r}\right) + \frac{1}{r^2\sin\theta}\frac{\partial}{\partial\theta}\left(\sin\theta\frac{\partial\psi}{\partial\theta}\right)$$
$$+\, \frac{1}{r^2\sin^2\theta}\frac{\partial^2\psi}{\partial\phi^2}. \tag{17}$$

These expressions are used when we discuss spherical waves in Section 2.4 and the earth's normal modes in Section 2.9.

A final point worth noting is that the elements of volume and surface used in integrals are different in spherical coordinates from rectangular coordinates. In spherical coordinates (Fig. A.7-5) there are several scale factors, so an element of surface area is

$$dS = r^2 \sin\theta\, d\theta d\phi, \tag{18}$$

and an element of volume is

$$dV = r^2 \sin\theta dr\, d\theta d\phi. \tag{19}$$

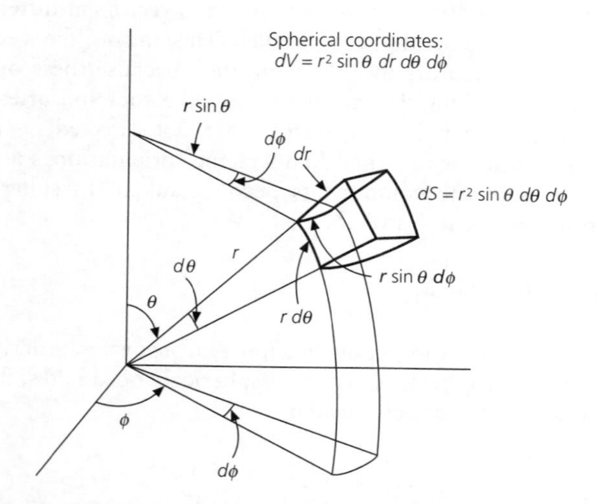

Fig. A.7-5 Definition of the element of volume in spherical coordinates. Unlike the case of Cartesian coordinates, the volume element in spherical coordinates in not a cube. (Marion, 1970. From *Classical Dynamics of Particles and Systems*, 2nd edn, copyright 1970 by Academic Press, reproduced by permission of the publisher.)

A.8 Scientific programming

Most seismological applications require computers, and these requirements, especially in exploration applications with very large data volumes, have spurred the development of computer software and hardware. Some remarks about the use of computers in seismology thus seem appropriate.

Computer usage in seismology includes several broad and overlapping categories:

- Computers are often used in data acquisition and recording systems.
- Data are initially displayed and manipulated using computers.
- Subsequent analysis is frequently done using computers. For example, seismograms can be filtered to enhance certain frequencies or combined to better resolve certain features.
- Theoretical, or *synthetic*, seismograms are often computed for a range of the parameters under study and compared to data to find the best fit.
- Computers are used to *invert* seismological data to determine the parameters of a model which best matches the data.
- Computer modeling is often used to draw geological inferences from seismological observations. For example, seismic velocity data are compared to the predictions of models for the velocity of rock as a function of composition, temperature, and pressure.

These applications often require *scientific programming*, a programming style used for essentially mathematical applications. Some problems in this book also require scientific programming. Although programming is a matter of personal style, this section discusses several points that may be helpful. The suggested reading provides some starting points for readers interested in pursuing these topics further.

A.8.1 *Example: synthetic seismogram calculation*

Consider a program to compute a synthetic seismogram for waves in a one-dimensional constant-velocity medium, a mathematically idealized string that illustrates features of wave behavior. The program is based on $u(x, t)$, the displacement as a function of position x and time t. The displacement is zero at the fixed ends of the string, $x = 0$ and $x = L$, between which waves travel at speed v. As in Section 2.2.5, the displacement can be written as the sum of the normal modes of the string, each of which is a standing wave with n half wavelengths along the string,

$$u_n(x, t) = \sin(n\pi x/L)\cos(\omega_n t), \tag{1}$$

and vibrates at a characteristic frequency, or *eigenfrequency*,

$$\omega_n = n\pi v/L. \tag{2}$$

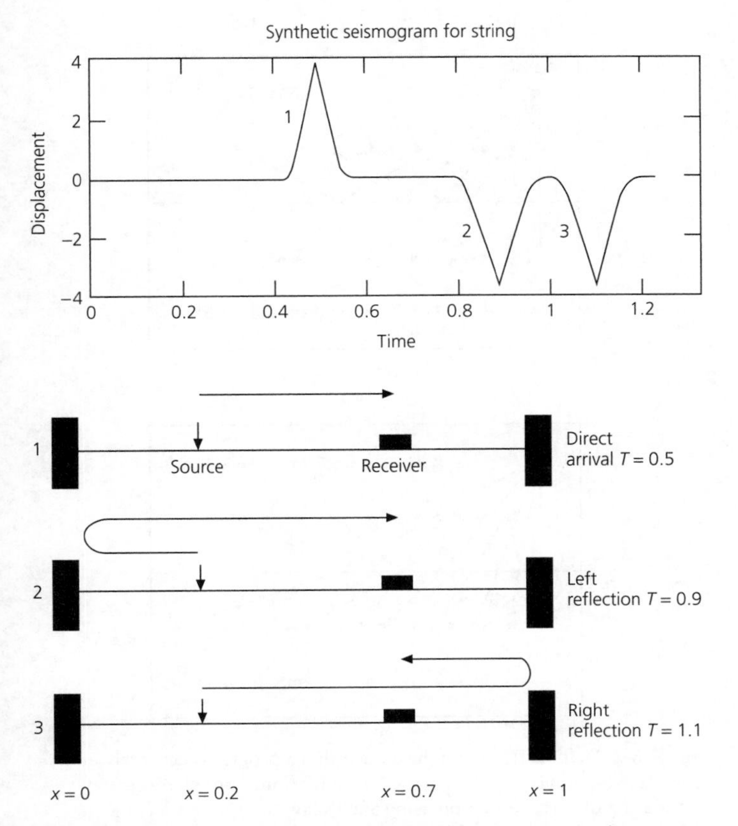

Fig. A.8-1 *Top*: Synthetic seismogram for a string showing the direct wave arrival (1) and reflections (2, 3) from both ends. *Bottom*: Geometry showing source and receiver positions, and the times of the direct and reflected arrivals.

If a source at position x_s generates a pulse at time zero with duration τ, the propagating waves are described by a weighted sum of the modes

$$u(x, t) = \sum_{n=1}^{\infty} \sin(n\pi x/L) \sin(n\pi x_s/L) \cos(\omega_n t) \exp[-(\omega_n \tau)^2/4]. \tag{3}$$

Given the displacement $u(x, t)$ for any position and time, a seismogram ("stringogram") giving the displacement versus time at a receiver position x_r is $u(x_r, t)$. Alternatively, a "snapshot" of the displacement everywhere on the string at time t_0 is $u(x, t_0)$.

Consider a program to evaluate a synthetic seismogram using this sum. For simplicity, we use a string of length 1 m[1] with a wave speed 1 m/s, a source at $x_s = 0.2$ m and a receiver at $x_r = 0.7$ m. To approximate the infinite sum, the program adds up 200 modes. The seismogram (Fig. A.8-1, *top*) is calculated at 50 time steps, covering 1.25 s. This program is written in Fortran, a language that is especially suitable for scientific programming and is therefore commonly used in seismology (and thus in this book). The program could be also written in other languages, but the general points would still apply.

```
C SYNTHETIC SEISMOGRAM FOR HOMOGENEOUS STRING
C DISPLACEMENT U AS FUNCTION OF TIME T
C CALCULATED BY NORMAL MODE SUMMATION
      DIMENSION U(200)
      PI = 3.1415927
C
C PARAMETERS (NORMALLY WOULD COME FROM INPUT)
C STRING LENGTH (M)
      ALNGTH = 1.0
C VELOCITY (M/S)
      C = 1.0
C NUMBER OF MODES
      NMODE = 200
C SOURCE POSITION (M)
      XSRC = 0.2
C RECEIVER POSITION (M)
      XRCVR = 0.7
C SEISMOGRAM TIME DURATION (S)
      TDURAT = 1.25
C NUMBER TIME STEPS
      NTSTEP = 50
C TIME STEP (S)
      DT = TDURAT/NTSTEP
C SOURCE SHAPE TERM
      TAU = .02
C
C LIST PARAMETERS
      WRITE (6,3000)
3000  FORMAT('SYNTHETIC SEISMOGRAM FOR STRING')
      WRITE (6,3001) NMODE
3001  FORMAT('NUMBER OF MODES', I6)
      WRITE (6,3002) ALNGTH, C
3002  FORMAT ('LENGTH (M)' F7.3, 'VELOCITY,
     X (M/S)', F7.3)
      WRITE (6,3003) XSRC, XRCVR
3003  FORMAT ('POSITION (M): SOURCE', F7.3,
     X 'RECEIVER', F7.3)
      WRITE (6,3004) TDURAT, NTSTEP
3004  FORMAT ('SEISMOGRAM DURATION (S)', F7.3,
     X I6, 'TIME STEPS')
      WRITE (6,3005) TAU
3005  FORMAT ('SOURCE SHAPE TERM', F7.3)
C
C INITIALIZE DISPLACEMENT
         DO 5 I = 1, NTSTEP
         U(I) = 0.0
5        CONTINUE
C
C OUTER LOOP OVER MODES
      DO 10 N = 1, NMODE
         ANPIAL = N*PI/ALNGTH
```

[1] It is easy to use arbitrary values on a computer; we could also use 1 km or 1 furlong. Finding a physical 1 km string is another matter . . .

```
C SPACE TERMS: SOURCE AND RECEIVER
          SXS = SIN(ANPIAL*XSRC)
          SXR = SIN(ANPIAL*XRCVR)
C MODE FREQUENCY
          WN = N*PI*C/ALNGTH
C TIME INDEPENDENT TERMS
          DMP = (TAU*WN)**2
          SCALE = EXP(-DMP/4.)
          SPACE = SXS*SXR*SCALE
C
C INNER LOOP OVER TIME STEPS
          DO 15 J = 1, NTSTEP
            T = DT*(J - 1)
            CWT = COS(WN*T)
C COMPUTE DISPLACEMENT
            U(J) = U(J) + CWT*SPACE
15        CONTINUE
10     CONTINUE
C
C OUTPUT SEISMOGRAM FOR LATER PLOTTING
       WRITE (6, 3101)(U(J), J = 1, NTSTEP)
3101   FORMAT (7F10.4)
       STOP
       END
```

This example brings out several points:

- *Is the answer correct?* Two different types of error occur in scientific programs. First, the *program* may be wrong. In this case, the mathematical formulation correctly describes the physical problem, but the program incorrectly implements this formulation. This is the usual situation, in which "bugs" are identified and corrected. Second, the *formulation* may be wrong, so the program correctly implements an incorrect mathematical model. This could occur because of a mathematical error, like an attempt to sum a divergent series, or a physical error, such as an equation that does not correctly describe waves on a string. An incorrect formulation is particularly disturbing because it cannot be detected by checking the program. For example, Fig. A.8-2 shows two computer simulations for waves bending as they pass from one medium into another with higher velocities. Figure A.8-2 (*top*) uses the correct formulation of Snell's law (Section 2.5), whereas Fig. A.8-2 (*bottom*) looks equally convincing but is wrong because the equation which the program illustrates is incorrect.

Programmers check for both types of errors by choosing cases for which the results can be predicted analytically and comparing the results to those of the program. Several tests are easily done for the string. The wave following the shortest (direct) path appears at the expected time, 0.5 s (Fig. A.8-1, *bottom*), because the source and the receiver are 0.5 m apart. The next two arrivals, reflections from the ends of the string, also occur at the expected times. Moreover, these arrivals have polarities opposite that of the initial pulse, as should occur (Section 2.2.3) upon reflection at the string's fixed ends. The program can also be checked for different string lengths, speeds, and source and receiver positions. Similarly, in addition to synthetic seismograms, displacements along the string at fixed times could be computed. Such tests are important, because if the mathematical model is not appropriate for the physical situation, then time spent debugging, documenting, and optimizing the program is wasted.

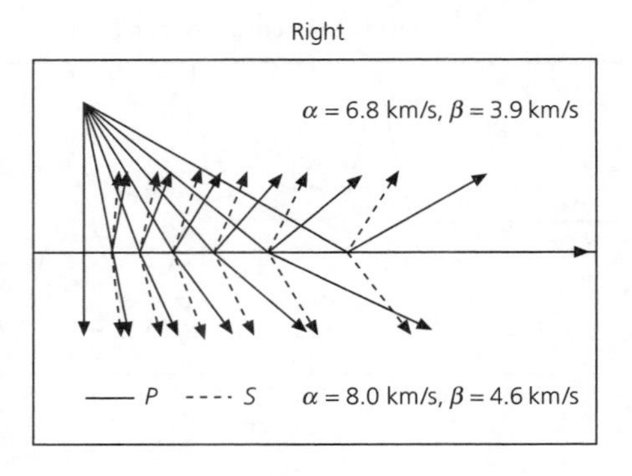

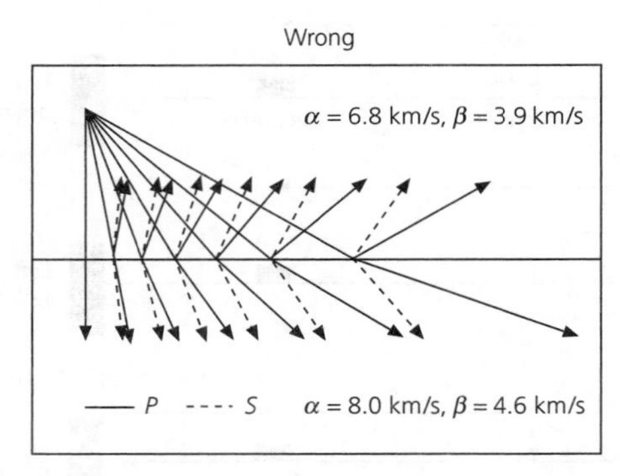

Fig. A.8-2 Demonstration of the danger that a program accurately computes an incorrect mathematical formulation. *Top*: A correct simulation of wave refraction using Snell's law, $\sin i_1/v_1 = \sin i_2/v_2$. *Bottom*: The same simulation using a wrong formula for Snell's law, $i_1/v_2 = i_2/v_2$.

- *The program is reasonably comprehensible.* Several features help clarify the program. The program's purpose and method are stated. Variable names somewhat resemble those in the equation: "SXS" is $\sin x_s$, and so on. Comments identify the functions of portions of the program.

- *The program uses loops and arrays.* The seismogram is described by the array U(J), and its values at successive times are calculated by looping. Using an array, rather than discrete variables UT1, UT2, etc., makes the program clearer, closer to the mathematical formulation, and simplifies output. The loop structure also makes the program clearer and allows the number of time steps to be changed simply by changing the parameter NTSTEP. Similarly, the number of modes is easily changed.

- *The output is labeled.* The seismogram was placed in an output file for later plotting. The parameters used to compute the seismogram are included, so examination of the output

```
C OUTER LOOP OVER MODES
            DO 10 N = 1, NMODE

                 terms for each mode
                 that do not depend on time

C INNER LOOP OVER TIME STEPS
                 DO 15 J = 1, NTSTEP

                 terms that depend on time

C COMPUTE DISPLACEMENT

15               CONTINUE
10       CONTINUE
```

Fig. A.8-3 Structure of the loops for the string synthetic seismogram calculation.

shows how it was computed. This helps avoid the common situation where, given a large collection of computer output, cases are rerun because it is unclear what parameters were used. Moreover, subsequent "improved" versions of the program can be checked to see whether they give the same results.

- *The program is somewhat efficient.* Some thought is generally put into *optimizing* scientific programs to make them run rapidly. The program could find the displacement by looping over time and summing all the modes at each time step. However, consideration of the equation shows that three terms, $\sin(n\pi x/l)$, $\sin(n\pi x_s/l)$, and $\exp[-(\omega_n\tau)^2/4]$ are evaluated only once for each mode, whereas only $\cos(\omega_n t)$ is evaluated for each time step. It is thus more efficient to loop over the modes and evaluate each at all times (Fig. A.8-3). Because the outer (mode) loop is executed 200 times, whereas the inner (time) loop is executed $200 \times 50 = 10{,}000$ times, the inner loop should be as efficient as possible. The program would run more slowly if the loops were reversed. The difference, though not significant for this calculation, might be significant for much larger numbers of time steps and modes.

Further improvements could be made to fully optimize the program. Optimization is not an end in itself, because the programmer's time and the intelligibility of the program are also important. Programmers typically try to write reasonably optimized programs without making them impossible to understand and debug. Once fully tested, a program that will be used heavily may be worth further optimization if the computer time savings justify the effort required. There is no point in "getting the wrong answer as fast as possible."[2] Certain computers, such as those using parallel processors, may require specialized optimization.

A.8.2 *Programming style*

The style in which programs are written can make them easier to develop, debug and use. A few suggestions, though not absolute rules, may be useful.

[2] Kernighan and Plauger (1978).

- *Document the program.* Computer programs can be almost useless without adequate documentation. Stonehenge has been described as "the world's largest undocumented computer system."[3] Failure to document is often justified by the assumption that the program will not be used again. This rationalization is self-fulfilling, because even the author may find an undocumented program difficult to reuse once the details are forgotten.

Documentation should state the program's goals and method. The input and output variables, their units, and how they are defined should be listed. Implicit assumptions and restrictions are worth noting. Comments should identify major portions of the program and describe their functions.

Documentation is best written when writing a program because it can aid in debugging. Moreover, once a program is fully written, it is harder to remember how it works. Documentation included in the program is less prone to be lost than that written separately.

Finally, documentation helps scientists exchange programs and work in collaboration. This can be useful, except in the apocryphal cases of programmers writing gigantic undocumented programs to maximize their job security.

- *Use modular programming.* Large programs can generally be divided into smaller subroutines or functions, which can be used like the functions (e.g., sine, square root) supplied by many computer languages. Each subroutine can be tested separately and then used in various programs. Subroutines can handle applications that frequently recur, such as reading or plotting data or carrying out a mathematical operation. This approach saves the time needed to write and debug portions of a program similar to one already available. Moreover, the overall structure of a program containing a set of calls to subroutines is generally easier to understand, because many complexities are isolated into subroutines.

- *Make programs comprehensible.* It is helpful to be able to understand programs once written. Clear documentation and modular programming help. In addition, it should be easy to tell what portions will be executed under which circumstances. For this purpose, portions of a program should be executed sequentially, rather than jumping backwards and forwards within a program.

Similarly, the statements themselves can be written clearly. The use of mnemonic variable names and natural groupings of variables can help. For example, it is somewhat unclear that

```
X=0.23873*A/(Y*Y*Y)
```

gives the average density X of a planet with mass A and radius Y, whereas

```
RHO=AMASS/((4.0/3.0)*PI*(RADIUS**3))
```

[3] Brooks (1975).

is clearer. For clarity, the latter expression is more verbose than required, has π previously defined, and is slightly less efficient.

- *Don't be clever.* Sometimes the shortest, "cleverest" way of programming something can be the worst. In addition to giving rise to lack of clarity, some shortcuts make it difficult to transfer programs between computers. This is especially true of programs that exploit specific properties of an individual computer or compiler, such as local variants of a standard programming language.

- *Keep a perspective on precision.* The program calculates and manipulates numbers that, at least in theory, correspond to physical entities. It is worth keeping track of the precision associated with the data and other quantities, and of that required to compute the desired results.

- *Organize programs and data.* Related programs and the associated files can be grouped into directories which include files listing and explaining the directory's contents. Data files can be organized similarly. Often seismograms, for example, go through multiple processing stages carried out by different programs. A common practice is to use specific types of file names to indicate various intermediate stages. In addition, the data files begin with *headers*, information identifying the data and recording the operations applied to it. The headers and file names should be updated by the programs themselves, rather than "by hand" at each stage. The output, whether text or graphic, should contain the parameters required to replicate the result. This can be especially important for interactive data processing because input files are not kept.

A.8.3 Representation of numbers

Several simple concepts about numerical calculations on a computer are worth bearing in mind. One is the consequences of the way in which numbers are represented and manipulated. Because computers use binary (base 2) arithmetic, numbers are written as sets of *bits*, single binary digits, grouped into *words*. Some general ideas about these representations can be illustrated without going into the schemes used by various computers.

Integers are represented by their binary equivalent. Thus 46 (decimal) is 101110, because

$$46 = 1\times 2^5 + 0\times 2^4 + 1\times 2^3 + 1\times 2^2 + 1\times 2^1 + 0\times 2^0.$$

Many computers represent integers by 16- or 32-bit words. The word length governs the range of possible integers. For example, using 16 bits, one of which indicates the sign, the largest positive integer that can be represented is

$$111\ 1111\ 1111\ 1111 \text{ (binary)} = 2^{15} - 1 = 32{,}767.$$

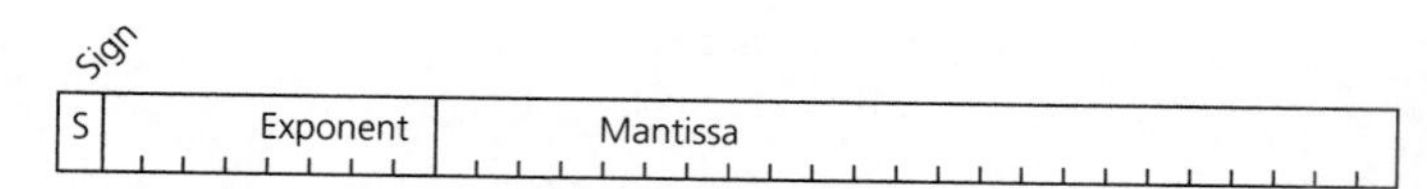

Fig. A.8-4 Representation of a floating point number using 32 bits.

Because a greater range is needed for scientific computation, floating point numbers are used:

$$\text{number} = (\text{mantissa}) \times 2^{\text{exponent}}.$$

Floating point numbers can accommodate fractions, with digits to the right of the binary point representing negative powers of two, just as digits to the left of the point represent positive powers of two. For example,

$$46.625 \text{ (decimal)} = 1\times 2^5 + 0\times 2^4 + 1\times 2^3 + 1\times 2^2 + 1\times 2^1 + 0\times 2^0 + 1\times 2^{-1} + 0\times 2^{-2} + 1\times 2^{-3}$$
$$= 101{,}110.101 \text{ (binary)} = 0.101110101\times 2^6.$$

To represent binary floating point numbers on a computer, a certain number of bits are assigned to the mantissa and the exponent. Figure A.8-4 shows one way in which a single precision floating point number might be represented by a 32-bit word. One bit is reserved for the sign of the mantissa, 8 bits are used for the exponent including its sign, and the remaining 23 bits contain the mantissa. The number of bits available for the exponent determines the *range* of the floating point numbers. Because $2^8 = 256$, the exponent can represent numbers between approximately 2^{127} and 2^{-128} or approximately 10^{38} to 10^{-39}. The number of bits in the mantissa determines the *precision* or number of significant digits. Because 2^{-23} is approximately 10^{-7}, the maximum number of significant decimal digits is about seven. Further precision can be obtained using *double precision* numbers with additional bits for the mantissa. The precise values of the range and the precision depend on details of the implementation.

The range and precision in use are worth bearing in mind because computers do not always issue "overflow" or "underflow" warnings. The computer may assign a value, such as the largest floating point number, and proceed. It can be frustrating to find that the peculiar answers produced by a program result from numbers outside the computer's range.

A related malady is *round-off error*, the loss of computational precision due to the limited number of significant digits. To illustrate the concept, suppose that a computer used six bits for the mantissa. The decimal addition

$$0.65625 + 0.96875 = 1.625$$

would, in binary, be

$0.10101 + 0.11111 = 1.10100,$

which, because no precision was lost, equals the exact answer. Now, consider the decimal addition

$5.25 + 0.96875 = 6.21875,$

which, in binary, becomes

$0.101010 \times 2^3 + 0.111110 \times 2^0.$

To carry out the binary addition, because the numbers have different exponents, the mantissa of the smaller number is shifted to produce a common exponent. If some of the bits representing the smaller number are lost, inaccuracy may result. For example, in this case,

$$0.101010 \times 2^3 + 0.000111 \times 2^3 = 0.110001 \times 2^3$$
$$= 6.125 \text{ (decimal)}.$$

The precision available on a computer is generally adequate to avoid significant round-off error. Nonetheless, it is a potential problem to keep in mind, especially in long calculations or in those such as a series sum where the answer is the difference between large numbers.

A.8.4 *A few pitfalls*

Difficulties often can be avoided by considering how various statements in the program will be executed. This is especially the case when using compilers that provide little error checking and few helpful warning and error messages. The computer, following its explicit rules, may yield results differing from those expected. The foibles here are for Fortran, but similar ones often appear in other computer languages.

- *Statement execution*. Problems often stem from the distinction between integers and floating point numbers. For example, if I and J are integer variables,

```
J=5
I=1/J
```

yields zero, because integer division yields an integer. This problem is not cured by setting the result equal to a floating point variable, or performing a floating point operation on the integer result:

```
X=1/J
Z=1.0*(1/J)
```

yield zero, because division is done as an integer operation, and the result (0) is converted to floating point (0.0). On the other hand, most compilers give 0.2 as the result of

```
X=1.0/J,
```

although a conservative policy is to explicitly convert the integer to floating point

```
X=1.0/FLOAT(J).
```

A second class of problems can result from the order in which operations are performed. For example, it may be unclear whether

```
-1.0**2
```

should be interpreted as $(-1.0)^2 = 1.0$ or $-(1.0)^2 = -1.0$. Although the computer language rules are explicit, it may be wise to use parentheses, e.g.,

```
(-1.0)**2
```

to ensure that operations are carried out as desired. The additional parentheses can also make the program more comprehensible.

- *Subroutines*. Subroutines are heavily used in writing scientific programs. As a result, problems can result while using computer languages like Fortran in which what appear to be arguments passed to a subroutine are actually the locations in memory of these arguments.

A common error is exemplified by the following program

```
CALL SUB(1.0)
X = 1.0
WRITE (6,*) 'X = ', X
STOP
END

SUBROUTINE SUB(Y)
Y = 5.0
RETURN
END
```

which, when executed, yields "X = 5.0." Because Y, a parameter in the subroutine definition, was set equal to 5.0, the value of the corresponding parameter in the subroutine call, "1.0" has been redefined as 5.0. This situation, which sometimes underlies inexplicable behavior by programs, can be avoided by not passing numerical values of an argument explicitly to a subroutine if the argument will be redefined. For example, had the first statements been

```
Z = 1.0
CALL SUB(Z)
```

the variable Z would equal 5.0, but "1.0" would not be affected.

Other errors occur when either the type or number of arguments in a call to a subroutine do not match those in its definition. For example, calling a subroutine with an integer variable may yield unexpected results if the definition is in terms of a real variable.

• *Arrays*. Scientific computing often involves dealing with *arrays*, groups of data addressed by their indices. For example, a seismogram giving a single component (e.g., vertical) of ground motion can be written as an array (U(1), U(2) . . .) of displacement versus time. Similarly, a seismogram giving all three (vertical, north–south, east–west) ground motion components can be written as a two-dimensional array

```
U(1, 1), U(1, 2), U(1, 3), U(1, 4) . . .
U(2, 1), U(2, 2), U(2, 3), U(2, 4) . . .
U(3, 1), U(3, 2), U(3, 3), U(3, 4) . . .
```

whose first index gives the component, and second index indicates the time.

Arrays are defined initially by statements giving their dimensions, i.e.,

```
DIMENSION A(N, M)
```

or

```
REAL A(N, M).
```

Typically, the computer selects a memory location for the first element in A and reserves N × M successive locations. Similarly, N×M×R locations are reserved for a three-dimensional array dimensioned (N, M, R). In Fortran, regardless of the number of dimensions, an array is stored as one-dimensional with the first index varying the most rapidly, then the second, and so on. In other words, if A is dimensioned (2, 3), the storage order is

```
A(1,1), A(2,1), A(1,2), A(2,2), A(1,3), A(2,3).
```

For two-dimensional arrays, this can be thought of as storing the array by columns. An individual array element is found by calculating its location relative to that of the first element. Thus, for an array dimensioned (N, M), with element (1, 1) at location 1, element (I, J) is found at location

```
1 + (I - 1) + (J - 1) × N.
```

Several computational difficulties can arise in dealing with arrays. A common set of errors involve being "off by one," either by starting or ending on the wrong element. This is especially easy because some computer languages (e.g., Fortran) start with the first element in an array being "1," whereas others (e.g., "C") start with the first array element as "0." Thus one needs to make sure that the array elements correspond to the expected variable values, such as seismic record times. Often, when an array index is computed by the program, an error yields an index outside the bounds dimensioned for the array. Because many compilers do not check for such errors unless specifically requested, a statement like

```
A(9) = 4.0
```

will usually be executed even for an array dimensioned

```
DIMENSION A(5).
```

Typically, the computer places 4.0 in whatever is 8 locations in memory beyond A(1). This location may contain some other variable, or a portion of the program itself. Often the program continues until it requires the contents of the overwritten location, at which point several things may occur. At best, the program "crashes"; at worst, it continues the calculation with erroneous values that propagate. Array element out-of-bounds problems are among the most common and most frustrating difficulties in scientific programming. When a compiler provides array bounds checking, it is worth using.

The nature of array storage can also lead to inefficient programs. On many computers, data which are actually on disk can be treated as resident in memory, and are automatically "swapped" into physical memory when needed. For efficiency, large adjacent regions of the disk are often swapped into physical memory together. Efficient programs minimize swapping by making the most possible use of data that reside in physical memory. By contrast, inefficient programs can produce "thrashing," a situation in which much of the computer's time is spent swapping rather than computing.

For example, consider[4]

```
DIMENSION A(1000, 1000)
DO 10 I = 1,1000
DO 10 J = 1,1000
10 A(I, J) = I + J
```

Because the elements of A are stored in column order, A(1, 1) and A(1, 2) are a thousand locations apart. It would be more efficient to reverse the loops

```
10 A(J, I) = I + J
```

so that adjacent locations (A(1, 1), A(2, 1) . . .) were used successively.

• *Uninitialized variables*. Problems frequently result from *uninitialized variables*: those used in calculation without their values being set. A common example, summing an array

```
DO 10 I = 1, N
10 SUM = SUM + A(I)
```

can give strange results unless the compiler initializes SUM as zero. Because this is not always the case, it is thus wise to explicitly initialize, e.g.,

```
SUM = 0.0
```

before executing the loop. Proper initialization also helps to ensure that programs do not give different results on different computers.

• *The computer may be wrong*. Although most problems result from programming errors, a very small fraction of the time the error may be the computer's. Compilers have been known to contain "bugs" in common routines such as square root, tangent, or complex arithmetic. This tempting explanation for the failure of a long and intricate program can generally be rejected unless a test program that carries out only the suspect operation yields the wrong answer.

A.8.5 Some philosophical points

To close our discussion, a few general thoughts are worth considering. Historically, computers were considered a scarce and valuable resource. Currently, as computer power increases and costs fall, it is increasingly practical to carry out investi-

[4] Hatton (1983c).

gations numerically. One example is the change, both in exploration and in global seismology, from earth models whose properties vary only with depth, to three-dimensional models that are evaluated numerically.

The role of analytic solutions is also changing. In addition to the traditional goal of providing exact solutions to simplified problems, analytic solutions provide test cases for numerical solutions of more complex problems. Analytic solutions can also yield the insight needed to evaluate numerical results.

Along with the increase in the complexity of problems that can be solved computationally comes an increase in the volume of output. Fortunately, a parallel development has been the increasing role of graphic output, often in color. The proverb "A picture is worth a thousand words" may be unduly conservative in this context. A thousand words on a computer might be 32,000 bits; graphic output often makes it possible to visualize data with millions of bits.

Finally, software such as spreadsheets or programs with sophisticated general mathematical capabilities often eliminates the need to write programs for a specific application. In this book, we do not assume that such software will be used for the problems, although many could be done this way. We think that programming without using such software gives a deeper understanding of the underlying principles. Hence, in educational applications, we strongly favor programming, even if in non-educational applications ease of use may favor sophisticated software.

Further reading

Many texts cover portions of the mathematical material summarized here. Feynman (1982) discusses general issues of the relations between mathematics and science. Butkov (1968) and Menke and Abbott (1990) provide introductions to many of these topics. Fung (1969), Hay (1953), Jeffreys and Jeffreys (1950), and Marion (1970) treat vectors, vector transformations, and vector differential operators. Applied linear algebra texts such as Franklin (1968) and Noble (1969) deal with the range of the subject including numerical methods.

Articles by Hatton (1983a–d, 1984a,b, 1985) provide a broad and witty introduction to computer science for geophysicists. Eckhouse and Morris (1979) and Sloan (1980) cover topics in computer software, including the representation of numbers and arithmetic operations. Kernighan and Plauger (1976, 1978) discuss topics in programming style. Brooks (1975) treats issues in the development and organization of computer software. Numerical analysis texts like Froberg (1969) cover round-off and other sources of error in numerical computations. Harkrider (1988) gives an entertaining anecdotal account of early (*c*.1960) computer usage in seismology.

The application of spherical geometry to the paths between an earthquake and a receiver, including the effects of the earth's ellipticity are discussed by Ben-Menahem and Singh (1981) and Bullen and Bolt (1985). The theory of the earth's shape is treated by Cook (1973) and Jeffreys (1976).

Problems

1. Find the angle between the vectors (1, 4, 2) and (2, 3, 1).
2. Show, using index notation, that for the three-dimensional vectors **a**, **b**, **c**:
   (a) $\mathbf{a} \times \mathbf{b}$ is perpendicular to both **a** and **b**.
   (b) $|\mathbf{a} \times \mathbf{b}| = |\mathbf{a}||\mathbf{b}| \sin\theta$, where θ is the angle between the two vectors.
   (c) $\mathbf{a} \cdot (\mathbf{b} + \mathbf{c}) = \mathbf{a} \cdot \mathbf{b} + \mathbf{a} \cdot \mathbf{c}$.
   (d) $\mathbf{a} \times (\mathbf{b} + \mathbf{c}) = \mathbf{a} \times \mathbf{b} + \mathbf{a} \times \mathbf{c}$.
   (e) $\mathbf{a} \cdot (\mathbf{b} \times \mathbf{c}) = \mathbf{b} \cdot (\mathbf{c} \times \mathbf{a}) = \mathbf{c} \cdot (\mathbf{a} \times \mathbf{b})$.
   (f) $\mathbf{a} \times (\mathbf{b} \times \mathbf{c}) = \mathbf{b}(\mathbf{a} \cdot \mathbf{c}) - \mathbf{c}(\mathbf{a} \cdot \mathbf{b})$.
3. Show that for arbitrary matrices A, B, and C:
   (a) $(AB)^T = B^T A^T$.
   (b) $(ABC)^T = C^T B^T A^T$.
4. Prove the following properties of determinants for the case of a 2×2 matrix:
   (a) The determinant of a matrix equals the determinant of its transpose.
   (b) If two rows or columns of a matrix are interchanged, the determinant has the same absolute value, but its sign changes.
   (c) If a multiple of one row (or column) of a matrix is added to another row (or column), the determinant is unchanged.
   (d) If two rows or columns of a matrix are the same, the determinant is zero.
5. Express the determinant of a 3×3 matrix using the definition in Eqn A.4.17.
6. Prove that if A has an inverse, the two solutions **x** and **y** satisfying $A\mathbf{x} = \mathbf{b}$ and $A\mathbf{y} = \mathbf{b}$ are equal.
7. Find the inverse of the matrix

$$\begin{pmatrix} 1 & 2 \\ 5 & 4 \end{pmatrix}$$

   both by the cofactor method and by row operations. Check that the solution is in fact the inverse.
8. Show that the inverse of a 2×2 matrix A is given by

$$A^{-1} = \frac{1}{|A|} \begin{pmatrix} a_{22} & -a_{12} \\ -a_{21} & a_{11} \end{pmatrix}.$$

9. Show that A, the transformation matrix for a rotation about the $\hat{\mathbf{e}}_3$ axis (Eqn A.5.9) satisfies $A^T A = I$ and is thus orthogonal.
10. Prove that the magnitude of a vector is preserved by an orthogonal transformation.
11. Expand the determinant that give the eigenvalues of a 3×3 matrix (Eqn A.5.19) and verify that the invariants (Eqn A.5.21) are the coefficients of the characteristic polynomial.
12. Prove the following vector identities using index notation:
    (a) For any vector field $\mathbf{u}(\mathbf{x})$, $\nabla \cdot (\nabla \times \mathbf{u}) = 0$.
    (b) For any scalar function $\phi(\mathbf{x})$, $\nabla \times \nabla\phi = 0$.
13. For the vector field $\mathbf{u}(x, y, z) = (3x^2y^2 + z, 2x^3y + 2y, x)$, find:
    (a) $\nabla \cdot \mathbf{u}$.
    (b) $\nabla \times \mathbf{u}$.
    (c) $\nabla^2 \mathbf{u}$.
    (d) A scalar field $\phi(x, y, z)$ such that $\mathbf{u} = \nabla\phi$.

14. Use index notation to show that the Laplacian in Cartesian coordinates of any vector field $\mathbf{u}(\mathbf{x})$ satisfies

$$\nabla^2 \mathbf{u} = \nabla(\nabla \cdot \mathbf{u}) - \nabla \times \nabla \times \mathbf{u}.$$

15. Show that at any point in a spherical coordinate system, the spherical basis vectors $(\hat{\mathbf{e}}_r, \hat{\mathbf{e}}_\theta, \hat{\mathbf{e}}_\phi)$ form an orthonormal set.

16. Use Eqn A.7.6 to derive the angular distance Δ between the locations of an earthquake and a seismic station as given in Eqn A.7.7.

Computer problems

The solutions may be useful for other problems in this and other chapters.

C-1. Find the largest integer your computer allows by starting with "2," "2 × 2," "2 × 2 × 2," and doing successive multiplication by 2. What happens when you exceed this number? Do the same for floating point numbers using "10.0" instead of "2" in both single and double precision. Does double precision allow larger floating point numbers?

C-2. Find when your computer starts to show round-off error by starting with "10.0" and doing successive multiplications by 10.0. At each step, add 1.0 to the result and subtract the two numbers. When does the difference become zero? Do the same in double precision.

C-3. Write subroutines to do the following operations on an input vector in three dimensions:

(a) Find the magnitude of a vector.
(b) Find the sum of two vectors.
(c) Find the scalar product of two vectors.
(d) Find the vector product of two vectors.

Your subroutines should include comment lines explaining the purpose of the routine and the various inputs and outputs.

C-4. Write a subroutine using the necessary subroutines from problem C-3 to find the angle between two vectors.

C-5. Use the solutions to problems C-3 and C-4 to find the magnitude, sum, scalar product, and vector product of the vectors (1, 4, 2) and (2, 3, 1), and the angle between the two vectors.

C-6. (a) Write a subroutine to multiply an $n \times m$ matrix by an m-element vector.

(b) Write a subroutine to multiply an $n \times m$ matrix by an $m \times r$ matrix.

(c) Write a subroutine to find the determinant of a 3×3 matrix.

C-7. (a) Write a subroutine that uses Gaussian elimination with partial pivoting to solve the system of equations $A\mathbf{x} = \mathbf{b}$. The routine should take an arbitrary 3×3 matrix A and 3-element vector $\mathbf{b}$ as inputs. The program should test the solution by multiplying $A\mathbf{x}$ and subtracting $\mathbf{b}$ from the result. The subroutines from C-6 may be helpful.

(b) Use the subroutine to solve

$$\begin{pmatrix} 10 & -7 & 0 \\ -3 & 2 & 6 \\ 5 & -1 & 5 \end{pmatrix} \begin{pmatrix} x_1 \\ x_2 \\ x_3 \end{pmatrix} = \begin{pmatrix} 7 \\ 4 \\ 6 \end{pmatrix}.$$

C-8. (a) Write functions that return the values of the δ_{ij} and ε_{ijk} symbols given the indices as arguments. Test the functions and show that they give the correct values.

(b) Write a program that uses these two functions to prove the identity

$$\varepsilon_{ijk}\varepsilon_{ist} = \delta_{js}\delta_{kt} - \delta_{jt}\delta_{ks}$$

by testing all possible combinations of indices.

C-9. (a) Write a subroutine to invert a 3×3 matrix using elementary row operations. The subroutine should first check to see if the matrix is singular. It should test the result by multiplying by the original matrix.

(b) Use this routine to invert

$$\begin{pmatrix} 1 & -1 & -1 \\ 3 & -1 & 2 \\ 2 & 2 & 3 \end{pmatrix}.$$

C-10. (a) Write a program to solve a 3×3 system of equations $A\mathbf{x} = \mathbf{b}$ using the matrix inversion routine from the previous problem. The program should test the solution by multiplying $A\mathbf{x}$ and subtracting $\mathbf{b}$ from the result. The subroutines from C-6 may be helpful.

(b) Use the program to solve the system of equations in C-7.

C-11. (a) Write a subroutine to find the roots of a general cubic equation using the method given below.[1]

A cubic equation $y^3 + py^2 + qy + r = 0$ may be converted to

$$x^3 + ax + b = 0$$

by defining

$$y = x - p/3, \quad a = (3q - p^2)/3, \quad b = (2p^3 - 9pq + 27r)/27.$$

If p, q, and r are real, the quantity

$$c = b^2/4 + a^3/27$$

characterizes the roots: if $c > 0$, there is one real root and two conjugate imaginary roots; if $c = 0$, there are three real roots, of which two are equal; and if $c < 0$, there are three real and unequal roots. Using

$$A = (-b/2 + c^{1/2})^{1/3}, \quad B = (-b/2 - c^{1/2})^{1/3},$$

the values of x given by

$$x = A + B, \quad [-(A + B) + (A - B)\sqrt{-3}]/2,$$
$$-[(A + B) + (A - B)\sqrt{-3}]/2$$

are the roots.

The subroutine requires complex arithmetic and should test the roots by substituting back into the equation.

(b) Use the result to solve

$$y^3 - 8y^2 + 19y - 12 = 0.$$

C-12. (a) Write a subroutine to find the eigenvalues and eigenvectors of a real, symmetric 3×3 matrix, using the results of C-11. The program should check that the eigenvectors and eigenvalues satisfy their definition. Be careful to avoid dividing by zero.

(b) Use this subroutine to find the eigenvalues and eigenvectors of

$$\begin{pmatrix} 1 & 2 & 3 \\ 2 & 4 & 5 \\ 3 & 5 & 6 \end{pmatrix}.$$

[1] Beyer (1984).

C-13. **(a)** Write a program that accepts the latitude and longitude of two points on the earth's surface and finds the angular distance and distance along the earth's surface between them, and the azimuth and back azimuth.

(b) Use your program to find the distances and azimuths between:

(i) Cairo, Illinois (37°N, 89°W) and Cairo, Egypt (30°N, 32°E).

(ii) Berlin, New Hampshire (44.5°N, 71.5°W) and Berlin, Germany (52.5°N, 13.5°E).

(iii) Montevideo, Minnesota (45°N, 95.5°W) and Montevideo, Uruguay (35°S, 56°W).

(iv) Mexico, Maine (44.5°N, 70.5°W) and Mexico City, Mexico (19°N, 99°W).

References

Agnew, D. C., B. Berger, R. Buland, W. Farrell, and F. Gilbert (1976) International deployment of accelerometers: A network for very long period seismology, *Eos. Trans. Am. Geophys. Un.*, *57*, 180–8.

Agnew, D. C., *et al.* (1988) *Probabilities of Large Earthquakes Occurring in California on the San Andreas Fault*, US Geol. Survey, Open-File Rep.

Ahrens, T. J. (ed.) (1995a) *Global Earth Physics: a handbook of physical constants*, Am. Geophys. Un., Washington, DC.

Ahrens, T. J. (ed.) (1995b) *Mineral Physics and Crystallography: a handbook of physical constants*, Am. Geophys. Un., Washington, DC.

Ahrens, T. J. (ed.) (1995c) *Rock Physics and Phase Relations: a handbook of physical constants*, Am. Geophys. Un., Washington, DC.

Aki, K. (1980) Presidential address to the Seismological Society of America, *Bull. Seism. Soc. Am.*, *70*, 1969–76.

Aki, K., and P. G. Richards (1980) *Quantitative Seismology: theory and methods*, W. H. Freeman, San Francisco.

Al-eqabi, G. I., K. Koper, and M. E. Wysession (2001) Source characterization of Nevada test site explosions and western United States earthquakes using Lg waves, with implications for regional discrimination, *Bull. Seism. Soc. Am.*, *91*, 140–53.

Alexander, D. (1993) *Natural Disasters*, Chapman and Hall, New York.

Ambraseys, N. (1989) Studies begin on Armenian quake, *Eos Trans. Am. Geophys. Un.*, *70*, 17.

Anderson, D. L. (1989) *Theory of the Earth*, Blackwell, Oxford.

Ando, M. (1975) Source mechanisms and tectonic significance of historical earthquakes along the Nankai Trough, Japan, *Tectonophysics*, *27*, 119–40.

Ando, M., Y. Ishikawa, and F. Yamazaki (1983) Shear wave polarization anisotropy in the upper mantle beneath Honshu, Japan, *J. Geophys. Res.*, *88*, 5850–64.

Argus, D. F., R. G. Gordon, C. DeMets, and S. Stein (1989) Closure of the Africa-Eurasia-North America plate motion circuit and tectonics of the Gloria fault, *J. Geophys. Res.*, *94*, 5585–602.

Astiz, L., P. S. Earle, and P. Shearer (1996) Global stacking of broadband seismograms, *Seism. Res. Lett.*, *67*, 8–18.

Atkinson, G. M., and I. Beresnev (1997) Don't call it stress drop, *Seism. Res. Lett.*, *68*, 3–4.

Atkinson, G. M., and D. M. Boore (1995) Ground-motion relations for Eastern North America, *Bull. Seism. Soc. Am.*, *85*, 17–30.

Babuska, V., and M. Cara (1991) *Seismic Anisotropy in the Earth*, Kluwer Academic Publishers, Boston.

Baker, B. B., and E. T. Copson (1950) *The Mathematical Theory of Huygens' Principle*, Clarendon Press, Oxford.

Batchelor, G. (1967) *An Introduction to Fluid Dynamics*, Cambridge University Press, Cambridge.

Bath, M., and A. J. Berkhout (1984) *Mathematical Aspects of Seismology*, Geophysical Press, London.

Bebout, G. E., D. W. Scholl, S. H. Kirby, and J. P. Platt (1996) *Subduction: top to bottom*; Geophysical Monograph 96, Am. Geophys. Un., Washington, DC.

Benioff, H. (1955) Seismic evidence for crustal structure and tectonic activity, in A. Poldervaart (ed.), *Crust of the Earth*, Geol. Soc. Amer. Spec. Pap. 62, pp. 61–74.

Ben-Menahem, A., and S. J. Singh (1981) *Seismic Waves and Sources*, Springer-Verlag, New York.

Bennett, R. A., J. L. Davis, and B. P. Wernicke (1999) Present-day pattern of Cordilleran deformation in the Western United States, *Geology*, *27*, 371–4.

Bent, A. (1995) A complex double-couple source mechanism for the M_s 7.2 1929 Grand Banks earthquake, *Bull. Seism. Soc. Am.*, *85*, 1003–20.

Benz, H. M., and J. E. Vidale (1993) Sharpness of upper-mantle discontinuities determined from high-frequency reflections, *Nature*, *365*, 147–50.

Bevington, P. R., and D. K. Robinson (1992) *Data Reduction and Error Analysis for the Physical Sciences*, McGraw-Hill, New York.

Beyer, W. H. (1984) *CRC Standard Mathematical Tables*, CRC Press, Boca Raton, FL.

Bina, C. R. (1997) Patterns of deep seismicity reflect buoyancy stresses due to phase transitions, *Geophys. Res. Lett.*, *24*, 3301–4.

Bina, C. R., and M. Liu (1995) A note on the sensitivity of mantle convection models to composition-dependent phase relations, *Geophys. Res. Lett.*, *22*, 2565–8.

Bina, C. R., and B. J. Wood (1987) The olivine-spinel transitions: Experimental and thermodynamic constraints and implications for the nature of the 400 km seismic discontinuity, *J. Geophys. Res.*, *92*, 4853–66.

Birch, F. (1952) Elasticity and constitution of the Earth's interior, *J. Geophys. Res.*, *57*, 227–86.

Birch, F. (1954) The earth's mantle: Elasticity and constitution, *Trans. Am. Geophys. Un.*, *35*, 79–85.

Birch, F. (1968) On the possibility of large changes in the earth's volume, *Phys. Earth Planet. Inter.*, *1*, 141–7.

Bland, D. R. (1988) *Wave Theory and Applications*, Oxford University Press, New York.

Bodine, J. H., M. S. Steckler, and A. B. Watts (1981) Observations of flexure and the rheology of the oceanic lithosphere, *J. Geophys. Res.*, *86*, 3695–707.

Boehler, R. (1996) Melting temperature of the Earth's mantle and core: Earth's thermal structure, *Ann. Rev. Earth Planet. Sci.*, *24*, 15–40.

Bolt, B. A. (1976) *Nuclear Explosions and Earthquakes: the parted veil*, W. H. Freeman, San Francisco.

Bolt, B. A. (1982) *Inside the Earth*, W. H. Freeman, San Francisco.

Bolt, B. A. (1999) *Earthquakes*, 4th edn, W. H. Freeman, San Francisco.

Bonini, W. E., and R. R. Bonini (1979) Andrija Mohorovičić: Seventy years ago an earthquake shook Zagreb, *Eos Trans. Am. Geophys. Un.*, *60*, 699–701.

Boore, D. M. (1977) Strong-motion recordings of the California earthquake of April 18, 1906, *Bull. Seism. Soc. Am.*, *67*, 561–77.

Boschi, E., G. Ekstrom, and A. Morelli (eds) (1996) *Seismic Modelling of Earth Structure*, Editrice Compositori, Rome.

Bott, M. H. P. (1982) *The Interior of the Earth: its structure, constitution and evolution*, Elsevier Science Publishing Co., Inc., New York.

Bott, M. H. P., A. P. Holder, R. E. Long, and A. L. Lucas (1970) Crustal structure beneath the granites of south-west England, in G. Newall and N. Rast (eds), *Mechanism of Igneous Intrusion*, Geol. J. Special Issue, 2, pp. 93–102.

Brace, W. F., and J. D. Byerlee (1970) California earthquakes: Why only shallow focus?, *Science*, *168*, 1573–5.

Brace, W. F., and D. L. Kohlstedt (1980) Limits on lithospheric stress imposed by laboratory experiments, *J. Geophys. Res.*, *85*, 6248–52.

Bracewell, R. (1978) *The Fourier Transform and its Applications*, McGraw-Hill, New York.

Braile, L. W., and C. S. Chiang (1986) The continental Mohorovičić discontinuity: Results from near vertical and wide angle seismic reflection studies, in M. Barazangi and L. Brown (eds), *Reflection Seismology: a global perspective*, Geodynamics Series, 13, Am. Geophys. Un., Washington, DC, pp. 257–72.

Braile, L. W., and R. B. Smith (1975) Guide to the interpretation of crustal refraction profiles, *Geophys. J. R. Astron. Soc.*, *40*, 145.

Braile, L. W., W. J. Hinze, R. G. Keller, E. G. Lidiak, and J. L. Sexton (1986) Tectonic development of the New Madrid rift complex, Mississippi embayment, North America, *Tectonophysics*, *131*, 1–21.

Braile, L. W., W. J. Hinze, R. R. B. von Frese, and G. Randy Keller (1989) Seismic properties of the crust and uppermost mantle of the conterminous United States and Canada, in L. C. Pakiser and W. D. Mooney (eds), *Geophysical Framework of the Continental United States*, Geol. Soc. Amer. Mem. 172, Boulder, CO., pp. 655–79.

Bray, J. D. (1995) Geotechnical earthquake engineering, in W. F. Chen (ed.), *The Civil Engineering Handbook*, CRC Press, Boca Raton, FL.

Brennan, B. J., and D. E. Smylie (1981) Linear viscoelasticity and dispersion in seismic wave propagation, *Rev. Geophys.*, *19*, 233–46.

Brigham, E. O. (1974) *The Fast Fourier Transform*, Prentice-Hall, Englewood Cliffs, NJ.

Brooks, F. P. (1975) *The Mythical Man-Month*, Addison-Wesley, Reading, MA.

Brown, G. C., and A. E. Mussett (1993) *The Inaccessible Earth*, Chapman and Hall, London.

Brumbaugh, D. (1999) *Earthquakes: science and society*, Prentice-Hall, Upper Saddle River, NJ.

Brune, J. N., W. M. Ewing, and J. T. F. Kuo (1961) Group and phase velocities for Rayleigh waves of period greater than 380 seconds, *Science*, *133*, 757–8.

Bullen, K. E. (1975) *The Earth's Density*, Chapman and Hall, London.

Bullen, K. E., and B. A. Bolt (1985) *An Introduction to the Theory of Seismology*, 4th edn, Cambridge University Press, Cambridge.

Bürgmann, R., P. A. Rosen, and E. J. Fielding (2000) Synthetic aperture radar interferometry to measure earth's surface topography and its deformation, *Ann. Rev. Earth Planet. Sci.*, *28*, 169–209.

Butkov, E. (1968) *Mathematical Physics*, Addison-Wesley, Reading, MA.

Byerlee, J. D. (1978) Friction of rocks, *Pure Appl. Geophys.*, *116*, 615–26.

Capon, J. (1969) Investigation of long-period noise at the large aperture seismic array, *J. Geophys. Res.*, *74*, 3182–94.

Chase, C. G. (1972) The *n*-plate problem of plate tectonics, *Geophys. J. R. Astron. Soc.*, *29*, 117–22.

Chase, C. G. (1978) Plate kinematics: The Americas, East Africa, and the rest of the world, *Earth Planet. Sci. Lett.*, *37*, 355–68.

Chave, A. D. (1979) Lithospheric structure of the Walvis Ridge from Rayleigh wave dispersion, *J. Geophys. Res.*, *84*, 6840–8.

Chen, W.-P., and P. Molnar (1983) Focal depths of intracontinental and intraplate earthquakes and their implications for the thermal and mechanical properties of the lithosphere, *J. Geophys. Res.*, *88*, 4183–214.

Chinnery, M. A. (1961) The deformation of the ground around surface faults, *Bull. Seism. Soc. Am.*, *51*, 355–72.

Chopra, A. K. (1995) *Dynamics of Structures: theory and applications to earthquake engineering*, Prentice-Hall, Upper Saddle River, NJ.

Choy, G., and P. G. Richards (1975) Pulse distortion and Hilbert transformation in multiply reflected and refracted body waves, *Bull. Seism. Soc. Am.*, *65*, 55–70.

Christensen, U. R. (1995) Effects of phase transitions on mantle convection, *Ann. Rev. Earth Planet. Sci.*, *23*, 65–87.

Chu, D., and R. G. Gordon (1998) Current plate motions across the Red Sea, *Geophys. J. Int.*, *135*, 313–28.

Chu, D., and R. G. Gordon (1999) Evidence for motion between Nubia and Somalia along the Southwest Indian ridge, *Nature*, *398*, 64–7.

Chung, W.-Y., and H. Kanamori (1980) Variation of seismic source parameters and stress drops within a descending slab and its implications in plate mechanics, *Phys. Earth Planet. Inter.*, *23*, 134–59.

Claerbout, J. F. (1976) *Fundamentals of Geophysical Data Processing*, McGraw-Hill, New York.

Claerbout, J. F. (1985) *Imaging the Earth's Interior*, Blackwell, Oxford.

Cloetingh, S., and R. Wortel (1985) Regional stress field of the Indian plate, *Geophys. Res. Lett.*, *12*, 77–80.

Coburn, A. W., and R. J. S. Spence (1992) *Earthquake Protection*, Wiley, New York.

Cook, A. H. (1973) *Physics of the Earth and Planets*, Wiley, New York.

Cox, A. (1973) *Plate Tectonics and Geomagnetic Reversals*, W. H. Freeman, San Francisco.

Cox, A., and R. B. Hart (1986) *Plate Tectonics: how it works*, Blackwell, Palo Alto, CA.

Creager, K. C. (1992) Anisotropy of the inner core from differential travel times of the phases PKP and PKIKP, *Nature*, *356*, 309–14.

Crossley, D. J. (ed.) (1997) *Earth's Deep Interior*, Gordon and Breach, Amsterdam.

Dahlen, F. A., and J. Tromp (1998) *Theoretical Global Seismology*, Princeton University Press, Princeton, NJ.

Davies, G. F. (1999) *Dynamic Earth: plates, plumes and mantle convection*, Cambridge University Press, Cambridge.

Davis, P., D. Jackson, and Y. Kagan (1989) The longer it has been since the last earthquake, the longer the expected time till the next, *Bull. Seism. Soc. Am.*, *79*, 1439–56.

DeMets, C., R. G. Gordon, D. F. Argus, and S. Stein (1990) Current plate motions, *Geophys. J. Int.*, *101*, 425–78.

DeMets, C., R. G. Gordon, D. F. Argus, and S. Stein (1994) Effect of recent revisions to the geomagnetic reversal time scale on estimates of current plate motion, *Geophys. Res. Lett.*, *21*, 2191–4.

Dewey, J. W. (1987) Instrumental seismicity of Central Idaho, *Bull. Seism. Soc. Am.*, *77*, 819–36.

Diebold, J. B., and P. L. Stoffa (1981) The travel time equation, tau-p mapping and inversion of common midpoint data, *Geophysics*, *46*, 238–54.

Dixon, T. H. (1991) An introduction to the global positioning system and some geological applications, *Rev. Geophys.*, *29*, 249–76.

Dobrin, M. B. (1976) *Introduction to Geophysical Prospecting*, McGraw-Hill, New York.

Dobrin, M. B., and C. H. Savit (1988) *Introduction to Geophysical Prospecting*, 4th edn, McGraw-Hill, New York.

Douglas, A., J. A. Hudson, and R. G. Pearce (1988) Directivity and the Doppler effect, *Bull. Seism. Soc. Am.*, *78*, 1367–72.

Doyle, H. (1995) *Seismology*, John Wiley & Sons, Chichester.

Dziewonski, A. M., and D. L. Anderson (1981) Preliminary reference Earth model, *Phys. Earth Planet. Inter.*, *25*, 297–356.

Dziewonski, A. M., T.-A. Chou, and J. H. Woodhouse (1981) Determination of earthquake source parameters from waveform data for studies of global and regional seismicity, *J. Geophys. Res.*, *86*, 2825–52.

Eakins, P. R. (1987) Faults and faulting, in C. K. Seyfert (ed.), *Encyclopedia of Structural Geology and Plate Tectonics*, Van Nostrand Reinhold, New York, pp. 228–39.

Eaton, J. P., D. H. Richter, and W. U. Ault (1961) The tsunami of May 23, 1960 on the island of Hawaii, *Bull. Seism. Soc. Am.*, *51*, 135–57.

Eckhouse, R. E., and L. R. Morris (1979) *Minicomputer Systems*, Prentice-Hall, Englewood Cliffs, NJ.

Ekeland, I. (1993) *The Broken Dice*, University of Chicago Press, Chicago.

Engeln, J. F., and S. Stein (1984) Tectonics of the Easter plate, *Earth Planet. Sci. Lett.*, *68*, 259–70.

Engeln, J. F., D. A. Wiens, and S. Stein (1986) Mechanisms and depths of Atlantic transform earthquakes, *J. Geophys. Res.*, *91*, 548–77.

Engeln, J. F., S. Stein, J. Werner, and R. Gordon (1988) Microplate and shear zone models for oceanic spreading center reorganizations, *J. Geophys. Res.*, *93*, 2839–56.

England, P., and J. Jackson (1989) Active deformation of the continents, *Ann. Rev. Earth Planet. Sci.*, *17*, 197–226.

Evans, B., and T.-F. Wong (1992) *Fault Mechanics and Transport Properties of Rocks*, Academic Press, San Diego.

Evans, R. (1997) Assessment of schemes for earthquake prediction, *Geophys. J. Int.*, *131*, 413–20.

Ewing, W. M., W. S. Jardetsky, and F. Press (1957) *Elastic Waves in Layered Media*, McGraw-Hill, New York.

Few, A. A. (1980) Thunder, in *Atmospheric Phenomena*, W. H. Freeman, San Francisco, pp. 111–21.

Feynman, R. P. (1982) *The Character of Physical Law*, MIT Press, Cambridge MA.

Feynman, R. P. (1988) *What Do You Care What Other People Think?*, W. W. Norton, New York.

Feynman, R. P., R. B. Leighton, and M. Sands (1963) *The Feynman Lectures on Physics*, Addison-Wesley, Reading, MA.

Finlayson, D. M., J. H. Leven, and K. D. Wake-Dyster (1989) Large-scale lenticles in the lower crust under an intra-continental basin in eastern Australia, in R. F. Mereu, S. Mueller and D. M. Fountain (eds), *Properties and Processes of Earth's Lower Crust*, IUGG 6, Am. Geophys. Un., Washington, DC, pp. 1–16.

Fischman, J. (1992) Falling into the gap, *Discover*, Oct., 58–63.

Flesch, L. M., W. E. Holt, A. J. Haines, and B. Shen-Tu (2000) Dynamics of the Pacific-North American plate boundary zone in the western United States, *Science*, *287*, 834–6.

Forsyth, D. W. (1975) The early structural evolution and anisotropy of the oceanic upper mantle, *Geophys. J. R. Astron. Soc.*, *43*, 103–62.

Forsyth, D. W., and S. Uyeda (1975) On the relative importance of the driving forces of plate motion, *Geophys. J. R. Astron. Soc.*, *43*, 162–200.

Forsyth, D. W., *et al.* (1998) Imaging the deep seismic structure beneath a mid-ocean ridge: The MELT experiment, *Science*, *280*, 1215–18.

Fouch, M. J., K. M. Fischer, E. M. Parmentier, M. E. Wysession, and T. J. Clarke (2000) Shear wave splitting, continental keels, and patterns of mantle flow, *J. Geophys. Res.*, *105*, 6255–76.

Foulger, G. R. *et al.* (2001) Seismic tomography shows that upwelling beneath Iceland is confined to the upper mantle, *Geophys. J. Int.*, 146, 504–30.

Fountain, D. M., and N. I. Christensen (1989) Composition of the crust and upper mantle: a review, in L. C. Pakiser and W. D. Mooney (eds), *Geophysical Framework of the Continental United States*, Geol. Soc. Amer. Mem. 172, Boulder, Co, pp. 711–41.

Fowler, C. M. R. (1990) *The Solid Earth: an introduction to global geophysics*, Cambridge University Press, Cambridge.

Frankel, A., C. Mueller, T. Barnhard, D. Perkins, E. Leyendecker, N. Dickman, S. Hanson, and M. Hopper (1996) *National Seismic Hazard Maps Documentation*, US Geol. Survey, Open-File Rep. 96–532, US Government Printing Office, Washington, DC.

Franklin, J. N. (1968) *Matrix Theory*, Prentice-Hall, Englewood Cliffs, NJ.

Freedman, D., R. Pisani, R. Purves, and A. Adhikari (1991) *Statistics*, W. W. Norton, New York.

French, A. P. (1971) *Vibrations and Waves*, W. W. Norton, New York.

Freymueller, J. T., S. C. Cohen, and H. J. Fletcher (2000) Spatial variations in present-day deformation, Kenai Peninsula, Alaska, and their implications, *J. Geophys. Res.*, *105*, 8079–101.

Froberg, C. E. (1969) *Introduction to Numerical Analysis*, 2nd edn, Addison-Wesley, Reading, MA.

Frohlich, C. (1989) The nature of deep focus earthquakes, *Ann. Rev. Earth Planet. Sci.*, *17*, 227–54.

Fung, Y. C. (1965) *Foundations of Solid Mechanics*, Prentice-Hall, Englewood Cliffs, NJ.

Fung, Y. C. (1969) *A First Course in Continuum Mechanics*, Prentice-Hall, Englewood Cliffs, NJ.

Garnero, E. (2000) Heterogeneity of the lowermost mantle, *Ann. Rev. Earth Planet. Sci.*, *28*, 509–37.

Geller, R. J. (1976) Scaling relations for earthquake source parameters and magnitudes, *Bull. Seism. Soc. Am.*, *66*, 1501–23.

Geller, R. J. (1997) Earthquake prediction: A critical review, *Geophys. J. Int.*, *131*, 425–50.

Geller, R. J., and H. Kanamori (1977) Magnitudes of great shallow earthquakes from 1904 to 1952, *Bull. Seism. Soc. Am.*, *67*, 587–98.

Geller, R. J., and S. Stein (1977) Split free oscillation amplitudes for the 1960 Chilean and 1964 Alaskan earthquakes, *Bull. Seism. Soc. Am.*, *67*, 651–60.

Geller, R. J., and S. Stein (1978) Normal modes of a laterally heterogeneous body: A one dimensional example, *Bull. Seism. Soc. Am.*, *68*, 103–16.

Geller, R. J., D. D. Jackson, Y. Kagan, and F. Mulargia (1997) Earthquakes cannot be predicted, *Science*, *275*, 1616–17.

Gere, J. M., and H. C. Shah (1984) *Terra Non Firma: understanding and preparing for earthquakes*, W. H. Freeman, New York.

Geschwind, C.-H. (2001) *California Earthquakes: science, risk, and the politics of hazard mitigation*, Johns Hopkins University Press, Baltimore.

Gibson, R. L., and A. R. Levander (1988) Lower crustal reflectivity patterns in wide-angle seismic recordings, *Geophys. Res. Lett.*, *15*, 617–20.

Gledhill, K., and D. Gubbins (1996) SKS splitting and the seismic anisotropy of the mantle beneath the Hikurangi subduction zone, New Zealand, *Phys. Earth Planet. Inter.*, *95*, 227–36.

Gordon, R. G. (1998) The plate tectonic approximation: Plate non-rigidity, diffuse plate boundaries, and global reconstructions, *Ann. Rev. Earth Planet. Sci.*, *26*, 615–42.

Gordon, R. G., and S. Stein (1992) Global tectonics and space geodesy, *Science*, *256*, 333–42.

Green, H. W., II, and H. Houston (1995) The mechanics of deep earthquakes, *Ann. Rev. Earth Planet. Sci.*, *23*, 169–213.

Gregersen, S., and P. Basham (1989) *Earthquakes at North-Atlantic Passive Margins: neotectonics and post-glacial rebound*, Kluwer, Dordrecht.

Griffiths, D. H., and R. F. King (1981) *Applied Geophysics for Geologists and Engineers; the elements of geophysical prospecting*, Pergamon, Oxford.

Gubbins, D. (1990) *Seismology and Plate Tectonics*, Cambridge University Press, Cambridge.

Gurnis, M., M. E. Wysession, E. Knittle, and B. Buffett (eds) (1998) *The Core–Mantle Boundary Region*, Am. Geophys. Un., Washington, DC.

Gutenberg, Beno (1959) *Physics of the Earth's Interior*, Academic Press, London.

Hale, L. D., and G. A. Thompson (1982) The seismic reflection character of the continental Mohorovicic discontinuity, *J. Geophys. Res.*, *87*, 4625–35.

Hanks, T. C. (1997) Imperfect science: Uncertainty, diversity, and experts, *Eos Trans. Am. Geophys. Un.*, *78*, 369–77.

Hanks, T. C., and C. A. Cornell (1994) Probabilistic seismic hazard analysis: A beginner's guide, in *Proceedings of the Fifth Symposium on Current Issues Related to Nuclear Power Plant Structures, Equipment, and Piping*, I/1-1 to I/1-17, North Carolina State University, Raleigh, NC.

Hanks, T. C., and H. Kanamori (1979) A moment magnitude scale, *J. Geophys. Res.*, *84*, 2348–50.

Hannay, J. H. (1986) Intensity fluctuations from a one-dimensional random wavefront, in B. J. Uscinski (ed.), *Wave Propagation and Scattering*, Oxford University Press, Oxford, pp. 37–48.

Harkrider, D. G. (1988) The early years of computational seismology at Caltech, *Bull. Seism. Soc. Am.*, *78*, 2105–9.

Hasegawa, A., N. Umino, and A. Takagi (1978) Double-planed structure of the deep seismic zone in the north-eastern Japan arc, *Tectonophysics*, *47*, 43–58.

Hasegawa, H. S., and H. Kanamori (1987) Source mechanism of the magnitude 7.2 Grand Banks earthquake of November 1929; double couple or submarine landslide?, *Bull. Seism. Soc. Am.*, *77*, 1984–2004.

Hatton, L. (1983a) Computer science for geophysicists, part I: Elements of a seismic data processing system, *First Break*, *1*, June, 18–24.

Hatton, L. (1983b) Computer science for geophysicists, part II: Seismic computer system architecture, *First Break*, *1*, Sept., 18–22.

Hatton, L. (1983c) Computer science for geophysicists, part III: Operating systems, I/O and the interrupt, *First Break*, *1*, Oct., 13–19.

Hatton, L. (1983d) Computer science for geophysicists, part IV: The user-interface, *First Break*, *1*, Nov., 18–23.

Hatton, L. (1984a) Computer science for geophysicists, part V: Databases and expert systems, *First Break*, *2*, Jan., 9–15.

Hatton, L. (1984b) Computer science for geophysicists, part VI: Communications and networks, *First Break*, *2*, Sept., 9–17.

Hatton, L. (1985) Computer science for geophysicists, part VII: Form and structure in programming, *First Break*, *3*, April, 9–19.

Hatton, L., M. H. Worthington, and J. Makin (1986) *Seismic Data Processing*, Blackwell, Oxford.

Hay, G. E. (1953) *Vector and Tensor Analysis*, Dover, New York.

Heaton, T. H., and S. H. Hartzell (1988) Earthquake ground motions, *Ann. Rev. Earth Planet. Sci.*, *16*, 121–45.

Hedlin, M. A., P. M. Shearer, and P. S. Earle (1997) Seismic evidence for small-scale heterogeneity throughout the Earth's mantle, *Nature*, *387*, 145–50.

Helffrich, G., S. Stein, and B. Wood (1989) Subduction zone thermal structure and mineralogy and their relation to seismic wave reflections and conversions at the slab/mantle interface, *J. Geophys. Res.*, *94*, 753–63.

Helmberger, D. V., and L. J. Burdick (1979) Synthetic seismograms, *Ann. Rev. Earth Planet. Sci.*, *7*, 417–42.

Henrion, M., and B. Fischhoff (1986) Assessing uncertainty in physical constants, *Am. J. Phys.*, *54*, 791–8.

Hernandez, B., F. Cotton, M. Campillo, and D. Massonet (1997) A comparison between short-term (coseismic) and long-term (1 year) slip for the Landers earthquake: Measurements from strong motion and SAR interferometry, *Geophys. Res. Lett.*, *24*, 1579–82.

Hill, D. P. (1998) Science, geologic hazards, and the public in a large, restless caldera, *Seism. Res. Lett.*, *69*, 400–2.

Hindle, D., J. Kley, E. Klosko, S. Stein, T. Dixon, and E. Norabuena (2002) Consistency of geologic and geodetic displacements during Andean orogenesis, *Geophys. Res. Lett.*, *29(7)*, 10.1029/2001GL013757, 2002.

Hough, S., J. G. Armbruster, L. Seeber, and J. F. Hough (2000) On the Modified Mercalli intensities and magnitudes of the 1811/1812 New Madrid, central United States, earthquakes, *J. Geophys. Res.*, *105*, 23,839–64.

Howell, B. F., Jr. (1985) On the effect of too small a data base on earthquake frequency diagrams, *Bull. Seism. Soc. Am.*, *75*, 1205–7.

Huang, P. Y., and S. C. Solomon (1988) Centroid depths of mid-ocean ridge earthquakes: Dependence on spreading rate, *J. Geophys. Res.*, *93*, 13,445–77.

Hubbard, W. (1984) *Planetary Interiors*, Van Nostrand, New York.

Hudnut, K., *et al.* (1996) Coseismic displacements of the 1994 Northridge, California, earthquake, *Bull. Seism. Soc. Am.*, *86*, S19–36.

Hudson, J. A. (1980) *The Excitation and Propagation of Elastic Waves*, Cambridge University Press, Cambridge.

Humphreys, E., and R. Clayton (1988) Adaptation of back projection tomography to seismic travel time problems, *J. Geophys. Res.*, *93*, 1073–86.

Huygens, C. (1962 [1690]) *Treatise on Light*, trans. S. P. Thompson, Dover, New York.

Igarashi, G., S. Saeki, N. Takahata, K. Sumikawa, S. Tasaka, Y. Sasaki, M. Takahashi, and Y. Sano (1995) Ground-water radon anomaly before the Kobe earthquake in Japan, *Science*, *269*, 60–1.

Isacks, B., and M. Barazangi (1977) Geometry of Benioff zones: Lateral segmentation and downwards bending of the subducted lithosphere, in M. Talwani and W. C. Pitman, III (eds), *Island Arcs, Deep Sea Trenches and Back Arc Basins*, Maurice Ewing Ser., 1, Am. Geophys. Un., Washington, DC, pp. 99–114.

Jackson, D. D. (1972) Interpretation of inaccurate, insufficient, and inconsistent data, *Geophys. J. R. Astron. Soc.*, *28*, 97–109.

Jackson, D. D., and Y. Y. Kagan (1993) Reply, *J. Geophys. Res.*, *98*, 9919–20.

Jackson, I. (1993) Progress in the experimental study of seismic wave attenuation, *Ann. Rev. Earth Planet. Sci.*, *21*, 375–406.

Jackson, J., and D. McKenzie (1988) The relationship between plate motions and seismic moment tensors, and the rates of active deformation in the Mediterranean and Middle East, *Geophys. J. R. Astron. Soc.*, *93*, 45–73.

Jacobs, J. A. (1987) *The Earth's Core*, 2nd edn, Academic Press, London.

Jaeger, J. C. (1970) *Elasticity, Fracture and Flow, with Engineering and Geological Applications*, 3rd edn, Barnes & Noble, New York.

Jaeger, J. C., and N. G. W. Cook (1976) *Fundametals of Rock Mechanics*, Chapman and Hall, London.

Jarchow, C. M., and G. A. Thompson (1989) The nature of the Mohorovičić discontinuity, *Ann. Rev. Earth Planet. Sci.*, *17*, 475–506.

Jarosch, H., and E. Aboodi (1970) Towards a unified notation of source parameters, *Geophys. J. R. Astron. Soc.*, *21*, 513–29.

Jeanloz, R. (1990) The nature of the earth's core, *Ann. Rev. Earth Planet. Sci.*, *18*, 357–86.

Jeffreys, H. (1976) *The Earth: its origin, history, and physical constitution*, 6th edn, Cambridge University Press, Cambridge.

Jeffreys, H., and K. E. Bullen (1940) *Seismological Tables*, British Association Seismological Committee, London.

Jeffreys, H., and B. S. Jeffreys (1950) *Methods of Mathematical Physics*, Cambridge University Press, Cambridge.

Jost, M. L., and R. B. Hermann (1989) A student's guide to and review of moment tensors, *Seism. Res. Lett.*, *60*, 37–57.

Julian, B. R., and S. A. Sipkin (1985) Earthquake processes in the Long Valley Caldera area, California, *J. Geophys. Res.*, *90*, 11,155–70.

Kagan, Y. Y., and D. D. Jackson (1991) Seismic gap hypothesis: Ten years after, *J. Geophys. Res.*, *96*, 21,419–31.

Kanamori, H. (1970a) Synthesis of long-period surface waves and its application to earthquake source studies — Kurile Islands earthquake of October 13, 1963, *J. Geophys. Res.*, *75*, 5011–27.

Kanamori, H. (1970b) The Alaska earthquake of 1964: Radiation of long-period surface waves and source mechanism, *J. Geophys. Res.*, *75*, 5029–40.

Kanamori, H. (1977a) The energy release in great earthquakes, *J. Geophys. Res.*, *82*, 2981–7.

Kanamori, H. (1977b) Seismic and aseismic slip along subduction zones and their tectonic implications, in M. Talwani and W. C. Pitman, III (eds),

Island Arcs, Deep Sea Trenches and Back Arc Basins, Maurice Ewing Ser., 1, Am. Geophys. Un., Washington, DC, pp. 163–74.
Kanamori, H. (1978) Quantification of earthquakes, *Nature*, *271*, 411–14.
Kanamori, H. (1986) Rupture process of subduction-zone earthquakes, *Ann. Rev. Earth Planet. Sci.*, *14*, 293–322.
Kanamori, H. (1988) Importance of historical seismograms for geophysical research, in W. H. K. Lee, H. Meyers and K. Shimizaki (eds), *Historical Seismograms and Earthquakes of the World*, Academic Press, San Diego, pp. 16–36.
Kanamori, H. (1994) Mechanics of earthquakes, *Ann. Rev. Earth Planet. Sci.*, *22*, 207–37.
Kanamori, H., and K. Abe (1968) Digital processing of surface waves and structure of island arcs, *J. Phys. Earth*, *16*, 137–40.
Kanamori, H., and D. L. Anderson (1975) Theoretical basis of some empirical relations in seismology, *Bull. Seism. Soc. Am.*, *65*, 1073–95.
Kanamori, H., and D. L. Anderson (1977) Importance of physical dispersion in surface wave and free oscillation problems: Review, *Rev. Geophys. Space Phys.*, *15*, 105–12.
Kanamori, H., and E. Boschi (1983) *Earthquakes: observation, theory, and interpretation*, Proc. Int. Sch. Phys. "Enrico Fermi", Course 85, North Holland, Amsterdam.
Kanamori, H., and J. J. Cipar (1974) Focal process of the great Chilean earthquake May 22, 1960, *Phys. Earth Planet. Inter.*, *9*, 128–36.
Kanamori, H., and J. W. Given (1981) Use of long-period surface waves for rapid determination of earthquake-source parameters, *Phys. Earth Planet. Inter.*, *27*, 8–31.
Kanamori, H., and J. W. Given (1982) Analysis of long-period seismic waves excited by the May 18, 1980 eruption of Mount St. Helens – a terrestrial monopole?, *J. Geophys. Res.*, *87*, 5422–32.
Kanamori, H., and G. S. Stewart (1976) Mode of the strain release along the Gibbs Fracture Zone, Mid-Atlantic Ridge, *Phys. Earth Planet. Inter.*, *11*, 312–32.
Kanamori, H., E. Hauksson, and T. H. Heaton (1997) Real-time seismology and earthquake hazard mitigation, *Nature*, *390*, 461–4.
Kanasewich, E. R. (1981) *Time Sequence Analysis in Geophysics*, University of Alberta Press, Edmonton.
Karato, S., and H. A. Spetzler (1990) Defect microdynamics in minerals and solid state mechanisms of seismic wave attenuation and velocity dispersion in the mantle, *Rev. Geophys.*, *28*, 399–421.
Kaula, W. M. (1975) The seven ages of a planet, *Icarus*, *26*, 1–15.
Kearey, P., and M. Brooks (1984) *An Introduction to Geophysical Exploration*, Blackwell, Oxford.
Kearey, P., and F. Vine (1990) *Global Tectonics*, Blackwell, Oxford.
Keller, E., and N. Pinter (1996) *Active Tectonics: earthquakes, uplift, and the landscape*, Prentice-Hall, Upper Saddle River, NJ.
Kendall, J. M., and P. G. Silver (1996) Constraints from seismic anisotropy on the nature of the lowermost mantle, *Nature*, *381*, 409–12.
Kennett, B. L. N. (1977) Towards a more detailed seismic picture of the oceanic crust and mantle, *Mar. Geophys. Res.*, *3*, 7–42.
Kennett, B. L. N. (1983) *Seismic Wave Propagation in Stratified Media*, Cambridge University Press, Cambridge.
Kennett, B. L. N., and E. R. Engdahl (1991) Traveltimes for global earthquake location and phase identification, *Geophys. J. Int.*, *105*, 429–65.
Kennett, B. L. N., E. R. Engdahl, and R. Buland (1995) Constraints on seismic velocities in the Earth from travel times, *Geophys. J. Int.*, *122*, 108–24.
Kernighan, B. W., and P. J. Plauger (1976) *Software Tools*, Addison-Wesley, Reading, MA.
Kernighan, B. W., and P. J. Plauger (1978) *The Elements of Programming Style*, McGraw-Hill, New York.
Kikuchi, M., and H. Kanamori (1991) Inversion of complex body waves – III, *Bull. Seism. Soc. Am.*, *81*, 2335–50.
Kirby, S. H. (1980) Tectonic stresses in the lithosphere: Constraints provided by the experimental deformation of rocks, *J. Geophys. Res.*, *85*, 6353–63.
Kirby, S. H. (1983) Rheology of the lithosphere, *Rev. Geophys. Space Phys.*, *21*, 1458–87.
Kirby, S. H., and A. K. Kronenberg (1987) Rheology of the lithosphere: Selected topics, *Rev. Geophys.*, *25*, 1219–44.
Kirby, S. H., E. R. Engdahl, and R. Denlinger (1996a) Intermediate-depth intraslab earthquakes and arc volcanism as physical expressions of crustal and uppermost mantle metamorphism in subducting slabs, in G. E. Bebout, D. W. Scholl, S. H. Kirby and J. P. Platt (eds), *Subduction: Top to Bottom*, Am. Geophys. Un., Washington, DC, pp. 195–214.
Kirby, S. H., S. Stein, E. A. Okal, and D. C. Rubie (1996b) Metastable phase transformations and deep earthquakes in subducting oceanic lithosphere, *Rev. Geophys.*, *34*, 261–306.
Klein, C., and C. Hurlbut, Jr. (1985) *Manual of Mineralogy*, John Wiley & Sons, Inc., New York.
Klein, M. V., and T. E. Furtak (1986) *Optics*, 2nd edn, John Wiley & Sons, Inc., New York.
Klosko, E., J. DeLaughter, and S. Stein (2000) Technology in introductory geophysics: The high-low mix, *Comp. Geosci.*, *26*, 693–8.
Kovach, R. L. (1995) *Earth's Fury: an introduction to natural hazards and disasters*, Prentice-Hall, Englewood Cliffs, NJ.
Krinitzsky, E. L., J. P. Gould, and P. H. Edinger (1993) *Fundamentals of Earthquake Resistant Construction*, John Wiley & Sons, New York.
Kuhn, T. (1962) *The Structure of Scientific Revolutions*, University of Chicago Press, Chicago.
Kuo, B. Y., D. W. Forsyth, and M. W. Wysession (1987) Lateral heterogeneity and azimuthal anisotropy in the North Atlantic determined from SS-S differential travel times, *J. Geophys. Res.*, *92*, 6421–36.
Lachenbruch, A. H., and J. H. Sass (1988) The stress-heat flow paradox and thermal results from Cajon Pass, *Geophys. Res. Lett.*, *15*, 981–4.
Lambeck, K. (1988) *Geophysical Geodesy: the slow deformations of the earth*, Clarendon Press, Oxford.
Lanczos, C. (1961) *Linear Differential Operators*, Van Nostrand, London.
Langston, C. A. (1978) Moments, corner frequencies, and the free surface, *J. Geophys. Res.*, *83*, 3422–6.
Lapwood, E. R., and T. Usami (1981) *Free Oscillations of the Earth*, Cambridge University Press, Cambridge.
Larson, K., R. Bürgmann, R. Bilham, and J. T. Freymueller (1999) Kinematics of the India-Eurasia collision zone from GPS measurements, *J. Geophys. Res.*, *104*, 1077–94.
Lay, T. (1992) Nuclear testing and seismology, in *The Encyclopedia of Earth System Science*, vol. 3, Academic Press, New York, pp. 333–51.
Lay, T. (1994) *Structure and Fate of Subducting Slabs*, Academic Press, New York.
Lay, T., and T. C. Wallace (1995) *Modern Global Seismology*, Academic Press, New York.
Lewis, B. T. R. (1978) Evolution of ocean crust seismic velocities, *Ann. Rev. Earth Planet. Sci.*, *6*, 377–404.
Liu, H.-P., D. L. Anderson, and H. Kanamori (1976) Velocity dispersion due to anelasticity: Implications for seismology and mantle composition, *Geophys. J. R. Astron. Soc.*, *47*, 41–58.
Liu, M., Y. Zhu, S. Stein, Y. Yang, and J. Engeln (2000) Crustal shortening in the Andes: Why do GPS rates differ from geological rates?, *Geophys. Res. Lett.*, *18*, 3005–8.
Lomnitz, C. (1989) Comment on "temporal and magnitude dependence in earthquake recurrence models" by C. A. Cornell and S. R. Winterstein, *Bull. Seism. Soc. Am.*, *79*, 1662.
Lomnitz, C. (1994) *Fundamentals of Earthquake Prediction*, Wiley, New York.
Lorenz, E. (1993) *The Essence of Chaos*, University of Washington Press, Seattle.

Lowrie, W. (1997) *Fundamentals of Geophysics*, Cambridge University Press, Cambridge.

Lundgren, P. R., and D. Giardini (1994) Isolated deep earthquakes and the fate of subduction in the mantle, *J. Geophys. Res.*, *99*, 15,833–42.

Madariaga, R. I. (1972) Toroidal free oscillations of the laterally heterogeneous Earth, *Geophys. J. R. Astron. Soc.*, *27*, 81–100.

Main, I. G. (1978) *Vibrations and Waves in Physics*, Cambridge University Press, Cambridge.

Main, I. (1996) Statistical physics, seismogenesis, and seismic hazard, *Rev. Geophys.*, *34*, 433–62.

Malvern, L. E. (1969) *Introduction to the Mechanics of a Continuous Medium*, Prentice-Hall, Englewood Cliffs, NJ.

Marion, J. B. (1970) *Classical Dynamics of Particles and Systems*, 2nd edn, Academic Press, New York.

Marone, C. (1998) Laboratory-derived friction laws and their application to seismic faulting, *Ann. Rev. Earth Planet. Sci.*, *26*, 643–96.

Mavko, G. M. (1981) Mechanics of motion on major faults, *Ann. Rev. Earth Planet. Sci.*, *9*, 81–111.

McCann, W. R., S. P. Nishenko, L. R. Sykes, and J. Krause (1979) Seismic gaps and plate tectonics: Seismic potential for major plate boundaries, *Pure Appl. Geophys.*, *117*, 1082–147.

McClusky, S., *et al.* (2000) Global positioning system constraints on plate kinematics and dynamics in the eastern Mediterranean and Caucasus, *J. Geophys. Res.*, *105*, 5695–719.

McElhinny, M. W. (ed.) (1979) *The Earth, Its Origin, Structure and Evolution*, Academic Press Inc., New York.

McKenzie, D. P. (1969) Speculations on the consequences and causes of plate motions, *Geophys. J. R. Astron. Soc.*, *18*, 1–32.

McKenzie, D. P., and F. M. Richter (1978) Simple plate models of mantle convection, *J. Geophys.*, *44*, 441–71.

Medawar, P. (1979) *Advice to a Young Scientist*, Basic Books, New York.

Meissner, R. (1986) *The Continental Crust*, Academic Press, Inc., San Diego.

Melchior, P. (1986) *The Physics of the Earth's Core*, Pergamon Press, Oxford.

Meltzer, A. S., A. R. Levander, and W. D. Mooney (1987) Upper crustal structure in the Livermore valley and vicinity, *Bull. Seism. Soc. Am.*, *77*, 1655–73.

Menard, H. W. (1986) *The Ocean of Truth: a personal history of global tectonics*, Princeton Series in Geology and Paleontology, ed. A. G. Fischer, Princeton University Press, Princeton, NJ.

Mendiguren, J. A. (1973) High resolution spectroscopy of the Earth's free oscillations knowing the earthquake source mechanism, *Science*, *179*, 179–80.

Menke, W. (1984) *Geophysical Data Analysis: discrete inverse theory*, Academic Press, Inc., Orlando, FL.

Menke, W., and D. Abbott (1990) *Geophysical Theory*, Columbia University Press, New York.

Michael, A. J., and R. J. Geller (1984) Linear moment tensor inversion for shallow thrust earthquakes combining first motion and surface wave data, *J. Geophys. Res.*, *89*, 1889–97.

Michaels, A., D. Malmquist, A. Knap, and A. Close (1997) Climate science and insurance risk, *Nature*, *389*, 225–7.

Minster, J. B., and T. H. Jordan (1978) Present-day plate motions, *J. Geophys. Res.*, *83*, 5331–54.

Minster, J. B., T. H. Jordan, P. Molnar, and E. Haines (1974) Numerical modeling of instantaneous plate tectonics, *Geophys. J. R. Astron. Soc.*, *36*, 541–76.

Mitchell, B. J. (1995) Anelastic structure and evolution of the continental crust and upper mantle from seismic surface wave attenuation, *Rev. Geophys.*, *33*, 441–62.

Mitchell, B. J., J. Xie, and S. Baqer (1997) *Lg Excitation, Attenuation, and Source Spectral Scaling in Central and Eastern North America*, Report to the Nuclear Regulatory Commission, NUREG/CR-6563, Washington, DC.

Molnar, P. (1988) Continental tectonics in the aftermath of plate tectonics, *Nature*, *335*, 131–7.

Mooney, W. D., and C. S. Weaver (1989) Regional crustal structure and tectonics of the Pacific coastal states: California, Oregon, and Washington, in L. C. Pakiser and W. D. Mooney (eds), *Geophysical Framework of the Continental United States*, Geol. Soc. Amer. Mem. 172, Boulder, Co, pp. 129–61.

Mooney, W. D., G. Laske, and T. G. Masters (1998) CRUST 5.1; a global crustal model at 5 degrees X5 degrees, *J. Geophys. Res.*, *103*, 727–47.

Moores, E. M., and R. J. Twiss (1995) *Tectonics*, W. H. Freeman, New York.

Morris, G. B., R. W. Raitt, and G. G. Shor, Jr. (1969) Velocity anisotropy and delay-time maps of the mantle near Hawaii, *J. Geophys. Res.*, *74*, 4300–16.

Morse, P. M., and H. Feshbach (1953) *Methods of Theoretical Physics*, McGraw-Hill, New York.

Muller, B., J. Reinecker, B. Sperner, and K. Fuchs (2000) The 2000 release of the World Stress Map.

Nakamura, Y. (1983) Seismic velocity structure of the moon's upper mantle, *J. Geophys. Res.*, *88*, 677–86.

Nataf, H. C. (2000) Seismic imaging of mantle plumes, *Ann. Rev. Earth Planet. Sci.*, *28*, 391–417.

Nettles, M., and G. Ekström (1998) Faulting mechanism of anomalous earthquakes near Bardarbunga Volcano, Iceland, *J. Geophys. Res.*, *103*, 17,973–83.

Newman, A., S. Stein, J. Weber, J. Engeln, A. Mao, and T. Dixon (1999) Slow deformation and lower seismic hazard at the New Madrid Seismic Zone, *Science*, *284*, 619–21.

Newman, A., J. Schneider, S. Stein, and A. Mendez (2001) Uncertainties in seismic hazard maps for the New Madrid Seismic Zone and implications for seismic hazard communication, *Seism. Res. Lett.*, *72*, 653–67.

Ni, J., and M. Barazangi (1984) Seismotectonics of the Himalayan continental collision zone: geometry of the underthrusting Indian plate beneath the Himalayas, *J. Geophys. Res.*, *89*, 1147–64.

Nicolas, A. (1995) *The Mid-Oceanic Ridges*, Springer-Verlag, Berlin.

Nishenko, S. P., and L. R. Sykes (1993) Comment on "Seismic gap hypothesis: ten years after" by Kagan and Jackson, *J. Geophys. Res.*, *98*, 9909–16.

Nishimura, C., and D. Forsyth (1989) The anisotropic structure of the upper mantle in the Pacific, *Geophys. J. R. Astron. Soc.*, *96*, 203–26.

Noble, B. (1969) *Applied Linear Algebra*, Prentice-Hall, Englewood Cliffs, NJ.

Nolet, G. (1987) *Seismic Tomography*, D. Riedel, Dordrecht.

Norabuena, E., L. Leffler-Griffin, A. Mao, T. Dixon, S. Stein, I. S. Sacks, L. Ocala, and M. Ellis (1998) Space geodetic observations of Nazca-South America convergence along the Central Andes, *Science*, *279*, 358–62.

Officer, C. B. (1958) *Introduction to the Theory of Sound Transmission, with Application to the Ocean*, McGraw-Hill, New York.

Okada, Y. (1985) Surface deformation due to shear and tensile faults in a half-space, *Bull. Seism. Soc. Am.*, *75*, 1135–54.

Okal, E. A. (1992) A student's guide to teleseismic body wave amplitudes, *Seism. Res. Lett.*, *63*, 169–80.

Okal, E. A., and R. J. Geller (1979) On the observability of isotropic seismic sources: The July 31, 1970 Colombian earthquake, *Phys. Earth Planet. Inter.*, *18*, 176–96.

Okal, E. A., and B. Romanowicz (1994) On the variation of b-values with earthquake size, *Phys. Earth Planet. Inter.*, *87*, 55–76.

Oliver, J., and B. Isacks (1967) Deep earthquake zones, anomalous structures in the upper mantle, and the lithosphere, *J. Geophys. Res.*, *72*, 4259–75.

Owens, T. J., S. R. Taylor, and G. Zandt (1987) Crustal structure at regional seismic test network stations determined from the inversion of broadband teleseismic P waveforms, *Bull. Seism. Soc. Am.*, *77*, 631–62.

Pacheco, J., L. R. Sykes, and C. H. Scholz (1993) Nature of seismic coupling along simple plate boundaries of the subduction type, *J. Geophys. Res.*, *98*, 14,133–59.

Pakiser, L. C., and W. D. Mooney (eds.) (1989) *Geophysical Framework of the Continental United States*, Geol. Soc. Amer. Mem. 172, Boulder, Co.

Parker, R. L. (1977) Understanding inverse theory, *Ann. Rev. Earth Planet. Sci.*, *5*, 35–64.

Parsons, B., and F. M. Richter (1980) A relation between the driving force and geoid anomaly associated with mid-ocean ridges, *Earth Planet. Sci. Lett.*, *51*, 445–50.

Parsons, B., and J. G. Sclater (1977) An analysis of the variation of ocean floor bathymetry and heat flow with age, *J. Geophys. Res.*, *82*, 803–27.

Pearce, R. G. (1977) Fault plane solutions using the relative amplitudes of P and pP, *Geophys. J.*, *50*, 381–94.

Pearce, R. G. (1980) Fault plane solutions using the relative amplitudes of P and surface reflections: further studies, *Geophys. J.*, *60*, 459–87.

Peltier, W. R. (ed.) (1989) *Mantle Convection*, Gordon and Breach, New York.

Pho, H.-T., and L. Behe (1972) Extended distances and angles of incidence of P waves, *Bull. Seism. Soc. Am.*, *62*, 885–902.

Poirier, J.-P. (2000) *Introduction to the Physics of the Earth's Interior*, 2nd edn, Cambridge University Press, Cambridge.

Press, F., and R. Siever (1982) *Earth*, 3rd edn, W. H. Freeman, San Francisco.

Rabiner, L. R., and C. M. Rader (1972) *Digital Signal Processing*, IEEE Press, New York.

Ragan, D. M. (1968) *Structural Geology*, Wiley, New York.

Ranalli, G. (1987) *Rheology of the Earth*, Allen and Unwin, Boston.

Rebollar, C. J., L. Quintanar, J. Yamamoto, and A. Uribe (1999) Source process of the Chiapas, Mexico, intermediate-depth earthquake, *Bull. Seism. Soc. Am.*, *89*, 348–58.

Reiter, L. (1990) *Earthquake Hazard Analysis*, Columbia University Press, New York.

Reynolds, J. M. (1997) *An Introduction to Applied and Environmental Geophysics*, John Wiley & Sons, Chichester.

Rial, J. A., and V. F. Cormier (1980) Seismic waves at the epicenter's antipode, *J. Geophys. Res.*, *85*, 2661–8.

Richards, P. G., and J. Zavales (1990) Seismic discrimination of nuclear explosions, *Ann. Rev. Earth Planet. Sci.*, *18*, 257–86.

Richardson, W. P., S. Stein, C. Stein, and M. T. Zuber (1995) Geoid data and the thermal structure of the oceanic lithosphere, *Geophys. Res. Lett.*, *22*, 1913–16.

Richter, C. F. (1958) *Elementary Seismology*, W. H. Freeman, San Francisco.

Ringwood, A. E. (1975) *Composition and Petrology of the Earth's Mantle*, McGraw-Hill, New York.

Ringwood, A. E. (1979) Composition and origin of the Earth, in M. W. McElhinny (ed.), *The Earth, Its Origin, Structure and Evolution*, Academic Press Inc., New York, pp. 1–58.

Robbins, J. W., D. E. Smith, and C. Ma (1993) Horizontal crustal deformation and large scale plate motions inferred from space geodetic techniques, in D. E. Smith and D. L. Turcotte (eds), *Contributions of Space Geodesy to Geodynamics: crustal dynamics*, Geodynamics Series 23, Am. Geophys. Un., Washington, DC, pp. 21–36.

Robinson, E. A. (1983) *Migration of Geophysical Data*, International Human Resources Development Corp., Boston.

Robinson, E. A., and S. Treitel (1980) *Geophysical Signal Analysis*, Prentice-Hall, Englewood Cliffs, NJ.

Roeloffs, E. A., and J. Langbein (1994) The earthquake prediction experiment at Parkfield, California, *Rev. Geophys.*, *32*, 315–36.

Romanowicz, B. (1991) Seismic tomography of the earth's mantle, *Ann. Rev. Earth Planet. Sci.*, *19*, 77–99.

Romanowicz, B. (1992) Strike-slip earthquakes on quasi-vertical transcurrent faults: Inferences for general scaling relations, *Geophys. Res. Lett.*, *19*, 481–4.

Romanowicz, B. (1995) A global tomographic model of shear attenuation in the upper mantle, *J. Geophys. Res.*, *100*, 12,375–94.

Romanowicz, B. (1998) Attenuation tomography of the Earth's mantle: A review of current status, *Pure App. Geophys.*, *153*, 257–72.

Romanowicz, B., and P. Guillemant (1984) An experiment in the retrieval of depth and source mechanism of large earthquakes using very long period Rayleigh-wave data, *Bull. Seism. Soc. Am.*, *74*, 417–37.

Rosendahl, B. R. (1987) Architecture of continental rifts with special reference to East Africa, *Ann. Rev. Earth Planet. Sci.*, *15*, 445–503.

Roth, E. G., D. A. Wiens, L. M. Dorman, J. Hildebrand, and S. C. Webb (1999) Seismic attenuation tomography of the Tonga-Fiji region using phase pair methods, *J. Geophys. Res.*, *104*, 4795–809.

Ruff, L., and H. Kanamori (1980) Seismicity and the subduction process, *Phys. Earth Planet. Inter.*, *23*, 240–52.

Sadigh, K., C.-Y. Chang, J. A. Egan, F. Makdisi, and R. R. Youngs (1997) Attenuation relationships for shallow crustal earthquakes based on California strong motion data, *Seism. Res. Lett.*, *68*, 180–9.

Sangree, J. B., and J. M. Widmier (1979) Interpretation of depositional facies from seismic data, *Geophysics*, *44*, 131–60.

Sarewitz, D., and R. Pielke, Jr. (2000) Breaking the global-warming gridlock, *Atlantic Monthly*, July, 56–64.

Sarewitz, D., R. Pielke, Jr., and R. Byerly, Jr. (2000) *Prediction: science, decision making, and the future of nature*, Island Press, Washington, DC.

Sato, H., and M. C. Fehler (1998) *Seismic Wave Propagation and Scattering in the Heterogeneous Earth*, Springer-Verlag, New York.

Savage, J. C. (1983) A dislocation model of strain accumulation and release at a subduction zone, *J. Geophys. Res.*, *88*, 4984–96.

Savage, J. C. (1991) Criticism of some forecasts of the national earthquake prediction council, *Bull. Seism. Soc. Am.*, *81*, 862–81.

Savage, J. C. (1993) The Parkfield prediction fallacy, *Bull. Seism. Soc. Am.*, *83*, 1–6.

Scherbaum, F. (1996) *Of Poles and Zeros*, Kluwer, Dordrecht.

Schneider, W. A. (1971) Developments in seismic data processing and analysis (1968–1970), *Geophysics*, *36*, 1043–73.

Scholz, C. H. (1990) *The Mechanics of Earthquakes and Faulting*, Cambridge University Press, Cambridge.

Segall, P., and J. Davis (1997) GPS applications for geodynamics and earthquake studies, *Ann. Rev. Earth Planet. Sci.*, *25*, 301–36.

Shearer, P. M. (1994) Imaging Earth's seismic response at long periods, *Eos Trans. Am. Geophys. Un.*, *75*, 449, 451, 452.

Shearer, P. M. (1996) Transition zone velocity gradients and the 520-km discontinuity, *J. Geophys. Res.*, *101*, 3053–66.

Shearer, P. M. (1999) *Introduction to Seismology*, Cambridge University Press, Cambridge.

Shedlock, K., D. Giardini, G. Grunthal, and P. Zhang (2000) The GSHAP global seismic hazard map, *Seism. Res. Lett.*, *71*, 679–86.

Sheriff, R. E., and L. P. Geldart (1982) *Exploration Seismology*, Cambridge University Press, Cambridge.

Shimazaki, K., and T. Nakata (1980) Time-predictable recurrence model for large earthquakes, *Geophys. Res. Lett.*, *7*, 279–82.

Sieh, K., and S. LeVay (1998) *The Earth in Turmoil: earthquakes, volcanos, and their impact on humankind*, W. H. Freeman, New York.

Sieh, K., M. Stuiver, and D. Brillinger (1989) A more precise chronology of earthquakes produced by the San Andreas fault in southern California, *J. Geophys. Res.*, *94*, 603–24.

Silver, P. G. (1996) Seismic anisotropy beneath the continents: Probing the depths of geology, *Ann. Rev. Earth Planet. Sci.*, *24*, 385–432.

Silver, P. G., R. W. Carlson, and P. Olson (1988) Deep slabs, geochemical heterogeneity, and the large-scale structure of mantle convection, *Ann. Rev. Earth Planet. Sci.*, *16*, 477–541.

Simon, R. B. (1981) *Earthquake Interpretations*, William Kaufmann, Inc., Los Altos, CA.

Sipkin, S. A., and T. H. Jordan (1979) Frequency dependence of Q_{ScS}, *Bull. Seism. Soc. Am.*, *69*, 1055–79.

Sleep, N. H. (1990) Hotspots and mantle plumes: Some phenomenology, *J. Geophys. Res.*, *95*, 6715–36.

Sleep, N. H. (1992) Hotspots and mantle plumes, *Ann. Rev. Earth Planet. Sci.*, *20*, 19–43.

Sleep, N. H., and K. Fujita (1997) *Principles of Geophysics*, Blackwell, Malden, MA.

Sleep, N. H., and B. R. Rosendahl (1979) Topography and tectonics of mid-oceanic ridge axes, *J. Geophys. Res.*, *84*, 6831–9.

Sloan, M. E. (1980) *Introduction to Minicomputers and Microcomputers*, Addison-Wesley, Reading, MA.

Smith, D. E., and D. L. Turcotte (1993) *Contributions of Space Geodesy to Geodynamics*, Geodynamics Ser. 23, Am. Geophys. Un., Washington, DC.

Smith, R. B., and L. W. Braile (1994) The Yellowstone hotspot, *J. Volcan. Geotherm. Res.*, *61*, 121–87.

Smithson, S. B. (1989) Contrasting types of lower crust, in R. F. Mereu, S. Mueller and D. M. Fountain (eds), *Properties and Processes of Earth's Lower Crust*, IUGG 6, Am. Geophys. Un., Washington, DC, pp. 53–63.

Snelson, C. M., T. J. Henstock, G. R. Keller, K. C. Miller, and A. Levander (1998) Crustal and uppermost mantle structure along the Deep Probe seismic profile, *Rocky Mountain Geology*, *33*, 181–98.

Snieder, R. K. (2001) *A Guided Tour of Mathematical Physics*, Cambridge University Press, Cambridge.

Snoke, J. A., I. S. Sacks, and D. E. James (1979) Subduction beneath western South America: Evidence from converted phases, *Geophys. J. R. Astron. Soc.*, *59*, 219–25.

Solomon, S. C., and N. C. Burr (1979) The relationship of source parameters of ridge-crest and transform earthquakes to the thermal structure of oceanic lithosphere, *Tectonophysics*, *55*, 107–26.

Solomon, S. C., and D. R. Toomey (1992) The structure of mid-ocean ridges, *Ann. Rev. Earth Planet. Sci.*, *20*, 329–64.

Song, X., and D. V. Helmberger (1993) Anisotropy of Earth's inner core, *Geophys. Res. Lett.*, *20*, 2591–4.

Spakman, W., and G. Nolet (1988) Imaging algorithms, accuracy and resolution in delay time tomography, in N. J. Vlaar, G. Nolet, M. J. R. Wortel and S. A. P. L. Cloetingh (eds), *Mathematical Geophysics*, Reidel, Dordrecht, pp. 155–87.

Spakman, W., N. J. Vlaar, and M. J. R. Wortel (1988) The Hellenic subduction zone: A tomographic image and its geodynamic implications, *Geophys. Res. Lett.*, *15*, 60–3.

Spakman, W., S. Stein, R. van der Hilst, and R. Wortel (1989) Resolution experiments for NW Pacific subduction zone tomography, *Geophys. Res. Lett.*, *16*, 1097–110.

Spudich, P., and J. Orcutt (1980) A new look at the seismic velocity structure of the oceanic crust, *Rev. Geophys. Space Phys.*, *18*, 627–45.

Stacey, F. D. (1992) *Physics of the Earth*, 3rd edn, Brookfield Press, Kenmore, Brisbane.

Stein, C. A., and S. Stein (1992) A model for the global variation in oceanic depth and heat flow with lithospheric age, *Nature*, *359*, 123–9.

Stein, C. A. and S. Stein (1993) Constraints on Pacific midplate swells from global depth-age and heat flow-age models, in M. Pringle, W. Sager, W. Sliter, and S. Stein (eds), *The Mesozoic Pacific*, Geophysical Monog. 77, Am. Geophys. Un., Washington, DC, pp. 53–76.

Stein, C. A., S. Stein, and A. M. Pelayo (1995) Heat flow and hydrothermal circulation, in S. Humphris, L. Mullineaux, R. Zierenberg and R. Thomson (eds), *Seafloor Hydrothermal Systems, Physical, Chemical, Biological, and Geological Interactions*, Geophys. Mono. 91, Am. Geophys. Un., Washington, DC, pp. 425–45.

Stein, R. S. (1999) The role of stress transfer in earthquake occurrence, *Nature*, *402*, 605–9.

Stein, R. S., and R. S. Yeats (1989) Hidden earthquakes, *Sci. Am.*, *260*, 48–57.

Stein, R. S., G. C. P. King, and J. Lin (1994) Stress triggering of the 1994 M = 6.7 Northridge, California, earthquake by its predecessors, *Science*, *265*, 1432–5.

Stein, S. (1978) An earthquake swarm on the Chagos-Laccadive Ridge and its tectonic implications. *Geophys. J. R. Astron. Soc.*, *55*, 577–88.

Stein, S. (1992) Seismic gaps and grizzly bears, *Nature*, *356*, 387–8.

Stein, S. (1993) Space geodesy and plate motions, in D. E. Smith and D. L. Turcotte (eds), *Contributions of Space Geodesy to Geodynamics: crustal dynamics*, Geodynamics Series 23, Am. Geophys. Un., Washington, DC, pp. 5–20.

Stein, S., and R. J. Geller (1978) Time-domain observation and synthesis of split spheroidal and torsional free oscillations of the 1960 Chilean earthquake: Preliminary results, *Bull. Seism. Soc. Am.*, *68*, 325–32.

Stein, S., and R. G. Gordon (1984) Statistical tests of additional plate boundaries from plate motion inversions, *Earth Planet. Sci. Lett.*, *69*, 401–12.

Stein, S., and E. R. Klosko (2002) Earthquake mechanisms and plate tectonics, in R. A. Meyers (ed.), *The Encyclopedia of Physical Science and Technology*, Academic Press, San Diego.

Stein, S., and G. C. Kroeger (1980) Estimating earthquake source parameters from seismological data, in S. Nemat-Nasser (ed.), *Solid Earth Geophysics and Geotechnology*, AMD Symp. Ser., 42, Amer. Soc. Mech. Engin., New York, pp. 61–71.

Stein, S., and A. Pelayo (1991) Seismological constraints on stress in the oceanic lithosphere, *Phil. Trans. R. Soc. London Ser. A*, *337*, 53–72.

Stein, S., and D. C. Rubie (1999) Deep earthquakes in real slabs, *Science*, *286*, 909–10.

Stein, S., and C. A. Stein (1996) Thermo-mechanical evolution of oceanic lithosphere: Implications for the subduction process and deep earthquakes, in G. E. Bebout, D. W. Scholl, S. H. Kirby and J. P. Platt (eds), *Subduction: top to bottom*, Am. Geophys. Un., Washington, DC, pp. 1–17.

Stein, S., and D. Wiens (1986) Depth determination for shallow teleseismic earthquakes: Methods and results, *Rev. Geophys. Space Phys.*, *24*, 806–32.

Stein, S., and D. F. Woods (1989) Seismicity: Midocean ridge, in D. E. James (ed.), *The Encyclopedia of Solid Earth Geophysics*, Van Nostrand Reinhold, New York, pp. 1050–4.

Stein, S., N. H. Sleep, R. J. Geller, S. C. Wang, and G. C. Kroeger (1979) Earthquakes along the passive margin of eastern Canada, *Geophys. Res. Lett.*, *6*, 537–40.

Stein, S., J. M. Mills, Jr., and R. J. Geller (1981) Q^{-1} models from data space inversion of fundamental spheroidal mode attenuation measurements, in Stacey *et al.* (eds), *Anelasticity in the Earth*, Geodynamics Series, 4, Am. Geophys. Un., Washington, DC, pp. 39–53.

Stein, S., J. F. Engeln, C. DeMets, R. G. Gordon, D. Woods, P. Lundgren, D. Agrus, C. Stein, and D. A. Wiens (1986) The Nazca–South America convergence rate and the recurrence of the great 1960 Chilean earthquake, *Geophys. Res. Lett.*, *13*, 713–16.

Stein, S., S. Cloetingh, N. Sleep, and R. Wortel (1989) Passive margin earthquakes, stresses, and rheology, in S. Gregerson and P. Basham (eds), *Earthquakes at North-Atlantic Passive Margins: neotectonics and postglacial rebound*, Kluwer, Dordrecht, pp. 231–60.

Stixrude, L. (1998) Elastic constants and anisotropy of $MgSiO_3$ perovskite, periclase, and SiO_2 at high pressure, in M. Gurnis, M. E. Wysession, E. Knittle and B. Buffett (eds), *The Core-Mantle Boundary Region*, Am. Geophys. Un., Washington, DC, pp. 83–96.

Stixrude, L., and R. E. Cohen (1995) High-pressure elasticity of iron and anisotropy of Earth's inner core, *Science*, *267*, 1972–5.

Strahler, A. N. (1969) *Physical Geography*, John Wiley, New York.

Su, W., R. L. Woodward, and A. M. Dziewonski (1994) Degree 12 model of shear velocity heterogeneity in the mantle, *J. Geophys. Res.*, *99*, 6945–80.

Sykes, L. R., and D. M. Davis (1987) The yields of Soviet strategic weapons, *Sci. Am.*, *256*, 29–37.

Sykes, L. R., and S. P. Nishenko (1984) Probabilities of occurrence of large plate rupturing earthquakes for the San Andreas, San Jacinto, and Imperial Faults, California, 1983–2003, *J. Geophys. Res.*, *89*, 5905–27.

Sykes, L. R., B. E. Shaw, and C. H. Scholz (1999) Rethinking earthquake prediction, *Pure Appl. Geophys.*, *155*, 207–32.

Talandier, J., and E. A. Okal (1979) Human perception of T waves: The June 22, 1977 Tonga earthquake felt on Tahiti, *Bull. Seism. Soc. Am.*, *69*, 1475–86.

Taner, M. T., and F. Kohler (1969) Velocity spectra — digital computer derivation and application of velocity spectra, *Geophysics*, *34*, 859–81.

Tappin, D., *et al.* (1999) Sediment slump likely caused Papua New Guinea tsunami, *Eos Trans. Am. Geophys. Un.*, *80*, 329–34.

Tapponnier, P., G. Peltzer, A. Le Dain, R. Armijo, and P. Cobbold (1982) Propagating extrusion tectonics in Asia: New insights from simple experiments with plasticine, *Geology*, *10*, 611–16.

Tatham, R. (1989) Tau-p filtering, in P. L. Stoffa (ed.), *Tau-p, a Plane Wave Approach to the Analysis of Seismic Data*, Kluwer, Dordrecht, pp. 35–70.

Telford, W. M., L. P. Geldart, R. E. Sheriff, and D. A. Keys (1976) *Applied Geophysics*, Cambridge University Press, Cambridge.

Thatcher, W. (1983) Nonlinear strain buildup and the earthquake cycle on San Andreas fault, *J. Geophys. Res.*, *88*, 5893–902.

Thio, H. K., and H. Kanamori (1996) Source complexity of the 1994 Northridge, California, earthquake and its relation to aftershock mechanisms, *Bull. Seism. Soc. Am.*, *86*, S84–92.

Thurber, C. H., and K. Aki (1987) Three dimensional seismic imaging, *Ann. Rev. Earth Planet. Sci.*, *15*, 115–39.

Toomey, D. R., S. C. Solomon, and G. M. Purdy (1988) Microearthquakes beneath the median valley of the Mid-Atlantic Ridge near 23°N: Tomography and tectonics, *J. Geophys. Res.*, *93*, 9093–112.

Torge, W. (1991) *Geodesy*, de Gruyter, Berlin.

Triep, E. G., and L. R. Sykes (1997) Frequency of occurrence of moderate to great earthquakes in intracontinental regions: Implications for changes in stress, earthquake prediction, and hazard assessments, *J. Geophys. Res.*, *102*, 9923–48.

Tromp, J. (1993) Support for anisotropy of the Earth's inner core from free oscillations, *Nature*, *366*, 678–81.

Tsai, Y.-B., and K. Aki (1970) Precise focal depth determination from amplitude spectra of surface waves, *J. Geophys. Res.*, *75*, 5729–43.

Tse, S. T., and J. R. Rice (1986) Crustal earthquake instability in relation to the depth variation of frictional slip properties, *J. Geophys. Res.*, *91*, 9452–72.

Turcotte, D. L. (1991) Earthquake prediction, *Ann. Rev. Earth Planet. Sci.*, *19*, 263–82.

Turcotte, D. L. (1992) *Fractals and Chaos in Geology and Geophysics*, Cambridge University Press, Cambridge.

Turcotte, D. L., and G. Schubert (1982) *Geodynamics: applications of continuum physics to geological problems*, John Wiley, New York.

Udias, A. (1999) *Principles of Seismology*, Cambridge University Press, Cambridge.

Usselman, T. N. (1975) Experimental approach to the state of the core, I: The liquidus relations of the Fe-rich portion of the Fe-Ni-S system from 30 to 100 kb, *Am. J. Sci.*, *275*, 291–303.

Uyeda, S. (1978) *The New View of the Earth*, W. H. Freeman, San Francisco.

van der Hilst, R. D., and W. Spakman (1989) Importance of the reference model in linearized tomography and images of subduction below the Caribbean plate, *Geophys. Res. Lett.*, *16*, 1093–6.

van der Hilst, R. D., S. Widiyantoro, K. C. Creager, and T. J. McSweeney (1998) Deep subduction and aspherical variations in P-wavespeed at the base of earth's mantle, in M. Gurnis, M. E. Wysession, E. Knittle, and B. Buffett (eds), *The Core-Mantle Boundary Region*, Am. Geophys. Un., Washington, DC, pp. 5–20.

van der Lee, S., and G. Nolet (1997) Upper mantle S velocity structure of North America, *J. Geophys. Res.*, *102*, 22,815–38.

Vassiliou, M. S. (1984) The state of stress in subducting slabs as revealed by earthquakes analysed by moment tensor inversion, *Earth Planet. Sci. Lett.*, *69*, 195–202.

Vassiliou, M. S., B. H. Hager, and A. Raefsky (1984) The distribution of earthquakes with depth and stress in subducting slabs, *J. Geodynam.*, *1*, 11–28.

Vera, E. E., J. C. Mutter, P. Buhl, J. A. Orcutt, A. J. Harding, M. E. Kappus, R. S. Detrick, and T. M. Brocher (1990) The structure of 0–0.2-m.y.-old oceanic crust at 9°N on the East Pacific Rise from expanded spread profiles, *J. Geophys. Res.*, *95*, 15,529–56.

Verhoogen, J. (1980) *Energetics of the Earth*, National Academy of Sciences, Washington, DC.

Vidale, J. E., and H. M. Benz (1992) Upper-mantle seismic discontinuities and the thermal structure of subduction zones, *Nature*, *356*, 678–83.

Von Huene, R., L. D. Kulm, and J. Miller (1985) Structure of the frontal part of the Andean continental margin, *J. Geophys. Res.*, *90*, 5429–42.

Walck, M. C. (1984) The P-wave upper mantle structure beneath an active spreading center: The Gulf of California, *Geophys. J. R. Astron. Soc.*, *76*, 697–723.

Wald, D. J., H. Kanamori, D. V. Helmberger, and T. H. Heaton (1993) Source study of the 1906 San Francisco earthquake, *Bull. Seism. Soc. Am.*, *83*, 981–1019.

Wald, D. J., T. H. Heaton, and K. Hudnut (1996) The slip history of the 1994 Northridge, California, earthquake determined from strong-motion, teleseismic, GPS, and leveling data, *Bull. Seism. Soc. Am.*, *86*, S49–70.

Wallace, T. C. (1985) A reexamination of the moment tensor solutions of the 1980 Mammoth Lakes earthquakes, *J. Geophys. Res.*, *90*, 11,171–6.

Wallace, T. C., A. Velasco, J. Zhang, and T. Lay (1991) Broadband seismological investigation of the 1989 Loma Prieta earthquake, *Bull. Seism. Soc. Am.*, *81*, 1622–46.

Ward, P. D., and D. Brownlee (2000) *Rare Earth*, Copernicus Press, New York.

Ward, S. (1989) Tsunamis, in D. E. James (ed.), *The Encyclopedia of Solid Earth Geophysics*, Van Nostrand-Reinhold, New York, pp. 1279–92.

Waters, K. H. (1981) *Reflection Seismology: a tool for energy resource exploration*, John Wiley, New York.

Weber, J., S. Stein, and J. Engeln (1998) Estimation of strain accumulation in the New Madrid seismic zone from repeat Global Positioning System surveys, *Tectonics*, *17*, 250–66.

Weertman, J., and J. R. Weertman (1975) High temperature creep of rock and mantle viscosity, *Ann. Rev. Earth Planet. Sci.*, *3*, 293–315.

Weidner, D. J. (1986) Mantle model based on measured physical properties of minerals, in S. K. Saxena (ed.), *Chemistry and Physics of Terrestrial Planets*, Advances in Physical Geochemistry, 6, Springer-Verlag, New York, pp. 251–74.

Wells, D. L., and K. J. Coppersmith (1994) New empirical relations among magnitude, rupture length, rupture width, rupture area and surface displacement, *Bull. Seism. Soc. Am.*, *84*, 974–1002.

Widmer, R., G. Masters, and F. Gilbert (1992) Observably-split multiplets — data analysis and interpretation in terms of large-scale aspherical structure, *Geophys. J. R. Astron. Soc.*, *111*, 559–76.

Wiegel, R. L. (ed.) (1970) *Earthquake Engineering*, Prentice-Hall, Englewood Cliffs, NJ.

Wiemer, S., and M. Wyss (1997) Mapping the frequency-magnitude distribution in asperities: An improved technique to calculate recurrence times?, *J. Geophys. Res.*, *102*, 15,115–28.

Wiens, D. A., and S. Stein (1983) Age dependence of oceanic intraplate seismicity and implications for lithospheric evolution, *J. Geophys. Res.*, *88*, 6455–68.

Wiens, D. A., and S. Stein (1984) Intraplate seismicity and stresses in young oceanic lithosphere, *J. Geophys. Res.*, *89*, 11,442–64.

Wiens, D. A., and S. Stein (1985) Implications of oceanic intraplate seismicity for plate stresses, driving forces and rheology, *Tectonophysics*, *116*, 143–62.

Wiens, D. A., *et al.* (1985) A diffuse plate boundary model for Indian Ocean tectonics, *Geophys. Res. Lett.*, *12*, 429–32.

Wiggins, R. A. (1972) The general linear inverse problem: Implication of surface waves and free oscillations for Earth structure, *Rev. Geophys. Space Phys.*, *10*, 251–85.

Wilson, J. T. (1976) *Continents Adrift and Continental Aground*, W. H. Freeman, San Francisco.

Wong, J., N. Bregman, G. West, and P. Hurley (1987) Cross-hole seismic scanning and tomography, *Leading Edge*, *6*, 36–41.

Wood, B. J., and D. G. Fraser (1977) *Elementary Thermodynamics for Geologists*, Oxford University Press, Oxford.

Woodhouse, J. H., and A. M. Dziewonski (1984) Mapping the upper mantle: Three dimensional modeling of Earth structure by inversion of seismic waveforms, *J. Geophys. Res.*, *89*, 5953–86.

Woods, M. T., and E. A. Okal (1987) Effect of variable bathymetry on the amplitude of teleseismic tsunamis: A ray-tracing experiment, *Geophys. Res. Lett.*, *14*, 765–8.

Wyllie, P. J. (1971) *The Dynamic Earth: textbook in geosciences*, John Wiley, New York.

Wysession, M. E. (1996a) Imaging cold rock at the base of slabs: The sometimes fate of slabs?, in G. E. Bebout, D. W. Scholl, S. H. Kirby, and J. P. Platt (eds), *Subduction: top to bottom*, Am. Geophys. Un., Washington, DC, pp. 369–84.

Wysession, M. E. (1996b) Large-scale structure at the core–mantle boundary from core-diffracted waves, *Nature*, *382*, 244–8.

Wysession, M. E., and P. J. Shore (1994) Visualization of whole mantle propagation of seismic shear energy using normal mode summation, *Pure Appl. Geophys.*, *142*, 295–310.

Wysession, M. E., J. Wilson, L. Bartko, and R. Sakata (1995) Intraplate seismicity in the Atlantic Ocean basin: A teleseismic catalog, *Bull. Seism. Soc. Am.*, *85*, 755–74.

Wysession, M. E., B. C. Hicks, D. A. Wiens, and P. J. Shore (1996) Determining the frequency–magnitude–depth relations of seismicity in the Tonga subduction zone. *Seism. Res. Lett.*, *67*, 62.

Wysession, M. E., T. Lay, J. Revenaugh, Q. Williams, E. J. Garnero, R. Jeanloz, and L. H. Kellogg (1998) Implications of the D″ discontinuity, in M. Gurnis, M. E. Wysession, E. Knittle, and B. Buffett (eds), *The Core–Mantle Boundary Region*, Am. Geophys. Un., Washington, DC, pp. 273–97.

Wyss, M., and R. Koyanagi (1992) Seismic gaps in Hawaii, *Bull. Seism. Soc. Am.*, *82*, 1373–87.

Wyss, M., R. K. Aceves, and S. K. Park (1997) Cannot earthquakes be predicted?, *Science*, *278*, 487–90.

Yeats, R. S., K. Sieh, and C. R. Allen (1997) *The Geology of Earthquakes*, Oxford University Press, New York.

Yilmaz, O. (1987) *Seismic Data Processing*, Society of Exploration Geophysicists, Tulsa, OK.

Young, C. J., and T. Lay (1990) Multiple phase analysis of the shear velocity structure in the D″ region beneath Alaska, *J. Geophys. Res.*, *95*, 17,385–402.

Young, G. B., and L. W. Braile (1976) A computer program for the application of Zoeppritz's amplitude equations and Knott's energy equations, *Bull. Seism. Soc. Am.*, *66*, 1881–6.

Youngs, R. R., and K. J. Coppersmith (1985) Implications of fault slip rates and earthquake recurrence models to probabilistic seismic hazard estimates, *Bull. Seism. Soc. Am.*, *75*, 939–64.

Yu, G.-K., and B. Mitchell (1979) Regionalized shear velocity models of the Pacific upper mantle from observed Rayleigh and Love wave dispersion, *Geophys. J. R. Astron. Soc.*, *57*, 311–42.

Zhao, L., T. Jordan, and C. Chapman (2000) Three-dimensional Frechet differential kernels for seismic delay times, *Geophys. J. Int.*, *141*, 558–76.

Zoback, M. L. (1992) First and second order patterns of stress in the lithosphere: The world stress map project, *J. Geophys. Res.*, *97*, 11,703–2104.

Solutions to selected odd-numbered problems

Note that in many problems (as in reality), the solution varies depending on the interpretation of the data or the assumptions used.

Chapter 2

(1) $R_{12}=0, T_{12}=1$.

(3a) $(3, -2, 5)$.

(3b) $(2, 1, 3)$.

(3c) $(5, 3, 0)/\sqrt{14}$.

(5a) $\sigma_1=2, \sigma_2=0, \sigma_3=-2$; $\mathbf{n}^{(1)}=(1, 1, 0)/\sqrt{2}$, $\mathbf{n}^{(2)}=(0, 0, 1)$, $\mathbf{n}^{(3)}=(1, -1, 0)/\sqrt{2}$.

(5b) $\tau=2$. Planes have normals $(1, 0, 0)$ and $(0, 1, 0)$.

(7b) -150 kbar.

(7c) $D=\begin{bmatrix} 0 & -2 & 1 \\ -2 & -5 & 3 \\ 1 & 3 & 5 \end{bmatrix}$.

(7d) 450 km.

(9) 2%.

(11a) $\begin{bmatrix} 6\lambda+6\mu & 0 & 0 \\ 0 & 6\lambda+2\mu & 2\mu \\ 0 & 2\mu & 6\lambda+4\mu \end{bmatrix}$.

(11b) $18\lambda+16\mu$.

(13a) $[(\lambda+2\mu)/\rho]^{1/2}$.

(13b) $(\mu/\rho)^{1/2}$.

(15) $\sqrt{3}$.

(17a) $\bar{\alpha}=11.25$ km/s; $\bar{\beta}=6.18$ km/s.

(17b) $\bar{\alpha}/\bar{\beta}=1.82$.

(19a) 0.8 km, 8 km, 800 km.

(19b) 0.000125 s, 0.125 s, 12.5 s; 8000 Hz, 8 Hz, 0.08 Hz.

(21) $i_2=13°$; $i_3=17°$; $i_c=37°$.

(23a) For the $i_1=0°$ wave: $i_2=0°$, $l_1=2$ km, $l_2=2$ km, $T=3.3$ s. For the $i_1=30°$ wave: $i_2=49°$, $l_1=2.3$ km, $l_2=3.0$ km, $T=4.3$ s.

(23b) For the $i_1=0°$ wave: $\mathbf{s}_1=(0, 1)$ s/km, $|\mathbf{s}_1|=1/v_1$, $\mathbf{s}_2=(0, 2/3)$ s/km, $|\mathbf{s}_2|=1/v_2$. For the $i_1=30°$ wave: $\mathbf{s}_1=(0.5, \sqrt{3}/2)$ s/km, $|\mathbf{s}_1|=1/v_1$, $\mathbf{s}_2=(0.5, 0.44)$ s/km, $|\mathbf{s}_2|=1/v_2$.

(25a) $\Psi_I=B_1 \exp[i(\omega t-k_x x+k_x r_\beta z)]$,
$\Psi_R=B_2 \exp[i(\omega t-k_x x-k_x r_\beta z)]$,
$\Phi_R=A_2 \exp[i(\omega t-k_x x-k_x r_\alpha z)]$.

(25b) $$\sigma_{xz}=\mu\left(2\frac{\partial^2\Phi}{\partial z\partial x}+\frac{\partial^2\Psi}{\partial x^2}-\frac{\partial^2\Psi}{\partial^2 z}\right)=0,$$
$$\sigma_{zz}=\lambda\left(\frac{\partial^2\Phi}{\partial x^2}+\frac{\partial^2\Phi}{\partial z^2}\right)+2\mu\left(\frac{\partial^2\Phi}{\partial z^2}+\frac{\partial^2\Psi}{\partial x\partial z}\right)=0.$$
$2r_\alpha A_2+(1-r_\beta{}^2)B_1+(1-r_\beta{}^2)B_2=0$,
$A_2(\lambda+\lambda r_\alpha^2+2\mu r_\alpha^2)-2\mu r_\beta B_1+2\mu r_\beta B_2=0$.

(25c) $B_2/B_1=-1$, $A_2/B_1=0$.

(25d) $\dfrac{|\mathbf{u}|_{SR}}{|\mathbf{u}|_{SI}}=\dfrac{|B_2|}{|B_1|}$; $\dfrac{|\mathbf{u}|_{PR}}{|\mathbf{u}|_{SI}}=\dfrac{\beta|A_2|}{\alpha|B_1|}$. $\dfrac{\dot{E}_{SR}}{\dot{E}_{SI}}=\dfrac{B_2^2}{B_1^2}$, $\dfrac{\dot{E}_{PR}}{\dot{E}_{SI}}=\dfrac{\eta_\alpha A_2^2}{\eta_\beta B_1^2}$.

(27a) For *ScS*: $j_{slab}=26°$, $i_{surf}=4°$. For *ScSp*: $i_{slab}=50°$, $i_{surf}=20°$.

(27b) 50 km.

(29) $\omega_1=0.58$, $\omega_2=1.16$.

(33) $(3/2)(5\cos^2\theta\sin\theta-\sin\theta)$.

(35a) 2591 s.

(35b) 4.4 km/s.

(35c) 8.3 km/s, 13.8 km/s, 74.5 km/s.

(35d) $c=5.36$ km/s, 5.01 km/s, 4.05 km/s; $\lambda=11{,}440$ km, 1312 km, 307 km.

(37a) $\Delta\omega/\omega$ observed: 0.059.

(37b) $\Delta\omega/\omega$ predicted: 0.037.

(C-3) For *P* waves at the CMB, $T_{mc}=0.975$, $R_{mc}=-0.025$, $T_{cm}=1.025$, $R_{cm}=0.025$, $\dot{E}_R/\dot{E}_I=0.0006$, $\dot{E}_T/\dot{E}_I=0.9994$. For *S* waves at the CMB, $T_{mc}=2$, $R_{mc}=1$, $\dot{E}_R/\dot{E}_I=1$, $\dot{E}_T/\dot{E}_I=0$.

Chapter 3

(1) $\alpha_0=5.7$ km/s, $\alpha_1=7.8$ km/s, $h_0=23$ km.

(3a) $\alpha_c=6.7$ km/s, $\alpha_m=8.2$ km/s.

(3b) 3.1 km.

(3c) 5.4 km.

(5) $\alpha_c=6.5$ km/s, $\alpha_m=8$ km/s, dip $=4°$, $h_u=50$ km, $h_d=30$ km.

(11) 24,000,000.

(13a) 9.34 km/s.

(13b) 11.24 km/s.

(15) For $D = 0$ km: $p_{40} = 8.3$ s/degree, $p_{60} = 6.9$ s/degree, $i_{40} = 26°$, $i_{60} = 21°$. For $D = 600$ km: $p_{40} = 7.9$ s/degree, $p_{60} = 6.6$ s/degree, $i_{40} = 52°$, $i_{60} = 41°$.

(17a) 4.5 s/degree.

(17b) 13.4 km/s.

(17c) *SKKS*.

(19a) $t_{1/e}({}_0T_2) = 58.3$ hr, $t_{1/e}({}_0T_{30}) = 3.0$ hr, $t_{1/e}({}_0S_{30}) = 4.2$ hr.

(19b) 54,300 km for ${}_0T_{30}$, 76,450 km for ${}_0S_{30}$.

(21a) 3%.

(21b) 0.3%.

(23) $M_c = 1.94 \times 10^{24}$ kg, $\bar{\rho}_c = 11$ g/cm^3.

Chapter 4

(3) Earthquake a: $(\phi, \delta, \lambda)_1 = (310°, 65°, 90°)$ (thrust); $(\phi, \delta, \lambda)_2 = (130°, 25°, 90°)$ (thrust); P axis (azimuth, plunge) = (40°, 20°); T axis = (220°, 70°); B axis = (130°, 0°).
Earthquake b: $(\phi, \delta, \lambda)_1 = (176°, 80°, 195°)$ (right-lateral strike-slip); $(\phi, \delta, \lambda)_2 = (83°, 75°, 350°)$ (left-lateral strike-slip); P axis (azimuth, plunge) = (40°, 18°); T axis = (309°, 3°); B axis = (209°, 72°).
Earthquake c: $(\phi, \delta, \lambda)_1 = (9°, 90°, 180°)$ (right-lateral strike-slip); $(\phi, \delta, \lambda)_2 = (99°, 90°, 0°)$ (left-lateral strike-slip); P axis (azimuth, plunge) = (234°, 0°); T axis = (144°, 0°); B axis = (undefined, 90°).
Earthquake d: First solution: $(\phi, \delta, \lambda)_1 = (16°, 85°, 90°)$ (dip slip); $(\phi, \delta, \lambda)_2 = (196°, 5°, 90°)$ (thrust); P axis (azimuth, plunge) = (106°, 40°); T axis = (286°, 50°); B axis = (196°, 0°). Second solution: $(\phi, \delta, \lambda)_1 = (78°, 66°, 25°)$ (left-lateral strike-slip); $(\phi, \delta, \lambda)_2 = (337°, 67°, 154°)$ (right-lateral strike-slip); P axis (azimuth, plunge) = (28°, 1°); T axis = (297°, 34°); B axis = (119°, 56°).

(7) 0.

(9a)
$$\begin{pmatrix} 0 & 0 & 0 \\ 0 & 0 & 0 \\ 0 & 0 & M_{zz} \end{pmatrix} = \begin{pmatrix} M_{zz}/3 & 0 & 0 \\ 0 & M_{zz}/3 & 0 \\ 0 & 0 & M_{zz}/3 \end{pmatrix} + \begin{pmatrix} -M_{zz}/3 & 0 & 0 \\ 0 & -M_{zz}/3 & 0 \\ 0 & 0 & 2M_{zz}/3 \end{pmatrix}.$$

(9b)
$$\begin{pmatrix} -2.14 & 0 & 0 \\ 0 & 2.01 & 0 \\ 0 & 0 & 0.13 \end{pmatrix} = \begin{pmatrix} -2.075 & 0 & 0 \\ 0 & 2.075 & 0 \\ 0 & 0 & 0 \end{pmatrix} + \begin{pmatrix} -0.065 & 0 & 0 \\ 0 & -0.065 & 0 \\ 0 & 0 & 0.13 \end{pmatrix}.$$

(Double-couple scalar moment)/(original scalar moment) = 0.999.
(CLVD scalar moment)/(original scalar moment) = 0.054.

(9c) There are two solutions:

Solution 1:

$$\begin{pmatrix} -2.14 & 0 & 0 \\ 0 & 2.01 & 0 \\ 0 & 0 & 0.13 \end{pmatrix} = \begin{pmatrix} -1.135 & 0 & 0 \\ 0 & 0 & 0 \\ 0 & 0 & 1.135 \end{pmatrix} + \begin{pmatrix} -1.005 & 0 & 0 \\ 0 & 2.01 & 0 \\ 0 & 0 & -1.005 \end{pmatrix}.$$

(Double-couple scalar moment)/(original scalar moment) = 0.546.
(CLVD scalar moment)/(original scalar moment) = 0.838.

Solution 2:

$$\begin{pmatrix} -2.14 & 0 & 0 \\ 0 & 2.01 & 0 \\ 0 & 0 & 0.13 \end{pmatrix} = \begin{pmatrix} 0 & 0 & 0 \\ 0 & 0.94 & 0 \\ 0 & 0 & -0.94 \end{pmatrix} + \begin{pmatrix} -2.14 & 0 & 0 \\ 0 & 1.07 & 0 \\ 0 & 0 & 1.07 \end{pmatrix}.$$

(Double-couple scalar moment)/(original scalar moment) = 0.452.
(CLVD scalar moment)/(original scalar moment) = 0.892.

(11) 4.24 mm/yr, 0.85 mm/yr, 0.42 mm/yr.

(13) $M_S = 5.2$.

(15) assuming $\mu = 3 \times 10^{11}$; 200,000 km; 43,333 km.

(17) ~0.04 Hz.

(19b) 0.003–0.03.

(21) Japan: 8 mo. ($M \geq 6$); 7 yr ($M \geq 7$); 65 yr ($M \geq 8$). S. California: 1 yr ($M \geq 6$); 8 yr ($M \geq 7$); ≈ 100 yr ($M \geq 8$). New Madrid: 92 yr ($M \geq 6$); 920 yr ($M \geq 7$); 9200 yr ($M \geq 8$).

(C-1) Earthquake a: $\hat{\mathbf{n}} = (0.453, -0.785, 0.423)$; $\hat{\mathbf{d}} = (0.098, 0.515, 0.852)$.
Earthquake b: $\hat{\mathbf{n}} = (0.853, -0.150, 0.500)$; $\hat{\mathbf{d}} = (0.492, -0.087, -0.866)$.
Earthquake c: $\hat{\mathbf{n}} = (0.853, -0.150, 0.500)$; $\hat{\mathbf{d}} = (-0.492, 0.087, 0.866)$.
Earthquake d: $\hat{\mathbf{n}} = (-0.633, -0.754, 0.174)$; $\hat{\mathbf{d}} = (0.758, -0.559, 0.337)$.
Earthquake e: $\hat{\mathbf{n}} = (-0.633, -0.754, 0.173)$; $\hat{\mathbf{d}} = (-0.758, 0.559, -0.337)$.

(C-3) Earthquake a: $\begin{pmatrix} 0.088 & 0.157 & 0.427 \\ 0.157 & -0.808 & -0.451 \\ -0.427 & -0.451 & 0.720 \end{pmatrix}$.

Earthquake b: $\begin{pmatrix} 0.840 & -0.148 & -0.492 \\ -0.148 & 0.026 & 0.087 \\ -0.492 & 0.087 & -0.866 \end{pmatrix}$.

Earthquake c: $\begin{pmatrix} -0.840 & 0.148 & 0.492 \\ 0.148 & -0.026 & -0.087 \\ 0.492 & -0.087 & 0.866 \end{pmatrix}$.

Earthquake d: $\begin{pmatrix} -0.960 & -0.218 & -0.082 \\ -0.218 & 0.843 & -0.351 \\ -0.082 & -0.351 & 0.117 \end{pmatrix}$.

Earthquake e: $\begin{pmatrix} 0.960 & 0.218 & 0.082 \\ 0.218 & -0.843 & 0.351 \\ 0.082 & 0.351 & -0.117 \end{pmatrix}$.

(C-7b) 0.95.

(C-9a) Gaussian: 0.1%; Poisson: 4%.

(C-9b) Gaussian: 0.3%; Poisson: 3%.

(C-9c) Gaussian: 0.5%; Poisson: 2%.

(C-11a) $\tau = 21.8$ yr; $\sigma = 7.2$ yr; Poisson: $p = 37\%$; Gaussian: C(1993, 1985) = 64%.

(C-11b) $\tau = 21.8$ yr; $\sigma = 1.5$ yr; Poisson: $p = 37\%$; Gaussian: C(1993, 1985) = 99%.

(C-11c) $\tau = 25.5$ yr; $\sigma = 11.1$ yr; Poisson: $p = 31\%$; Gaussian: C(2018, 2010) = 82%.

(C-11d) $\tau = 27.2$ yr; $\sigma = 14.7$ yr; Poisson: $p = 29\%$; Gaussian: C(2028, 2020) = 74%.

Chapter 5

(1a) 0.77 m; $M_w = 6.8$, length = 31 km.

(1b) 4.62 m; $M_w = 7.8$, length = 240 km.

(3a) 40 mm/yr.

(3b) 125, 250, 500 yr.

(3c) For 25%: 500, 1000, 2000 yr; for 50%: 250, 500, 1000 yr.

(3e) $M_w \approx 8.4$; $M_0 = 5 \times 10^{28}$ dyn-cm.

(5) 6×10^{31} dyn-cm; M_w 10.5.

(7a) 47 mW/m^2.

(7b) 33 mW/m^2.

(7c) 84 mW/m^2.

(9) ~1 Ga.

(11) −21.5 bar/°C.

(13) $vL^3/(24\kappa^2 t)$; 28 (for $v = 10$ cm/yr and $t = 150$ Ma).

(17) 58°; 251 MPa.

(C-1b) San Andreas: 46 mm/yr at 324°; Aleutian: 53 mm/yr at 346°.

(C-3b) $(\theta, \phi, |\omega|) = (-63.0°, 107.4°, 0.641$ °/my).

(C-3c) Hawaii: 66 mm/yr at 299°.

Chapter 6

(1a) $a_0 = 0, a_k = 0, b_k = (2/k\pi)(1 - \cos(k\pi))$.

(1b) $a_0 = 0, a_k = 0, b_k = -\cos(k\pi)/k\pi = (-1)^{k+1}/k\pi$.

(3a) −1.

(3b) $4i$.

(3c) $-i$.

(3d) $1.5 + 2.6i$.

(7a) $\pi e^{-i\pi/2}[\delta(\omega - \omega_0) - \delta(\omega + \omega_0)]$.

(9a) $a^2\sigma_u^2 + b^2\sigma_v^2 + 2ab\sigma_{uv}^2$.

(9b) $a^2v^2\sigma_u^2 + a^2u^2\sigma_v^2 + 2a^2uv\sigma_{uv}^2$.

(9c) $(a^2/v^2)\sigma_u^2 + (a^2u^2/v^4)\sigma_v^2 - 2(a^2u/v^3)\sigma_{uv}^2$.

(9d) $a^2b^2u^{(2b-2)}\sigma_u^2$.

(11a) $v\Delta t/(2\cos i)$.

(11b) 10 km.

(11c) $(\sigma_v^2\Delta t^2 + \sigma_{\Delta t}^2 v^2 + \sigma_i^2 v^2\Delta t^2 \tan^2 i)/4\cos^2 i$.

(11d) 4 km.

Appendix

(1) 21°.

(5) $a_{11}a_{22}a_{33} - a_{11}a_{23}a_{32} - a_{12}a_{21}a_{33} + a_{12}a_{23}a_{31} + a_{13}a_{21}a_{32} - a_{13}a_{22}a_{31}$.

(7) $A^{-1} = \begin{pmatrix} -2/3 & 1/3 \\ 5/6 & -1/6 \end{pmatrix}$.

(13a) $6xy^2 + 2x^3 + 2$.

(13b) (0, 0, 0).

(13c) $(6y^2 + 6x^2, 12xy, 0)$.

(13d) $xz + x^3y^2 + y^2 +$ constant.

(15) Hint: use Eqn A.7.4.

(C-5) $|(1, 4, 2)| = \sqrt{21}$; $|(2, 3, 1)| = \sqrt{14}$; sum = (3, 7, 3); $\mathbf{a} \cdot \mathbf{b} = 16$; $\mathbf{a} \times \mathbf{b} = (-2, 3, -5)$; $\theta = 21.1°$.

(C-7b) (0, −1, 1).

(C-9b) $\begin{pmatrix} 0.7 & -0.1 & 0.3 \\ 0.5 & -0.5 & 0.5 \\ -0.8 & 0.4 & -0.2 \end{pmatrix}$.

(C-11b) 1, 3, 4.

(C-13b) i: $\Delta = 93°, \zeta = 48°$.
ii: $\Delta = 54°, \zeta = 49°$.
iii: $\Delta = 87°, \zeta = 148°$.
iv: $\Delta = 35°, \zeta = 232°$.

Index